TEMPERATURE CONVERSIONS

Fahrenheit-Celsius

°F	°C		Some Conversions of Common Interest
	230	110	
	220		
212°F	210	100	100°C — Water boils at standard temperature and pressure
	200		
	190	90	
	180	80	
	170		
160°F	160	70	71°C — Flash pasteurization of milk
	150		
	140	60	
131°F	130		55°C — Many enzymes inactivated
	120	50	
	110		
	100	40	
98.6°F	100		37°C — Human body temperature
	90	30	
	80		
68°F	70	20	20°C — Standard room temperature
	60		
	50	10	
	40		
32°F	30	0	0°C — Water freezes at standard temperature and pressure
	20		
	10	-10	
	0		
	-10	-20	
	-20	-30	
	-30		
	-40	-40	-40°, F° = C°

To convert temperature scales:

Fahrenheit to Celsius $°C = \dfrac{5}{9}(°F - 32)$

Celsius to Fahrenheit $°F = \dfrac{9}{5}(°C) + 32$

LENGTH CONVERSIONS

Centimeter-Inches

Centimeters	Inches
15	6
14	
13	5
12	
11	
10	4
9	
8	3
7	
6	
5	2
4	
3	1
2	
1	
0	0

THE NETWORK OF LIFE

Biology

SECOND EDITION

Sponsoring Editor: Liz Covello
Developmental Editor: Thom Moore, Jennifer Eddy
Project Editor: Anthony Calcara
Design Administrator: Jess Schaal
Text and Cover Design: Terri W. Ellerbach
Cover Photograph: John Giustina/The Wildlife Collection
Photo Research: Carol Parden
Production Administrator: Randee Wire
Project Coordination and Composition: Electronic Publishing Services Inc.
Printer and Binder: R.R. Donnelley & Sons Company
Cover Printer: Phoenix Color Corporation

On the Cover: The Slow Lorises depicted have been photographed in captivity. Once viewed as zoological gardens for the public display of animals, zoos are now recognized for helping to preserve species threatened with extinction. In addition to their involvement in conserving biodiversity, zoos are also contributing to the preservation of natural habitats of endangered species.

Biology: The Network of Life, Second Edition

HarperCollins® and ⬛® are registered trademarks of HarperCollins Publishers Inc.

Library of Congress Cataloging-in-Publication Data
Mix, Michael C.
   Biology : the network of life / Michael C. Mix, Paul Farber, Keith I. King. ~ 2nd ed.
   p. cm.
   Includes index.
   ISBN 0-673-52353-5
   1. Biology I. Farber, Paul Lawrence, 1944- . II. King, Keith I. III. Title.
   QH308.2M58   1996
574~dc20   95-35185
CIP

95 96 97 98   9 8 7 6 5 4 3 2 1

For information about any HarperCollins title, product, or resource, please visit our World Wide Web site at http://www.harpercollins.com/college

Dedication
The students we have had the good fortune to have in classes
during the past 25 years were the source of inspiration for this project.
To all biology students—past, present, and future—we dedicate this book.

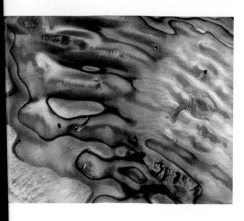

Contents in Detail

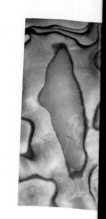

5   The Diversity and Classification of Life: Kingdoms Prokaryotae, Protoctista, and Fungi   80

6   The Diversity and Classification of Life: Kingdoms Plantae and Animalia   99

Unit II   The Interactions of Life   122

7   The Biosphere   124

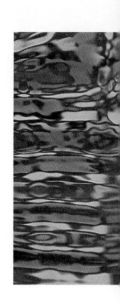

35 The Respiratory, Circulatory, Digestive, and Urinary Systems   656

36 Control and Regulation: The Nervous System   676

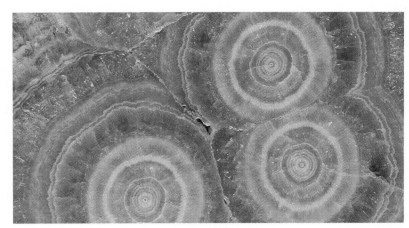

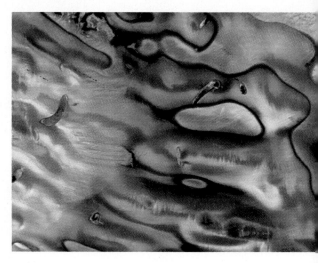

39 Defense Systems 737

40 Human Diseases 763

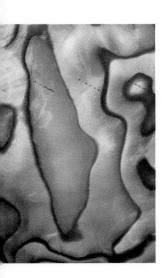

Preface

The study of biology continues to undergo profound changes as we approach the twenty-first century. Courses and textbooks that emphasize facts, vocabulary, isolated concepts, and content without context are being supplanted by new curricula designed to provide students with an understanding of the scientific enterprise and the excitement that characterizes studies of the living world. We wrote *Biology: The Network of Life* as a textbook to compliment this new way of teaching and learning. Since the publication of the first edition we have communicated with hundreds of colleagues. Their comments and suggestions validated the relevancy of our approach and aided us considerably in refining the second edition.

GOALS OF THE SECOND EDITION

We have pursued six goals in writing the second edition: They are:

■ To convey *fundamental knowledge* of the major principles, concepts, and ideas of modern biology.

■ To explain principles, concepts, and ideas related to biology within a framework of the *process of science*. Accurate historical information is an essential component of the approach.

■ To promote an understanding of *scientific inquiry*, the methods, approaches, and techniques scientists use to answer interesting questions.

■ To present *content in context* to facilitate learning.

■ To concentrate on *relevant information* that has significance, now and in the future.

■ To provide students with a background that will encourage *lifetime learning* and ongoing interest in biology.

THE PROCESS OF SCIENCE

The framework of *Biology: The Network of Life* is that scientific information is presented within a context of the *process of science*. The process of science considers the total dynamic enterprise of science. It encompasses questions that biologists address, the origins of those questions, the methods used to tackle the questions, the answers achieved, the resulting effect on the general picture of nature, the directions of new research, the institutions in which all the work is performed and the social context in which questions arise, are investigated, and are funded.

Central to the process of science are questions that motivate *scientific inquiry*. Scientists work hard to answer important questions, using diverse methods, techniques, and technologies. They also raise new questions, often as a consequence of answers obtained to old questions, or as a result of new, related observations, technological advances, or relevant information from another field. In some cases, questions that drive research are answered quickly, within a few months or years. Some questions, however, were posed by our ancient ancestors and, to this day, have not been answered completely. Early Greek philosophers first asked, "Why do offspring resemble their parents?" Although we now have a reasonably solid answer to that question, it is not yet complete because molecular geneticists continue to make new discoveries that apply to the original, ancient question. Finally, some questions, after driving research for a period of time, turn out to be dead ends. For example, in 1628, William Harvey published a book in which he accurately described the motion of the heart and the circulation of blood in humans. Soon after this a new question arose: "Do plants have hearts?" Although this question stimulated research for many years and led to a better understanding of plant structure, no one found a plant heart, and scientists turned to other questions. Thus science can be thought of as a *lineage of questions* in which the search for answers to one question or a set of questions leads to a new question or even a whole new set.

Lineage of Questions

The concept of lineage of questions reflects the dynamic nature of biology. It is often difficult to grasp the ever-changing nature of biology because of the enormous amount of information being generated. However, historians have shown that science is constantly changing. This feature of science is often confusing to students, who may believe that change means that earlier conceptions were wrong. In fact, change is consistent with the dynamic nature of science. We emphasize the concept of the lineage of questions so that it is integral to the framework of the second edition.

Each unit of *Biology: The Network of Life* 2nd edition opens with a lineage of questions that serves as an organizational structure to enhance teaching and learning. Within each unit we examine the origin of significant questions, describe how they are derived from earlier questions, consider various methods used to answer them, and discuss the social context in which they were studied. This approach allows students to understand that the search for knowledge in science is an interesting, dynamic process, not a rigid procedure as described by the mythical scientific method.

CONTENT IN CONTEXT

To maintain our focus on scientific process, facilitate learning of fundamental principles and ideas, and ensure our concentration on relevant information, all content is presented *in context*. This can mean different things within the organization of the text. Within chapters, specific topics are generally introduced where there is a *need to know*. For example, detailed discussion of proteins, enzymes, and nucleic acids appears in the fourth chapter of our genetics unit, Chapter 17, where we consider the discovery of DNA as being the genetic material. For five of the six units in the book, material is presented in the framework of scientific process. However, Unit I differs from the other units because time is used as the context for presenting information about origins of the universe, Earth, and life.

COVERAGE AND ORGANIZATION

Biology: The Network of Life 2nd edition consists of six units, each of which deals with major areas of contemporary importance. We present information, ideas, and concepts in the approximate sequence in which they originated or became known to scientists. This sequence reflects the actual development of science and allows students to follow the progression of scientific ideas that have led to what we currently know. It also gives a sense of the direction of future research in the life sciences. By emphasizing the historical development of

knowledge, we can better understand the process of science and social context in which science operates.

Unit I, *The Sphere of Life*, deals with the origins of the universe, Earth, and the life that came to exist on Earth. Starting with the mythical world views of our ancient ancestors, this unit traces scientific thought on the origins of the universe and Earth. Current hypotheses on the origin of life and the early evolution of cells are then discussed. The nature and functions of cells and their ability to transform energy set the stage for a survey of the tremendous diversity of life that exists on Earth.

Unit II, *The Interactions of Life*, focuses on our environment, knowledge of the global distribution of life, the interactions that occur among plants and animals, and the impact that humans have had and are having on our environment. It begins by explaining the physical forces that act on earth, world climate, and the development of biomes. The unit then traces the roots of modern ecology and explains its major concepts. Separate chapters focus on our knowledge of terrestrial and aquatic ecosystems. The growth and impact of the human population and an up-to-date chapter on global climate change end the unit.

The detailed level of understanding we now have in genetics represents one of the most stunning breakthroughs in biology. New genetic technologies are revolutionizing medicine and agriculture, and these changes raise controversial social questions. In Unit III, *The Language of Life*, we follow the entire development of the field of genetics from Aristotle to the present. While exploring some of the elegant experiments that led to our current knowledge, students will experience the ingenuity and power of critical thinking applied to answering interesting scientific questions.

Unit IV, *The Evolution of Life*, explains the central unifying idea of modern biology. We trace the origin of the theory of evolution and show how the theory arose to answer a set of interesting scientific questions. The nature of scientific theories is discussed along with the relationship between scientific theory and religious thought. Following the development of the theory of evolution to the present, students can appreciate how the original theory has expanded and how it relates to the various fields of biology. A chapter on human evolution discusses current hypotheses of human origins, and two chapters on animal behavior explore the evolutionary significance of this subject.

Plants are discussed throughout the entire text, but Unit V, *Plant Systems of Life*, examines specific features of plants. We explain why, until the nineteenth century, the study of plants was limited to identification and description. Modern knowledge of plant life is described by following the general life history of a plant from seed to adult. Current research on how plants handle environmental stress and on plant adaptation is included.

The concluding unit, Unit VI, *Human Systems of Life*, explores the area of biology that currently commands the greatest amount of research funding worldwide. This unit has increased coverage of human body systems, especially in Chapters 34 and 35. Individual chapters trace the growth of our knowledge of different human organ systems and describe the frontiers of research. A chapter on human defense systems leads into an examination of human disease, both infectious and noninfectious, and discusses current ideas about cancer, vascular disease, AIDS, risk factors, and lifestyle management, along with the future of biology.

Distinguishing Features of the Second Edition

Besides the content changes to the second edition, there are new features to help you learn biology.

Lineage of Questions and Unit Conclusions. The lineage of questions summarize the significant questions scientists have investigated over time; they appear on the first page of each unit to set the stage for the topics to be studied. The unit conclusions appear on the final page of each unit and relate what was studied to the process of science.

Chapter Outlines open each chapter to establish a conceptual framework for that chapter.

Chemistry in Context. An important part of our approach is to integrate the coverage of chemistry exactly when and where it is needed to understand biology concepts. To that end, basic chemistry concepts are introduced in Chapter 2 in a setting of chemical element origin and universe evolution. Chapter 3 introduces energy transformation processes in relation to cell evolution. A new Chapter 4 expands on the background given in Chapter 3 by including such topics as forms of energy in the physical and biological realms, energy-managing molecules, photosynthesis, glycolysis and cellular respiration, the process of chemiosmosis, and the significance of biological energy transformations in the history of life on Earth. A new appendix, Chemistry in the Biological Context, is a self-contained resource and reference tool for students.

"Focus on Scientific Process" features are stimulating essays within each chapter. These essays might describe a classic experiment, recount the development of a major new theory or the rejection of an old one, explain the use of technology in answering old and new questions, describe complex ecological research, or interpret the significance of certain biological processes in understanding the history of life. An index of all the Focus features follows the table of contents.

"Before You Go On" is an in-text summary that highlights key ideas; it allows students to review their comprehension of the main points of the chapters as they read the text.

The Illustration Program is revised to enhance the visual aspect of learning biology. Nearly 60 percent of the line art and photographs are new to this second edition. Color is used consistently to indicate chemical, anatomical, and physiological detail.

Figures with Questions. Many of the figures in the text include a question or two about the figure. These questions allow students to test themselves and use their critical thinking skills to draw conclusions. Answers to the Figure Questions are at the end of the chapter for convenient reference.

Chapter-End Material includes the following elements to help students study the text:

Summaries recap the salient principles and concepts presented in each chapter. **Working Vocabulary** lists key terms new to the text discussion. **Review, Essay, and Discussion Questions** help students synthesize the information discussed. **Reference and Recommended Readings** indicate where students can look for more information on chapter topics.

The Supplement Package for Biology: The Network of Life Second Edition

The text is supported with an extensive package of supplements for both the instructor and the student.

For the Instructor

The ***Instructor's Manual with Activities*** offers instructors valuable support in preparing lectures for their introductory biology course. For each chapter there are tips on teaching science, discussion questions, problem-solving exercises, suggested references for print and media, and demonstration ideas. The manual also contains 40 laboratory activities that are linked to the text.

The printed ***Test Bank*** by Lesley Blair of Oregon State University, consists of 3000 multiple-choice, true or false, matching, short answer, and essay questions that are new to the second edition.

TestMaster software in IBM and Macintosh formats is available to adopters who prefer a computerized testing system. *TestMaster* enables instructors to select problems for any chapter, scramble them as desired, or create new questions. New with this edition, ***QuizMaster*** coordinates with the *TestMaster* program. *QuizMaster* allows students to take tests at the computer. Upon completion, a student can evaluate his or her test score and view or print a diagnostic report that lists the topics or objectives that need further study. With network access to *QuizMaster*, student scores are saved on disk and instructors can use the utility program to view records and print reports for individual students, class sections, and entire courses.

Overhead Transparencies of over 250 four-color figures cover key topics in the introductory biology course. These full-color illustrations are taken from the art program of the text as well as from other sources; the transparencies are available to adopters.

The ***HarperCollins Encyclopedia of Biology*** videodisc is an exciting way to integrate multimedia presentations into your classroom. Loaded with still images, motion footage, and animations of key biological concepts, the disc allows students to see complex biological topics in ways that are impossible to present with two-dimensional illustrations. A bar code manual accompanies the disc along with software drivers for both IBM and Macintosh computers. A guide for integrating the videodisc with the text is also available to adopters.

For the Student

The *Study Guide*, by Amy Hacker of Oregon State University and Mette Hansen, includes chapter outlines, a summary, and key terms and concepts. Chapter Challenge sections give student study tips for tough topics in biology. There are practice tests with short-answer, multiple-choice, and critical thinking questions so students can test themselves before exam time. To order use ISBN 0-673-52355-1.

Essays on Wellness, by Barbara A. Brehm of Smith College, is a collection of 29 essays that cover a variety of human health issues. These essays can supplement Unit VI of the text. To order use ISBN 0-06-501549-5.

The *Harper Dictionary of Biology* is a valuable reference tool. By W. G. Hale and J. P. Margham, both professors of biology at the Liverpool Polytechnical Institute, it covers all the major subjects—anatomy, biochemistry, ecology, evolutionary theories—plus has biographies of important biologists. It contains 5600 entries that provide in-depth explanations and examples. Diagrams illustrate such concepts as genetic organization, plant structure, and human physiology. To order use ISBN 0-06-461015-2.

Writing About Biology, by Jan A. Pechenik of Tufts University, is a brief guide that prepares students to meet the demands of writing at all levels of biology. Every aspect is covered: laboratory reports, research proposals, research papers, essay exams, oral presentations, and applications for jobs or graduate schools. To order use ISBN 0-673-52128-1.

Studying for Biology, by Anton E. Lawson of Arizona State University, details ways to improve basic study skills. Several chapters cover the thinking patterns that biologists use to answer questions and formulate hypotheses and theories. The book also introduces the basic postulates of the major theories in biology. To order use ISBN 0-06-50065-X.

The Biology Coloring Book is part of new approach to learning biology. Each page shows a biologic process or concept. You create the color key and color in each part of the plate. As you color, you can read an explanation about each element in the accompanying text. When completed, the colored plates are an excellent tool for reviewing what you've learned. To order use ISBN 0-06-460307-5. Other books in the series include the *Anatomy Coloring Book* (0-06-455016-8), *Physiology Coloring Book* (0-06-043479-1), *Botany Coloring Book* (0-06-460302-4), and *Zoology Coloring Book* (0-06-460301-6).

The Illustrated Five Kingdoms: A Guide to the Diversity of Life on Earth is by Lynn Margulis, University of Massachusetts, Amherst; Karlene V. Schwartz, University of Massachusetts, Boston; and Michael Dolan, University of Massachusetts, Amherst. It contains full-page, unlabeled drawings of examples of each of the major groups (phyla) illustrated in their natural habitats. Each full-page drawing is shown in a reduced size with labels for reference. Accompanying text introduces and describes both familiar and unfamiliar plants, animals, and microorganisms. To order use ISBN 0-06-500843-X.

Acknowledgments

This book grew out of our 25 years of teaching university biology and history of science courses. Along the way we have been aided by numerous people from many institutions. A comprehensive list of these friends would be unreasonably long and unavoidably incomplete; nevertheless, we acknowledge the contributions of all of our esteemed teaching colleagues.

Bonnie Roesch was the first to recognize the novel dimensions of our project, and she has been a tireless supporter. She has always been there when we needed support and her continuous encouragement has inspired us; we will always feel indebted to her. Glyn Davies made the first edition of this book possible and played an important role in producing this second edition. He expanded our concept of what a text can and should be, and he helped define our dream for the second edition. We are also grateful to Ed Moura and Liz Covello for their advice and suggestions for improving this book. The efforts of Rebecca Strehlow, our developmental editor in the first edition, led to positive results that are still evident in this second edition. Thom Moore was developmental editor for the second edition, and he was a superb taskmaster. The meticulous attention he has given the manuscript is evident in every page. He also brought a vision, a wealth of new ideas, and a remarkable sense of humor that were incredibly stimulating to all of us.

Many other people made significant contributions to the second edition. Molly Bloomfield, Joel Hagen, Richard Halse, John Matsui, and Patricia Muir all wrote initial drafts of material used in various chapters. Jacqueline Webb, Villanova University, provided invaluable assistance in helping us evaluate art and figures. Administrators at Oregon State have strongly supported our ideas for developing a contemporary introductory biology course. Fred Horne, Dean of the College of Science, continues to be especially helpful and encouraging. Laura Mix Kohut generously provided us with legal assistance.

Several distinguished scientists and scholars graciously answered our questions about how they would like to see biology develop in the coming decades. We thank Ernst Mayr, Ledyard Stebbins, David Hull, E. O. Wilson, Marvin Druger, and the late Linus Pauling for their time and thoughts. We especially appreciate the time Professor Pauling spent reviewing our first edition.

We recognize our spouses, Marilyn Henderson, Vreneli Farber, and Roberta King. They tolerated unreasonably long hours of our absence, yet were unfailing in their encouragement and support.

Finally, we are particularly indebted to the numerous reviewers whose comments, suggestions, and insights have substantially enhanced the revision of this text. Many instructors reviewed segments of the manuscript. Others completed surveys that helped us shape the revision. We sincerely thank them for their outstanding contributions to the second edition of *Biology: The Network of Life*: Douglas Allchin, University of Texas, El Paso; David W. Aldridge, North Carolina A&T University; Jane M. Beiswenger, University of Wyoming; Robert Bergad, University of St. Thomas; Robert Barkman, Springfield College; Susan H. Brawley, University of Maine; George Damoff, Liberty University; Roger M. Davis, University of Maryland, Baltimore County; Nathan Dubowsky, Westchester Community College; Bruce W. Elliot, Jr., Boston University; Richard T. Fraga, Lane Community College; Bernard Goldstein, San Francisco State University; Elliot S. Goldstein, Arizona State University; Joel B. Hagen, Radford University; Heather Hall, Montgomery College-Takoma Park; Richard D. Harrington, Rivier College; Linda Hunt, University of Notre Dame; Murray Jensen, University of Minnesota; Florence L. Juillerat, Indiana University-Purdue University at Indianapolis; LeAnn Kirkpatrick, Glendale Community College; Tyjuanna LeBennet, Ball State University; Peter Lardner, Flagler College; Jane Maienschein, Arizona State University; Peter May, Stetson University; James Mickles, North Carolina State University; Michael O'Donnel, Trinity College; Joel Ostroff, Brevard Community College; Roland Prudhon, Shelby State Community College; Nancy Raffetto, University of Wisconsin, Madison; Michael D. Rourke, Bakersfield College; Paul Sager, University of Wisconsin, Green Bay; J. S. Shipman, Providence College; Curtis Swanson, Wayne State University; Robin W. Tyser, University of Wisconsin, LaCrosse; Lance Urven, University of Wisconsin, Whitewater; Steven P. Vives, Georgia Southern University; Lauren E. Wentz, University of Wisconsin, Whitewater; Ann Zayaitz, Kutztown University.

Through interviews and telephone calls, we canvassed professors about their opinions about introductory biology courses and about our manuscript. Thank you for sharing your thoughts and time with us. **HarperCollins Partnership Focus Groups**: Frank Awbrey, San Diego State University; Moonyearn Brower, Armstrong State College; Virginia Buckner, Johnson County Community College; Rush Buskirk, University of Texas at Austin; Rebecca Cook, University of Tennessee; Richard T. Fraga, Lane Community College; David Futch, San Diego State University; Durwynne Hsieh, Los Medanos College; Tom Kane, University of Cincinnati; LeAnn Kirkpatrick, Glendale Community College; Derek Maden, Modesto Junior College; Keith Morrill, South Dakota State University; Becky Moulton, Georgia Southern University; Jal Parakh, Western Washington University; Gary Peterson, South Dakota State University; Oscar Pung, Georgia Southern University; Edward Southwick, SUNY College at Brockport; Bill Thwaites, San Diego State University; Nels Troelstrup, South Dakota State University; Steven Vivas, Georgia Southern University; Edward Weiss, Christopher Newport University.

We also thank the following instructors who responded to a market survey for the revision of this second edition. **Survey Respondents**: Stephen Bernstein, University of Colorado-Boulder; Virginia Buckner, Johnson County Community College; Roger Christianson, Southern Oregon State University; Kevin Conway, The American University; Claire Cronmiller, University of Virginia; Michael Cummings, University of Illinois at Chicago; Rowland Fergusson, University of the District of Columbia; Bill Frase, University of Cincinnati; Joseph Glass, Camden County College; William Glider, University of Nebraska at Lincoln; Donna Harpold, Virginia Western Community College; Jean Helgeson, Collin County Community College; Kent Hodson, Santa Monica College; Ann Hooke, Miami University of Ohio; Elizabeth Johnson, Evergreen Valley College; Tom Kantz, California State University-Sacramento; Douglas Morrison, Rutgers University-Newark; Lee Neary, Joliet Junior College; Francis Raleigh, Saint Peter's College; Alison Roberts, University of Rhode Island; Katya Yarosevich, Butte College.

We sincerely hope that all the readers of this textbook have a rewarding experience. Please send us your comments or questions about *Biology: The Network of Life* 2nd edition. Send us letters in care of HarperCollins*CollegePublishers* or feel free to connect with us through the Internet.

Michael C. Mix   mixm@bcc.orst.edu
Paul Farber   farberp@cla.orst.edu
Keith I. King   kingk@css.orst.edu

Studying the Living World

Chapter Outline

In 1995, the U.S. government spent over $10 billion to support research in the biological sciences, and more than 200,000 biological scientists were employed in the United States. You and hundreds of thousands of other students are required to study biology in school. Why all of these big numbers? The most fundamental reason is that biology is concerned with important questions about the living world that have immediate and practical implications for solving major problems that we currently face. What can we do about environmental degradation? AIDS? Threatened species? Overpopulation? Genetic disorders (see Figure 1.1, page 2)? Rarely does a day go by that you do not encounter some discussion of a biological issue in the media. Biology is a subject that concerns all of us, and it is relevant for understanding ourselves and making decisions about our lives.

THE PROCESS OF SCIENCE

What is *science?* General descriptions of the nature of science often begin with the so-called scientific method, a standardized procedure (observe, hypothesize, test, conclude) that is supposed to explain how scientists investigate problems. However, the picture portrayed by the scientific method is misleading. Careful assessments of the activities of scientists reveal that there is no single scientific method in biology or in any other science. Today, philosophers, scientists, and historians generally agree that the scientific method model not only fails to portray science accurately but actually gives the false impression that scientific research is based on a plodding step-by-step procedure designed to generate new "facts." It also fails to convey the excitement that characterizes science and scientific studies (see the **Focus on Scientific Process**: Past and Future).

Science as a Lineage of Questions

A more fruitful way of depicting the nature of biology is to consider the total dynamic enterprise of science, an approach known as the *process of science.* The process of science encompasses questions that biologists address, the origins of those questions, the techniques and methods used to tackle the questions, the answers achieved, the resulting effect on the general picture of nature, the directions of new research, the institutions in which all the work is performed, and the social context in which questions arise, are investigated, and are funded.

Central to scientific process are questions that motivate scientific inquiry. Scientists work hard to answer important questions, and they use diverse methods, techniques, and technologies to do so. They also raise new questions, often as a consequence of answers obtained to old questions or as a result of new and related observations, technological advances, or relevant information from another field. In some cases, questions

Figure 1.1   Biology is an exciting subject that is concerned with important questions about the living world. These questions—and their answers—have immediate and practical implications for solving major problems that we currently face. (A) Worker's cleaning up an oil spill on Texas' San Jacinto River. (B) AIDS quilts in Washington, D.C. (C) An endangered species, the spotted owl. (D) Densely populated Bangalore, India. (E) An individual with Down Syndrome.

that drive research are answered quickly, within months or a few years. However, some questions—even some that were posed by our ancient ancestors—have not yet been answered completely. The question "Why do offspring resemble their parents?" was first asked by the early Greeks. Although we now have a solid answer to that question, it is not yet complete because molecular geneticists continue to make new discoveries that apply to the original, ancient question. Finally, some questions, after driving research for a period of time, turn out to be dead ends. For example, in 1628, William Harvey published a famous book in which he accurately described the motion of the heart and the circulation of blood in humans (dis-

cussed in Chapter 35). Soon after, a new question arose—"Do plants have a heart?"—that stimulated research for years and led to a better understanding of plant structure. However, no plant heart was found, and scientists turned to more fruitful questions. Thus science can be thought of as a *lineage of questions* in which the search for answers to one question or set of questions leads to a new question or set.

The lineage-of-questions concept also reflects the dynamic nature of biology. It is often difficult to appreciate the changing nature of biology because of the enormous amount of information being generated. However, historians have shown that science is constantly changing. This feature of science is often

Past and Future

One of the major features of *Biology: The Network of Life* is that scientific information is presented within the context of the process of science. We examine the origin of significant questions, trace how they grew out of earlier questions, explore various methods used to answer them, and consider the social context in which they were studied. This approach reflects the dynamic nature of the search for knowledge in science.

Many other biology textbooks have been criticized for presenting information in a way that fails to teach effectively about science, the work of scientists, and the nature of the scientific enterprise. There are significant differences between learning about the living world from a process-of-science perspective and from older, more restrictive, less accurate approaches. This problem is associated with the two issues that are discussed here.

The "Scientific Method"

Countless textbooks describe what has become known as **the** *scientific method.* In presenting this method, authors usually imply that scientists solve problems according to some variation of the following sequence: observe, formulate a hypothesis, test the hypothesis, and conclude or generalize. But do biologists actually follow such a rigid procedure in answering questions or generating new knowledge? Although scientists certainly make use of observations, hypotheses, and tests, the scientific method does not accurately describe science, scientific research, or the creative endeavors of scientists.

To understand how biologists approach problems, it is important to recognize that *no single method* is used by scientists to obtain knowledge of the natural world. As you will discover in reading this text, scientists use a variety of methods to explore nature. They choose methods that are appropriate to the question they are asking. Indeed, one of the most important aspects of scientific research is determining the best method for investigating a specific problem. Scientists make observations, compare them, create hypotheses, conduct experiments, generalize, and formulate laws and theories. They also use a variety of mathematical data analyses, work on computers to generate images, and construct various types of models. But there is no single scientific method that can be used in cookbook fashion to tell us about the natural world. Science is a much more creative enterprise.

Why is the scientific method included in so many science textbooks? Early in the twentieth century, a group of philosophers, known today as *logical empiricists,* believed that scientific knowledge was fundamentally different from other types of knowledge. These philosophers concentrated mostly on analyzing the logical structure of scientific explanations, and they devised numerous abstract formulations to describe those arrangements. In the 1960s, several philosophers attempted to codify a procedure that scientists could use to generate scientific explanations. The most famous of these analysts were Carl Hempel of Princeton University and Karl Popper of the London School of Economics. Both believed that the formulating and testing of hypotheses was at the heart of scientific investigation. Hypotheses attempt to explain interesting observations or answer scientific questions. The method of science, in the view of these philosophers, was to test predictions made by a hypothesis. For example, if we observed that after eating in the same restaurant, several individuals became infected with a bacterium that causes severe intestinal distress, we might hypothesize that the restaurant had served them food contaminated with the bacterium. We might then predict that bacteria-contaminated food could be found in the restaurant that day. If contaminated food was indeed found, we would probably conclude that the hypothesis was correct or had a high level of certainty.

Formulating and testing hypotheses is an important activity of scientists, but it is not the only method used in investigating nature. Some scientists spend a lifetime compiling observations, others compare observations and draw conclusions, and still others apply knowledge from one area to another.

Scientists and philosophers have agreed for many years that a picture of formulating and testing hypotheses does not accurately describe science or the work of scientists. Unfortunately, the scientific method formula was simple and easy to memorize, and it was used by science writers who were looking for an elementary way to characterize science. The "hypothetico-deductive model," as it was called, soon took on a life of its own. Because it contained elements of methods used by scientists, it had a certain appeal. However, it has not been effective in teaching students about the work of scientists or the scientific enterprise. In the worst case, the scientific method conveys an impression that scientists use a simple recipe to concoct new facts for students to memorize.

As described in this chapter, a more accurate and more useful approach for learning about science is identified as the *process of science,* which considers questions that scientists ask, the origins

box continues

FOCUS ON SCIENTIFIC PROCESS

of those questions, methods used to answer the questions, the answers achieved, the resulting effect on the general picture of nature, the directions of new research, the institutions in which all the work is performed, and the social context in which questions arise, are investigated, and are funded.

Interest in the process of science also dates from the 1960s. Thomas Kuhn, a science historian now at the Massachusetts Institute of Technology, played a key role by calling attention to the importance of understanding that *science changes.* Historians and philosophers came to recognize the underlying complexities inherent in the scientific enterprise and began to refer to scientific change and the factors responsible for change as the "process of science." During the past two decades, detailed research has convinced historians, philosophers, science educators, and scientists the value of understanding how and why science changes. Obviously, the process of science is not a simple, step-by-step method used to describe science, but it presents a far more accurate and more exciting picture of the scientific enterprise.

"Textbook History of Science"

Since at least the eighteenth century, science textbook writers have traditionally introduced a subject with a short "historical sketch." Professional historians usually cringe when they read these introductory essays and refer to them contemptuously as "textbook history." What are the principal flaws of textbook history? It is often inaccurate and usually irrelevant, overlooking what was truly of historical significance. Textbook history typically suffers two major flaws: descriptions of precursors and constructions of scientific myths.

Textbooks often contain a few brief vignettes about *precursors*—individuals who are alleged to have had ideas similar to those we hold today. However, when the actual writings of these individuals are analyzed carefully, they are found to be superficially similar but quite different from ideas that followed. Moreover, there are usually no actual links between precursors and later scientists; that is, the purported forerunners had no real influence on the development of significant ideas. Further, a description of precursors does not illuminate the process of science; instead, it clouds our understanding of how science progresses.

A typical example of a set of precursors is often included in "historical" discussions of the theory of evolution. In such accounts, a few early individuals are described, such as the fifth-century B.C. Greek philosopher Empedocles, the early-eighteenth-century French naturalist Buffon, and the late-eighteenth-century naturalist Lamarck. Of these three, only Lamarck had a theory of evolution, but his incorrect theory was significantly different from Darwin's, and his ideas had little impact—in fact, Darwin strongly resented any association of his ideas with those of Lamarck! The history of the theory of evolution is more complicated than such a simplistic and misleading set of precursor sketches implies.

Equally problematic is the creation of *scientific myths:* the presentation, in chronological order, of a group of historical figures—often famous individuals sprinkled among a few obscure names—as parts of a continuing tradition. Careful examination reveals that these stories actually represent bogus constructions. They often turn out to be characters from different stories that have been artificially grafted together to create a sort of fable. Historians object strongly to these creative versions of history because they miss the true and very interesting story of how science operates and changes. "Histories" of genetics that link Aristotle, Albert the Great, Leeuwenhoek, Spallanzani, Mendel, and Pasteur are typical of this genre. All of these individuals wrote about heredity in one way or another, but they do not form a tradition, and the context of their ideas was totally different.

Biology: The Network of Life surveys knowledge of the living world within a new framework—the one that historians and philosophers of science refer to as the process of science. Instead of sustaining old ideas about science or creating myths, we trace the process by which we arrived at our present understanding of life and attempt to explain where that process may take us as we move into the twenty-first century.

confusing to students, who may believe that change means that earlier conceptions were wrong. In fact, change is consistent with the dynamic nature of science.

Scientific Inquiry

How do biologists answer interesting questions? Like all scientists, biologists use techniques and methods that are appropriate for answering the question at hand. However, certain general features of scientific research constitute a powerful approach for learning about the natural world.

At the most basic level, biologists *observe* nature, directly or with instruments, in order to answer certain questions (see Figure 1.2). Early naturalists (people who studied nature) simply wanted to know which plants and animals were present in different parts of the world. They were also interested in similarities that existed among plants and among animals, and this led to some careful *comparative* studies.

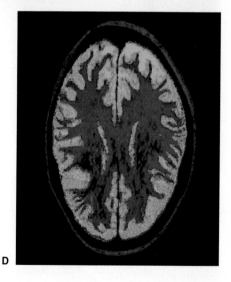

Figure 1.2 Progress in biology has been increasing rapidly, partly because advanced technologies have led to the development of new, powerful tools. (A) An electron microscope is used to magnify materials over 100,000 times. Electron micrographs (EMs) are photographs of such magnified objects, and many of them appear in this book. (B) Huge machines are now used in making products from genes, bits of genetic material that dictate structure and function in all organisms. (C) Special techniques allow biologists to determine the effects of pollutants on plants in their natural environment; here we see a normal leaf (left) and two leaves damaged by air pollution (right). (D) Sophisticated instruments such as a magnetic resonance imaging (MRI) scanner are used to create vivid images of human organs. MRI images of the brain are used to evaluate normal and abnormal functions of this incredibly complex organ.

Today, biologists ask more complex questions: What would happen to plants and animals in the Appalachian Mountains if the temperature increased by a few degrees? Will a person eventually develop Alzheimer's disease if his or her parents or grandparents had it? Can AIDS be cured? To answer such questions, scientists study nature in search of regularities, mechanisms, and relationships. They use logical arguments to integrate particular phenomena into more general patterns. Sometimes, when scientists cannot explain their observations or do not understand or cannot interpret well-established patterns or changes in established patterns, they develop **hypotheses**—tentative explanations derived from observations and sup-

ported by results from experiments or other evidence—that may lead to an answer. Hypotheses are important in science because of two key features: (1) they can be tested in some way to determine whether or not they are accurate, and (2) they can be used to make predictions.

Testing a hypothesis may be done by conducting a simple experiment or it may involve multifaceted studies carried out over a long period of time. For example, in recent years, many scientists have voiced concern over what they believe to be a worldwide decline in amphibian populations (see Figure 1.3). This has been a difficult hypothesis to test because there is little information on amphibian population sizes around the world. Studies are underway in some areas to determine the distribution and abundance of amphibian populations. The data collected will be compared with earlier data sets (where available) and will also provide a baseline for future comparisons. In some cases, results indicate that amphibian populations are indeed declining, and a new question has been asked: what is causing the population to decline? Some data suggest that many of the sites where amphibian declines may be occurring are at high altitudes; scientists have hypothesized that exposure to increased levels of ultraviolet (UV) light may be responsible for local population declines. Ultraviolet light is harmful to living organisms, and UV radiation is more intense at higher elevations than at lower elevations. Recent data from field studies have also shown that amphibian eggs exposed to UV light were less likely to survive and develop normally compared to unexposed eggs. Much research will be required to test the accuracy of this UV hypothesis.

Unifying Ideas in Biology

Ultimately, most hypotheses are confirmed or rejected after appropriate research. In either case, scientists usually ask new questions as a result of their investigations. Equally important, biologists use their answers to relate individual bits of information about organisms to larger bodies of information and to more general ideas about life. Regularities that are observed continually are called **laws**—invariable relationships among objects, given a specific set of conditions. On the broadest level, biologists formulate scientific **theories**—systematic sets of concepts that relate data, explain why the world is as it is, and serve to guide future research.

The *theory of evolution* constitutes the most general framework in which all life can be understood. Many similarities among modern organisms are reflections of a shared ancestry. So, too, are the comparable physical characteristics of living organisms and fossil remains of extinct organisms (see Figure 1.4). The astonishing diversity of life on Earth today is a product of evolutionary forces that molded the appearance of populations throughout millions of years. Even the patterns of geographical distribution among plants and animals can be understood through knowledge of their evolutionary history.

Other unifying ideas in biology relate large bodies of information, and we will explore them in later chapters. Ecology,

A   B

Figure 1.3 In recent years, scientists have voiced concern over what they believe to be a worldwide decline in amphibian populations, including certain species of frog (A) and salamanders (B). This hypothesis has been difficult to test because there is relatively little information on amphibian populations around the world.

Figure 1.4 In various places can be found the remains of extinct fossils that are similar to contemporary forms. In this photo, it can be seen that fossil scallops resemble modern scallops.

genetics, cell biology, and molecular biology all focus on fundamental aspects of life. These fields are based on broad unifying ideas that are compatible with evolution and can be interpreted in evolutionary terms. Nevertheless, biology is far from a completely systematic and totally integrated science. It is not yet clear how certain bodies of knowledge relate exactly to other collections of information about life. In this sense, biology is no different from most other sciences, such as physics or chemistry.

Institutions of Biology

The process of science also involves the institutional settings in which scientists work. Because of the wide range of subjects studied, biologists are today found everywhere in the world and in all social institutions. Yet less than a century ago, before biology became a significant field of science and before governments began to support biological research on a large scale, examinations of the living world were undertaken by a relatively small number of people, and those few who did investigate the living world were typically found in one of three places: the museum, the field, or the laboratory.

Natural history was primarily a museum subject in the eighteenth and nineteenth centuries. Preserved specimens were examined with the goal of classifying and describing them. Relatively few individuals studied organisms in their natural habitats. They either collected specimens to be sent back to museums or, more rarely, made field observations of organisms' behavior, life stages, or distribution. Governments supported such work on a limited scale, mostly in cases where it held potential commercial value. For example, in the nineteenth century, the British were interested in making observations on plants that could be raised on plantations and then exported from British colonies in Asia. For the most part, however, natural history was an amateur undertaking; that is, it was not done for pay but rather for love of the subject.

Laboratory work in the nineteenth century was mostly performed by medical researchers. Because it was impractical to use human subjects, animals were often studied. Claude Bernard, one of the greatest researchers of the nineteenth century, conducted experiments on animals. At that time, society was less sensitive to animal suffering, and many of Bernard's experiments were, from a modern point of view, rather appalling and probably could have been done with considerably less discomfort to the experimental subjects. Nevertheless, such studies permitted Bernard and other researchers to make great progress in learning how the animal body functions. Agricultural research also yielded knowledge about both plants and animals, especially in the areas of heredity, growth, and nutrition.

Given the large number of biologists today, it is difficult to appreciate that until recently, a comparatively small group of people studied biological sciences. For example, in the eighteenth century, individuals who were seriously interested in natural history were able to correspond with every other such individual in the world! Today, it is almost impossible for a biologist to maintain communication with all specialists in any given subdiscipline, such as AIDS research, let alone with all other biologists.

The explosion of biological research is a twentieth-century phenomenon. Universities and other centers of higher learning have extensive research faculties; the U.S. government maintains large research operations (the National Institutes of Health alone employ more biological scientists than the entire English-speaking world had at any one time in the nineteenth century); and governmental bodies, such as the Environmental Protection Agency, fund extensive research programs. Private foundations, such as the Salk Institute and numerous corporations and businesses, are also major contributors to our growing knowledge of the living world.

Social Context of Science

In considering the institutions of science, it is apparent that extensive funding has been necessary to acquire the knowledge of biology that we have today. Science is obviously influenced by society and exerts an influence on it. For example, two of the most compelling questions that drive research today, in terms of money spent and numbers of researchers involved, are linked with problems that concern many citizens: how can complex diseases such as AIDS and cancer be treated or cured, and how can the quality of our environment be maintained or improved? These two issues drive gigantic research programs in the United States and in other countries.

In later chapters, we will see that the influence of society on science may be subtle or not so subtle. For example, nineteenth-century theories about human intelligence stressed the inferiority of females, children, and "nonwhite races" compared to white males. Serious scientists of that period were convinced of the merit of these ideas and collected vast amounts of data on skull size, behavioral "observations," and reports from explorers. We now recognize that the social attitudes and values of that period influenced the observations and interpretations related to these ideas. Although scientists try to be completely objective, social opinions may color the ways in which they formulate questions and interpret information.

For this reason, many biologists today are eager to promote the increasing number of women and people of color in science (see Figure 1.5). A broader social range of scientists may help all of us become more aware of the social biases we bring to research.

THE ORIGIN OF BIOLOGY

The word *biology* combines the Greek words *bios* "life" and *logos* "word" or "study"); it literally means the "study of life." First used around 1800, the term referred to the study of func-

tions of living organisms, a field we now call *physiology.* **Biology** has since come to mean the "scientific study of life." Today, biologists study the functions of all living organisms, their appearance, their habitats, their interactions with one another and with the environment, and their changes over time. Although extremely broad, the study of biology does not incorporate everything known or written about living organisms because life can be viewed from numerous perspectives beyond the scientific. For example, artists and writers can express personal ideas about the beauty of organisms or about a landscape that are quite different from the ideas that a scientist considers.

How did biology arise as a scientific discipline? The general study of living beings traces its roots back to the ancient Greeks, when it was known as *natural history*. Aristotle, a famous Greek philosopher in the fourth century B.C., wrote a book titled *History of Animals* in which he described 500 types of animals and what he knew of them. Aristotle's book was one of the first classics of natural history.

The goals of natural history changed over many centuries. The Romans, whose civilization succeeded the Greeks in the second century B.C., were a practical people. Roman natural history writers like Pliny (see Figure 1.6), concentrated on such practical issues as the proper maintenance of horses, the recognition of medically valuable plants, and the treatment of domestic animals. In the Middle Ages—the period between the collapse of the Roman Empire in the fifth century A.D. and the Renaissance nine centuries later—natural history reflected the moral lessons of religious writers, as can be seen in Figure 1.7, which shows a page from a famous medieval work on animals.

A

B

Figure 1.5 Compare these pictures of scientific laboratories from (A) the 1950s and (B) the 1990s. Although white males have been responsible for a great deal of good science, there can be no doubt that science has benefited enormously from the ever-increasing diversity of the scientific community.

Figure 1.6   Pliny at work, as illustrated in a fifteenth-century manuscript.

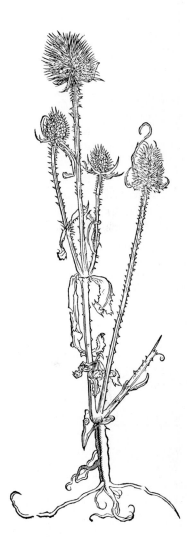

Figure 1.7   This woodcut, from a Renaissance-era edition (1587) of the *Physiologus,* a medieval collection of morality tales using animals to examine human behavior, shows vultures descending on a lone dead wolf in an open field. Wolves were particularly despised in early Europe. The moral lesson of this illustration revolves around the lonely and unattractive end that awaits humans who exhibit "wolfish" behavior.

After the Renaissance, the goals of natural history were to describe and classify all the "natural" products on Earth—all plants, animals, and minerals. Starting in the Renaissance, natural history books were illustrated, sometimes with magnificent and often fanciful woodcuts and later with beautiful and reasonably accurate ones such as that shown in Figure 1.8.

Although the ambitions of natural scientists to describe and classify natural products were important, modern biology focuses on many other fascinating questions about organisms. However, until recently, scientists in what were then separate disciplines, such as medicine and agriculture, made efforts to answer many of these questions. For example, until the nine-

Figure 1.8   Renaissance natural history books had beautiful and accurate illustrations, like this one of teasel (*Dipsacus fullonum*) from Brunfels's *Herbarum vivae eicones* (1530).

teenth century, questions about the structure and function of organisms were studied in medicine, not natural history. Early physicians made great contributions to what we now know as biology. Similarly, detailed examination of hereditary traits was of considerable interest to those who raised plants and animals for food or pleasure. These same people were also interested in plant and animal diseases and nutrition. Agriculture is thus also an important ancestor of modern biology.

BIOLOGY: THE NETWORK OF LIFE

Biology: The Network of Life consists of six units, each of which deals with major areas of contemporary importance. The general approach is to present information, ideas, and concepts in the approximate sequence in which they originated or became known to scientists. This sequence reflects the actual development of science and allows you to follow the progression of scientific ideas that have led to our current state of knowledge. It will also give you a sense of directions to be taken in future research. By emphasizing the historical development of knowledge, we can better understand the process of science and social context in which science operates.

Content

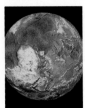

Unit I, "The Sphere of Life," deals with the origins of the universe, Earth, and the life that came to exist on Earth. Starting with the early mythical worldviews of our ancient ancestors, Unit I traces scientific thought on the origins of the universe and Earth. We then discuss current hypotheses on the origin of life and the early evolution of cells. The nature and functions of cells and their ability to transform energy set the stage for a survey of the tremendous diversity of life that exists on Earth. The state of our environment has become a critical issue in the past two decades. International leaders agree that something must be done. Although there are complex political dimensions to the current discussions, any action taken will have to be based on our knowledge of the global distribution of life, the interactions that occur among plants and animals, and the impact that humans have had and are having on our environment.

Unit II, "The Interactions of Life," begins by explaining the physical forces that act on Earth, world climates, and the development of *biomes*—large regions of Earth with characteristic assemblages of plants and animals. We then trace the roots of modern ecology and explain its major concepts. Separate chapters focus on our knowledge of terrestrial and aquatic ecosystems. The impacts of the human population are discussed in the last two chapters of Unit II. You will learn why many scientists believe that human population growth is the most critical factor involved in environmental deterioration. An up-to-date chapter on global climate change explains why scientists have become so concerned about this complex and critical contemporary issue.

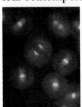

The detailed level of understanding we now have in genetics represents one of the most stunning breakthroughs in biology. New genetic technologies are now revolutionizing medicine and agriculture, and these changes are associated with explosive social questions. In Unit III, "The Language of Life," we follow the entire development of the field of genet-

ics, from Aristotle to yesterday. In exploring some of the elegant experiments that led to current knowledge, you will see the ingenuity and power of critical thinking applied to answering interesting scientific questions. By learning about modern ideas of human genetics and their applications, you will better appreciate the technical dimensions of choices that our society now faces.

Unit IV, "The Evolution of Life," explains the central unifying idea of modern biology. We trace the origin of the theory of evolution and show how the theory arose to answer a set of difficult scientific questions. The nature of scientific theories is also discussed in this unit, along with the relationship between scientific theory and religious thought. By following the development of the theory of evolution to the present, you can better appreciate how the original theory has been expanded and how it relates to the various fields of biology. A chapter on human evolution discusses current hypotheses on human origins, and two chapters on animal behavior explore the evolutionary significance of this subject.

Plants are discussed throughout the entire text, but Unit V, "Plant Systems of Life," examines some specific features of plants. We explain why, until the nineteenth century, the study of plants was limited to identifications and descriptions. Modern knowledge of plant life is described by following the general life history of a plant from seed to adult. Special attention is given to current research on how plants handle environmental stress and on plant adaptation.

The concluding unit, Unit VI, "Human Systems of Life," explores the area of biology that currently commands the greatest amount of research funding worldwide. Individual chapters trace the growth of our knowledge of different human organ systems and describe the frontiers of research now being approached. A chapter on human defense systems leads to an examination of human disease, both infectious and noninfectious. Current ideas about cancer, vascular disease, AIDS, risk factors, and lifestyle management are also discussed. The final chapter of the text considers the future of biology.

Goals

We have pursued six goals in writing *Biology: The Network of Life:*

1. To convey *fundamental knowledge* of the major principles, concepts, and ideas of modern biology.

2. To explain principles, concepts, and ideas related to biology and their social contexts, within a framework of the *process of science.* Accurate historical information is an essential component of this approach.

3. To promote an understanding of *scientific inquiry,* the methods and approaches used by scientists to answer interesting questions.

4. To present *content in context* to facilitate learning. Different contexts are used, depending on scale. Specific topics are generally introduced where there is a "need to know." For units, material is usually presented in the framework of scientific process. However, in Unit I, time is used as the context for presenting information about origins of the universe, Earth, and life.

5. To concentrate on *relevant information* that has significance now and in the future.

6. To provide students with a background that will encourage *lifetime learning* and interest in biology.

By the time you finish this book, you will have come to view science as a way of knowing and will understand contemporary biology and its relevance and importance in modern society and to your life. We hope that the new knowledge you have gained will keep you interested in biology for the rest of your life.

The Sphere of Life

2000

Present

1900

How have cells evolved?

When did cells originate?

What is the basic unit of life?

1800

1700

When did life originate on Earth?

How old is Earth?

A.D. 1600

How old is the universe?

Greek
Culture
5000 B.C.
to
2000 B.C.

Earth is composed of what basic elements?

How did life on Earth originate?

Ancient
Cultures

How did Earth originate?

How did the universe originate?

What is the structure of the universe?

2

Origins: The Universe, Earth, and Life

Chapter Outline

Reading Questions

1. Why did ideas about the origin of the universe change through time?

2. How does the big bang theory explain the origin of the universe?

3. How may the universe have evolved during its existence?

4. How was the planet Earth formed?

5. How was Earth transformed from a barren planet into one that could support life?

From the earliest written records, it is evident that our ancestors were curious about the universe they observed (see Figure 2.1). How could they explain stars in the night sky? What was the relationship between Earth, the sun, other planets, and the stars? How did these heavenly bodies originate? Ancient peoples tried to answer these questions, and their answers reflected worldviews that dominated different periods of the past.

ANCIENT WORLDVIEWS

The earliest conceptions of the world were mythical, and phenomena were organized into stories. All known ancient cultures developed cosmological myths that were grounded in beliefs of supernatural spirits, divine beings, or heroes. For example, an early Greek myth includes stories about gods, humans, animals, and monsters being changed into stars or star clusters.

Figure 2.1 Throughout history humans have been fascinated by the stars seen in the evening sky. What were they? How were they created? Most of the stars we see are in the Milky Way galaxy, one of billions in the universe and the home of our planet, Earth.

The mythical worldviews were not simplistic. Individuals from many great early civilizations—Babylonian, Egyptian, Mayan, Indian, and Chinese—made rigorous studies of the heavens and noted regularities in the celestial bodies they observed. Such efforts were related to the view that celestial bodies had a direct influence on their lives; further, they believed that certain patterns were signs that could be interpreted to forecast unusual events such as invasions or floods. These ideas contributed to the development of *astrology,* a pseudoscience based on a belief that human affairs and people's personalities are influenced by the positions of the planets.

Greeks in the fifth century B.C. were the first people known to stop using myths to explain the cosmos. Although the Greeks were interested in the astrological implications of their observations, their attempts to explain observed regularities of the heavens in physical terms reflect a more rational worldview and comprise an early form of modern *astronomy,* the field of science that studies the natural world beyond Earth. They applied mathematics to their observations and created a sophisticated model that predicted motions of the sun, the moon, and the planets. **Models** are simplified representations of a system, or abstractions of reality. They are widely used in science and are often expressed as mathematical representations. However, as described in Figure 2.2, the Greek model was consistent with their belief that a stationary Earth occupied the center of the universe and that the sun, moon, planets, and stars all revolved

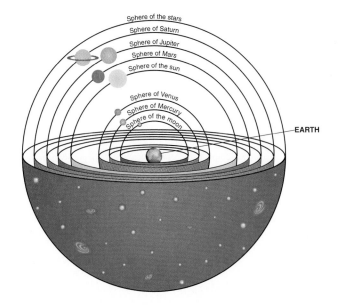

Figure 2.2 The early Greeks created a model of the universe in which a stationary Earth occupied the center. According to this model the sun, moon, planets, and stars orbited Earth. Although this model now seems foolish, it was consistent with observations and worldviews of that period.

Question: *With what development in the history of science was the change in this model associated?*

around Earth in almost perfect spherical orbits. According to this model, nothing existed beyond the outer sphere of stars. This picture of the universe, although incorrect, explained simple observations like the sun's rising and setting and the lack of any feeling of Earth's actually moving. The Greek model served as the basis for detailed astrological predictions, and it was a view that lasted almost 2,000 years.

THE SCIENTIFIC REVOLUTION

Not until the sixteenth century was the erroneous Earth-centered picture of the universe overthrown by a small group of thinkers who revived the early Greek tradition of rigorous astronomical observation. They also became interested in reevaluating certain assumptions of the early Greek model of universe structure. In 1543, a member of this group, the Polish astronomer, Nicolaus Copernicus, claimed that a more accurate model could be created if one accepted the idea that the sun, not Earth, was at the center of the known universe. It took over a century before this theory became accepted; it was during that time that what is now known as the *scientific revolution* began. The **scientific revolution** of the seventeenth century emphasized the importance of considering the world as a physical system and explaining observed regularities in terms of laws of nature. It also asserted that the most effective methods for obtaining knowledge of the natural world were careful observation, comparisons with existing information, and experimentation, whenever possible. Thus myth and religion were no longer considered as valid information sources about the natural world, but they still had a role in interpreting the significance of what exists in the natural world. Astrology also came to be rejected as a method to inquire about nature or to predict the future of human events. **Cosmology,** the science concerned with the structure of the universe as a whole, came to replace the earlier, nonscientific approaches.

The scientific revolution and the laws it generated provided the fertile ground from which modern cosmological sciences have sprung and matured. Cosmology is still in its infancy, and the sheer scale of the subject makes it one of the most challenging sciences today.

UNDERSTANDING ORIGINS

Why study origins of the physical universe and Earth? To understand life, it helps to appreciate that living organisms on Earth are the result of evolutionary processes that began with the origin of the universe. From a starting point of approximately 15 billion years ago, scientists hypothesize that matter first originated, followed in sequence by the synthesis of simple chemicals, the origin of our solar system with its planets (including Earth) about 4.5 billion years ago, the synthesis of complex chemicals, and finally, the origin of life on Earth between 3.5 and 3.8 billion years ago. We explore what is known about the origins of the universe, Earth, and life in the rest of this chapter.

The scientific hypotheses and theories about origins of the universe, Earth, and life have varying levels of certainty (see Figure 2.3). A complete understanding of the ideas examined in this chapter requires knowledge of principles from the physical sciences that is beyond the reach of this book. However, basic information from those areas and a lively imagination should allow you to grasp current ideas about origins of the universe, Earth, and life. Figure 2.4 describes the relationships in the universe.

BEFORE YOU GO ON   Our ancient ancestors tried to explain stars and other heavenly bodies in terms of creation myths and supernatural beings. The Greeks, reflecting a more rational worldview, created a model in which Earth occupied the center of the universe. In the sixteenth century, Copernicus and others demonstrated that the sun, not Earth, occupied the center of the known universe. The scientific revolution of the seventeenth century gave rise to logical approaches that could be used in trying to understand the natural world.

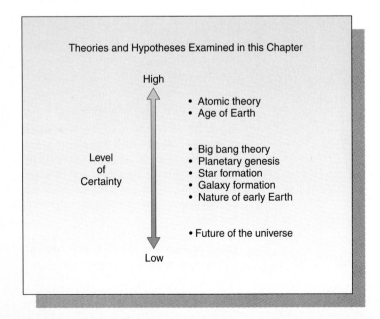

Figure 2.3 Scientific theories and hypotheses are associated with different levels of certainty, depending on the nature and strength of supporting evidence. Those described in this chapter have different levels of certainty, as indicated in this figure. Some theories, such as the big bang theory, are not regarded as having the highest level of certainty because they are very broad and inclusive and may be modified as new observations and experimental results are obtained.

Question: *What might increase or decrease the level of certainty for some of these hypotheses or theories?*

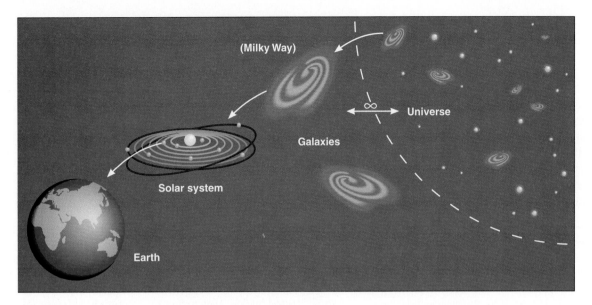

Figure 2.4 The universe is infinite space with no known boundaries. Within the universe are billions of galaxies, such as the Milky Way, and within galaxies are dust, gases, stars, solar systems, and planets.

THE ORIGIN OF THE UNIVERSE

Our distant ancestors were not only limited in their understanding of the universe and its organization, but they also had imperfect knowledge about the age and size of the universe and of Earth. Some of their ideas about the age of our planet were derived from the Old Testament. For example, in the fourth century A.D., Saint Augustine, based on his careful studies of biblical family trees, estimated that Earth was 6,000 years old. Later, in 1611, an edition of the King James Bible included a calculation by James Ussher, an archbishop and biblical scholar, indicating that time began on Sunday, October 23, 4004 B.C. However, scientists began to use methods that led to more accurate assessments of Earth's age. By the nineteenth century, *geologists* (scientists who study Earth's history) had formulated a significant conclusion: The planet Earth was far older than the biblical 6,000 years.

Early astronomers could study the heavens only with their naked eyes. Then, early in the seventeenth century, a new viewing device—the telescope—became available to observers. Larger and more efficient telescopes continued to be developed during the next three centuries. By the nineteenth century, distances from Earth to the sun and to other planets had been accurately determined, and by the early twentieth century, astronomers had discovered that the universe was filled with enormous numbers of galaxies, stars, and smaller bodies arrayed in vast areas of space. New questions arose and old questions could be reconsidered: How old is the universe? How did it arise? How old is Earth? How was it formed?

A Key Observation

Contemporary ideas on cosmology date from less than a century ago. In the late 1920s, Edwin P. Hubble, using giant telescopes at Mount Wilson and Mount Palomar in California, made observations of distant galaxies. In 1929, he announced a remarkable conclusion: All galaxies in the universe were moving away from each other at incredible speed! What did Hubble's stunning observations and conclusion mean? Most scientists felt they indicated that the universe was expanding from a highly concentrated initial state that existed somewhere in space billions of years ago.

The Big Bang Theory

Cosmologists continued to study the expansion phenomenon for many years. The rate of expansion was determined, and by extrapolating backward, it was possible to estimate the approximate time when all matter and energy were originally present at a single point in space, smaller than a dust particle. In the late 1940s and early 1950s, a major theory about the origin of the universe emerged from these and other calculations. According to the **big bang theory,** the universe originated from an infinitely dense, infinitely hot point in space that exploded some 13 to 18 billion years ago. Following this explosion, an expansion of space began, matter soon formed, and ultimately, galaxies, stars, and solar systems were created. **Matter** is anything that has mass or weight and occupies space; it may exist as a solid, liquid, or gas.

By understanding the properties and behavior of matter and knowing that light travels at a definite speed, scientists have been able to create a picture of the universe that begins immediately after the big bang.

The First Second of the Universe

In the first fraction of a second after the big bang, the universe was contained within a space smaller than a dust particle. Matter did not exist, only energy. **Energy** is the capacity to do work or cause change; for example, moving a car and reading this text both require energy. By that moment, the embryonic universe had begun to expand, and its temperature, initially billions of degrees, began to decrease.

Within the next fraction of a second, the universe increased to the size of a grapefruit, but its temperature was still too high for matter to exist. A split second later, the universe became filled with exotic classes of newborn *particles,* the fundamental units of matter, that were created from energy. These first particles were the precursors of more familiar particles, *neutrons* and *protons,* that came into existence a moment later. Within minutes after the big bang, simple *atoms* of hydrogen and helium appeared in the young universe.

> **BEFORE YOU GO ON** According to the big bang theory, the universe originated at time zero, 13 to 18 billion years ago. In the first second of its existence, the universe underwent a rapid expansion, and energy was converted into matter that later gave rise to atoms. Since the big bang, the universe has expanded continuously.

EVOLUTION OF THE EARLY UNIVERSE

Much of the support for the big bang theory, and for related theories concerned with universe evolution, can be traced to the nineteenth century when people were exploring questions about matter and energy. In 1808, John Dalton, an English chemist, proposed the **atomic theory,** which stated that matter is composed of small indivisible units called *atoms* (from Greek for "cannot be cut"). That theory was fully developed over the next 150 years, and even today it continues to be extended as it explains new observations. Where, when, and how did atoms originate in the universe? We will need to review some basic chemistry before these questions can be addressed.

Elements and Atoms

Today, matter exists in many forms and is not limited only to atoms of hydrogen and helium. An **element** is a single type of matter, such as oxygen, gold, or lead, that cannot be broken down into simpler substances by ordinary chemical processes. The smallest unit of an element that retains its chemical identity is called an **atom.** There are 92 naturally-occurring elements and each has a unique name and a one- or two-letter symbol, for example, hydrogen (H), oxygen (O), nitrogen (N), calcium (Ca), and iron (Fe). When two or more atoms of different elements are bonded together, a **compound** is formed. For example, one sodium (Na) atom will combine with one chlorine (Cl) atom to form common table salt, NaCl. Similarly, two hydrogen atoms will combine with one oxygen atom to form the compound, water. Water is described by the chemical formula H_2O, which indicates that it is made of two hydrogen atoms and one oxygen atom. A unit that contains two or more atoms of the same or different elements is called a **mol-**ecule. The formula O_2 represents one oxygen molecule, and $6CO_2$ represents six molecules of carbon dioxide. Molecules formed by living organisms can contain thousands of atoms; such molecules are known as **macromolecules.**

Atomic Structure

Research has shown that an atom is not indivisible, as Dalton originally thought. Rather, it consists of many smaller *subatomic particles,* the most important of which are **protons, neutrons,** and **electrons** (see Table 2.1). *Atomic mass* is an arbitrary unit of measure that allows comparisons of the relative masses of subatomic particles and elements. Protons and neutrons both have an atomic mass of 1 unit, whereas electrons have a much smaller relative mass, 1/1837 (that is, the atomic mass of an electron is 1/1837 of a proton or neutron). A proton has a positive electrical charge, an electron has a negative charge, and a neutron has no electrical charge. An atom has a small, central nucleus that contains both protons and neutrons. The greater the number of protons and neutrons, the heavier the atom. An atom's electrons are found around the nucleus in regions or "clouds" called *energy levels* (see Figure 2.5). The number and arrangement of electrons give an atom (and an element) its chemical properties.

Table 2.1 The Subatomic Particles of an Atom and Some of Their Properties

Name of Particle	Location in the Atom	Electrical Charge	Atomic Mass
Proton	Nucleus	Positive (1+)	1
Neutron	Nucleus	Neutral (0)	1
Electron	Outside the nucleus	Negative (1−)	$1/1837$

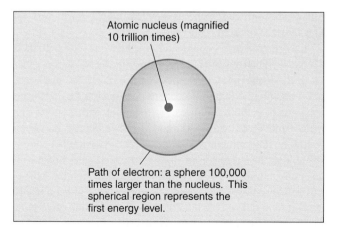

Figure 2.5 A diagrammatic structure of hydrogen, the simplest atom. The hydrogen nucleus contains a single proton, and one electron occupies the first energy level around the nucleus.

An element is defined by its **atomic number,** which is the number of protons in the nucleus of one of its atoms. For example, a hydrogen atom has one proton in its nucleus, so its atomic number is 1; oxygen has eight protons, so its atomic number is 8. In electrically neutral atoms, the number of protons in the nucleus equals the number of electrons around the nucleus. For example, carbon has an atomic number of 6; a neutral carbon atom has six protons in the nucleus and six electrons around the nucleus.

All atoms of the same element have an identical number of protons, but they may contain varying numbers of neutrons in their nuclei. Atoms of the same element that have different numbers of neutrons in their nuclei are called **isotopes.** Some isotopes, called *radioisotopes,* are unstable. To become more stable, they "decay" or undergo nuclear changes that result in the emission of subatomic particles and the release of energy from the atom; these emitted particles and energy are forms of *radiation.* The length of time required for one-half of a radioisotope's atoms in a given sample to decay is known as its *half-life.* For example, a radioisotope of hydrogen called tritium (hydrogen-3 or ^3H) has a half-life of 12.36 years, and the half-life of the radioisotope called carbon-14 (^{14}C) is 5730 years.

Electrons around the nucleus of a single atom occupy different energy levels. As shown in Table 2.2, there are seven energy levels; each level can hold only a certain maximum number of electrons. The first energy level is closest to the nucleus; higher-numbered energy levels are located farther from the nucleus, and the electrons in them contain more energy. Electrons can move from one energy level to another by gaining or losing energy. In some chemical reactions inner-energy-level electrons may "capture" energy and move to a higher energy level, thus "saving" that energy for later use. If, at a later time, that electron moves from the higher energy level to a lower energy level, the energy is released.

Table 2.2   Energy Levels in Atoms

Energy Level Number	1	2	3	4	5	6	7
Maximum Number of Electrons	2	8	18	32	32	18	8

Hydrogen atoms have only one electron, so only the first energy level is occupied by an electron. Oxygen atoms have 8 electrons: 2 in the first energy level and 6 in the second energy level. Larger atoms have electrons in more energy levels. When atoms react to form chemical bonds, however, they have a tendency to acquire 8 electrons in their outermost energy level, regardless of their possible maximum number.

Chemical Bonds

Chemical bonds are forces of attraction that hold two or more atoms together. Chemical bond formation between two atoms generally requires the rearrangement of electrons in their outermost energy level. Electrons in the outermost energy level of an atom are called *valence electrons* (see Table 2.3). When atoms react to form chemical bonds, most tend to reach a stable configuration of eight valence electrons, regardless of the possible maximum number of electrons that may occupy their outer energy level. That is, unstable atoms will enter into chemical reactions that involve the gain, loss, or sharing of electrons in order to reach a stable configuration of eight valence electrons. The stable configuration for the smallest atoms, hydrogen and helium, is two valence electrons.

Table 2.3   Electron Dot Diagrams for Atoms of Various Elements

Element (Symbol)	Number of Valence Electrons	Electron Dot Diagram
Hydrogen (H)	1	H·
Sodium (Na)	1	Na·
Calcium (Ca)	2	Ċa·
Carbon (C)	4	·Ċ·
Nitrogen (N)	5	·N̈·
Oxygen (O)	6	·Ö·
Chlorine (Cl)	7	·C̈l·
Neon (Ne)	8	:N̈e:

Electron dot diagrams are used to show the number of valence electrons in an atom of an element. When forming chemical bonds, most elements shown here need 8 valence electrons to become stable; hydrogen needs 2 valence electrons. Which atoms shown in this table is naturally stable?

Atoms of elements that are very different from one another may transfer electrons from one atom to another to become more stable. For example, common table salt, sodium chloride (NaCl), is formed when atoms of sodium lose one electron to atoms of chlorine. A sodium (Na) atom contains one valence electron. Would it be easier for Na to give up its one valence electron, in which case its new outermost energy level would contain eight electrons, or to gain seven more electrons? The same question can apply to chlorine (Cl), which has seven electrons in its outer shell. As shown in Figure 2.6, an atom of sodium gives up its valence electron to an atom of chlorine, forming the compound NaCl. When this occurs, the Na atom has one more proton than electron and thus has a positive charge (indicated as Na^+). The Cl atom now has one more electron than proton and has a negative charge (Cl^-). Atoms that are positively or negatively charged are called **ions.** The force of attraction between the oppositely charged Na^+ and Cl^- ions results in the formation of an **ionic bond.**

Elements that have similar numbers of valence electrons share electrons in order to have eight valence electrons and become stable. Sharing electrons between two atoms results in an attractive force called a **covalent bond.** This attractive force is stronger than the attraction between ions in an ionic bond, and it takes a large amount of energy to break a covalent bond.

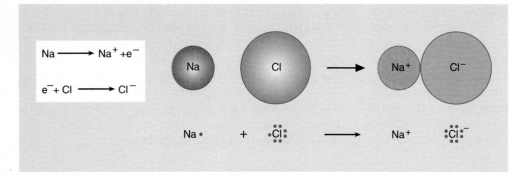

Figure 2.6   Ionic bonds form between sodium (Na) and chlorine (Cl) to create table salt (sodium chloride). To form such a bond, a single but unstable Na atom first donates its outer electron to a single unstable chlorine atom. After transfer and acceptance of the electron, Na$^+$ and Cl$^-$ ions are formed, and the force of attraction between the positive and negative electrical charges results in the creation of an ionic bond.

Question: *Why are single Na and Cl atoms unstable?*

As shown in Figure 2.7, some atoms share a pair of electrons, thus forming a single covalent bond; others share two pairs of electrons, forming a double covalent bond. A few elements, such as nitrogen, form molecules in which three pairs of electrons are shared between two atoms which results in a triple covalent bond.

Hydrogen bonding is a third type of chemical bonding. Unlike ionic and covalent bonds, which join atoms together, a **hydrogen bond** links molecules that contain atoms with different electrical charges. Hydrogen bonds are extremely important for living organisms because they determine the shapes and properties of many molecules associated with life. These bonds are fragile, having about one-tenth the strength of a covalent bond. When a hydrogen atom is covalently bonded to an oxygen, nitrogen, or fluorine atom, a positive region forms around the hydrogen on the molecule. As described in Figure 2.8, this positive region will be attracted to the negative region of the oxygen, nitrogen, or fluorine on another molecule. Hydrogen bonds determine the shapes of many large biological molecules and give water its unusual properties, such as having higher freezing and boiling points than other molecules of the same size. Water is discussed further in a later section of this chapter.

BEFORE YOU GO ON   Elements are different types of matter. Each element is composed of atoms that have specific numbers of protons, neutrons, and electrons. Protons and neutrons are found in an atom's nucleus, whereas electrons exist in regions known as energy levels that surround the nucleus. Atoms become more stable by gaining, losing, or sharing electrons and forming chemical bonds with other atoms.

Atom	Molecule	
H •	H⦂H Shared pair of electrons	H – H Single bond
⦂O⦂	O O Two shared pairs of electrons	O = O Double bond

Figure 2.7   Hydrogen atoms share a pair of electrons in forming a single covalent bond to make a hydrogen molecule (H_2). In a molecule of oxygen (O_2), oxygen atoms share two pairs of electrons in forming a double covalent bond, which results in eight electrons being present around each atom.

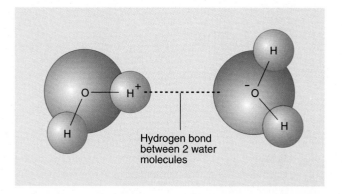

Figure 2.8   Water molecules (H_2O) are held together by hydrogen bonds (shown here as dotted lines) that form between the hydrogen atom of one molecule and the oxygen atom of another molecule.

The Origin of Chemical Elements

What is starlight? Why does the sun shine? By the dawn of the twentieth century, researchers had hypothesized that light and heat emitted by stars, including the sun, were generated from some form of atomic energy. What are the relationships between atoms, energy, and universe evolution? How did chemical elements originate in the history of the universe? The *theory of galaxy evolution,* which is part of the big bang theory, is summarized in the discussion that follows and addresses these questions.

Galaxy Formation

As temperatures declined after the big bang, protons, neutrons, and electrons combined to make atoms of the lightest elements, primarily hydrogen (H) and helium (He). By one million years after the big bang, the universe was composed of a uniform mixture of hydrogen and helium atoms. Expansion of this primordial gas continued through the next 2 billion years. During that time, the gas cooled and in a few regions of space, H and He atoms became more concentrated than elsewhere. These consolidations ultimately formed vast clouds of matter that gave birth to *galaxies,* large aggregations of dust and gas, or clusters of galaxies, in which the first generation of stars was created (see Figure 2.9).

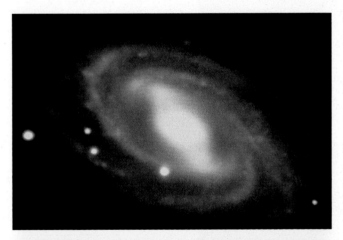

Figure 2.9 A spiral galaxy contains billions of stars. Our galaxy, the Milky Way, probably looks like the one on the top, a barred spiral galaxy (M109, NGC 3992). The one on the bottom is a standard spiral (M83).

Star Birth

Star birth begins when a concentration of gas becomes very dense and starts to contract. A true star comes into existence after *nuclear fusion* begins in the compressed gas as a result of incredibly high interior pressure and temperature, in the range of 10 million °C. **Fusion** is a process in which small atomic nuclei combine to form larger, more stable nuclei. When trillions of nuclei are involved, this results in the release of enormous amounts of energy. Within young first-generation stars, fusion reactions primarily involve hydrogen nuclei combining to form helium nuclei, with an accompanying release of energy in the form of light and heat.

Star Death

The length of a star's life depends on the amount of hydrogen "fuel" present and the rate at which this fuel is consumed in fusion reactions; the life span of a star ranges from 50 million years to an estimated 200 billion years. As a star ages, its temperature increases and additional fusion reactions occur that produce heavier, larger elements such as carbon, oxygen, magnesium, and iron. In the final phase of its life, the star's core suddenly collapses to a super-dense state, which may trigger a colossal explosion called a *supernova.*

Supernovas played two key roles in early universe evolution. First, energy generated by supernovas was so great that a final round of complex nuclear reactions created elements heavier than iron (for example, lead, gold, copper, and silver). Second, shattered star fragments, including the heavy elements formed during a star's life and death, were blown into space where they became thoroughly blended into interstellar gas clouds. Later generations of stars were formed by the compression of these gas clouds, so they are richer in heavy elements than first-generation stars.

What kind of evidence supports the galaxy evolution theory? One prediction of the theory is that later generations of stars and planets (including those in our solar system) will contain a greater percentage of heavier elements, such as carbon, oxygen, silicon, iron, and gold, than first-generation stars. That prediction has now been confirmed, which offers strong support for the theory.

Thus we can conclude that we are truly children of the stars. The calcium in our teeth, iron in our red blood cells, sulfur in our hair—indeed, all atoms that make up Earth and its inhabitants—were forged within stars billions of years before we were born and before the sun and Earth even existed. The early hypothesis about some form of atomic energy creating starlight and sunlight was correct; light and heat are released from fusion reactions within stars.

BEFORE YOU GO ON The theory of galaxy evolution explains that the first galaxies and their stars originated from dense concentrations of hydrogen and helium gases in scattered regions of space. After these gas clusters became highly compressed, forming first-generation stars, fusion reactions that liberated energy—light and heat—began and generated new chemical elements such as carbon and iron. Heavier elements such as lead and gold were created in supernovas that ended stars' lives. Later generations of stars and planets were made from elements created in first-generation stars.

THE ORIGIN OF OUR SOLAR SYSTEM

Throughout the history of the universe, small concentrations of matter formed seedlike clusters that ultimately ballooned into vast galaxies or clusters of galaxies. Evidence indicates that our own galaxy, the *Milky Way galaxy,* which contains about 300 billion stars and developed several billion years ago, is one of billions of galaxies that now fill the universe.

How did our solar system come into existence? Cosmologists believe that the formation of our sun and the system of planets that encircles it followed a developmental path that may not be uncommon in the universe.

Our solar system is hypothesized to have originated 5 to 10 billion years ago in a fog of gas and dust that would have appeared as a tiny speck in one arm of the Milky Way galaxy. The proposed sequence of events in the creation of our solar system is traced in Figure 2.10. The sun began its life when an ancient cloud of hydrogen, helium, and a sprinkling of heavier elements from first-generation stars became large enough to be condensed and compressed by gravity (*gravity* is the uni-

versal, mutual attraction of all massive objects to one another). As condensation and compression continued, the sun began to get hotter. As the young sun continued to increase in size, it accumulated more matter from the galactic haze. Eventually, great quantities of these substances were spewed into surrounding orbits, and a slowly rotating disk was formed. This huge disk, billions of miles across, was the forerunner of our solar system.

The raw materials necessary for forming planets—hydrogen and helium gases and dust containing calcium, silicon, uranium, aluminum, iron, carbon, sulfur, phosphorus, and oxygen—were present in the disk surrounding the young sun. At some point, these substances started to coalesce into solids that grew larger and larger. As the temperature of the sun rose, a few of these developing bodies increased in size until they formed embryonic planets that continued to accumulate additional materials by gravitational attraction.

About 5 billion years ago, our solar system entered its final phase of development. At that time, the sun began emitting scouring blasts, blowing gases out into space. This reduced the

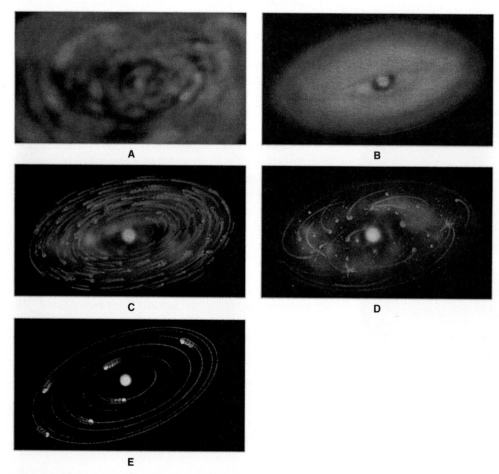

Figure 2.10 The major events of planetary genesis probably occurred in this sequence (A–E).

smaller inner planets essentially to rock with no surrounding gases. The giant outer planets, farther from the sun, retained more of their primordial gases. This gave them greater size. The sun contracted into its final, stable state approximately 4.6 billion years ago.

The sun, shown in Figure 2.11, is the major driving force of our solar system. It is a star that derives its enormous energy from the fusion of hydrogen into helium. We recognize this energy as sunshine, which illuminates and warms the planet we inhabit. At 4.6 billion years of age, the sun has now consumed about 50 percent of its fuel, so its life is nearly half over.

DEVELOPMENT OF THE PLANET EARTH

The beautiful blue planet known as Earth, shown in Figure 2.12, changed dramatically after its origin nearly 4.6 billion

---

BEFORE YOU GO ON The Milky Way galaxy was formed several billion years after the big bang. Our solar system—the sun and planets—arose in one arm of the Milky Way galaxy 5 to 10 billion years ago. The sun was forged from a gaseous cloud containing hydrogen, helium, and heavier elements created by first-generation stars. Later, planets were formed within the disk of gases and other materials that surrounded the sun.

---

years ago. Starting as a barren, rocky planet, it evolved in less than a billion years to a stage when the first forms of life arose. Knowledge from several fields of science, including geology, atmospheric science, chemistry, and biology, has been used in forming and testing scientific hypotheses about the maturation and development of Earth into its present form. Most of the hypotheses are supported by techniques and data that have been used to determine the age and atomic and chemical characteristics of various objects (for example, meteorites, rocks, and fossils). Nevertheless, much remains to be learned about the transformation of early Earth.

Early Earth

Geologists hypothesize that the surface of early Earth was bare and cold, with no oceans or atmosphere. However, its interior is now thought to have been initially hot. Consequently, interior, molten layers of different densities were formed. More than 4 billion years ago, as Earth began to cool, this density differentiation resulted in the establishment of internal layers as very heavy materials sank toward the planet's center and lighter elements accumulated near the surface. This process produced the major divisions of Earth's inner space—the core, mantle, and crust—that are shown in Figure 2.13A (see page 24).

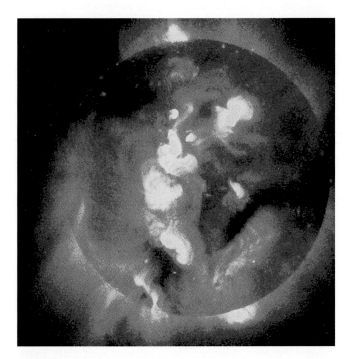

Figure 2.11 The sun shown in this X-ray image is the source of radiant energy and heat, two forms of energy that maintain Earth's environments and the life that inhabits this planet.

Question: *How are light and heat created in the sun? What will happen on Earth when the sun's fuel is eventually exhausted?*

Figure 2.12 The blue planet—Earth—as it appears from outer space.

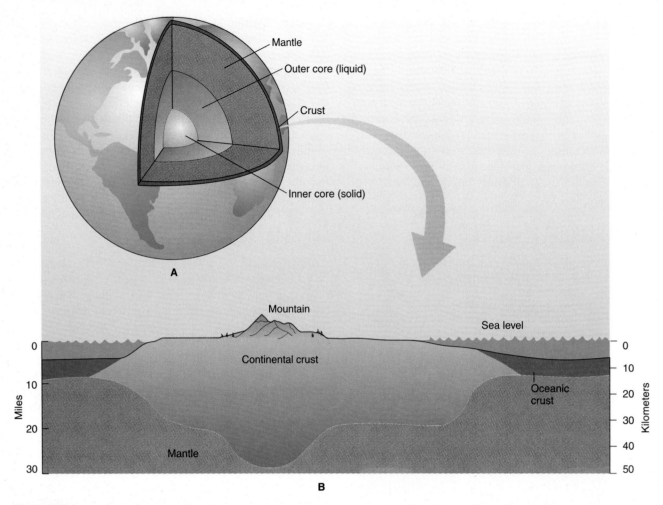

Figure 2.13   (A) A cutaway view of Earth showing its interior and exterior structures. The solid inner core and the hot, molten outer core are surrounded by a thick mantle. The lower mantle and upper mantle are 1300 and 300 miles deep, respectively. Overlaying the mantle is a thin outer crust. (B) The structure of Earth's outer crust. A thick continental crust supports landmasses; a much thinner oceanic crust supports Earth's oceans.

Question: *What appears to be the relationship between continental crust thickness and the presence of overlying mountains?*

Earth's Physical Structure

Geological studies now indicate that Earth's *inner core* is solid, being composed mostly of iron. The surrounding *outer core* is liquid and consists of hot, molten metals, primarily iron and nickel, with smaller amounts of other chemical elements. Above the core is the planet's thick, rocky *mantle,* which constitutes about 80 percent of its volume. Finally, the thin upper *crust* is composed of less dense rocky material and varies in thickness. As shown in Figure 2.13B, the continental crust that supports large landmasses is much thicker than the oceanic crust, which supports overlying oceans.

Early Landmasses

Early continental landmasses were not fixed at specific locations for great lengths of time (see Figure 2.14). Rather, they moved restlessly on shifting continental *plates*—immense slabs of continental crust that support major landmasses and

oceans. It is now known that slow plate movements have occurred throughout Earth's history, from the time when the first plates formed to the present. The ceaseless movements and interactions of the plates resulted in earthquakes, mountain building, and volcanoes. Movements of these plates continue to have an impact on geographical areas now occupied by humans (see Figure 2.15).

The Planet Matures

Several factors influenced the development of early Earth. Foremost among these were the sun and geological changes.

Nature of the Sun's Energy

The sun emits *electromagnetic radiation,* which consists of particles of light, or **photons,** that are composed of different wavelengths, as shown in Figure 2.16 (see page 26). The wave-

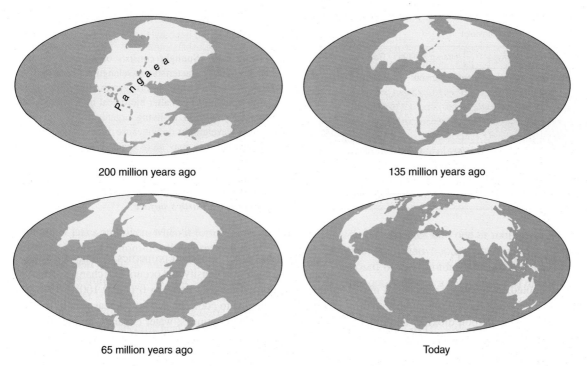

200 million years ago

135 million years ago

65 million years ago

Today

Figure 2.14 Continental and oceanic plates have moved slowly throughout Earth's history. Since existing in a single colossal landmass called Pangaea 200 million years ago, the major continental plates have migrated to the positions shown in this figure.

length of a photon is related to its energy: the shorter the wavelength, the greater the energy of the photon. The high-energy photons, called gamma rays and X rays, are very harmful to living organisms because they can alter biological molecules. However, these types of solar radiation rarely reach Earth's surface. Ultraviolet (UV) radiation is also damaging to biological molecules, and this type of radiation penetrated to the surface of early Earth. Infrared radiation emitted by the sun, and also from young Earth as it cooled, has low energy levels, as do microwaves and radio waves. The shortest visible light wave-

lengths are seen as violet or blue colors, and the longest wavelengths appear red. Radiation of wavelengths longer than those of visible light does not directly alter biological molecules because it does not contain significant amounts of energy.

Formation of the Early Atmosphere

Between 4 and 4.5 billion years ago, the first volcanoes penetrated Earth's crust and vented steam and gases that formed the primordial atmosphere. This early atmosphere is now hypothesized to have consisted primarily of water vapor (H_2O), carbon monoxide (CO), carbon dioxide (CO_2), and nitrogen (N_2) gases. Little or no oxygen gas (O_2) is thought to have been present at this time. Thus the gaseous composition of early Earth's atmosphere was dramatically different from our present atmosphere, which consists primarily of N_2 and O_2. In addition, lightning and intense concentrations of UV radiation—significant energy sources—may have played important roles in the later origin of life forms. Characterized by erupting volcanoes; high temperatures (about 85°C); lightning; vicious winds and heat; the absence of solid landmasses, oceans, and lakes; and noxious gases in the atmosphere, early Earth would be described as an extremely hostile environment by modern observers.

Figure 2.15 When continental or oceanic plates shift position, major changes can occur in areas supported by the affected plates. The eruption of Pacaya Volcano in Guatemala was a consequence of plate movements.

BEFORE YOU GO ON After Earth formed about 4.6 billion years ago, it slowly changed during a billion-year period. From a bare, cold, rocky planet, it was transformed into one that was warm, and had solid landmasses, water, and a gaseous atmosphere.

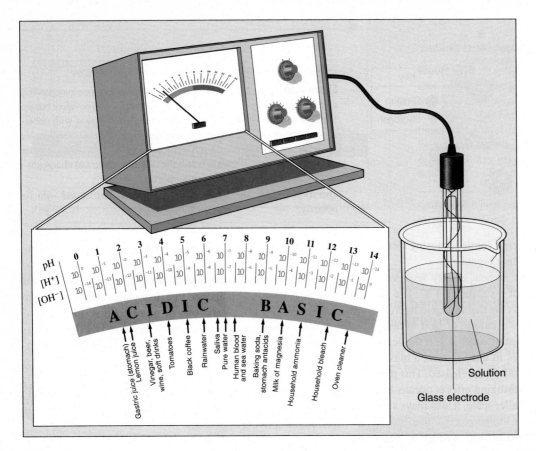

Figure 2.20   The pH scale is used as a measure of H^+ concentrations in a solution. Mathematically, pH = $-\log(H^+)$. pH values range from 0 to 14; the difference between each increment represents a tenfold change in the H^+ concentration. For example, a solution with a pH of 4 has 10 times *more* H^+ than a solution with a pH of 5 and 1000 times more H^+ than a solution with a pH of 6. A solution with a pH of 9 has 10 times *less* H^+ than a solution with a pH of 8. Pure water is neutral, having equal concentrations of H^+ and OH^-. Acidic solutions have greater concentrations of H^+ than OH^-, and basic solutions have greater concentrations of OH^- than H^+. The greater the difference in concentrations, the lower or higher the pH.

used in generating the first life forms on Earth (see the Focus on Scientific Process, "The Origin of Life").

Carbon is the key element in forming diverse organic compounds because it has four electrons in its outer (valence) shell. Consequently, it is very reactive and can form up to four covalent bonds with other atoms in order to achieve an outer electron shell that contains eight electrons. Carbon can combine with other carbon atoms to make short and long straight chains, branched chains, and rings; carbon also bonds with hydrogen and oxygen in forming an enormous variety of organic compounds. All life is characterized by organic compounds—living organisms synthesize organic molecules and they are composed of organic compounds. There are four major classes of organic compounds: carbohydrates, lipids, proteins, and nucleic acids. Each has various functions in living organisms. Following is a summary of the general structures and functions of organic compounds. These subjects are treated in greater detail in later chapters.

Carbohydrates

Carbohydrates are composed of carbon, hydrogen, and oxygen. Carbohydrates are found primarily in plants and account

for about 75 percent of a plant's body weight, compared to only 2 or 3 percent of your body weight. There are two major types of carbohydrates: (1) *monosaccharides,* also known as simple sugars, are used primarily as energy sources, and (2) *polysaccharides,* which are composed of long chains of monosaccharides, serve as carbohydrate storage molecules and as structural components in plants.

Glucose, a simple sugar, is a critically important energy source derived from plants (see Chapter 4). The chemical formula of glucose is $C_6H_{12}O_6$, and its structural formula is shown in Figure 2.21A (see page 32). In this text, simple structural formulas are used to represent most chemicals. *Starch,* a polysaccharide, is a major storage molecule composed of hundreds of glucose molecules and is described in Figure 2.21B (see page 32).

Lipids

Lipids include a great assortment of biological compounds including fats, oils, and some vitamins and hormones. Lipids are important structural components of biological membranes; they also serve as energy storage molecules. The simplest and most abundant lipids in living organisms are commonly called *triglyc-*

The Origin of Life

How did life on Earth begin? Most early cultures accepted the idea of *spontaneous generation,* in which living organisms could be created from non-living substances. This idea prevailed until the nineteenth century. Why did our ancestors believe in spontaneous generation? Primarily, it was based on such common observations as frogs seeming to emerge from mud, mosquitos from swamp water, or maggots from rotting meat. No scientist seriously questioned spontaneous generation until 1688, when Francesco Redi, a famous Italian physician, showed that maggots in decaying meat were actually larvae that hatched from fly eggs. He conducted a set of simple but ingenious experiments in which he placed pieces of meat into covered and uncovered containers. In the covered containers no maggots appeared, even after many days, whereas in the open containers, maggots were soon found, and these developed into flies.

Redi's experiments convinced most educated people that flies and frogs were not spontaneously generated. However, the invention of the microscope in the seventeenth century reopened the issue. One of the startling observations made with the new instrument was that microorganisms, such as bacteria, were extremely abundant in a drop of pond water. Where had these microorganisms come from? Many naturalists were convinced by their observations that microorganisms were spontaneously generated from organic material present in pond water or substances made from organic matter (beef broth, for example). Others were not persuaded, believing instead that microorganisms either reproduced sexually or that they came from "seeds" that had existed since Earth was formed.

Experiments conducted by Louis Pasteur in the 1860s finally ended the debate. As shown in Figure 1 on page 30, Pasteur filled flasks with different substances known to "generate" microorganisms, boiled the flasks to kill anything within, and then sealed them. No microorganisms were found in the flasks as long as they remained sealed. However, after opening a set of flasks, Pasteur observed that some of them soon contained microorganisms. He concluded that it must be "something" in the air, not the air itself, that gave rise to the microorganisms. Pasteur then conducted experiments in which he demonstrated that microorganisms existed on dust particles in the air. He used "swan-necked" flasks that were bent in such a way that they remained open but could trap any dust particles floating in the air. He placed sterilized medium into the flasks and found they always remained sterile. For Pasteur and most of his colleagues, these experiments proved that spontaneous generation did not give rise to microorganisms. But had spontaneous generation ever occurred on Earth?

Beginning in the 1920s, relevant information from many fields of science was used in formulating hypotheses about the probable course of events that took place on Earth during the previous 4 billion years. Some of the hypotheses have been supported by research conducted during the past 40 years. Nevertheless, the antiquity of events and uncertainties about early Earth's environmental conditions have made it necessary to make debatable assumptions in designing related experiments.

Three major scientific hypotheses deal with the origin of life on our planet. One involves the transport of life to Earth from somewhere in outer space and is unsatisfying to scientists because it does not explain the origin of *those* living organisms. The other two hypotheses are concerned with changes in chemicals assumed to have been present on early Earth.

The Chemical Evolution Hypothesis

The **chemical evolution hypothesis** states that beginning about 4 billion years ago, life evolved gradually in oceans through a series of continuous changes in chemical systems that existed on the primitive planet. The purported sequence of changes in the chemical evolution hypothesis can be summarized as follows: (1) creation of complex organic molecules from simple chemicals that existed on Earth at that time; (2) formation of an "organic soup" in the early oceans; (3) development of complex chemical systems capable of arranging and using organic soup molecules for life processes such as self-replication and energy capture; and (4) evolution of the first prelife system.

The basic ideas about these events were clearly formulated in 1938 by A. I. Oparin of the Soviet Union, although J. B. S. Haldane of Great Britain expressed some similar ideas in 1929. It was not until the 1950s, however, that scientists began to conduct experiments designed to test the feasibility of the chemical evolution hypothesis. Logical questions provided a starting point for designing experiments. How were complex chemicals formed from simple chemicals assumed to have existed in the early atmosphere? Could existing energy sources—assumed to be lightning, ultraviolet light, and heat—together or separately, have been capable of causing the formation of complex molecules?

box continues

FOCUS ON SCIENTIFIC PROCESS

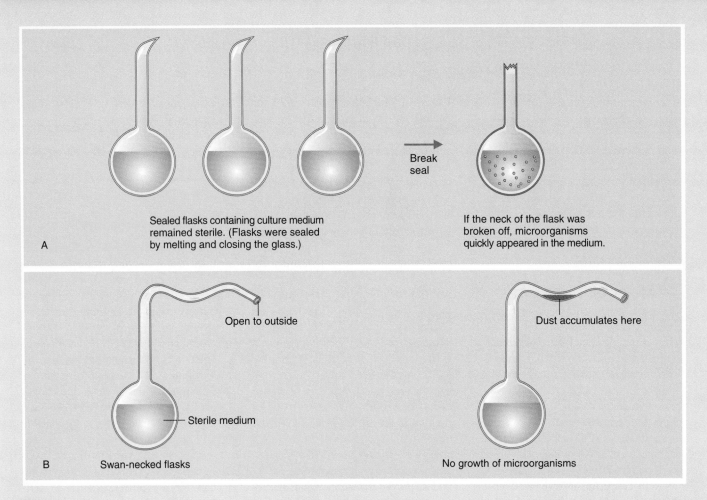

A   Sealed flasks containing culture medium remained sterile. (Flasks were sealed by melting and closing the glass.)

Break seal

If the neck of the flask was broken off, microorganisms quickly appeared in the medium.

Open to outside

Dust accumulates here

Sterile medium

B   Swan-necked flasks

No growth of microorganisms

Figure 1   Louis Pasteur conducted two experiments in the 1860s that addressed spontaneous generation. (A) In the first experiment, flasks were filled with culture medium, boiled, and then sealed by melting and closing the glass. Flasks that were later opened by breaking the glass quickly became filled with microorganisms. (B) In the second experiment, Pasteur used swan-necked flasks that were open to the air but which trapped microorganisms or dust particles within the flask neck. No microorganisms were ever found in these flasks.

In 1953, Stanley L. Miller conducted a famous experiment that attempted to simulate conditions on early Earth. He set up an airtight apparatus in which warmed water vapor containing a mixture of hydrogen (H_2), methane (CH_4), and ammonia (NH_3)—hypothesized gases in the early atmosphere—was circulated past electrical discharges (used to mimic lightning) (see Figure 2). These gases were circulated for one week; then the contents of the experimental system were removed and analyzed. An astonishing number of organic compounds had been formed.

Miller's experiment provided the first evidence that one of the stages of the Oparin–Haldane hypothesis—organic compound synthesis—could have occurred. However, his results have been questioned by researchers who doubt that early Earth's atmosphere contained concentrations of the specific gases used in his study. Nevertheless, since 1953, many additional experiments have been carried out using different mixtures of gases and other

energy sources, such as UV light and heat. Hundreds of different organic compounds have been formed in these primitive Earth simulation experiments, including representatives of all important molecules found in cells today. The assortment of experimental conditions in which atmospheric chemical synthesis has been demonstrated has led scientists to conclude that the first step in chemical evolution has a relatively high level of certainty.

In the last 20 years, organic molecules have been identified, using new

Figure 2   This figure shows the type of apparatus Miller used, superimposed on a drawing depicting the conditions on primitive Earth, which were assumed to be similar to those in the apparatus. The results of Miller's experiments supported one step in the chemical evolution hypothesis.

technologies, in the atmospheres of Jupiter and Saturn, on meteorites, and in gas clouds throughout space. Could such molecules have been carried to Earth by comets? Many scientists feel that it is unlikely because such molecules would probably be destroyed upon entering Earth's atmosphere. Others, however, have not rejected the idea and feel extraterrestrial organic matter may have played a role in the origin of life on Earth.

The second stage of the chemical evolution hypothesis—formation of an organic soup—assumes that over millions of years newly formed organic molecules accumulated in early Earth's oceans. Although this may have occurred, there is little solid evidence to support this idea.

The Vent Origin of Life Hypothesis

A second hypothesis about the origin of organic compounds and life on Earth emerged in the 1980s. After viewing and studying deep underwater vents along ocean floor plate seams, where superheated water, gases, and magma blend with cool salty water (as they probably did 4 billion years ago), some geologists and oceanographers felt that those sites may have been the major source of early organic molecules. They postulated the following sequence of events: (1) water seeped into cracks and came into contact with molten rock lying under the plates; (2) the water became superheated, reacted with the rocks, and extracted carbon, nitrogen, oxygen, hydrogen, and sulfur—the chemicals required to make organic molecules; (3) organic molecules were formed; and (4) primitive life forms developed. Further, they felt that life could have originated at ancient vents where hydrogen-rich gases escaping from Earth's interior reacted with carbon-rich gases in the water.

What type of evidence would be required to support this "vent origin of life" hypothesis? What questions could be formulated to guide research on this hypothesis? What types of organisms exist at such sites today? Two key observations have been that hydrogen sulfide gas (H_2S) streams from the vents and that bacteria presently exist near these sites that grow in H_2S in laboratory experiments. Such bacteria are considered to be ancient. Could they have evolved at these vents billions of years ago? What about the water temperatures at the vents? They have been reported to be extremely hot, up to 350°C. In 1988, Stanley L. Miller (yes, the same Stanley Miller) and a colleague reported results from an experiment that indicated that organic molecules are unstable and decompose rapidly at temperatures of 350°C. From this, they concluded that it was highly unlikely that life originated at vent sites. A counterargument to this conclusion is that newly formed organic molecules could have moved rapidly into cooler water immediately after their synthesis. But then how could organisms be created in the absence of an energy source to promote further chemical reactions?

This hypothesis is also supported by the presence of unique bacteria at similar sites today. One related finding of considerable interest is that organisms existing at the vent sites do not depend on the sun as their source of energy because these sites are shrouded in complete darkness. Energy that currently supports the assemblage of life at these sites originates from the heat and chemicals emitted from Earth's interior.

Debates about early Earth's environmental chemistry will undoubtedly continue well into the future. It seems likely that complete answers to all questions may remain elusive given the antiquity and complexity of events associated with the origins of life on Earth.

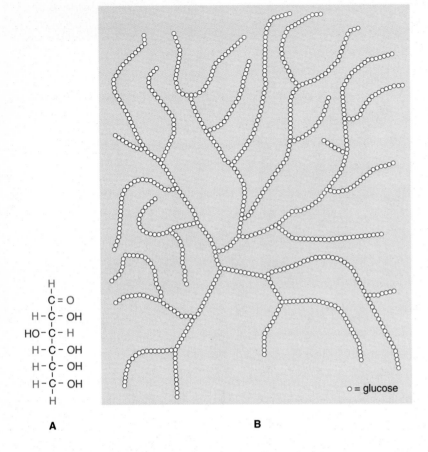

A

B

○ = glucose

Figure 2.21 (A) The structural formula of glucose (its chemical formula is $C_6H_{12}O_6$). Each line connecting the atoms represents a single covalent bond. (B) The general structure of starch, a polysaccharide composed of hundreds of glucose molecules (each bead represents a glucose molecule).

Question: *In (A), what does the double line between the carbon and oxygen atom represent?*

erides, which contain only carbon, hydrogen, and oxygen. The basic structure of a triglyceride is shown in Figure 2.22. Triglycerides consist of a 3-carbon *glycerol* backbone to which three *fatty acids* are attached. Fatty acids are long, straight chains; their lengths may vary in a triglyceride molecule.

Phospholipids, described in Figure 2.23A, have a similar structure except that one of the fatty acids is replaced by a negatively charged phosphate group (PO_4^{3-}) linked with a positively charged nitrogen-containing compound. As a result, phospholipid molecules have a polar head that is attracted to

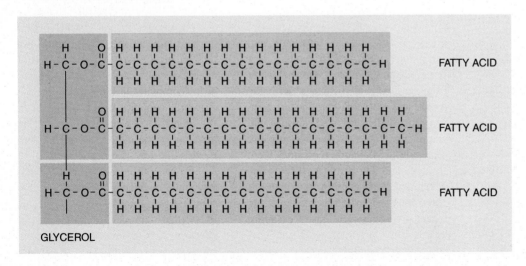

FATTY ACID

FATTY ACID

FATTY ACID

GLYCEROL

Figure 2.22 Triglycerides are the simplest and most common lipid in living organisms. They are composed of one glycerol molecule joined to three fatty acid molecules that may be of different lengths.

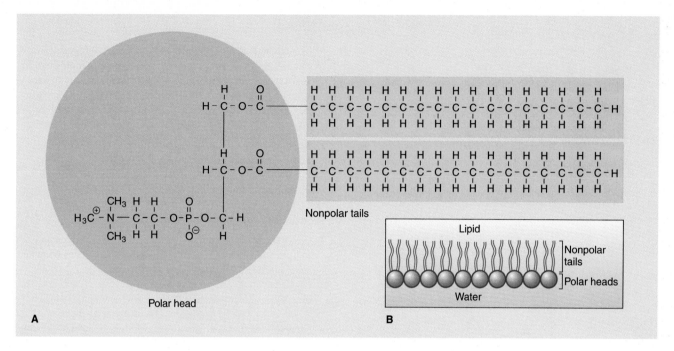

Figure 2.23 (A) Phospholipids have a structure similar to a triglyceride except that a negatively charged phosphate group (PO_4^{3-}) linked with a positively charged nitrogen-containing compound replaces one of the fatty acids attached to the glycerol molecule. The electrical charges make the "head" of a phospholipid molecule polar, but the neutral fatty acids' "tails" are nonpolar. (B) The polar heads of a phospholipid molecule are soluble in water, but the nonpolar tails are not.

water and two nonpolar tails that are repelled by water (see Figure 2.23B). This feature is especially important in membrane structure (discussed in Chapter 3).

Proteins

Proteins are the most diverse and complex group of organic molecules found in living organisms. They have various functions related to structure, storage, transport, regulation, communication, and protection, which are described in Units III and VI. Proteins are constructed of smaller building-block molecules called *amino acids,* which are composed of carbon, hydrogen, oxygen, nitrogen, and, in some cases, sulfur. There are 20 different amino acids, each with the same general structure shown in Figure 2.24A. Protein molecules are composed of varying numbers of amino acids linked to one another by a special covalent bond called a *peptide bond,* which is described in Figure 2.24C. The structure and synthesis of proteins are discussed further in Chapter 18. *Enzymes,* which are protein molecules that function as biological catalysts in chemical reactions, are described in Chapter 4.

Nucleic Acids

Nucleic acids are the largest organic molecules in living organisms. There are two principal types of nucleic acid: *deoxyribonucleic acid* (*DNA*) and *ribonucleic acid* (*RNA*). DNA is the repository of hereditary or genetic information in most living organisms. RNA functions primarily in protein synthesis. The structure and function of DNA and RNA are the subjects of Chapters 17 and 18.

THE CHARACTERISTICS OF LIFE

What is life? This is one of the oldest and most complex questions in biology, and there is no precise answer. Rather, biologists recognize certain features that characterize living systems; they include:

- A high level of organization
- An ability to exchange materials and energy with the surrounding environment
- A capacity for self-regulation and control, which usually results in maintenance of a "steady state"
- A capacity to respond and adapt to the surrounding environment
- An ability to reproduce
- A capacity to evolve

Today complex organic molecules—carbohydrates, lipids, proteins, and nucleic acids—carry out these activities in living organisms. Scientists, therefore, hypothesize that simple, building-block organic molecules (for example, monosaccharides and amino acids) must have been present early in Earth's

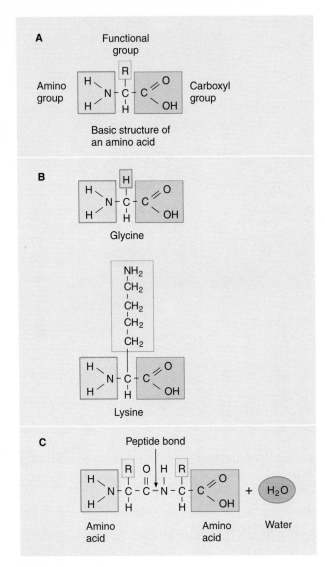

Figure 2.24 (A) All amino acids have a similar basic structure: an amino group (NH_2), a carboxyl group (COOH), and a functional group (R) that varies in structure. (B) Two amino acids are shown here. The functional group of the simplest amino acid, glycine, is a single hydrogen atom; lysine has a more complex functional group. (C) Proteins are composed of tens or hundreds of amino acids held together by peptide bonds that form between the amino group of one amino acid and the carboxyl group of another.

Question: *If a protein were made of 500 amino acids, how many peptide bonds would it have?*

history, before life originated. More complex organic molecules then formed, perhaps through activities connected with the first forms of life. Although hypotheses have been developed that attempt to explain how such compounds may have evolved on Earth, there is great uncertainty about the steps between chemical evolution and the appearance of life. Exactly when and how did life arise on Earth, after organic molecules were present? We'll examine this question in Chapter 3.

SUMMARY

1. The big bang theory proposes that the universe began 13 to 18 billion years ago with a colossal explosion at a single point in space, where all matter and energy originally existed in one form. During the first 500,000 years after the big bang, particles were created from energy, hydrogen and helium atoms were formed, and galaxies began to develop in the universe.

2. Matter is composed of atoms. An atom consists of a central nucleus containing neutrons and protons, and electrons that occupy different energy levels in orbit around the nucleus. Two or more atoms can be joined by chemical bonds. Three types of chemical bonds—covalent, ionic, and hydrogen—are important in biology.

3. According to the theory of galaxy evolution, hydrogen and helium nuclei were created by the big bang and became mixed into galactic gas clouds that eventually condensed, forming first-generation stars. Within these stars, fusion reactions gave rise to heavy elements that later became scattered through space by supernova explosions. These heavy elements were incorporated into later generations of stars, including our sun, and other celestial bodies such as Earth.

4. Our solar system, consisting of the sun and its orbiting planets, was formed 5 to 10 billion years ago as a result of matter's becoming condensed and compressed. Planets formed when very small bodies began to accumulate vast amounts of solar material that later coalesced. The physical development of the planet Earth was completed between 4 and 5 billion years ago.

5. From its original rocky state, Earth matured into its present state through the effects of several processes that led to the formation and trapping of heat, the separation of internal substances into layers, the creation of landmasses, the development of an atmosphere, and the formation of warm, shallow oceans.

6. It has been hypothesized that about 4 billion years ago, a chemical evolution process began on Earth in which inorganic chemicals reacted to give rise to the first simple organic chemicals. Subsequently, the major classes of complex organic macromolecules—carbohydrates, proteins, lipids, and nucleic acids—were formed, perhaps by the first systems that had characteristics of life.

WORKING VOCABULARY

atom (p. 18)
carbohydrate (p. 28)
chemical bond (p. 19)
compound (p. 18)
covalent bond (p. 19)
electron (p. 18)
element (p. 18)
hydrogen bond (p. 20)
ion (p. 19)
ionic bond (p. 19)
lipid (p. 28)

macromolecule (p. 18)
matter (p. 17)
model (p. 15)
molecule (p. 18)
neutron (p. 18)
nucleic acid (p. 33)
organic compound (p. 27)
pH scale (p. 27)
protein (p. 33)
proton (p. 18)

REVIEW QUESTIONS

1. How does the big bang theory explain the origin of the universe?

2. What is matter? What are elements?

3. What are atoms? Describe their structure.

4. Describe the major types of chemical bonds. How are atoms joined together by chemical bonds?

5. Describe a set of hypotheses, best supported by current evidence, that explains the evolution of the universe after the big bang.

6. Where and how were atoms and elements formed?

7. How was our solar system created?

8. Describe the sequence of events that led to the formation of Earth.

9. What factors influenced the early development of Earth 4 to 4.5 billion years ago?

10. What are the four major classes of organic compounds? What is the general structure and function of each?

ESSAY AND DISCUSSION QUESTIONS

1. Why doesn't science today use supernatural events or processes to explain or interpret natural phenomena?

2. What additional types of evidence would *increase* the level of certainty for the big bang theory? What types of findings could *reduce* its level of certainty?

3. Is it possible that life could have developed on Earth if the planet's chemistry were different? What would we need to know in order to decide?

REFERENCES AND RECOMMENDED READING

Clark, S. 1995. *Stars and Atoms: From the Big Bang to the Solar System.* New York: Oxford Univ. Press.

Coles, P. 1995. *Cosmology: The Origin and Evolution of Cosmic Structure.* New York: Wiley.

Gribbin, J. 1993. *In the Beginning: After COBE and Before the Big Bang.* Boston: Little, Brown.

Hartquist, T. W. 1995. *The Chemically-Controlled Cosmos: Astronomical Molecules From the Big Bang to Exploding Stars.* New York: Cambridge Univ. Press.

Hawking, S. W. 1988. *A Brief History of Time: From the Big Bang to Black Holes.* Toronto: Bantam Books.

Manchester, W. 1993. *A World Lit Only by Fire: The Medieval Mind and their Renaissance.* Boston: Little, Brown.

National Geographic. 1983. The once and future universe. 163: 704–749.

Osterbrock, P. E., J. A. Gwinn, and R. S. Brashear. 1993. Edwin Hubble and the expanding universe. *Scientific American,* 269: 84–89.

Peoples, P. J. E., D. N. Schramm, E. L. Turner, and R. G. Kron. 1992. The case for the relativistic hot big bang cosmology. *Nature,* 352: 769–776.

Rebek, J. 1994. Synthetic self-replicating molecules. *Scientific American,* 271: 48–57.

Ronan, C. A. 1991. *The Natural History of the Universe: From the Big Bang to the End of Time.* New York: Macmillan.

Scientific American. 1994. Special Issue: Life in the Universe. *Scientific American,* 271: 44–126.

Silk, J. 1989. *The Big Bang.* rev. ed. New York: Freeman.

Singer, S. F. 1990. *The Universe and Its Origins.* New York: Paragon House.

Turner, M. S. 1993. Why is the temperature of the universe 2.726 Kelvin? *Science,* 262: 861–867.

van den Bergh, S., and J. E. Hesser. 1993. How the Milky Way formed. *Scientific American,* 268: 72–78.

York, D. 1993. The earliest history of the earth. *Scientific American,* 268: 90–96.

ANSWERS TO FIGURE AND TABLE QUESTIONS

Figure 2.2   The scientific revolution of the seventeenth century.

Figure 2.3   New observations or additional evidence that support (increase) or new observations and new evidence that contradict (decrease) existing hypotheses.

Figure 2.6   They do not have eight (valence) electrons in their outer electron shell.

Figure 2.11   By nuclear fusion reactions. It will once again become a bare, rocky, lifeless planet.

Figure 2.13   Continental crust regions that support mountains are thicker than in zones that do not have overlying mountain ranges.

Figure 2.16   High-energy wavelengths—gamma rays and X rays.

Figure 2.18   Its ability to dissolve most substances.

Figure 2.21   A double covalent bond (the sharing of two pairs of electrons).

Figure 2.24   499.

Table 2.3   Neon.

3

Origins: Cells

Chapter Outline

Reading Questions

1. What does the cell theory explain?

2. What are the major eukaryotic cell structures and functions?

3. What are the characteristics of prokaryotic cells?

4. How may cells have originated on Earth?

5. What factors influenced cell evolution?

In Chapter 2, we described hypotheses indicating that the origin of the first forms of life on Earth may have been linked to chemical evolution, a process that began approximately 4 billion years ago. Questions about the nature and evolution of such life forms have received increased attention from scientists during the last two decades, and these subjects are considered in later sections of this chapter. However, long before hypotheses were formulated about the first organisms on Earth, biologists were interested in identifying the fundamental structural units of life. Whereas seventeenth-century astronomers trained their telescopes on the heavens, other inves-

tigators interested in smaller worlds focused their lenses on earthbound objects and discovered that all living organisms are composed of cells. Thus **cells** are defined as the fundamental structural and functional units of all living organisms on Earth.

CELLS

The **cell theory** is a major unifying set of ideas in biology. It states that (1) living organisms consist of individual cells and that the basic processes defining life are carried out by cells or groups of cells; (2) cells carry out activities that enable organisms to develop, grow, survive, and reproduce; and (3) all living organisms are made up of one or more microscopic cells.

Two Cell Types

Two basic cell types occur in the biological world; they are described in Figure 3.1. **Prokaryotic** ("prenuclear") **cells,** most of which are bacteria, are of the simplest cell type. Bacteria have neither a *nucleus* nor any other membrane-bound *organelles*. A **nucleus** is a discrete, membrane-bound structure inside a cell that contains DNA, the genetic material. **Organelles** are microscopic structures within the cell that are responsible for carrying out specific functions. **Eukaryotic** ("true nucleus") **cells** have a more complex internal structure and are able to carry out a large range of cellular activities. In contrast to prokaryotes, eukaryotic cells have a nucleus and many specialized membrane-bound organelles. Prokaryotic cells are usually 1 to 10 micrometers in size, whereas typical eukaryotic cells generally range from 10 to 100 micrometers. The relative sizes of various physical and biological structures, including eukaryotic and prokaryotic cells, are described in Figure 3.2 (see page 38).

The Cell Theory

The origin of the cell theory demonstrates the complexity that often characterizes the birth of new scientific ideas. An understanding of the fundamental importance of cells, how they are organized, and how they function began to develop about 150 years ago. To a large extent, this knowledge grew out of medical studies of the human body, but results from this research were extended to include all living organisms. The theory also developed as a result of advances in technology. Because most cells cannot be seen with the naked eye, it was not until the invention of the microscope in the seventeenth century that the cell theory could begin to take shape. Even then early microscopes were imperfect instruments that often greatly distorted any object being examined. Seventeenth-century microscopists, such as England's Robert Hooke, observed structures such as the cell walls shown in Figure 3.3 (see page 38), but they did not appreciate the significance of such observations. Although many interesting discoveries were made with these early microscopes, development of the cell theory was stimulated by the invention of high-quality lenses in the 1820s and 1830s.

The origin of the cell theory involved more than making observations with a good microscope. The theory was formulated because many scientists in the first half of the nineteenth century were trying to find something like the cell. Several French and German medical researchers were convinced that the key to understanding the human body was to consider it as a large, organized chemical machine. This seems to be an unusual idea, for when we look in the mirror, we certainly don't see a machinelike entity, nor do we imagine ourselves and our friends to be bags of chemicals. However, powerful ideas in science have not always been intuitively obvious. Scientists such as Henri Dutrochet in France and Theodor Schwann in Germany also felt that a new way of looking at how bodies function was necessary if they were to understand them. Not only did they think that living organisms were made of organized chemicals, but they also recognized that these chemicals must be contained within some type of small basic unit. Dutrochet's microscope was not adequate to see these units, but two German scientists, Matthias Schleiden (working with plants) and Schwann (working with animals), saw many different kinds of cells through their microscopes. In 1838 and 1839 Schleiden and Schwann published separately their idea that the bodies of plants and animals were composed of cells. Thus what was to become the cell theory was first articulated.

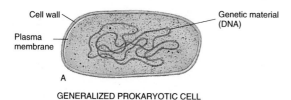

Cell wall
Plasma membrane
Genetic material (DNA)

A

GENERALIZED PROKARYOTIC CELL

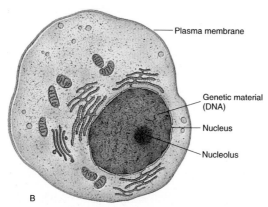

Plasma membrane

Genetic material (DNA)

Nucleus

Nucleolus

B

GENERALIZED EUKARYOTIC CELL

Figure 3.1 All organisms on Earth are composed of one of two cell types. (A) Prokaryotic cells are the simplest cell type. They do not have a nucleus or any membranous organelles. (B) Eukaryotic cells are more complex. They have a nucleus and numerous organelles.

Question: *Which cell type probably evolved first?*

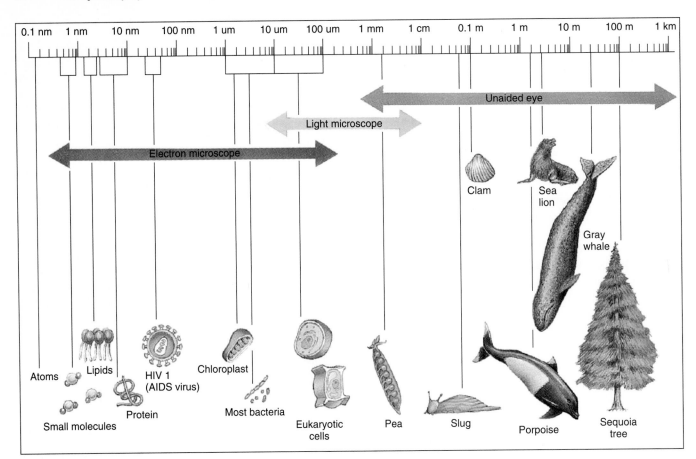

Figure 3.2 Physical and biological structures vary enormously in size. Note that prokaryotic cells, including bacteria such as *Escherichia coli* (*E. coli*), are generally 1 to 10 micrometers in size, whereas most eukaryotic cells, such as human red blood cells and nerve cells, are between 10 and 100 micrometers.

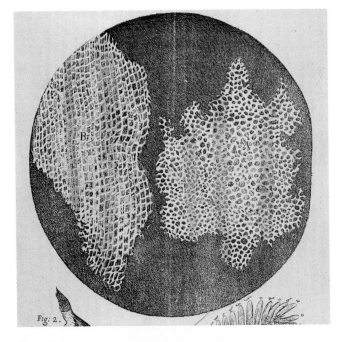

Figure 3.3 Using a simple microscope seventeenth-century scientist Robert Hooke observed plant cell walls and made these drawings.

Maturation of the Cell Theory

Throughout the rest of the nineteenth century, new discoveries and theoretical advances contributed to an extension of the cell theory. In 1855, German physician Rudolf Virchow stated that new cells could originate only by the division of preexisting cells and not from nonliving substances. The corollary statement—that all existing cells can trace their origins to ancestral cells—was added in 1880. Throughout this period, descriptions were published of some of the structures observed inside of cells and the wondrous events that occurred when a cell divided into two apparently identical cells.

By the end of the nineteenth century, there was a broad but shallow understanding of cell structure and function as a result of the earlier descriptive studies. Biologists began to penetrate ever further into the cell's interior to discover the ultimate secrets of its internal structures and functions.

Modern Cell Theory

Today the cell theory has been extended and modified to reflect advances in knowledge gained during recent decades. It is now clear that all cells share many basic similarities. They are alike in their chemical composition and in several structures and functions. They also share similar mechanisms for using energy, building cell structures, and synthesizing genetic material. Modern cell theory acknowledges these similarities and considers activities of organisms to be a function of the properties and interactions of cells.

The cell theory has had a long and rich history that is not yet over. Technological innovations have taken the initial ideas far beyond what was originally described, but not beyond the original inspiration to understand the functions of living organisms in terms of their chemical and physical components. The pioneers of the cell theory believed that it would ultimately unify all our knowledge of living organisms. However, the theory of evolution, which emerged a few decades later, was even more comprehensive and actually incorporated the cell theory into its framework. We will discuss the theory of evolution in detail in Unit IV of this book.

BEFORE YOU GO ON The concept that all organisms are composed of cells originated as a result of research using microscopes and the formulation of powerful new ideas derived primarily from medical studies of the human body. These activities led to the formation of the cell theory—a broad, unifying set of statements about the nature of cells and of life.

EUKARYOTIC CELLS

Most familiar organisms on Earth—plants and animals—are composed of eukaryotic cells and are commonly identified as **eukaryotes.** An examination of cell types from different eukaryotes reveals a bewildering array of shapes and sizes. However, there is a basic pattern in the structure of all living cells that is apparent when viewed through a microscope. Although many cells in multicellular plants and animals have the potential to operate independently, they do not do so under normal circumstances. Rather, most cells are combined with other cells to form tissues and organs that perform specific functions within the body of an organism.

Visualizing Cells

The *compound microscope* (light microscope), which consists of objective and ocular lenses and a lighting system, made it clear that eukaryotic cells are discrete units bound by a *plasma membrane* that encloses a prominent *nucleus,* which is surrounded by *cytoplasm* (see Figure 3.4). What exactly exists in the nucleus and the cytoplasm? Because a compound microscope can magnify objects clearly only 1,000 times, that question could not be answered until the 1950s, when the *transmission electron microscope* (*TEM* or *EM*), which could magnify objects over 100,000 times, revealed a level of structural complexity that was previously unknown. Later, the *scanning electron microscope* (*SEM*) made it possible to view a cell three-dimensionally at these high magnifications.

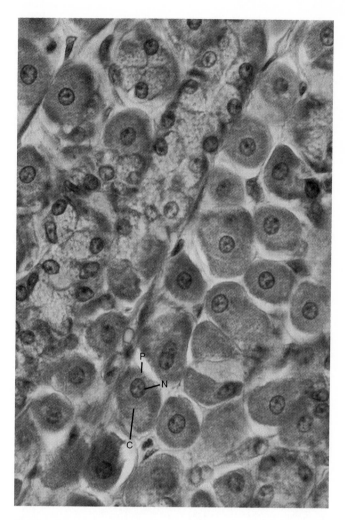

Figure 3.4 Most eukaryotic cells, such as these cells of a cat's stomach, consist of a nucleus (N) and cytoplasm (C) surrounded by a plasma membrane (P).

For cells and tissues to be visualized with most microscopes, including the electron microscopes, they must first be treated and processed by special techniques. Basically, biological materials are fixed for viewing by first placing them in chemicals that preserve their structure. Next, they are embedded in a medium (for example, wax or plastic) that can be sliced into ultrathin sections. Finally, these sections are stained or impregnated with some substance that makes their structure visible. Cells and tissues prepared for viewing through a compound microscope are stained with biological dyes that impart different colors to various structures. Those to be viewed through an electron microscope are impregnated with the salts of heavy metals (for example, lead or uranium) that act to reflect electrons, which leads to the creation of a black and white image. Hence, electron micrographs (EMs), which are photos taken through an electron microscope, are normally black and white. However, new computer technologies allow EMs to be color-enhanced to better show specific structures. Photographs of cells and cell parts, using different microscopes and technologies, including many color-enhanced EMs, appear in this chapter and later parts of the text.

General Cell Structure

Figures 3.5 and 3.6 illustrate the detailed structure of animal and plant cells. Once the EM was developed, it became possible to examine the internal architecture of the cell. Within the internal matrix of each cell, a remarkable assemblage of fibers, collectively known as the *cytoskeleton,* and organelles were observed.

Complex Cell Structure

Eukaryotic cells consist of a plasma membrane that encloses cytoplasm and a nucleus. Within the cytoplasm are various

organelles that constitute compartments defined by a membrane or a membrane system. Organelle compartmentalization allowed eukaryotic cells to become much more specialized because each organelle carries out its function in an isolated environment. All eukaryotic cells contain organelles, although there are variations in type and number among the cell types. Figure 3.6B reveals that plant cells differ from animal cells in that each plant cell is enclosed by a **cell wall,** which is a tough, fibrous material composed of cellulose, a polysaccharide, and other substances. Plant cells also have chloroplasts and vacuoles (which are described later in the chapter).

Plasma Membrane

All cells are surrounded by a plasma membrane. As shown in Figure 3.7, **plasma membranes** have a complex lipid bilayer

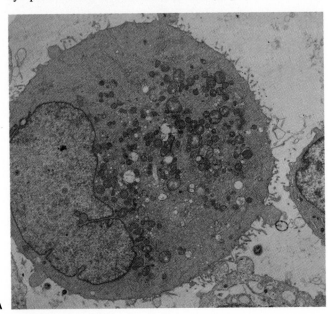
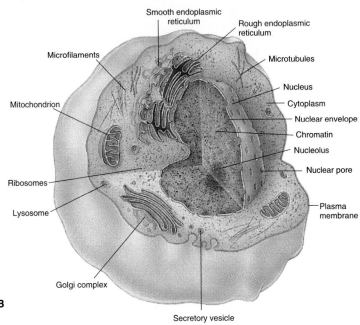

Figure 3.5   (A) An electron micrograph (EM) of an animal cell showing the nucleus and cytoplasm. (B) Animal cells contain some or all of the structures shown in this figure, depending on their function.

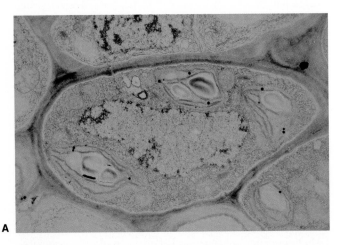
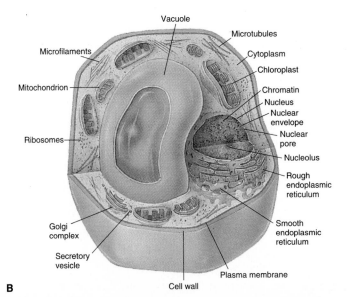

Figure 3.6   (A) An EM of a plant cell. (B) Plant cells contain some or all of the structures found in animal cells. In addition, they also have cell walls, chloroplasts, and vacuoles.

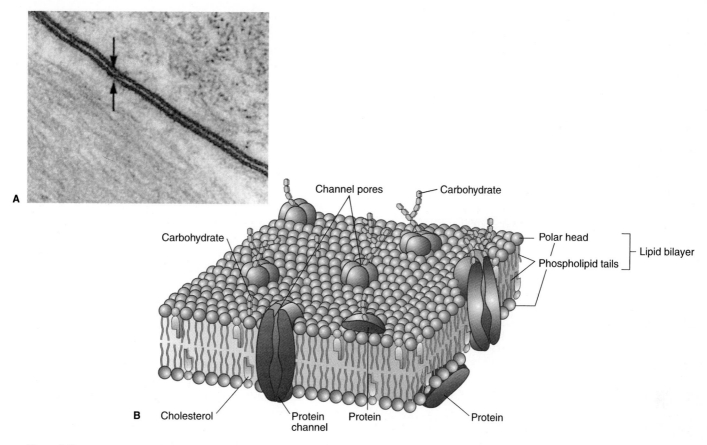

A

B

Channel pores   Carbohydrate

Carbohydrate

Polar head

Phospholipid tails

Lipid bilayer

Cholesterol

Protein channel   Protein   Protein

Figure 3.7 (A) An EM showing a plasma membrane (between the arrows). (B) All cells are enclosed by a plasma membrane, a complex structure that carries out many activities critical for cell function. Plasma membranes are composed of two layers of phospholipid molecules in which various proteins are embedded. The polar phospholipid heads are oriented toward areas of high water concentration, whereas the tails occupy the middle area between the heads where water concentrations are lower. Cholesterol molecules are found in the plasma membranes of most animal cells, and carbohydrate molecules are attached to both lipid and protein molecules. As described by the fluid mosaic model, the plasma membrane can be viewed as a fluid medium composed of lipid molecules in which proteins float.

Question: *What is the primary function of the plasma membrane?*

structure with two layers of highly organized phospholipid molecules in which various proteins are embedded. Small carbohydrate molecules are attached to some of the lipid and protein molecules.

Recall from Chapter 2 that each phospholipid has a polar phosphate head that is attracted to water and two nonpolar fatty acid tails that avoid water. Because the internal and external cellular environments are mostly water, the phosphate heads lie closest to the inside and outside of the cell; the fatty acid tails occupy the middle area, away from the highest water concentrations. Consequently, the plasma membrane can be viewed as a fluid medium composed of phospholipid molecules in which proteins "float." This description of the plasma membrane is known as the *fluid mosaic model.* Most animal cells, but not plant cells, also have cholesterol molecules in the plasma membrane; these add to the membrane's fluid properties. Figure 3.8 (see page 42), shows that the plasma membrane is responsible

for a variety of operations critical to the well-being of the cell it surrounds. Its most important function is to regulate the flow of chemical substances—water and various molecules and ions—into or out of the cell. Because each cell type has exact requirements for its internal chemical composition, precise regulation by the plasma membrane is required for maintaining stable operating conditions within the cell. In later chapters, we describe specific functions of the plasma membrane. In the next section we examine two processes related to the regulation of substances transported through the plasma membrane.

Cells exist in environments containing hundreds or thousands of different molecules and ions. Plasma membranes are *selectively permeable,* which means that in regulating the movements of various substances, some can pass directly into the cell but others cannot. Whether or not molecules can pass through a plasma membrane depends on their size, polarity, and electrical charge. Molecules and ions move across plasma

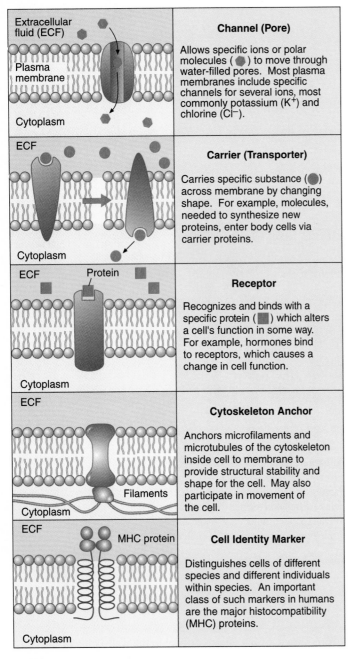

	Channel (Pore) Allows specific ions or polar molecules (⬡) to move through water-filled pores. Most plasma membranes include specific channels for several ions, most commonly potassium (K⁺) and chlorine (Cl⁻).
	Carrier (Transporter) Carries specific substance (●) across membrane by changing shape. For example, molecules needed to synthesize new proteins, enter body cells via carrier proteins.
	Receptor Recognizes and binds with a specific protein (■) which alters a cell's function in some way. For example, hormones bind to receptors, which causes a change in cell function.
	Cytoskeleton Anchor Anchors microfilaments and microtubules of the cytoskeleton inside cell to membrane to provide structural stability and shape for the cell. May also participate in movement of the cell.
	Cell Identity Marker Distinguishes cells of different species and different individuals within species. An important class of such markers in humans are the major histocompatibility (MHC) proteins.

Figure 3.8 The protein molecules embedded in a plasma membrane are responsible for the various functions described in this figure. Many activities involve regulating materials and coordinating events between the cell's interior (cytoplasm) and its external environment (extracellular fluid, or ECF).

membranes by two general mechanisms: *passive transport,* which does not require the cell to expend energy, and *active transport,* which does.

Passive Transport Molecules at temperatures above freezing are constantly in motion, which means they continually collide with each other and with other molecules. In the absence of any other energy, such collisions ultimately result in molecules being distributed randomly in any particular environment, whether as solids, liquids, or gases. In cells, however, molecules and ions are rarely distributed randomly; rather, they tend to be more concentrated in certain areas than in others. In passive transport, the movement of a substance is related to its *concentration gradient*—the difference in the number of molecules, atoms, or ions between two points or areas. **Simple diffusion** is the movement of molecules down a concentration gradient—that is, from an area where they are highly concentrated to an area where they are less concentrated (see Figure 3.9). Gases such as carbon dioxide and oxygen diffuse through plasma membranes until the concentrations on both sides are equal, a point that represents the *equilibrium level.* As described in Figure 3.10, essential molecules, such as glucose, that are too large to diffuse through the plasma membrane combine with *carrier proteins* within the membrane and are then transported through the membrane and released on the other side. This process is known as **facilitated diffusion**, and it is used to move molecules down their concentration gradient, in both directions, across plasma membranes.

Osmosis is the diffusion of water molecules across the plasma membrane, from the side where it is more concentrated to the side where it is less concentrated. A major task of the plasma membrane is to prevent too much water from entering the cell. What determines relative water concentrations on either side of a plasma membrane? The answer to this question sheds light on a problem cells face in maintaining a steady state. A *solution* is the mixture formed when some substance is dissolved in water or another liquid. The water or liquid in which the substance is dissolved is called the *solvent,* and the dissolved substance is called the *solute.* Cytoplasm can be seen as a complex solution in which various solutes (molecules, salts, and ions) are dissolved in water (the solvent). Table 3.1 summarizes the general types of solutions and their characteristics relative to a cell's interior environment; Figure 3.11 (see page 44) describes the effects of different solutions. Ultimately, water concentrations are determined by solute concentrations. An *isotonic solution* has the same solute concentration

Table 3.1 Types of Solutions and Their Effects on Water Movement Into and Out of Cells

Type of Solution	Solvent (H₂O) Concentration	Solute Concentration	Water Movement
Isotonic	Same as in the cell	Same as in the cell	None
Hypertonic	Less than in the cell	Greater than in the cell	Out of the cell
Hypotonic	Greater than in the cell	Less than in the cell	Into the cell

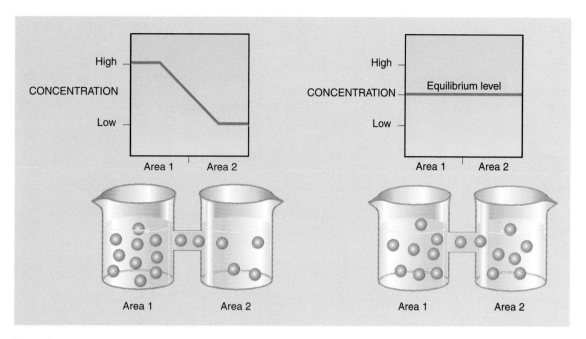

Figure 3.9 Diffusion is the movement of a substance down a concentration gradient—from an area of high concentration to an area of low concentration. When concentrations of the substance become equal in the two regions, the equilibrium level is reached and diffusion stops.

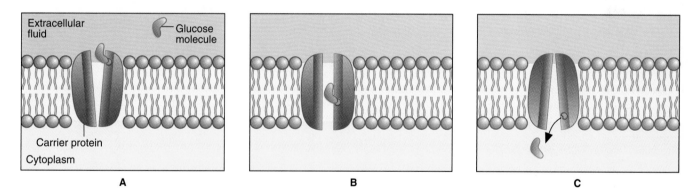

Figure 3.10 In facilitated diffusion, glucose molecules bind with carrier proteins (binding indicated by green dot) in the plasma membrane. This induces a conformational change in the carrier protein that transports glucose into the cell.

Question: *Does facilitated diffusion occur against a concentration gradient? Does it require energy?*

as the inside of a cell, and no osmosis occurs. A *hypertonic solution* has a higher solute concentration (and therefore, a lower solvent concentration), and water diffuses out of the cell in response. A *hypotonic solution* has a lower solvent concentration, and water diffuses into the cell, which may cause it to rupture. Cells are constantly subjected to osmotic changes, and most organisms have mechanisms that help regulate concentration gradients of various molecules and ions.

Active Transport   Many molecules and ions involved in cellular functions cannot diffuse into or out of a cell because they are normally more concentrated on the other side of the plasma membrane. In these cases, cells use energy to transport important ions or molecules against their concentration gradient, a process known as **active transport**. In general, active transport is used to transport needed molecules into the cell and waste molecules out of the cell. Carrier proteins and membrane channel proteins also participate in active transport. Specific cases involving various molecules and ions are described in later chapters.

Cytoplasm

The interior of a cell contains a gelatinous material that consists primarily of different molecules dissolved in water. Within this medium exists a variety of specialized cellular

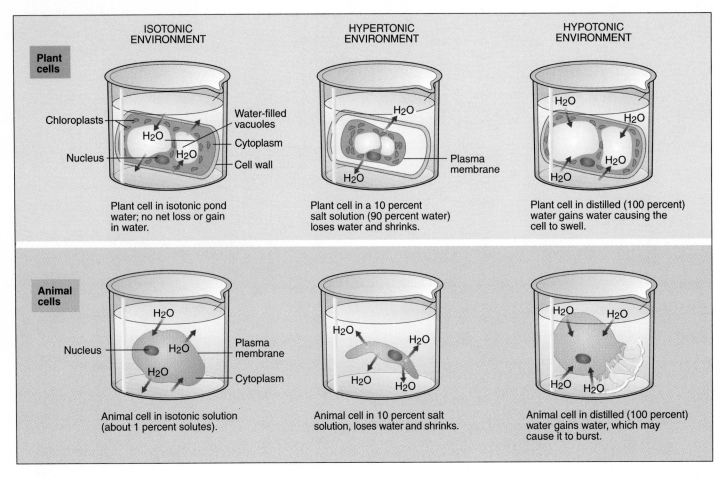

ISOTONIC ENVIRONMENT — HYPERTONIC ENVIRONMENT — HYPOTONIC ENVIRONMENT

Plant cells

Chloroplasts — Water-filled vacuoles — Cytoplasm — Nucleus — Cell wall — H_2O

Plant cell in isotonic pond water; no net loss or gain in water.

H_2O — Plasma membrane

Plant cell in a 10 percent salt solution (90 percent water) loses water and shrinks.

H_2O

Plant cell in distilled (100 percent) water gains water causing the cell to swell.

Animal cells

H_2O — Nucleus — Plasma membrane — Cytoplasm

Animal cell in isotonic solution (about 1 percent solutes).

H_2O

Animal cell in 10 percent salt solution, loses water and shrinks.

H_2O

Animal cell in distilled (100 percent) water gains water, which may cause it to burst.

Figure 3.11   The changes in plant and animal cells when placed in different solutions. In isotonic solutions, there is no net gain or loss of water from the cells. In hypertonic solutions, water diffuses out of the cell, causing it to shrink. In hypotonic solutions, water diffuses into the cell and may cause it to rupture.

Question: *Why does water diffuse out of a cell if it is placed in a hypertonic solution?*

organelles. **Cytoplasm** refers to all of the substances inside the plasma membrane, except for the organelles.

Nucleus

The **nucleus** is the largest organelle in most cells. It is a spherical structure that is surrounded by two membranes called the *nuclear envelope,* which separate it from the cytoplasm (see Figure 3.12, page 45). The inner membrane encloses the nucleus itself. In many cells, the outer membrane branches and connects with a membrane system called the *endoplasmic reticulum.* Each cell usually contains a single nucleus, although there are some interesting exceptions. For example, mature human red blood cells do not have a nucleus; each muscle cell in your biceps contains many nuclei. The nuclear envelope, like the plasma membrane, regulates the movement of materials into and out of the nucleus. Large *nuclear pores* are apparent in the surface of the membrane; they are involved in transporting substances between the nucleus and the cytoplasm.

The interior of the nucleus is referred to as *nucleoplasm.* The genetic material DNA is found in the nucleus; it is the sub-

ject of several chapters in Unit III. A prominent structure seen in the nucleus of most cells is a *nucleolus* (meaning "little nucleus"). Its primary function is related to protein synthesis, which is described in Chapter 18. Mechanisms for the control and regulation of most cell functions reside within the nucleus. The underlying basis for most of these regulatory activities lies in the production of unique proteins that cause specific effects in different cells.

Other Organelles

Eukaryotic cells have certain organelles in common. Both plant and animal cells possess a nucleus, endoplasmic reticulum, Golgi complexes, and mitochondria. However, only plant cells also contain chloroplasts and, often, vacuoles.

The **endoplasmic reticulum** (**ER**) is a large network of interconnected membranes where proteins and lipids are synthesized in the cell (see Figure 3.13). **Rough ER** is studded with small organelles called **ribosomes** that play a major role in the synthesis of proteins. Some proteins synthesized by the cell are required for building and maintaining organelles or for

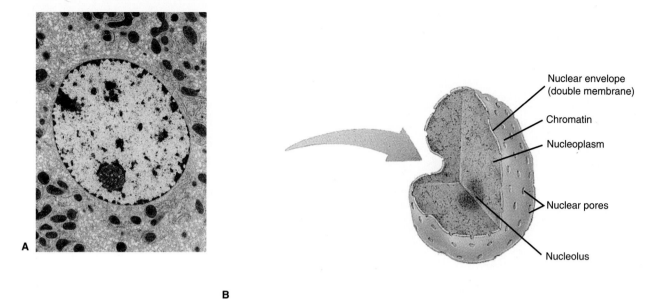

Figure 3.12   (A) An EM of a nucleus. (B) The nucleus is a complex organelle enclosed by a nuclear envelope (membrane) that is riddled with nuclear pores. These large openings regulate the movement of substances in and out of the nucleoplasm. Chromatin is scattered genetic material. The nucleolus is a specialized structure involved in protein synthesis.

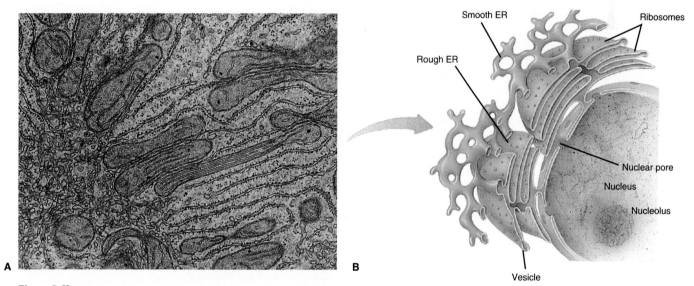

Figure 3.13   (A) An EM of endoplasmic reticulum (ER). (B) The ER is a network of membranes that branch off the nuclear envelope but are continuous with it. Ribosomes are scattered on the surface of the rough ER, where they participate in protein synthesis. The smooth ER has no ribosomes, and it functions in lipid synthesis.

participating in activities within the cell, but others are transported to other cells in the organism. Proteins produced in the rough ER are incorporated into small membrane-bound sacs called *vesicles* before they are distributed within the cell or exported. **Smooth ER,** which does not have ribosomes on its surface, is the site where lipids are synthesized.

Proteins synthesized on ribosomes attached to the rough ER are transported by vesicles from the ER to a **Golgi complex,** an organelle that consists of a series of flattened membranes (see Figure 3.14, page 46). It is generally believed that Golgi

complexes are responsible for modifying, sorting, and transporting proteins to different compartments within the cell. Proteins that are to be exported from the cell are packaged in *secretory vesicles* that move to the plasma membrane, where their contents are expelled to the cell's exterior. *Lysosomes* are small, single-membrane-bound organelles that contain enzymes used to digest macromolecules in certain situations. Enzymes used by lysosomes are replenished by small, enzyme-bearing vesicles that bud from Golgi complex. Figure 3.15 (see page 46) describes the relationships among membranous organelles.

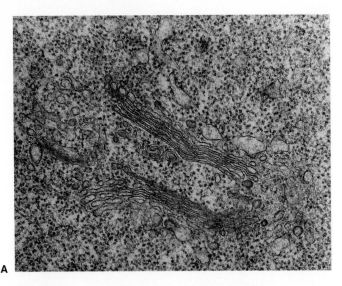
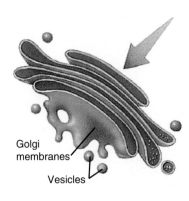

Figure 3.14   (A) An EM of a Golgi complex. (B) Golgi complexes are networks of flattened membranes. Their primary function is to process and move proteins produced within the cell. Transported proteins are enclosed in membranous vesicles.

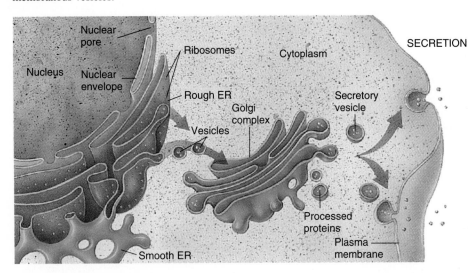

Figure 3.15   The structural and functional relationships among various membranous organelles in eukaryotic cells. Molecules responsible for initiating protein synthesis move from the nucleus to the rough ER, where synthesis occurs. From there, newly produced proteins are bound into vesicles and transported to the Golgi complex. Proteins to be exported from the cell are placed into new secretory vesicles that move to the plasma membrane where they are secreted. Lipids are synthesized in the smooth ER.

Cells obtain energy by converting simple sugars into another energy-rich substance that can be used by the cell (described in Chapter 4). This transformation of sugars occurs primarily in **mitochondria,** which are large organelles that are encircled by a double membrane (see Figure 3.16). The outer membrane surrounds the mitochondrion, and the inner membrane has numerous infoldings that are the sites of energy generation. Most cells contain numerous mitochondria; the more active the cell, the larger the number. For example, muscle cells are filled with mitochondria, whereas certain types of fat cells may have only one mitochondrion.

Chloroplasts, shown in Figure 3.17, usually contain a green pigment called *chlorophyll* that is critical in carrying out **photosynthesis,** the process whereby radiant energy is used to produce sugars (explained in a later section). Chloroplasts have a double outer membrane and an extensive internal membrane system, which is where some of the reactions of photosynthesis take place.

Most plant cells contain at least one large *vacuole,* a cavity that is surrounded by a single membrane. Vacuoles store substances such as water, sugar, proteins, and sometimes pigments and may also act as reservoirs for harmful waste products created by the plant cell.

The Cytoskeleton

Within the cytoplasm of eukaryotic cells is a delicate, organized array of protein filaments called the **cytoskeleton** (see Figure 3.18). This microscopic internal skeleton gives animal cells their shape, anchors organelles and in some cases transports them from one area of the cell to another, and permits movement in certain cells (muscle cells, for example). The cytoskeleton of plant cells performs similar functions. The principal components of the cytoskeleton are microtubules and microfilaments. *Microtubules* are made of globular, beadlike subunits organized into long, hollow, cylindrical tubes that form a skeletal network of fibers in the cytoplasm of many cells. Microtubules have a variety of roles in cells of living organisms, such as the control of cell shape and the movement of different materials and organelles within the cell. *Microfil-*

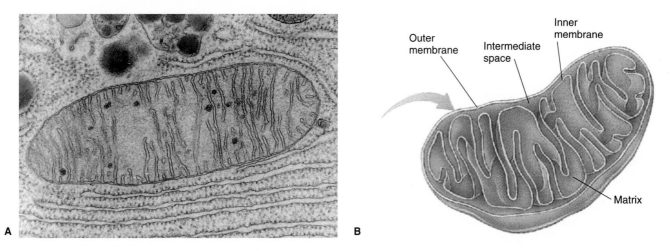

Figure 3.16   (A) An EM of mitochondria. (B) Mitochondria are membranous organelles that provide the cell with energy.

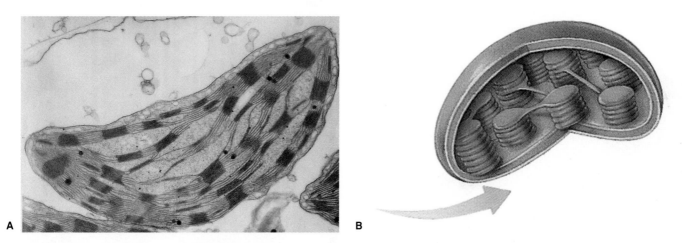

Figure 3.17   (A) An EM of a chloroplast. (B) Plant cells have chloroplasts (large organelles that contain chlorophyll, which is a pigment capable of transforming radiant energy into organic molecules during photosynthesis).

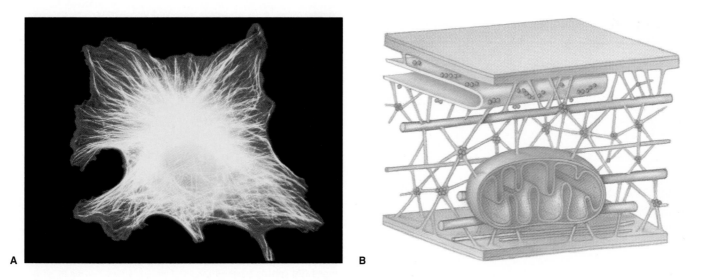

Figure 3.18   (A) The cytoskeleton is an internal support system composed of microtubules and microfilaments. In this photograph of a cell, microtubules appear as green fibers; microfilaments are red. (B) Microtubules have a hollow tubular structure. They control cell shape and various materials around the cell. Microfilaments are long, thin rodlike fibers that transport organelles from one part of the cell to another and assist in the movement of cells specialized for that function.

aments are long, thin, rodlike fibers composed of globular protein subunits. They also function in moving organelles around the cell and in contraction movements in cells specialized for that activity.

PROKARYOTIC CELLS

Simple organisms, mostly bacteria, that are composed of a single prokaryotic cell are commonly known as *prokaryotes. Bacteria,* such as those shown in Figure 3.19, are spherical, rod-shaped, or spiral-shaped prokaryotes. Bacteria are divided into two groups: *eubacteria,* which live in soil, water, and larger living organisms, and *archaebacteria,* which live in more hostile environments such as hot springs, salt marshes, bogs, and deep ocean depths. The general organization of a bacterial cell (in this case, *Escherichia coli,* or *E. coli*) is shown in Figure 3.20. Bacteria are enclosed in a plasma membrane surrounded by a rigid cell wall, which has a different chemical composition than a plant cell wall. The cell

wall protects the cell and is responsible for the characteristic shapes of different bacteria. Within the plasma membrane is a single cytoplasmic compartment that contains neither a nucleus nor membranous organelles, such as mitochondria or endoplasmic reticulum. Ribosomes, the only organelle commonly found in prokaryotic cells, are scattered throughout the cytoplasm. Bacterial genetic material, DNA, consists of a single circular molecule located in a *nucleoid* region within the cytoplasm.

BEFORE YOU GO ON Eukaryotic cells have a basic structure consisting of an outer plasma membrane that encloses a nucleus and cytoplasm. Within the cytoplasm are various membrane-bound organelles—endoplasmic reticulum (ER), Golgi complexes, mitochondria, chloroplasts, and vacuoles—that carry out specific cellular functions. The numbers and kinds of organelles may vary in different cells of the same organism and between cells of organisms. Prokaryotic cells are much simpler; they have no nucleus, and ribosomes are the only organelle commonly found in their cytoplasm.

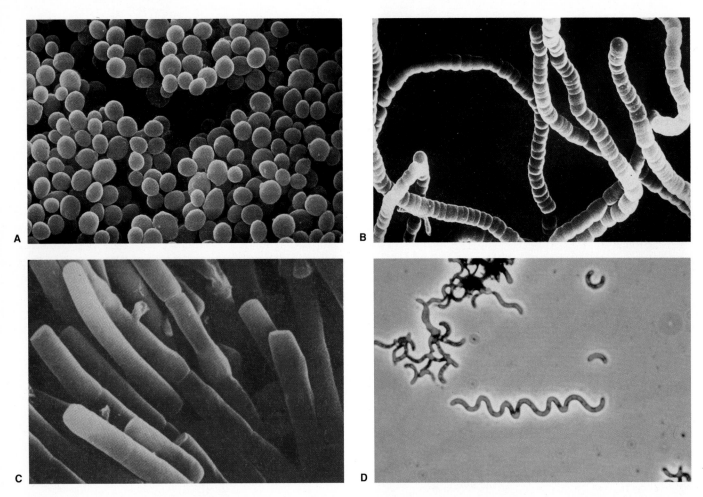

A B C D

Figure 3.19 Bacteria are prokaryotic cells that have diverse shapes. Cocci (round) bacteria grow (A) separately or in pairs or (B) as long chains, (C) bacilli bacteria are rod-shaped, and (D) spirilla bacteria are corkscrew-shaped.

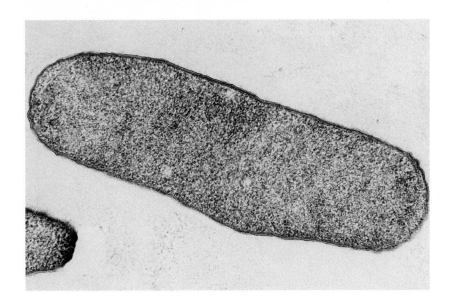

Figure 3.20   A bacterium with a typical prokary- otic cell structure: an outer cell wall, a plasma membrane, and cytoplasm containing ribosomes and a nucleoid that contains the genetic material.

Question: *What are the structural differences between a prokaryotic cell and a normal eukary- otic cell?*

THE ORIGIN OF CELLS

When did living systems and cells evolve on Earth? The *geological time scale,* described in Figure 3.21 (see page 50), organizes the entire history of Earth into different time intervals. *Precambrian* time includes Earth's history between 570 million years ago and 4.6 billion years ago, and it consists of two eons: the *Archean eon,* from 2.5 to 4.6 billion years ago, and the *Algonkian eon,* from 570 million to 2.4 billion years ago. Life and the first prokaryotic cells are thought to have originated in the Archean eon whereas fossil evidence indicates that eukaryotic cells evolved early in the Algonkian Eon. *Fossils* are remains or traces of ancient organisms that are embedded in rocks or other substances in the Earth's crust; *microfossils* are the remains of single-celled or relatively simple organisms. The subject and significance of fossils is explored fully in Chapter 27. Here we note that fossils constitute a permanent record of history; when a fossil can be identified and its age determined, it relates what existed in the distant past—thousands, millions, or even billions of years ago.

Geological evidence indicates that about 4 billion years ago, Earth's crust had begun to solidify and it may have cooled sufficiently for naturally formed, stable, complex organic molecules to exist in the early oceans. The first organisms for which microfossils exist are nearly 3.5 billion years old. These microfossils have been interpreted as being the remains of ancient bacteria that lived in marine (ocean) environments. It can therefore, be concluded that the time required for the formation of organic molecules and the appearance of the first living systems was rather short, approximately 500 million years. However, the difference in complexity between organic molecules and even the simplest cell is enormous. How were living systems first formed from organic molecules?

Development of Living Systems

Precellular systems are structures that may have originated as aggregates of organic molecules and become capable of arranging and using available organic chemicals for processes which we associate with life. These hypothesized systems, as described in Figure 3.22 (see page 51), may have evolved through the following sequence of events: development of a membrane or similar barrier that separated the interior and exterior environments of the "individual;" structural organization; an ability to maintain an internal environment that differed from the surrounding external environment; competence in converting chemical energy available in organic molecules into energy that could be used for maintenance, and, finally, self-replication, or the ability to reproduce.

The development of a mechanism for self-replication was a critical step in the origin of cells because the ability to reproduce is a hallmark of life. Once self-replicating, ordered precellular systems had developed, rapid progress in a continuous evolution toward a more highly ordered cell became possible. However, the point at which an actual cell evolved from a precellular form remains speculative. The escalation from a self-replicating, precellular system to even the simplest cell is immense and must have included many small developmental steps and possibly one or more precellular stages.

How did prokaryotic and eukaryotic cells originate after the first precellular system appeared? What is the nature of the evidence used to support hypotheses on the origin and evolution of cells? Much of what follows is supported by results from experiments conducted during the past decade.

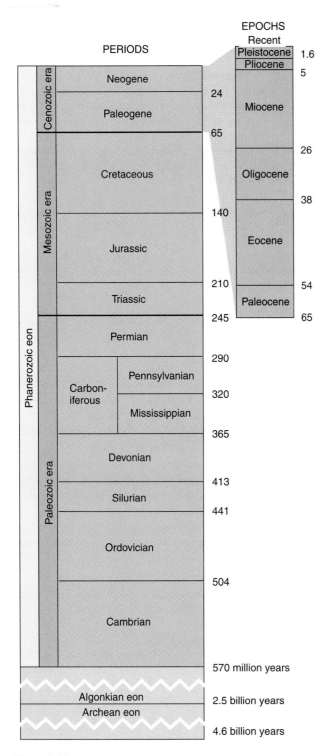

Figure 3.21 The geologic time scale chronicles the past eons, eras, periods, and epochs of Earth's history. Most of these divisions are based on the ages that have been determined for different rock layers. Except for Precambrian time, the time scale is in millions of years ago.

BEFORE YOU GO ON he first precellular system is hypothesized to have evolved through organizing complex organic chemicals present in early Earth's environment. It was probably surrounded by a membrane that separated its internal environment from the external environment around it, and it must have been able to use energy and to reproduce. Precellular forms gave rise to the first cells, which appeared in early marine environments approximately 3.5 billion years ago.

The Origin of Prokaryotic Cells

Biological evolution, the process of changes that occur in populations of living organisms over a long time, and the beginning of definite *lines of descent* are thought to have begun during the Archean eon in Earth's history. A line of descent may have involved one type of cell that gave rise to a slightly different cell type. With the process continuing over thousands or millions of years, a new cell type may have eventually evolved.

Figure 3.23 illustrates a current hypothesis about how different cells and organisms arose and evolved on Earth. According to this hypothesis, both prokaryotic and eukaryotic cells arose independently from a precellular system called a *progenote* (after the term *progenitor*). Two separate ancestral lines of prokaryotic cells evolved from the progenote between 3.5 and 3.8 billion years ago. One branch gave rise to the *archaebacteria,* primitive prokaryotes that inhabited harsh environments where little or no oxygen was present. Archaebacteria that exist on Earth today are found in similar environments—extremely salty, hot, or acidic environments where little or no oxygen is present. The second branch gave rise to *eubacteria,* which are more complicated prokaryotes that could live in diverse environments. Most modern prokaryotes are eubacteria. The oldest microfossils, which are about 3.5 billion years old, resemble modern archaebacteria.

A critical function of all living cells is the ability to obtain energy by converting chemical substances from one form to another. High-energy forms are converted to low-energy forms, and the energy liberated by these transformations is used by the cell to carry out its activities. What were the possible sources of energy used by these first prokaryotes? As described in Chapter 2, organic molecules are hypothesized to have been present in marine environments of early Earth. Thus the first primitive cells may have had a "free lunch" that provided them with the energy they required. In time, however, these organic molecules would have been depleted as "output" would have greatly exceeded "input"; that is, the number of organic molecules needed to make new cells and to serve as an energy source for those cells would have been far greater than the few molecules being formed. Life on Earth might have entered the abyss of extinction were it not for the evolution, at least 2.5 to 3.5 billion years ago, of a new chemical process in certain types of bacteria that provided a fresh and continual source of high-energy molecules.

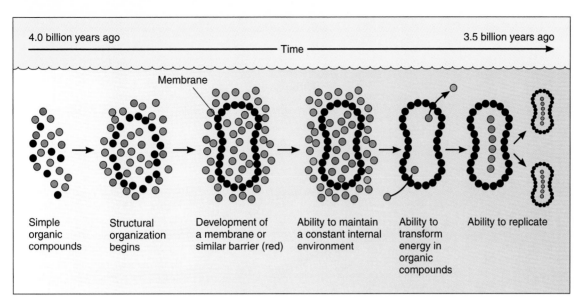

Figure 3.22 A summary of events hypothesized to have led to the development of the first living system on Earth, beginning 4 billion years ago.

Question: *What was the source of the simple organic compounds?*

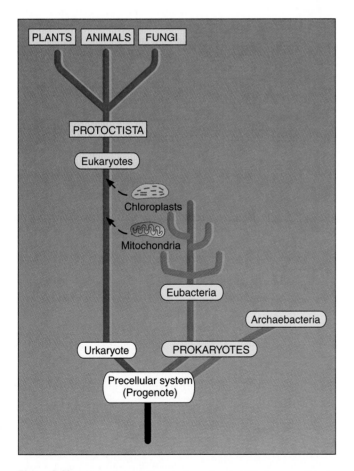

Figure 3.23 The origin and evolution of major life forms on Earth.

Question: *When is each of the various cells and organisms hypothesized to have first appeared on Earth?*

A New Supply of Energy

The new process, *photosynthesis,* has been appraised by scientists as being one of the most important evolutionary developments in Earth's history. The steps of photosynthesis are briefly outlined in Figure 3.24 (see page 52) and are described in greater detail in Chapter 4. **Photosynthesis** yields a continual source of energy because photosynthetic cells use radiant energy from the sun to convert the raw materials of carbon dioxide (CO_2) and water (H_2O) into energy-rich sugars—the source of energy for most organisms. Oxygen (O_2), a byproduct of photosynthesis, played a central role in shaping the life forms that were to appear on Earth later. The sugars from photosynthesis replaced the depleted organic chemicals and were used by cells that continued to appear and evolve during the next 2 billion years.

The Importance of Oxygen

As the O_2 produced during photosynthesis began to accumulate, it had a major impact on existing organisms and on Earth itself. Free O_2 was probably toxic to many of the prokaryotes that first inhabited the planet. Thus as the concentration of O_2 increased from 0.0001 percent to about 2 percent 1.5 to 2 billion years ago, most oxygen-intolerant prokaryotes generally became extinct. New prokaryotes, however, evolved that were not only able to tolerate O_2 but which could actually use it in reactions that yielded great quantities of energy (described in Chapter 4). Another significant development occurred as O_2 diffused into the outer atmosphere, where a portion of it was converted into the gas ozone (O_3). *Ozone* forms a layer in the outer atmosphere that absorbs UV radiation emitted by the sun and protects life from the harmful effects of this highly reac-

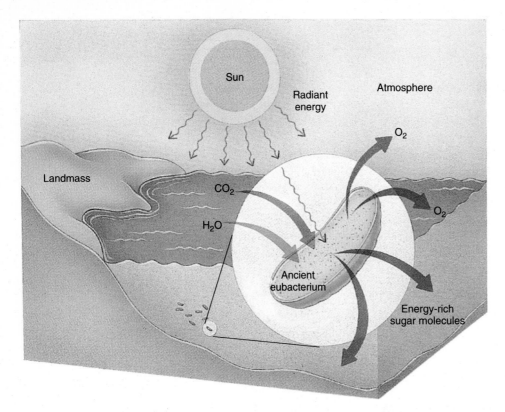

Figure 3.24 Photosynthesis is a chemical process in which radiant energy from the sun is used to convert carbon dioxide (CO_2) and water (H_2O) into energy-rich sugars with oxygen (O_2) released as a byproduct. Scientists believe that photosynthesis evolved in eubacteria more than 2.5 billion years ago, although it may have developed much earlier in Earth's history.

Question: *How did photosynthesis change Earth's early environment?*

tive energy source. Thus prokaryotic photosynthesis provided two of the original conditions thought essential for modern cells to arise on ancient Earth: free O_2 and protection from UV radiation.

Despite its primitiveness in comparison with modern eukaryotic cells, prokaryotic life must be judged a marvelous success. Early prokaryotes helped transform Earth from a violent, barren landscape into a fertile planet capable of accommodating eukaryotic cells that began to appear 2 billion years ago.

The Origin of Eukaryotic Cells

Most organisms present on Earth today are composed of eukaryotic cells. These cells all have a similar structure, use the same systems for carrying on chemical reactions necessary for normal life, and employ identical mechanisms for processing and preparing their genetic material for transmission to the next generation. Because of these similarities, biologists believe that eukaryotic cells evolved from a single, common precursor cell. But what was the nature of this ancestral cell, and how did the complicated eukaryotic cell, which first appeared about 2 billion years ago, evolve?

In the 1960s, new ideas were formulated that traced the emergence of the eukaryotic cell line from the progenote (see again Figure 3.23). According to current hypotheses, the first cell type to evolve from the progenote was the *urkaryote* (*ur* is German for "primordial" or "ancient"). Just as for the progenote, we have no direct evidence of the urkaryote. However, analyses of the molecular structure of prokaryotic and eukaryotic cells from hundreds of different organisms suggest that the urkaryote lineage appeared at about the same time as the prokaryote lineages.

The **endosymbiosis hypothesis** envisions eukaryotic cells arising about 2 billion years ago as a result of unions between at least two types of eubacteria that became to live within ancestral eukaryotic cells that had evolved from the urkaryote. Two organelles found in modern eukaryotic cells are of particular importance in this hypothesis: mitochondria and chloroplasts. Mitochondria are found in all eukaryotic cells and function to provide energy for the cell. Chloroplasts function in photosynthesis and are found only in eukaryotic plant and algal cells. Both mitochondria and chloroplasts contain ribosomes and nucleic acids. Molecular biologists have found that the appearance and structure of eukaryotic mitochondria and chloroplasts are very similar to ancient and existing prokaryotic cells. In other words, today's mitochondria and chloroplasts were once free-living prokaryotic cells that were engulfed by ancestral eukaryotes. Once inside ancestral eukaryotic cells, a mutually beneficial (symbiotic) relationship was established, which led to the evolution of modern eukaryotic cells. What were the benefits for each cell type? Mitochondria require oxygen for carrying out their energy-releasing reactions; therefore, ancient ancestral eukaryotic cells containing "mitochondria" (symbiotic prokaryotic cells) would have had the means for creating abundant energy for use in complex activities. Likewise, eukaryotic cells with "chloroplasts" (another type of symbiotic prokaryotic cell) would have been able to create their own energy supply. The prokaryotic symbionts would have benefited by being sheltered and by having access to nutrients in the eukaryotic cell host.

The Origin of Multicellular Organisms

Multicellular forms of life were abundant on Earth by 600 million years ago. Most major groups of animals without backbones that are present on Earth today are represented in the fossil record from that time. However, because of the general absence of fossils from about 600 million and 1.5 billion years ago, little is known about how multicellular organisms arose from unicellular (single-cell) eukaryotes. It is clear, though, that an explosive increase in animal size and diversity occurred. The great diversification of plants occurred later, after life became established in terrestrial environments. Chapters 5 and 6 describe the diversity of life on Earth.

Multicellular organisms, constructed with eukaryotic cells, appeared on the planet less than a billion years ago and now seem to dominate it. However, that is an illusion related to size—microscopic prokaryotes outnumber eukaryotic organisms by at least 10^{100} to 1. The prokaryotes, although no longer the sole inhabitants of our planet, continue to play a critical role in maintaining life on Earth.

BEFORE YOU GO ON   Prokaryotic and eukaryotic cells are thought to have evolved from a single ancestral life form called a progenote. Prokaryotic cells evolved and became widespread first, and they changed conditions on Earth in a way that favored the development and evolution of more complex eukaryotic cells. Eukaryotic cells first appeared about 2 billion years ago. According to the endosymbiosis hypothesis, modern eukaryotic cells evolved through symbiotic relationships formed between ancestral eukaryotic cells and certain prokaryotic cells.

THE CURRENT PICTURE

During the last three decades new geological discoveries have provided data related to cell evolution (see the Focus on Scientific Process, "Cell Evolution"). The oldest prokaryotic microfossils are approximately 3.5 billion years old. Eukaryotic microfossils date back to almost 2 billion years ago, although eukaryotic cells did not become abundant until about 1 billion years ago. Consequently, knowledge of life in Precambrian time has increased dramatically in the last decade. Although while much remains to be learned, the following tentative picture has emerged from research results described in the previous section (see Figure 3.25).

- Liquid water existed early in Earth's history, by at least 3.8 billion years ago.

- Prokaryotic cells existed in early marine environment at least 3.5 billion years ago.

- Photosynthesis evolved between 2.7 and 3.5 billion years ago.

- Atmospheric O_2 levels increased significantly about 2 billion years ago.

- The first eukaryotic cells appeared about 2 billion years ago, and they may have been present even earlier.

- Eukaryotic cells were common in marine environment 1.7 to 1.9 billion years ago and were widespread by 1 billion years ago.

- The first prokaryotic cells appeared on land, beginning about 1 billion years ago.

- Multicellular organisms were abundant on Earth 600 million years ago.

Figure 3.25   Could Earth have looked like this 3 or 3.5 billion years ago? Evidence suggests that during the first billion years after its formation, Earth was characterized by erupting volcanoes, lightning, poisonous gases in the atmosphere, and little, if any, water on its surface. As the planet matured, it cooled and became covered with water, gases in the atmosphere changed, and the first life forms (perhaps stromatolites) developed.

Cell Evolution

Research on the origin and evolution of cells has greatly expanded during the last ten years, primarily because of new data and knowledge from geology and molecular biology. Recall the hypothesized appearance of cells in early marine environments on Earth, from first to most recent: precellular form → progenote → prokaryotic cells → modern eukaryotic cells that arose through symbiotic relationships between ancestral eukaryotes and certain prokaryotes that today are mitochondria and chloroplasts. What types of evidence support this hypothesized progression of cell evolution?

The Archean Microfossil Record

Fossils constitute highly convincing evidence of biological history. It is not surprising, therefore, that Archean fossils have been sought since the 1800s; however, several obstacles have hindered success. Most notably, few existing rock formations on Earth, which might contain Archean fossils, are billions of years old. Also, older methods for dating rock formations and, hence, fossils, were relatively crude, and it was hard to identify and classify "candidate" Archean microfossils with confidence.

It was not until the 1960s that new technologies and discoveries generated success in the search for authentic Archean microfossils. What changed the situation? First, methods were developed that could accurately determine the age of geological rock formations. Second, two well-preserved rock formations were discovered—one in eastern South Africa and another in the remote outback region of Western Australia—that were shown to be 3.3 to 3.5 billion years old. Third, microfossils were found in these two ancient rock formations. An important related finding was the discovery of the oldest known rock formation on Earth, determined to be 3.8 billion years old, in southwestern Greenland. However, during geological history, the Greenland rocks had been greatly modified by heat and high pressure, which would have destroyed any microfossils even if they had existed. Nevertheless, the Greenland rock formation has structural characteristics indicating that liquid water and carbon dioxide existed on Earth when they were formed. Finally, the microfossils were subjected to careful study and analysis.

Stromatolites

Stromatolites are mound-shaped, finely layered rock structures that form reefs in shallow seas. The existence of these curious structures was recognized in the fossil record in the 1800s, but little was known about their formation (see Figure 1A). That changed, however, when *living* stromatolites were discovered in the 1960s (see Figure 1B). Living stromatolites consist primarily of calcium-containing rock structures that are formed by surface coverings of complex communities composed of different prokaryotes.

The oldest known stromatolites have been found in the South African and Australian rock formations and are, therefore, 3.3 to 3.5 billion years old. They appear to be like modern stromatolites, both living and fossilized, yet no prokaryotic microfossils have been found in these ancient structures. Because the occurrence of microfossils would support the hypothesis that prokaryotes existed on Earth 3.4 to 3.5 billion years ago; what does their absence indicate? Basically, that whereas the Archean stromatolites are suggestive of prokaryotic life, they raise the level of certainty for the hypothesis only slightly.

A   B

Figure 1   (A) Fossilized stromatolites appear as rocky laminated mounds or columns. (B) The discovery of living stromatolites revealed that the layers are produced by prokaryotic cells that live in mats on the rocky surface. When the cells die, they become fossilized and are replaced by new cells at the surface. Fossilized stromatolite reefs may represent the oldest record of life on Earth.

Microfossils

Microfossils, shown in Figure 2, were found at both the South African and Australian sites in the last ten years. They are filamentlike and, in most cases, their appearance is nearly identical to modern prokaryotes known as *cyanobacteria.* These discoveries were significant for two reasons: They indicate that prokaryotes existed on Earth 3.5 billion years ago, and, since cyanobacteria are photosynthetic, this oxygen-producing process may have already evolved by this time! This latter point is especially intriguing because the raw materials of photosynthesis—carbon dioxide and water—were apparently present on Earth by 3.8 billion years ago, as indicated by the Greenland geological deposits. Analysis of geologic rock records, however, indicates that oxygen did not increase in the atmosphere until approximately 2 billion years ago. If photosynthesis evolved over 3 billion years ago, why was oxygen not abundant until 2 billion years ago? Most of the oxygen produced by photosynthesis between 2 and 3 billion years ago is thought to have oxidized (reacted with) iron, which was extremely abundant in the early seas, forming molecules that sank and accumulated in ocean basins. By 2 billion years ago all of the "free" iron had been oxidized, and oxygen began to accumulate in the atmosphere. The iron deposits formed during the Archean eon occur widely on Earth, and today these iron-rich ores are mined, processed, and used to make steel. Is it possible that prokaryotic cells evolved earlier than 3.5 billion years ago? Because no other rocks with ages greater than 3.5 billion years are known, questions concerned with possible earlier Archean cells remain out of reach for the present time.

The Algonkian Microfossil Record

The Algonkian fossil record is considerably more abundant than the Archean record, which is rather sparse. Algonkian microfossils have provided rich details about the history of life on Earth beginning about 2 billion years ago, the time at which oxygen was increasing in the atmosphere.

Many prokaryotes have been found in the Algonkian fossil record. A remarkable feature of these fossils is that they bear a striking resemblance to modern prokaryotes. Thus it can be reasonably concluded that this group has hardly evolved or changed in the last 2 to 3 billion years. (Why might that be the case?) Recent evidence also suggests that prokaryotic cells began to inhabit terrestrial environments about 1 billion years ago.

What does the Algonkian microfossil record reveal about eukaryotic cells? How could the earliest eukaryotic microfossils be distinguished from prokaryotic microfossils? Because most eukaryotic cells require oxygen, it is unlikely that they could have appeared much before 2 billion years ago, when oxygen became more abundant in the atmosphere. The difference in size between prokaryotic and eukaryotic cells is the most useful distinguishing feature in identifying microfossils. Prokaryotes are generally less than 10 micrometer in diameter, whereas single-celled eukaryotes tend to be larger. Microfossils greater than 60 micrometers in size would be classified as "definite" eukaryotic cells; microfossils between 10 and 60 micrometers would be "probable" eukaryotic cells. Microfossils that satisfy the probable criteria have been found in China and are estimated to be 1.8 to 1.9 billion years old. Recently, large spiral-shaped microfossils resembling eukaryotic algae have been discovered and are estimated to be 2.1 billion years old. Other geological (and biochemical) evidence suggests that eukaryotic cells were common, if not abundant, in existing marine ecosystems as early as 1.7 to 1.9 billion years ago.

In the last decade, much has been learned about cell evolution. New discoveries seem to be pushing the dates at which prokaryotic and eukaryotic cells arose on Earth back further into history. Although exact dates and details about cell origin and evolution may remain elusive, the general picture seems to be coming into sharper focus as a result of studies such as those described in this chapter.

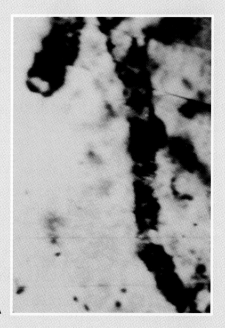

A

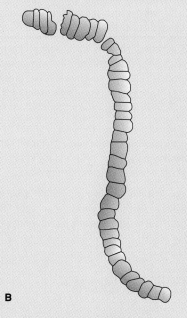

B

Figure 2 Microfossils found in South Africa and Australia that are approximately 3.5 billion years old. They are similar to modern prokaryotic cyanobacteria.

As always, new questions or modified questions have been raised as a result of fresh research results. For example, is it possible that cells evolved earlier than 3.5 billion years ago. Because prokaryotes existed by 3.5 billion years ago, how much earlier did precellular forms evolve? Can you think of other new questions? The answers now guide research in the field of origins. The future conclusions should be very exciting.

SUMMARY

1. During the nineteenth century, the cell theory proposed that all living organisms were composed of cells and that all cells arose from preexisting cells.

2. The structure and function of cells became an important area of research. New technologies, such as the electron microscope (EM), provided scientists with the opportunity to examine cells at extremely high powers of magnification. As a result, previously unknown internal cell structures were identified, and their functions were determined using other techniques.

3. Eukaryotic cells consist of a nucleus and cytoplasm enclosed by a plasma membrane. Within the cell cytoplasm are small membrane-bound organelles that carry out specific activities and a cytoskeleton that provides support and allows for movement. Major organelles found in plant and animal cells include endoplasmic reticulum (ER), Golgi complexes, and mitochondria. Plant cells also contain chloroplasts and vacuoles.

4. Prokaryotic cells do not have a nucleus, membrane-bound organelles, or a cytoskeleton. All of their cellular activities occur in the cytoplasm rather than in the organelles.

5. Life on Earth is hypothesized to trace its origins to organic chemicals that were present in the early oceans. These chemicals became organized into self-replicating systems that later evolved into precellular systems.

6. It is now hypothesized that both prokaryotic and eukaryotic cells evolved from the progenote, a precellular stage of organization that existed between 3.5 and 4 billion years ago.

7. The earliest cells on Earth were prokaryotes. Over millions of years, they changed Earth's environment and evolved a new process called photosynthesis. The products of photosynthesis—sugars and oxygen—ultimately led to the evolution of new, more complex eukaryote cells about 2 billion years ago.

8. The endosymbiosis hypothesis states that modern eukaryotic cells arose from ancestral eukaryotic cells that came to shelter certain prokaryotic cells—today's mitochondria and chloroplasts—in a symbiotic relationship.

9. The first multicellular organisms arose from unicellular eukaryotes about 1 billion years ago. These early life forms gave rise to all the plants and animals on Earth today.

WORKING VOCABULARY

active transport (p. 43)
cell (p. 36)
cell theory (p. 36)

cell wall (p. 40)
chloroplast (p. 46)
cytoskeleton (p. 46)

simple diffusion (p. 42)
endoplasmic reticulum (ER) (p. 44)
endosymbiosis hypothesis (p. 52)
eukaryotic cell (p. 37)
facilitated diffusion (p. 42)
Golgi complex (p. 45)

mitochondrion (p. 46)
nucleus (p. 44)
organelle (p. 37)
osmosis (p. 42)
photosynthesis (p. 51)
plasma membrane (p. 40)
prokaryotic cell (p. 37)
ribosome (p. 44)

REVIEW QUESTIONS

1. What are the major ideas of the cell theory?

2. What are the general functions or roles of the plasma membrane, the nucleus, and the cytoplasm?

3. What are the specific functions of the following cellular organelles: rough and smooth endoplasmic reticulum, Golgi complexes, mitochondria, and chloroplasts? Describe the structure and function of the cytoskeleton.

4. What are the differences between prokaryotic cells and eukaryotic cells?

5. What events may have led to the formation of the first precellular system?

6. What is a progenote? What is its hypothesized relationship with prokaryotic and eukaryotic cells?

7. How did photosynthesis and changes in atmospheric oxygen concentrations influence the evolution of life on Earth?

8. How does the endosymbiosis hypothesis explain the origin of eukaryotic cells?

ESSAY AND DISCUSSION QUESTIONS

1. Why is the cell theory classified as a unifying scientific theory with broad application?

2. One of the key ideas in the early cell theory was that the difference between living and nonliving objects was in the material organization of living beings. Is this an adequate distinction today? Why or why not?

3. How do biologists explain the common structures and organelles that are present in the cells of all eukaryotic organisms?

4. Is it possible that a biological system, such as a precellular form or progenote, could arise from nonliving materials on Earth today? Explain.

5. If life is discovered on a distant planet, what features would you expect it to share with Earth's living beings? What differences would you expect?

REFERENCES AND RECOMMENDED READING

Alberts, B., D. Bray, J. Lewis, M. Raff, K. Roberts, and J. D. Watson. 1994. *Molecular Biology of the Cell.* 3d ed. New York: Garland.

Baker, J. R. 1948–1955. The cell-theory: A restatement, history, and critique. A five-part series published over seven years in the *Quarterly Journal of Microscopical Sciences.*

Cherif, A. H., and G. E. Adams. 1994. Planet Earth: Can other planets tell us where we are going? *American Biology Teacher,* 56: 26–37.

Dyer, B. D. and R. A. Obar. 1994. *Tracing the History of Eukaryotic Cells.* New York: Columbia Univ. Press.

Fenchel, T. 1995. *Ecology and Evolution in Anoxic Worlds.* New York: Oxford Univ. Press.

Fredrick, J. F. (ed.). 1981. Origins and evolution of eukaryotic intracellular organelles. *Annals of the New York Academy of Sciences,* 361: 1–512.

Horodyski, R. J., and L. P. Knauth. 1994. Life on land in the Precambrium. *Science,* 263: 494–498.

Kasting, J. F. 1993. Earth's early atmosphere. *Science,* 259: 920–925.

Knoll, A. H. 1992. The early evolution of eukaryotes: A geological perspective. *Science,* 256: 622–627.

Lipps, J. J. 1993. *Fossil Prokaryotes and Protists.* Cambridge, Mass.: Blackwell Scientific.

Margulis, L. 1970. *Origin of Eukaryotic Cells.* New Haven, Conn.: Yale University Press.

Schopf, J. W. 1992. *Major Events in the History of Life.* Boston: Jones and Bartlett.

Schopf, J. W. 1993. Microfossils of the early Archean Apex chert: New evidence for the antiquity of life. *Science,* 260: 640–646.

ANSWERS TO FIGURE QUESTIONS

Figure 3.1 Prokaryotic cell.

Figure 3.7 It regulates the movement of ions and molecules into and out of the cell.

Figure 3.10 No. No.

Figure 3.11 The water concentration is greater inside the cell (99 percent) than outside the cell (90 percent) and water diffuses down a concentration gradient.

Figure 3.20 Eukaryotic cells contain a true nucleus and various types of membrane-bound organelles whereas prokaryotic cells do not contain these structures.

Figure 3.22 It has been hypothesized that they were formed in ancient seas beginning about 4 billion years ago (see Chapter 2).

Figure 3.23 Prokaryotic cells (archaebacteria and eubacteria) appeared about 3.5 billion years ago; eukaryotic cells were present 2 billion years ago but may have evolved much earlier; protoctista and multicellular fungi, plants, and animals were probably present 1 billion years ago (± 3 million years).

Figure 3.24 It resulted in an atmosphere rich in oxygen, and it was a new source of organic molecules.

4

Energy Transformations in Cells

Chapter Outline

Reading Questions

1. What are the different forms of energy that are important for sustaining life on Earth?

2. What is photosynthesis, and why is it important?

3. What are glycolysis and cellular respiration, and why are they important?

4. How is ATP generated in cells?

5. How can life on Earth be explained in terms of energy and energy transformation processes?

In Chapter 3, we examined various hypotheses concerning cell evolution. Available evidence tells us that both the origin of cells (between 3.5 and 4 billion years ago) and the subsequent evolution of cells were influenced by an ability to obtain energy. Early in Earth's history, the original supply of organic chemicals, which had served as an ener-gy source for precellular systems, became depleted, and the first life forms faced extinction. A new process—*photosynthesis*—evolved in early prokaryotic cells about 3 billion years ago and, quite literally, saved the day. Photosynthesis consists of chemical reactions in which energy from the sun is converted into organic chemical products and oxygen. The chemical products provided evolving cells with a new source of energy, and the released oxygen changed Earth's environment and led to the evolution of eukaryotic cells. In this chapter, we describe how cells transform energy and how that energy sustains life on Earth.

ENERGY IN THE BIOLOGICAL REALM

There exist on Earth complex levels of organization that are supported by a continuous input of light (radiant) energy from the sun (see Figure 4.1). Enormous quantities of light energy are essential for maintaining temperatures suitable for life on Earth and for driving the water cycle that distributes moisture over its landscapes (described in Chapter 9).

At simpler levels of organization, trees, humans, frogs, worms, and bacteria carry on activities that require an uninterrupted supply of energy. The sun is also the ultimate source of energy for almost all of these biological systems. Only in deep-ocean hydrothermal vent systems, and perhaps a few anaerobic hot springs, are different sources of energy—hydrogen sulfide and methane—used to support life (see Chapter 2).

Plants are able to capture light energy emitted by the sun and convert it into chemical energy. Almost all organisms depend on plants to provide the chemical forms of energy required for carrying out life-supporting functions. In this chapter, we examine the energy-transforming pathways used by organisms and the importance of energy-transforming reactions for sustaining life on Earth.

ENERGY IN THE PHYSICAL REALM

What is energy? Most people have an intuitive sense about energy; it is something associated with activity, movement, or power—the ability to cause change. Scientists describe energy in more technical terms. **Energy** is the ability to do work, and it occurs in two general states. **Potential energy** is stored or inactive energy that is capable of doing work later. A stick of dynamite, a liter of gasoline, and the vast quantities of snow deposited on mountains during winter have potential energy. **Kinetic energy** is the energy of action or motion. A match applied to the dynamite or gasoline will explosively convert the potential energy into kinetic energy. When the warm spring sun melts snow, the water flowing downhill represents kinetic energy (see Figure 4.2A, page 60). Dams harness the kinetic energy of falling water to produce electricity (see Figure 4.2B, page 60).

Forms of Energy

Several forms of energy exist on Earth. **Chemical energy** is the potential energy contained in chemical bonds within compounds such as gasoline, sugar, and fat. Recall that *chemical bonds* are forces of attraction that hold two or more atoms together. When these bonds are broken, energy is released. **Thermal energy** (heat) is the kinetic energy of molecular motion that can be measured with a thermometer. With appropriate mechanical systems, heat can be harnessed to do certain types of work, such as power an automobile or produce electricity. Within the biosphere, the movement of heat from regions of higher to lower temperature creates winds and powers the water cycle. **Electrical energy** is produced by a flow of electrons and may occur in diverse objects such as batteries and the cells of certain organisms. Speeding race cars and flexed muscles reflect **mechanical energy,** which is expressed as motion. Energy emitted from the sun travels to Earth as **radiant energy** in the form of light waves.

Figure 4.1 Almost all organisms on Earth depend on the sun for the energy to sustain their lives.

Question: *How is the sun's energy converted to energy that can be used by organisms on Earth?*

The Laws of Thermodynamics

Where does energy come from? What are the relationships between the various forms of energy? All forms of energy are interrelated and can be converted from one form to another.

Figure 4.2   (A) Snow constitutes a source of potential energy. When snow melts and flows into streams and rivers, it has been transformed into kinetic energy. (B) Dams transform the kinetic energy of moving water into electrical energy.

Energy transformations—changing one form of energy to another—is the basis for sustaining life on Earth. This transformation, however, is not very efficient because some energy is lost during each change. For example, a liter of gasoline (chemical energy) transformed to move (mechanical energy) a car 8 kilometers could actually move it 24 kilometers if the energy transformation were 100 percent efficient. Why are energy transformations so wasteful? Two laws of thermodynamics formulated by physicists help answer questions like this. Recall that laws are the most powerful scientific concepts; they describe regularities that we think are invariable.

The First Law of Thermodynamics

Thermodynamics describes the relationships between heat and other forms of energy. The **first law of thermodynamics** (also called the *law of conservation of energy*) states that energy can be converted from one form to another, but it can never be created or destroyed.

The first law offers this startling perspective: no organism can create the energy it requires for survival. The significance of the first law is that radiant energy from a distant source, the sun, must be converted into chemical energy that can be used by living organisms.

The Second Law of Thermodynamics

The **second law of thermodynamics** states that every energy transformation results in a reduction in the total usable energy of the system. To maintain order, organisms must transform energy every second of their lives. All usable forms of energy are ultimately dissipated as heat through energy transformations. As a result, **entropy**—the disorder (that is, "useless energy") of any system, whether the universe or a cell—continuously increases while order, or organization, decreases.

The ultimate biological consequences of the second law are harsh. Life on Earth will end in several billion years when the sun's energy is finally used up in accordance with the sec-

ond law. Also, all organisms eventually lose their battle against entropy, biological order fades, and the organism dies.

Picture a romantic setting in which dinner is being served by candlelight. To open (and quite possibly close) the conversation, how would you explain the burning of the candles in terms of the first and second laws of thermodynamics?

BEFORE YOU GO ON   An unending supply of energy is required to sustain life on Earth. This supply is provided by radiant energy emitted by the sun, which is captured by plants and converted into chemical energy, a form of potential energy. Organisms transform chemical energy into different forms of kinetic energy that they use for carrying out their activities. The laws of thermodynamics explain the primary consequence of energy transformations—that is, an increase in entropy.

BIOLOGICAL ENERGY TRANSFORMATION PROCESSES

The energy transformation process by which solar energy is captured by plants and converted into high-energy organic compounds, such as the sugar glucose, is called **photosynthesis.** Almost all organisms can convert the chemical energy in these compounds into other forms of energy necessary for their activities through two sets of reactions, termed **glycolysis** and **cellular respiration** (the latter term should not be confused with breathing, or respiring, which refers to the exchange of gases between air and lungs). Figure 4.3A describes the fundamental relationships of the energy-transforming pathways and reviews concepts explored further in Chapter 8. Using light energy, green plants are able to manufacture complex organic compounds ("food") from carbon dioxide and water during photosynthesis.

Photosynthetic organisms, including most plants, are able to use the food they synthesize to support their activities and

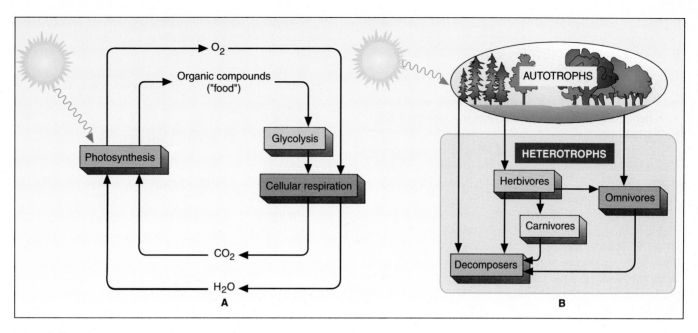

Figure 4.3 (A) Radiant energy from the sun is transformed by plant cells into organic compounds used as food through the process of photosynthesis. Plants and animals carry out the reactions of glycolysis and cellular respiration to convert food (chemical energy) into energy that can be used in the cell. Oxygen, carbon dioxide, and water are recycled through these three biological energy transformation processes. (B) Autotrophs produce food for themselves and for various heterotrophs.

Question: *What ultimately happens to most of the energy in food?*

are classified as **autotrophs** (in ecology, *producers*). **Heterotrophs** (the *consumers* of ecology) are organisms that must obtain energy and nutrients from the food molecules produced and stored by autotrophs. Heterotrophic organisms are further classified according to how they accomplish this task. *Herbivores* obtain energy-rich, carbon-containing compounds directly by eating plants or plant parts. *Carnivores* acquire products manufactured by autotrophs indirectly by consuming herbivores and other animals. *Omnivores,* including humans, obtain energy from either plants or animals. *Decomposers* obtain food by breaking down dead organic materials from any source (see Figure 4.3B).

The reactions of photosynthesis, glycolysis, and cellular respiration are linked through an interdependence of products formed during each process. Heterotrophs and autotrophs both carry out cellular respiration, which results in the production of carbon dioxide and water. These are the raw materials of photosynthesis, which can be cycled back to plants for the synthesis of more sugar. Oxygen, a byproduct of photosynthesis, is required for the reactions of cellular respiration.

The chemical reactions of photosynthesis, glycolysis, and cellular respiration are complex and numerous. We shall limit our descriptions of chemical reactions to a few and concentrate on broader concepts related to these questions:

1. What enables chemical reactions to take place in living cells?

2. How do plants harness the sun's energy?

3. What are the major products of the energy transformation pathways?

4. What are the relationships between the energy-generating and the energy-consuming pathways?

5. How is ATP generated?

ENZYMES

Chemical reactions involve changes in the basic chemical composition of the substances involved. They occur when molecules, which are always in motion, collide with one another randomly and chemical bonds between atoms or molecules are formed or broken. Thousands of chemical reactions normally occur within cells of living organisms; they are required for obtaining energy, manufacturing proteins, maintaining proper temperature, and so on. However, life would end quickly (or would never have evolved) if chemical reactions were limited to infrequent random collisions between molecules within cells. To frame the problem, consider that a rare collision necessary for chemical reaction *X* might occur a few times in the lifetime of a cell, or more likely not at all, but the cell requires 10,000 or more *X* reactions per second! How have organisms solved this problem?

Catalysts are substances that greatly increase the rate at which a chemical reaction occurs. Catalysts are active in small quantities, and they are neither destroyed nor modified during the reaction. **Enzymes** are biological catalysts, composed mostly of protein, that participate in most chemical reactions within cells. Like other catalysts, enzymes greatly accelerate the rate at which chemical reactions occur, but they are not used up during the course of the reaction. A single enzyme molecule may enter into thousands of reactions every second without being modified or degraded. The general concept of enzyme action is shown in Figure 4.4.

Enzymes are highly specific. Most enzymes catalyze only one chemical reaction within cells; they usually react with a single *substrate*—a molecule on which an enzyme acts—changing it into *products*.

Each enzyme has an **active site,** which is a small surface region of the molecule that recognizes and reacts with a substrate. The classic *lock-and-key hypothesis,* first developed in 1884, explains enzyme specificity in terms of the active site's being the "lock" and the substrate's being the "key." According to a more recent hypothesis proposed in 1973, the *induced fit model,* active sites have a flexibility that enables them to close tightly around a substrate molecule so that molecule becomes properly aligned for the reaction to occur.

Once the substrate becomes engaged at the active site, an *enzyme-substrate complex* is formed. The substrate molecule is then altered, broken down, or combined with another substrate molecule to form reaction *products.* After the reaction is completed, enzyme and products separate from one another.

Enzymes are remarkably efficient. Substrate molecules are commonly converted to products at a rate of 1 to 10,000 per second.

Enzymes are regulated by a variety of cellular control mechanisms. Some of these are described in later chapters.

ENERGY-MANAGING MOLECULES

For energy to be produced and used in an orderly manner, certain molecules in cells are employed to manage the energy as it becomes available. Essentially they serve either to store energy or to transport it between reactions until it can finally be released to the cell.

Adenosine Triphosphate

The energy used by plant and animal cells is provided by **adenosine triphosphate** (**ATP**). Figure 4.5 shows the simplified molecular structure and the reactions of ATP. Adenosine triphosphate is composed of the molecule adenosine and three linked phosphate (PO_4) groups identified as ⓟ. *Energy is*

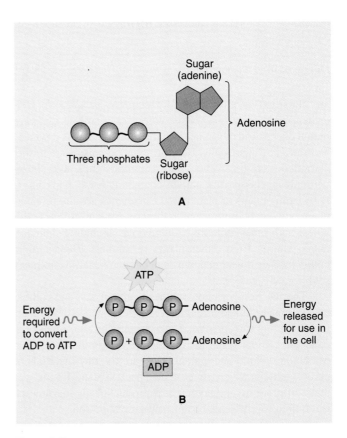

Figure 4.5 (A) Adenosine triphosphate (ATP) is composed of an adenine base, ribose, and three phosphates. The wavy line between phosphates signifies a high-energy chemical bond. (B) ATP is able to both store and release energy at different times. Energy is stored for future use when adenosine diphosphate (ADP) combines with a phosphate (P) to form ATP. Energy is released to the cell when a phosphate is removed from ATP, converting it back to ADP.

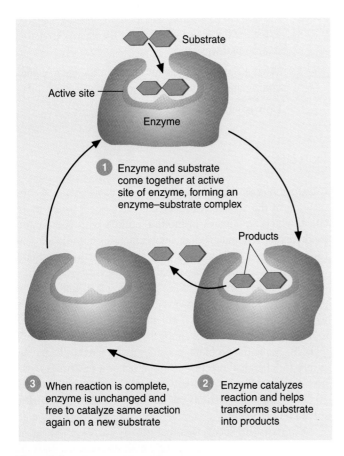

Figure 4.4 How an enzyme works.

released in the cell when the terminal phosphate is removed by a specific enzyme to produce *adenosine diphosphate* (*ADP*) and a phosphate group. This reaction is described as:

$$ATP \rightarrow ADP + (P) + energy$$

Energy is stored when ADP combines with a phosphate to regenerate ATP, which is shown as:

$$ADP + (P) + energy \rightarrow ATP$$

The ATP–ADP system is very dynamic because the cell is a cauldron of constant chemical activities, with some of the reactions releasing energy and some storing energy. To illustrate this flux, consider that an average ATP molecule is consumed within 60 seconds of its formation and that each typical plant or animal cell contains several million molecules of ATP–ADP. Even on the quietest day of your life, your cells require the generation of about 45 kilograms (99 pounds) of ATP simply to keep you alive!

Oxidation–Reduction Reactions

Oxidation–reduction reactions involve the transfer of electrons (e^-) between atoms. Hydrogen atoms are involved in many energy-transforming reactions. A hydrogen atom is composed of one electron and one proton; the proton in this case is identified as H^+. In many oxidation reactions that occur in energy transformations, two electrons and two protons are removed together, which is equal to two hydrogen atoms ($2e^- + 2H^+ = 2H$). Thus **oxidation** occurs when electrons or hydrogen atoms are removed from a molecule. Oxidation reactions cause a decrease in the energy content of a molecule, which means that energy is liberated. **Reduction** occurs when electrons or hydrogen atoms are added to a molecule. Reduction reactions increase the energy content of a molecule, which means that energy is stored. As described in Figure 4.6, oxidation–reduc-

tion reactions are always *coupled*—whenever one molecule is oxidized, electrons and energy are transferred to another molecule that is then reduced.

Electron Carriers

During the chemical reactions of photosynthesis, glycolysis, and cellular respiration, a continuous series of molecular degradation and rearrangement results in the release of energetic electrons that must be captured to prevent damage to the cell. How is this accomplished? **Coenzymes** are organic molecules that are required for the operations of enzymes. Many coenzymes are able to accept the energetic electrons, thereby serving as *electron carriers*. Three coenzymes play important roles in managing energetic electrons produced during biological energy transformations. In this chapter, we will use their common names. When not carrying electrons—that is, when they are in their oxidized form—they are referred to as *NADP+*, *NAD+*, and *FAD*. When they are carrying electrons in their reduced form, they are identified as *NADPH, NADH,* and *FADH$_2$*. Adenosine triphosphates are produced when energy released from electrons carried by NADPH, NADH, and FADH$_2$ is captured in a different chemical form as the electrons are transferred to other electron carriers in later reactions (see Table 4.1).

BEFORE YOU GO ON   There are different biological energy transformation processes, but the organisms involved in each are interrelated. Autotrophs convert radiant energy into chemical energy by photosynthesis. Both autotrophs and heterotrophs transform this chemical energy into forms of kinetic energy during glycolysis and cellular respiration. Several molecules—the ATP–ADP system and various coenzymes—participate in the controlled release of kinetic energy in the cell through oxidation–reduction reactions.

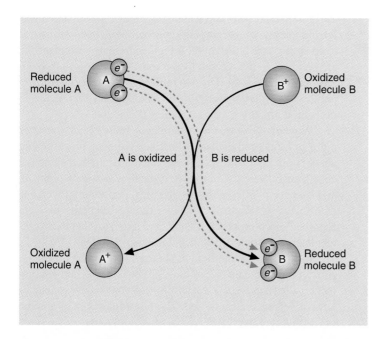

Figure 4.6   Initially, molecule A contains two electrons (e^-) and is reduced, and molecule B, lacking 2 e^- is oxidized. When A is oxidized, its 2 e^- are transferred to B, which then becomes reduced. Oxidized molecules have a positive charge.

Table 4.1 Summary of Energy Reactions That Occur in Cells

Chemical Reaction		This Reaction Occurs During		
Reactions–Products	Outcome	Photosynthesis	Glycolysis	Cellular Respiration
$ATP \rightarrow ADP + \text{P} + energy$	Energy released for use in the cell	✓	✓	✓
$ADP + \text{P} + energy \rightarrow ATP$	Energy stored for later use in the cell	✓	✓	✓
$NADP^+ \rightarrow NADPH + H^+$	Energetic electrons captured by coenzyme	✓		
$NADH + H^+ \rightarrow NADP^+$	Energy from electrons released	✓		
$NAD^+ \rightarrow NADH + H^+$	Energetic electrons captured by coenzyme		✓	✓
$NADH + H^+ \rightarrow NAD^+$	Energy from electrons released			✓
$FAD \rightarrow FADH_2$	Energetic electrons captured by coenzyme			✓
$FADH_2 \rightarrow FAD$	Energy from electrons released			✓

All l iving organisms transform energy at the cellular level. **Plants** use all of these reactions because they carry out photosynthesis, glycolysis, and cellular respiration. **Animals** use only the reactions of glycolysis and cellular respiration.

PHOTOSYNTHESIS

Photosynthesis consists of a series of complex chemical reactions driven by energy originally supplied by the sun. What is the significance of this process for organisms inhabiting Earth? All plants and animals require food and oxygen for their survival. Plants manufacture their own food through photosynthesis. However, animals cannot make food and are ultimately dependent on the energy-rich, organic compounds produced by plants. Thus each of the millions of carbon molecules in every one of your billions of cells came from a molecule of carbon dioxide that was once present in the atmosphere. The warmth of a fire on a cold winter night represents energy that originally came from the sun tens, hundreds, or millions of years ago. Finally, most organisms require the oxygen that is released during photosynthesis. According to contemporary scientific hypotheses, if the remarkable process called photosynthesis had not evolved, there would be no oxygen and no food supply. The most familiar life forms—plants and animals—would never have appeared on Earth.

Photosynthesis occurs primarily in the cells of green plants, algae, and cyanobacteria. The process is commonly summarized as follows: Radiant energy from the sun is used to convert carbon dioxide (CO_2) and water (H_2O) into glucose ($C_6H_{12}O_6$) and oxygen (O_2) according to the following simplified equation. Equations like this summarize the number and types of molecules *used* (the "reactants") to the left of the arrow and the number and types *produced* (the "products") to the right of the arrow.

$$6CO_2 + 6H_2O + \text{radiant energy} \rightarrow C_6H_{12}O_6 + 6O_2$$

Radiant Energy

As described in Chapter 2, the sun emits radiant energy consisting of various *photons* that are each composed of a different wavelength and energy level (refer to Figure 2.16).

The shorter the wavelength, the greater the energy of the photon. High-energy photons called gamma rays and X rays are very harmful to biological systems because they can cause mutations in DNA, but they rarely reach Earth's surface. Ultraviolet radiation is also harmful, but it but is mostly blocked by gases (especially ozone) present in Earth's upper atmosphere.

Only photons that fall within the narrow *visible light spectrum* possess energy that can be captured and transformed through biological processes without causing harm. Photons with characteristic wavelengths in the visible light spectrum are perceived as distinct colors. The shortest wavelengths are seen as violet or blue colors, whereas the longest visible light photons appear to be red.

Light-trapping Pigments

How do plants capture energy from sunlight? Plant *pigments* are protein molecules capable of absorbing photons and transferring energy by rearranging their own molecular structures to create an "energized state." A pigment remains energized for only a fraction of a second; then it transfers its energy through a variety of processes. The two most important mechanisms in photosynthesis are transmitting the energy to another molecule or using it directly to drive a chemical reaction.

Why are the leaves of most plants green? Many types of pigments are found in plant cells. Each pigment, however, can absorb only light with specific wavelengths; those they do not absorb are reflected. *Chlorophyll* is the most common pigment, and it exists in different forms, the most important of which is *chlorophyll a*. Figure 4.7 shows the wavelengths of light absorbed by chlorophylls *a* and *b* and indicates that these pigments absorb primarily blue and red wavelengths.

In addition to chlorophyll, photosynthetic cells (or organisms) usually contain one or more yellow, orange, and red pigments called *carotenoids* that absorb violet and blue wavelengths but not yellow, orange, or red. The beautiful colors of autumn

Figure 4.12 During anaerobic exercise, fermentation reactions lead to the production of lactate, a compound that causes muscle cramps unless sufficient oxygen becomes available for its breakdown.

colysis cannot occur, and the cell will not be able to obtain any energy. Second, large amounts of potential energy remain in the products of fermentation. Lactate cannot be stored and is broken down when adequate oxygen becomes available. Ethanol, however, can either be broken down or converted into fat, a storage molecule.

CELLULAR RESPIRATION

Microorganisms, certain simple plants and animals, and even some specialized cells, such as human red blood cells, can function with the meager amount of energy obtained through glycolysis. However, most cells and tissues of multicellular organisms require large amounts of ATP that are generated by the complete oxidation of glucose during the aerobic reactions of cellular respiration. Cellular respiration consists of two sets of reactions that result in the release and capture of energy in small steps: *Krebs cycle reactions* and *electron transport chain reactions.*

In eukaryotic cells, the many reactions of cellular respiration take place in specialized organelles called *mitochondria* (described in Chapter 3). Mitochondria consist of a complex system of folded membranes that are organized to carry out aerobic respiration reactions (see Figure 4.13). All of the enzymes, coenzymes, and other molecules necessary for Krebs cycle reactions are found within the matrix of mitochondria. The cascade of events that occurs in electron transport chain reactions takes place in the inner mitochondrial membrane.

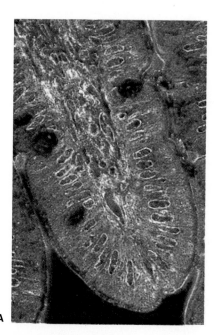

A

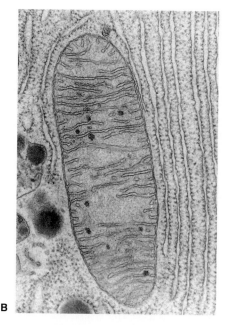

B

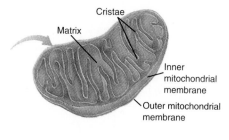

C

Figure 4.13 (A) An animal cell contains numerous mitochondria. (B) Mitochondria are highly specialized organelles composed primarily of membranes in which the reactions of cellular respiration are carried out. The inner membrane of mitochondria is folded into cristae that project into the internal matrix. (C) Krebs cycle reactions take place in the matrix, and electron transport chain reactions take place in the inner membrane.

Krebs Cycle Reactions

Following glycolysis, molecules of pyruvate move into the mitochondrial matrix (see Figure 4.14). In a preparatory step pyruvate is converted to a two-carbon fragment called acetyl-coenzyme A, or *acetyl-CoA,* for each molecule of glucose:

$$2 \text{ pyruvate} + 2NAD^+ + 2 \text{ coenzyme A} \rightarrow$$
$$2 \text{ acetyl-CoA} + 2NADH + 2H^+ + 2CO_2$$

As described in Figure 4.15, each acetyl-CoA molecule then proceeds through the **Krebs cycle reactions** (named after Hans Krebs, who described them in 1937 and later won a Nobel Prize for his research), where it is completely oxidized, yielding a variety of products:

$$2 \text{ acetyl-CoA} + 6NAD^+ + 2FAD + 2ADP + 2\text{\textcircled{P}} \rightarrow$$
$$6NADH + 6H^+ + 2FADH_2 + 2ATP + 4CO_2 + 2 \text{ coenzyme A}$$

Note that two turns of the Krebs cycle occur for each glucose molecule because two pyruvate molecules are produced from each glucose molecule.

After completing the Krebs cycle, virtually nothing remains of the original glucose molecule. Its individual atoms have met various fates. Some of the oxygen and hydrogen atoms were used to form a water molecule; all carbon and some oxygen atoms were exhaled as carbon dioxide, some of which recycles back to the plant realm for use in photosynthesis; and the high-energy electrons released were used to reduce NAD^+ or FAD. Coenzyme A molecules are also recycled. The high yield of ATP finally occurs in the next set of reactions, when the electron carriers, NADH and $FADH_2$, are finally relieved of their cargo.

ATP Production

Thus far the amount of ATP produced from various reactions in glycolysis and the Krebs cycle has been limited. Glycolysis reactions yields only two ATP, and Krebs cycle reactions also nets two ATP. Yet much of the energy liberated during these reactions has been stored in the form of high-energy electrons carried by NADH and $FADH_2$. How is this energy finally converted to ATP?

The Electron Transport Chain

During electron transport chain reactions, NADH and $FADH_2$ pass their high-energy electrons on to a series of more than 15 different electron carriers located in the inner mitochondrial membranes. The various electron carriers are organized into three large *respiratory enzyme complexes* in the mitochondrial membrane. Figure 4.16 (see page 72) illustrates events in the electron transport chain. Initially, two hydrogen atoms are removed from NADH and $FADH_2$ and separated into protons (H^+) and electrons (e^-). Electrons from NADH enter the first respiratory enzyme complex while those from $FADH_2$ enter at the second complex. Electrons are passed through carriers of one complex and then transferred sequentially from one complex to another by two mobile electron carriers denoted as Q and cytochrome *c*. Each electron carrier is alternately reduced as it picks up electrons; it is then oxidized as it passes the electron to the next carrier. When a pair of electrons finally completes its movement along the chain, it combines with oxygen, the last electron acceptor, and protons to form a water molecule. About 90 percent of the oxygen we inhale from the air is used for this process.

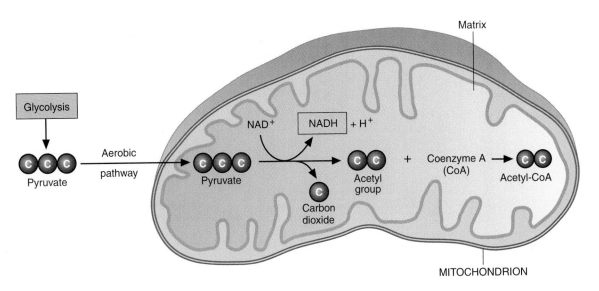

Figure 4.14   When abundant oxygen is present, pyruvate enters into the aerobic reactions of cellular respiration. Pyruvate enters the mitochondrial matrix where it is oxidized to form an acetyl group, which then combines with coenzyme A to form acetyl coenzyme A (acetyl-CoA).

Question: *Which molecule shown in this figure becomes reduced?*

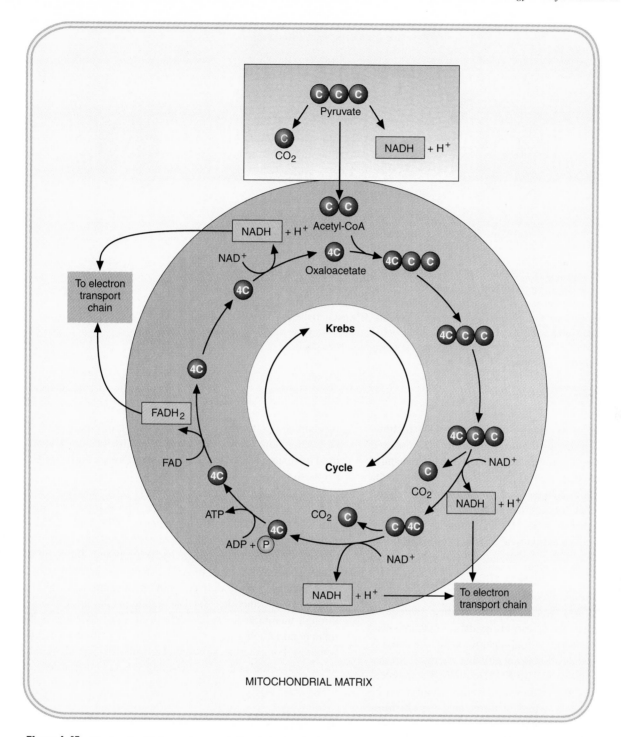

MITOCHONDRIAL MATRIX

Figure 4.15   During the Krebs cycle, acetyl-CoA initially combines with a four-carbon compound called oxaloacetate that "escorts" it through the series of reactions. Each arrow represents one or more reactions in the Krebs cycle. During the Krebs cycle the two-carbon acetyl-CoA undergoes various oxidation reactions, which lead to its being completely degraded. The energy and electrons from the oxidation reactions reduce electron carriers (3 NAD$^+$ → 3 NADH + H$^+$ and FAD → FADH$_2$) and generate 1 ATP. The four-carbon oxaloacetate undergoes various modifications during the Krebs cycle but is reformed by the end of the cycle.

Question: *What happens to the CO$_2$ given off during the Krebs cycle?*

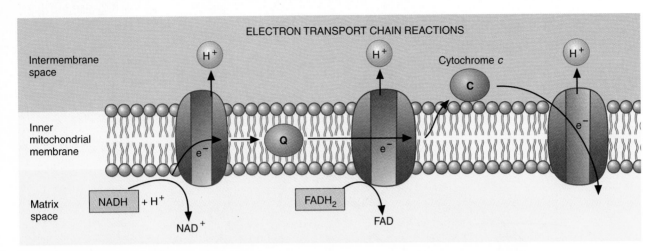

Figure 4.16 In electron transport chain reactions, energetic electrons (*e*−) carried by NADH and FADH₂ are released to one of three large respiratory enzyme complexes in the inner mitochondrial membrane. Each complex contains numerous electron carriers. After entering a complex, *e*− react sequentially with carriers in each complex, releasing energy in a stepwise fashion that is used to pump protons (H⁺) out into the intermembrane space. Electron carriers designated as Q and cytochrome *c* are capable of moving electrons between the respiratory enzyme complexes. At the end of the electron transport chain, the *e*−, which have been depleted of energy, combine with protons (H⁺) and oxygen to form water.

Question: *Why do we need oxygen?*

What happens to the energy released during oxidation–reduction reactions involving the electron carriers? It is not used to generate ATP directly; rather, it is used by the three respiratory enzyme complexes to pump H⁺ from the matrix through the inner mitochondrial membrane to the inner membrane space. Consequently, the H⁺ concentration outside the membrane becomes much higher than the H⁺ concentration inside. This concentration difference is finally used to generate ATP.

Chemiosmosis

Before 1961, various hypotheses had been advanced to explain how ATP was generated. The leading hypothesis explained that the energy of various oxidation reactions was somehow used to generate a high-energy bond between a Ⓟ and unknown molecules. Subsequently, the energy from that bond was used to convert ADP to ATP. However, despite intensive research efforts, the unknown molecules could never be identified. In 1961, Peter Mitchell formulated the **chemiosmosis hypothesis,** which proposed that ATP was generated through the diffusion of H⁺ back into the mitochondrial matrix through a special channel in the inner membrane; this process is known as **chemiosmosis.** At the time, the hypothesis seemed highly speculative and lacked experimental support. However, experimental supporting evidence soon became available, the hypothesis became widely accepted, and Mitchell received a Nobel Prize for his work.

As shown in Figure 4.17, H⁺ concentrations in the inner membrane space are much higher than in the matrix; consequently H⁺ diffuses rapidly back across the inner membrane through a large transmembrane enzyme complex called *ATP synthetase.* ATP synthetase converts the energy of H⁺ flow into

ATP, much like turbines in the dam shown in Figure 4.2 convert water flow energy into electricity. Chemiosmosis is also used to generate ATP in chloroplast membranes during photosynthesis.

The complete processing of one NADH molecule yields two (those from glycolysis) or three (those from the Krebs cycle) ATP, whereas one FADH₂ yields two ATP. Table 4.2 summarizes the theoretical maximum quantity (net) of ATP that can be produced from the complete oxidation of one glucose molecule. The actual number of ATP generated is probably closer to 25 because energy derived from chemiosmosis is used for other cellular functions, such as moving different molecules through membranes. Once synthesized, most ATP leaves the mitochondria and enters the cell, where it is used in various activities. For example, substantial amounts of ATP are used by the cytoskeleton and in synthesizing DNA. Figure 4.18 (see page 74) summarizes energy transformation reactions in cells.

BEFORE YOU GO ON Glucose is ultimately converted to ATP through oxidation–reduction reactions within the cell. During glycolysis, glucose is partially broken down into two molecules of pyruvate. In the absence of oxygen, pyruvate enters into fermentation reactions, where it is converted to ethanol or lactate, depending on the cell type. If oxygen is present in the cell, pyruvate moves into Krebs cycle reactions, where it is completely oxidized. Electrons liberated during oxidation reactions are captured by coenzymes and then processed in the electron transport chain, where their energy is used to pump protons into the intermembrane space. Through the process of chemiosmosis, the flow of protons back into the mitochondrial matrix results in the production of ATP.

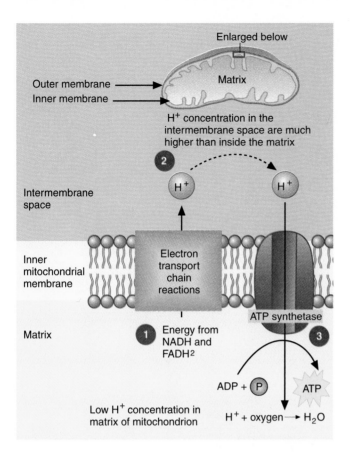

Figure 4.17 As a result of electron transport chain reactions (1), proton (H^+) concentrations in the intermembrane space are much higher than inside the matrix (2). Consequently, H^+ rapidly diffuses back through the mitochondrial membrane through specialized protein channels called ATP synthetase. ATP synthetase is able to use the energy from the H^+ flow to synthesize ATP in much the same way that a dam converts water flow energy into electricity (3). This process is known as chemiosmosis.

THE SIGNIFICANCE OF PHOTOSYNTHESIS AND CELLULAR RESPIRATION

Scientists have developed hypotheses to address questions about the evolution of biological energy transformation processes and the relationships among those reactions. Because photosynthesis, glycolysis, fermentation, and cellular respiration evolved several billion years ago, the following hypothesized events cannot be proven. Rather, as with all good hypotheses, they are consistent with the facts now available. (See the **Focus on Scientific Process**: The Road to Understanding Photosynthesis.)

Recall that the first forms of life on Earth are hypothesized to have evolved between 3.5 and 4 billion years ago in primordial oceans (see Chapter 2). These first life forms may have obtained energy from organic molecules present in the early oceans. But how may we assume that ATP was present in these early oceans? Every organism present on Earth today uses ATP for energy. This suggests that ancestral prokaryotic cells came to possess metabolic machinery for processing ATP, rather than some other molecule, that this capability had adaptive significance, and that it was passed to all descendant cells. As ATP and other organic chemicals became depleted, natural selection may have favored cells that could create ATP from available molecules. Thus glycolysis and fermentation reactions are thought to have been the first biological energy-transformation systems on Earth. These ancient energy-releasing pathways probably evolved between 3 and 4 billion years ago, when little oxygen was present in the atmosphere (see Chapter 3). Some modern organisms, such as certain bacteria, still depend entirely on glycolysis and fermentation for energy.

Nearly all organisms have retained glycolysis as a first step in the reactions they use to release energy. The major disadvantage of a total dependency on those reactions is that only

Table 4.2 Breakdown of Glucose

Reaction	Net ATP Gain	Other Products Formed
Glycosis	2	2NADH
Pyruvate to acetyl-CoA	0	2NADH
Krebs cycle	2	$2FADH_2$ + 6NADH
Electron transport chain		
2 NADH (from glycolysis)	4	
2 NADH (from pyruvate → acetyl-CoA)	6	
$2FADH_2$ (from Krebs cycle)	4	
6NADH (from Krebs cycle)	18	
Total ATP	**36**	

This table summarizes ATP production from the complete breakdown of one molecule of glucose during cellular respiration.

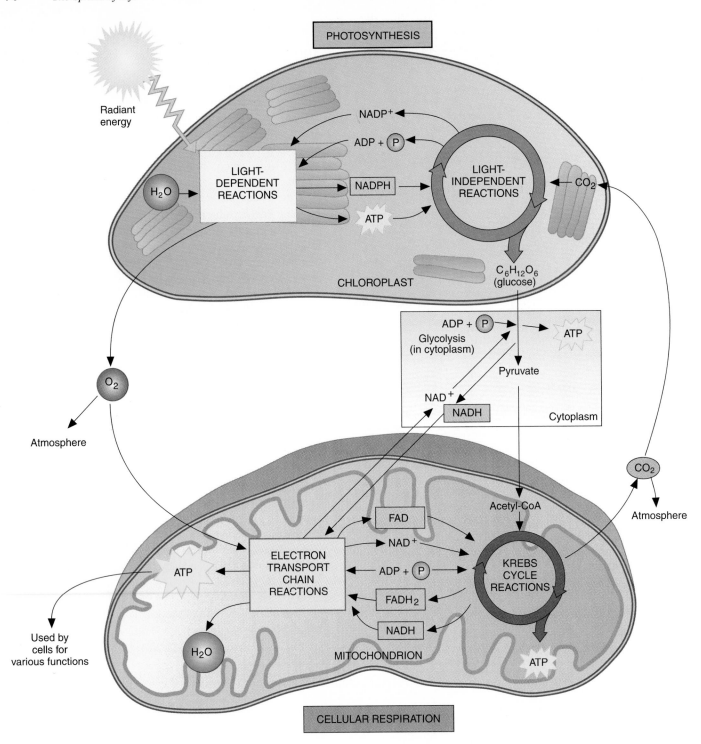

Figure 4.18 A summary of cellar energy transformations. In photosynthesis, radiant energy is captured in light-dependent reactions used to generate NADPH and ATP, which are then used in light-independent reactions to produce glucose. During glycolysis, glucose is converted to pyruvate in the cytoplasm of cells. Subsequently, pyruvate molecules enter mitochondria, where they are completely broken down during Krebs cycle reactions to yield various products, including NADH and $FADH_2$, which carry energetic electrons. The energy from these electrons is converted to significant quantities of ATP during electron transport chain reactions. Note that most of the chemicals involved in cellular energy transformations—ATP+(P), $NADP^+$-NADH, and $FADH$-$FADH_2$—are recycled continuously in the cell. Carbon dioxide (CO_2) and oxygen (O_2) molecules enter the atmosphere and may take part in subsequent cellular energy transformation reactions.

The Road to Understanding Photosynthesis

In science, modern understandings of many fundamental mechanisms have been derived, in part, from experiments conducted by earlier investigators. Today, the molecular events of photosynthesis are known in great detail. However, to understand the advancement of scientific knowledge, it is instructive to review some of the early, classical experiments that shed the first light on this critical biological process. An examination of these works will also reveal how results of experiments are evaluated in an existing theoretical context. Scientific studies are based on ideas, hypotheses, and theories that prevail at the time they are undertaken. Investigations relating to photosynthesis began early in the seventeenth century. At that time, naturalists were asking two questions: What accounts for plant growth? Where does the matter used to make the branches, roots, and leaves of plants actually come from? Until about 1650, it was assumed that plant growth resulted from the uptake and accumulation of materials removed from soil. Jan van Helmont conducted a simple experiment designed to test this hypothesis.

concluded that the 164 pounds of new leaves, wood, bark, and roots were "derived from water alone."

How did he arrive at this conclusion? Van Helmont evaluated his results on the basis of a popular theory that has since been discredited, the *transmutation theory*. According to this theory, one substance could be transformed into another. A familiar example was the erroneous belief that lead or some other metals could be converted into gold. Those who studied or practiced such transmutations were called *alchemists*. Even though it was incorrect, the transmutation theory, like others that were previously held but eventually discredited, was important in the development of modern science. The experiments and observations of alchemists prepared the ground from which the modern discipline of chemistry emerged. Some of the ideas of alchemists were more fruitful than others, however, and some, given historical hindsight, can be seen as hindering the advancement of knowledge. Van Helmont's conclusion that new plant matter

was derived from water clearly belongs in the latter category.

Hales's Experiments

Eighty years after van Helmont's experiment, Stephen Hales, an English clergyman, conducted an experiment to test his idea that plants somehow interacted with air, a possibility not considered by van Helmont. What was the basis for this hypothesis? In part, Hales was led to his experiments because of new information about plants revealed by the microscope. Using this instrument, he became aware of the existence of the very tiny openings (stomata) on the surfaces of leaves. What function did they serve? Did they function as pores in the skin of animals? Also, by that time, new knowledge had shifted the focus of chemistry. Alchemy had become chemistry, and amateurs such as Hales supported a theory that assumed the world was made up of small, hard "corpuscles of matter" that differed in size and shape. Some proponents of this *corpuscular theory* also believed that cer-

Van Helmont's Experiment

Figure 1 illustrates van Helmont's experiment. He planted a 5-pound willow tree in a vessel containing 200 pounds of dried soil. For five years, he added only rainwater to the tub of soil. At the end of that period, the tree weighed a little more than 169 pounds, whereas the weight of the soil was almost unchanged, having lost only 2 ounces. Clearly, the new growth was not accounted for by materials removed from the soil. What was the source of the new plant matter? Van Helmont

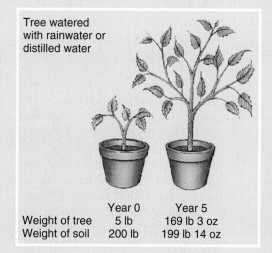

Tree watered with rainwater or distilled water

	Year 0	Year 5
Weight of tree	5 lb	169 lb 3 oz
Weight of soil	200 lb	199 lb 14 oz

Figure 1 Jan van Helmont conducted a study to determine the source of substances used in the growth of new plant tissue. Was it water, soil, or something else? He thought that the transmutation theory held the answer.

box continues

FOCUS ON SCIENTIFIC PROCESS

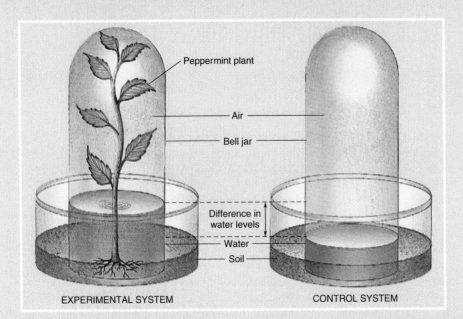

EXPERIMENTAL SYSTEM

CONTROL SYSTEM

Labels: Peppermint plant, Air, Bell jar, Difference in water levels, Water, Soil

Figure 2   Stephen Hales carried out a sophisticated experiment that provided new information on the identity of substances used in plant growth. He showed that plants interacted with the air and changed its composition. He based his analyses on the corpuscular theory.

tain particles of matter were more "elastic" or active than others. Hales approached his studies on plants within the context of this corpuscular view of matter.

As shown in Figure 2, Hales's experimental system consisted of two glass containers that were filled with water and a layer of soil at the base. He added a peppermint plant to one and then covered it with a glass bell jar. The other he simply covered with the jar and used as an *experimental control,* a standard for comparison. In general, a control consists of the same components as the experimental apparatus except for the variable being tested (in this case, the peppermint plant).

Hales observed and recorded changes in the water levels of the two systems. The control was essential because changes observed in both vessels could not be associated with the plant, but those unique to the experimental container could be considered as caused by the plant. Note that van Helmont did not have a control in his experiment. What could have served as van Helmont's experimental control?

Hales found that the water levels fluctuated as a function of barometric pressure (the pressure or "weight" exerted by the atmosphere) and temperature. Because they fluctuated similarly in both systems, he knew that the water levels were not associated with the plant's activities. However, after several months, Hales found that the water level in the experimental system rose significantly higher than that in the control. What could have caused this difference? Hales felt that the plant had "imbibed" or absorbed the air, thus creating a void in the jar that was "filled" by the water. He also found that there was no further reduction in air volume after two or three months. What is more, when he removed the plant from the experimental system and replaced it with a new plant, the new plant died after a few days. When a new plant, which had been confined for the same period, was placed in the control system, however, the plant remained healthy. What could account for these complicated observations? Hales was unable to offer a precise interpretation of his findings. Nevertheless the signif-

icance of his experiment was clear: plants *did* interact with the atmosphere and, in some unexplained way, modified the surrounding air.

Priestley's Experiments

The Reverend Joseph Priestley was a British chemist who conducted experiments in 1771 that significantly advanced what was then known about the relationship between plants and air (see Figure 3). Priestley established experimentally that both a burning candle and a mouse placed in a system closed to air would soon expire. From these observations, he concluded that the air had been "rendered noxious." On a broader level, he recognized that some large-scale process must exist for "rendering it fit for breathing again." Without such a mechanism, he speculated that "the whole mass of [Earth's] atmosphere would, in time, become unfit for the purpose of animal life." But how was this reconditioning process accomplished?

Priestley conducted a further series of experiments that led him to conclude that he had discovered "at least one of

FOCUS ON SCIENTIFIC PROCESS

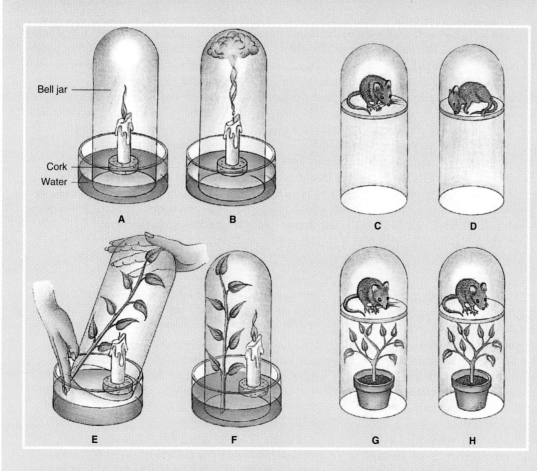

Figure 3 Joseph Priestley demonstrated that plants somehow purified air and made it fit for animals to breathe. His experiments showed that when a burning candle (A) was placed in a closed system, it went out within minutes (B). Likewise, when a mouse (C) was placed in the closed system it died within minutes (D). However, when both an unlit candle and a plant (E) were placed in the closed system, the candle burned when lit a few days later (F). And when a plant accompanied the mouse (G), the mouse was still alive a few days later (H). Priestley based his analysis of these experiments on the phlogiston theory.

the restoratives which nature employs for . . . restoring air which has been injured by the burning of candles. . . . It is vegetation." To test his hypothesis, Priestley added plants to his closed experimental systems. He found that candles would burn again and mice could live in such a system. He concluded that plants served to purify the atmosphere by reversing the harmful effects caused by an animal's breathing or burning materials.

What happened chemically when the air was rendered "noxious" or when it was improved? Priestley, like most of his generation, interpreted his experiments in light of the popular chemical theory of his day, the *phlogiston theory.* Proponents of the theory held that when a substance was burned, something called *phlogiston,* "the material and principle of fire," was expelled during

combustion. Conclusions formulated in using this theory seemed reasonable and consistent with existing knowledge. For example, charcoal was thought to consist of nearly pure phlogiston because it almost disappeared when burned. A candle burning in a sealed bell-shaped jar went out because the air became "phlogisticated" or "fixed" and was no longer able to support combustion or a mouse placed in the jar. "Dephlogisticated" air was capable of supporting the burning of a candle or the life of a small animal placed in the jar.

Priestley, an advocate of the phlogiston theory, very naturally interpreted his results from that perspective. Plants, according to this view, altered the air by removing phlogiston. Thus like Hales, Priestley recognized that plants change the air, but he focused on qualitative

changes rather than the quantitative changes that Hales described.

Ingenhousz's Experiments

Jan Ingenhousz, a Dutch physician, confirmed and extended Priestley's experimental findings eight years later. He, too, explained his results in terms of the phlogiston theory. According to Ingenhousz, plants removed phlogiston introduced into the air by animals and burning candles and changed it into dephlogisticated air. Ingenhousz did more than just agree with Priestley; he made some other, truly significant observations. First, he noted that plants performed their "beneficial operation . . . only after the sun [appeared] above the horizon." He also observed that only the green parts of the plant performed

box continues

"this office" (photosynthesis) and "that all plants contaminate the surrounding air by night, and even in the daytime in shaded places."

Soon after Ingenhousz's studies in 1779, the phlogiston theory exited from the stage of science, primarily as a result of an article published in 1786 by Antoine-Laurent Lavoisier titled "Réflexions sur le phlogistique," in which he described a new theory of combustion that accurately identified the gases present in the atmosphere. Moreover, Lavoisier demonstrated that combustion and respiration were similar chemical processes.

Subsequently, a number of remaining questions about photosynthesis were soon answered. In 1796, Ingenhousz proclaimed that photosynthesis was the primary process for producing new organic materials in the biosphere. By 1845, the underlying mechanism of photosynthesis—the conversion of light energy to chemical energy—was understood. Studies conducted during the 1980s largely completed the journey that led to an understanding of photosynthesis, a trip that began nearly 300 years earlier.

The modern understanding of photosynthesis draws on many scientific hypotheses that were developed in the past century and are strongly supported by research results. Like the early ideas on plants, our current knowledge of photosynthesis is rooted in broader theories of energy, chemical interactions, and biological processes.

a small amount of energy (two ATP) can be obtained from each glucose molecule, and the energy-rich products formed (pyruvate, ethanol, and lactate) cannot be broken down directly to yield ATP even though they contain a great amount of additional potential energy; ethanol and lactate are also toxic if they are not removed. Consequently, organisms that remained dependent on glycolysis and fermentation for obtaining energy did not evolve into more complex organisms which require greater quantities of energy.

Between 3 and 4 billion years ago, photosynthesis evolved in primitive bacteria and, subsequently, in all plants. As a result, a new source of food became available on Earth, and great quantities of oxygen were released into the atmosphere and seas. Following these developments, a new energy-releasing set of reactions (aerobic cellular respiration) evolved in certain early prokaryotic cells. Such cells would have had an enormous advantage over those that remained dependent on glycolysis and fermentation because they were able to generate large amounts of ATP. Consequently, they flourished and eventually gave rise to all the multicellular plants and animals that came to inhabit the Earth. The diversity of life on Earth is the subject of Chapters 5 and 6.

SUMMARY

1. Different forms of energy—chemical, thermal, electrical, mechanical, and radiant—are used by organisms that inhabit Earth. All of these are interrelated and can be converted from one form to another. The first and second laws of thermodynamics help explain the behavior of energy in the universe.

2. Photosynthesis is a process whereby radiant energy from the sun is transformed by plants into chemical energy that is used by almost all organisms for maintaining life.

3. In photosynthesis, beginning products (carbon dioxide and water) are converted into final products (glucose and oxygen) by means of energy provided by the sun. Pigments, such as chlorophyll, are unique light-trapping protein molecules that are able to capture radiant energy, transform it, and use it in driving the reactions of photosynthesis.

4. Cells obtain energy through the systematic breakdown of glucose into carbon dioxide and water in a series of chemical reactions. During glycolysis—a process that does not require oxygen—glucose is broken down into two smaller molecules of pyruvate. If no oxygen is present in the cell, pyruvate enters into fermentation reactions from which little energy is made available to the cell. If oxygen is present, pyruvate in a cell is completely degraded to carbon dioxide and water by the reactions of cellular respiration. Cellular respiration results in the production of significant amounts of ATP that can be used by the cell.

5. During electron transport chain reactions, high-energy electrons from NADH and $FADH_2$ are passed through a series of electron carriers located in the inner mitochondrial membrane. These electron transfers result in the release of energy that is used by respiratory enzyme complexes to pump protons (H^+) into the intermembrane space. Spent electrons combine with H^+ and oxygen to form water.

6. Adenosine triphosphate (ATP) is generated through chemiosmosis. Protons rapidly flow from the high H^+ concentrations in the intermembrane space to low H^+ concentrations in the mitochondrial matrix, through specialized transmembrane protein channels known as ATP synthetase. The strong flow of H^+ pro-

vides the energy required for ATP synthetase to synthesize ATP from ADP (adenosine diphosphate) +ⓅP.

7. Available evidence indicates that the first organisms to evolve on Earth used glycolysis and fermentation to obtain energy. Subsequently, certain bacteria and, later, eukaryotic cells acquired the mechanisms necessary to transform energy using the reactions of photosynthesis. Cellular respiration was the last biological energy transformation process to evolve, appearing in later prokaryotes and then eukaryotes that depended on the glucose and oxygen made available by photosynthesis.

WORKING VOCABULARY

adenosine triphosphate (ATP) (p. 62)
cellular respiration (p. 60)
chemiosmosis (p. 72)
electron transport chain (p. 70)
energy (p. 59)
energy transformation (p. 60)
entropy (p. 60)
glycolysis (p. 60)
kinetic energy (p. 59)
oxidation (p. 63)
photosynthesis (p. 60)
potential energy (p. 59)
reduction (p. 63)

REVIEW QUESTIONS

1. What is energy? What are the two general energy states?
2. List and define the different forms of energy.
3. What do the first and second laws of thermodynamics tell us about energy transformations?
4. What are the major biological energy-transforming processes? How are they related?
5. What are ATP, ADP, and coenzymes? What is the role of each in energy-transforming reactions?
6. What are photons? Pigments? Describe their functions in photosynthesis.
7. Describe the light-dependent and light-independent reactions of photosynthesis. What initial products are used and what end products are generated during each?
8. Describe the reactions of glycolysis and the different types of fermentation. What are the end products of each?
9. In which organisms do each of the following take place: glycolysis, alcoholic fermentation, and lactate formation?
10. Where in a cell does cellular respiration occur?
11. What are the end products of Krebs cycle reactions? Electron transport chain reactions?
12. How is ATP generated by chemiosmosis?

ESSAY AND DISCUSSION QUESTIONS

1. Is it possible that unique forms of energy not present on Earth would exist on other planets in other galaxies? Why?
2. Assume that life exists on other planets in the universe. Is it reasonable to believe that they use the same energy-transforming systems as their counterparts on Earth? Why? Describe a different set of transforming reactions that might have evolved.
3. Is it probable that during the next 2 billion years on Earth, one of the existing energy-transforming processes will become "extinct"? Which process might be the primary candidate for elimination? Why?

REFERENCES AND RECOMMENDED READING

Bassham, J. A. 1962. The path of carbon in photosynthesis. *Scientific American,* 206: 88–100.
Bennett, J. 1979. The protein that harvests sunlight. *Trends in Biochemical Science,* 4: 268–271.
Calvin, M. 1962. *The Photosynthesis of Carbon Compounds.* New York: Benjamin.
Govindjee, and W. J. Coleman. 1990. How plants make oxygen. *Scientific American,* 262: 50–58.
Hendry, G. 1988. Where does all the green go? *New Scientist,* 120: 38–42.
Hinkle, P. C., and R. E. McCarty. 1978. How cells make ATP. *Scientific American,* 238: 104–112.
Kay, J., and P. D. J. Weitzman. 1987. *Krebs Citric Acid Cycle: Half a Century and Still Turning.* London: Biochemical Society.
Krebs, H. A. 1970. The history of the tricarboxylic cycle. *Perspectives in Biology and Medicine,* 14: 154–170.
Mitchell, P. 1979. Keilin's respiratory chain and its chemiosmotic consequences. *Science,* 206: 1148–1159.
Morowitz, H. J. 1970. *Entropy for Biologists.* Orlando, Fla.: Academic Press.
Walker, D. 1992. *Energy, Plants, and Man.* Brighton: Oxygraphics.

ANSWERS TO FIGURE QUESTIONS

Figure 4.1 By the reactions of photosynthesis.
Figure 4.3 It is transformed to heat.
Figure 4.7 They reflect green light, which is what we see as a result.
Figure 4.9 It is used in cellular respiration reactions. (We cannot live without it!)
Figure 4.10 Carbon dioxide.
Figure 4.11 Yeast cells.
Figure 4.14 NAD^+.
Figure 4.15 In humans, it is exhaled during breathing.
Figure 4.16 To combine with the spent electrons from electron transport reactions and remove them from the cell.

5

The Diversity and Classification of Life: Kingdoms Prokaryotae, Protoctista, and Fungi

Chapter Outline

Reading Questions

1. How great is the diversity of life on Earth?

2. What hierarchies are used to organize the various life forms on Earth?

3. What are the five kingdoms of life, and how do they differ?

4. How is each species assigned a unique name?

The enormous diversity of species present in tropical rain forests has recently been verified by a number of studies. A **species** consists of all the organisms of a particular kind that have the ability to interbreed and produce offspring. Alwyn H. Gentrey of the Missouri Botanical Gardens found nearly 300 species of trees per hectare in a forest near Iquitos, Peru, and stated:

> I conclude that the ever-wet forests of Upper Amazonia may be the world's richest in tree species. Indeed, it is hard to

imagine a more diverse forest than at Yanamono where there are only twice as many individuals as species in a 1-hectare patch of forest, with 63% of species represented by single individuals and only 15% of species represented by more than two individuals.

Other recent studies in tropical Amazonia have determined that the world's largest inventories of birds, butterflies, amphibians, reptiles, and mammals occur in these rain forests.

The extent of species diversity on a worldwide basis remains unknown, but Edward O. Wilson of the Museum of Comparative Zoology at Harvard University has concluded that about 1.7 million species have been described, including approximately 250,000 flowering plants, 47,000 vertebrates, and over 750,000 insects. Estimates of total potential diversity, considering all *biota*—the plants and animals—of Earth, range from 5 to 30 million species!

SYSTEMATIC ORGANIZATION OF THE DIVERSITY OF LIFE

The systematic organization of this complex diversity of life was given its modern form in the eighteenth century, when Swedish naturalist Carolus Linnaeus (1707–1778) developed a hierarchical classification system. He divided all known living organisms into two **kingdoms,** Plantae (plants) and Animalia (animals). Linnaeus subdivided each kingdom into sequentially less inclusive groups, from larger to smaller, which he named **class, order, genus,** and **species** (see the Focus on Scientific Process, "Classification"). Here is an analogy to illustrate the increasing specificity of information in this system. If a similar classification scheme were to be used to locate the place where a certain student lives, we might expect the following organization: *kingdom,* Earth; *class,* country (say, the United States); *order,* state (California); *genus,* town (Eureka); *species,* address (555 North Student Lane).

Other levels have since been added to the Linnaean system. These include the **phylum** between the kingdom and class and the **family** between the order and genus. The use of terms or names from the two smallest groups—genus and species—constitutes a binomial system that provides an individual name for each species.

The Binomial System

The word *nomenclature* is from the Latin for name (*nomen*) and caller or crier (*calator*) and refers to the act or system of giving names to things. The **binomial** (literally "two-name") **system** of nomenclature is used today for naming all species. For example, the genus *Pinus* contains all the various pine trees of the world; the species *contorta* is limited to the lodgepole pine (see Figure 5.1). By convention, genus names always begin with a capital letter and species names begin with a lowercase letter. Both are italicized or underlined when used as a scientific name, as in, for example, *Pinus contorta.*

Figure 5.1 The lodgepole pine has the scientific (genus and species) name *Pinus contorta.*

Question: *In what way might the scientific name of the eastern white pine differ from that of the lodgepole pine?*

Linnaeus, who invented the binomial system, obtained the names he gave to organisms from Latin, a practice that continues today. Usually, the names refer to some characteristic of the species. The surname of an individual, who has identified or studied a new species, however, is sometimes used in naming the species. For example, the scientific name of the cutthroat trout is linked with the famous Lewis and Clark expedition that first explored the western United States (see Figure 5.2). As

Figure 5.2 There are four cutthroat trout subspecies within the species, *Oncorhynchus clarki.*

Question: *How does a subspecies differ from a species?*

Classification

Figure 1 This beautiful water lily graced Brunfels's famous herbal, published in 1530. It was one of the first of many Renaissance herbals that contained plants drawn from nature.

Classification is the orderly arrangement of information. It has many practical applications because it permits the retrieval of information and can organize it to reflect relationships that exist among the objects classified.

A survey of different cultures shows that all peoples have constructed classification systems of natural objects. Some are complex, and others are simple. Distinctions are made in some languages that are not possible in others. Attempts to classify organisms date back to the beginning of science. The earliest scientific classifications were fairly simple and reflected the relatively small number of plants and animals known at the time. Aristotle, for example, described only 500 animals in his writings.

As a response to new information during the Renaissance, many "herbals" were published. These books, which were beautifully illustrated, as shown in Figure 1, attempted to describe all the plants and animals then known. They were arranged for the most part in a simple style, usually alphabetically, although some herbalists attempted to devise a system of classification.

During the seventeenth century, classification became the central issue among naturalists, and two general types of classification systems were developed: *artificial* and *natural*. An **artificial system** is a classification system that is used primarily for retrieving information and makes no claims about the relationships among the objects classified. By contrast, the main purpose of a **natural system** is the construction of a classification that reflects an order existing in nature. Natural systems can also be used for information retrieval, but often they are not as convenient as artificial systems. In the seventeenth century, Joseph Pitton de Tournefort, one of France's most famous botanists, organized 10,146 species of plants into 698 genera, many of which are still rec-

ognized today. Tournefort believed that God created plants according to a plan and that by careful observation, a trained scientist could "intuitively" come to know the plan. His system was a natural classification system, and for him it was a picture of God's arrangement.

An English contemporary of Tournefort, John Ray, did not believe that a natural system was possible. He didn't doubt that God had a plan, but he thought that people had no access to it. Ray described 18,000 species of plants and went on to classify animals as well. He was able to arrange these numerous organisms into simple systems, called keys, that allowed someone to identify a specific plant or animal. Ray did not claim that his keys reflected anything

other than convenient systems for identification and organization.

As might be expected, there has been considerable debate over the decades between proponents of artificial and natural systems. In the eighteenth century, Linnaeus constructed an artificial system of classification that became widely used throughout the world. He also proposed using a binomial nomenclature, which brought some order into the then-chaotic realm of naming organisms. Until Linnaeus, there was no universal method of naming organisms, and one plant might have 25 different names given at different times by different writers!

In Paris, Linnaeus's contemporary and rival, Georges-Louis Leclerc de

FOCUS ON SCIENTIFIC PROCESS

Buffon, constructed a natural system of classification of animals and encouraged others to work on a natural system for plants. Buffon's system was based on more than 20 years of research, but it encountered a problem that plagued most of the writers who constructed natural classifications—disagreement over the criteria. Tournefort had earlier said that naturalists could intuitively see the relationships among organisms. Although many agreed with this point of view, when it came to elaborating a system, many naturalists saw different patterns of organization, resulting in no common agreement over what constituted the natural system.

Debates persisted throughout the nineteenth century. Some experts hoped that comparative anatomy of animals would uncover the natural criteria; others felt that animal behavior would provide the key. In botany, a wide variety of plant parts were proposed as determinants.

Darwin's theory of evolution (discussed in Chapter 23) provided a new perspective from which to approach classification. He had stated in his writings that species of animals or plants often resembled other species because they descended from a common ancestor. He went on to suggest that classification should be based on evolutionary relationships.

Darwin's idea quickly caught on; in practice it has often been problematic because virtually no historical record remains for many plants and animals. Moreover, organisms that share similarities are not always related directly; rather, they may have evolved the similarities independently.

Today many different tools are used to assess the evolutionary relationships among organisms. In addition to adult and embryo morphology, scientists use information from biochemical analysis, biogeography, fossils, geology, and behavior. Scientists who attempt to develop evolutionary classifications still disagree on the basic differences that should be emphasized. Two distinct approaches have emerged: cladistic classification and evolutionary systematics.

Cladistic classification stresses the sequential order in which groups, known as *clades* arise, but the degree of difference is largely irrelevant. The root word for *cladistic* and *clades* is the Greek *klados,* which refers to a branch or a shoot.

Thus, appropriately, a cladistic classification diagram—called a *cladogram*—is typically represented as a tree consisting of a series of dichotomous branches. Each branching point is defined by a new trait that is not present in earlier members of the tree. Figure 2, for example, shows a cladogram for five common vertebrates: lizards, cows, seals, dogs, and cats. They all have four limbs, so the number of limbs cannot be used to define a new branching point. Hair, however, is an evolutionary innovation that did not involve lizards, and so hair becomes the first branching point. A different type of tooth, called an "involuted cheek tooth," occurs in seals, dogs, and cats, but not in cows; this tooth, therefore, becomes the second branching point. Likewise, cats and dogs—but not seals—possess carnassial teeth (teeth specialized for slicing), which becomes the third branching point. Finally, at the fourth branching point, cats possess retractable claws, and dogs (thank goodness!) do not.

Whereas cladistic classification does not stress the degree of difference among groups, *evolutionary systematics* emphasizes the usefulness and significance of

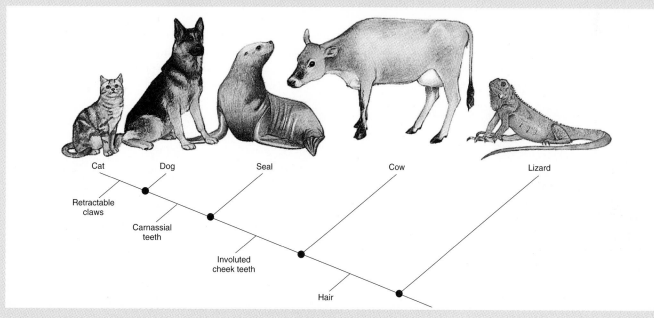

Figure 2   This cladogram shows certain features (clades) that characterize each branching point. Note that cladograms can be much more complex than this one!

box continues

FOCUS ON SCIENTIFIC PROCESS

evolutionary innovations in defining different groups. The term *evolutionary systematics* is somewhat misleading because cladistics is also evolutionary. Evolutionary systematics, however, attempts to more fully reflect evolutionary history in its classification relationships. For example, an evolutionary innovation that led to the successful exploitation of a new environment (say, a particular type of nail or claw) is given greater value than an innovation that had less influence (say, a certain type of tooth).

When information on evolutionary relationships is insufficient, taxonomists resort to simple keys for part or all of the classification system. There are also scientists who recognize the intellectual value of evolutionary classification systems but nonetheless construct artificial keys for practical purposes. For example, a field guide to birds is much more convenient for the early morning bird-watcher than a multivolume treatise that attempts to organize the avian world into a phylogenetic scheme.

Still other scientists reject evolutionary classifications and believe that a more rigorous system can be devised by using the number of similarities that exist among organisms. These **phenetic systems** tabulate traits, and through computer manipulation scientists can generate lists of groups that share the most traits. Generally, those who have been attracted by phenetic classification work with organisms with uncertain evolutionary histories.

described by Meriwether Lewis in his *Journals of the Lewis and Clark Expedition* on June 13, 1805:

> These trout are from 16 to 23 inches in length, precisely resemble our mountain or speckled trout in form and the position of their fins, but the specks on these are of a deep black instead of the red or gold of those common in the United States. These are furnished with long teeth on the pallet and tongue and have generally a small dash of red on each side behind the front ventral fins; the flesh is of a pale yellowish red, or when in good order, of a rose red.

The scientific name given to this trout, which was first described by Lewis, is *Oncorhynchus clarki* after William Clark. Cutthroat trout with slightly different characteristics inhabit a number of distinct habitats throughout the West. Although there is some disagreement among scientists, each of the 14 different forms is designated by a different **subspecies** name. For example, the cutthroat trout *Oncorhynchus clarki clarki* lives in western coastal streams and migrates to the ocean. *O. clarki lewisi* inhabits areas of the inland Pacific Northwest, *O. clarki bouvieri* is the Yellowstone species, and *O. clarki pleuriticus* inhabits the Colorado River.

Both genus and species names are required to identify each species. A common term can be used in naming two different species. For example, *Bison bison* is the North American bison, but *Enophrys bison* is the buffalo sculpin, a small (38-millimeter) marine fish living over shallow reefs along the Pacific coast (see Figure 5.3).

THE KINGDOMS OF LIFE

For a time, Linnaeus's two-kingdom system worked well enough to classify the organisms that had been identified in the early eighteenth century. However, as better microscopes brought the world of microorganisms into view, the two kingdoms, Plantae and Animalia, were no longer sufficient for classifying the huge number of diverse and unique microscopic organisms (see Figure 5.4). Many microscopic species were found to have characteristics of both plants and animals, and it became increasingly difficult to place them comfortably in either of the two kingdoms. This problem prompted a number of scientists to suggest the creation of a third kingdom.

A

B

Figure 5.3 The same name has been used for more than one species. For example, the term *bison* appears in the scientific name of (A) the North American buffalo, *Bison bison,* and (B) the small marine buffalo sculpin, *Enophrys bison.*

Question: *Can the same name be used for different genera within a phylum? Why or why not?*

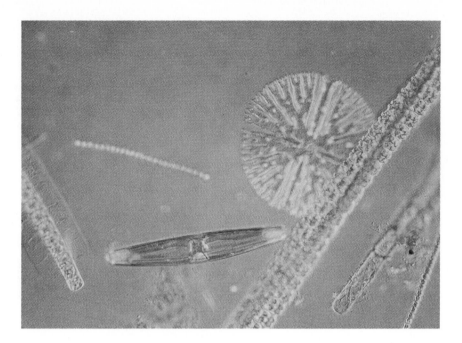

Figure 5.4 The compound microscope enabled early naturalists to study a new universe of life forms that could not be seen with the unaided eye. Consequently, a third kingdom, named Protoctista by John Hogg in 1861 and Protista by Ernst Haeckel in 1866, was created.

In 1861, John Hogg proposed the kingdom **Protoctista** for microscopic organisms, and in 1866, Ernst Haeckel suggested creating the kingdom **Protista.** The classification system advanced by Haeckel separated the blue-green algae and bacteria (prokaryotes) from the nucleated protists (eukaryotes, such as the amoeba) and placed them in a special group he called *Monera,* within the kingdom *Protista.* Other classification systems were proposed during the late-nineteenth and early-twentieth centuries, but Haeckel's three-kingdom system predominated.

In 1938, Herbert Copeland proposed a four-kingdom system. In addition to the plants and animals, he elevated the prokaryotes to kingdom status and in 1956 accepted Hogg's term *Protoctista* as the kingdom name for the eukaryotic protists. The names applied to these two kingdoms—Monera and Protoctista—have varied over the past half century. However, the conceptual separation of life into four kingdoms remained unchanged until 1957, when Robert Whittaker proposed elevating the fungi to kingdom rank, thus creating a fifth kingdom. Whittaker's five kingdoms were Monera, Protista, Fungi, Plantae, and Animalia, and this system has been adopted by most biologists since the 1960s.

In this book, we use Whittaker's five-kingdom system as modified by Lynn Margulis and Karlene Schwartz in their 1988 edition of *Five Kingdoms: An Illustrated Guide to the Phyla of Life on Earth* and in their more recent book (with co-author Michael Dolan), *The Illustrated Five Kingdoms: A Guide to the Diversity of Life on Earth* (1994). Their five kingdoms are Prokaryotae, Protoctista, Fungi, Plantae, and Animalia (see Figure 5.5, page 86). However, they concede that their system, like all its predecessors, is imperfect, stating in the earlier book: "Our system has the advantage of defining the three multicellular kingdoms precisely, but the disadvantage of grouping together as protoctists amoebae, kelps, water molds, and other eukaryotes that have little in common with one another."

Taxonomy

From Linnaeus's time, taxonomy has been an important component of descriptive biological sciences. **Taxonomy** (or systematics) is the science of arranging organisms, plants, and animals into related groups based on certain factors common to each. The different levels within the taxonomic hierarchy are referred to as **taxa.** The classification groups (taxa) biologists now use are kingdom, phylum, class, order, family, genus, and species; each of these groups may be further broken down into subcategories (subkingdoms, subphyla, and so on). For the kingdom Prokaryotae only, *division* is used between the subkingdom and the phylum. In Table 5.1, you can see that humans share the same taxa with sheep up to the level of order. You are classified in the order Primate, family Hominoidae, genus *Homo,* and species *sapiens.* In the binomial system of nomenclature, a sheep is called *Ovis musimon,* whereas you are called *Homo sapiens.* Of course, you can clearly see that neither you nor the sheep

Table 5.1 A Comparison of the Taxonomic Classification of the Black Oak, the Domestic Sheep, and the Human

	Black Oak	**Sheep**	**Human**
Kingdom	Plantae	Animalia	Animalia
Phylum	Tracheophyta	Chordata	Chordata
Subphylum	Spermatophyta	Vertebrata	Vertebrata
Class	Angiospermae	Mammalia	Mammalia
Subclass	Dichotyledoneae	Theria	Theria
Order	Spindales	Artiodactyla	Primate
Family	Fagaceae	Bovidae	Hominoidea
Genus	*Quercus*	*Ovis*	*Homo*
Species	*velutina*	*musimon*	*sapiens*

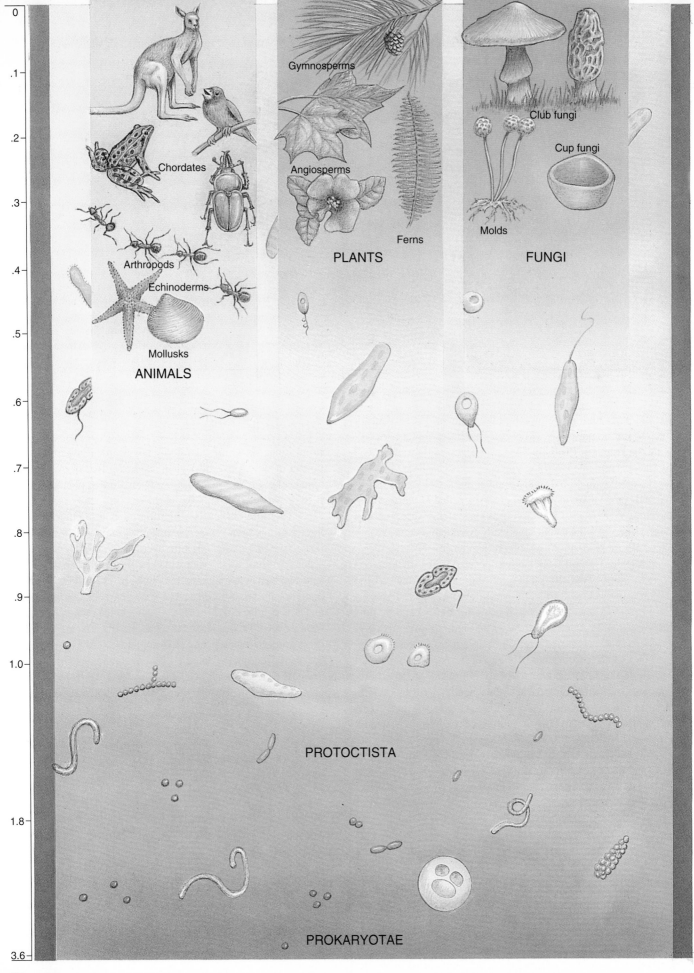

Figure 5.5 An overview of the five-kingdom system of classification, after Margulis and Schwartz.

Question: *What is the main disadvantage of the five-kingdom system?*

share any taxa with the plant (black oak, or *Quercus velutina*), whose taxonomy is listed in the column at left.

Figure 5.6 traces the taxonomic classification of *Homo sapiens* starting with kingdoms and working its way through phyla, subphyla, classes, and so on to species. Note the large number of phyla and families that exist. Also bear in mind that the taxonomic classification of newly discovered organisms and modification of the status of previously classified species continues today as new information becomes available.

In this book we will concentrate on the major groups of living organisms in the five kingdoms that contribute significantly to our understanding of biological science.

Classification Terminology and Tradition

Biologists today refer to organisms as "higher" or "lower," "simple" or "advanced," and "primitive" or "complex." These are not value-laden terms and are not used to degrade the humble sponge and glorify the proud peacock. In part, the use of such terms is traditional, dating back to a time when it was believed that all living organisms could be arranged on a single scale called the **chain of being** (see Figure 5.7 on page 88). This scale began with minerals, moved to plants and animals, and finally reached what seemed to people at the time to be the most "perfect" animals, humans. Although biologists no longer

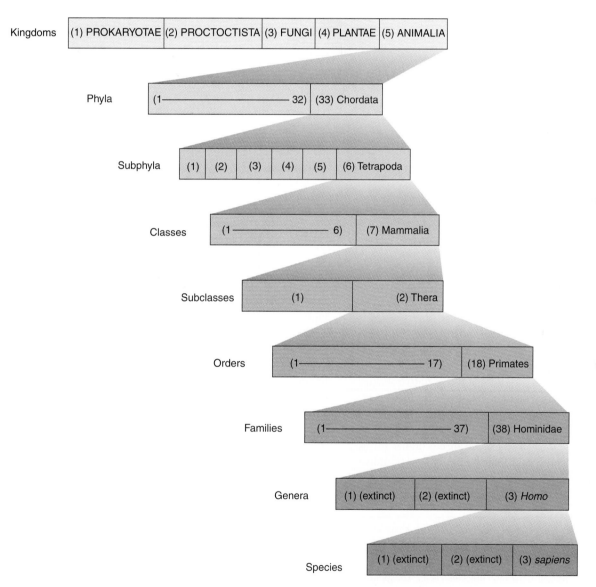

Figure 5.6 The taxonomic classification of humans begins in the kingdom Animalia. This kingdom is composed of 33 phyla. Our phylum, Chordata, is divided into six subphyla, including our subphylum Tetrapoda ("four feet"). Tetrapoda has four classes, including mammals. Our class, Mammalia, is split into two subclasses, including Thera (true mammals). Primates are only one of 18 orders in our subclass, and there are 38 families in the Primate order, including the family of humans, Hominidae. There were three genera in our family, but two became extinct. Within the genus *Homo*, there were three species, but only ours, *sapiens*, survives today.

Question: *How many taxa do we share with chimpanzees?*

Figure 5.7 As depicted in the eighteenth century, all beings on Earth were linked in a scale that went from most perfect (humans) to nonliving minerals.

think in terms of a scale between "imperfect" and "perfect," they continue to use terms like *simple* and *advanced* to distinguish among organisms that have relatively simple organizations and those that have many complicated organ systems. *Primitive* and *complex* do not mean that the primitive species have given rise through evolution to advanced forms but, rather, that their organization is less developed. *Higher* and *lower* similarly reflect levels of organization. Since in the evolution of organisms, more complex, advanced, higher forms generally came into being later than simple, lower, primitive ones (not the ones alive today, but their evolutionary ancestors), these terms also reflect some aspects of ancestry.

> **BEFORE YOU GO ON** | Life on Earth is classified into the kingdoms Prokaryotae, Protoctista, Fungi, Plantae, and Animalia. Other hierarchies include the phylum, class, order, family, genus, and species, each of which is less inclusive than the preceding level. The genus and species names combine to give each kind of organism a unique scientific name.

KINGDOM PROKARYOTAE

The kingdom **Prokaryotae** includes all known organisms constructed of cells that have neither an organized nucleus nor, with few exceptions, specialized cellular organelles (intracel-

lular structures). It contains some species with characteristics similar to the earliest life forms that are thought to have occupied Earth. Today they continue to participate in the cycling of many mineral elements that are required by other organisms within the biosphere. Some prokaryotes however, are detrimental because they cause diseases in eukaryotic organisms from the other four kingdoms of life.

Prokaryotes are either *autotrophs* (self-feeders) or *heterotrophs* (other-feeders). **Autotrophic bacteria** can synthesize complex organic molecules from simple molecules. Some can use light as an energy source for this synthesis and are known as **photosynthetic bacteria.** Others can use energy-rich molecules such as hydrogen sulfide or methane; these are known as **chemosynthetic bacteria. Heterotrophic bacteria** obtain their nourishment from organic molecules formed by autotrophs and other heterotrophs. Bacteria participate in the decay of organic matter from organisms in all five kingdoms.

Research completed in the early 1980s indicates that there are two distinct types of prokaryotes. Consequently, Margulis and Schwartz separate the prokaryotes into two *subkingdoms:* the Archaebacteria and the Eubacteria.

Subkingdom Archaebacteria

The subkingdom **Archaebacteria** includes only two phyla, one containing *methanogenic bacteria* and the other **halophilic** (salt-loving) and **thermoacidophilic** (heat [and] acid-loving) **bacteria. Methanogenic bacteria** are unusual organisms that use simple organic molecules other than carbohydrates or proteins as food sources. In Chapter 9, we will see that methanogenic bacteria are very important in the global "cycling" of carbon because they chemically reduce carbon dioxide and oxidize hydrogen to produce methane, a colorless, odorless, flammable gas. Recall that oxidation is a process in which an atom or molecule loses one or more electrons, and reduction is the process whereby an atom or molecule gains one or more electrons.

Halophilic and thermoacidophilic bacteria can exist in environments that extend their habitable ranges to nearly unbelievable limits. Some live in hot springs where temperatures may approach 90°C, and others live in environments where extreme pH ranges of 1 or 2 are common. Representatives of the thermoacidophilic archaebacteria are found in the hot springs of Yellowstone National Park (see Figure 5.8).

Subkingdom Eubacteria

The subkingdom **Eubacteria** includes all "true" bacteria (eubacteria) and is partitioned into three different *divisions* based on the type of cell wall present in the bacteria. The first division contains only one phylum, which consists of species lacking rigid cell walls. The bacteria in this phylum cause some types of pneumonia in mammals, including humans.

The other two divisions have been distinguished historically by a contrasting affinity for a specific biological staining agent known as *Gram stain* (named after the Danish microbiologist Hans Christian Gram). *Gram-negative bacteria* have a cell wall that retains little of the Gram stain during the staining process,

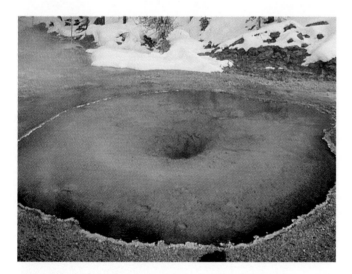

Figure 5.8 Yellowstone National Park is located on a geological "hot spot." Surface water seeps into the ground, where it becomes superheated. As it rises back to the surface, the heated water carries mineral compounds that have dissolved in it. Halophilic and thermoacidophilic bacteria can inhabit environments like this.

Question: *Could eukaryotic organisms live in such a geological hot spot? Why or why not?*

causing them to appear light pink. In contrast, *gram-positive bacteria* have cell walls that retain greater amounts of the stain, causing the cells to develop a deep purple color.

Gram-negative bacteria include many of the photosynthetic cyanobacteria (blue-green algae); the spirochetes that cause the sexually transmitted disease known as syphilis, a skin disease transmitted by eyeflies in the tropics called yaws, and other pathogenic diseases; and one group of nitrogen-fixing bacteria, which are able to convert atmospheric nitrogen into a form that plants can use (discussed in Chapter 9), including species in the genus *Rhizobium* found in the root nodules of legumes such as alfalfa, beans, and lupine.

Gram-positive species include the fermenting bacteria that emit lactic acids (which impart unique flavors to yogurt, some cheeses, and other fermented foods) or hydrogen sulfide (marsh or rotten-egg gas) as end products of their chemical activities. Also among the gram-positive bacteria are species capable of nitrogen fixation and some species in the genus *Streptomyces*. These bacteria are important because they are used in the production of antibiotics such as *streptomycin*, a powerful drug that interferes with protein synthesis in several bacteria harmful to humans.

> **BEFORE YOU GO ON** The kingdom Prokaryotae contains all organisms composed of prokaryotic cells. This kingdom has two subkingdoms: the Archaebacteria and the Eubacteria. Each of these subkingdoms has one or more divisions, and each division contains a number of phyla.

KINGDOM PROTOCTISTA

The kingdom **Protoctista** contains four major groups of eukaryotic organisms: single-celled protozoans, unicellular algae, multicellular algae, and slime molds. Table 5.2 (see page 90) shows that this kingdom is defined by exclusion. All members have characteristics that exclude them from the other four kingdoms, but they are constructed of nucleated cells, they respire aerobically (that is, they require oxygen), and most have **flagella** (specialized structures used in motility) at some stage of their life cycle. Unlike plants and animals, however, they do not develop from a **blastula** (the hollow ball stage of animal development) or an **embryo.** It is a kingdom of extreme body forms, for it encompasses organisms as diverse as protozoans and giant kelps (see Figure 5.9, page 90). This diversity is reflected in the myriad methods of reproduction and development, from the simple cell division of an amoeba to the sexual reproduction of kelp, in which the fertilized egg germinates in response to light, rootlike rhizoids grow downward to become the *holdfast* (anchoring structures), and a leaflike blade grows upward toward the surface of the ocean.

Some scientists have recently suggested that various members of the kingdom Protoctista are so distinctive that they could justify the creation of at least 20 separate kingdoms. Others have suggested that 40 to 50 phyla should be developed for all of the different varieties of organisms. Margulis and Schwartz list 27 phyla to accommodate this diverse assemblage of organisms.

Table 5.2   Key to the Five Kingdoms

		Kingdom
1. (a)	Cells without nucleus ⟶	Prokaryotae
	or	
(b)	Cells with nucleus ⟶	2 (go to number 2)
2. (a)	Flagella absent, absorptive nutrition, form spores ⟶	Fungi
	or	
(b)	Flagella usually present, photosynthetic or ingestive nutrition ⟶	3
3. (a)	Ingestive nutrition, blastula present during development ⟶	Animalia
	or	
(b)	Ingestive or photosynthetic nutrition, blastula absent ⟶	4
4. (a)	Photosynthetic nutrition, embryo present during development ⟶	Plantae
	or	
(b)	Ingestive or photosynthetic nutrition, embryo absent ⟶	Protoctista

Keys designed to classify living organisms provide for choices between sets of characteristics. In this key, the absence of a nucleus (choice 1a) separates the kingdom Prokaryotae from the other four. The absence of flagella, absorptive nutrition, and the presence of spores (choice 2a) separates the kingdom Fungi. Ingestive nutrition and blastula formation are characteristics of organisms in the kingdom Animalia (choice 3a). The presence of an embryo (choice 4b) separates the kingdom Plantae from the kingdom Protoctista.

The Giant Amoeba

The very primitive giant amoeba, *Pelomyxa palustris,* is the only species in the first phylum of the kingdom Protoctista (see Figure 5.10). These amoebas are large, multinucleated cells with no chromosomes or specialized cellular organelles. They obtain energy from methanogenic bacteria, which reside inside them. Giant amoebas inhabit mud at the bottom of freshwater ponds, where they contribute to the degradation of organic molecules and the cycling of carbon within the biosphere.

Protozoans

The protozoans classified by Haeckel in his kingdom Protista are placed in the kingdom Protoctista by Margulis and Schwartz, who have created seven phyla that include these "animal-like," generally unicellular organisms. The four major phyla contain the following general types of eukaryotic organisms: complex, single-celled amoebas; flagellated protozoans; ciliated protozoans; and foraminifera. In addition to flagella, protozoans also feature other hairlike structures called **cilia,** which are used in mobility.

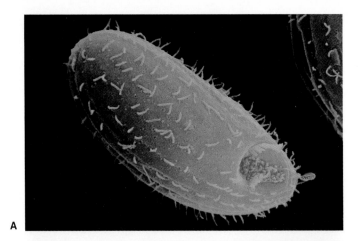

A

B

Figure 5.9   The kingdom Protoctista includes such diverse species as (A) single-celled ciliated protozoans and (B) kelps. Kelps are giant brown algae that live in coastal waters.

Single-celled Amoebas

This phylum includes all free-living freshwater, marine (ocean water), and soil amoebas, as well as those that are parasites of animals. A **parasite** is any organism that lives on or within and at the expense of another organism, referred to as a **host.**

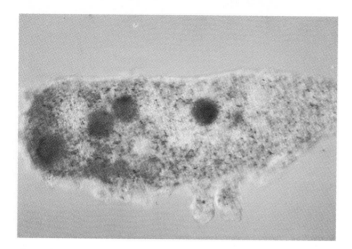

Figure 5.10 The giant amoeba *Pelomyxa palustris* may be the most primitive of all eukaryote life forms. This species has multiple membrane-bound nuclei but none of the other organelles found in all other eukaryotes.

Question: *What would have to happen in order for this life form to become multicellular?*

Amoebas lack flagella and move by forming specialized, temporary membrane extensions called **pseudopodia** ("false feet"), as shown in Figure 5.11. They do not reproduce sexually, rather, they produce new cells through the asexual process of simple cell division, all of which we will consider in more depth in Chapter 15. Many members of this phylum form tough, enclosed, resting stages called **cysts** at some stage of their life cycle. This allows them to withstand adverse conditions for long periods of time. The intestinal parasites that cause amoebic dysentery (an infection of the intestinal tract) germinate from such resistant cysts within the digestive tracts of their mammalian hosts, including humans.

Flagellated Protozoans

Members of this phylum are free-living, *mutualistic,* or *parasitic* species. **Mutualism** refers to a type of symbiosis in which two dissimilar organisms live together in an intimate association that benefits both. By contrast, in **parasitism,** only one of the organisms benefits while the other is generally affected in a negative way. An example of a mutualistic flagellate is *Trichonympha,* which inhabits the digestive systems of dry-wood termites (see Figure 5.12). Bacteria associated with these flagellates digest the wood eaten by the termites. Both the termites and the flagellates benefit by deriving energy from the digested wood products. Some flagellated protozoans cause severe human parasitic diseases, including a form of sleeping sickness in Africa and Chagas's disease, which affects millions of people in Central and South America (see Figure 5.13). In temperate North America, the parasitic flagellates are represented by the genus *Giardia,* which includes a number of forms

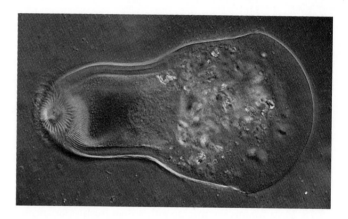

Figure 5.12 The flagellate *Trichonympha* sp. has up to 14 flagella. Trichonymphs live as symbionts in a mutualistic association in the gut of dry-wood termites, where they digest cellulose in the wood fibers consumed by the termites.

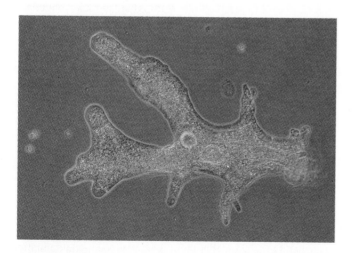

Figure 5.11 The amoeba *Amoeba proteus* moves by using pseudopodia. These "false feet" form from temporary extensions of the cell body, allowing the amoeba to engulf its food.

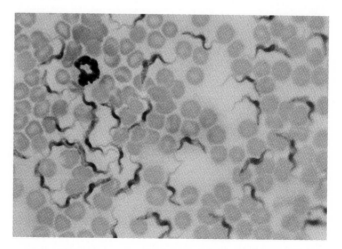

Figure 5.13 Some flagellated protozoans, including these trypanosomes, are parasitic and cause human diseases such as African sleeping sickness.

that inhabit many seemingly pristine mountain lakes and streams. *Giardia* causes severe digestive tract problems when taken in by a suitable mammalian host, such as a human. The potential presence of this flagellate in all wilderness areas has made it necessary for campers routinely to boil all water used for personal consumption.

Ciliated Protozoans

This phylum contains species that are among the most complicated single-celled organisms. Members of this phylum have many specialized structures, including cilia for mobility, **trichocysts** for capturing food and defense, **micronuclei** and **macronuclei,** and **contractile vacuoles** for expelling excess water from their cells (see Figure 5.14). They are also distinguished by their use of **conjugation,** a variety of sexual reproduction that involves the "mating" of two cells, with an actual exchange of nuclear material, to produce new individuals. Members of this phylum live in aquatic environments, and more than 8,000 freshwater and marine species have been described. Most samples of water collected near the bottom of any pond will contain ciliates. They are very efficient "swimmers" that feed on bacteria and other microscopic creatures as they move through the water column. Few ciliates cause human diseases, but they have been of interest for decades to cell biologists because they are large, complex, and easy to grow in laboratory environments.

Foraminifera

Some species of foraminifera, also called forams, are the most aesthetically pleasing protoctists, as you can see in Figure 5.15. Most forams are marine rather than freshwater species, and most live attached to various surfaces or to other organisms. However, some are free-living members of the plankton community (floating or drifting microscopic organisms in oceans, ponds, and lakes) and are important members of

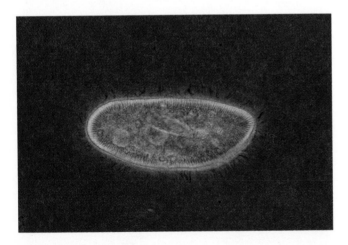

Figure 5.14 When living *Paramecium* species are stained, their trichocysts explode. In living paramecia, trichocysts function as darts that can be used to capture the small prey on which they feed.

Figure 5.15 Foraminifera species have unique tests with beautiful geometric patterns.

aquatic food chains. Forams are known as the "shelled protoctists" because they all have **tests**—external, hard coverings composed of mineral and organic molecular complexes. Great numbers of **filopodia** (needlelike pseudopodia) protrude from pores in the tests and are employed for mobility and feeding on algae, ciliates, and other microscopic forms of life.

Other Protozoan Phyla

The three remaining phyla of protozoans cover one group of small marine plankton known as *actinopodes* and two groups—*sporozoans* and *microsporidians*—that are parasites of higher vertebrates. One sporozoan genus is *Plasmodium,* which contains the parasites that cause malaria.

Unicellular Algae

Unicellular algae are separated into a number of protoctist phyla, the most important being the *dinoflagellates,* many of which are a bright red color. Microscopic dinoflagellates have nuclei bound to intercellular membranes and two flagella, a longitudinal one and a traverse one that circumscribes the cell within a groove in the cell wall or test. The tests of most dinoflagellates consist of **cellulose** (structural carbohydrate) plates embedded in the cell membrane and encrusted with cilia. Dinoflagellates often undergo an explosive increase in numbers during the summer months, resulting in "blooms" that cause *red tides* in near-shore marine waters. The dinoflagellate responsible for red tides, *Gonyaulax tamarensis,* produces a neurotoxin that is concentrated by shellfish that feed by constantly filtering and ingesting whatever edible organisms are in the water. Because this toxin affects humans who eat the shellfish (the effects range from a tingling of the ears to respiratory paralysis and death), the harvesting of clams, mussels, and oysters in affected areas is suspended until the red tide dissipates and tests show that there is no persistent contamination (see Figure 5.16).

The amount of photosynthesis carried out by dinoflagellates is surpassed only by the marine **diatoms.** The phylum of

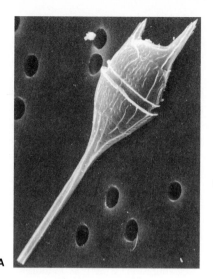

Figure 5.16 The dinoflagellates (A) *Ceratium* and (B) *Gonyaulax tamarensis.* Summer blooms of the dinoflagellate *Gonyaulax tamarensis* can produce (C) red tides in near-shore ocean waters. When filter-feeding shellfish feed on *Gonyaulax,* they may accumulate high levels of the neurotoxin concentrated in the unicellular algae. If humans then eat the shell fish, the neurotoxin they contain can be lethal.

Question: *Would you expect the fish taken from red tide areas to be as toxic as the shellfish? Why or why not?*

the diatoms includes marine and freshwater species that are very important in aquatic food chains. Each of the nearly 10,000 species of diatoms displays a unique pattern in its test, an extraordinary example of the variety present in nature (see Figure 5.17).

More unicellular algae are found in two other phyla. *Euglenoid flagellates* possess flagella for motility. *Euglena gracilis,* a representative species, is often used in studies of cell structure and function (see Figure 5.18). *Golden algae* have

flagella and tend to have tests made primarily of silicon. Golden algae can also exist in aggregates of cells, which suggests that they may have given rise to the true multicellular algae that arose later.

Multicellular Algae

The greatest degree of *organized cell aggregation* (a cellular organization that approaches the level of a multicellular organism) in the kingdom Protoctista is seen in the phyla of brown, red, and green algae. These phyla include the major aquatic *producers* of freshwater and marine environments. By "producers" we mean organisms that are capable of capturing light

Figure 5.17 Every diatom species has a unique test pattern. Photosynthetic diatoms are very important in aquatic systems because they are able to convert sunlight into a form of chemical energy that is stored in organic molecules.

Figure 5.18 Flagellated unicellular algae, including *Euglena,* are important photosynthetic organisms in aquatic systems.

energy and making complex organic substances from simple inorganic molecules.

Brown Algae

About 1,500 species of brown algae are found within and below the intertidal zone of rocky coasts in temperate continental regions. The larger forms, some attaining a length of nearly 100 meters, are known as **kelp** (see Figure 5.19A). Kelp dominates these intertidal habitats and forms the base of some food chains, described in Chapter 9.

Brown algae are a varied group, ranging in size from microscopic forms composed of single filaments to the large kelps. They also have a number of characteristics that enable them to survive in harsh intertidal areas. Some species have hollow sacs that are filled with water, which helps to prevent drying during exposure to air at low tides. In other species, these sacs are inflated with air and serve as float bladders, bringing the algae up near the surface and into the light during high tides; this allows the algae to carry on photosynthesis throughout the entire tidal cycle. Moreover, kelps have anchoring structures called **holdfasts** that anchor them to the rocky shore and allow them to survive in the pounding surf (see Figure 5.19B).

Red Algae

Red algae also reside in rocky intertidal areas along temperate coasts; however, their distribution is more cosmopolitan because they can occupy both sandy and rocky tropical beaches (see Figure 5.19C). There are about 4,000 species of red algae, making them the most diverse multicellular algae phylum. Because some become encrusted with calcareous materials composed of calcium carbonate, these algae add a splash of color and texture to many intertidal pools exposed at low tide along rocky, temperate coastlines.

Green Algae

The green algae are contained in two phyla: the gamophytes and the chlorophytes. *Gamophytic green algae* reproduce sexually when their amoeboid gametes (sex cells), which do not have flagella, fuse during conjugation, which, as mentioned earlier, is a form of sexual reproduction involving the "mating" of two cells, with an actual exchange of nuclear material, to produce new individuals. These algae are important members of freshwater communities, and we study them further in Chapter 11. Microscopic gamophytes include *desmids* and some common *filamentous algae* that produce "pond scums" during summer blooms (see Figure 5.20).

Figure 5.19 (A) Bullwhip kelp (*Nereocystis luetkeana*) forms dense beds in offshore ocean waters up to 80 meters deep. Its holdfasts often give way in the pounding surf, and masses of kelp wash up on Pacific beaches from Alaska to Baja California. *N. luetkeana's* float bladders and long stalks, up to 40 meters in length, resemble the whips used by bull teamsters, hence their common name, bullwhip kelp. (B) Sea palms are common brown algae that inhabit many outer rocky coasts. Their strong holdfasts anchor them to rocky shorelines despite the pounding surf. (C) Coralline red algae give both color and texture to tidepools found in the rocky intertidal zones of most oceans.

Figure 5.20 The scums that form on the surfaces of ponds during summer usually result from the growth of green algae.

Chlorophytic green algae live in both freshwater and marine environments and are distinguished from gamophytes by their flagellated gametes, which fuse during conjugation. Freshwater chlorophytes, along with gamophytic green algae and cyanobacteria (blue-green algae), are the floating producers present in lakes and ponds. In marine environments, chlorophytes add bright green colors that contrast with the drab brown algae and deep red algae also present in the intertidal areas. Many unique chlorophytes occur in these marine habitats, characterized by the following distinguishing forms: stiff or flexible filaments similar to those of freshwater species; thin, bright green sheets, as in the sea lettuce *Ulva;* and thick, spongy growths of dark green "fingers," as in the genus *Codium.*

Other Protoctists

Myxomycota and the other remaining protoctist phyla include the peculiar group of organisms collectively known as *slime molds* (see Figure 5.21). They are separated into different phyla

based on their body types. Some tend to be cellular, some form nets, and some form large cellular masses at certain stages in their life cycles. A few of these unusual organisms share many characteristics with species found in the kingdom Fungi, but they still produce flagellated cells, thereby retaining their membership in the kingdom Protoctista.

> **BEFORE YOU GO ON** The kingdom Protoctista contains single-celled protozoans, unicellular and multicellular algae, and slime molds. This kingdom includes the more primitive eukaryote life forms that have in common aerobic respiration and flagella during some life stage.

KINGDOM FUNGI

The kingdom **Fungi** encompasses eukaryotic, spore-forming organisms that lack flagella. It consists of five phyla that contain more than 100,000 species. Most fungi are **saprophytic,** obtaining their nutrients through absorption of breakdown products from dead plants and animals. A number of parasitic species cause many diseases in plants and a lesser number in animals.

Most new fungi arise directly from spores produced asexually by mature fungi in structures called *sporangia.* Fungal spores give rise to structures called **hyphae,** which usually become multicellular when connecting "cross walls," or **septa,** join neighboring hyphae (see Figure 5.22). Hyphae form mats of tissue called **mycelia** that constitute the body of a fungus. In the life cycles of most fungi, sexual reproduction also occurs by conjugation, which results in the production of other spore types from which new hyphae develop.

Figure 5.21 Slime molds are protoctists that have some fungal characteristics during certain stages of their life cycle. For example, they may develop spore-forming reproductive structures that mature and release spores. After dispersal, these spores germinate, forming a new generation of slime molds like the one shown here growing on a dead Douglas fir tree.

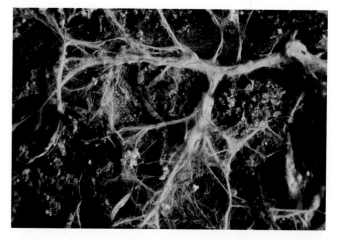

Figure 5.22 Most fungi consist of hyphae that form mats of mycelia within the soil, rotting logs, and vegetation in which they grow. Microscopic magnification of fungal hyphae reveals networks such as those shown here.

Question: *The saprophytic life-style of fungi would be most prevalent in which of the kingdoms discussed so far?*

Zygomycotes

The phylum of the zygomycotic fungi includes black bread molds in the genus *Rhizopus*. These molds were a problem in food preservation before mold inhibitors were routinely added to bakery products. Some of the zygomycotic fungi are parasitic, but most are saprophytic (see Figure 5.23).

Ascomycotes

In this phylum, bladder fungi are distinguished by microscopic reproductive structures called **asci** ("little sacs") that appear as tubular spore sacs filled with spores. This phylum includes certain edible products such as morels and truffles, as well as the yeasts used in many fermenting processes, such as those used for producing beer and wine and leavening bread (see Figure 5.24). There are a few parasitic forms in this phylum, including the pathogenic species that causes Dutch elm disease, which has decimated most populations of North American elm trees.

Basidiomycotes

The most familiar fungi are basidiomycotes. This phylum contains smuts, puffballs, and mushrooms (see Figure 5.25). Most of a living basidiomycote actually exists underground or within the rotting log on which it may live. The mushroom protruding aboveground is the spore-producing, reproductive structure.

Many species of zygomycotes, ascomycotes, and basidiomycotes form mutualistic associations with the roots of trees and other plants that are called *mycorrhizae.* The mycorrhizal relationship between roots and fungi is beneficial to the tree species because it increases the rate at which mineral nutrients liberated by the fungi are transported from the soil into the roots. Fungi benefit by receiving carbohydrates produced by the tree.

Figure 5.24 Ascomycotes are a diverse group of fungi that include the edible yellow morels shown here (*Morchella esculenta*).

Deuteromycotes

The fourth phylum of fungi includes species that have lost the ability to reproduce sexually or that never were sexual; for this reason, they have often been classified as the *fungi imperfecti.* This group is defined as a phylum on the basis of structural form. Deuteromycotes include the genus *Penicillium,* from which the important antibiotic penicillin is obtained, as well as a number of parasitic genera including those that cause root rot in wheat and athlete's foot in humans (see Figure 5.26). Most scientists consider fungi in this phylum to have evolved from the ascomycotes or basidiomycotes.

Lichens

Lichens are classic *symbionts* (participants in a close living association); their bodies consist of two different species: one a fungus and the other a species of cyanobacteria (Kingdom Prokaryotae) or chlorophytic algae (Kingdom Protoctista). Currently, lichens are classified according to their fungal component. Most lichens consist of a fungus in the phylum ascomycotes; therefore, such lichens are also classified as ascomycotes. Lichens play important roles in many natural communities (see Figure 5.27).

Figure 5.23 Black bread mold, belonging in the genus *Rhizopus,* forms stalks on which sporangia develop spores. If dispersed to suitable substrates, these spores will germinate and grow into new fungi.

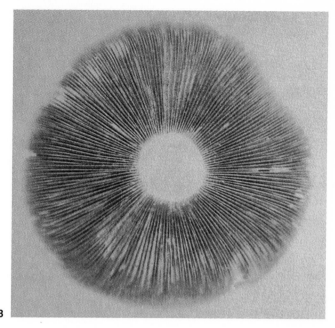

Figure 5.25 (A) Basidiomycotes include familiar puffballs that produce spores. (B) The spore print of a mushroom is often used to identify a species (*Agaricus campestris*). Spore prints are made by placing mushrooms on white, colored, or black paper and allowing them to dry. As they dry out, the mushrooms release their spores onto the paper in a characteristic pattern.

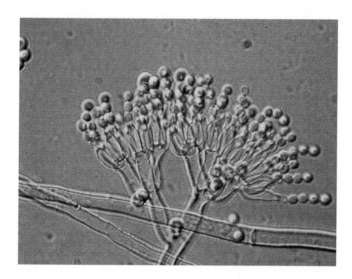

Figure 5.26 A photomicrograph of *Penicillium*, magnified × 2000. This fungus yields the important antibiotic penicillin.

Figure 5.27 Lichens are classified with fungi because of their growth form. However, they are composed of both a fungus and a cyanobacterium, or green alga, forming a symbiotic relationship that is beneficial to both organisms.

BEFORE YOU GO ON The kingdom Fungi includes eukaryotic organisms that have no flagella. Most fungi are saprophytic, but some are important parasites in both plants and animals, including humans. They all produce reproductive spores, some through asexual reproduction and others through sexual processes. Usually only the spore-forming structure is visible because the bulk of a fungus grows within the soil or decaying vegetation.

SUMMARY

1. Estimates of the total species diversity in the biosphere range from 5 to 30 million, but only 1.7 million species have been described. Linnaeus developed a binomial system of nomenclature that uses a unique genus and species name for identifying each organism.

2. All the biota of Earth are classified into five kingdoms: Prokaryotae, Protoctista, Fungi, Plantae, and Animalia.

3. The prokaryotes are separated into two subkingdoms, the Archaebacteria and the Eubacteria, which are further partitioned into a number of divisions based on the staining properties of their cell walls. Each division is broken down into a number of phyla. Archaebacteria phyla include methanogenic, halophilic, and thermoacidophilic forms. Eubacteria include autotrophs, heterotrophs, important nitrogen-fixing forms, and pathogenic forms.

4. Organisms classified in the kingdom Protoctista are aerobic eukaryotes that have flagella during some stage of their life cycle. They include both single-celled and multicellular forms that range in size from microscopic protozoans and algae to large marine kelps.

5. The kingdom Fungi contains saprophytic and parasitic organisms composed of hyphae that form mats of mycelia and asexual reproductive structures that produce spores. Most fungi can also reproduce sexually.

WORKING VOCABULARY

Archaebacteria (p. 88)
class (p. 81)
Eubacteria (p. 89)
family (p. 81)
Fungi (p. 95)
genus (p. 81)

kingdom (p. 81)
phylum (p. 81)
Prokaryotae (p. 88)
Protoctista (p. 85)
species (p. 80)

REVIEW QUESTIONS

1. What is a scientific name? How does it differ from the common name of an organism?

2. How is the hierarchy of a division used in the kingdom Prokaryotae?

3. Why is the kingdom Protoctista said to be defined by exclusion?

4. Describe the various types of flagella and how they are used in the classification of protoctists.

5. Which types of algae are you likely to find in the intertidal zone along rocky coasts?

6. How do cyanobacteria differ from green algae?

7. How do fungi obtain their food?

8. Which phyla of fungi are most useful to humans, and in what ways?

9. What are lichens? Why are they classified with the fungi?

ESSAY AND DISCUSSION QUESTIONS

1. Scientists believe that large-scale destruction or alteration of natural environments (for example, tropical rain forests) threatens the diversity of life. Why might the loss of species diversity be a serious problem?

2. Why are some species in the kingdom Prokaryotae thought to be most similar to the first organisms to evolve on Earth?

3. Construct an artificial system of classification.

REFERENCES AND RECOMMENDED READING

Ahmadjian, V. 1982. The nature of lichens. *Natural History,* 91: 30–37.

Corliss, J. O. 1984. The protista kingdom and its 45 phyla. *BioSystems,* 17: 87–126.

Gentry, A. H. 1988. Tree species richness of upper Amazonian forests. *Proceedings of the National Academy of Science,* USA, 85: 156-159.

Krogmann, D. 1981. Cyanobacteria (blue-green algae): Their evolution and relation to other photosynthetic organisms. *Bio-Science,* 31(2): 121–124.

Margulis, L., and K. Schwartz. 1988. *Five Kingdoms: An Illustrated Guide to the Phyla of Life on Earth.* New York: Freeman.

Margulis, L., K. Schwartz, and M. Dolan. 1994. *The Illustrated Five Kingdoms: A Guide to the Diversity of Life on Earth.* New York: HarperCollins.

Smith, A. 1980. *The Mushroom Hunter's Field Guide.* Ann Arbor: University of Michigan Press.

Vidal, G. 1984. The oldest eukaryotic cells. *Scientific American,* 250: 48–57.

Whittaker, R., and L. Margulis. 1978. Protist classification and the kingdoms of organisms. *BioSystems,* 10: 3–18.

Woese, C. 1981. Archaebacteria. *Scientific American,* 244: 92–122.

ANSWERS TO FIGURE QUESTIONS

Figure 5.1   It would differ only in the species name; eastern white pine is *Pinus strobus.*

Figure 5.2   A species consists of a population (or populations) that cannot interbreed successfully with another species. A subspecies consists of a group within a species that is usually geographically isolated from other subspecies. If ranges of different subspecies do overlap, successful interbreeding can occur.

Figure 5.3   No. However, different species within a phylum can have the same name.

Figure 5.5   It groups together as protoctists amoebae, kelps, water molds, and other eukaryotes that actually have little in common.

Figure 5.6   Six: We are both in the kingdom Animalia, phylum Chordata, subphylum Tetrapoda, class Mammalia, subclass Thera, and order Primate.

Figure 5.8   No. Environmental conditions far exceed those in which any eukaryotic organisms could survive.

Figure 5.10   Membranes would have to form, isolating the nuclei into separate cells.

Figure 5.16   No. Shellfish eat by continuously filtering whatever is in the water, whereas fish eat selectively. Thus, fish would not concentrate the toxin as filter-feeding shellfish do.

Figure 5.22   Most bacteria of the kingdom Prokaryotae absorb their nutrients in a saprophytic manner. (We will also see that a few organisms in the kingdom Plantae are also saprophytic.)

6

The Diversity and Classification of Life: Kingdoms Plantae and Animalia

Chapter Outline

Reading Questions

1. What characteristics are used in classifying plants?

2. What characteristics are used in classifying animals?

3. What is the basis for classifying tetrapod vertebrates?

4. What is a taxonomic classification system? A functional classification system? How do they differ?

KINGDOM PLANTAE

Plants have been the major occupants of terrestrial landscapes for more than 400 million years, and today nearly 500,000 species are known to inhabit Earth. The actual number is certainly much larger. Plants are eukaryotic organisms that are distinguished by a life cycle featuring an **alternation of generations,** in which two different structural forms—*gametophytes* and *sporophytes*—alternatively produce one another through time. The **gametophyte** is the gamete-producing stage of plants, and the **sporophyte** is the spore-producing

phase, shown in Figure 6.1 (see page 100). Through the process of *mitosis,* gametophytes produce gametes (sperm and egg cells) that unite during fertilization, develop into multicellular embryos, and ultimately mature to become sporophytes. Through the process of *meiosis,* sporophytes produce spores that germinate and develop into gametophytes. Details of these processes are presented in Chapter 15.

All plants are classified into one of two major groups: **tracheophytes** or **bryophytes. Tracheophytes** (vascular plants) have vascular tissues—vessels and tubes—that transport water and minerals obtained from the environment, and sugar produced during

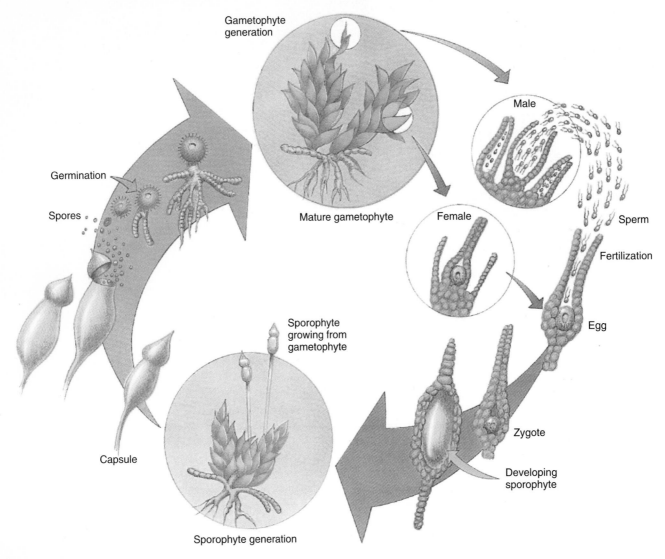

Figure 6.1 Plant life cycles have an alternation of generations. The sporophyte generation of a moss produces spores. Spores germinate and develop into the gametophyte generation, which produces sperm and eggs. These gametes unite during fertilization and form a zygote, which will grow into a mature sporophyte.

Question: *Which is the more prominent generation in mosses?*

photosynthesis, from one region of the plant to another. **Bryophytes** (nonvascular plants) have no vascular tissues, and required substances simply diffuse from the environment into their tissues.

Bryophytes

The bryophyte phylum includes about 24,000 species that are separated into three classes with somewhat unappealing names: *liverworts, hornworts,* and *mosses* (see Figure 6.2). Nonvascular bryophytes are considered to be the most primitive plants because the sporophyte is inconspicuous and nutritionally dependent on the gametophyte, which dominates the life cycle. The gametophyte stage develops from spores and appears as a small leafy structure with rootlike **rhizoids** that are composed of elongated single cells or multicellular filaments. Female gametophytes produce eggs that are fertilized by flagellated sperm furnished by male gametophytes. The sperm must swim

to the egg, an awkward requirement for land plants that limits bryophyte distribution to environments that are wet during some part of the year. Zygotes result from the union of egg and sperm and develop into embryos that grow and develop into the sporophyte phase of the bryophyte life cycle.

Most bryophytes are terrestrial organisms, although some are aquatic. A few species inhabit the splash zone of intertidal habitats along rocky coasts, but none are considered to be strictly marine organisms. They are important members of many different terrestrial communities.

BEFORE YOU GO ON Bryophytes are photosynthetic nonvascular plants. This phylum includes mosses, liverworts, and hornworts. All plants, including the bryophytes, have an alternation of generations. In this type of life cycle, a gametophyte generation produces gametes that, after fertilization, develop into a sporophyte generation, which produces spores that germinate and grow into gametophytes.

Figure 6.2 (A) Liverworts and (B) mosses have no vascular tissue. This accounts for their having a low growth form.

Question: *Why does the lack of vascular tissue result in low growth forms?*

Tracheophytes

Tracheophytes include the most advanced species of plant life. The more conspicuous sporophyte generation is characterized by the presence of specialized groups of cells and intercellular substances that form vascular conducting tissues.

Spore-forming Tracheophytes

Four phyla (whisk ferns, clubmosses, horsetails, and true ferns; see Figure 6.3) include primitive tracheophytes that produce both spores and gametes during their life cycles. Ferns are by far the most widespread of these four phyla on Earth at the present time.

Ferns have flagellated sperm, and like the bryophytes, they are restricted to habitats that occasionally have sufficient moisture to allow fertilization. Such moist conditions occur in a number of different biomes, permitting ferns to become members of many plant communities. A few aquatic ferns form symbiotic associations with nitrogen-fixing cyanobacteria, which leads to their involvement in the nutrient cycles of many aquatic systems.

Seed-forming Tracheophytes

There are two types of seed-forming tracheophytes: those that have a "naked" seed, the *gymnosperms,* and those with seeds enclosed within certain plant structures, the *angiosperms.* Seed-

Figure 6.3 (A) Clubmosses, (B) horsetails, and (C) true ferns are tracheophytes that do not produce seeds.

Question: *If these plants produce no seeds, how do they reproduce?*

A   B   C

Figure 6.4 Living gymnosperms include (A) cycads, small shrubby plants that have a cursory resemblance to pineapples; (B) gnetophytes, like the cone-bearing, seed-producing *Welwitschia mirabilis* that grows in the deserts of southern Africa; and (C) conifers, including the largest plant known, the giant sequoia tree.

forming tracheophytes are the dominant plant forms on Earth today.

Gymnosperms **Gymnosperms** are separated into four different phyla: *cycads, ginkgos, gnetophytes,* and *conifers* (see Figure 6.4, page 102). **Conifers** (cone producers) are the most common gymnosperms and are presently represented by about 50 genera of *softwood* species, including those that are harvested by the timber industries in North America. **Softwood** is a common term applied to lumber from conifers. Commercially important species include pines, Douglas fir, true fir, larch, spruce, and redwoods.

Angiosperms All plants that develop a fruit are classified as **angiosperms.** Fruits usually develop from *ovaries,* a female part of a flower, but in some angiosperms other flower parts may also contribute to fruit development. The seeds of angiosperms develop within the ovaries after eggs are fertilized by sperm contained in pollen. More than 300 families of angiosperms are separated into two major classes: monocots and dicots.

Monocots develop from germinated seeds that have a single embryonic leaf (called a cotyledon) and grow into plants with parallel leaf venation and floral parts that occur in threes or multiples of three, as shown in Figure 6.5. The vascular tissues in cross sections of monocot stems appear as scattered bundles of cells.

Monocots include most species that are cultivated as food crops by humans. The grass family is by far the largest and most important; it includes rice, wheat, barley, oats, other grains, and the grasses on which most animal *herbivores* feed (see Figure 6.6). **Herbivores** are organisms that feed on plants or other primary producers. Horses, sheep, cows, and many other animals are herbivores. Larger members of the grass family include corn, sugarcane, and bamboo. The largest monocot species belong to the pineapple, banana, and palm families.

Many spring wildflowers are monocots in the lily, iris, and orchid families (see Figure 6.7). The orchid family has the largest number of flowering plants—over 20,000 species—most of which grow in tropical or subtropical regions of the world.

Dicots develop from germinated seeds that have two cotyledons, grow into plants with netted leaf venation, and produce flowers with parts in fours or fives or multiples of four or five. The vascular tissue in the dicot stem forms a circular pattern when viewed in cross section.

Nearly 70 percent of the 250,000 known species of flowering plants are dicots. They vary in form and habitat, ranging from deciduous trees, such as maples and elms, to herbaceous (nonwoody) ground cover species found on the floor of old-growth forests, to the giant saguaro cactus, a hallmark species in the Sonoran Desert of southern Arizona (see Figure 6.8, page 104).

Dicot species that have been adapted for human cultivation include peas, beans, alfalfa, and clover—all members of the pea family; lemons, oranges, and grapefruit in the citrus family; and the diverse rose family, which includes roses, berries, cherries, peaches, almonds, pears, and apples among its 3,000 species.

The aster family features over 19,000 species, some of which are among the most evolutionarily specialized plants. This largest, most diverse dicot family includes sunflowers, ragweed, zinnias, asters, sagebrush, lettuce, and dandelions.

Appendix B at the end of the book lists the phyla in the kingdom Plantae and includes an example of a genus from each.

BEFORE YOU GO ON Tracheophytes include whisk ferns, clubmosses, horsetails, true ferns, gymnosperms, and angiosperms. They all produce spores during their life cycle, but only gymnosperms and angiosperms form seeds. Cycads, ginkgos, gnetophytes, and conifers comprise four phyla of gymnosperms. All angiosperms produce true flowers. They are separated into two classes: monocots and dicots.

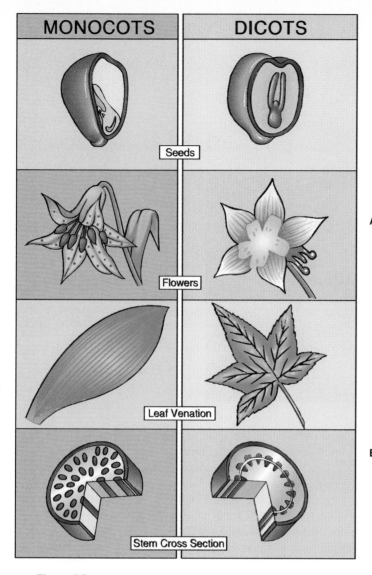

A

B

C

Figure 6.5 Monocots and dicots can be distinguished in a number of ways. Monocots have one cotyledon and dicots have two. The floral parts of monocots occur in threes or multiples of three, those of dicots in fours or fives or multiples of four or five. Vascular bundles in monocots form a parallel pattern of leaf veins; dicot bundles form a netted pattern. In monocot stems, vascular tissue forms a scattered pattern in cross section; in dicots it forms a circular pattern.

Question: *Are there more monocot or dicot tree species in eastern North America?*

Figure 6.6 (A) Rice is a monocot in the grass family. Grains make up an important part of the human food supply throughout the world. (B) Bananas and (C) date palms are among the largest monocots. They are an important human food source in some regions.

Figure 6.7 Many early wildflowers, including these yellow fawn lilies, are monocots.

Figure 6.8   The dicots include more species than any other terrestrial plant class. Their diversity ranges from (A) the saguaro cactus that lives in the hot Southwest deserts to (B) many of our fruits, such as the peaches shown here.

Question: *Which characteristic of a walnut could be used to identify it as a dicot or monocot?*

KINGDOM ANIMALIA

The classification of the animal kingdom is based to a large extent on patterns of early embryonic development. The kingdom Animalia includes all heterotrophic eukaryotes in which a **zygote** is produced from the fertilization of an egg by a sperm. Animal zygotes undergo successive cellular divisions to form a hollow ball of cells, the *blastula,* shown in Figure 6.9. Further development of the blastula proceeds when certain surface cells begin to move inward, forming the **blastopore,** an opening that will develop into the mouth in some animal species and into the anus in others. These migrating surface cells become a cell layer known as the **endoderm.** As they continue to migrate inward, some cells move to the area between the developing endoderm and the outer layer of cells, the **ectoderm.** These "middle cells" form the **mesoderm.** At this stage, the original zygote has reached the embryonic stage of development known as the **gastrula,** shown in Figure 6.10. The inward-moving cylinder of endoderm eventually unites with the ectoderm on the other pole of the gastrula, thus forming the animal's primitive gut (see Figure 6.11).

One of the major characteristics that is used to separate animal phyla is the fate of the blastopore. Animal phyla in which the blastopore becomes the mouth are known as **protostomes** ("first mouth"), and those in which it becomes the anus are called the **deuterostomes** ("second mouth"). Figure 6.11 shows some of the developmental differences between these two types of animal phyla.

The symmetry of developing animal embryos follows one of two general patterns, as described in Figure 6.12 (see page 106): a primitive radial type or the more advanced bilateral form. Phyla in which animals develop bilateral symmetry are separated into three groups on the basis of whether or not a body cavity is present, and if so, where it develops, as described in Figure 6.13 (see page 106). Animals in the most advanced bilateral phyla form a body cavity known as a

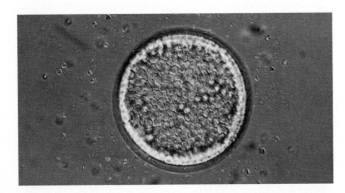

Figure 6.9   The formation of a blastula, like that of a sea urchin shown here, is an important characteristic used in classifying animals.

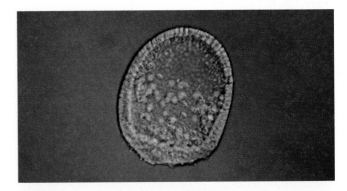

Figure 6.10   During sea urchin development, a blastopore appears as the blastula develops into a gastrula, shown here. In echinoderms and chordates, this opening is destined to become the anus, but in all the other animals in which it forms, it becomes the mouth.

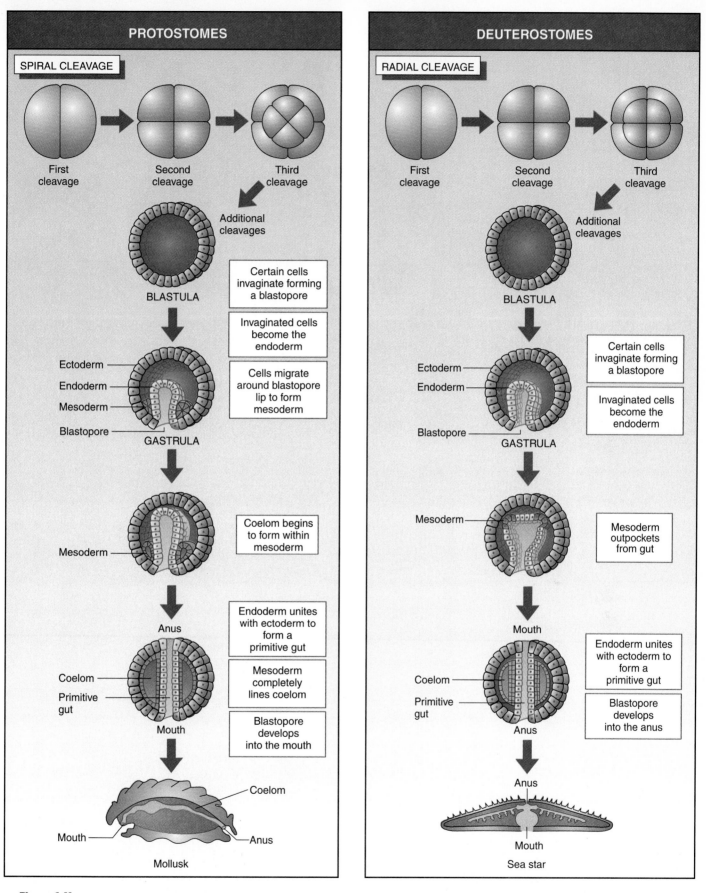

Figure 6.11 A protosome (mollusk) develops as follows: the fertilized egg divides three times, producing uniform cells, and the forth cleavage (not shown) results in cells arranged in a spiral pattern. A blastula (a hollow ball of cells) develops, followed by a gastrula. Mesoderm, formed from cells that migrate to the blastopore, then gives rise to the coelom, the body cavity in which the internal organs will develop. Finally, the blastopore develops in the mouth. Deuterostome (sea star) development differs in that cells after the forth cleavage have a radial pattern, mesoderm arises from an outpocketing of gut endoderm cells, and the blastopore becomes the anus.

Figure 6.12   Radial symmetry in adult animals can be seen in (A) the green anemone (*Anthopleura xanthogram-mica*). This male anemone, with the shell of a shore crab in its tentacles, is releasing its gametes (sperm). Both its internal and external structures are arranged around a central axis; these structures radiate outward, creating a pattern much like that of a freshly cut pie. (B) The tree frog (*Hula cinerea*) is a bilateral animal because only one plane can pass from its anterior end ("head") to its posterior end ("tail"), forming equal right and left halves.

Question: *What kind of symmetry do you have?*

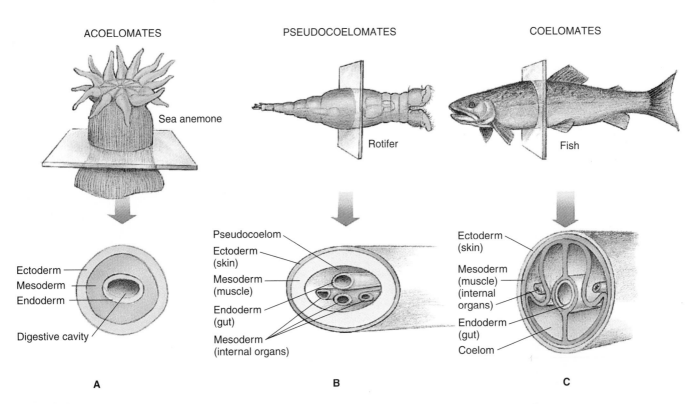

Figure 6.13   Internal body structures and organ development vary among the animal phyla. In acoelomates such as (A) the sea anemone, no internal cavity develops between the tissue layers. In (B) rotifers and other pseudo-coelomates, a cavity referred to as a pseudocoelom ("false coelom") develops between the endoderm (gut) and the mesoderm (muscles) in which the internal organs develop. In (C) the fish and other coelomates, a coelom develops within the mesoderm. The internal organ systems of the fish develop and are suspended within this cavity.

coelom within the mesoderm and are called **coelomates.** Most internal organ systems (for example, the digestive system) of coelomates develop and are suspended within the coelom. The most primitive bilateral phyla do not have a coelom and are known as the **acoelomates.** The other group of animal phyla develop a body cavity between the mesoderm and the endoderm that is commonly called a *pseudocoelom* ("false" coelom); hence they are classified as **pseudocoelomates.**

Table 6.1 indicates that the 33 animal phyla are separated according to their pattern of embryonic development and the presence or absence of organ systems. Appendix B at the end of the book lists the phyla in the kingdom Animalia and includes an example of a genus from each.

Primitive Animal Phyla

The most primitive animal phyla are those that have only a tissue level of organization (that is, they have no true organs). They are represented by living animals in two phyla: *Placozoa,* which has only a single species, *Trichoplax adhaerens,* a microscopic animal found in seawater and on the walls of marine tanks in laboratories and aquariums; and *Porifera,* which includes all the sponges. The sponges shown in Figure 6.14 have tissues that are composed of two cell layers separated by **mesenchyme,** a gelatinous layer in which *spicules* and *amoebocytes* can be found.

Figure 6.14   The yellow candle sponge is in the phylum Porifera, one of the most primitive of the animal phyla.

Spicules are skeletal elements composed of calcium carbonate or silicates produced by **amoebocytes,** which are cells that can move through the cell layers with their pseudopodia. Sponges are filter feeders that remove particles and *plankton*—small marine or freshwater organisms inhabiting the water's surface layer—from the water.

Table 6.1   Key to the Phyla in the Kingdom Animalia

	Phylum
1. (a) Tissue level of organization ⟶	Placozoa / Porifera
or	
(b) Tissue and organ level of organization ⟶	2
2. (a) Radial symmetry ⟶	Cnidaria / Ctenophora
or	
(b) Bilateral Symmetry ⟶	3
3. (a) Coelom absent (acoelomates) ⟶	Mesozoa / Platyhelminthes / Nemertina / Gnathostomulida / Gastrotricha
or	
(b) Coelom present ⟶	4
4. (a) Partial coelom present (pseudocoelmomates) ⟶	Rotifera / Kinorhyncha / Loricifera / Acanthocephala / Entoprocta / Nematoda / Nematomorpha
or	
(b) Coelom present ⟶	5
5. (a) Blastopore becomes the mouth (protostomes) ⟶	Ectoprocta / Phoronida / Brachiopoda / Mollusca / Priapulida / Spinucula / Echiura / Annelida / Targigrada / Pentastoma / Onychophora / Arthropoda
or	
(b) Blastopore becomes the anus (deuterostomes) ⟶	Pogonophora / Echinodermata / Chaetognatha / Hemichordata / Chordata

As explained in Chapter 5, keys designed to classify taxa alllow choices to be made between two different characteristics. In this key to the phyla in the kingdom of Animalia, the first characteristic (1) is the animal's level of organization, either tissue only (a) or tissue and organ level (b). The second characteristic (2) is symmetry, either radial (a) or bilateral (b). Other choices include the presence or absence of a coelom (3), the type of coelom, when present (4), and the fate of the blastopore (5).

Radially Symmetrical Phyla

The two phyla of animals with radial symmetry are the *Cnidaria* and the *Ctenophora*. Most cnidarians are marine and include the familiar true jellyfish, sea anemones, and corals. Hydras and a few other species live in fresh water. The marine corals are responsible for the production of some of the largest structures produced by any organism, the coral reefs. The phylum Ctenophora contains about 90 species of marine organisms known as comb jellies (see Figure 6.15).

Acoelomate Phyla

Acoelomates are separated into five phyla of wormlike animals that are either parasitic or free-living. Nemertine ribbon worms found in intertidal zones and the brown planaria from freshwater streams are among the most common free-living species. Some of the parasitic flatworms, such as liver flukes, blood flukes like the schistosomas, and tapeworms, are very detrimental in many developing tropical and subtropical coun-

tries because of the human and other mammalian diseases they cause. Members of the other acoelomate phyla are for the most part small freshwater or marine animals seldom observed in nature.

Pseudocoelomate Phyla

Important pseudocoelomate phyla include *Rotifera* and *Nematoda* (see Figure 6.16). Rotifers are common aquatic organisms in freshwater ponds and lakes. Of the 2,000 known species, only 50 are marine. Free-living nematodes occur in all habitats, where they are often the most numerous of all animals. There are also a number of parasitic nematodes that cause diseases in both plants and animals, resulting in significant economic losses of some agricultural crops, wildlife, and domestic animals. The number of known nematode species is approaching 100,000, but it is considered likely that there will be more than 1 million at the final accounting.

Loricifera is the most recently described (1983) animal phylum. It includes microscopic, spiny-headed, wormlike pseudocoelomates that live in clean, coarse marine sands or fine gravel.

A

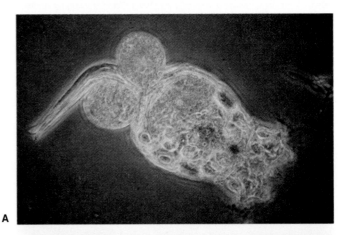

A

B

Figure 6.15   (A) The hydra in the phylum Cnidaria and (B) the comb jelly in the phylum Ctenophora represent the two radially symmetrical animal phyla.

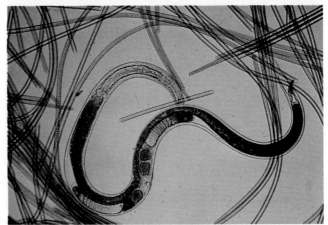

B

Figure 6.16   (A) Rotifers live in most aquatic environments. (B) Nematodes are found in most environments on Earth.

Coelomate Protostome Phyla

The most primitive coelomate animals occupy 12 protostome phyla. In these animals, the blastopore forms the mouth of mature adults. Many of these phyla include marine, wormlike organisms rarely seen by most biologists. *Mollusca* and *Arthropoda* are the two protostome phyla that contain organisms of the greatest economic importance.

Annelids

Segmented worms in the phylum *Annelida* are among the best-known protostomes—the most familiar being the common earthworm or night crawler. Species in this phylum are distributed from the tropics to the Antarctic seas, where they exist as scavengers or predators in terrestrial, marine, and freshwater habitats (see Figure 6.17).

Mollusks

The phylum *Mollusca* is composed of about 110,000 species separated into eight classes, of which three (Gastropoda, Bivalvia, and Cephalopoda) are the best known and of the greatest importance to humans (see Figure 6.18). *Gastropods* occur

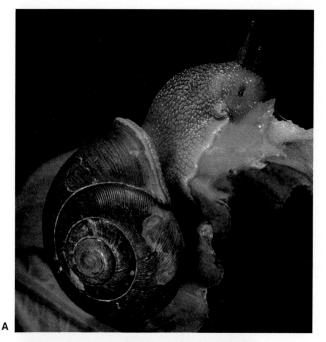

A

A

B

B

C

Figure 6.17 Annelid worms live in both marine and terrestrial areas. (A) This marine worm, *Nereis vexillosa,* reaches about 15 centimeters in length and lives in mussel beds along the rocky Pacific coast. (B) Earthworms, or night crawlers, are terrestrial annelids.

Figure 6.18 Mollusks are a diverse coelomate protostome phylum. (A) Garden snails, (B) mussels, and (C) octopuses represent three of the eight classes in this phylum.

in aquatic and terrestrial environments and include the land snails and slugs familiar to gardeners in moist climates, both temperate and tropical. **Bivalves** include most of the marine and freshwater organisms known as "shellfish." Although they are not actually fish, they do include many economically important species of clams, oysters, scallops, and mussels that are cosmopolitan in both their distribution and their appeal as food to humans. **Cephalopods** are a class of mollusks that includes the largest and most complex protostome coelomates. Giant squids are known to reach lengths greater than 18 meters and weights

in excess of 1,000 kilograms. Other cephalopods include the cuttlefish, the octopus, and the nautilus.

Arthropods

The phylum *Arthropoda* is the largest animal phylum, containing 18 classes distributed among three subphyla: *Crustacea, Chelicerata,* and *Uniramia.* As a group, the crustaceans predominate in many aquatic environments. Crayfish, barnacles, shrimp, crabs, and lobsters contribute to the bounty that humans harvest from rivers and oceans (see Figure 6.19). The chelicerates include

Figure 6.19 Arthropoda, the largest animal phylum, is divided into three subphyla: Crustacea, Chelicerata, and Uniramia. Of the arthropods shown here, the (A) ghost crab and (B) crayfish are in the first subphylum, (C) marine horseshoe crabs and (D) spider are in the second, and (E) centipedes and (F) insects represented by the dragon fly are in the third.

Question: *Which insect taxa represent commonly used terms like flies, beetles, butterflies, bees, and others?*

marine horseshoe crabs, sea spiders, and the class of terrestrial arachnids containing mites, spiders, and scorpions. Uniramians include centipedes, millipedes, and insects. Insects are the most numerous terrestrial animals. Many insect species have aquatic stages in their life cycle. The class *Insecta* is partitioned into more orders than any other animal class, and each order contains numerous families, many with vast numbers of genera and species. About 750,000 species of insects have been described, but intensive collections in tropical forests indicate that their total number, on a world basis, may exceed 30 million!

> **BEFORE YOU GO ON** Sponges have only a tissue level of organization, jellyfish and comb jellies have radial symmetry, and the other animal phyla are distinguished by their coelom development and the fate of the blastopore.

Coelomate Deuterostome Phyla

Five phyla of deuterostomes contain the most familiar animals that inhabit Earth. They all have a true coelom, and the blastopore of developing embryos becomes the anus in adults. *Echinodermata* is by far the best known of the four primitive deuterostome phyla. It includes sea stars (starfish) and sea urchins, shown in Figure 6.20, which live in the intertidal areas

Figure 6.20 (A) Sea stars and (B) red urchins are in the phylum Echinodermata. Adult echinoderms have radial symmetry, but they develop from bilateral larvae.

Question: *What does the blastopore become in adult sea stars and urchins?*

along rocky coasts. This phylum is unique among the advanced phyla because the adults have radial symmetry, a trait that is characteristic of more primitive jellyfish and other cnidarians.

The phylum *Chordata* contains the most advanced animals. Members of this phylum all have a **dorsal** hollow nerve cord (located near the back), **pharyngeal gill slits** (openings that allow water to flow through the mouth, over the gills, and out the gill slits) during some stage of their life, and a **notochord** (an internal, cartilaginous rod). The chordates are divided into four major subphyla: *Tunicata, Cephalochordata, Agnatha,* and *Gnathostomata* (see Table 6.2, page 112).

Tunicates and Cephalochordates

Tunicates have a notochord only during their motile, larval stage, and adults live as sessile (attached) organisms in marine environments. *Lancelets* are cephalochordates that have a notochord and dorsal nerve cord in both their larval and adult stages. They are primitive, marine, fishlike chordates that live buried in sand. They feed on plankton delivered to their mouth by cilia-induced currents (see Figure 6.21).

Agnaths

The parasitic lampreys are living relics of the primitive agnaths, the jawless fishes. Their notochord and pharyngeal gill slits persist throughout life, and their dorsal nervous system includes a very simple brain. By parasitizing and killing large numbers of fish, lampreys have had a great negative impact on fisheries in the Great Lakes.

Figure 6.21 (A) Tunicates and (B) lancelets are members of primitive chordate phyla.

Table 6.2 Key to the Phylum Chordata

1.	(a)	Brain Absent	⟶	2
		or		
	(b)	Brain present	⟶	3
2.	(a)	Notochord present in larvae only	⟶	Subphylum Tunicata (tunicates)
		or		
	(b)	Notochord present in larvae and adults	⟶	Subphylum Cephalochordata (lancelets)
3.	(a)	Jaws absent	⟶	Subphylum Agnatha (lampreys)
		or		
	(b)	Jaws presesent	⟶	Subphylum Gnathostoma, 4
4.	(a)	Paired jointed appendages absent	⟶	Superclass Pisces, 5
		or		
	(b)	Paired jointed appendages apresent	⟶	Superclass Tetrapoda, 6
5.	(a)	Internal skeleton (composed of cartilage)	⟶	Class Chondrichthyes (sharks and rays)
		or		
	(b)	Internal skeleton (composed of bone)	⟶	Class Osteichthyes (bony fish)
6.	(a)	Poikilothermic	⟶	7
		or		
	(b)	Homeothermic	⟶	8
7.	(a)	Scales absent	⟶	Class Amphibia (frogs and salamanders)
		or		
	(b)	Scales present	⟶	Class Reptilia (lizards, snakes, and turtles)
8.	(a)	Teeth absent, feathers present, hard-shelled egg	⟶	Class Aves (birds)
		or		
	(b)	Teeth present, feathers absent, young nourished with milk	⟶	Class Mammalia (mammals)

This key to the phylum Chordata separates the taxa by the presence or absence of the following characteristics: brain (1), notochord (2), jaws (3), paired jointed appendages (4), internal sekelton (5), temperature control (6), scales (7), and teeth, feathers, and nourishment of young (8).

Gnathostomates

Superclass Pisces   The superclass *Pisces* includes the two classes of true fish: *Chondrichthyes,* the cartilaginous sharks and rays that live in seawater (marine) environments, and *Osteichthyes,* the bony marine and freshwater fishes. In adult sharks, the notochord is replaced by vertebrae that are composed of cartilage. The pharyngeal gill slits persist throughout life, serving as openings for water that flows through the mouth and over the gills. Sharks have paired *pectoral fins* located behind the head, paired pelvic fins on the underside near the anus, and an asymmetrical tail (see Figure 6.22A).

Members of the class Osteichthyes have an ossified (bony) skeleton composed of a skull, a vertebral column, and ribs. About 25,000 species of bony fish live in marine and fresh-

water systems. Bony fish have several adaptations for living in aquatic environments, including a swim bladder that functions in depth control, paired pectoral and pelvic fins, and a symmetrical tail fin (see Figure 6.22B). The gill slits of bony fish consist of a single opening covered by an **operculum** that protects the gills and allows oxygen-containing water to enter the mouth and flow over the gills in a controlled fashion.

Superclass Tetrapoda   Tetrapoda is composed of the other four familiar vertebrate classes: Amphibia, Reptilia, Aves, and Mammalia. Organisms in these classes are characterized by the presence of bilateral symmetry, increasing cephalization (the concentration of specialized nervous tissues in the head), and four limbs used for locomotion. The amphibians and reptiles are poikilothermic in that they have no internal

Figure 6.22 (A) The leopard shark is a member of the class Chondrichthyes, the cartilaginous fishes that inhabit marine environments. (B) Members of the class Osteichthyes (bony fish), live in both fresh and marine water. Some class members, such as these sturgeon, resemble ancestors that swam in rivers 400 million years ago.

Question: *If* ichthy *is Greek for* fish, *what might* chondr *and* oste *mean?*

mechanisms that control their body temperature, but the birds and mammals are homeothermic because they have such mechanisms.

The transition of amphibians from aquatic to terrestrial life was never completed, and they all must reproduce in moist or aquatic environments, like the bullfrog in Figure 6.23. Amphibians include salamanders, newts, frogs, and toads, many of which have existed with few changes for over 150 million years. They all have smooth, moist skin and gills at some life stage. Most species develop lungs as adults.

Figure 6.23 The life cycle of an amphibian such as this bullfrog starts with sexually mature adults. Frog eggs are fertilized within the female after she picks up a sperm packet that was deposited by a male on a stick, a leaf, or the ground. After internal fertilization, (A) eggs are laid in fresh water (B) where they develop. (C) Young larvae (tadpoles) grow to resemble their parents and mature to (D) adult size through successive metamorphosis. The adult bullfrog is the largest frog species in the New World. As tadpoles, bullfrogs live in ponds and lakes and feed on algae; as adults they are carnivores, feeding mainly on insects.

Reptiles were the first truly terrestrial vertebrates. Certain characteristics acquired over millions of years of evolution enabled them to inhabit terrestrial environments. These characteristics included skin with protective scales, internal fertilization, and the *amniote egg*, with a waterproof, leathery shell and a fluid-filled membrane, shown in Figure 6.24. The last two developments freed reptiles from any dependence on water to complete their reproduction. Today reptiles include lizards, snakes, turtles, alligators, and crocodiles.

Birds, which make up the Aves vertebrate class, evolved from reptiles millions of years ago. All birds retain reptilian-like scales on their feet, and their feathers are actually highly modified scales. Unlike reptiles, however, modern birds produce amniote eggs that have hard, waterproof, calcium-impregnated shells; they have no teeth; and their forelimbs have evolved into wings. There are 19 orders of birds containing more than 8,600 species (see Figure 6.25). With their highly evolved structures enabling them to fly, they dominate the airspace they share with arthropod insects and mammalian bats.

Unique characteristics of the class Mammalia include the presence of hair, functional red blood cells without nuclei, and

A

B

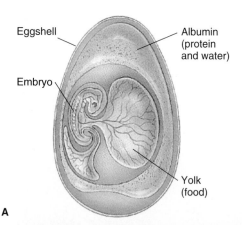

A

Eggshell

Albumin (protein and water)

Embryo

Yolk (food)

B

Figure 6.24 Reptiles are terrestrial poikilotherms that have been liberated from aquatic environments because of their amniote (shelled) egg (A). With a waterproof covering and an internal water supply, reptiles, such as this pilot blacksnake with the eggs she just laid, have been able to inhabit most regions of Earth (B).

Figure 6.25 Unique features of birds include feathers, wings, and calcareous shelled amniote eggs. Each of these birds, (A) young screech owls and (B) a parrot, represents one of the 19 bird orders.

female mammary glands. There are two subclasses of mammals: the primitive *Prototheria,* consisting of the egg-laying, duck-billed platypus and the spiny anteater, and *Theria* ("true mammals"), which contains all other species (see Figure 6.26). The subclass Theria includes both *marsupials* and *placental mammals* and is divided into 17 orders containing 4,600 species that inhabit most aquatic, terrestrial, and aerial environments (see Figure 6.27). The *marsupials* have internal fertilization and early development, but later development occurs in an external pouch, where the embryo feeds on milk provided through the nipple of a mammary gland. *Placental mammals* complete their development within the female uterus, nourished by the *placenta,* an organ jointly produced by the mother and the developing embryo.

Figure 6.26 The class Mammalia includes species with diverse reproductive strategies that range from (A) the primitive egg-laying platypus to (B) the marsupial opossum to (C) the more advanced placental rodent, a beaver.

Figure 6.27 The 17 mammalian orders are represented by the four shown here. (A) Dolphins are marine mammals, (B) eastern gray kangaroos are marsupials, (C) elephants are the largest terrestrial mammals, (D) and the pig-tailed macaque represents the primates.

Question: *What do you think are the largest and smallest (adult) mammals?*

BEFORE YOU GO ON   With the exception of the echinoderms, the most complex lower animals are coelomate protostomes. The echinoderms and chordates are coelomate deuterostomes. Reptiles, birds, and mammals are the only truly terrestrial chordate phyla, and the latter two are the only homeothermic ones.

A FUNCTIONAL CLASSIFICATION SYSTEM

The five kingdoms of life can also be classified by an alternative *functional classification system* (see the Focus on Scientific Process, "Functional Classification"). The basis for such a system is the mode of nutrient acquisition in each of the major groups as described in Figure 6.28. **Functional classification systems** emphasize processes that occur among organisms, and they are becoming important in both terrestrial and aquatic ecology.

Producers

Cyanobacteria in the kingdom Prokaryotae, algae in the kingdom Protoctista, and the entire kingdom Plantae are all classi-

fied as primary **producers.** Through photosynthesis, they combine the energy-poor molecules of carbon dioxide and water and convert them into energy-rich organic molecules (sugars) using sunlight as the energy source and releasing oxygen as a by-product. Most other organisms on Earth depend on the products formed during photosynthesis by these primary producers.

Decomposers

Archaebacteria and eubacteria in the kingdom Prokaryotae and the entire kingdom Fungi can be classified as **decomposers.** They all produce digestive enzymes that reduce complex organic molecules into a form that the organisms can absorb. The free-living species of this group are saprophytic; they digest nonliving organic matter. Pathogenic forms have the ability to digest and destroy living cells, and in some ways they might be considered consumers. For convenience, however, all of these organisms are classified as decomposers because they do not ingest nutrients; rather, they absorb essential energy-rich organic molecules.

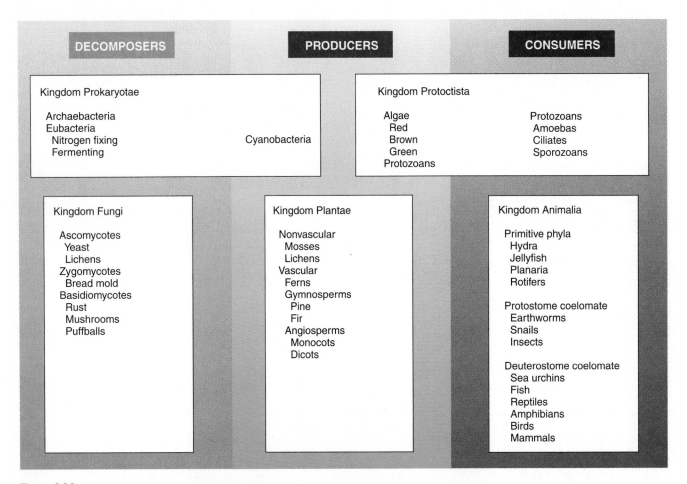

Figure 6.28   This classification system distinguishes organisms in the five kingdoms in two different ways. First, organisms are grouped with respect to their taxonomy within their kingdom. Second, in the functional system, organisms of a kingdom are regrouped depending on their mode of nutrient acquisition. Decomposers include all of the fungi and some prokaryotes; producers include all of the plants and the prokaryote cyanobacteria, as well as protoctist algae; and consumers include all of the animals and the nonphotosynthetic protoctists.

FOCUS ON SCIENTIFIC PROCESS

Functional Classification

Any system that attempts to classify organisms represents a compromise between potentially conflicting requirements. Biologists often prefer a *natural classification system,* one that reflects a real order in nature. Since Darwin's theory of evolution was published in 1859, this has meant that taxonomic categories and their arrangement must reflect known evolutionary relationships. A natural system must also be inclusive and relatively easy to use because biologists from diverse fields depend on it for information about the living world. The taxonomic system described in Chapter 5 basically satisfies these two requirements. It is recognized as the universal classification system used by biologists and students around the world to identify and name organisms. Nevertheless, if different relationships are of interest, then alternative *special-purpose systems* of classification are useful.

Some biologists are more interested in particular functional relationships among organisms than in their general evolutionary relationships. In such cases, biologists may group together unrelated species that all have specific characteristics or adaptations. Such groupings comprise a *functional classification system* and may be particularly appropriate for certain purposes. For example, consider the three plant species shown in Figure 1. Although they all appear cactuslike, only one belongs to the cactus family (Cactaceae). Why do they all look like cacti? Basically because these unrelated species independently evolved the protective spines, fleshy tissues for water storage, reduced leaves that prevent water loss, and thick epidermis necessary to survive and reproduce in a desert environment. Plants that share these adaptations for life in dry habitats are sometimes described as *succulents,* a

A

B

C

Figure 1   Cactus-like plants in the spurge family, *Euphobiaceae* (A); milkweed family, *Aesclepiadaceae* (B); and cactus family, *Cactaceae* (C).

box continues

term that does not correspond to any Linnean taxonomic group. It is a useful artificial category, however, for discussing functional similarities among desert plants.

The early twentieth century was a period of rapid growth and specialization in biology. Biologists were drawn to many new areas of research that had scarcely been considered by earlier naturalists. Not surprisingly, many special-purpose classification systems were proposed after 1900. Most of these were later abandoned, but a few have become widely adopted. For instance, in 1908 Danish botanist Christian Raunkiaer proposed a new functional classification system for plants. He identified five basic groups of plants, which he referred to as *life forms,* based on the position of their buds (see Figure 2). Because *buds* contain the embryonic leaves that will emerge during the next growing season, they must be adapted for protecting dormant cells during cold, dry conditions. Raunkiaer described the following plant groups:

- *Group 1:* Plant species with exposed, relatively unprotected buds.
- *Group 2:* Plants with buds restricted to the sheltered lower branches of the plant.
- *Group 3:* Plants with buds located on runners or stolons that extend along the surface of the ground.
- *Group 4:* Plants with well-protected, underground buds.
- *Group 5:* Plants (not shown in Figure 2) that died at the end of each growing season; these plants propagated from year to year solely by seeds.

All of Raunkiaer's life forms are artificial groups and, like the succulents described earlier, they do not correspond to any Linnean taxa. In Raunkiaer's artificial system two very closely related species may be classified as different life forms, and each life form includes many unrelated species.

Why did Raunkiaer create such a system to classify plants? He reasoned that bud position was strongly correlated with a plant's ability to survive cold periods. If that was true, then the life forms could be used as "biological indicators" of climate. Using this classification system, Raunkiaer and his followers completed detailed surveys of the distribution and abundance of vegetation in various parts of the world. For example, when he compared plants in his native Denmark with those growing on two Caribbean islands (St. Thomas and St. John), Raunkiaer found that the two areas had approximately the same numbers of species (1,084 versus 904), but the distribution of life forms was strikingly different. Nearly 60 percent of the plants on the two tropical islands had exposed, unprotected buds, compared to only 7 percent in Denmark. In contrast, 22 percent of Danish species had well-protected underground buds, whereas only 4 percent of tropical species exhibited this life form. Similar trends could be found when plants at different altitudes were compared, or when moist tropical areas were compared with deserts. Thus Raunkiaer's

system of life forms became an important tool for doing quantitative studies of plant geography.

Raunkiaer and other early ecologists sometimes suggested that their functional classification systems were more scientific than the traditional natural system used by Linnaen taxonomists. Some biologists even felt that the traditional taxa might be replaced by more rigorously defined functional groups. This led to heated arguments over the merits of various classification systems. Such disputes are not uncommon when new ideas are presented to the scientific community, and they often prove useful in focusing attention on previously neglected areas of research. By the middle of the twentieth century, most ecologists and taxonomists had grown weary of the controversies. Raunkiaer's life forms gave rise to a useful system of classification and, like some other functional systems of classification, it continues to play an important role in modern biology. Because it is a special-purpose system applicable only in a narrow area of research, however, it cannot replace the general Linnean taxonomic system.

Figure 2   Raukiaer's plant life forms.

Consumers

The protozoan species in the kingdom Protoctista and all members of the kingdom Animalia are considered to be **consumers** because they ingest complex energy-rich organic molecules. Herbivores consume primary producers, and carnivores feed on other animals or protozoans.

SUMMARY

1. All species in the kingdom Plantae exhibit alternation of generations. Sporophyte plants produce spores that germinate and grow into gametophytes, which in turn produce gametes that unite during fertilization to produce new sporophytes.

2. The bryophytes are liverworts, hornworts, and mosses; the tracheophytes include all of the higher forms of plant life. Ferns have flagellated sperm that restrict them to environments that are moist during their period of sexual reproduction.

3. Gymnosperms and angiosperms are two major types of living tracheophytes. Four phyla of gymnosperms contain the cycads, ginkgos, conifers, and gnetophytes, respectively, whereas all true flowering plants—those that produce seeds enclosed within fruits—are in a single phylum.

4. Angiosperms are divided into two classes: monocots, which have a single cotyledon, leaves with parallel venation, and floral parts in threes or multiples of three; and dicots, which have two cotyledons, netted leaf venation, and floral parts in fours or fives or multiples of four or five.

5. The phyla in the kingdom Animalia are segregated according to their pattern of embryonic development and the presence or absence of organ systems. Sponges have tissues composed of two cell layers separated by mesenchyme, whereas jellyfish and sea anemones are characterized by radial symmetry.

6. Acoelomates are represented by the planaria and ribbon worms; pseudocoelomates are represented by rotifers and nematodes. Two developmental forms of coelomates—the protostomes and the deuterostomes—make up the remaining animal phyla. Annelids, mollusks, and arthropods are characteristic protostomes, and echinoderms and chordates are representative deuterostomes.

7. The presence of a brain, notochord, jaws, and paired jointed appendages differentiate organisms in the phylum Chordata. The four classes of tetrapods are amphibians, reptiles, birds, and mammals.

8. Producers, decomposers, and consumers are the major categories in a functional classification system of all the organisms in the five kingdoms of life. The basis of this system is the mode of nutrient acquisition.

WORKING VOCABULARY

angiosperm (p. 102)
blastopore (p. 104)
blastula (p. 104)
bryophyte (p. 99)
consumer (p. 119)
decomposer (p. 116)
deuterostome (p. 104)
dicot (p. 102)
ectoderm (p. 104)
endoderm (p. 104)
gastrula (p. 104)
gymnosperm (p. 101)
herbivore (p. 102)
mesoderm (p. 104)
monocot (p. 102)
producer (p. 116)
protostome (p. 104)
taxonomy (p. 117)
tracheophyte (p. 101)
zygote (p. 104)

REVIEW QUESTIONS

1. What are the major components in the life cycle of a moss?

2. What are the two major types of tracheophytes, and how are they different?

3. How do ferns and gymnosperms differ?

4. What characteristics are used to separate monocots from dicots?

5. How do sponges differ from members of the radially symmetrical phyla?

6. How does a pseudocoelom differ from a coelom?

7. What characteristics are typical of annelids, mollusks, and arthropods?

8. Which phyla are classified as coelomate deuterostomes? What characteristics result in their inclusion in this group?

9. What are the characteristics that separate classes in the superclass Tetrapoda?

10. What is a functional classification system?

ESSAY AND DISCUSSION QUESTIONS

1. What are some of the characteristics that distinguish plants from animals?

2. How do plant and animal life cycles differ?

3. What are three reasons for classifying organisms?

4. How do taxonomic and functional classification systems differ?

REFERENCES AND RECOMMENDED READING

Gordon, M.S. 1995. *Invasions of the Land: The Transitions of Organisms from Aquatic to Terrestrial Life*. New York: Columbia Univ. Press.

Hickman, C. P. 1995. *Animal Diversity*. Dubuque: W. C. Brown.

Little, L. 1980. *The Audubon Society Field Guide to North American Trees: Western Region*. New York: Knopf. Also see the other volumes in this series, including *Birds: Eastern Region* and *Western Region; Butterflies, Fish, Whales and Dolphins; Mammals; Reptiles and Amphibians; Seashore Creatures; Trees: Eastern Region;* and *Wildflowers: Eastern Region* and *Western Region.*

Pearson, L. C. 1995. *The Diversity and Evolution of Plants*. Boca Raton: CRC Press.

Raven, P., R. Evert, and S. Eichhorn. 1986. *Biology of Plants*. 4th ed. New York: Worth.

Walters, D. R. 1996. *Vascular Plant Taxonomy.* 4th ed. Dubuque: Kendall/Hunt.

Wilson, E. O. 1992. *The Biodiversity of Life.* Cambridge: The Belknap Press.

ANSWERS TO FIGURE QUESTIONS

Figure 6.1   The gametophyte.

Figure 6.2   Vascular tissue not only transports sugars and water, but it gives strength to the stems and roots of higher plants. Plants without vascular tissue lack the resources to grow vertically.

Figure 6.3   They produce spores that germinate and develop into gametophytes that produce gametes.

Figure 6.5   Dicot species far outnumber monocot species.

Figure 6.8   When you crack a walnut open, you can see two cotyledons; thus, it is a dicot.

Figure 6.12   Humans have bilateral symmetry.

Figure 6.19   Orders.

Figure 6.20   Because Sea stars and urchins are deuterostomes, the blastospore becomes the anus.

Figure 6.22   The Greek prefix *chondre* means "cartilage;" the Greek prefix *oste* means "bone." Thus Chondrichthyes are "cartilaginous fish" and Osteichthyes are "bony fish."

Figure 6.27   Blue whales are the largest mammals, and some of the shrews are the smallest.

Unit One Conclusion
The Sphere of Life and the Process of Science

Questions related to the origin of the universe, Earth, and life on Earth were first asked in early human cultures, and some of those questions have still not been completely answered. We began by considering the mythical worldviews of our ancient ancestors and how those beliefs shaped their answers to questions about the structure and organization of the universe. Our understanding of the natural world fundamentally changed during the scientific revolution of the seventeenth century. At that time, scientists began to develop a picture of the universe that was based on basic laws of nature. Technological advances, such as the development of powerful telescopes, allowed scientists to make more precise observations that led to new, more rigorous ideas about the universe. Early ideas about the origin of life on Earth changed after Redi and Pasteur demonstrated that living organisms could not be created by spontaneous generation from nonliving matter.

The universe, Earth, and life on Earth are composed of chemicals that were created as a result of galaxy evolution, which began over 10 billion years ago. In 1808, the atomic theory explained that matter was composed of atoms, and subsequently, details of atomic structure were investigated. In the twentieth century, geologists established the age of Earth and conducted studies that supported hypotheses about the nature of early Earth. It also became clear that the evolution of photosynthesis and cellular respiration played critical roles in Earth's maturation and the appearance of life on Earth. Consequently, new hypotheses were formulated about chemical evolution, the origin of life on Earth, and the origin and evolution of prokaryotic and eukaryotic cells. Through the last half of the twentieth century, these hypotheses have been supported by research using new technologies, such as the electron microscope, but they continue to be modified as more information becomes available. Today, questions related to the origins of life are currently the subjects of intense research efforts. Finally, the big bang theory emerged in the 1940s and 1950s and has since guided investigations on questions about the origin of the universe, matter, and Earth.

In Unit I, we described the remarkable progress made in our understanding of the natural world. Some of the major questions—What are the basic elements that make up the universe and Earth? How old is Earth? What is the basic unit of life?—have been answered. Attractive hypotheses and powerful theories have been formulated that provide partial answers to the other questions: What is the structure of the universe? How did the universe originate? How did Earth originate? How did life on Earth originate? How old is the universe? When did life originate on Earth? When did cells originate? How have cells evolved? These questions continue to guide research today.

The Interactions of Life

2000	
Present	
1980	Is the biosphere maintained in equilibrium by life?
	Is anthropogenic global climate change underway?
1960	
	How do organisms modify their environment?
	What is the relationship between energy flow and ecosystem function?
1940	
	What limits the number of trophic levels in an ecosystem?
	How are biogeochemical cycles regulated?
1920	
1900	
	Do communities reach a predictable, final stage of maturation?
	How do different kinds of life interact within communities?
1880	What is a biological community?
1860	
	What environmental factors affect plant growth?
1840	
1820	
	What affects the distribution of plants and animals?
A.D.1800	Can the human population outgrow its food supply?
Greek Culture 5000 B.C. to 2000 B.C.	How can organisms best be classified?

7

The Biosphere

Chapter Outline

Reading Questions

1. What factors led to the formation of the biomes on Earth?

2. How is the curvature of Earth related to its climates?

3. Why do climates differ across North America?

4. What are some predictive models of biome distribution?

The **biosphere** consists of a thin envelope of air, water, and land in which all known life forms on Earth exist. It provides the physical and chemical conditions necessary to support living organisms that interact with the physical environment. Through these interactions, the physical environment is maintained, yet it becomes modified over time.

In this chapter, we will explore how energy from deep within the planet caused the continents to form numerous alignments in past eons, how sculpting forces have created the present surface features of the biosphere, and how forces in our solar system contribute to the march of the seasons and the development of world climates. Integrating these developments will help you understand why distinct biomes exist in different areas of Earth. **Biomes** are large regions with characteristic assemblages of plants and animals that change from the polar to the equatorial latitudes.

CONCEPTUAL BEGINNINGS

The term *biosphere* dates from 1875, when Austrian geologist Edward Suess briefly described the interactions of several "envelopes" of Earth in his book on the development of the Alps mountain range. Fifty years later, the famous Russian scientist Vladimir Ivanovetch Vernadsky shaped the biosphere concept into its modern form. Vernadsky's biosphere was divided into the **hydrosphere** (all the water on Earth's surface), the **lithosphere** (Earth's crust), and the **atmosphere** (the gases surrounding Earth). These domains are characterized by three conditions necessary for life. First, the range of temperatures allows water to exist in liquid form. Second, the optimal quantity and quality of solar energy allows photosynthesis to occur, and the intensity of harmful ultraviolet radiation from the sun is not so great as to damage biological systems. Third, atmospheric gases, particularly oxygen and carbon dioxide, are found in concentrations necessary to sustain life processes.

Unlike other geologists and chemists of his day, Vernadsky stressed the importance of living organisms as agents of change. According to Vernadsky, organisms not only are sustained by the biosphere, they also actively maintain and modify it. He thought this to be especially true for the human species. Prophetically, in an article written shortly before his death in 1945, Vernadsky warned that humans were becoming a "large-scale geological force" capable of causing irreversible changes in the biosphere. This article was published in the United States less than six months before the first atomic bomb was detonated in the desert near Alamogorado, New Mexico.

Evolution of the Biosphere

Earth has changed continuously during its 4.6 billion years of existence. With the emergence of life between 3.5 and 4 billion years ago, the biosphere began to evolve. Some biosphere transformations occur regularly, on an annual cycle, whereas others take place irregularly, over thousands of years.

Energy Sources

Long-term changes in the biosphere through time resulted from two energy sources. Tectonic movements of the continental plates were driven by energy derived from deep within Earth before and during the formation of Pangaea, 230 to 280 million years ago. These forces created the present alignment of the continents and will continue to rearrange them throughout the geologic future. Refer back to Chapter 3 for the geological time scale.

Different assemblages of organisms inhabited the continents as they drifted apart following the breakup of Pangaea 180 million years ago. Throughout this long period, most life forms have depended on energy derived from the sun. However, as shown in Figure 7.1, this energy is not evenly distributed over the face of the planet. This is because Earth is a sphere whose curvature permits the sun's rays to spread over a greater area near the poles than near the equator. Thus energy—light and heat—is more concentrated at the equator and becomes more diffuse toward the poles. This difference is the primary factor responsible for the development of world climates and for major differences in the distribution of life within the biosphere.

Sculpting Forces

The topography (surface features) of Earth consists of coastal plains, elevated plateaus, canyons, and mountain ranges of varying age. These surface features have resulted from the interactions of major lithospheric, hydrospheric, and atmospheric forces throughout the vast expanse of geologic time. When the continental blocks of Pangaea broke apart during the Mesozoic era, they drifted up and over the oceanic blocks, producing massive uplifting forces that created many of the major continental mountain ranges (see Figure 7.2, page 126). Climatic conditions varied throughout Earth's history, and for extended periods of time, large areas of the continents in the Northern Hemisphere were covered with massive sheets of ice. These periodic ice ages

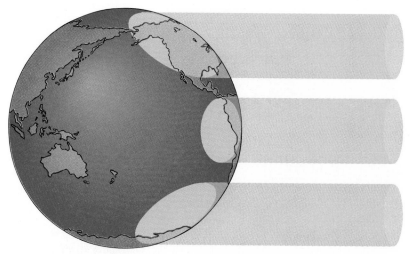

Variation in area covered by solar beams carrying equal amounts of energy

Figure 7.1 Solar energy strikes Earth directly at the equator and at oblique angles as latitude increases, which results in less energy per unit surface area.

Question: *Which location would receive the greater amount of solar energy each year: Fairbanks, Alaska, or Miami, Florida?*

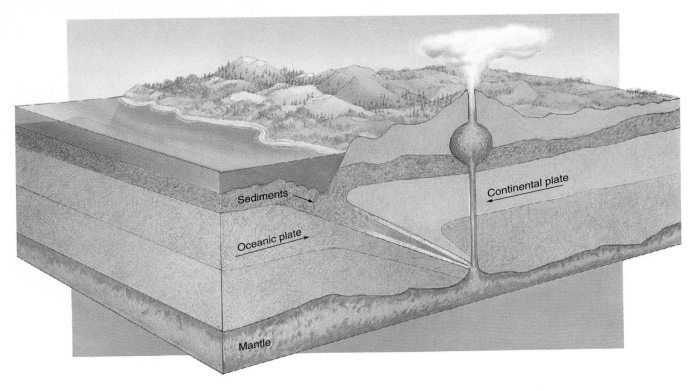

Figure 7.2   When continental and oceanic plates collide, organic sediments are dragged beneath the continental plate along with the subducting oceanic plate. This results in massive uplifting forces that can create new mountain ranges and volcanoes.

Question: *Since the North American continental block is moving west, subduction and its effects should occur at which location: New York City or San Francisco?*

had profound effects on the structure of mountain terrains. The last Great Ice Age occurred during the Pleistocene epoch, and lasted from 20,000 to 10,000 years ago. As seen in Figure 7.3, this ice age caused dramatic changes in Earth's surface. In North America, Pleistocene glaciers left many reminders of their colossal forces, including the Great Lakes in the Midwest and the alpine valleys and lakes in the Rocky Mountain and Sierra Nevada ranges (see Figure 7.4A). The receding glaciers even changed the direction of some river flows. When massive glacial dams melted, the enormous deluges of water they released scarred large landscapes in certain areas (see the Focus on Scientific Process, "Glacial Lake Missoula and the Spokane Floods").

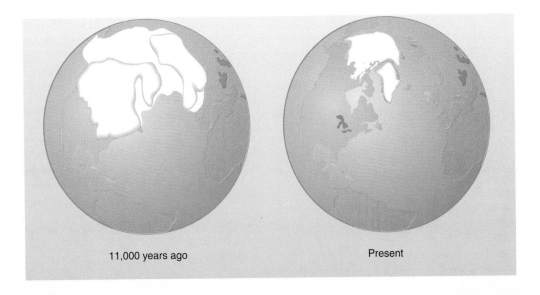

11,000 years ago   Present

Figure 7.3   Comparing the present world map with a map of 11,000 years ago dramatizes the extent of glaciation (white) at the end of the last ice age.

Question: *Where would you rather have lived 11,000 years ago: Portland, Oregon or Portland, Maine?*

A

B

C

Figure 7.4 (A) This lake was formed in the depression made by mountain glaciers during the last ice age. (B) Rock strata of the Grand Canyon represent all of the major eons, eras, periods, and epochs listed on the geologic time scale from the Early Precambrian to the present. (C) Ancient terrains in the Badlands of South Dakota show wind and water erosion.

Not all areas of North America were affected by the ebb and flow of the glaciers, but other forces have shaped vast expanses of ancient terrains. The Colorado River, and others before it, carved canyons over a kilometer and a half deep, exposing rocks that are 1.7 billion years old (see Figure 7.4B). In other areas, wind and water combined in an erosion process that sculpted unique forms in varied landscapes (see Figure 7.4C).

March of the Seasons

Living organisms inhabit environments from polar regions to the equator, from below sea level in Death Valley to about 6 kilometers in the highest mountains, and from the oceans' surfaces to their deep trenches. The inhabitants of these different regions have adapted to tremendous variations in day length, temperature, and available moisture that result from Earth's planetary relationships within the solar system. The term *adaptation* can be used in both an ecological and an evolutionary context. In ecology it is the short-term process of adjustment of an individual organism to environmental stresses. In its evolutionary context, adaptation can result from the long-term processes that modify species over time, improving their reproductive and survival efficiencies.

Earth orbits the sun once each year and makes one complete rotation on its axis each day. Because the axis of its rotation is inclined 23½° from its orbital plane, the Northern Hemisphere tips toward the sun during summer (see Figure 7.5). The

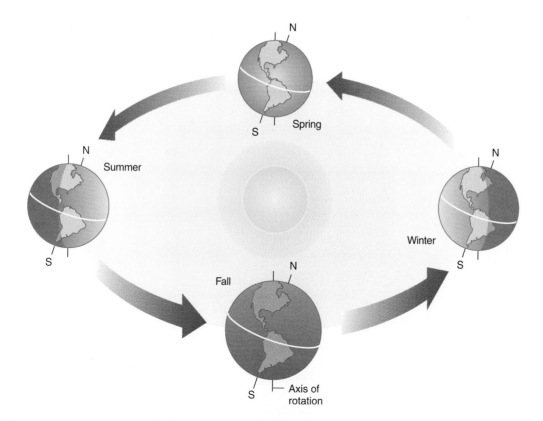

Figure 7.5 The "march of the seasons" results from the 23½° tilt of Earth's axis from the plane of its orbit around the sun. In the fall, the oblique angle of sunlight increases in the Northern Hemisphere as the North Pole tips away from the sun. This reduces the amount of solar energy received per unit area and causes the cooler temperatures that lead to winter. As Earth's axis tips toward the sun, the oblique angle of light decreases, which increases the amount of solar energy per unit area and leads to spring and summer.

Question: *How would the "march of the seasons" be affected if Earth were tilted 5° from its axis of rotation?*

FOCUS ON SCIENTIFIC PROCESS

Glacial Lake Missoula and the Spokane Floods

The Roaring Twenties produced many innovations in the science and culture of the Western world. New hypotheses were being advanced in many fields of science, and great controversies often developed as their merits were debated in the scientific literature.

Early in the nineteenth century, the science of geology was anchored in the **theory of uniformitarianism,** as developed by Charles Lyell in his *Principles of Geology,* which was published in three volumes between 1830 and 1833. Lyell had revived ideas first explained by James Hutton in his 1788 book, *Theory of Earth.* Hutton argued that Earth's sedimentary rocks (rocks formed by condensing loose sediments or by settling from a watery solution) were not, as many people believed, the result of a single 40-day and 40-night flood. Rather, he believed that over vast periods of time they developed layer upon layer in ancient seas that had once covered the land. He also proposed that mountains were continuously being eroded by water flows, which produced sediments that accumulated in ancient seas. Hutton's ideas received some attention at the end of the eighteenth century, but they were soon overshadowed by a doctrine known as *catastrophism.*

Catastrophism grew out of fossil studies and the realization that the living world had been dramatically different in the past. Some scientists felt that fossils occupied distinct layers of rock and that each rock layer represented the remains of a single catastrophic event. Such events were thought to have regularly changed the surface of Earth, and many geologists believed that the "catastrophes" were caused by supernatural forces.

For most geologists, uniformitarianism replaced catastrophism by the 1840s, largely because of Lyell's works. Lyell reinstated Hutton's uniformitarian doctrine by showing clearly that "catastrophic" events in the geological record could be explained by gradual changes observed on Earth today. Consequently, many geologists were stunned when J. Harlen Bretz at the University of Chicago published a paper in the *Journal of Geology* titled "The Channeled Scablands of the Columbia Plateau." In this 1923 paper, Bretz advanced the hypoth-

esis that a large region (40,000 square kilometers) of eastern Washington had been eroded by a catastrophic flood, creating what he termed "channeled scablands" (see Figure 1). Bretz's "Spokane flood hypothesis" was disputed by most prominent geologists of the 1920s and 1930s because it seemed to revert to catastrophism, which for them had long been displaced by uniformitarianism.

Bretz had solid geological evidence—erosion features and depositional patterns—that the channeled scablands were the result of an immense flood (see Figure 2). But what was the original water source? In later papers, he presented two possible hypotheses: (1) a period of very rapid climatic warming had accelerated snow and ice melting, which led to floods; and (2) volcanic activity beneath a large glacier resulted in a massive flood of meltwater. Bretz, however, criticized both hypotheses, noting that 42 cubic miles of ice would have to melt each day to account for his calculated flow rates during the hypothesized flood.

Meanwhile, prominent geologists proposed other hypotheses to explain the channeled scablands. One, for example,

Figure 1   The Dry Falls, shown here, in the Grand Coulee of the Columbia River were formed during the Spokane floods. When these great floods occurred, water over 70 meters deep poured across these falls.

FOCUS ON SCIENTIFIC PROCESS

proposed that ice dams had once formed in the Columbia River gorge, blocking upstream water and causing it to accumulate in a reservoir that spread over eastern Washington. Another suggested that complicated alternations of glacial ice advances had changed water drainage systems throughout the area, which had led to floods. Bretz used field evidence to discount all the alternative hypotheses. Most geologists found this evidence to be convincing but, clinging to the dogma of uniformitarianism, they still refused to accept Bretz's hypothesis.

Dr. Bretz finally explained the location of his water source in 1930—it was Glacial Lake Missoula, which is described in Figure 3 on page 130. This enormous lake had formed during the regression of an enormous ice sheet that once covered southern British Columbia, eastern Washington, northern Idaho, and western Montana. During its retreat, the ice sheet came to rest against the Bitterroot Mountains in the Idaho panhandle region. This resulted in an ice dam, over 600 meters high, that blocked the Clark Fork of the Columbia River. The water behind the ice dam inundated valleys of western Montana,

ultimately creating an inland lake of glacial meltwater that was about half the size of Lake Michigan. It is now known that the ice dam moved back and forth several times, and each time it moved, massive flooding occurred into eastern Washington, creating the channeled scablands. Hence, a series of floods had occurred, not just one.

One curious element in the history of Bretz's Spokane flood hypothesis, and its hostile reception by geologists, was related to the lack of an adequate water source. In 1910, 20 years *before* Bretz identified his water source, J. T. Pardee of the U.S. Geological Survey published a paper in the *Journal of Geology* titled "The Glacial Lake Missoula." Pardee described in detail the location, size, and extent of the lake impounded behind the ice dam on the Clark Fork River. However, because of geologists' reluctance to accept "catastrophic" events, Pardee's superiors apparently did not allow him to study the lake further or to describe its possible connection with the channeled scablands. Not until 1942 did Pardee finally publish a paper in which he presented field evidence that strongly supported Bretz's Spokane flood hypothesis.

Evidence supporting Bretz's hypothesis can still be found today. On mountains east of Missoula, Montana, 240 kilometers upriver from the site of the glacial ice dam, strandlines of Glacial Lake Missoula can be observed at an elevation of 1,280 meters (see Figure 3). It is estimated that 2,100 cubic kilometers of water were released when the ice sheet made its final retreat 11,000 years ago. This last massive flood performed the final sculpting of the eastern Washington channeled scablands, terraced the Columbia River Gorge, and filled, to a depth of 20 meters, the 520-square-kilometer Willamette Valley of western Oregon with soil from eastern Washington, northern Idaho, and western Montana.

Bretz's Spokane flood hypothesis is a good example of a controversial hypothesis that rises in level of certainty as new evidence becomes available. There is also a lesson to be learned from events associated with this hypothesis. Part of geologists' resistance to the idea of a massive flood was related to its clashing with conventional concepts about geological processes. Final acceptance of the Spokane flood hypothesis

Figure 2 The channeled scablands, shown here, were created by a series of enormous floods that swept over a vast geographic area. These floods removed most soil, leaving bedrock, rock rubble, and sand. Today, only hardy grasses, sagebrush, and a few pine trees grow in the channeled scablands.

box continues

Figure 3   The great Spokane floods occurred whenever the ice dam holding back Glacial Lake Missoula melted. It is now known that this occurred about every 50 years or so over a 2,000-year period.

was one of several factors that led geologists to modify their assumptions about how the face of Earth has been modified over time. Geologist's initial adherence to a strict uniformitarian doctrine delayed acceptance of Bretz's hypothesis and demonstrates how dogmatic loyalty to a doctrine can be unproductive. What dogmas do we hold today that prevent us from accepting new ideas? It is often difficult to know what represents "dogma" at any particular time. Thus one of the lessons to be learned from studying the process of science is the importance of keeping an open mind, even when considering what appears to be an "outrageous hypothesis."

first day of summer (the summer solstice) occurs on June 21, when the vertical rays of the sun strike Earth at the Tropic of Cancer, resulting in 24 hours of continual sunlight in the far northern latitudes. As summer progresses to autumn, the days become shorter in the Northern Hemisphere, and fall begins with the autumnal equinox on September 23. On this date, the vertical rays of the sun fall on the equator, and throughout the entire biosphere, day and night last 12 hours each.

As fall slips into winter in the northern latitudes, summer arrives in the Southern Hemisphere. On December 22, the winter solstice occurs, when the sun's rays are vertical on the Tropic of Capricorn. Far northern Arctic latitudes are cast into continual darkness, but the southern latitudes bask in the warmth of full summer. While Alaskans brace for subzero temperatures and blizzards move over the northern Great Plains, revelers spend New Year's Day on sunny beaches in Sydney and Rio de Janeiro. Spring occurs on March 21 (the vernal equinox), when once more the vertical rays of the sun fall on the equator, and day and night are again equal.

The source of the seasons is related to the tilt of the planet. The differential heating from the equator to the polar latitudes creates forces with far-reaching effects over the entire biosphere. This variation in energy generates both the climates of the world and the major currents that circulate in the great oceans.

BEFORE YOU GO ON   Life on Earth was hypothesized in the 1920s to play a role in producing and maintaining habitable environments. Sunlight and energy from deep within the planet affect both the development and the maintenance of the biosphere. Topography, the tilt of the planet, and Earth's orbital variations all interact in creating climates and seasons.

WORLD CLIMATE

The variations in energy received in different areas of the globe cause winds in the atmosphere that deliver moisture to the continents and establish the major circulation patterns in the oceans. The polar latitudes receive only about one-fifth of the total annual sunlight received in the tropics. Temperate latitudes receive about twice the solar energy at the summer solstice as they do at the winter solstice. Why doesn't this uneven distribution of heat lead to even greater temperature extremes than are observed? Do global mechanisms redistribute this uneven concentration of energy?

Atmospheric Cells

In 1735, George Hadley, an English scientist, attempted to explain the prevailing wind patterns of the Atlantic Ocean. He knew that the tropics receive more energy from the sun than the polar regions and also, that as air warms, it rises. From these observations, he developed the concept of an **atmospheric cell,** a circular atmospheric zone in which warm air rises in one area, cools, and then descends in another area. Hadley's original idea has since been extended; today, Earth's wind patterns are explained using the three-cell model described in Figure 7.6. Air movement within the multiple atmospheric cells accounts for all major wind patterns on Earth.

Ocean Circulation

Atmospheric cells and their associated winds are responsible for the large surface circulation patterns in the oceans that are described in Figure 7.7 (see page 132). **Coriolis forces** are associated with Earth's rotation; they cause circular currents in

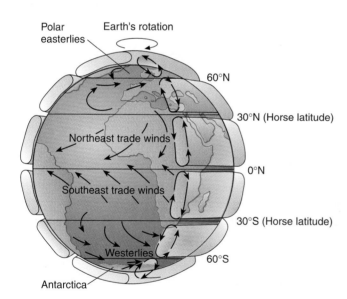

Figure 7.6 In the three-atmospheric-cell model, a combination of the decreasing circumference of Earth from equator to pole and its rotation from west to east (counterclockwise) results in the formation of three atmospheric cells in each hemisphere. After air rises near the equator, creating the low pressure of the doldrums, it descends at about 30° north or south latitude, creating the high pressure that causes surface winds which blow toward the equator. This pressure gradient causes the northeast and southeast trade winds. The descending air between 30° and 35° north or south latitude produces little, if any, surface wind in an area known as the horse latitudes. A second cell forms when air rises once again, causing low pressure at about 60° north or south latitude. This air descends at the horse latitudes and produces a surface wind known as the westerlies. Polar easterlies form from the cycle of the third atmospheric cell.

large ocean basins in the Northern Hemisphere to rotate clockwise and those in the Southern Hemisphere to rotate counterclockwise. These ocean currents transport about 30 percent of the excess equatorial heat north or south, while the atmosphere transports 70 percent. Consequently, the oceanic currents, in combination with solar heating and surface winds, contribute to continental climates.

> **BEFORE YOU GO ON** Three atmospheric cells in each hemisphere produce wind patterns that vary with latitude. These winds and Coriolis forces create the major ocean currents. Both the winds and the ocean currents transport excess equatorial heat poleward.

CLIMATES OF NORTH AMERICA

Climates vary greatly throughout the world, and many of them are represented somewhere in North America. North American climates include the western maritime (ocean-influenced), mountain, and central and eastern continental climates.

Western Maritime Climates

The high capacity of oceans to store heat and the circulation patterns of currents combine to produce profound climatic effects on the western coast of the major continents. From about 40° north or 40° south latitude, the circulation patterns in the oceans produce onshore flows of warm, moist air that moderate these climates, causing them to be milder than their latitude alone would predict. These maritime climates owe their special characteristics to the unique properties of water (see Chapter 2 and Appendix A).

In North America, maritime climates sustain forests of ancient origin. The coastal redwoods of northern California, shown in Figure 7.8A (see page 132), contained species of trees that covered much of North America in past geologic eras, when the continental climate was warmer and moister than it is today. The cool onshore breezes of summer bring moisture in the form of fog, but storm cells of winter become water-laden as they sweep over the warm waters of the Pacific Ocean, bearing heavy rains to these ancient forests.

A short distance north of the California–Oregon border, the effect of increasing latitude lowers the average annual temperature below the tolerance limit of redwoods. Here they are replaced by Sitka spruce, a tree that dominates the coastal fringe from southern Oregon to southern Alaska (see Figure 7.8B). The spruce forests of coastal Alaska persist to 60° north latitude. They could not exist that far north were it not for the moderating effects of the ocean.

Mountain Climates

Maritime coastal forests of the West would extend far inland were it not for the result of tectonic forces operating in the past and the present. These forces have been responsible for the

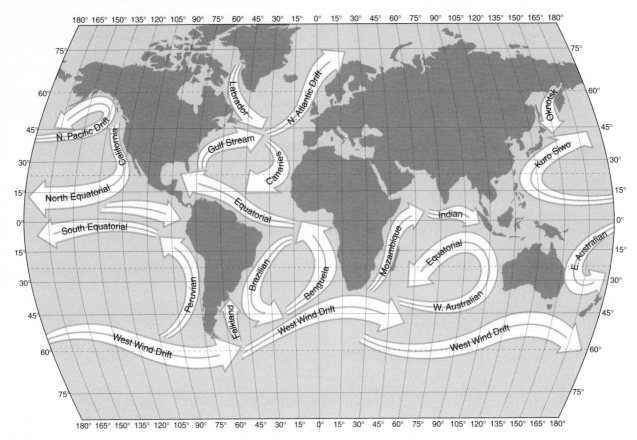

Figure 7.7   The North Pacific gyre (a spiral system of ocean currents) brings cold water in the California current that moderates climates of the west coast of North America. The Gulf Stream and the North Atlantic drift of the North Atlantic gyre tempers climates of the eastern seaboard and western Europe.

Question: *What would be the climatic effect if the directions of flow were reversed for the Gulf Stream and California currents?*

Figure 7.8   (A) Redwood forests exist where cool, moist marine air flows onshore from the Pacific Ocean. (B) Sitka spruce forests grow from southern Oregon to coastal Alaska, replacing the redwoods in these higher latitudes.

development of major western mountain chains. The Coast, Cascade, and Sierra Nevada ranges and the Rocky Mountains all have major modifying effects on the climate of central North America (see Figure 7.9).

When winter storm cells surge over the West Coast, they encounter these mountain ranges, and air is forced to rise. As air flows up and over mountain summits, cooling occurs, moisture condenses, and clouds release tremendous volumes of water in the

A

Figure 7.9 (A) Mount Whitney is the highest mountain in the Sierra Nevada range. (B) The eastward-flowing air from the Pacific Ocean loses much of its moisture as it flows up and over the west-facing slopes of the Sierra Nevada. Deserts occur in rainshadows east of the Sierra Nevada range.

Question: *Is this a photograph of the east face or the west face of Mount Whitney? How do you know?*

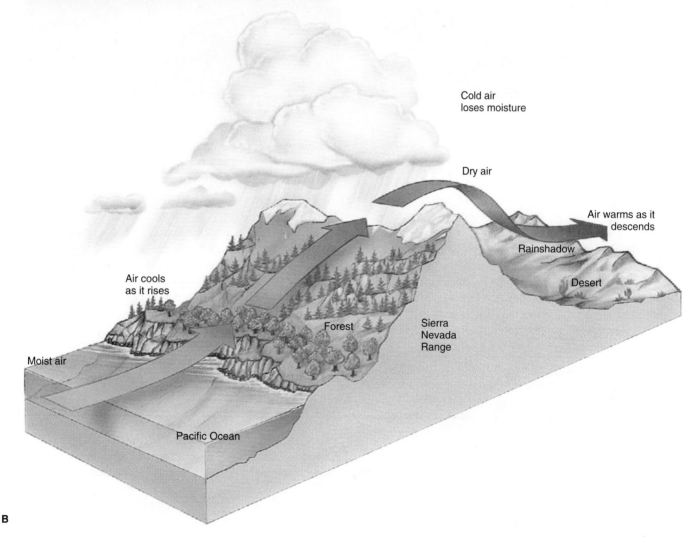

B

form of rain or snow, depending on elevation and season (see Figure 7.9B). When rain forms, heat energy is released, warming the air mass, and as it descends the eastern mountain slopes, compression and further warming occur. This warm, descending air mass tends to absorb moisture. As a result precipitation decreases, forming a **rainshadow** east of these major mountain ranges. As moisture decreases in the rainshadow, forests are replaced by open woodlands, which give way to grasslands and deserts.

Central and Eastern Climates

The climates of central and eastern North America are often modified by maritime influences, but to a lesser extent than western and mountain regions. Typically, for example, summer rains in the Midwest are the result of weather systems spawned in the Gulf of Mexico. Most major winter storms in the eastern half of North America, however, originate in the Arctic, then travel south and east across the Great Plains, creating a continental climate with much greater fluctuations in temperature and moisture than those found in true maritime climates. The grasses of the North American continent's tallgrass prairie have developed certain strategies for dealing with such fluctuations. With widespread and finely branched root systems, these grasses are able to absorb sufficient water from even the lightest rainfalls during dry periods. They can also lie dormant for lengthy periods of harsh weather conditions and then quickly spring back to life (see Figure 7.10A, page 134).

Figure 7.10   (A) Most original tallgrass prairies have been altered by human activities. However, remnants of such prairies can still be seen in a few areas such as Flint Hills National Grasslands, Kansas, which is shown here. (B) Weather fluctuations in New Hampshire mountains, shown here, are less extreme than those of western mountain ranges

Such continental climatic conditions are tempered on the eastern seaboard, where the waters of the Gulf Stream, a warm Atlantic Ocean current, moderate some of the extremes. Thus temperature and other weather fluctuations tend to be less extreme in the mountains of, say, New Hampshire (see Figure 7.10B) than in the middle western prairie or the mountains of Colorado. The Gulf Stream extends its moderating effects beyond North America to the island provinces of eastern Canada, southern Greenland, Iceland, the isles of Great Britain, and much of northwestern Europe.

> **BEFORE YOU GO ON**   Western maritime, mountain, and central and eastern continental climates occur west to east across North America. Proximity to oceans, high mountain ranges, and latitude all contribute to climate.

DISTRIBUTION OF WORLD BIOMES

Because the biosphere is not uniform, great diversity exists in the types of plants and animals that inhabit different parts of the world. Thus variations among climates of the world have resulted in many different biomes, each with a unique assemblage of organisms.

Biome Formation

Variations in solar energy and moisture have led to the formation of the general latitudinal pattern of biome distribution that is described in Figure 7.11. This pattern began to be recognized in maps drawn during expeditions early in the sixteenth century. It is verified today by photographs from space, which show six major biome types: tropical forests, deserts, temperate forests, grasslands, taiga, and tundra.

Tropical forests are found in the equatorial regions of continents, between the Tropic of Cancer and the Tropic of Capricorn. **Deserts** tend to be located in the doldrums of 30° north and south latitude, where air in the Hadley cells descends. **Temperate forests** occur in latitudes of the westerlies, between 5° and 10° north and south of the fortieth parallel. **Grasslands** develop in these temperate regions in areas devoid of major mountain systems and where distinct wet and dry seasons occur. The **taiga** begins at about 50° north latitude and ends at the lower limit of the permafrost, where the **tundra** begins. The taiga and tundra are characteristic of only the Northern Hemisphere, because there are no landmasses of sufficient size in the Southern Hemisphere where these types of biomes could develop.

It becomes evident on inspection of a world biome map that the general pattern of latitudinal biome distribution is modified by topography, proximity to oceans, and continental size. These modifying factors change the distribution of some major biomes and lead to the development of other minor biomes. Minor biomes include **scrub forests** and **savannas** in the tropics and **chaparral** in the Mediterranean climates. We will consider the characteristics of the major and minor biomes in greater detail in Chapter 8.

Different biomes result from different climates, which are in turn caused by variations in the amount of available solar energy and moisture, which vary with latitude. Changes in **topography** associated with increased elevation, however, can produce similar differences in the general distribution of life forms. This fact became evident to European explorers when they first visited tropical regions.

Altitudinal Zonation

The Personal Narrative of the Travels to the Equinoctial Regions of America (1799–1804), by Alexander von Humboldt and Aimé Bonpland, includes numerous descriptions of the

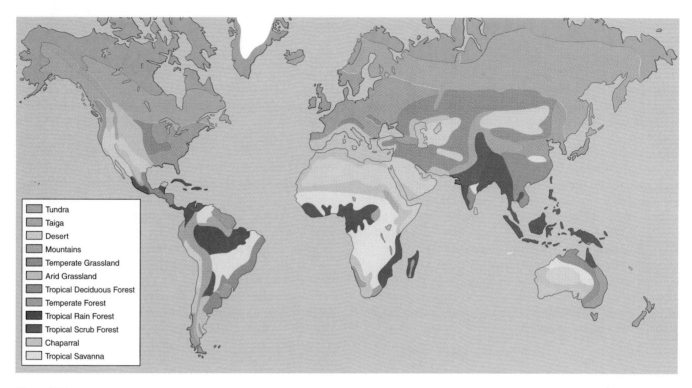

Figure 7.11 World biomes developed in response to latitude, proximity to oceans, topography, and other factors that generate world climates. Greater expanses of grasslands, boreal forests, and tundra occur in the Northern Hemisphere than in the Southern Hemisphere because there is insufficient landmass in southern latitudes for the development of those biomes.

authors' ascents of several high mountains, most in the Andes of Venezuela. From his journals, it is evident that Humboldt developed an understanding of how the pattern of life was related to altitude. Early in his journey with Bonpland, while still in the Canary Islands off the west coast of Africa, Humboldt described the ascent of the volcano Tenerife: "We beheld the plants divided by zones, as the temperature of the atmosphere diminished with the elevation of the site." At the top of equatorial mountains, Humboldt discovered **alpine zones** that were similar to tundra biomes. Humboldt's concept of **altitudinal zonation** can be defined as a change in community structure as a function of changing elevation. That is, the types of organisms found at sea level will be much different from those found on a mountain 2,000 meters above sea level.

Biogeographical Models

Humboldt's idea was developed further in the 1890s by C. Hart Merriam while he was studying the distribution of vegetation on San Francisco Peaks in Arizona (see Figure 7.12). Merriam believed that he could predict which organisms would be present at specific latitudes or elevations by applying what he called the *Law of Temperature Control of Geographic Distribution*. This law stated that "animals and plants are restricted in southward distribution by the mean temperature of a brief period covering the hottest part of the year" and that their northern distribution "is determined by the total quantity of heat." Merriam used this law in developing different **life zones** that were defined in terms of elevation.

Merriam's life zone concept represents a model. A *model* is a simplified representation of some natural phenomenon. A model can be expressed in different forms: a written description, a picture, a graph, a mathematical equation, or a computer program. By focusing on important factors (*variables*) and by eliminating extraneous details, a good model can make accurate predictions about cause-and-effect relationships. Simple models that account for only one or two variables can sometimes be remarkably predictive. Increasing predictive power is

Figure 7.12 Early life zone studies were conducted by C. Hart Merriam in the San Francisco Peaks of Arizona, shown here.

generally achieved by including more variables in the model; however, as the complexity of a model increases, the model becomes more difficult to understand or apply. Therefore, models created by biologists are often compromises between simplicity and predictive power.

In Merriam's model, a single variable—temperature—was used to predict which life zones would be present at any particular latitude or elevation. But there is no simple cause-and-effect relationship between temperature and the distribution of life on Earth, and Merriam could not always accurately predict the location of his life zones. Could Merriam's model be made more predictive by adding a second variable? During the 1960s several plant ecologists combined two variables—average temperature and annual precipitation—to predict types of vegetation (see Figure 7.13). These models are better predictors of climatic conditions in the major biomes. Their predictive value, however, decreases at poorly defined boundaries between biomes. Ecologists are continually trying to build models that include several variables in order to adequately account for the global distribution of vegetation.

Distribution of Biome Vegetation: A New Predictive Model

At present, a new model with high predictive value is emerging. Examination of this model offers insight into the type of research now being conducted on biomes. The first part of this model states that vegetation mass, structure, and *leaf area index* can be predicted from the *hydrological balance* in different biomes. **Leaf area index (LAI)** is the ratio of total leaf surface area to the total ground surface area covered by the leaves (that is, LAI = total leaf surface area ÷ total ground surface area covered). For example, if 1 square meter of ground is covered with 1 square meter of leaf surface area, LAI = 1, but if the same area of ground is covered with 3 square meters of leaf surface area, LAI = 3. The greater the amount of foliage covering the ground, the larger the LAI.

The **hydrological balance (HB)** is computed using a variety of data, including the amount of precipitation that penetrates the canopy and reaches the ground surface as well as the total loss of water from evaporation and by the foliage itself. In general, there is a direct correlation between HB and LAI. A high HB is predictive of a high LAI, and a low HB is predictive of a low LAI.

Another component of the model uses minimum yearly temperatures to predict the *physiognomy* in each biome as described in Figure 7.14A. **Physiognomy** refers to the types of vegetation present in an area (see Table 7.1, page 138). Physiognomy is affected by low temperatures associated with both freezing and **supercooling,** which occurs when water cools rapidly and fails to freeze at 0°C. When water cools slowly, ice crystals form, and freezing occurs at 0°C. These ice crystals may damage the cells of plants. However, when supercooling occurs, crystals do not form, and plants are not harmed at temperatures below freezing.

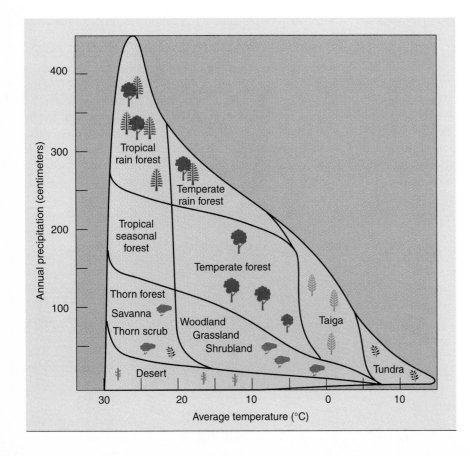

Figure 7.13   This model shows vegetation type predicted by using two variables: annual precipitation and average temperature.

Question: *What vegetation type does this model predict for the following locations: Yuma, Arizona (20°C, 28 cm); Boston, Massachusetts (9°C, 113 cm); and Pierre, South Dakota (8°C, 43 cm)?*

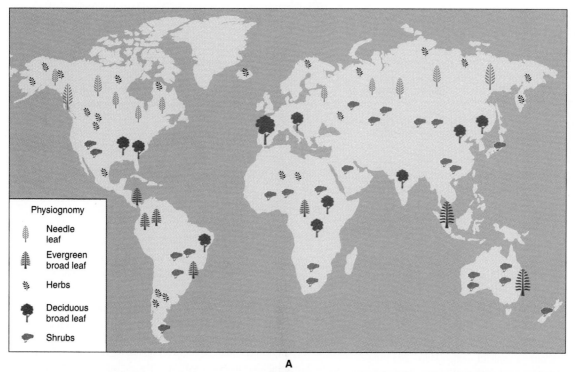

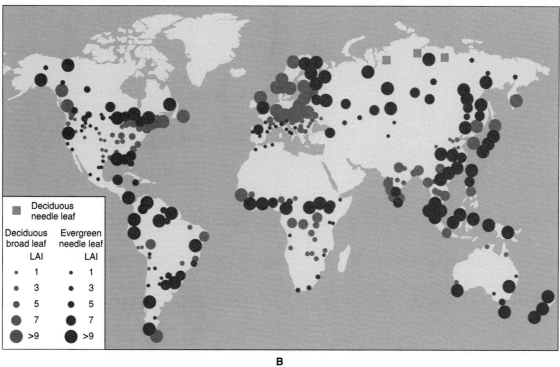

Figure 7.14   (A) Average minimum yearly temperatures were used to create this map of vegetation physiognomy for the world. (B) Hydrological balances were used to predict the leaf area index (LAI) within world vegetation physiognomy regions. The LAI varies from 1 to more than 9.

Table 7.1   Minimum Temperatures and Expected Physiognomy

Temperature Range (°C)	Phenomenon	Expected Physiognomy
>15	Temperature not limiting	Broad-leaved evergreen when rainfall inadequate
−1 to 15	Chilling	Broad-leaved evergreen when rainfall adequate
−15 to 0	Freezing and supercooling	Broad-leaved evergreen
−40 to −15	Freezing and supercooling	Broad-leaved deciduous
<−40	Freezing and supercooling	Evergreen and deciduous needle-leaved (coniferous)

Predicted dominant physiognomy is based on minimum yearly temperatures. Broad-leaved trees can be evergreen, as in the tropics, or deciduous, as in the temperate zones. Most conifers are evergreen, but some species, such as larch, are deciduous.

Source: After Woodward (1987), pp. 97–107.

When both physiognomy and the LAI were computed for the biomes of Earth, the map shown in Figure 7.14B resulted. Compare the maps in Figure 7.14 with Figure 7.11. How accurate are the predictions made by this model?

BEFORE YOU GO ON   Variations in sunlight, temperature, and moisture produce latitudinal patterns of biome distribution. An increase in altitude results in changes in life zones. A highly predictive model that explains changes in vegetation between the biomes uses minimum temperatures and hydrological balance to predict leaf area index (LAI) and physiognomy.

The biomes present on Earth result from the interaction of altitude, latitude, and proximity to large oceans because they all affect both temperature and water balance. In Chapter 8, we will consider further the unique characteristics of different biomes, along with representative organisms that have become adapted to live within them.

SUMMARY

1. The biosphere is composed of the hydrosphere, the lithosphere, and the atmosphere. Living organisms have contributed to the creation and maintenance of habitable environments on Earth.

2. Tectonic forces deep within the Earth have been responsible for changing continental alignments throughout the planet's history, and they will continue to produce new alignments in the future.

3. Solar energy is distributed unevenly over Earth's surface. This results in differential heating that is responsible for world climates and ocean circulation and the development of major world biomes.

4. The tilt of the Earth on its axis and differential global heating are responsible for the seasons. The movement of air within atmospheric cells transports excess energy from equatorial regions north and south to temperate and polar regions. This atmospheric circulation, coupled with Earth's rotation, causes major ocean currents.

5. Moisture-laden air flowing onshore from the Pacific Ocean loses much of its water to precipitation as it rises up and over the major mountain ranges of western North America; as a result rainshadows develop on their eastern slopes. Climates in central and eastern North America are influenced by weather systems spawned in the Gulf of Mexico and the Arctic and by the Gulf Stream.

6. Major world biomes include tropical and temperate forests, deserts, grasslands, taiga, and tundra. Latitude has a major effect on the location of world biomes. Life zones similar to those of the major biomes develop in areas with increasing topographic elevation.

7. Older life zone models are not as strongly predictive as the current model using leaf area index (LAI) and vegetation physiognomy.

REVIEW QUESTIONS

1. What are the three major components of the biosphere?

2. How are tectonic forces related to the present continental alignments on Earth?

3. How is the uneven distribution of solar energy over Earth's surface related to its climate and ocean circulation?

4. Which orbital properties of Earth are related to the "march of the seasons," and how does each season affect the Earth?

5. How do Hadley's cells explain the existence of the horse latitudes, trade winds, and the doldrums?

6. Which topographic features are related to rainshadows?

7. Which factors are responsible for the development of the major biomes of Earth?

8. What is altitudinal zonation, and what are its causes?

9. How is a leaf area index (LAI) computed?

WORKING VOCABULARY

adaptation (p. 127)
alpine zones (p. 135)
altitudinal zonation (p. 135)
atmosphere (p. 125)
atmospheric cell (p. 131)
biomes (p. 124)
biosphere (p. 124)
catastrophism (p. 128)
chaparral (p. 134)
Coriolis force (p. 131)
deserts (p. 134)
grasslands (p. 134)
hydrological balance (p. 136)
hydrosphere (p. 125)

leaf area index (p. 136)
life zones (p. 135)
lithosphere (p. 125)
model (p. 135)
physiognomy (p. 136)
savannas (p. 134)
scrub forests (p. 134)
supercooling (p. 136)
taiga (p. 134)
temperate forests (p. 134)
theory of uniformitarianism (p. 128)
tropical forests (p. 134)
tundra (p. 134)

ESSAY AND DISCUSSION QUESTIONS

1. How have energy sources, sculpting forces, and the march of the seasons caused the development of each major biome present on Earth?

2. How do oceans affect maritime climates in North America?

3. What factors combine to produce altitudinal life zones on mountains similar to the different biomes associated with increasing latitude?

4. What is the value of predictive models such as the LAI physiognomy model? How might they be used by scientists?

REFERENCES AND RECOMMENDED READING

Ager, D. V. 1995. *The New Castastrophism: The Importance of the Rare Event in Geological History.* New York: Cambridge Univ. Press.

Allen, J. E., and B. Burns. 1986. *Cataclysms on the Columbia.* Portland, Oreg.: Timber Press.

Baker, V. R., G. Benito, and A. N. Rudoy. 1993. Paleohydrology of late pleistocene superflooding, Altay Mountains, Siberia. *Science,* 259: 348–350.

Balandin, R. K. 1982. *Vladimir Vernadsky, Outstanding Soviet Scientist.* Moscow: MIR Publishers.

Briggs, J. C. 1995. *Global Biogeography.* New York: Elsevier.

COHMAP (Cooperative Holocene Mapping Project) Members. 1988. Climatic changes of the last 18,000 years: Observations and model simulations. *Science,* 241: 1043–1052.

Gohav, G. 1991. *A History of Geology.* New Brunswick, NJ: Rutgers University Press.

McGowan, B. 1992. Fifty million years ago. *American Scientist,* 78: 30–39.

Merriam, C. H. 1898. Life zones and crop zones of the United States. *Bureau Biological Survey,* 10: 1–79.

Murphy, J. B., and R. D. Nance. 1992. Mountain belts and the supercontinent cycle. *Scientific American,* 266: 84–91.

Rowell, G. 1989. The John Muir Trail: Along the high, wild Sierra. *National Geographic,* 175: 467–493.

Smith, R. S. 1992. *Elements of Ecology.* 3d ed. New York: HarperCollins.

Woodward, F. I. 1987. *Climate and Plant Distribution.* Cambridge: Cambridge University Press.

ANSWERS TO FIGURE QUESTIONS

Figure 7.1 Fairbanks would receive less solar energy than Miami. In Fairbanks, both the angle of incidence of sunlight and the total number of daylight hours per year are less than they are in Miami.

Figure 7.2 San Francisco, as well as along the entire West Coast of the United States.

Figure 7.3 Depending on how much you like winter sports—although good skates and skis were probably hard to come by 11,000 years ago—Portland, Oregon, would have been the preferred habitat because it was ice free. Portland, Maine, was buried under the continental ice sheet.

Figure 7.5 The "march of the seasons" would be much less dramatic in the southern and northern latitudes.

Figure 7.7 Cold water from the North Atlantic and the Arctic Oceans would flow down along the East Coast of North America, making the climate much colder as far south as Florida and the Caribbean. Also, Great Britain and northern Europe would no longer be warmed by the Gulf Stream. On the West Coast, warm water would flow as far north as Alaska, bringing a wetter climate with warmer temperatures to the entire western coast of the continent.

Figure 7.9 The desert in the foreground and the sparse vegetation of the mountain indicate that this is the east-facing slope of Mount Whitney, because deserts develop only in rainshadows of the mountain range.

Figure 7.13 The predicted vegetation types would be as follows: Yuma, Arizona: desert; Boston, Massachusetts: temperate forest; and Pierre, South Dakota: grassland.

8

World Biomes

Chapter Outline

Reading Questions

1. What environmental factors give rise to Torrid Zone forests?

2. Why do western coniferous forests exist at the same latitude as eastern deciduous forests?

3. Which climatic conditions are responsible for growth of grasslands?

4. Why do most deserts occur at 30° north or 30° south latitudes?

5. In what ways do savannas differ from temperate grasslands?

The journals written by intrepid naturalists of past centuries provide insightful descriptions of the environments they encountered during expeditions to regions of the world previously unknown to Europeans. These expeditions penetrated all major biomes of Earth, from the torrid closeness of the tropical forests to the frozen expanses of the arctic tundra. In order to give you a sense of the impressions that these naturalists had when they first explored unique regions of Earth, we have used excerpts from a few of their journals as introductions to the different biomes.

FORESTS OF THE TORRID ZONE

We descended slightly from an elevated, dry, and sandy area to a low and swampy one; a cool air breathed on our faces, and a mouldy smell of rotting vegetation greeted us. The trees were now taller, the underwood less dense, and we could obtain glimpses in to the wilderness on all sides. The leafy crowns of the trees, scarcely two of which could be seen together of the same kind, were now far away above us, in another world as it were.

This description of a tropical rain forest at the mouth of the Para River is from *The Naturalist on the River Amazons,* by Henry Walter Bates, which describes the explorations of Amazonia made by Bates and Alfred Russel Wallace in the late 1840s.

The **Torrid Zone** is Earth's surface between the Tropic of Cancer and the Tropic of Capricorn, divided by the equator. It receives a greater amount of solar radiation than any other part of the planet. Tropical and subtropical forests are found in this zone.

Tropical Forests

Tropical forests (see Figure 8.1) occur in the equatorial portion of the Torrid Zone, where more than 240 centimeters of annual rainfall combine with an average annual temperature greater than 17°C to create the most productive forests on Earth. Many lowland areas, below 1,000 meters in elevation and within 10° to 15° of the equator, receive up to 450 centimeters of rain a year, creating environmental conditions that support a dense, stratified rain forest characterized by high species diversity among both plants and animals. It is thought that over 50 percent of all species on Earth are native to tropical forests. An example of this great diversity can be found in the book *Forest Environments in Tropical Life Zones: A Pilot Study,* by L. R. Holdridge, et al. Costa Rica has an area of 51,100 square kilometers, which makes it about 20 percent smaller than the state of West Virginia (which has an area of 62,629 square kilometers). Whereas 14 different species of trees were found in the forest canopy of West Virginia's Allegheny Mountains, a typical section of Costa Rican forest yielded an astonishing 860 different tree species!

The distribution of animal life mirrors the stratification in the tropical rain forests. There is an aerial community above the forest, which includes birds and bats; a canopy community with birds, fruit bats, squirrels, and monkeys; a subcanopy community of climbing animals; and a community on the ground, which includes large and small mammals, reptiles, and birds. Insects are present in all strata throughout the forest.

Subtropical Forests

Subtropical forests tend to develop in northern and southern portions of the Torrid Zone and at higher elevations on mountains within rain forests. They result from a decrease in total available moisture or a seasonal distribution of rain. Tropical seasonal forests include semievergreen rain forests and deciduous seasonal forests.

Forest Nutrients

Forests of the Torrid Zone have soils with unusual properties. The interactions of excessive rain, warm humid conditions, and the biota (all organisms present in a given area) combine either to recycle soluble nutrients quickly or to remove them from the system. In these forests, most of the nutrients reside within living organisms, not in the soil as they do in other biomes. Nutrients in the soil are the "life blood" of temperate biomes, but in tropical forest biomes, high rates of decomposition and nutrient cycling are the driving forces that contribute to the forest stability through time.

BEFORE YOU GO ON Torrid Zone forests occur between the Tropic of Cancer and the Tropic of Capricorn. They include tropical rain forests, semievergreen rain forests, and seasonal deciduous forests. Most nutrients are found within living organisms present in Torrid Zone forests rather than in the soils.

Figure 8.1 Tropical forests include the greatest number of different species per unit area of any terrestrial biome, and they probably contain more than half the species living on Earth today.

Question: *Why is the clearing of vast expanses of tropical forests of great concern to biologists who study plant and animal diversity?*

TEMPERATE FORESTS

Virginia doth afford many excellent vegitables and living Creatures, yet grasse there is little or none, but that groweth in Iowe Marishes: for all the Countrey is overgrowne with trees, whose dropings continually turneth their grasses to weedes, by reason of the ranckness of the ground which would soone be amended by good husbandry. The wood that is most common is Oke and Walnut, many of their Okes are so tall and straight, that they will beare two foote and a halfe square of good timber for 20 yards long. . . . There is some Elme, some black walnut tree, some Ash: of Ash and Elme they make sope Ashes.

This description of the forests that greeted the first European settlers in Virginia is from *A Map of Virginia,* written by Captain John Smith in 1612.

The **temperate forest biome** occurs in Earth's Temperate Zones, where increasing latitude results in greater seasonal extremes, with lower average temperatures and less precipitation than in biomes of the Torrid Zone. In North America, these environmental conditions have produced three major forest types: *deciduous forests* of the eastern region, *coniferous forests* of the Pacific Northwest, and *western montane forests* in mountainous areas of the West.

Deciduous Forests

Extensive **deciduous forests** have developed in only three temperate regions of Earth, all in the Northern Hemisphere. They are found in central Europe, eastern Asia, and eastern North America, where variations in temperature and precipitation correlate directly with increasing latitude and altitude.

In deciduous forests of North America, precipitation ranges from 75 to 250 centimeters per year, and temperatures vary from summer highs of 38°C to winter lows of −30°C. These extremes create a temperate forest biome characterized by four distinctive seasons, as shown in Figure 8.2. This seasonality results in large variations in the amount of available moisture. There is a high probability that water will be unavailable to the plant community for more than half the year, during the cold months, in northern regions of deciduous forests. Thus growth is limited to late spring, summer, and early fall,

Figure 8.2   Leaves of deciduous trees emerge in spring (A); growth continues through summer (B) and into early fall, when the leaves become ablaze with color, die, and fall to the forest floor (C). Fallen leaves enrich the soil and prepare trees for a dormant period in winter (D).

Question: *Why does soil accumulate in temperate deciduous forests but not in tropical forests, where there are more trees per any given area?*

and the total amount of forest growth is far less than that of tropical forests.

When the lengthening days of early spring release the frozen grip of winter, the floor of a deciduous forest blooms with a profusion of flowers. To survive, these flowering plants must complete most of their annual growth and reproduce before oak, hickory, maple, beech, and other tree species grow leaves, closing the forest canopy and reducing light intensities on the forest floor by 95 to 99 percent. The growth of these tree species is supported to a large extent by rains spawned in the Gulf of Mexico and carried northeast by the many storms and hurricanes of summer.

As the days begin to shorten in late summer, growth slows in deciduous forests, and the deep green summer foliage is replaced by the brilliant colors of fall. As fall gives way to winter, the trees lose their leaves, and the northern forests become snow-covered and dormant through the cold months of winter.

Leaves and other litter accumulate and decompose on the forest floor, producing rich soils. These soils contain far greater concentrations of nutrients than soils in the tropics. However, the rates of decomposition and nutrient cycling are much lower because of dormant periods during winter when they lie frozen.

The extensive deciduous forests that greeted colonists in the early settlement of eastern North America have been reduced in size by more than 80 percent. During the westward migration in the Ohio Valley, an increasing human population cleared the land of trees and converted most of this magnificent forest biome into land for agricultural production, leaving only a few ancient stands undisturbed. The forested areas of the southeastern coastal plain were converted to agricultural lands in the late seventeenth and early eighteenth centuries, and only

about 3000 non-native Americans (that is, European immigrants) were in the Ohio Valley by 1790. By 1800, however, the Ohio Valley had an immigrant population of 45,000. After the War of 1812, this population rose sharply to 2 million by 1850 and to 4.1 million by 1900.

Coniferous Forests

Evergreen rain forests, composed of a variety of conifer (cone-bearing) species of spruce, fir, pine, and hemlock, blanket the coastal plains and mountains of the Pacific Northwest. They range from sea level to an elevation of about 1,800 meters in the Coast Range of Oregon and extend north along the Pacific coast to southern Alaska, where their upper elevation limit drops to about 900 meters. The southern portion of this **coniferous forest biome** occurs at the same latitude occupied by deciduous forests in eastern North America. Eastern deciduous forests result from summer rains and cold, frozen winters. In contrast, coniferous forests in the Pacific Northwest are influenced by onshore flows of winter storms that bring up to 400 centimeters of rain per year. As a result, western conifers have extended growth periods into the wet winter season, in which temperatures are often mild.

Redwood and Sitka spruce forests, characteristic of the fog zone along coastal lowlands, are replaced by forests of Douglas fir and hemlock as elevation increases in the Coast and Cascade ranges of Oregon, Washington, and British Columbia (see Figure 8.3A). These forests are replaced at even higher elevations by mountain hemlock and the true fir (see Figure 8.3B).

The rainshadow on eastern slopes of the Cascade Range is associated with a sharp decline in moisture that results in a

Figure 8.3 (A) In coniferous forests of the Pacific Northwest, inland from the coastal plain, the fog zone is left behind as the elevation increases and the spruce forest gives way to western hemlock. (B) At higher elevations, hemlock is replaced by mountain hemlock and true firs.

Grasslands exist where droughts occur in 30- to 50-year cycles. The degree of moisture decline varies in different areas of the globe, leading to the development of various types of grasslands. Average annual rainfall can vary from 100 centimeters in the tallgrass prairies of Illinois to 25 centimeters in shortgrass prairies in the rainshadows of the Rocky Mountains. Some trees may live and grow for a time, but forests do not develop in these grasslands because in the long run a drought will occur and trees will die.

Trees do occur in grasslands, but only in drainage basins at the bottoms of hills and along streams where there is sufficient moisture to support them through periods of drought. Some drought-resistant tree species will occur singly or in scattered clumps on the tropical savannas of eastern Africa and northern Australia and in the velds of southern Africa and northern South America (*veld* is an old Dutch word adopted into Afrikaans meaning "field"). In these areas, secondary disturbances—immense herds of grazing animals, wildfires, and activities of humans— also affect tree density and distribution (see Figure 8.7).

North American Grasslands

The major grasslands of North America occur in areas without large mountain ranges. On these rolling plains, persistent winds develop during hot summer months and cause an increase in the evapotranspiration rate far above that found in deciduous forests. This high rate of water loss reduces the amount of precipitation that percolates through the soil and thus reduces the amount lost to groundwater. As a result, the quantity of soil nutrients removed is small, and grassland soils develop a layer of rich organic matter.

North American Prairies

Before the westward migrations of European immigrants in the nineteenth century, **tallgrass prairies** covered most of southern Minnesota, much of Iowa, and the eastern edges of North and South Dakota, Nebraska, and Kansas. It also extended east

of the Mississipi River through central Illinois, northwestern Indiana, and into southern Michigan. Today this region constitutes the major portion of the "corn belt," where corn and soybeans form the base of an agricultural economy.

Vast expanses of **mixed** and **shortgrass prairies** originally extended from Texas north to Saskatchewan and west to the eastern slopes of the Rocky Mountains. Over the millennia, these prairies withstood the advance and retreat of at least three major glaciations, extensive seasonal fires, droughts, and grazing by herbivores, including migrating herds of up to 50 million bison (see Figure 8.8A). After the westward migrations of the nineteenth century, however, farmers converted these grassland biomes into the breadbasket of the nation and, to some degree, the entire world. Today combines are used to harvest vast crops of wheat, and cattle graze in areas where herds of deer, elk, pronghorn antelope, and bison once migrated in uncounted numbers (see Figure 8.8B).

BEFORE YOU GO ON Grasslands occur in areas where droughts occur in 30- to 50-year cycles, which limits the establishment of forests. Average annual rainfall varies from 100 centimeters in tallgrass prairies to 25 centimeters in shortgrass prairies. Tallgrass, mixed, and shortgrass prairies have rich soils in which much of the world's food supply is now grown.

THE TAIGA

August 17th. Rainy weather, made East coming on the River about $3\frac{1}{2}$ Miles, horrid Roads. . . . The mountains confining this Valley are here on both sides lofty spacious & majestic bellying into promontories alternately advancing into the valley on both sides & their Red sides (for they are here partly composed of Redish stone & earth) appearing through a Thick Coat of wood & when viewed as we now see them through a Summer shower droping from a black pending cloud with all the phenomenon of the beautiful Variations of the Rain Bow, is realy Grand & Beautiful—here we find some sizeable Birch

Figure 8.7 The most extensive savannas on Earth are found in Africa; they support numerous species of herbivores, as well as the carnivores that feed on the herbivores.

A

B

Figure 8.8 Extensive grasslands of the prairie states, associated today with cattle, once supported vast numbers of large herbivores including bison, deer, elk, and (A) pronghorn antelope. Vast areas of shortgrass prairie biomes have been plowed and modified for producing wheat and other grains harvested with modern combines (B).

Question: *How was the soil base of this Great Plains agricultural system created?*

Trees, but not big enough to make a small Canoe—two kinds of the aspin Poplars & Liards but of stinted Growth.

This passage is from *A Journal of a Voyage from Rocky Mountain Portage in Peace River to the Sources of Finlays Branch and North West Ward in Summer 1824* by Samuel Black for Hudson's Bay Company. It describes a valley in the northern forest of north-central British Columbia, Canada.

The **taiga** is the most northern coniferous forest biome. It forms a continuous belt across Eurasia and North America, beginning at the southern limit of the winter arctic front, near 50° north latitude, and extending to the southern limit of the summer arctic front, between 60° and 70° north latitude. The taiga is a relatively young biome, formed in the wake of receding glaciers of the last ice age. Species diversity in the taiga is the lowest of any forest biome, with dominant species being four types of coniferous trees: spruce, pine, fir, and larch. Deciduous species of alder, birch, and poplar are interspersed within these coniferous forests (see Figure 8.9).

Many lakes and rivers are found in the valleys across this great expanse of taiga. These aquatic environments provide habitats for many migratory and resident species of birds and mammals. The large northern bears, wolves, and moose are mammals adapted to this unique biome (see Figure 8.10, page 148).

Cool and cold northern **boreal forests** of the North American taiga biome are composed of two general forest types: *closed-canopy forests* and *open woodlands*. **Closed-canopy forests,** where spruce and larch shade understories of small shrubs and moss-covered ground, extend from Alaska to Newfoundland (see Figure 8.11, page 148). These forests are seldom limited by moisture, and water shortages are rare, occurring only on some open, south-facing slopes.

Figure 8.9 Boreal forests are the most extensive forest type in North America. They include woodland areas that form the habitat for woodland caribou, shown here.

Question: *Which caribou predator would you expect to find in boreal forests?*

Figure 8.10   Moose, the largest species in the deer family, have also adapted to the taiga.

Figure 8.11   Black spruce covers great expanses of the taiga.

Extreme variations in solar radiation over the course of a year create a climate of cool summers and extremely cold winters. At some sites, seasonal average monthly temperatures can vary by 90°C. Within the taiga, the number of days with an average temperature of 10°C or greater ranges from 120 days per year in the south to 30 days in the north and seems to set a limit on the amount of vegetation present.

Permafrost, the subsurface layer of permanently frozen ground, is a major factor defining the distribution of trees in boreal forests. In summer, only a thin surface layer of soil thaws, allowing growth of grasses and shrubs. Permafrost, however, forms a barrier that prevents the growth of tree roots. Permafrost may also be responsible for the development of treeless areas within **boreal woodlands** of the northern taiga. In these woodland areas, white and black spruce grow where permafrost does not occur, and deciduous trees are restricted to margins along watercourses where deeper soils have developed from past sedimentation. The ground cover of these open woodlands includes mosses and lichens, which provide food for woodland caribou, the dominant herbivores inhabiting this biome.

> **BEFORE YOU GO ON**   The taiga of North America consists of both closed-canopy forests and open woodlands. The number of days with average temperatures of 10°C or greater sets a limit on the amount of taiga vegetation. Permafrost may be responsible for treeless areas within the boreal woodlands.

THE TUNDRA

On the 9th . . . a large musk bull was shot, and his flesh was found excellent—the skeleton will be preserved. A short time after midday on the 10th we arrived here having been five days coming from the coast, during some of which we were 14

hours on foot and continually wading through ice cold water or wet snow which was too deep to allow our Exquimaus boots to be of any use.

The latter part of our journey if not the most fatiguing was by far the most disagreeable. . . . Our principal food was geese, partridges, and lemmings. The latter being very fat and large were very fine when roasted before the fire or between two stones. These little animals were migrating northward and were so numerous that our dogs as they trotted on, killed as many as supported them, without any other food.

This passage is from a letter to Sir George Simpson, governor-in-chief of Hudson's Bay Company, by John Rae at Kendall River Provision Station, describing part of Rea's exploration of the High Arctic tundra in the northern Northwest Territories, Canada, in June 1851.

The **tundra,** treeless arctic plains, like the taiga, is a biome of the Northern Hemisphere. It does not exist in Antarctica, where terrestrial life is limited to a few meadows that develop during the short southern summer. The Northern Hemisphere has an extensive area of arctic tundra that is commonly divided into the *Low Arctic* and *High Arctic,* as shown in Figure 8.12.

The Low Arctic

The **Low Arctic** begins as a treeless zone at the edge of boreal woodlands and continues north to areas where low temperature and available moisture limit the growth of vegetation (see Figure 8.13). Plants growing on the tundra are well adapted to low temperatures, and during the continuous light of summer, their growth rates approach those of similar temperate species. The growing season, however, is short, beginning with the melting of snow in spring and ending with the first frost in late July or early August. With 24 hours of light each day in early summer, the total amount of solar energy approaches that of temperate regions, but the limited length of the arctic summer

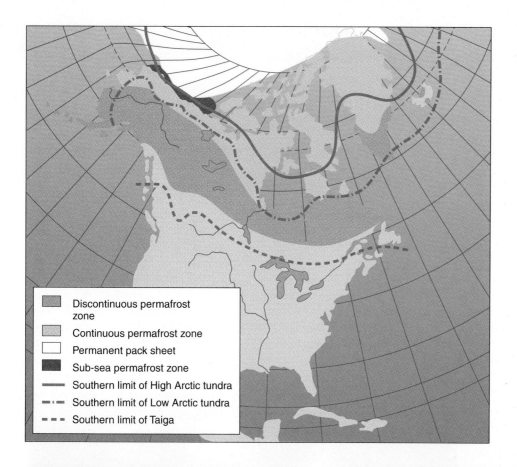

Discontinuous permafrost zone

Continuous permafrost zone

Permanent pack sheet

Sub-sea permafrost zone

—— Southern limit of High Arctic tundra

—·— Southern limit of Low Arctic tundra

--- Southern limit of Taiga

Figure 8.12 The Low Arctic tundra includes the tundra from just north of the boreal woodlands, where the dominant plant species are woody shrubs, to the southern edge of the High Arctic tundra, where nonwoody plants predominate. The High Arctic tundra ends at the edge of the permanent ice sheet.

Question: *Was tundra in the same location 10,000 years ago as it is now? If not, where was it?*

Figure 8.13 Low Arctic plants such as those shown here (A), are eaten by woodland caribou (*Rangifer caribou*) as they move north during their annual migration (B).

FOCUS ON SCIENTIFIC PROCESS

The Gaia Hypothesis

The *Gaia hypothesis* is one of the most important and controversial ideas in biology today. According to this hypothesis, the entire biosphere constitutes a self-regulating or homeostatic system that is capable of maintaining Earth's environment within the narrow range of conditions required for life. Furthermore, the hypothesis explains that life itself plays the central role in this regulatory system.

Ironically, the Gaia hypothesis did not originate from studies of life on Earth, but from the search for life on other planets. During the 1960s the U.S. National Aeronautics and Space Administration (NASA) became interested in the possibility that primitive life forms exist on Mars. With travel to the moon imminent and interplanetary travel being discussed seriously by scientists and politicians, the search for extraterrestrial life had important practical and theoretical implications. NASA's research efforts climaxed when two unmanned *Viking* space probes landed on Mars in 1976 (see Figure 1). Aboard these space probes were instruments to detect the presence of amino acids and other organic molecules; these probes also conducted simple experiments designed to detect possible by-products from chemical reactions carried out by organisms. The data collected from these investigations provided no evidence of life existing on Mars.

Before the *Viking* missions, at least one scientist had argued that the question of life on Mars could be answered from observations made from Earth. According to James Lovelock, a chemist and engineer who had worked for NASA, the composition of the atmosphere is a good indicator of whether or not life exists on a planet. The Martian atmosphere, which could be analyzed by observations using special infrared tele-

scopes, is composed largely of carbon dioxide and contains very small amounts of oxygen and nitrogen gases. According to Lovelock, this type of atmosphere would be expected on a lifeless planet. In contrast, Lovelock claimed that Earth's atmosphere, which contains high oxygen and nitrogen concentrations and significant amounts of other gases such as methane, could only occur on a planet inhabited by an abundance of living organisms. Thus according to Lovelock, one need only to study the atmosphere of a distant planet to

determine whether or not life exists there. These atmospheric studies led to the Gaia hypothesis.

According to the Gaia hypothesis, the improbable composition of Earth's atmosphere resulted from complex interactions between organisms and their environments. The fact that atmospheric gas concentrations have remained relatively constant for millions of years suggests that life actively regulates the physical environment. Cellular energy transformations—photosynthesis and cellular respiration—carried out on a

Figure 1 The surface of Mars taken by *Viking Lander 1* (above), August, 1976.

FOCUS ON SCIENTIFIC PROCESS

global scale, are responsible for maintaining an exquisite balance of oxygen and carbon dioxide. As Lovelock pointed out, if the concentration of oxygen in the atmosphere were to increase by only a few percent, even the wettest rain forests would be consumed by uncontrollable fires. Because this has never occurred during Earth's history, supporters argue that this constitutes evidence in support of Gaian-type regulation.

Lovelock publicized his ideas in a popular book, *Gaia: A New Look at Life on Earth* (1979), and together with biologist Lynn Margulis, he published several technical articles about the Gaia hypothesis. Margulis was the leading supporter of the endosymbiotic hypothesis of cellular evolution, an idea that was just beginning to gain support during the 1970s (see Chapter 3). Her knowledge of the biology of microorganisms provided authoritative support for Lovelock's ideas, but her unorthodox theories have also contributed to the controversy over the Gaia hypothesis.

The Gaia hypothesis has become very popular among non-scientists, particularly those in the environmental movement. It has also generated interest in the scientific community, although most scientists remain quite skeptical of the broader claims made by Lovelock and Margulis. Three general types of criticism have been leveled at the Gaia hypothesis. Many physical scientists have argued that the hypothesis is unnecessary because more traditional chemical and physical explanations can adequately account for most of the unique characteristics of the biosphere. Although they grant that life is important, these critics dismiss the claim that living organisms play the central role in controlling the biosphere. Many biologists have argued that a complex homeostatic control system involving all living things could not evolve according to known evolutionary principles. Scientists, in general, have criticized the

Figure 2 James Lovelock (left) and Lynn Murgulis are principal advocates of the Gaia hypothesis.

hypothesis as a vague metaphor that provides no testable predictions. In part, this is owing to the fact that Lovelock and Margulis have written for a general audience. The name "Gaia" refers to the ancient Greek goddess of Earth, and Lovelock and Margulis have often described the biosphere as if it were an entity with both purpose and foresight. As they are quick to admit today, this is not the case, but some critics continue to see Gaia as nothing more than a romantic fiction.

The controversy over the Gaia hypothesis is instructive for understanding the process of science. Like many ideas on the fringes of science, this hypothesis started out as a vague and speculative idea. Supporters of the hypothesis overstated their case for the hypothesis, and they were frequently careless in the language they used to describe complex, natural phenomena. Through the give-and-take of scientific discussions carried out at conferences and in journals, Lovelock and Margulis were forced to modify the hypothesis and clarify their claims about the biosphere. They, and other scientists, also conducted research designed to support

or refute claims made by the hypothesis. Although it remains very controversial, many scientists now take the Gaia hypothesis seriously and are willing to consider it a source of important scientific questions. Perhaps more significant, Lovelock and Margulis have forced earth scientists to take biology seriously. Prior to the 1960s, it was not uncommon for physical scientists to explain the geological characteristics of Earth without reference to living organisms, but very few scientists would do so today.

Whether or not life regulates Earth's climate and atmosphere as Lovelock and Margulis claim, most scientists now accept a view of the biosphere as a complex, interactive system of organisms and physical environments. This breaking down of disciplinary boundaries between the biological and physical sciences has been a major goal of Lovelock and Margulis. As they have often pointed out, solving complex environmental problems such as global warming and acid rain will require the cooperation of scientists from many disciplines. Perhaps such interdisciplinary teams of scientific researchers will become more common in the future.

allows only a few centimeters of upper permafrost to thaw. This limited amount of soil is usually saturated with water in the Low Arctic summer, and tundra species develop shallow root systems.

The High Arctic

The **High Arctic** tundra is found at higher latitudes and in some areas is considered a desert, for the annual precipitation may be as low as 5 centimeters. These far northern latitudes are areas of high atmospheric pressure produced by descending air from the northern air cell. In many areas, if moisture does not limit vegetation, the permafrost does. This leads to a scattered distribution of a few hardy species of moss, lichens, herbs, and dwarf willows (see Figure 8.14).

Permanent High Arctic residents include a few species of birds, including the ptarmigan (see Figure 8.15A) and the snowy owl. Mammals include the musk-ox (see Figure 8.15B), arctic hare, arctic fox, and lemming. These resident species are joined by large herds of migrating mammals and immense flocks of migratory birds, including many species of waterfowl that breed and rear their young in the long days of arctic summer. Caribou migrate from their winter ranges in boreal forests and woodlands to graze on spring and summer foilage of the tundra. During these migrations of several hundred kilometers, cows give birth to young that grow and develop rapidly during the High Arctic summer.

BEFORE YOU GO ON Taiga and tundra biomes exist only in the Northern Hemisphere. The tundra consists of Low and High Arctic regions. Animals of the tundra consist of both resident and migratory species.

Figure 8.14 As shown in this figure, dwarf willows (*Salix arctica*) have a low growth form.

Question: *What are some environmental factors that may account for this type of growth form?*

A

B

Figure 8.15 The tundra supports resident populations of (A) willow ptarmigan and (B) musk-oxen.

THE DESERTS

January 28.—We were favored with another charming morning, mild and without a breeze. Following an Indian trail down White Cliff valley, we soon came to a projecting rock, beneath which were walled partitions, with remnants of fires, showing signs of having been recently occupied. . . . The valley was covered with dense groves of cotton-wood, beneath which flowed the prettiest brook we have found since leaving Pueblo creek. . . . The stream, turning southerly, appeared a short distance below to join a wide arroyo from the north, called Big Sandy. . . . The weather is spring-like. Vegetation begins to conform to that of Rio Gila. Canotias are mingled with cedars upon the dry arroyos, and mezquites with cotton-wood upon the flowing streams. Numerous varieties of cacti also abound, and from the huge Echino cactus of Wislizenus, to humbler mammillaria.

This descriptive passage is from *Reports of Explorations and Surveys, to Ascertain the Most Practicable and Economical Route for a Railroad,* written in 1853 and 1854 by A. W. Whipple of the Corps of Topographical Engineers, on the route near the thirty-fifth parallel in the desert of Arizona (see Figure 8.16).

The **deserts** of the world are found in arid regions, where total annual precipitation is less than 25 centimeters. These

Figure 8.16 The campsite depicted in the lithograph *Bivouac January 28,* from A. W. Whipple's report of his surveys, shows vegetation types which he encountered near the thirty-fifth parallel in the Arizona desert.

deserts occur in western regions of the major continents, near 30° north and south latitude, where the descending air of equatorial and temperate air cells creates high atmospheric pressures and low humidity.

The general circulation patterns of ocean currents contribute to the formation of deserts along western continental margins, where cold water originating in polar seas flows toward the equator, whereas eastern coasts are influenced by warm water originating in equatorial seas and flowing toward the poles. An example of this can be seen in South America. On the west coast, the Atacama Desert lies on the Tropic of Capricorn; in Brazil, a tropical rain forest occurs at this latitude. The same conditions prevail in Australia, where extensive deserts occur on the western edge of the continent, but at the same latitude on the east coast, subtropical forests predominate (refer to Figure 7.11, p. 135).

The largest deserts occur in the Northern Hemisphere because greater continental landmasses exist near the 30° latitude.

The Sahara in Africa is the largest desert on Earth. It begins on the west coast and traverses the entire continent and the Arabian Peninsula.

Deserts also occur in western North America, where descending air from the equatorial and temperate air cells combines with the climatic effects produced by the cold offshore California current, resulting in an arid climate. At the same latitude on the east coast, subtropical climates in Georgia and Florida result from the warm waters of the Gulf Stream.

Rainshadows of the Sierra Nevada and Cascade ranges extend western deserts beyond 45° north latitude, creating both cold and hot deserts that are distinguished by average yearly temperatures (see Figure 8.17A). **Cold deserts** are found within the Great Basin of southeastern Oregon, southern Idaho, western Utah, and the northern three-fourths of Nevada. The elevation of this desert region is about 1,220 meters, with numerous mountain ranges reaching 2,440 meters (see Figure 8.17B).

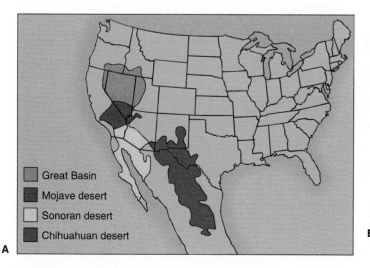

Great Basin
Mojave desert
Sonoran desert
Chihuahuan desert

A

B

Figure 8.17 (A) Deserts of the western United States include the northernmost cold Great Basin Desert and the hot Mojave, Sonoran, and Chihuahuan Deserts of the Southwest. (B) The Steens Mountains rise from the Great Basin Desert in southeastern Oregon to an elevation of 2967 meters.

Question: *Which deserts receive winter moisture?*

The Great Basin Desert covers an area of about 410,000 square kilometers, making it the largest desert in the United States. Most drainage basins in this desert area have no outlet; all precipitation is stored as surface water, enters the groundwater, is transpired by plants, or evaporates from the surface. This results in a continuous increase in salt concentration in these desert soils, which causes many desert lakes to have very high salinities (see Figure 8.18).

The Great Basin Desert can be divided into two different portions based on the major vegetation present in each. Sagebrush dominates the northern portion, and in the southern part, shadescale predominates. Other plants occur within these two desert areas, but trees are limited to areas of drainage at the base of rimrocks and along intermittent streams.

Hot deserts include the Mojave, Sonoran, and Chihuahuan, each differing in latitude and vegetation type. The Mojave Desert covers the south-central portion of California and the northwestern quarter of Arizona. The Sonoran Desert is in the southeastern corner of California and southwestern Arizona and extends into Baja California and western Mexico, enclosing most of the Sea of Cortez. The Chihuahuan Desert occurs in central Mexico, with a northern extension across southern New Mexico and into western Texas.

The moisture regime is quite different among these four deserts. The Great Basin Desert receives moisture from winter storms that move in from the Pacific Ocean. Some of this winter moisture also falls in the Mojave and Sonoran deserts but does not reach far enough east to affect the Chihuahuan Desert.

The southern deserts receive summer moisture from thundershowers spawned in storm systems that develop in the Gulf of Mexico. These summer thundershowers reach north into the Chihuahuan and Sonoran Deserts but not into the Mojave or the Great Basin. Thus only the Sonoran Desert receives both summer and winter moisture, creating a unique assemblage of desert plants (see Figure 8.19).

A

B

Figure 8.19 (A) The Sonoran Desert, characterized by creosote bush, ocotillo, and the giant saguaro cactus, receives moisture in both summer and winter. (B) Agavaceae, yucca, and Joshua trees grow in the Mojave Desert, where most of the annual moisture falls in the winter.

BEFORE YOU GO ON Both cold and hot deserts occur in western North America. The Great Basin, Mojave, and Sonoran deserts receive winter moisture, and the Sonoran and Chihuahuan deserts receive summer moisture. Thus only the Sonoran receives both summer and winter moisture, which results in a distinctive plant community.

THE MINOR BIOMES

Tropical Savannas

Savannas of the world occur in a tract across Africa south of the Sahara, in areas in northern Australia, and in regions

Figure 8.18 Many areas in western deserts occupy drainage basins that have no outlets to oceans. As water evaporates from lakes in these areas, salts are left behind, which causes the lakes to increase in salinity.

Question: *What is the source of salts in lakes such as these?*

south of the Amazon basin and east of the Andes in South America. **Savannas** are modified tropical grasslands characterized by alternating wet and dry seasons that produce areas typical of grasslands. During the wet season, grasses develop rapidly, but the dry season reduces them to brown expanses. This type of climate supports variable densities of scattered thorn trees and brush, creating a habitat quite different from that of typical grasslands found in temperate regions.

Seasonal moisture cycles in savannas are reflected in a cyclic appearance of vegetation, which causes herbivores to follow predictable patterns of migration. Wildebeests of the Serengeti Plain are typical of this cycle of life (see Figure 8.20). Giraffes, gazelles, zebras, and other herbivores inhabit these grasslands, accompanied by a variety of carnivores, including jackals, hyenas, and lions.

Chaparral

Chaparral areas receive significant winter rains, but the summers are typical of the desert latitudes in which they occur. Most chaparral biomes are located within arid zones, near 30° north or south latitude. They occur in regions near the Mediterranean Sea, near the tip of South Africa, and on the western coast of North and South America. The proximity of these areas to major oceans subjects them to maritime climatic effects that modify the desert, producing a typical chaparral biome.

Vegetation in the chaparral typically consists of a number of shrub species, with a few species of grass interspersed among them. The shrubs produce a considerable amount of living mass that is subjected to frequent burning, an environmental characteristic that many homeowners in southern California can verify (see Figure 8.21).

Figure 8.20 Vast herds of wildebeests migrate across the Serengeti Plain of East Africa in response to wet and dry seasonal cycles that are characteristic of tropical savannas.

Question: *Prior to the European settlement of North America, which species would have been analogous to the African wildebeest?*

Figure 8.21 Chaparrals occur in latitudes where deserts are usually found as well as where maritime climates moderate the effects of high temperature. Extensive foliage of the chaparral may become tinder dry in the hot summers and can burn readily, destroying homes and other structures.

Question: *In which language would you expect to find the root word for* chaparral?

SUMMARY

1. Tropical forest biomes may contain more than 50 percent of all species on Earth. Tropical forest soils are usually nutrient-poor, and most nutrients are stored in organisms living in these biomes.

2. Deciduous forests occur in eastern North America, where wet summers are prevalent and the growing season is limited. In the west, coniferous forests predominate because summer precipitation is minimal, but growth occurs all year. Most of the original ancient deciduous forests in eastern North America were cleared, and the arable lands was converted for agricultural production.

3. Species of plants change with increasing elevation in western coniferous forests. Coastal plains and western slopes are carpeted with redwood, pine, spruce, and Douglas fir trees. True fir, hemlock, and cedar predominate at higher elevations; on the eastern slopes, pine and juniper form open woodlands.

4. Quantities of moisture define the transition from forests to grasslands and between short- and tall-grass prairies. Grazing animals, wildfires, and human activities are secondary disturbances that contribute to the continued existence of grasslands.

5. The taiga is the youngest and most northern coniferous forest biome. Its species diversity is the lowest of any forest biome. The presence of permafrost is a major factor in defining tree distribution in the northern forest.

6. The tundra biome occurs only in the Northern Hemisphere and is composed of both Low Arctic and High Arctic areas. During the continuous light of summer, growth rates in some Low Arctic species approach those of similar temperate species. Much of the High Arctic area often receives so little precipitation that many experts consider it a desert. The tundra provides many migratory species with habitats conducive to their reproduction and resources necessary for the early growth of their young.

7. Major deserts occur in areas near 30° north and 30° south latitudes, where descending air of the equatorial air cells creates high pressure and low humidity. The four major North American deserts are the Great Basin, Mojave, Sonoran, and Chihuahuan. Variations in total sunlight and the frequency of precipitation are major factors contributing to the development of deserts.

8. Minor world biomes include tropical savannas and chaparrals. Savannas are modified grasslands that have cyclic wet and dry seasons. They are characterized by the presence of widely dispersed trees, and in Africa they support vast numbers of animals. Chaparrals occur in arid zones, but they are influenced by their proximity to oceans and are characterized by shrub species.

WORKING VOCABULARY

chaparral (p. 155)
coniferous forest (p. 143)
deciduous forest biome (p. 142)
desert (p. 152)
grassland biome (p. 145)
montane forest (p. 144)
permafrost (p. 148)
savannas (p. 155)

subtropical forest (p. 141)
taiga (p. 147)
temperate forest biome (p. 142)
Torrid Zone (p. 141)
transpiration (p. 145)
tropical forest (p. 141)
tundra (p. 148)

REVIEW QUESTIONS

1. How do tropical rain forests differ from temperate deciduous forests in terms of important nutrients?

2. Why are coniferous forests found in the Pacific Northwest at the same latitude where deciduous forests occur in the northeastern states?

3. What factors are responsible for the rich soils formed in grasslands?

4. What factors are related to the formation of grassland biomes?

5. Why are taiga and tundra biomes not found in the Southern Hemisphere?

6. How does permafrost affect the development of Low and High Arctic areas in the tundra biome?

7. How do growth rates of plants in arctic biomes compare with rates for similar species in the temperate biomes? What accounts for the differences?

8. Why do major deserts occur near 30° north and 30° south latitudes?

9. What are some minor biomes? Where do they occur? What factors influence their development?

ESSAY AND DISCUSSION QUESTIONS

1. What factors may have been responsible for the great species diversity that characterizes tropical rain forest biomes?

2. As you travel north from the Torrid Zone to tundra biomes, what changes would you expect in species diversity? Why?

3. Why are most grassland biomes located in interior areas of continents?

4. If weather patterns are disturbed or change naturally, what will happen to the current distribution of biomes? Construct some possibilities.

REFERENCES AND RECOMMENDED READING

Barbour, P. L. (ed.). 1986. *The Complete Works of Captain John Smith* (1580–1631). Vol. 1. Chapel Hill: The University of North Carolina Press.

Brown, J., and B. Maurer. 1989. Macroecology: The division of food and space among species on continents. *Science*, 243: 1145–1150.

Brown, L. 1985. *Grasslands*. New York: Knopf. Also see other volumes in the Audubon Society Nature Guide series, including *Atlantic and Gulf Coasts, Deserts, Eastern Forests, Pacific Coast,* and *Western Forests.*

Holdridge, L. R., W. C. Grenke, W. H. Hatheway, T. Liang, and J. A. Tosi. 1971. *Forest Environments in Tropical Life Zones: A Pilot Study.* Elmsford, N.Y.: Pergamon Press.

Joern, A. and K. H. Keeler. 1995. *The Changing Prairie: North American Grasslands.* New York: Oxford Univ. Press.

Knowlton, J. 1995. *The Deserts of the World.* New York: Harper Collins.

Lovelock, J. E. 1995. *Gaia, A New Look at Life on Earth.* Oxford: Oxford University Press.

McClaran, M. P. and T. R. VanDevender (eds.). 1995. *The Desert Grassland.* Tucson: Univ. Arizona Press.

McNaughton, S., R. Ruess, and S. Seagle. 1988. Large mammals and process dynamics in African ecosystems. *BioScience,* 38: 794–800.

National Geographic Society. 1976. *Our Continent: A Natural History of North America.* Washington, D.C.

Nichol, J., and C. Newton. 1990. *The Mighty Rain Forest—In Association with Worldforest 1990.* London: Newton Abbot.

Norse, E. 1990. *Ancient Forests of the Pacific Northwest.* Washington, D.C.: Wilderness Society/Island Press.

Richards, P. W. and P. Westmacott. 1995. *Tropical Rainforests: An Ecological Study.* 2nd ed. Cambridge: Cambridge Univ. Press.

Special Issue Articles on Stability and Change in the Tropics. 1992. *BioScience,* 42: 818–869.

Sinclair, A., and M. Norton-Griffiths. 1979. *Serengeti: Dynamics of an Ecosystem.* Chicago: University of Chicago Press.

Trimble, S. 1989. *The Sagebrush Ocean: A Natural History of the Great Basin.* Reno: University of Nevada Press.

ANSWERS TO FIGURE QUESTIONS

Figure 8.1 Tropical forests contain the greatest diversity of animal and plant species on Earth.

Figure 8.2 Cold winters retard bacterial decomposition of the organic matter produced by temperate deciduous forests, thus organic matter accumulates as soil.

Figure 8.4 Cold temperatures result in a short growing season. Also, water is often frozen and therefore unavailable to plants.

Figure 8.6 The dramatic increase in the number of Europeans and consequent modification or outright destruction of native bison habitats resulted in a dramatic reduction in the bison population. An important secondary factor was the widespread hunting of bison. This practice had already greatly depleted bison herds by the time railroads were constructed, which gave hunters virtually unlimited access to the remaining herds of the Great Plains.

Figure 8.8 From organic matter produced in natural prairie systems.

Figure 8.9 Wolves.

Figure 8.12 No. It would have been further south, below the continental glaciers.

Figure 8.14 Cold temperatures, high winds, short growing seasons, permafrost, and limited amounts of water.

Figure 8.17 The Great Basin, Mojave, and Sonoran deserts receive most of their moisture during the winter.

Figure 8.18 Salts dissolve in snow melt and rain water as it flows through soils and moves with water into the salt lakes.

Figure 8.20 The North American bison (buffalo).

Figure 8.21 Spanish. The Spanish word *chaparral* describes an evergreen oak thicket of shrubs and thorny bushes. The chaparral biome is similar to that of southern Spain, especially along the Mediterranean Sea.

9

Ecology

Chapter Outline

Reading Questions

1. What discoveries led to the development of early ecological concepts?

2. What causes ecological succession?

3. What are the components of an ecosystem?

4. In what ways are trophic levels related to ecological pyramids?

5. Which environmental factors move through biogeochemical cycles?

6. Which factors regulate the abundance and distribution of populations in natural communities?

cology is defined as the scientific study of interrelationships that exist between organisms and their environments. Within an environment, organisms are greatly affected by both **biotic factors,** such as other living plants, animals, and microbes, and **abiotic factors,** various physical, chemical, and temporal (time) components (see Figure 9.1). Major *physical factors* are the force of gravity and energy in the form of light and heat. Important *chemical factors* are water and all the nutritional elements required by living organisms. *Temporal factors* consist of normal changes that occur throughout the life of an organism and gradual environmental changes over long periods of time.

ORIGINS OF ECOLOGY

Although the term *ecology* was used infrequently before 1900, the roots of ecological science were formed much earlier. For over a century before Charles Darwin's work in the 1850s, naturalists believed not only that God had created each species but also that

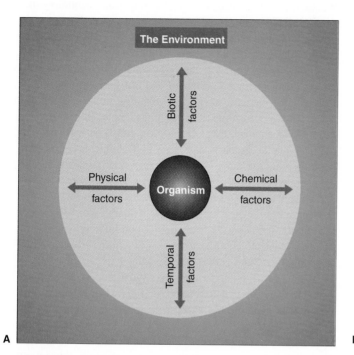

Figure 9.1 (A) All organisms are affected by factors present in their environment. Physical factors include energy and gravity, chemical factors include water and all essential nutrients, biotic factors include all other living organisms, and temporal factors consist of changes in environments that occur during the life of the organism and those that have occurred over very long time periods. (B) The organisms living in this lake are affected by energy (sunlight), gravity (which will pull them to the bottom if they are not mobile), chemicals in the water, and changes that occur as the lake slowly fills with sediments.

Question: *What are some biotic factors that may affect organisms living in this lake?*

He had created a "balance of nature." In other words, plant and animal numbers remained roughly constant owing to the creator's plan of having just the right balance between plants and animals, predator and prey, and so on. Linnaeus, in his overview of classification (see Chapter 5) popularized this view; he referred to a natural balance in which the forces of reproduction and destruction maintained stable numbers of plants and animals.

Darwin presented a much more informed account of how biotic interactions preserved the balance of nature. In his famous treatise, *On the Origin of Species* (1859), Darwin described how clover growing in fields near his home in rural England was regulated by the number of bees that could pollinate the flowers. In turn, the number of bees was limited by numbers of mice that destroyed their underground hives. Finally, mice were regulated by the number of neighborhood cats. Thus Darwin reasoned that the clover population was indirectly regulated by the cat population! Further, he felt that such complex interactions between predators and prey were common in nature and that such interactions, rather than divine foresight, explained the balance of nature.

Other early scientists recognized that abiotic factors play important roles in maintaining the balance of nature. In 1840, an agricultural chemist, Justice von Liebig conducted studies designed to evaluate successful plant growth. Liebig discovered that if there were insufficient quantities of one essential chemi-

cal element in the environment (for example, magnesium), plant growth would be limited regardless of the concentrations of other essential elements. These conclusions led him to formulate Liebig's "law of the minimum," which stated that plant growth was limited by the scarcest nutrient. Consequently, in nature, the distribution and abundance of plant species would be determined by the availability of essential nutrients in the soil.

> **BEFORE YOU GO ON** The science of ecology traces its roots to the eighteenth and nineteenth centuries when there was great interest in understanding the "balance of nature." Early naturalists held to the prevailing belief that a balance of nature was part of God's plan. Subsequently, Darwin and other scientists explained that biotic and abiotic factors accounted for the relatively constant numbers of organisms that lived in an environment.

FUNDAMENTAL ECOLOGICAL CONCEPTS

Early ecological concepts emerged from studies of relationships among organisms and their environment. Many of the ideas generated by this research have become basic concepts in modern ecology.

Limiting Factors

Liebig's original ideas were extended and modified by scientists interested in understanding how plants and animals adapted to their physical environments. Some of the first ecologists recognized that Liebig's law also applied to other **limiting factors** such as temperature and light. In 1913, Victor Shelford expanded the concept of limiting factors to incorporate his belief that there were maximum, optimum, and minimum levels of all abiotic factors present in the environment. He concluded that if organisms were to survive, they must be able to function within a range of environmental conditions (see Figure 9.2). Further, this range of environmental conditions played a major part in determining the distribution of each species. Shelford's principle became known as the **law of toleration** (also called the law of tolerance).

Studies of limiting factors were important in the early history of ecology. The law of toleration is a useful, general concept that explains why organisms live where they do. Organisms, however, do not always live under optimal conditions because biotic interactions, such as competition, may force them to exist in less than ideal environments.

Succession

Near the end of the nineteenth century, two American plant ecologists, F. E. Clements and H. C. Cowles, developed the major ecological principle of *succession,* which explained more clearly why communities change over time. They postulated that natural systems resulted from a gradual **succession** of different plant–animal associations through time. If an area of "new" ground came into existence (for example, through volcanic outflow or sand dune formation), **primary succession** would occur. Clements believed that a progression of different species would come to inhabit the new site. The sequence would begin with early **pioneer community** that would slowly change the original microenvironmental (small local habitat) conditions. Changes in concentrations of specific elements or the amount of shade available are examples of microenvironmental alterations. These first modifications would then enable other species with different environmental requirements to invade the original site. Sequential succession would continue until a permanent community of organisms called the **climax community** existed. This stable climax community would continue to perpetuate itself unless some major disturbance occurred and the environment–organism equilibrium was disrupted. Such disturbances often occurred as a result of fire, disease, insect infestations, or logging. If disturbed,

the site would revert back to an earlier stage, and **secondary succession** would begin. The communities present on the site would progress again through a number of predictable types and in time would culminate in a new climax community.

In Clements's opinion, the community developed much like an organism—progressing from an immature, pioneer stage through intermediate stages to a mature, climax community stage. However, in some of his writings he overemphasized the progressive changes that often occur during succession. His ideas about the balance of nature and about an equilibrium between climax community and physical environment were also exaggerated. Cowles shared many of Clements's ideas about succession, but Cowles felt that succession is often variable and that communities can retrogress as well as advance toward a stable climax state. Cowles wrote that "a condition of equilibrium is never reached" (see the Focus on Scientific Process, "Succession on the Sand Dunes of Lake Michigan").

The Clementsian "school of organismic ecology" prevailed in many early American ecological writings, but a number of ecologists did not agree with its principal concepts. In 1913, W. S. Cooper, a student of Cowles, described results from a study he conducted on Isle Royale in Lake Superior. He observed that succession did not occur uniformly throughout a community. Rather, it resembled a series of "braided streams" that, when woven together, would appear as a "mosaic" of variously aged stands of trees, each representing a different stage of continuous change. He felt that such mosaics existed because of small, localized disturbances within the forest that were caused by diseases, insects, or windfalls.

Cooper's argument was made even more forcefully by H. A. Gleason, who thought that succession could be explained in terms of individual species adapting to different environmental conditions. Gleason made the following argument: (1) every species had a unique set of limiting factors; (2) these limiting factors changed over time; (3) therefore, the distribution and abundance of species in a community would also change. According to Gleason, succession was simply a statistical change in which some species increased over time while others decreased.

Many articles in modern ecological literature include elements of both Clements's "organismal" view and Gleason's "individualistic" view of succession. Some ecologists still view succession as a type of developmental process because communities often change in fairly predictable ways. However, the process of succession is much different than the development of an organism. Most modern ecologists agree with Cooper and Gleason that all communities are mosaics formed by constant changes in the distribution and abundance of individual species.

Figure 9.2 In Shelford's law of toleration, organisms were thought to function most effectively and be most abundant when abiotic factors were present in their optimum range. Decreases in function and abundance would occur if higher or lower concentrations of the abiotic elements were present in the organisms' environment.

Question: *Which factors affecting crayfish in Figure 9.3 pertain to Shelford's law?*

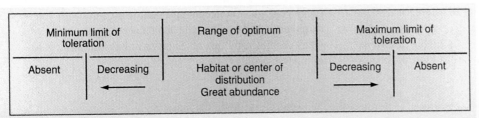

Minimum limit of toleration		Range of optimum	Maximum limit of toleration	
Absent	Decreasing	Habitat or center of distribution Great abundance	Decreasing	Absent

FOCUS ON SCIENTIFIC PROCESS

Succession on the Sand Dunes of Lake Michigan

Figure 1 Henry C. Cowles developed some of the early concepts of ecological succession from studies he conducted on the south shore Lake of Michigan.

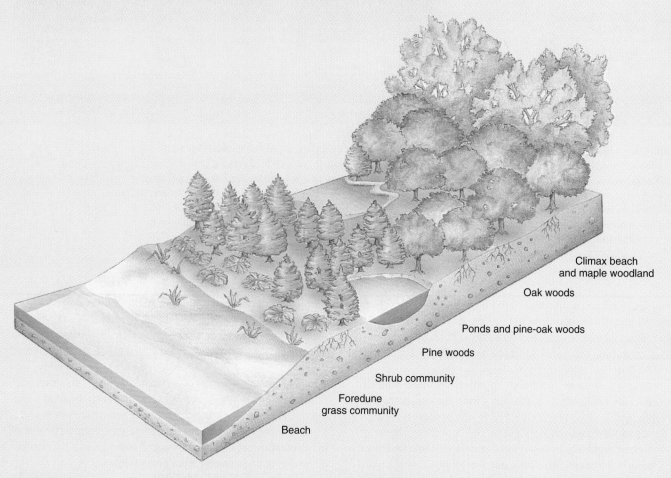

Climax beach
and maple woodland

Oak woods

Ponds and pine-oak woods

Pine woods

Shrub community

Foredune
grass community

Beach

Figure 2 Succession starts with the formation of new ground when sand is deposited on the beach and the wind forms a foredune. Larger dunes develop inland from the foredune where the pioneer community develops. Grasses and small shrubs develop on the larger dunes. Farther inland, pine trees invade the grass and shrub community as conditions change. Pine forests develop, and the environmental conditions change once again. In these shaded forests, acorns germinate, and oak trees grow and compete for available light. Small depressions in the dunes fill with water, forming ponds in which aquatic communities develop. Some areas have enough moisture to support cedar trees. The climax community develops when beech trees replace the pine and oak forests. By walking from the beech forest to the open beach, you can, in effect, walk back through time because the stages through which you pass will all occur in the present beach with the passage of enough time.

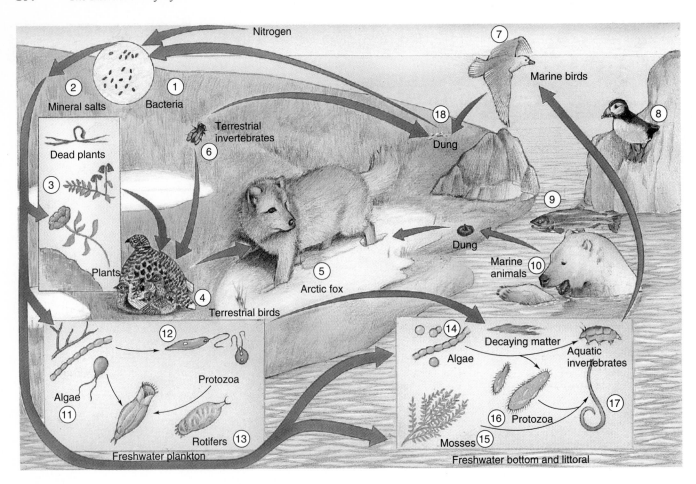

Figure 9.4 Energy entered Elton's food cycle through photosynthesis by aquatic algae and mosses (11, 14, and 15) and terrestrial plants (3). Algae become food for plankton (12, 13, and 16), which feed aquatic invertebrates (17); these are eaten by fishes (9), which in turn are eaten by birds and bears (7, 8, and 10). Terrestrial plants are eaten by insects (6) and birds such as the ptarmigan (4), which are then eaten by arctic foxes (5). Dung (18) and other wastes feed bacteria (1), which release mineral salts and fix nitrogen (2) used by the plants and algae. As the last few patches of snow melt, both the arctic fox and the ptarmigan replace their white winter color with brown fur or feathers that will be more adaptive during the coming summer.

Question: *Which represents the greatest living biomass in this community: polar bears or fishes?*

diagrams such as the one shown in Figure 9.5. Odum's system refined the traditional descriptive, nonquantitative way of describing energy flow within ecosystems because it incorporated quantitative estimates of the fate of energy as it flowed among different trophic levels.

Odum's aquatic ecosystem in Silver Springs, Florida, received abundant amounts of energy from the sun. A fraction of this radiant energy was captured by the producers—plants and algae—through photosynthesis and stored as sugars or other plant materials in the form of **photosynthetic biomass.** Biomass is the quantitative estimate of the total mass of organisms within a given area at a given time. However, not all of the photosynthetic biomass was available to the *herbivores.* The producers' **gross production**—the total amount of new plant substances generated—was reduced by a quantity called the **plant respiratory biomass,** which represented the amount of energy required by the plants to conduct their normal activ-

ities. Consequently, only the **net plant production** remained available to the herbivores. In a similar fashion, energy received by the herbivores was reduced by their respiration. As a result, only a small fraction of the original energy that entered the system was available to the **carnivores,** and even less was available to the **top carnivores.** Ultimately, **decomposers** broke down the biomass that was not passed up the food chain. All of the remaining energy finally departed the system in the form of low-grade heat created during metabolic activities of the decomposers, or it was exported as organisms that drifted downstream and entered other ecosystems.

Ecological Pyramids

In his 1937 book, Elton devised "pyramids of numbers" that were based on his observations of an oak forest and a small pond. He wrote:

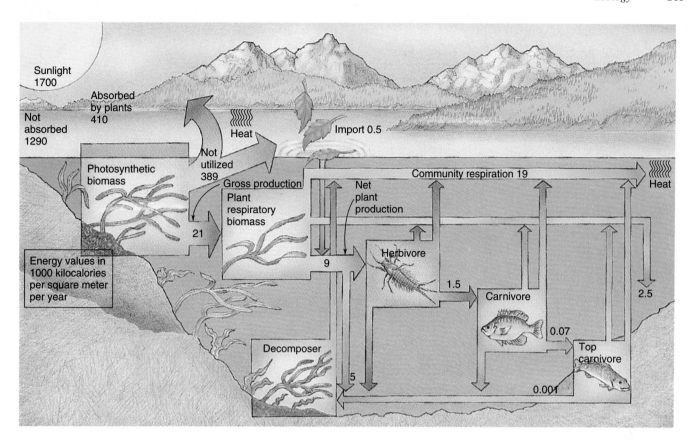

Figure 9.5   Howard T. Odum attempted to show both the direction and the amount of energy moving among trophic levels in Silver Springs, Florida. Of the 410,000 kilocalories per square meter per year absorbed by plants, only 6 kilocalories moved through the trophic levels and into the top carnivores. Ultimately, most of the energy left the ecosystem in the form of heat.

Question: *Why is there less energy available to each successive trophic level?*

The animals at the base of a food-chain are relatively abundant, while those at the end are relatively few in numbers, and there is a progressive decrease in between the two extremes. . . . This arrangement of numbers in a community, the relative decrease in numbers at each stage in a food-chain, is characteristically found in animal communities all over the world, and to it we have applied the term "pyramid of numbers."

Howard Odum used Elton's pyramid concept in structuring his energy flow diagrams for Silver Springs, and Eugene Odum applied it to pyramids of numbers, biomass, and energy flow, as shown in Figure 9.6 (see page 166). **Pyramids of energy** are used to describe rates of energy flow and can be graphically presented as kilocalories per area (square meters) per year. When energy flows from one trophic level to another, there is always less available to the next highest level because of the inescapable loss of some energy in the form of heat when organisms transform energy from one form to another (discussed in Chapter 4). As a result, energy pyramids are never inverted.

Pyramids of numbers are graphic representations of the density of organisms in each trophic level, expressed as numbers per area or volume. **Pyramids of biomass** depict measurements of the amount of living mass in each trophic level and are usually expressed as grams of dry weight per area or volume. Pyramids of numbers and biomass represent the **standing crop** (the number or amount of organisms present at any time) in each trophic level of an ecosystem. The standing crop is usually expressed in grams of dry weight per area or volume, or it can be converted into energy equivalents and expressed as kilocalories per area or volume. Because there is no time interval involved, pyramids of numbers and biomass can be inverted. For example, in a pyramid of numbers representing a population of bark beetles infesting a Douglas fir, the tree represents a single producer (P), but the herbivorous beetles (C_i) may number in the thousands (see Figure 9.7, page 166).

Biogeochemical Cycles

The early food cycle of Elton (see again Figure 9.4) illustrates several food webs and traces the flow of energy (food) through an arctic ecosystem. It also shows, in an uncomplicated way, the cycling of important "mineral salts," nitrogen, and "dung"

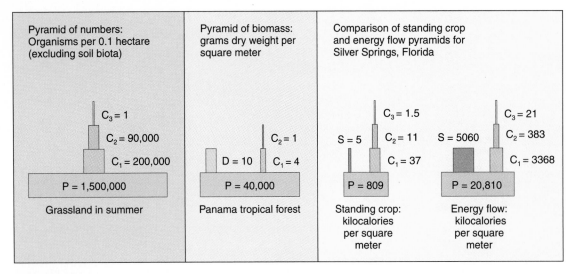

Pyramid of numbers: Organisms per 0.1 hectare (excluding soil biota)	Pyramid of biomass: grams dry weight per square meter	Comparison of standing crop and energy flow pyramids for Silver Springs, Florida	
$C_3 = 1$ $C_2 = 90,000$ $C_1 = 200,000$ $P = 1,500,000$	$C_2 = 1$ $D = 10$ $C_1 = 4$ $P = 40,000$	$S = 5$ $\begin{array}{l} C_3 = 1.5 \\ C_2 = 11 \\ C_1 = 37 \end{array}$ $P = 809$	$S = 5060$ $\begin{array}{l} C_3 = 21 \\ C_2 = 383 \\ C_1 = 3368 \end{array}$ $P = 20,810$
Grassland in summer	Panama tropical forest	Standing crop: kilocalories per square meter	Energy flow: kilocalories per square meter

Figure 9.6   Eltonian pyramids of numbers, biomass, and energy were created by Eugene Odum and show the relative amounts of each present in different trophic levels. P = producers, C_1 = primary consumers (herbivores), C_2 = secondary consumers (carnivores), C_3 = tertiary consumers (carnivores), D = decomposers, and S = saprotrophs (decomposers). The energy flow data from Silver Springs show that only 16 percent of the energy available in producers was transferred to primary consumers and that less than 12 percent of the energy flowed to secondary consumers. Tertiary consumers represented only 5 percent of the energy present in the secondary consumers. Similar decreases occur in pyramids of numbers for grasslands in summer, biomass in tropical forests, and the standing crop of Silver Springs.

Question: *What percentage of the energy present in producers is available to secomdary consumers in Silver Springs?*

Figure 9.7   Douglas firs can be attacked by Douglas fir bark beetles when growing under stressful conditions. Thousands of beetles can infest a single tree and lay eggs under the bark. Hatched larva feed on nutrient-rich tree tissues and leave a network of tracks, shown here. A pyramid of numbers representing these two trophic levels would be inverted.

by bacteria. Within Elton's Bear Island ecosystem, energy could flow through food webs, but only in one direction and in ever-decreasing amounts. In contrast, bacteria could release the "mineral salts" in decaying matter, making them available for the future growth of new plants and algae. Thus Elton described a recycling of minerals among producers, consumers, and decomposers.

It is now known that many biogeochemicals are cycled through ecosystems—carbon, calcium, potassium, phosphorus, oxygen, and other elements required by the biota. In general, cycling of these nutrients occurs through one of three types of **biogeochemical cycles.** Oxygen and nitrogen are cycled rather

quickly through **gaseous cycles** in which the atmosphere and hydrosphere serve as the primary *reservoirs,* places where large amounts of an element are found. In contrast, cycling of common elements such as phosphorus, calcium, sulfur, magnesium, and potassium occurs slowly through complicated **sedimentary cycles** in which rocks, soil, or sediments act as reservoirs. Movement through sedimentary cycles may involve hundreds, thousands, or millions of years. The **hydrologic cycle** (water cycle) connects and drives the various sedimentary biogeochemical cycles that are critical to functioning ecosystems. That is, crucial elements are transported by water as it moves through its global cycle as shown in Figure 9.8.

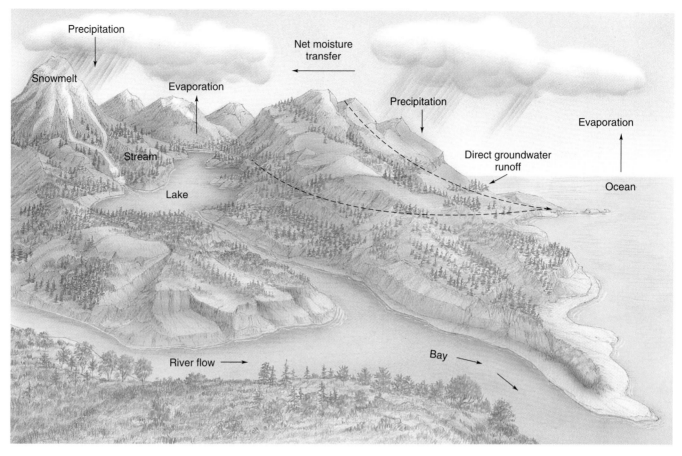

Figure 9.8 The hydrologic cycle describes the movement of water between oceans, atmosphere, and landmasses. Solar energy causes evaporation from the water surface. Onshore flows of moist air result in a net moisture transfer landward, where the increased elevation of mountains causes air to cool and lose its moisture as precipitation. Water deposited over land evaporates or it is stored in snowfields, glaciers, lakes, and the biota or it is transported back to the ocean in groundwater or surface streams and rivers.

Question: *What is the major reservoir for water in the hydrologic cycle?*

Phosphorus Cycle

Most sedimentary nutrients move through similar cycles, so the movements of phosphorus can serve as a model to illustrate a sedimentary cycle. Phosphorus is particularly important because insufficient quantities of this element often limit the growth of producers in many aquatic and some terrestrial ecosystems (see Figure 9.9, page 168). Phosphorus contained in rocks is liberated by erosion as phosphate (PO_4^{-3}). This inorganic phosphate (IP) becomes available to plants and algae in ecosystems. These producers incorporate PO_4^{-3} into an organic molecule to create organic phosphate (OP), the form present in organisms. Phosphate can cycle through numerous food webs, moving from plants to animals, ultimately to decomposers, and then back to other plants.

Phosphates and other biogeochemicals can be routed through many different cycles. As ecological succession occurs in terrestrial environments, varying amounts of phosphates (important chemical forms of phosphorus) are removed by surface runoff and transported to rivers and lakes, where they cycle through aquatic food webs. Eventually, these phosphates are transported downstream and enter oceans. Once there, phosphates may be cycled through marine food webs, returned to land in the form of bird guano, or even conveyed back to headwater streams in the bodies of salmon returning from the ocean to spawn (see Figure 9.10, page 168). Once spawning is completed, the salmon die, and through decomposition they replenish the stream or river system with phosphates and other nutrients. Phosphates may also accumulate in the bodies of marine organisms, which sink to the ocean floor when they die. If this path is followed, deep ocean sediments serve as a reservoir from which the phosphates may not escape for millions of years.

Carbon–Oxygen Cycle

All biological systems, cells, and tissues that have ever existed on Earth have been primarily constructed with carbon

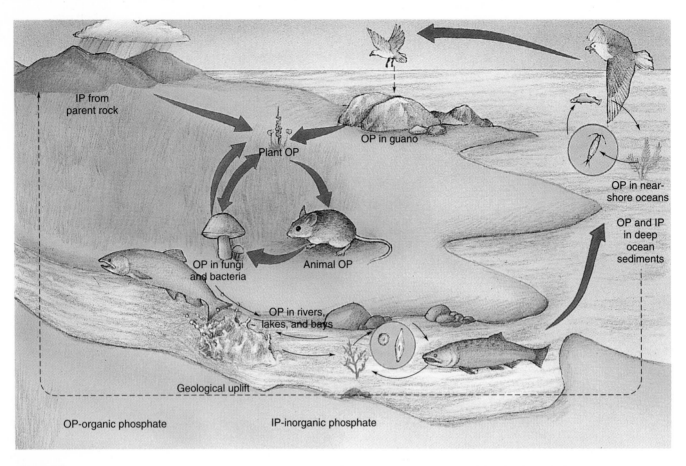

Figure 9.9 Inorganic phosphate (IP) enters terrestrial ecosystems when parent (sedimentary) rock breaks down during weathering. Once in soil, plants convert IP into organic phosphate (OP) in their growing tissues. Plants are either eaten by herbivores or are decomposed by fungi and bacteria when they die. OP from animals is also degraded by decomposers. Once returned to the soil, OP can be taken up once again by plants or enter aquatic systems, where it cycles through both freshwater and marine communities. Some OP can move from oceans landward in migrating fish, such as salmon, or in the droppings (guano) of birds that feed in marine ecosystems. Most OP in the ocean, however, is lost to deep ocean sediments when marine organisms die and sink to the bottom. There the OP is destined to be compressed and uplifted during some future eon to become "new" parent rock.

Question: *Why are there many fossils of marine organisms in sedimentary rock?*

Figure 9.10 Phosphates present in ocean food chains are concentrated in the bodies of adult salmon as they feed in offshore marine communities. When they return to their home rivers and streams, they mate, spawn, and die. Bacterial and fungal decomposition of these large fish release phosphates that can once again cycle in the food webs of these freshwater ecosystems.

Question: *Why is death essential for life in salmon life cycles?*

atoms. Oxygen (O_2) is mainly confined to a gaseous cycle, and carbon is involved in both gaseous and sedimentary cycles as shown in Figure 9.11A. In the gaseous phase, carbon contained in carbon dioxide (CO_2) is cycled through the hydrosphere and atmosphere. However, the major carbon reservoir is calcium carbonates of sedimentary rock strata. Erosion of rocks allows "new" carbon to enter ecosystems, where it can be used by living organisms.

The two major metabolic processes of life—photosynthesis and cellular respiration (discussed in Chapter 4)—link the

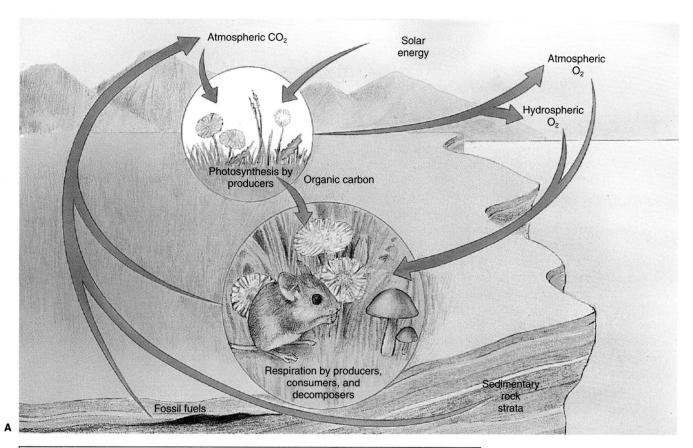

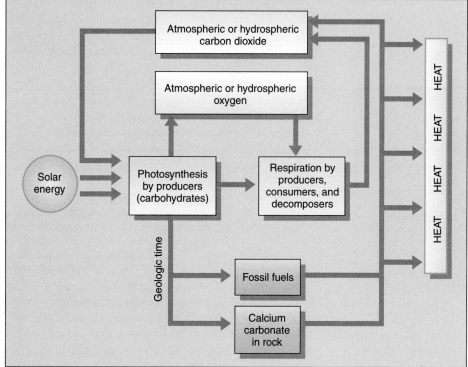

Figure 9.11   (A) Carbon and oxygen move through ecosystems in both short- and long-term cycles. Short-term cycles involve the extraction of carbon dioxide (CO_2) by plants during photosynthesis. Complex carbon compounds formed by photosynthesis are eaten by animals or degraded by decomposers, releasing CO_2 back into the atmosphere. Oxygen (O_2) is released during photosynthesis and used during the respiratory activities of plants, animals, and decomposers. Long-term cycles involve fossil fuels and sedimentary rock strata containing calcium carbonate. (B) The energy that drives all these cycles comes from the sun, and most of it is ultimately lost as heat.

Question: *Under what conditions might carbon be stored for long (geological) time periods?*

hydrologic and carbon–oxygen cycles as indicated in the following equation:

$$CO_2 + water + energy \underset{\longleftarrow\ cellular\ respiration}{\overset{photosynthesis\ \longrightarrow}{= = = = = = = = = = = =}} sugars + O_2$$

The reactions of photosynthesis, which are dependent on solar energy, drive the equation to the right and lead to the production of energy-rich carbohydrate molecules (sugars), as shown in Figure 9.11B. *Aerobic cellular respiration* reactions release the energy stored in the sugar so that it can be used by the organism, driving the equation to the left. Relative to carbon cycling, plants are able to extract carbon molecules, as CO_2, from the hydrosphere and atmosphere during photosynthesis. Cellular respiration occurs in all plants, animals, and decomposers and results in the return of CO_2 to the environment. Similarly, cellular respiration requires O_2 and photosynthesis releases O_2.

During some past geological periods (for example, the Triassic period, about 200 million years ago), cellular respiration by all of the existing biota occurred at lower rates than the photosynthetic reactions of plants. As a result, great quantities of carbon were extracted by plants from the hydrosphere and atmosphere. Eventually the masses of resulting plant material became stored in fossil fuel "sinks" such as coal, natural gas, oil, and oil shales through complex geological processes (see Figure 9.12). Burning of these fossil fuels by humans is returning much of this carbon to the atmosphere, where it once again is available to plants (discussed in Chapter 12). Consequently, some of the carbon atoms present in the breakfast cereal you most recently consumed may have last cycled through biota that existed on Earth during the Triassic period!

Nitrogen Cycle

The nitrogen cycle is probably the most complex biogeochemical cycle. In the biosphere, the element nitrogen occurs in three different states: dinitrogen molecules, nitrogen oxides, and reduced nitrogen. The atmosphere is composed of about 79 percent *dinitrogen molecules* (N_2), the most stable form of nitrogen. A number of different nitrogen oxides exist, but those of greatest biological importance are nitrite (NO_2^-) and nitrate (NO_3^-). Forms of reduced nitrogen include ammonia (NH_3), ammonium (NH_4^+), and various forms of organic nitrogen (ON), including that in organic chemicals which make up living organisms and different nitrogenous products excreted by animals.

Most plants absorb nitrogen from soil as NO_3^-, but once within their tissues, it changes to NH_4^+, which is then used by rapidly growing younger cells for building new plant tissues (see Figure 9.13). A few plant species are able to assimilate NH_4^+ directly from the soil, which effectively shortens the cycling time for nitrogen.

Organic nitrogen in plant tissues is consumed by animals and converted to animal ON. When death occurs, both plant and animal ON enters decomposer food chains, where fungi and bacteria convert it to ammonia through the process of *ammonification*. Soil bacteria in the genus *Nitrosomonas* convert NH_4^+ to NO_2^-, and *Nitrobacter* species convert NO_2^- into NO_3^-, thus making nitrogen available once again for new plant growth.

Within oxygen-limited terrestrial and aquatic environments, about 25 different genera of anaerobic bacteria convert NO_3^- into inert forms of nitrogen, including the atmospheric form N_2, through a process known as *denitrification*. More than 70 bacterial genera, however, including some free-living forms and those like *Rhizobium* that inhabit the root nodules of plants in the pea family (Leguminosae), convert atmospheric N_2 into ammonia by a process called *nitrogen fixation*.

In terrestrial ecosystems, nitrogen-fixing organisms are critically important because they change biologically inert nitrogen into ammonia, thus making nitrogen available to producers. Eukaryotic cells are not capable of carrying out nitrogen fixation; therefore, most eukaryotic life forms on Earth are dependent on the nitrogen-fixing reactions carried on by free-living and nodule-associated prokaryotes. Exceptions are crops grown using modern nitrogen-based fertilizers, which are produced by using fossil fuels, usually natural gas.

When water is present in a soil environment, NH_3 is converted into NH_4^+, which becomes bound to soil particles. This serves to retard nitrogen loss by runoff or through groundwater movement. However, some soil bacteria can convert NH_4^+ into NO_2^- and NO_3^- in a process known as *nitrification*. Both of these oxide forms of nitrogen are very soluble in water. Consequently, they can be removed from the ecosystem during surface runoff or groundwater transport. These processes tend to reduce the amount of available nitrites and nitrates, often causing them to become major limiting macronutrients in terrestrial ecosystems.

Figure 9.12 Carbon was extracted from the atmosphere during the Triassic period and stored in fossil fuel "sinks" like the coal in the mine shown here. When we burn this coal, the carbon is released, which increases the concentration of carbon dioxide in our atmosphere.

BEFORE YOU GO ON Lindeman and the Odum brothers applied concepts from physics to quantify the flow of energy and material cycling within ecosystems. Organic life forms actively participate in the cycling of materials such as carbon, oxygen, nitrogen, and phosphorus. The solar energy that drives the ecosystems of Earth is converted into many useful forms, but ultimately it is all dissipated as heat.

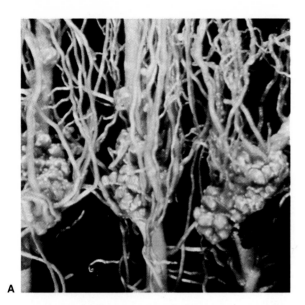

A

Figure 9.13 (A) Inert atmospheric nitrogen (N$_2$) enters terrestrial ecosystems when nitrogen-fixing bacteria convert it to ammonium (NH$_4^+$). Some of these bacteria are free-living; others live in the root nodules of legumes such as beans, shown here. (B) Ammonium and ammonia (NH$_3$) can be converted to nitrite (NO$_2^-$) and nitrate (NO$_3^-$) forms of nitrogen by the process of nitrification. This process is accomplished by bacteria such as *Nitrosomonas* and *Nitrobacter*, which live free in the soil, not in nodules of legumes. Many plants can use both the ammonia and nitrate forms of nitrogen for their growth processes, by converting the nitrogen compounds into organic nitrogen (ON), which is used by animals for their growth and body maintenance. Nitrogenous waste products from both plants and animals are processed by decomposers, and through ammonification, the modified nitrogen enters the ammonia pool in the soil. Other bacteria present in the environment can convert nitrate forms of nitrogen into N$_2$ by the process of denitrification, thus completing the nitrogen cycle.

Question: *Where is the greatest amount of nitrogen found in the biosphere?*

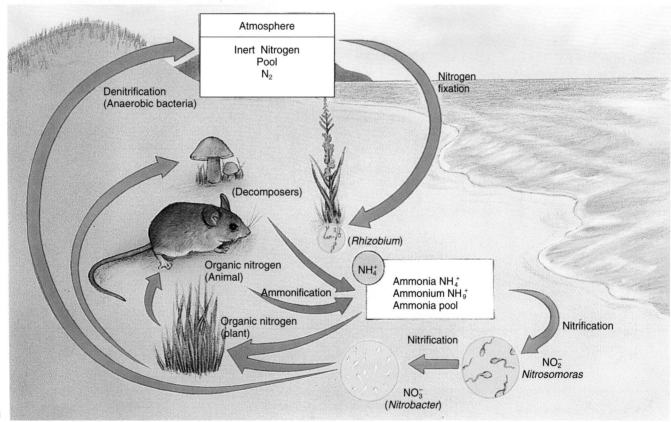

B

POPULATION BIOLOGY

The ecosystem concept is not the only idea used to describe ways in which organisms live together. Another fruitful approach is to consider the distinct populations that make up a community. A **population** is a group of individual organisms of the same species that occupy the same area and interact with one another. From antiquity to the present, many individuals have been interested in biotic and abiotic factors that affect the distribution and abundance of populations. Aristotle (384–322 B.C.) was curi-

ous about reproduction in populations, and he speculated on the relationship between an organism's size and the number of its offspring. In 1798, Thomas Robert Malthus wrote *An Essay on the Principle of Population* in which he contended that natural checks would limit the size of human populations. The ideas presented in Malthus's essay were very important in the development of nineteenth-century biology. For example, Charles Darwin used them as part of the foundation for his theory of evolution. In *Animal Ecology* (1927), Charles Elton noted that all populations have the capacity to grow very rapidly. **Exponen-**

tial growth (also called *geometric growth*) occurs when the growth rate of a population can be described mathematically by the formula 10^x, where the exponent x is some number that defines the rate of growth. Some natural populations may increase exponentially, but such growth can be sustained for only short periods of time. Elton reasoned that if the balance of nature was to be maintained, factors such as competition for resources, predation, or disease (*environmental resistance*) must act to reduce birth rates and increase death rates. Thus he believed that over long periods of time most populations tend to fluctuate around some optimum size. This population size is often referred to as the **carrying capacity** of the environment (see Figure 9.14). Elton, however realized that the carrying capacity is never constant—that it fluctuates with changes in the abiotic and biotic factors in the environment.

While Elton was writing his famous book on ecology, a number of his contemporaries were applying mathematics in studying population growth and regulation. These scientists—Peal, Lotka, Volterra, and Gause—created a rich body of mathematical theory on which modern population ecology developed. In Chapters 10 and 11 we will consider many of their ideas about animal populations in greater depth. Chapter 12 is devoted to human population issues.

BEFORE YOU GO ON Population biology is an important field of study in modern ecology. Mathematical descriptions of population growth enable ecologists to measure the influence of various factors on environmental carrying capacity. Population sizes tend to be related to changes in abiotic and biotic environmental factors.

THE ECOLOGICAL NICHE

The ecosystem is not the only concept used to describe ways in which communities of organisms live together. Another fruitful approach is to consider a population as the fundamental ecological unit and determine which biotic and abiotic factors affect its abundance and distribution. Joseph Grinnell at the University of California was one of the first ecologists to use the term *niche,* In a 1917 paper titled *The Niche-Relationships of the California Thrasher,* Grinnell summarized the factors that defined the distribution of this secretive bird (see Figure 9.15):

These various circumstances, which emphasize dependence upon cover, and adaptation in physical structure and temperament, thereto, go to demonstrate the nature of the ultimate associational niche occupied by the California Thrasher. This

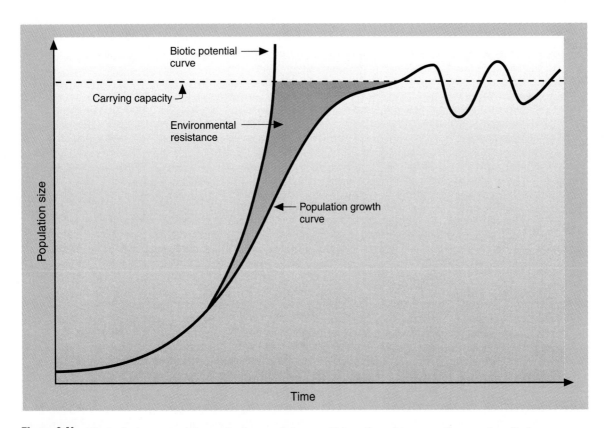

Figure 9.14 Unchecked exponential growth of a population would be reflected in a growth curve described as its biotic potential. As environmental factors check exponential growth, the population growth curve increases more slowly until it reaches the environmental carrying capacity. *Environmental resistance* refers to all environmental factors that affect the growth of a population; it is represented by the area between the biotic potential curve and the population growth curve.

Question: *Why does the population growth curve fluctuate once it reaches the environmental carrying capacity?*

Figure 9.15 California thrashers live in chaparral foothills and dense shrubs in parks and gardens west of the Sierra Nevada and south into Baja California. Grinnell first used the term *niche* to describe the factors that defined the distribution of this bird species.

is one of the minor niches which with their occupants all together make up the chaparral association. It is, of course, axiomatic that no two species regularly established in a single fauna have precisely the same niche relationship.

Grinnell conceptualized the thrasher's niche as a unit defined by its collective behaviors in association with elements from its physical structure.

In *Animal Ecology* (1927) Elton expanded the niche concept to include other factors present in the community. He reasoned, "Animals have all manner of external factors acting upon them—chemical, physical, and biotic—and the niche of an animal means its place in the biotic environment, *its relations to food and enemies.*"

The emerging concept of the species niche was subjected to mathematical interpretations during the 1920s and 1930s by Volterra, Lotka, Gause, and numerous other ecologists. Grinnell's general idea that only one species can occupy a single niche was further expanded by Volterra and Gause, and it became known as the Gause–Volterra principle or, more descriptively, the *competitive exclusion principle.* Today, a **niche** is considered to be the ecological role a species plays within a community. Competition between two similar species at niche boundaries became the focal point for one school of ecology, and it continues today as a focus for many scientific studies.

> **BEFORE YOU GO ON** A niche is defined as the role a species plays in a community. It includes the behaviors of individuals in a species and the influences of other biotic and abiotic factors that affect its life.

MODERN ECOLOGY

Two major approaches for studying ecology have emerged in the twentieth century. One group of ecologists views populations as the fundamental unit of study. The **population–community school** has been strongly influenced by Darwin's theory of natural selection, as the population is the unit of evolution. Competition, predation, and other biotic interactions affect individual survival and reproductive success and, consequently, influence population evolution (discussed further in Chapter 24).

Another group of ecologists views the ecosystem as the fundamental unit of study. The **ecosystem school** has stressed the importance of energy flow, biogeochemical cycling, and other interactions between abiotic and biotic components of ecosystems. From their perspective, populations are often considered to be less important than trophic levels.

There have often been sharp disagreements between the two schools, but both perspectives may contribute to a more complete understanding of ecology. Conflicts between the two schools have led to interesting new questions about the study of ecology. For example, as discussed previously, both schools have been strongly influenced by a belief in the "balance of nature." Today, however, many ecologists question the extent or even the existence of such a balance. In the next four chapters we will encounter many examples of environmental disturbances and discover what processes lead to environmental instability and change. We will also consider whether the balance of nature does exist, and to what degree natural processes or human interference can disrupt it—and once disrupted, whether the balance of nature can be restored.

SUMMARY

1. Ecology is the scientific study of interrelationships between organisms and other biotic and abiotic factors in their environment. Early ecological concepts included the idea of a *balance of nature.* Darwin and other naturalists proposed mechanisms to help explain this perceived balance.

2. Liebig discovered that an insufficient concentration of only one required salt could limit the growth of plants in an area. This concept was later expanded into the law of toleration.

3. Clements and Cowles presented different views of how ecological succession occurs. Clements thought that terrestrial succession led to a climax community with a developmental pattern similar to that of an organism. Cowles thought that succession could both advance and retrogress and that natural

communities never reached equilibrium. Cooper and Gleason later described communities in terms of mosaics formed by continuous changes in species present.

4. Forbes and later Elton described patterns of feeding relationships called trophic levels. Plants, herbivores, and carnivores represent three different trophic levels. The amount of energy present at a given trophic level is always less than that present in the previous (lower) trophic level but greater than the energy that will be available to the next (higher) trophic level. Food chains represent one possible route of energy flow through an ecosystem; food webs represent all possible routes.

5. The ecosystem concept was developed during the 1930s and 1940s to include both the biotic community and its abiotic environment. Lindeman and the Odum brothers attempted to quantify ecosystem functions by measuring the energy flow between trophic levels. They determined basic biological efficiencies by relating this flow to the growth that occurred within each trophic level. They also applied the concepts of trophic dynamics to changes that occur during succession.

6. Ecological pyramids of energy, numbers, and biomass are graphical representations of the trophic structure of an ecosystem.

7. Nutrients and chemical elements are cycled through ecosystems. Gaseous, sedimentary, and hydrologic biogeochemical cycles provide a continuous supply of these materials to organisms in the environment.

8. A population consists of all members of a species in an environment. Population growth can be defined in mathematical terms. The effect of environmental factors will be reflected by population growth curves.

9. The niche is the ecological role that a species plays within a community.

10. Both the population–community school and the ecosystem school have contributed to our understanding of ecology.

WORKING VOCABULARY

abiotic factors (p. 158)
balance of nature (p. 159)
biogeochemical cycles (p. 165)
biotic factors (p. 158)
carnivores (p. 162)
carrying capacity (p. 172)
climax community (p. 160)
ecological niche (p. 173)
ecological pyramids (p. 164)
ecosystem (p. 162)
energetics (p. 162)
exponential growth (p. 171)

food chain (p. 162)
food web (p. 162)
herbivores (p. 162)
limiting factors (p. 160)
pioneer community (p. 160)
plant respiratory biomass (p. 164)
population (p. 171)
producers (p. 162)
succession (p. 160)
trophic level (p. 162)

REVIEW QUESTIONS

1. What contribution did Darwin make to early ecological concepts?

2. What is the significance of the law of limiting factors?

3. How did Clements and Cowles differ in their views on ecological succession?

4. What contributions did Shelford and Elton make to early animal ecology?

5. What are trophic levels, and how do they affect the flow of energy and material cycling in ecosystems?

6. Which environmental factor can be accurately represented by ecological pyramids?

7. What is the hydrologic cycle, and how does it operate?

8. What are the major sinks for carbon, oxygen, nitrogen, and phosphorus?

9. Which processes in the nitrogen cycle produce nitrogen compounds that are useful to plants?

10. What determines the carrying capacity of an environment?

11. What is a species' ecological niche?

ESSAY AND DISCUSSION QUESTIONS

1. Energy flows through an ecosystem only once, whereas materials can cycle through numerous times. What will happen if Earth's energy source is extinguished? Why? What human activities might disrupt biogeochemical cycling?

2. What might be some reasons why early concepts of plant ecology were developed before those of animal ecology?

3. Which concept described in this chapter do you feel was probably the most important in advancing the field of ecology? Why?

REFERENCES AND RECOMMENDED READING

Butlin, R. A. (ed.). 1995. *Ecological Relations in Historical Times: Human Impact and Adaptation.* Cambridge: Oxford Univ. Press.

Cowles, H. 1899. The ecological relations of the vegetation of the sand dunes of Lake Michigan. *Botanical Gazette,* 27: 95–117, 167–202, 281–308, 361–391.

Darwin, C. 1859. *On the Origin of Species.* London: Murray.

Elton, C. 1927. *Animal Ecology.* London: Sidgwick & Jackson.

Hagen, J. B. 1992. *An Entangled Bank: The Origins of Ecosystem Ecology.* New Brunswick: Rutgers Univ. Press.

Hoekstra, T. W., T. F. H. Allen, and C. H. Flather. 1991. Implicit scaling in ecological research. *BioScience,* 41: 148–154.

Kingsland, S. E. 1995. *Modeling Nature: Episodes in the History of Population Ecology.* Chicago: Univ. Chicago Press.

McIntosh, R. 1987. *The Background of Ecology.* Cambridge: Cambridge University Press.

Odum, E. 1971. *Fundamentals of Ecology.* 3d ed. Philadelphia: Saunders.

Odum, H. 1957. Trophic structure and productivity of Silver Springs, Florida. *Ecological Monographs,* 27: 55–112.

O'Neill, R., D. DeAngelis, J. Waide, and T. Allen. 1986. *A Hierarchical Concept of Ecosystems.* Princeton, N.J.: Princeton University Press.

Smith, L. 1986. *Elements of Ecology.* 2d ed. New York: Harper-Collins.

Stauffer, R. 1957. Haeckel, Darwin, and ecology. *Quarterly Review of Biology,* 32: 138–144.

ANSWERS TO FIGURE QUESTIONS

Figure 9.1   Biotic factors could include organic material transferred from land areas around the lake when snow melt and precipitation entered the lake as surface and ground water. Interactions among organisms living in the lake are also biotic factors.

Figure 9.2   An increase in lake surface temperature is one abiotic factor that could limit the range of the crayfish, forcing them to live in deeper, cooler water.

Figure 9.3   Materials can cycle numerous times within communities, but a given unit of energy can flow only once through a community.

Figure 9.4   Total fish biomass far exceeds total polar bear biomass. Great numbers of salmon feed throughout the vast oceans. Bears feed on relatively small continental margins, where their number is far smaller than that of fishes.

Figure 9.5   Each successive trophic level has less energy available because each energy transformation results in a loss of energy, usually in the form of heat.

Figure 9.6   Only 1.84 percent of the energy present in producers is available to secondary consumers (383 kilocalories per square meter present in secondary consumers, $\div$ by 20,810 kilocalories per square meter in producers $\times$ 100).

Figure 9.8   The oceans of Earth are major water reservoirs for the hydrologic cycle.

Figure 9.9   In past eons, marine organisms died and sank to the ocean floor to become fossilized in sedimentary rocks. Some of these fossils are present in sedimentary rocks that are exposed on continents today.

Figure 9.10   Nutrients present in the decaying bodies of adult salmon "fertilize" food chains in the streams, rivers, and estuaries in which their offspring spend two or three years after hatching.

Figure 9.11   Carbon will be stored for long, geological time periods if (1) it accumulates in habitats that are devoid of oxygen and (2) it is covered with sediments. Pete bogs are examples of such habitats.

Figure 9.13   The atmosphere.

Figure 9.14   The effects of environmental resistance usually vary over time. This variance results in population density fluctuations above and below the carrying capacity of the environment.

Terrestrial Ecosystems

Chapter Outline

Reading Questions

1. What are the basic components of biotic ecosystem models?

2. What are the basic components of functional ecosystem models?

3. How do major terrestrial ecosystems differ in their patterns of energy flow, material cycling, population densities, and succession?

4. What types of factors regulate these ecosystem processes?

In the last chapter we discussed some basic concepts in ecology and saw how the concept of the *ecosystem* provides a useful framework for understanding the relationships and interactions between living organisms and the abiotic factors that affect those organisms. In this chapter we discuss terrestrial ecosystems. The root word for *terrestrial* is *terra,* which is the Latin word referring to "earth or land," as distinguished from water (*aquatic* ecosystems are discussed in the next chapter).

Concepts for terrestrial ecosystems were originally developed by integrating results from various environmental studies that focused on different space and time scales. It was found that terrestrial ecosystems undergo succession, and that environmental regulatory factors often affect different trophic levels during each successional stage. Early successional stages are usually dominated by producers, and their growth is largely dependent on available amounts of water and nutrients. In later successional stages, herbivores play a major role by regulating the growth of producers through their grazing activities. As succession continues, the complexity of the entire system increases. The degree of complexity, as indicated by factors such as *species diversity*—the number of different species present in an area—is considered to be of major importance in sustaining an ecosystem when external disturbances occur. However, species diversity may decrease late in succession, which would make the ecosystem more vulnerable to natural or human-caused disturbances.

ECOSYSTEMS AND COMMUNITIES

Ecosystems have long been described in *biotic* terms by considering each species to be a member of a specific trophic level in a community. The community includes producers, consumers (herbivores and carnivores), and decomposers (see Figure 10.1). Traditional **biotic ecosystem models** have emphasized the *amount* of energy and materials moving up through different trophic levels. The more recently conceived **functional ecosystem models** are concerned with the *rate* at which energy and materials move through the system. We will briefly consider both types of models here.

Biotic Ecosystem Models

Until recently, most descriptions of terrestrial ecosystems have had a biotic emphasis: species populations are considered to operate in fixed trophic levels within constraints imposed by the abiotic environment. Under this system of organization, producers, herbivores, carnivores, and decomposers take center stage, and

populations belonging to the various trophic levels become the focal point of most studies. Following Lindeman's original ideas, published in 1942, it has been assumed that energy flows sequentially from producers to herbivores to carnivores to decomposers (as described in Chapter 9). Some scientists questioned Lindeman's simple energy flow model in the 1940s and 1950s; contemporary ecological studies (as we will see in the next section) have uncovered certain conceptual problems in this model.

Trophic Levels

Producers

Producers in terrestrial ecosystems are all the plants that carry on photosynthesis. In most plants, such as the maple tree, only a relatively small fraction of living biomass (usually leaves and the surface layer of stems) actually participates in photosynthetic activities; that is, only a small fraction of the plant is actually a producer. The nonphotosynthetic plant biomass—woody stems and roots—interacts with the soil and serves as a reservoir for key nutrients in terrestrial ecosystems.

Figure 10.1 Ecosystems can be described by both biotic and functional models. The biotic ecosystem model, shown here, is based on the flow of energy through trophic levels. Energy enters the ecosystem through photosynthesis by producers; it flows to herbivores, carnivores, and decomposers. Ultimately it is lost from the ecosystem as heat.

Herbivores, Carnivores, and Decomposers

Energy and nutrients generally move from producers to herbivores, carnivores, and decomposers (fungi, bacteria). However, assigning a species to a single trophic level may not always be an accurate way to describe its role in an ecosystem. Consider the case of the North American grizzly bears (see Figure 10.2). To the relief of most nature lovers and wilderness hikers, grizzlies usually feed as herbivores, consuming plant species in the genus *Equisetum* (horsetails) throughout the bears' entire North American range. Various species of horsetails populate different sections of their range, and grizzlies feed on them during all seasons. Horsetails are usually found in association with grasses and sedges, which the bears also consume. In coastal and mountain habitats, cow parsnip, clover, and dandelions supplement the bear's spring and early summer diet. The animals also consume a variety of other plants, including shrub genera that produce huckleberries and blackberries, during late summer and fall.

The question of trophic level placement for grizzlies becomes more complex because they are also carnivores. Grizzly bears are opportunistic and will feed on any living fish or mammal they are able to kill, as well as on most carrion (dead, decaying flesh). In the northern portion of their range, they have been observed to steal carcasses of animals killed by wolf packs or by other predators. Herbivores such as caribou, elk, moose, and small rodents (ground squirrels and lemmings, for example) make up significant portions of grizzly bears' spring diet. In the coastal areas of their range, they consume large quantities of migrating salmon, and here they may also feed on the beached carcasses of marine mammals (whales, walruses, and seals), some of them herbivores and others carnivores. In which conventional trophic level does a grizzly bear belong?

The term **omnivore**—a consumer of both plants and animals, living or dead—is often used, but food webs describing energy flow in far northern ecosystems become quite confusing when they include grizzly bears.

Trophic Guilds

In developing his concepts of trophic dynamics, Lindeman used the community of Cedar Bog Lake in central Minnesota as a model. His trophic concepts had their roots in an unpublished paper written by his mentor, G. E. Hutchinson. Hutchinson considered the productivity of any trophic level to be the rate at which energy (food) was advanced from the previous trophic level; Hutchison felt that trophic structure could be defined in terms of a linear chain consisting of eaters and the eaten. However, during the half century of ecological research since these concepts originated, scientists have come to realize that few, if any, heterotrophic organisms feed exclusively at any one trophic level.

Currently, trophic levels are no longer considered to be connected in a series to form a chain, and species are not necessarily assigned to a single trophic level. Instead, species are assigned to **trophic guilds,** which are defined as groups of species that exploit the same class of trophic resources in a similar way. Trophic guilds are comparable to the traditional concept of trophic levels; that is, they consist of producers, herbivores, carnivores, and decomposers. However, membership in trophic guilds is fluid, and most organisms, including grizzly bears, are thought to belong to more than one guild, depending on their feeding habits. Use of the trophic guild concept allows modern ecologists to define and quantify more precisely the flow of energy and nutrient cycling in the ecosystems they study.

A

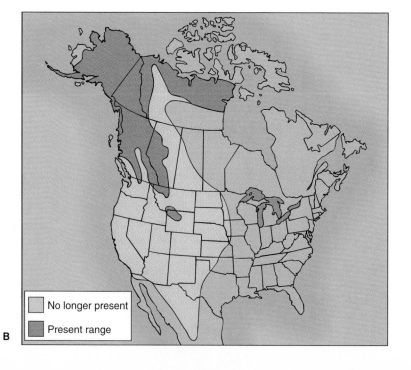

No longer present

Present range

B

Figure 10.2   (A) The grizzly bear *(Ursus arctos)* is classified as an omnivore. In the coastal regions of the grizzly bears' range, salmon become an important component of their diet. (B) Grizzly bear distribution in North America.

Functional Ecosystem Models

When animals such as grizzly bears are partitioned into different trophic guilds, we can begin to think of them as elements in a *functional system,* where they are classified as *consumers.* As consumers, the bears affect both the *rate* of energy flow and the *flux,* or movement, of materials (see Figure 10.3). You may wish to review the discussion of "A Functional Classification System" at the end of Chapter 6 as well as that chapter's Focus on Scientific Process, "Functional Classification."

Functional ecosystem models of energy flow represent a modern, alternative conceptual approach to the more traditional concepts of the biotic model. The central assumption in functional ecosystem theory is that the existing structure of complex ecosystems results from different rates of energy flow and material cycling during each successional stage. Thus the major controlling factors that operate during each successional stage affect the availability of nutrients and energy for a certain period of time. However, modifications of those factors can occur through the successional history of a site and must also be recognized and factored into the functional ecosystem model.

When viewing ecosystems from a functional perspective, organisms are placed into one of two general categories: (1) energy capture and nutrient retention and (2) rate regulation (see Figure 10.3). Consumers and decomposers are important because they regulate rates of energy flow through the system. Thus ground squirrels, lemmings, snowshoe hare, elk, moose, lynx, grizzly bears, fungi, and bacteria are important in "functional ecology" because they all influence the *rate* of energy flow and the flux of nutrients and other materials. Functional models also incorporate population-regulating factors derived from studies grounded in the biotic ecosystem model.

> **BEFORE YOU GO ON** The biotic model of an ecosystem categorizes organisms into different trophic levels identified as producers, herbivores, carnivores, and decomposers. Because most animal species can belong to more than one trophic level, the new concept of trophic guilds was formulated. By focusing on trophic guilds, researchers can more accurately measure energy flow and material cycling in ecosystems.

Figure 10.3 In the functional ecosystem model, the hierarchy includes components of energy capture, nutrient retention, and regulation of the rate at which energy flows to consumers, decomposers, and ultimately out of the ecosystem as heat. Material cycling is not shown.

Question: *Where would you place grizzly bears in the models described in Figure 10.1 and Figure 10.3?*

POPULATIONS

The study of populations has been a key element in expanding our knowledge of ecosystems, particularly in understanding succession. Currently, there is great emphasis on identifying factors that influence the size of populations within ecosystems.

Population Parameters

Populations of organisms affect one another over time. Within each community, however, only a small portion of each species population is present at any given time. These local interbreeding populations represent a lower tier within the *biotic hierarchy*—a level of organization within the biosphere—called a **deme,** which is described in Figure 10.4.

Research has shown that changes within demes constitute a major driving force of succession. This observation has led to a number of important questions that are now guiding research, which could lead to a better understanding of terrestrial ecosystems. What are the relationships between different demes and other important components of an ecosystem? What factors regulate the demes in a population? Are those factors biotic, abiotic, or both? Finding the answers to these and other pertinent questions presents major challenges for ecologists today and tomorrow. These questions and some of the partial answers already obtained are addressed in the remainder of this chapter.

Deme Density

Deme density is usually expressed as the number of individuals of the same species in a given area. Examination of a specific case provides an opportunity for some interesting analyses relative to deme density. Table 10.1 summarizes results from a trapping study conducted in Point Pelee National Park, in southern Ontario, Canada. From that study researchers determined the density of white-footed mice *(Peromyscus leucopus)* and meadow voles *(Microtus pennsylvanicus)* in four different *microhabitats* (small,

Table 10.1 Deme Density of Two Species in Different Microhabitats

| Species | Year | Microhabitat | | | |
		Grassland	Old Field	Sumac	Forest
Peromyscus	1978	10	11	31	59
Leucopus	1979	11	39	84	121
(White-footed mouse)					
Microtus	1978	97	26	0	5
Pennsylvanicus	1979	147	30	0	2
(Meadow vole)					

The number of different individuals of *Peromyscus* and *Microtus* captured in four adjacent microhabitats, each 3.4 hectares in size, in Point Pelee National Park, Canada, during 1978 and 1979.
Source: After Morris (1987).

specialized habitats) during 1978 and 1979. A trapping system covered an area of about 3.4 hectares in each of the four microhabitats. Thus in the grassland microhabitat, the density of white-footed mice in 1978 is expressed as 10 per 3.4 hectares.

The data in Table 10.1 show that white-footed mice are adapted to a wider range of microhabitats than the meadow voles. Why can this be concluded? In general, each of the four microhabitats represents a sequential stage in a successional series that begins with the grassland and climaxes with the forest. It appears, therefore, that meadow voles may not be adapted for surviving in the later successional stages in Point Pelee National Park.

Terrestrial succession can be understood in terms of changes in deme densities over time. It is important, therefore, to identify the factors that regulate the *rate of change* in the density of a deme. Densities of a deme can increase, decrease, or remain the same over a given time period. What causes deme densities to change? Four factors have been identified, as illustrated in Figure 10.5: (1) changes in the numbers of live births

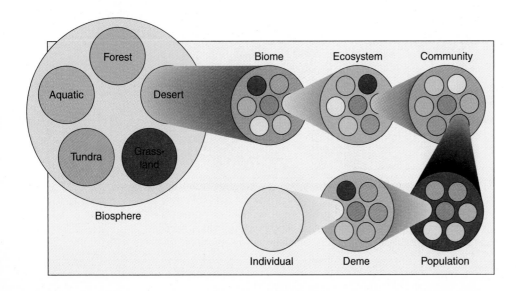

Figure 10.4 In the biotic ecosystem model, an ecological hierarchy exists in which there are different levels of organization. The biosphere is the highest level, followed by biomes, ecosystems, communities, populations, demes, and individuals.

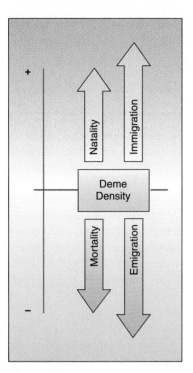

Figure 10.5 Natality and immigration of individuals increase deme density, whereas mortality and emigration of individuals decrease deme density.

per female per unit of time (**natality**); (2) changes in the numbers of new individuals entering from demes in other areas (**immigration**); (3) changes in the numbers of deaths per unit of time (**mortality**); and (4) changes in the numbers of individuals that move away from an area (**emigration**). The rates at which deme densities change are shown in Figure 10.6. Elements that influence these rates of change include environ-mental factors that are either dependent on deme density or independent of deme density.

Density-dependent Factors

Density-dependent factors vary directly with changes in the deme density (that is, as deme density increases, so does the factor). These factors include nutrients, food, cover, *intraspecies interactions* (among individuals of a species within a deme or among members of the same species from other demes), and *interspecies interactions* (among demes of different species). Important intraspecies interactions that affect the densities of animal demes are direct competition for space, food, and mates; those that affect plant densities are nutrients, moisture, space, and light. Critical interspecies, density-dependent interactions include parasitism, predation, and competition.

Density-independent Factors

Density-independent factors are usually abiotic. Severe climatic conditions, including droughts, extended periods of unusual freezing temperatures, flash floods, or volcanic eruptions, can alter the rate at which a deme's density changes (see Figure 10.7, page 182). However, the amount of change in these cases would not depend on the deme's density.

Population Density Controls

It has been difficult for ecologists to measure the separate effects of density-dependent and density-independent factors in changing animal population densities. To illustrate the problems, consider the following classic study, based on an examination of historic records and data of the Hudson's Bay Company. From these data, Charles Elton graphed the yearly totals of fur pelts taken from arctic snowshoe hare and their major predator, the

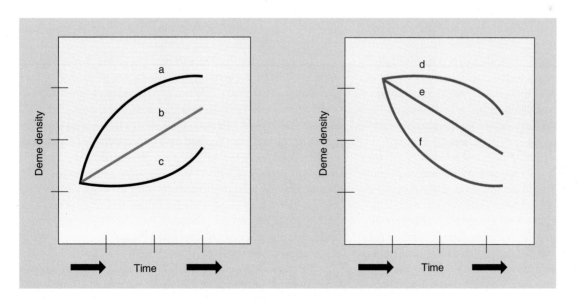

Figure 10.6 Deme densities can expand at rates that are (a) decreasing, (b) constant, or (c) increasing, or they can be reduced at (d) increasing, (e) constant, or (f) decreasing rates.

Figure 10.7 Deme densities can be affected by abiotic factors that range from small, localized changes in temperature or available moisture to catastrophic regional events such as the eruption of Mount St. Helens in southwestern Washington. Not only are demes decimated, but entire communities, both terrestrial and aquatic, can be destroyed.

Canadian lynx, over a 100-year period (see Figure 10.8). He made two assumptions about the data. First, the geographical area from which the pelt data were collected remained constant. Second, the graph represented the approximate densities of the two species over the 100-year period. Inspection of the graph revealed an apparent cyclic oscillation in lynx and hare densities that was a function of time. Hare numbers peaked at intervals of about ten years, and each peak was followed by a rapid decline—the result of massive mortalities. The graph of the lynx pelt data revealed similar oscillations, but in most decades the lynx peak numbers lagged a few years behind those of the hare. What would be a logical hypothesis to explain this pattern?

Elton's observations led him to form a concept of **intrinsic control** between these two species. When hare numbers were high, lynx numbers would also increase. Ultimately, after heavy predation, hare numbers crashed, followed by a dramatic decline in the numbers of lynx. This attractive, simple hypothesis was accepted by ecologists for more than 30 years. Like any hypothesis, accumulating corroborative evidence increases confidence in its validity, but dissenting observations can undermine this confidence. Such contradictory observations lay ahead, and as a result, the "hare-lynx intrinsic control hypothesis" entered a phase of scientific limbo.

Throughout most of Canada, lynx are major predators of snowshoe hares. In the 1960s, hare populations studied on Anticosti Island, Quebec, and more recently on the coast of British Columbia, fluctuated in ten-year cycles in the absence of lynx! Thus Elton's explanations about oscillations in lynx and hare numbers came under suspicion. The fluctuations apparently were not caused exclusively by intrinsic interactions of predator and prey populations, at least not in Anticosti Island or British Columbia hare populations. Factors other than lynx predation must also limit the size of hare populations. What were they?

Research reported in 1988 in areas populated by both lynx and hare has indicated that when hare numbers increased, a shift occurred in their diet. Their preferred winter food is normally older twigs from swamp birch. During times of high hare density, however, these twigs became scarce because of heavy browsing, so hares ate willow twigs instead. When the available willows were consumed, the hare turned to spruce and buffalo berry as a supply of food.

What were the consequences of these changes from their normal diet? Birch was found to have the highest nutritional value for hares, the other twig species lower value. As high-quality food became scarce, hares were forced to spend more time feeding away from protective cover. This altered behavior had two related negative effects on hares. Both the search for and the effects of lower-quality food made them more vulnerable to predation. Most of the mortality observed in the recent studies occurred because of predation, not starvation.

Thus it appeared that a modification of Elton's original hypothesis was in order. It could not be rejected entirely because the oscillations in numbers of hare seemed to influence the numbers of lynx. Nevertheless, the decline in hare numbers in the 1988 studies resulted directly from changes in the quantity and quality of the animals' food. Increased lynx predation occurred only because of the poor nutritional state of the hare. If the lynx had not been present, as was the case in the island and coastal studies, hare mortality would still have occurred at the same level owing to starvation, but it would have been spread over a longer period of time.

In 1993, Canadian researchers proposed a complicated new hypothesis to explain the cyclic nature of snowshoe hare populations in Canada. They collected and analyzed historical data on the following factors: (1) hare population sizes; (2) dark ("scar") marks left on small spruce trees—a poor nutrient source—that hares feed on only during periods of high population when more desirable food sources become depleted; (3) solar (sunspot) activity; and (4) climate changes. The researchers found a positive correlation between all four factors; moreover, all factors tend to follow ten-year cycles. That is, large hare populations, abundant dark marks on spruce trees, elevated solar activity, and increased climate changes all occur together approximately every ten years.

Researcher do not yet fully understand the mechanisms that relate all of the factors. Here is one possible model: Elevated solar activity → increases the amount of available moisture (related to climate change) → spurs the growth of pre-

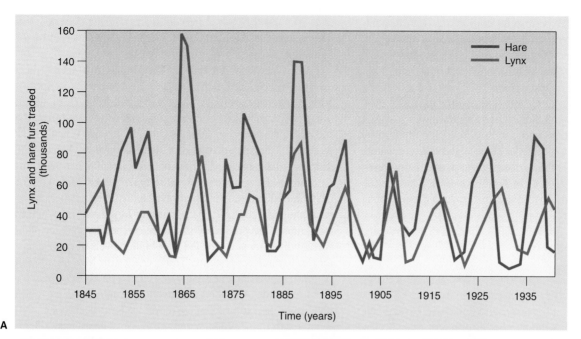

A

B

C

Figure 10.8 (A) Graph drawn from Elton's data suggested to him that the density of prey such as (B) hares allowed the density of predators such as (C) lynx to reach such high levels that increased predation caused a crash in prey density, which in turn led to a subsequent crash in the density of predators. These cyclic oscillations occurred about every 10 years. Recent research has shown that Elton's explanation was incomplete.

Question: *What other factors may regulate hare populations?*

ferred hare food → leads to increased hare population size → leads to increased dark marks on spruce trees as hares begin to feed on them when their preferred food is depleted. This intriguing hypothesis needs further exploration, and efforts are now underway to gather more data, particularly climatic data.

The hare-lynx story is a fascinating example of how scientific ideas change. Elton realized that population regulation was a complex process. He suggested a simple hypothesis in the hope of discovering the underlying causes of population cycles. He and other scientists then tested this hypothesis by studying other natural populations. This research demonstrated that Elton's tentative explanation was inadequate, so the original hypothesis was modified. This resulted in a better understanding of natural population regulation. The new hypothesis may further illuminate the ultimate underlying factors responsible for the cyclic fluctuations of hare populations in boreal forests across Canada.

BEFORE YOU GO ON Deme density is determined by both density-dependent and density-independent factors. Ecological succession results in modifications of habitats that are populated by animal demes and leads to further changes in their density. The hare-lynx case study illustrates clearly the scientific process.

MAJOR TEMPERATE TERRESTRIAL ECOSYSTEMS

The application of functional perspectives to ecosystem research is now guiding studies of terrestrial and aquatic ecosystems. A primary objective of most modern ecological studies is to obtain abundant, precise numerical data that can be analyzed and used to explain different aspects of ecosystem processes. In this section, we will investigate some specific case studies in order to: (1) illustrate how components of the functional ecosystem model are influenced by regulating factors and (2) highlight experimental approaches used in contemporary studies of terrestrial ecosystems. For now, we will explore the application of functional ecosystem models to terrestrial ecosystems. In Chapter 11, we will see how such models are applied to aquatic ecosystems.

Different terrestrial ecosystems exist within the major biomes described in Chapter 8. For each biome, the number of distinct terrestrial ecosystems varies with latitude, longitude, elevation, and proximity to oceans or other large bodies of water. Forest biomes dominate the northern, eastern, and many western and mountainous regions of North America. The central portion of the continent is covered with grasslands, and deserts occur in the southwest.

Temperate Forest Ecosystems

In North America, a large-scale biogeographical pattern exists for total organism diversity. For example, in Figure 10.9, transects (lines used to chart organism distribution across a given area) A and B show that the number of different tree species decreases from a maximum of 180 in the high plateau south of the Appalachian Mountains in North Carolina to only 10 species near the timberline in northern Canada and Alaska.

As you can see on the transects, change in species diversity is largely associated with increasing longitude from central North America to the Appalachian plateau (transect A), whereas from central Canada to the northern timberline,

increasing latitude has a greater effect (transect B). The complex patterns in the west are related to the presence of major mountain ranges.

What factors might account for this pattern of tree species diversity? Recent research correlates the number of tree species with the **average annual evapotranspiration (AAET),** which is defined as the average amount of water evaporated from the soil plus that transpired from vegetation, per year. When AAET increases, tree species diversity decreases, and when AAET decreases, tree species diversity increases (remember that summer rainfall increases from west to east). About 76 percent of the variation in tree species numbers across North America can be explained by AAET; the remaining 24 percent is related to topography and proximity to oceans. Generally, greater plant species diversity is correlated with higher levels of energy capture in natural ecosystems. Thus AAET can be a significant factor in regulating energy capture in certain ecosystems.

Longitude, latitude, climate, topography, and elevation determine the type of forest—coniferous or deciduous—but within each forest type, local modifications among microhabitats determine the characteristics of a specific forest. For example, the forest biome east of the Cascade Mountains in Oregon includes a number of different species of evergreen trees that are affected by various factors. However, elevation is a primary regulating element, and at higher elevations, ponderosa pine is replaced by lodgepole pine, which can thrive at higher altitudes.

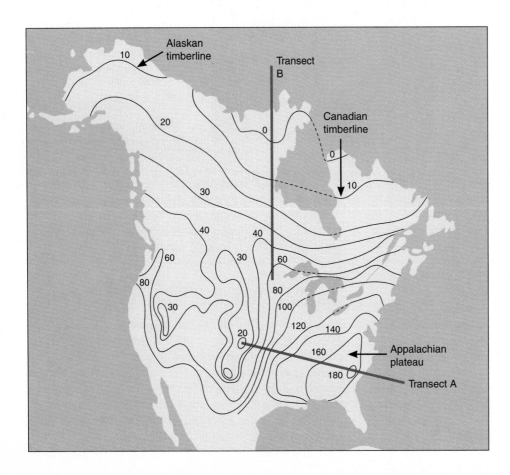

Figure 10.9 Tree species diversity in North America decreases from east to west and from south to north. Along transect A, the change in species diversity is primarily associated with increasing longitude, whereas along transect B, species diversity decreases with increasing latitude. The red lines deliniate areas, each labeled with the number of species that grow there.

Question: *Which transect would be most influenced by a temperature gradient? Why?*

A different controlling factor regulates parasitic bark beetle populations in pine forests. Ponderosa and lodgepole pine forests are often infested with species of bark beetles. The larvae of these beetles consume nutrient-transporting tissues located under the bark, and if their density reaches high levels, they can kill the trees. One important species, mountain pine beetles *(Dendroctonus ponderosae),* infests both lodgepole and ponderosa pine trees (see Figure 10.10A). One major problem ecologists face is identifying the factors that trigger large increases in bark beetle populations, which result in massive pine tree mortalities over large forested tracts.

In 1985, forest ecologists conducted a study of an epidemic population of beetles infesting these lodgepole pine in eastern Oregon. First, they established experimental plots and added nitrogen fertilizers to the forest floor. Then they compared the growth rates of trees on nitrogen-supplemented plots with growth rates on plots that received no additional nitrogen. The results indicated that nitrogen fertilizers stimulated the growth of trees and that greater numbers of beetles were required to kill these more vigorous trees. In areas where tree growth exceeded 100 grams of new wood per square meter of foliage, the beetles were unable to penetrate the bark because they were "pitched out" (see Figure 10.10B). Adult beetles are "pitched out" when the flow of tree sap exceeds their ability to penetrate past the sap flow and into the nutrient layers under the bark. When the sap is exposed to air, it dries to form pitch and entombs the encroaching insect.

What about the amount of available nitrogen on the unfertilized plots? Was it adequate to support growth efficiencies capable of retarding beetle attacks? What conditions in the pine forest ecosystem would influence levels of available nitrogen? What options are available to forest managers that would help control epidemics of bark beetles? Such questions are typical of those raised by ecologists who address modern problems involving forest ecosystems. The answers can only be determined by carefully designed experiments in pine forest ecosystems. Some studies have been completed on bark beetles, and they provide insights into the interesting and complex relationships of ecosystem components that may initially seem to be unrelated.

Under natural conditions, low available nitrogen levels most likely result from extended periods of drought. During drought years, trees are stressed in a number of ways. Most significant, soil moisture decreases while soil temperature increases. These conditions cause a reduction in rates of nitrogen fixation and ammonification by soil bacteria and cyanobacteria. The drought conditions also place added stress on pine trees by reducing the amount of water available for photosynthesis, cell activities, and growth. The combined effects of these processes reduce the overall vigor of pine trees. As a result, they can no longer produce sufficient quantities of sap to protect themselves from beetle attacks.

In turn, prolonged drought has a positive effect on the parasitic beetles. Local beetle demes have access to vastly increased food supplies in the form of stressed trees. As a result, they are released from the major density-dependent factor, food, that normally controls the size of their regional population. Thus during drought years, their numbers can explode and lead to epidemics in pine forests.

Results of these studies appear to offer limited options for forest managers interested in maximizing tree growth while preventing bark beetle epidemics in pine forests. The best approach may be to increase the amount of nitrogen available to trees during drought years. This could be accomplished by fertilizing forests with expensive commercial fertilizers or by thinning the affected stands, which would reduce the competition for nitrogen among trees. By removing some trees, those remaining might be vigorous enough to withstand beetle attacks. However, thinned stands are more vulnerable to damage by high winds.

Figure 10.10 Trees in both lodgepole and ponderosa pine forests can be attacked by beetles. (A) When sufficient numbers of mountain pine beetles infest a pine tree, it dies. However, if environmental conditions are optimal for the tree, they can produce enough sap to "pitch out" invading beetles. (B) When red turpentine beetles infest ponderosa pines, whitish pitch tubes form where the beetles have bored through the thick bark and into the sap.

These studies of interacting functional ecosystem components—beetles and trees, consumers and energy capturers—have related numbers of bark beetles to concentrations of available nitrogen at the ecosystem level. In drought years, beetles play a major role in regulating the rate of total energy flow and material cycling in the forest by increasing the mortality of pine trees.

> **BEFORE YOU GO ON** Species diversity in forest ecosystems is strongly affected by AAET. Characteristics of forest microhabitats are also influenced by abiotic factors such as elevation and biotic factors such as parasites. Studying the relationships between parasitic bark beetle populations and pine species offers insights into the complexity of regulating factors in forest ecosystems.

Grassland Ecosystems

Primary production, or rates of energy capture, of the North American central grassland region differs between *shortgrass* and *tallgrass* communities. Research has shown that the average **aboveground net primary production (ANPP)** in both the shortgrass and tallgrass prairies is correlated directly with the amount and distribution of precipitation (see Figure 10.11).

Major controlling factors affecting ANPP include water, carbon dioxide, nitrogen, and sunlight. Increased solar radiation leads to accelerated growth in the spring, when light levels and day lengths are maximal. Plant growth rates change with longitude from east to west, but the total sunlight energy required for growth decreases with increasing latitude, as shown in Figure 10.12. For example, an ANPP **isopleth** (a line on a graph or map that connects points of equal or corresponding values) of 300 grams per square meter extends from northeastern North Dakota to southwestern Texas in years of average precipitation. During years of below-normal precipitation, the isopleth shifts eastward and extends from north-central Minnesota to southeastern Texas. In what ways does this range of annual precipitation affect community structure in these prairie ecosystems? Do other abiotic factors contribute substantially to community structure and function?

Prairie grassland ecosystems are largely dependent on adequate moisture, but soil type and nitrogen availability also influence grassland ecology. In fact, all three factors are interrelated. One hypothesis suggests that ANPP in dry prairie ecosystems is greater for fine-textured soils with high water-holding capacity (loamy soils containing clay, silt, sand, and organic matter) than for coarse-textured soils with low water-holding capacity (sandy soils). Also rates of ammonification and nitrification are dependent on specific soil moisture and temperature combinations.

Variations in prairie community structure are influenced not only by annual precipitation and soil texture but also by factors that affect the amount of water available to the plant community over time. One such factor is the *rate of moisture loss*. Surface runoff, groundwater export, surface evaporation, and transpiration from vegetation all contribute to moisture loss from prairie communities.

Grazing animals and wildfires are other factors that can affect the structure and function of grassland communities. By consuming grass and carrying out their normal activities, grazing animals influence rates of energy flow and material cycling in the ecosystem. Wildfires remove surface litter and dead foliage, allowing increased amounts of light to reach the surface. Although some nitrogen is lost in smoke when wildfires sweep the prairies, the loss of surface litter results in elevated soil temperatures that stimulate nitrogen fixation and ammonification in the critical weeks of early spring, when moisture is present.

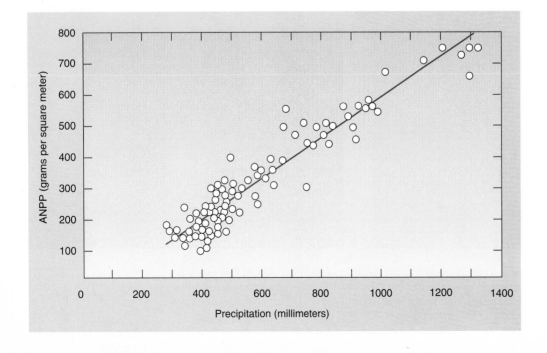

Figure 10.11 The relationship between aboveground net primary production (ANPP) and average annual precipitation was determined by taking measurements from 100 major land resource areas across the central grassland region of North America. The graph shows that locations with greater precipitation have a higher ANPP.

Question: *What is the average increase in ANPP per centimeter precipitation?*

Normal Years

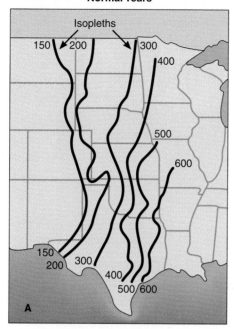

Drought Years

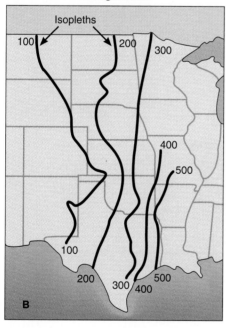

Figure 10.12 (A) Isopleths of ANPP range from 150 to 600 grams per square meter in the central grassland region of the United States during years of average precipitation. (B) During drought years production decreases to ANPP isopleths of 100 to 500 grams per square meter.

Question: *What is the difference in ANPP range between normal and drought years in Oklahoma?*

Recent research considers prairie wildfires as a positive secondary factor in established grasslands. The major impact of such fires is to increase the demand for materials essential for perpetuating prairie grassland communities. In effect, fires

serve to accelerate the recycling of important nutrients and minerals. This type of research results has led some ecologists to advocate "let it burn" policies that have proved to be controversial (see the Focus on Scientific Process "Yellowstone: Drought, Fires, and Carrying Capacity").

Grassland ecosystems illustrate a general ecological principle that is related to *scale of analysis*. As the scale becomes smaller, in moving from biome to ecosystem to community to population to deme to individual, the total number of variables that may influence an observed pattern becomes greater. Thus when the scale reaches a single plant on a prairie, an enormous number of variables can affect it. Each species' niche can be influenced by several levels in both biotic and functional ecosystem models: those that determine the type of biome, those that specify the ecosystem, those that regulate the rate at which energy flows and materials cycle within the community, those that determine the density of populations, and ultimately those that control the growth and reproduction of individual prairie plants.

BEFORE YOU GO ON A key measure of the condition of grassland ecosystems is ANPP, which is determined primarily by precipitation. ANPP isopleths are lines that describe a specific level of grassland productivity. In low-rainfall years, they shift from west to east, reflecting the influence of moisture on ANPP. Prairie grassland ecosystems are also affected by soil types, nitrogen availability, and the rate of moisture loss. All of these may be strongly influenced by grazing animals and wildfires.

Desert Ecosystems

To anyone who has visited a desert, it comes as no surprise that water availability is the chief factor that influences desert ecology. The amount of moisture and its geographical distribution throughout the year play leading roles in regulating the rates of energy flow and material cycling in desert ecosystems.

In the vast, hot deserts of southwestern North America, two species of plants are most common. The creosote bush *(Larrea divaricata)* and the white burr sage *(Ambrosia dumosa)* compete in both space and time in many desert communities (see Figure 10.13, page 191). Research has shown that these two shrubs, both *perennials* (plants that live for many successive years), respond to the harsh environment in different ways over the course of time. Burr sage is deciduous and loses its leaves each year, but the creosote bush is evergreen. Burr sage usually forms a continuous pattern of distribution; it is spread more or less evenly over the desert. In contrast, creosote bushes form a dispersed or clumped pattern. The dispersed pattern of the long-lived creosote bush results from competition for water among individuals that occurs during the critical period when they become established. Over time, certain individual creosote bushes will have a competitive advantage in obtaining water. This leads to the elimination of some plants in the area, and the

Yellowstone: Drought, Fires, and Carrying Capacity

During the summer of 1988, dramatic events in Yellowstone, our nation's oldest national park, resulted in ecological changes on a scale never before observed in the region. Hot, dry weather during the summer, coupled with a low snowpack formed during the previous winter, had significantly reduced the grassland productivity in low-elevation areas of the park. Precipitation during June, July, and August was only 36 percent of normal levels, and temperatures in June were 5°C higher than normal, leading to the development of high winds. Together these conditions caused a 50 percent reduction in grass production on the summer wildlife range.

Historical factors also played a role in the story that was to unfold. From its establishment in 1872 until 1972, the National Park Service's forest management policy was to extinguish all natural fires within park boundaries. This management strategy, in conjunction with natural forest succession, resulted in a century during which the park experienced no fires of significant size. Consequently, extensive tracts of mature lodgepole pine forest characterized the Greater Yellowstone Area (GYA), and great quantities of accumulated deadwood lay on much of the forest floor.

In August, relative humidities were as low as 6 percent, and winds up to 96 kilometers (60 miles) per hour swept through the park. Thus the stage was set for the most extensive forest fire in the park's history (see Figure 1). As shown in Figure 2, the August–September infernos burned about 45 percent of the area within Yellowstone Park. Figure 3 reveals that after these extensive fires, the park appeared as a mosaic of black, severely burned areas and brown, less burned area, interspersed with green, unburned forests.

What effect did these fires have on the park's famous wildlife? Predators such as black bears, grizzly bears, and

Figure 1   Extensive fires burned huge areas in Yellowstone National Park during the summer of 1988.

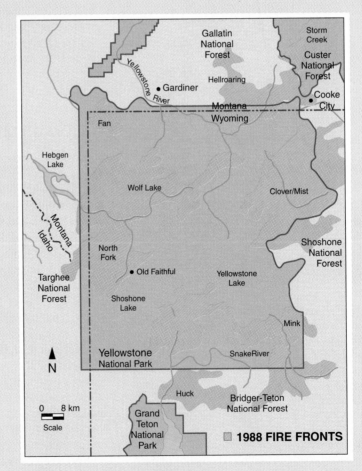

Figure 2   The major fire fronts in Yellowstone National Park during the summer of 1988.

FOCUS ON SCIENTIFIC PROCESS

Figure 3   The extensive areas that were burned during the 1988 Yellowstone fires.

The 1988 drought and fires, a deep snowpack during the winter of 1988–1989, and bitterly cold temperatures during the last two weeks of February 1989 resulted in high mortalities in the northern elk herd. The National Park Service estimates that 4,400 to 5,500 elk died of "natural" causes during the winter of 1988–1989. This mortality, added to the 2,350 elk harvested during a late season hunt near Gardiner, Montana, reduced the size of the population by about 35 percent.

Wildlife biologists carefully analyze mortality data to determine answers to a number of important questions. For example, was mortality evenly distributed among all age classes, or were certain age classes of elk more

mountain lions are dependent on populations of elk, deer, and other herbivores. Ultimately, then, the effects of the fire on predator populations are linked to the fire's impact on herbivore populations, with elk being the most important. During the summer of 1988, some 31,000 elk inhabited the park. After the fires, surveys of burned areas within the park revealed that less than 1 percent of the elk had been killed. The fires, however, combined with existing drought conditions and an extremely harsh winter that lay ahead, set the stage for a massive winterkill of the northern Yellowstone elk herd, the largest migratory elk herd in the world.

The northern herd, consisting of about 22,500 elk in 1988, typically spends most of the year grazing within Yellowstone Park. When winter comes, the population migrates northward to a range that straddles the park boundary, as seen in Figure 4. The 1988 fires burned extensive, overlapping regions of the northern summer and winter ranges within park boundaries. Because of this loss, about 50 percent of the northern elk herd was forced to migrate out of the park 4 to 6 weeks earlier than usual in search of food.

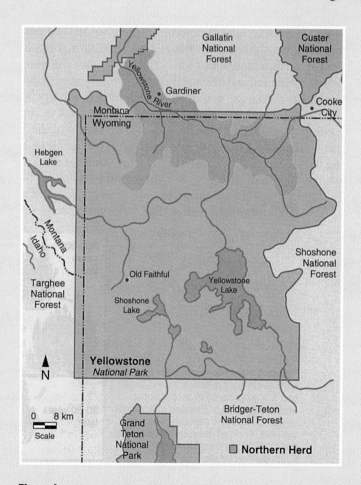

Figure 4   The winter range of the northern elk herd includes parts of Yellowstone National Park and regions north of the park boundary.

box continues

FOCUS ON SCIENTIFIC PROCESS

likely to die? Did the loss resulting from hunting affect the age classes in the same way as the winterkill mortalities? What age classes are important for reproduction? Hunters generally harvest elk from the 3- to 8-year-old age class, the group most responsible for reproduction in the population. In contrast, surveys of dead elk on the northern winter range indicated that only about 17 percent of the carcasses were in this age class. Thus the northern elk herd appeared to have the potential for a full recovery. However, new troublesome questions emerged.

The poor condition of the northern elk herd during the winter of 1988–1989 resulted in a universal clamor for feeding the stressed elk. This humanitarian request seems to constitute a simple and effective solution, but how does it mesh with realities imposed by the natural environment? Ecologists recognize a number of problems related to artificial winter feeding programs for wild big game animals such as elk, including these:

1. Winter feeding can sustain excessive elk population sizes that will crash if feeding is terminated—that is, if the populations reach a size that cannot be supported by the natural environment.

2. Elk, at excessive densities, degrade the natural plant community, which can in turn affect the entire ecosystem.

3. Winter feeding can create a long-term dependency on feeding locations that may alter or stop natural migrations to normal winter ranges.

4. The concentration of large numbers of animals in a limited area results in a high risk of disease in the herd.

5. The usual pattern of winter mortality—the loss of a certain percentage of young, old, and diseased animals—that normally regulates the population size does not occur, and excessive elk numbers result.

Figure 5   After the major fire in the Greater Yellowstone Area, secondary successions will begin, and the progression of successional stages will result in changes in the environmental carrying capacity for elk in the area. Food will be more abundant in the early stages, when the forest canopy closes.

For these reasons, the Montana Fish and Game Commission decided not to feed elk from the northern Yellowstone herd.

Can the natural summer and winter ranges provide long-term support for an elk population of the size that existed in the summer of 1988? Will the Yellowstone fires of 1988 result in a greater or smaller summer range in the future? After this major fire, natural succession is expected to take place over the next 300 years and result in a mature forest of lodgepole pine, subalpine fir, and Engelmann spruce, depending on elevation. Previous research indicates that during early successional stages, wildlife range increases greatly as grasses and shrubs become established after the fire. However, as the forest begins to grow and mature, the canopy closes, and the amount of food for herbivores decreases (see Figure 5). Consequently, the **environmental carrying capacity (ECC)** of Yellowstone Park can be expected to change over time. If we consider the ECC as being the density of an animal deme that can be sustained during each stage of succession, then the total number of elk that both the summer and winter range can support will change as the forest undergoes natural succession. In the short term, the ECC of the summer range is expected to be high. As a result wildlife ecologists are now considering the answers to some difficult questions: Will the ECC of the summer range within the park greatly exceed the ECC of the winter range? If so, what will be the consequences for the northern elk populations? Will winterkill continue to be the major density-dependent regulating factor for the northern elk herd in Yellowstone Park in the foreseeable future? By 1995, studies indicated that fire effects on the winter range are likely to persist for only a few years. Decades of further research will be required to provide final answers to these and other questions.

resultant dispersed pattern eventually appears. In areas of older, established creosote bushes, competition for water is reduced because of the spacing arrangement of the surviving plants.

In contrast, burr sage is a short-lived deciduous shrub that can respond quickly to short-term climatic changes. When water is available, it grows close together, but as water becomes limited, the plants compete, and only a few will survive. They are able to propagate numerous offspring because they produce many seeds under favorable conditions. During periods of water stress in areas where both burr sage and creosote bush exist, competition occurs between the two species and among burr sage individuals, but not among the older creosote bushes.

In areas of the Sonoran Desert, many cacti cannot survive until after early successional plants, including species of palo verde, become established. Why is it essential for palo verde to become established in an area before cacti can survive? Basically, cacti and other desert perennials cannot germinate and grow except in the shade of mature palo verde or other pioneer plant species. The microenvironment under these "nurse plants" is enriched with leaf and twig litter, and the palo verde also provides some protection from the hot sun (see Figure 10.14, page 192).

Surprisingly, it has been shown that plant productivity in hot deserts can approach that of a young coniferous forest if the

A

B

Figure 10.13 The growth pattern of (A) the creosote bush is quite different from that of (B) the burr sage. Competition for water results in the uniform, evenly distanced distribution shown here for the creosote bush. By comparison, burr sage typically grows much closer together if water is available.

Ricklefs, R. E. and D. Schluter (eds.). 1994. *Species Diversity in Ecological Communities: Historical and Geographical Perspectives.* Chicago: Univ. Chicago Press.

Roughgarden, J., R. May, and S. Levin (eds.). 1989. *Perspectives in Ecological Theory.* Princeton, N.J.: Princeton University Press.

Sala, O., W. Parton, L. Joyce, and W. Lauenroth. 1988. Primary production of the central grassland region of the United States. *Ecology* 69: 40–45.

Sinclair, A. R. E., J. M. Gosline, G. Holdsworth, C. J. Krebs, S. Boutin, J. N. M. Smith, R. Boonstra, and M. Dale. 1993. Can the solar cycle and climate synchronize the snowshoe hare cycle in Canada? Evidence from tree rings and ice coves. *The American Naturalist,* 141(2): 173–198.

Van Cleve, K., F. S. Chapin III, C. T. Dryness, and L. A. Viereck. 1991. Element cycling in taiga forests: State-factor control. *BioScience,* 41: 78–88.

ANSWERS TO FIGURE QUESTIONS

Figure 10.3 In the biotic ecosystem model, the grizzly bear would be in both the herbivore and the carnivore trophic levels. In the functional ecosystem model, the bear would be a consumer that regulates the rate of energy flow through the ecosystem.

Figure 10.8 Recent research indicates that sun spot activity, food quality, and climate changes may all be involved in the cyclic population densities of snowshoe hares.

Figure 10.9 Transect B would be more influenced by a temperature gradient because it covers a greater range of latitudes.

Figure 10.11 The average increase in ANPP would be 6.25 grams per square meter for every centimeter of precipitation.

Figure 10.12 In Oklahoma, the ANPP ranges between 200 and 600 grams per square meter in normal years, but in drought years it drops to a range of between less than 100 to 400 grams per square meter.

11

Aquatic Ecosystems

Chapter Outline

Reading Questions

1. What are the principal types of freshwater ecosystems?

2. What is the nature of regulatory mechanisms that operate in different freshwater ecosystems?

3. What are the major classes of marine ecosystems?

4. What factors account for the differences among various marine ecosystems in their capacities for supporting life?

Aquatic ecosystems are found within both freshwater and marine bodies of water. **Freshwater ecosystems** are composed of *lotic* systems, characterized by free-flowing streams and rivers, and *lentic* systems, which are ponds and lakes. **Marine ecosystems** include the oceans of Earth, bays, and estuaries (areas where fresh water from rivers mixes with the saltwater of the ocean). In this chapter we will see that although the two types of systems have an aquatic—water-based—environment in common, there are significant differences between freshwater and marine ecosystems.

FRESHWATER ECOSYSTEMS

Classical ecological concepts were extended from the study of terrestrial ecosystems to the study of aquatic ecosystems over several decades. These concepts include Forbes's microcosms, Shelford's law of toleration, Lindeman's trophic dynamics, and the Odums' ecosystem energetics (see Chapter 9). More recently, biologists have developed helpful new approaches for the study and understanding of freshwater ecosystems.

Lotic Ecosystems

Major environmental factors that influence free-flowing **lotic ecosystems** vary greatly between small headwater streams (those at the highest elevations in drainage basins) and huge rivers such as the Columbia and the Mississippi (see Figure 11.1, page 196). Water volumes and **gradient variations** resulting from the slope of the drainage basin create corresponding changes in the communities that exist within streams and rivers of different size.

A

B

Figure 11.1 Streams and rivers are classified by orders ranging from 1 to 12. First-order streams originate at the top of drainage basins, usually from melting snow. They flow into larger streams (A). The order of streams and rivers continues to increase until they reach the size of huge rivers, such as the Columbia River shown here (B).

Classification of Lotic Ecosystems by Size

In the stream and river classification system now widely used, small headwater streams are categorized as *first-order streams. Second-order streams* originate when two first-order streams meet. *Third-order streams* arise at the junction of two second-order streams, *fourth-order streams* at the junction of two third-order streams, and so on up to *twelfth-order rivers,* which are the size of the Columbia or the Mississippi by the time they reach the ocean.

> **BEFORE YOU GO ON** Aquatic ecosystems exist in both freshwater and marine environments. Although different ecological principles have been used to explain terrestrial and aquatic ecosystems, some concepts, such as functional ecosystem models and trophic guilds, have been applied to both. Free-flowing lotic ecosystems are classified according to gradient variation and size.

A Functional Model of Lotic Ecosystems

What approaches do modern ecologists use to study aquatic ecosystems? Figure 11.2 diagrams a useful conceptual model for describing energy flow in lotic systems. Developed in 1977 by K. W. Cummins, this model allows aquatic ecologists to identify and investigate the major routes through which nutrients move in biota that inhabit lotic ecosystems. It helped provide answers to the question, How do different organisms survive in lotic ecosystems? Cummins's model consists of a functional classification of the community that is based on the niches of organisms within the trophic structure of streams and rivers. It includes the light energy used in primary production by *microproducers* (diatoms) and *macroproducers*

(mosses and algae) living in the stream, as well as terrestrial products (leaves, for example) and plant debris that enter the stream. Rate-regulating decomposers are bacteria and fungi, and the consumers consist of aquatic insect shredders, collectors, scrapers, and numerous species of insect predators and fish.

Two different components form the base of food chains in lotic systems of small and large sizes. Small headwater streams usually occur in steep terrain, near the head of drainage basins. Their **riparian zones** (the areas along the banks that are influenced by water flow) are bordered by terrestrial producers such as deciduous willows, birch or alder trees, or conifers such as pine or fir trees. These trees often grow out over the stream and provide deep shade on the stream's surface. In shaded streams, photosynthesis by aquatic plants and algae is minimal, so primary production in these aquatic environments is low. However, the overhanging foliage contributes substantial quantities of organic matter to the streams in the form of leaves, bark, twigs, and occasionally larger parts of trees. Thus in headwater streams, most aquatic food chains begin with this *coarse particulate organic matter (CPOM).*

Insects observed by anyone who has walked along a stream include mayflies, stoneflies, caddisflies, and dragonflies. These are adults that have emerged to breed; in the stream they will lay eggs that will hatch and become the next generation, as shown in Figure 11.3 (see page 198). Most stages in the life cycle of aquatic insects occur in streams or rivers.

Cummins classified the aquatic insects that consume CPOM as **shredders.** Shredders generally obtain most of their nutrients from digesting the bacteria and fungi that first colonize leaves and other organic debris as it enters the stream.

Figure 11.2 Functional models of lotic ecosystems have been useful for designing ecological studies. Aquatic insects are classified on the basis of their mode of nutrient acquisition. Fallen leaves and other terrestrial materials enter small-order streams, where bacterial and fungal degradation begin. Shredders feed on the decomposing leaves, a major component of coarse particulate organic matter (CPOM). Insect fecal pellets and other organic matter form fine particulate organic matter (FPOM), which flows downstream. Collectors feed on the FPOM, and predators feed on both shredders and collectors. Farther downstream, as light penetration increases, diatoms and other microproducers and mosses, algae, and aquatic plants—the macroproducers—grow and become food for increasing numbers of scrapers and other grazers.

Question: *Would first-order streams have a higher percentage of shredders or grazers? Why?*

Shredder fecal pellets (undigested foods that pass through their digestive system) constitute much of the *fine particulate organic matter (FPOM)* that is taken up by insect **collectors.** The importance of other biotic elements in Cummins's model varies in lotic systems of different sizes. The changes in energy and nutrient flow that occur as water moves toward the ocean are described in Figure 11.4, beginning with streams and ending with large-order rivers.

Small-order Streams   One important indication of how factors influence energy flow in aquatic systems is the **production to respiration ratio (P/R).** A P/R of greater than 1 indicates that productivity—in this case, the amount of new organic material produced by photosynthesis—exceeds the amount of organic material used in the activities (respiration) of all organisms in the community. A P/R of 1 symbolizes a precise balance; a P/R of less than 1 indicates that the level of community production alone is inadequate to sustain the total ecosystem.

The community structure of small streams is influenced to a large extent by the presence of CPOM. Because of low light levels resulting from overhanging plants, primary production within the stream provides little food for invertebrate **grazers,** a group composed primarily of insect nymphs or larvae that consume algae on rocks and other surfaces. The herbivore trophic guild, consisting of all shredders, collectors, and grazers, provides enough energy for only a small biomass of predators, as shown in Figure 11.4 (see page 199). Because little photosynthesis occurs in the stream and there are high levels of community respiration owing to CPOM received from terrestrial riparian plants, the P/R would be less than 1.

Mid-order Rivers   As smaller streams combine to form larger fourth-, fifth-, and sixth-order rivers, width and depth increase, but gradient and current velocity decrease. Rivers are usually distinguished from streams between the third and fourth orders.

Figure 11.3 Mayfly life cycles are typical of aquatic insect species. Adult females deposit their eggs at the surface of the water after mating. The eggs hatch into tiny nymphs that begin to feed and grow. Mayflies progress through a number of successively larger nymphal stages. After about a year, mature nymphs crawl out of the water onto rocks along the shore, and adults emerge from the last nymphal stage. The adults form mating pairs that reproduce and complete the life cycle.

Question: *In which trophic level would you place the colorful rainbow trout?*

Riparian vegetation does not cover entire expanses of mid-order rivers, so greater amounts of sunlight are received at their surfaces. Decomposition activities in small-order streams farther up a drainage basin result in increased concentrations of carbon dioxide, nitrogen, phosphorus, and other nutrients that move downstream. Consequently, the conditions required for plant growth and photosynthesis are often ideal in fourth-, fifth-, and sixth-order rivers. Because of small quantities of terrestrial-generated CPOM and increased levels of primary production by vascular plants and **periphyton** (plants or algae adhering to the rock substrate in the bottom of the stream), significant changes occur in the proportions of herbivore types. The relative number of collectors remains about the same as in small-order streams, but grazers increase and shredders decrease in number and importance. The herbivore community in mid-order rivers supports about the same percentage of

predators. With increased photosynthesis, the P/R increases and becomes greater than 1. Thus organisms in these mid-order rivers generate more biomass than they consume, and the excess organic products can be exported to the larger rivers into which they flow.

The most diverse biological communities in lotic systems are normally found in mid-order rivers. These larger lotic systems provide a wide range of habitats for different forms of collectors, grazers, a few shredders, and insect predators. In turn, this abundance of insect fauna can support a greater variety of fish, including many warm-water species in rivers at low elevations, and sculpin and trout species in rivers at higher elevations. Mid-order streams and rivers are the most highly productive for trout species. Many "free stone rivers," those flowing over gravel and rocks, like the river shown in Figure 11.5 (see page 200), provide the best trout fishing

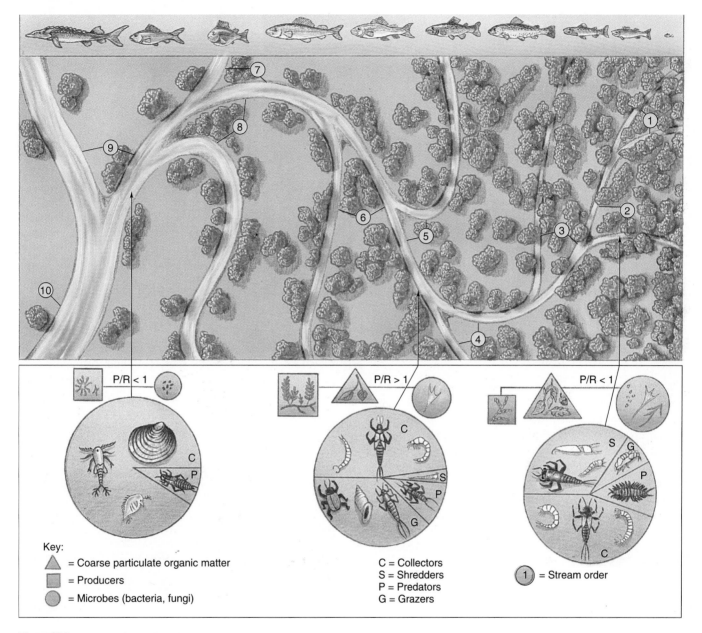

Key:

△ = Coarse particulate organic matter

▢ = Producers

● = Microbes (bacteria, fungi)

C = Collectors
S = Shredders
P = Predators
G = Grazers

① = Stream order

Figure 11.4 When first-order streams flow together, they form second-order streams, which combine to form third-order streams. This pattern continues, eventually forming tenth-order or larger rivers. The insect community changes as the stream order changes. Tree canopies cover many first-, second-, and third-order streams. In these streams, the insect community is represented by large proportions of shredders and collectors, few grazers, and only about 10 percent predators. Farther down the drainage basin, the streams become rivers, and the insect community changes. Here grazers increase and shredders decrease in abundance, but the collectors and predators remain in about the same proportions as in the smaller streams farther up the drainage basin. In tenth-order and larger rivers, the insects are usually represented by collectors and about 10 percent predators. The production-to-respiration ratios (P/R) are usually less than 1 in small-order streams because the tree canopy restricts light penetration to the water surface, limiting in-stream photosynthesis. As the canopy opens up in mid-order streams and rivers, P/R becomes greater than 1 owing to increasing light penetration, which results in greater amounts of in-stream photosynthesis. The depth of large rivers prevents light penetration to where aquatic plants and benthic algae could live, which results in a P/R of less than 1.

Question: *How would fish abundance be related to P/R?*

Figure 11.5   The rock substrate of a free stone river maximizes the habitat for aquatic producers and insect life, which can support large fish populations.

Question: *What is the likely stream order for this free stone river?*

because they include areas of maximum primary productivity. This leads to greater insect diversity and numbers and, hence, to larger trout populations.

Large-order Rivers   Large rivers, in the orders from 7 through 12, become deep and wide. Because of depth and an increased load of light-filtering organic matter and inorganic sediments received from upstream, the light energy required for photosynthesis does not penetrate the entire water column in these systems (see Figure 11.6). Because primary production is limited as a result of insufficient light, the consumer community, now composed of only collectors and predators, uses energy at rates higher than it is fixed during the total primary production. This results in a P/R of less than 1 in most large-order rivers. Consequently, large-order rivers are not productive areas for aquatic algae or plants; the insect population consists mainly of collectors, few predators thrive, and the fish tend to be species like sturgeon or catfish that feed on the organic sediments present on the river bottom.

Drift

A unique feature of lotic systems is that water always flows toward the sea. The moving water carries dissolved materials, suspended particulates—including CPOM and FPOM—and living insects. Collectively these constitute the **drift.**

The life cycle of most aquatic insects begins when the adults emerge, mate, and then fly upstream to lay their eggs. The eggs hatch in the stream and develop into nymphs. They drift downstream as they grow and increase in size. This drift usually occurs at night, and it is related to a number of self-sustaining features of the stream community. Normal spring floods that occur in most drainage systems without dams scour the bottoms of streams and remove significant portions of the flora and fauna. However, since adult insects fly upstream to lay their eggs, these flood-stressed systems are repopulated on an annual basis. Most trout species also swim

upstream to spawn. Thus when the eggs hatch, young fish occupy an area populated with small nymphs, which are their most important insect food source. As both fish and insects grow larger, they tend to move downstream together.

Drift not removed during its seaward movement will ultimately settle and be deposited in estuaries near the mouth of the river system. This organic material adds to the biological richness of these unique estuarine ecosystems.

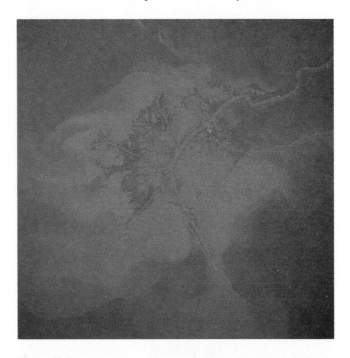

Figure 11.6   This classic view of the Mississippi River shows an example of a "crow foot" delta. The distribution of riverborne sediments is clearly evident in the Gulf of Mexico.

Question: *What might have been the source of sediments forming the delta?*

Lentic Ecosystems

Lentic ecosystems consist of standing bodies of fresh water, including all lakes and ponds. Four major zones of life exist in lentic ecosystems, as shown in Figure 11.7. The **littoral zone** includes the area of transition between the riparian zone along the shore and water extending to a depth of about 11 meters. It is usually the most productive zone and contains the greatest diversity of aquatic life. The **limnetic zone** consists of the portion of the lake, excluding the littoral zone, in which sufficient light energy penetrates to depths where photosynthesis can occur. The **profundal zone** is the deeper portion of the lake, where light does not penetrate, and the **benthic zone** is the bottom of the lake, where sediments accumulate and most bacterial decomposition occurs.

Lentic systems differ fundamentally from lotic systems because the water does not move in a constant-elevation gradient-induced flow. However, water in lentic systems can move, depending on such factors as the amounts of water received from the drainage basin and the rates of discharge to groundwater, streams, or rivers. In addition, wind-generated waves can circulate surface water to variable depths, depending on the wind velocity and duration. Finally, surface water actually sinks to the bottom of a pond or lake when it reaches its maximum density at 4°C. This occurs in lakes where water temperatures cool in the fall before ice forms and again in the spring when the water warms after the ice melts.

Primary production in temperate lakes has a marked seasonal aspect, with photosynthesis being highest in the late spring, after the surface water reaches its greatest density and sinks. As the cold surface water sinks, it is replaced by deep water from near the lake bottom in a process called **turnover.** The water brought from the depths is rich in nitrates, phosphates, and other nutrients because of bacterial decomposition of organic material that occurred on the lake bottom throughout the winter. The nutrient-laden bottom waters circulate to the surface in late spring, and sunlight continues to increase until the summer solstice.

During the summer, primary productivity will often become limited by either decreased concentrations of available nitrogen or phosphorus. However, the total amount of primary productivity is dependent on the growth of algae and aquatic plants, which in turn is related to the quantities of available nutrients and a number of biotic factors.

Trophic Classification of Lakes

In temperate regions, lakes are often classified on the basis of their relative *net productivity* (P_n), the amount of organic material produced by plant photosynthesis minus the amount used by plant respiration per year. Figure 11.8A (see page 202) describes this classification system. **Oligotrophic** (*oligo* is Greek for "few" or "small") **lakes** like the one shown in Figure 11.8B (see page 202) are the least productive, with P_n ranging from less than 0.1 to about 10 grams of carbon per square meter per year (gC/m²/y). **Mesotrophic** (*mes* is Greek

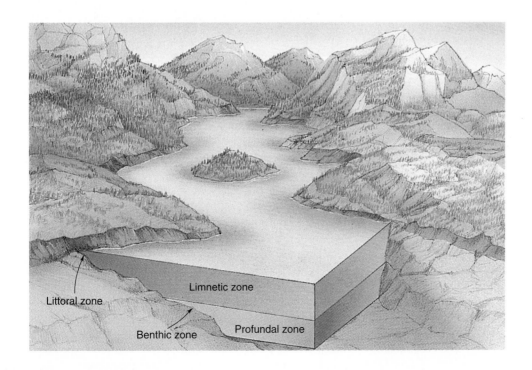

Figure 11.7 Life zones present in lakes and other lentic systems include the near-shore littoral zone; the limnetic zone, through which light will penetrate; the benthic zone, or bottom of the lake; and the profundal zone, which is too deep for light to penetrate.

Question: *Which zone would contain the greatest abundance of producers?*

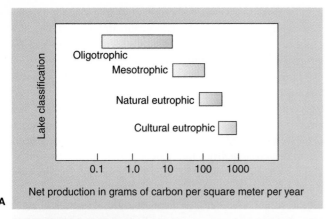

Figure 11.8   Lakes are often classified on the basis of net production (A), measured in grams of carbon per square meter per year. Oligotrophic lakes (B) are the least productive, ranging from about 0.1 to 10 grams, mesotrophic lakes (C) range from 10 to 70 grams, natural eutrophic lakes (D) from about 70 to 400 grams, and cultural eutrophic lakes from 400 to 1000 grams or more.

Question: *How does lake trophic classification relate to its stage of succession?*

for "middle") **lakes** (shown in Figure 11.8C) range from about 10 to 70 gC/m²/y, and **natural eutrophic lakes** (shown in Figure 11.8D) range from 70 to 400 gC/m²/y. **Cultural eutrophic lakes** can produce 1,000 or more gC/m²/y because excessive nutrients generated by human activity are introduced into the system.

The productivity of oligotrophic lakes is usually nutrient-limited. Mesotrophic lakes are more productive and often contain larger populations of fish. Natural eutrophic lakes can accommodate the growth of algae and aquatic plants to such levels that the herbivore community cannot consume them all. The excess eventually becomes organic matter that is decomposed by bacteria. This decomposition is an aerobic process that may lead to depletion of oxygen in the lake water. Occasionally, oxygen levels become so low that fish can no longer survive. Massive fish mortalities caused by oxygen depletion can occur in natural eutrophic lakes, and they are very common in cultural eutrophic lakes (see Figure 11.9).

Biotic Regulation of Lentic Ecosystems

Recent research in the field of **limnology** (the study of freshwater ecosystems) has revealed some factors that profoundly affect lake productivity, material cycling, and food web interactions. In most lakes, phosphate (PO_4^{-3}) concentrations limit primary production. Thus factors that regulate PO_4^{-3} availability control and shape the lentic community. Low nitrate concentrations can also be limiting, but if sufficient PO_4^{-3} is present, nitrogen-fixing cyanobacteria such as *Oscillatoria* and *Anabaena* will often "bloom," which leads to increases in the available nitrate concentrations.

A functional model of the organisms in a lake food web can be described using five trophic guilds, as illustrated in Figure 11.10. The **producers** (P) are first, consisting of nanoplankton (microscopic algae between 2 and 20 micrometers in diameter) and larger edible and inedible phytoplankton. **First-order consumers** (C1), which are the herbivores, include rotifers and small and large crustacean zooplankton. Because large crustacean zoo-

Figure 11.9   When eutrophication occurs in lentic ecosystems, the concentration of oxygen in the water may decrease, resulting in the death of many fish such as those shown here.

Question: *What reduces oxygen concentrations in eutrophic lakes?*

plankton also feed on phytoplankton, small crustacean zooplankton, and the rotifers, they must be placed in two trophic guilds: C1 and **second-order consumers** (C2). **Third-order consumers** in C3 trophic guilds include insect nymphs and other small arthropods, small minnows, and young trout, bass, and pike that feed on the C1 and C2 trophic guilds. Finally, these minnows and small fish are eaten by **fourth-order consumers** (C4), including larger trout, bass, or pike, depending on the lake system.

Cascading Trophic Interactions   A number of studies have shown that changes in the relative abundance of organisms in different trophic guilds can regulate the rate of PO_4^{-3} cycling, which in turn serves to regulate the entire community through a series of interrelated reactions. The discovery of these related effects has given rise to a theory of cascading trophic interactions (CTI). Like all theories, it allows scientists to make predictions that can be tested in carefully designed experiments.

To illustrate the theory, assume that the number of trout, bass, or pike is depleted by human harvest with nets, traps, or hook and line. What will happen? Why? If such a reduction

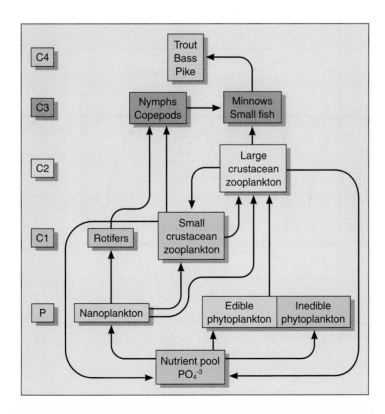

Figure 11.10   Five trophic guilds can be used as a functional ecosystem model for the organisms comprising the food web of a lake. Nutrients and energy are fixed during photosynthesis by the producers, which are consumed by first-order consumers, including rotifers, and both small and large crustacean zooplankton. Large zooplankton (second-order consumers) feed on small zooplankton as well, but when the large zooplankton are young, they can be eaten by mature small zooplankton species. Demes at the C1 and C2 levels become food for insect nymphs, copepods, minnows, and other small fish (including the young of larger fish species). Third-order consumers in level C3 are consumed by mature trout, bass, and pike.

Question: *If large adult fish were removed from this ecosystem, what would be the affect on the phosphate nutrient pool?*

occurs, the density of minnows will increase. The increased minnow population will lead to a decrease in large crustacean zooplankton, insect nymphs, and copepods (aquatic gill-breathing crustaceans). These changes will have corresponding effects on the small herbivores, and this in turn will have an impact on the phytoplankton.

The effects of such interactions are both direct and indirect. Changes in herbivore densities directly affect the density of phytoplankton. An indirect effect also occurs because most of the nutrient cycling within the limnetic zone is dependent on the decomposition of crustacean waste products. Thus as shown in Figure 11.10 (see page 203), the densities of the crustacean populations control the amount of the nutrients available for photosynthesis by the phytoplankton.

Figure 11.10 (see page 203) and the three graphs in Figure 11.11 summarize results from several studies concerned with the CTI theory. They indicate that maximum production rates for the minnows, herbivores, and phytoplankton occur when levels of trout, bass, or pike biomass are intermediate. It is of great significance that the biomass of the lower trophic guilds was found to change in concurrence with predictions derived from the CTI theory. This affirmation indicates that the theory has power and can be used to guide aquatic ecologists in designing further experiments to learn more about lentic systems.

BEFORE YOU GO ON Lentic ecosystems consist of different zones that are defined in terms of light energy received and productivity. These factors are greatly influenced by depth, temperature, and available nutrients. A functional model consisting of five trophic guilds describes energy flow in lentic ecosystems. The theory of cascading trophic interactions (CTI) explains the system of effects that follow if one of the trophic guilds changes significantly.

Eutrophication Abatement   The CTI theory and an examination of Figure 11.10 (see page 203) suggest a possible management strategy for lakes suffering from eutrophication. What might be done to prevent the explosive growth of aquatic algae and plant populations? If excessive algae blooms occur, the addition of large predatory fish (trout, bass, or pike) may offer a viable solution. Why? Because increasing their density would lead to a reduction in the number of minnows, and that effect would cascade through the food web. The most desired effect would be increased densities of herbivore species that would consume much of the excess plankton productivity. This strategy, however, would be appropriate only for lentic systems that contained reasonable levels of PO_4^{-3}. In cultural eutrophic lakes, the excessive amounts of nutrients would place the system beyond the regulatory capacity offered by the cascading trophic interactions.

Beavers and Lake Succession   Ironically, the ultimate fate of most lakes is that one day they will cease to exist. A number of distinct successional patterns occur in different natural lentic systems. For example, one type of succession may occur in a lake formed in the wake of a receding glacier (see Figure 11.12). With the passage of time, the combination of an accumulation of organic matter entering the lake from the watershed along with an excess of lentic biomass production will result in the lake's eventually filling with sediments.

A different type of succession can occur in lakes, like the one shown in Figure 11.13A. The outline of the original lake can be identified in the photograph. Trees now grow in areas that once were part of the lake. If you were to walk from these trees toward the lake, you would simulate passing through time, from the oldest shrub communities to a younger boggy wetland area to a still younger marshy area and finally to the

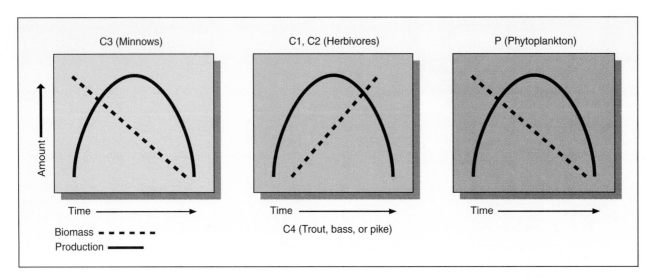

Figure 11.11   When the biomass of C4 predatory fishes (trout, bass, and pike) increases (as indicated by the time arrow along the horizontal axis), the biomass of C3 organisms (minnows) decreases and the biomass of C2 and C1 herbivores increases, which results in a P (producer) biomass decrease. Production (rate of growth) in each trophic level is greatest when C4 predators are at intermediate biomass levels. The effects of increasing the biomass of C4 predators cascade down through the entire food web.

Figure 11.12　Many lakes were formed in the wake of receding alpine glaciers as the last ice age ended. These lakes are usually oligotrophic.

Question: *Would the rate of succession in this lake be more rapid or slower than in a eutrophic lake? Why?*

present lake. The lake system in the photograph is in the last successional stages of its existence. In time, the lake will disappear, and terrestrial succession will begin. In the distant future, the area will be covered by a climax forest that is characteristic of its elevation and biome.

The time line of lake succession is not always predictable. Unanticipated forces in the environment can lead to a surprising alteration in the predicted progression of succession. In 1981, the lake shown in Figure 11.13A became occupied by a family of beavers, which constructed a dam that they continued to expand during the next few years. The dam blocked the outflow of water from the lake, and rather quickly, the lake enlarged and water flooded the riparian habitats, killing many

of the shrubs and trees, as shown in Figure 11.13B. What would you predict will be the ultimate fate of this "new" lake?

MARINE ECOSYSTEMS

Marine and freshwater ecosystems differ in a number of significant ways. Most notably, a high concentration of dissolved salts (salinity) is the major parameter that affects the occurrence, distribution, and abundance of organisms in marine environments. Also, the great ocean depths act as "sinks" for important nutrients. In most areas of the oceans, nutrients locked in sediments in these sinks are slowly cycled through geological time scales.

Figure 11.13　(A) By 1977 lower Crabtree Lake in the Cascade Mountains of western Oregon was nearing the end of its aquatic succession because it was nearly filled with sediments. (B) In 1981 a family of beavers moved into the Crabtree Lake area and constructed a dam. Consequently, succession was reversed, and a "new" lake was formed.

Question: *Given sufficient time, what do you expect will occupy this site?*

Marine systems are categorized as either *neritic* or *oceanic* (see Figure 11.14). **Neritic waters** include a *littoral zone* that occurs in near-shore waters adjacent to the coasts, bays, and estuaries as well as waters extending over a portion of the continental shelf. **Oceanic waters** include all the other areas of marine ecosystems.

Oceanic Ecosystems

Oceans are classified into four vertical zones because environmental conditions vary so greatly with increasing depth that distinct ecosystems with very different structures and functions exist. The **euphotic zone** is defined by the depth of light penetration and occurs from the surface down to a maximum depth of about 200 meters. The **bathyal zone** occurs between 200 and 1,500 meters, the **abyssal zone** from 1,500 to 6,000

meters, and the **hadal zone** from 6,000 down to 11,000 meters in deep ocean trenches.

Upwelling

The topography of the ocean floor directly influences both the movement of deep ocean currents and nutrient circulation. Because of gravity, organic nutrients are deposited in the depths of the oceans. Only in specific areas do conditions allow these nutrients to be transported back to the surface waters. The vertical transport of nutrients from ocean depths to surface waters is called **upwelling.**

In general, three different current systems are capable of causing upwelling, as shown in Figure 11.15. In areas where divergent currents separate surface waters, deeper ocean waters move to the surface. The presence of mid-oceanic ridges on the ocean floor causes deep ocean currents to be forced up and over

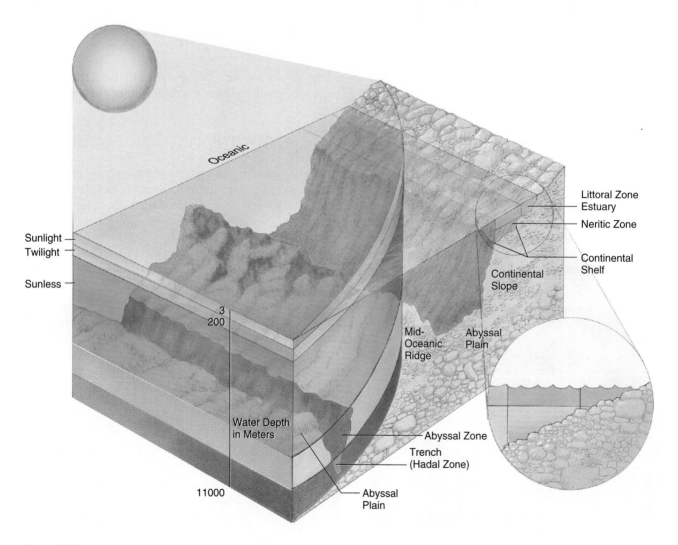

Figure 11.14   Marine life zones vary with distance from shore and depth. The zones include near-shore neritic waters over continental shelves and littoral zones along the shore, bays, and estuaries. Oceanic life zones consist of the upper regions where light will penetrate and the sunless bathyal, abyssal, and hadal zones of the deep oceans. Ocean topography includes abyssal plains, midoceanic ridges, the continental slope, and the continental shelf.

Question: *In which zones would you expect to find producers?*

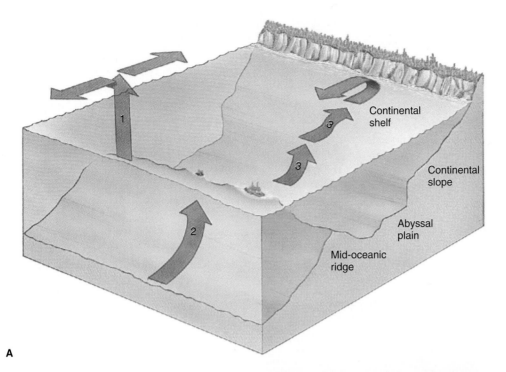

A

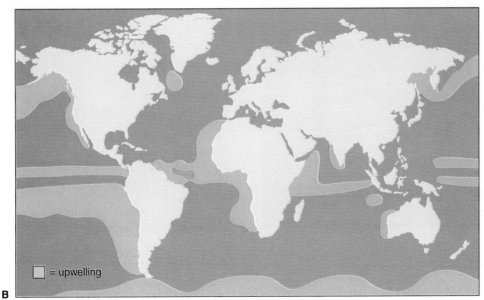

= upwelling

B

Figure 11.15 (A) Upwelling occurs as a result of (1) divergent surface currents, (2) underwater currents reflected off midoceanic ridges, and (3) the offshore movement of surface water. This is then replaced by deep water from the abyssal plain, which moves up the continental slope and along the continental shelf to the near-shore surface areas. Upwelled water enriches the surface zones with important nutrients released by bacterial degradation of organic matter in the ocean benthic environments. (B) The locations of global upwelling and other major regions of marine community productivity.

Question: *Where would you expect to find the greatest abundance of whales?*

them, which brings deeper waters to the surface. Some oceanic ridges emerge above the surface and create islands, such as the Hawaiian chain. Finally, **continental upwelling** occurs where the combined effect of surface winds and Earth's rotation toward the east causes a net offshore movement of surface water on many western continental coasts. As this water moves offshore, it is replaced by cold, deep ocean water that moves up the continental slope, across the continental shelf, and into the near-shore surface waters. All these types of upwelling result in the transfer of nutrient-rich deeper waters to the surface layers of the oceans.

In addition to the three types of upwelling, a mechanism exists for the vertical transport of nutrients in far northern and southern ocean areas that is similar to turnover in temperate lakes. In the cold polar regions, ocean surface water reaches its maximum density just before it freezes. The temperature of greatest density varies with salinity, but when it is most dense, the seawater sinks and is replaced by deep, nutrient-laden water from the ocean floor.

The global regions where upwelling and vertical transport of deep ocean water occur are the areas of highest marine productivity. These are the sites of major modern marine fisheries and the whaling industry of the past. The salmon, tuna, and halibut fisheries off the western coast of North America are located in areas of continental upwelling, as are the anchovy fisheries off western South America and major fisheries along both coasts of Africa.

Nutrient Deserts and Coral Reefs

Most areas of the deep oceans are not in upwelling zones. These expanses are nutrient-poor, for there is no mechanism to cycle any nutrients present on the bottom back into the euphotic zone. Such areas are identified as *nutrient deserts* where few organisms live.

Coral reefs in shallow tropical and subtropical seas are also unique marine systems. They have an enormous diversity of life that is dependent on the nutrients cycled with continental upwelling or wave action circulation (see Figure 11.16).

A Marine Food Web

Much of the whaling industry of past decades was centered in the upwelling regions. Heavily hunted areas included the North Pacific from Alaska to the northern islands of Japan, south and west of Greenland in the North Atlantic, and the nutrient-rich waters north of Antarctica in the southern oceans.

Most large whales feed on **krill,** which are small crustaceans that are present in unimaginably large numbers in some marine systems. Krill feed on zooplankton, which feed on phytoplankton, whose growth is dependent on nutrients brought to the surface in these areas of upwelling (see Figure 11.17A). As shown in Figure 11.17B and C, excessive harvests of blue whales led to decreases in total catch and average length of females, an increase in the proportion of sexually immature whales, and a dramatic decrease in catch per unit effort, a measure of how long it takes to find and kill each whale. The sharp declines in populations of the great blue, the fin, and other large whales have led most nations to adhere to international whaling moratoriums. However, because of the overharvesting practices of the past, the ultimate fate of these marine mammals, (continues page 210)

A

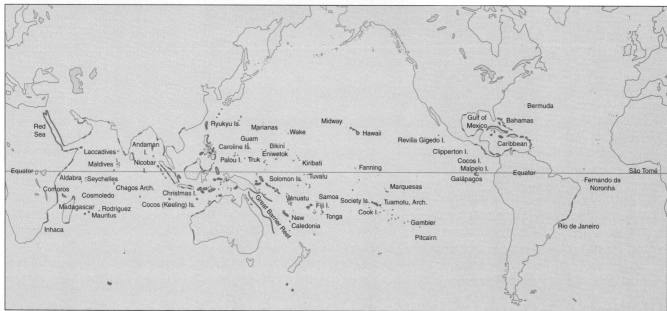

B

Figure 11.16 (A) Coral reef communities are important ecosystems in warm, shallow ocean areas. (B) Most coral reefs are found in tropical and subtropical ocean areas where they are usually associated with the continental shelf or with islands of archipelago systems.

Question: *Which abiotic factors might explain the great diversity of life around coral reefs?*

Table 11.1 Changes in Marine Mammal Populations Since Mid-century

Species	Population at Mid-century	Current Population[a]
Declines		
Sei Whale	200,000	25,000
Fin Whale	470,000	110,000
Blue Whale	200,000	2,000
Humpback Whale	125,000	10,000
Right Whale	200,000	3,000
Bowhead Whale	120,000	6,000
Northern Sea Lion	154,000	66,000
Juan Fernandez Fur Seal	4,000,000	600
Hawaiian monk seal	2,500	1,000
Recoveries		
Gray Whale	10,000	21,000
Galápagos fur seal	near extinction	30,000
Antarctic fur seal	near extinction	1,530,000
Walrus	50,000	280,000
Dugong	30,000	55,000

[a]1980s and early 1990s.

Source:From Brown. L. R., H. Kane, and E. Ayres. 1993. *Vital Signs 1993,* p.110. World Watch Institute, W.N. Norton & Company. New York.

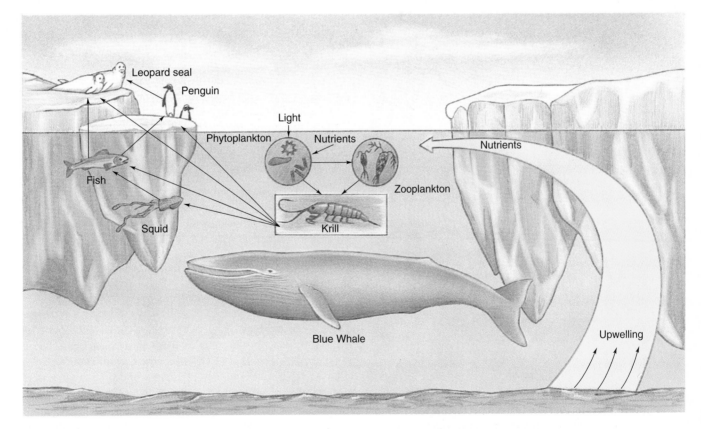

Figure 11.17   (A) In a food web that occurs in southern oceans, upwelling enriches the surface waters with important nutrients used by phytoplankton that supports a thriving zooplankton herbivore guild. They, in turn, sustain immense numbers of krill, the main food for the great blue whale and other large whales. Other guilds in the food web include fishes, penguins, and seal populations. (B) Between 1934 and 1965 the blue whale catch varied, but the catch per unit effort declined. (C) Between 1934 and 1964 a greater percentage of the more abundant younger whales were killed when adults became scarce after excessive harvesting. No data were available for 1942–1947 because of World War II and its aftermath.

figure continues

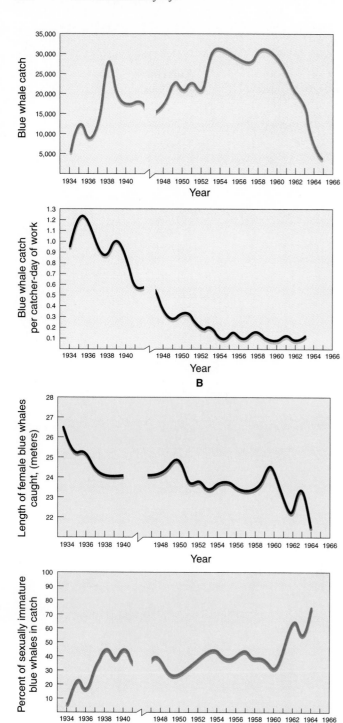

Figure 11.17   *continued*

fornia, and Mexico (see Figure 11.18). All other whale populations appear to be declining. Some sea lion and seal populations are recovering, but others are not, and the prospect seems dim for these endangered species. In the meantime, despite a global ban on whaling, some countries still maintain or allow largely unregulated harvesting of whales. Japan, for example, continues to take whales "for scientific study," and Norway announced in December of 1993 that it would resume its whaling program because the country's economy depended on it.

Neritic Ecosystems

Neritic waters include the nutrient-rich areas over the continental shelf, littoral zones from areas reached by high tide down to about 30 meters, and bays and estuaries where rivers enter the ocean. Many of these areas are enriched from two major sources, continental upwelling and nutrient outflow from rivers and estuaries.

The Littoral Zone

The littoral zone on both coasts of North America hosts a number of ecological systems that support rich diversities of marine life (see Figure 11.19). They include the subtidal areas offshore, open beaches, and the rocky intertidal region around headlands.

The communities established in rocky intertidal areas are stratified into different zones based on their aspect (the direction they face) and tidal depth. Organisms on north-facing rocks are subjected to less solar drying during low tide than those on south-facing rocks. As a result, there is a greater diversity of life in the moist, shaded, north-facing microhabitats. The areas where water remains in pools at low tide also support a greater diversity of life than is found on adjacent, open rock faces.

Red, green, and brown algae constitute the base of the food chain in the intertidal ecosystems as well as the subtidal areas offshore. These algae support a variety of marine organisms, culminating with the top carnivores represented by marine otters and harbor seals (see Figure 11.20).

Bays and Estuaries

Bays differ from estuaries in many ways. A bay is a quiet arm of the ocean extending landward, where the water characteristics are basically the same as those in the near-shore ocean waters (see Figure 11.21). In contrast, **estuaries** are areas where fresh water from rivers mixes with saline ocean waters that flow upbay during high tides. Estuarine waters vary in salinity according to depth because the high concentration of salts in ocean water makes it more dense (and thus heavier) than fresh water. The entering ocean water flows as a *salt wedge* along the bottom of the estuary on the incoming tide, whereas fresh water from the river flows seaward on the surface, as described in Figure 11.22 (see page 212).

the largest species on Earth, remains uncertain at the present time; some data suggest that marine mammal populations continue to decline today. Data in Table 11.1 show that only gray whale populations seem to be recovering. These whales migrate annually from the near-freezing waters of the North Pacific to the tropical waters of the Gulf of California, between Baja, Cali-

Figure 11.18 Although most countries adhere to the present global whaling ban, the populations of great blue, fin, and other large whales continue to decline. This gray whale and her calf are members of the only whale population that seems to be recovering right now.

Figure 11.19 Common intertidal organisms are sea stars and anemones.

Figure 11.20 The marine otter shown here represents mammals that feed at the top of marine food chains. They also eat crabs, sea urchins, scallops, abalone, and fish.

Figure 11.21 Marine bays are quiet landward extensions of the ocean where the water characteristics are very close to those of the near-shore ocean waters. They are important nursery areas for marine organisms, and many, like this one in Frenchman Bay, near Bar Harbor, Maine, have resident populations of fish, crabs, lobsters, and shellfish.

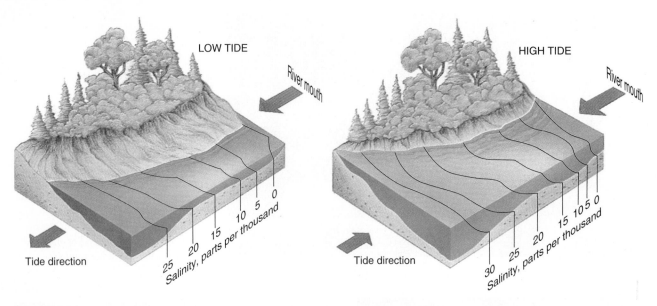

Figure 11.22 The salt water that enters a marine bay or estuary during an incoming tide is more dense than the fresh water flowing seaward from the river. This results in the formation of salinity gradients from the bottom to the surface and from the ocean to the river mouth. Thus the salinity of bays and estuaries varies during the day between high tide and low tide and seasonally with fluctuations in the volume of river flow.

Question: *What might account for the rich abundance of life in bays and estuaries?*

Marine bays are common along the western coast of North America, and their estuarine portions occur far inland. The location of the major area where fresh water and saltwater mix in these estuaries varies seasonally, depending on the volume of river flow. East-coast rivers develop more extensive estuaries and wetlands of coastal marshes, mangrove swamps, and deltas along major rivers in the southern states (see Figure 11.23).

BEFORE YOU GO ON Marine ecosystems are incredibly diverse with respect to size and productivity. Depth and upwelling are two major factors that determine the capacity of a marine ecosystem to support life. Neritic ecosystems are generally the richest biologically of marine ecosystems because they receive an abundance of light and nutrients. They support some of the most diverse communities on Earth.

Figure 11.23 Mangrove swamps are shallow tidal ecosystems found in tropical and subtropical regions.

HUMAN INTERVENTIONS

Bays, estuaries, and wetlands are important feeding habitats for many species of birds and major nursery areas for numerous species of marine life (see Figure 11.24A). Because they are close to major areas of human habitation, these fragile ecosystems have been seriously affected by disturbances, including agricultural, industrial, and municipal wastes, and draining and filling for housing, industrial, and recreational uses, as shown in Figure 11.24B (see also the Focus on Scientific Process, "Wetlands"). The scope of the loss was described in a federal survey, which determined that about 11 million acres of wetlands were converted to other uses in the lower 48 states between the mid-1950s and the mid-

1970s. Most of this loss resulted from drainage or diking, the conversion of tidelands into agricultural lands. What will be the ultimate cost of these alterations to the affected estuarine systems?

In the next two chapters, we turn our attention away from ecosystem structure and function and attempt to assess some of the effects human activities are having on natural systems. Increasing environmental impact from human activities has been correlated directly with the increase in human population density. Parameters that influence human population growth over time are the major topics in Chapter 12. In Chapter 13, we will consider the extent to which human activity is changing the global climate and how such changes will likely affect the biota present in all biomes of Earth.

A

B

Figure 11.24   (A) Great blue herons are important predators that feed on fish, shrimp, and other aquatic life present on or in exposed mudflats during low tide in marine bays and estuaries. (B) Biologically rich areas of marine bays and estuaries have been lost in the past because of filling. The "new land" is often very valuable real estate because of its proximity and access to the water. Uses of such filled areas are varied, but they can include sites for oceanfront development, as shown here in Ocean County, New Jersey.

Wetlands

In past decades, swamps, marshes, and bogs often called up visions of mud, mosquitoes, snakes, and other unpleasant objects. Today, however, such environments have become the subject of fierce debates concerning their importance in the lives of many species, including humans. **Wetlands** are transitional environments that develop between terrestrial and aquatic ecosystems. They are characterized by emergent plants growing in soils that are normally or periodically saturated with water.

Wetland Classification

Wetlands are difficult to classify. There are great variations in wetland characteristics because they have features of surrounding terrestrial and aquatic habitats. Many classification systems have been proposed. Some are based on the amount of available water and type of vegetation, whereas others are based on energy flow. A more exact system classifies wetlands on the basis of their location (coastal or inland), salinity (saltwater or freshwater), and the dominant type of vegetation. Based on these features, a wetland is identified as a marsh, bog, swamp, riparian zone, or wet tundra. As shown in Figure 1, marshes, bogs, swamps, and riparian zones exist at most latitudes, but tundra wetlands are found only in the Arctic region.

Freshwater and saltwater **marshes** develop where land is covered by 15 to 300 centimeters of water. Marsh plants include grasses, sedges, rushes, cattails, reeds, and some aquatic plants such as water lilies that are found on top of the water surface (see Figure 2, page 216).

Bogs are characterized by acidic waters, insectivorous ("insect-eating") plants, peat mosses, and peat accumulations. *Peat* forms when bunches of dead mosses and other plants fail to decay completely because acidic conditions and the absence of oxygen inhibit the growth and action of decomposers. Growth of nitrogen-fixing nodule bacteria is also restricted in acidic bogs where pH values fall below 3.5. Such conditions favor the growth of insectivorous plants such as round leaf sundew, cobra lilies, and pitcher plants (see Figure 3, page 216). These plants satisfy their nitrogen requirements by capturing and digesting insects.

Shrub and wooded **swamps** often occur along sluggish lotic ecosystems, or within their flood plains, in upland flat areas containing depressions that fill with water, or around shallow lentic systems. Swamp soil is saturated with water during much of the year, and trees are commonly seen standing in 30 to 60 centimeters of water. Northern wooded swamps harbor red maple, white cedar, black spruce, black ash, and a few other tree species, whereas in the southern states, cypress, black tupelo, black gum, and water oak are common (see Figure 4, page 217).

The Arctic **tundra** contains vast wetland expanses, including nearly 60 percent of Alaska, and significant areas of Canada, Greenland, northern Europe, and Asia (see Chapter 7). Tundra wetlands are usually dominated by lichens, peat mosses, some grasses and sedges, and dwarf woody plants. Permafrost restricts root growth and also prevents wetland drainage because water cannot seep through permanently frozen ground.

Riparian zones occur along the edges of most rivers, and in North America they often consist of areas with dense vegetation. "Bottom lands" are riparian zones found along most rivers that cross flat lowlands in the southern United States. More than 100 species of woody shrubs and trees grow in the bottom lands along rivers from eastern Texas to Virginia.

Other Wetlands

Coastal saltwater and freshwater *tidal marshes* are found in most eastern and southern coastal states. The largest tidal marshes occur in the mid-Atlantic region and along the Texas and Louisiana Gulf coasts. In the mid-Atlantic states, freshwater wetlands known as "pocosins" and "Carolina bays" are composed of boglike vegetation communities that grow on waterlogged soils. Gulf Coast wetlands are characterized by extensive marshes and forested swamps in Louisiana and Mississippi, and a saltwater lagoon marsh system along the southern Texas coast.

Mangrove swamps cover about 14 million hectares (1 hectare = 2.47 acres) of the world's coastal zones located between 25° north and 25° south latitudes. Over 80 species of mangrove trees and shrubs form a variety of different communities in these wetlands. In North America, mangrove swamps are confined to Florida, the Caribbean, and the Gulf Coast of Mexico. Those in Florida are near the northern limit of their climatic range; these wetlands are usually occupied by small shrubs that grow in discrete, widely separated clumps.

Wetland Values and Functions

Wetlands have intrinsic value for many people. In heavily populated areas, wetlands are often the only natural habitats that retain a wilderness character, and their abundant flora and fauna provide extensive recreational opportunities for city dwellers. Wetlands also provide important ecological services. Man-

FOCUS ON SCIENTIFIC PROCESS

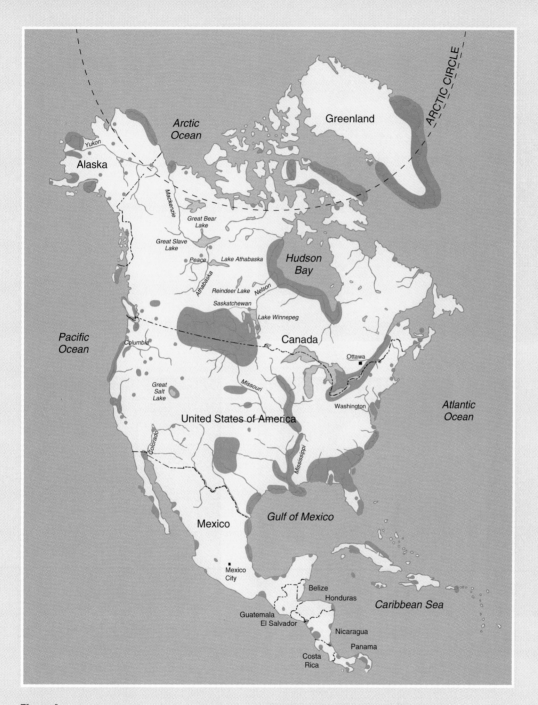

Figure 1   The dark regions of the map indicate wetland areas in Greenland, North America, Central America, and the Caribbean. The geographic areas of the United States that have the most extensive wetlands are the "pothole" area along the north central border, the Playa Lake area of Texas and Oklahoma, and the bottom lands along rivers in southeastern states. (From Finlayson and Moser, 1991.)

box continues

FOCUS ON SCIENTIFIC PROCESS

Figure 2 Rushes and water lilies are common marsh plants.

A

B

C

Figure 3 Bog plants such as round leaved sundrews *(Dorsera rotundafolia)* (A), cobra lilies *(Darlingtonia californica)* (B), and pitcher plants *(Sarracenis purpurea)* (C), all have different structural adaptations used in capturing insects, which provide the plants with nitrogen after they are digested.

grove swamps prevent coastal erosion, cypress and other swamps act as sinks for various pollutants, river bottom lands trap sediments and control flooding, and peat bogs act as global carbon dioxide sinks. In addition, wetlands are critically important ecosystems in the life cycles of many species. For example, coastal marshes provide habitats at some stage of life for over two-thirds of the commercial fish and shellfish harvested in the United States. Also, northern and prairie marshes provide nesting and feeding habitats for the majority of North American shore birds and water fowl.

Wetland Conversions

When the U.S. Declaration of Independence was signed in 1776, there were nearly 90 million hectares of wetlands in the area that was to become the contiguous 48 states. Now only about 40 million hectares remain undeveloped.

FOCUS ON SCIENTIFIC PROCESS

Figure 4   Bald cypress *(Taxodium distinctum)* grows in swamps that were once far more extensive than they are today in many southern and southeastern states.

Most wetland losses involved conversions to agricultural uses. However, human-caused alterations have not been uniform, and some states now have only small remnants of their original wetlands: For example, California has only 9 percent of its original wetlands; Ohio, 10 percent; Iowa, 11 percent; and Indiana and Missouri, each 13 percent. In addition, more than half of Florida's everglades and half of Connecticut's coastal marshes have disappeared. Each year, in the continental United States, 122,000 hectares of wetlands are converted to agricultural, industrial, commercial, and residential uses. If this conversion rate continues, an additional area of wetlands greater in size than the state of Delaware will be lost in less than five years.

On a global scale, wetland conversions are often associated with increasing human population size. At present, about one-third of the world's urban population lives within 60 kilometers of a coast. Between 1980 and 2000, this coastal urban population is expected to grow by 380 million people, an increase about equal to the total North American population in 1995.

Wetland Regulation

What will prevent the continual loss of biologically important wetlands in the United States? Wetland conversions to agricultural and other uses are now regulated by both state and federal legislation. State laws vary greatly among the 50 states, but federal laws apply to every state in the nation. Federal laws fuel controversies when their implementation appears to interfere with private property rights. Section 404 of the 1972 Federal Water Pollution Control Act, "swampbuster" provisions of the 1985 Food Security Act, and the 1990 Food, Agriculture, Conservation and Trade Act all restrict wetland conversions on both public and agricultural private lands. The U.S. Army Corps of Engineers is authorized by Section 404 to regulate the disposal of both dredged and fill materials into all waters of the United States, including wetlands. The implementation of Section 404 has reduced wetland loss each year by 24 to 39 percent. Under the 1990 swampbuster statutes, farmers can be fined from $750 to $10,000 if they drain protected wetlands. Swampbuster provisions may also deny farm program benefits to farmers who convert any wetlands to dry agricultural land. Such restrictions may apply to loans from the Commodity Credit Corporation, federal crop insurance coverage, deficiency payments, and other federal programs. In addition, the Title IV Tax Reform Act of 1986 reduced tax liability deductions of drainage expenses and taxed gains on sales from converted wetlands at higher rates.

With tough federal and state laws now in place, why are wetlands still being converted to other uses? Wetlands owned by federal and state governments do have a high degree of protection but, because most wetlands in the United States are privately owned, vast areas are not covered by drainage protection laws. Should society's right to an uninterrupted flow of benefits from wetlands outweigh the private property rights of wetlands owners? Should new laws be passed that regulate private wetlands? If so, should the landowners be compensated? How much? Answers to these and other pertinent questions will not be based exclusively on scientific research. Wetlands not only have ecological value, but they also represent important social, commercial, and political resources. It seems likely that the future of wetlands may depend more on social and political decisions than on issues raised through ecological studies.

SUMMARY

1. Aquatic ecology includes studies of lotic and lentic freshwater systems and estuarine, neritic, and open-ocean marine systems. The amount of light energy and availability of the nutrients are the most critical limiting factors in aquatic systems.

2. The order of a stream increases by one unit when two streams of the same order meet. Insect shredders in headwater stream communities use CPOM; insect collectors use FPOM as major energy sources. As the stream order increases, shredders are replaced by grazers that feed on periphyton. Only mid-order streams and rivers have a P/R greater than 1. Drift not removed during its seaward movement will settle in estuarine sediments, adding to their biological richness.

3. Turnover increases the productivity of temperate lakes by recharging surface waters with nutrients twice each year. Lakes are classified as oligotrophic, mesotrophic, natural eutrophic, and cultural eutrophic, depending on community productivity.

4. A functional model of lake food webs assigns the organisms to a number of trophic guilds, depending on their food habits. Changes in these guilds can change the rate of PO_4^{-3} cycling, which can lead to a modification of the trophic structure and can regulate the function of the entire community. Most lakes are temporary because, given enough time, natural successional events will cause them to be replaced by terrestrial ecosystems.

5. Cold arctic and antarctic waters and areas along some continental margins are the most productive marine environments. In these areas, mechanisms recharge the euphotic zone with nutrients liberated by the decomposition of organic matter on the ocean floor. Most other ocean areas have no mechanism for this type of nutrient transport and are, therefore, considered to be nutrient deserts.

6. Estuaries are among the most productive aquatic areas, but their proximity to major areas of human habitation has resulted in the significant degradation of many of these important natural ecosystems.

WORKING VOCABULARY

aquatic ecosystems (p. 195)
cascading trophic interactions (CTI) (p. 203)
estuaries (p. 210)
eutrophic (p. 202)
lentic ecosystems (p. 201)

lotic ecosystems (p. 195)
production to respiration ratio (P/R) (p. 197)
upwelling (p. 206)
wetlands (p. 214)

REVIEW QUESTIONS

1. What are the classes of aquatic and marine ecosystems?

2. How are lotic ecosystems classified?

3. How does Cummins's functional model explain energy flow in lotic ecosystems of different sizes? What are the types and roles of different classes of insects in the model? What is the significance of P/R in the model?

4. What is drift? Why is it important?

5. What factors influence lentic ecosystems?

6. How are lakes classified? What are some important regulating factors in lentic ecosystems?

7. What does the theory of cascading trophic interactions explain?

8. How are marine ecosystems classified?

9. What is upwelling? Where does it occur, and why is it important?

10. Why are neritic ecosystems generally richer biologically than oceanic ecosystems?

ESSAY AND DISCUSSION QUESTIONS

1. Assume that you are an environmental consultant responsible for preparing an impact statement. Predict the direct impact of logging (removing trees from) watersheds of first- and second-order streams. Also project the effects on larger streams and rivers into which these streams flow.

2. How would you rate the vulnerability of the following aquatic ecosystems to disruption by human activities (from most sensitive to least sensitive): an second- order stream, a small bay, coral reefs, the Mississippi River, a small lake, Lake Superior, Chesapeake Bay, the open ocean between Hawaii and the mainland? What criteria would you use in making such a list?

3. Some ecologists believe that it is easier to conduct "good" studies of aquatic systems than of terrestrial ecosystems. Do you agree? What factors might be considered in coming to such a conclusion?

REFERENCES AND RECOMMENDED READING

Allan, J. D., and A. S. Flecker. 1993. Biodiversity conservation in running waters. *BioScience,* 43: 32–43.

Baker, M., and W. Wolff (eds.). 1987. *Estuarine and Brackish-Water Sciences Association Handbook.* Biological Surveys of Estuaries and Coasts. Cambridge: Cambridge University Press.

Barnes, R., and R. Hughes. 1988. *An Introduction to Marine Ecology.* 2d ed. Oxford: Blackwell.

BioScience. 1995. Special Section Articles: Ecology of large rivers. *BioScience,* 45: 134–205.

Carpenter, S., and J. Kitchell. 1988. Consumer control of lake productivity. *BioScience,* 38: 764–769.

Carpenter, S., J. Kitchell, and R. Hodgson. 1985. Cascading trophic interactions and lake productivity. *BioScience,* 35: 634-639.

Cummins, K. W. 1977. From headwater streams to rivers. *American Biology Teacher,* 5: 305–312.

Cummins, K. W., M. A. Wilzbach, D. M. Gates, J. B. Perry, and W. B. Taliaferro. 1989. Shredders and riparian vegetation. *BioScience* 39: 24–30.

Cushing, C. E., Cummins, K. W., and G. W. Minshall. 1995. *River and Stream Ecosystems*. New York: Elsevier.

Dennison, W. C., R. Orth, K. Moore, J. Steverson, V. Carter, S. Kollar, P. Bergstron, and R. Batiuk. 1993. Assessing water quality with submerged aquatic vegetation. *Bioscience,* 43: 86–94.

Findlayson, M., and M. Moser. 1991. *Wetlands.* New York: Oxford.

Horton, T., and R. W. Madden. 1993. Chesapeake Bay—hanging in the balance. *National Geographic,* 183: 2–35.

Mitchell, J. 1992. Our disappearing wetlands. *National Geographic,* 182: 3–45.

Nicol, S., and W. Mare. 1993. Ecosystem management and the antarctic krill. *American Scientist,* 81: 36–47.

Northcote, T. 1988. Fish in the structure and function of freshwater ecosystems: A "top-down" view. *Canadian Journal of Fisheries and Aquatic Science,* 45: 361–379.

Spencer, C. N., B. R. McClelland, and J. A. Stanford. 1990. Shrimp stocking, salmon collapse, and eagle displacement—cascading interactions in the food web of a large aquatic ecosystem, *Bio-Science,* 41: 14–21.

Turner, R. E., and N. N. Rabal. 1991. Changes in Mississippi River water quality this century. *BioScience* 41: 140–147.

World Bank. 1995. *Restoring and Protecting the World's Lakes and Reservoirs.* Washington, D. C.

ANSWERS TO FIGURE QUESTIONS

Figure 11.2 Because light is limited and CPOM is maximum in first-order streams, there would be many shredders and few, if any, grazers.

Figure 11.3 Rainbow trout would be second- or third-order consumers in river ecosystems. Top carnivores—such as bull trout, river otter and other mammals; and eagles, osprey, and other birds—would feed on trout.

Figure 11.4 Fish abundance would be greatest in areas where the P/R is greater than 1 (mid-order rivers).

Figure 11.5 This is most likely a seventh- or eighth-order river.

Figure 11.6 These sediments would have come from eroded soils along the Mississippi River and its tributaries.

Figure 11.7 The littoral zone would have the greatest amount of nutrients and available light; thus it would contain the greatest abundance of producers.

Figure 11.8 As lake succession occurs, the trophic classification changes from oligotrophic to mesotrophic to natural eutrophic, then to a swamp or marsh, and finally to a meadow.

Figure 11.9 Bacterial decomposition of excessive primary productivity reduces the oxygen concentration in eutrophic lakes.

Figure 11.10 If large adult fish were removed, the minnows and other small fish would increase in density. This would lead to a decrease in density of crustaceans and their waste products, which would decrease the rate of phosphate cycling in the limnetic zone.

Figure 11.12 This is an oligotrophic lake that has a very low concentration of nutrients. The rate of succession in this lake would be much slower than in a eutrophic lake.

Figure 11.13 In time the site may become a meadow and, later, a forest.

Figure 11.14 Photosynthesis occurs only in the presence of light; therefore, producers could live only in the littoral zone, sunlight zones, and in some twilight zones of the neritic and oceanic regions.

Figure 11.15 Whales would be most abundant in regions where upwelling occurs because only in these locations are sufficient nutrients transported into the sunlight zone to support their food webs.

Figure 11.16 Coral reefs occur in shallow seas where there is abundant sunlight and wave actions can circulate nutrients from the sea floor back into the sunlight zone.

Figure 11.22 Nutrients are transported from terrestrial and lotic ecosystems to bays and estuaries, where they settle out as water flow decreases. The concentrated nutrients allow abundant growth to occur in these unique habitats.

Human Populations and the Environment

Chapter Outline

Reading Questions

1. What factors have influenced the size of human populations throughout history?

2. What factors may act to limit human population growth?

3. In what ways have human populations affected the environment?

4. What types of technological solutions to biological problems have resulted in new environmental problems?

The emergence of early human populations from the Rift Valley in East Africa set the stage for the development of a species that would radiate into all of the major biomes of Earth. We humans, who evolved out of the dim past and were "fruitful and multiplied," now exceed 5.7 billion. That number will likely double in the next four decades. Numerous factors have contributed to this phenomenal growth, including the development of agriculture, the industrial revolution, advances in medicine and public health, the green revolution of the 1960s, and the genetic revolution occurring today. The effect of these factors on the population growth rate is illustrated in Figure 12.1.

EARLY HUMAN SETTLEMENT

For millennia, the environmental impact of *Homo sapiens* was related to the increasing densities of various human populations that inhabited Earth. The success of the human species at producing offspring often resulted in population densities that

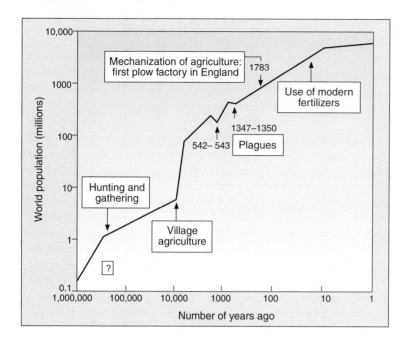

Figure 12.1 The history of human population growth expressed as a logarithmic plot.

exceeded the food-producing capacity of the area in which humans lived. This led to famine or the emigration of part or all of the affected population to less used locales.

Certain ancient human populations led an existence that enabled them to obtain food from a dependable source. They did this by correlating their movements with those of certain animals. Many species of large herbivores, such as the bison and caribou of North American and the wildebeests of Africa, evolved a survival strategy that allowed their population densities to exceed what any given habitat could long support. When an area became devoid of important nutrients, the herds migrated, following the "march of the seasons" to greener fields (see Figure 12.2). Early human populations often developed a similar nomadic lifestyle. They followed these migratory herds on their annual treks or settled along their routes of

migration and exploited the animals as food resources during specific periods of each year.

The oldest known permanent human settlement sites date from approximately 11,500 years ago. They appeared between the end of the Pleistocene glaciation and the early interglacial Holocene epoch, a time when the climate began to warm. What factors led humans to exchange their nomadic existence for life in a village? How did survival become dependent on cultivating food resources as opposed to harvesting available food? Was the development of agriculture a cause or an effect of settlement?

These questions cannot be answered completely because information is lacking. The story is complicated by the fact that human history has been influenced by cultural as well as biological factors. A unique attribute of our species' evolution was the development and application of human culture. In the

Figure 12.2 Many species of large herbivores have evolved strategies that include extensive migrations on an annual basis. The wildebeests in East Africa move in such patterns, following more favorable forage conditions as the seasons change.

Question: *What changes would you expect in the wildebeest population if their migration were prevented?*

broadest sense, **culture** includes languages, social structures, value systems, and the development of tools and their use in agriculture. Human culture can be viewed as an adaptive human trait, but many anthropologists are skeptical that adaptation is all that is involved in culture. In recent years, human culture has been treated as a partly independent factor of human evolution constrained, but not fully determined, by biological limits.

In this text, we are concerned primarily with the physical and biological aspects of human population effects. It is important to understand, however, that human populations at different times have chosen to limit or expand their size for reasons that are not strictly related to biological concerns. Regardless of the choices they make, human populations are still limited by biological factors. For example, a population may decide to quadruple in size, but if it lacks food or other necessary resources, its decision cannot be implemented. Similarly, if millions of people want to drive to work every day, at some point they may have to face the realization that the air they breathe is contaminated by this action.

One of the physical factors hypothesized to have been important in the development of permanent human settlements is global climate change. Warming climates developed after the continental glaciers retreated. Consequently, native flora and fauna flourished in many areas, which may have induced humans to settle in villages there. As the villages' population densities increased, environmental damage often began to occur in the surrounding areas used for hunting and gathering food. With the food supplies diminished, human options were probably limited: the population could starve, emigrate, or attempt to increase the production of some important plants and animals that were used for food. This last choice may have been the impetus for the development of agriculture. Early farming allowed settlements to exist for long periods of time, but available evidence indicates that decreased soil fertility and changing climate, coupled with ever-increasing population densities, led to the collapse of most prehistoric villages.

> **BEFORE YOU GO ON** Early human populations were nomadic because humans obtained much of their food from killing and eating migratory animals. Approximately 11,000 years ago, humans began to settle into villages. The reasons for this life-style change are not known with certainty, but cultural, biological, and physical factors may have been involved. Expanding populations led to environmental damage that ultimately forced these early villages to be abandoned.

HUMAN POPULATION GROWTH THROUGH TIME

Growth of global human populations was very slow for more than 15,000 years. Then, beginning in the seventeenth century, the population began to grow at a very rapid rate. If this high growth rate continues, the global human population could reach 10 billion—almost double what it is now—within your lifetime.

Prehistoric Human Population Growth

By the end of the Pleistocene glaciation, the human species had radiated from its cradle of evolution in eastern Africa to inhabit most of the major continents of Earth. The estimated world population numbered about 5 million people, who generally subsisted in small hunter–gatherer tribes. Because of the application of primitive agricultural techniques, humans were able to multiply, and the world population increased slowly for the next 9,500 years. About 2,000 years ago, the rate of population growth slowed and remained relatively constant for the following 1,000 years. During this millennium, few major advances occurred in agriculture. Periodic famines during times of drought were common. Although very impressive "high" civilizations (such as those of the Greeks, Romans, and Arabs) existed during this time, most people lived in appalling conditions. Poor sanitary practices resulted in epidemics of typhus, smallpox, plagues, and other serious diseases. These diseases caused an increase in the mortality rate, and as it approached the birth rate, world population growth approached zero.

Historic Human Population Growth

Figure 12.3 shows population data graphed on an arithmetic scale, but this approach fails to divulge details about the nature of human population growth over long periods of time. However, if the arithmetic data are converted to logarithmic numbers, the resultant graph, as already shown in Figure 12.1 (see page 221), is much more revealing. The logarithm of a number is the power to which 10 must be raised in order to produce the number; for example, $100 = 10^2$, so $\log (100) = 2$. The logarithmic plot in Figure 12.1 shows clearly many of the major events that occurred throughout the history of human population growth.

Village agriculture contributed to the growth of early human populations by increasing the amount of available food (see Figure 12.4). Figure 12.1 describes this general trend and also reflects some other significant historical events, including the effects of two major disease plagues. The plagues of 542–543 A.D. and 1347–1350 A.D. caused a temporary decrease in the density of the world's human population, but they had little long-term effect on population growth.

The total human population stood at about 500 million by the year 1650, when the rate of growth began to increase. A number of factors are likely to have contributed to this increase. European populations began to emigrate, especially to Africa, Oceania (Australia and the Pacific islands of New Zealand, Melanesia, Micronesia, and Polynesia), and the New World (North and South America). Nutrition began to improve as a result of new developments in agriculture. The death rate began to decrease—not yet because of medical advances, but most likely because of declines in the severity of diseases. Public health and sanitation advances in the nineteenth century extended the human life span, and by that century's end, medicine had developed into a scientific art that began to alter life expectancy dramatically.

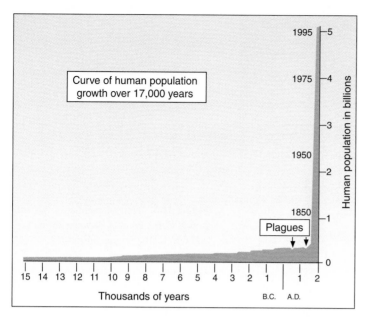

Figure 12.3   An arithmetic graph of human population growth during the past 17,000 years.

Figure 12.4   As indicated by this ancient manuscript, early village life may have developed as a result of advances in crop production and animal husbandry.

Question: *What factors may have led to the failure of prehistoric villages?*

Table 12.1   Human Population Trends, 1900–2025

	Population (Millions)			
	1900	**1950**	**1995**	**2025**
Developing Regions (Total)	1070	1682	4527	7140
Africa	133	222	747	1597
Asia[a]	867	1294	3287	4785
Latin American	70	166	493	758
Developed Regions (Total)	560	835	1243	1365
Europe, Japan, former Soviet Union, Oceania[b]	478	669	957	1033
Canada, United States	82	166	286	332
World (Total)	1630	2517	5770	8505

[a]Excludes Japan
[b]Includes Australia and New Zealand

Sources: Merrick, T. W. (1986), p. 16; *World Resources* (1992–93), pp. 246–247.

BEFORE YOU GO ON   Around 11,000 to 12,000 years ago, the world held about 5 million people. Through the next 10,000 years, global human population increased slowly, limited largely by food and disease. By 1650, some 500 million humans inhabited Earth, and at that time the population began to increase sharply because of an increased food supply and a decline in mortality caused by disease. This accelerated rate of growth continued until recently.

The global human population increased to 1.6 billion by 1900, and the rapid, exponential nature of its growth continued worldwide into the late 1960s. **Exponential growth** (also called *geometric growth*) occurs when the growth rate of a population can be described mathematically by the formula 10^x, where x (the exponent) is some computed number that defines the rate of growth. Exponential growth still characterizes populations of many developing regions, and it is expected to continue into the foreseeable future (see Table 12.1). The data from Table 12.1, when graphed, indicate a great disparity in growth rates between the developing and developed regions of Earth (see Figure 12.5, page 224).

Modern Human Population Growth

In the late 1960s, the world rate of human population growth began to decrease (see Figure 12.6, page 224). However, many countries in developing regions of the world continue

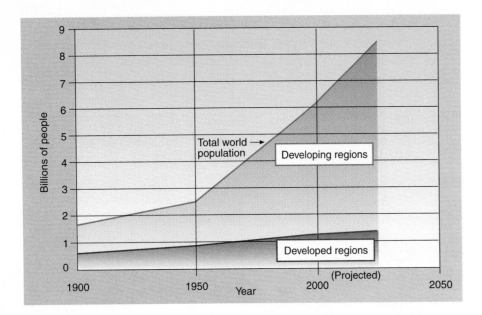

Figure 12.5 Human population growth trends differ greatly between developed and developing regions of Earth.

Question: *Is there a limit to the size of human population that can be supported by the biosphere?*

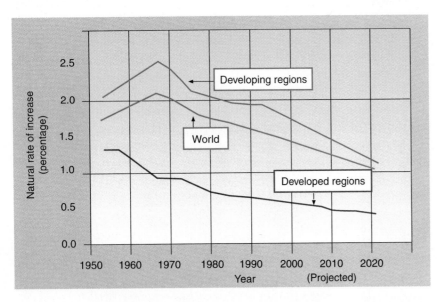

Figure 12.6 The average annual rate of human population growth for developing and developed regions and the entire world is expected to decrease.

to increase their populations at high rates. For example, the growth rate throughout most of Africa is expected to exceed 3 percent per year until the end of this century. A growth rate of 3 percent translates into a **doubling time** (the number of years it takes a population growing at a given rate to double in size) of 23 years. Thus the total population in Africa is projected to double between 1995 and 2018 (see Table 12.1). Other developing areas where high population growth rates persist include ten Middle Eastern Arab nations and the Central American countries Belize, Guatemala, Honduras, and Nicaragua.

Differences in the densities of humans in developing and developed regions of Earth have also increased in relation to their changing populations. Past, present, and expected densities are shown in Table 12.2. The land area will remain constant at about 95 million square kilometers for the developing regions and at about 61 million square kilometers for the developed regions. If the rates of population increase remain constant in the developing regions, their density will become

Table 12.2 Human Population Density in Developing and Developed Regions, 1900–2025

	Population Density[a]			
	1900	**1950**	**1995**	**2025**
Developing Regions[b]	11	18	48	75
Developed Regions[c]	9	14	20	22

[a]Values are expressed as numbers of people per square kilometer
[b]*Developing regions* are Africa, Asia, and Latin American (total area = 95 million square kilometers).
[c]*Developed regions* are Europe, the former Soviet Union, Japan, Oceania, Canada, and the United States (total area = 61 million square kilometers).

greater than 35 people per square kilometer by the year 2025 (see Figure 12.7). Is this likely to occur? Will some factor intercede to reduce this potential population explosion?

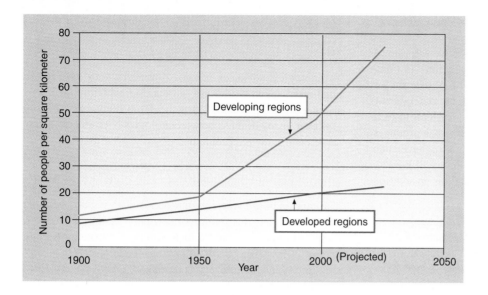

Figure 12.7 The density of humans is expected to continue increasing in the developing regions of Earth but to plateau in the developed regions. (Data are from Table 12.2.)

In an even more extreme case, the rate of natural increase now approaches 4 percent in some developing countries, the most prominent being Kenya in eastern Africa. In mid-1992, the growth rate in Kenya was 3.7 percent, which computes to a doubling time of 19 years. By contrast, in mid-1992, France, a developed country covering an area similar to that of Kenya, had a rate of natural increase of 0.4 percent and a doubling time of 175 years. The expected change in density between these two countries is shown in Table 12.3. In 1992, the population density of France was greater than twice that of Kenya, by 2017 they will be equal, and by 2040 the density in Kenya will be twice that in France. Are there factors that will prevent this extreme population growth from occurring in Kenya?

Table 12.3 Projected Population Densities in Kenya and France, 1992–2049

	Population Density[a]			
	1992[b]	**2011**	**2030**	**2049**
Kenya	45	98	181	361
France	104	114	123	134

[a]Values are expressed as numbers of people per square kilometer.
[b]In 1992 the rate of increase, r, was 3.7% in Kenya and 0.4% in France. If these rates remain constant, Kenya's population density will double every 19 years, whereas the population of France will double in 175 years, or by the year 2167.

Malthusian Limits to Human Population Growth

In 1798, the English clergyman Thomas Robert Malthus published the first edition of *An Essay on the Principle of Population.* He based his essay on two assumptions:

> First, That food is necessary to the existence of man. Secondly, That the passion between the sexes is necessary and will remain nearly in its present state. . . . Assuming then, my postulata as granted, I say, that the power of the population is

indefinitely greater than the power in the earth to produce subsistence for man.

Malthus was aware of the disparity between the rate of population growth and the rate at which the food supply could be increased. "Population, when unchecked, increases in a geometrical ratio. Subsistence increases only in an arithmetical ratio." He concluded:

> The only true criterion of a real and permanent increase in the population of any country is the increase of the means of subsistence. . . . Famine seems to be the last, the most dreadful resource of nature. The power of population is so superior to the power in the earth to produce subsistence for man, that premature death must in some shape or other visit the human race.

Malthus assumed that further increases in human populations would greatly exceed the ability of agriculture to feed all the people. However, he did not anticipate the impact of the industrial revolution that was dawning throughout Europe. Mechanization during the nineteenth and twentieth centuries allowed agricultural production to keep pace with the increasing human population in most Western countries. Nevertheless, numerous famines did occur in less developed regions, but migrations of significant numbers of people to the Western Hemisphere often reduced their impact.

The Green and Genetic Revolutions and Malthus

The **green revolution** of the 1960s was built on the selective propagation of high-yielding varieties of cereal grains such as wheat and rice. These new varieties helped especially in feeding the burgeoning populations in Asia and Latin America. When treated with modern chemical fertilizers and pesticides, crop yields often increased two- or threefold. In many respects, the green revolution of the past three decades has allowed the world population to surpass 5.7 billion people. We have now entered the next revolution, the **genetic revolution,** in which advanced techniques of modern science and technology are further amplifying the production of plant and animal varieties.

Malthus, therefore, was wrong in his predictions, not only because of his assumptions about the rate of population growth but also because of his assumptions about the growth rate of food production. Will the "power of the population," in time, still exceed the ability of agriculture to supply subsistence for humans? That is, will our human population ultimately grow to a size that exceeds the ability of Earth's ecosystems to feed us? Of equal importance, why does a world that produces enough food to feed the entire human population have such extensive famine? The answers to this question will have to come through cultural, political, and social actions, in addition to biological research.

BEFORE YOU GO ON   In 1798, Malthus described a numerical relationship that showed a great difference between the rate at which human populations grew and the rate at which food was produced. He felt that unless human population growth was slowed, mass starvation was inevitable. However, during the nineteenth and twentieth centuries, technologies were developed that increased rates of food production in the world to a level that kept pace with population growth.

HUMAN IMPACTS ON DRINKING WATER

Early environmental problems caused by our ancestors were related primarily to the degradation of agricultural areas adjacent to developing villages, towns, and cities and to the effects of inadequate disposal of human wastes. Human feces and kitchen wastes accumulated in or near the inhabited areas, creating squalid conditions. Such sites were ideal breeding grounds for vermin, which transmitted infectious diseases. These rats and fleas were infected with the bacteria that caused the famous plagues of the sixth and fourteenth centuries. However, in attempting to rid their villages of garbage, early city dwellers often exchanged one environmental problem for another. In general, they dumped their wastes into the nearest aquatic system, usually the river on which the city was located. This often led to reduced fish populations and water that became unsafe for drinking. Contaminated waters were also the major sources of typhus, cholera, and many diarrheal diseases. With a few notable exceptions—such as the Minoans (ca. 3000 – 1100 B.C.), the Romans (ca. 100 B.C. – 400 A.D.), and Arab society of the tenth and eleventh centuries, all of whom developed workable sewage removal systems—most early planners did not triumph in their attempts to create cleaner, safer villages and cities.

In developed regions, water pollution continued to be a major environmental problem well into the twentieth century, and it remains a critical problem in many developing regions. With the application of modern sewage treatment and disposal, however, most developed countries have now succeeded in reducing the degrading effects of human municipal and industrial wastes in their aquatic systems.

TECHNOLOGY AND MODERN ENVIRONMENTAL PROBLEMS

Technological advances in agricultural practices have led to huge increases in world crop production, which in turn have sustained the exponential growth of the human population. However, many of these new technologies have created a new set of environmental problems.

Environmental Consequences of the Green Revolution

The success of the green revolution resulted, in part, from the development and distribution of new plant varieties. However, these new crops required the expenditure of considerably more energy in preparing fields, pumping irrigation water, manufacturing synthetic fertilizers and pesticides, and applying them to fields and forests. These new agricultural methods allowed vast areas to be planted with single-species crops known as **monocultures.** These areas, with their uniform crops, not only provided food for humans, but they also became expansive tracts of unlimited food supplies for agricultural pests such as insects, soil nematodes, and other destructive species (see Figure 12.8).

Pesticides are substances that kill unwanted or harmful organisms. Pesticides developed before and during World War II were adapted for agricultural use and significantly reduced the loss of crops attributed to many pests. One class of **insecticides** (substances that kill insects)—the chlorinated hydrocarbons that include the well-known DDT—proved to be nearly 100 percent effective against many insect pests when first used. Also, the application of **herbicides** (substances that kill plants) reduced competition from weeds. The liberal application of these pesticides provided the foundation for increasing crop production and managing pests. During the 1960s, however, it suddenly became apparent that the concentrations of pesticides required for acceptable pest control had severe environmental consequences.

Pesticide Resistance

After a decade or so of use, stronger pesticide concentrations had to be used because a small number of plant and animal pests survived each application. These survivors then gave rise to new populations that could tolerate pesticide applications in ever-increasing numbers. Eventually, certain pest populations developed **pesticide resistance** to a number of pesticides, some of which were used at remarkably high concentrations. Thus the ability to manage their numbers slipped from control (see Figure 12.9).

Pesticide Effects in Nontarget Species

DDT and other chlorinated hydrocarbon insecticides were effective because they had two important characteristics. They were *broad-spectrum insecticides,* which means that they kill most insect species, and they had very low water solubilities

Figure 12.8 Modern agriculture in developed regions often uses large areas for monoculturing more valuable crops, such as the wheat being harvested here.

Question: *What happened to natural species diversity in areas such as this?*

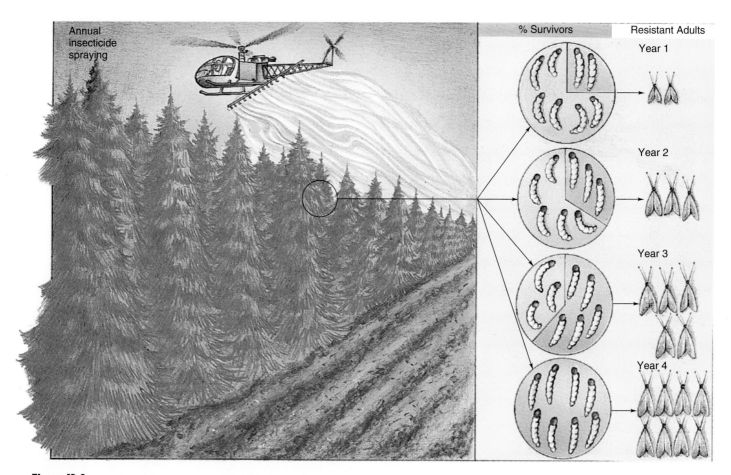

Figure 12.9 Spruce budworm larvae feed on the new foliage of many coniferous tree species. Spraying forests with DDT caused high budworm mortality, represented by the curved larvae. However, a few individuals that were resistant to the insecticide, represented by the straight larvae, survived. After several generations, some budworm populations were highly resistant to the insecticide, and either much larger doses or different insecticides had to be used if desired insect mortality was to be achieved.

Question: *How may nontarget organisms be affected when larger pesticide doses are required to kill resistant target species?*

(that is, very little DDT will dissolve in water). The low water solubility of DDT allowed it to remain effective for a long time because rains would not wash it from the plant foliage and it was less likely to be broken down by natural mechanisms. Thus it would often continue to kill many insects, including those that were not pests, for days or weeks after application.

In the absence of chemical insecticides, insect pests are normally subjected to the assaults of insect predators, parasites, and disease organisms. Unfortunately, many of these beneficial insects are more sensitive than crop pests to synthetic insecticides. For example, the wasps that parasitize spruce budworm are much more sensitive to insecticides than the budworms themselves. Consequently, when chlorinated hydrocarbons such as DDT were used, these natural systems of control were squandered.

Biological Magnification

Bird populations make up another important element of natural insect pest control. By the early 1960s, evidence was accumulating that several bird populations were being adversely affected by high concentrations of DDT. Terrestrial raptors (birds of

prey), including owls, hawks, ospreys, and eagles, as well as many marine predators, such as cormorants and pelicans, were among the species being affected by increasing environmental DDT concentrations (see Figure 12.10). Why were birds so vulnerable to DDT?

In large-scale spraying operations, DDT was applied to crops and forests. Scientists now know that once sprayed into the environment, DDT moved from one trophic guild to the next, being stored in the lipids (fats and oils) of each organism. As a result, body burdens of DDT increased within each successive trophic guild in the affected food chains, a process known as **biological magnification** which is described in Figure 12.11. The highest concentrations generally occurred in birds feeding at the top of DDT-fouled food chains. These birds, most notably the peregrine falcon, developed a condition known as the thin-eggshell syndrome. When body burdens of DDT reached high levels, females produced eggs with abnormally thin shells. Consequently, these fragile eggs were often broken by the parent birds during nesting. This resulted in low hatching success and sharp declines in the population densities of many birds of prey.

Figure 12.10   (A) Ospreys and (B) brown pelicans feed at the top of aquatic food chains by eating fish. They were vulnerable to biological magnification of DDT and other chlorinated hydrocarbon insecticides. Consequently, their population densities declined largely because of reproductive failures caused by insecticides.

Question: *Why would predators at the top of food chains be more affected by DDT than species lower in food chains?*

A

B

Figure 12.11 Food chain concentrations of insecticides like DDT occur as a result of its low water solubility and its high affinity for oils and fats. The DDT concentrations in aquatic plants may be very low, but when the insecticide passes to each successive trophic guild, concentrations increase until, at the top of a food chain, they may be high enough to harm top-level carnivores.

Question: *If DDT were highly soluble in water would it be biomagnified through food chains?*

Finally, it was discovered that DDT had been distributed over the entire globe by wind transport mechanisms. Thus after massive spraying operations conducted over two decades, even polar bears in pristine arctic regions were found to have measurable quantities of DDT in their flesh.

As a result of all these negative environmental effects—biological magnification as well as persistence, toxicity to nontarget species, and worldwide transport—the U.S. Environmental Protection Agency banned the use of DDT in the United States in 1972 and prohibited the use of other chlorinated hydrocarbon insecticides by 1975. Nevertheless, DDT continues to be used in tropical regions of the world to control malaria-spreading mosquitoes. For example, 19 million kilograms of DDT are now used for malaria control in India each year. About 80 percent of it is used for mosquito control in domestic houses and cattle sheds in rural areas. Concern is building about significant DDT contamination of stored grains. By eating these stored, contaminated foods, people may raise their tissue DDT levels to nearly twice the currently acceptable U.S. standard. Administrators in these countries, however, have decided that the severe economic and personal costs of malaria override the environmental and human consequences of the DDT contamination.

BEFORE YOU GO ON Contamination of water with human wastes and related effects were early environmental problems associated with population growth. In the twentieth century, technological solutions to biological problems resulted in new environmental impacts. Many of these environmental consequences were related to the extensive use of chemical pesticides, especially DDT.

The Future of Agriculture

The success of the green revolution has altered world agriculture in many positive ways. Millions of people have avoided famine, and new varieties of rice and wheat remain major staples throughout much of the developing world. In many areas, including China and India, more than half of the available cropland consists of monocultures of these two grain crops. However, too much of a good thing can often have negative economic and environmental consequences. For example, global rice production now often exceeds demand. As a result, prices have fallen to very low levels, and many farmers in both developed and developing countries are left with excess crops that have market values less than their costs of production.

Monoculturing of grains leads to a dependency on high-energy agriculture, including synthetic fertilizers and pesticides, irrigation, and high transportation costs. The potential for crop failure is especially high in drought-prone regions, and in some areas, the use of monocultures may have already reached the point of diminishing returns.

Research in many developing countries is once again focusing on **multiple cropping,** an approach that offers several economic and environmental advantages. By growing a number of different interspersed crop species, the demand for pesticides often decreases, and farmers are placed in a more stable economic position should the price of a single crop fall on the world market. Other innovative agricultural technologies hold the promise of reducing energy requirements and dependence on expensive chemical pesticides and synthetic fertilizers.

Malthusian Population Limits and the Future

Many important questions about the quality of human life in the future remain unanswered (see Focus on Scientific Process: "Is Overpopulation the Cause of Environmental Problems?"). Will modern agricultural practices and the emerging genetic revolution contribute adequate food resources for the ever-increasing human population density, or will the predictions of Malthus ultimately prevail? What will be the social, cultural, and environmental costs of a continuously expanding human population? Will fertility control be applied on a worldwide basis, and if so, will it be successful in slowing human population growth?

Modern methods of human fertility control include procedures that prevent fertilization or implantation, and in some countries, abortion if implantation has occurred. Fertilization prevention can be achieved through the use of mechanical, chemical, surgical, and behavioral contraceptive techniques (discussed in Chapter 38). Mechanical and chemical techniques are also used to prevent implantation. These methods have ethical considerations and thus are applied in varying degrees among the different human cultures.

Within the developing regions of the world, contraceptive use is highest in eastern Asia and portions of Latin America, where up to 75 percent of the women report using some method. In general, the burden of fertility reduction rests with the women of the world. It is widely believed that significant control will result when more women in developing regions employ some method of contraception, particularly since men in many cultures throughout the world seldom use contraceptive measures. This will not occur, however, until women in the developing areas are liberated from their heavy workload and given a form of security other than that derived from their children. It has been estimated that women do as much as two-

In 1975, the global human population reached 4 billion persons. By 1987, it increased another billion—a gain of 1 billion people in about 12 years. As we saw in Figure 12.3, reproduced again here as Figure 1, it took from the beginning of human time—over 1 million years ago—until A.D. 1850 to reach our first billion. Less than 150 years later, we have exceeded 5 billion. What conclusions can be drawn about the human population growth rate? What are the consequences of such growth? Each new person requires food, water, clothing, and shelter. To accommodate our expanding population, more people cut forests for land to raise food, strain the productive capacity of existing land, and deplete water supplies. In industrial societies, more people use fossil fuel–based energy to travel, heat and light homes, and produce goods. Air pollutants from these fossil fuel–based technologies affect air quality in urban and remote areas, and may be modifying the global climate (see Chapter 13).

In part, rapid human population growth is related to various technological innovations (such as changes in agriculture and medicine) that have reduced death rates. Technological innovations have also led to greater quantities of resources being used by each person. The increase in per capita resource use has been most obvious in industrially developed nations, but it is also escalating in lesser developed nations. Between 1930 and 1990, average per capita energy use, on a global basis, more than doubled. Extraction and expanding use of natural resources have been related to environmental degradation.

Pressures imposed on the environment by the increasing human population and its use of resources are reflected in various ways, including the following.

Is Overpopulation the Cause of Environmental Problems?

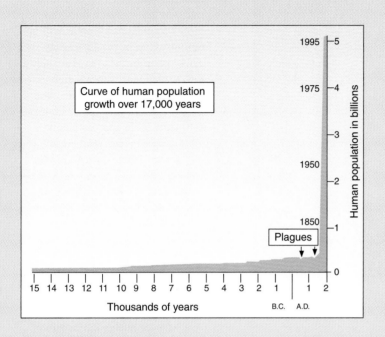

Figure 1 An arithmetic graph of human population growth during the past 17,000 years.

■ Global per capita food production increased rapidly between 1950 and 1985, but since then per capita production of basic foods such as grain and soybeans has either leveled off or declined.

■ In a related development, the amount of land devoted to raising grain per person, on a global basis, has been declining since the late 1950s. This decrease has been fueled directly and indirectly by the expanding human population. In the direct case, dividing a constant acreage by an increasing population base decreases per capita acreage. However, the amount of acreage has not been constant; rather, it has been decreasing, largely because of changes indirectly related to the increased population size. These changes include conversion of cropland to nonfarm uses, such as cities and highways, and abandonment of land that was degraded by overuse or inappropri-

ate use. Consequently, more people must be fed from less land.

■ Not only is less land available per person, but water used for irrigation, which made farming productive in dry areas, is also disappearing. Irrigated land area per person decreased 6 percent between 1978 and 1989. The increasing population figures directly into this change. The decrease, however, has also been caused by competing demands for water and by increased energy costs associated with pumping irrigation water. In addition, irrigation uses of water are often not sustainable over the long term. In many areas of the United States and other nations, water is pumped for irrigation use faster than it can be replenished, and excessive water withdrawals have adverse effects on other systems that depend on the water. For example, the Aral Sea in the Commonwealth of Independent States (formerly the

Soviet Union) is now 40 percent smaller than it was in 1960, largely because of excessive withdrawal of irrigation water from the rivers that flow into it (see Figure 2, page 232).

■ Habitat destruction and the related extinctions of species continue at a rapid rate. Such destruction is driven by demands of an increasing population for agricultural land and for resources from newly opened lands. For example, the world's tropical rain forests hold about half of the world's species, yet between 1981 and 1990, approximately 4.6 million hectares (11.4 million acres) per year of these forests were cleared (see Figure 3, page 232). The rate of tropical deforestation is much higher when other forest types (such as moist and dry deciduous forests) are included. The annual rate of deforestation of all tropical forest types, between 1981 and 1990, was approximately

box continues

FOCUS ON SCIENTIFIC PROCESS

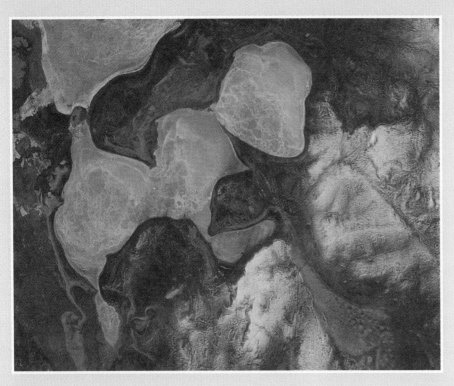

Figure 2   This satellite photograph of the Aral Sea, taken in 1990, clearly show its vast reduction in size in a very short period of time. Whereas the Aral Sea is now separated into the "Lesser Sea" in the north (upper left) and the "Greater Sea" in the south (upper middle), and has a surface area of about 30,000 km^2, in 1960, it covered approximately two-thirds of the area shown in the photo and had a surface area of 68,000 km^2. The main reason for the decline was the withdrawal of irrigation water from rivers that flow into the Aral.

Figure 3   If the present rate of deforestation in tropical forests—about 15.4 million hectares (38 million acres) per year—continues, more than half of all global tropical forests will be deforested in less than 90 years.

15.4 million hectares (38 million acres)—an area slightly greater than the state of Georgia! If that rate continues, over half of the global tropical forest areas will be deforested in less than 90 years. This loss in forest area also leads to the extinction of species that depend on this habitat. The current global rate of species extinctions is estimated to be approximately 20,000 species per year, which means that 1 out of every 1,000 species becomes extinct each year. What is the cause of deforestation? The most important one, in terms of acreage, is not commercial timber production or large corporate conversion of forest to rangeland, but the quest for land, fuel wood, and fodder by members—usually poor—of the growing population.

■ Although air quality has improved over much of the United States in recent decades as a result of strong air pollution regulations, this is not true in much of the world. Emissions of pollutants that contribute to forming acid deposition and photochemical smog are increasing rapidly in many lesser developed nations as their populations increase and they undergo industrial development. For example, sulfur and nitrogen oxides are pollutants that contribute to acid deposition and, in the case of nitrogen oxides, to ozone pollution (see Chapter 13). Emissions of these pollutants from Asia, Africa, and South America are projected to increase dramatically over the next several decades. Thus despite the declining emissions of these pollutants in the more developed nations of the world, global emissions may be greater in the future, owing to substantial increases from less developed nations.

■ Emissions of "greenhouse gases"— such as methane (CH_4)—that help

FOCUS ON SCIENTIFIC PROCESS

trap heat in the atmosphere have also been increasing rapidly over the past 250 years because of human activities (see Chapter 13). Two of the most important sources of the CH_4 being added to the atmosphere are livestock such as cattle (bacteria that live in the guts of these animals produce CH_4, which the animals then release by belching) and bacteria living in rice paddies. Because human population has increased, the number of livestock and the area cultivated for rice have also increased (the area devoted to rice fields has nearly tripled in the last century). As a result, there has been increased atmospheric CH_4 and the greater potential for global climate change. Carbon dioxide (CO_2) is another major greenhouse gas showing a rapid increase in atmospheric concentration. The most important human-related sources of CO_2 include fossil fuel combustion and deforestation of tropical forests, both of which are related to an expanding human population.

The related phenomena of increased population, increased resource use, and the technologies that make this resource use possible have contributed to a vigorous debate among environmental scientists about the root causes of our current environmental problems. Some scientists argue that our large and rapidly growing population is the fundamental cause of environmental degradation and resource depletion, while recognizing that increased per capita resource use and ill-conceived technologies also contribute to the problem. This school of thought dates back at least to the writings of Thomas Malthus, whose book *An Essay on the Principle of Population* is described in this chapter. Contemporary scientists, led by Paul Ehrlich and Garrett Hardin, also share this view. Other scientists, however, such as Barry

Commoner, argue that the root cause of resource depletion and environmental degradation is inappropriate technology, rather than population size and growth. They believe that population *contributes* to the problems of resource depletion and environmental degradation but that technological change is the fundamental cause of the problem.

Is there a single "right" or "wrong" answer in this debate? Perhaps an evaluation of the examples listed here may shed light on the answer to this question. For some problems, such as the conversion and degradation of agricultural land, population is clearly the fundamental force. The land base is diminishing because there are a great many people who need land for various purposes. For other problems, such as air pollution, inadequate technologies may be more important. This point of view is supported by the observation that emissions of sulfur and nitrogen oxides and carbon dioxide have decreased (or at least stabilized) in the United States during the last 10 to 20 years despite continued growth in population and in industrial output. A combination of technological improvements and conservation ethics has been responsible for the declining emissions.

A fruitful approach for evaluating the relative importance of population and technology in resource depletion and environmental degradation was suggested by Paul Ehrlich and John Holdren. They argue that in any society, whether agricultural or industrial, each human has some negative impact on the environment. These impacts result from the utilization of renewable and nonrenewable resources. The impact *(I)* of any population can be expressed as a product of three characteristics: the population's size *(P)*, its affluence or per capita consumption *(A)*, and the environmental damage *(T)* inflicted by the technologies used to supply each unit of consumption:

$$I = PAT$$

In many situations, it is difficult to quantify each of these terms, but even in these cases the model has heuristic value. By attempting to estimate the terms, we are forced to recognize the interdependencies among them, and to recognize that their relative importance differs depending on the impact being addressed. For example, Commoner and others who believe that problematic technologies are the most important cause of environmental degradation tend to focus on problems related to *pollution in industrial nations.* For other types of problems, however, exemplified by deforestation in tropical regions, technology is obviously far less important than the sheer number of people seeking land.

The simple formula also clarifies the disproportionate environmental impact of people in industrialized nations of the world, particularly for those activities associated with energy-intensive technologies. Whereas less developed nations hold nearly four-fifths of the world's population and are growing rapidly, these nations used only 24 percent of the world's oil, coal, natural gas and electricity in 1991. The average person in an industrial country used about nine times more commercial energy than the average person in a developing nation in 1990!

Clearly, there is no simple answer to the question, What is the root cause of environmental degradation and resource depletion? The most important cause varies among geographical regions of the world and from situation to situation. In each case, however, population size is implicated as being an important—if not always *the* most important—contributing factor. Two conclusions seem inescapable: (1) If there weren't so many of us, our impacts would be much smaller, and (2) the resources of Earth are finite, and increasing numbers of people are straining those resources.

thirds of the physical work and produce 60 to 80 percent of the food in many developing countries, including Asia, Africa, and parts of Latin America.

Moreover, women and children frequently represent the largest proportion of the very poor, landless peoples of the world. Often their only security is in numbers. Unless this cycle is broken—through the education of females, agrarian reforms, and the improved status of women—population growth rates will likely continue at high levels. Until the females among the poor are educated and allowed to participate in the security of land ownership, their dependence on large numbers of children will not likely diminish.

> **BEFORE YOU GO ON** Many important questions about human population growth in the future cannot be answered at the present time. Major uncertainties concern food production, the quality of human life, and social and cultural dilemmas related to controlling population growth.

Malthus's postulate "that the passion between the sexes is necessary and will remain nearly in its present state" retains some validity today. Although we now have the technical means to prevent conception, the question remains if we will have the determination, on a global basis, to apply them. What is the upper limit of human population density that the ecosystems of Earth can support? Will new food-producing revolutions unfold in the future? Can we, as modern humans, control our fertility in time? What will be the economic and environmental consequences if we do not?

SUMMARY

1. The development of agriculture by early human cultures allowed the human population to expand from 5 million to over 5 billion in only 10,000 years. Major human migrations resulted in the habitation of all the major biomes on most continents of Earth.

2. The world population increased exponentially from the late 1800s until the late 1960s, when the rate of growth began to decrease. However, the total population continues to increase and will double in about 40 years if growth continues at the present rate.

3. Malthus assumed that the need for food and the continuing passion between the sexes would lead populations to outstrip the food supply, and famine would visit the human race. However, the agricultural, industrial, and genetic revolutions have allowed human population densities to exceed the limits that Malthus predicted.

4. The propagation of monocultures has led to a reliance on pesticides, fertilizers, and other high-energy processes in agriculture. Pesticide resistance, biological magnification, and nontarget species mortality are some negative side effects of pesticide use.

5. If significant population control is to occur, females in the developing regions must have greater security beyond that derived from their children. This security could include increased access to education and participation in land ownership.

WORKING VOCABULARY

biological magnification (p. 228)
doubling time (p. 224)
exponential growth (p. 223)
genetic revolution (p. 225)
green revolution (p. 225)

herbicide (p. 226)
insecticide (p. 226)
monoculture (p. 226)
multiple cropping (p. 230)
pesticide resistance (p. 226)

REVIEW QUESTIONS

1. What was the relationship between early human culture and food supply?

2. Describe the pattern of prehistoric human population growth.

3. What factors affected human population growth until the modern period?

4. Describe the current pattern of human population growth.

5. What is the significance of Malthus's ideas on human population growth, published in 1798?

6. What was the green revolution? How did it affect human population growth?

7. Describe some of the consequences of using DDT to control insects.

8. What changes may be in store for agriculture in the future?

9. What do experts predict about human population growth in the future?

ESSAY AND DISCUSSION QUESTIONS

1. What predictions can you make about the size of the world's human population in the year 2500? What factors might be considered in making such predictions?

2. What types of technological "revolutions" might occur in the future that would allow for expanded human population growth? What might be the consequences of such revolutions?

3. What are the lessons to be learned from the DDT experience?

REFERENCES AND RECOMMENDED READING

BioScience. 1995. Special Issue: Science and Diversity Policy. (Supplement).

Caldwell, J. C., and P. Caldwell. 1990. High fertility in sub-Saharan Africa. *Scientific American,* 262: 118–125.

Caputo, R. 1993. Tragedy stalks the Horn of Africa. *National Geographic,* 184: 88–122.

Cherif, A. H., and G. E. Adams. 1994. Planet Earth. Can other planets tell us where we are going? *The American Biology Teacher,* 56: 26–37

Cohen, J.E. 1995. *How Many Peaple Can the Earth Support?* New York: Norton

Daily, G. C., and P. R. Ehrlich. 1992. Population, sustainability, and Earth's carrying capacity. *BioScience* 42: 761–771.

Ehrlich, P., and A. Ehrlich. 1990. *The Population Explosion.* New York: Simon & Schuster.

Lewin, R. 1988. A revolution of ideas in agricultural origins. *Science,* 240: 984–986.

Ludwig, D., R. Hilborn, and C. Waters. 1993. Uncertainty, resource exploitation, and conservation: Lessons from history. *Science,* 260: 17.

Malthus, T. 1798. *An Essay on the Principle of Population.* London: J. Johnson.

Managing Planet Earth (special issue). 1989. *Scientific American,* 261 (3).

Stanley, D. J. and A. G. Warne. 1993. Nile Delta: Recent geological evolution and human impact. *Science,* 260: 628–634.

Stranahan. S. 1993. Empowering women. *International Wildlife,* 23 (5): 12–19.

Torrey, B. B., and W. W. Kingkade. 1990. Population dynamics of the United States and the Soviet Union. *Science,* 247: 1548–1552.

Weiss, H., M.-A. Courty, W. Wetterstrom, F. Guichard, L. Senior, R. Meadow, and A. Curnow. 1993. The genesis and collapse of the third millennium North Mesopotamian Civilization. *Science,* 261: 995–1003.

World Resources Institute, International Institute for Environment and Development, and United Nations Environment Programme. 1990–1995. *World Resources,* New York, Oxford.

World Resources Institute. 1994–95. *People and the Enviroment.* New York: Oxford Univ. Press.

Zuckerman, B. and D. Jefferson (eds.). 1995. *Human Population and the Environmental Crises.* Boston: Jones and Bartlett.

ANSWERS TO FIGURE QUESTIONS

Figure 12.2 If their migration were prevented, wildebeests would drastically reduce their food supply in any given region. Consequently, their population density would decline sharply because of starvation.

Figure 12.4 Excessive numbers of people may have resulted in surrounding lands being overworked. Eventually, the continuous removal of nutrients from the soil may have led to crop failure, famine, and migration.

Figure 12.5 Because global primary productivity has a finite limit, there must be a finite human population density that the biosphere can support.

Figure 12.8 Agricultural monocultures decrease biodiversity by destroying the habitats of native organisms.

Figure 12.9 Many nontarget insect species are more sensitive to pesticides than target species; thus, they would die before the target species were controlled.

Figure 12.10 Biological magnification results in increases in DDT concentrations at each successive trophic level. Because birds of prey are the top consumer, their food would have the largest DDT concentration.

Figure 12.11 If DDT were highly soluble in water, it would be excreted in waste products at each trophic level, and it would also be degraded more quickly.

13

Global Climate Change

Chapter Outline

Reading Questions

1. How has Earth's climate changed throughout its history? What factors account for past changes?

2. What factors are influencing global climate at the present time?

3. What are the hypothesized consequences of global climate change?

4. How have various types of pollution affected forest biomes?

5. What types of human activities have affected global carbon sinks? What are the predicted consequences?

As described in Chapter 12, environmental problems have resulted from **anthropogenic** (human-caused) activities and are now known to affect the biosphere at local, regional, and global levels. Many of these problems are related to human population growth. The effects that lead to disruptions in terrestrial and aquatic biomes on a global scale are anticipated to have the greatest impacts in the future. Most environmental scientists believe that changes in world climate will be the most important environmental problem facing Earth's inhabitants in the twenty-first century.

Are anthropogenic activities causing changes in climate on a global scale? To comprehend present and future global climate changes, it is first necessary to understand how similar changes occurred since Earth's atmosphere formed about 3 billion years ago. Records from the distant past have allowed scientists to study the natural sequences of climate change that have occurred throughout geologic history. From resulting data they have been able to formulate hypotheses about anthropogenic effects and to make predictions about what the future may hold. In this chapter we concentrate on those hypotheses and what they tell us about the future of our planet. Although there are many uncertainties, the predicted effects are diverse and generally serious. It is a complex but critically important story for all of us to understand.

TIME SCALES AND CLIMATE CHANGES

Global climate has changed over the past decades, centuries, thousands of years, and millions of years. These changes have resulted in major biome shifts over Earth's surface as glacial and interglacial periods followed one another.

Hundred-Million-Year Intervals

Earth's climate has fluctuated throughout its history. **Paleoclimatic indicators** (records that indicate types of ancient climates) contained within rock strata, lake sediments, ancient coral reefs, and continental ice accumulations have revealed that ice age cycles began 2.5 billion years ago. Other major ice ages occurred 1 billion years ago, 700 million years ago, 450 million years ago, and from 300 million to 250 million years

ago. The most recent glacial period started about 2 million years ago, at the beginning of the Pleistocene epoch. There have also been numerous minor ice ages. What caused these irregular glacial cycles? Did similar factors lead to the different ice ages? An abundance of scientific data has been used in developing hypotheses about global climate changes in the past and in the future.

Million-Year Intervals

A large body of data dating back to the ice ages documents a series of glacial cycles that began about 30 million years ago. At that time, a number of factors intensified a general cooling trend. Those factors, in combination with tectonic forces centered in Antarctica, caused the formation of a massive ice cap over the South Pole. The ninth and last Pleistocene ice age started about 100,000 years ago, and maximum ice accumulations occurred about 18,000 years ago.

Orbital Factors

Three properties of Earth's orbit are thought to contribute to the alternating climatic cycles that are responsible for very long glacial and interglacial periods (see Figure 13.1). First, the present tilt of Earth's axis of rotation is 23.5°. Evidence, however suggests that that the axis of rotation shifts every 41,000 years to a new tilt, between 22° and 24.5° (see Figure 7.5 in Chapter 7). Second, during every 100,000-year period, Earth's orbit around the sun changes from more circular to more elliptical and back to circular. This variation has little effect at the equator, but the impacts are amplified with increasing latitude and are maximal at the poles. Also, tilt variation is not constant because as Earth rotates, it wobbles like a spinning top. The

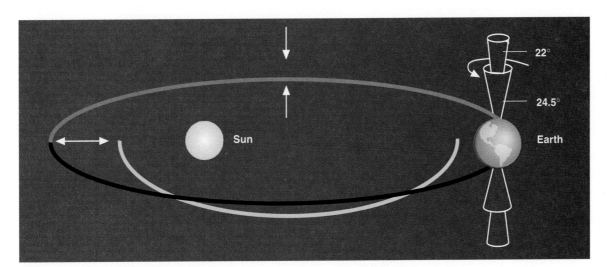

Figure 13.1   Regular changes in Earth's orbital properties are responsible for long-term changes in climatic cycles. These changes include variations from the present 23.5° tilt of Earth's axis of rotation to between 22° and 24.5°, a shift in Earth's orbit from a more circular to a more elliptical orbit around the sun within a period of about 100,000 years, and a variation in the time of year when the Earth is closest to the sun.

Question: *What combination of orbital properties would result in the greatest amount of solar energy reaching Earth's surface?*

third orbital factor is a change in the time of year when Earth comes closest to the sun. Figure 7.5 shows that at present, on the northern summer solstice (June 21), Earth is at its greatest distance from the sun. However, 11,000 years ago it was at its closest point on this date. All of these factors have affected Earth's climatic history.

Tectonic Factors

Plate tectonics, the movements of Earth's continental blocks, constitute another factor that may have influenced long-term cooling and warming patterns in the distant past. One recent hypothesis addresses the effects of uplifting of two major plateaus in the Northern Hemisphere. The mountainous plateaus of western North America and the Tibetan Plateau in southern Asia were thrust up during the past 45 million years. These rising landmasses created large weather cells that deflected both the prevailing west-to-east flows of surface air and the jet stream northward. The deflected jet stream would have developed an arc, with a north-to-south orientation that caused cold arctic air to flow into the continental interiors of both Asia and North America. These changes may have led to a general period of cooling that culminated in the last Pleistocene glaciation.

Other aspects of tectonic activity are also thought to be implicated in the nine shorter glacial cycles of the Pleistocene. Atmospheric warming probably occurred when volcanic activity, in association with the continental plate movements responsible for creating plateaus, injected massive quantities of gaseous and particulate matter into the atmosphere. These volcanic gases and ash could have altered Earth's energy balance by increasing the concentration of carbon dioxide (CO_2) and amplifying the atmospheric *greenhouse effect,* as described in Figure 13.2. The **greenhouse effect** occurs when concentrations of CO_2, water vapor, and other gases increase and absorb more of the sun's longer (infrared) wavelengths that are radiated from Earth's surface. The effect of this process is a positive forcing that warms the atmosphere. **Forcing** refers to the tendency of a specific gas to increase (positive forcing) or decrease (negative forcing) temperature. Carbon dioxide, methane (CH_4), and water vapor are positive forcing, whereas gases such as chlorofluorocarbons (CFCs) are negative forcing. Sulfur dioxide (SO_2) and ozone (O_3) can be either positive or negative forcing depending on where and how they react within the atmosphere. The source (where these substances come from), effects (how they alter atmospheric chemistry), sinks (biospheric areas capable of storing large amounts of them), and forcing differ for each of these gases, but they all contribute to the greenhouse effect (see Table 13.1). For example, erosion increases following periods of mountain formation and continental uplifting, which exposes various materials that were previously covered. Newly exposed silica combines with atmospheric CO_2 and is then transported to oceans through the hydrologic cycle. This process decreases atmospheric CO_2 con-

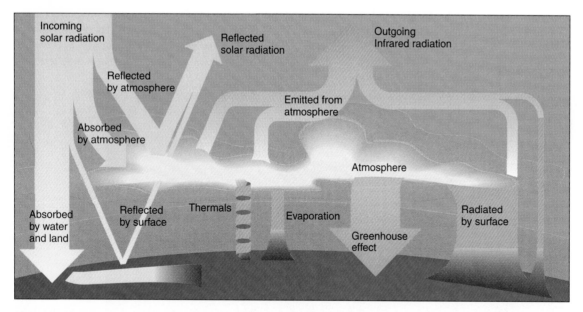

Figure 13.2 Water vapor, carbon dioxide, and other greenhouse gases generally do not inhibit incoming visible and near-infrared wavelengths of sunlight from reaching Earth's surface. The hydrosphere and lithosphere absorb some of this incoming energy, but they also reflect part of it. The reflected wavelengths are radiated back into the atmosphere as longer-wavelength infrared energy. Greenhouse gases absorb infrared energy more efficiently than visible or near-infrared wavelengths. The result is in an increase in atmospheric warming. Thus, as the concentration of greenhouse gases increases, atmospheric warming also increases. Widths of arrows indicate relative amounts of energy.

Question: *What might happen to Earth's biomes if incoming solar energy exceeded the total of reflected solar radiation and outgoing infrared radiation?*

Table 13.1 Sources, Sinks, Effects, and Forcing of Greenhouse Gases

Pollutant	Primary Source	Sink	Effect	Forcing
Sulfur dioxide (SO_2)	Burning of fossil fuel, volcanoes	Atmosphere; removed by acid dispositions	Soil and water acidification, increase atmospheric reflection	+ and −
Nitric oxides (NO_X)	Burning of fossil fuel and agricultural effluent	Atmosphere removed by acid dispositions	Soil and water, affects O_3 concentration	?
Nitrous oxides (N_2O)	Cultivated soils, biomass burning, oceans, and wet forests	Stratospheric (130 years lifetime)	Controls O_3 concentration	?
Methane (CH_4)	Wetlands; termites and cows; coal, natural gas, and petroleum industry	Atmospheric breakdown, soil removal	Breakdown products (CO_2 and H_2O) increase greenhouse warming	+
Ozone (O_3)	Results from conversions of NO_X and hydrocarbons	Stratospheric, depleted by CFCs	Stratospheric loss results in increased UV light; tropospheric damage to vegetation	+ and −
Chlorofluorocarbons (CFCs)	Foam expansion, refrigerants, and solvents	Stratosphere	Breakdown of O_3	−
Carbon dioxide (CO_2)	Fossil fuel combustion, land use changes, and cement production	Global atmosphere and hydrosphere (strong in Northern, but weak in Southern Hemisphere) and global vegetation	Increase global atmospheric temperature	+

centrations, (because the carbon would be added to ocean sediments) and results in atmospheric cooling. Thus CO_2 releases during tectonic activity could eventually be balanced by the carbon that is removed by the CO_2-silica-sediment mechanism.

As the atmosphere warmed during an interglacial cycle, more surface water evaporated, and the water vapor load of the atmosphere increased. This atmospheric water vapor would also have absorbed more heat energy radiating from Earth's surface, thus augmenting the general warming trend. However, more rain would lead to increased erosion, and in time the warming trend would end as a result of declining atmospheric CO_2 concentration.

Volcanoes may have been one factor that combined with others to bring the last ice age to an end. Volcanoes have also been implicated in atmospheric cooling trends that have varied in length from a few years or decades to a century or two. Cooling resulted from injections of SO_2 into the stratosphere during volcanic eruptions. Once present in the atmosphere, SO_2 was converted into tiny droplets of sulfuric acid. Such droplets are very shiny; as their numbers increased, more incoming solar energy was reflected by Earth's atmosphere. This reduction in incoming solar energy would have resulted in a global cooling trend. Recent research indicates that certain aerosols, such as SO_2, can be involved in either atmospheric heating or cooling. Particles larger than 1 micron in diameter increase greenhouse warming by absorbing more infrared energy being radiated from Earth's surface than they reflect from incoming sunlight. Particles smaller than 1 micron reflect more energy than they absorb, which tends to have a cooling effect.

Disappearances of numerous early human cultures have been correlated with periods of intense volcanic activity. If very fine volcanic aerosols, injected into and suspended in the upper atmosphere, led to precipitous drops in the average global temperature, they may have contributed to the demise of certain human cultures by shortening the growing seasons, reducing rainfall, or lowering winter temperatures.

BEFORE YOU GO ON In order to comprehend current changes in global climate, it is helpful to understand natural patterns of change that have occurred in the past. Evidence indicates that Earth's climate has changed regularly during the past 2.5 billion years. These regularities have been primarily related to Earth's orbital properties and tectonic factors associated with wind, volcanoes, and atmospheric changes.

The Present

Calculations of orbital relationships between Earth, the sun, and other planets in our solar system indicate that the present tilt and orbital position of Earth should result in the smallest average difference between summer and winter temperatures in Earth's recent history. On the basis of past records, scientists would expect that summers would now be rather cool and winters quite mild. During cool summers, the accumulated snow and ice of winter should not melt totally. If this occurred, an enlargement of existing icefields and snowfields over the globe would herald the end of the present interglacial warm period. The last four interglacial warm periods of the 2-million-year Pleistocene glacial epoch lasted 8,000 to 12,000 years. The present intergalacial warm period started 10,000 years ago and peaked about 6,000 years ago. Thus scientists expect ice to begin accumulating again within the next 2,000 years, with a glacial maximum following in about 23,000 years. However, all available evidence indicates that alpine glaciers and arctic and antarctic icefields are retreating, not advancing (see Figure 13.3). Are there errors in the orbital calculations or in scientists' interpretations, or are there other explanations?

HUMAN-INDUCED CLIMATE CHANGES

Human activities, particularly during the past century, have modified the atmosphere by increasing the relative concentrations of greenhouse gases. The burning of fossil fuels, depletion of temperate and tropical forests, increased desertification, and certain modern agricultural and forestry practices have combined to increase significantly the atmospheric concentrations of CO_2 (see Figure 13.4). Because these activities are expected to continue, this trend is likely to persist into the next century.

Figure 13.3 Alpine glaciers are presently retreating, as are arctic and antarctic ice caps. As the ice melts, the volume of water on Earth increases, resulting in a general rise in sea level.

The photosynthetic activities of living plants and algae extract CO_2 from the atmosphere or hydrosphere, and the activities of all organisms return CO_2 to the air or water through both aerobic and anaerobic cellular respiration (see Chapters 4, 9, and 30). Minor temperature variations have a relatively limited effect on the photosynthetic rate, but they cause major changes in respiration rates. Thus cellular respiration of fungi and bacteria—the major decomposers—varies directly with changes in temperature. With increasing environmental temperatures, the rate of organic decomposition is expected to increase while the global photosynthetic rate remains relatively constant. What effect will this have on the atmospheric concentration of CO_2?

Anaerobic decomposition takes place in areas where oxygen is limited. Methane is a major by-product of this type of decay. If temperatures rise, the amount of atmospheric methane should increase because of a general increase in the rate of decomposition. Are atmospheric methane concentrations increasing? Yes, but the growth rate of atmospheric methane decreased from 1.3 percent per year in the early 1980s to 0.6 percent per year by 1992. No convincing explanation accounts for the rate decrease of this important greenhouse gas. Methane seems to be the only gas that is now increasing at a slower rate, but decreased emission rates from anthropogenic or natural sources could be involved.

Analyses of polar ice cores from Vostok station, Antarctica, and others from Greenland indicate that atmospheric methane concentrations began increasing significantly in the 1750s, resulting in a doubling of the atmospheric load of this gas. These data correlate well with increasing CO_2 concentrations in the atmosphere. When coupled with the effects of increasing CFC concentrations, these changes in atmospheric chemistry are now predicted to cause an increase in average global temperature between 1.5°C and 4.5°C by 2030.

General circulation models (GCMs) are computer models used to predict long-term changes in global conditions. GCMs that incorporate greenhouse warming data predict that the general rise in temperature will not be uniform over Earth's surface. Rather, temperatures in the subpolar and temperate regions may rise between 5°C and 10°C, and in the polar regions they may increase by as much as 20°C. How significant would these changes be in the next 35 years? What would be the effects of a 20°C temperature rise in far northern and southern latitudes? Is the greenhouse effect induced by human activities inhibiting the general cooling expected? Is Earth entering its next predicted ice age or not?

BEFORE YOU GO ON During the past century, human activities have caused changes in concentrations of greenhouse gases in the atmosphere. As a result, normal biological cycles related to photosynthesis, respiration, and decomposition have been modified quantitatively. Scientific analyses indicate that the long-term global consequences of these changes will include depletion of stratospheric ozone and increased warming.

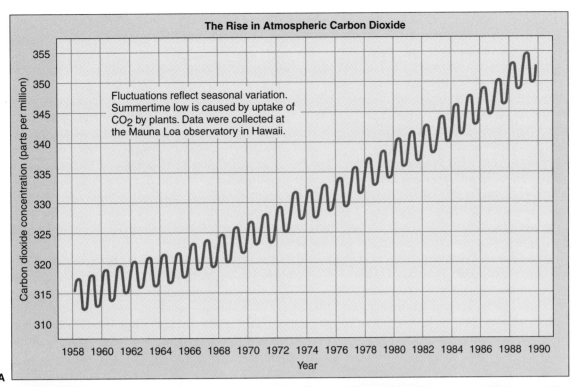

Figure 13.4   (A) Data collected at the Mauna Loa climate observatory in Hawaii indicate that the atmospheric concentration of carbon dioxide (CO_2) has increased by about 40 parts per million since 1957. Seasonal variations in these data reflect the increased photosynthesis by producers during Northern Hemisphere summers. (B) The burning of Douglas fir forest debris after logging, (C) tropical forest clearing and burning, and (D) industrial pollution from a carbon black plant in Transylvania, Romania, all contribute to the ever-increasing concentration of atmospheric CO_2.

Question: *Which major life processes on Earth increase atmospheric CO_2 concentrations? Which decrease them?*

THE ECOLOGICAL EFFECTS OF GLOBAL TEMPERATURE INCREASE

During the peak of the last ice age 18,000 years ago, average surface temperatures for both land and oceans are estimated to have been 5°C cooler than at present. The interglacial warming period ended about 6,000 years ago, when summer temperatures throughout the continental interiors of Europe, Asia, and North America were an average of 2.7°C higher than at present. If Earth has indeed entered its next cooling period, and it took 6,000 years for the temperature to decline 2°C, what might happen if global average surface temperatures increase 4.5°C in the next 38 years? Major predicted effects include an increase in the average global sea level and severe negative impacts on Earth's biota.

The Rising Tide

A global rise in sea level has occurred during the past century. Ocean levels are 16 centimeters higher over Earth's surface, and they continue to rise at an average rate in excess of 1.75 millimeter per year. Much of this sea-level rise can be attributed to thermal expansion of the ocean waters caused by global warming, because when water is heated, its density decreases but its volume increases. The remaining portion is thought to be owing to the melting of terrestrial glaciers and floating icebergs. If the greenhouse effect continues unabated, Earth's oceans are expected to continue to rise, and salt water will flood important estuarine habitats and many low-lying terrestrial areas.

Recent ice core data indicate that most, if not all, of the Greenland ice sheet melted during a major interglacial period more than 100,000 years ago. This evidence correlates closely with data from studies of ancient coral reefs, which show that during the same period, the average sea level was 6 meters higher than at present. Much more of Greenland's ice sheet is thought to have melted during that earlier interglacial period than during the present phase. It is postulated that this was owing to a number of combined orbital and tectonic factors that resulted in greater insulation and warmer temperatures.

What effects would an ever-increasing rise in sea levels have on major centers of human population that are located along the coasts of most continents? The effect will likely vary with latitude, but one recent computer model predicts that in Florida and in other Gulf states, vast areas will be flooded and beaches will disappear or reform inland. In addition, there will be an increase in the frequency and severity of hurricanes, which will cause further direct damage and extensive saltwater intrusions into many freshwater systems.

Warming Effects on Biomes

The increasing thermal effects linked to accumulating atmospheric greenhouse gases are elevating the average surface temperatures of both marine and terrestrial areas of Earth. This warming is expected to produce changes in general circulation patterns of the atmosphere and oceans (see Chapter 7). These changes could produce markedly different rainfall distributions that may result in variations in global soil moisture (see Figure 13.5). When coupled with increases in average annual temperatures, major climatic conditions will likely be altered within and between the biomes of Earth.

The response of communities within different biomes will vary with increasing latitude. Major effects, however, will probably occur in all biomes as each species responds to rapidly changing environmental conditions. Plant communities will suffer the greatest effects because plants are affected directly by changes in available moisture, and unlike animals, plants cannot move.

Shifts in Forests

Temperate and boreal forests will probably be affected to a greater degree than tropical and subtropical forests. One computer model that assumes a doubling of CO_2 levels predicts that today's environmental conditions in temperate forests of beech, birch, hemlock, and sugar maple in eastern North America will change and cause the optimal conditions for these forests to

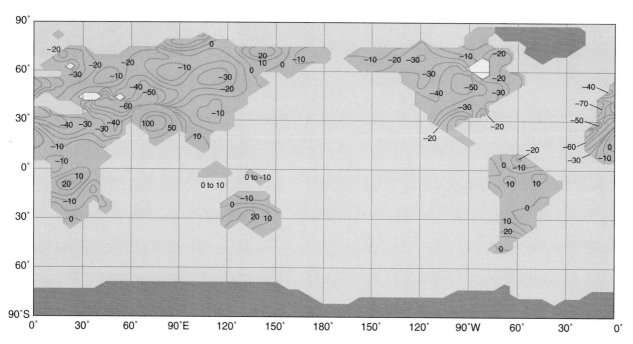

Figure 13.5   This map, generated from a computer model, shows the percentage change in soil moisture that could result from a doubling of atmospheric CO_2. The variations correlated with changing latitudes have important ecological ramifications.

shift north as far as 500 kilometers. In other words, if the eastern beech forests now growing from southern Canada to Florida are to survive, they would need to "migrate" (colonize a new area) to the area between northern New England and the shores of Hudson Bay. Can beech trees migrate this far this fast?

With the retreat of Pleistocene glaciers, forests in eastern North America developed at a rate of about 20 kilometers per century, as temperatures rose an average of 3°C to 5°C over a 10,000-year period. It is clear that if the temperature rises by an equal amount in only one century, the eastern deciduous forests cannot develop at an equal rate, and this forest type may undergo a rapid decline. The old trees could survive for a time, but the production of new saplings would be reduced from current levels.

The coniferous forests of western North America will also undergo significant changes as their temperature and moisture regimes respond to greenhouse warming. Some species may migrate to higher elevations, but the available area in the higher reaches of the western mountains is limited, and for some trees, the proper climatic conditions may exist only farther north in the boreal forest and tundra regions of Alaska and Canada.

Forest migrations are expected to occur throughout most of the Northern Hemisphere. Forests now growing in northern Europe and the great expanses of boreal forest in the former Soviet Union will likely join this parade.

Shifts in Grasslands and Deserts

With moisture and temperature maximums shifting northward, major environmental changes are also expected in the shortgrass and tallgrass prairie regions of central North America. The growing season in the corn belt will likely become drier and hotter, and normal summer rains may fail to fall. The vast areas of wheat now in the original shortgrass prairie regions may shift from the northern Great Plains into north-central Canada. The deserts of the western United States are also likely to expand northward.

The amounts of moisture now captured and stored in the deep winter snowpacks of the Cascade, Sierra Nevada, and Rocky Mountain ranges may decrease. Such a decline would severely limit the amount of water available to the major population centers in the southwestern states. What will be the fate of Los Angeles, Phoenix, and other large southwestern cities as their sources of water decrease during shorter winters and longer, even hotter summers? Will California's Central Valley continue to provide its agricultural bounty? Will the aqueducts of Arizona run dry?

BEFORE YOU GO ON Slight increases in global temperature over time may result in significant changes in global sea levels. Increasing temperatures and sea level as well as atmospheric changes are expected to have major impacts on Earth's biomes. The species composition of forests and grasslands will change, and the geographical locations of most biomes will shift. Some biomes will shrink while others expand.

Global Warming and Biodiversity

A consensus is emerging within the scientific community that biological diversity (numbers of different species) will diminish globally in response to the continued global warming resulting from anthropogenic activities. This loss in diversity will be in addition to that which is caused by the continued loss of habitat related to growth of the global human population. The decrease in species diversity resulting from global warming will not be uniform over the surface of Earth; major effects are predicted for the arctic biomes.

In his book *Biodiversity* (1988), E. O. Wilson offers numerous reasons why the loss of biodiversity is important to all humans. Among them he states: "We have come to depend completely on less than one percent of the living species for our existence, the remainder waiting untested and fallow." What potential treasures will be lost when these uncounted organisms enter the abyss of extinction? Will a new important food source have been among them? Will we lose a potential cure for some devastating human disease? Do the biota with which we share the biosphere have any "right" to continued existence? At what point will the loss of biodiversity begin to affect human populations directly?

Changes in Arctic Biomes

The warming trend will be greater at higher latitudes, where extreme effects on numerous arctic tundra species are predicted. Many species of migratory birds could be affected when rising sea levels inundate their nesting sites. The massive insect blooms that occur in the Arctic each spring may appear much earlier, long before the birds arrive for their nesting season. Numerous species depend on these insects for feeding their young. Will migratory bird species be able to adjust their life cycles to these new conditions?

Higher average temperatures, earlier seasonal changes, and the melting of the permafrost will have major effects on many resident and migratory animal species and on seasonal progression in the growth of different plant species. If these predicted scenarios occur in the future, some species may not have the genetic, behavioral, or physiological flexibility to accommodate such radical changes. If that is the case, will they face extinction?

Changes in Temperate and Tropical Biomes

Global warming will also affect species in temperate and tropical regions. Tropical diseases from equatorial regions may spread into the subtropical and southern temperate regions. Parasitic diseases such as malaria, schistosomiasis (snail fever), and onchocerciasis (river blindness) may be able to enter large human population centers that are located north and south of the areas in which they now occur. Malaria still kills about 5 million people per year worldwide. How many will die if global warming brings more rain, warmer temperatures, and malaria-carrying mosquitoes into today's disease-free subtropical regions?

With increasing insolation, the hot winds blowing out of East Africa will increase the rate of water evaporation as they cross the Indian Ocean. As global warming continues, monsoons sweeping across south-central India may carry twice the amount of rainfall of current levels. What impact will these torrential rains have on the immense human populations of India and Pakistan? Will their effects on rain forests increase or decrease species diversity in these tropical regions?

Changes in Aquatic Systems

In the Great Lakes Basin, the effects of a global temperature rise are predicted to include warming habitats in streams and lakes, shrinking areas acceptable to salmon and trout species, and increasing habitats for less valuable warm-water fish species. Habitat changes will include decreases in wetland and littoral areas as less precipitation occurs in the basin and more water is lost to evaporation. These changes will adversely affect shallow fish-spawning and nursery areas. The general effects will probably include geographical shifts of entire fish associations that will result in changes in the relative abundance of different species. Salmon and trout species will move farther north within the basin.

The impact that these changes will have on the southern Great Lakes basin sport fishery will likely be significant. Will warm-water species capture the same interest of the sport fishing industry that the cold-water salmon and trout species do at present? If not, what will be the total economic consequences in affected communities?

> **BEFORE YOU GO ON** Scientists hypothesize that biodiversity will be reduced on a global scale if abnormal global warming occurs. The effects of this loss cannot be predicted with certainty, but they will likely be significant for the human population. Although all terrestrial biomes will be disrupted, arctic biomes will probably undergo the greatest change. Alterations in aquatic ecosystems will ultimately change the species composition of most communities.

REGIONAL EFFECTS OF HUMAN ACTIVITY

The environmental effects of human activities are not uniform over all inhabited areas of Earth. Factors that contribute to these variations in effects include differences in population density, variable levels of technological development, and climatic variations.

Acid Deposition

Acid deposition is a process in which acidic substances are delivered from the atmosphere to Earth's surface. The major precursors of these acids enter the atmosphere as emissions of SO_2 and NO_x. Chemical reactions in the atmosphere, or after deposition, convert these precursors into sulfuric (H_2SO_4) and nitric (HNO_3) acids. As described in Figure 13.6, these acids arrive on surfaces in three forms: **wet deposition** as acid precipitation, **occult deposition** directly from fog or cloud droplets, and **dry deposition** of particles.

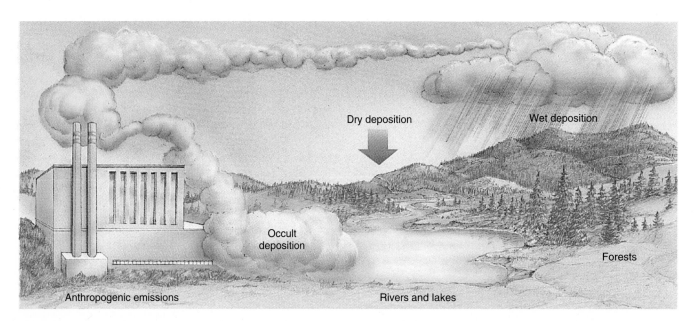

Figure 13.6 Major pathways of acidic compounds from anthropogenic emissions include occult deposition directly from fog or cloud droplets, dry deposition of acidic particles, and wet deposition as acidic precipitation. Both terrestrial and aquatic ecosystems have been seriously affected by acid deposition.

Question: *Which forms of energy do you use that contribute to acid depositions?*

Emission Sources

Substances that contribute to total atmospheric acidity originate from both natural and anthropogenic sources. Globally, SO_2 emissions from industrial processes are estimated to be about equal to those from natural sources. Biogenic (natural) sulfur emissions are greater in the tropics, whereas anthropogenic sulfur is greatest in Europe and North America. Major biogenic sulfur emissions include large quantities of dimethylsulfide produced by marine phytoplankton. In addition, hydrogen sulfides and organic sulfur compounds are emitted from swamps, mud flats, and other anaerobic environments.

Major anthropogenic sources of both SO_2 and NO_x include coal-fired electrical generating plants and, for NO_x especially, exhaust from automobile engines. By far the greatest amounts of SO_2 come from burning coal that has a high sulfur content (see Figure 13.7). These emissions are associated with high atmospheric acid concentrations on a regional basis. In North America the regions of highest atmospheric acid concentrations are downwind from major power-generating plants located in midwestern and northeastern states and the southeastern Canadian provinces, as shown in Figure 13.8 (see page 246). More than 75 percent of the United States' SO_2 emissions originate east of the Mississippi River.

Ecological Effects

Ecological effects of acidic depositions include measurable increases in the acidity of lakes and streams, observed deterioration in the quality of coastal waters, and increased groundwater contamination. Acidic depositions have also resulted in major declines in aquatic communities. Many affected lakes no longer contain fish. Finally, acid deposition has tentatively been identified as one of the contributing factors in forest diebacks (discussed in the next section) in affected regions of the United States and Europe.

Surface Water Acidification The degree to which acid deposition results in the acidification of surface waters depends to a large extent on the **buffering capacity** (ability to neutralize acids) of the water. Alkalinity (the amount of bicarbonate ion, HCO_3^-) is the most useful measure of the **acid neutralizing capacity (ANC)** of receiving waters because alkaline substances in the water will neutralize a certain quantity of acid. The geology of a watershed and its hydrologic characteristics (flow rates, flow paths, and storage capacity) determine the buffering capacity of its lentic systems.

When acidic depositions fall on a watershed, the magnitude of their effects is generally determined by the ANC of the system. Lakes with low alkalinity are more sensitive to acidification than lakes with higher alkalinity. If wet, occult, and dry acidic depositions from anthropogenic sources exceed the ANC of susceptible lakes, they become acidic.

Biotic Effects When the acidity of surface waters increases, biotic effects often include declines in the populations of salmon, trout, and other cold-water fish species. Changes in species diversity are among the first effects of aquatic acidification. Studies of the effect of increasing acidity suggest that mortality during early life stages is responsible for declines in fish population in acidified lakes.

Recent studies estimate that nearly 500 lakes (18.7 percent of the total number) in the Adirondack Mountains in northern New York State have become acidified as a result of acidic depositions from anthropogenic sources. These lakes once supported large recreational fisheries, but because of increasing acidification, the golden days have passed. The demise of these fish populations has resulted in projected losses in excess of

Figure 13.7 An example of a coal-fired, electrical-generating plant.

Question: *What might be the effect of releasing gases from tall smoke stacks at these power plants?*

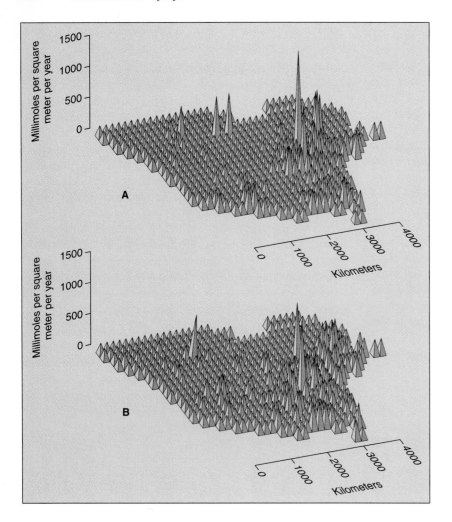

Figure 13.8 These maps show portions of Canada and the United States south from 60° north latitude. They have a grid structure of 1° longitude by 1° latitude per cell, and each cell represents 9435 square kilometers. The first map (A) describes emissions of sulfur dioxide; the second map (B), nitrogen oxides. At a finer resolution, the major concentration peaks indicate that emissions come from point sources, such as coal-fired electricity-generating plants.

$13 million per year, excluding the "loss of amenity" costs to the surrounding communities.

Acidification of surface waters has also resulted in losses or declines of Atlantic salmon in rivers in Nova Scotia and other areas in eastern Canada, Norway, and Sweden. The total dollar cost of these losses is unknown, but with continued input of SO_2 and NO_x from coal-fired power plants and other anthropogenic sources, it will continue to increase.

BEFORE YOU GO ON Acid depositions trace their origin to gaseous sulfur and nitrogen compounds emitted into the atmosphere from various natural and industrial processes. Lakes that receive acid precipitation may or may not become acidic, depending on their buffering capacity. Fish populations have been eliminated from lakes that have become acidified.

Air Pollution and Forest Decline

In some temperate forests of Europe and North America, a number of tree species in stands adjacent to or downwind from major industrial centers began showing symptoms of decline and dieback in the late 1970s and early 1980s. In **forest decline,** trees develop yellow leaves or needles, and tree growth per unit of ground area decreases. **Forest dieback** is the mortality of select species within a forest stand or even of an entire stand.

Forest decline and dieback are rapidly becoming global in scope. Major forest declines have occurred in New Hampshire and New York. In Vermont, a significant dieback of mature red spruce has occurred on the upper slopes of an area called Camels Hump. Profound declines and diebacks have occurred in the forests surrounding the Mexico City Basin, in the Bavarian forests in Germany, in numerous other forests in central Europe, and in the subalpine forest of central Japan.

In general, sites of forest decline and dieback contain trees that are dying for reasons other than normal successional or density-related mortality. Studies in Germany suggest that a network of effects, described in Figure 13.9, are combining to produce forest decline and dieback. This hypothesis indicts numerous compounds common to the atmospheric emissions generated by many industrial activities. These include SO_2, NO_x ammonia (NH_3), and O_3. For some of these atmospheric pollutants, the pathways leading to reduced tree growth have been well established; those of the others are at present only speculative. For example, surface-level concentrations of O_3 are known to cause damage to the needles in many coniferous tree species.

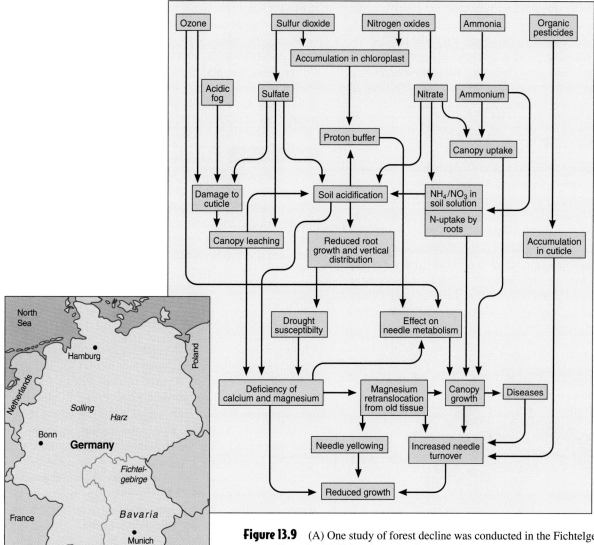

Figure 13.9 (A) One study of forest decline was conducted in the Fichtelgebirge in northeastern Bavaria. (B) This schematic diagram relates important air pollutants to their effects in soil and plants. The total effect of the pollutants is to reduce plant growth, which can lead to forest decline.

Question: *How might forest declines on a global scale affect the greenhouse effect?*

Effects on Soil

Certain atmospheric pollutants are thought to damage forest stands principally by their effects on soil. Depositions of sulfur, nitrate, and ammonium have been shown to alter soil chemistry. In these chemically altered soil environments, tree roots absorb ammonium rather than nitrate, a process that diminishes the uptake of magnesium (Mg). Trees suffering from reduced Mg uptake often compensate by mobilizing Mg reserves in old needles, causing those needles to turn yellow. If normal tree growth is to continue, Mg concentrations must somehow be maintained because Mg is an essential element for all plants. Acidification also alters a tree root's ability to obtain the necessary ratios of other required nutrients.

Studies in forests of the Camels Hump region of Vermont support the results found in Germany. They also indicate that existing atmospheric concentrations of nitrogen-containing compounds can *overfertilize* the forest because trees can take up these substances from the air through their leaves or needles in a process called **foliar uptake.** These compounds include nitrate, ammonium, and, at some sites, gaseous nitric acid (HNO_3). Foliar uptake of excess nitrogen compounds induces trees to grow rapidly, at rates that cause Mg to become limiting. Liebig's law of the minimum (see Chapter 9) seems to apply in these circumstances because the lack of one required element (Mg) seems to limit tree growth. Thus uptake of atmospheric nitrogen-based pollutants can promote canopy growth, which in turn exhausts the tree's reserves of Mg. To compound the problem, essential minerals like Mg become more soluble in some acidic soils. They are then often lost when percolating water leaches them from the root zones of trees and other plants.

Although other atmospheric pollutants are involved, SO_2 and NO_x are primarily responsible for causing soil acidification. Acidification induces various nutrient imbalances that create stress on trees. Also, nitrates are removed from the soil and transferred to groundwater, where they are often inaccessible to plants. The resultant nutrient imbalance leads to reduced growth, forest decline, and eventually forest dieback.

Ozone Effects on Forests

Increasing concentrations of O_3 in the lower portion of the troposphere (the near-surface, weather-generating layer of the atmosphere) are adding a stress to many temperate forests, and, as we will see later, stratospheric **ozone** concentrations are related to the amount of ultraviolet (UV) light reaching Earth's surface (see Figure 13.10). Recent laboratory studies indicate that photosynthesis and growth rates are reduced in many tree species when they are exposed to O_3 levels now common in many areas of the United States. These studies were conducted on seedlings, and the legitimacy of extrapolating the results

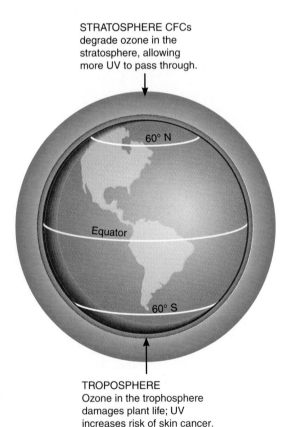

STRATOSPHERE CFCs
degrade ozone in the
stratosphere, allowing
more UV to pass through.

60° N

Equator

60° S

TROPOSPHERE
Ozone in the trophosphere
damages plant life; UV
increases risk of skin cancer.

Figure 13.10   The presence of ozone in different parts of the atmosphere is associated with different effects. In the stratosphere, ozone is beneficial because it absorbs harmful ultraviolet (UV) radiation emitted by the sun. At present, chlorofluorocarbons (CFCs) are degrading stratospheric ozone, which allows more UV energy to reach Earth's surface. Increased exposure to UV energy has been correlated with an increased incidence of skin cancer. Ozone in the troposphere, however, is harmful because it can damage plants.

Question: *Is stratospheric ozone depletion uniform over Earth's surface?*

to a living forest is debatable. Nevertheless, the following questions are of great economic importance: Does O_3 preferentially damage the taller, more valuable trees in the forest? Are faster-growing species affected to a greater extent by O_3 than slower-growing types? Do the effects differ between multispecies and single-species stands? These and other questions will require more field studies before the complete environmental impact of O_3 can be assessed.

BEFORE YOU GO ON   The forests of the world have proved to be especially susceptible to various forms of pollution. Increased atmospheric concentrations of sulfur- and nitrogen-containing gases have been correlated with tree mortalities in several parts of the world. Acidified soils disrupt normal tree activities and diminish chances for long-term survival. Increased ozone levels are suspected to have negative effects on trees, and this problem is being carefully studied at present.

HUMAN IMPACTS ON CARBON SINKS

Changes in natural and anthropogenic carbon sources are causing increases in atmospheric concentrations of CO_2. The effect of this increase, however, is partly reduced by the global carbon cycle in which dynamic mechanisms extract CO_2 and store it in a number of sinks (see Chapter 9). Global photosynthetic activities of terrestrial plants and aquatic algae involve extraction of CO_2 from the atmosphere and the hydrosphere. This CO_2 is converted to sugars and can be stored in or converted to tissues by all organisms. In past ages these processes maintained a balance within the global carbon cycle. As a result, excessive global warming resulting from elevated concentrations of atmospheric CO_2 did not occur.

However, numerous human activities have had detrimental effects on these natural **carbon sinks.** The great forests of temperate and tropic biomes have been greatly reduced through harvesting practices that exceeded rates of replacement. This reduction has decreased the number of living trees that through photosynthesis removed carbon as CO_2 from the atmosphere and through their growth stored it as wood for long periods of time (see Figure 13.11).

Tropical Deforestation

One major global carbon sink is composed of tropical forests. Large-scale tropical **deforestation** continues to accelerate in the Amazon basin south of the Amazon River because the construction of all-weather roads has drastically increased human access to formerly remote parts of the tropical rain forest. Nations with jurisdiction over this tropical forest biome include Brazil, Venezuela, Colombia, Ecuador, and Peru. The major areas of deforestation have occurred in the Brazilian states of Acre, Mato Grosso, and Rondônia. For example, the state of Rondônia is at present undergoing the greatest amount of deforestation in its tropical forests (see Figure 13.12). Rondônian deforestation has been occurring since the early 1970s, but with increased access provided by new all-weath-

Figure 13.11 (A) Students in Kenya, Africa, planting trees to reclaim deforested areas. (B) An "urban forest" created on Earth Day 1990.

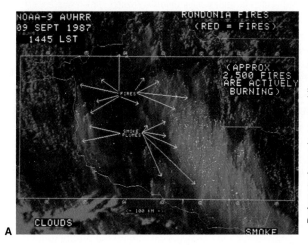

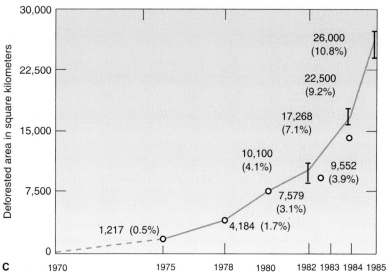

Figure 13.12 (A) This view from space was generated from data transmitted by the NOAA-9 weather satellite as it passed over the Brazilian state of Rondônia. Clouds are labeled at the lower left, and fires and smoke plumes can be seen in the center of the picture. These fires are the result of massive efforts to clear and burn tropical rain forests in this region. (B) This map of the northern states of Brazil shows highway BR-365, which runs through Rondônia. The access provided by this all-weather road opened up the tropical forest to new population migrations, which led to the forest's being cleared and burned. (C) The rate of deforestation in Rondônia from 1970 to 1985. Note the rapid increase since the early 1980s. The brackets at the data points for 1982, 1984, and 1985 indicate a range of more or less than 1000 square kilometers.

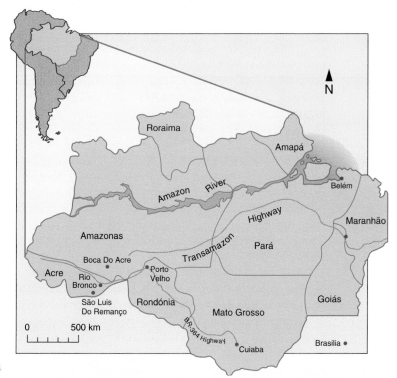

er roads, it has now reached devastating proportions, as indicated by Figure 13.12C (see page 249). To a large extent, this deforestation correlates directly with new frontier expansion policies. These policies have resulted in a surge of human migrations from overpopulated cities in Brazil into these rain forests, which has led to the clearance of vast areas of the forest for planting subsistence crops and grazing cattle.

The loss of these vast tracts of tropical forests will obviously reduce this global carbon sink to some degree. Amazonia, however, is not the only tropical area where deforestation is occurring. Areas of Africa, India, and Oceania are also undergoing major deforestation. It has been estimated that on a global basis, tropical forests are disappearing at an annual rate of 11 million hectares, an area larger than the state of Ohio. Because of the acceleration in deforestation rates, the extent of denuded areas is expected to increase rapidly in the near future.

Marine Carbon Sinks

The other major global carbon sink consists of marine algae and diatoms that extract CO_2 from ocean waters. When these organisms die, the carbon they contain eventually sinks, forming carbonate deposits on the ocean floor. Human activities may also be decreasing the amount of carbon entering this sink as a result of marine pollution and stratospheric ozone depletion.

The pollution of near-shore waters with municipal garbage, industrial effluent, and major oil spills is likely to have an adverse effect on the primary producers in these aquatic life zones. Worldwide, the magnitude of marine pollution is unknown; thus the extent of decrease in CO_2 extraction rates is also unknown. Some scientists believe that this loss is significant because deposits from past ages contain immense amounts of carbon. Hence in the past, major quantities of CO_2 must have been removed by primary producers inhabiting these areas.

Stratospheric Ozone Depletion and CO_2 Sinks

Depletion of stratospheric ozone has resulted from increases in stratospheric CFC concentrations and concentrations of some other anthropogenic and naturally occurring aerosols. Ozone depletion has become especially apparent in the Southern Hemisphere. A hole in the ozone layer there appears seasonally over Antarctica and seems to be enlarging. However, concern over ozone depletion in the northern hemisphere is increasing (see Figure 13.13). As O_3 concentrations decrease, greater quantities of high-energy UV light, consisting of wavelengths normally blocked by stratospheric O_3, enter the troposphere and penetrate to the surface of Earth. These higher surface levels of UV energy are expected to lead to an increase in the incidence of skin cancer because this high-energy form of sunlight is known to induce damaging changes in skin cells.

Increased stratospheric concentrations of CFC and other aerosol compounds may also indirectly affect marine phytoplankton because as concentrations decrease, the amount of UV light reaching the ocean surfaces will increase. These increased levels of UV light may have a negative impact on the phytoplankton community. If elevated levels of UV light reduce the photosynthetic rates of marine phytoplankton, their efficiency as a major carbon sink could be reduced as a result of decreasing the rate at which CO_2 is removed from the hydrosphere and, ultimately, the atmosphere as well.

> **BEFORE YOU GO ON** In the past, atmospheric levels of CO^2 and O^3 remained relatively constant. As a result of human activities during the past century, atmospheric CO^2 concentrations have been increasing and global carbon sinks have been reduced. Stratospheric O^3 concentrations have also decreased, which has apparently led to elevated levels of harmful UV radiation reaching Earth's surface.

IS GLOBAL GREENHOUSE WARMING UPON US?

Are the effects of anthropogenic atmospheric changes now being reflected in a change in average global temperature (see Figure 13.14)? Is global greenhouse warming now occurring? These questions may seem rhetorical, but if the answers are affirmative, as most researchers believe, they will have enormous importance for the policies formulated by governments throughout the world (see the Focus on Scientific Process, "Global Warming"). Because the major air pollutants involved in global climate change are associated with the production and uses of energy sources (oil, gas, coal), any regulations aimed at their reduction would require changes in our patterns of energy use and lifestyles. What would be the effects of mandated reductions in energy use? How will our industrial societies accommodate these reductions? Will people stop driving cars?

Throughout the geologic past, the best predictor of climate variations has been the global sea level. When warming occurred, the sea level rose, and when cooling resulted in glaciation, the sea level dropped. At present, the sea level is rising. In the past, when global sea levels rose, concentrations of CO_2 increased. The atmospheric level of CO_2 is now also increasing.

It is unclear whether or not Earth's temperature is rising or if the recent temperature increases are within the normal range of natural variations. What will be the consequences if the implementation of new energy production and use policies is delayed until the temperature rise clearly exceeds this "variation noise"? Will it be too late to alter the total global greenhouse warming effects?

When one considers the potential environmental problems likely to be of major importance in the twenty-first century, none compares to the potential effects of continued global warming. How will the biota of Earth and our ever-increasing human population cope with the effects brought on by global warming?

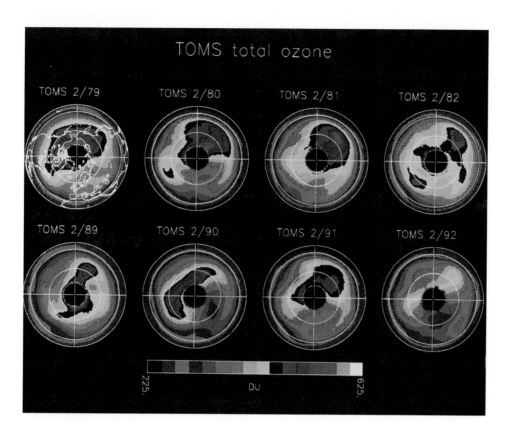

Figure 13.13 In this photograph, eight Total Ozone Mapping Spectrometer (TOMS) images show the February monthly average in two four-year periods in the Northern Hemisphere (1979–82, top panels; and 1989–1992, bottom panels). The color scale (bottom) indicates low total ozone as blues/purples and high total ozone as red/green. These observations indicate that in February, 1992, the total amount of Northern Hemisphere ozone was as low or lower than any previous year. Usually, there is a substantial build-up of total ozone in northern mid-to-high-latitude during late winter and early spring. However, the 1992 accumulation of ozone was 10 to 15 percent lower compared to previous years. Reduced ozone in 1992 (bottom right) is indicated by the scarcity of the relatively high ozone values 50 to 70 degrees North (approximately Canadian latitudes). This contrasts with the higher ozone values (reds and greens) shown in earlier years. The black region in the center of each image is the polar night, a period when TOMS does not make ozone measurements.

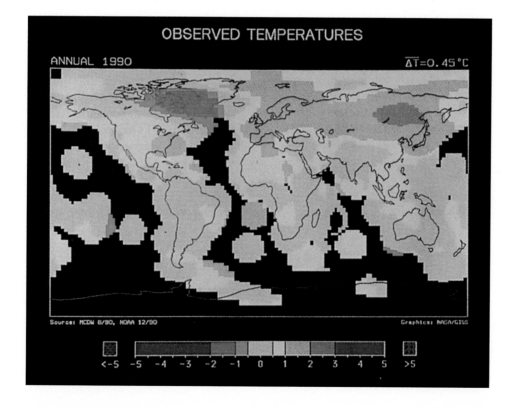

Figure 13.14 Recent data from 2000 weather stations show patterns of temperature change. Average temperatures in 1990 were compared with average temperatures from the reference period of 1951–1980. Colors represent deviations from the reference temperatures, which are shown in green. Except for two areas, most regional temperatures increased. The global average temperature in 1990 was 0.45°C higher than the 30-year reference average.

Question: *Is this evidence of global warming related to natural temperature variations?*

Global Warming

Scientific questions related to global warming are numerous and complex. Hypotheses related to global warming have relatively low levels of certainty. Perhaps the most fundamental question is, How can it be determined whether or not global climate change is actually occurring? The simplest idea—measure global temperature change over time—is unlikely to be successful because natural, yearly temperature variations would probably conceal any true increase caused by global warming. Many scientists now feel that some combination of measurements associated with a single variable may be more useful in answering the key question. One model, which we will refer to as the *ocean level change model*, is receiving increased attention. It proposes that ocean level changes—the single variable—are affected by atmospheric CO_2 (and other greenhouse gas) concentrations that lead to changes in atmospheric temperature and glacial processes (melting and freezing). What does the best available evidence tell us about past, present, and projected future conditions on Earth? What do the data tell us about the model? What do the conclusions drawn from these data mean for society?

Atmospheric CO_2 concentration has increased by about 30 percent since the early 1800s. Figure 13.4 indicates that the rate of increase from 1960 to 1990 has been nearly 0.4 percent per year. During the 1980s, this greater rate resulted in 70 trillion kilograms of CO_2 being added to the global atmosphere. Most of this gain was the result of two factors: increased emissions from burning fossil fuels (77 percent) and a decrease in CO_2 being extracted by plants as a result of global deforestation and other land use changes that reduced vegetation (23 percent). In 1990 scientists estimated that of this 70 trillion kilograms of CO_2, 49 percent remained in the atmosphere and 29 percent was taken up by the oceans—both major carbon sinks. The location of the remaining 22 percent is unknown; CO_2 in this category is known as *missing carbon*. Where is the missing carbon? Recent research indicates that its most likely locations are mangrove swamp peat and the vegetation and sediments of other coastal wetland ecosystems.

What are the likely consequences of increasing CO_2 concentrations? In the past, long-term changes in atmospheric CO_2 concentrations have been linked to cyclic changes in temperature, ice ages, and sea levels. The relationships are direct and can be described simply as:

- Increases in CO_2 means higher temperature, glacial melting, and thermal expansion of ocean water means higher ocean level.

- Decreases in CO_2 means lower temperature, glaciers form, and thermal contraction of ocean water means lower ocean level.

These relationships are the basis of the ocean level change model.

Temperature fluctuations during the last 150 years are shown in Figure 1 (compare with Figure 13.14). Average global sea level changes during the last century are described in Figure 2. Global temperatures and sea levels were essentially stable through the early part of this century. In the 1930s, however, they both began to increase. The zero line on the sea level graph (Figure 2) represents the average sea level between 1950 and 1970. The five year average (thick line) has been greater than zero since the mid-1960s, and it continued to increase through the 1980s and into the 1990s.

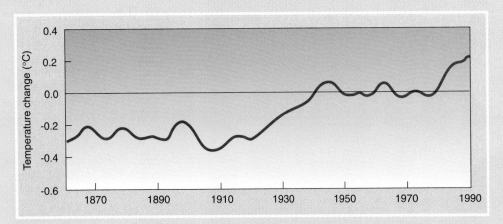

Figure 1   Average global yearly temperatures (land-air and sea surface) for the period 1861–1989. Temperature values were calculated relative to average temperatures during 1951–1980.

FOCUS ON SCIENTIFIC PROCESS

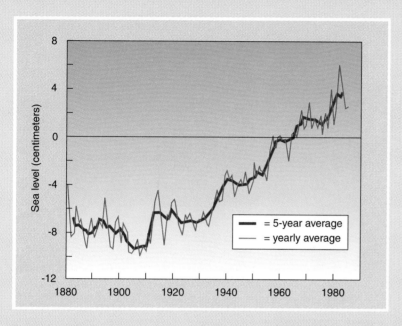

Figure 2 Global average sea level rise from 1880 to mid-1980. The zero baseline is the average for 1951–1970. The thick line represents five-year running averages; the thin line shows the average for each year.

The impact of a rising sea level varies in different geographical regions. Some continental margins are rising because of plate tectonic activity and reductions in ice mass from the last ice age. Overall, average sea level is estimated to have increased by 1.75 millimeter per year since the early 1900s, which means that it has risen by 16 centimeters during this century. It is assumed that if the present rate of CO_2 emissions, deforestation, and other land use changes continue, then both the present atmospheric temperature rate increase and sea level increases will continue.

Recent studies have provided other useful measurements and explained some interesting connections. Elevated global temperatures have caused more ocean water to evaporate. A substantial amount of this water has been deposited by rain and snow on Greenland and Antarctica, which has led to increases in their ice fields. Such increased ice masses are usually associated with decreasing global ocean volume. However, higher temperatures have also caused more rapid melt-

ing of ice along their ocean margins, which has added water to global oceans. The net result of these two processes is about equal. Research has also revealed an irregular decrease in the size of North-

ern Hemisphere snowfields since the early 1970s and all continental glaciers are currently melting, which adds water to the oceans (see Figure 3). All of these data seem to support the ocean level change model. The basic prediction of the model is that ocean levels will continue to increase if global warming is occurring. What does this mean for society? What does the future hold if the prediction is correct?

A 1-meter rise in global sea level would cause flooding of 36,269 square kilometers of dry land and coastal wetlands in the United States. Fifty percent of the population lives within 80 kilometers of a coastline. By the year 2030, sea levels are projected to be 0.6 meters higher than at present. What affects will this have on important coastal cities in the United States?

Are there other consequences of global warming? Current general circulation models indicate that atmospheric CO_2 will double by 2030. Consequently, global average temperature is expected to increase between 1.5°C and 4.5°C and rainfall to increase between 7 and 16 percent, depending on latitude and topography.

Figure 3 Emmons Glacier, on Mount Rainier (4392 meters) in Washington State's Rainier National Park, is in retreat, as are all the other glaciers on the mountain. In the wake of the retreating glacier, rock rubble remains where thick ice once existed.

box continues

FOCUS ON SCIENTIFIC PROCESS

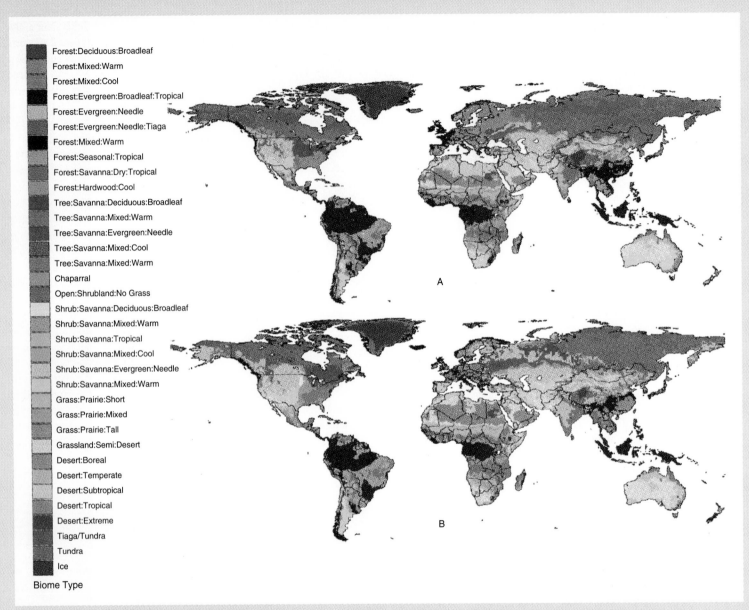

Biome Type:
- Forest:Deciduous:Broadleaf
- Forest:Mixed:Warm
- Forest:Mixed:Cool
- Forest:Evergreen:Broadleaf:Tropical
- Forest:Evergreen:Needle
- Forest:Evergreen:Needle:Tiaga
- Forest:Mixed:Warm
- Forest:Seasonal:Tropical
- Forest:Savanna:Dry:Tropical
- Forest:Hardwood:Cool
- Tree:Savanna:Deciduous:Broadleaf
- Tree:Savanna:Mixed:Warm
- Tree:Savanna:Evergreen:Needle
- Tree:Savanna:Mixed:Cool
- Tree:Savanna:Mixed:Warm
- Chaparral
- Open:Shrubland:No Grass
- Shrub:Savanna:Deciduous:Broadleaf
- Shrub:Savanna:Mixed:Warm
- Shrub:Savanna:Tropical
- Shrub:Savanna:Mixed:Cool
- Shrub:Savanna:Evergreen:Needle
- Shrub:Savanna:Mixed:Warm
- Grass:Prairie:Short
- Grass:Prairie:Mixed
- Grass:Prairie:Tall
- Grassland:Semi:Desert
- Desert:Boreal
- Desert:Temperate
- Desert:Subtropical
- Desert:Tropical
- Desert:Extreme
- Tiaga/Tundra
- Tundra
- Ice

Figure 4   This figure compares current global biome distributions (A) with those predicted if the atmospheric carbon dioxide concentration were to double (B). These predictions are the outcome of super computer models used by the U. S. Forest Service and the Geophysical Dynamics Laboratory. By comparing any region in these two maps, the biome change can be predicted in a double carbon dioxide world.

How might these climatic changes affect global terrestrial ecosystems? North American ecosystems? Figure 4 compares present global biomes with those predicted for a *double CO$_2$* world in 2030, and Figure 5 shows the changes predicted for North America. Major life zones, described in Chapter 7, are expected to develop hundreds of kilometers north or up 1000 meters in altitude. In a double CO$_2$ world, higher temperatures would increase both the growing season and the rate of evapotranspiration. Such changes would cause drought stress in 80 to 85 percent of global terrestrial biomes and 85 to 90 percent of temperate and boreal forests. What does Figure 5 tell us about changes expected in the grasslands of central North America? Will the deserts of the Southwest increase or decrease in size? How will Rocky Mountain and Pacific Northwest forests change? Where will optimal conditions exist for eastern hardwood forests? What changes are expected in your region?

It is estimated that 60 percent of global warming is related to atmospheric CO$_2$ concentrations. Given the uncertainties associated with global warming hypotheses and models, should society

FOCUS ON SCIENTIFIC PROCESS

now take steps to reduce CO_2 emissions and initiate programs to enhance carbon uptake and storage? If so, who should pay for these programs? Will they work, or is it already too late? Carbon emissions could be reduced by decreasing quantities of fossil fuels burned in electrical power–generating plants, automobiles, and other machines. Carbon uptake and storage could be increased by expanding global forests and coastal wetlands. To balance current fossil fuel carbon emis-

sions, however, it would be necessary to double the size of global forests. Are either of these strategies possible given the magnitude of population increases in the developing world and the pattern of overconsumption in the developed world?

Should we "wait and see" what will happen before we act? Is waiting a reasonable position to take given the predicted consequences of global warming by 2030? The predicted consequences for

the United States include frequent catastrophic forest fires; erosion of western mountain ranges owing to increased rainfall; coastal erosion from more frequent storms; increased sedimentation in streams, rivers, and estuaries; a shift of the midwestern grain belt north into central Canada; and diebacks of eastern hardwood forests. If we elect to do something now, what will it cost? If we choose to do nothing, what will it cost our children and grandchildren?

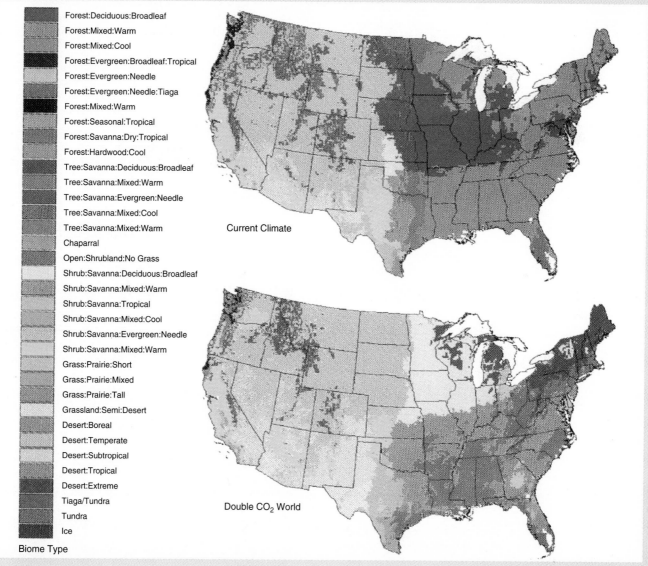

Figure 5 This figure compares predictions for the United States in a double carbon dioxide world. Biome shifts in the eastern half of the country are much greater than those predicted for the western region. (What changes are predicted for your state?)

SUMMARY

1. Anthropogenic activities affect the biosphere at the local, regional, and global levels. Earth's climate has fluctuated on different time scales as a result of both orbital and tectonic factors.

2. The greenhouse effect, resulting from atmospheric increases of carbon dioxide, methane, and compounds of nitrogen and sulfur, may be causing global rises in both temperature and sea levels that may lead to major biome changes. Scientists predict that these effects will occur in variable magnitudes that increase with latitude. The northern biomes will be affected to the greatest degree.

3. Anthropogenic activities contributing to global warming include the direct addition of greenhouse gases to the atmosphere, forest destruction, and marine pollution.

4. Acids arrive on surfaces from the atmosphere in wet, occult, and dry depositions. The sources of these depositions are related to fossil fuels burned for transportation and energy production. Acid rain degrades aquatic systems and is implicated in significant forest declines and diebacks in many areas downwind from major industrial centers.

5. Increased stratospheric CFC concentrations are thought to be destroying Earth's protective ozone shield, which will result in an increased intensity of harmful UV light received at the surface.

WORKING VOCABULARY

acid deposition (p. 244)
acid neutralizing capacity (ANC) (p. 245)
anthropogenic (p. 236)
buffering capacity (p. 245)
deforestation (p. 248)
dry deposition (p. 244)
foliar uptake (p. 247)
forest decline (p. 246)

forest dieback (p. 246)
general circulation models (GCMs) (p. 240)
greenhouse effect (p. 238)
occult deposition (p. 244)
paleoclimatic indicators (p. 237)
wet deposition (p. 244)

REVIEW QUESTIONS

1. What are paleoclimatic indicators? How are they used in studies of global climate change?

2. What types of long-term regularities characterized global climate changes of the past? What factors are thought to have been responsible for these cycles?

3. What is the greenhouse effect? What causes it?

4. How should natural factors be changing global climate at the present time?

5. What types of human activities have modified the normal patterns of global climate change? How has the expected pattern been modified?

6. What types of ecological effects are expected to occur if global temperatures increase?

7. How are biomes expected to be affected by increasing global temperatures?

8. How will biodiversity be affected by global warming?

9. What is acid deposition? What are its causes? What are its effects?

10. How has air pollution affected forests throughout the world?

11. What effects have soil acidification and abnormal ozone concentrations had on forests?

12. Why have vast tracts of tropical forests disappeared?

13. How have marine pollution and stratospheric ozone depletion affected carbon sinks? What are some of the possible consequences of these actions?

ESSAY AND DISCUSSION QUESTIONS

1. Some ecologists believe that changes in temperature serve as a single integrative measure of anthropogenic change. What might be the basis for coming to this conclusion? Do you agree? Explain.

2. Assume that you are a political leader (for example, the head of the U.S. Environmental Protection Agency) responsible for taking steps to reduce global warming. What recommendations would you and your task force make? Do you think they would be implemented? Why?

3. If predicted global temperatures increase in the future, how might different biomes of Earth be changed after 100 years? After 1,000 years?

REFERENCES AND RECOMMENDED READING

Barker, T. 1995. *Global Warming and Energy Demand.* New York: Routledge.

Blaustein, A. R., P. D. Hoffman, D. G. Hokit, J. M. Kiesecker, S. C. Wall, and J. B. Hays. 1994. UV repair and resistant to UV-B in amphibian eggs: A link to population decline? *Proceedings of the National Academy of Science,* 91: 1791–1795.

Bridgham, S. D. , C, A. Johnson, J. Pastor, and K. Undegraff. 1995. Potential feedback of northern wetlands on climate change. *Bio-Science,* 45: 262–274

Corson, W. 1990. *The Global Ecology Handbook: What You Can Do About the Environmental Crisis.* Boston: Beacon Press.

Douglas, B. 1991. Global sea level rise. *Journal of Geophysical Research,* 96: 6981–6992.

Fletcher, C. H. 1992. Sea-level trends and physical consequences: Applications to the U.S. shore. *Earth-Science Reviews,* 33: 73–109.

Graham, N. E. 1995. Simulation of Recent Global Temperature Trends. Science, 267: 666–671.

Houghton, R., and G. Woodwell. 1989. Global climate change. *Scientific American,* 260: 36–34.

Howells, G. P. 1995 *Acid Rain and Acid Waters.* 2nd ed. New York: E. Harwood..

Issar, A. S. 1995. *Climate Change and the History of the Middle East. American Scientist , 83: 350–355.*

Jones, P., and T. Wigley. 1990. Global warming trends. *Scientific American,* 264: 84–91.

Matthews, S., and J. Sugar. 1990. Under the sun: Is our world warming? *National Geographic,* 178 (4): 66–99.

Mooney, H. A., B. G. Drake, R. J. Luxmoore, W. C. Oechel, and L. F. Pitelka. 1991. Predicting ecosystem responses to elevated CO_2 concentrations. *BioScience,* 41: 96–104.

NATO ISI Series. 1995. Series I, Global Envirormental Change, volume 32. *Atmospheric Ozone as a Climate Gas: General Circulation Model Simulation. New York: Springer.*

Peltier, W., and A. Tushingham. 1989. Global sea level rise and the greenhouse effect: Might they be connected? *Science,* 244: 806–810.

Pimm, S. L., and A. M. Snyder. 1994. Tropical diversity and global change. *Science,* 263: 933–934.

Repetto, R. 1990. Deforestation in the tropics. *Scientific American,* 262: 36–47.

Schneider, S. H. 1994. Detecting climatic change signals: Are there any "fingerprints"? *Science,*263: 341–347.

Schwartz, S. 1989. Acid deposition: Unraveling a regional phenomenon. *Science,* 243: 753–762.

Silver, C. S., and R. S. DeFries. 1990. *One Earth, One Future: Our Changing Global Environment.* Washington, D.C.: National Academy Press.

Skole, D., and C. Tucker. 1993. Tropical deforestation and habitat fragmentation in the Amazon: Satellite data from 1978 to 1988. *Science,* 260: 1905–1909.

White, R. 1990. The great climate debate. *Scientific American,* 263: 36–43.

Wilson, E. O. 1992. *The Diversity of Life.* New York: Norton.

Wittwer, S. H. 1995. *Food, climate, and Carbon Dioxide: the Global Environment and World Food Production.*Boca Raton: Lewis.

World Resources, 1992–1993. Oxford: Oxford University Press.Special issue on regional climate change. 1991. G. H. Orians (ed.). *The Northwest Environmental Journal,* 7: 169–201.

ANSWERS TO FIGURE QUESTIONS

Figure 13.1 When Earth's orbit is closest to the sun (that is, most circular), more solar energy would reach the surface.

Figure 13.2 Heat would accumulate on Earth and increase average temperatures.

Figure 13.4 Global respiration raises the atmospheric CO_2 concentration, whereas global photosynthesis reduces it.

Figure 13.6 Coal-fired electrical power generation is responsible for most acid deposition.

Figure 13.7 They would lead to acid deposition downwind from the plants.

Figure 13.9 Decreases in the amount of global forests would reduce an important carbon sink, causing an increase in atmospheric CO_2 concentrations. This would lead to increased greenhouse warming.

Figure 13.10 No. Stratospheric ozone depletion is maximal in the polar regions.

Figure 13.14 Increases in observed global temperatures are still within the "noise" (normal variations) of values established from the past. Thus it is not yet certain that increased global warming is occurring if only temperature is considered.

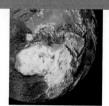

The Interactions of Life and the Process of Science

Early scientific explorers observed interesting patterns in the distribution of organism on Earth's continents. Von Humboldt and other scientists asked questions based on their observations. What affects the distribution of plants and animals? Many answers were proposed, but the most fruitful hypotheses focused on average temperature and water balance. Von Liebig and other nineteenth-century naturalists began to measure the effects of various environmental factors on plant growth. Shelford extended this idea to animals when he developed his law of toleration and food relations diagrams.

Ecology emerged as a new field of science when more biologists began to concern themselves with the interesting and practical questions generated form studies of interactions among organisms and their relationship to the surrounding environment. What is a biological community? How do communities change over time? Ecological research during the early part of the twentieth century came to be dominated by questions about communities. Clements and Cowles differed in their views of how communities change over time. For Clements, community change, or ecological succession, was constant and resulted in climax communities that had characteristics of living organisms. Cowles thought that the climax was never reached and that ecological succession was a variable, not a constant, because it could retrogress under some conditions.

The term *ecosystem* was introduced in 1935 by Tansley, but it was neglected until Lindeman integrated it with his trophic–dynamic concepts of food cycles in lakes. Howard and Eugene Odum extended the ecosystem concept and found that certain laws of physics determined the number of trophic levels in an ecosystem. New ways of looking at ecosystems were reflected in studies of biogeochemical cycling rates and energy flow and how they are related to succession. The early work of Cowles and other scientists gave rise to new questions about how organisms modify their environment and how environmental factors affect the density and distribution of life. Most important ecology questions raised in the past 100 years still guide research today.

The world of 2030 is likely to be much different from the world of today. If current growth rates continue, the human population will reach nearly 9 billion by that time. Scientists in universities, government laboratories, and government agencies are conducting research on the future consequences of such growth for ecological systems. Predictions vary, but ever-expanding development, and the use of new technologies will alter natural ecosystems. Answers to questions about global climate change, habitat destruction, species extinctions, and food shortages remain open, but most debates currently center on the timing of events and the degree of damage, no on the likelihood of their occurrence. It remains to be determined if effective solutions to the pending problems identified by environmental scientists can be found and whether or not they will be acceptable to society.

The Language of Life

III

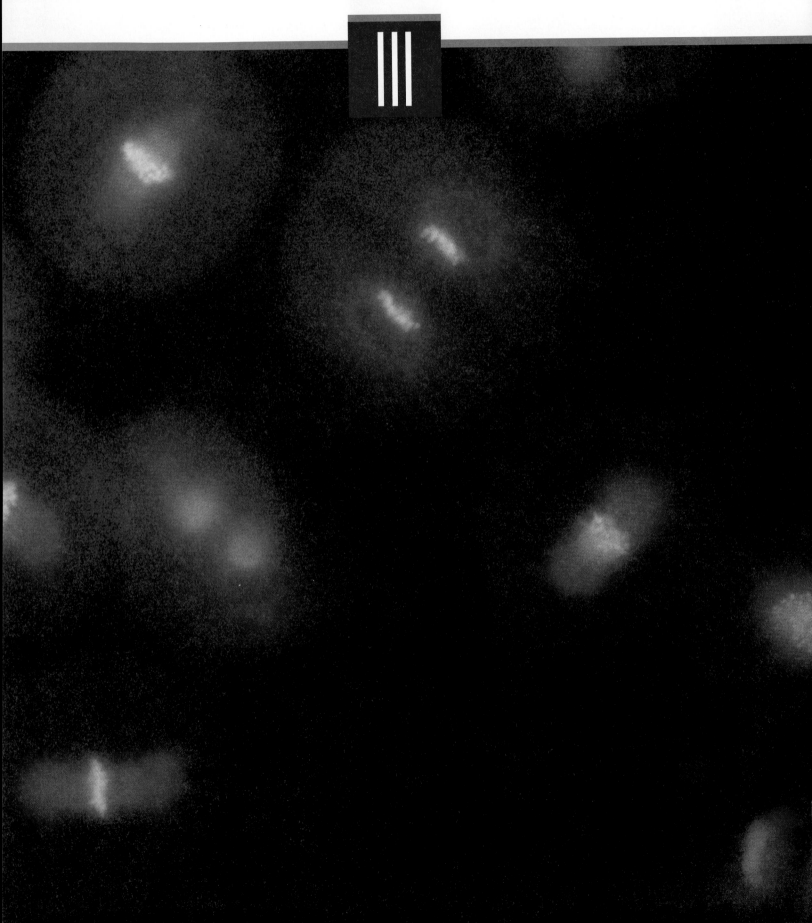

Greek Ideas About Reproduction

In the fourth century B.C., Aristotle, the greatest naturalist of ancient times wrote an extensive treatise on reproduction. Like most scientists before the eighteenth century, he thought that only animals reproduce sexually. He concentrated primarily on animals in his studies because of their importance for understanding the biology of humans. Therefore, he described heredity in terms of sexual reproduction.

In his treatise *On the Generation of Animals,* Aristotle stated that males and females made different contributions to what become their offspring. Reflecting the social values of his day, Aristotle claimed that the male's contribution was more important than the female's. He held that women were inferior to men and should not be allowed any place in politics, nor should they be given the same education as men. A woman's place, from this view, was basically in the household. In reproduction, according to Aristotle, the female primarily supplied "matter." This was acted on by an "active spirit" contained in the male's semen, which molded and formed the future offspring. If all went well in the process, the normal result would be a healthy male child that resembled his father. If conditions were sub-optimal (due to the male's not being vigorous), the offspring would be a male that displayed some of his mother's traits. In the worst case, where the male was inadequate, the result was a "monstrous," deformed offspring who lacked the ability to reason clearly, had less virtue, and did not even have "proper" sexual organs. In other words, it was a female.

How could such a great mind hold such disparaging and ridiculous views on the female role in reproduction? Aristotle merely stated ideas that seemed obvious to him; they were so widely held at the time that apparently no one stopped to consider if they were true. Although other Greek, and later Roman, writers occasionally differed with Aristotle's views, for the most part his theory of heredity lasted for almost 2,000 years.

> **BEFORE YOU GO ON** Aristotle's theory of heredity reflected the culture of his time. He held that the female contributed matter and the male contributed an active spirit that molded the matter into the future offspring.

The Scientific Revolution and Sex

Aristotle's view was finally challenged in the 1600s because of a major shift in the way nature was conceptualized and because of new scientific discoveries. The seventeenth century was the age of scientific revolution, and with it came the notion that nature could be regarded as a wondrous mechanism created by God. The study of nature was an expression of piety, and contemplation of God's creation was thought by scientists to lead to a greater appreciation of his power. Central to the scientific revolution was the metaphor of the world as a machine. Philosophers and scientists tried to explain all phenomena in terms of pieces of matter following the laws of physics, a worldview known as **mechanical philosophy.**

From this new perspective, Aristotle's theory of reproduction (indeed, most of Aristotle's scientific writings) looked hopelessly out of date and misguided. What could an "active spirit" possibly mean in a world that was like a big windup clock?

A new set of discoveries also called into question Aristotle's position on reproduction. The seventeenth century saw the invention of numerous scientific instruments, including the *microscope.* It created a sensation, for when it was trained on even the most mundane objects (moldy cheese, pond water, onion skin, fleas, the ever-present body lice), an entirely new world of microstructure was revealed (see Figure 14.1).

Researchers soon focused their microscopes on plants and animals and exposed their microstructure. A study of the reproductive organs of mammals led to the discovery of what appeared to be eggs in the female ovaries. Since eggs were known to be one of the main reproductive products in other

A

B

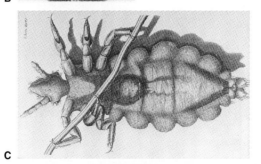

C

Figure 14.1   These illustrations are from Robert Hooke's book *Micrographia,* published in 1665, which popularized the microscope by reproducing a set of interesting observations. Hooke used this compound microscope (A) for his observations. Among the many objects he viewed were the head of an insect (B) and the body louse (C), an all-too-frequent companion in the seventeenth century.

vertebrates, the discovery of mammalian eggs led some scientists to believe that all animal life began in eggs.

Even more exciting was what careful observations with the microscope revealed about the metamorphosis of caterpillars into butterflies. To a number of microscopists, it appeared that all parts of a butterfly were actually already present in the caterpillar stage, merely folded up and ready to expand. These and similar observations led to the formulation of the idea of **preformation,** which states that successive generations are all preformed and encapsulated in previous generations like nested boxes or Russian matrioshka dolls (see Figure 14.2).

Today, this idea seems rather strange. Why was the theory of preformation attractive to major thinkers of the scientific revolution? One reason was that it resolved a number of issues associated with viewing the world as a machine. For example, although machines are marvelous, they were never seen reproducing. Put a pair of gerbils in a cage for a period of time, and one will soon see baby gerbils; put two watches in a box, and no baby watches will issue forth, not even cheap imitations. The preformation theory, by conceiving of all generations as being preformed, explained how a mechanism like an animal could produce similar mechanisms. The theory also fit nicely with religious ideas of the day. Because scientists who took part in the scientific revolution believed that God had created the world, they easily accepted the idea that God had encapsulated all generations of humankind within Eve and generations of all animals and plants within the first creations. They held that slowly, over hundreds of years, the entire course of history unfolded according to a divine plan. Preformation therefore not only explained a complex biological problem but also tied together science and religion.

Preformation was a highly popular scientific position. However, some seventeenth-century naturalists expressed dissatisfaction over the emphasis on the female's role in reproduction.

If all generations of living organisms are encapsulated in ovaries or their equivalents, then the male's role was reduced, at best, to merely stimulating the egg to begin development. A few scientists presented an alternative idea that each sperm contained a *homunculus,* a "little candidate for life" that consisted of a partly or fully formed miniature human being that grew and developed after being introduced into the uterus. This view, which seems to us today a rather odd attempt to rescue male pride, did not catch on, and for almost a century, the classic preformation theory based on the unfolding of eggs (called *ovism*) dominated writings on reproduction.

BEFORE YOU GO ON   Preformation reflected the mechanical philosophy of science. It explained heredity as part of an overall divine plan wherein successive generations were preformed and encapsulated in previous generations like a set of nested boxes. The site generally thought to contain future generations was the female ovary.

The Beginning of Modern Genetics

Belief in the preformation theory broke down in the late eighteenth century for several reasons. The most important was that many scientists came to believe that although science and religion were compatible, science should not rely on the supernatural to explain the physical or biological world. Instead, they thought it more fruitful to attempt to understand the world by searching for physical and biological laws of nature and then seeking to test and verify them. Because religious and supernatural concepts cannot be physically tested, they were considered inappropriate for scientific investigation. Although religious ideas may provide an overview of the significance of the world, they have no role in designing scientific investigations.

Figure 14.2 These Russian matrioshka dolls can fit inside one another (smaller into the next larger) to become nested into a single unit. Similarly, the idea of preformation described all future generations as being contained in the first living beings.

Preformation ran into observational problems as well. The late eighteenth century was a time of great reform in agriculture and animal husbandry. Part of that movement consisted of vast breeding programs intended to improve the quality of cattle, sheep, corn, potatoes, and fruit. The information collected by these practical efforts showed farmers and scientists that traits from *both* the male and the female were passed on to their descendants.

An additional problem for preformation was a general recognition in the eighteenth century that plants reproduce sexually. Studies of plant reproduction during this time clearly showed that this process is very complex. The view that the embryo was somehow preformed in the egg made no sense in light of such studies.

By the nineteenth century, scientists had concluded that preformation was not an adequate explanation for inheritance. Unfortunately, they did not all agree on one good working theory. Most scientists believed that both the male and the female contributed to the offspring and that the hereditary material was blended. This view, illustrated in Figure 14.3, was called the **blending theory of inheritance.** Although scientists were unable to describe rigorous, predictable laws that governed inheritance, they researched this area carefully because it offered considerable practical applications. The impressive research on animal and plant breeding had literally revolutionized farming, and in countries like Great Britain, the threat of famine was eliminated for the first time in history. Scientists now realized that if predictable laws of heredity could be discovered, this agricultural revolution could be extended to other organisms. In addition, Darwin's theory of evolution had stimulated research on inheritance (discussed in Chapter 23).

Around the end of the nineteenth century, a new way of looking at inheritance and a set of careful experiments fundamentally altered existing concepts of heredity. The central figure in this turning point was a Dutch botanist named Hugo de Vries (see Figure 14.4). From his studies with the evening primrose (see Figure 14.5), he collected evidence that a blending of hereditary material did *not* occur. Instead, it appeared to him that individuals inherited from each parent specific traits that were then passed on to later generations in unblended, discrete units. He also claimed that he had discovered a law that described and predicted the inheritance of these specific traits. De Vries was actually engaged in a project larger than genetics. He hoped to reformulate Darwin's theory of evolution in such a way as to make it a more experimental theory.

Figure 14.4 Hugo de Vries was a Dutch botanist whose research on genetics led to wide acceptance of the idea that genetic traits were inherited in discrete units rather than through blending in the offspring. His work was the first to call attention to Mendel's famous 1866 publication.

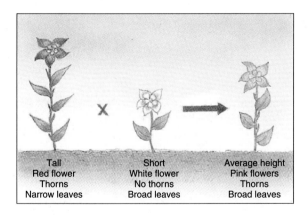

Tall
Red flower
Thorns
Narrow leaves

Short
White flower
No thorns
Broad leaves

Average height
Pink flowers
Thorns
Broad leaves

Figure 14.3 These imaginary plants illustrate the blending theory of inheritance. Hereditary material from each parent blended or mixed to produce intermediate offspring. The blending of a particular trait is seen when red and white parents produce pink offspring and when tall and short parents blend to produce an offspring of average height. The mixing of different traits is illustrated by the thorns of one parent and the wide leaf shape of the other parent, resulting in an offspring that has both thorns and wide leaves.

Question: *What would you expect if plants of the offspring of the illustrated cross were self-pollinated?*

Figure 14.5 The evening primrose was one of Hugo de Vries's most important experimental plants.

Figure 14.6 (A) Gregor Mendel, whose simple and perceptive experiments gave rise to the first genetic principles that later became his laws of segregation and independent assortment. (B) Mendel's laboratory was a simple garden at his monastery.

The publication in 1900 of de Vries's experiments marked a turning point in genetics between the idea of blending inheritance and the idea of discrete units of inheritance. Unfortunately, de Vries is usually not credited with the innovation; instead, the honor goes to Gregor Mendel (1822–1884), a relatively obscure church figure who lived in Brno, in the Czech Republic (then part of the Austro-Hungarian Empire). Thirty-four years before de Vries published the results of his experiments, Mendel (see Figure 14.6) had published in the proceedings of the Natural History Society of Brno a paper that clearly stated the law that de Vries had "discovered." In fact, de Vries may have used Mendel's paper in formulating his ideas. In any case, de Vries's publication called attention to Mendel's paper, which had been ignored until then. The recovery of what came to be known as *Mendel's laws* eventually overshadowed de Vries's work, and today few people recognize his importance in reorienting our understanding of heredity.

BEFORE YOU GO ON   The decline of preformation theory resulted from new observations and new criteria for studying living organisms. In 1900, Hugo de Vries popularized the idea of discrete units of inheritance and rediscovered Mendel's laws of heredity.

MENDEL'S LAWS

Although Mendel's research had little influence in its own day, his experiments, conducted with 10,000 plants over a period of eight years, were so carefully designed, his data so intelligently analyzed, and his conclusions so well expressed that his 1866 paper has received much belated honor. It is, indeed, a model scientific paper. What did Mendel's research tell us about heredity?

Mendel's Experiments and the First Law

Mendel was searching for laws that regulated the transmission of characteristics from generation to generation, and he chose as his experimental subject the garden pea plant, *Pisum sativum,* which has easily recognizable traits. Figure 14.7 illus-

trates the seven traits he selected for his study: (1) shape of ripe seeds (round or wrinkled); (2) color of cotyledon or seed leaf (yellow or green); (3) color of seed coat (gray or white), which is always associated with flower color (violet or white); (4) shape of ripe pod (inflated or constricted); (5) color of unripe pod (green or yellow); (6) flower position (axial or terminal); and (7) stem length (long or short).

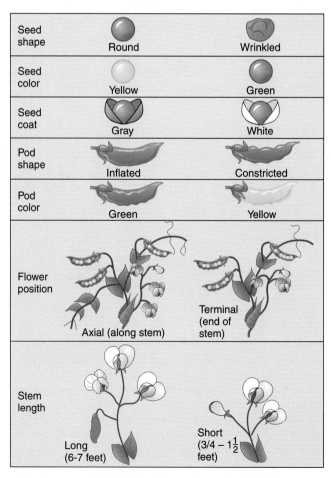

Figure 14.7   The seven pairs of pea plant traits that Mendel studied.

Not only did the pea plant that Mendel chose have easily recognizable traits, but its flower structure also permitted him to control pollination and hence fertilization (see Chapter 32). The plant has male and female organs on the same flower, and both the structure of the flower and the development of its male and female sex cells, or *gametes* (sperm and eggs), are such that the plant normally fertilizes itself. These features, shown in Figure 14.8, allowed Mendel to control mating reliably. He could allow the plant to self-pollinate, or he could artificially cross-pollinate by removing the male pollen-bearing part of the flower (anthers) and then dusting the end of the female part (stigma) with pollen from a chosen plant.

The self-pollinating nature of the plant also made it possible for Mendel to acquire and use **pure-breeding lines**, plants that when self-pollinated or cross-pollinated with a plant from the same line produce only offspring that are identical to the parent or parents. He could therefore plan experiments with individual plants that were known to display particular traits.

Mendel's Monohybrid Crosses

Mendel set out to determine what happened to traits when various crosses were made. He began his work by focusing on pure-breeding lines that differed by only a single trait. Such crosses, involving two forms of a trait (for example, a long stem and a short stem), are called **monohybrid crosses** (see Figure 14.9, page 268). The offspring of the crosses between parents that have different forms of the trait he called **hybrids**.

Mendel found that his hybrids did *not* show blended inheritance. Offspring of a plant with a long stem crossed with a short-stemmed plant did not have medium stem length. Instead, one of the two traits, in each of the seven pairs of traits, was always expressed. He called these **dominant traits** because they seem to dominate in the cross. The traits that did not show up in the first generation of offspring—referred to as the first *filial generation* (*filius* means "son" in Latin) and represented as F_1—he called **recessive traits** because they appeared to recede or disappear in the hybrid. For example, as shown in Figure 14.9, when Mendel crossed a plant that had white seed coats with a plant that had gray seed coats, the resulting F_1 plants did *not* yield seeds with light gray seed coats; they all had gray seed coats. Thus the trait for gray seed coats was dominant over the recessive trait, white seed coats. Later researchers would discover that dominance of one trait over another is not universal and that traits do not always exist in pairs. Our modern understanding of Mendel's traits will be discussed in Chapter 17.

Were there any other patterns that could be uncovered? Mendel's monohybrid crosses revealed even more interesting results than the existence of dominant and recessive traits. When he allowed his F_1 hybrids to self-pollinate and then examined the resulting second filial generation (F_2), he found some with the dominant trait and some with the recessive trait. By keeping careful records of the number of each, he showed that the traits appeared in a ratio of about three dominant to one recessive. Clearly, the hereditary material had not been blended together in the F_1 hybrids. Rather, Mendel concluded, the F_1 hybrids received **hereditary factors** (by "factor" Mendel meant hereditary material responsible for a particular trait) from each of its parents, but the expression of the recessive factor was masked by the dominant factor. The recessive factor, although not expressed in the F_1, had not ceased to exist and could be passed on to the F_2.

Mendel conducted experiments to determine the hereditary makeup of these F_2-generation plants. Plants that showed the recessive trait produced only offspring with the recessive trait when allowed to self-pollinate. The plants with a dominant trait in the F_2, however, were not all the same. When he allowed them to self-pollinate, Mendel discovered that one third produced offspring with the dominant trait only, and subsequent generations raised from their seeds continued to do the same. However, the other two thirds of the F_2 that displayed the dominant trait gave rise to dominant and recessive in the same ratio as the offspring of the F_1 plants, 3:1. How could Mendel explain this result? Mendel reasoned that they must be like the F_1 hybrids. The F_2 generation, then, was complicated. The F_2 plants exhibiting the ratio of three dominant to one recessive were actually one recessive to one dominant to two hybrid.

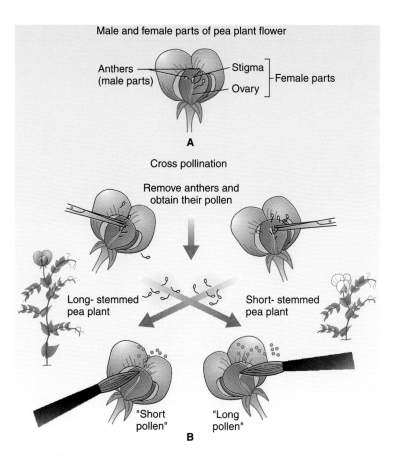

Male and female parts of pea plant flower

Anthers (male parts) — Stigma
— Ovary] Female parts

A

Cross pollination

Remove anthers and obtain their pollen

Long- stemmed pea plant   Short- stemmed pea plant

"Short pollen"   "Long pollen"

B

Figure 14.8 (A) A section of the pea plant flower shows the location of its male and female parts. (B) By removing the anthers of flowers, one can cross-pollinate pea plants. Sperm develop in pollen grains.

Trait	Dominant form of trait in one parent	Recessive form of trait in one parent	Proportion of traits in F₁ hybrids (first generation)	Proportion of traits in F₂ hybrids (second generation) on average

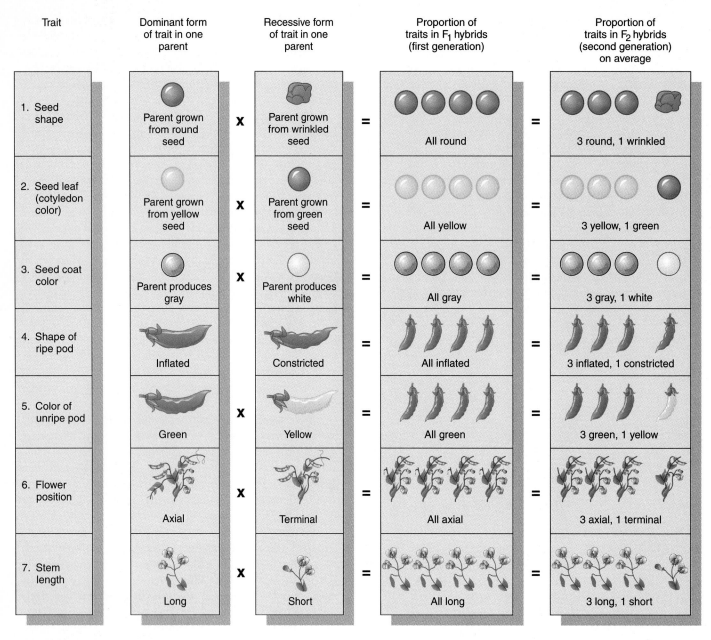

Figure 14.9   The results of Mendel's first monohybrid crosses.

Question: *Which traits show up in the F₁ hybrids?*

Mendel's First Law: The Law of Segregation

Mendel concluded from his results that the two alternative forms of a trait were determined by a pair of hereditary factors. He held that each individual pea plant had a pair of factors for each trait. The pair of factors could be the same, a state we now call *homozygous,* or the pair could consist of the two alternative factors, a state we now call *heterozygous.* In future chapters, we will discuss some of the things that scientists have discovered about these factors and how much more complicated the situation is than it first appeared to Mendel.

The generalizations that Mendel drew from his set of monohybrid crosses are still considered valid. He explained that the appearance of the F₁ generation of hybrids was due to the combination of a dominant factor inherited from one of the parent plants and a recessive factor inherited from the other parent plant. It made no difference which gamete, egg or pollen, carried the factor. But why were these regularly observed ratios of dominant and recessive traits not present in the F₂? A brief look at the possible combinations of gametes will explain why.

The checkerboard-like diagram in Figure 14.10 is called a **Punnett square** and is a convenient, modern way to illustrate the possible combinations of gametes. By convention, each pair of factors can be represented by a pair of letters, a capital letter standing for the dominant factor and a lowercase letter standing for the recessive factor. Figure 14.11 illustrates Mendel's interpretation of his data. He explained that the dominant and recessive factors in the F₁ plant separated in the course of egg

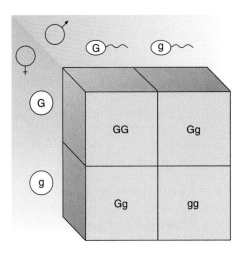

Figure 14.10 The Punnett square is a convenient modern method used to tabulate all the possible combinations of gametes in fertilization. The larger circles represent eggs, and the little circles with tails represent sperm or pollen. The letters inside the circles depict factors for the trait being studied. Capital letters stand for dominant; lowercase letters stand for recessive. Each square represents one of the possible combinations.

and pollen formation and that these factors randomly reunited in fertilization to produce the F_2. The separation of parental factors is known as Mendel's **law of segregation.** This law directly contradicted the then-current idea of blending inheritance.

BEFORE YOU GO ON Mendel experimented with pairs of traits in pea plants. He found that for each pair of traits, one was dominant and one was recessive. He believed that each form of the trait was caused by a pair of factors, one from each parent. Mendel's law of segregation states that during gamete production, these factors separate, and only one member of the pair enters a particular gamete. Fertilization randomly brings the pairs of factors together again and determines the type of trait in offspring.

Mendel's Second Law

Mendel's experimental organism had seven true-breeding traits, and his first law was discovered by considering plants that differed in a single trait. What occurred when two or more traits were considered? Mendel did extensive experiments and found that not only did his law of segregation always hold when more

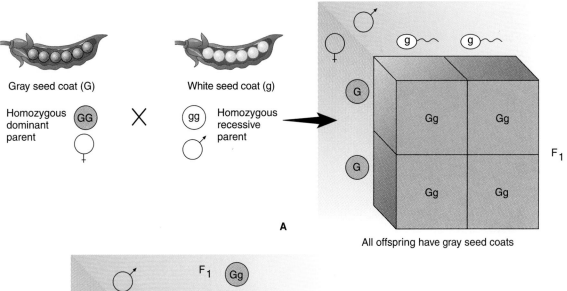

Figure 14.11 (A) A Punnett square illustrating the ratio Mendel obtained in a monohybrid cross, involving gray seed coat and white seed coat plants. (B) The F_1 heterozygote yields a 3:1 ratio in the F_2.

Question: *What would F_2 individuals look like according to the ideas of blending inheritance?*

than one trait was studied, but the traits were transmitted independently of one another—a discovery now referred to as Mendel's second law, the **law of independent assortment.**

The law of independent assortment can be illustrated by considering one of Mendel's crosses involving individuals with two different pairs of traits, a **dihybrid cross.** Figure 14.12 describes a cross involving Mendel's pea plants with two pairs of traits: long stem versus short stem and gray seed coat versus white seed coat. As can be seen in the figure, long stems and gray seed coats are dominant, and in the F_1, the plants have long stems and gray seed coats.

These F_1 dihybrids, when self-pollinated, produced pollen and eggs in four different combinations: LG, Lg, lG, and lg. Figure 14.13 shows the eggs and pollen produced by the F_2 and their possible combinations. Nine different combinations are possible. Simple observation, however, will not distinguish all nine. Because of dominance, only four different combinations of factors are observed, as indicated in Figure 14.14.

Mendel was able to determine that the factors were sorting independently by making a **test cross,** in which he mated F_1

dihybrid pea plants with plants that were known to be homozygous recessive for both traits. Figure 14.15 (see page 272) illustrates the results of his test cross. Since the homozygous test-cross parent could contribute only double recessive factors to the eggs and pollen, the ratio of traits seen in the offspring reflects the ratio of different pairs of factors in the eggs and pollen produced by the dihybrid parent. If the two traits were assorted independently, then four combinations are possible and would be observed in a ratio of 1:1:1:1. If the traits were not independent, some other ratio would have been observed.

Another way of showing independent assortment was to consider the ratios of each trait in the F_2 dihybrids separately (that is, long stem with gray seed coats and short stem with white seed coats). Ratios for each trait turned out to be 3:1. Students who are familiar with statistics may remember that the multiplication law of mathematical probability (which Mendel had learned at Vienna University) predicts that for traits that assort independently from each other, the proportions from a cross involving two or more traits can be obtained by multiplying the proportions expected for each trait alone. So in this

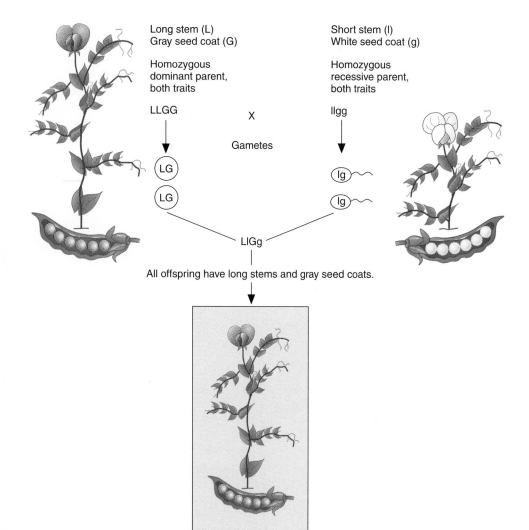

Figure 14.12 One of Mendel's crosses that illustrates the law of independent assortment. This cross involves two traits: long stem (L) versus short stem (l) and gray seed coat (G) versus white seed coat (g).

LIGg X LIGg

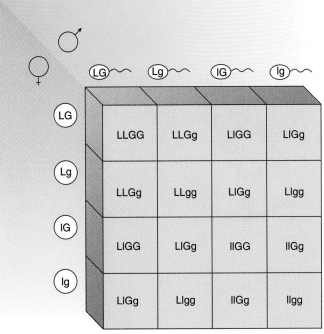

Dihybrid cross

Figure 14.13 The dihybrid cross offers nine possible combinations of factors in the F$_2$.

Question: *Can all nine be observed? Why?*

case, the 3:1 ratios (3 long stem, 1 short stem) × (3 gray seed coats, 1 white seed coat) = 9 long stem with gray seed coats to 3 long stem with white seed coats to 3 short stem with gray seed coats to 1 short stem with white seed coat.

BEFORE YOU GO ON Mendel's law of independent assortment states that the factors responsible for two or more traits are inherited independently.

MENDEL AND LATER WORK IN GENETICS

When we read Mendel's paper today, it is difficult to imagine that it was generally ignored in the years after 1866. Why was such an important paper overlooked for more than 30 years? In part, it went against the common wisdom that hereditary material was blended. It is as difficult for scientists as it is for other professionals to give up a widely held position unless compelling evidence is brought forth. In Mendel's case, that evidence did not appear until the turn of the century, when de Vries and two other biologists, Carl Correns and Erich von Tschermak, rediscovered and confirmed the laws that now bear Mendel's name.

Another reason Mendel's laws did not receive a better reception was that the use of mathematics—ratios, statistics, and counting—was not widely practiced in biology at that time. In the 1860s, biology was largely a descriptive science. Even when experiments were conducted, as in physiology, the work was largely qualitative rather than quantitative (see the Focus on Scientific Process, "Understanding Nature").

Mendel's laws were not accepted for many years for a third reason. When other scientists attempted to verify Mendel's results by testing other organisms, they did not always get the simple ratios that Mendel's laws predict—nor did Mendel when he experimented with certain other plants. Although Mendel's laws are valid, they are not observed for all traits in all crosses in all species. We will see why in Chapters 17 and 18.

LIGg x LIGg

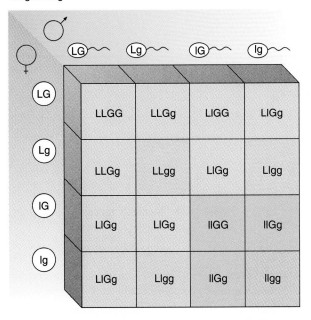

9	Long-stemmed with gray seed coat
3	Long-stemmed with white seed coat
3	Short-stemmed with gray seed coat
1	Short-stemmed with white seed coat

Figure 14.14 Of the nine possible combinations in the dihybrid cross, only four are actually observed.

Question: *Which are they?*

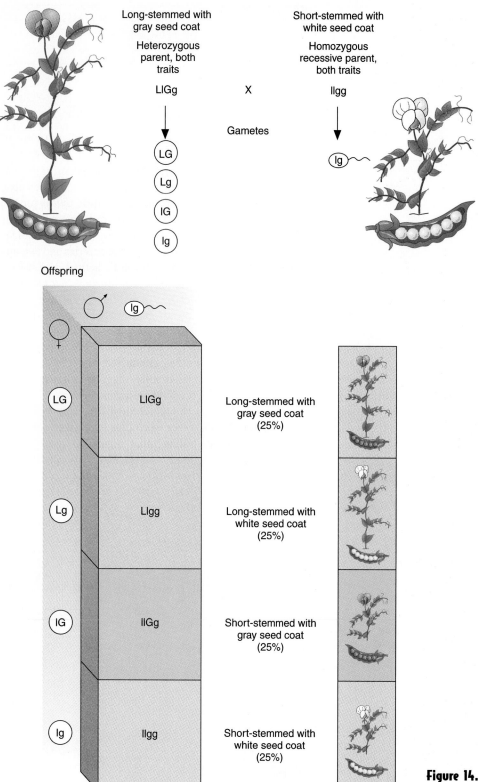

Long-stemmed with gray seed coat

Heterozygous parent, both traits

LlGg   X   llgg

Short-stemmed with white seed coat

Homozygous recessive parent, both traits

Gametes

LG

Lg

lG

lg

lg

Offspring

LlGg	Long-stemmed with gray seed coat (25%)
Llgg	Long-stemmed with white seed coat (25%)
llGg	Short-stemmed with gray seed coat (25%)
llgg	Short-stemmed with white seed coat (25%)

Figure 14.15   Mendel's testcross demonstrated the law of independent assortment.

After finishing his now-famous experiments, Mendel's life became increasingly busy with professional duties (he was elected the abbot of his monastery), and he was not able to pursue his own research. In 1900, however, after the rediscovery of his work, a very extensive research program in genetics began. Soon researchers would not only verify Mendel's laws but also expand the knowledge of genetics beyond the early pioneers' wildest dreams.

Understanding Nature

Biologists use different methods to identify the laws of nature. For hundreds of years, studies of the biological world focused on describing and naming organisms. *Natural history,* as the subject was called, had as its goal creating a catalog of all living beings. Some natural historians also hoped that when all plants and animals had been discovered, described, and cataloged, a natural order would reveal itself. The assumption was that God had a plan and that the plan would be recognizable once all the pieces were assembled. Other natural historians were skeptical that humans had the ability to discern God's plan. They held that the best we could hope for would be to devise some pattern to organize the vast amount of information (a brief discussion of the history of classification can be found in Chapter 5).

A method that proved very helpful in extending the knowledge gained from the description of organisms was the **comparative method.** By comparing the anatomy (structure) of various animals, scientists were able to discover similarities and differences that led them to classify organisms in new ways (see Figure 1). It gave them a sense of what was fundamental to particular groups, and it suggested relationships among various life processes. Georges Cuvier, perhaps the greatest comparative anatomist who ever lived, was able to construct a complete classification system based on the comparative anatomy of different organ systems.

The **experimental method** did not enter the biological sciences until quite late (experimental methods are described in the Chapter 15 Focus on Scientific Process). Although William Harvey performed some astonishingly insightful experiments in the seventeenth century on the circulation of

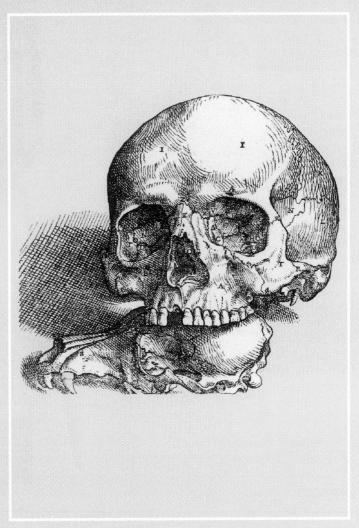

Figure 1   This illustration from a sixteenth-century anatomy book is an example of one of the first uses of the comparative method. A human skull is compared with a dog skull.

blood in animals (described in Chapter 35), it was not until the late eighteenth century that experiments began to be used as a tool of investigation in biology. When first introduced, experimentation was employed primarily for the investigation of the function of the body, that is, physiology. The method was so powerful, however, that soon it was employed in most scientific areas.

Today, experimentation is almost synonymous with the biological sciences.

With the experimental method came the use of **quantitative methods.** Experimenters measured, counted, and calculated. For example, the famous French scientists Antoine Lavoisier and Simon Laplace showed that the amount of CO_2 given off by both a candle burning and a mouse breathing was proportional

box continues

FOCUS ON SCIENTIFIC PROCESS

tional to the amount of heat produced by each (see Figure 2). From that experiment they concluded that breathing and combustion involve the same chemical process.

Mendel was one of the first to employ **statistical methods** to analyze experimental results. Statistics refers to the mathematical analyses and interpretations of data (numerical facts) that enable a scientist to make an objective appraisal about the reliability of the conclusions based on the data obtained. The key to generating powerful conclusions is to make as many observations as possible and to obtain a large data set. For example, Mendel could be relatively confident about his 3:1 ratios, not because in every cross he obtained exactly a 3:1 ratio but because he made many crosses and had hundreds of offspring data that always resulted in about a 3:1 ratio. Modern biological investigations are carefully designed to provide data that can be organized and subjected to specific statistical tests that indicate whether or not a given hypothesis should be rejected.

Figure 2   Lavoisier and Laplace used an ice calorimeter to measure the amount of heat given off in respiration and combustion. This engraving from their paper of 1783 shows a cutaway illustration of their basic apparatus.

SUMMARY

1. Genetics is the scientific study of heredity.

2. Theories of reproduction have changed dramatically throughout history. Aristotle held that females contributed matter and that males contributed an active spirit that molded the matter into the future offspring.

3. The mechanical philosophy of the scientific revolution replaced Aristotle's worldview. The invention of the microscope made possible new observations that, combined with the mechanical view of nature, led to the formation of the theory

of preformation, which stated that all successive generations are encapsulated in previous generations.

4. Preformation theory was discarded because scientists chose to explain nature in terms of physical and biological laws and because observations seemed to contradict it. A blending theory of inheritance replaced the preformation theory.

5. Blending inheritance, however, did not give rise to rigorous, predictable laws that governed heredity. Dissatisfaction with the blending theory led to a new way of looking at heredity, which was the belief that traits are transmitted in discrete, unblended units.

6. Hugo de Vries carried out experiments to demonstrate the idea of discrete units of inheritance. His discoveries, published in 1900, called attention to the earlier work of Gregor Mendel.

7. Mendel's now-famous paper of 1866 describes his experiments on pea plants. He chose true-breeding lines and selected seven traits to study. His monohybrid crosses yielded hybrids that displayed dominant traits in the F_1 generation, but the recessive traits reappeared in the F_2. Mendel's experiments also showed that the traits were inherited in the F_2 in a ratio of three dominant to one recessive. However, two of the dominants were hybrids, so he concluded that the F_2 consisted of one dominant, two hybrids, and one recessive.

8. Mendel postulated that a pair of factors existed for each trait. If an individual has a pair of like factors, it is called *homozygous*. If the individual has the two alternative factors, it is called *heterozygous*.

9. In the production of eggs and pollen, the two factors that determine a trait in the parent separate and then recombine in the F_2. The separation of factors is known as Mendel's first law, the law of segregation.

10. Mendel discovered that the traits he studied were transmitted independent of one another. This independence is known as Mendel's second law, the law of independent assortment.

11. Mendel's work was not appreciated in his own lifetime because it did not fit with accepted ideas, it was not properly verified using other species, and the use of quantitative methods was novel in biology. After the rediscovery of his laws, however, active research in genetics began.

WORKING VOCABULARY

blending theory of inheritance (p. 265)
dominant trait (p. 267)
dihybrid cross (p. 270)
hybrid (p. 267)
law of independent assortment (p. 270)

law of segregation p. (268)
monohybrid cross p. (267)
Punnett square p. (268)
recessive trait p. (267)

REVIEW QUESTIONS

1. What is genetics?

2. What was Aristotle's theory of reproduction?

3. Why was preformation theory ultimately rejected?

4. Hugo de Vries believed that the idea of blending inheritance should be replaced by what?

5. What is a monohybrid cross? A dihybrid cross?

6. Why did Mendel conclude that the two alternative forms of the traits he was studying were determined by a pair of hereditary factors?

7. What is the difference between a homozygous pair and a heterozygous pair of factors?

8. What is the law of segregation?

9. What is the law of independent assortment?

10. Why was Mendel's work neglected for so long?

PROBLEMS

1. (a) What ratios among the F_1 generation would you predict would result from the following monohybrid crosses with Mendel's pea plants: yellow pod (YY) × green pod (yy)? (b) What ratios would you predict if you allowed the F_1 plants to selfpollinate?

2. Construct a Punnett square that illustrates the possible gamete combination in a dihybrid cross using two of the seven traits Mendel studied.

3. What sort of test cross could Mendel have used to determine that the F_1 plants in his monohybrid crosses were heterozygous?

ESSAY AND DISCUSSION QUESTIONS

1. Aristotle was unaware of the value judgments he brought to his theory of reproduction. What assumptions that we hold today might influence our picture of nature?

2. What might you have thought if Mendel had written to you describing his experiments with *Pisum* and describing other experiments with a different plant that did not yield the same ratios? What advice might you have given him? Why?

REFERENCES AND RECOMMENDED READING

Allen, G. E. 1975. *Life Science in the Twentieth Century.* New York: Wiley.

Bowler, P. J. 1989. *The Mendelian Revolution.* Baltimore: Johns Hopkins University Press.

Corcos, A., and F. Monaghan. 1985. Role of de Vries in the recovery of Mendel's work. *Journal of Heredity,* 76: 187–190.

Dunn, L. C. 1965. *A Short History of Genetics.* New York: McGraw-Hill.

Gasking, E. 1967, *Investigations into Generation, 1651–1828.* Baltimore: Johns Hopkins University Press.

Olby, R. 1966. *Origins of Mendelism.* New York: Schocken Books.

Roger, J. 1996. *The Life Sciences in the Age of Enlightenment.* Stanford University Press.

Stern, C., and E. R. Sherwood, eds. 1966. *The Origin of Genetics: A Mendel Source Book.* New York: Freeman.

ANSWERS TO FIGURE QUESTIONS

Figure 14.3 They would look like their parents: average height, pink flower, thorns and wide leaves.

Figure 14.9. The dominant traits.

Figure 14.11 The F_2 would look the same as the F_1.

Figure 14.13 No. Because of dominance.

Figure 14.14 (1) Long-stemmed, gray seed coat; (2) long-stemmed, white seed coat; (3) short-stemmed, gray seed coat; (4) short-stemmed, white seed coat.

15

Cell Reproduction

Chapter Outline

Reading Questions

1. How do prokaryotic cells reproduce?

2. What is mitosis? What does it accomplish?

3. What is meiosis? What does it accomplish?

4. What are the basic differences between plant and animal life cycles?

5. How do meiosis and fertilization explain Mendel's laws?

By the end of the nineteenth century, scientists had developed two fundamental principles that were crucial for understanding biological inheritance. First, the cell theory explained that all living organisms were composed of cells and that all cells arose from preexisting cells. Early microscopic studies had revealed how this was accomplished: certain plant and animal cells could divide into two apparently identical cells. Second, as explained in Chapter 14, Hugo de Vries advanced the concept that inherited traits were transmitted from generation to generation by discrete units. As microscope technology improved in the late 1800s, more became known about interesting cellular structures called *chromosomes* that could be seen after being stained with dyes. Observations of chromosome structure and behavior during cell division were related to the following questions: How are preexisting cells able to divide and give rise to new cells? What is the physical basis of Mendel's laws?

CHROMOSOMES

Chromosomes are specialized structures that contain DNA (deoxyribonucleic acid), the genetic material of a cell. Evidence indicates that chromosomes in both prokaryotic (bacterial) cells and eukaryotic cells contain only one long DNA molecule.

Bacterial cells, and cells of other prokaryotic species, contain a single circular chromosome that consists of one tightly coiled DNA molecule. As described in Chapter 3, a bacterial chromosome is located in a nucleoid region within the cytoplasm and is attached to the plasma membrane.

As shown in Figure 15.1, eukaryotic chromosomes are compact, filamentous structures composed of protein and tightly coiled DNA. Eukaryotic chromosomes are approximately 1,000 times larger than prokaryotic chromosomes and contain much more genetic information. However, these chromosomes can be seen through a microscope only when a cell is dividing into two cells. What happens to chromosomes at other times? During most of a cell's life, when it is not dividing, its chromosomes exist as a dispersed substance called **chromatin,** which is made up of uncoiled DNA and associated proteins. Viewed with an electron microscope, chromatin appears as small fibers or granules scattered throughout the nucleus (see Figure 15.2).

Cells of each eukaryotic species contain a specific number of paired chromosomes, and each pair is similar in size and genetic content. As depicted conceptually in Figure 15.3, each chromosome pair consists of two **homologs,** one contributed by the female parent (the maternal chromosome) and one by the male parent (the paternal chromosome); together the two homologs make up a **homologous chromosome pair.** Different pairs of homologous chromosomes in a cell vary in size and genetic content. The number of homologous chromosome pairs varies among species. For example, fruit flies have 4 pairs, corn

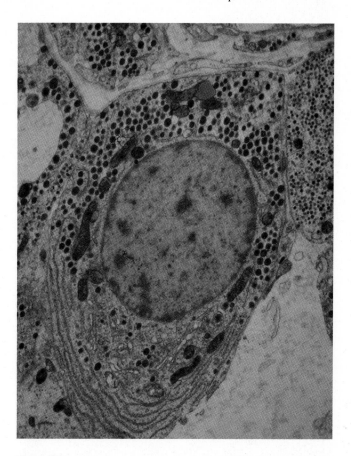

Figure 15.2 When they are not dividing, chromosomes are dispersed as a substance called chromatin (dark patches inside nucleus in center of cell).

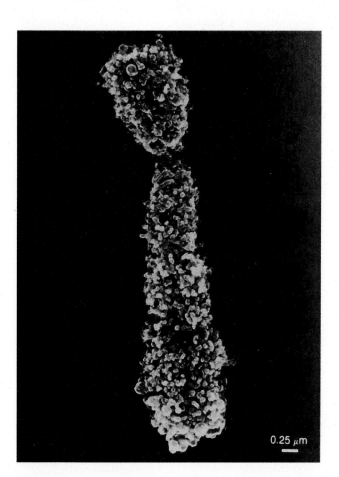

Figure 15.1 A scanning electron micrograph of a eukaryotic chromosome taken during cell division.

Question: *Where are the chromosomes located in a eukaryotic cell?*

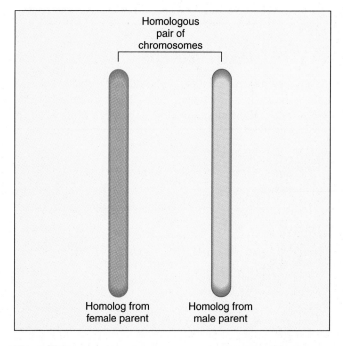

Homologous
pair of
chromosomes

Homolog from
female parent

Homolog from
male parent

Figure 15.3 This conceptual figure indicates that each homologous pair of chromosomes consists of two homologs (chromosomes), one contributed by the female parent and one by the male parent. Each homolog is similar in size and genetic content.

has 10 pairs, humans have 23 pairs, ducks and pumpkins each have 40 pairs, and crayfish have 50 pairs. Obviously, complexity is not related to the number of chromosomes a species possesses. The relationship between the two members of each homologous chromosome pair is interesting and fundamental in conveying hereditary information between generations. One member of each pair of homologous chromosomes came from each parent. Therefore, each parent contributed exactly one half of the chromosomes present in cells of the offspring. In humans, 23 chromosomes came from the mother and 23 from the father.

A cell with one complete set of homologous chromosome pairs is called **diploid,** abbreviated as $2n$ where n = the number of different homologs in one complete chromosome set. **Gametes,** sperm and egg cells, or other cells with only one homolog of each type, are called **haploid** and are signified as n. To extend this example to humans, each of our body cells is diploid ($2n$) and contains 23 homologous pairs of chromosomes, or 46 chromosomes in total; one homolog of each pair came from each parent. Our sperm or egg cells are haploid (n) and contain only 23 chromosomes, one of each type of homolog. Prokaryotic cells, with their single circular chromosome, are considered haploid.

BEFORE YOU GO ON   Questions related to cell division and inheritance guided research in the late 1800s. Much of the interest in these questions was generated by microscopic observations of chromosomes during cell division. Prokaryotic cells contain a single circular chromosome. Cells of each eukaryotic species contain a specific number of homologous chromosome pairs. Diploid cells contain a complete set of homologous chromosome pairs, whereas haploid cells contain only one set of homologs.

TYPES OF CELL REPRODUCTION

As expressed in the cell theory, a fundamental characteristic of cells is an ability to reproduce themselves by a process called *cell division.* **Reproduction** is the origination of new cells or organisms from preexisting ones. Cell division is a type of **asexual reproduction,** a process of replication in which a single *parent cell* gives rise to two or more identical (or nearly identical) *daughter cells.* To produce identical daughter cells, two cell division processes—*binary fission* and *mitosis*—evolved, which ensured that each new cell received the same genetic information as the original parent cell as well as a full complement of cellular organelles. A third type of cell division—*meiosis*—gives rise to haploid daughter cells that differ from the parent cell. Table 15.1 compares the three types of cell division.

Binary Fission

Recall that prokaryotic cells have neither a nucleus nor any membranous organelles, although they do have ribosomes. Bacterial cells reproduce by **binary fission,** a form of asexual reproduction that consists of DNA replication followed by cell division (see Figure 15.4). Prior to cell division, the single circular chromosome is duplicated through replication of its DNA molecule (DNA replication is described in Chapter 17). After replication is completed, the two bacterial chromosomes remain attached, at different points, to the plasma membrane of the parent cell. **Cytokinesis,** or cytoplasmic division that results in two cells, follows immediately as the plasma membrane wedges inward at the center of the cell between the two chromosomes, cell walls form along the same plane, and the cell divides into two identical daughter cells that each contain a chromosome and ribosomes obtained from the parent cell. The entire process usually takes less than one hour. Binary fission is a bacterium's means of reproducing asexually, and the two daughter cells represent independent new offspring.

Mitosis

Eukaryotic cells have a nucleus; numerous other membranous organelles, such as endoplasmic reticulum, mitochondria, Golgi complexes, and, in plants, chloroplasts; and ribosomes.

Table 15.1   Comparison of Types of Cell Division

Feature	Type of Cell Division		
	Binary Fission	**Mitosis**	**Meiosis**
Cell type	Prokaryotic cells	Eukaryotic cells	Eukaryotic cells
Parent cells	Haploid (n)	Diploid ($2n$)	Diploid ($2n$)
Number of chromosome replications	1	1	1
Number of nuclear divisions	1	1	2
Number of cytoplasmic divisions	1	1	2
Number of daughter cells produced	2	2	4
Characteristics of daughter cells produced	Haploid: identical to parent cell	Diploid: identical to parent cell	Haploid: genetically different from the parent cell: gametes (animals) or spores (plants)

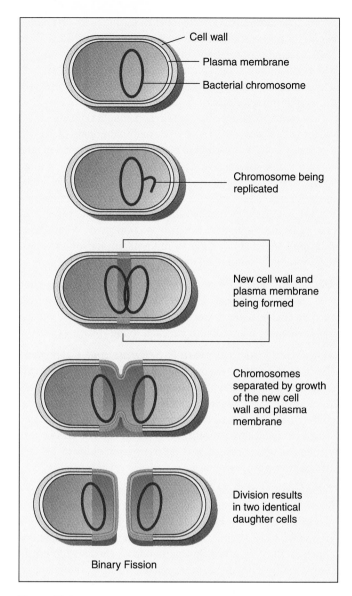

Figure 15.4 Bacteria reproduce by binary fission. The process begins when the bacterial chromosome is duplicated; the resulting chromosomes are attached to the plasma membrane at a different point. Following chromosome duplication, new cell wall and plasma membrane synthesis begins around the cell's center. Binary fission is completed when the cell divides into two identical daughter cells.

Question: *If binary fission takes one hour and all new daughter cells continue to reproduce asexually, how many bacteria would be produced in 24 hours?*

Many, but not all, eukaryotic cells may divide one or more times during their lifetime. Such cells engage in a repetitive process in which they grow and divide. This process is called the *cell cycle.*

The Cell Cycle

The **cell cycle** of dividing eukaryotic cells is the sequence of events that make up one complete cell division. The length of a cell cycle varies among species and cell types, but 24 hours is the common standard. As shown in Figure 15.5, the cell cycle consists of four phases, labeled G_1, S, G_2, and M.

Immediately after cell division, **interphase** begins with G_1 ("first gap"), a period in which the new cell grows rapidly and generates new membranous organelles. Interphase continues through the S and G_2 phases.

The *S* ("synthesis") phase begins when DNA replication starts and ends when all of the chromosomes have been duplicated. Chromosomes are not the only structures duplicated during S phase. Microtubules play a major role in moving chromosomes about during cell division. In most animal cells, microtubules are generated and organized in a cytoplasmic area called the **centrosome,** a structure that appears as a small cytoplasmic cloud containing two **centrioles** from which the microtubules seem to emerge. Both the centrosome and its centrioles are duplicated during the S phase. Note that plant cells also have centrosomes but no centrioles.

As synthesis of the various structures is completed, the cell enters the G_2 ("second gap") phase, in which preparations take place for chromosome condensation and construction of the microtubule framework required for cell division. Interphase ends when M ("mitosis") begins.

Phases of Mitosis

Mitosis, or nuclear division that results in two nuclei, consists of a remarkable sequence of dynamic events involving chromosomes and microtubules. Mitosis is divided into four stages

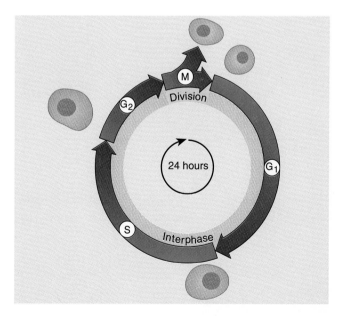

Figure 15.5 The eukaryotic cell cycle consists of four phases and typically takes about 24 hours. Interphase begins after cell division is completed. The G_1 phase is characterized by rapid cell growth. In the S phase, chromosomes, microtubules, centrosomes, and centrioles are all duplicated. During the G_2 phase, the cell prepares to divide. Cell division occurs during the M phase.

for convenience in describing the process: prophase, metaphase, anaphase, and telophase (see Figure 15.6). However, the process is actually continuous, and the phases have no distinct borders. Also, there are many variations in the mitotic process in eukaryotic organisms from different kingdoms. The entire cell division process is described in Figure 15.7.

Prophase: Chromosomes Appear and the Mitotic Spindle Forms During **prophase,** chromatin condenses into distinct duplicated chromosomes that can be seen with a light microscope. Each duplicated chromosome consists of two identical **sister chromatids** that are held together at their **centromere,** a shared, constricted region of DNA (see Figure 15.8, page 282). As the sister chromatids become visible, the *mitotic spindle*—an elaborate system of cytoplasmic microtubules that will

coordinate chromosome movements during cell division—begins to appear outside the nucleus (see Figure 15.9, page 282). During prophase, the duplicated centrosomes and centrioles (in animal cells) split and move to the ends of the cell, where they form the two *spindle poles.* Each centrosome produces microtubules that form the mitotic spindle.

Soon after the mitotic spindle appears, the nuclear envelope breaks down into *vesicles,* or small membrane pieces. At this stage, spindle microtubules enter the nuclear region, and some become attached to each centromere at newly formed specialized protein structures called *kinetochores.* As described in Figure 15.10 (see page 282), each sister chromatid has one kinetochore, and its attached microtubules extend toward the pole opposite the one to which it is connected. When living cells are viewed at this stage of mitosis, chromosomes appear to be moving around in an excited fashion.

Metaphase: The Chromosomes Align After prophase, the cell enters **metaphase,** a stage in which sister chromatids become aligned at the center of the cell—the *metaphase plate*—by kinetochore microtubules attached to the spindle poles.

Anaphase: Sister Chromatids Separate and Cytokinesis Begins **Anaphase** begins when paired sister chromatids are suddenly separated and pulled toward opposite poles as their kinetochore microtubules shorten. This stage marks the point at which the duplicated homologs become separated into two sets of identical chromosomes received by each of the two new identical daughter cells. Cytokinesis, division of the cytoplasm, also begins during anaphase.

Telophase: Daughter Cells Form During the last stage of mitosis, **telophase,** daughter cell chromosomes reach the poles, and kinetochore microtubules fade. Nuclear envelopes form around each set of chromosomes, the chromosomes then disappear as the condensed chromatin uncoils, and nucleoli reappear. Finally, cytokinesis is completed, forming two identical daughter cells.

Cytokinesis occurs differently in animal and plant cells. In animal cells, the plasma membrane in the center of the cell begins to fold inward between the two daughter nuclei near a ring of microfilaments that lies just beneath the plasma membrane (see Figure 15.11, page 282). When the infolding is completed, the cell divides into two cells.

Plant cell cytokinesis is not based on membrane infolding as in animal cells. Rather, beginning in late anaphase, small membranous vesicles begin to form in the cell where the metaphase plate once existed (see Figure 15.12, page 283). These vesicles grow outward, eventually merge, and form a double-layered membrane, or *cell plate,* that extends across the entire cell. Cell walls then begin to form between the two membrane layers. When construction of the cell wall is completed, the cell divides along the cell plate into two daughter cells.

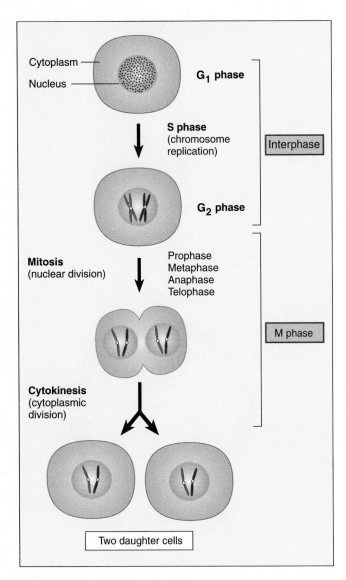

Figure 15.6 Following interphase, mitosis—nuclear division—occurs. The events of mitosis are categorized in four phases: prophase, metaphase, anaphase, and telophase. The M (mitosis) phase ends when cytokinesis occurs and the two daughter cells enter into a new cell cycle.

Mitosis and the Division of Organelles

The remarkable process by which daughter cells receive a nucleus and chromosomes from a parent cell was understood

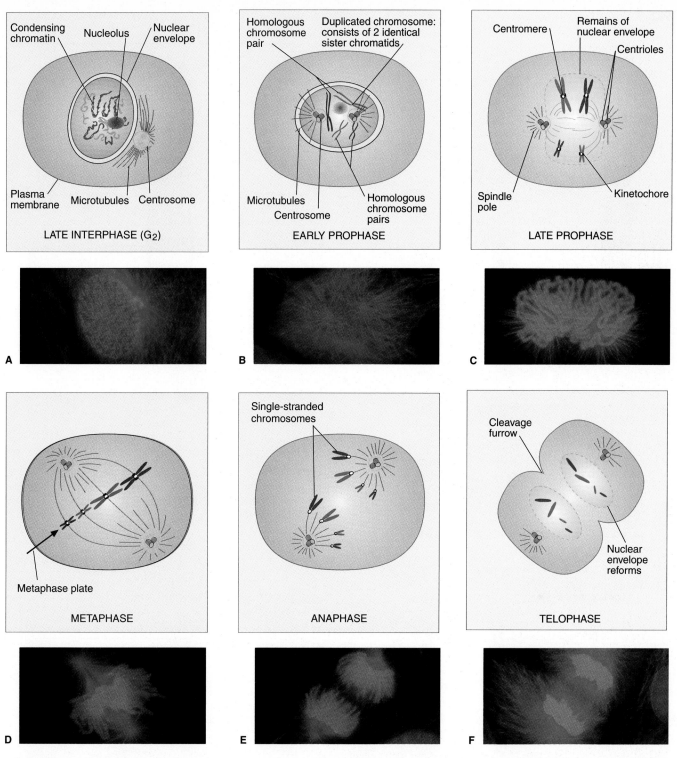

Figure 15.7 The events of mitosis are fully described in this figure. The photomicrographs are of dividing animal cells. the schematic drawings are included to depict structures that cannot be seen clearly in the photos. (A) LATE INTERPHASE (G2). Chromosomes have been duplicated and chromatin (blue) starts to condense. Centrosomes (yellow) have also been replicated, but they have not yet separated, and microtubules (green) generated and organized by centrosomes become visible. The nuclear envelope and nucleolus remain intact during interphase. (B) EARLY PROPHASE. Chromatin condenses into duplicated chromosomes, each of which is composed of 2 identical sister chromatids. In this drawing, there are two homologous chromosome pairs that can be distinguished by length (that is, the two long chromosomes are members of one homologous chromosome pair and the two short chromosomes are members of another pair). The color differences between homologous indicate different parental origin (let's assume that the light blue homologous came from the female parent and the dark blue homologous came from the male parent). These diagrammatic chromosomes will be used to highlight the movement and separation of chromosomes during the different phases of mitosis. Note that the centrosomes have separated and microtubules are easily observed. (C) LATE PROPHASE. Sister chromatids, joined at their centromere, are evident. Both the nuclear envelope centrioles, have migrated to opposite ends of the cell, where they form spindle poles, which are part of the mitotic spindle (shown in green). Microtubules emanating from the spindle poles become attached to kinetochores, which are part of the centromeres. (D) METAPHASE. All sister chromatid pairs become aligned at the metaphase plate in the middle of the cell. (E) ANAPHASE. Sister chromatids separate into two identical sets of daughter chromosomes, and each set is pulled toward a spindle pole in two regions of the cell. (F) TELOPHASE. Daughter chromosomes reach the poles, and microtubules fade. Nuclear envelopes form around each chromosome set, and the cell undergoes division as the membrane folds inward, creating a cleavage furrow. Cytokinesis concludes when division is completed and two new daughter cells, each identical to the original parent cell, have been formed.

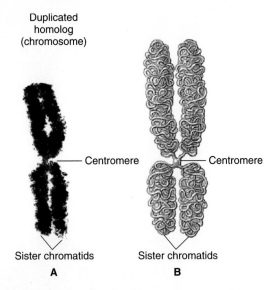

Duplicated
homolog
(chromosome)

Centromere   Centromere

Sister chromatids   Sister chromatids
      A   B

Figure 15.8   (A) The appearance of one duplicated homolog during mitosis. Each duplicated homolog consists of two identical sister chromatids that are joined at their centromere. (B) The centromere is a short region of DNA that is shared by two sister chromatids.

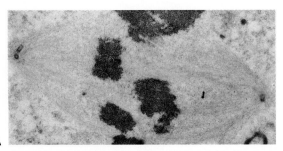

A

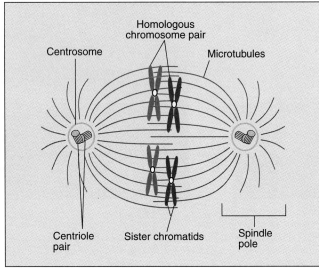

Homologous
chromosome pair

Centrosome   Microtubules

Centriole   Sister chromatids   Spindle
pair   pole

B

Figure 15.9   (A) An electron micrograph showing a mitotic spindle. (B) The mitotic spindle found in animal cells is an elaborate system of microtubules, centrosomes, and centrioles, anchored at spindle poles at each end of the cell. The mitotic spindle coordinates the movements of sister chromatids during mitosis. Mitotic spindles of plant cells are similar but have no centrioles.

Question: *When does the mitotic spindle appear and disappear in mitosis?*

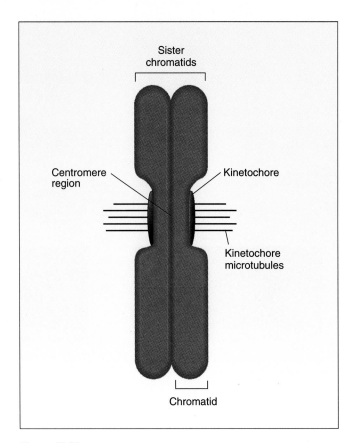

Sister
chromatids

Centromere   Kinetochore
region

                            Kinetochore
                            microtubules

Chromatid

Figure 15.10   A drawing showing a duplicated homolog at metaphase. Each sister chromatid has a kinetochore to which microtubules become attached. At the beginning of anaphase, the kinetochore microtubules pull the chromatids toward opposite spindle poles.

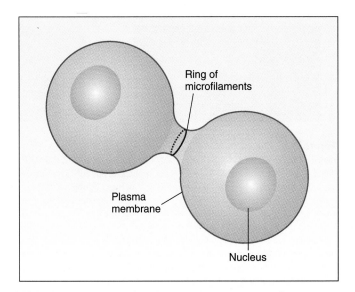

Ring of
microfilaments

Plasma
membrane

Nucleus

Figure 15.11   During cytokinesis, animal cells are drawn inward by a ring of microfilaments. Eventually, the parent cell pinches apart, forming two daughter cells.

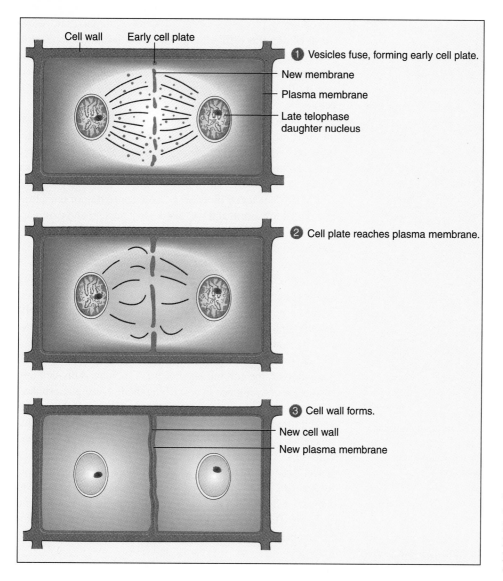

1 Vesicles fuse, forming early cell plate.

Cell wall   Early cell plate

New membrane

Plasma membrane

Late telophase
daughter nucleus

2 Cell plate reaches plasma membrane.

3 Cell wall forms.

New cell wall

New plasma membrane

Figure 15.12 Cytokinesis in plant cells involves the formation of a cell plate along which a new cell wall and plasma membrane form. When the cell plate is completely developed, the cell divides into two daughter cells.

over 100 years ago. Decades later, new technologies, such as electron microscopes, led to an interesting new question about mitosis. How does each daughter cell obtain the other organelles necessary for it to survive and function? The endosymbiotic theory (see page 52 in Chapter 3) explains that modern eukaryotic mitochondria and chloroplasts evolved about 2 billion years ago from prokaryotic cells that came to inhabit early eukaryotic cells. Although these organelles are functionally integrated with the nucleus, they retain their own DNA and divide independently by a process similar to binary fission (see Figure 15.13, page 284).

Like bacteria, chromosomes of mitochondria and chloroplasts are composed of circular DNA strands, but there are multiple copies of the chromosome in each organelle. Evidence now indicates that new mitochondria and chloroplasts arise from the growth and fission of preexisting mitochondria and chloroplasts during interphase. They are then somehow distributed to each daughter cell prior to cytokinesis. Other membranous organelles such as Golgi complexes and the endoplasmic reticulum (ER) simply break up into many small fragments and vesicles during interphase and become distributed throughout the cell. This ensures that each daughter cell

receives some of the membranous materials that can be used in rebuilding Golgi complexes and ER during the G_1 phase of the next cell cycle.

Although mitosis has been carefully studied for over a century, challenging new questions continue to arise. For example, what factors are responsible for construction of the mitotic spindle? For initiating DNA synthesis? What regulates the rates of mitosis in normal cell populations? What causes a loss of mitotic regulation that can lead to cancer? The complexity of such questions indicates that mitosis will likely continue to be a fruitful field of research for many more decades.

Why Mitosis?

What does mitosis accomplish? Basically, it preserves the genetic content of cells from generation to generation. Mitosis produces daughter cells that are genetically identical to a parent cell. What kinds of cells divide by mitosis? As described in a later section, certain haploid plant cells use mitosis to produce gametes, or sperm and egg cells. Mitosis also occurs in all eukaryotic cells involved in growth, repair, development, and cell replacement. Thus it has tremendous importance in the

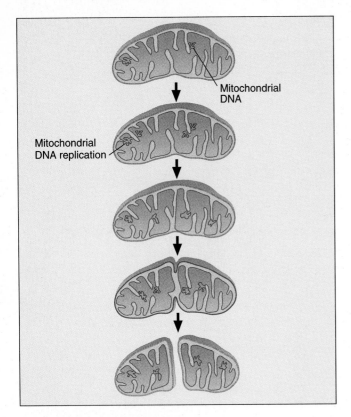

Mitochondrial DNA

Mitochondrial DNA replication

Figure 15.13   Mitochondria in eukaryotic cells are thought to have evolved from prokaryotic cells. During cell division, they divide by a process similar to binary fission. Daughter mitochondria become distributed into each daughter during cell division.

Question: *What other eukaryotic organelle is also capable of dividing independently of the nucleus?*

lives of all eukaryotic organisms. Let us consider what this means to human beings. In adult humans, it is estimated that about 1 trillion mitoses occur every 24 hours to provide new cells to replace those that normally die in performing their functions! An average red blood cell, for example, lives approximately 120 days, and millions of these cells die every hour and are replaced by mitosis. Similarly, wounds and injuries are repaired, and a fertilized egg (one cell) becomes a baby (trillions of cells) in about nine months.

BEFORE YOU GO ON   Many cells have the ability to divide by some type of asexual cell division. In bacteria, binary fission gives rise to new individuals. Eukaryotic cells divide by mitosis and cytokinesis, a process characterized by one chromosomal duplication, one nuclear division, and one cytoplasmic division. The resulting daughter cells are identical to the original parent cell.

Meiosis

Sexual reproduction is not a term that usually sends people in search of a dictionary. However, since hundreds of different

reproductive processes occur in plants, animals, and other forms of life, it is difficult to formulate a definition applicable to all organisms. In eukaryotic multicellular organisms, **sexual reproduction** is the generation of a new cell or a new organism by the fusion of two haploid gametes—sperm and egg cells.

How are haploid cells produced from diploid cells? That question arose in the late 1800s when microscopic investigations revealed that a species's sperm and egg cells contained only half the number of chromosomes found in their other cells.

Meiosis is a cell division process consisting of one chromosome duplication and two successive nuclear and cytoplasmic divisions. Consequently, one diploid parent cell produces four haploid daughter cells. In animal species, the four haploid cells produced are gametes. In plants, meiosis also results in four haploid cells, but they are not gametes. In some plant species, these cells are **spores,** structures that may grow directly into a new plant. In other plant species, the haploid cells develop into structures that contain gametes (for example, pollen grains).

Meiosis is illustrated in Figure 15.14 (see pages 286-287). Like mitosis, the continuous process of meiosis is divided into phases for convenience in describing them. During interphase, before meiosis begins, parent diploid cells grow and develop, and chromosomes exist only as diffuse chromatin. While in this phase, each homolog is duplicated, and thus each chromosome consists of two sister chromatids. Meiosis is now ready to take place. Because there are two nuclear divisions during meiosis instead of one, as in mitosis, roman numerals are appended to indicate whether the events occur during the first or second division. Though there are general similarities between mitosis and meiosis, the latter has many special features and is more complex.

Meiosis I

The first stage, **prophase I,** is by far the longest phase and usually account for 90 percent or more of the time spent in meiosis. Early prophase begins when chromatin starts to condense. Although each duplicated homolog consists of two sister chromatids, separate chromatids do not become visible until late prophase. In early prophase, duplicated homologs appear as long stringlike structures. Duplicated homologs next undergo **synapsis,** a process in which the duplicated maternal and paternal homologs of each homologous chromosome pair are joined together (synapsed) and become aligned lengthwise. A synapsed pair of duplicated homologs is known as a **bivalent.** During synapsis, homologs are connected with one another by a long zipperlike formation called the **synaptonemal complex,** which is made primarily of proteins (see Figure 15.15A, page 288). When synapsis is completed at mid-prophase, chromosomal exchanges known as **crossing-over** occur between homologs (see Figure 15.15B). During crossing-over, DNA strands in both the maternal and paternal homologs break and reattach to DNA in the other homolog. Sites on the homologs where these exchanges occur remain attached to one another and can be observed later in prophase. Crossing-over occurs during every meiosis, and in human chromosomes, two or three crossovers usually occur between each pair of homologs. In late prophase, the synaptonemal complex dissolves, and sister chromatids, connected at their centromere, finally become visible. Nonsister chromatids that crossed over appear as X-shaped connections called *chiasmata.* The nuclear envelope and nucleoli disappear, a spindle is formed, and microtubules become attached to the kinetochore of each duplicated homolog.

The continuing meiotic process advances to **metaphase I,** when bivalents become aligned at the center of the cell. Their orientation is random, and there is a 50–50 chance that one duplicated homolog will go to a particular pole during the next phase. Metaphase I ends when the spindle fibers begin to shorten and separate the homologs of each bivalent from one another.

During **anaphase I,** spindle fibers pull the separated homologs toward opposite poles of the cell. During this phase of meiosis, the sister chromatids, which form the duplicated homolog, remain attached and move together as a single unit. Anaphase I is a key event in creating new genetic combinations in sexually reproducing species. New combinations of genetic material or chromosomes, created by any process (including crossing-over), is termed **genetic recombination.** Because the original pairs of maternal and paternal homologs are *independently assorted* in anaphase I, each duplicated homolog has a 50–50 chance of ending up in one of the new daughter cells. Consequently, the gametes produced can have any possible combination of maternal and paternal chromosomes (the actual number of combinations possible is 2^n where n = the haploid number of chromosomes. For humans, this number is $2^{23} = 8.4$ million).

The first meiotic division concludes during **telophase I,** when the separated homologous chromosomes cluster at each pole and the first cytokinesis takes place. Each homolog is still composed of two sister chromatids sharing a single centromere, but the total chromosome number has been reduced by half. The two daughter cells harbor an equal number of homologs, but each contains different maternal and paternal homologs because they separated independently during this first division.

Interkinesis

In most species, daughter cells created during the first meiotic division enter **interkinesis,** a period when the homologs briefly fade and new nuclear envelopes form around daughter cell nuclei. The amount of time cells spend in this state varies with the species and the sex of the reproducing organism, but for most cells, it is a relatively short period. No chromosomal duplication occurs during interkinesis.

Meiosis II

The second meiotic division resembles a normal mitotic cell division. It begins with **prophase II,** when the newly formed nuclear envelopes of the two daughter cells disappear and sister chromatids become visible. This second prophase ends when sister chromatid pairs move to the center of each daughter cell. During **metaphase II,** sister chromatids are aligned at the center of each daughter cell. The attached spindle fibers begin to contract, and metaphase II ends. In **anaphase II,** the sister chromatids are separated as kinetochore microtubules pull them toward opposite poles. The result is that each of the four daughter cells receive, in **telophase II,** one set of homologs. A second cytokinesis concludes the meiotic process.

Why Meiosis?

Meiosis produces daughter cells that differ from parent cells in three significant ways. First, the daughter cells are haploid rather than diploid. Second, independent assortment during anaphase I creates daughter cells that contain different assortments of maternal and paternal chromosomes. Third, crossing-over during meiosis I also results in daughter cells having unique genetic variations. Consequently, daughter cells resulting from meiosis are genetic originals, and offspring formed from these cells will be genetically distinctive. However, new questions have been raised about the fundamental importance of meiosis in sexual reproduction (see the Focus on Scientific Process: "The Significance of Meiosis and Sexual Reproduction").

Fertilization

Fertilization is the process in which haploid gametes unite to form a diploid *zygote* that develops into a new offspring. The outcome of fertilization is the restoration of the normal diploid number of chromosomes $(n + n = 2n)$. Thus fertilization restores the normal chromosome number in a species.

BEFORE YOU GO ON Meiosis is a mechanism used by diploid eukaryotic organisms to create haploid cells. It consists of one chromosomal duplication in a parent cell followed by two nuclear and cytoplasmic divisions. In completing meiosis, a single diploid ($2n$) parent cell will produce four haploid (n) cells. Because of crossing-over between homologs and the random distribution of sister chromatids during anaphase I of meiosis, each new haploid cell represents a unique genetic combination. Fertilization involving two haploid gametes reestablishes the diploid condition in a zygote.

LIFE CYCLES

Eukaryotic species have evolved countless variations on the theme of sexual reproduction to produce offspring in their life cycle. Both animal and plant life cycles generally involve meiosis and fertilization; however, plants also use mitosis.

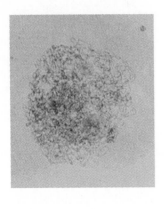
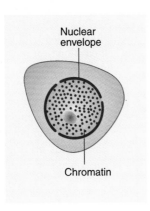

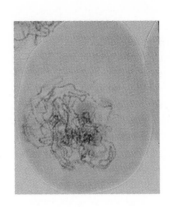
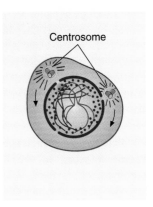

Nuclear envelope

Chromatin

Centrosome

EARLY PROPHASE I

MIDDLE PROPHASE I

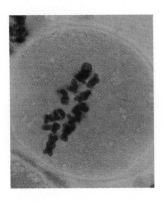

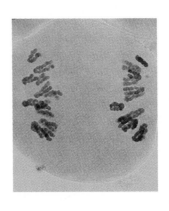
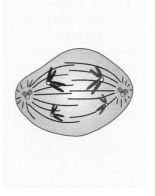

METAPHASE I

ANAPHASE I

METAPHASE II

ANAPHASE II

Figure 15.14 Meiosis consists of one chromosome duplication and two nuclear divisions and results in the formation of four haploid cells. The events of meiosis are described in this figure. Photos and drawings of events in meiosis I are from animal cells. Only drawings are shown for meiosis II.

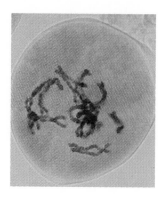
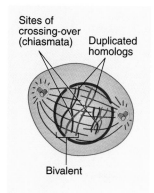

Sites of
crossing-over
(chiasmata)   Duplicated
homologs

Bivalent

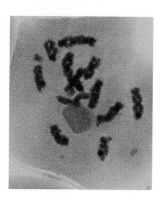
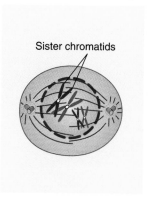

Sister chromatids

MIDDLE PROPHASE I   **LATE PROPHASE I**

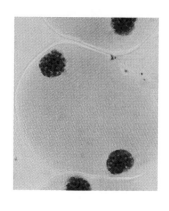
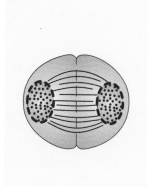

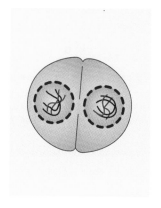

TELOPHASE I   **PROPHASE II**

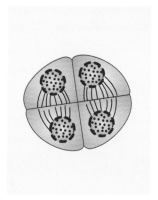

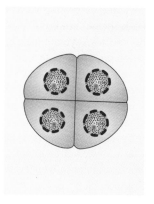

TELOPHASE II   **ONSET OF
CYTOKINESIS**

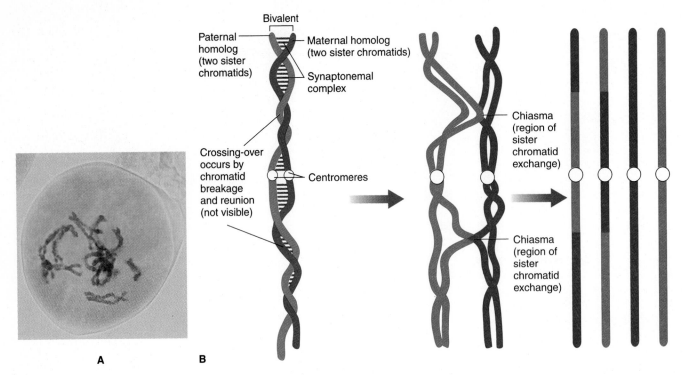

Figure 15.15 (A) The zipperlike synaptonemal complex joins duplicated homologs together during prophase I. (B) Crossing-over occurs when bivalent homologs become intertwined, sister chromatid segments break, and each becomes joined to a sister chromatid from the other homolog. When chromosomes become visible during late prophase I, crossover sites can be identified by the presence of chiasmata.

Question: *What is the importance of crossing-over in genetics?*

Animals

The basic animal life cycle, shown in Figure 15.16 (see page 290), involves sexual reproduction in which haploid sperm and egg cells from two distinct sexual types, male and female, unite in fertilization to produce a diploid zygote that develops into a single offspring. Although most animal life cycles follow the basic model, there are some interesting variations. For example, a certain crustacean species employs "gonad transplants" in its life cycle! In this species, a young male crab uses a form of hypodermic injection to insert a few of its cells into a female. The injected cells eventually form a *testis,* the male gonad, which then produces the sperm used by the female to fertilize her eggs. Thus the female produces both types of sex cells, yet technically the sperm cells came from a male. The list of life cycle variations used by animals is almost endless if all deviations from the basic bisexual form are included. Most, however, follow a general pattern in which meiosis and fertilization are of central importance.

Plants

Plants have life cycles that are fundamentally different from those of animals (see Figure 15.17, page 290). Two distinct phases, haploid and diploid, occur during the life cycle of a plant species. In a single species, each phase results in the production of a life form that is often remarkably different from the other. The diploid plant form, known as a *sporophyte*, produces haploid cells by meiosis. In some plant species, these haploid cells are spores that can germinate directly into one form of the plant. Because fertilization does not occur, this form is haploid. The haploid plant, or *gametophyte*, produces haploid cells by mitosis, and some of them mature into gametes. Under proper conditions, fertilization will occur and lead to the development of the diploid sporophyte form of the plant.

Mosses, familiar to many of us as the green carpeting on trees, logs, forest floors, and roofs, have a conspicuous gametophyte form and an obscure sporophyte phase. In contrast, the fern that most of us recognize is a diploid sporophyte form; the haploid fern gametophyte is inconspicuous and generally overlooked. All other plants have reproductive strategies that involve variations of the same haploid–diploid alternation theme.

MENDEL REVISITED

For convenience in discussing certain concepts, biologists have developed terms to describe chromosomes and their relationships with one another. These are presented in summary form

The Significance of Meiosis and Sexual Reproduction

Why have sex? Understanding the ultimate nature of sex and sexual reproduction requires, first, full comprehension of the process of meiosis and the significance of its final accomplishment, the production of discrete reproductive cells containing new arrangements of genetic information. Second, it requires an appreciation of the significance of fertilization, the process that restores the normal chromosome number for the species being perpetuated. Though somewhat abstract, the following is an intriguing example of the types of inquiries scientists make about biological processes.

Given the complexities involved, biologists have posed a number of new and interesting questions about the adaptive value of a dependence on meiosis and sexual reproduction for producing offspring. There are a number of obvious *disadvantages* in these operations, among them reliance on an extremely complex mechanism, meiosis, for producing haploid cells; the frequent occurrence of errors (for example, inaccurate duplication of chromosomes or separation of chromosomes); the need to maintain complex sex organs; and the effort required to locate a mate and establish circumstances that can bring about fertilization. These costs cause scientists to ask, what are the possible *advantages* of using this elaborate system of reproduction?

Conventional reasoning has run along the following lines. Together, meiosis and sexual reproduction, including fertilization, create opportunities for generating novel genetic combinations and producing offspring that are genetic originals. Variation can be introduced in two ways. Genetic recombination is any process that generates new gene or chromosome combinations. In most organisms, this occurs in meiosis I, when the homologs are randomly assorted; consequently, each gamete is almost certain to be distinctive in the type of genetic information it contains. It also occurs when crossing-over during meiosis increases the variety of genetic information in cells produced. In *outbreeding,* offspring are produced from gametes contributed by two different parents.

Genetic recombination due to crossing-over and independent assortment of homologs in meiosis I is thought to be the more fundamental aspect of sex because many different plant and animal life cycles retain this feature but do not preserve outbreeding. However, certain questions are very troublesome for this line of reasoning. For example, why is it advantageous to break up genetic combinations once they are established and presumably successful?

The traditional view is now being reexamined by scientists who have studied new evidence about sex and sexual reproduction in the biological world. The relatively recent discovery that genetic recombination is virtually universal, occurring in all organisms including those that usually reproduce asexually (such as bacteria), has great significance. For example, it is now known that genetic recombination occurs at astonishing rates in bacteria (through mechanisms that we will discuss in future chapters). Recombination also tends to randomize genetic information in all organisms; the greater the randomization, the greater the variety of offspring that can be produced. For bacteria, this may be an advantage that is related to creating new variations quickly. Because of recombination, bacteria, which may produce a new generation in minutes or hours, can quickly adjust to new environmental conditions. Recombination is what allows bacteria to become resistant to certain drugs in relatively short periods of time.

However, can recombination and the resulting randomization of genetic material convey such marked advantages to eukaryotic plants and animals? In comparison with bacteria, genetic recombination must have less importance in long-lived organisms that are dependent on meiosis and sexual reproduction because rapid adjustments to changing conditions are not possible. Although it is clear that meiosis and sexual reproduction do produce genetic variation, and this may be important in the long run, many biologists now believe that their preeminent effect is to serve as stabilizing mechanisms. According to this *stabilization hypothesis,* the randomization of genetic information is actually minimized as a consequence of the elaborate, time-consuming events required for the successful completion of meiosis and sexual reproduction. These biologists therefore believe that the long-term genetic stability of a sexually reproducing species is of greater importance than any short-term advantage arising from variations resulting from recombination.

An intriguing extension of this stabilization hypothesis has recently been published. It is called the *DNA repair hypothesis,* and, in simplified form, it proposes that for eukaryotes, recombination and outbreeding provide opportunities for replacing, eliminating, or repairing DNA that has become altered in gamete-producing parental cells. Because transmitting damaged DNA or deficient chromosomes to a new generation would probably be very harmful, these mechanisms counteract that risk by removing damaged genetic materials. By combining DNA from two parents, it is much less likely that the offspring will be affected by abnormal gene for one parent will probably contribute a normal allele.

The true significance of meiosis and fertilization to sexually reproducing eukaryotic organisms is not yet clear. However, the old view that they conveyed unique advantages for creating new variations in offspring has been questioned. Reexamining accepted ideas like these is a characteristic of all good science.

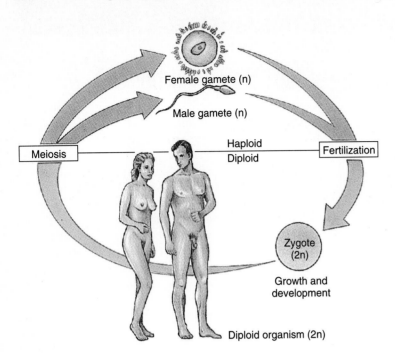

Figure 15.16   In the basic animal reproductive strategy, haploid (*n*) gametes are produced by two distinct sexes, male and female. When gametes unite in fertilization, a diploid (2*n*) zygote is formed that develops into an offspring resembling the adult.

Question: *Where does mitosis fit into this figure?*

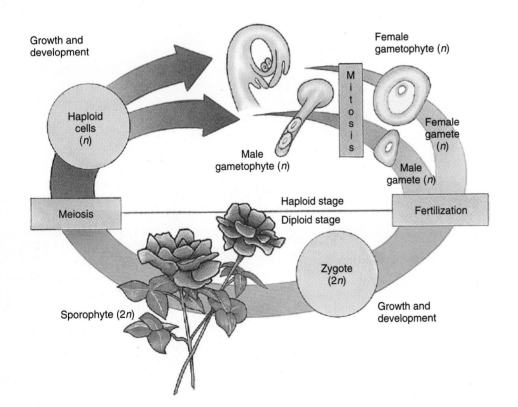

Figure 15.17   A plant life cycle consists of two distinctly different forms. The haploid gametophyte stage produces gametes by mitosis, and after fertilization, the resulting diploid zygote develops into a sporophyte plant. The diploid sporophyte stage produces spores by meiosis. Spores grow into gametophyte plants. This type of life cycle is known as *alternation of generations*.

in Figure 15.18, along with related terms used in genetics, using one pair of homologous chromosomes. Recall that each chromosome pair consists of two homologs, one contributed by the female parent (the maternal chromosome) and one by the male parent (the paternal chromosome), and that the two homologs make up a homologous chromosome pair.

Homologous pairs of human chromosomes contain about 50,000 bits of genetic information, called **genes,** each composed of a DNA segment that determines specific traits. The

term *gene* was introduced in 1909 to replace *hereditary character* or *discrete factor* as the name of the fundamental hereditary unit. Each homolog of the homologous chromosome pair contains one or more genes for the same trait (for example, flower color) at a specific **locus,** or identical location, along their length. A single gene may have one or more alternate forms that are called **alleles.** For example, a gene for flower color, a single trait, may have different alleles that result in red, white, or purple colors, which are alternative forms of the trait.

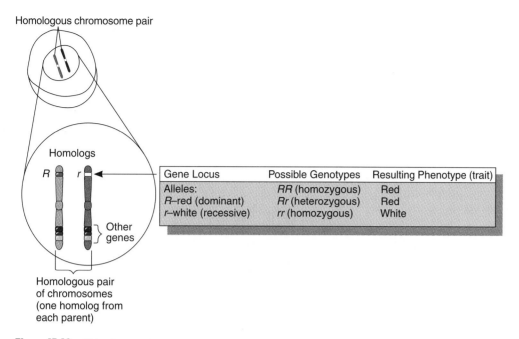

Homologous chromosome pair

Homologs

R   r

Other genes

Homologous pair
of chromosomes
(one homolog from
each parent)

Gene Locus	Possible Genotypes	Resulting Phenotype (trait)
Alleles:	*RR* (homozygous)	Red
R–red (dominant)	*Rr* (heterozygous)	Red
r–white (recessive)	*rr* (homozygous)	White

Figure 15.18   This diagram describes how various terms in genetics are used in describing the relationships between chromosomes and also chromosomal structures and locations. Chromosomes are not visible until after they have been duplicated prior to cell division. The duplicated chromosome is not shown in this figure. Eukaryotic chromosomes contain thousands of *genes* that determine all traits of an organism. Each gene exists at a specific location called a *gene locus*. In this example, a gene for flower color occurs at a locus in the upper arm of the chromosome. A diploid plant will contain two flower color genes, one on each homolog. The gene for flower color has two *alleles*, identified as *R* (the dominant allele) and *r* (the recessive allele). The parent plant *genotype* for flower color is *Rr*, and its *phenotype* is red. Haploid gametes produced by a parent will contain one flower color allele; when fertilization restores the diploid number, offspring plants will each have two alleles. The possible genotypes of offspring plants are *RR*, *Rr*, and *rr*. Since *R* is dominant, offspring phenotypes are red if their genotype is *RR* or *Rr*, white if their genotype is *rr*. *Homozygous* refers to genotypes with two identical alleles (*RR* or *rr*); *heterozygous* indicates two different alleles (*Rr*).

The specific alleles contained in a cell (or an individual) are called its **genotype.** The physical trait that occurs as a result of a specific genotype combined with any environmental influence is referred to as the **phenotype.**

Understanding meiosis and fertilization makes it possible to look back and reevaluate Mendel's studies and the laws of genetics that bear his name. Using pea plants and mathematical analyses, but with no knowledge of genes or chromosomes, Mendel was able to describe regular patterns of inheritance that had not been identified previously. By luck or design, he chose traits that occurred in one of two phenotypic forms (for example, stems were either long or short, and unripe pods were either yellow or green). When his experiments were completed, he was able to make the following inferences:

1. Pea plants possessed two discrete hereditary factors (alleles) for each trait.

2. One member of the allele pair expressed itself (was dominant) in the presence of the other, recessive allele.

3. When gametes were formed, the two alleles separated independently, and the resulting sex cells contained only one allele for each trait.

4. Alleles for each of the traits studied were sorted randomly into the gametes.

5. The alleles were recombined when fertilization occurred.

When Mendel's work was rediscovered in the early twentieth century, his conclusions were formally summarized as laws and named in his honor.

The mechanisms of meiosis and fertilization are now understood, and hence it is possible to correlate Mendel's observations and laws with events that occur during the two processes. His discrete pairs of factors, now called *alleles,* are carried on separate chromosomes, with each homolog of a homologous pair carrying one allele. The two alleles separate as a function of chromosome separation during meiosis (anaphase I) and return to the diploid state during fertilization. All the axioms from Mendelian genetics are derived from these two core principles. Thus Mendel's law of segregation is based on the fact that the two alleles determining a trait separate, or segregate, during meiosis and each gamete contains only one of the alleles. Mendel's law of independent assortment states that the alleles of one gene pair separate independently, or randomly, from alleles of another gene pair during meiosis. In Mendel's pea plants, for example, the alleles for plant height sorted independently from the alleles for pod color because they were on separate chromosomes. Once these central principles were defined and understood, they provided a starting point for the golden age of genetics that was to follow.

SUMMARY

1. During the nineteenth century, the cell theory proposed that all living organisms were composed of cells and that all cells arose from preexisting cells. Improved microscopes provided biologists with an opportunity to learn how this was accomplished.

2. Chromosomes in cells contain genetic information in the form of single, long DNA molecules.

3. Asexual reproduction gives rise to daughter cells or offspring that are identical to the parent.

4. Bacteria reproduce asexually by binary fission, a process in which the single circular bacterial chromosome is duplicated and then the parent cell divides into two daughter cells.

5. Eukaryotic cells that reproduce asexually follow a cell cycle that consists of cell growth, DNA and microtubule replication, and mitosis.

6. Mitosis has four phases—prophase, metaphase, anaphase, and telophase—and results in chromosomes being distributed precisely to two daughter cells that form after nuclear division and cytokinesis. Each of the two daughter cells is identical to the parent cell.

7. Sexual reproduction generates offspring that differ from their parents. It is accomplished by the fusion of two haploid gametes. Each sexually reproducing species has a characteristic diploid number of homologous chromosome pairs. Offspring receive one half of their homologs from each parent.

8. Meiosis is a process that gives rise to haploid gametes in animals and spores or haploid cells in plants. In meiosis, there is one chromosome duplication in the parent cell followed by two nuclear and cytoplasmic divisions, which results in four daughter cells. The daughter cells contain a haploid number of chromosomes and vary in their genetic content; thus they are different from the parent cell.

9. Two events in meiosis, independent assortment and crossing-over, are the source of much of the genetic variation that occurs in sexually reproducing species. Fertilization involves the fusion of two haploid gametes and restores the diploid chromosome number in the zygote.

10. Most animals have a basic life cycle in which two parents produce haploid gametes that fuse during fertilization to produce a diploid zygote that develops into a new offspring.

11. Plant life cycles include unique sporophyte and gametophyte generations. The sporophyte generation produces spores by meiosis that develop directly into the gametophyte generation. The gametophyte produces gametes by mitosis, and after fertilization, a new sporophyte generation develops.

12. Meiosis and fertilization can be used to explain Mendel's laws of segregation and independent assortment.

WORKING VOCABULARY

allele (p. 290)
asexual reproduction (p. 278)
cell cycle (p. 279)
chromosome (p. 276)
crossing-over (p. 284)
cytokinesis (p. 278)
diploid (p. 278)
fertilization (p. 285)
gametes (p. 278)
gametophyte (p. 290)
gene (p. 290)
genotype (p. 290)
haploid (p. 278)
homolog (p. 277)
meiosis (p. 284)
mitosis (p. 279)
phenotype (p. 290)
sexual reproduction (p. 284)

REVIEW QUESTIONS

1. What are chromosomes? Describe their structure.

2. When are chromosomes visible through the microscope? Why?

3. What are homologous chromosome pairs? Why is this an important concept in genetics?

4. What is the fundamental difference between haploid and diploid cells?

5. What is asexual reproduction?

6. How do bacteria reproduce?

7. Describe events in the cell cycle of dividing eukaryotic cells.

8. Explain how mitosis and cytokinesis allows one eukaryotic cell to produce two identical daughter cells. What is the importance of mitosis?

9. What is sexual reproduction?

10. Explain how meiosis allows a single eukaryotic cell to produce four animal gametes or plant spores.

11. What is accomplished by fertilization?

12. How does the basic animal life cycle differ from a plant life cycle?

13. What is the difference between a gene and an allele? Between a genotype and a phenotype?

14. How do meiosis and fertilization explain Mendel's law of segregation? His law of independent assortment?

ESSAY AND DISCUSSION QUESTIONS

1. Why are chromosomes visible only during cell division? What are some possible adaptive or survival advantages to the cell?

2. What might be some advantages to organisms that reproduce by asexual reproduction rather than by sexual reproduction? What are the advantages of sexual over asexual reproduction?

3. Students sometimes confuse mitosis and meiosis. What do the processes have in common? In what ways are they different?

4. Would you expect two cells (or individuals) with the same genotype ever to have different phenotypes? If so, under what conditions?

REFERENCES AND RECOMMENDED READING

Alberts, B., D. Bray, J. Lewis, M. Raff, K. Roberts, and J. D. Watson. 1994. *Molecular Biology of the Cell.* 3rd ed. New York: Garland.

American Society of Zoologists. 1989. Science as a way of knowing: VI. Cell and molecular biology. *American Zoologist,* 29: 483–817.

Baker, J. R. 1948–1955. The Cell-theory: A restatement, history, and critique. Five parts, published over seven years in the *Quarterly Journal of Microscopical Sciences.*

Blackwelder, R. E., and B. A. Shepard. 1981. *The Diversity of Animal Reproduction.* Boca Raton, Fla.: CRC Press.

Earnshaw, W. C., and A. F. Pluta. 1994. Mitosis. *BioEssays,* 16: 639–643.

Glover, D. M., C. Gonzales, and J. W. Raff. 1993. The centrosome. *Scientific American,* 268: 62–69.

Hyams, J. S., and B. R. Brinkley (eds.). 1989. *Mitosis: Molecules and Mechanisms.* London: Academic Press.

John, B. 1990. *Meiosis.* New York: Cambridge University Press.

McIntosh, J. R., and M. P. Koonce. 1989. Mitosis. *Science,* 246 622–628.

Murray, A. W., and M. W. Kirschner. 1991. What controls the cell cycle? *Scientific American,* 264: 56–63.

Penny, D. 1985. The evolution of meiosis and sexual reproduction. *Biological Journal of the Linnaean Society,* 25: 209–220.

Sagan (Margulis), L. 1967. On the origin of mitosing cells. *Journal of Theoretical Biology,* 14: 225–274.

Sawin, K. E., and S. A. Endow. 1993. Meiosis, mitosis, and microtubule motors. *BioEssays,* 15: 399–407.

ANSWERS TO FIGURE QUESTIONS

Figure 15.1   In the nucleus.
Figure 15.5   16,777,216.
Figure 15.10 It appears during prophase and disappears during telophase.
Figure 15.14 Chloroplasts.
Figure 15.16 Crossing-over generates new genetic combinations in gametes.
Figure 15.17 Mitosis provides cells for growth and development.

16

Chromosomes and Heredity

Chapter Outline

Reading Questions

1. What kinds of knowledge led to the formation of the chromosome theory?

2. How does the chromosome theory explain genetic inheritance?

3. What kinds of experiments proved the chromosome theory to be correct?

4. What are the bases for non-Mendelian patterns of inheritance?

5. What are the genetic consequences of chromosomal alterations?

 y the dawn of the twentieth century, it had become clear that the studies of Mendel and De Vries provided a theoretical foundation for understanding the heredity of certain traits in plants and animals. Within this framework, it was possible to make and test predictions about the inheritance of at least some traits in some organisms. Where did Mendelian "factors," later known as *genes* (the term we will use in most of this chapter), reside? Since gametes are the only direct link between generations, a partial answer to that question was evident, but the more difficult part of the question remained unanswered. Where in the sperm and egg cells were genes found? What structure or structures were they associated with?

ORIGIN OF THE CHROMOSOME THEORY OF HEREDITY

As microscope technology improved in the late 1800s, a new science, **cytology,** the study of cells, generated much information that led to a greater understanding of cell structure and function. Microscopes had revealed that sperm and egg cells consisted of a nucleus and cytoplasm. Were genes found in one or both of these cell parts? The cytoplasm seemed an unlikely location. There was very little of this substance in sperm cells compared to egg cells, and by then it was believed that both sexes made equal contributions in producing offspring. The nucleus, however, was approximately the same size in both gametes; thus most attention became focused on this structure. Where in the nucleus were genes found?

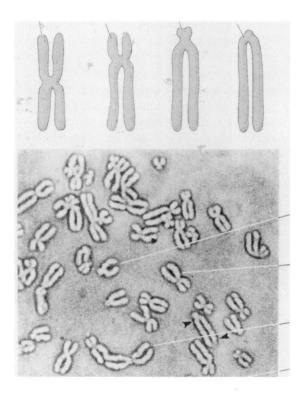

Figure 16.1 By the end of the nineteenth century, improved microscopes had made it possible to study chromosomes. Early observers were especially interested in the structure, behavior, and function of chromosomes.

Question: *What was the hypothesized function of chromosomes in heredity?*

Learning About Chromosomes

Several discoveries during the second half of the nineteenth century were relevant to this unanswered question. Certain threadlike structures within the nucleus were found strongly to absorb special dyes or stains that made them easily visible for microscopic studies (see Figure 16.1). These structures, dubbed *chromosomes,* could now be examined by placing cells on a glass slide, treating them with certain preservative chemicals, staining them, and observing them through the microscope. Using these methods, cytologists learned about the behavior of chromosomes during meiosis. Although the exact connection between meiotic events and the hereditary transmission of traits remained undefined, there seemed to be some relationship between chromosomes and genes.

Formulation of the Chromosome Theory

In 1884 and 1885, four German biologists independently postulated that chromosomes carried the units of heredity, but this concept required further elaboration before it could be considered a theory. By the first years of the twentieth century, scientists were in hot pursuit of the biological basis of heredity. Many people were closing in on the answer, but the entire matter was suddenly resolved by two precise and detailed papers, one written in 1902 and the other in 1903, by Walter S. Sutton. The earlier paper, which described results from experiments on grasshopper chromosomes, included the first clear demonstration that chromosomes exist within the cell as sets of distinguishably different pairs of like (homologous) chromosomes, one having come from each parent. Sutton concludes the paper with this statement: "I may finally call attention to the probability that the association of paternal and maternal chromosomes in pairs and their subsequent separation during the reducing division [meiosis] . . . may constitute the physical basis of the Mendelian law of heredity." The 1903 paper expanded on this hypothesis and included additional data on the random orientation of paired chromosomes on meiotic spindles, thus accounting for the independent segregation of separate pairs of the genes ("factors") previously described by Mendel. Sutton also proposed that Mendel's results could be explained if genes were located on chromosomes. Further, he speculated that each chromosome must carry many genes.

Sutton's brilliant conceptual breakthrough explained the hereditary basis of segregation and independent assortment and their relation to chromosomes. However, several years would pass before Sutton's hypothesis became widely accepted. The main obstacle for many scientists was lack of experimental evidence that proved that genes were located on chromosomes or that chromosomes contained numerous genes. Ultimately, the formalized statement that inheritance patterns may be generally explained by assuming that genes are located on chromosomes became known as the **chromosome theory** of heredity.

Thus by 1903, a mechanism that explained Mendelian inheritance had been proposed, and the field of **genetics** (named in 1900) was ready to explode. Sutton's hypothesis served as a guide for much of the genetics research that occurred during the next two decades.

BEFORE YOU GO ON Following Mendel's classic experiments, there was great interest in learning where genes were located in the cell. Based on a brilliant analysis of results from his own studies, Sutton advanced the idea that numerous genes were located on chromosomes.

CONFIRMATION OF THE CHROMOSOME THEORY

Sutton's papers concluded one phase of genetics history, but additional pieces were necessary to complete the puzzle of chromosomes and heredity. In fact, the field of genetics was in a general state of confusion during the first decade of the 1900s. Many scientists were either not aware of Sutton's papers or did not understand them. Also, many investigators had difficulty moving beyond Mendel's simple model of inheritance. Although Sutton had formally associated chromosomes with genes, chromosomes could not be related to specific traits in a one-to-one fashion, as inferred from Mendel's studies.

Questions About Chromosomes and Heredity

Recall that Mendel studied seven specific hereditary traits, each determined by one pair of genes. His experiments with peas always resulted in the offspring ratios expected if independent assortment and segregation occurred. From his work, it could reasonably be concluded that each chromosome carried a gene for only one trait. However, when scientists conducted new investigations with additional traits or with other plants or animals, the offspring ratios obtained often failed to conform to the common Mendelian ratios.

The primary difficulty, recognized by Sutton and others, was accounting for the fact that in most organisms, there were obviously more hereditary traits than there were chromosomes. It became essential to conduct experiments that answered certain questions. Specifically, what exactly did chromosomes have to do with heredity? Were genes contained within chromosomes, and were several genes located on each chromosome, as Sutton hypothesized? How could the inheritance of hundreds of traits be explained? Did Mendelian laws apply only to pea plants? Definitive answers to these questions emerged from a small laboratory smelling of bananas and ether and jammed with half-pint milk bottles full of flies.

Fruit Flies Yield Their Secrets

The keepers of these milk bottles and their unusual contents were Thomas Hunt Morgan (see Figure 16.2) and his brilliant group of students (Alfred Sturtevant, Calvin Bridges, and Hermann Muller), who shared a laboratory at Columbia University in New York City. In 1904, they began the research in the "Fly Room" that eventually defined the relationships between chromosomes and heredity.

For their experiments, Morgan selected the tiny, red-eyed fruit fly *Drosophila melanogaster,* which proved an ideal research animal and has since been widely studied in laboratories around the world (see Figure 16.3). The use of fruit flies offered many advantages over using other animals: they are small, only 2 to 3 millimeters in length; they are cheap to raise and feed (banana gelatin); they are easy to maintain in small bottles and simple to examine (after being anesthetized with ether); they reproduce frequently, furnishing a new generation about every two weeks; and hundreds of offspring are produced in each generation. Experiments that would have taken Mendel years with his pea plants could be completed within months using fruit flies. Another major advantage is that *Drosophila* has only four pairs of chromosomes.

Drosophila Chromosomes and Sex Determination

There are two basic types of chromosomes in the cells of fruit flies and most other animals. A single pair of **sex chromosomes** plays a role in determining the sex of an individual; all other chromosomes are known as **autosomes.** Examination of *Drosophila* chromosomes reveals the presence of one pair of sex chromosomes and three pairs of autosomes. The relationship between the sex chromosomes and sex type had been described for many insects by 1909. Sex determination in *Drosophila* was determined a year later.

X–Y sex determination occurs in many organisms, including humans. In 1910, scientists described the model for sex determination illustrated in Figure 16.4 (see page 298). Females contain two X chromosomes, but males have one X and one Y; thus with respect to sex chromosomes, females are XX and males are XY. During meiosis, the male produces two types of sperm, one type bearing an X chromosome and the other type bearing a Y chromosome. The female produces eggs that all have an X chromosome. Fertilization by an X-bearing sperm produces a female (XX); by a Y-bearing sperm, a male (XY). Equal numbers of males and females are produced because of the segregation of X and Y chromosomes during meiosis. This documentation of the link between sex determination and chromosomes offered strong support for Sutton's hypothesis that chromosomes were the basis of inheritance.

Abnormal Fruit Flies

Morgan's group was enthusiastic about detecting fruit flies with unusual traits. Such traits, when caused by a **mutation,** occur when a gene undergoes a permanent structural change that results in a new allele that gives rise to a different form of an expressed trait (recall that a gene may have several different alleles). Mutant alleles are almost always recessive to the normal trait, and an individual showing the recessive trait is called a **mutant.** Analyses of mutants played an important part in the development of many fundamental concepts in classical genetics. Early in 1910, one such mutant appeared in the form

Figure 16.2 Thomas Hunt Morgan, shown here late in his career, and his students confirmed the chromosome theory of heredity.

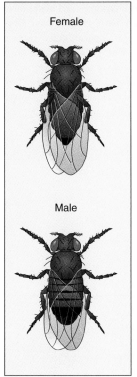

Female

Male

A

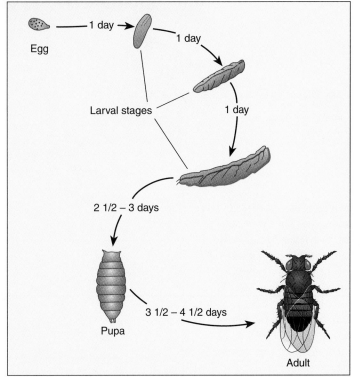

Egg

1 day

1 day

Larval stages

1 day

2 1/2 – 3 days

Pupa

3 1/2 – 4 1/2 days

Adult

B

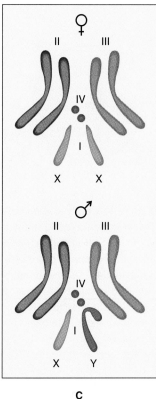

♀

II   III

IV

I

X   X

♂

II   III

IV

I

X   Y

C

Figure 16.3 Morgan's research group used fruit flies (*Drosophila melanogaster*) in its genetics studies. (A) The natural eye and body coloration and the relative size of males and females are shown. These flies proved ideal experimental animals because of their small size, the ease of maintaining them in the laboratory, and their ability to produce many generations of offspring in a short period of time. (B) The life cycle of *Drosophila*. (C) The insect's cells have four pairs of homologous chromosomes—three pairs of autosomes (designated as Group II, III, and IV) and one pair of sex chromosomes (Group I, XX, or XY).

of a white-eyed male fruit fly. This fly became the most famous fruit fly in the history of science, and the subsequent studies it generated constitute a classic story in the history of genetics (see the first Focus on Scientific Process, "The White-eyed Fruit Fly").

Linkage

Experiments with white-eyed male fruit flies yielded a critical finding: the gene for a specific trait (eye color) was shown to be carried on the X chromosome. Other unusual mutant traits,

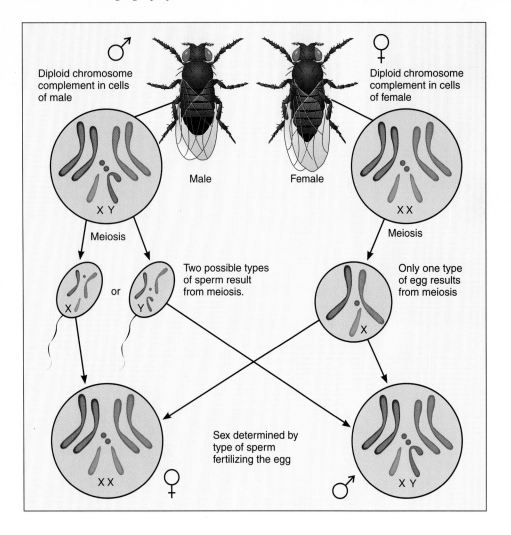

Figure 16.4 In organisms with X–Y sex determination, females have two X chromosomes and males have one X and one Y chromosome. During meiosis, sperm cells receive either an X or a Y sex chromosome, while all eggs produced carry an X chromosome. The sex of the resulting offspring depends on whether the egg is fertilized by an X- or a Y-bearing sperm.

Question: *What is the chance that an offspring will be female?*

such as yellow body and miniature wings, were soon discovered also to be linked to the X chromosome. Thus it became apparent, as Sutton had hypothesized, that several traits were determined by genes located on a single, specific chromosome. Sutton had also postulated that "all of the [genes] represented by any one chromosome must be inherited together." Could this hypothesis now be tested in the Fly Room?

Morgan and his students continued their search for mutant flies, and by 1915, they had identified 85 different mutant traits. Further, as Sutton had suggested, many of these traits, and the alleles that caused them, tended to be inherited together. **Linkage** refers to the concept that specific traits tend to be inherited together because their associated genes are arranged in linear fashion along the same chromosome (see Figure 16.5). All genes present on a single chromosome are referred to as a **linkage group.** How many linkage groups did Morgan's group find for *Drosophila?* Not surprisingly, they found four, equal to the haploid number of chromosomes.

Many of the traits described by Morgan were linked to the X chromosome, but others fell into three separate linkage groups, and numerous experiments showed that the traits of each group were usually inherited together. For example, as shown in Figure 16.6, if flies with three mutant recessive traits

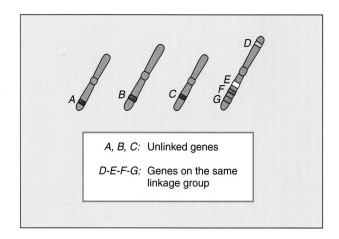

A, B, C: Unlinked genes

D-E-F-G: Genes on the same linkage group

Figure 16.5 Genes located on different chromosomes (*A, B,* and *C*) are not linked. Genes located on the same chromosome (*D, E, F,* and *G*) belong to a linkage group.

Question: *Which genes will usually be inherited together?*

FOCUS ON SCIENTIFIC PROCESS I

The White-eyed Fruit Fly

"In a pedigree culture of *Drosophila* which had been running for nearly a year through a considerable number of generations, a male appeared with white eyes. The normal flies have brilliant red eyes" (see Figure 1). So reads the first paragraph in T. H. Morgan's classic paper, "Sex-limited Inheritance in *Drosophila*," which was published in 1910. He speculated that a mutation within a single egg resulted in elimination of the factor (gene) for red eye color. The new character, white eyes, and the "sport" (Morgan's term for the white-eyed male) that carried it, were exhaustively analyzed in different genetic experiments.

Like Mendel, Morgan made standard crosses involving P, F_1, and F_2 flies. In the first parental cross, the white-eyed male was mated with his red-eyed sisters. The F_1 consisted of 1,237 red-eyed offspring along with 3 white-eyed males. The unexpected appearance of white-eyed males in the F_1 was thought to have been "due evidently to further sporting," and these flies were ignored for the rest of the paper (they may have been early-emerging F_2 flies contained in the same bottles). The red-eyed F_1 hybrids were then crossed, and the F_2 consisted of the following offspring: 2,459 red-eyed fe-

males, 1,011 red-eyed males, 782 white-eyed males, and no white-eyed females. Finally, the white-eyed male was crossed with some of his F_1 daughters, producing 129 red-eyed females, 132 red-eyed males, 88 white-eyed females, and 86 white-eyed males ("the four classes of individuals occur in approximately equal numbers"). How could these experimental results be interpreted?

Morgan described a hypothesis to account for the results. Note that his hypothesis was formulated *after* the first experiments were completed.

The results just described can be accounted for by the following hypothesis. Assume that all of the spermatozoa of the white-eyed male carry the "factor" for white eyes *W* and that half of the spermatozoa carry a sex factor *X* that the other half lacks; that is, the male is heterozygous for sex. Thus the symbol for the male is *WWX* and for his two kinds of spermatozoa *WX* and *W*. Assume that all of the eggs of the red-eyed female carry the red-eyed factor *R* and that all of her eggs (after reduction) carry one *X* each; the symbol for the red-eyed female will therefore be *RRXX,* and that for her eggs will be *RX* and *RX*.

The relevant phenotypes, genotypes, and gametes produced, using Morgan's ter-

minology in his paper, are shown in Table 1 (see page 300). Because a hypothesis makes predictions, Morgan then conducted several additional crosses to test predictions made by his hypothesis. The predictions and results of the experiments are summarized in Table 2 (see page 300). In each experiment, his prediction was confirmed.

This famous paper—the description of a white-eyed male, the results of various test crosses, the formulation and testing of a hypothesis, and its contribution of convincing support for Sutton's hypothesis that genes were carried by chromosomes—occupies a prominent spot in the library of genetics historians. Yet it had major deficiencies, since one of its primary assumptions was incorrect. Can you identify the problems with the hypothesis set forth by Morgan?

Recall that sex in *Drosophila* is determined by X and Y chromosomes (females are XX and males are XY). Morgan incorrectly assumed that females were XX and males simply had one X chromosome (expressed as X or X0). Further, even though he considered eye color alleles to be "associated" with the X chromosome, his hypothesis and experimental interpretations were based

Figure 1   Two fruit flies, one with normal red eyes, the other a white-eyed mutant.

box continues

FOCUS ON SCIENTIFIC PROCESS I

Table 1 Phenotypes, Genotypes, and Possible Gametes of Fruit Flies Used in Morgan's 1910 Experiments

Generation	Phenotype	Genotype	Possible Gametes
P	Red-eyed female	*RRXX*	*RX*
	White-eyed male	*WWX[1]*	*WX, W*
F$_1$	Red-eyed female	*RWXX*	*RX, WX*
	Red-eyed male	*RWX*	*RX, W[2]*
F$_2$	Red-eyed female	*RRXX or RWXX*	
	Red-eyed male	*RWX*	
	White-eyed male	*WWX*	
Other	White-eyed female	*WWXX*	

[1]Males were assumed to have only one sex chromosome, an X.
[2]Morgan assumed that only two types of sperm were produced by the red-eyed males, *RX* or W; "otherwise the results will not follow."

Morgan's symbols are used in this table: *R* = red eye factor (allele), *W* = white eye allele, and *X* = sex chromosome. The *R* and *W* alleles are located at the same locus.

Table 2 Results of Morgan's 1910 Experiments

Prediction 1: Offspring from a cross between a white-eyed female and a white-eyed male should be "white [eyed], and male and female in equal numbers."
Result of Experiment 1: All offspring were white-eyed.

Prediction 2: If two types of red-eyed F$_2$ females are crossed with white-eyed males, "there should be four classes of individuals in equal numbers."
Result of Experiment 2: The predicted ratios (25 percent each of red-eyed females, white-eyed females, red-eyed males, and white-eyed males) were obtained.

Prediction 3: If F$_1$ females heterozygous for white eyes are crossed with white-eyed males, "the four combinations [of offspring] last described" will occur.
Result of Experiment 3: Offspring consisted of approximately 25 percent each of red-eyed females, white-eyed females, red-eyed males, and white-eyed males.

Prediction 4: If F$_1$ red-eyed males (*RWX*) are crossed with white-eyed females, "all the female offspring should be red-eyed, and all the male offspring white-eyed."
Result of Experiment 4: All females were red-eyed and all males were white eyed.

Morgan's experiments were designed to test predictions derived from his hypothesis that the factor for white eye color is associated with the X chromosome.

on the idea that eye color genes and sex chromosomes were inherited independently. This led to complications in explaining the results of some additional experiments not discussed in this essay.

Within months, it became clear to Morgan that all his experimental results made sense if he assumed that the eye color alleles were actually transmitted *as part of the X chromosome.* Also, by then he had become aware that male fruit flies have an X and a Y chromosome. Thus in 1911, he formulated and published a modification of his original hypothesis that proved to be entirely correct.

The experiments on *Drosophila* have led me to two principal conclusions: First, that sex-limited [now called *sex-linked*] inheritance is explicable on the assumption that one of the material factors of a sex-limited character is carried by the same chromosomes that carry the maternal factor for femaleness [that is, the X chromosome]. Second, that the "association" of certain characters in inheritance is due to the proximity in the chromosomes of the chemical substances [genes] that are essential for the production of those characters.

Figure 2 summarizes key aspects of Morgan's experiments in light of his modified hypothesis.

Based on his data, Morgan hypothesized that the allele for white eyes was recessive and was carried on the X chromosome. Figure 2A shows that the gene for eye color occurs at a specific locus on the X chromosome but that there is no corresponding gene on the Y. Since male fruit flies have only one X chromosome, they carry only one allele for eye color. The two alleles for eye color, carried by the X chromosome, are shown as *R* for the dominant red color and *w* for the recessive white color. According to Morgan's modified hypothesis, gametes of the P$_1$ male contained either an X chromosome with a white-eye allele or a Y chromosome, which had no eye color allele. Gametes from red-eyed P$_1$ females all contained an X chromosome with an allele for red eye color. Genotypes of the F$_1$ are shown in Figure 2B. All F$_1$ flies were red-eyed, and all females would have been expected to be heterozygous, or carriers of the white-eye allele.

The possible gametes formed by F$_1$ flies are shown in Figure 2C. The F$_1$ flies were mated, and the results are shown in the Punnett square in Figure 2D. The F$_2$ females were all red-eyed, as predicted by the hypothesis. Morgan also predicted that an equal number of males would be produced, with approximately 50 percent having red eyes and

FOCUS ON SCIENTIFIC PROCESS I

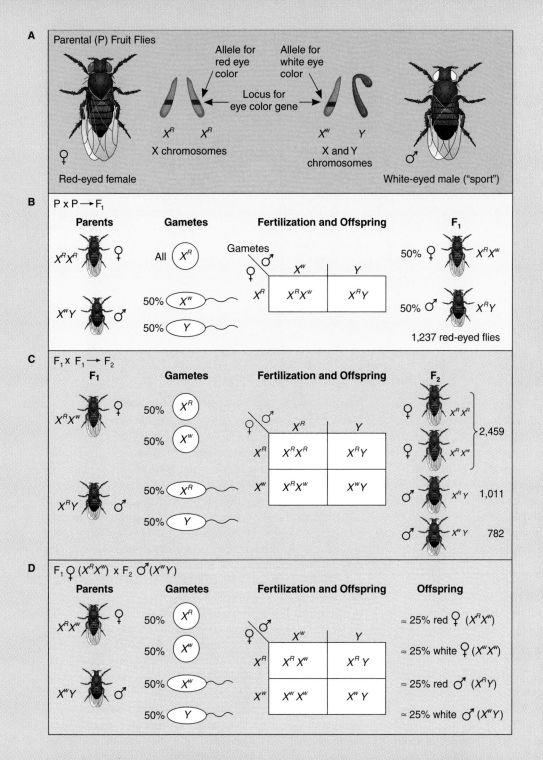

Figure 2 A summary of some of Morgan's experiments and results. X and Y = sex chromosomes, R = the dominant allele for red eye color, w = the recessive allele for white eye color; X^R indicates that the X chromosome carries the allele for red eyes; X^W indicates that the X chromosome carries the allele for white eyes. The Y chromosome has no locus for eye color genes. (A) These are the phenotypes, sex chromosomes, and alleles for eye color in Morgan's experimental fruit flies in the P generation. (B) In Morgan's first preliminary experiment ($P \times P$), homozygous red-eyed females were mated with the white-eyed male. The females could produce gametes containing only X^R. The sperm of the male could contain either a Y chromosome or an X^W. A Punnett square is used to describe the offspring produced by fertilization of the parental gametes. The phenotypes and genotypes (and ratios) of the F_1 are also shown. (C) In Morgan's second preliminary experiment, F_1 females were mated with F_1 males. (D) The results of Morgan's third prediction in Table 2 were confirmed by this experiment.

50 percent white eyes. As can be seen, the actual numbers were considerably lower, especially the number of white-eyed males. It is now known that in any such experiment, there is always a lower number of white-eyed offspring produced than would be predicted. This is due to the death of a significant number of flies with the white-eye allele during embryonic development, which suggests that some other (unknown) characteristic may also be affected. The results of a test cross between F_1 females and F_2 white-eyed males are shown in Figure 2D.

box continues

These experiments were repeated by numerous investigators, with the same results. They provided strong support for the hypothesis that the gene for this trait was carried on the X chromosome and thus for Sutton's genes-on-chromosomes hypothesis. A valid hypothesis enables scientists to test its predictions. In this case, Morgan tested many predictions by conducting experiments summarized in Table 2. (Students are encouraged to construct diagrams similar to those of Figure 2 for the crosses described in Table 1.) These experiments served to verify his hypothesis, since no other interpretation for the white-eyed phenomenon could be seriously considered after a careful examination of the results.

The relatively simple picture Morgan had of white and red eyes in his fruit flies has since become exceedingly complex as a result of modern studies using new experimental tools. It is now known that genes at 18 separate loci on different chromosomes act together to produce the normal red eye color. If any of these genes undergoes a mutation, different eye colors, including the classic white eyes, will result. Some of the other eye colors are coffee, carrot, buff, apricot, honey, purple, and spotted. Sex determination in *Drosophila* has also been found to depart from the original X–Y model. Sex-determining genes are present on both the X chromosome *(X)* and the three pairs of autosomes *(A)* but not on the Y chromosome, although it is required for male fertility. It is the *balance* between *X* and *A* that determines the sex of any fruit fly. In normal flies, an *XX* genotype has an *X:A* ratio of 2:2 = 1 (that is, both are present in diploid numbers; it will have two X chromosomes and two of each autosome) and will be a female. An *XY* genotype has an *X:A* ratio of 1:2 = 0.5 and will be a male.

Morgan's first experiments initially led to the formulation of an interesting hypothesis. Subsequently, new experiments provided convincing evidence to support his modified hypothesis. His studies on the white-eyed fruit flies, though conducted far in the past, are still considered an elegant example of scientific experimental methods.

(black bodies, purple eyes, and dumpy wings) were mated with standard flies heterozygous for dominant traits (brown bodies, red eyes, and normal wings), all three mutant traits generally appeared together in some of the offspring. Thus the concept of linkage was extended to include the three fruit fly autosomes.

The results of Morgan's studies showing that a large number of traits were associated with specific chromosomes and that the number of linkage groups was equal to the haploid number of fruit fly chromosomes were very important. They added substantial credibility to the evolving chromosome theory.

Crossing-over and Chromosome Mapping

As we just saw, all genes carried on one fruit fly chromosome constitute a linkage group, and Morgan's early experiments showed that the traits they determine were usually inherited together. Within a short time, however, further studies demonstrated that linked genes did not always remain together, and flies with traits connected with different homologous chromosomes *did* appear in experimental crosses; such flies were called **recombinants.** This finding could be explained by the effects of chromosomes crossing over, breaking, and rejoining at some point during meiosis (see the second Focus on Scientific Process: "Linkage, Linkage Groups, and Chromosome Mapping").

Morgan's group hypothesized that a precise correlation must exist between recombination frequency and the linear distance separating the linked genes; that is, genes separated by long distances would tend to have a higher probability of crossing over than genes that were close neighbors on their chromosome. After conducting further experiments with its fruit flies and obtaining additional data, Morgan's group was able to create **chromosome maps,** which showed both the location of linked genes and their exact linear order along the length of the chromosome. Final confirmation of the *Drosophila* maps required more sophisticated techniques that did not become available until much later. A remarkable aspect of these chromosome maps, which were constructed using an indirect, mathematical approach, was that their gene sequences proved to be entirely correct, although the distances they proposed turned out to be somewhat less precise.

Sturtevant's mapping study (see the second Focus on Scientific Process), in particular, offered solid evidence that genes were arranged in a linear sequence along the chromosome. It also provided the theoretical foundation for constructing chromosome maps for many species besides *Drosophila*. It remained for another of Morgan's students, Calvin Bridges, to conduct experiments that overwhelmingly convinced the scientific community of the accuracy of the chromosome theory. His studies also centered on crosses of red- and white-eyed fruit flies, but they involved extremely complex aspects of abnormal chromosome behavior ("nondisjunction") that are beyond the scope of this book.

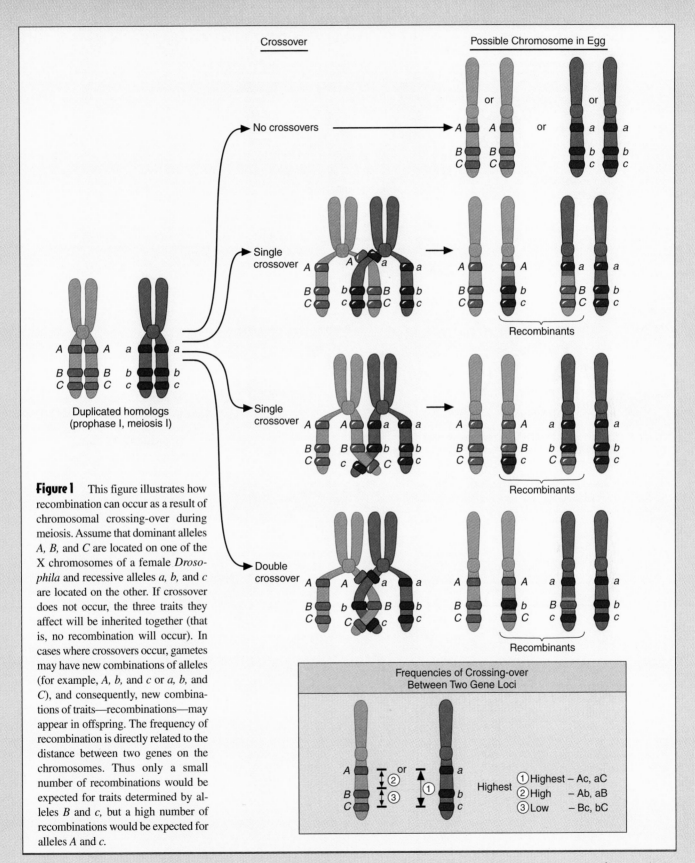

Figure 1 This figure illustrates how recombination can occur as a result of chromosomal crossing-over during meiosis. Assume that dominant alleles *A, B,* and *C* are located on one of the X chromosomes of a female *Drosophila* and recessive alleles *a, b,* and *c* are located on the other. If crossover does not occur, the three traits they affect will be inherited together (that is, no recombination will occur). In cases where crossovers occur, gametes may have new combinations of alleles (for example, *A, b,* and *c* or *a, b,* and *C*), and consequently, new combinations of traits—recombinations—may appear in offspring. The frequency of recombination is directly related to the distance between two genes on the chromosomes. Thus only a small number of recombinations would be expected for traits determined by alleles *B* and *c,* but a high number of recombinations would be expected for alleles *A* and *c.*

FOCUS ON SCIENTIFIC PROCESS II

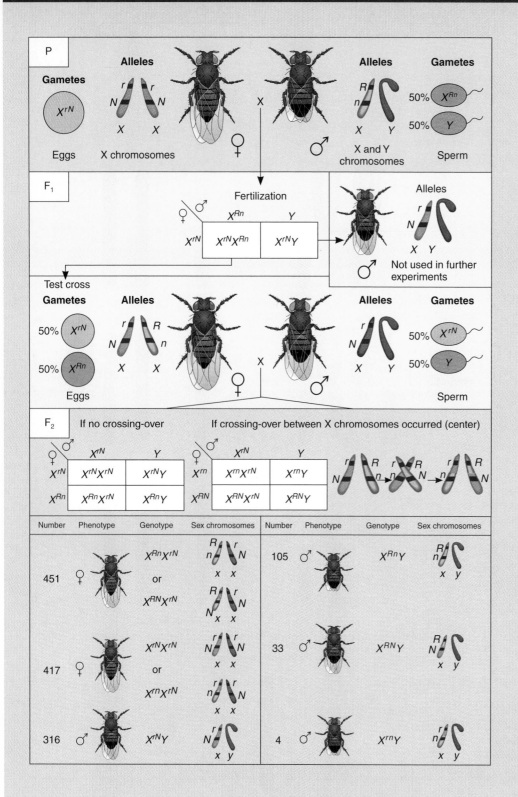

Figure 2 Sturtevant carried out many experiments in which pairs of recessive traits linked to the X chromosome in *Drosophila* were studied. His basic approach was first to cross females (P) homozygous for one normal dominant trait (in this example, normal long wings) and one for one recessive trait (vermilion eye color) with males (P) that had the opposite combination of traits on their single X chromosome (in this example, rudimentary wings and red eyes). The resulting F_1 females, which were heterozygous for both traits being studied, were test-crossed with a different set of males that had normal wings and vermilion eyes to produce the F_2. A simplified diagram describes crossing-over in the X chromosomes of F_1 females. Numbers of F_2 flies with the different traits were then counted (N = normal wing; n = rudimentary wing; R = red eyes; r = vermilion eyes). Those data are included in Table 1.

FOCUS ON SCIENTIFIC PROCESS II

Table 2 Sturtevant's Reasoning

(a) The distance between *y* and *r* = 32.2 (the percentage of crossovers between *y* and *r*). These can be placed arbitrarily as the first genes on the X chromosome.

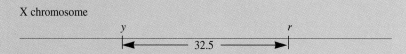

X chromosome

(b) The distance between *y* and *n* = 35.5. But is *n* to the right or to the left of *y*? If it lies to the left of *y*, the distance between *n* and *r* would be approximately 32.5 + 35.5 = 68. If *n* is to the right of *y*, the distance between *r* and *n* would be about 35.5 − 32.5 = 3. The data in Table 1 show that the actual distance, as indicated by the percentage of crossovers, between *r* and *n* is 3. Therefore *n* lies to the right of *y*.

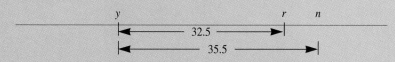

(c) This is Sturdevant's complete map, published in his paper in 1913.

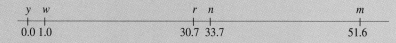

Sturtevant used this type of reasoning in analyzing his data and mapping the X chromosome for the five genes studied in his experiments. *y* = yellow body, *w* = white eyes, *r* = vermilion eyes, *m* = miniature wings, and *n* = rudimentary wings

linear series at least mathematically." In fact, this study, and others conducted by members of Morgan's group during the next three years, completed the picture showing the relationships of chromosomes, genes, and inheritance.

IMPORTANCE OF THE CHROMOSOME THEORY

The period that began with Hugo de Vries and ended with the time-honored studies of Morgan and his students was full of remarkable accomplishments. Beginning with only a dim outline of how traits were transmitted from parent to offspring, scientists established these three fundamental principles:

1. Genes are the fundamental hereditary units that determine specific traits in an organism.

2. Chromosomes are the carriers of genes.

3. The basis of Mendel's laws—segregation and independent assortment—lies in the behavior of chromosomes during the events of meiosis.

In the process of establishing these concepts through hypothesis generation, scientific experiments, and data analysis, scientists laid to rest many misconceptions about heredity. Chromosomes were shown to sort independently during meiosis, whereas genes, except when separated by crossing-over or unless they occur on different chromosomes, do not separate independently during meiosis. This is because genes on the same chromosome constitute a linkage group, and they normally segregate into the same gamete. Because they are on the same chromosome, they do not sort independently. The old hypothesis that a chromosome and a gene (Mendel's hereditary "factor") were one and the same was modified to state that a chromosome contains several genes and that these genes are located in a linear order along the length of the chromosome. This picture of genes as parts of chromosomes provided a logical basis for understanding the normal mechanism of genetic transmission of different traits in sexually reproducing organisms. Equally important, the chromosome theory united the disciplines of genetics and cell biology. Now that many of the questions about chromosomal inheritance had been answered, scientists focused their efforts on understanding the mechanisms and roles of chromosomes and genes in inheritance.

OTHER TYPES OF NORMAL INHERITANCE

The most familiar and most easily understood patterns of gene transmission fall into the category of *Mendelian inheritance,* in which traits normally exist in one of two forms (tall versus short, red versus white) and are determined by a single pair of alleles (*Tt* and *Rr*). We discussed Mendelian inheritance at the end of Chapter 15, in the section titled "Mendel Revisited," where we noted that homologous pairs of human chromosomes contain about 50,000 bits of genetic information called *genes.* Each gene is composed of a DNA segment that determines a specific trait. We also saw that each homolog of a homologous chromosome pair contains one or more genes for the same trait (for example, flower color) at a specific *locus* (an identical location) along their length. *Alleles* refer to alternative forms of a single gene. For example, a gene for flower color, a single trait, may have different alleles—*R* and *r*—that result, for example, in red or white color (*Rr*), which are alternative forms of the trait. The specific alleles contained in a cell (or an individual) are called its *genotype.* The physical trait that occurs as a result of a specific genotype and any environmental influence is referred to as the *phenotype.*

Although patterns of Mendelian inheritance provide useful classical models that help students begin to understand genetics, even a cursory examination of common plants and animals, including humans, makes it clear that surprisingly few traits appear to be inherited in this way. Most flowers come in a variety of colors, and humans are not simply either tall or short. In fact, a remarkable range of shapes, colors, heights, and other features characterizes the inheritable traits of all species. Let us therefore examine some types of inheritance that differ from the single-gene, all-or-none Mendelian concepts.

Dominance Models

In genetics, *dominant* and *recessive* refer to the appearance of phenotypic traits. Different dominance models can be defined by the phenotype of a heterozygous genotype. Mendel's F_1-generation peas always looked like one of the homozygous parents for two reasons: (1) all seven phenotypic traits were determined by one gene pair with two alleles, and (2) **complete dominance,** in which one allele in a heterozygote determined the phenotypic trait. For example, in Mendel's peas, homozygous parental genotypes *GG* and *gg* resulted in gray and white seed coats, respectively; F_1 peas with the heterozygous genotype *Gg* always had gray seed coats. Thus in a heterozygote, allele *G* determined the phenotype (gray seed coat), and allele *g* had no affect on seed coat color.

A different type of dominance model is revealed by four-o'clock flowers. As shown in Figure 16.7, when red and white four-o'clocks are crossed, the F_1 heterozygotes are all various

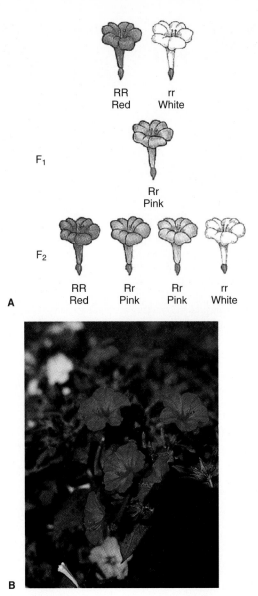

Figure 16.7 (A) The color of four-o'clock flowers is determined by a gene pair with two alleles, *R* and *r,* that have incomplete dominance. Flowers with genotype *RR* are red and *rr* are white, but flowers with genotype *Rr* are some shade of pink. (B) Photo of four-o'clocks.

shades of pink. When F_1 pinks are crossed, the F_2 consists of an approximate 1:2:1 ratio of three phenotypes—one red to two pink to one white. This type of inheritance, in which heterozygotes show a range of intermediate phenotypes between homozygous parents, is called **incomplete dominance.**

Codominance occurs when a heterozygote reflects the phenotypes of both parental homozygotes. For example, humans have an M–N blood group that is determined by a single gene pair with two alleles (*M* and *N*). Homozygous genotypes *MM* and *NN* have blood types M and N, whereas the heterozygote, *MN,* has blood type MN. For many traits that show variable phenotypes, it is difficult to determine whether codominance or incomplete dominance is involved.

Multiple Alleles

Genetic phenotypic traits we have studied so far have been determined by genes with two alleles. However, a single gene may have **multiple alleles** for a particular trait, not just two. In such cases, the phenotype of an individual is still determined by two alleles, but within a population, there are many possible phenotypes because there are many different allelic combinations. For example, eye color in fruit flies (*Drosophila*) is determined by one gene pair, but there are many different eye colors because the eye color gene has several different alleles. The common eye color in "wild-type" (normal) flies is dominant and therefore results in red eyes in both homozygotes and heterozygotes. This is now represented as w^+, where w refers to the gene for eye color and $+$ indicates the allele for normal red eye color. Because w^+ is completely dominant, other eye colors are only observed in fruit flies without an w^+ allele. The allele for white eye color, for example, is designated w; fruit flies with genotype ww will thus have white eyes, while those with genotype w^+w will have red eyes. Other eye color genotypes and alleles are w^aw^a (apricot-colored eyes), w^ew^e (eosin), $c^{ch}c^{ch}$ (cherry), w^ww^w (wine), and $w^{co}w^{co}$ (coral). These represent only a few of the multiple alleles for eye color in *Drosophila*.

Polygenic Inheritance

Many traits show a great range of phenotypic variation. Such traits include human height, the color of wheat kernels, and possibly the weights of whales (not easily determined!). If you recorded height in a random sample of 100 students, a typical result might show that the shortest measured 5 feet and the tallest 6 feet 4 inches. The heights of the other 98 would form a continuum between the two extremes. A variable phenotypic trait, such as height, that is determined by many genes is said to be governed by *polygenic inheritance.*

Polygenic inheritance involves several genes that have a small but cumulative effect, with or without dominance or with incomplete dominance, in determining a trait. A classic example of this concept was demonstrated for kernel color in wheat by Herman Nilsson-Ehle in 1909 (see Figure 16.8). In a certain wheat variety, kernel color is determined by the cumulative action of two pairs of genes, each with two possible alleles: an incompletely dominant allele for red and another for white. Kernels with four red alleles are dark red in color, and those with four white alleles are white. Kernels with different combinations, such as two white and two red or three red and one white, are intermediate in color.

BEFORE YOU GO ON Following confirmation of the chromosome theory, new patterns of inheritance were identified. Incomplete dominance results in phenotypic traits in the F_1 that are intermediate between homozygous parents. Codominance results in offspring that reflect equally the phenotypes of homozygous parents. Many traits involve multiple alleles, which represent many variations of a single gene. Polygenic inheritance indicates that a trait is determined by several genes.

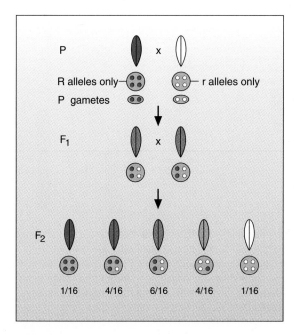
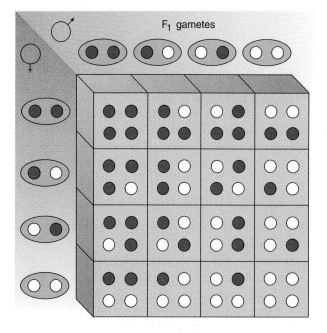

Figure 16.8 Nilsson-Ehle found that kernel color in a certain strain of wheat was determined by two gene pairs, each with two alleles. One gene pair came from each parent. Wheat with dark red kernels had four alleles for red color (designated *R*), and white kernels were found in wheat in which all four alleles were *r*. If dark red-kernel wheat was crossed with white-kernel wheat, the F_1 were all medium red. $F_1 \times F_1$—diagrammed in the Punnett square— yielded an F_2 with kernel colors in the ratios indicated in the figure.

SOME TYPES OF ABNORMAL INHERITANCE

As we discussed in Chapter 15, the normal mechanisms of inheritance in sexually reproducing organisms follow a precise trail that begins with the events of meiosis and ends with fertilization, the formation of a zygote, and the development of new offspring. This pathway leads to the orderly, largely predictable transmission of chromosomes, genes, and genetic traits between generations.

Occasionally, however, deviations in normal genetic processes lead to abnormal combinations or a modification in the structure of genes or chromosomes within the cells that produce gametes. A permanent structural change within the hereditary material is called a *mutation.* The end result of a mutation—the effect on the offspring—can vary dramatically: there may be no effect, new variants (offspring with new forms of a trait) may appear, or premature death may occur before or after birth. A change that results in death is classified as a **lethal mutation.** Although most mutations are considered harmful, the creation of new genetic combinations and new variations is also a vital process in the evolution of organisms. Two basic levels of mutation are recognized: chromosomal mutations and gene mutations.

Chromosomal Mutations

Chromosomal mutations occur when segments of chromosomes, entire chromosomes, or even complete sets of chromosomes are involved in genetic change. The effects of chromosomal mutations are caused primarily by new arrangements of the chromosomes. They are generally harmful in animals but may produce interesting new variations in plants.

Most chromosomal mutations can be traced to their behavior during prophase I of meiosis because at that time, homologous regions of chromosomes have a powerful pairing affinity. During this phase, chromosomes often break, and the broken ends of such chromosomes are very likely to attach to other broken ends. The consequences of such behavior are loss, duplication, or rearrangement of chromosomes or segments of chromosomes. These chromosomal changes are described in Figure 16.9.

Deletions

The loss of a segment of a chromosome after breakage is called a **deletion.** The loss can range from a small section of a gene to a portion of a chromosome that contains several genes. If the same deletion occurs in both members of a homologous chromosome pair, it is almost always lethal. Studies of deletions indicate that most of the genetic information carried on chromosomes is required for an organism to develop and function normally. Even single deletions are frequently harmful because for some traits, this upsets a specific balance of genes. Deletions of certain regions of a chromosome often create a characteristic phenotype. For example, in *Drosophila,* notched wings result from one particular deletion.

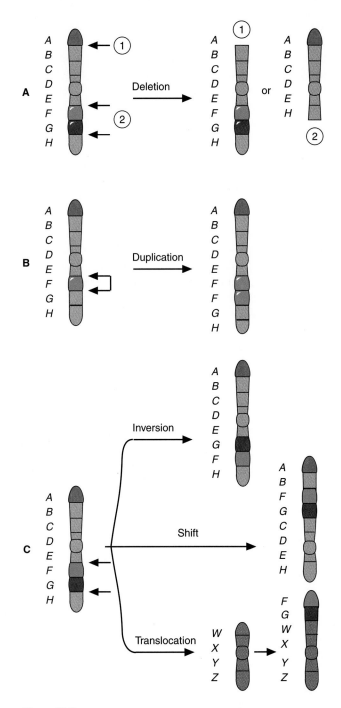

Figure 16.9 Several types of changes in chromosome structure may occur during meiosis. In this figure, changes in one homolog from a homologous pair are described. Capital letters indicate different regions of the chromosome. (A) Deletions occur when breaks develop (at sites indicated by the arrows) and portions of the chromosome are detached. (B) Duplications (arrows) result in extra copies of genes or larger regions of chromosomes. (C) If a section of a chromosome breaks (arrows), it may become inverted in the same part of the chromosome (inversion), become incorporated into a different part of the same chromosome (shift), or become integrated into a different chromosome (translocation).

Question: *Why are such changes generally harmful to animals?*

Duplications

Duplications result in an increase in the amount of genetic material carried by a chromosome. Unless a large number of genes or a major segment of the chromosome is duplicated, the potential for a negative effect is apparently quite small. Indeed, there are some indications that duplications may, in some instances, actually supply additional genetic material that leads to an increase in the diversity of gene functions. Thus duplications can be a source of new genetic variation.

Inversions, Shifts, and Translocations

When a portion of a chromosome breaks, it may rotate 180° and become reinserted at the same position in the chromosome (**inversion**), become inserted in a different region of the same chromosome (**shift**), or become inserted in a nonhomologous chromosome (**translocation**). These types of mutations can also occur during the complex chromosomal gymnastics connected with crossing-over (discussed in Chapter 15). Chro-

mosomal inversions, shifts, and translocations all have the potential to generate gametes with abnormal chromosomes and consequently are usually harmful to any zygote produced.

Changes in Chromosome Number

In the normal flow of events during meiosis, paired homologous chromosomes separate during anaphase of meiosis I, and paired chromatids separate during anaphase of meiosis II. Occasionally, **nondisjunction,** the failure of certain chromosomes to separate during either of these phases, results in the production of gametes with abnormal numbers of chromosomes (see Figure 16.10). Changes of this type, which involve a fraction of the chromosomes in a complete chromosome set (one chromosome in Figure 16.10), result in offspring that have cells with an abnormal number of chromosomes. Such cells or individuals are called **aneuploid.** You can see in Figure 16.10 that the gametes produced will contain one extra chromosome or lack one chromosome. Assuming fertilization of these abnormal gametes with a

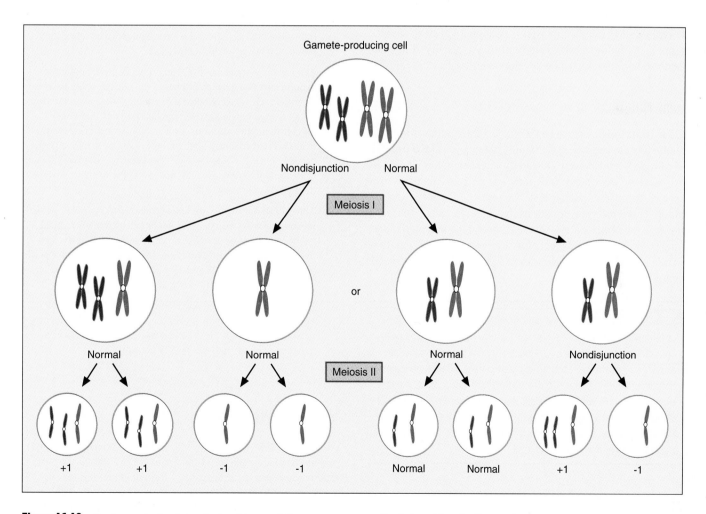

Figure 16.10   During meiosis, sister chromatids may fail to separate normally during either the first or the second division. As a result of this nondisjunction, gametes produced will have either one more or one less than the normal chromosome number. If one of these irregular gametes participates in fertilization, the resulting zygote will also have the same abnormal chromosome number as the gamete (+1 or −1).

normal sperm or egg, the resulting cells of the aneuploid zygote will contain one chromosome too many (in which case it is called *trisomic*) or one too few (*monosomic*), compared to the normal diploid number. In humans and other animals that have been carefully studied, aneuploidy usually results in spontaneous (natural) abortion of the fetus. If the offspring survives, it usually suffers from a variety of symptoms caused by the abnormal concentration of genetic material.

Normal diploid organisms have two sets ($2n$) of the monoploid (n) number of chromosomes in their cells (technically, the term *haploid* refers to the n number of chromosomes found in a gamete; *monoploid* refers to the n number of chromosomes in any cell). However, organisms—mostly plants—can be produced that have $3n$, $4n$, or even a greater number of complete sets of chromosomes. Such organisms are called **polyploids,** and they usually result from nondisjunction.

Breeders of plants have a long history of developing new varieties and species of flowers, fruits, and grains using methods that involve the creation of polyploid forms. For example, New World cotton has 26 pairs of chromosomes and is a polyploid of the original Old World cotton, which has 13 pairs. Polyploidy has minor importance in animal genetics, as possession of multiple sets of chromosomes usually results in death soon after embryonic development begins.

Gene Mutations

Gene mutations occur when a single gene changes from one allelic form to another. For example, white eyes of the fruit fly in Morgan's studies were caused by a gene mutation of the red eye alleles. Most gene mutations are probably harmful to a species because they alter a stable, successful genetic constitution. However, gene mutations that occur in gamete-producing cells of animals are a source of new genetic material and may have a positive aspect, for they can increase variation within a population. We will discuss gene mutations in greater detail in later chapters.

BEFORE YOU GO ON Abnormal inheritance often involves changes in chromosomes or genes. Segments of a chromosome may be deleted, duplicated, or broken and relocated to another site on the chromosome or to another chromosome altogether. When nondisjunction occurs during meiosis, gametes that have an abnormal number of chromosomes may be formed. If a gamete carries an extra set of chromosomes, polyploid offspring may be produced. Gene mutations occur when one allelic form of a gene changes to another and may result in a new phenotypic trait.

CLOSE OF AN ERA

The first third of the twentieth century saw enormous progress in the study of genetics, progress that ushered out the age of classical genetics. The reality of genes, their location on chromosomes, and their vital roles in determining hereditary traits became clearly established. T. H. Morgan and many others successfully employed multicellular organisms in complex breeding experiments and mathematical analyses to derive early genetic principles. With these achievements to build on, scientists extended classic concepts of Mendelian inheritance as they made new discoveries about other patterns of inheritance. By the 1930s, scientists were asking exciting new questions, and a new age of genetics research had begun.

SUMMARY

1. Following the rediscovery of Mendel's work, there was great interest in establishing the cellular location of Mendelian factors (genes) and the mechanisms that accounted for the determination of genetic traits.

2. The chromosome theory of heredity, articulated early in the twentieth century primarily by Walter S. Sutton, stated that genes were located on chromosomes. Mendel's laws and the transmission of genetic traits were thought to be related to events that occurred during meiosis.

3. Thomas H. Morgan and his group of students conducted experiments that eventually proved the chromosome theory to be correct. By using an ideal experimental organism, the fruit fly *Drosophila melanogaster,* they carried out studies of mutants with unusual traits such as white eyes. Their research demonstrated conclusively that these traits were due to genes carried on the X chromosome. Thus certain genes were shown to be located on specific chromosomes, and the chromosome theory was confirmed.

4. Through studies involving linked traits, crossing-over of chromosomes during meiosis, and uncomplicated mathematical analyses, it became possible to create maps of fruit fly chromosomes. Such maps serve to identify the locations of specific genes on chromosomes and also to determine their sequence and relative distances from one another.

5. Different types of inheritance were identified in the first third of the twentieth century. It also became recognized that chromosomes and genes undergo changes that give rise to new genetic variations that are usually harmful.

WORKING VOCABULARY

autosome (p. 296)
chromosomal mutation (p. 310)
chromosome map (p. 302)
chromosome theory (p. 295)
codominance (p. 308)
complete dominance (p. 308)
deletion (p. 308)
gene mutation (p. 310)

incomplete dominance (p. 308)
linkage (p. 298)
linkage group (p. 298)
multiple alleles (p. 309)
nondisjunction (p. 311)
polygenic inheritance (p. 309)
sex chromosome (p. 296)
translocation (p. 311)

REVIEW QUESTIONS

1. Why did early cytologists believe that there was a link between meiosis and chromosomes and genes?

2. How does the chromosome theory explain genetic inheritance?

3. Why are fruit flies (*Drosophila*) excellent experimental animals for genetics studies?

4. How was it shown that genes for certain traits of *Drosophila* were located on the X chromosome?

5. What types of experiments and results confirmed the chromosome theory?

6. What is meant by Mendelian inheritance?

7. What is meant by non-Mendelian inheritance? Give three examples, and explain why they are non-Mendelian.

8. Why are chromosomal mutations more likely to be harmful than gene mutations?

PROBLEMS

1. Using Nilsson-Ehle's wheat variety, what offspring ratios would be expected if one of the two parent plants had three alleles for red kernels and one for white and the other had two red alleles and two white alleles?

2. Assume that a gene pair for wing length in *Drosophila* is found on one of the autosomes and that the allele for long wings (*L*) is dominant over the allele for short wings (*l*). What genotype and phenotype ratios would you expect in the F_1 generation if both parents were heterozygous for wing length? If one was heterozygous and the other had short wings?

3. From Morgan's eye color experiments, what would you expect in the F_1 if a white-eyed female was crossed with a red-eyed male?

4. Construct a chromosome map for alleles *A, B, C, D,* and *E,* given the following percentage rate of crossovers: *A* and *B* = 12, *A* and *C* = 18, *A* and *D* = 2, *A* and *E* = 22, *B* and *C* = 30, *B* and *D* = 14, *B* and *E* = 10, *C* and *D* = 16, *C* and *E* = 40, *D* and *E* = 24.

ESSAY AND DISCUSSION QUESTIONS

1. When the chromosome theory was proposed, no one had actually seen a gene. Genes were a theoretical concept that were hypothesized to exist and to determine specific traits in organisms. Later research verified the existence of genes and their structure. Were scientists justified in postulating the existence of genes and a hereditary role for these unseen factors? Why?

2. Mendel proposed two simple laws of genetics. Subsequent investigation showed the situation to be much more complex for most traits in most organisms. To what extent are we justi-

fied in believing in the validity of so-called Mendelian inheritance?

3. The chromosome theory developed through theoretical ideas, experimental investigations, and careful observations. Evaluate the relative importance and roles of each of these elements in creating hypotheses or theories.

4. Confirmation of the chromosome theory depended on a group of scientists who had laboratory facilities, graduate students who did research, and time. What were the possible sources of funds for all of the space, equipment, supplies, and investigator salaries? Discuss possible reasons why funds were made available for supporting the type of research described in this chapter.

REFERENCES AND RECOMMENDED READING

Allen, G. E. 1978. *Thomas Hunt Morgan: The Man and His Science.* Princeton, N.J.: Princeton University Press.

Crowe, J. F. 1991. Anecdotal, historical and critical commentaries on genetics. Our diamond anniversary. *Genetics,* 127: 1–3.

Dunn, L. C. 1965. *A Short History of Genetics: The Development of Some of the Main Lines of Thought, 1864–1939.* New York: McGraw-Hill.

Griffiths, A. J. F., J. H. Miller, D. T. Suzuki, R. C. Lewontin, and W. G. Gelbart. 1993. *An Introduction to Genetic Analysis.* New York: Freeman.

Kohler, R. E. 1994. *Lords of the Fly.* Drosophila *Genetics and Experimental Practice.* Chicago: University of Chicago Press.

Lederman, M. 1989. Research note: Genes on chromosomes: The conversion of Thomas Hunt Morgan. *Journal of the History of Biology,* 22: 163–176.

Moore, J. A. 1986. Science as a way of knowing: Genetics. *American Zoologist,* 26: 583–747.

Morgan, T. H. 1910. Sex-limited inheritance in *Drosophila. Science,* 32: 120–122.

Morgan, T. H., A. M. Sturtevant, H. J. Muller, and C. B. Bridges. 1915. *The Mechanism of Mendelian Heredity.* New York: Holt.

Sturtevant, A. H. 1913. The linear arrangement of six sex-linked factors in *Drosophila,* as shown by their mode of association. *Journal of Experimental Zoology,* 14: 43–59.

Sturtevant, A. H. 1965. *A History of Genetics.* New York: Harper-Collins.

Sutton, W. S. 1902. On the morphology of the chromosome group in *Brachystola magna. Biological Bulletin,* 4: 24–39.

Sutton, W. S. 1903. The chromosomes in heredity. *Biological Bulletin,* 4: 231–251.

ANSWERS TO FIGURE QUESTIONS

Figure 16.1 They carry genes.
Figure 16.4 0.5 or $\frac{1}{2}$.
Figure 16.5 Those on the same chromosome (D, E, F, G).
Figure 16.9 They lead to abnormal development and death during the embryonic stage.

17

The Birth of Modern Genetics

Chapter Outline

Reading Questions

1. What questions guided genetics research between 1920 and 1955?

2. What types of organisms were used to answer questions about DNA?

3. What is molecular biology?

4. What were the key studies in identifying the genetic material?

5. How did Watson and Crick determine the structure of DNA?

6. How is DNA replicated?

After the first third of the twentieth century, genetics research turned to new problems as more was learned about the elemental mechanisms of genetics. Mendel and de Vries had established that discrete hereditary factors (genes) were responsible for determining specific traits. Later, Sutton postulated that genes were located on chromosomes, a theory that was verified by T. H. Morgan and his students.

NEW FRONTIERS

Following these achievements, new questions emerged that defined the direction of genetics research. The basis of many of these questions concerned the chemical identity of the hereditary material. Exactly what molecules in the cell constituted genes and chromosomes? The research conducted to answer this question and the knowledge uncovered by numerous studies are described in this chapter and the next. During the 1930s, several developments changed the science of genetics. In genetics, progress has often been accelerated when scientists turned to new organisms for their experiments. Most of the early work, including Mendel's, was performed on plants. Later, Morgan and his colleagues advanced the field of genetics by using their famous fruit flies.

By the 1930s, finer degrees of genetic resolution became the goal of scientists. To accomplish this goal, bacteria and later viruses were recruited as experimental subjects (see Figure 17.1). Bacteria and viruses have major advantages over the eukaryotic plants and animals that had been used in earlier studies. Their genetic materials are simpler and more easily accessible, culturing expenses are lower, and they multiply at extraordinary rates, giving rise to large populations in a short period of time. Perhaps the greatest gain was that experiments could be completed within days or even hours instead of years (Mendel's peas) or months (Morgan's *Drosophila*).

Although the most obvious advantage of using microorganisms was that they were simple and easy to work with, an underlying rationale for using such organisms in modern studies is that all species evolved from a common ancestral cell line (see Chapter 3) and have the same mechanisms for genetic processes. Thus the discovery and understanding of a genetic molecule or process in a plant, fruit fly, or bacterium leads to at least a tentative perception of how similar mechanisms operate in more complex organisms, such as humans.

Other developments that aided progress in genetics included new methods, techniques, and scientific instruments. These advances made it possible to conduct more complex experiments and to analyze results with greater precision. The development that had the greatest impact on genetics—indeed, it resulted in the transformation of biology—was the merging of physics, biochemistry, microbiology (the study of microorganisms, such as bacteria), and genetics to create a new field of science that came to be known as **molecular biology.** During the 1930s, the great interest in comprehending genes at the molecular level required experimental approaches, methods, and expertise that were within the domains of physicists and chemists. Fortunately, many of these scientists had become interested in understanding biological processes at the level of atoms and molecules and had immigrated to biology. Their classical studies on the molecules involved in heredity provided a basis for creating the field of **molecular genetics.**

BEFORE YOU GO ON | In the period following Morgan's verification of the chromosome theory, geneticists turned to microorganisms as experimental subjects. A new scientific discipline, molecular biology, arose from an influx of scientists from the fields of physics, biochemistry, microbiology, and genetics who had an interest in understanding life processes.

CELLS AS CHEMICAL FACTORIES

To understand the investigations described in this chapter, it will be helpful to review some basic concepts about the nature of chemicals discussed in Chapter 2. The incredible variety of

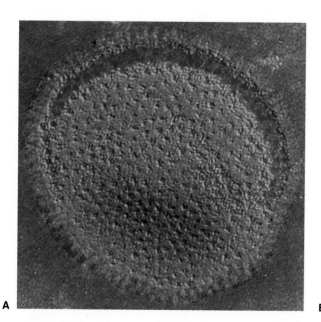
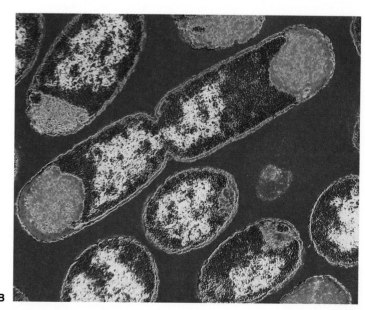

Figure 17.1 Studies using viruses (A) and bacteria, especially *Escherichia coli*, shown here (B) accelerated progress in genetics research during the twentieth century. (Note: the *E coli* in the center is dividing by binary fission.)

Question: *What were the advantages of using viruses and bacteria in genetics studies?*

activities carried out by different cell types can ultimately be understood as the result of numerous chemical processes that take place within the cell. To sustain their functions, cells are always engaged in producing, using, and recycling chemicals.

Cells contain hundreds of small, inorganic chemicals. These include water; elements such as phosphorus, iron, sodium, potassium, and calcium; and various types of salts. Cells also contain thousands of large, organic **macromolecules,** which are formed by linking specific types of small molecules into long chains. Most of the macromolecules involved in chemical processes occupy one of four categories: **carbohydrates** (sugars), **lipids** (fats), **proteins,** and **nucleic acids.**

Proteins

Proteins of every organism on Earth are composed from an alphabet of 20 different small, building-block molecules called **amino acids** (see Table 17.1). Amino acids are attached to each other by a form of chemical linkage called a **peptide bond** (see Figure 17.2A); two amino acids are joined by a single peptide bond. Proteins (such as insulin, diagramed in Figure 17.2B) differ from one another because each has a distinctive sequence of amino acids. There are no predictable patterns or rules in the amino acid sequence of proteins; the 20 amino acids never occur in equal amounts and are rarely all found in any single protein. Most proteins range in size from about 50 amino acids (insulin) to 1,250 amino acids (gamma globulin, an antibody protein).

Table 17.1 The 20 Amino Acids Used to Construct Proteins

Alanine (Ala)	Glycine (Gly)	Proline (Pro)
Arginine (Arg)	Histidine (His)	Serine (Ser)
Asparagine (Asn)	Isoleucine(Ile)	Threonine (Thr)
Aspartic acid (Asp)	Leucine (Leu)	Tryptophan (Trp)
Cysteine (Cys)	Lysine (Lys)	Tyrosine (Tyr)
Glutamic acid (Glu)	Methionine (Met)	Valine (Val)
Glutamine (Gln)	Phenylalanine (Phe)	

The proteins in all living organisms on Earth are constructed of combinations of these 20 amino acids.

A central theme in biology is that the specific structures and functions of each cell type are determined by the proteins it contains. There are more than 50,000 different kinds of protein in the cells and tissues of your body, and individual cells may contain hundreds of different proteins. Proteins are the most abundant macromolecules, accounting for over 50 percent of the dry weight of living cells. Proteins have enormous structural and functional diversity and play a wide variety of biological roles.

In living organisms, proteins serve as structural units, carry out specific biochemical reactions within cells (enzymes), transport substances (for example, hemoglobin in human blood,

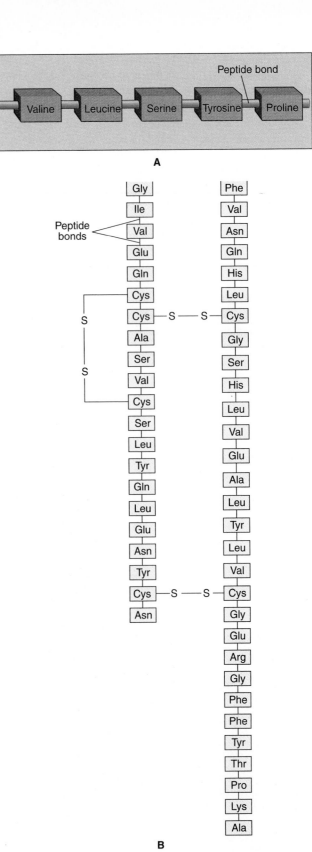

Figure 17.2 (A) Proteins are made of chains of amino acids attached to each other by peptide bonds. Most individual proteins contain 50 to over 1,250 amino acids. (B) Insulin is a protein molecule composed of two chains, 21 and 30 amino acids long, held together by bonds between sulfur atoms.

which transports oxygen), defend cells (antibodies), and regulate processes (hormones). In addition, diverse biological molecules and products are made of proteins, including muscle fibers, feathers, antibiotics, and mushroom poisons.

Enzymes

Recall from Chapter 4 that *chemical reactions* involve changes in the basic chemical composition of the substances involved. They occur when molecules, which are always in motion, collide randomly and chemical bonds between atoms or molecules are formed or broken. Thousands of chemical reactions normally occur within cells of living organisms; they are required for obtaining energy, manufacturing proteins, maintaining proper temperature, and all other life processes.

Enzymes are biological catalysts, composed mostly of protein, that participate in most chemical reactions within cells. Like other catalysts, enzymes greatly accelerate the rate at which chemical reactions occur, but they are not used up during the course of the reaction. A single enzyme molecule may enter into thousands of reactions every second without being modified or degraded.

Nucleic Acids

Two types of nucleic acids, **DNA** (deoxyribonucleic acid) and **RNA** (ribonucleic acid), exist in cells, and both play central roles in heredity. These roles are described here and in following chapters. Both DNA and RNA have the same fundamental structure; each is a linear unit composed of varying numbers of four different basic molecules called nucleotides. A **nucleotide** is a molecule containing a nitrogen base, a phosphate, and a five-carbon sugar—ribose in RNA and deoxyribose in DNA. The nitrogen bases found in the nucleotides of DNA are adenine (A), guanine (G), cytosine (C), and thymine (T). In RNA, adenine, cytosine, and guanine are present, but thymine is replaced by uracil (U). Figure 17.3 shows structur-

al diagrams of the five DNA bases. Nucleotides with A and G are classified as **purines** because of their double-ring structure; those with C, T, and U are **pyrimidines,** having single rings.

> **BEFORE YOU GO ON** A wide variety of chemicals are found inside cells. Two types of macromolecules—proteins and nucleic acids—are especially important in cellular activities. Proteins are composed of amino acids and have a great diversity of structure. They are used in building cellular structures and performing various functions, depending on the type of cell. Nucleic acids (RNA and DNA) are linear molecules constructed of nucleotides. They have important functions in heredity.

THE DISCOVERY OF DNA AND ITS ROLE IN LIVING ORGANISMS

The quest to uncover the repository of hereditary information spanned a century, from the 1860s to the 1960s. The road traveled by scientists on this journey was long and relatively straight except for one major detour.

Several experiments conducted during this period were extraordinarily beautiful in their design and outcome. We devote the rest of this chapter to describing some of these elegant experiments because they vividly illustrate not only the progress of biochemistry, molecular biology, and genetics but also the process of science.

A Chemical in the Nucleus

The revolution in molecular biology traces its origins back to the time when chemists were trying to identify the chemicals present in the nuclei of cells. The discovery of DNA is credited to a Swiss biochemist named Johann Friedrich Miescher, who worked on the problem from the 1860s until 1874 and returned to it shortly before his death in 1895. His source of

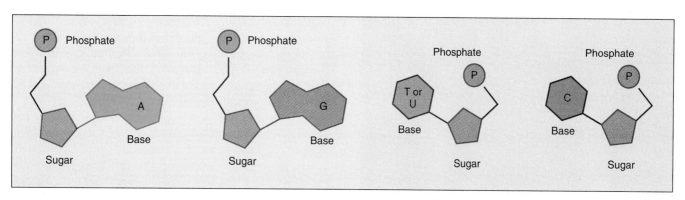

Figure 17.3 The primary structural units of nucleic acids are five nucleotides that each consist of a unique nitrogen-containing base. The bases used in the construction of DNA are adenine (A), guanine (G), cytosine (C), and thymine (T). RNA is composed of A, G, C, and uracil (U) in place of T. Each nucleotide contains one base, a phosphate molecule (P), and a sugar molecule. In DNA molecules, the sugar is deoxyribose; in RNA molecules, it is ribose.

experimental material would not have been the first choice of many scientists: Miescher analyzed pus, largely composed of white blood cells, which he obtained by washing out bandages taken from surgical patients. After analyzing these cells, Miescher found that the nuclei contained an unknown organic chemical with an unusually high amount of phosphorus. He and his colleagues named the new chemical *nuclein* (see Figure 17.4). He later continued his studies using sperm from salmon. Not only was this material probably more pleasant to work with, but it was also an especially rich source of nuclein because few other chemicals are present. Miescher published his results on nuclein, but they were vague and were severely criticized. Also, he was unable to offer a viable hypothesis about its role in the cell.

Based on work by Miescher and others, nineteenth-century scientists formed opinions about the possible nature of nuclein. Some felt, with little supporting evidence, that nuclein and chromatin (known to be the substance of chromosomes) were identical.

Further advances were made during the next three decades as a result of new research in biochemistry. The previously controversial suggestion that nuclein and chromatin were the same substance was shown to be true. This finding, coupled with the assumption from genetics that chromosomes carried the hereditary material, made the chemical analysis of nuclein highly significant. Through the use of new biochemical techniques, nuclein was found to consist of DNA and protein. This critical discovery can be seen as a promising beginning toward understanding the ultimate chemical nature of chromosomes. The number of possible answers for the original question had been reduced to two: the genetic material was either DNA or protein—but which? The trail was now to grow cold as a significant detour, in the form of the tetranucleotide hypothesis, delayed progress for many years.

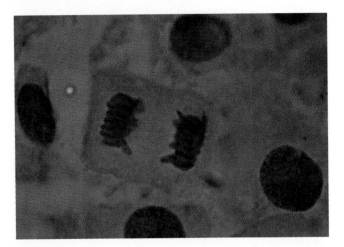

Figure 17.4 In Miescher's early studies to identify chemicals inside cells, special stains revealed a previously unknown chemical he called *nuclein*. In this photo, the red substance is Miescher's nuclein.

Question: *What chemical did nuclein turn out to be?*

The Tetranucleotide Hypothesis

Phoebus Levene and his colleagues at the Rockefeller Institute for Medical Research investigated the chemistry of nucleic acids from the early 1900s until 1940. During the first decade of this work, they were responsible for many of the chief discoveries that helped form a clearer picture of nucleic acid structure.

By his great successes and distinguished reputation, Levene came to dominate the field of nucleic acid research. His analyses suggested to him that equal amounts of the four nucleotides were present in the DNA of all organisms. The implication of this **tetranucleotide hypothesis** was that DNA was simple and repetitive. Given the complexity and great diversity of genetic traits, it seemed obvious that a static DNA molecule could not be the hereditary blueprint. Instead, scientists assumed that proteins, because of their remarkably varied structure, must somehow be responsible for defining the exceptional variations of hereditary traits. Because of Levene's venerable reputation, this hypothesis became an unchallenged paradigm of biochemistry even though it had no support from experimental studies. Consequently, little headway was made until this hypothesis was finally abandoned in the face of results from later experiments.

In hindsight, it is easy to understand that the uncritical acceptance of the tetranucleotide hypothesis led to a period in the 1920s and 1930s when little was accomplished in determining the chemical identity of the genetic material. This case illustrates how premature acceptance of a hypothesis retards the normal processes of questioning (modifying, rejecting, or forming new hypotheses) and experimenting that generate progress in science.

Nevertheless, the coming years would see several brilliant experiments whose results would increasingly point to DNA as the genetic material. We shall trace the progress of these important experiments.

The "Transforming Principle"

It often happens in science that research conducted in one area provides unexpected insights into a seemingly unrelated field. One such event revolved around work with certain strains of a bacterium, *Diplococcus pneumoniae* (commonly called *pneumococcus*), that causes pneumonia in humans.

In studying bacteria, individual cells are spread in a shallow dish or a test tube that is partly filled with a culture medium, a gelatinous substance that contains the nutrients required for bacterial growth. Once placed in culture, bacterial cells divide continuously. Within a day or two, one cell will give rise to millions of identical cells that appear as a spot or clump on the medium. These solitary bacterial masses are called *colonies*.

In the 1920s, it was shown that two basic *Diplococcus* varieties existed. The first was an encapsulated form that was surrounded by a carbohydrate coat and produced smooth, shiny, rounded colonies when grown in culture; this became known as the *S form*. A second type did not have a carbohydrate coat, and its colonies appeared rough rather than smooth; this was called the *R form*. The form that causes pneumonia is the S form. Exposure to the R form causes no negative effects.

The Griffith Experiment with *Diplococcus* Bacteria

Fred Griffith was a microbiologist interested in learning more about the effects of different strains of pneumococcus on humans. In 1928, he published a paper describing some interesting observations that puzzled scientists for many years.

Griffith's experiments are described in Figure 17.5. Standard results were first obtained when he injected mice with the two forms of *Diplococcus:* mice injected with the S form developed pneumonia and usually died, while those injected with the R form suffered no ill effects. In a second set of experiments, he first inactivated ("killed") the S-form bacteria by applying gentle heat. Not surprisingly, the heat-killed S bacteria did not cause pneumonia. However, a startling result occurred when he mixed the live R form with the heat-killed S form. Mice exposed to this combination contracted pneu-

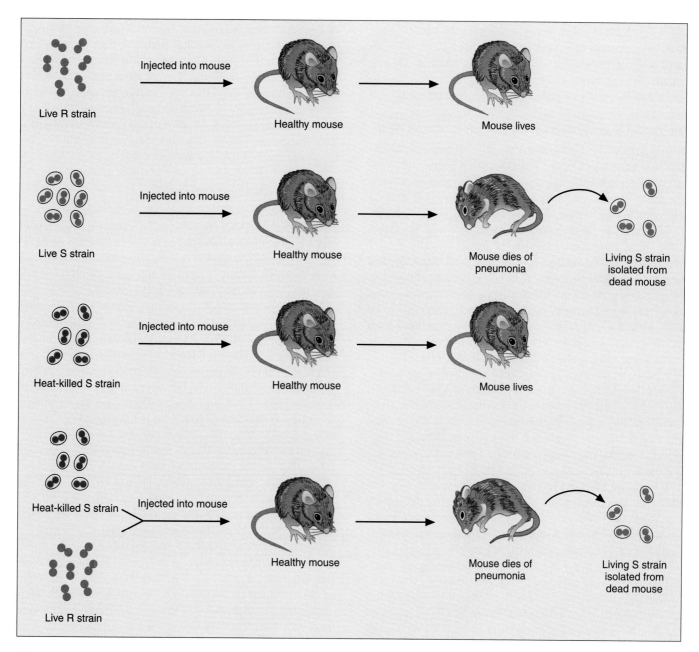

Figure 17.5   Beginning in the 1920s, key experiments were conducted using two strains of *Diplococcus* bacteria, a smooth (S) form and a rough (R) form. In Griffith's first experiment, mice were injected with either the S or R *Diplococcus.* Those injected with the R strain were unaffected, but those receiving the S strain died of pneumonia. In later experiments, mice injected with heat-killed S bacteria lived. However, if those S forms were first mixed with live R bacteria and then injected into a mouse, it died and was found to contain live S forms.

Question: *What conclusion was drawn from these experiments?*

monia. Curiously, blood samples collected and analyzed from those mice revealed the presence of live S bacteria that had caused the disease! When those bacteria were grown in culture, they produced the smooth colony characteristic of the S variety. What had happened? How could these results be explained? To Griffith, the results indicated that the R bacteria had mysteriously obtained something that became known as a **transforming principle** from the heat-killed S form. The harmless R form had somehow been transformed into the deadly S type of pneumococcus.

Griffith did not attempt any further experiments, and there is no indication that he related his findings to a transfer of hereditary material between the dead and living bacteria. He was killed in his lab during a German bombing raid over Britain in 1941, and the puzzle he created was not solved until 1944.

The Studies of Avery, Macleod, and McCarty

Soon after Griffith's publication in 1928, several other scientists confirmed his results. Also, transformations were produced under normal bacterial cell culture conditions without the use of mice. In the cell culture system, living R bacteria placed in test tubes containing the heat-killed S form became transformed into the encapsulated S pneumococcus. In a clever 1933 study, a method was devised to shatter the cells of heat-killed S bacteria by alternately subjecting them to freezing and heating. The different chemicals obtained from inside the cells were then separated from the cell debris using a special filtration technique. The transforming substance was found to be contained in the extracted group of chemicals isolated from inside the cell. In other words, this extract of chemicals alone could transform harmless R bacteria to the virulent S form.

Much of the work just described was done in the laboratory of Oswald T. Avery, where the transforming principle was the center of attention for over a decade. Beginning in 1935, Avery and his colleagues initiated experiments to try to isolate the transforming principle and to identify its chemical composition. A long period elapsed before a positive outcome was achieved. Finally, in a famous 1944 publication, Avery, Colin Macleod, and Maclyn McCarty combined results from several studies they conducted over many years using the smooth and rough pneumococcus strains. The complete set of results seemed to provide a clear answer about the chemical nature—and identity—of the transforming principle.

Their initial tactic involved an indirect process of elimination as described in Figure 17.6A. First, proteins, RNA, and sugars were eliminated as the active material through studies in which the chemical extract from inside S cells was attacked by specific enzymes. The researchers found that even though proteins were chopped apart, transformation still occurred. They obtained the same result for RNA and sugars. Hence none of these chemicals could be the transforming principle. However, when an enzyme that split DNA apart was introduced into the extract, transformation no longer took place! What did this suggest?

The second approach that Avery and his colleagues used, as described in Figure 17.6B, involved direct attempts to identify the transforming material by using chemical analytical methods. In this technique, R cells were exposed to purified protein, RNA, sugar, and DNA extracted from S cells. Only cells exposed to the DNA were transformed. The researchers stated that the active material was "a highly polymerized and viscous form of sodium deoxyribonucleate," that is, a form of DNA.

The obvious conclusion from these powerful studies was that DNA was the transforming principle. However, the impact of this report was delayed because of the absence of a solid connection between biochemical studies on bacteria and genetics at that time. For example, it was unclear whether or not bacteria even had chromosomes. Nevertheless, many researchers no longer doubted that DNA was, in fact, the genetic material. More conservative scientists adopted a cautious attitude and awaited further evidence. To some of them, it still seemed possible that proteins might play a hereditary role in other organisms or in other circumstances.

The Hershey and Chase Experiment

By 1952, many molecular biologists were using viruses to learn more about DNA. Viruses that infect bacteria, called **bacteriophages** or simply **phages,** were special favorites (see Figure 17.7). Viruses occupy a twilight zone between living and nonliving forms. They depend on host cells for all phases of their life cycle. Phages have an intricate structure, but chemically they are very simple, consisting primarily of a protein coat surrounding an inner core of nucleic acid (either DNA or RNA, but not both). When a phage encounters a bacterium, it attaches itself and injects DNA into the bacterial cell. The protein portion of the phage remains outside the bacterium, where it appears as a "ghost." The bacterium, unable to distinguish between phage DNA and its own DNA, reproduces phages according to instructions contained in the phage DNA. Eventually, the new phages cause the bacterial cell to rupture, releasing 100 to 200 progeny viruses that can then infect other bacteria. This entire cycle, from DNA injection to bacterial rupture, takes only a few minutes. Human viruses operate in much the same way when they infect our cells.

In a brilliant experiment, the results of which were published in 1952, Alfred Hershey and Martha Chase used T2 phages to establish beyond any doubt the identity of the genetic material. A T2 phage is a type of phage that infects the common bacterium *Escherichia coli,* which is a normal inhabitant of the intestines of mammals, including humans. The Hershey–Chase experiments are illustrated in Figure 17.8 (see page 322. First, bacteria were grown on culture media containing radioactive isotopes of either phosphorus (^{32}P) or sulfur (^{35}S). These two isotopes were used because DNA contains phosphorus but not sulfur and proteins contain sulfur but not phosphorus. If ^{32}P radioactivity were detected in a bacterial chemical, the source must have been DNA; ^{35}S radioactivity would indicate that it came from a protein.

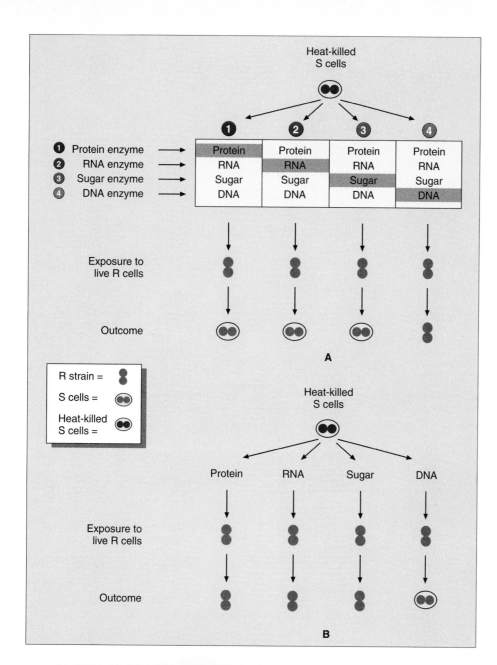

A

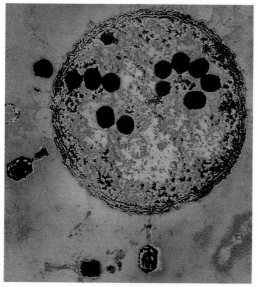

A

Figure 17.6 In a classic scientific paper published in 1944, Avery, Macleod, and McCarty summarized the results of experiments they conducted with *Diplococcus*. (A) Using suitable enzymes, they sequentially inactivated proteins (1), RNA (2), sugars (3), and DNA (4) from heat-killed smooth (S) cells, while leaving the other molecules intact. All groups of molecules, each containing one inactive type, were then introduced separately into live rough (R) cells; only those receiving inactive DNA failed to become transformed into S cells. (B) In a second approach, live R cells were exposed to purified molecules extracted from heat-killed S cells. Only R cells receiving DNA became transformed to S cells. The results of these experiments seemed to answer the question about the chemical identity of the genetic material.

Question: *What was the genetic material? What result would have been expected if* protein *were the genetic material?*

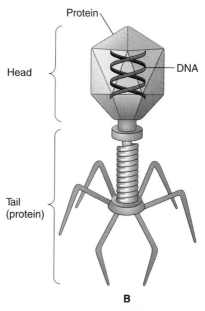

B

Figure 17.7 (A) Bacteriophages are viruses that infect and kill bacteria. In this electron micrograph, T2 phages appear as black ovals outside and inside an infected bacterium. (B) A diagram of a T2 phage.

Question: *Why is it necessary for phages to infect bacteria to complete their life cycle?*

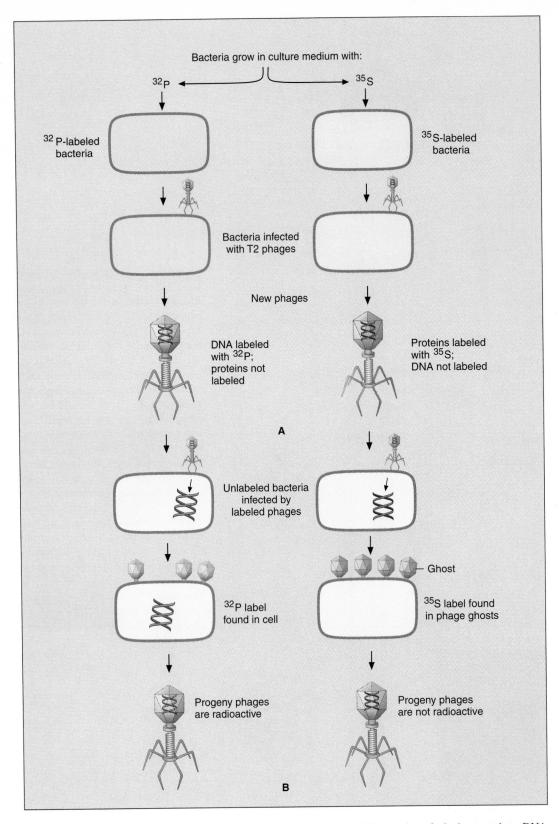

Figure 17.8 Hershey and Chase's famous phage experiment finally settled the question of whether protein or DNA was the genetic material. (A) *E. coli* bacteria were first grown in culture medium containing either radioactive ^{32}P or ^{35}S and allowed to reproduce. Bacterial offspring containing either ^{32}P or ^{35}S were then infected with T2 phages. Phages from ^{32}P-labeled bacteria contained radioactive DNA and no radioactive proteins, whereas phages produced from ^{35}S-labeled bacteria had radioactive proteins but no radioactive DNA. (B) These radioactive phages then infected nonradioactive bacteria. All of the ^{35}S radioactivity was found in the phage ghosts that remained attached to the bacterial cell wall. ^{32}P radioactivity was first found inside the bacterial cell and later in the DNA of progeny viruses.

Question: *Why did this experiment demonstrate conclusively that DNA is the genetic material?*

After labeling, phages were introduced into the two bacterial cultures, and the cycle of infection and phage reproduction was allowed to follow its normal course. As shown in Figure 17.8A, new phages generated from *E. coli* labeled with ^{32}P had radioactive DNA, and those propagated from bacteria labeled with ^{35}S had radioactive proteins. In the second part of the experiment (see Figure 17.8B), the two types of radioactive phages infected unlabeled bacteria. By determining which chemical was present in the bacteria after a replication cycle was completed, the true identity of the genetic material would be revealed. Unfortunately, a slight snag developed when at first both ^{32}P and ^{35}S seemed to be incorporated into the bacteria. The researchers then discovered that the phage ghosts (protein coats) remained attached to the bacteria, so a method to remove the ghosts had to be devised. Gentle agitation in a blender was found to detach the ghosts from the bacteria. The resulting mixture was then centrifuged, forcing the bacteria to the bottom while the phage ghosts remained at the top. The ^{32}P label, indicating DNA, was found inside the bacterial cells, whereas the ghost layer contained protein, as indicated by the presence of ^{35}S. It was also found that phage progeny produced by bacteria originally infected by ^{32}P-labeled viruses contained 30 percent or more of the parent-phage phosphorus but almost no radioactive sulfur.

This elegant experiment marked the end of scientific skepticism about DNA as the genetic material (at least in this group of viruses). For this and other studies on phages, Hershey received the Nobel Prize in 1969.

> **BEFORE YOU GO ON** A central question had guided research in molecular biology during the first half of the twentieth century: what was the chemical identity of the genetic material? The two leading candidates were proteins and DNA. Due primarily to the flawed tetranucleotide hypothesis, proteins were assumed to be the hereditary substance in the 1920s and 1930s. Then, through a series of classic experiments, DNA was finally identified as the chemical responsible for transmitting genetic information.

THE STRUCTURE OF DNA

The experiments just described answered the question about whether the genetic material was protein or DNA (see the Focus on Scientific Process, "Experimental Methods in Biology"). It was now clear that DNA was the source of genetic information transmitted between generations or, in the case of viruses, between the virus and the host cell.

Demise of the Tetranucleotide Hypothesis

By the 1940s, molecular biologists were aggressively pursuing answers to questions about molecules associated with life. Initially, proteins were of greater interest, but as evidence began to accumulate in favor of DNA as the hereditary molecule, emphasis shifted toward learning more about this complex chemical. What kind of structure could account for all of the genetically determined traits and functions with which it was associated?

The strength of the tetranucleotide hypothesis was seriously weakened as results of experiments on the transforming principle emerged. Edwin Chargaff, studying lipids in cells at Columbia University, was deeply impressed with the findings presented by Avery, Macleod, and McCarty in 1944, and he and his team began studies on the structure of nucleic acids. They made careful analyses of the nucleotide composition of DNA obtained from many organisms. Their results, published during the late 1940s, are summarized in Table 17.2. Before reading any further, review the table, examine the data, and formulate any conclusions that are supported by this extremely interesting data set. Do the data support or contradict the tetranucleotide hypothesis? Why?

Two conclusions that were drawn from Chargaff's data struck at the heart of the tetranucleotide hypothesis. First, the DNA in all species does not contain equal amounts of the four bases, and second, the base composition of DNA differs from one species to another. These conclusions effectively demolished whatever support remained for the obsolete tetranucleotide hypothesis. It had became obvious that DNA is complex and highly variable in structure.

It was also apparent that the base composition of DNA from different cells of the same species was characteristic and constant. Finally, the data indicated a consistency in the base composition of all DNAs that became known as **Chargaff's rule**: the amount of adenine is always approximately equal to the amount of thymine (A = T), and the amount of guanine equals the amount of cytosine (G = C). At the time, no hypothesis could explain this observation. (In Table 17.2, the A-T and G-C amounts are not exactly equal because the techniques in use at that time did not permit perfect measurements.)

In Pursuit of DNA's Structure

By the early 1950s, interest in DNA was in full blossom. No one doubted the genetic role of DNA any longer, and attention became directed toward determining the mechanism whereby DNA accomplished the transfer of information. Exactly how did DNA operate? To answer this question, the fundamental structure of the molecule would have to be determined.

Scientists from many fields, including many members of the first generation of molecular biologists, were drawn to the challenge of accomplishing this difficult task. The cast of characters that became involved in this venture included many celebrated scientists of the period, located primarily in the United States and England. Success was finally achieved by two young, ambitious scientists, James D. Watson and Francis Crick, who had not yet acquired strong reputations. The story of their discovery of the structure of DNA in 1953 is fascinating and has been the subject of numerous books and at least one film.

Watson, a young Ph.D. from the United States, and Crick, a brilliant British Ph.D. candidate, came to share a lab at Cam-

Table 17.2 Base Composition of DNA in Several Species

Organism	Cells Analyzed	Component Nucleotide Bases (Percent)			
		Adenine	Thymine	Guanine	Cytosine
E coli	—	26.0	23.9	24.9	25.2
Diplococcus	—	29.8	31.6	20.5	18.0
Yeast	—	31.3	32.9	18.7	17.1
Wheat germ	—	27.3	27.1	22.7	22.8
Sea Urchin	Sperm	32.8	32.1	17.7	18.4
Herring	Sperm	27.8	27.5	22.2	22.6
Rat	Bone Marrow	28.6	28.4	21.4	21.5
Human	Thymus	30.9	29.4	19.9	19.8
	Liver	30.3	30.3	19.5	19.9
	Sperm	30.7	31.2	19.3	18.8

Summary of results from Chargaff's experiments measuring the nucleotide base composition of DNA from several species.

bridge University in Cambridge, England, in 1951 (see Figure 17.9). Both were interested in determining the structure of DNA, and each brought with him unique skills that were useful in attacking the problem. Crick's first college degree, which he received in 1938, was in physics, and after World War II, his interests turned toward biology. Watson began his studies in zoology but later worked with scientists studying DNA in viruses. Crick and Watson did not conduct conventional experiments. Rather, they used a combination of strategies, including an application of existing principles and assumptions from quantum physics; the construction of models with cardboard cutouts, wire, and sheet metal; and guesswork. In addition, they exploited vital information obtained from others.

Uncovering the three-dimensional structure of DNA presented substantial technical challenges. Not even large molecules, such as DNA, could be seen in detail through existing microscopes. Special instruments and techniques were used to acquire data that could provide hints about the structure of DNA. The most important clues were obtained using a technique called **X-ray crystallography.** In this method, a crystal of the purified chemical of interest is exposed to X rays. The X rays are diffracted in specific ways characteristic of the crystal's structure. The resulting pattern is then captured on photographic film (see Figure 17.10). Examination and analysis of the photograph by expert crystallographers provided details about the precise structure of the crystal DNA. (Today, com-

Figure 17.9 In 1953, using unconventional methods, Watson (left) and Crick (right) identified the fundamental molecular structure of DNA, one of the most important accomplishments in the history of biology.

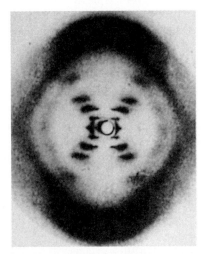

Figure 17.10 X-ray crystallography was a powerful tool used in identifying the structure of DNA. This X-ray diffraction photograph of DNA, taken by Rosalind Franklin, was apparently used by Watson and Crick in determining the structure of DNA. It suggested that DNA had a helical structure.

puters are used in creating molecular models from X-ray crystallography data.)

Neither Watson nor Crick had exceptional expertise in crystallography, yet they fully realized the importance of obtaining X-ray diffraction pictures of DNA. Meanwhile, a major effort to obtain such pictures was being made at King's College in London by Maurice Wilkins and Rosalind Franklin, both acknowledged crystallography experts, who were also interested in determining the structure of DNA (see Figure 17.11). In early 1953, Watson obtained a print of one of Franklin's best photographs (reproduced in Figure 17.10). The events surrounding that acquisition, the subsequent use Watson and Crick made of the picture, and their initial failure to acknowledge its critical importance in their success remain controversial. It later became apparent, however, that Franklin and her coworkers had been very close to determining the structure of DNA.

After obtaining Franklin's picture, Watson and Crick began a feverish month of model building that ended on March 7, 1953, with a full DNA model. It conformed perfectly with the X-ray measurements and the requirements imposed by quantum physics and chemical principles. Their accomplishment, elucidating the three-dimensional structure of DNA, has been described as one of the greatest achievements of twentieth-century biology.

In 1962, Watson, Crick, and Wilkins received the Nobel Prize for their work on DNA. Franklin would probably have shared in this honor had she not died of cancer before the award was bestowed, as this prize is awarded only to living scientists.

The Watson–Crick Model of DNA

The DNA molecule described by Watson and Crick in their historic paper is shown in Figure 17.12A (see page 328). It consists of two strands twisted into the shape of a **double helix,** much like a ladder twisted about its long axis. The backbone of each strand (the upright of the twisted ladder) consists of long chains of nucleotides, bonded sugar and phosphate groups each with one of the four bases (A, T, G, or C) attached and projecting inward. The two strands are held together (appearing as rungs of the twisted ladder) by hydrogen bonds, weak chemical bonds that join the bases from one strand to the bases from the other strand. The strands are *antiparallel* because they run in opposite directions relative to their sequence of bases (see Figure 17.12B). The key to the Watson–Crick model is that the bases always pair in the same way, A with T (or T with A) and C with G (or G with C), partly because of hydrogen-bonding requirements. Three hydrogen bonds can form between G and C but only two between A and T. These bases (A–T, G–C) are called **complementary base pairs.** The underlying basis for Chargaff's rule—equal amounts of A and T and of C and G in DNA—is now evident.

Several features of the Watson–Crick DNA model are especially interesting. The two antiparallel strands are not identical to each other either in the base sequence or in composition. Rather, they are complementary to each other with A always opposite T and G always opposite C. Any linear sequence of bases can exist in a DNA molecule; there are no restrictions except that the composition of one strand is determined by the sequence of the other because of complementary base pairing. This last point is related to the most fascinating feature of the Watson–Crick model: it hints at how chromosomes—and DNA—become duplicated before cell division. Indeed, near the end of their paper is a sentence, apparently put in at the insistence of Crick, that is considered one of the greatest understatements in the literature of science: "It has not escaped our notice that the specific pairing we have postulated immediately suggests a possible copying mechanism for the genetic material." In other words, not only did Watson and Crick describe the structure of DNA, but they also understood, at least in part, how it was replicated.

> **BEFORE YOU GO ON** After DNA was identified as the genetic material, the next challenge facing molecular biologists was to determine its structure. After an exciting race, James Watson and Francis Crick described the chemical structure of DNA in 1953. The DNA molecule has the shape of a double helix, with two chains of nucleotides held together by hydrogen bonds formed between complementary base pairs.

Figure 17.11 Rosalind Franklin was an expert in X-ray crystallography. One of her diffraction photographs of DNA (see Figure 17.10) proved crucial in determining its structure.

DNA REPLICATION

Recall that prior to mitosis and meiosis, chromosomes are duplicated during interphase. The basis of chromosome duplication is the **replication,** or synthesis, of DNA. A new question now occupied researchers: how was DNA replicated in the nucleus?

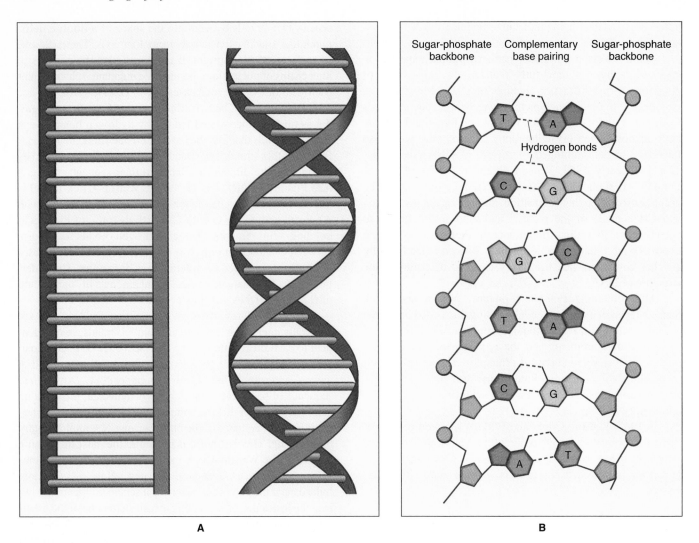

A   **B**

Figure 17.12   (A) As described in Watson and Crick's historic 1953 paper, the basic structure of DNA is a double helix in which two strands (rails of the ladder), connected by hydrogen bonds (steps of the ladder), are twisted around each other. (B) Each DNA strand consists of a long chain of different nucleotides that are paired in specific ways (A–T, T–A, G–C, C–G). The two strands are held together by weak hydrogen bonds that exist between complementary base pairs of each strand.

Question: *Why did the specific pairing "suggest a possible copying mechanism for the genetic material"?*

The Watson–Crick Hypothesis

In a second paper that followed on the heels of the first, Watson and Crick proposed that by breaking the weak hydrogen bonds, the two parent (original) strands of the double helix could unwind and serve as templates for constructing new daughter strands. The two new strands of DNA, each consisting of a parent combined with its daughter, would be identical. This type of replication, shown in Figure 17.13, is called **semiconservative replication** because each new DNA molecule contains one parent strand and one newly synthesized daughter strand. If replication were fully conservative, the entire DNA molecule would be copied, and one daugh-

ter cell would get the original and the other would get the copy.

Shortly after this paper appeared in 1953, Watson and Crick developed new interests, and their celebrated collaboration ended. In the years since, Crick has become established as a dominant intellectual force in the field of theoretical molecular biology. Watson became director of the prestigious Cold Spring Harbor laboratories on Long Island, New York, and has also written several important books on molecular biology (see also Chapter 22). Both have written delightful books that are recommended for anyone interested in their accounts of the DNA story (Watson published *The Double Helix* in 1968, Crick *What Mad Pursuit* in 1988).

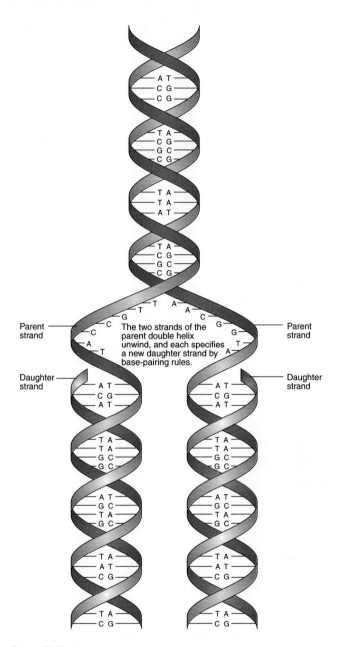

Figure 17.13 This simple model describes the basic process of DNA replication. The two parent strands separate, and each serves as a template for forming a new daughter strand. This type of replication is called semiconservative because each new DNA molecule consists of one "old" (parent) strand and one "new" (daughter) strand.

The Replication Process

In 1958, scientists confirmed the essential correctness of Watson and Crick's 1953 outline for semiconservative DNA replication. Further studies have since provided full details about how DNA is synthesized.

DNA replication is illustrated in Figure 17.14. It begins when hydrogen bonds linking the base pairs are broken and the two parent strands of DNA partially unwind. **DNA poly-**

merases, enzymes that copy DNA by joining complementary bases, then move along each opened strand, bonding appropriate base pairs together. The parent base sequence serves as a template for determining the precise arrangement of the new daughter chain. Regular complementary base pairing occurs as the replication proceeds; A pairs with T, and C pairs with G.

Several features of the replication process are noteworthy. Prior to replication, the cell synthesizes the nucleotides that will be required in the construction of new DNA strands. Replication begins at specific positions, or regions, in the inner portion of the DNA chain (never at the end) and then proceeds in both directions away from the **initiation site.** The number of initiation sites ranges from one in small bacterial chromosomes to hundreds or thousands in the comparatively huge eukaryotic chromosomes. During replication, DNA polymerases race along the two DNA strands in opposite directions. Along one parent strand, the *leading strand,* the chain is copied continuously and smoothly. Replication proceeds in the opposite direction along the other, the *lagging strand.* Here numerous short segments of the new DNA strand are produced simultaneously by many polymerases and then pieced together and joined to the template strand by hydrogen bonds. In eukaryotes, replication progresses at a rate of about 50 base pairs per second.

Complementary base-pairing rules dictate the organization of the new strand formed from the parent template. Each daughter strand is identical to the original parent strand that did *not* serve as its template. Consequently, the two new double helices are identical to each other and also to the original parent DNA molecule. Because DNA is the intergenerational genetic link, the integrity of the base sequence must be maintained through correct base pairing during replication. To ensure the accuracy of newly formed DNA molecules, cells of eukaryotes and prokaryotes have "proofreading" molecular mechanisms that pinpoint and correct errors in base pairing. In *E. coli,* replication errors, after correction, occur at an amazingly low frequency—one mistake per every billion pairs bonded.

BEFORE YOU GO ON DNA replication occurs before cell division and the transmission of genetic information between generations. The basic process begins when the two parent strands of DNA open, followed by the construction of two daughter strands, by complementary base pairing, using the parent strands as templates. The end products of this semiconservative replication process are two identical DNA molecules, each consisting of one parent strand and one new daughter strand.

Finally, most of what is known about DNA replication has been derived from studies of bacteria. Considerably less is understood about DNA synthesis in eukaryotes. Nevertheless, cells of all organisms appear to produce copies of their DNA in essentially the same manner (see Figure 17.15). The universal possession of such a critically important biochemical mechanism is consistent with the concept of unity in biology: that all life holds in common features that developed early in the history of life on Earth.

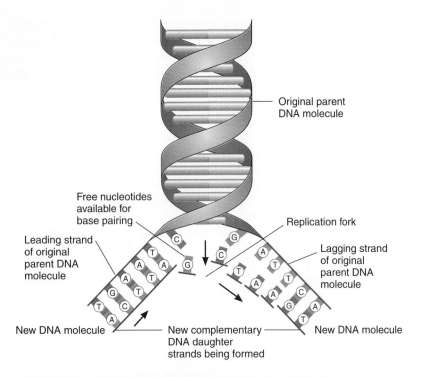

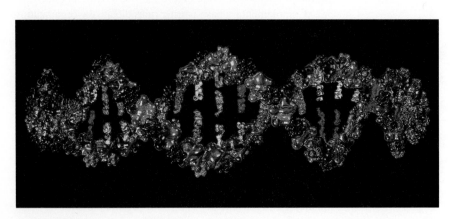

Figure 17.14 DNA replication begins when DNA parent strands partly unwind and the initiation site is exposed. DNA polymerases then add complementary bases to each parent strand, creating a daughter strand. Along the leading strand, bases are added continuously until completion; along the lagging strand, short segments are added at several points and then pieced together. The newly replicated DNA molecule consists of one parent strand and one daughter strand.

Figure 17.15 A computer-generated model of DNA.

SUMMARY

1. Following confirmation of the chromosome theory, the next fundamental problem to be addressed in genetics was determining the molecular substance associated with genes and chromosomes. The leading candidates were protein and DNA.

2. In experiments using bacteria that cause pneumonia, Griffith discovered that a "transforming principle" could convert one strain of bacteria to another.

3. Several ingenious experiments were conducted to identify the substance that could transform bacteria. Ultimately, it was determined that DNA, not protein, was the genetic material.

4. The next problem that generated intense interest was determining the structure of DNA. Using nontraditional scientific methods, Watson and Crick discovered that DNA had the structure of a double helix. Attention then shifted to the process of DNA replication.

5. Watson and Crick proposed that parent DNA strands served as templates for producing new daughter strands, a process termed semiconservative replication. That hypothesis was later confirmed.

WORKING VOCABULARY

amino acid (p. 316)
bacteriophage (p. 320)
complementary base pairs (p. 327)
DNA (p. 317)
DNA polymerase (p. 329)

enzyme (p. 317)
nucleic acid (p. 316)
nucleotide (p. 317)
protein (p. 316)
replication (p. 327)
RNA (p. 317)

REVIEW QUESTIONS

1. Why are microorganisms used in genetics studies?

2. What are the roles of proteins in cells?

3. What was the significance of Miescher's studies?

4. What is the tetranucleotide hypothesis? What did it contribute to identification of the genetic material?

5. What was concluded from Griffith's *Diplococcus* experiments?

6. How did the studies by Avery, Macleod, and McCarty and by Hershey and Chase prove that DNA was the genetic material?

7. What was the significance of Chargaff's experiments? Of Franklin's work?

8. What methods did Watson and Crick use in determining the structure of DNA?

9. How is DNA constructed? What features of DNA's structure are important in its replication?

ESSAY AND DISCUSSION QUESTIONS

1. Why is the Hershey–Chase experiment considered a classic in science?

2. Is it possible that another tetranucleotide-type hypothesis—one that restrains progress in a field—could appear today? Why?

3. What might explain the emergence of the field of molecular biology in the first half of the twentieth century?

REFERENCES AND RECOMMENDED READING

Avery, O. T., C. M. Macleod, and M. McCarty. 1944. Studies on the nature of the substance inducing transformation of pneumococcal types. *Journal of Experimental Medicine,* 79: 137–158.

Crick, F. 1957. *Nucleic Acids.* San Francisco: Freeman.

Crick, F. H. C. 1988. *What Mad Pursuit: A Personal View of Scientific Discovery.* New York: Basic Books.

Gribbon, J. 1985. *In Search of the Double Helix: Quantum Physics and Life.* New York: McGraw-Hill.

Griffith, F. 1928. The significance of pneumococcal types. *Journal of Hygiene,* 27: 113–159.

Hershey, A. D., and M. Chase, 1952. Independent functions of viral protein and nucleic acid in growth of bacteriophage. *Journal of General Physiology,* 36: 39–56.

Judson, H. F. 1979. *The Eighth Day of Creation: The Makers of the Revolution in Biology.* New York: Simon & Schuster.

Magasanik, B. 1988. Research on bacteria in the mainstream of biology. *Science,* 240: 1435–1438.

McCarty, M. 1985. *The Transforming Principle: Discovering That Genes Are Made of DNA.* New York: Norton.

Moore, J. 1986. Science as a way of knowing: Genetics. *American Zoologist,* 26: 583–747.

Nelkin, D. 1995. *The DNA Mystique: The Gene as a Cultural Icon.* New York: Freeman.

Sayre, A. 1975. *Rosalind Franklin and DNA.* New York: Norton.

Stent, G. S. (ed.). 1980. *The Double Helix.* New York: Norton.

Watson, J. D. 1968. *The Double Helix: A Personal Account of the Discovery of the Structure of DNA.* New York: Atheneum.

Watson, J. D., and F. H. C. Crick. 1953. Genetical implications of the structure of deoxyribonucleic acid. *Nature,* 171: 964–967.

Watson, J. D., and F. H. C. Crick. 1953. Molecular structure of nucleic acids: A structure for deoxyribose nucleic acid. *Nature,* 171: 737–738.

ANSWERS TO FIGURE QUESTIONS

Figure 17.1 Single bacterial cells or viruses are much simpler to manipulate than multicellular organisms; have less complex and more easily obtained DNA (genetic material); have rapid life cycles, producing abundant offspring in a short time period; and are relatively inexpensive and easy to maintain in a laboratory.

Figure 17.4 DNA.

Figure 17.5 That something was able to "transform" living R-strain bacteria into living S-strain bacteria.

Figure 17.6 DNA. Live R cells would have become transformed to live S cells after being exposed to *Diplococcus* proteins, and such a transformation would not occur after exposure to DNA.

Figure 17.7 They lack the cellular structures or molecules necessary for reproducing (or even to be classified as living).

Figure 17.8 By definition, genetic material is transmitted from one generation to the next. Labeled DNA was the only molecule that was found in both parent and progeny viruses.

Figure 17.12 If the two strands were separated and complementary base pairs were then added to the two single strands, the two resulting DNA molecules ("copies") would be identical.

18

The Nature of Genetic Information

Chapter Outline

Reading Questions

1. Which early studies provided insights into the genetic role of DNA?

2. What is the relationship between metabolic pathways and genes?

3. How are proteins synthesized in eukaryotic cells?

4. What is the genetic code?

The wondrous discoveries described in Chapter 17 opened up vast new areas of scientific research in molecular biology. In the 1950s and 1960s, questions about DNA and genes were always at the heart of these investigations. The answer to the question "What is a gene?" has changed continually during the past 60 years. Even as the final answer to that question awaits further research, scientific process continues to generate many new questions. How are genes connected with the formation of a new individual, with all of the characteristics of its parents, and with the properties of its species? How do genes regulate the universe of activities carried on in cells that are necessary for life? In this chapter, we will consider current ideas on these questions.

THE GENETIC ROLE OF DNA

Even before DNA's structure and replication mechanism were described in the 1950s, earlier research had provided scientists with major clues about the nature of DNA and how it functioned in the cell.

Garrod's Concept of Chemical Individuality

In 1901, Archibald Garrod, an English physician, published a paper in which he stated that a human disease, **alkaptonuria**— a disorder characterized by excretion in the urine of large amounts of a chemical called *homogentistic acid,* which causes the urine to turn black when exposed to air—was inherited, not acquired later in life. His hypothesis was based on observations that symptoms of alkaptonuria were evident in newborn children. Garrod also studied albinism and phenylketonuria. **Albinism** is the condition of having no pigment in certain cells, which results in milky-colored skin, white hair, and pinkish eyes (see Figure 18.1). **Phenylketonuria** (**PKU**) is distinguished by abnormally high levels of an amino acid, phenylalanine, in the blood of newborn babies and can result in severe mental retardation. In 1908, the results of his studies were published in the classic book *Inborn Errors of Metabolism.* Concluding that these diseases were inherited, Garrod hypothesized that each condition resulted from a deficiency involving a specific enzyme. William Bateson and R. C. Punnett, two prominent Mendelian geneticists of the early 1900s, had noted that alkaptonuria was an example of recessive inheritance because it occurred in children whose parents did not have the disease. Garrod made reference to their conclusion in his book.

Garrod is often credited with proposing a direct relationship between a *gene* and an *enzyme.* However, there is no evidence in his writings that he ever made such a clear connection. Rather, he argued for the concept of **chemical individuality**— the idea that every species and every individual within a species, including humans, had different proteins. In a 1923 lecture, he

Figure 18.1   Albinos lack pigments in their hair, eyes, and skin. This phenotypic effect is caused by an "inborn error of metabolism" associated with a recessive gene.

Question: *What is the cause of albinism?*

stated that "chromosomes . . . obviously include factors [genes] which determine both the forms and metabolic peculiarities of the organism which originate from them." Throughout his life, Garrod asserted that inherited chemical individuality, in the form of variable enzymes, could explain certain diseases in humans. Because of these hypotheses (chemical individuality and its connection with inherited diseases), Garrod is often recognized as the father of biochemical genetics. Note that his conceptual breakthrough was based primarily on observation and analysis, not on experimental studies.

We now know that Garrod's pioneering hypotheses were basically correct, yet they were not widely accepted, or even recognized, by the time of his death in 1936. Why were his important ideas overlooked for so long? First, because he did not state his hypotheses in terms of Mendelian inheritance, Garrod's ideas may not have come to the attention of classical geneticists of the 1920s and 1930s. Also, studies of human genetics during that period had turned toward eugenics (as you will see in the Focus on Scientific Process in Chapter 19). Second, Garrod's work was not easily integrated into any established scientific disciplines of the period. For example, medicine at that time was concerned primarily with the treatment of diseases; there was nothing in Garrod's writings that related to treatments. Besides, rare, relatively harmless diseases such as albinism and alkaptonuria were of much less consequence to medicine and society as a whole when compared with the overwhelming problems caused by infectious diseases. The field of genetics itself had not yet developed to the point where genes were associated with chemicals such as proteins and enzymes. Biochemists probably paid the most attention to Garrod's work, but their interest was generally confined to questions about protein and enzyme structure and function, not to links with genes. Third, Garrod's hypotheses could not be tested because technologies did not yet exist for measuring proteins, amino acids, or enzyme activities in individuals affected by the diseases Garrod worked with. Thus although Garrod established the theoretical groundwork for a one-gene, one-enzyme relationship, this hypothesis was neither clearly articulated nor confirmed until after his death.

The One-Gene, One-Enzyme Hypothesis

In the late 1930s, George Beadle and Edward Tatum conducted classic experiments that confirmed the idea that genes gave rise to enzymes. They used a fungus, the pink bread mold *Neurospora* shown in Figure 18.2A (see page 334), that could be cultured (grown) easily and rapidly in the laboratory on a liquid *minimal growth medium* containing only a few simple substances. The fungal cells could synthesize all required amino acids, vitamins, and other molecules from the basic chemicals contained in the minimal medium. In their experiments, partly illustrated in Figure 18.2B (see page 334), Beadle and Tatum first exposed this microorganism to X rays and ultraviolet light, agents known to induce mutations in genes. After exposure, *Neurospora* cells were allowed to grow on a *complete growth medium* containing amino acids, vitamins, and other chemicals normally found in the minimal growth medium. New fungal

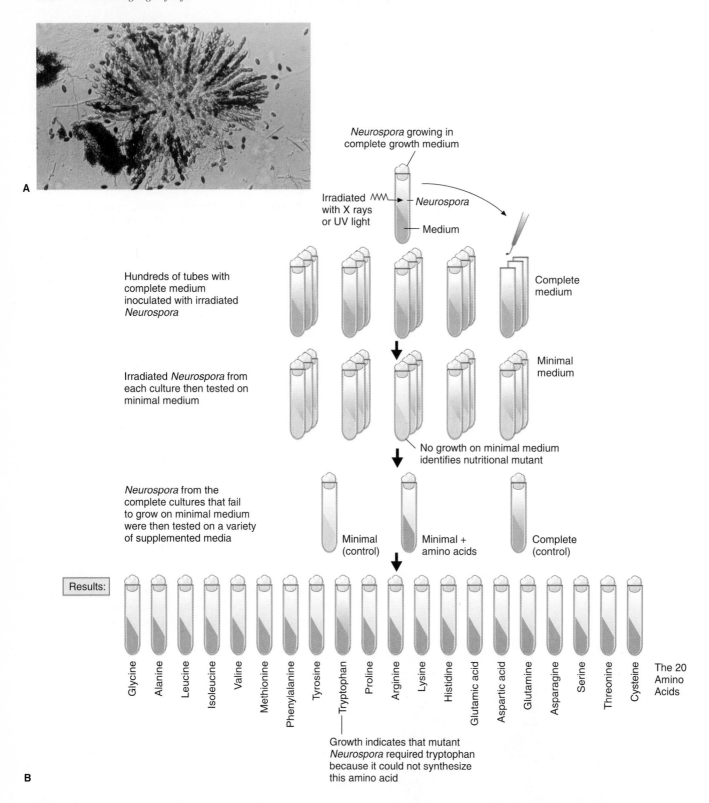

Figure 18.2 (A) *Neurospora* is a simple fungus that proved useful in early genetics studies designed to identify the role of genes in organisms. (B) Beadle and Tatum induced mutations in *Neurospora* by exposing it to radiation. In analyzing the effects of these mutations, they found that affected *Neurospora* often required that a specific amino acid (or vitamin) be added to the minimal medium for growth to occur. Beadle and Tatum concluded that mutants lacked a necessary enzyme for synthesizing the required substances and that genes were responsible for producing these enzymes.

Question: *What conclusion did Beadle and Tatum draw from the experiment described in this figure?*

cells growing on the complete medium were then removed and placed on the minimal medium. Some of these cells were unable to grow on the minimal medium. What did this indicate? Beadle and Tatum concluded that one or more mutations prevented the nongrowing *Neurospora* from synthesizing either an amino acid or a vitamin (necessary for their growth). How could this hypothesis be tested? The mutant, nongrowing *Neurospora* strains were placed on minimal medium supplemented with single amino acids (or vitamins) to see if they grew. Beadle and Tatum reasoned that if a mutant could grow on a minimal medium supplemented with a specific amino acid, the mutation must have involved an enzyme necessary for producing that amino acid. In their studies, they showed that the need for a specific amino acid or vitamin was due to a missing enzyme that, if present, would have promoted synthesis of the required substance. In 1941, Beadle and Tatum concluded that each metabolic reaction in *Neurospora* was catalyzed by a single enzyme and that each enzyme was specified by one gene. This concept became known as the **one-gene, one-enzyme hypothesis.** For their discoveries, they received the Nobel Prize in 1958.

> **BEFORE YOU GO ON** Two early investigations provided important information about the possible role of genes in cells. Between 1901 and 1908, Garrod reported that certain human diseases appeared to be inherited. He hypothesized that inherited variations that affected normal enzyme function were responsible. In the late 1930s, Beadle and Tatum studied the effects of gene mutations in *Neurospora.* They found that mutations resulted in missing enzymes. This finding led to the one-gene, one-enzyme hypothesis.

Metabolism

As the title of Archibald Garrod's 1908 book, *Inborn Errors of Metabolism,* indicates, we need to know a little more about metabolism to understand the full significance not only of Garrod's work with human diseases but also of Beadle and Tatum's work with *Neurospora.* Within a living cell, thousands of different molecules are continuously synthesized from simple precursor molecules, while others are being broken down or modified. **Metabolism** refers to *all* chemical reactions involved in the synthesis and degradation of molecules within a living cell. This is the process by which living organisms assimilate energy and use it for maintenance and growth. Metabolic processes that involve the synthesis (construction) of molecules—for example, the synthesis of proteins from amino acids—are referred to as **anabolism.** Metabolic processes by which a cell breaks down molecules—for example, sugar broken down into water and carbon dioxide—are referred to as **catabolism.** A wide variety of biochemical reactions enable cells to carry on activities that include the breakdown of energy-rich molecules and the synthesis of cellular components such as structural proteins, enzymes, hormones, and DNA. Cellular metabolism is involved in every aspect of an organism's structure and function.

A **metabolic pathway** is a complete sequence of chemical reactions in which a molecule becomes progressively modified. Some metabolic pathways consist of only a few reactions; oth-

ers may involve hundreds of stepwise reactions. Each reaction is specifically catalyzed by a single enzyme. The basic types of metabolic reactions include **synthesis** (simple substances A, B, C, and so on are combined to form a final product, $A + B + C \rightarrow X$), stepwise **modification** of a chemical ($M \rightarrow N \rightarrow O \rightarrow P$ and so on), or **breakdown** into different products ($X \rightarrow Y + Z$); each arrow represents a chemical modification. Each reaction typically requires the presence of a specific enzyme (see Figure 18.3, page 336). Given this information about metabolism, how can Garrod's hypotheses, which centered on missing enzymes, be interpreted? How can this information be applied to Beadle and Tatum's studies?

A small part of a metabolic pathway is shown in Figure 18.4 (see page 336). Phenylalanine normally undergoes a series of modifications that ultimately lead to much of it being broken down into carbon dioxide and water. As indicated in the figure, some of it is converted into tyrosine, another amino acid, which is then transformed into **melanin,** the pigment that gives color to our skin, hair, and eyes. If one of the enzymes of this pathway is not present or not functional, three things may happen:

1. A substance would increase in concentration because it would not be broken down by an enzyme (resulting in alkaptonuria).

2. A needed compound would not be produced in the absence of a required enzyme (resulting in albinism).

3. Excesses of unwanted substances would be produced from compounds that would accumulate if no enzyme were present (resulting in PKU).

In PKU, the major pathway—phenylalanine converted to tyrosine—is blocked, and a minor pathway—phenylalanine converted to phenylpyruvic acid—becomes dominant. This leads to elevated levels of phenylalanine in the blood and phenylpyruvic acid in the urine; both of these substances have been implicated in causing mental retardation in untreated individuals. Since 1966, screening tests for detecting phenylalanine and its **metabolites** (products of metabolism) in the urine of newborn babies have been required for American hospitals. An infant who tested positive was treated by being placed for a few years on a special diet low in phenylalanine to aid in intellectual development. Recently, however, studies have shown that phenylalanine-restricted diets should be continued well into the adult years to prevent loss of intellectual abilities. Also, it is strongly recommended that women with PKU remain on the diet during reproductive years because high phenylalanine concentrations in a mother's blood can cause mental retardation in her offspring.

> **BEFORE YOU GO ON** Metabolism refers to the breakdown and synthesis of chemicals within the cell. A metabolic pathway includes all of the reactions in which a single chemical is modified. Each step in a metabolic pathway usually requires a specific enzyme. If all components of a pathway are known, it is sometimes possible to determine the underlying cause of a genetic disorder and also to devise a treatment.

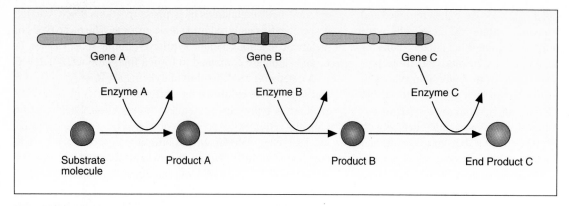

Figure 18.3   A hypothetical metabolic pathway in which a substrate molecule becomes modified by the action of different enzymes.

Question: *What would happen in this pathway if gene A mutated?*

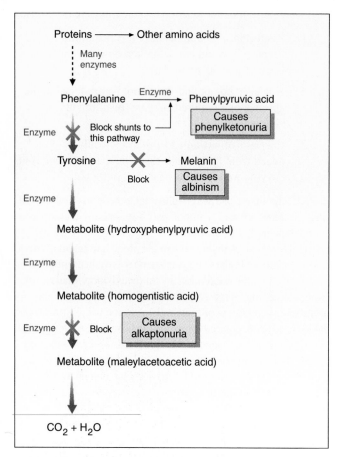

Figure 18.4   Metabolic pathways describe the sequence of chemical reactions in which a molecule becomes modified. This partial metabolic pathway describes both the normal breakdown of the amino acid phenylalanine and the phenotypic effects caused if specific enzymes are missing at different parts of the pathway.

Question: *What is the cause of phenylketonuria (PKU)?*

The One-Gene, One-Polypeptide Hypothesis

As a result of the pioneering metabolic studies of Garrod, Beadle and Tatum, and many others, the one-gene, one-enzyme hypothesis came to be widely accepted. As more became known about genes and proteins and their structure, the original hypothesis was extended. It later became known as the **one-gene, one-polypeptide hypothesis.** The reason for the modification was that some proteins contain more than one **polypeptide,** which is a chain of amino acids that may be smaller than a complete protein, and each polypeptide is determined by a single gene.

We will see later in this chapter that because of the recent discovery of *introns* and *exons* in DNA, the concept of a gene was again modified. A **gene** is now defined as all of the DNA sequences necessary to produce a single protein or RNA product. Thus the one-gene, one-polypeptide hypothesis has itself been modified and can now be restated as the **one-gene, one-protein** (or **one-RNA-product**) **hypothesis.** Most references to this hypothesis simply state "one-gene, one-protein."

By the early 1950s, it was generally assumed that genes were small segments of DNA and that they gave rise to specific proteins in the cell. How was the genetic information in genes turned into proteins?

Watson, Crick, and the "General Idea"

Watson and Crick had ushered in a new era of molecular biology with their description of DNA's structure in 1953. They, and many other scientists, now entered a decade of fruitful research that focused on the central questions of genes and proteins.

By 1953, scientists understood, first, that DNA is located on chromosomes within the nucleus and, second, that protein synthesis occurs outside the nucleus, in the cytoplasm. Scientists had also identified and described RNA, the other nucleic acid. Whereas DNA molecules are large and double-stranded, RNA molecules are much smaller and single-stranded (see Figure 18.5). Since there could be no direct transfer of information from DNA confined to the nucleus to proteins in the cytoplasm, scientists postulated that some intermediate molecule must be involved in the transfer of genetic information from DNA to protein.

From the time the structure of DNA was described in 1953, the leading candidate for this intermediate information transfer molecule was RNA. Results from several studies supported this view. The cytoplasm of cells actively engaged in protein synthesis was found to contain large amounts of RNA. Other experiments, such as the one summarized in Figure 18.6 (see page 338), demonstrated that RNA was synthesized in the nucleus and hours later appeared in the cytoplasm. Also, it was easy to visualize the possible synthesis of single-stranded RNA with one strand of a DNA molecule serving as a template (take another look at Figure 18.5).

For historical flavor on how answers to the gene–protein question were pieced together, consider a comment that Francis Crick made many years after the question had been answered: "Jim [Watson], you might say, had it first. DNA makes RNA makes protein. That became then the *general idea.*" In a famous 1958 article, Crick elaborated on the general idea, which basically consisted of two principles, neither of which was strongly supported by experimental evidence at the time. The first was the **sequence hypothesis,** the concept that the sequence of bases in DNA and RNA molecules specified the sequence of amino acids in proteins. The second principle, which Crick called the **central dogma,** is summarized

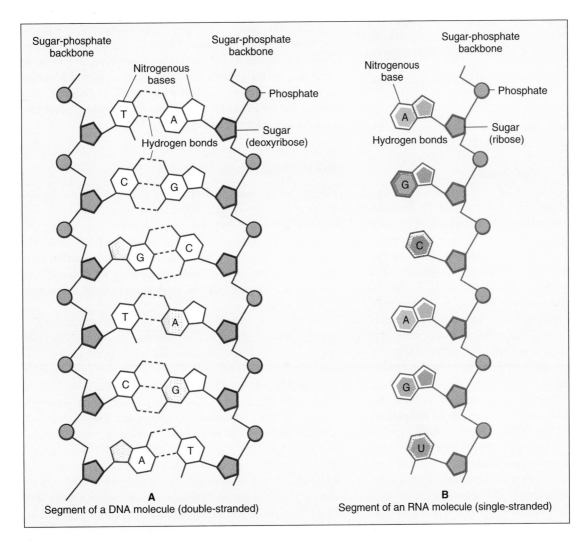

Figure 18.5 A comparison of DNA (A) and RNA (B) shows basic similarities. They are both composed of linear chains of nitrogenous bases, sugars, and phosphate groups (P). There are three principal differences in their structure: the sugar molecule is deoxyribose (D) in DNA and ribose (R) in RNA; DNA is double-stranded, but RNA is single-stranded; and both contain four nitrogenous bases, but RNA has uracil (U) instead of the thymine (T) found in DNA.

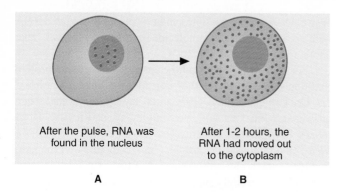

After the pulse, RNA was found in the nucleus

After 1-2 hours, the RNA had moved out to the cytoplasm

A   B

Figure 18.6   (A) To show that RNA is synthesized in the nucleus and then moves into the cytoplasm, cells were first exposed to a "pulse" of RNA nucleotides labeled with radioactive molecules. The labeled molecules were incorporated into RNA molecules within 15 minutes, and special sensitive film, which is exposed by radiation emitted from the labeled nucleotides, was used to reveal their location. For a brief time the black dots, indicative of exposed film from radioactive RNA, were found only in the nucleus. (B) One to two hours later, they were found in the cytoplasm, indicating that they moved from the nucleus to the cytoplasm.

Question: *Why does RNA move from the nucleus to the cytoplasm?*

in Figure 18.7. Essentially, the central dogma described the flow of genetic information as DNA to RNA to protein. It also asserted that information flow in the reverse direction, from protein to RNA to DNA, was not possible.

You may be surprised to see the word *dogma* used in relation to a scientific principle, since a dogma usually describes a belief or an idea that is not open to discussion or speculation—an idea that is "written in stone." Crick later explained that he used the term *dogma* for two reasons. First, he had already used the term *hypothesis* in reference to the sequence hypothesis, and he did not wish to reuse that word. Second, he wanted to indicate that he considered the central dogma to be a more powerful theoretical idea for which there was no reasonable supporting evidence. His colleagues were somewhat appalled by his application of *dogma* to an untested hypothesis, and Crick has never again used the term in reference to any of his later theoretical ideas.

As conceived in 1958, the central dogma made these predictions: first, DNA strands serve as templates for the production of either complementary DNA molecules (*replication*) or RNA molecules (*transcription*); next, the RNA molecules move from the nucleus to the cytoplasm, where they serve as templates for the sequence of amino acids in proteins (*translation*).

The general idea turned out to be mostly accurate. Two decades later, however, it was discovered that certain virus-

es called *retroviruses* could direct the synthesis of DNA from RNA. Thus the concept of a one-way flow of genetic information from DNA to RNA to protein proved to be only part correct. By the mid-1960s, most of the gaps in the general idea had been filled in, and molecular biologists finally understood how one gene could give rise to a protein or RNA product.

PROTEIN SYNTHESIS

DNA constitutes the information that is transferred between generations of organisms. Proteins make up the active labor force carrying out the biological functions of the cell. To maintain the integrity of a cell's activities, mechanisms are necessary for ensuring the continuous synthesis of proteins, each with an exact sequence of amino acids. As described by the one-gene, one-protein hypothesis, a gene encodes instructions for building a specific protein. This is analogous to a recipe that encodes instructions for making a certain type of cookie. As predicted by the general idea, the linear sequence of amino acids in each protein is specified by DNA, and translation is carried out by RNA. How does this remarkable process take place inside eukaryotic cells?

Transcription

Figure 18.8 provides an overview of the events in protein synthesis in a eukaryotic cell. When a cell requires a specific protein, the DNA segment that constitutes the gene serves as a template for synthesizing a single-stranded RNA molecule. This is the initial step in the transfer of information from DNA. Essentially, the information-bearing RNA serves as a "messenger" from DNA and is therefore called **messenger RNA (mRNA)**.

RNA synthesis from DNA is called **transcription** because information coded in the sequence of DNA nucleotides is copied ("transcribed") into RNA nucleotides. As described in Figure 18.9, transcription begins when **RNA polymerase,** an enzyme that catalyzes the synthesis of RNA molecules from a DNA template, attaches to a specific site on the DNA molecule and causes the double strands to separate partially into two strands, exposing a small number of bases. Note that a gene consists of a DNA *sense strand,* which encodes the protein or RNA product, and a *template* (or *nonsense) strand.* Messenger RNA is assembled as RNA polymerase moves along the DNA strand, adding RNA nucleotide bases—according to complementary base–pairing rules G–C,

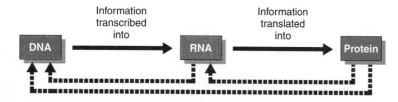

Information transcribed into   Information translated into

DNA   RNA   Protein

Figure 18.7   The central dogma postulated that the flow of genetic information proceeded from DNA to RNA to protein but not in the reverse direction indicated by the dashed lines.

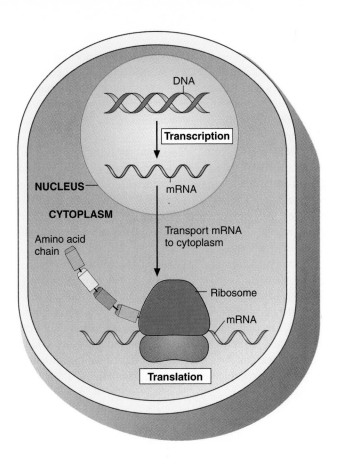

Figure 18.8   Genes in eukaryotes are usually expressed at some point through the synthesis of proteins. DNA is transcribed, in the nucleus, into messenger RNA (mRNA), which then moves out into the cytoplasm through nuclear pores. Once in the cytoplasm, mRNA becomes associated with ribosomes and is translated through the production of amino acid chains.

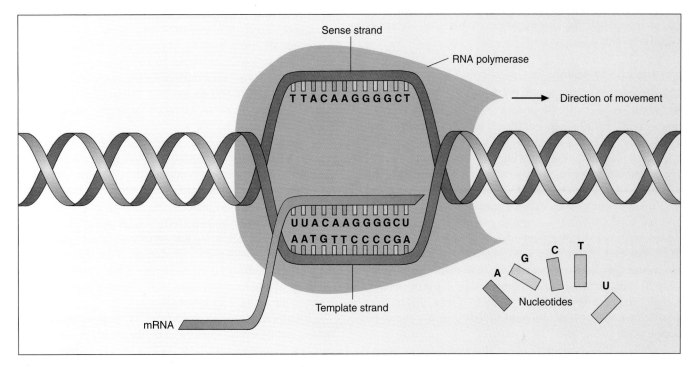

Figure 18.9   A gene is a section of DNA that consists of two strands, a sense strand that encodes the protein or polypeptide to be synthesized and a template strand that is used in synthesizing RNA molecules. DNA is transcribed into RNA primarily through the actions of an enzyme called RNA polymerase. This enzyme first separates the double-stranded DNA; the gene is then transcribed into a strand of RNA by adding base pairs that are complementary to the DNA template strand. Thus the base sequence of the messenger RNA strand shown in this figure is identical to the base sequence of the DNA sense strand, except that T (thymine) in DNA is replaced by U (uracil) in RNA molecules.

C–G, T–A, A–U (recall that uracil replaces thymine in RNA)—to the DNA template strand being transcribed. Hence the base sequence of the mRNA molecule formed is *complementary* to that of the template strand but *identical* to that of the sense strand (except that U replaces T). Nucleotides continue to be added one at a time until the RNA polymerase reaches a specific DNA nucleotide sequence that acts as a signal to terminate the transcription. The single strand of mRNA is then released from the DNA, and the two DNA strands are rejoined.

In addition to mRNA, two other types of RNA that participate in protein synthesis are transcribed from DNA. These are called **transfer RNA (tRNA)** and **ribosomal RNA (rRNA)**, and their roles will be described shortly. The three types of RNA are constructed using the same complementary base–pairing procedure. Once synthesized, the RNA molecules leave the nucleus and enter into activities related to protein synthesis. Thus there are two general types of genes associated with protein synthesis. **Structural genes** are those that code for proteins, and **nonstructural genes** code for tRNA and rRNA.

Translation

Translation is the process of synthesizing a protein whose linear amino acid sequence is directed by the mRNA sequence specified by a gene. In this operation, the genetic information of the nucleotides is translated into the language of amino acids and proteins. These proteins are synthesized on small organelles called ribosomes.

Messenger RNA Codons

The discovery that a sequence of DNA nucleotides coded for the synthesis of a specific protein, with its precise array of amino acids, led to a new question: How were sequences of DNA nucleotides translated into sequences of amino acids? Clearly, it was not possible for one DNA nucleotide to specify one amino acid in a one-to-one fashion because there are 20 amino acids but only four DNA nucleotides (A, T, G, and C). A pair of nucleotides would expand the coding possibilities but would still be inadequate, as only 16 (4^2) different amino acids could be accommodated. Thus scientists postulated that each amino acid was dictated by a code of three nucleotides, a prediction that was accurate. Although there are only 20 amino acids in proteins produced by living organisms, a triplet code can specify 64 (4^3) unique "words." In fact, it is now known that several different triplets code for the same amino acid.

Each mRNA nucleotide triplet that codes for an amino acid is called a **codon.** For example, the DNA template sequences TAC and TCC stipulate, via complementary base pairing, mRNA codons reading AUG and AGG. Because most proteins contain between 50 and 1,250 amino acids, each mRNA strand consists of between 50 and 1,250 codons.

Transfer RNA

Individual amino acids are brought to ribosomes by specific tRNA molecules with which they are chemically linked. These tRNA "adapter" molecules must recognize the information in the mRNA codon so that the appropriate amino acid is added to the protein at the correct location in the linear sequence. How might this be accomplished?

Two attributes of tRNA account for its ability to serve as the adapter molecule in protein synthesis: each tRNA recognizes one amino acid and one specific codon. Transfer RNA molecules are short (70 to 80 nucleotides long) and have a complex, three-dimensional structure (see Figure 18.10A). The characteristic shape of tRNA molecules is due to the formation of hydrogen bonds between short nucleotide sequences that contain complementary base pairs. Several sites exist that are associated with specific functions (see Figure 18.10B). The amino acid attachment site has the same base sequence, CCA, in all tRNAs, but the precise mechanism by which a particular tRNA recognizes its amino acid is currently the subject of intense study. The tRNA site that can recognize the complementary base pairs of a specific codon is called the **anticodon;** its bases are complementary and antiparallel to those of the codon.

Ribosomes

Ribosomes appear as small particles in the cytoplasm or attached to rough endoplasmic reticulum. They are composed of complex aggregations of rRNA and proteins. During protein synthesis, each ribosome is composed of a large subunit and a small subunit (see Figure 18.11A, page 342). The larger subunit contains three rRNA molecules, ranging in size from about 120 to 4,500 nucleotides. The smaller subunit contains a single long rRNA molecule of approximately 1,800 nucleotides.

Ribosomes have a grooved structure that enables them to provide a temporary abode for mRNA as its message is being deciphered during protein synthesis. In cells actively engaged in protein synthesis, clusters of several ribosomes may read the same mRNA strand to increase the amount of protein being manufactured. Ribosomes perform several critical functions in protein synthesis. These include correct alignment and reading of the mRNA strand, proper orientation of the tRNAs bearing their amino acids, accurate handling of the growing protein chain as synthesis progresses, and forming peptide bonds between the amino acid molecules.

Ribosomes involved in protein synthesis contain three binding sites for RNA molecules on the small ribosomal subunit (see Figure 18.11B, page 342). The small subunit binds mRNA and tRNAs whereas peptide bonds are formed between amino acids on the large subunit. As the ribosome moves along the mRNA, successive codons are brought into position for ordering the respective amino acids of the protein being synthesized. As each tRNA brings its amino acid to the site of the

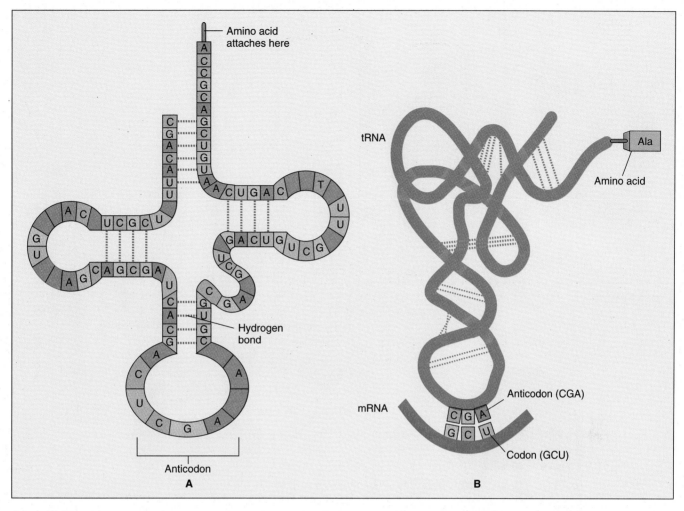

Figure 18.10   (A) Transfer RNA molecules have a two-dimensional cloverleaf shape but are actually considerably more complex, as indicated by their three-dimensional structure. Weak hydrogen bonds formed between complementary base pairs account for the shape of tRNA molecules. (B) Transfer RNA molecules have two functions, indicated by their structure and nucleotide composition. The tRNA anticodon at one end pairs with the complementary mRNA codon, and each tRNA molecule can carry one specific amino acid at the other end.

Question: *Where are tRNA molecules synthesized?*

growing protein chain on the ribosome, it binds to the complementary mRNA codon and transfers its amino acid to the growing chain.

Constructing a Protein

Figure 18.12 (see page 343) describes the synthesis of proteins inside a cell. All protein chains begin with a "start" codon, AUG, which also specifies a form of the amino acid methionine if it is located elsewhere in the mRNA strand. In addition, several other interacting protein molecules called *initiation factors* must be present before synthesis can begin.

Once mRNA is positioned and protein synthesis begins, the ribosome moves along its length one codon at a time in a three-step cycle. First, as each codon is exposed on the small subunit, a tRNA with the appropriate anticodon moves into place. Second, the amino acid it carries is linked, by a peptide bond, to the growing protein chain on the large subunit. Third, the tRNA is ejected from the ribosome. This three-step cycle is repeated until the complete protein has been synthesized. At that point, one of three codons, UAA, UAG, or UGA, serves as a "stop" signal. Thus the sequence of codons specifies a precise linear sequence of amino acids in a protein and also contains signals indicating where synthesis is to begin and end.

When the genetic message has been completely translated, the finished protein is released from the ribosome. The synthesis of a protein takes 20 to 60 seconds. The new protein is then usually modified by enzymes and assumes a distinctive three-

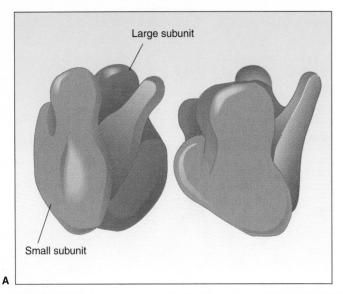

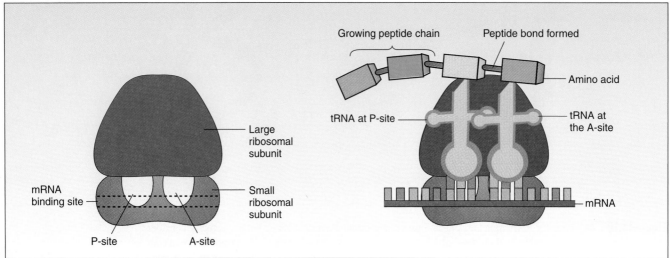

Figure 18.11   (A) During protein synthesis, mRNA is translated on ribosomes that consist of a small and a large subunit with a specific three-dimensional structure. (B) Ribosomes contain three binding sites for RNAs on the small ribosomal unit: one mRNA site and two tRNA binding sites known as the *A-site* and the *P-site*. During protein synthesis, the ribosome moves quickly along the mRNA, and amino acids are added to the growing protein chain in a three-step cycle: (1) a tRNA with its amino acid moves into the A-site, (2) its amino acid is joined with one previously added by a peptide bond at the P-site, and (3) the tRNA is ejected from the ribosome. The cycle continues until the protein is synthesized.

dimensional configuration, like the one shown in Figure 18.13 (see page 344), that has relevance to its function. The rules that govern the precise bending and folding of amino acid chains into functional proteins are not yet understood, although significant progress has been made, especially for small proteins. Through the use of new technologies, it has also become possible to view and study precise "pictures" of protein structure.

Eventually, the mRNA that encoded the genetic message for producing the protein is destroyed by appropriate enzymes. This mechanism guards against the overproduction of a protein, which could have harmful effects on cell activities. The average life span of an mRNA molecule in eukaryotic cells is 6 to 24 hours, during which time it can direct the synthesis of thousands of protein molecules.

Fate of Synthesized Proteins

Protein synthesis occurs continually in most cells, and each day millions of protein molecules are produced by multicellular organisms. What happens to these protein molecules? A protein may carry out its function in the cytoplasm; in a specific cell compartment (organelle), such as the nucleus, a mitochondrion, or a chloroplast; or outside the cell, at some other site in the body. Proteins that function in a specific organelle are routed to that compartment, using complex genetic mechanisms, after being synthesized on ribosomes in the cytoplasm. Proteins that function outside the cell where they are produced are synthesized on ribosomes attached to membranes of the rough endoplasmic reticulum (ER) and then secreted from the

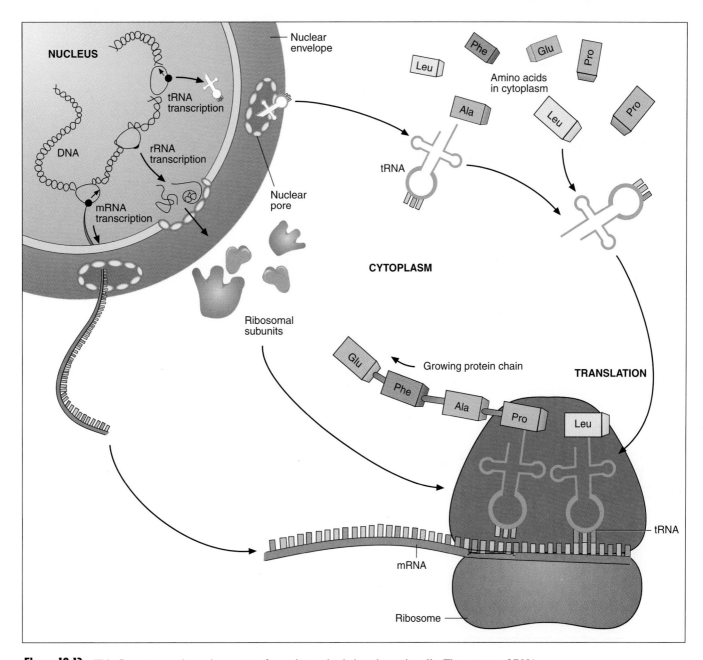

Figure 18.12 This figure traces the major events of protein synthesis in eukaryotic cells. Three types of RNA—mRNA, tRNA, and rRNA—are synthesized from genes (DNA) in the nucleus and enter the cytoplasm. Ribosomal RNAs becomes ribosomal structural subunits, and tRNAs transport amino acids to ribosomes, the site of protein synthesis. A protein or polypeptide encoded by a gene is transcribed from DNA to mRNA, which becomes attached to ribosomes and is then translated into the protein. Amino acids are brought to the ribosome by tRNA and placed in the appropriate sequence specified by the mRNA. When the protein is assembled, it is released from the ribosome.

cell. As described in Figure 18.14 (see page 344), proteins to be secreted are transported from the ER to the Golgi apparatus, where they are processed and packaged into vesicles. Secretion from the cell occurs when a vesicle fuses with the plasma membrane, opens, and releases its protein.

Proteins perform many functions, and their synthesis is carefully regulated according to need. Basically, the life cycle of a protein can be described as synthesis → function → degradation. Some specific examples of protein-regulating mechanisms are considered in Unit VI, "Human Systems of Life."

BEFORE YOU GO ON Synthesis of a protein involves a sequence of reactions that begins in the nucleus, where a segment of DNA—a gene—is transcribed into mRNA. Two other types of RNA, tRNA and rRNA, are also synthesized from DNA in the nucleus. Messenger RNA migrates to the cytoplasm, where translation occurs on ribosomes. Messenger RNA codons specify the exact type and sequence of amino acids that are used to build the protein defined by the gene. Synthesized proteins are localized in specific organelles or secreted from the cell for use elsewhere.

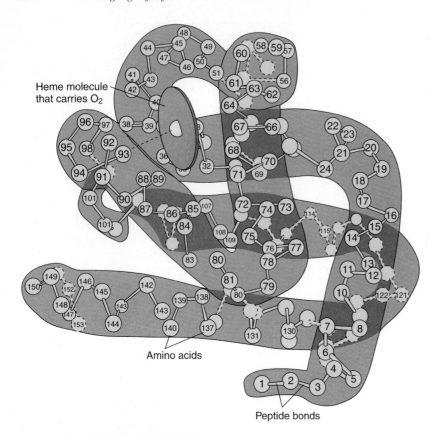

Heme molecule that carries O₂

Amino acids

Peptide bonds

Figure 18.13   All proteins have a specific structure that is related to their amino acid sequence. The shape of the protein is usually related to its function. This complex molecule is myoglobin, a protein that is responsible for carrying oxygen to muscle cells. Numbered spheres represent amino acids, which are joined by peptide bonds.

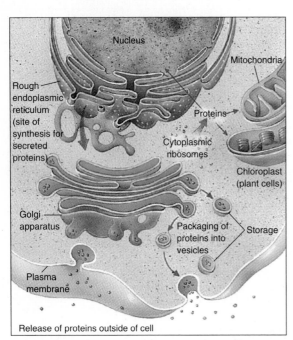

Nucleus

Mitochondria

Rough endoplasmic reticulum (site of synthesis for secreted proteins)

Proteins

Cytoplasmic ribosomes

Chloroplast (plant cells)

Golgi apparatus

Packaging of proteins into vesicles

Storage

Plasma membrane

Release of proteins outside of cell

Figure 18.14   Proteins required by cellular organelles such as mitochondria, chloroplasts, and the nucleus are synthesized on cytoplasmic ribosomes and then routed to the proper organelle. Proteins to be secreted from the cell are synthesized on ribosomes attached to the rough endoplasmic reticulum, processed and packaged into vesicles in the Golgi apparatus, and then released through the plasma membrane.

A CLOSER LOOK AT PROTEIN SYNTHESIS

The description of protein synthesis in the preceding section provided a general review of the process. Molecular biologists have expended great amounts of time and money learning more about all aspects of protein synthesis and the molecules involved. Here we describe some of the results of their studies.

Breaking the Genetic Code

After mRNA had been found to serve as the interpreter between DNA and the sequence of amino acids in specific proteins, there was great interest in determining how the nucleotide message (64 possible codons) was translated. What was the **genetic code** by which DNA, through mRNA base triplets, dictated specific amino acids? Breaking this genetic code is considered to be one of the greatest achievements of modern biochemistry.

The problem was solved by employing all of the usual molecules necessary for normal protein synthesis (tRNAs, amino acids, and ribosomes) along with mRNAs that were produced synthetically. The base sequence of synthetic mRNAs could be specified, and analysis of experimental results in which they were used enabled investigators to determine the amino acid indicated by each codon.

In 1961, Marshall Nirenberg and Heinrich Mathaei conducted the first experiment, described in Figure 18.15, that con-

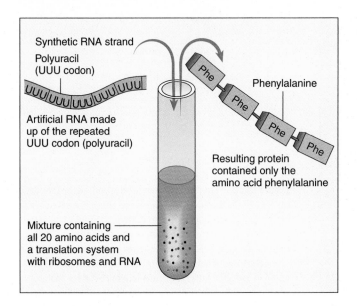

Synthetic RNA strand
Polyuracil
(UUU codon)

Artificial RNA made
up of the repeated
UUU codon (polyuracil)

Phenylalanine

Resulting protein
contained only the
amino acid phenylalanine

Mixture containing
all 20 amino acids and
a translation system
with ribosomes and RNA

Figure 18.15 To determine the amino acid called for by a specific codon, an early experiment was conducted that used an artificial mRNA with a codon reading UUU. The only amino acid found in the protein produced was phenylalanine.

Question: *What did the results of this experiment show?*

sisted of using mRNAs composed only of uracil ("poly-U"); thus the codon read UUU. When incorporated into the experimental system, a protein was produced from poly-U that contained only phenylalanine! What did this mean? Clearly, it meant that UUU codes for phenylalanine. Codons consisting of pure strings of the other three DNA bases (AAA, GGG, and CCC) and their specified amino acids were also quickly determined.

The next series of experiments, using codons of more than one nucleotide, were obviously more difficult. Basically, synthetic mRNAs consisting of alternating bases (for example, . . . ACACAC . . .) were used. A second experiment used a different repeat of the two bases (for example, . . . AACAAC . . .). By logical analysis of the end product (strings of certain amino acids), it was possible to assign amino acids to specific codons. Most of this work was done by groups working in Nirenberg's lab at the National Institutes of Health in Bethesda, Maryland, or with Severo Ochoa at New York University.

Other experiments eventually led to the complete deciphering of the genetic code, which is shown in Figure 18.16 (see page 346), by 1964. Results from all this research produced the following information:

1. The code is nonoverlapping, and there are no internal punctuation marks.

2. All codons consist of three nucleotides.

3. Of the 64 codons, 61 code for specific amino acids.

4. The code is "degenerate" in that many amino acids may be encoded by more than one codon. However,

one codon never specifies more than one amino acid.

5. One of the codons, AUG, codes for methionine but may also act as a start signal, depending on its location in the mRNA strand, and three of the codons—UAA, UAG, and UGA—are stop signals that terminate synthesis of the protein chain.

Regulating Gene Expression

We now understand that genes are expressed through the production of specific proteins. **Gene expression** is defined broadly to include all the steps necessary to transpose a genotype to a phenotype. However, many proteins are rarely needed, some are synthesized only by certain types of cells, and others must be available only for a brief time in an organism's life. What happens when the protein product of gene expression is not needed? How does a cell "know" when a certain protein is required? How do cells regulate the expression of their genes?

Prokaryotes

A variety of genetic control mechanisms have been fully described in prokaryotes. In general, they are much simpler to describe than control mechanisms in eukaryotic cells. Therefore, we will focus our consideration of these complex activities on one of the best-understood models to illustrate the concept of gene regulation.

The bacterial lifestyle often requires the sudden production of a certain enzyme in response to a rapid change in the environment. If, for example, a new food source (say, a specific sugar) is encountered, it may be beneficial to a bacterium to be able to respond immediately. It does this by producing enzymes capable of converting the sugar molecules into forms that can be metabolized. However, because of an extremely small amount of space within its cell and limited cellular resources, a bacterium does not profit from having extra enzymes that it may never use. Thus scientists faced a challenging question: Is there some mechanism by which bacteria control the production of enzymes (or other proteins) by using some sort of on–off switch? Two French scientists, Jacques Monod and François Jacob, received the Nobel Prize in 1965 for their classic studies, conducted in the 1950s, that helped answer this question.

What worthwhile information was available that could help put the puzzle together? From earlier studies, Monod and Jacob knew that the common bacterium *E. coli* produces a specific enzyme, β-galactosidase (β-gal), at high rates when the sugar lactose is present. For lactose to be used by the bacterium, it must first be broken down by β-gal into two smaller sugars, glucose and galactose. They also knew that the presence of lactose greatly accelerates the rate at which RNA polymerase binds to the gene (DNA) that codes for β-gal and initiates the synthesis of mRNA. The increased amount of the enzyme-coding mRNA leads to an expanded supply of β-gal.

First letter of code	Second letter of code				Third letter of code
	U	C	A	G	
U	Phenylalanine	Serine	Tyrosine	Cysteine	U
	Phenylalanine	Serine	Tyrosine	Cysteine	C
	Leucine	Serine	STOP	STOP	A
	Leucine	Serine	STOP	Tryptophan	G
C	Leucine	Proline	Histidine	Arginine	U
	Leucine	Proline	Histidine	Arginine	C
	Leucine	Proline	Glutamine	Arginine	A
	Leucine	Proline	Glutamine	Arginine	G
A	Isoleucine	Threonine	Asparagine	Serine	U
	Isoleucine	Threonine	Asparagine	Serine	C
	Isoleucine	Threonine	Lysine	Arginine	A
	START or methionine	Threonine	Lysine	Arginine	G
G	Valine	Alanine	Aspartic acid	Glycine	U
	Valine	Alanine	Aspartic acid	Glycine	C
	Valine	Alanine	Glutamic acid	Glycine	A
	Valine	Alanine	Glutamic acid	Glycine	G

Figure 18.16 This table of the genetic code—the language of life—is in the form of a 64-word dictionary with each amino acid, start code, or stop code identified by a specific mRNA triplet codon. To determine the amino acid specified by a codon, read the three-letter sequence of any codon and determine where they intersect. For example, C (first letter) A (second letter) A (third letter) codes for the amino acid glutamine; UGG codes for tryptophan.

Question: *Of the 64 codons, 61 encode amino acids, one (AUG) specifies an amino acid and is also used to initiate protein synthesis, and three are stop codons, which do not specify an amino acid. Which codon or codons specify the amino acid serine?*

From their own experiments and the work of others, a picture of the system emerged (see Figure 18.17). The gene for β-gal, *Z,* lies next to two other genes that code for enzymes that also have a role in *E. coli* lactose metabolism. One of the genes, *Y,* codes for permease, an enzyme that helps lactose enter the bacterium. The other gene, *A,* codes for transacetylase, an enzyme that removes modified lactose fragments that are not useful to the bacterium. A single strand of mRNA contains codons for expressing all three genes, and when lactose is available, equivalent amounts of the three enzymes are produced (from here on, β-gal refers to this three-gene complex). Now the central question could be addressed: How is the synthesis of β-gal turned on and off?

From their studies, Monod and Jacob developed the following hypotheses:

1. A certain molecule acted as a β-gal **repressor** when it was bound to a specific site on the DNA called the **operator (O)**.

2. When the repressor was bound to the operator, RNA polymerase was prevented from attaching to the DNA and synthesizing β-gal mRNA.

3. The repressor protein molecule was encoded by a gene referred to as *I* (because it controlled *inducibility*).

4. Lactose would act as an **inducer** of β-gal synthesis by binding to the repressor molecule and preventing it from becoming attached to the operator.

5. Once the repressor was removed from the operator, the structural genes would be turned on, and mRNA and β-gal would be synthesized.

6. β-gal mRNA synthesis began at a specific site called the **promoter (P)**.

7. When the lactose was broken down and eliminated from the cell, the repressor would once again bind to the operator, and the β-gal genes would be turned off.

Through further investigations, all seven hypotheses were verified. This *E. coli* genetic system became known as the *lac* operon. An **operon** is a set of adjacent structural genes whose mRNA is synthesized in one piece, plus the adjacent regulating signals that affect transcription of the structural genes. Thus the *lac* **operon** includes three consecutive regions of DNA consisting of structural genes that code for enzymes needed to metabolize lactose (*Z, Y,* and *A*) and the adjacent **control site,** which features a promoter and an operator. A separate **regulatory gene,** *I,* located nearby on the chromosome, encodes a repressor protein that controls inducibility. The work that led to an understanding of the *lac* operon and the mechanisms associated with the regulation of the operon illustrate the efficiency of the systems that control prokaryotic gene expression. Many similar on–off operons have since been identified in other prokaryotes.

Eukaryotes

Gene regulation in multicellular organisms involves a complex collection of signals, many of which are not fully understood. Mechanisms of eukaryotic gene control apparently exist at each stage of protein synthesis from transcription through translation. Experiments have revealed that multiple control processes exist at the transcriptional level. Available evidence also indicates that regulation of gene expression occurs after

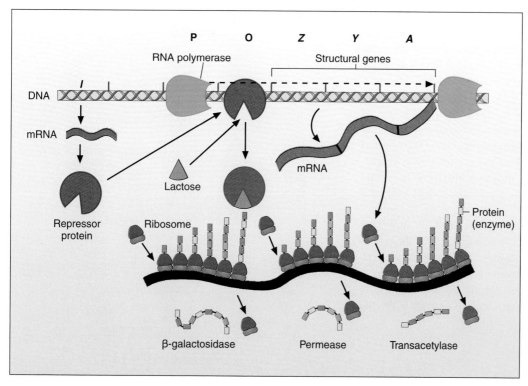

Figure 18.17 The *E. coli* genetic system responsible for breaking down the sugar lactose is called the *lac* oper-on. It is composed of three major components, a gene (*I*) that controls regulation; control sites (promoter, P, and operator, O), acted on by products encoded by the regulatory gene; and three structural genes (*Z, Y,* and *A*) that encode enzymes necessary for metabolizing lactose. Regulation of the *lac* operon system is controlled by a repres-sor protein encoded by the *I* gene that interacts with the operator region. In the presence of lactose, the repres-sor–lactose complex cannot bind to the operator; structural genes *Z, Y,* and *A* are turned on; transcription occurs; enzymes are produced; and lactose can be broken down.

Question: *When is this system turned on in* E. coli?

transcription in eukaryotes. However, large gaps exist in defin-ing the mechanisms responsible for gene control at this level, and it remains an exciting research frontier.

> **BEFORE YOU GO ON** In the early 1960s, the genetic code was deter-mined through ingenious experiments. Meanwhile, a deep question began to drive genetics research: How is gene translation regulat-ed in the cell? Monod and Jacob discovered how the *lac* operon reg-ulates the production of enzymes involved in a specific metabolic pathway in the bacterium *E. coli.* Subsequently, many other prokary-otic gene regulation mechanisms have been described. Less is known about gene regulatory mechanisms in eukaryotic cells.

Mutations

Mutations, as we saw in Chapter 16, are permanent alterations in DNA. They occur spontaneously, with no apparent cause, or after DNA is exposed to radiation (for example, ultraviolet light or X rays) and certain chemicals and viruses. Mutations have always been the subject of considerable interest in genet-ics. If they occur in gamete-producing cells, they can be a source of new alleles and, through translation, new hereditary

traits (remember the white-eyed fruit fly?). Understanding gene expression allows us to cast a new light on certain types of gene mutations.

Gene mutations are generally described as changes in the normal sequence of nucleotide bases within a single gene. When one nucleotide or base pair is replaced by another—for example, a T replaces an A or a T–A base pair becomes a G–C pair—the change is called a **base substitution.** A **point muta-tion** is a change in the sequence of nucleotide bases in DNA or RNA due to a base substitution or to the addition or dele-tion of a single base pair. Because of redundancy in the genet-ic code, base substitutions and point mutations may have no effect—no change in the amino acid sequence—or they may lead to the synthesis of an altered protein if there is a change in the amino acid sequence. Sickle-cell anemia, described in Chapter 20, is an example of a genetic disorder caused by a sin-gle base substitution.

Mutations can also result when one or more base pairs are added to or deleted from a DNA strand. If the gene affected by this mutation codes for a protein, a **frameshift mutation** occurs, and the translation process is disrupted. Because the genetic code involves "reading" codons, the addition or dele-tion of any number that is not a multiple of three results in an inability to read the codons correctly (that is, the "reading frame" is shifted, hence the name *frameshift mutation*). For example, assume that codons represent three-letter words in the following sentence: THE FAT CAT ATE THE BIG RAT. If the first F were deleted, the sentence would read THE ATC ATA TET HEB IGR AT and would no longer make sense.

Another class of mutation involves "jumping genes" or **transposons,** which are DNA sequences capable of migrating and inserting into a different chromosomal site. Transposons in eukaryotic organisms may lead to harmful effects characteristic of mutations. However, they may also be an important force in evolution by creating new combinations of DNA and hence new proteins.

Because few mutations are beneficial, it is not surprising that cellular mechanisms have evolved that reduce the rate of mutation but do not eliminate it entirely. DNA polymerases, the enzymes primarily responsible for DNA synthesis, play a major role in identifying and deleting erroneous bases or base sequences in DNA and also in repairing any damage at the site involved.

Exons and Introns

Many important discoveries have occurred since the breaking of the genetic code and the elucidation of prokaryotic gene regulatory systems. Prior to the 1970s, it was hypothesized that the basic structure of genes was the same in prokaryotes and eukaryotes. However, that hypothesis was proved incorrect. A eukaryotic gene is now known to be a functional unit composed of a specific DNA segment that contains discrete islands of transcribed nucleotide sequences scattered within stretches of DNA that are not transcribed. **Exons** are all nucleotide sequences of a gene that are transcribed as RNA. The nontranscribed sequences between exons are called **introns** (see Figure 18.18). After transcription, introns are removed from the mRNA by enzymes before the mRNA migrates to the cytoplasm to be translated.

The discovery of exons and introns in 1977 caused a bit of a shock to molecular biologists. Until that time, it was assumed that a gene was a single, continuous interval of DNA that would code for a single protein. Most eukaryotic genes encoding a polypeptide have at least one intron and usually more. It is now known that in human and other mammalian cells, over 75 percent of "gene DNA" consists of introns. In light of this new information, the gene could no longer be thought of as a single stretch of DNA. Further, it became known that a small number of genes code for tRNAs and rRNAs, not proteins. This discovery led to the revised definition of a gene given earlier in this chapter.

Molecular Biology and the History of DNA

Research in molecular biology has also yielded some fascinating details about DNA and its history (see the Focus on Scientific Process: "Ancestors"). The chromosomes of prokaryotes do not contain collections of introns. Rather, many of

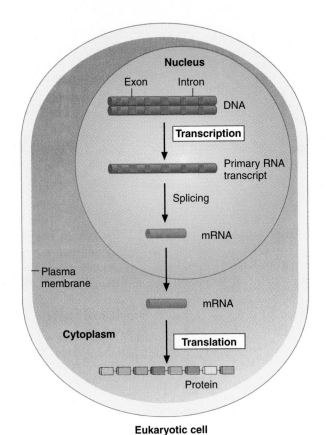

Eukaryotic cell

Figure 18.18 Eukaryotic genes consist of exons, expressed as part of a protein encoded by the gene, and introns, intervening DNA sequences that are not expressed. For the encoded protein to be synthesized, introns are removed from the primary RNA transcript after transcription occurs. Messenger RNA then moves into the cytoplasm where translation takes place.

Question: *How do prokaryotic genes differ from eukaryotic genes?*

their protein-coding genes are arranged in operons, and little nontranslated DNA is found within or between these operons.

The simplest eukaryotes, such as yeast, also contain little excess DNA. However, chromosomes of complex plants and animals contain an enormous load of "junk" DNA in the form of introns and other apparently nontranslated segments. Protein-coding portions of genes (exons) occupy only 1 to 20 percent of the total DNA in the cells of higher organisms, and active transcription units are usually separated by great distances. What is the significance of this great difference in exon-to-intron DNA ration between lower and higher organisms?

Hypotheses concerned with the origins of life on Earth offer an explanation for the disparity in amounts of nonfunctional DNA in prokaryotes and eukaryotes. As explained in Chapter 3, scientists now believe that three cell lines—archaebacteria, eubacteria ("common" bacteria), and eukaryotes—arose several billion years ago from a single ancestral cell type called the progenote.

FOCUS ON SCIENTIFIC PROCESS

Ancestors

For over a century, biologists have been aware that there are great similarities in structure and function among widely dissimilar organisms. This view is expressed in the phrase "unity in diversity." Diversity refers to the millions of species that have inhabited Earth since life evolved.

Unity was first observed in the common anatomical structures of different plants and the similar structures present among animals. Recently, as the highly sophisticated techniques of molecular biology have been applied to studying lower organisms, it has become clear that common molecular mechanisms are found throughout nature. For example, all organisms synthesize proteins and transmit genetic information in essentially the same way; such fundamental mechanisms are known as *biological generalizations.* Some scientists hypothesize that molecular mechanisms that exist in all organisms trace their origins to the progenote, the hypothesized ancestor of all living species. The molecules and processes involved have been highly "conserved" (that is, they have not changed significantly in species over time), and they are widely shared by prokaryotes and eukaryotes, including humans, at the cellular and molecular levels. Two specific examples illustrate this point. A molecule used by yeast cells for "mating" has likewise been found to be a component of certain sex hormones of higher organisms. Also, an insulinlike molecule has been found in fungi and yeasts that functions as a cellular signal, much as insulin does in mammals.

Interest in this area increased dramatically during the 1980s, primarily because of two developments. First, in a 1985 report from the National Academy of Sciences, *Models for Biomedical Research: A New Perspective,* a sugges-

tion was made to enlarge the scope of biological generalizations by creating a computerized information system that incorporates new data and information gained from studies of all organisms. Analysis of such information may reveal fundamental biological patterns that exist throughout the living world.

At the heart of this approach is the widely held belief that undiscovered organizational principles can be demonstrated from studies of much different or simpler organisms. We have examined a number of such cases in Chapters 15, 16, and 17. The fruit fly *Drosophila* was used to derive many important genetic concepts that are now known to be universal. The universal genetic code was discovered first in bacteria and viruses but is now known to be shared by all living organisms on Earth.

Second, the results of recent studies on DNA, RNA, and protein synthesis have revolutionized thinking about the beginning of life. Consequently, new hypotheses have been formulated on the origin and evolution of the first life forms that may have existed on Earth some 4 billion years ago. The discovery of a unique type of RNA revived interest in this area.

The 1989 Nobel Prize in chemistry was awarded to Sidney Altman of Yale University and Thomas Cech of the University of Colorado for their independent discovery, in 1983 and 1984, that "catalytic" RNA molecules can act as their own enzymes in certain microorganisms. Specifically, catalytic RNA segments termed *ribozymes* snip the introns out of various pieces of tRNA and rRNA and rejoin the remaining pieces before they become functional in protein synthesis. This finding was especially interesting in that it had been widely assumed that only proteins carried out metabolic activities within the

cells. The ribozyme finding implied that RNA may have been the first molecule to evolve that possessed metabolic capabilities. Further, it immediately suggested that RNA was the first nucleic acid to evolve; if that is true, introns must have existed in the first genetic molecules because a mechanism exists for their removal. Results from these studies led to the *RNA world hypothesis,* which states that all modern life forms descended from a primordial ancestral system that used RNA both in self-replication and in its metabolic activities.

To be classified as living, an organism must be able to perform two primary functions: catalyze chemical reactions required for life ("metabolism") and store and transmit "genetic information." As we have seen in this chapter, modern organisms accomplish these tasks through interactions among DNA, RNA, and proteins. How did the first life forms solve the two problems? Since the discovery of the self-splicing RNA, many molecular biologists have come to believe that the first living systems must have used RNA for all of their chemical activities. Some extremely interesting models of this *RNA world* have recently been developed, based on analyses of metabolic pathways, genetic organization, chemical structure, and enzymes that exist in modern organisms.

The most widely accepted model is described in Figure 1 (see page 350). It indicates that the first organism, or "riboorganism," used only RNA for catalyzing reactions and encoding genetic information; DNA and proteins did not exist in the RNA world. In the second stage of the origin of life, the "breakthrough organism" appeared, with the following characteristics: it obtained energy and carried out other complex reactions that were catalyzed by proteins, and it continued to store genetic

box continues

information in RNA. Though the breakthrough organism is not hypothesized to have had standard protein-synthesizing capabilities, it may have possessed pieces of the translation machinery. The presence of a complete protein-synthesizing system and the use of DNA as a repository for genetic information first appeared in the progenote, the common ancestor for all forms of existing life. Thus, according to the model, the sequence in which the fundamental molecules of life originated, from first to last, was RNA, then protein, then DNA.

The intriguing RNA-world hypothesis promises to occupy molecular biologists and theoreticians for some time. Not all scientists agree that events occurred so quickly (over the course of half a billion years or so). However, evidence in support of the RNA-world hypothesis continues to build. For example, it was recently discovered that peptide bond formation on the ribosome can be catalyzed solely by ribosomal RNA during protein synthesis. It will be intriguing to follow future developments and perhaps learn when and how we came to live in a DNA world.

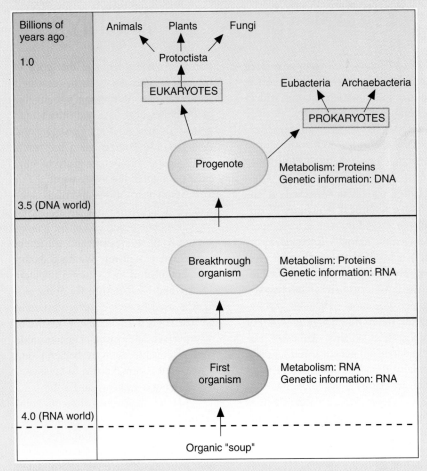

Figure 1   The latest hypotheses on the origin and nature of the first living organisms to have appeared on Earth and the evolution of life are reflected in this diagram. The hypothesized first organism to develop from the organic "soup" was capable of both carrying out metabolic activities and reproducing using RNA molecules. This RNA-world inhabitant gave rise to the breakthrough organism, which used proteins for its metabolic activities and RNA as its genetic material. Finally, the progenote evolved, which used DNA as its genetic material; this organism is thought to have been the ancestral cell to all modern life forms on Earth (discussed in Chapter 3).

Based on results from many studies, an interesting hypothesis has emerged (see Figure 18.19). The progenote is now postulated to have contained intron-rich DNA. Thus the earliest cells that descended from the progenote would also have had DNA that was rich with introns. However, since bacteria reproduce very quickly, they are thought to have evolved more rapidly. As a result, they developed into streamlined protein-synthesizing units by eliminating the noncoding ("useless") regions of their DNA. In contrast, eukaryotic cells have evolved much more slowly and still retain the introns inherited eons ago from the ancestral progenote. This hypothesis has not gone unchallenged. Some scientists argue that introns evolved later, which is why they are common in eukaryotes but not in prokaryotes.

Recently, a new idea has stimulated great interest. According to the *intron–RNA–gene regulation hypothesis,* some introns encode RNAs that help regulate the expression of certain genes in eukaryotic cells. They apparently carry out this function by way of binding to either DNA or RNA products transcribed from DNA. A few such regulatory RNAs have been identified in eukaryotic cells, and they bind either with DNA or with messenger RNA. This new hypothesis also suggests that introns would be harmful to prokaryotes because transcription of DNA into mRNA and translation of mRNA into protein occur simultaneously in these organisms. Thus in prokaryotic cells, introns could not be removed before a protein was synthesized, and the outcome would be a nonfunc-

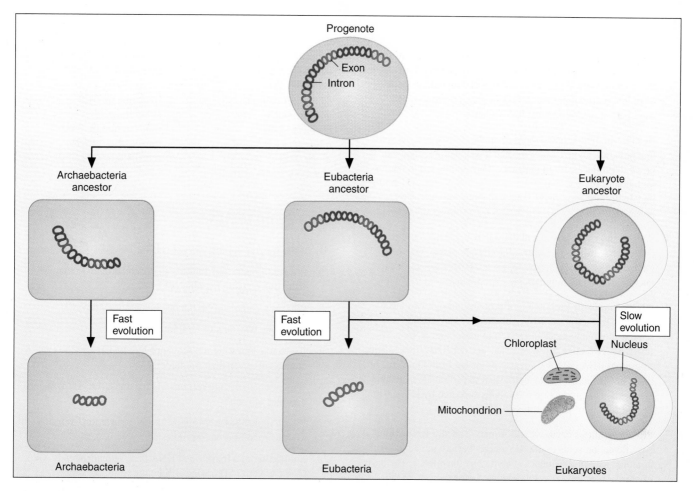

Figure 18.19 All life is hypothesized to have originated from a cell type called the progenote. The progenote gave rise to three cell lines that evolved into primitive bacteria (archaebacteria), modern bacteria (eubacteria), and modern plants and animals (eukaryotes). The progenote probably contained intron-rich sequences. Because of their short life cycles and rapid evolution, archaebacteria and eubacteria have eliminated the "useless" introns, whereas such DNA remains in eukaryotic cells.

tional ("nonsense") protein. The intron–RNA–gene regulation hypothesis has generated many new questions, and it is currently the subject of intense research activities. It has also led to new definitions of a chromosome. Rather than being simply a collection of genes scattered within junk DNA, a chromosome is now viewed by some scientists as a complex "information organelle" containing both genes and sophisticated maintenance and regulatory systems.

BEFORE YOU GO ON Research on DNA has revealed that genes are composed of exons and introns. Prokaryotic genes are composed mostly of exons, whereas the genes of eukaryotes have an abundance of introns. Differences in the exon-to-intron balance between DNA of prokaryotes and eukaryotes are remarkable. Scientists have formulated hypotheses that attempt to explain the contrast in terms of differential evolution of the two cell types from the ancestral progenote.

THE LANGUAGE OF LIFE

The path from Mendel and de Vries to the present traverses an expanse of knowledge that began with the concept of hereditary factors and ended with the mechanism of gene expression. Despite overlying complexities, the genetic language of DNA turned out to be rather basic, considering the possibilities: three nucleotides = one codon = one amino acid. Genes, first defined in terms of enzymes, are now defined in terms of proteins and RNA molecules and also structural features such as introns and exons.

The genetic information transmitted from parent to offspring is basically a set of instructions on how to make specific proteins and RNAs. This process underlies many aspects of biology—human genetic traits, genetic disorders, high-technology uses of DNA, evolution, human physiology, cancer, AIDS—that we will consider in later chapters of this book.

SUMMARY

1. A desire to learn about the nature of genetic information guided research through much of the twentieth century. Early in the century, Garrod hypothesized that certain human diseases were inherited through variations in enzyme structure that were part of one's chemical individuality.

2. Beadle and Tatum, using the pink mold *Neurospora,* showed that radiation altered genes in such a way that the organism failed to produce certain enzymes. This finding led to the formulation of the one-gene, one-enzyme hypothesis, which ultimately became modified to one-gene, one protein or RNA product.

3. Watson and Crick postulated that DNA in the nucleus produced RNA, which then moved to the cytoplasm, where it guided the synthesis of specific proteins. This scheme was later confirmed by innumerable experiments.

4. DNA controls the production of proteins through two mechanisms. In transcription, RNA polymerase converts the sequence of nucleotides in DNA into messenger RNA (mRNA). Two other types of RNA, transfer RNA (tRNA) and ribosomal RNA (rRNA), are also produced from DNA. All of these RNAs then enter the cytoplasm to participate in protein synthesis. The production of proteins specified by DNA is a central feature of gene expression.

5. Protein synthesis occurs on ribosomes. Transfer RNAs bring specific amino acids to the site where they are joined to another amino acid in a sequence specified by the mRNA. The basis of correct matching is complementary base pairing. In the genetic code, an mRNA three-base sequence called a codon is read by a corresponding anticodon on the tRNA. Amino acids continue to be added in a three-step cycle until the protein encoded by the DNA is completed.

6. Proteins required by cytoplasmic organelles, such as a true nucleus, are synthesized on cytoplasmic ribosomes and then routed to a specific organelle. Proteins to be secreted are synthesized in the rough endoplasmic reticulum, then routed to the Golgi apparatus, where they are processed before being released through the plasma membrane.

7. Gene expression in prokaryotes involves regulation by complex systems of genes, proteins, and other factors called operons. The well-understood *lac* operon enables bacteria to respond to the presence of lactose in the surrounding environment.

8. Gene mutations cause changes in the normal nucleotide sequence of DNA. As a result, alterations in protein production that can be harmful to the organism may occur.

9. Eukaryotic genes contain exons and introns, whereas prokaryotic genes generally contain only exons. The difference between the two cell types is thought to be related to their evolutionary history.

WORKING VOCABULARY

anticodon (p. 340)
codon (p. 340)
exon (p. 348)
gene expression (p. 345)
genetic code (p. 344)
intron (p. 348)
messenger RNA (mRNA) (p. 338)
metabolic pathway (p. 335)

metabolism (p. 335)
ribosomal RNA (rRNA) (p. 340)
ribosome (p. 340)
structural gene (p. 340)
transcription (p. 338)
transfer RNA (tRNA) (p. 340)
translation (p. 340)

REVIEW QUESTIONS

1. What was the significance of Garrod's studies of metabolism?

2. How did Beadle and Tatum's studies lead them to the one-gene, one-enzyme hypothesis?

3. What is metabolism? What is a metabolic pathway?

4. What was Watson's general idea?

5. To what extent were the sequence hypothesis and the central dogma correct in their predictions?

6. Which molecules are responsible for transcription? What are the roles of each?

7. Which molecules are responsible for translation? What are the roles of each in protein synthesis?

8. If a portion of a transcribed DNA base sequence reads AAA-CACGCATCGATC, what will be the resulting amino acid sequence?

9. How does the genetic code work?

10. Describe the operation and function of the *lac* operon.

11. Explain the relationships between mutations and the genetic code.

12. What are exons and introns? In which types of cells are they found—prokaryotes, eukaryotes, or both? What explains this finding?

ESSAY AND DISCUSSION QUESTIONS

1. Why would scientists in the 1950s be driven to determine how DNA was turned into proteins?

2. Why might the central dogma have come to be labeled a dogma? Would you expect dogmas to be common in biological science? Why?

3. Which experiment, idea, or hypothesis described in this chapter do you think might have been the most important in understanding gene action? Why?

4. Why has the definition of a gene changed repeatedly during the past century?

REFERENCES AND RECOMMENDED READING

Alberts, B., D. Bray, J. Lewis, M. Raff, K. Roberts, and J. D. Watson. 1994. *Molecular Biology of the Cell.* New York: Garland.

Beadle, G. W. 1959. Genes and chemical reactions in *Neurospora* (Nobel Prize lecture). *Science,* 129: 1715–1726.

Bearn, A. G. 1993. *Archibald Garrod and the Individuality of Man.* New York: Oxford University Press.

Benner, S. A., A. D. Ellington, and A. Tauer. 1989. Modern metabolism as a palimpsest of the RNA world. *Proceedings of the National Academy of Science,* 86: 7054–7058.

Crick, F. H. C. 1958. On protein synthesis. *Symposium of the Society for Experimental Biology,* 12: 138–163.

Crick, F. H. C. 1970. Central dogma of molecular biology. *Nature,* 227: 561–563.

Doolittle, R. F. 1985. Proteins. *Scientific American,* 253: 88–99.

Garrod, A. 1908. *Inborn Errors of Metabolism.* Oxford: Oxford University Press.

Gesteland, R. F., and J. F. Atkins (eds.). 1993. *The RNA World.* New York: Cold Spring Harbor Laboratory Press.

Jacob, F., and J. Monod. 1961. Genetic regulatory mechanisms in the synthesis of proteins. *Journal of Molecular Biology,* 3: 318–356.

National Academy of Sciences. 1985. *Models for Biomedical Research: A New Perspective.* Washington, D. C.: National Academy Press.

Nowack, R. 1994. Mining treasures from "junk DNA." *Science,* 263: 608–610.

Osawa, S. 1995. *Evolution of the Genetic Code.* New York: Oxford University Press.

Schwartz, A. W. 1995. The RNA world and its origins. *Planetary and Space Science,* 43:161–167.

ANSWERS TO FIGURE QUESTIONS

Figure 18.1   Lack of the enzyme that converts the amino acid tyrosine into melanin, a pigment protein that normally gives color to skin, hair, and eyes.

Figure 18.2   That each metabolic reaction was catalyzed by a single enzyme and that one gene specified the structure of one enzyme.

Figure 18.3   The substrate molecule would not be converted to Product A.

Figure 18.4   Lack of an enzyme causes phenylalanine to be converted to phenylpyruvic acid instead of tyrosine; accumulation of this acid results in PKU.

Figure 18.6   To take part in protein synthesis, which takes place in the cytoplasm.

Figure 18.10   In the cytoplasm or on membranes of the rough endoplasmic reticulum.

Figure 18.15   That the codon UUU specified the amino acid phenylalanine.

Figure 18.16   UCU, UCC, UCA, UCG, AGU, and AGC.

Figure 18.17   When lactose is present in the bacterium's environment (usually in the absence of other sugars).

Figure 18.18   Most prokaryotic genes do not contain introns.

19

Human Genetics and Patterns of Normal Inheritance

Chapter Outline

Reading Questions

1. What is known about the human karyotype?

2. What types of genes and DNA are found in human chromosomes?

3. Which human traits are determined by a single gene pair? By multiple alleles? By polygenes?

4. How is human sex determined?

In earlier chapters, we saw how the studies of Mendel, de Vries, Morgan, and others provided scientists with a general understanding of how traits are passed between generations and of the roles that chromosomes and genes play in inheritance. The work of these early researchers was and remains fascinating and significant in the history and practice of science. Using relatively simple organisms, they were able to describe many fundamental genetic principles, including segregation, independent assortment, sex-linked inheritance, linkage groups, and chromosome mapping.

HUMAN HEREDITY

These early studies gave hope to scientists who had long believed that the scientific investigation of human heredity was potentially of great importance. As early as 1869, Francis Galton, a leading British scientist (and first cousin of Charles Darwin), argued that hereditary factors played a major role in determining intelligence. He had conducted his research by using mathematical analyses to evaluate information about human "genius," and although his conclusions were controversial, they provided a starting point for the study of human genetics. Galton's results suggested to him that certain human traits were inherited, but they did not provide any knowledge about *how* these traits were inherited.

Despite other stimulating observations and hypotheses, such as those of Archibald Garrod (discussed in Chapter 18), human genetics did not enter the mainstream of scientific research until the middle of the twentieth century. Scientists interested in studying human genetics faced biological and ethical impediments. Because humans have a long generation time span and few offspring, this species does not lend itself to the same experimental manipulations as fruit flies and bacteria. Relative to ethical considerations, there is virtually universal agreement that informed consent is required for any research involving humans and that humans may not be treated as experimental animals. Furthermore, genetics breeding experiments, routinely conducted using plants and animals, are morally unacceptable with humans. Finally, the Mendelian model of inheritance—a single gene pair determining a specific trait with complete dominance—was generally inadequate for explaining inheritance in humans.

Mendelian patterns of inheritance do, of course, occur in humans. Traits *are* determined by genes carried on chromosomes that segregate and undergo independent assortment. Genes are expressed in terms of proteins synthesized, which are in turn responsible for determining phenotypic characters such as blood type and eye color. Deeper levels of understanding have resulted from the explosive growth of knowledge in human genetics during the past decade. It has been convenient for us to use the classical genetic principles as a departure point for understanding human genetics. However, we now turn to new information that makes it possible to obtain a more sophisticated appreciation of the enormous complexities involved in human genetics.

The Human Karyotype

For over a century, biologists have recognized the existence of chromosomes in cell nuclei. All species have a unique **karyotype,** which is defined as a complete set of chromosomes that is visible during metaphase of mitosis. A species's karyotype includes details on the specific number of chromosomes present and their sizes, shapes, and staining patterns. By the 1920s, karyotypes had been determined for familiar experimental organisms, such as Mendel's pea plants and Morgan's fruit flies. However, correct determination of the human karyotype proved to be troublesome. In 1907, a German cytologist reported that human cells contained 47 chromosomes, a finding that was accepted for

over a decade. In 1921, an American cytologist, T. S. Painter, using different techniques, reported that human cells held 48 chromosomes, or 24 pairs. This conclusion was confirmed by other investigators and became generally accepted dogma until the 1950s. Finally, in 1956, Swedish researchers who were interested in inherited diseases established that the human karyotype consisted of 46 chromosomes (23 pairs).

The primary difficulty in accurately determining a karyotype is that chromosomes are not ordinarily visible in the cell except during cell division, either meiosis or mitosis (see Chapter 15), when they are in a condensed form. Cytologists use special techniques—illustrated in Figure 19.1 (see page 356)—to reveal condensed chromosomes microscopically in certain cells. Appropriate cells (usually white blood cells) are removed from the body and grown in culture dishes containing a substance that stimulates cell division. After about 72 hours, the cells are exposed to *colchicine,* a chemical that stops cell division at metaphase. At this time, the cells are placed in a solution that causes them to swell and allows the chromatids to separate. After swelling, the cells are gently squashed and split open on a microscope slide, and the chromosomes are then stained and allowed to dry. After completing this procedure, the "chromosome spread" can be examined under a microscope. To prepare a karyotype, photographs are taken, enlargements are made, and pictures of metaphase chromosomes are cut out with scissors and placed in a conventional sequence according to their size. The amount of genetic information each chromosome contains is roughly related to its dimensions.

Human male and female karyotypes have 23 pairs of chromosomes. The first 22 pairs of the chromosomes are identical in males and females and are called **autosomes.** The unnumbered X and Y chromosomes are called **sex chromosomes;** cells of males contain an X and a Y chromosome, whereas those of females have two X chromosomes. By convention, the karyotype of a normal male is represented as 46,XY (46 total chromosomes of which one is an X and one a Y), a female as 46,XX.

BEFORE YOU GO ON   Research on human genetics lagged far behind studies of the genetics of simpler organisms. The human karyotype was not accurately described until 1956, when it was determined that human cells contain 23 pairs of chromosomes. Normal males and females each have 22 pairs of autosomes and one pair of sex chromosomes. The autosomes are identical in both sexes, but females have two X sex chromosomes, males one X and one Y.

Chromosome Nomenclature

To facilitate communication, scientists studying genetics worldwide have developed standard terminology to classify human chromosomes.

Position of the Centromere

Before 1970, stains used in karyotyping gave chromosomes a uniform dark color. As a result, all human chromosomes were

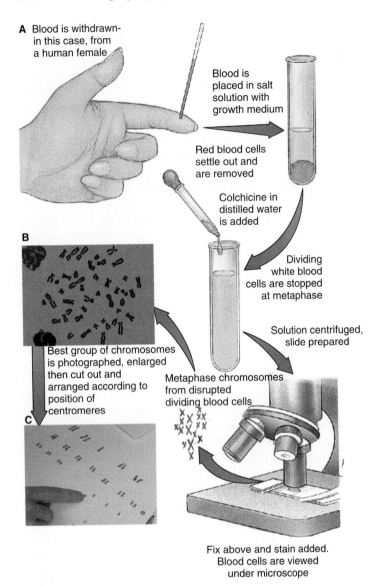

Figure 19.1 A human karyotype is commonly determined using the following procedure. (A) Blood is removed from the finger; white blood cells are isolated and exposed to colchicine, a chemical that halts cell division (mitosis) at metaphase. (B) Dividing white blood cells that were arrested in metaphase are gently broken, causing the chromosomes to be released from the cell. The chromosome set from a single cell is then photographed and enlarged. (C) The best group of chromosomes is cut out and arranged according to size and position of the centromere.

classified according to two variables, length and the position of the **centromere** (middle, off-center, or near the end of a chromosome). Based on the position of the centromere, the short (upper) arm of a chromosome is designated by the letter *p* and the long (lower) arm by the letter *q*.

Chromosome Bands

After 1970, a number of special staining techniques were developed that significantly advanced the study of human chromo-

somes. Now each human chromosome can be further distinguished by a unique series of bands along its length after being stained with different dyes (see Figure 19.2). **Chromosome bands** are alternating light- or dark-staining sections along a chromosome that are visible using a compound microscope. These bands are revealed by applying dyes called quinacrine mustard or related stains (Q bands) and Giemsa stain (G bands) or by heat pretreatment of chromosomes followed by Giemsa staining that results in a banding pattern that is the reverse (R bands) of Q or G bands.

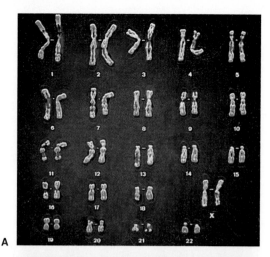

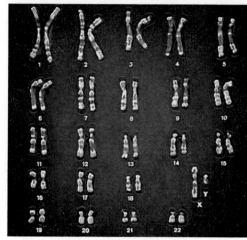

Figure 19.2 Two full metaphase sets of stained chromosomes taken from white blood cells of a male (A) and a female (B). Both karyotypes have 22 pairs of autosomes and one pair of sex chromosomes, which in the male is XY and in the female is XX.

The chromosome arms are divided into regions separated by **landmarks,** defined as consistent features—position of the centromere, position of major bands, or the ends of chromosome arms—that can be used in identifying a specific chromosome. As shown in Figure 19.3 (see page 358), geneticists divide each arm of a chromosome into large areas or regions that are numbered consecutively from the centromere (for example, p1 designates the short or upper arm, region 1) and then into smaller subareas identified by band numbers. Use of this procedure allows for the precise location of an area of interest in a chromosome. For example, 1p22 refers to chromosome 1, the short (upper) arm, region 2, band 2, and 4q43 refers to chromosome 4, the long (lower) arm, region 4, band 3. Recent improvements in banding techniques now permit using *subbands* for even more precise chromosomal location; subbands are added as a decimal to the band number (for example, 1p22.3). Distinctive banding patterns have allowed geneticists to make standardized maps of bands of human chromosomes that are used widely in modern genetics studies.

Chromosome Structure

Each of the 46 chromosomes in a human cell contains a single long strand of DNA, and the total DNA in all chromosomes within a single cell is about 2 meters long. It has been estimated that the DNA from all cells in your body would extend from Earth to the sun and back again about 50 times! What allows 46 DNA molecules, which are several centimeters long, to be packed into a nucleus that is only several micrometers in diameter?

Recall that eukaryotic chromosomes are composed of chromatin—a mixture of DNA and protein molecules—that looks like beads on a string when viewed through an electron microscope (see Figure 19.4A, page 359). In the 1970s, it became clear that different levels of chromatin packing occur in nuclei of interphase cells and in chromosomes of cells that are undergoing mitosis or meiosis. The basic chromatin structural unit in eukaryotic chromosomes is a *nucleosome* (the "bead"), as illustrated in Figure 19.4B. A **nucleosome** consists of a central core of eight *histones*—structural protein molecules—around which the DNA "string" is wrapped twice. An additional histone molecule, *H1 histone,* helps bind DNA to the central core. As shown in Figure 19.4C, during cell division, nucleosomes are packed into coils that are then twisted into highly compressed loops, forming supercoils that make up chromosomes. Once cell division is completed, the supercoils dissipate, and chromatin within interphase cells once again takes on its characteristic beads-on-a-string appearance. Both DNA replication and transcription take place when chromatin exists in this state. It is not yet fully understood how replication and transcription machinery can pass through the nucleosomes and gain access to the DNA.

THE HUMAN GENOME

A **genome** is the complete collection of genes in one organism or within a diploid chromosome set of a eukaryotic species. In the 1960s, molecular tools made it possible to obtain rough estimates of genome sizes, and some interesting results appeared from the first studies. For example, the genome of *E. coli,* a prokaryotic bacterium, was estimated to contain about 4.5 million base pairs. Because *E. coli* contains about 4,000 genes and a typical gene consists of 1,000 base pairs, the genome calculations seemed consistent with the prevailing view that chromosomes contained linear collections of genes laying next to one another, much like the words on this page. At that time, it was estimated that eukaryotic organisms contained up to 50,000 genes and therefore that their genomes would consist of approximately 50 million base pairs. However, eukaryotic genomes were found to contain 3 to 4 *billion* different base pairs. If only 6 to 8 percent of eukaryotic chromosomes are occupied by functional genes, what accounts for the remaining 92 to 94 percent, which apparently consists of noncoding "surplus" DNA? The complete answer to that puzzle is still being pursued, although major pieces are now in place.

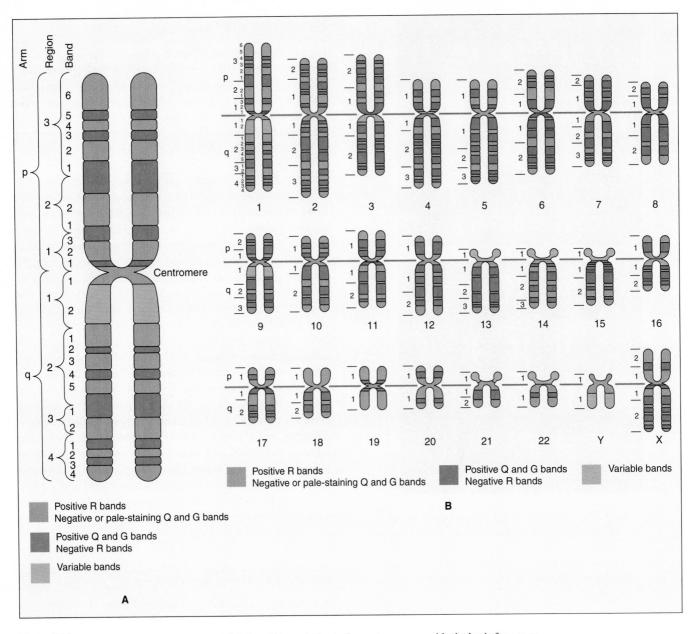

Figure 19.3   The bands of a chromosome and their position relative to the centromere provide the basis for a numbering system that is used to create a map of the chromosome. The details about the bands and the numbering system for human chromosome 1 (A) apply to the band maps for all human autosomes and sex chromosomes (B).

The human genome is immense and complex. Its 6 to 7 billion DNA nucleotide base pairs are distributed among 46 chromosomes and include approximately 100,000 genes. Because homologous chromosome pairs of diploid organisms each contain copies of the same genes (but not necessarily the same allele), the term *human genome* commonly refers to the DNA or genes within the 22 autosomes and the human X and Y chromosomes. The latter definition will be used in this chapter because it reflects the actual number of different genes. According to this definition, the human genome consists of 3.0 to 3.5 billion base pairs and includes about 50,000 different genes. What is known about these 50,000 genes?

Types of Genes

Various numbers and arrangements of genes exist in eukaryotic genomes (see Figure 19.5, page 360). There are *single genes* that each encode a specific protein or RNA. There are also *multiple copies* of a specific gene, which allow a cell to synthesize protein or RNA more rapidly.

In addition to single genes and multiple copies of a specific gene, there are **gene families** consisting of two or more genes that encode either proteins with similar functions or proteins that are expressed at different times of development or in different tissues. The genes in any given family are often clus-

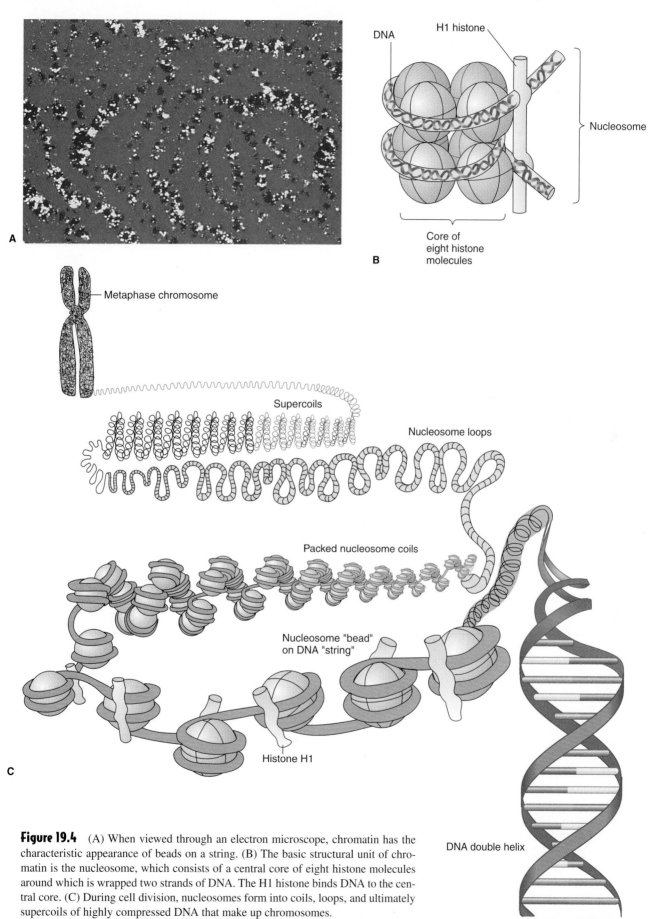

Figure 19.4 (A) When viewed through an electron microscope, chromatin has the characteristic appearance of beads on a string. (B) The basic structural unit of chromatin is the nucleosome, which consists of a central core of eight histone molecules around which is wrapped two strands of DNA. The H1 histone binds DNA to the central core. (C) During cell division, nucleosomes form into coils, loops, and ultimately supercoils of highly compressed DNA that make up chromosomes.

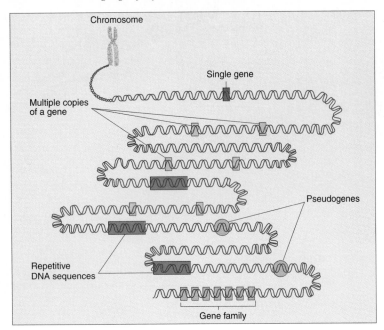

Figure 19.5 Chromosomes of eukaryotes contain single genes, multiple copies of single genes, gene families, and pseudogenes. However, most of the DNA consists of repetitive sequences whose functions are poorly understood.

Question: *What are pseudogenes?*

tered together on the same chromosome, although in some cases they are widely distributed throughout the genome. The genes in any given family are thought to have descended from one ancestral gene. According to this hypothesis, some ancestral genes were replicated but never separated during cell division processes, and they accumulated continuously over time. In modern eukaryotes, they are represented as gene families that are similar to the ancestral gene. Color vision is hypothesized to have evolved through this process.

Gene families also contain **pseudogenes,** genes descended from functional genes that are no longer expressed due to a mutation or deletion at some point in their evolutionary history. There are usually many pseudogenes for each functional gene in a gene family. Even though they are no longer expressed, pseudogenes remain in the genome and account for part of the noncoding surplus DNA.

Types of DNA Sequences

Different classes of DNA have been described in the genome. DNA sequences that have from one to several copies per genome are composed of **unique-sequence DNA;** these include single genes, certain multiple genes and pseudogenes, and, most commonly, sequences with no known function. Eukaryotic chromosomes also contain vast stretches of **repetitive DNA sequences,** most of which have no known function. **Moderately repetitive DNA sequences** are repeated between 1,000 and 100,000 times in the genome. A few transfer RNA and ribosomal RNA nonstructural genes and several structural genes are found in this DNA class. **Highly repetitive DNA sequences** are repeated 100,000 to 10 million times in the genome, and little is known about their function.

Research conducted on various mammals indicates that 50 to 70 percent of mammalian genomes consist of unique-sequence DNA, 20 to 40 percent of moderately repetitive DNA sequences, and 1 to 10 percent of highly repetitive DNA

sequences. The functions of most repetitive DNA sequences have not been identified; they apparently account for much of the surplus DNA found in the human genome.

BEFORE YOU GO ON After staining with special dyes, each human chromosome appears under a microscope with a unique series of bands and subbands. These bands, together with centromere position, are used to create a standard blueprint of each chromosome. The human genome is thought to contain about 50,000 genes. Within the genome are single genes, multiple copies of single genes, gene families, and pseudogenes. Three DNA classes—unique-sequence DNA, moderately repetitive DNA sequences, and highly repetitive DNA sequences—have been defined to reflect the number of times they occur in a genome.

MENDELIAN INHERITANCE IN HUMANS

Mendelian inheritance refers to traits that are inherited in accordance with Mendel's laws of segregation and independent assortment and also show dominance or recessiveness as Mendel defined them. Such traits are generally determined by a limited number of genes, most commonly one pair.

In 1966, Victor A. McKusick published the first edition of a historic book titled *Mendelian Inheritance in Man: Catalogs of Autosomal Dominant, Autosomal Recessive, and X-linked Phenotypes.* Since then, ten updated editions of the book have appeared, the most recent in 1994. These massive volumes represents a gold mine for geneticists or any person interested in learning more about human heredity. *Mendelian Inheritance in Man* is an encyclopedia of human genetic *loci*—specific sites where genes are located on certain chromosomes. It is also a valuable catalog of human traits (phenotypes associated with a single locus), their history, and their genetic basis. The first edition identified 1,487 human phenotypes; the most recent number obtained from the World Wide Web Mendelian Inheritance in Man (OMIM) home page in September, 1995, is 6,634. Of this latter figure, McKusick considers that for 4,238 traits, the mode of inheritance is proven. Of this number, 3,302 are listed as autosomal dominant (caused by a dominant gene located on an autosome), 677 as autosomal recessive, and 239 as X-linked (caused by genes carried on the X chromosome), and 20 as Y-linked (caused by genes carried on the Y chromosome). For 2,396 traits (1,196 autosomal dominant, 1,023 autosomal recessive, and 173 X-linked, and 4 Y-linked), the evidence concerning their inheritance is not proven but strong enough to include them in the list.

Scientists are making rapid advances in identifying human genes, discovering their functions, and determining their exact locations on specific chromosomes. For example, our largest chromosome, number 1, is thought to represent about 6 percent of the total human genome. At present, 353 gene loci (thought to be about 3 to 5 percent of the total genes on this chromosome) have been identified and mapped. A very small human autosome, chromosome 21, contains about 2 percent of the human DNA. So far, 38 gene loci have been assigned to this chromosome.

Autosomal Dominant and Recessive Traits

The question of whether a trait (or an allele) is dominant or recessive is generally much more complex in humans than for the simple alternative phenotypes (for example, round or wrinkled peas) first studied by Mendel. In those cases, the dominant allele is expressed in the phenotype of individuals with either the homozygous or heterozygous genotype. A recessive allele is expressed only when it is homozygous. However, in many carefully studied human traits, the distinction between dominant *(H)* and recessive *(h)* alleles is not this perfect. Complications arise because in heterozygotes *(Hh),* each allele frequently forms a gene product, usually a structural protein or an enzyme. In these cases, there may be either incomplete dominance or *codominance* (equal expression of both alleles), or some other mechanism may be involved. As a result, different levels of the protein may be present in each of the three genotypes *(HH, Hh,* or *hh),* and this variation may result in three different phenotypes. For traits determined by more than one gene pair, it is easy to see how complex the genetic analysis can become.

Some Human Traits Determined by a Single Gene Pair

What sorts of human traits are determined by a single gene pair? We invite you to check out your phenotype as you proceed through this section.

A relatively limited number of observable human traits are clearly determined by a single gene pair, and attempts to list them are not without peril. For example, *tongue rolling,* the ability to curl the tongue into a longitudinal trough, has long been cited in biology texts as an example of a dominant Mendelian trait in humans since it was first described in 1940 by A. H. Sturtevant (of fruit fly fame). However, several stud-

ies of human groups have shown conclusively that it is not due to a single gene pair but must involve a variety of complex factors. To illustrate this point, nonrolling parents ("homozygous recessive," according to the old model) often produce tongue-rolling offspring, and people have been known to learn tongue rolling through practice. Nevertheless, tongue rolling remains in many biology and genetics texts as an example of a human trait determined by a single gene pair.

Figure 19.6 shows some human traits commonly assumed to be determined by a single gene pair. Some caution is recommended in evaluating these traits, since most are not simply expressed as two alternative forms. There is often **variable expressivity,** which means that the trait appears differently in people that apparently have the same genotype. There may also be **incomplete penetrance,** meaning that the trait is not evident in all individuals who have the relevant genotype. Final-

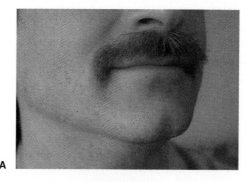

A

B

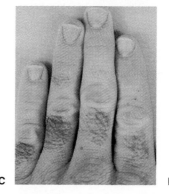

C

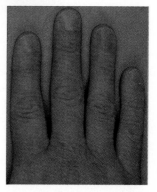

D

Figure 19.6 A few observable human traits are thought to be determined by a single gene pair including (A) chin fissure, (B) male-pattern baldness, (C) mid-digital hair, and (D) bent little finger.

Question: *Why are there relatively few bald women?*

ly, more than a single gene pair may eventually prove to be involved in determining a trait, or environmental or sexual influences may also affect the gene's expression.

For some traits, reference is made to a particular population. The underlying reason for this is that geneticists have found that many traits, particularly genetic disorders, are strongly associated with certain populations. To identify and evaluate such associations, data collected on hereditary traits are categorized according to this standard. In many cases, there are extensive data sets for white Americans but incomplete data for other populations. Thus reference to white Americans reflects the reliability of the data and does not indicate that the traits are not found in individuals of other populations.

The inheritance of a *free earlobe* is thought by some scientists to be due to a dominant gene. People with the homozygous recessive genotype have earlobes attached directly to their heads. Because several ear-related traits are known to be controlled by many genes, some geneticists doubt that lobe attachment is controlled by a single gene pair. Nevertheless, inheritance of earlobes seems to best fit a simple Mendelian recessive–dominant model involving a single gene pair.

The presence of *hair on the middle segment* of the four fingers may be due to a dominant allele. Homozygotes for the recessive allele have no hair on these finger segments, even though they may have abundant hair elsewhere.

The *ability to taste PTC* (phenylthiocarbamide) has long interested human geneticists. It also illustrates that genes can influence some rather bizarre sensory traits in humans, since PTC is a chemical that is never normally encountered in substances consumed by humans. To most people (the 70 percent with a dominant allele), PTC tastes quite awful; to the rest, it is tasteless. Evidence suggests that the gene for PTC tasting is located on chromosome number 7. In the white American population, approximately 20 percent are homozygous tasters (genotype *TT*), 50 percent are heterozygous tasters *(Tt)*, and 30 percent are homozygous nontasters *(tt)*. However, the simple tasting–nontasting classification may be misleading, since among tasters there is considerable variation in taste quality reported, indicating that there may be more than one gene pair involved.

A *bent little finger* (camptodactyly) is a minor hand variation characterized by an inward curvature of the little finger. It is due to a dominant allele and has both incomplete penetrance and variable expressivity.

A *mid-chin fissure* is the most common type of chin dimple and is caused by a dominant allele. It is a nice example of how a single gene pair can affect skeletal development. A number of features associated with this trait illustrate the complexities of "simple" inherited human traits:

1. It shows variable expressivity, ranging from a small dimple to a deep fissure.

2. It has nearly complete penetrance in males, but only about 50 percent of females with the allele have a fissure. Thus expression of the trait is *sex-influenced*—affected by the gender of the individual bearing the allele.

3. It has a *variable age of onset,* most commonly appearing in childhood or early adulthood.

4. It is most common among people of German heritage, with 20 percent of males and 10 percent of females having the trait.

Male-pattern baldness is the common form of hair loss and is another example of a sex-influenced trait caused by a gene that is fully expressed in males but less so in females. The hairline begins a progressive recession in young adult males as hair begins to fall out at the sides near the front of the scalp. Eventually, a bald spot appears at the rear of the crown, and finally, in about a third of affected males, only a fringe of hair remains along the lower crown of the head. It is estimated that at least 50 percent of white males and a much smaller number of white females are affected by such hair loss. Females experience much less hair loss than males and do not become bald unless they are homozygous for this trait. Thus, as with chin fissures, male-pattern baldness is sex-influenced, shows variable expressivity, and has a variable age of onset.

Some Human Traits Determined by Multiple Alleles and Polygenes

Most familiar human traits are determined by *multiple alleles*—the case in which there are alternative forms of a gene (alleles) that map to a single locus—or by *polygenes*—many genes at different loci. In *polygenic inheritance,* the number, chromosomal locations, and degree of expression of all the different genes are usually not known. Also, external environmental factors often influence the expression of such genes. Some of these traits are listed in Table 19.1.

Table 19.1   Some Human Traits Determined by Multiple Alleles or by Polygenic Inheritance

Trait	Pattern of Inheritance
Blood type	Multiple alleles with codominance
Eye color	Polygenic with incomplete dominance
Fingerprints	Polygenic, influenced by external factors
Ear structure	Many features polygenic, development related to gender

Blood Type

The ABO blood type system was discovered in 1900 and was proved to be genetically determined in 1911. Its early discovery was probably connected with its clear pattern of Mendelian inheritance. Blood type is one of the most completely understood human traits associated with multiple alleles; it is controlled by a single gene locus with three alleles. Even the site of the gene locus is known; it is located on chromosome 9q34. A, B, and O refer to the presence of specific molecules found on the surfaces of red blood cells. Blood types are linked to three alleles, designated I^A, I^B, and i. All humans have two alleles that determine which types of molecules will be present on their red blood cells. Humans with genotype $I^A I^A$ or $I^A i$ have blood type A, while genotypes $I^B I^B$ and $I^B i$ are type B; genotype $I^A I^B$ results in type AB, and individuals with genotype ii have blood type O. Figure 19.7 shows a cross between a female with blood type AB and a male with blood type O.

Eye Color

Human eye color is a trait that was poorly understood by classical geneticists. In 1907, an interesting research paper by Gertrude and Charles Davenport, titled "Heredity of Eye-Color in Man," appeared in the journal *Science*. The appeal of a simple Mendelian model had led to an assumption that eye color "is inherited as an alternative character," meaning that there were two primary human eye colors, brown and blue. The intent of the 1907 study was to test the hypothesis that human eye colors were inherited according to Mendelian predictions. The basic approach used was to study eye color in individuals from a number of families. After examining the data, the authors concluded that "blue eye-color is recessive to brown" and that "when two recessive individuals are mated," they produce "only the recessive type." There was also a consideration of "imperfect [blue] owing to the presence of specks or patches of pigment—the 'gray' or 'hazel' color." However, the major conclusions of the paper—two primary eye colors, with brown being dominant over blue—became incorporated into textbooks and was accepted as correct until relatively recently. The inadequacy of this simple model becomes obvious upon examining any large group of humans: you can see all sorts of black, brown, green, and blue eye colors.

Recent studies have led to the conclusion that eye color is a polygenic trait. There are no true blue, green, or black pigments in the human eye. Rather, eye color is mostly determined by the amount of *melanin* (the same brown pigment that determines skin and hair color) present in the iris. A current model, illustrated in Figure 19.8 (see page 364), proposes that at least two separate genes, each with two incompletely dominant alleles, control eye color. Each dominant allele results in the production of a certain amount of melanin. If there is little pigment (no dominant alleles), the eyes appear blue because of the way light is scattered by the sparse amount of pigment. The amount of pigment increases progressively with an increase in the number of dominant alleles. As a result, humans have eye colors that may be blue (no dominant allele), shades of green or gray (one dominant allele), light brown (two), brown (three), and dark brown or black (four). This model, along with allele segregation and independent assortment that occur during meiosis, also explains how two parents with light-colored eyes can have children with dark eyes.

Fingerprints

One of the most carefully studied human traits determined by polygenic inheritance is fingerprints. Three basic types of fingerprints occur in humans (see Figure 19.9, page 364). All humans have unique sets of permanent fingerprints, which are formed by the twelfth week of embryonic development. The three basic patterns are thought to be determined by a complicated polygenic system of dominant, incompletely dominant, and recessive genes. However, since not even identical twins have the same fingerprints, external factors in the uterine environment must play some role in their formation during pregnancy.

Ear Structure

A final example of an odd trait determined by polygenic inheritance is ear structure. Like fingerprints, every human has a unique pair of ears. The number of genes and chromosomes involved in the creation of an ear is unknown but is thought to be large, given the impressive variety of ear shapes and sizes. Also, it is clear that many structural features of the human ear are under different genetic control and developmental regulation in males and females. The net effect of these differences is that males tend to have bigger ears than females, even though their heads are of similar size. The significance of the ear size difference is unknown.

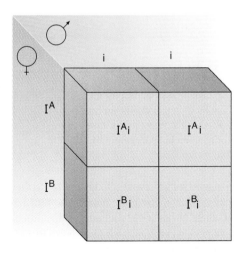

Figure 19.7 If a woman with type AB blood and a man with type O blood had children, each child would have a 50–50 chance of having type A or type B blood.

Question: *Which blood types would you expect to find in the children of a mother with type AB blood and a father with type A blood?*

BEFORE YOU GO ON The genetic basis and pattern of Mendelian inheritance are now known for over 6,500 human traits. A majority are due to autosomal dominant genes, about one third can be traced to autosomal recessive genes, and 5 to 10 percent are caused by X-linked genes. Several human traits, including hair on fingers, free earlobes, and a mid-chin fissure, involve a single gene pair. Many single-gene traits can be modified due to variable expressivity, sex, age, and other factors. Blood type is determined by multiple alleles, and eye color and fingerprints are examples of traits determined by polygenic inheritance.

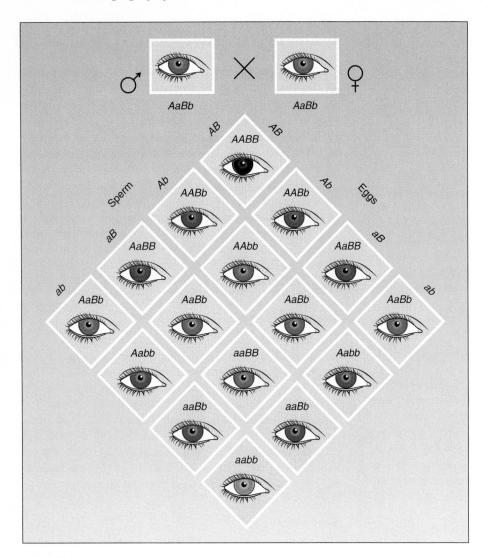

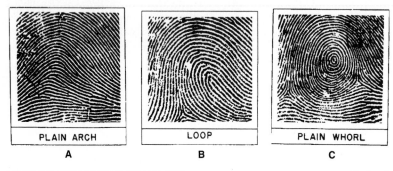

Figure 19.8 Human eye color is thought to be determined by polygenic inheritance. One model proposes that at least two genes, each with two incompletely dominant alleles, are responsible for producing melanin. Eye color is determined by the quantities of this pigment produced.

Question: *If one parent was brown-eyed with genotype* AaBb *and the other was green-eyed with genotype* aaBb, *which eye colors could their children have?*

Figure 19.9 All humans have unique fingerprints that are determined by complicated interactions of polygenes and environmental factors during embryonic development. The three basic types of fingerprints—arch (A), loop (B), and whorl (C)—are shown.

HUMAN SEX CHROMOSOMES

The karyotypes of human females and males are identical for the 22 pairs of autosomes but differ with respect to the sex chromosomes. Recall that cells of females contain two X chromosomes and those of males an X and a Y chromosome. The differences in size and gene content between the two sex chromosomes are striking.

Traits Carried on Sex Chromosomes

Many early investigators interested in human genetics noted a peculiar pattern in the inheritance of certain traits. A Swiss oph-thalmologist (physician who studies eyes) noted in the 1870s that in families that he had studied, many more males than females were red–green color-blind. What could account for this difference? In 1911, E. B. Wilson, a cytologist, made certain assumptions and came to an interesting conclusion about the inheritance of color blindness. If the gene for color blindness is recessive and males are XY, then the gene must be located on the X chromosome. A faulty gene on the X chromosome would always be expressed in males because they had only the one X chromosome. In contrast, females were likely to have a normal gene on one of their X chromosomes and would therefore rarely be color-blind.

The X chromosome has been studied more intensively than any other human chromosome. These efforts are related to the unique inheritance patterns of the X chromosome—males have only one X chromosome, which comes from the mother, and hence an X-linked recessive allele is always expressed in a male—and to the occurrence of more than 100 genetic disorders caused by genes carried on this chromosome.

X Chromosome Inactivation

In 1949, M. L. Barr, a Canadian cytologist, observed a dark-staining structure in the nucleus of cells from female cats that was not present in cells from male cats. Barr referred to the structure as "sex chromatin," and it is now called the **Barr body** (see Figure 19.10). Subsequently, Barr bodies were found in nuclei from human females, and it was also shown, in studies of individuals with abnormal numbers of sex chromosomes, that the number of Barr bodies present in a cell nucleus was always one less than the number of X chromosomes. Thus cells of normal females (46,XX) have one Barr body, males (46,XY) have none, 47,XXX females have two, and 47,XXY males have one. What could explain this peculiar but consistent finding?

Figure 19.11   Mary F. Lyon, a British cytologist, formulated a powerful hypothesis in 1961 that explained the basis of X chromosome inactivation. The Lyon hypothesis still guides research.

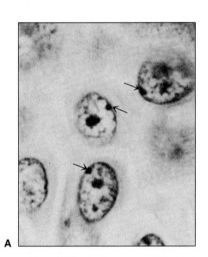

A

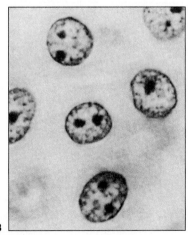

B

Figure 19.10   Barr bodies appear as dark-staining objects in the nucleus. (A) Female cells (XX) have one barr body. (B) Male cells (XY) have no barr body.

Question: *How are X chromosomes inactivated?*

In 1959, it was proposed that the Barr body was an X chromosome with highly condensed, dark-staining DNA, but no significant insights were offered about its development or relevance. Two years later, Mary F. Lyon, a British cytologist (see Figure 19.11), published a classic paper in which she consolidated diverse observations and information in formulating a coherent hypothesis that explained the nature of Barr bodies. In her beautiful, short (one page!) publication, Lyon described a model for **X chromosome inactivation.** The **Lyon hypothesis** explained that (1) the dark-staining X chromosome (Barr body) was genetically inactive; (2) either of the X chromosomes (that is, the one contributed by either the mother, X_m, or the father, X_p) could be randomly inactivated in any given cell; (3) inactivation of the X chromosome took place early in embryonic development; and (4) any descendants of the cell would have the same inactivated X chromosome as the parent cell.

What was the genetic significance of the Lyon hypothesis? How could it be verified for humans? First, it explained the phenomenon of **gene dosage compensation,** in which both males and females produce the same amount of X gene protein products. The major prediction of the hypothesis was that cells of females are mosaics for X-linked genes; that is, cells with an inactivated X_m chromosome would express genes from X_p, and vice versa. This prediction was verified in 1963 through experiments on an X-linked enzyme called G6PD, which has two forms, G6PD-F and G6PD-S. The results showed that in females heterozygous for the enzyme, cells produce either G6PD-F or G6PD-S but not both.

Another Lyon hypothesis prediction is now the focus of intense research. Lyon predicted the existence of a gene locus, now called the **X inactivation center (XIC),** where events occur that regulate the X chromosome inactivation process. Recent studies have given rise to a model, shown in Figure

19.12, that may soon explain how an X chromosome becomes inactivated. The key to the model appeared in 1992, when a new gene called *XIST* (for "X-inactive-specific-transcript gene") was found in the XIC. The *XIST* gene encodes an apparently unique type of RNA, which is large, never leaves the nucleus, and concentrates in the region of the inactivated X chromosome. How is *XIST* RNA involved with X chromosome inactivation? There are several hypotheses, and they are described in Figure 19.12.

The Lyon hypothesis continues to be extended as new studies provide more detailed information. It is now known that inactivation is not entirely random if one of the X chromosomes is structurally abnormal (for example, if part of it has been deleted). In these cases, the defective X chromosome is always inactivated. Also, there are several genes on inactivated X chromosomes that are still expressed. Results from studies of mice have shown that inactivation may not be a permanent condition. For unknown reasons, the proportion of cells with reactivated X chromosomes seems to increase with age. This finding raises many questions, the most important of which may be whether reactivation is an example of a general deterioration of gene control that is related to the aging process.

Genes Carried by the X Chromosome

It is well known that many genetic disorders trace their origins to defective genes carried on the X chromosome. Although we know much less about normal traits carried by this chromosome, we can make certain inferences from the effects of X-linked genetic disorders. For example, various forms of muscular dystrophy occur as a result of mutated genes on the X chromosome; therefore, some genes on that chromosome must be responsible for the *normal* development or functioning of muscle cells in humans.

In common with all chromosomes, the X chromosome contains genes that encode structural proteins and enzymes that are important for normal operations of the organism. Genes for blood-clotting factors and normal red–green color vision are also carried by the X chromosome.

Genes Carried by the Y Chromosome

The X–Y basis of sex determination in humans and other mammals has been known since 1923; females have two X chromosomes (XX), and males have one X and one Y (XY). However, until the late 1950s, an open question remained unanswered: Which sex chromosome, the X or the Y, possesses the gene or genes responsible for governing the sexual destiny of a developing embryo? Some discoveries in the late 1950s and early 1960s seemed to point to the answer. Studies of individuals with abnormal numbers of sex chromosomes revealed that 45,X0 humans developed ovaries (female reproductive organs) and 47,XXY developed testes (male reproductive organs). What did this indicate? Consistent with these

findings, it was hypothesized that the Y chromosome possessed a gene that dictated development of a male.

The search for the sex-determining gene on the Y chromosome led down several blind alleys. In 1987, the answer finally seemed near when researchers reported the results of studies that led them to conclude that they had likely found the testis-determining factor *(TDF)* gene. However, further work soon showed that their proposed *TDF* gene did not determine sex. Finally, in 1991, a research team reported the discovery of a gene called *SRY* (sex-determining region Y). The *SRY* gene has since been proved to be the gene that governs sex determination in all mammals studied, including humans.

The following picture of human sex determination has now emerged. The development of male and female embryos proceeds identically until about six or seven weeks after fertilization, when sexual differentiation occurs. At that time, a genetic signal initiates a cascade of biochemical events that leads to the development of a male. The genetic signal is hypothesized to be a protein encoded by *SRY* that regulates the expression of other genes. If the *SRY* gene is absent, the embryo develops into a female. The next research goal is to identify the genes regulated by *SRY,* for we know nothing yet about genes involved in the *SRY*-initiated cascade. Thus in spite of the great excitement created by the discovery of the *SRY* gene, the complete story of sex determination in humans (and other mammals) has not yet been deciphered, and research on this exciting question continues.

A few other structural genes are located on the Y chromosome. Two or three Y genes encode specific proteins found on cell surfaces. Also, available evidence indicates that one or more genes affecting sperm production, stature (height), tooth size, and hair on ears are located on the Y chromosome.

BEFORE YOU GO ON The X chromosome has been studied longer than any other human chromosome. Many genes on the X chromosome determine traits that can be studied because of their pattern of inheritance in males and females. In females, only one of the X chromosomes remains active in cells, and the other is inactivated during embryonic development. Relatively few genes are known to be carried by the Y chromosome, the most significant being the *SRY* gene, which determines the sex of offspring.

THE EMERGENCE OF HUMAN GENETICS

It seems strange that the discipline of human genetics lagged far behind *Drosophila* genetics, microbial genetics, and the genetics of agricultural plants and animals. What were the reasons for this delay? A simple but incomplete answer is that human genetics was a much more complicated field. This, along with ethical and biological considerations, meant that Mendelian geneticists interested in studying human heredity could not expect to make rapid progress. In addition, many human geneticists turned their attention to *eugenics,* with the goal of improving the human race (see the Focus on Scientific Process, "Human Genetics and the Abuse of Science").

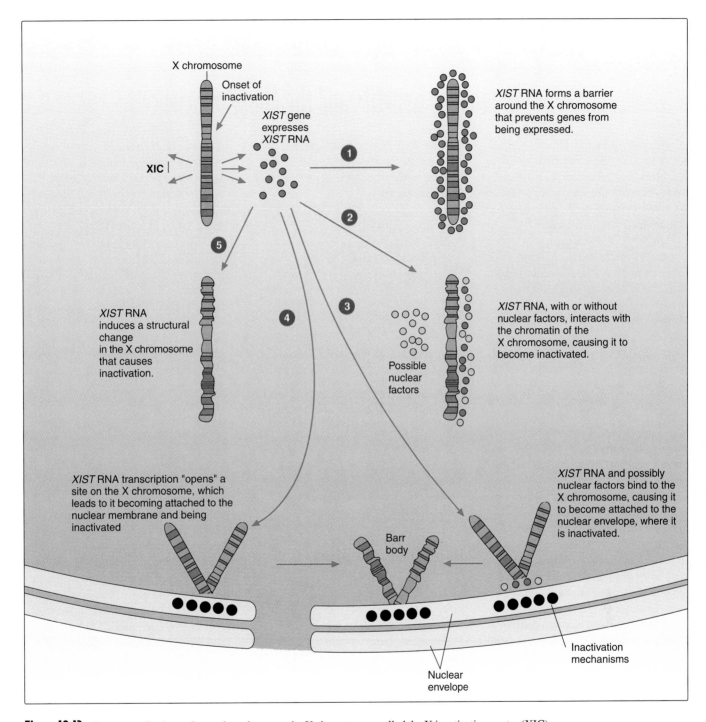

Figure 19.12   Recent studies have shown that a locus on the X chromosome called the X inactivation center (XIC) contains a gene that may regulate X chromosome inactivation. The gene is known as *XIST* (X-inactive-specific-transcript), and it encodes a unique type of RNA. Several hypotheses have been proposed to explain how the *XIST* gene may inactivate an X chromosome. (1) *XIST* RNA forms a barrier around the DNA that prevents transcription. (2) It interacts with the chromatin, either with or without the help of unknown nuclear factors (proteins), which leads to the X chromosome becoming inactivated. (3) *XIST* RNA, perhaps along with nuclear factors, binds to a site on the X chromosome, causing it to become attached to the nuclear envelope, where the actual inactivation mechanisms exist. (4) The act of transcribing *XIST* RNA exposes a site on the X chromosome, which leads it to become attached to the nuclear envelope, where it is inactivated. (5) The act of transcribing *XIST* RNA induces a structural change in the X chromosome, which leads to inactivation.

Question: *Is it possible that none of these hypotheses will prove to be correct? Why?*

Human Genetics and the Abuse of Science

A review of the history of human genetics shows how science and human values can become intertwined. It also demonstrates how scientific research results can be abused by flawed analyses and unwarranted extrapolations.

In 1900, at the time of the rediscovery of Mendel's laws of segregation and independent assortment, biologists throughout Europe and the United States were debating the nature of human intelligence and the comparative influences of heredity and the environment. To many of these biologists, Mendelian genetics appeared to provide a valuable additional perspective in understanding human heredity. Some of these scientists were also concerned with implications of evolution for human heredity and believed that knowledge from both genetics and evolution could be combined to "improve the human race." These individuals called themselves **eugenicists,** and they conducted different types of research. For example, the American eugenicist (and famous geneticist) Charles Davenport established a well-funded station at Cold Spring Harbor on Long Island, New York, for "the experimental study of evolution." There he studied different families with the aim of improving humankind.

Davenport and others throughout the worldwide scientific community were genuinely concerned with what they perceived to be disturbing implications of evolution and genetics for the human population. They argued that harmful variations, normally removed by natural selection in other species, were "artificially" preserved by human society. In 1914, W. Grant Hague, in *The Eugenic Marriage,* expanded on this theme. Two quotations will serve to convey the flavor of eugenicists' concerns. Hague wrote that unless something was done immediately,

every second child born in this country, in fifty years, will be unfit; and, in one hundred years, the American race will have ceased to exist. We mean by this that every second child born will be born to die in infancy, or, if it lives, will be incapable of self-support during its life, because either of mental degeneracy or physical inefficiency.

In calling attention to the potential harm of existing conditions, he wrote,

any condition that fundamentally means race deterioration must be rendered intolerable. The prevalent dancing craze is an anti-eugenic institution, as is the popularity of the delicatessen store. No sane person can regard with complacency the vicious environment in which the future mothers of the race "tango" their time, their morals, and their vitality away. We do not assume to pass judgment on the merits of the dance; we do, however, emphatically condemn the surroundings.

In retrospect, we can easily see that these early researchers had a very incomplete view of human genetics. There is, of course, no such group as the "American race" that Hague wrote about, and it is highly doubtful that the tango or the deli had much to do with degeneracy. These scientists did not appreciate that a deleterious allele could remain in a population in a heterozygous state. Even more misguided and of greater importance, the eugenicists also mixed morality with genetics. They did this by assuming that all physical, mental, and moral traits were inherited in accordance with simple Mendelian ratios. Therefore, insanity, epilepsy, alcoholism, criminality, and feeblemindedness were treated in the same fashion as eye color and albinism. Feeblemindedness in particular drew the attention of early eugenicists, as indicated by Figures 1 and 2.

The early eugenicists also mixed their political opinions into their scientific discussions. Since the post–Civil War years, the United States had been receiving millions of immigrants from southern and eastern Europe. These people came from different ethnic groups than the predominately Anglo-Saxon people who originally colonized the United States and who still represented the majority of the population. Many of the early eugenicists considered these immigrants, as well as people of African origin already in the United States, racially inferior and therefore worried about the "degeneration" of the American people. More enlightened eugenicists argued that only individuals with a family history of undesirable elements (mental retardation, insanity, criminality, alcoholism, and so on) should be barred from entry to the United States. Still, many eugenicists completely opposed the immigration of "inferior races." Similar racial and ethnic prejudices existed in European countries.

Today, we can look back with disbelief on some of the excesses of these naive views. However, it is important to realize that they had significant influence at the time. Eugenicists in the United States were active in campaigning for sterilization laws. Although most of the geneticists considered these laws premature, eugenicists considered them an important part of "racial hygiene." By 1931, as many as 30 of the

FOCUS ON SCIENTIFIC PROCESS

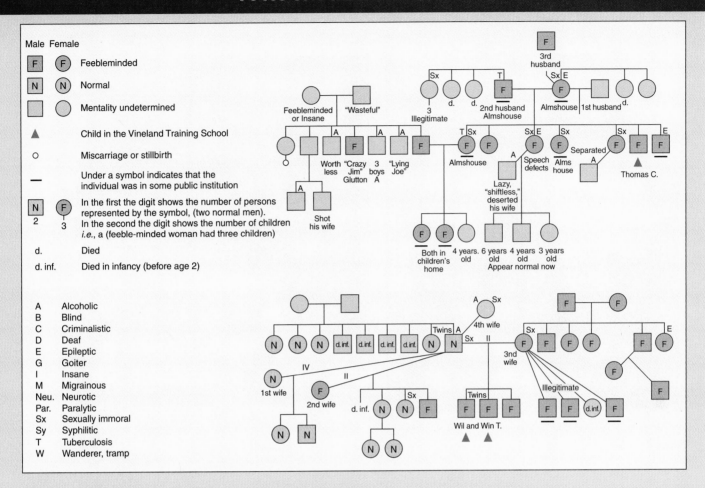

Figure 1   Early eugenicists described the inheritance of feeblemindedness and other "undesirable" phenotypes in pedigrees such as this.

48 states had laws that permitted sterilization of the feebleminded, the insane, alcoholics, and rapists. By 1958, over 60,000 sterilizations had been performed for eugenic reasons. Although in later years many of these laws were modified to protect individual rights and to reflect more "enlightened" genetics, most are still on the books today.

Like sterilization laws, immigration laws passed early in the twentieth century were strongly influenced by the eugenics movement. The Immigration Restriction Act of 1924 limited the annual immigration from each European nation to 2 percent of that country's natives living in the United States at the time of the 1890 census. That act was an attempt to keep the Anglo-Saxon Protestant majority from being "diluted," and it served as the template for future immigration legislation.

By 1930, American geneticists had come to realize that the early eugenics program was seriously flawed. They understood that politics and morality had been mixed with incomplete genetic knowledge, and for the most part, they repudiated their earlier program. They argued for a more "scientific" approach that distinguished cultural from biological factors and stressed the need for research on human genetics. In spite of repudiation by leading geneticists, the early American eugenics movement persisted. Not until the worldwide reaction of outrage to Nazi extermination of millions of people in the name of "racial hygiene" was the early American eugenics movement finally extinguished.

box continues

FOCUS ON SCIENTIFIC PROCESS

Figure 2   Eugenicists felt that the type of house a person occupied was related to qualities of the mind.

Another factor that deferred progress in human genetics was that no compelling reason was advanced for society to support research in this field when there were other, more important biological problems, such as infectious diseases, to be addressed. Despite the beautiful, pioneering work of Garrod described in Chapter 18, few researchers in the medical sciences saw a connection between genetics and human disorders. During the first half of the twentieth century, human genetics was not taught in medical schools because it was not considered relevant to medicine. Hence most geneticists worked in biology departments or agricultural experimental stations where the organisms of interest were not human. What changed this situation? Events described in Chapters 17 and 18 played a role, but the development of improved cytologic techniques and the accurate determination of the human karyotype in 1956 were especially significant. Soon after this, cytogeneticists discovered numerous abnormalities in human karyotypes (for example, 45,X0, 47,XXY, and 47,XXX) that were clearly the causes of specific disorders. Thus beginning in the 1960s, the central issue of medical implications, combined with increased funding for research, ushered in an era of explosive growth in human genetics, which is now a preeminent discipline in biology.

During the 1980s and 1990s, scientists have achieved much deeper knowledge of human genetics. The organizational features of our chromosomes are now well known. Detailed knowledge of chromosome structure has permitted cytologists to draw precise maps. Researchers have made amazing progress in unraveling the structure, types, numbers, and locations of genes. However, it is humbling to consider that the complete human genome can be compared to a library holding millions of books, of which only about 50,000 contain pages that we know how to read!

SUMMARY

1. The field of human genetics developed slowly because humans are complex, have long life spans, and could not be used in controlled genetics experiments for ethical reasons. Until the 1960s, genetics researchers were much more interested in using organisms such as fruit flies, bacteria, and viruses that could help them answer questions in a relatively brief period of time.

2. The human karyotype was not correctly determined until 1956. Cells of normal females have 22 pairs of autosomes and one

pair of X sex chromosomes; males have 22 pairs of autosomes and one X and one Y sex chromosome.

3. Chromosomes have a consistent structure. When stained with certain dyes, each chromosome has a unique set of bands and subbands that serves as landmarks that can be used to distinguish it from other chromosomes. These bands are also used as maps in determining the precise locations of genes on a chromosome.

4. The human genome consists of 3.0 to 3.5 billion different base pairs and includes about 50,000 genes. Single genes, multiple copies of single genes, and gene families are all found on human chromosomes.

5. Human and other eukaryotic chromosomes are characterized by different classes of DNA. They contain unique-sequence DNA, moderately repetitive DNA sequences, and highly repetitive DNA sequences. The functions of most repetitive DNA sequences are not understood.

6. To date, more than 6,500 human phenotypic traits can be explained in terms of Mendelian inheritance. The genes that determine over 4,200 of these traits have been identified, and their locations on specific chromosomes are known.

7. Human traits thought to be determined by a single gene pair with complete dominance are free earlobes, hair on the mid-joint of fingers, the ability to taste PTC, a bent little finger, a mid-chin fissure, and male-pattern baldness. A trait determined by multiple alleles is blood type. Eye color, fingerprints, and ear structure are examples of polygenic traits.

8. Many genes are carried on the X chromosome, but only a few have been found on the Y chromosome; one, the *SRY* gene, determines sex of offspring.

9. Only one of two X chromosomes is genetically active in humans and other mammals. The second X chromosome of females is randomly inactivated during embryonic development. Consequently, the amount of gene products derived from the X chromosome is the same in both males and females.

WORKING VOCABULARY

chromosome bands (p. 356)
gene dosage compensation. (p. 365)
genome (p. 357)
incomplete penetrance (p. 361)
karyotype (p. 355)
repetitive DNA sequences (p. 360)
sex-influenced trait (p. 362)
variable expressivity (p. 361)
X chromosome inactivation (p.365)

REVIEW QUESTIONS

1. What is a karyotype? How are karyotypes determined?
2. Describe the human karyotype.
3. What are chromosome bands? How are they used in human genetics?
4. Describe the human genome.
5. What different classes of DNA are found in the chromosomes? What is known about the function of each?

6. Explain differences between the genetic bases of these human traits: free earlobe, blood type, eye color, and color blindness.
7. Explain the pattern of inheritance for sex-linked traits in humans.
8. What is X chromosome inactivation? Why is it significant?
9. What is the *SRY* gene? Where is it located? What does it do?

ESSAY AND DISCUSSION QUESTIONS

1. Using ideas from hypotheses related to the origin of life on Earth (Chapter 2) and the RNA world (Chapter 18), how might the existence of different DNAs in eukaryotic chromosomes be explained?
2. Why might studies of "normal" genetically determined human traits be expected to be less fruitful than studies of human genetic disorders? Can results in one area have applications in the other? Explain.
3. Which areas of human genetics research do you feel should be emphasized (and hence receive funding) in the next decade? Why?
4. A woman with blood type A files a paternity suit against a man with blood type B, claiming that he is the father of her child. The child has blood type O. Could the alleged father be the biological father? Discuss.

REFERENCES AND RECOMMENDED READING

Beckwith, J. 1993. A historical review of social responsibility in genetics. *BioScience,* 43: 327–333.

Bulger, R. E., E. Heitman, and S. J. Reiser. 1993. *The Ethical Dimensions of the Biological Sciences.* New York: Cambridge University Press.

Carson, H. L. 1986. Patterns of inheritance. *American Zoologist,* 26: 797–809.

Charlesworth, B. 1991. The evolution of sex chromosomes. *Science,* 251: 1030–1033.

Cummings, M. R. 1994. *Human Heredity: Principles and Issues.* 3rd ed. St. Paul: West. Publ.

Daniel, W. L. 1995. *Genetics and Human Variation.* Champaign: Stipes.

Davenport, G. D., and C. B. Davenport. 1907. Heredity of eye-color in man. *Science,* 26: 589–592.

Hague, W. G. 1914. *The Eugenic Marriage: A Parental Guide to the New Science of Better Living and Better Babies.* New York: Review of Reviews.

Kevles, D. J. 1985. *In the Name of Eugenics: Genetics and the Uses of Human Heredity.* Berkeley: University of California Press.

Lyon, M. F. 1961. Gene action in the X-chromosome of the mouse (*Mus musculus* L.). *Nature,* 190: 372–373.

McKusick, V. A. 1994. *Mendelian Inheritance in Man: Catalogs of Autosomal Dominant, Autosomal Recessive, and X-linked Phenotypes.* 11th ed. Baltimore: Johns Hopkins University Press.

Morell, V. 1994. Rise and fall of the Y chromosome. *Science,* 263: 171–172.

Novak, R. 1993. Curious X-inactivation facts about calico cats. *Journal of NIH Research,* 5: 60–65.

Powledge, T. M. 1993. The genetic fabric of human behavior. *BioScience,* 43: 362–367.

Russell, P. J. 1996. *Genetics.* 4th ed. New York: HarperCollins.

Thurman, E., and M. Susman. 1993. *Human Chromosomes: Structure, Behavior, and Effects.* New York: Springer-Verlag.

ANSWERS TO FIGURE QUESTIONS

Figure 19.5   Pseudogenes are decended from functional genes but, because of a mutation or deletion, they are no longer expressed.

Figure 19.6   Females must be homozygous for the male-pattern baldness gene, whereas affected males can be heterozygous. Also, because the trait is sex-influenced, it is less likely to develop in females.

Figure 19.7   If the father is has genotype *AA,* there would be a .50 chance of a child having blood type A or type AB. If the father's genotype is *Ai,* the chances would be 0.50 for blood type A (.25 *AA* and .25 *Ai*), .25 type AB, and .25 type B (*Bi*).

Figure 19.8   Children's chances of having the following genotypes are: .25 *AaBb,* .25 *aaBb,* and .125 *AaBB, Aabb, aaBB,* and *aabb.* Therefore, there would be a .125 chance for a child to have three dominant alleles and dark brown eyes, .375 for two dominant alleles and brown eyes, .375 for one dominant allele and green or gray eyes, and .125 for no dominant alleles and blue eyes.

Figure 19.10   It has been hypothesized that genetic events in the X inactivation center regulate X chromosome inactivation. A new gene, called the *XIST* gene, encodes a unique type of RNA that may be directly responsible. The precise mechanism remains to be identified.

Figure 19.12   That no current hypothesis will prove to be correct is always possible. However, the proposed hypotheses are consistent with observations on the appearance and behavior expected of *XIST* RNA if it regulates X chromosome inactivation. It seems likely that one of the proposed mechanisms will prove to be accurate or nearly so.

20

Human Genetic Disorders

Chapter Outline

Reading Questions

1. Why is research on human genetic disorders important?

2. What are pedigrees? How are they used in the study of human genetics?

3. What is the genetic basis of sickle-cell anemia? What are the effects of this genetic disorder?

4. What are the genetic nature and effects of Huntington's disease? Of familial hypercholesterolemia?

5. Under what circumstances might prenatal diagnosis and genetic counseling be advisable?

K nowledge about human genetics has been expanding at a phenomenal rate. Methods and experimental approaches used in the *new genetics,* a term introduced in 1979, allowed scientists to probe human chromosomes at a level not previously imaginable. One area of intense interest for the new genetics is human genetic disorders. A **genetic disorder** is characterized by a predictable consequence or set of consequences caused by mutations in genes and chromosomes.

Research designed to identify the causes of serious genetic disorders now occupies a mainstream position in the field of human genetics. There are several important reasons for this: genetic disorders occur in a significant number of individuals, they are not curable and often not even treatable, they cause great emotional and financial stress on the families of affected individuals, and they place a considerable burden on health, social, and other community services. In addition to contributing to understanding the disease process, detailed knowledge about an abnormal inherited condition will frequently provide insights into the underlying, but often poorly understood, normal processes that have been disrupted.

THE NATURE OF GENETIC DISORDERS

The causes of genetic disorders are rooted in mutations of normal DNA or chromosomes (or both) in cells that produce gametes (sperm or egg cells). Some **congenital malformations** (deformities that occur during fetal development) seem to have a strong inherited element and may also be traced to altered genetic material. The genetic basis of most human genetic disorders is illustrated in Figure 20.1. In offspring produced by fertilization involving a defective gamete, the altered DNA or chromosomes may result in abnormal concentrations of normal proteins, the absence of critical proteins, or the production of abnormal proteins. Such mutations can then lead to biochemical deficiencies or structural abnormalities of cells, tissues, or organs.

Two general categories of genetic changes are recognized: *gene mutations* occur when a single gene changes from one allelic form to another, and *chromosomal mutations* occur with changes in segments of chromosomes, whole chromosomes, or entire sets of chromosomes (see Chapter 16). By definition, chromosomal mutations affect more than one gene.

Genetic disorders caused by all mutations and their associated abnormalities occur in 2 to 5 percent of all live births and are an important cause of death in children under the age of 15 years. About 10 percent of the adult population may be affected by chronic disorders with a significant genetic component (for example, certain types of heart disease and cancer).

Gene Mutations

Single-gene mutations (mutations of one gene) cause a number of inherited diseases that can be followed through many generations of an affected family. Genetic disorders caused by single-gene mutations are often classified as dominant or recessive, autosomal or sex-linked. Recessive disorders such as phenylketonuria (PKU), discussed in Chapter 18, usually involve a defective enzyme or hormone, whereas dominant disorders are frequently characterized by an abnormality in a structural protein. All human organ systems (for example, the nervous, circulatory, skeletal, and immune systems) are affected by one or more genetic disorders. Most, if not all, metabolic pathways that depend on a variety of normal enzymes for processing DNA, RNA, lipids, carbohydrates, or proteins can also be disrupted by gene mutations. More than 3,000 human genetic disorders caused by single-gene mutations have now been identified.

Chromosomal Mutations

A second group of inherited abnormalities are caused by chromosomal mutations that arise primarily as a result of imperfect meiosis during gamete formation, as described in Chapter 15. There are two principal types of chromosomal mutations: structural rearrangements and numerical changes.

Structural rearrangements occur when there is a loss or relocation of genes or sets of genes along specific chromosomes. As shown in Figure 20.2, chromosome segments containing many genes may break and do one of the following: (1) be removed from the chromosome (an instance of *deletion*), (2) be *inverted* and incorporated in reverse orientation in the same chromosome, (3) be reattached elsewhere on the same chromosome (a type of *translocation*), or (4) be reattached on a different chromosome (another type of translocation). A *duplication* occurs when multiple copies of genes are replicated. Structural rearrangements generally have harmful effects; duplications are usually also harmful but may sometimes have

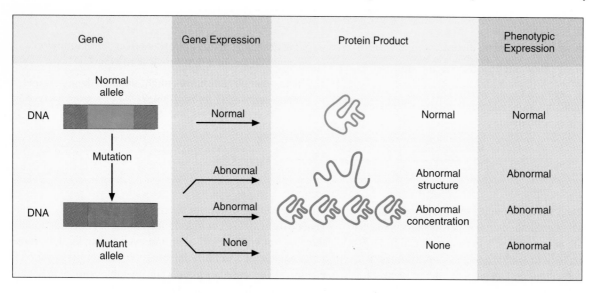

Figure 20.1 The basis of most human genetic disorders lies in the alteration or elimination of normal proteins, which occur because mutated genes are not expressed normally.

Question: *Why are abnormal proteins produced from mutant alleles?*

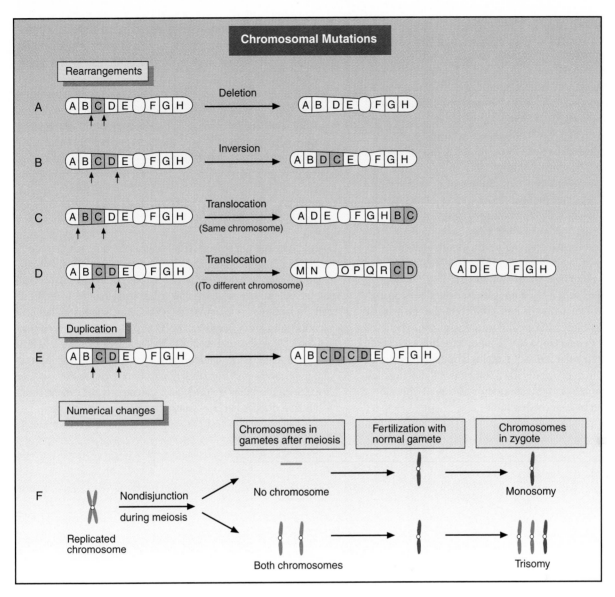

Figure 20.2 Various types of chromosomal mutations cause human genetic disorders. Rearrangements involve chromosome segments (indicated by ↑) that are detached from the chromosome and either deleted (A), inverted (B), or translocated elsewhere in the same chromosome (C) or to another chromosome (D) during meiosis. Genes or sets of genes may be duplicated an abnormal number of times, which can result in multiple gene copies (E). If nondisjunction occurs during meiosis, a gamete may have zero or two copies of a chromosome instead of one. If that gamete participates in fertilization, cells of the zygote will contain one copy (monosomy) or three copies (trisomy) of the chromosome (F).

a positive effect by creating new genetic combinations capable of new functions.

Numerical changes increase or decrease the number of whole chromosomes without changing the structure of the individual chromosomes. The primary cause of numerical mutations is *nondisjunction,* the failure of replicated chromosomes to separate during meiosis, resulting in one gamete receiving both and the other gamete receiving neither of the affected chromosomes. If the defective gamete participates in fertilization, an abnormal number of chromosomes will be found in every cell of the resulting embryo. **Trisomy** occurs when a

third copy of a homologous chromosome is present in cells. The condition in which cells lack one chromosome from a homologous pair is called **monosomy.**

Scientists estimate that chromosomal mutations occur in approximately one of every ten conceptions. About 90 percent of these cause such serious abnormalities that development ceases after a short period and a spontaneous abortion (miscarriage) occurs, often before the mother is even aware that she is pregnant. Offspring that are born usually have distinctive structural and functional symptoms, a **syndrome** that is quite characteristic for each specific chromosome mutation. The biological consequences of such effects include premature mortality, infertility, and physical and mental handicaps.

For live births, the most frequent structural rearrangements involve chromosome numbers 13, 14, 15, 21, and 22. Numerical changes are generally limited to trisomies of autosomes 13 (Patau syndrome: one in 5,000 live births), 18 (Edwards syndrome: one in 10,000), 21 (Down syndrome: one in 700), and abnormal numbers of sex chromosomes, usually 47,XXY (Klinefelter syndrome: one in 2,000), 47,XXX (female: one in 1,000), and 45,X0 (Turner syndrome: one in 3,000), this last being the only human monosomy capable of developing and surviving. Trisomies and monosomies occur in other chromosomes, but their devastating effects are not compatible with survival and so are rarely seen in live births. Children with Patau or Edwards syndrome have many severe organ abnormalities and usually die within one year after birth. Children with Down syndrome have numerous problems but are often capable of leading productive, relatively long lives (see the Focus on Scientific Process, "Down Syndrome"). X0 females are usually short, have many body abnormalities, and are sterile because their ovaries fail to develop. XXX females have limited fertility and may be slightly retarded. XYY males have underdeveloped testes, are infertile, and usually have some mental deficiencies. Figure 20.3 summarizes the quantitative effects of chromosomal mutations on human conceptions.

BEFORE YOU GO ON Human genetic disorders are caused by gene mutations and chromosomal mutations and affect a considerable number of people. Mutant genes associated with more than 3,000 human genetic disorders have now been identified. Chromosomal mutations are relatively common, and their effects on the developing embryo are usually extremely harmful. At present, knowledge about the underlying genetic and molecular nature of genetic disorders is increasing rapidly, but there has been limited success in treating or curing most conditions.

HUMAN PEDIGREE ANALYSIS

People who are unfamiliar with human genetic research tend to associate a pedigree with dogs, horses, cats, and other domesticated animals. However, pedigree analysis is a useful tool for tracing the inheritance of genetic disorders in human families. Basically, a **pedigree** is a simple diagram constructed from detailed records for as many family members as possible over several generations. Extensive and complete records permit construction of pedigrees that may reveal informative patterns in the way that certain traits are passed from individuals in one generation to those in the next. Pedigrees are often used in research and genetic counseling; we discuss genetic counseling later in this chapter.

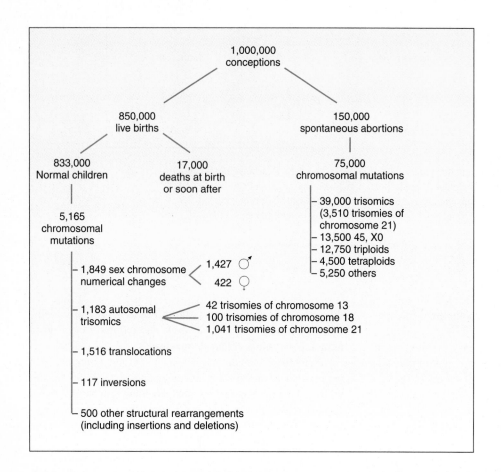

Figure 20.3 A summary of the outcome of 1 million human conceptions and the influence of chromosomal mutations.

Question: *Why do spontaneous abortions often occur as a result of chromosomal mutations?*

FOCUS ON SCIENTIFIC PROCESS

Down Syndrome

The most common chromosomal mutation, occurring in one out of every 700 to 800 live births in the general population, is a trisomy of chromosome 21 that causes **Down syndrome (DS)** (see Figure 1). The occurrence of trisomic DS is most often related to the age of the mother: the risk of having a DS child is one in 50 in women over age 40. Recent evidence also suggests that older men are more likely to father DS offspring. Down syndrome is the leading cause of mental retardation in the United States. Affected children suffer from multiple physical and mental problems.

Reviewing DS offers an opportunity to examine how advances in scientific knowledge and scientific technology are used to extend the understanding of a genetic disorder. As more becomes known, particularly at the molecular level, it may become feasible to design methods for treating the effects of the genetic disorder. Ultimately, it may even become possible to prevent the effects caused by such disorders. Scientists are now investigating the molecular implica-

tions of having an extra copy of chromosome 21 and are trying to correlate these molecular effects with the DS phenotype.

Various records indicate that DS has occurred in children of all racial and ethnic groups throughout human history. The syndrome was formally recognized in 1866 when an English physician, John L. Down, published a comprehensive description of physical symptoms in certain mentally retarded patients. However, the cause of DS was not discovered until the late 1950s, when scientists observed that affected patients had three copies of chromosome 21 in their cells. Individuals with DS suffer from many anatomical and biochemical abnormalities. They have flattened facial features, abnormal (epicanthic) folds of the eyelids, unusual creases on the palm, muscular flaccidity, and short stature. In addition to mental retardation, 40 percent of DS patients have congenital heart defects, many develop cataracts or other eye problems because of lens defects, and they are more susceptible to infections

and are much more likely to develop leukemia than normal people. Most individuals with DS live a maximum of 30 to 50 years.

Recently, the number of families raising DS children at home, rather than committing them to an institution at birth, has increased. Studies indicate that DS children benefit significantly when allowed to remain in a home environment with caring and loving family members along with formal occupational and physical therapy. They show enhanced intellectual and physical development, achieve a higher degree of independence, and have an increased life expectancy (see Figure 2, page 378).

Most chromosome 21 trisomies are a result of nondisjunction during meiosis; however, 3 to 5 percent of DS individuals are a result of trisomies caused by chromosomal translocations. This latter process occurs when a major portion of chromosome 21 is translocated to another chromosome, usually 13, 14, 15, 22, or, in rare cases, the other 21. In translocation DS, one of the parents is

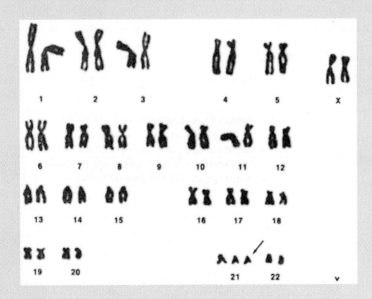

Figure 1 Down syndrome is most often caused by a trisomy of chromosome 21.

box continues

Figure 2 Down individuals participating in athletic competition.

generally a translocation carrier that is phenotypically normal but at increased risk of producing a DS offspring. Chromosome 21 is estimated to contain roughly 1,500 of the 50,000 genes in the complete human genome. Of that number, only a few have been precisely located and identified using high-resolution genetic-mapping procedures. Molecular biologists are particularly interested in answering the following questions: Which genes on chromosome 21 are implicated in producing the abnormal phenotype? Which genes are responsible for the harmful effects? What proteins are encoded by these genes? Why does having three copies of these genes lead to DS?

Investigations have now established that the DS phenotype is linked to a number of genes that lie within the 21q22 band of chromosome 21, as shown in Figure 3. Although speculative at the present time, ideas about the relationships among the genes identified, the proteins they produce, and the effects they cause have recently emerged.

The high risk of leukemia is thought to be associated with the *ets-2* gene,

which is known to be a cancer-causing gene (oncogene). Studies are under way to learn more about the expression of this gene in DS and normal individuals.

The *Gart* gene codes for three different enzymes that play a role in the synthesis of purines (recall that two purine bases, adenine and guanine, are found in DNA molecules). Individuals with DS have elevated purine levels in their blood serum. High purine levels are linked with a variety of problems, including mental retardation. Thus it has been hypothesized that the presence of a third *Gart* gene leads to a higher dosage of purines, which accounts for many of the problems that are characteristic of DS. An alternative hypothesis has recently been proposed. A gene has been described that codes for S100 protein, which is found in the nervous systems of all vertebrates. The highest levels of S100 are found in the brain, where it interacts with other brain proteins. It is possible that abnormal concentrations of S100 protein may also be responsible for the adverse nervous system effects of DS. One scenario for therapy in the future involves early detection of DS in

the fetus followed by immediate attempts to regulate levels of the purines or proteins ultimately found to be responsible for the severe symptoms.

The α-*A-crystalline* gene codes for a protein of the same name that is a structural component of the lens of the eye. The abnormal expression of that protein may be related to the increased risk of cataracts and lens defects in DS.

Finally, the gene *SOD-1* codes for a protective enzyme that prevents damage to cells from reactive chemicals produced in certain metabolic pathways. Alterations in SOD-1 levels may account for the accelerated rate of aging in DS and may also contribute to mental retardation. What might explain the mental retardation? Recently, researchers discovered the SOD-1 mutations are responsible for causing inherited amyotrophic lateral sclerosis (Lou Gehrig's disease), a disease characterized by the death of nerve cells.

These hypotheses remain to be verified. Nevertheless, the supporting evidence is impressive, and these efforts offer a detailed view of ongoing research into an important human genetic disorder.

FOCUS ON SCIENTIFIC PROCESS

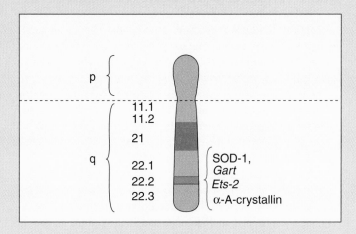

Figure 3 New techniques have been used to identify genes associated with Down syndrome. Scientists are now trying to determine how these genes cause the effects leading to the Down syndrome phenotype.

Geneticists use standard symbols and conventions in diagraming pedigrees, as shown in Figure 20.4A (see page 380). A pedigree can often determine the type of gene mutation responsible for the disorder being studied. What patterns of inheritance would be expected for each type of gene mutation? Researchers use established criteria to distinguish recessive, dominant, and sex-linked traits in families. *Autosomal dominant* genes cause genetic disorders, such as Huntington's disease, that can be transmitted through continuous generations in a family, can affect both males and females, and can be transmitted by both males and females to male and female offspring. Disorders caused by *autosomal recessive* genes tend to appear in alternate generations because affected individuals often do not have children; otherwise, hereditary patterns are similar to those of autosomal dominants. A pedigree with many affected males in two or more family units connected through female relatives is evidence of a disorder caused by an *X-linked recessive* gene; Duchenne muscular dystrophy is an example of such a disorder. If a trait is due to an *X-linked dominant* gene, all the daughters, but no sons, of an affected father will have the disorder; few of these disorders occur in humans. Figure 20.4B shows typical pedigrees that demonstrate the forms of inheritance involving various gene mutations.

BEFORE YOU GO ON Pedigrees are charts of family generations that show individuals affected by genetic disorders. Through logical analysis of pedigrees, it is usually possible to determine the type of gene mutation—dominant or recessive, autosomal or sex-linked—that is responsible for the disorder.

GENETIC DISORDERS OF HUMANS

Human pedigree analyses have been used successfully for decades in establishing the genetic basis of human genetic disorders. However, until the age of the new genetics, there were few major successes in learning about the molecular causes of a specific genetic disorder. Now we have reached a point where human genetics researchers are frequently able to identify not only the genes responsible for causing genetic disorders but also the underlying biological process that is disrupted. We will analyze four human genetic disorders that provide a picture of the rapid progress being made in this field. As you will discover, there are interesting deviations from the classic Mendelian dominant–recessive model of inheritance.

The relative frequency of most genetic disorders varies significantly from one human population to another. For example, sickle-cell anemia occurs in 1 to 2 percent of offspring born to black Africans but is extremely rare in most other populations. The reasons for these differences are not always clear, but geneticists have speculated that they may often be related to some survival advantage conveyed by the allele in individuals who are heterozygous for the condition. Thus heterozygotes may be able to withstand some environmental factor (for example, a disease) better than individuals without the allele. Table 20.1 (see page 381) provides some information on the genetic disorders discussed in this chapter.

Sickle-Cell Anemia

Sickle-cell anemia has great significance in the history of human genetic disorders. It is now known that there are two forms of sickle-cell anemia, and they are distinguished by name. **Sickle-cell disorder** occurs in individuals who are homozygous for the mutant allele; **sickle-cell trait** occurs in heterozygous individuals. People with the sickle-cell trait generally have milder anemia and are not as seriously affected as those with sickle-cell disorder. Sickle-cell disorder was the first disorder in which clear relationships were established involving a simple Mendelian trait, a heritable gene mutation, an abnormal gene product, and an abnormal molecular structure.

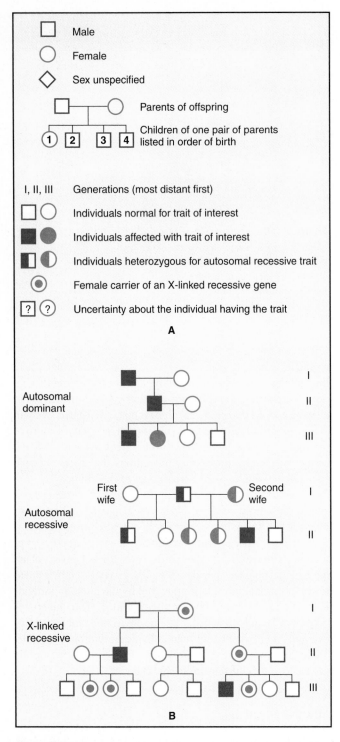

□ Male

○ Female

◇ Sex unspecified

□—○ Parents of offspring

①②③④ Children of one pair of parents listed in order of birth

I, II, III Generations (most distant first)

□ ○ Individuals normal for trait of interest

■ ● Individuals affected with trait of interest

◧ ◐ Individuals heterozygous for autosomal recessive trait

⊙ Female carrier of an X-linked recessive gene

? ? Uncertainty about the individual having the trait

A

Autosomal dominant

I
II
III

Autosomal recessive

First wife   Second wife

I
II

X-linked recessive

I
II
III

B

Figure 20.4 (A) Human pedigrees are constructed using these symbols. (B) These pedigrees show typical patterns of inheritance associated with different types of gene mutations.

A syndrome referred to as *sickle cell disease* was first described in 1910, but scientists at the time knew only that it affected both male and female blacks; it often caused serious symptoms, including fever and extreme pain in bones and joints; and some of the red blood cells of affected individuals

were sickle-shaped (see Figure 20.5). Severe *anemia* (low number of red blood cells) often led to premature death. By examining pedigrees, scientists hypothesized that sickle-cell disease was transmitted as an autosomal recessive trait.

Subsequently, scientists discovered that red blood cells of affected individuals underwent a reversible change in shape called *sickling* when internal oxygen concentrations decreased in the body. Some suspected that this effect might be due to a structural flaw in the *hemoglobin (Hb) molecule,* an iron-containing protein of red blood cells that carries oxygen from the lungs to the internal tissues. What were the biochemical and genetic bases of such a defect? In 1949, Linus Pauling and his colleagues found that the sickling characteristic was in fact due to a structural aberration in hemoglobin molecules.

Since Pauling's pioneering work, sickle-cell anemia has become one of the best-understood human genetic disorders. The underlying cause of this hemoglobin structural abnormality was determined in 1957. Normal hemoglobin, known as *Hb-A,* is described in Figure 20.6. It is composed of two pairs of polypeptide chains, termed α (alpha) and β (beta), that are almost equal in size; α chains have 141 amino acids, and β chains 146 amino acids. The genes that encode the two polypeptide chains are located on the *p* arm of chromosome 11. As shown in Figure 20.7 (see page 382), sickle-cell hemoglobin, *Hb-S,* differs from Hb-A by only one amino acid in the β chain, yet this single change alters the molecule to such an

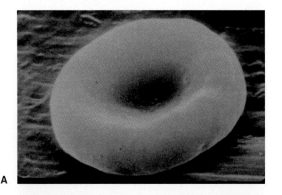

A

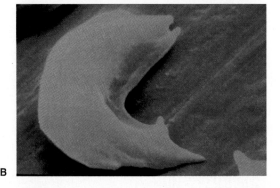

B

Figure 20.5 (A) Photo of a normal red blood cell. (B) Sickle-cell anemia is characterized by red blood cells that develop an abnormal sickled shape when oxygen concentrations in the body are reduced.

Table 20.1 Genetic Disorders Described in This Chapter

Disorder	Genetic Basis	Site of Locus
Sickle-cell anemia	Single gene locus on chromosome 11 at 11p15.5; codominant alleles	p — Sickle-cell anemia, q, Chromosome 11
Sickle-cell disorder	Individuals are homozygous	p — Huntington disease, q, Chromosome 4
Sickle-cell trait	Individuals are heterozygous	
Huntington disease	Single gene locus on chromosome 4 at 4p16.3; mutant dominant alleles	p — Familial hypercholesterolemia, q, Chromosome 19
Familial hypercholesterolemia	Single gene locus on chromosome 19 at 19p13.2; codominant alleles	p — Duchenne muscular dystrophy, q, X chromosome
Duchenne muscular dystrophy	Single gene locus on the X chromosome at Xp21.2; one or more sex-linked recessive alleles	

extent that it can become nonfunctional if both β chains are affected. In Hb-A, the DNA triplet CTT (cytosine–thymine–thymine) codes for glutamic acid, but in Hb-S, the corresponding sequence reads CAT (cytosine–*adenine*–thymine), which codes for valine. Thus for hemoglobin β chains, there are at two alleles, $β^A$, which encodes a normal β chain, and $β^S$, which encodes a β chain containing valine. There are many other β chain alleles and hemoglobin variants in the human population.

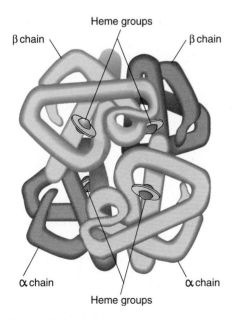

Heme groups

β chain   β chain

α chain   α chain

Heme groups

Figure 20.6 Hemoglobin, which carries oxygen attached to the four heme groups, is composed of two α chains and two β chains. The complex structure of hemoglobin is determined by the amino acid sequences of the α and β chains.

Question: *Which chains are altered in sickle-cell anemia?*

Individuals with sickle-cell disorder are homozygous, which means that both alleles, $β^Sβ^S$, encode abnormal β chains and all hemoglobin molecules are Hb-S. Heterozygous individuals with sickle-cell trait have alleles $β^Aβ^S$ and produce equal amounts of Hb-A and Hb-S. Therefore, $β^A$ is codominant with $β^S$ because both are expressed in the heterozygote.

This would seem to represent the ultimate level to which the cause of a genetic disorder can be traced. The single change in one DNA base creates a defective hemoglobin molecule that under low-oxygen conditions forms a spiral, rigid, or fiberlike structure that causes the affected red blood cell to lose its normal shape and assume a sickled form. Figure 20.8 (see page 382) shows the cascade of effects that results from sickle-cell anemia. Anemia occurs because sickled cells are susceptible to damage during circulation through blood vessels and are likely to be destroyed as a result.

Recently, a new therapy has been developed that is effective in treating adults with sickle-cell anemia. Patients receive doses of hydroxyurea (a simple organic chemical), which reduces the frequency of severe episodes caused by the cascade of effects. The underlying mechanism responsible for this positive result is now the subject of intense sutdy.

About 8 percent of African Americans have the heterozygous sickle-cell trait. This frequency is apparently decreasing slowly. As indicated in Figure 20.9 (see page 383), the mutant allele is also quite common in inhabitants of Africa, India, Arabia, and the Mediterranean area. Since alleles that cause such negative biological effects are relatively rare in human populations, why do certain human populations have these high allelic frequencies?

Scientists now know that heterozygotes with the sickle-cell trait are more resistant to malaria than those with normal hemoglobin. Thus in areas where malaria is present, there is a survival advantage associated with the allele that apparently outweighs the disadvantage related to the early deaths of a small number of homozygous individuals.

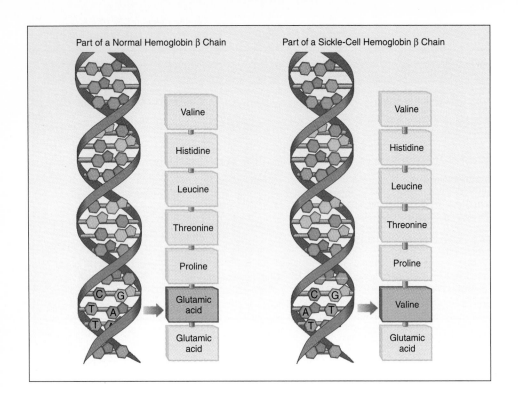

Part of a Normal Hemoglobin β Chain

Valine
Histidine
Leucine
Threonine
Proline
Glutamic acid
Glutamic acid

Part of a Sickle-Cell Hemoglobin β Chain

Valine
Histidine
Leucine
Threonine
Proline
Valine
Glutamic acid

Figure 20.7 Sickle-cell anemia is caused by a gene mutation that results in the synthesis of abnormal β chains in Hb-S proteins. The abnormal β chain in Hb-S differs from the normal Hb-A β chain by only one amino acid, having valine instead of glutamic acid.

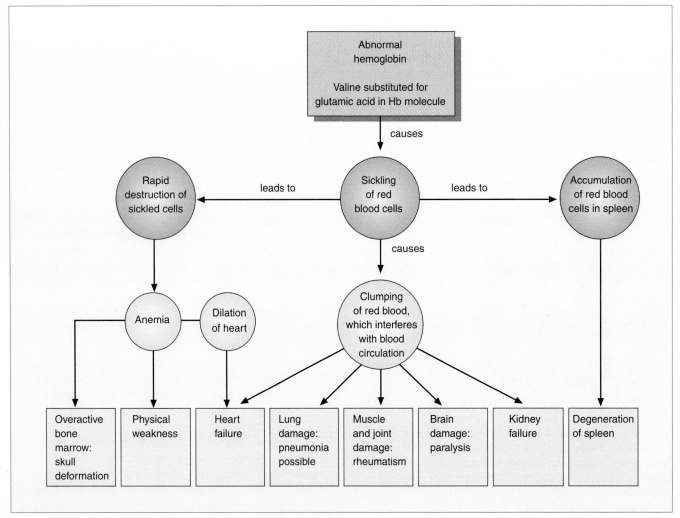

Abnormal hemoglobin

Valine substituted for glutamic acid in Hb molecule

causes

Rapid destruction of sickled cells ← leads to — Sickling of red blood cells — leads to → Accumulation of red blood cells in spleen

causes

Anemia — Dilation of heart

Clumping of red blood, which interferes with blood circulation

Overactive bone marrow: skull deformation

Physical weakness

Heart failure

Lung damage: pneumonia possible

Muscle and joint damage: rheumatism

Brain damage: paralysis

Kidney failure

Degeneration of spleen

Figure 20.8 The effects of sickle-cell anemia are complex and numerous. Many different tissues and organs are affected.

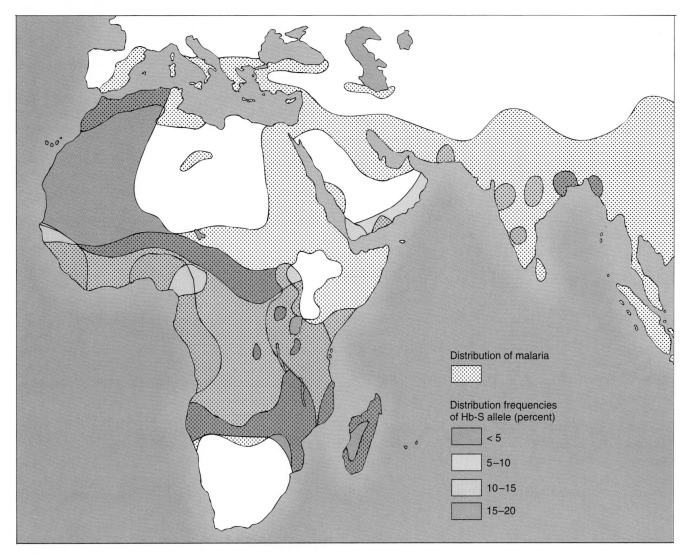

Figure 20.9   A higher percentage of individuals from tropical regions of the world have sickle-cell anemia than people who live elsewhere in the world. This difference is thought to be related to its advantage in conveying resistance to malaria. As shown here, the distribution of malaria coincides with the frequencies of the Hb-S allele.

BEFORE YOU GO ON   Sickle-cell anemia is a genetic disorder associated with two codominant alleles. Homozygous individuals with two mutant alleles develop sickle-cell disorder, a dangerous condition that has direct effects on red blood cells and indirect effects on many other tissues and functions. Sickle-cell disorder results from a flaw in the structure of hemoglobin due to the replacement of a single amino acid in the protein chains that make up part of this molecule. Heterozygotes suffer from sickle-cell trait, a less serious form of this hemoglobin disorder.

Huntington's Disease

Huntington's disease (HD) was first described in 1872 by George Huntington, an American physician. In the 100 years after its description, much was learned about HD, even though it occurred in a relatively small number of individuals.

Characteristics of HD

Huntington disease generally conforms to an autosomal dominant inheritance pattern; individuals with genotypes *HH* or *Hh,* where *H* indicates the dominant allele and *h* the recessive allele, will develop Huntington disease. Huntington disease is considered one of the cruelest human genetic disorders for several reasons: the first symptoms usually do not appear until the person is between 30 and 40 years of age; it affects the nervous system, causing the death of brain cells, which leads to a long decline and, ultimately, total incapacitation and death 10 to 20 years after initial onset; and there is no treatment or cure at the present time.

The early symptoms of HD vary but often include clumsiness, forgetfulness, and depression. The effects soon escalate, with **chorea**—constant and uncontrollable body movements—being the most obvious. Gradually, an affected individual begins to stagger and fall, compulsively clench and unclench the fists, and experience rapid, uncontrolled movements of the arms and legs. In the later stages of HD, there is loss of memory and intellectual functions and a general physiological deterioration leading to death. These effects are all caused by the death of nerve cells in the brain.

Until the 1970s, opportunities for studying HD on a large scale were limited because of its low incidence in people of European ancestry, the primary group that suffers from HD. However, in 1972, a surprising discovery was reported: an extraordinary number of people living in a remote village on Lake Maracaibo, Venezuela, had HD. In 1976, the U.S. Congress established a national commission on HD and chose Nancy Wexler, now of Columbia University, as its executive director (see Figure 20.10). Wexler is an expert on HD on both a professional and a personal level; her mother died of the disorder. The commission recommended in 1977 that a genetic study of the Venezuelan community be undertaken, and investigations began in 1979 under Wexler's direction. An extensive pedigree, consisting of over 7,000 individuals, was traced out, and blood samples were obtained from about 1,500 of these people. Of the 7,000, more than 100 have HD and another 1,100 are at risk because one of their parents had the disease. Why does such a large proportion of this population have HD? Apparently, five generations ago, when the population of the village was small, one woman had several children, including some who developed HD. These affected individuals may have constituted 5 to 10 percent of the population at that time. As the village population expanded, the HD gene continued to appear in each succeeding generation at approximately the same relatively high percentage. It is interesting to note that the mother of the original HD children in Venezuela did not develop HD herself; thus it would be interesting to know something about the father of these children. Some scientists have speculated that he may have been a British sailor.

Figure 20.10 Nancy Wexler and colleagues have studied Huntington's disease in members of a Venezuelan village for over 20 years. Here she is shown evaluating an individual from the village with HD.

Question: *Why are people in such a remote location being studied?*

Discovery of the HD Gene

Analysis of blood samples obtained from the affected Venezuelan population led to important developments. In 1983, using a new type of linkage analysis, James Gusella of Massachusetts General Hospital and his colleagues (including Nancy Wexler) discovered a **gene marker** for HD—a distinctive allele or DNA segment that was always present in HD patients but not in unaffected people from the same family and was therefore assumed to be very close to the HD gene locus on the chromosome. The general principle of identifying gene markers traces its roots to the early fruit fly experiments described in Chapter 16. Recall from Sturtevant's crossover experiments that the closer two genes lie together on a chromosome, the more likely they will be inherited together. Today, gene markers used in human genetics studies generally consist of unique alleles or DNA segments that occur once in the genome and are therefore found only on one chromosome. They include single alleles that encode specific enzymes or blood factors and segments of unique DNA that are not expressed as proteins, which was the case with the HD gene marker. Analyses showed that the HD gene marker was found only on chromosome 4, and further studies revealed that it was present at the terminal band of the short arm of chromosome 4 (4p16), which meant that the HD gene itself was located nearby.

After discovery of the gene marker, there was great optimism that the HD gene would be found quickly, as its approximate location was now known. However, despite intensive efforts of research teams from many countries, the problem turned out to be far more difficult than originally imagined. Success was not realized until 1993, when Gusella's research team and five other groups from the United States, England, and Wales (the Huntington's Disease Collaborative Research Group) published results of their studies comparing chromosomes of 150 HD patients with those from individuals without HD. They described both the location (4p16.3) and the base sequence of the HD gene, which they identified as Huntingtin. The normal protein product of *Huntingtin* has not yet been identified.

The Genetic Basis of HD

Huntingtin is a "giant" gene spanning more than 10,000 nucleotide bases. The predicted size of the *Huntingtin* protein is 3,144 amino acids. What distinguishes a normal allele from the mutant allele of the gene that causes HD? As described in Figure 20.11, the answer turned out to be rather uncomplicated but quite surprising: numerous repeats of a single trinucleotide sequence, cytosine–adenine–guanine (CAG), occur near the start of the *Huntingtin* gene (that is, CAG is repeated many times). The position of the CAG repeats is the site of the mutation that leads to HD.

Research has revealed that most normal *Huntingtin* alleles have 9 to 36 CAG repeats. Alleles from people with HD generally had between 37 and 66 repeats, but some had 70 to 90 and a few had over 100. Individuals with 33 to 36 CAG repeats are said to have a *premutation,* which means that they do not show

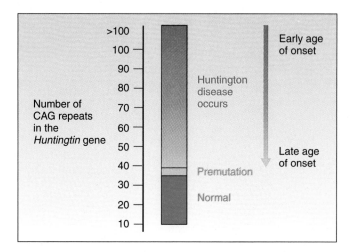

Figure 20.11 The *Huntingtin* gene contains multiple consecutive repeats of the nucleotide base sequence CAG (cytosine–adenine–guanine). Normal individuals have two *Huntingtin* alleles with 9 to 36 CAG repeats. Individuals with Huntington disease have one (heterozygotes) or two (homozygotes) mutant alleles with 37 or more CAG repeats. Individuals with alleles having 33 to 36 repeats are normal but are likely to have children who develop HD, a condition known as *premutation*. The number of CAG repeats is hypothesized to be inversely related to the age of HD onset.

Question: *What might explain the development of HD in a 2-year-old child?*

clear symptoms of HD but are more likely to have children who will have HD than are parents with fewer than 33 repeats. Further, there is a correlation between the number of CAG repeats and the age of onset for HD. Persons with the highest number of repeats experienced HD as juveniles, while those with lower numbers developed HD in their fifties. Individuals with over 100 repeats developed HD as early as 2 years of age.

Why is a parent with an HD premutation at greater risk for having a child who will develop HD? Recent observations indicate that the number of CAG repeats tends to expand in the next generation; that is, children with a parent who has HD or a premutation number of CAG repeats will often have a greater number of repeats than the parent. Repeat expansions are more likely when the HD gene is transmitted from the father.

HD is an autosomal dominant disorder. Homozygous individuals with two mutant alleles, each encoding an abnormal number of CAG repeats, develop typical HD. Heterozygotes with one normal allele and one mutant allele, both of which may be expressed, also develop HD even though there may be some normal *Huntingtin* protein present. How many *Huntingtin* alleles exist? The actual number is not known, but it was determined that in people without HD, each CAG repeat number (9 to 36) represents a different allele. Therefore, it is reasonable to assume that there is also a large number of mutant alleles.

The discovery of *Huntingtin* and the mutation responsible for HD has led to many new questions and hypotheses. What is the structure and function of the normal *Huntingtin* product?

Why does the mutation lead to HD? The Huntington's Disease Collaborative Research Group hypothesized that the CAG repeat expansion causes HD by disrupting the expression or structure of the normal *Huntingtin* protein product. Alternatively, the expression of other genes may be affected by the abnormal *Huntingtin* protein. These questions will be difficult to answer. It may be years before the biochemical basis of HD is understood and effective therapies can be developed. Nevertheless, the discovery of *Huntingtin* represents an important first step in this direction.

BEFORE YOU GO ON Huntington disease usually appears in middle-aged individuals and causes a gradual decline in nervous system function that leads to death. HD is an autosomal dominant disorder caused by a mutant Huntingtin allele that has an abnormal number of CAG base repeats. The underlying molecular cause of HD has not yet been determined, and there is no treatment for this genetic disorder.

Familial Hypercholesterolemia

Whereas sickle-cell anemia and Huntington disease damage specific cells, some genetic disorders disrupt a major metabolic pathway. Collectively, heart diseases are the most common disorders resulting from known genetic mutations. And it is a metabolic disorder, **familial hypercholesterolemia (FH),** that is the most frequent cause of genetic heart disease. *Hypercholesterolemia* means "too much cholesterol in the blood." The underlying genetic cause of FH is now understood in great detail, largely as a result of research conducted in the 1980s. Like many genetic disorders, it involves an inborn error of metabolism, but of a special variety that is not due to an absent or deficient enzyme. To understand FH, it is necessary to answer this question: How is cholesterol normally processed in the body?

Cholesterol, a type of lipid molecule, is required by every cell for synthesizing membranes. Under normal circumstances, the liver synthesizes most of the cholesterol required, with the remainder provided by the food we eat. Cholesterol is transported through the blood in particles known as **low-density lipoproteins (LDL).** As illustrated in Figure 20.12 (see page 386), each LDL particle holds about 1,500 cholesterol molecules enclosed within a lipid layer, which contains a single large protein molecule responsible for organizing the particle. How do LDL particles enter the cell? Cells in need of cholesterol make *receptor* molecules that are inserted into the plasma membrane. A **receptor,** usually a membrane-bound protein, recognizes and binds with a specific molecule and helps it enter the cell. Figure 20.13 (see page 386) shows that LDL receptors are concentrated in *coated pits,* where they bind with LDL. Once LDL becomes linked with a membrane-bound LDL receptor, the entire complex pinches off to form a vesicle that enters the cell; this process is known as **receptor-mediated endocytosis.** Once inside the cell, receptors are detached from the vesicle and are recycled to the plasma membrane. The LDL particle then fuses

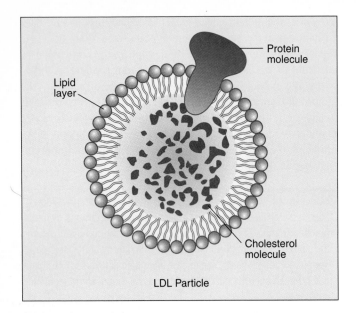

Figure 20.12 Cholesterol molecules are packaged and transported through the blood in low-density lipoprotein (LDL) particles. Each LDL particle contains about 1,500 cholesterol molecules surrounded by a lipid layer in which a protein is embedded. The protein's primary function is to organize construction of the LDL particle.

Question: *What other role does the protein portion of the LDL particle play in the cholesterol transport system?*

with a lysosome and is broken down to release cholesterol molecules that are used for building new membranes.

LDL receptors are encoded by a single gene locus that has four comon alleles, three of which are mutants, with incomplete dominance. The normal allele, *C,* encodes normal LDL receptors. Of the three mutant alleles, *c,* one encodes a receptor with a reduced binding capacity, one encodes a receptor that cannot bind to LDL, and one encodes a receptor that cannot function in endocytosis. Homozygous individuals with genotype *CC* have normal LDL uptake systems. Heterozygotes (*Cc*) have one normal allele and one of the three mutant alleles. The net result of all heterozygous conditions is that only a limited amount of cholesterol can enter the cell. There are two negative consequences of this deficiency: cells deprived of cholesterol, including liver cells, begin to make their own, and LDL and cholesterol levels in the blood increase because the excess is not taken up by affected cells. The body copes with the surplus by depositing cholesterol in the walls of blood vessels, which leads to **atherosclerosis,** a blockage of arteries that supply the heart (see Figure 20.14). Thus heterozygotes frequently develop early *coronary heart disease* and commonly suffer heart attacks in their forties or fifties. For homozygous (*cc*) individuals, heart attacks occur in childhood and usually result in death.

In most human populations, the frequency of heterozygotes born is at least one in 500, making FH the most prevalent genetic disorder and one of the most serious. Homozygotes occur with a frequency of one in 1 million. Figure 20.15 shows two hypothetical pedigrees to illustrate how FH is transmitted in humans.

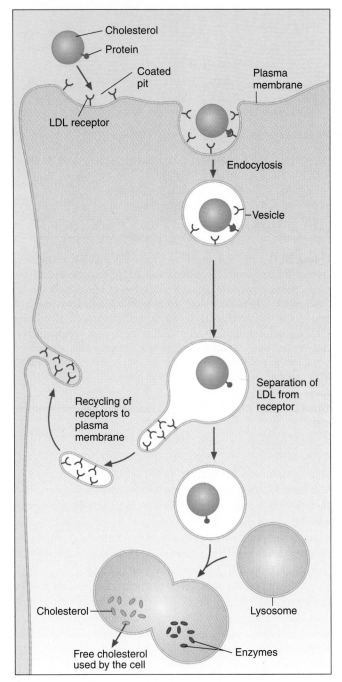

Figure 20.13 LDL particles interact with LDL membrane receptors that are concentrated in coated pits of cells requiring cholesterol. Once engaged, the LDL–LDL receptor–membrane complex pinches off, forms a vesicle, and enters the cell. Once inside the cell, LDL receptors are separated from the vesicle and recycled to the plasma membrane. The vesicle then fuses with a lysosome containing enzymes that digest the LDL particle, releasing free cholesterol molecules needed by the cell in membrane synthesis.

Question: *Which part of this system fails to function in individuals with familial hypercholesterolemia?*

Unlike sickle-cell anemia and many other genetic disorders, FH does not occur more frequently in specific ethnic groups. Also, scientists know the precise molecular effects of

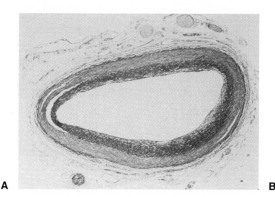 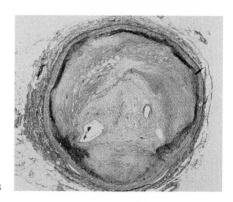

Figure 20.14 Continuous high levels of cholesterol in the blood eventually result in the formation of deposits in arteries that can block or inhibit normal blood flow and lead to a heart attack. (A) This is a normal artery from a 100-year-old woman who never had evidence of heart disease in her life. (B) This clogged artery has three small openings but is otherwise blocked.

be traced to the X chromosome by their distinctive pattern of inheritance in families; they typically appear in males but only very rarely in females, unless dominant. More than 300 human genetic disorders are thought to be caused by defective alleles on the X chromosome.

Most cases of **Duchenne muscular dystrophy (DMD)** are caused by an X-linked recessive allele that appears in at least one of every 3,500 males born. By comparison, hemophilia, perhaps the best-known X-linked disorder, occurs once in every 10,000 males. DMD is one of ten forms of muscular dystrophy but is by far the most common and one of the most severe.

DMD is a fatal genetic disorder caused by the degeneration of all types of muscle cells. The disease may sometimes first appear in adolescents, young adults, and even middle-aged people, with varying degrees of severity. However, in the classic, childhood-onset form of DMD, the first signs are evident at about 2 years of age. By early adolescence, affected children are usually wheelchair-bound. DMD patients rarely live beyond young adulthood, and death is commonly caused by respiratory or heart failure.

Dystrophin and DMD

The gene associated with DMD encodes a protein called **dystrophin.** The *dystrophin* gene is "mammoth" in size, spanning more than 2.4 million base pairs on the X chromosome (at Xp21), with at least 70 exons and 60 introns (recall that exons are parts of genes that are expressed and introns are nonexpressed regions of genes). What causes DMD? The structure of the normal *dystrophin* allele was determined in 1988. Subsequently, scientists studied different *dystrophin* alleles from muscular dystrophy patients and proposed a model for explaining the cause of DMD. Exon deletions from the normal *dystrophin* allele cause DMD, and the severity of the muscular dystrophy is related to the quantity and quality of the dystrophin produced. If some or most of the key portions of the normal *dystrophin* allele were present, a milder form of muscular dystrophy occurred because some dystrophin (or a simpler form of it) was produced. If major proportions of the allele were deleted, little or no dystrophin was produced, and severe DMD resulted.

A New Model for DMD

Based on new observations, the model describing the underlying cause of DMD continues to be extended. The specific

---

Figure 20.15 Two hypothetical pedigrees illustrate how FH is transmitted in humans.

the mutant alleles. As a result, they have devised methods of treatment for heterozygous individuals who are at risk for premature heart diseases associated with FH. Today, drugs are used that accelerate excretion of excess cholesterol from the body. These drugs do have side effects such as nausea and diarrhea. However, the medication, along with lifestyle modifications involving dietary changes to lower total body fat and weight, and getting increased regular exercise, has allowed heterozygotes to improve their chances of survival beyond middle age. New ideas for treating FH are described in Chapter 22.

> **BEFORE YOU GO ON** Familial hypercholesterolemia is the most common human genetic disorder and the leading cause of genetic heart disease. FH is caused by codominant mutant alleles that encode abnormal membrane receptors required for normal cholesterol processing. Homozygous individuals rarely reach their teens. Heterozygotes commonly suffer from coronary heart disease by middle age. Drugs and lifestyle modifications are used to treat FH.

Duchenne Muscular Dystrophy

Sickle-cell anemia, Huntington disease, and familial hypercholesterolemia are caused by autosomal genes. However, there are also many genetic disorders that are sex-linked. Different mutant alleles and the resulting affected phenotypes can

function of dystrophin has not yet been defined, though recent research has provided valuable clues about how it may interact with other proteins in normal muscle cells. Figure 20.16 illustrates the hypothesis that dystrophin binds to six specific protein molecules embedded in the muscle cell plasma membrane and links them to microfilaments of the intracellular cytoskeleton. The six membrane-associated protein molecules are known as **dystrophin-associated proteins (DAP).** What is the function of the coupled dystrophin–DAP complexes? While not yet verified, DAP may help regulate calcium ion channels; an **ion channel** is a protein channel complex that regulates the movement of a specific type of ion through the membrane. Figure 20.17 is a simple diagram that describes an ion channel.

The new model explains that calcium ions have a known role in normal muscle contraction processes, coupled dystrophin–DAP complexes may function in muscle cell contraction by regulating the flow of calcium ions between the external and internal cellular environment, and DMD may develop if the calcium-regulating mechanism is disrupted or destroyed. What sorts of observations and evidence support the model? Research has shown that dystrophin interacts with DAP and that if dystrophin is absent or greatly reduced, DAP is quickly degraded by enzymes. It is also known that DAP levels are very low or absent in DMD patients. New studies will tell whether this model is correct or if it will need modification.

What causes DMD? Despite rapid progress, the question remains open, and the model suggests that there may be more

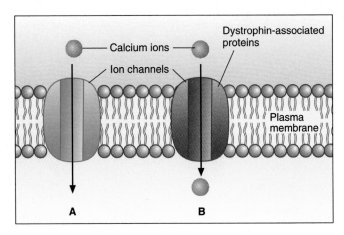

Figure 20.17 Dystrophin-associated proteins may form ion channels that allow calcium ions to diffuse into the cell (A). In some cases, ion channels have the ability to regulate the entry of ions into a cell (B).

than one answer. Studies indicate that 85 percent of DMD patients have mutations in the *dystrophin* gene. The other 15 percent may have undetected mutations in the *dystrophin* gene or in one or more of the genes encoding the six-protein DAP complex. Further research is required.

BEFORE YOU GO ON DMD is a muscle-wasting genetic disorder that usually affects males and commonly leads to death by young adulthood. DMD is associated with a mutant sex-linked gene that encodes a protein called dystrophin. Dystrophin is now hypothesized to interact with membrane-bound proteins that may regulate the movement of calcium ions into muscle cells. Mutations of genes encoding the membrane proteins may or may not also cause DMD.

PRENATAL DIAGNOSIS AND GENETIC COUNSELING

It is beyond the scope of this book to describe the methods used by sociologists and economists to measure the costs of genetic disorders in human populations. Genetic disorders and congenital malformations occur in 2 to 5 percent of live births. In developed countries, they account for up to 30 percent of children's admissions to hospitals and are a significant cause of childhood deaths. In the United States, it is estimated that over 1 million persons are hospitalized each year for hereditary or congenital disorders. Chronic disorders with a genetic cause are thought to occur in about 10 percent of the total adult population.

The costs of genetic disorders have two aspects. They place major financial strains on affected families and community health and social services. And because they primarily affect children, there is an emotional price that is incalculable. For these reasons, prospective parents at risk for a genetic disorder may often opt for **prenatal diagnosis,** a process used to determine the genetic status of the developing fetus. Genetic counselors explore various choices, including medical abortion, before such testing and assist in the evaluation of test

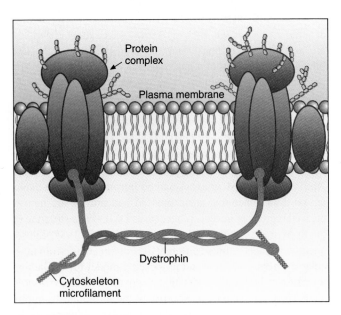

Figure 20.16 Dystrophin is a huge protein that is hypothesized to connect protein complexes embedded in muscle cell plasma membranes with cytoskeleton microfilaments inside the cell. According to a recently proposed model, Duchenne muscular dystrophy (DMD) results because of *dystrophin* gene mutations or mutations of genes encoding the membrane-bound proteins.

Question: *Why does DMD occur much more frequently in males than in females?*

results and any decision-making processes that follow. Parents unwilling to consider abortion may also choose to determine if the fetus has a genetic disorder. If the tests are positive, they can then become familiar with the care that will be required for the child.

Techniques of Prenatal Diagnosis

The field of prenatal diagnosis is growing rapidly, along with advances in molecular genetics. Many single-gene defects, such as Duchenne muscular dystrophy, and all chromosomal mutations can be identified in developing fetuses. In addition to classic genetic disorders, many common congenital malformations without known causes can be identified prenatally. Because these malformations frequently appear in certain families, they are hypothesized to involve multiple genes or various environmental factors (or combinations of both). Many common structural malformations are related to abnormal development of the neural tube, the structure that gives rise to the brain and spinal cord. These **neural tube defects** can result in severe abnormalities. Depending on the portion of the neural tube affected, effects range from a partial skull with no higher brain centers to abnormal spinal cord closure, a condition called **spina bifida,** which affects approximately one of every 500 children born in the United States. Neural tube defects are first indicated by the presence of abnormally high concentrations of a protein called alpha-fetoprotein in the amniotic fluid surrounding the fetus.

Visualization Techniques

Two methods are used routinely to visualize a developing embryo inside the womb. **Ultrasound scanning** employs high-frequency sound waves to create a *sonogram,* an image of the structural features of the fetus such as that seen in Figure 20.18. Ultrasound scanning can illustrate the orientation of the fetus

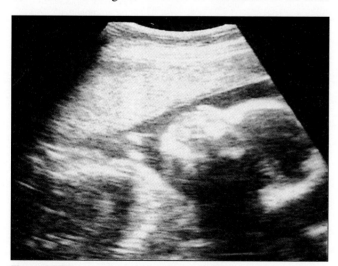

Figure 20.18 Ultrasound scanning is a relatively safe method for examining the structural features of the fetus before further tests are conducted. This sonogram shows the orientation of the fetus in the womb as well as specific developing structures.

and the location of the placenta prior to conducting further tests. It can also be used to confirm multiple fetuses and normal or abnormal brain development. Ultrasound scanning is generally considered safe for both the fetus and the mother.

Fetoscopy requires insertion of a *fetoscope,* a small instrument, into the amniotic sac through a small surgical opening in the mother's abdomen. Because it is an invasive surgical procedure, fetoscopy poses a significant risk of inducing a miscarriage, and its use is therefore limited. Its advantages over ultrasound scanning are that it can detect deformities of the limbs and can also be used to remove a fetal blood sample for evaluation.

Testing for Genetic Disorders: Amniocentesis and Chorionic Villus Sampling

Two principal methods are used in prenatal diagnoses for genetic disorders: amniocentesis and chorionic villus sampling (see Figure 20.19, page 390). **Amniocentesis** is a procedure for testing the fluid surrounding the fetus within the amniotic cavity. The amniotic fluid is produced by the fetus itself and contains various solutions, including fetal urine, and several kinds of fetal cells. A physician passes a long, thin needle through the mother's abdominal wall and into the amniotic cavity. A small amount of fluid is removed and tested to identify possible genetic and biochemical defects. The extracted fluid can be analyzed to detect enzyme abnormalities that suggest certain developmental defects or genetic disorders. Fetal cells in the fluid must first be cultured (grown in an artificial medium) and later analyzed for chromosomal abnormalities when enough cells have been produced.

Amniocentesis has revolutionized prenatal genetic counseling since 1980. It is usually performed during the fourteenth to sixteenth week of pregnancy and, when ultrasound is used to locate precisely the fetal head and the placenta, is considered to be reasonably safe for both the fetus and the mother. It is very accurate in detecting chromosomal mutations and about 50 genetic mutations expressed as inborn errors of metabolism. A disadvantage of amniocentesis is that a final analysis of the cells cannot be made until 17 to 20 weeks into the pregnancy.

Like amniocentesis, **chorionic villus sampling (CVS)** is an invasive procedure. Chorionic villi are hairlike projections of the membrane that surrounds the embryo early in pregnancy. This membrane is a rich source of fetal cells until the tenth week of pregnancy, when it is replaced by the placenta. CVS involves the removal of a plug of tissue from the villi with a small tube inserted through the cervix. Since there is an abundance of fetal cells in the collected tissue, cells do not need to be cultured, as in amniocentesis, and analysis can begin immediately.

CVS offers certain other advantages over amniocentesis. Most important is that the sample is taken six to nine weeks into the pregnancy, and the results of chromosomal analyses and biochemical tests are available within days. A major drawback of CVS, however, is that it is not suitable for determining neural tube defects, which cannot be detected until later in fetal development. Also, CVS has a 1 to 2 percent risk of inducing a spontaneous abortion—two to four times the risk of amniocentesis.

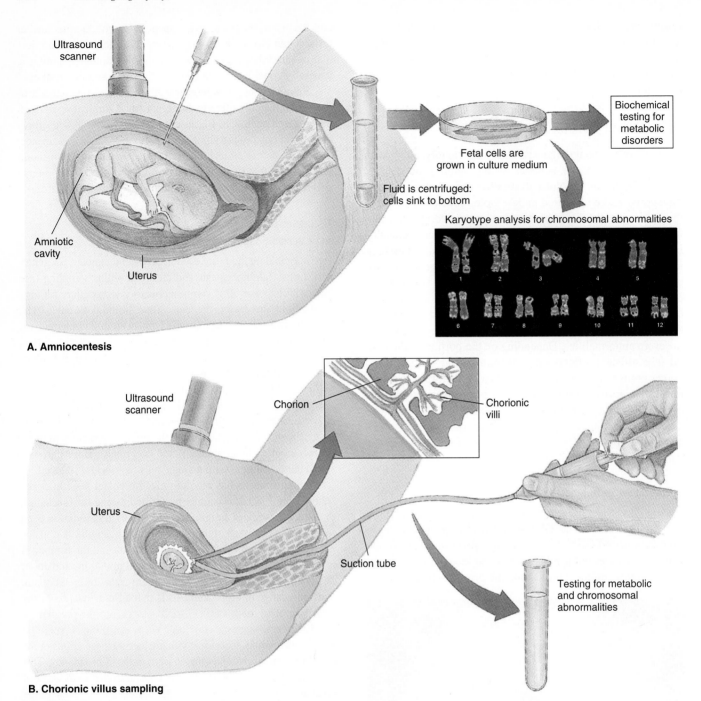

A. Amniocentesis

B. Chorionic villus sampling

Figure 20.19   Amniocentesis (A) and chorionic villus sampling (B) are procedures used in prenatal tests to determine the presence of chromosomal mutations or certain genetic disorders in a developing fetus.

BEFORE YOU GO ON Prenatal diagnosis involves various techniques that are used to examine a developing fetus and to analyze the content of its cells. Such procedures may enable a doctor to determine whether or not the fetus has or will develop a specific genetic disorder. Many, but not all, genetic disorders can be detected. Prenatal diagnosis and genetic counseling are usually recommended for parents who may be at risk for a specific genetic disorder.

Genetic Counseling

The American Society of Human Genetics offers the following definition of genetic counseling:

Genetic counseling is a communication process concerning the risks of occurrence of a genetic disorder in a family. It involves an attempt to help the person or family comprehend the medical facts, appreciate the hereditary nature and recurrence risks in specific relatives, understand the options for dealing with the risk, choose the most appropriate course of action, and make the best possible judgments.

Genetic counselors often use family pedigree analysis as a starting point to help people understand the nature and cause of genetic disorders that may be inherited in their family. It may also be possible to establish whether or not a person is heterozygous for a recessive mutation that causes a genetic disorder.

For people planning to have children, genetic counselors may be able to state the precise risk of having a child with genetic disorder. For example, if both parents are heterozygous for a recessive disorder, the chance of their offspring having the disorder would be 25 percent. In general, prenatal testing is recommended if one or both parents are at risk for a genetic disorder. The following criteria are used in determining risk:

1. The prospective mother is 35 or older or the father 50 or older, as increased age is correlated positively with chromosomal mutations.

2. The prospective mother is known or suspected to be a carrier of an X-linked genetic disorder.

3. The parents have previously had a child with a chromosomal mutation or a neural tube defect.

4. Both prospective parents are known to be carriers of an autosomal recessive allele that causes a genetic disorder in homozygotes.

5. One parent is known to carry a defective autosomal dominant allele.

Once the results of prenatal tests are known, genetic counselors review the tests, explain their meaning, and, if the tests reveal that the fetus will have a genetic disorder, discuss the consequences of having an affected child. Ultimately, it is left to the parents to make an informed decision about whether or not to continue the pregnancy. Science, scientists, and legitimate counselors do not abrogate the personal responsibilities of individuals in making their own reproductive decisions.

SCIENTIFIC PROCESS AND HUMAN GENETIC DISORDERS

As described in Chapter 18, Garrod's studies, conducted early in the twentieth century, provided the first clues about the underlying causes of human genetic disorders. Pauling's hypothesis that sickle-cell anemia was caused by a molecular structural change that disrupted the normal architecture of red blood cells represented a major conceptual breakthrough. Decades later, Pauling referred to this as a "pretty nice idea." Indeed, his "nice idea" opened the door to the future and has served as a guide for studies of human genetic disorders for the past 50 years.

The four genetic disorders examined in this chapter—sickle-cell anemia, Huntington disease, familial hypercholesterolemia, and Duchenne muscular dystrophy—represent different levels of understanding. Sickle-cell anemia seems to hold no secrets beyond its treatment or cure. Likewise, the molecular defects that cause FH are well known, although more mutant receptor alleles may be found and other genes may be implicated. The DMD model will continue to evolve as more becomes known about dystrophin and the dystrophin-associated proteins. Finally, after a long search, the *Huntingtin* gene has been found, but nothing is yet known about the biochemistry of HD.

These four case studies also illustrate the sequence of questions that guides contemporary investigations of human genetic disorders. The first step in this sequence typically involves attempts to identify the location of the gene causing the disorder. On which chromosome is the gene found? Exactly where on the chromosome is it located? What is the DNA sequence of the normal allele? The second step consists of trying diverse approaches that might lead to the identity of the protein encoded by the gene. What is the encoded protein? What is the amino acid sequence of the protein itself? Next, the function of the protein encoded by the normal allele is pursued if it is not already known, along with the effects of the modified or missing protein that results from the mutant allele. What is the role of the normal protein? Why does disease occur if the protein is modified or missing? The final step of the process leads to questions that have only rarely been answered. What can be done to treat the disorder? Can the disorder be prevented or cured? Ultimately, molecular biologists strive for answers to all of these questions for all genetic disorders. This sequence of questions generally reflects the degree of difficulty in finding answers. And only when scientists have the answers will it be possible to counteract the effect of gene mutations in the human population.

SUMMARY

1. In the age of the new genetics, much research on human genetics has been directed at determining the specific nature of inherited disorders. Genetic disorders are caused by gene mutations and chromosome mutations.

2. Gene mutations result in disorders that are often classified as dominant, codominant or recessive, autosomal, or sex-linked. Genetic disorders associated with recessive gene mutations are usually due to defective enzymes or hormones. Dominant disorders typically affect a structural protein. More than 3,000 genetic disorders are known to be caused by single-gene mutations.

3. Chromosomal mutations consist of structural changes in chromosomes or alterations of the normal number of chromosomes. Chromosomal mutations usually result in such serious abnormalities that development is arrested early in pregnancy and spontaneous abortion occurs.

4. Human pedigrees are used to determine the pattern of an inherited disorder. Evaluations of pedigrees may enable inves-

tigators to decide if a genetic disorder is caused by a recessive, dominant, or X-linked gene.

5. Sickle-cell anemia is caused by a codominant mutant allele of a gene that encodes polypeptide chains found in hemoglobin molecules carried by red blood cells. Consequently, hemoglobin structural abnormalities cause red blood cells to lose their normal shape, which result in severe anemia. Huntington disease is caused by an abnormal number of CAG base repeats in dominant, mutant *Huntingtin* alleles, but nothing is known about the molecular effects of the mutation. Familial hypercholesterolemia occurs because of flawed membrane receptors required for normal cholesterol processing. At least three codominant mutant alleles are involved in FH. Duchenne muscular dystrophy is an X-linked recessive disorder that occurs because of mutations to the *dystrophin* gene or to genes encoding membrane-bound proteins that may interact with dystrophin in regulating calcium in muscle cells.

6. Parents at risk for a genetic disorder have access to various prenatal diagnostic procedures that will enable them to learn whether or not the developing fetus has a specific disorder. These procedures include visualization techniques, amniocentesis, and chorionic villus sampling. All have advantages and disadvantages that must be considered before diagnosis proceeds. Genetic counselors are trained to interpret the results and inform parents of available options. Final decisions, such as continuation or termination of pregnancy, are left to the individuals at risk after they have considered all available options.

7. Describe the cause and effects of familial hypercholesterolemia. What is the molecular basis of this disorder? How can its harmful effects be reduced?

8. What is the genetic nature of Duchenne muscular dystrophy? Explain the model describing the mechanism responsible for its effects.

9. Describe the techniques used in prenatal diagnoses for detecting genetic disorders. What are the advantages and disadvantages of each?

10. What is genetic counseling?

ESSAY AND DISCUSSION QUESTIONS

1. Why did the discovery of the affected Venezuelan population constitute a breakthrough for researchers studying Huntington's disease?

2. What recommendations would you make to agencies that fund research on human genetic disorders regarding emphasis in the decade ahead? Which disorders should receive the most attention? Why? What are some criteria that might be used in making this decision?

3. Should insurance companies pay for prenatal diagnoses and genetic counseling? Why?

WORKING VOCABULARY

amniocentesis (p. 389)
chorionic villus sampling (p. 389)
Duchenne muscular dystrophy (p. 387)
familial hypercholesterolemia (p. 385)
gene marker (p. 384)

genetic disorder (p. 373)
Huntington's disease (p. 383)
ion channel (p. 388)
low-density lipoproteins (p. 385)
pedigree (p. 376)
receptor (p. 385)
sickle-cell anemia (p. 379)
syndrome (p. 375)

REVIEW QUESTIONS

1. What are genetic disorders? What is their importance in the human population?

2. What types of gene and chromosome mutations cause human genetic disorders?

3. Describe the pattern of inheritance that is typical for different classes of gene mutations.

4. What is the probable type of gene mutation that causes the genetic disorder present in the family pedigrees below?

5. What is the genetic cause of sickle-cell anemia? What is the molecular basis of this disorder? What accounts for its effects?

6. What is the genetic cause of Huntington's disease? What are the biological effects of this disorder?

REFERENCES AND RECOMMENDED READING

Cavalli-Sforza, L. L. 1994. *The History and Geography of Human Genes.* Princeton, N.J.: Princeton University Press.

Cooper, D. N., and M. Krawczak. 1993. *Human Gene Mutation.* McLean, Va.: Books International.

Darden, L. 1991. *Theory Change in Science: Strategies from Mendelian Genetics.* New York: Oxford University Press.

Eaton, W. A., and J. Hofrichter. 1995. The biophysics of sickle cell hydroxyurea therapy. *Science,* 268:1142–1143.

Folstein, S. E. 1989. *Huntington Disease: A Disorder of Families.* Baltimore: Johns Hopkins University Press.

Huntington's Disease Collaborative Research Group. 1993. A novel gene containing a trinucleotide repeat that is expanded and unstable in Huntington's disease chromosomes. *Cell,* 72: 971–983.

Lander, E. S., and N. J. Schork. 1994. Genetic dissection of complex traits. *Science,* 265: 2037–2048.

McKusick, V. A. 1994. *Mendelian Inheritance in Man: Catalogs of Autosomal Dominant, Autosomal Recessive, and X-linked Phenotypes.* 11th ed. Baltimore: Johns Hopkins University Press.

Myant, N. B. 1990. *Cholesterol Metabolism, LDL, and the LDL Receptor.* San Diego: Academic Press.

Patterson, D. 1987. The causes of Down syndrome. *Scientific American,* 259: 52–60.

Rasko, I. 1995. *Genes in Medicine.* London: Chapman & Hall.

Serjeant, G. R. 1992. *Sickle Cell Disease.* 2d ed. New York: Oxford University Press.

Weiss, K. M. 1993. *Genetic Variation and Human Disease: Principles and Evolutionary Approaches.* Cambridge: Cambridge University Press.

Wexler, N. 1992. *Mama Can't Remember Anymore: How to Manage the Care of Aging Parents.* Thousand Oaks, Calif.: Wein & Wein.

ANSWERS TO FIGURE QUESTIONS

Figure 20.1 The synthesis of a normal protein is dependent on the expression of an allele that has a specific DNA base sequence. Any change (mutation) in the normal allele of a gene may result in production of an abnormal protein because the "recipe" (base sequence) encoding the normal protein has been altered.

Figure 20.3 Generally because the presence of more or less than 46 chromosomes in cells of a human embryo seriously disrupts normal development, causing death of the embryo and spontaneous abortion.

Figure 20.6 The β chains are altered.

Figure 20.10 In such a large sample of affected individuals, pedigree analyses can be more detailed and revealing and more genes are available for analysis.

Figure 20.11 An abnormally large number of CAG repeats in a mutant *Huntingtin* allele.

Figure 20.12 It binds to LDL receptors in the plasma membrane.

Figure 20.13 Abnormal receptors prevent LDL particles from being brought into the cell.

Figure 20.16 Males have only one dystrophin allele; if it is mutated, DMD will develop. By contrast, females have two dystrophin alleles; if one is normal, it will be expressed (due to X chromosome inactivation and the selective inactivation of abnormal alleles).

Biotechnology

Chapter Outline

Reading Questions

1. What is gene cloning? How are genes cloned?
2. What events helped establish modern biotechnology?

3. How has genetic engineering been applied to plants and animals?
4. What factors must be considered in releasing genetically engineered organisms into the environment?

ncient records reveal that humans have long used microorganisms, most notably bacteria and yeasts, to make their diet more interesting. For example, Egyptian tomb paintings dating from 2000 B.C. show clearly that ancient people used a brewing process requiring microorganisms to produce wine and beer (see Figure 21.1). These people may have been the first to use a "biological technology"—the use of an organism to create something desired by humans. To this day, humans have continued to use microorganisms in brewing and baking. The technology used in these activities is **fermentation,** a biological process that occurs when microorganisms synthesize worthwhile products—alcohol and other substances—from

nutrients and raw materials in the absence of oxygen. Fermentation arts are also used in making olives, pickles, cheese, soy sauce, and other substances that increase dietary diversity.

ZYMOTECHNOLOGY AND THE RISE OF BIOTECHNOLOGY

By the nineteenth century, fermentation systems were used by large alcohol-producing industries throughout the world. Starting in the 1870s, industrial brewing technologies became

Figure 21.1 In this painting from an Egyptian tomb, workers at the right are picking grapes in an arbor, while workers at left are crushing the grapes with their feet. The worker in the middle of the panel collects the pressed juice. After fermentation, the wine was placed in large earthenware jars marked with the vintage and place of origin.

integrated with different scientific disciplines to form a dynamic field called **zymotechnology** (from the Greek root *zyme* meaning "leaven"). The general goal of zymotechnology was to improve all types of industrial fermentation, from the manufacture of lactic acid to curing leather, through engineering and applied chemistry, bacteriology, and botany. The zest for zymotechnology at the end of the nineteenth century is reflected in the following statement from the head of a Berlin brewing institute in 1884: "With the sword of science and the armor of practice German beer will encircle the world."

In the twentieth century, industrial fermentation systems continued to improve and enlarge. However, new and broader ideas about biological production systems emerged, and zymotechnology gave way to *biotechnology,* a concept that has had many different meanings, depending on social, cultural, and political context. The term *biotechnologie* was first used in 1917 by Karl Ereky, the owner of a huge pig-fattening station in Hungary, to describe "all such work by which products are produced from raw materials with the aid of living organisms." Ereky hoped that *biotechnologie* would help transform agriculture in Hungary. In the 1920s and 1930s, European philosophers envisioned a "biotethic" society in which harmless biological production systems replaced more unpleasant, polluting industrial factories.

After World War II, various industries used the term *biotechnology* to describe the manufacture of brewing products, pharmaceutical products (such as drugs and antibiotics), and agricultural products and the production of chemicals by microorganisms. Biotechnology was now described by one scientist to "embrace all aspects of the exploitation and control of biological systems and their activities." By the 1960s, biotechnology programs existed in major universities in the United States; the biotechnology concept was embraced by biologists, engineers, physiologists, and microbiologists; and there was a

journal of biotechnology and bioengineering. Idealists felt that biotechnology could be the solution to world hunger and energy demands and that it would not deplete nonrenewable resources or cause environmental pollution.

But the 1960s also saw the emergence of a powerful new idea: it might soon be possible to alter the genetic properties of biological systems, which could lead to enormous increases in biological production systems. What was the basis of this idea? What would be its effect on the concept and definition of biotechnology?

THE DNA REVOLUTION

The period following Watson and Crick's historic description of the DNA molecular structure in 1953 has often been described as the "age of the DNA revolution." A constant stream of new discoveries flowed from research laboratories throughout the 1950s, 1960s, and 1970s, providing information about the mechanism of DNA replication (1958), the complete genetic code used to specify proteins synthesized by cells (1964), and various enzymes that participate in reactions involving DNA molecules (1970s).

Molecular biologists also discovered unconventional processes for transferring genetic information through two different biological entities, *plasmids* (1965) and *bacteriophages (phages)* (1960s and 1970s), which we discuss in detail below and are defined in Table 21.1 (see page 396), along with other terms used in this chapter. Neither are considered to be true organisms in a biological sense because they cannot reproduce independently outside of the host cells they normally inhabit. Nevertheless, plasmids and bacteriophages are capable of transmitting genetic information from one species to another. A spectacular accomplishment occurred in 1973, when DNA removed from different bacterial cells was spliced together—the first successful experiments in which DNA from two different species was "recombined."

The molecular tools identified and developed during the two decades following 1953 gave rise to novel techniques collectively referred to as **genetic engineering** and to products known as **recombinant DNA** (see Table 21.1). It had become clear that DNA constitutes a complete set of instructions that, once expressed, regulate the development, growth, functions, and aging of organisms throughout their lives. The common feature that unites all life on Earth is the genetic code, which is virtually the same for all organisms, from bacteria to humans. All modern genetic technologies are derived from the reality that genes are interchangeable among species and that the universality of the genetic code allows for a gene obtained from one species to be expressed in a different species.

BEFORE YOU GO ON Humans have long used microorganisms to produce useful products. Successful fermentation technologies used for over 4,000 years led to the establishment of zymotechnology as a scientific enterprise in the nineteenth century. In the twentieth century, zymotechnology gave way to a broader field known as biotechnology. The DNA revolution that occurred after the description of DNA's structure in 1953 gave rise to new molecular tools and processes that could be used to modify organisms genetically.

Gene Cloning

How can genes be manipulated, transferred from one organism to another, and then replicated or expressed? **Gene cloning** is the central process of genetic engineering that scientists use to create large numbers of a specific gene of interest—for exam-

---

Table 21.1 Terms Used in Genetic Technology

Bacteria One-celled organisms (prokaryotes lacking a nucleus) capable of carrying on many chemical reactions (for example, protein synthesis) that are similar to those of higher organisms. *Strains* of bacteria are a group of organisms within a species that is characterized by some particular quality (for example, the rough and smooth strains of *Diplococcus pneumoniae* discussed in Chapter 17).

Bacteriophage (phage) A virus that infects bacteria.

Biotechnology A broad field that employs an industrial technology based on the biological synthesis of important chemical compounds, especially proteins (for example, insulin), genetic engineering of plants and animals, and other, related technologies.

Clone A large number of *cells* or *molecules* identical to an ancestral cell or molecule. Also, a specific gene sequence, isolated and replicated.

Cloning vectors Small *plasmid, phage,* or animal *virus* DNA molecules used to transfer a DNA fragment from a test tube into a living cell. Cloning vectors are capable of multiplying inside of living cells.

Complementary DNA (cDNA) DNA copied from a messenger RNA (mRNA) molecule using an enzyme called *reverse transcriptase*. The DNA sequence is thus *complementary* to that of the mRNA.

DNA insert A foreign DNA segment inserted into vector DNA.

DNA ligase An enzyme that connects two separate DNA molecules end to end.

Escherichia coli (E. coli) The bacterium most widely used in recombinant DNA research; commonly found in the digestive tracts of mammals

Expression vector A *plasmid* designed to permit expression (transcription) of a foreign gene inside a cell.

Genetic code The sequence of mRNA base triplets that specify the exact series of animo acids for a protein.

Genetic engineering The manipulation of genetic information (DNA or RNA) of an organism to alter the characteristics of that organism. The basic process consists of inserting foreign DNA carrying instructions for a valuable enzyme, hormone, or other protein in the DNA of some other organism so that the host organism makes the desired product at the same time as it makes its own proteins.

Genomic DNA All DNA sequences of an organism.

Host cell A cell (usually a bacterium) in which a *cloning vector* can be propagated.

λ (lambda) phage A particular *bacteriophage* used extensively in gene cloning; also referred to as *bacteriophage* λ.

Plasmids Small, circular DNA molecules found inside bacterial cells. Plasmids reproduce every time the bacterial cell reproduces.

Recombinant DNA A DNA molecule containing two or more regions of DNA from two different genes or species (for example, a fragment of human DNA spliced into plasmid DNA).

Restriction enzymes Enzymes that cut DNA molecules at specific nucleotide sequences.

Reverse transcriptase An enzyme purified from retroviruses that makes DNA from RNA.

Transformation The insertion of any foreign gene into any organism.

Vector Any *plasmid, phage,* or *DNA segment* that carries a DNA insert.

ple, one that encodes a "desirable" protein, such as human insulin. Genes from any organism, prokaryotic or eukaryotic, may now be cloned. Figure 21.2 illustrates a general procedure for gene cloning.

Obtaining Genes

The first step in the gene-cloning procedure is to obtain a desired gene for cloning, which is usually done using one of two methods. In one method, cells are subjected to a detergent that dissolves plasma and nuclear membranes, releasing the DNA. This liberated **genomic DNA** is then exposed to one or more **restriction enzymes** (see Table 21.1), obtained from bacteria, which cut DNA into small fragments in a precise and predictable way. The normal function of restriction enzymes in bacteria is to destroy ("restrict") foreign DNA that attempts to enter the cell. Through this mechanism, bacteria are protected against the insertion of DNA by bacteriophages that may infect

and kill them. Approximately 200 different restriction enzymes are now used by genetics researchers.

Each restriction enzyme recognizes a specific short nucleotide sequence, binds with that sequence, and then cleaves both strands of the double helix at or near the binding site. For example, as shown in Figure 21.3 (see page 398), *Eco*RI, a restriction enzyme from the bacterium *Escherichia coli,* specifically recognizes the sequence GAATTC and cleaves it at a restriction site between the guanine and adenine, G↓AATTC, leaving two DNA fragments with complementary G and AATTC ends. These single-strand complementary ends, known as "sticky ends," can bind easily with complementary DNA segments obtained from another organism, using the same restriction enzyme. The two end nucleotides of the same strand can be bonded together by the enzyme **DNA ligase** (see Table 21.1).

The other method for obtaining a gene of interest is to create a **complementary DNA (cDNA)** strand from a messenger RNA (mRNA) molecule. This procedure is used when specif-

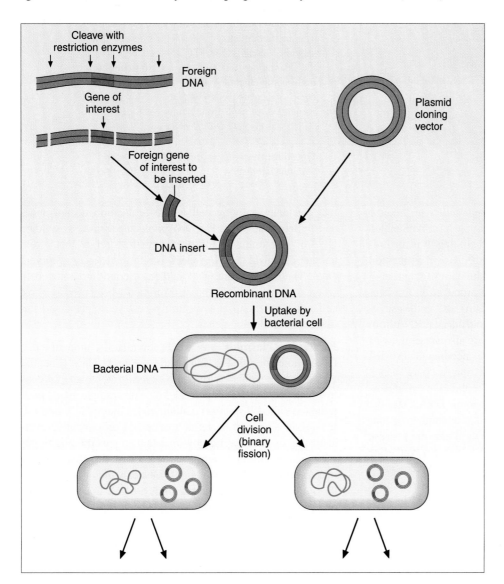

Figure 21.2 New genetic technologies are now used to make unlimited numbers of desired genes in a process called *gene cloning.* A gene of interest, most commonly one that encodes an important protein, can be identified and joined with the DNA of a cloning vector. The resulting product—a combination of DNA from two different species or organisms—is known as *recombinant DNA.* The recombinant DNA is introduced into a bacterial cell, and after a day or two of cell division, millions of bacteria and an equal or greater number of the gene of interest have been produced.

Question: *Why are cloning vectors required for gene cloning?*

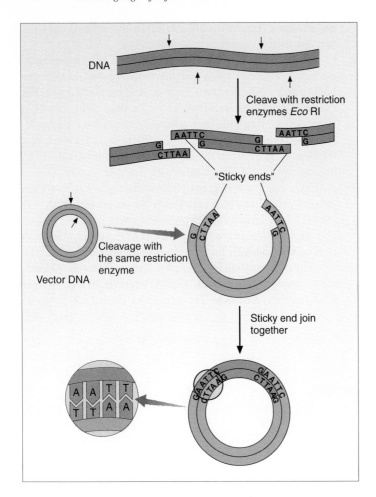

Figure 21.3 Restriction enzymes cut DNA at sites where specific nucleotide sequences occur. The restriction enzyme *Eco*RI recognizes a sequence reading GAATTC and cuts between the guanine and adenine. By making cuts in both complementary DNA strands, "sticky ends" are created that can join with complementary sticky ends from cloning vector DNA, which was also cut with *Eco*RI.

Question: *What allows sticky ends to combine with one another?*

ic mRNA molecules can be isolated from cells that produce high levels of a certain protein. For example, a protein called *factor VIII* is necessary for proper blood clotting. It is the protein that most hemophiliacs are missing because of a mutation in the relevant structural gene. From several clever studies, high levels of this protein were found to be produced in the liver along with the mRNA molecules directing its synthesis. By separating the factor VIII protein and the affiliated mRNA from other proteins and other mRNAs that were also present, it was possible to determine the DNA sequence encoding factor VIII. The major advantage of this method is that a "pure" DNA sequence can be isolated that does not contain the extraneous DNA sequences (introns) that occur in genomic DNA. Once the mRNA is isolated and purified, it is used as a template to construct a cDNA strand using the unique viral enzyme **reverse transcriptase** (see Table 21.1, page 396), which can synthesize DNA from RNA (see Figure 21.4). When the complementary DNA strand has been formed, the mRNA strand is removed, and a second DNA strand, which is complementary to the first, is added, resulting in a double-stranded cDNA molecule.

Cloning Vectors

Because isolated genes cannot replicate themselves, a gene to be cloned must be inserted into the DNA of a suitable **cloning vector** (see Table 21.1, page 396), usually a plasmid or a virus,

that can enter a living cell where replication occurs under appropriate conditions. The DNA molecule formed by splicing DNA of the cloning vector together with "foreign" DNA from another organism (a **DNA insert**) is referred to as *recombinant DNA*. The recombinant DNA is then inserted into a host cell, where it is replicated. *E. coli* has commonly been used as a host cell, although scientists now routinely use yeast, plant, and animal cells for cloning genes.

Plasmids **Plasmids** are small, circular, double-stranded DNA molecules that reside inside bacteria. They generally encode proteins that are of some benefit to the bacterial cells in which they live (see Figure 21.5A). For example, many plasmids (so-called R *factors*) contain genes that confer antibiotic resistance on their bacterial host cells. Consequently, several strains of bacteria are now resistant to specific antibiotics such as ampicillin and tetracycline because they harbor plasmids with genes that encode enzymes capable of neutralizing the antibiotic. The fact that plasmids are able to confer antibiotic resistance to a bacterial host has been exploited in identifying bacteria that contain such plasmids. Plasmids replicate independently of the bacterial chromosome, although they are dependent on the host cell's DNA-synthesizing machinery.

Figure 21.5B describes the insertion of DNA into a plasmid containing an antibiotic resistance gene. Plasmid DNA is cut by the same restriction enzyme used in preparing the DNA

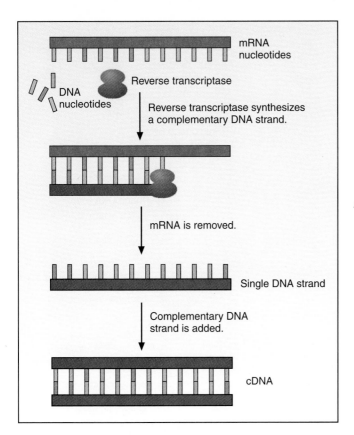

Figure 21.4 If the mRNA encoding a gene of interest can be isolated, a DNA strand that is complementary to this mRNA can be synthesized by the enzyme reverse transcriptase. The resulting DNA strand is then used as a template for adding the complementary DNA strand to create a double-stranded cDNA molecule, which can then be inserted into a vector.

Question: *Under what circumstances might cDNA be used in gene cloning instead of genomic DNA?*

insert. Plasmid and insert DNAs are then mixed together in the presence of DNA ligase, which binds some of them together to create a recombinant DNA molecule. The manipulations described so far are carried out in vitro (in a test tube). For the recombinant DNA molecules to be replicated, they must be placed in the presence of a suitable bacterial strain. Only a very small percentage of the bacterial cells undergo **transformation,** meaning that they take up plasmids containing a DNA insert. Bacterial cells that contain plasmid DNA must then be distinguished from those that do not. How might this be accomplished? Bacterial cells are grown in a medium containing the antibiotic whose resistance is encoded by the plasmid. Only transformed bacterial cells containing the plasmids survive; untransformed bacterial cells perish.

Bacteriophages **Bacteriophages,** commonly known as **phages,** are viruses that infect bacteria (see Figure 21.6A, page 400). They are more complicated than plasmids because they contain genes encoding proteins used in their own structures and for their replication. Like plasmids, however, they depend

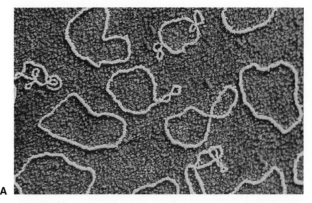

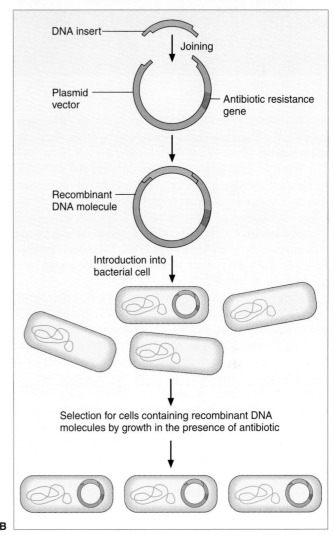

Figure 21.5 (A) Plasmids are small pieces of circular DNA that live in bacterial cells and replicate independently. (B) To clone a gene of interest, a DNA insert is joined with a plasmid cloning vector DNA to create a recombinant DNA molecule, which is then introduced into a bacterial cell. Both bacterial cells and recombinant DNA molecules are cloned as a result of cell division. Bacterial cells containing the plasmid vector–recombinant DNA system can be determined by exposing them to the antibiotic for which the plasmid conveys resistance. Bacteria with the plasmid vector will survive in the presence of the antibiotic, but those without the plasmid will die.

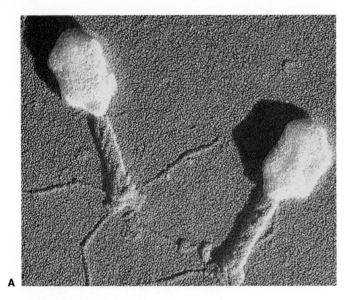

A

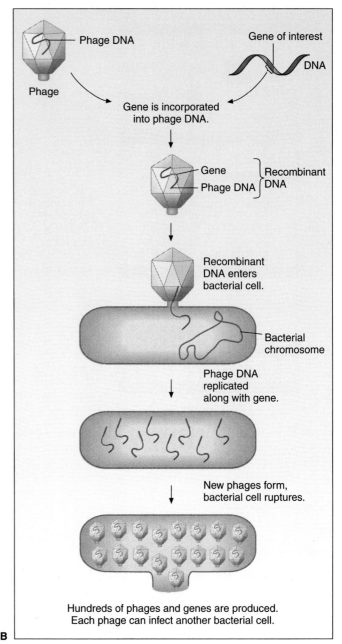

B

on a bacterial host cell to actually carry out their DNA replication. Each phage type can replicate only in a specific bacterial species or strain.

There are three major advantages in using phages instead of plasmids as cloning vectors: (1) phages are easier to introduce into bacteria, (2) larger foreign genes or gene combinations can be inserted into the phage DNA to create stable recombinant DNA molecules, and (3) phages have a greater efficiency for incorporating foreign DNA. The success for plasmids is usually 10 percent or less, whereas for phages it is greater than 50 percent.

Many phages have been used in gene-cloning operations, but scientists have most frequently used the λ (lambda) phage (bacteriophage λ) that infects *E. coli*. The procedures for incorporating foreign DNA into the λ phage genome follow the same lines as for plasmids and are shown in Figure 21.6B. The viral DNA is cut using specific restriction enzymes; then the DNA insert is introduced and the two DNAs are joined by DNA ligase to form a recombinant DNA molecule. The λ phage cloning vector then infects a bacterial cell, in which the recombinant DNA replicates rapidly to produce new phages; this ultimately kills the bacterial host cell but results in high yields of the cloned gene.

The Role of Bacteria

Recall that bacteria reproduce very rapidly by binary fission—dividing into two cells after a brief growth period (see Figure 21.7). They live in practically every environment, but the species of greatest use to molecular biologists are those that can be easily cultivated in the laboratory by placing them on a suitable culture medium contained in a small dish. Under optimal conditions, a single bacterial cell can give rise to millions of identical cells (clones) within a single day. When a vector containing recombinant DNA is inserted successfully into a bacterial cell, it is replicated, and hence the DNA insert is cloned.

Figure 21.6 (A) Bacteriophages, shown in this scanning electron micrograph, are viruses that infect bacteria. The "head" of the phage contains DNA used to direct the synthesis of proteins and nucleic acids necessary to complete its life cycle. The "tail" is used for attachment to the bacteria and for inserting the viral DNA. (B) Phages can be used as cloning vectors to introduce recombinant DNA into bacterial cells. Once inside a cell, the recombinant DNA may begin replicating, and new phages, each containing the gene of interest, are formed. The bacterial cellular machinery synthesizes the vector system proteins and DNA, but the bacterium is destroyed when the phages are released.

Question: *What are the advantages of using phages instead of plasmids?*

E. coli and some other bacteria are used to perform two functions in biotechnology. First, they are used to replicate unlimited numbers of genes. Second, pure cultures, in which

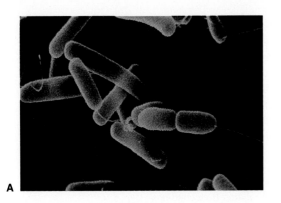

A   **B**

Figure 21.7 (A) Bacteria are single-celled prokaryotes that inhabit all types of environments. *E. coli,* shown here, is widely used in recombinant DNA technology for synthesizing foreign genes and proteins of interest. (B) One advantage of using bacteria in scientific studies is that enormous numbers can be grown quickly in culture dishes in the laboratory. Each spot in the culture dish represents a colony consisting of millions of bacteria that originated from a single cell.

every bacterium contains a gene encoding a desired protein, can be created and put to work manufacturing huge quantities of that protein (see Figure 21.8). For example, a bacterium containing a human insulin gene will synthesize human insulin. Plasmids that contain specific genes to direct protein synthesis are called **expression vectors** (see Table 21.1 on page 396).

BEFORE YOU GO ON Recombinant DNA is created when a gene from one species is combined with DNA from a different species. Genes to be cloned are combined with plasmid or phage DNA and introduced into a bacterial cell. In replicating the plasmid or phage, host bacteria also replicate the foreign gene carried by these cloning vectors, thus producing large numbers of the desired gene. Such genes may then be expressed in bacteria to produce important proteins.

MODERN BIOTECHNOLOGY

Today, **biotechnology** can be defined as a broad field that employs an industrial technology based on the biological synthesis of important chemical compounds, especially proteins, genetic engineering of plants and animals, and other, related technologies. Note that this definition accommodates genetic technologies. However, biotechnology is still defined differently by various countries, scientific disciplines, and companies.

As you have learned, the discovery of molecular tools, enzymes, and biological processes associated with bacterial genetic engineering and recombinant DNA spanned parts of three decades. By the 1970s, scientists understood that organisms could be genetically altered and endowed with new capabilities. Bacteria could be made to produce foreign proteins,

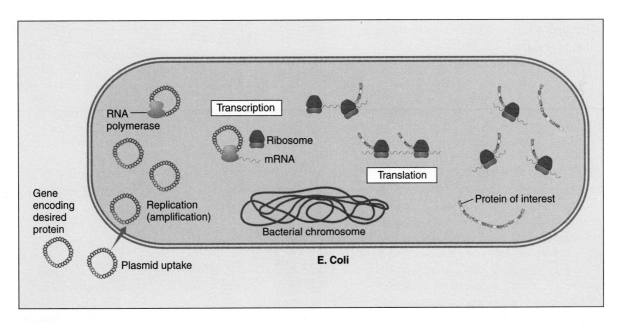

Figure 21.8 Bacteria such as *E. coli* are used as "protein-producing factories" in recombinant DNA technology. This schematic drawing shows the basic components of the system used for producing desired proteins encoded by a gene of interest that was obtained through gene cloning. The bacterium transcribes the gene carried by the plasmid vector, producing mRNA that is then translated on bacterial ribosomes. Under ideal conditions, the resulting foreign protein may account for 10 to 40 percent of the total protein synthesized by the bacterium.

Question: *Why can bacteria produce* human *proteins?*

even human proteins, on demand; this outcome of the DNA revolution had obvious significance for the old art of fermentation. Genetically modified plants and animals were envisioned that could have major human and commercial impacts. It seemed as though a new biological revolution was approaching, and the flame of biotechnology began to flare brightly.

Biotechnology Policies in the United States

Nevertheless, through the 1960s and into the 1970s, many scientists and politicians, reflecting public concern over creating new life forms, called for caution. The phrase "genetic engineering" had often been related to the idea of *human* genetic engineering; in this context, it brought back memories of the ill-fated eugenics movement (discussed in Chapter 19). There was also concern about releasing genetically modified organisms into the environment. Consequently, in February 1975, a group of distinguished molecular biologists met at Asilomar, California, to discuss the implications of genetic engineering. One important result of their discussions was the suspension of further studies on genetic engineering for 16 months, until government guidelines could be developed for regulating research. Other outcomes of the Asilomar meeting reflected concern for the influence of science on society. Scientists no longer casually applied the term *genetic engineering* to humans because there were no intentions at that time to do research involving humans. Also, benefits became a principal focus in public communications on genetic technologies. In June 1976, the U.S. National Institutes of Health (NIH) issued guidelines on good research practice (to be discussed shortly), and the 16-month suspension ended. Subsequently, concerns dissipated as benefits came into sharper focus, no harmful results occurred in any new experiments, and the public began to view biological technologies in the same positive light as the evolving electronic (computer) technologies.

The promise of using recombinant DNA and genetic engineering technologies to produce medicines, human proteins, and agricultural products soon led to the formation of new biotechnology companies. The first biotechnology company in the United States, Genentech, was founded in 1977, and there are now hundreds of such businesses throughout the world.

The research guidelines issued by the NIH in June 1976 defined the risks of certain kinds of experiments, described types of physical facilities required for various experiments, and listed experiments that were not to be performed. Regulatory policies have continued to evolve over the past 20 years to keep pace with the explosive progress in recombinant DNA technology. Whereas early concerns centered on human safety in using genetically engineered microorganisms in the laboratory, today the major issues are protection of the environment and human health.

In June 1986, the United States government published its *Coordinated Framework for the Registration of Biotechnology*, which assigned responsibilities for reviewing genetically engineered products and microorganisms to specific federal agencies. In most cases, the statutory authority comes from other, related acts (for example, the Environmental Protection Agency regulates under the Federal Insecticide, Fungicide and Rodenticide Act; the Toxic Substances Control Act; and a few other antipollution statutes). This has often led to confusion, controversy, and legal actions. Nevertheless, there is presently no strong movement for creating new legislation that specifically covers the products of biotechnology. The responsibilities of the key agencies are outlined in Table 21.2.

Table 21.2 Federal Agencies and Their Responsibilities in Regulating Biotechnology

Animal and Plant Health Inspection Service (APHIS) Part of the Department of Agriculture that oversees broad categories of genetically engineered plants and animals and microbes altered for agricultural purposes.

Environmental Protection Agency (EPA) The lead agency for monitoring all microbial pesticides and conducting required reviews of genetically engineered microorganisms prior to release into the environment. The primary criterion for approving a pesticide is that it will not cause "unreasonable adverse effects on human health or the environment." The data requirements for satisfying this criterion are detailed and complex.

Food and Drug Administration (FDA) Agency that has authority over human and animal drugs—"biologics," including any "virus, serum, toxin, antitoxin, vaccine, blood, blood component or derivative, allergenic product or analogous product . . . applicable to the prevention, treatment, or cure of diseases or injuries," food or color additives, and medical devices. Obtaining approval of a new drug is normally a long and expensive process, averaging six to eight years and tens of millions of dollars. In the case of recombinant DNA products, approval time is often reduced if information is already available.

National Institutes of Health (NIH) Agency whose Recombinant DNA Advisory Committee (RAC) has primary responsibility for recombinant DNA research. Its 1976 publication *Guidelines for Research Involving Recombinant DNA Molecules* assigned different categories of risk to certain types of experiments. Some types of studies were prohibited, and others had to employ special containment and safety procedures, depending on the level of risk. Since then, after the completion of millions of recombinant DNA experiments, it has become clear that the early assessments were overly conservative. Many of the research risks either did not exist or were exaggerated. Today, the guidelines have been modified, and only 10 percent of the experiments conducted fall under guideline restrictions

Occupational Safety and Health Administration (OSHA) Agency whose primary biotechnological role is monitoring genetically engineered microbes in the workplace.

In general, the regulatory thrust in the United States is adaptive rather than anticipatory. In other words, the emphasis is toward monitoring developments rather than trying to create complex regulations that attempt to anticipate all potential problems that may arise. However, there is still no complete agreement about the value of this regulatory strategy, and two opposing views have emerged. On one side are people who agree with the flexible approach. They tend to support strong governmental oversight as a way to secure public trust in an industry sometimes viewed as potentially hazardous. To them, overly rigid regulations would impede progress and delay the enormous benefits that biotechnology promises. On the other side are people who have criticized and legally contested the existing regulations, believing that the tightest possible restrictions are required to ensure that no unanticipated or undesirable outcome occurs. As with other scientific concerns that become political issues, the questions raised by genetic technologies are very complex and not always readily resolved on the basis of experimental data alone.

> **BEFORE YOU GO ON** Following the successes of the DNA revolution, there was a brief pause in research in the 1970s while scientists and politicians evaluated the implications of genetically modifying organisms and made decisions about how to proceed. Regulatory policies were issued in 1976, and within a year, new biotechnology companies were being established in the United States.

Protein Products

Biotechnology companies use complex, high-technology fermentation systems in which microorganisms are cultured in huge reactor vessels maintained under carefully controlled conditions (see Figure 21.9). So far, most commercial efforts have been directed at producing proteins with human medical appli-

Figure 21.9 Large reactor vessels are used in the industrial production of proteins by microorganisms carrying recombinant DNA.

cations. These include insulin and other hormones, vaccines, blood products, immune system molecules, and antibiotics.

Product development of proteins using genetic technology consists of three phases: creating the system to produce the gene, expression vector, or bacteria protein; large-scale production of the protein; and approval for marketing. The last phase requires extensive testing of the product, first using animals to establish its safety and then in clinical trials on human patients. Along the way, various applications must be submitted and approval obtained from various federal agencies responsible for product safety. In the United States, final approval for marketing comes from the Food and Drug Administration. In addition to pharmaceutical and medical companies, other industries involved in food processing, agriculture, chemicals, pollution control, and energy are exploiting these technologies to manufacture commercially important products.

Genetically Engineered Bacteria

The processes used in gene cloning are also used to create genetically engineered bacteria. These bacteria have great potential relative to environmental applications. For example, one of the fondest dreams of genetic engineers is to create bacteria that will provide commercially important plants with nitrogen. Plants require nitrogen for growth, yet they cannot absorb nitrogen directly from the atmosphere and transform it into chemical forms they can use (see Chapter 9). Therefore, industrial chemical processes are used to manufacture fertilizers containing nitrogen in a form required by plants.

Meanwhile, however, bacteria of the *Rhizobium* species can supply certain crop plants with nitrogen, the critical nutrient more commonly provided by expensive chemical nitrogen fertilizers in the United States and other developed countries (see Figure 21.10, page 404). Rhibozia form nodules in the roots of legume plants such as peas, beans, peanuts, soybeans, clover, and alfalfa, providing nitrogen to the plant and receiving nutrients in return. Rhibozia that live in the soil have a symbiotic relationship with a limited number of plant species and can transform nitrogen into useful forms. Since Roman times, farmers have exploited this relationship to increase soil fertility.

Now genetic engineers are identifying and manipulating *Rhizobium* genes to create strains that will be more efficient in producing nitrogen or that will form associations with other plants. Efforts will continue because farmers spend vast amounts of money each year on chemical nitrogen fertilizers.

Despite the potential benefits, the idea of applying genetically engineered microorganisms in agricultural systems has been controversial. The following questions provide a focus for considering the release of genetically engineered microorganisms:

Will the altered microbe survive in the natural environment?

Will it multiply?

Will it spread to areas outside the site of introduction?

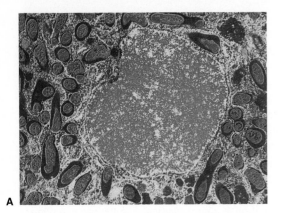

A

B

Figure 21.10 (A) An EM of *Rhizobium* (red) in the root cell of a pea plant. The yellow/green structure is the nucleus of the cell; the *Rhizobium* around the nucleus is enclosed within a membrane (blue) provided by the plant. *Rhizobium* species are able to convert nitrogen from the atmosphere into forms that can be used by plants. (B) These bacteria can live in root cells of legume plants such as peas and soybeans. When rhizobia infect a root cell, a nodule is formed in which the bacteria live and form nitrogen compounds that pass into the plant's tissues from the roots.

Question: *Why is there such great interest in genetically engineering plants to produce their own nitrogen?*

Can its altered genes be transferred to other species?

Will the altered microbe or any species receiving its genes prove harmful?

The general concern expressed by opponents of field testing with altered microbes is that dangerous new organisms might be released into the environment with a potential for causing harm to other crops or native plants or to the health of animals or humans. Proponents argue that a single or limited number of gene modifications are highly unlikely to result in the creation of a microorganism presenting any significant risk. No consensus on this subject has yet emerged among the public, although many scientists favor the latter view.

BEFORE YOU GO ON Recombinant DNA containing a gene of interest can be placed into bacteria, which then express the gene and produce significant quantities of the encoded protein. Using this technology, companies have succeeded in producing vaccines, hormones, and many other products. Issues surrounding the release of genetically engineered bacteria have not yet been fully resolved.

Genetic Engineering of Multicellular Organisms

Plants and animals have been manipulated reproductively since the dawn of the agricultural age. At the simplest level, agricultural plants with desirable production traits (juiciest fruits, highest yields, resistance to disease) have been bred for centuries in attempts to establish varieties that produce offspring with the valued feature. This approach has a long record of success. For example, yields of corn and wheat crops have continued to increase over a 50-year period. However, the conventional selection methods used by plant breeders typically require many years to achieve the desired result. Because DNA governs the expression of desired characteristics, genetic engi-

neering may offer powerful opportunities for creating or amplifying production traits.

Following the revolutionary developments in genetically engineering microorganisms, attention turned toward multicellular plants and animals. For two decades, a central question has been how to use genetic technologies to create new phenotypes with special capabilities that would benefit humans.

Because of the enormous potential benefits for humankind, efforts have focused mostly on research with agricultural applications. There are two approaches by which genetic technology can be used to improve agricultural products. First, new biochemicals can be created, using standard recombinant DNA technology, that will enhance the growth or survival of agricultural commodities. Vaccines, drugs, antiviral substances, food additives, and growth hormones are examples of such chemicals.

A second approach involves directly altering the genetic material of agriculturally important plants and animals. Though the potential value of such organisms is enormous, many technical difficulties remain to be resolved. Paramount among these problems are the complexities involved in identifying genes (or gene clusters) and gene products that determine traits of interest and creating stable changes in the DNA of genetically engineered plants and animals. Nevertheless, the past few years have seen rapid progress in developing the potential of genetic engineering not only in agriculture but also in research on disease, environmental science, and fundamental molecular processes in plants and animals that are not yet understood.

Plants

Two strategies are used in the genetic engineering of plants: to make genetic changes through cell fusion, and to insert or alter genes in cells using the recombinant DNA methods described for microorganisms, with one important conceptual distinction. For bacteria, genetic changes are made at the cellular level, while for plants, changes at the cellular level must usually be reproduced stably into every cell (and their offspring) if they are to be of commercial value.

Protoplasts Several methods have been developed to facilitate gene transfer in plants. A procedure for producing new plants with genes and chromosomes from different species is shown in Figure 21.11. **Protoplasts** are plant cells that have had their outer cell wall removed by digestion with certain enzymes. Protoplasts of some species have the potential to regenerate a whole plant from a single cell. In genetic engineering, foreign DNA can be introduced into these "naked" cells to create new plant varieties

or even species. Various methods are used to insert DNA into protoplasts: two protoplasts from two different plant species may be stimulated to fuse together in a process called **protoplast fusion**; DNA may be introduced by a disease-causing **vector** (see Table 21.1 on page 396); or, most commonly, DNA segments can be introduced into protoplasts by means of chemical, electrical, or mechanical procedures.

Genetically altered or unaltered protoplasts can be grown in media containing nutrients, vitamins, and plant growth hormones to form a mass of unspecialized cells called a **callus** (see Figure 21.11C). The callus is then induced to develop into an entire plant by exposing it to appropriate hormones. Unaltered

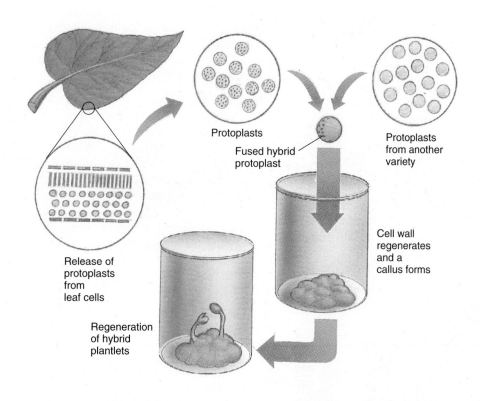

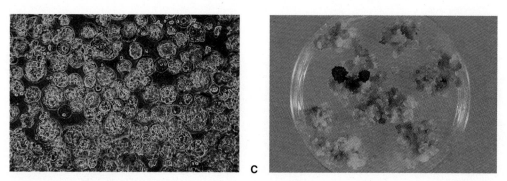

Figure 21.11 Protoplasts (plant cells without the surrounding cell wall) from different species can be isolated or fused with a protoplast from a different species and then cultured, as shown in the diagram (A). Protoplasts from separate species, like these tobacco leaf protoplasts (B), can be placed on culture medium and stimulated to grow into a callus (C) and ultimately into a plant.

protoplasts and calli of petunias, tobacco, tomatoes, and carrots have regenerated into whole plants. However, this process has only recently been used successfully for some of the most important crop species, such as legumes and cereals.

Agrobacterium tumefaciens   For successful genetic engineering, foreign genes must be inserted into an organism's genetic material. How can genes be inserted into a plant chromosome? Geneticists have made remarkable progress in this area. Recombinant DNA methods can be used for cloning plant genes, although it has often been difficult to identify specific DNA segments that harbor genes of importance. To introduce foreign DNA or genes, a unique, natural vector has been used. Most **transgenic plants** (those containing genes from a different species) produced so far have been created using an *Agrobacterium* system.

Agrobacterium tumefaciens is a bacterial species that normally lives in the soil. However, it can infect plant tissues exposed by wounds and cause a harmful tumorous enlargement called a *crown gall* that grows at the site of invasion (see Figure 21.12A). The agent responsible for inducing these tumors

is a large plasmid, called the **Ti** ("tumor-inducing") **plasmid** harbored by the bacterium. *A. tumefaciens* has the ability to transfer a DNA segment called **T-DNA** from the Ti plasmid directly into the plant cell's chromosome, as illustrated in Figure 21.12B. The T-DNA encodes enzymes for the production of different amino acids and hormones that redirect normal plant cell growth patterns and eventually cause the tumors. Species of *Agrobacterium* can infect and induce tumors in a large number of plants.

Molecular biologists have been successful in removing crown gall–inducing genes from plasmid T-DNA, inserting beneficial genes from other species in place of those genes, and transferring the new combination of genes into the chromosomes of different plants. In the first successful experiment with *A. tumefaciens,* a bacterial gene conveying resistance to an antibiotic was introduced into a petunia. The petunia became resistant to the antibiotic, and some of its offspring were also resistant. The importance of this finding is that an introduced gene became a *stable trait* that was inherited according to Mendelian expectations.

Desirable Plant Traits   A number of other experiments have been conducted to examine the prospects for successfully introducing genetically engineered traits into plants that would lead to improvements in producing crops. What types of genetically determined traits would be associated with increased crop production? To date, four traits have received considerable attention: insect resistance, disease resistance, biological efficiency (for example, increased photosynthesis or an ability to withstand drought), and weed control.

Plants are constantly under attack from herbivorous insects. Billions of dollars are spent yearly on chemical pesti-

A

Figure 21.12   (A) The bacterial species *Agrobacterium tumefaciens* can infect plant cells and cause them to grow at an uncontrolled rate, which leads to the formation of a tumor called a crown gall, shown here. (B) This effect is caused by a Ti plasmid in the bacterium that contains genes (T-DNA) that can be integrated into the plant's chromosomes, resulting in a transformed cell. When these plasmid genes are expressed, they cause abnormal plant growth.

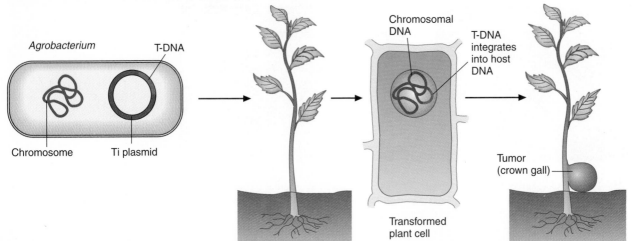

B

cides to manage agriculturally important insect pests. Might some other control strategy be possible? In the 1950s, a bacterium, *Bacillus thuringiensis,* was found to produce a protein that was toxic to the larvae of moths and butterflies and a few other insects that consume plants. This substance has no effect on other organisms, including beneficial adult insects, animals, or humans. The gene encoding the toxic protein was inserted into tomato, tobacco, and cotton plants using *Agrobacterium* as a vector (see Figure 21.13). Recent field tests of transgenic plants indicated a high level of insect control. These results indicate a promising commercial future for the *B. thuringiensis* system and perhaps for other insecticidal genes.

Genes for resistance against a broad spectrum of viruses have successfully been transferred into a range of plants. Transgenic tomato, tobacco, and potato plants have been created that are resistant to numerous viral plant diseases. Resistance to viruses could greatly increase the production of important crops such as wheat, corn, rice, and soybeans.

As all gardeners come to realize, weeds compete with economically important plants for space, nutrients, and other soil

resources. To combat undesirable plants ("weeds"), chemical herbicides (substances that kill plants) are widely used in agriculture. Unfortunately, producing herbicides that kill weeds but not desired plants is an expensive proposition. Spraying herbicides over fields with resistant crops would allow for more effective and less costly weed control but may also promote the use of chemical herbicides in agriculture, a prospect that is unwelcome to many. Although a few herbicide-resistant plants have already been developed, prospects for their production and use in the future remain clouded by the chemical herbicide question.

Future Research   The multibillion-dollar floral industry has started to use genetic engineering in attempts to create flowers with new characteristics (for example, blue roses), and commercial companies are expected to become more heavily involved in the future. However, it seems likely that most plant genetic engineering research will continue to emphasize improving agricultural crops, and much of the effort will be directed toward increasing resistance to disease.

New frontiers are also being explored. Transgenic plants have recently been created that can express genes from various species, including humans! For example, tobacco and potato plants have been genetically engineered to produce human albumin, a blood protein. *Arabidopsis thaliana* (mouse ear cress), shown in Figure 21.14, is a small plant (about 30 centimeters tall) that is often identified as a weed. However, it has become a model for plant genetics studies in the same way that *E. coli* and *Drosophila* are models for bacteria and animals. Recently, *A. thaliana* was genetically engineered to produce polyhydroxybutyrate (PHB), a biodegradable plastic, by inserting genes from a soil bacterium that encode enzymes used in synthesizing PHB. This marked the first time that a transgenic plant was engineered to produce a nonprotein. Might it be possible to create *Arabidopsis* that can synthesize other nonrenewable organic products, such as petroleum products?

Gene transfer in plants has become a fertile field of genetic engineering, and despite some formidable technical obsta-

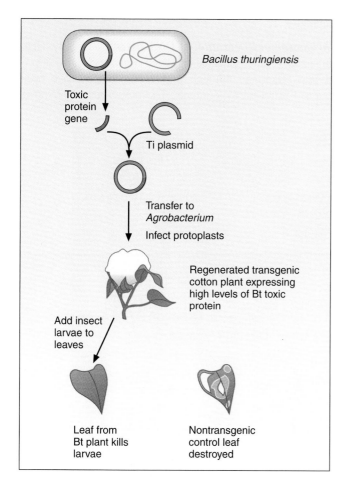

Figure 21.13   A *Bacillus thuringiensis* (Bt) gene encodes a protein that is toxic to certain insect larvae that feed on valuable plants. Scientists can use the Ti plasmid–*Agrobacterium* system to introduce the gene into protoplasts, which then grow into mature plants. If successfully transformed, the transgenic plant will express the toxic protein, and insect larvae that feed on the plant will die.

Figure 21.14   *Arabidopsis thaliana* is a small plant that is being intensely studied to learn about plant genetic mechanisms.

cles, the future appears very promising. However, complex scientific problems have arisen in attempts to create new plants or plants with enhanced capabilities through genetic engineering. For example, many of the desirable properties, such as the ability to fix nitrogen or to resist disease, heat, and cold are apparently determined by the interactions of many genes. For such traits, it is difficult to identify all of the relevant genes, or gene products, or to transfer these materials from one plant to another. Despite the problems, it seems certain that genetic engineering will be preeminent in the development of new or enhanced plants as we move toward the twenty-first century.

> **BEFORE YOU GO ON**   Plant protoplasts have been subjected to various genetic manipulations in efforts to create new traits. A vector system consisting of Agrobacterium tumefaciens, its Ti plasmid, and the T-DNA it contains has been created to introduce genes into plants. Plant traits of interest to genetic engineers include resistance to insects and disease, increases in biological efficiency, and weed control. In the future, plants may be used to synthesize valuable organic chemicals.

Animals

Commercially important animals in agriculture (cows, pigs, goats, sheep, and poultry), like plants, have benefited to some degree from the products of recombinant DNA technology, such as vaccines and growth hormones. Unlike plants, whole animals cannot be regenerated from the growth of a single somatic cell that has been extracted and manipulated genetically. Only a zygote, formed at fertilization by the fusion of gametes, normally has the capacity to develop into a whole animal. For a foreign gene to be incorporated into all cells of an animal (and their offspring), it must be introduced into a sperm or egg cell or the zygote. Although the esoteric genetic technologies have not yet been applied to animals on a broad scale, momentum is increasing, primarily because of potential medical applications.

Transgenic Animals   Scientists began to develop techniques for inserting foreign genes into higher animals in the late 1970s. The first successful gene transplant experiments were carried out several years later.

The introduction of foreign genes into the germ cell lines (fertilized eggs or embryos) of mammals is one of the most significant advances in the revolutionary field of genetic engineering. Embryos that integrate genes or DNA from a different species into their chromosomes may subsequently develop into **transgenic animals** (see the Focus on Scientific Process, "The Making of a Transgenic Mouse").

Transgenic mice have been used in a number of fruitful studies that have offered unprecedented insights into how DNA controls embryonic development and the functions of cells in the organism. For a transgenic experiment to be a success, the following events should occur: (1) genes inserted by the vector must be integrated into the host genome, (2) the encoded protein must be produced in specific tissues and be fully functional, and (3) any activities required to make the protein oper-

ate normally must take place (for example, molecular modifications or transport to a specific cellular site).

Introducing Genes into Animals   Two techniques have commonly been used for introducing genes into animals: virus vectors and microinjection. In using **virus vectors**, a foreign gene is incorporated into the viral genome, and the virus is injected into an embryo; if successful, the new gene becomes integrated into a host chromosome, and the encoded protein is produced in the resulting offspring. **Microinjection** is the technique of inserting cloned genes or DNA strands generated by recombinant DNA techniques into fertilized eggs or cells of an embryo using finely engineered instruments such as that shown in Figure 21.15. It has been the most widely used method for creating transgenic animals. Using this procedure, between 100 and 30,000 foreign gene copies are introduced into a sperm or egg nucleus (called **pronuclei** before they combine to form a zygote) of a fertilized egg or directly into cells of a developing embryo. In a typical experiment, the injected zygote or embryo is implanted into a foster mother and allowed to develop. After birth, the offspring are subjected to specialized analyses to determine whether or not the microinjected gene became integrated.

Chromosomal Integration of DNA Inserts   Usually, 10 to 30 percent of microinjected fertilized eggs survive and develop into offspring, although higher percentages have been obtained in recent experiments. The proportion of offspring

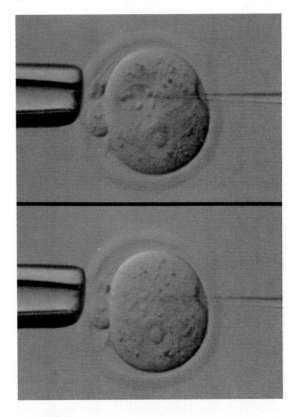

Figure 21.15   High-precision instruments, such as the microneedle in the photograph, are used to inject genes directly into sperm or egg pronuclei just before fertilization occurs.

FOCUS ON SCIENTIFIC PROCESS

The Making of a Transgenic Mouse

By the early 1980s, molecular biologists faced an interesting challenge with important implications for research on human and animal diseases. The molecular tools and experimental techniques had become available for transferring genes between cells of different animal species. Yet before the first experiments could begin, careful thought had to be given to two important questions. Which animals species would be the best candidates for use in the pioneering experiments? The answer to this question was relatively obvious. Rats and mice have long been used as experimental animals by biologists, and a wealth of relevant information on their reproductive processes, embryonic development, and genes was available. Which gene could be transferred? Perhaps the most important criterion for making this decision was ascertaining which gene would make its presence known if it was expressed in the recipient animal. Richard Palmiter of the Howard Hughes Medical Institute in Seattle and his colleagues selected the gene for growth hormone. *Growth hormone* is required for normal growth; if insufficient quantities are produced by the pituitary gland, dwarfism results. Excessive levels cause abnormally large individuals. If genes for rat growth hormone (rGH) could be successfully introduced into fertilized mice eggs, integrated into chromosomes, and later expressed, recipient mice would be expected to be larger than experimental controls that did not receive the rGH gene.

In what will probably become a classic experiment in the history of biology, Palmiter and his coworkers successfully microinjected copies of a *rat* gene encoding growth hormone into fertilized *mouse* eggs in 1982. The affected offspring were much larger than their normal littermates. The experimental procedures used are described in Figure 1. How could the rGH gene from a rat be introduced into the genome of a mouse?

To insert the rGH gene into fertilized mouse eggs, a *fusion gene* was constructed combining DNA segments from different organisms. First, recombinant DNA methods were used to clone the rGH gene, which consists of five introns and six exons. Second, an interesting question had to be addressed: how could the rGH gene be regulated during the experiment? Because little was known about normal regulation of the gene, was there some simple mechanism that could be devised that would allow researchers to turn the rGH gene on or off in mice that had the gene? An ingenious solution was found by fusing the regulatory sequence of a mouse metallothionine promoter to the rGH gene. **Regulatory sequences** are specific DNA sequences that control the production of a protein—in this case, the synthesis of metallothionine. *Metallothionine (MT)* is a protein that is expressed in all tissues but especially in the liver. It binds to metals and helps prevent metal poisoning. The MT gene is inducible (turned on) by exposure to heavy metals, such as cadmium or lead. Thus by injecting small quantities of such metals into experimental mice, the rGH gene would be expressed; when metal exposures were terminated, the MT gene would be turned off. Third, the rGH gene and the MT promoter were integrated into a special plasmid, pBR322, to create the fusion gene called an *MTrGH plasmid*.

About 600 copies of the fusion gene were microinjected into the mouse sperm nucleus after it entered the egg but before it fused with the egg nucleus

to form a zygote. The mouse zygotes were then inserted into foster mothers. From 170 zygotes, 21 mice developed, of which 7 were positive for the rGH gene. Six of the seven grew to an abnormal size.

In a second, even more spectacular study published in 1983, the same group of scientists extended their earlier research by using *human growth hormone (hGH)* genes! They used comparable procedures to create an MThGH fusion gene and followed a similar experimental protocol. The results were very interesting, and some will be reviewed here.

Scientists often formulate questions in designing and analyzing their research. In evaluating results, new questions often suggest themselves, providing direction for future investigations. The brief summaries that follow relate some of the relevant questions and resulting conclusions. We leave it to you to think of new experiments that would be of value for extending the results of this study.

1. How many transgenic mice were produced? A total of 101 mice developed from approximately 1,400 injected eggs, of which 33 were transgenic (that is, positive for hGH DNA). Why was the percentage of offspring produced so low? It was concluded that since the injection procedures involve trauma, implantation of the embryo into the placenta often failed to occur and therefore significant numbers of experimental eggs did not develop.

2. The number of MThGH genes in each transgenic ani-

box continues

FOCUS ON SCIENTIFIC PROCESS

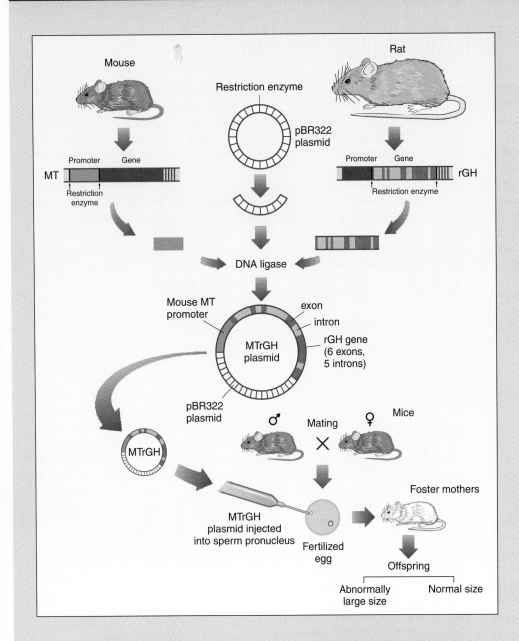

Figure 1 Summary of the genetic engineering experiment involving mice and rats. A rat growth hormone (rGH) gene and the promoter region for a mouse metallothionine (MT) gene were removed from chromosomes by restriction enzymes and combined with a pBR322 plasmid to create a fusion gene called an MTrGH plasmid. After experimental mice were mated, copies of this plasmid were injected into the sperm nucleus just before it combines with the egg nucleus to form a fertilized egg. The resulting zygotes were placed in foster mothers to develop. Several of the offspring grew to an abnormally large size, indicating that the rGH gene functioned in these mice.

mal varied from 0.9 to 455 copies per cell. From this it is obvious that there was great variability in the number of genes that became integrated.

3. Did transgenic mice grow to a greater size? Of the 33 transgenic mice, 23 were more than 18 percent larger than their littermates, and several were twice as large (see Figure 2).

Therefore, enhanced growth in transgenic mice does not depend on GH from a closely related species or a hormone with close structural similarities. Rat GH and human GH differ in 67 out of 191 amino acids, yet growth patterns in the two studies were similar.

4. Was it possible to derive a *predictive relationship* (that

is, to obtain a data set that is good enough to be able to predict the value for one variable, given a value for the other) between growth rate and the number of inserted hGH genes? There was no relationship between growth rate and the number of MThGH genes in the cell. Some of the largest animals had few copies, but of the ten

FOCUS ON SCIENTIFIC PROCESS

Figure 2 Photograph of a normal-sized mouse and an abnormally large littermate that incorporated and expressed the gene for human growth hormone.

that were not larger, most had fewer than three GH genes.

5. Was there a predictive relationship between the rate of growth and the amount of hGH hormone present in the blood? All of the larger transgenic mice had hGH in their serum, but only a rough relationship existed between growth rate and circulating hGH. Some animals with high hGH levels did not grow as well as others. One animal with substantial levels of hGH did not grow larger than normal at all. The largest transgenic mouse had only two integrated copies of the MThGH gene per cell and

had a moderate level of hGH. Therefore, the answer to the question was no. However, expression of the foreign MThGH gene was undoubtedly responsible for the increased growth. Further, there was only a poor relationship between gene dosage and level of gene expression. In these types of experiments, variability may be related in a general way to the number of integrated genes, but the site on the chromosome where they are integrated may have a much more profound effect.

6. Is the MThGH gene in transgenic mice heritable? One

transgenic male was bred to a nontransgenic female. The first litter contained seven pups, four of which had inherited the gene. A second litter showed the same results. Thus the trait could be inherited according to Mendelian expectations, since the genes were contained on only one of the two homologous chromosomes in the transgenic male. That is, half the offspring would be expected to have the trait.

There were many other significant aspects of this fascinating study. We have highlighted some of the results related to information considered in previous chapters.

that have integrated the foreign gene has been highly variable, ranging from a few percent to as high as 40 percent.

Some generalizations can be drawn from animal studies conducted to date: (1) the foreign gene is usually integrated into only one of the two homologous chromosomes, (2) the number of genes integrated varies from a few to hundreds, and (3) the foreign gene is integrated into both somatic cells and gametes (eggs or sperm) of the offspring. Consequently, the gene can be transmitted as a stable Mendelian trait.

Gene Expression in Transgenic Animals In an ideal experiment, the integrated gene would be expressed in a predictable and tissue-specific manner. For example, if a foreign gene from a rat encodes an enzyme normally produced in the liver, certain

levels of the same enzyme would be expected in a transgenic mouse receiving the gene. Furthermore, synthesis of the enzyme would be under host regulatory control and be produced by liver cells, not cells of the brain or muscle. That would indicate that the gene had been integrated and was operating normally. In the early studies, there was often very little evidence of regulated or tissue-specific expression of the inserted gene. However, these criteria have been satisfied in several transgenic mouse, rabbit, sheep, and pig experiments involving insulin, immune molecules, and certain enzymes and hormones.

Future Research What are some trends for research in animal genetic engineering? Genetically altered mice will continue to help answer questions about gene function. Two approaches are now widely used: producing "standard" transgenic mice, in which a gene of interest is incorporated into the genome, and creating *"knockout mice,"* in which genes are "knocked out" (made nonfunctional). Both methods can yield important information about genes from various organisms. For example, transgenic mice were used to confirm that the *sry* gene determines sex in mammals (discussed in Chapter 19). Mice with two X chromosomes and only the *sry* gene from a Y chromosome developed into males, whereas XX mice without the *sry* gene were normal females.

Knockout mice have recently become a hot item in genetics research. The basic idea is to determine the function of a specific gene by inactivating it during embryonic development and then identifying the effect in offspring produced. More than 150 knockout mice have been developed, and the number is expected to increase sharply in the next few years. So far, studies of knockout mice have yielded interesting results. In some cases, the expected effect was observed. For example, knocking out genes known to be important in preventing cancer resulted in tumor development in young knockout mice. In other studies, hypothesized results were not obtained. For example, two genes thought to be critical for normal muscle development—*myo-D* and *myf-5*—were knocked out. Would these knockout mice die before birth? If they survived to birth, would they have any muscles or would their muscles be abnormal? The results: mice lacking the *myo-D* gene were normal and those without *myf-5* had no ribs. Obviously, much remains to be learned about normal gene function in mice and other animals, but genetic engineering seems to offer remarkable opportunities in this area.

Transgenic methods have also been used successfully in experiments with other animal species. For example, the gene for the human blood clotting protein, factor IX, was engineered into sheep, and they secreted factor IX in their milk. Consequently, there is now great interest in genetically engineering mammals to become "bioreactors." As described in Figure 21.16, the concept is familiar: genes encoding human proteins are used to create transgenic pigs, cows, goats, or sheep; if the gene becomes established and is expressed, human proteins will be synthesized and, it is hoped, secreted in milk; and finally, the protein can be isolated and purified. Several companies are now attempting to develop transgenic mammals that will produce proteins, such as factor IX, that can be used by humans who may lack the normal protein. It remains to be seen how soon they will appear in the marketplace, as the Food and Drug Administration will require extensive testing to establish their efficacy and safety.

Several problems have impeded progress in animal research. A better understanding of the regulation and expression of genes is needed, and ways must be devised to control where genes are inserted into chromosomes. The complexity of traits determined by multiple genes (probably the rule for most animal characteristics of economic importance) must also be unraveled. Gene identification is a serious problem in the genetic engineering of mammals; of the 50,000 or so expressed genes, the protein products of only about 1,500 have been identified. Finally, the ability to insert genes and follow their integration and expression constitutes an extraordinarily powerful tool for learning more about fundamental biological processes. Nevertheless, such a capability has raised concerns among certain sectors of the general public about scientists trespassing into areas where they may not belong. We will investigate this subject in the next chapter.

> **BEFORE YOU GO ON** Transgenic animals, with genes from other species incorporated into their genome, have been produced. The technology used in creating such animals may be used for learning more about normal gene function, gene regulation, and genetic disorders and for improving production traits in livestock. Transgenic mammals may soon be used to produce significant quantities of human proteins in their milk.

RELEASE OF GENETICALLY ENGINEERED ORGANISMS

The fundamental question of whether or not to release new strains of genetically altered microorganisms, plants, and animals into the natural environment is still being debated. Potential problems with microorganisms were described previously. What about the need to regulate the introduction of modified plants and animals that could provide important new food resources for humankind?

Progress in the science of producing such organisms has exceeded our ability to establish an appropriate decision-making process. Several nontechnical obstacles remain that must be addressed before genetically engineered plants and animals will become commonplace. For example, biotechnology businesses feel that there is a critical need to create an approval process that is straightforward and allows for the rapid change from field tests to full-scale market production of a new food commodity. Also, companies consider patent protection—court decisions have ruled that genetically engineered organisms can be patented—essential if they are to recover their costs in developing new organisms. Problems in applying these legal decisions to all aspects of producing such organisms have not been resolved completely.

What about potential risks to the environment posed by genetically engineered organisms? In 1989, the Ecological Society of America published a special review article ("The Release of Genetically Engineered Organisms") in the journal *Ecology* that helped guide discussion of this question. Recommendations from this prestigious scientific society have influenced the development of regulations in this area. The major points expressed in the publication were these: (1) the "authors support the timely development of environmentally sound products, such as improved agricultural varieties, fertilizers, pest control agents, and microorganisms for waste

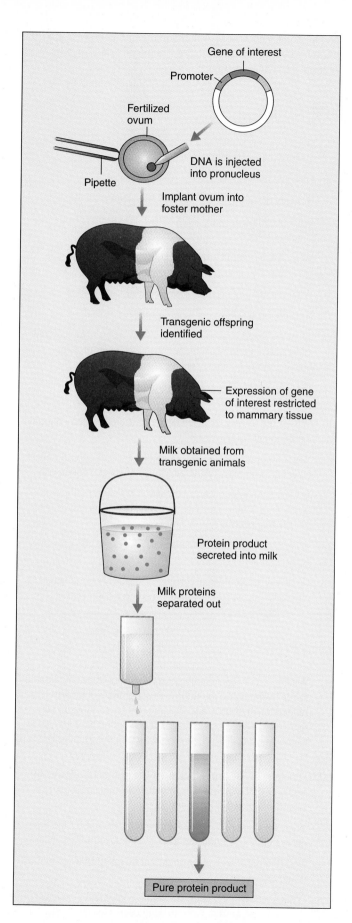

treatment . . . [that do not compromise] sound environmental management"; (2) "we support and will continue to assist in the development of methods for scaling the level of oversight needed for individual cases according to objective, scientific criteria, with a goal of minimizing unnecessary regulatory burdens"; (3) "genetically engineered organisms should be evaluated and regulated according to their biological properties (phenotypes), rather than according to the genetic techniques used to produce them"; and (4) the issues that must be considered before releasing such organisms into the environment "include survival and reproduction of the introduced organism, interactions with other organisms in the environment, and effects of the introduced organisms on ecosystem function." They also recommended procedures for regulatory approaches and suggested that instead of the present case-by-case review for each new product, it may become possible to use a generic approval process depending on the level of risk as determined by several criteria (for example, nature of the genetic alteration, genetic stability of the alteration, and survival under adverse conditions).

BEFORE YOU GO ON   Decisions and policies relative to the release of genetically engineered organisms into the environment have not yet been established. The Ecological Society of America published an important paper that helped focus the debate and moved the decision-making process forward.

Progress in creating genetically engineered organisms with special attributes is expected to continue at a swift pace. This subject is frequently discussed in editorials and science sections of most major newspapers and newsmagazines. You may wish to follow new developments in this field by reading such articles.

SUMMARY

1. Humans have long used fermentation to produce useful products. In the nineteenth century, zymotechnology improved industrial fermentation processes through science and engineering. In the twentieth century, zymotechnology gave way to a broader enterprise called biotechnology.

Figure 21.16 Experiments have shown that human genes encoding desired proteins can be injected into pronuclei in the fertilized eggs of cows, pigs, or sheep to create transgenic animals. If gene expression occurs only in the mammary glands of transgenic animals, the protein will be secreted in milk, from which it can be separated and purified.

Question: *In the United States, what regulatory activity would have to occur before such human proteins could become available in the marketplace?*

2. During the two decades following Watson and Crick's report on the structure of DNA, scientists learned much about how DNA was replicated, the enzymes involved, and mechanisms that could be used to combine DNA from different species. These processes gave rise to a new field of science called genetic engineering and new products produced through recombinant DNA technology.

3. Genes from any organism that encode important proteins can now be isolated and cloned using the tools of molecular biology. A gene is first inserted into a cloning vector, creating recombinant DNA. This recombinant DNA is then placed into a bacterial cell, where it is replicated to produce thousands or millions of copies of the gene.

4. After a gene has been isolated and amplified, copies can be introduced into bacterial cells, which then manufacture the protein specified. Industries now use this process to produce large quantities of commercially important proteins as well as genetically engineered bacteria.

5. Once techniques had been developed for manipulating DNA and genes, scientists became interested in creating genetically engineered organisms with unique properties that could be beneficial to humans. After governmental research regulations were formulated after a conference at Asilomar, biotechnology companies were formed that used genetic technologies to make new products.

6. Genetically engineered bacteria have been used to produce important proteins and for environmental applications.

7. Beneficial genes can be incorporated into a plant species genome by protoplast fusion, by an *Agrobacterium tumefaciens*–Ti plasmid vector system, or other protoplast transformation systems. Most research has been directed at improving economically important plants by conveying resistance to viral diseases and insect herbivores and at reducing competition from unwanted plant species. Transgenic plants may soon be used to synthesize valuable organic chemicals.

8. So far, application to animals of the genetic engineering techniques used for bacteria and plants has not been straightforward. However, transgenic animals have been developed successfully and are currently used in research on gene function and on human diseases. Transgenic mammals may eventually be used to produce important human proteins that can be secreted in their milk.

9. For both plants and animals, more knowledge is required about genes that control traits of interest, the products of these genes, the control of their expression, and how they can be stably integrated into the genome.

10. The debate about introducing transgenic organisms into the environment and how they should be tested and regulated has not yet been resolved.

WORKING VOCABULARY

bacteriophage (phage) (p. 399)
biotechnology (p. 401)
cloning vector (p. 398)
gene cloning (p. 396)
genetic engineering (p. 396)
plasmid (p. 398)
protoplast (p. 405)
recombinant DNA (p. 396)
restriction enzymes (p. 397)
transgenic animals (p. 408)
transgenic plants (p. 406)

REVIEW QUESTIONS

1. What types of "molecular tools" contributed to the development of recombinant DNA technology?

2. Describe the general process of gene cloning.

3. How are desired DNA sequences obtained for cloning?

4. What is complementary DNA? How does it differ from genomic DNA?

5. What are restriction enzymes? How are they used in gene cloning?

6. What are cloning vectors? How are they used in gene cloning?

7. How are bacteria used in gene cloning?

8. What are some important products now being manufactured using biotechnology? What are the advantages of using biotechnology to create such products?

9. How are genetically engineered microorganisms now being used?

10. Why is the release of genetically engineered bacteria into the environment controversial?

11. What are protoplasts? How are genes inserted into protoplasts? How has protoplast fusion been used in genetic engineering?

12. What are the relationships between *Agrobacterium tumefaciens,* Ti plasmids, and T-DNA? How is this system used in genetic engineering?

13. What criteria are used to determine success in creating transgenic animals?

14. What has the Ecological Society of America recommended about the release of genetically engineered organisms?

ESSAY AND DISCUSSION QUESTIONS

1. What types of biotechnology products should be emphasized for development in the future? Because the federal government funded much of the basic research that led to the creation of these technologies, should the general public have a voice in making such decisions, or should they be left to industry? Why or why not?

2. Should individuals or bioindustry companies be allowed to patent any DNA sequences or genetically engineered organisms they create? Why or why not?

3. What are some possible explanations to account for the frequent failure of engineered traits to be transmitted to offspring?

4. Design a transgenic plant or animal that might be economically important.

5. What would you recommend, relative to process, policies, or agency responsibilities, for making decisions about releasing genetically engineered organisms?

REFERENCES AND RECOMMENDED READING

Aldridge, S. 1995. *The Thread of Life: The Story of Genes and Genetic Engineering.* New York: Cambridge University Press.

Bud, R. 1993. *The Uses of Life: A History of Biotechnology.* New York: Cambridge University Press.

Chrispeels, M. J. 1994. *Plants, Genes, and Agriculture.* Boston: Jones and Bartlett.

Fransman, M., G. Junne, and A. J. M. Roobeek (eds.). 1995. *The Biotechnology Revolution.* Cambridge, Mass.: Blackwell.

Kornberg, A. 1995. *The Golden Helix: Inside Biotech Ventures.* Mill Valley, Calif.: University Science Books.

Krimsky, S. 1982. *Genetic Alchemy: The Social History of the Recombinant DNA Controversy.* Cambridge, Mass.: MIT Press.

Lebowitz, R. J. 1995. *Plant Biotechnology.* Oxford: Brown.

Mellor, A. 1994. Genetic engineering of animals. *Trends in Biotechnology,* 12: 335–349.

Palmiter, R. D., G. Norstedt, R. E. Gelinas, R. E. Hammer, and R. L. Brinster. 1983. Metallothionein–human GH fusion genes stimulate growth of mice. *Science,* 222: 809–814.

Pinholster, G. 1994. Debatable edibles: Bioengineered foods. *Environmental Health Perspectives,* 102: 636–639.

Rollin, B. E. 1995. *The Frankenstein Syndrome: Ethical and Social Issues in the Genetic Engineering of Animals.* New York: Cambridge University Press.

Science. 1995. Emerging plant science: Frontiers in Biotechnology (special section). *Science,* 267:654–696.

Tiedje, J. M., R. K. Colwell, Y. L. Grossman, R. E. Hodson, R. E. Lenski, R. N. Mack, and P. J. Regal. 1989. The release of genetically engineered organisms: A perspective from the Ecological Society of America. Special feature: The planned introduction of genetically engineered organisms: Ecological considerations and recommendations. *Ecology,* 70: 297–315.

Wang, K., A. Herrera-Estrella, and M. Van Montague (eds.). 1995. *Transformation of Plants and Soil Microorganisms.* New York: Cambridge University Press.

ANSWERS TO FIGURE QUESTIONS

Figure 21.2   Exposed genes or DNA cannot be replicated. To be cloned, they must be integrated in an intact DNA molecule that can be replicated in bacterial cells.

Figure 21.3   Hydrogen bonding between complementary base pairs and sealing of adjacent base pairs on the same strand by DNA ligase.

Figure 21.4   When the protein of interest and its amino acid sequence are known, a mDNA sequence can be deduced using the genetic code. It may be simpler to find an mRNA molecule (which contains only exons) than a specific gene, which contains exons and introns.

Figure 21.6   Phages are easier to place into a bacterium, larger DNA segments can be cloned, and phages are more efficient.

Figure 21.8   Because the genetic code is universal for all organisms.

Figure 21.10   To save enormous sums spent on chemical fertilizers and to reduce environmental problems (for example, groundwater contamination) associated with chemical fertilizers used in agriculture.

Figure 21.16   The U.S. Food and Drug Administration would first have to establish their effectiveness and safety and then approve the proteins for sale.

Humans and Genetic Technologies

Chapter Outline

Reading Questions

1. How can human genetic disorders be diagnosed?

2. What is the purpose of genetic screening programs?

3. What are some ethical issues associated with human genetic technologies?

4. How has somatic cell gene therapy been used in human studies?

5. What are the goals of the Human Genome Project?

The development of new molecular tools and genetic technologies described in Chapter 21 prompted an important scientific question: could they be applied to humans? They also posed a difficult question for society: *should* they be applied to humans? At first, it seemed unlikely that extending new genetic technologies to humans would proceed rapidly because of complex technical problems. However, by the late 1980s, unanticipated progress provided a dramatic answer to the first question: it *is* possible to manipulate human genes using the new molecular techniques.

DIAGNOSING HUMAN GENETIC DISORDERS

Recall that four human genetic disorders—sickle-cell anemia, Huntington disease, familial hypercholesterolemia, and

Duchenne muscular dystrophy—were described in Chapter 20. Scientists have intensively applied new molecular technologies in diagnosing these and other human genetic disorders. *Carriers*—people who are heterozygous for certain autosomal recessive disorders caused by a single gene and are therefore more likely to have affected children—can now be easily identified for an increasing number of genetic disorders. Also, for some late-onset genetic disorders caused by a dominant allele (for example, Huntington disease), it is possible to determine if an individual *at risk* (having a higher probability of developing a disease) will develop the disorder later in life. Finally, diagnostic techniques have been adapted to prenatal tests for disorders such as Duchenne muscular dystrophy, Huntington disease, and cystic fibrosis. How can these new diagnostic technologies best be used in our society?

Ethical Questions

Despite the potential benefits, the diagnostic technologies—which, in effect, convey an ability to predict the future—have severely challenged the ethics and values of individuals all over the world. Scientists, physicians, lawyers, politicians, clergy, parents and prospective parents, and many other concerned people, question whether we are ethically and morally prepared as a society for the problematic issues that could arise as genetic technologies are developed further and applied to human beings. *Ethics* in this context is defined as a system or set of moral principles and values. The problem of applying ethical standards to current biological issues is that in pluralistic societies, there is a wide spectrum of opinion about what is or is not acceptable (see Figure 22.1).

A host of ethical questions arises from being able to determine who has or who will develop a genetic disorder, especially late-onset disorders. Should individuals at risk for a genetic disorder be required to have a diagnostic test? For what purposes? Are there circumstances that make a difference? For example, individuals who are deficient for an enzyme called alpha-1-antitrypsin are more sensitive to certain substances (such as cotton dust) in the workplace than those who do not lack the enzyme. Do companies have a right to require testing

to reduce the risk of exposing susceptible employees and deny them higher-paying jobs associated with the hazard? Conversely, can companies justify not requiring tests if without that information, employees might be placed at increased risk?

Do insurance companies have a right to obtain the results of genetic tests that can determine whether or not an individual has a high probability of having a mutant allele associated with a disorder? If a family has a history of a genetic disorder, should insurance companies be able to mandate testing as a prerequisite for insuring individuals at risk? The relevance of these questions is associated with the high costs of medical treatment for individuals with genetic disorders and the greater probability that they will die at an early age. Is it reasonable to withhold insurance from carriers or people at risk or from people who test positive? Should people who have no genetic disorders be required to pay higher insurance rates to subsidize those who need or will need treatment for such disorders?

In the case of serious genetic disorders that appear in certain families and cause death soon after birth, should physicians be responsible for requiring prenatal tests in pregnancies involving parents at risk for a genetic disorder (see Figure 22.2, page 418)? If the offspring has a genetic disorder, should the parents be able to sue the doctor for negligence if the doctor did not recommend the test? Who should make those decisions?

Huntington Disease: A Model

What is an effective approach for using the new technologies to test individuals at risk for developing a serious genetic disorder? The ability to diagnose Huntington disease (HD), a late-onset neurological disorder that is progressively debilitating and invariably fatal, has resulted in the evolution of an approach for addressing complicated problems that arise from administering predictive genetic tests.

After discovery of the HD (*Huntingtin*) gene in 1993, a simple new blood test was developed to determine if an individual has a mutant *Huntingtin* allele and, if so, when the person is likely to develop the disease. The HD test can be conducted at any time during an individual's life. Obviously, the decision to be

Figure 22.1 Besides being genetically unique, individuals have differing systems of values that influence their thinking about the uses of genetic technologies.

Figure 22.2 New procedures can be used in prenatal testing for an increasing numbers of genetic disorders. Prospective parents who are at risk for a genetic disorder may or may not wish to take advantage of such tests.

tested or not is an enormously difficult one for a person to make. Such a highly personal choice can only be made by the individual at risk, usually in consultation with family members. An additional problem concerns children at risk for HD. Institutions offering the test have chosen to restrict administering it to people age 18 and older who can give informed consent and who have undergone extensive counseling.

Professional staffs conducting HD diagnoses have recognized the importance of including a person's family or support group at every level of the process. All test centers offer extensive counseling, both before and after testing, to enable individuals to deal with the results in the most constructive way possible. Results indicate that such a highly personal approach, with a sensitive professional staff, can serve as a model for other genetic testing programs.

Genetic Screening

Genetic screening refers to broad testing programs used to identify carriers for specific genetic disorders. Such programs typically target ethnic populations or groups at risk for a specific genetic disorder. Government-sponsored genetic screening programs were developed in the United States in the early 1970s for testing Ashkenazic (eastern European) Jews for Tay–Sachs disease and for testing African Americans for sickle-cell anemia (described in Chapter 20). Tay–Sachs disease is a single-gene disorder inherited as an autosomal recessive (that is, both parents must carry the mutant allele to produce a homozygous child with the disease). In the classical form of the disease, symptoms appear at 3 to 5 months of age. Affected infants progressively become severely retarded, paralyzed, and blind. Death occurs at age 2 or 3, and there is no effective treatment or cure. The responsible gene is identified as *HEXA*, located on chromosome 15, and over 30 different mutant *HEXA* alleles have been described. Tay–Sachs affects one of every 3,600 Ashkenazic Jews, and sickle-cell anemia affects

one of every 400 African Americans (both disorders also occur in other ethnic groups). Histories of the two early genetic screening programs represent contrasting models for creating successful policies.

The highly successful Tay–Sachs screening program was influenced and supervised by people of Ashkenazic descent. Efforts were made to educate Jews about the disease and its genetic basis. Diagnostic tests to identify carriers and prenatal tests for pregnant women were offered at convenient times and locations. Genetic counselors explained the implications of being a carrier and provided reliable information about results from prenatal tests that could be used in making an informed decision about having an abortion. As a result, the number of children born with Tay–Sachs decreased from 100 in 1970 to 13 in 1980. Many Jewish communities continue to support Tay–Sachs screening programs.

In comparison, the first sickle-cell screening program was not a success (see Figure 22.3). Little effort was made to educate people about the disease, the purpose of the tests, or the implications of being a carrier. Consequently, there was great confusion over the test results and their meaning. For example, many people identified as carriers thought that they would develop the severe form of the disease. Because no prenatal test was then available for sickle-cell anemia, some carriers were told not to have children, which resulted in charges of racism. In some cases, test results were not treated as confidential information and carriers were denied health insurance. No significant decrease in the incidence of sickle-cell anemia resulted from the original genetic screening program, whose legacy is how such a program should not be operated.

Because of the importance of sickle-cell disease in terms of costs and number of people affected, new and very successful testing programs were established in the mid-1980s by the U.S. Department of Health and Human Services. The key elements of these comprehensive programs, which now exist in more than 40 states, include screening newborn children to

HUMAN GENE THERAPY

Gene therapy, the substitution of a normal allele in place of a mutant allele, is a concept that emerged in the early 1970s when it became known that during some stage of their life cycles, tumor-causing viruses insert foreign genes into mammalian cells. When new laboratory techniques were developed during the age of recombinant DNA technology to perform this same feat, it seemed reasonable to assume that forms of gene therapy could be devised for treating human genetic disorders. In 1989 and 1990, the first human gene therapy experiments were conducted, and within just a few years, the dream of using human gene therapy to treat a genetic disorder had become a reality.

There are two general types of gene therapy (see Figure 22.4, page 420). **Somatic cell gene therapy** involves replacing genes in somatic (body) cells. **Germline gene therapy** consists of inserting genes in cells that produce sperm or eggs, in a fertilized egg, or in cells of early developing embryos. Whereas somatic cell gene therapy concerns only the person treated, germline gene therapy involves genetic alterations that can be passed on to future generations. Before turning to the processes and techniques of somatic cell gene therapy, we will discuss some of the ethical issues surrounding human gene therapy.

Ethical Issues

The policies of the U.S. National Institutes of Health (NIH) and the U.S. Food and Drug Administration (FDA) currently prohibit germline therapy experiments, but human somatic cell therapy studies are taking place throughout the world. Scientists and physicians generally accept somatic cell gene therapies as extensions of approved medical procedures, comparable to blood and marrow transfusions used for treating diseases. Consequently, few specific objections have been raised based on ethical considerations.

In contrast, most groups have indicated strongly that they are not yet ready to support research centered on human germline alterations. Nevertheless, because of the successes of somatic cell gene therapy studies, the public debate over germline gene therapy has been opened even though germline experiments will not take place for years. Many interesting and difficult medical, philosophical, and ethical issues are now subjects of lively discussion among individuals in science, government, medicine, law, ethics, and the general public.

There are many arguments for and against developing human germline gene therapies. People who favor developing these therapies argue that (1) health professions have an obligation to use the best methods available for preventing or treating genetic disorders, (2) parents have the right to use this technology if they wish to have a normal child and to prevent a genetic disorder from being transmitted to future generations, (3) onetime germline gene therapy is more efficient and cost-effective than the repeated use of somatic cell gene therapy

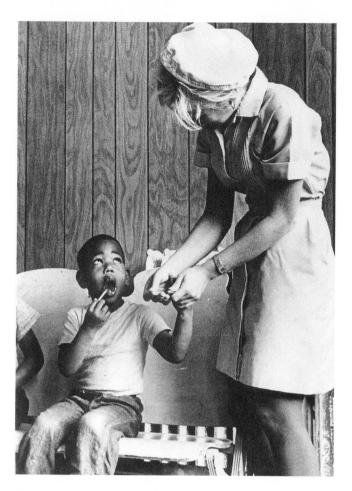

Figure 22.3 Government-sponsored genetic screening programs were developed in the United States in the early 1970s for testing Ashkenazic Jews for Tay–Sachs disease and African Americans for sickle-cell anemia. The first sickle-cell screening program, the source of this photo, was not a success because little effort was made to educate people about the disease, the purpose of the tests, or the significance of being a carrier.

identify those with sickle-cell disease, providing affected children with appropriate medical care, and offering parental education and genetic counseling. The most important medical treatment involves the administration of oral penicillin twice daily, which has been shown to reduce illness and mortality caused by pneumonia. The education and genetic counseling services provided to all parents of affected infants are modeled after the successful Tay–Sachs screening programs.

BEFORE YOU GO ON   New genetic technologies are being applied to humans. The ability to diagnose genetic disorders has already reached a high level. Counseling systems developed for individuals at risk for Huntington disease provide a valuable model for other diagnostic programs. The history of early genetic screening programs for Tay–Sachs disease and sickle-cell anemia provided valuable information for developing new programs.

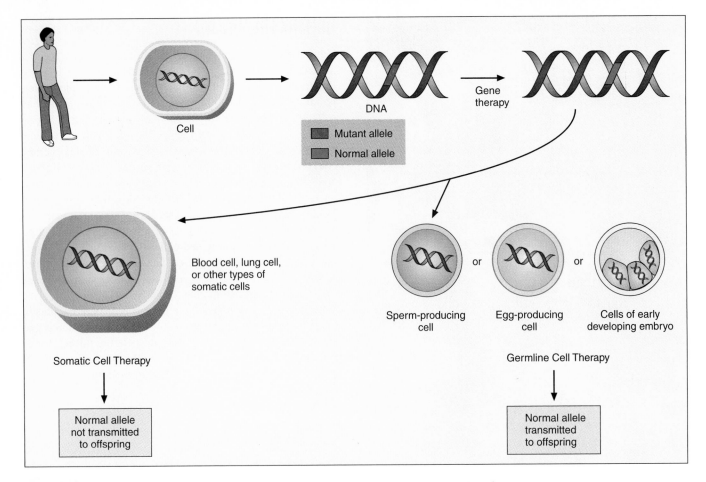

Figure 22.4   Gene therapy consists of various strategies that are used to substitute a normal allele for a mutant allele. If normal alleles are placed in a somatic cell, they may reduce or eliminate effects caused by a mutant allele but will not be transmitted to offspring. If inserted into cells that produce sperm or egg cells, a fertilized egg, or cells of an early developing embryo, the normal allele may be transmitted to offspring.

Question: *What explains society's general acceptance of somatic cell gene therapy?*

over several generations, and (4) science and medicine have the fundamental right to pursue knowledge that has intrinsic value. Opponents of further development of human germline gene therapies argue that (1) such therapies will be very expensive and potentially beneficial to a relatively small number of people; (2) they will always have unavoidable risks, and mistakes will be irreversible; and (3) pressures will arise to use germline gene therapy for "improving" offspring. A decision on whether or not to proceed with germline gene therapy will require a major public policy decision. Most people agree that this determination will best be made after intense, extended public debate and well before the new technologies reach the point where they can be applied to humans.

Somatic Cell Gene Therapy

Two basic requirements must be met before somatic cell gene therapy can be used in treating a particular genetic disorder. First, it is necessary to identify the normal allele of a gene—

its locus and structure—and the protein it encodes. Second, vectors must be created that can insert the normal allele into cells where it can become integrated into a chromosome and expressed. In human gene therapy, a virus usually serves as the vector that transmits an allele into a cell.

Vectors

Since the early 1970s, the use of mouse and bird retroviruses as vectors for transferring normal human alleles has been carefully investigated. **Retroviruses,** shown in Figure 22.5A, are unusual in that their genetic material is RNA, not DNA. Once they infect a host cell, retroviruses are replicated by the action of reverse transcriptase, the enzyme that can transcribe RNA into DNA. This, of course, is the opposite of the normal transcription of genetic information, DNA into RNA, hence the prefix *retro-*. Human retroviruses cause several important diseases, including some cancers and AIDS.

Retroviruses have several advantages as vectors compared to others that have been studied. They are easy to manipulate

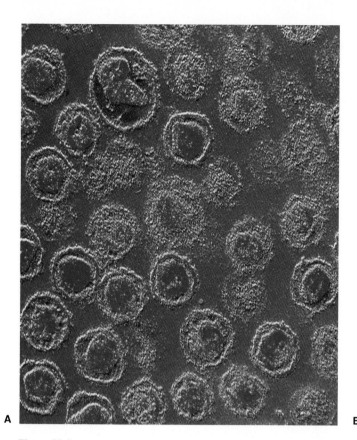
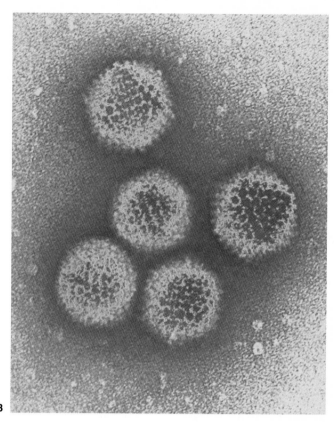

A   B

Figure 22.5   Retroviruses (A) and adenoviruses (B) have been used as vectors in gene therapy experiments.

and have a demonstrated ability to transfer foreign genes into cells and have the genes expressed, even after they are genetically engineered so that they cannot replicate. However, retroviruses also have disadvantages. They cannot carry large genes, they are sometimes unstable, it is not always possible to control where the genes they carry are inserted, and even if they do become integrated, the genes are not always expressed. Also, recipient human cells must be dividing before the gene carried by a retrovirus can become integrated in their chromosomes. Because of the cell division requirement, retrovirus vectors cannot be used for introducing genes into mature cells that do not divide, such as brain, nerve, and muscle cells. Despite these problems, retrovirus vectors have been used successfully in human studies.

Other viral vectors have also been created. For example, **adenoviruses,** shown in Figure 22.5B, which cause colds and viral pneumonia by infecting and killing cells of the respiratory system, have been genetically engineered and are now used as vectors for introducing genes into lung and nervous system cells. Unlike retroviruses, adenovirus vectors can insert genes into nondividing cells. Another advantage is that they can be delivered in an aerosol spray rather than being injected.

Types of Somatic Cell Therapy

As illustrated in Figure 22.6 (see page 422), two somatic cell gene therapy procedures have been developed. In the transfusion technique, normal alleles are inserted by vectors into cells that are removed from an affected individual; the *transformed cells*—those in which the allele becomes inserted—are then injected back into the person, where they can be expressed. A second technique involves the direct inoculation of vectors carrying the normal allele into an affected person, where they may become integrated into appropriate cells and produce the normal protein. Both techniques have been used successfully in human studies, as described in the next sections.

BEFORE YOU GO ON Gene therapy offers great promise for treating human genetic disorders. Somatic cell gene therapies are being developed worldwide. Germline gene therapy is now the subject of intense debate. To use somatic cell gene therapy in treating a genetic disorder, the structure and location of the normal allele must be known as well as the protein encoded by the gene. Vectors must also be created that can insert the normal allele into cells.

Developing Experimental Protocols

There are many steps leading to the eventual application of new genetic technologies to human beings. Ultimately, it is necessary to develop an approved experimental **protocol,** the detailed plan of a scientific experiment or medical treatment. However, for ethical and practical reasons, humans are never used to develop a new protocol. Given this constraint, how is

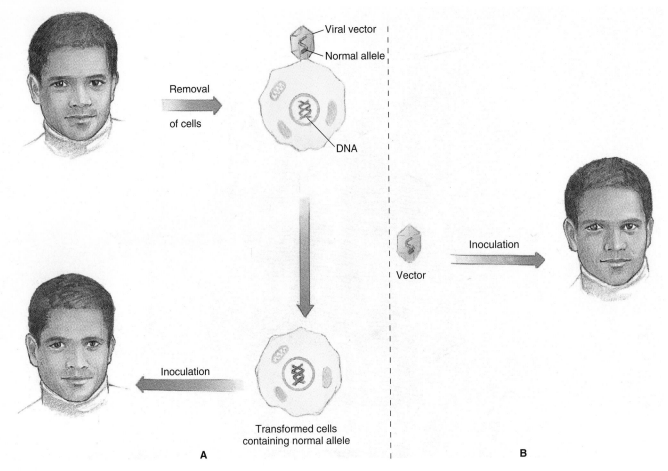

Figure 22.6   Two general types of somatic cell gene therapy have been developed. (A) Using a transfusion protocol, a normal allele is inserted by a vector into cells removed from a person with a genetic disorder. After being transformed, the cells are inoculated back into the person. (B) For some genetic disorders, vectors containing the normal allele may be inoculated directly into a patient. Using either approach, the goal is for transformed cells to express the normal allele and produce the desired protein.

Question: *Which type of somatic cell gene therapy was first used successfully for treating a human genetic disorder? What was the disorder?*

it possible to establish experimental protocols that are appropriate for treating humans?

Cell Studies

In general, new techniques are first developed and tested in the laboratory, using cells obtained from various animals, including humans, that are *cultured* (grown and amplified by cell division) in small containers. If studies using cultured cells are successful, the new methods are then usually tested on animals. The development of an adenovirus vector, shown in Figure 22.7, illustrates this process. The goal of the research was to create a vector system that could be used to infect brain neurons (nerve cells). The sequence of key events is as follows:

1. An adenovirus vector was created that could not replicate after it infected cells. This was accomplished by removing genes required for replication.

2. A *gene marker* (in this case, an allele not normally found in a vector's genome, encoding a protein that can be easily detected if the gene is expressed) for a bacterial enzyme, β-galactosidase (β-gal), was added to the viral vector DNA; the presence of β-gal in an infected cell would indicate that vector DNA had been inserted and was being expressed.

3. Rat neurons growing in cell cultures were exposed to the adenovirus/β-gal vectors. The neurons were transformed, expressed the β-gal gene (that is, β-gal protein was produced), and were not harmed by being infected.

4. Adenovirus/β-gal vectors were next injected directly into rats' brains. All injected animals expressed β-gal activity, and β-gal protein was detected in brain cells as early as 24 hours after inoculation and continued for two months. No harmful effects were

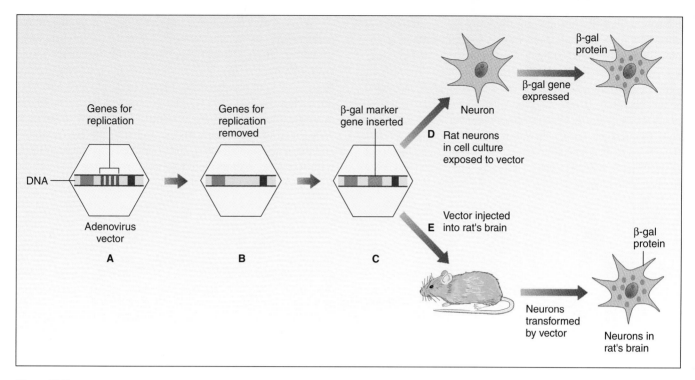

Figure 22.7   Human gene therapy protocols are usually developed and tested using cells grown in culture and various animals. This figure describes the sequence of events in creating and testing an adenovirus vector. (A) Adenovirus genes and their functions are determined. Some genes are necessary for normal replication of the virus. (B) Genes used in viral replication are removed. (C) A β-galactosidase (β-gal) gene marker is inserted into the viral DNA. This is an enzyme produced by bacteria and not normally produced by either viruses or animal cells. (D) Adenovirus vectors containing the β-gal gene are placed in cultures containing rat neurons (nerve cells). The neurons are transformed as indicated by the production of β-gal Protein in the cytoplasm. (E) The vector is then injected directly into rats' brains, and hours later, brain neurons express the β-gal gene, which indicates that they have been successfully transformed.

Question: *Why is it necessary to remove or neutralize viral replication genes?*

observed in injected animals, and they all recovered without any apparent abnormalities.

Animal Models

Once methods that may have useful applications in human gene therapy have been developed, they are tested and analyzed using animals, primarily mice, rats, or rabbits. For some disorders, animals have been discovered with the same mutation that causes a human genetic disorder. For example, golden retrievers were found with the same *dystrophin* gene defect that causes Duchenne muscular dystrophy in humans. Such a species is referred to as an **animal model;** its disease serves as a model for learning about the nature of the human disorder. Knockout mice, described in Chapter 21, that model human genetic disorders have also been created. The use of animal models for studying human diseases (for example, cancer) has a long and successful history. Only after exhaustive tests and demonstrations of safety are specific gene therapies ever used on humans.

BEFORE YOU GO ON   Gene therapy protocols are generally developed using cultured cells and various animals. Vectors are first created and tested by inserting them into cells to see if the gene of interest is inserted and expressed. If cell studies are successful, the vector is then tested on animals. In some cases, animals have been found that have the same genetic disorder as humans.

The First Human Gene Therapy Experiment

Before conducting an experiment related to human gene therapy, scientists must obtain legal approval from several regulatory agencies. Research plans must first be reviewed by a local ethics panel and institutional review boards and institutional biosafety committees. At the national level, final approval is required from the NIH and the FDA.

Early in 1988, a request to conduct the first human gene therapy experiment in the United States was submitted for approval. The proposal stemmed from earlier investigations on a promising new strategy for treating cancerous tumors. Specific cells—white blood cells called *lymphocytes*—that kill "foreign" cells, including cancer cells, normally exist in the body. In early studies, lymphocytes that had infiltrated the tumors of seriously ill patients were removed surgically and induced to grow and divide in the laboratory. These cells are called *tumor-infiltrating lymphocytes,* or *TIL cells.* When sufficient numbers of cultured TIL cells had been generated, they were reinfused into the patient. TIL therapy caused significant tumor regression in about 50 percent of the 25 patients tested who had advanced cancers and had failed to respond to other treatments.

Why was TIL therapy effective only in some of the patients? To try to answer this question, the investigators wanted to insert a single gene marker (for resistance to the antibiotic neomycin) into TIL cells using a retrovirus vector and follow their progress once placed back into a patient. Figure 22.8 illustrates the TIL marker protocol. The gene marker would allow researchers to identify TIL cells with the gene by collecting blood samples periodically during treatment and then exposing all blood cells removed to neomycin; those with the gene marker would survive, and those without the resistance gene would perish. The gene marker would have

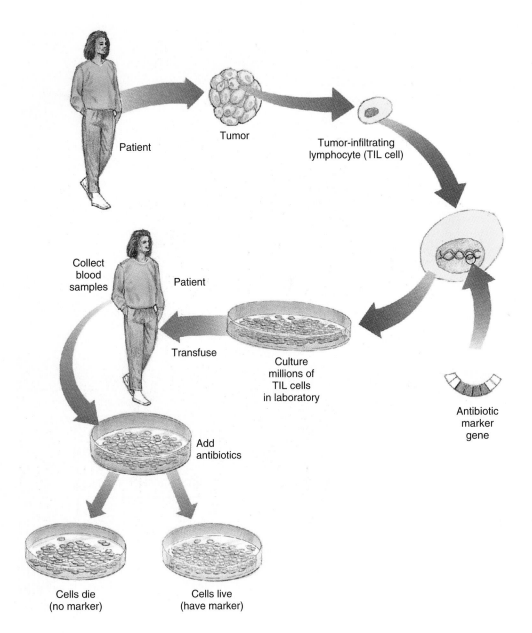

Patient

Tumor

Tumor-infiltrating
lymphocyte (TIL cell)

Collect
blood
samples

Patient

Transfuse

Culture
millions of
TIL cells
in laboratory

Antibiotic
marker
gene

Add
antibiotics

Cells die
(no marker)

Cells live
(have marker)

Figure 22.8 The first approved human gene therapy experiment was conducted to determine the fate of a type of lymphocyte that was known to react against tumor cells. Gene markers for antibiotic resistance were inserted into the lymphocytes, which were then transfused into the patient. By exposing lymphocytes that were collected periodically in blood samples to the antibiotic, it was possible to determine if they had survived and reacted against the tumor.

no effect on either the TIL cells or their capacity for attacking cancer cells.

The request was finally approved by the NIH and the FDA, and on May 22, 1989, a severely ill cancer patient received the first infusion of TIL cells that had been genetically altered to contain the neomycin gene. Additional patients were treated during the following weeks. Results from the experiment were similar to those from earlier studies—some patients showed dramatic regression of their tumors, and others did not. The significance of the research lies in what was learned about the marked TIL cells introduced into the patient. They survived in the body, apparently attacked and killed tumor cells, and were unchanged by the inserted gene.

Subsequent Gene Therapy Tests

The TIL experiment immediately opened the doors to further studies of human gene therapies. The same group of researchers received approval to test the effects of a *tumor necrosis factor* gene inserted into lymphocytes. Tumor necrosis factor is a protein that shrinks tumors by cutting off their blood supply. The first patients were treated in 1991, and there were no negative effects from the gene transfer or from the tumor necrosis factor secreted by the transformed cells. However, there was no clear indication that the procedure was effective as a cancer therapy.

In 1990 and 1991, two young girls with a rare genetic disorder called *severe combined immune deficiency* were treated using gene therapy. This disorder is caused by a missing enzyme known as ADA (adenosine deaminase). Retroviral vectors were used to insert normal ADA alleles into lymphocytes taken from the children. Transformed lymphocytes were injected into the first child, a 4-year-old, in 1990, and the second, a 9-year-old, in 1991. The immune systems of both children were at least partly restored, and the experiment was judged a success. Since then, the two girls have received gene-corrected cells weekly, along with injections of artificial ADA. The 1990 ADA treatment is generally considered the first use of human gene therapy because a normal human allele was used, not a gene marker.

BEFORE YOU GO ON Human gene therapy protocols must be approved by the NIH and the FDA. In the first human gene therapy experiment, conducted in 1989, a gene marker was inserted into lymphocytes removed from cancer patients. The cells were injected back into the patient, where they functioned and survived. In 1990 and 1991, two similar experiments were conducted. Tumor necrosis factor genes were used in a study involving cancer patients. In a second study, normal ADA alleles were inserted into the lymphocytes of two children with a severe immune disease. Transformed cells expressed ADA, and immune function was partly restored.

Familial Hypercholesterolemia: A Model

Following the successful TIL and ADA experiments, interest in human gene therapy accelerated. Which human genetic disorders might be treated using gene therapies? Disorders receiving the most attention were those with existing animal models.

Recall from Chapter 20 that familial hypercholesterolemia (FH), which is caused by codominant alleles, is a human genetic disorder characterized by high levels of blood cholesterol that often leads to heart disease and premature death. FH is caused by defects in the function or expression of low-density lipoprotein (LDL) receptors on liver cells that help remove cholesterol from the blood. In 1973, a Japanese researcher, Yoshio Watanabe, made a crucial observation in an experimental rabbit population: the feet of one rabbit he was studying were covered with fat-filled lumps. By breeding this rabbit, an FH animal model—the Watanabe rabbit—was developed that has high blood cholesterol concentrations caused by a shortage of LDL receptors (see Figure 22.9). The Watanabe rabbit became well known in the late 1970s and early 1980s when it was used in Nobel Prize–winning research that determined the underlying cause of FH.

Could an acceptable experimental protocol be developed and approved for treating humans with FH? In the early 1990s, researchers provided information relevant to this question by accomplishing the following:

1. Methods were devised for removing a small part of a rabbit's liver and establishing cell cultures. Thus pure liver cells could be modified outside the body if a suitable vector could be created.

2. A retroviral vector containing a normal rabbit LDL allele was developed.

3. The retroviral vector was used to insert normal LDL alleles into cultured rabbit liver cells. Once inside the cells, the LDL allele was expressed and produced functional receptors that removed cholesterol from the experimental system.

Figure 22.9 The Watanabe rabbit is an animal model used for studying familial hypercholesterolemia, the most common human genetic disorder.

Question: *Why are animal models considered so valuable for studying human genetic disorders?*

Following the successful cell studies, a gene therapy experiment was conducted using the Watanabe rabbit and the methods just described. Cultured liver cells were first transformed using retroviral vectors and then injected back into the donor rabbits. The results of this experiment were quite remarkable. Most injected cells quickly became established in the liver, blood cholesterol levels decreased 25 to 45 percent within 24 to 48 hours, and LDL gene expression continued for over six months, until the study was terminated.

Could a comparable experimental protocol now be used for treating humans? In 1992, the NIH and the FDA approved a similar somatic cell gene therapy protocol for treating five homozygous FH patients with very high cholesterol levels, and tests began in 1992. In 1994, researchers reported that the treatment had reduced one patient's blood cholesterol level and that the effect had lasted for almost two years. Data for the other patients were not released. The FH model—develop an appropriate vector, test using cell cultures, develop an experimental protocol using an animal model—will likely be duplicated for many genetic disorders in the future.

Promising New Therapeutic Approaches

Gene therapy seems to have great promise for treating genetic disorders in which mutant alleles are not expressed. Are there other possible treatments for mutant alleles that express an abnormal protein? One strategy now being aggressively pursued involves the creation of **antisense compounds**—molecules that can block expression of abnormal alleles (they are called "antisense" because they prevent expression of the "sense" DNA strand in a gene). An antisense compound is usually a short, single-stranded DNA molecule with a base sequence that is complementary to the mRNA encoding the abnormal protein (see Figure 22.10). When the antisense DNA strand binds with mRNA by complementary base pairing, it prevents the mRNA from being translated, and the encoded abnormal protein cannot be produced. Many biotechnology companies are now developing antisense compounds.

So far, gene therapy has been used only to *treat* certain genetic disorders. Might it become possible to *cure* such disorders? The protocols used so far require the continual introduction of transformed cells because these cells die and their alleles are lost after a few days or weeks. **Stem cells,** described in Figure 22.11, are a unique long-lived type of cell that divides continuously to produce daughter cells that either mature to become specialized functional cells or continue to act as stem cells. For example, stem cells in the bone marrow produce daughter cells that can mature into red blood cells or various types of white blood cells such as lymphocytes. If stem cells could be transformed with the normal allele required, every new lymphocyte produced would also carry the normal allele, and it would not be necessary to continue to introduce lymphocytes transformed outside the body (see Figure 22.12). In other words, the transformation of stem cells would constitute a cure, for no additional treatments would be required. The stem cell protocol has been proposed and approved by the FDA. However, blood stem cells are difficult to isolate, and for other organs, such as the lung, they have not yet been identified. It will be interesting to see how soon stem cell therapy will be applied and whether or not it will provide a cure for the disorder tested.

BEFORE YOU GO ON After successful gene therapy experiments were conducted in 1990 and 1991, a protocol was developed for treating familial hypercholesterolemia. A retroviral vector was created and tested using cells and an animal model, the Watanabe rabbit. In 1992, five humans with FH were treated using the newly developed gene therapy protocol; at least one showed improvement. In the future, human genetic disorders may be treated with antisense compounds developed by biotechnology companies. If stem cells can be transformed, gene therapy could provide a cure rather than just treatment for certain human genetic disorders.

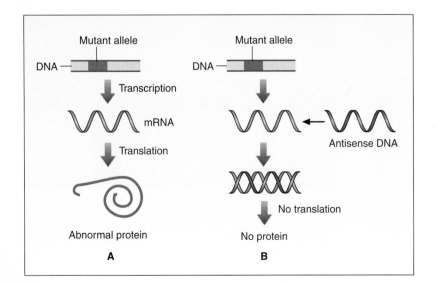

Figure 22.10   In some genetic disorders, a mutant allele is transcribed into mRNA, which is then translated to produce an abnormal protein. To prevent the expression of mutant alleles, a new strategy is being developed. Antisense DNA, a single-stranded DNA molecule with a nucleotide sequence that is complementary to the mRNA strand transcribed from a mutant allele, can bind with the mRNA by complementary base pairing, which prevents it from being transcribed. Consequently, no abnormal protein is synthesized.

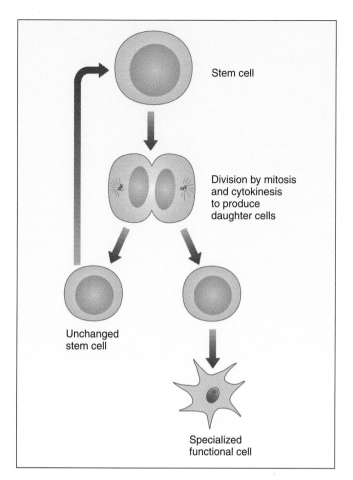

Figure 22.11 Stem cells divide by mitosis and cytokinesis, giving rise to daughter cells that can either remain stem cells or develop into specialized functional cells, such as blood cells or lung cells. Stem cells are long-lived, whereas specialized functional cells normally die after a few days or weeks of service.

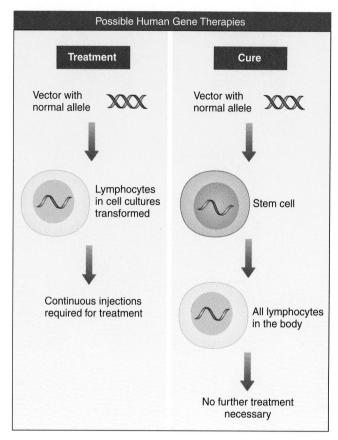

Figure 22.12 Human gene therapies developed so far have been used to treat genetic disorders. As in the example shown, transformed lymphocytes must be injected into a patient regularly for the treatment to succeed. If vectors could insert normal alleles into stem cells that produce lymphocytes, all lymphocytes produced would have the normal allele, and the patient would be cured if the allele is expressed.

THE HUMAN GENOME PROJECT

Progress in genetics research developed along many fronts during the 1980s. In a monumental achievement, the first partial map of the human genome was published in 1987. This first map was relatively primitive, but the blanks have continued to be filled in, and new genome maps are published every year. Figure 22.13 (see page 428) describes the type of information included for each chromosome in new genome maps.

A question on an even grander scale was raised during the mid-1980s: Could a high-resolution physical map consisting of the exact sequence of nucleotides in the entire human genome be drawn? A **physical map** shows the actual location of genes on a chromosome and also the physical distance, expressed in terms of base pair numbers, between genes as determined by constant, identifiable landmarks. The crudest physical map uses chromosomal bands as landmarks. The ultimate physical map will be the complete nucleotide sequence of the entire genome. In 1985, a massive project designed to elucidate the complete sequence of the 3 billion nucleotide bases and the approximately 50,000 genes in the human genome was formally proposed. The enterprise became known as the **Human Genome Project**, and the original projection for completion was 15 years, at a cost of $3 billion.

The initial reaction of the scientific community to the Human Genome Project was mixed. On the one hand, there was great enthusiasm about gaining access to the "Holy Grail of biology," as one molecular biologist referred to the human genome. On the other hand, there was concern about the magnitude and cost of what promised to be biology's first "big science" project—a huge project costing billions of dollars and involving large, organized, interdisciplinary teams of scientists—and about the possible reduction in resources for traditional "small science" projects. However, concerns abated, and in 1989, the proposal was approved and placed under control of the NIH. The official starting date was October 1, 1990, and large-scale sequencing trials began in 1991. The original projected completion date was September 30, 2005, approxi-

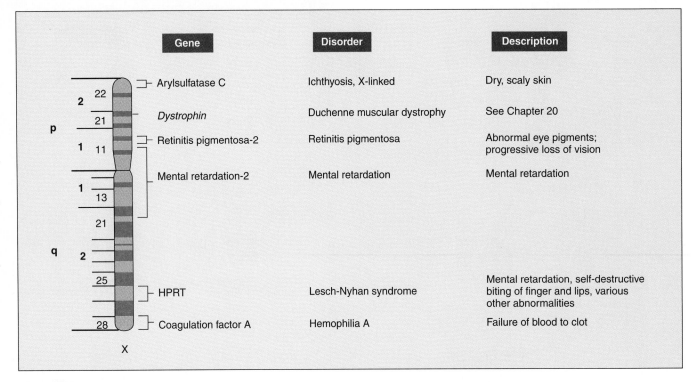

Figure 22.13 Modern genome maps describe the loci of tens or hundreds of genes. Over 160 genes are known to reside on the human X chromosome. This limited map shows the locations of six genes known to be associated with serious genetic disorders. The exact location of five of the six genes on the chromosomes is not known with certainty.

Question: *One gene locus is known with certainty; which one is it?*

mately 50 years after James Watson and Francis Crick published their seminal paper on the structure of DNA. The same James Watson served as director of the genome project from 1989 to 1992.

Goals of the Project

What are the goals of the Human Genome Project? From 1990 to 1994, emphasis was placed completing a high-density human physical map, a goal that was achieved. DNA sequencing (that is, determining the exact sequence of the nucleotide bases) was initiated for model organisms such as *E. coli,* yeast, *Drosophila,* the plant *Arabidopsis thaliana,* and mice. As a result of this phase of the project, the first complete genome sequence of a free-living organism—the bacterium, *Haemophiulis influenzae*—was achieved in 1995. Interesting regions of the human genome—for example, those known to be near genes that cause important genetic disorders such as Huntington disease—were also sequenced. Starting in 1995, some of the smaller human chromosomes are being completely sequenced. During the final phase of the project, sequencing of the entire genome

will be completed. The precise timetable for the project remains fluid. New technologies have already moved the timetable forward, and many researchers now believe that human genome sequencing will be completed before the year 2000.

Importance of the Project

The knowledge gained from the Human Genome Project will be of immediate importance in diagnosing single-gene disorders that affect our species. Progress can also be expected in learning about the underlying nature of enormously complex polygenic (or suspected polygenic) disorders—including alcoholism, mental illnesses, abnormal behavior, obesity, certain cancers, and heart disease—that affect millions of people but are poorly understood at present. Significant biotechnology spin-offs are also anticipated. Further down the road, the new information will also contribute to a greater understanding of gene regulation in eukaryotes, chromosome organization, human genome organization, normal and abnormal cellular growth and development, and evolution.

The genetic information in the genome represents the fundamental instructions for the origin, development, maintenance, and eventual demise of an organism. All questions related to these topics can conceivably be addressed from the conceptual framework provided by knowledge of the genome. However, deep knowledge of the genome does not guarantee that we will be able to understand complex processes such as the development of an organism from a fertilized egg or the underlying molecular basis of all genetic disorders. Even though all genes and their locations may be known, key details for understanding how all of this information is processed may not be available for years or decades.

Ethical, Legal, and Social Implications

Approximately 3 percent of the Human Genome Project budget has been used to fund a program for studying its ethical, legal, and social implications—a program informally identified by its acronym, *ELSI*. The ELSI program is the first federally funded research project in which scientists, social scientists, health care workers, legal experts, and philosophers have formed a working group to consider implications of the Human Genome Project. Nancy Wexler, president of the Huntington Disease Foundation, is the chairperson of the ELSI working group.

The broad goal of the ELSI program is to ensure that the new genetic information will be used to benefit society. The ELSI working group has recognized four primary issues related to the Human Genome Project: fairness, privacy, delivery of health care, and education. These issues are not new, but they will expand as vast amounts of new knowledge become available.

According to ELSI, *fairness* refers to freedom from discrimination on the basis of genotype—for example, not being denied employment or insurance coverage for having a genetic disorder or being a carrier. These problems were mentioned earlier, but both may become increasingly serious as more genetic information becomes available from the Human Genome Project. Fairness also relates to cultural and societal differences. For example, ethnic groups tend to respond to health problems in different ways, and such differences must be appreciated by educators and others describing new genetic information.

Privacy means that an individual controls the generation and disclosure of personal genetic information. Privacy also applies in cases where individuals do not want to have themselves tested for mutant alleles. For example, some people simply may not wish to know whether they will develop Huntington disease later in life.

Delivery concerns practices of the physicians, counselors, and laboratories that generate and supply information about genotypes. Prenatal tests can reveal if an embryo has a genotype associated with a serious genetic disorder, and results of such tests are often used in making decisions about abortion.

To make the best decision, pregnant women must have access to the most accurate information available. This will become increasingly difficult for counselors to provide as the Human Genome Project matures and potentially overwhelming amounts of information become available. Adequate training of genetic counselors is a related problem.

Education means helping policymakers, educators, health care professionals, biologists, and social scientists as well as the general public, become aware of the new knowledge and of the problems and opportunities that it creates. Ultimately, for society to receive the maximum benefits from the Human Genome Project, everyone must be able to understand both the science and the issues.

BEFORE YOU GO ON The Human Genome Project will enable scientists to learn the location and nucleotide sequence of all genes in the human genome. Genomes of other organisms used in genetics research will also be studied. New information from the genome project will provide an intellectual framework for studying the human organism, from conception to death. Although support for the project is widespread, concerns have been raised about ethical, legal, and social implications. Research on these topics is also part of the genome project.

THE FUTURE OF HUMAN GENETIC TECHNOLOGIES

In the 1990s, progress related to genetic technologies and humans has occurred with breathtaking speed (see the Focus on Scientific Process, "Cystic Fibrosis"). Human gene therapy came of age in a matter of years. In 1990, the question of whether or not to use somatic cell gene therapy was being debated; by 1993, gene therapies were being tested on humans with genetic disorders. Today, somatic cell gene therapy trials are being conducted worldwide, and researchers have flooded the field. The number of human diseases that may respond to gene therapy seems unlimited.

The Human Genome Project offers the supreme scientific challenge to both science and society. Scientists anticipate obtaining the complete set of instructions responsible for our conception, development, maturation, and death. As Nobel laureate James Watson stated, "A more important set of instruction books will never by found by human beings. When finally interpreted, the genetic messages encoded within our DNA molecules will provide the ultimate answers to the chemical underpinnings of human existence." Society must ultimately determine how this knowledge will be used.

Cystic Fibrosis

An old folktale holds that a midwife (a woman who assists another woman in childbirth) would lick the forehead of a newborn baby and, if the sweat tasted extremely salty, predict that the child would die of lung disease. **Cystic fibrosis** (CF)—the disease referred to in the tale—is the leading genetic cause of childhood deaths in Caucasian populations. Approximately 4 percent of Caucasians are carriers, and one child in 2,000 is born with CF. Significant numbers of African Americans are also affected, and CF occurs in other ethnic groups although incidence data are imprecise.

The pathway that led to an understanding of CF, from folklore to present, is a nice example of scientific process and the lineage-of-questions concept. The first detailed description of CF, published in 1938, emphasized effects on the pancreas, not the lung. Subsequently, it was recognized that CF disrupted functions of the lung, liver, pancreas, intestine, sweat glands, and male reproductive tract. Further, CF effects on various organs varied widely among affected individuals, as did the age at which they died. Infants whose lungs were affected frequently died soon after birth, but patients with mild forms of CF, with little lung involvement, lived to middle or old age. These differences led to confusion about whether CF was a single disease or several diseases.

What is the genetic basis of CF? Pedigree analyses of affected families indicated that about one-fourth of all CF children had unaffected parents. Based on these studies, autosomal recessive inheritance, involving a single gene, was first reported in 1949. During the next three decades, little was learned about the underlying cause of CF. Because it was caused by a single recessive allele, many researchers assumed that a single abnor-

mal enzyme was responsible. The effects of CF on various organs—pancreas, liver, intestine, sweat glands, reproductive system, and lungs—were carefully cataloged during this period. In 1953, careful studies confirmed that excessive salt loss occurs in the sweat of children with CF, and this observation led to the development of a diagnostic test in which sodium and chloride concentrations were measured in sweat.

In the 1960s and 1970s, moderately effective treatments were devised for children with CF, and further improvements have since been made. CF patients are given antibiotics to reduce infections, engage in physical therapy, and follow carefully planned diets. Because of these treatments, the life expectancy of CF patients has increased from less than 30 years to over 40 years at present. It was well understood that CF patients produce abnormally thick, sticky mucus, especially in the lungs, and that the primary cause of death involved restricted air flow and severe bacterial infections in accumulated mucus. Was there something in the secreted materials of CF patients that affected lung cells in some way? To answer this question, rabbit lung tissues and oyster and freshwater mussel gills were exposed to blood proteins obtained from CF patients. The results suggested to investigators that lung and gill cell function was inhibited after exposure to CF proteins. Subsequently, there was an intense effort to identify a hypothesized inhibitory "factor" that caused such an effect, but contradictory results stifled interest, and attention turned in different directions.

By the early 1980s, new genetic and biochemical technologies gave rise to new questions. Was there a single gene locus for CF? It was still not

known if CF was a single-gene disorder or a collection of diseases caused by mutations at different gene loci. Detailed pedigree studies conducted in 1983 suggested there may be two gene loci involved in CF. However, by 1985, most geneticists were convinced that CF and all its variable effects were caused by mutations of one gene. Recall that earlier influential studies of sickle-cell anemia had shown that mutations of a single gene could result in different phenotypes. Exactly where was the CF gene locus? Various chromosomes were hypothesized to be the site of the CF gene locus—chromosome 5 in 1968, chromosome 4 in 1980, chromosome 13 in 1984, and chromosome 7 in 1985, which turned out to be correct.

What was responsible for the increased salt concentration in the sweat of CF patients? In 1983, it was first hypothesized that water and chloride ions were unable to move normally through plasma membranes in CF patients. By 1986, this hypothesis had been extended to indicate that CF was caused by chloride ion channels that functioned abnormally (see Figure 1). An *ion channel* is simply a water-filled opening in a plasma membrane that, when open, allows ions to move passively across the membrane in both directions. In CF patients, channels remain closed, chloride ions cannot leave the cell, and water does not diffuse across the membrane. Consequently, mucus produced in the lung is thick and dry rather than watery, and sweat secreted from sweat glands is abnormally salty. An alternative hypothesis stated that a mutant ion channel regulatory protein, not the ion channel itself, was defective.

Sensational CF-related research results have been reported continually

FOCUS ON SCIENTIFIC PROCESS

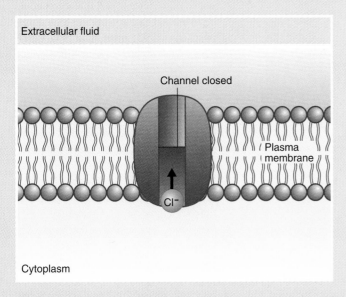

Extracellular fluid

Channel closed

Plasma membrane

Cl⁻

Cytoplasm

Figure 1   Chloride ion channels are water-filled openings in plasma membranes that allow ions to move between the inside and outside of a cell when open but not when closed.

since 1989. On September 8, 1989, scientists from Toronto's Hospital for Sick Children, together with colleagues from the Howard Hughes Medical Institute at the University of Michigan, published three papers on CF. They reported that the CF gene locus is on chromosome 7, band q31 (see Figure 2). The CF gene, now identified as *CFTR*, is categorized as "giant," is composed of 250,000 nucleotide base pairs with 24 exons. The protein product encoded by *CFTR* is named **cystic fibrosis transmembrane-conductance regulator (CFTR).** This *CFTR* protein consists of 1,480 amino acids, and 70 percent of CF cases are caused by deletion of a single codon specifying phenylalanine at amino acid position 508; these mutations are identified as Δ*F*508.

New questions quickly emerged: How can *CFTR* be cloned? If normal *CFTR* alleles were inserted into CFTR cells, would they be expressed? Would they function normally? In 1990, the nor-

mal *CFTR* allele was cloned using cDNA/retroviral vectors in bacteria. Normal *CFTR* alleles were then introduced into pancreatic cells and lung cells obtained from CF patients, using viral vectors. The alleles were inserted into chromosomes and expressed normal CFTR, which led to the establishment of functional chloride ion channels! What is the exact structure and function of the CFTR?

In 1991 and 1992, this question was at least partially answered. In one elegant experiment, normal *CFTR* alleles were inserted into cell types that normally do not have chloride ion channels (for example, certain human cancer cells). The alleles were expressed, and once CFTR was produced, functional chloride ion channels appeared in the cell's plasma membranes. The same effect was demonstrated when CFTR formed chloride channels in artificial lipid membranes. Thus CFTR was

shown to function as a chloride ion channel. Additional research led to the detailed structural model of the ion channel that is described in Figure 3 (see page 432). According to the model, the huge *CFTR* protein has five different regions, or "domains": two transmembrane domains that span the membrane; two nucleotide-binding folds (NBFs) that bind and cleave adenosine phosphate (ATP), which provides energy for closing and opening the channel by moving the R domain in and out of the channel pore. ATP is a compound that provides energy for many cell functions; its production and use are explained in Chapter 4. As shown in Figure 4 (see page 433), additional studies revealed that cells with mutant *CFTR* alleles

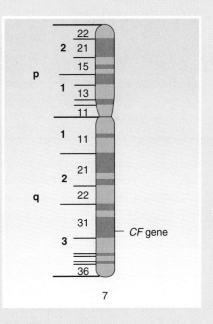

Figure 2   In 1989, the gene responsible for cystic fibrosis is identified and mapped to chromosome 7. The gene is identified as *CFTR* and its protein product is named cystic fibrosis transmembrane-conductance regulator, or CFTR.

box continues

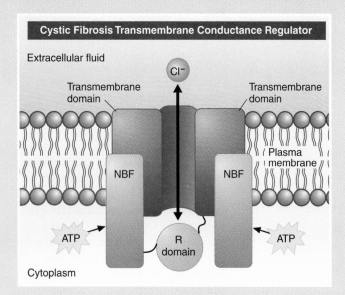

Figure 3   The huge protein encoded by *CFTR* is an enormously complex chloride channel that has five domains, or parts. Two are transmembrane domains that span the plasma membrane, and two are known as nucleotide-binding folds (NBFs), which can obtain energy from ATP and use it to move the R domain, which acts as the "gate" in opening and closing the ion channel.

encode a protein that either fails to move from the endoplasmic reticulum (ER) and Golgi apparatus, where it is synthesized and processed, to the plasma membrane where it functions (for example, the $\Delta F508$ mutation has this effect) or fails to open properly if it is transported to the membrane. How do these defects cause cystic fibrosis? That question is still in search of an answer. Does the loss of functional ion channels by itself explain all CF abnormalities, or does the extremely complex *CFTR* protein have additional functions that remain to be identified?

The successful cell studies completed in 1990 suggested that gene therapy protocols could be developed. In 1992, normal human *CFTR* alleles were inserted into adenovirus vectors that were sprayed directly into the lungs of cotton rats. A significant number of lung cells expressed *human* CFTR within two days, and expression continued for up to six weeks. Could a similar gene therapy protocol be used on

humans? A proposal was submitted, and approved somatic cell gene therapy tests began in 1993. Normal *CFTR* alleles carried by adenovirus vectors were placed in nasal cells of three people with CF, and the infected cells soon began to produce normal CFTR. Subsequently, experiments were conducted in which lung cells were infected, and the results were the same. However, no therapeutic effects have resulted from recent gene therapy trials.

Identification of the *CFTR* gene in 1989 led to accurate diagnostic tests that could be used for both prenatal testing and genetic screening. Because so many Caucasian and African Americans are carriers, the medical community expressed interest in developing a genetic screening program. In 1990, the following guidelines for CF genetic screening were established at the NIH: screening should be voluntary, and confidentiality must be assured; screening requires informed consent; test providers have an obligation to ensure adequate education and counseling; and

there should be equal access to testing. However, the screening test was postponed until more *CFTR* mutations could be identified and appropriate tests developed. By 1993, more than 300 *CFTR* gene mutations had been identified, which made it possible to detect 90 to 95 percent of CF carriers. Consequently, pilot CF screening programs were started in the United States, Canada, and Europe. Society will probably determine if CF screening will be extended to the general population in these countries.

A final interesting question has been asked about CF: why is the $\Delta F508$ mutation so common in Caucasians? Was there once an evolutionary advantage in being heterozygous for CF? (Recall that heterozygotes for sickle-cell anemia have some resistant to malaria.) *Cholera* is a bacterial disease characterized by severe diarrhea, which results in severe water loss that can lead to death. Until the nineteenth century, cholera outbreaks caused major mortalities in European populations. One hypothesis proposes that since the chloride ion channel defect prevents water loss, CF carriers might have had greater resistance to bacterial-caused intestinal diarrhea and would therefore have been more likely to survive cholera epidemics. A recent study, using a mouse animal model, showed that heterozygous mice carrying one mutant *CFTR* allele were less sensitive to cholera toxin than those with two normal alleles. Further studies will be necessary to confirm or reject this intriguing hypothesis.

The phenomenal pace of discoveries is expected to continue in CF research. The recent CF story—*CFTR* gene discovered, *CFTR* protein structure and function determined, normal *CFTR* allele cloned and inserted successfully into cells and animals, human gene therapy tests—will undoubtedly be repeated for other human genetic disorders. What new questions and challenges lay ahead for both science and society?

FOCUS ON SCIENTIFIC PROCESS

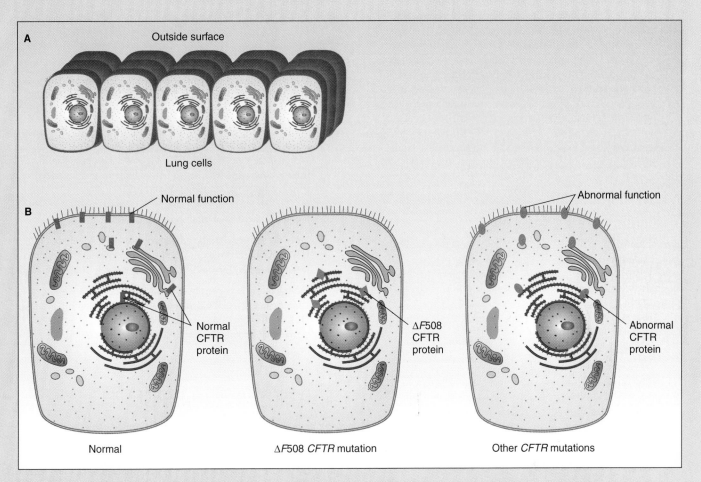

Figure 4   (A) A drawing of lung cells, showing their general organization. (B) In normal cells, CFTR protein is synthesized in the endoplasmic reticulum, is processed in the Golgi complex, and then moves into the plasma membrane, where it functions as a chloride ion channel. Cystic fibrosis patients with the ΔF508 *CF* mutation produce abnormal CFTR protein that remains in the ER, where it is degraded. Cystic fibrosis patients with other *CF* mutations produce abnormal CFTR protein that is transported to the plasma membrane but fails to function.

SUMMARY

1. New genetic technologies have emerged with the potential to benefit humans. It is already possible to diagnose a variety of genetic disorders through prenatal testing or analyzing individuals at risk for such disorders. In the future, the number of genetic disorders that can be diagnosed is expected to increase sharply.

2. The use, or potential use, of diagnostic technologies has led to a number of societal and ethical problems that interest and concern people from all walks of life.

3. Genetic screening programs may be possible for certain genetic disorders. However, such programs have had mixed success in the past, and careful planning and comprehensive education of participants are essential for future programs.

4. Gene therapy—the substitution of a normal allele for a mutant allele—holds great promise for treating certain genetic disorders and other important diseases. Somatic cell gene therapy experiments are now being conducted worldwide. Germline gene therapy studies have not yet been approved but are the subject of current debate.

5. Protocols used for human gene therapies require the development and testing of appropriate vectors using cultured cells, animals, and animal models. The first successful human gene therapy experiment was conducted in 1989, and two more were initiated in 1990 and 1991. Since then, gene therapy has been used in experimental tests to treat familial hypercholesterolemia and other human genetic disorders.

6. New strategies for treating genetic disorders are being pursued. Two of the brightest prospects are antisense com-

pounds, which prevent the expression of mutant alleles, and stem cell gene therapy, which may lead to cures for certain genetic disorders.

7. The Human Genome Project is the most ambitious biology project ever conceived. The ultimate goal of the project is to determine the exact nucleotide sequence of every gene in the human genome. Researchers from many disciplines began studies in 1990 to achieve intermediate objectives such as completing a higher-resolution human genetic map than presently exists and sequencing the genomes of other organisms such as *Drosophila,* and certain plants and animals. Completion of the definitive nucleotide map will open doors for rapid advances in both applied and basic sciences. Ethical, legal, and social implications of the project are also being studied.

WORKING VOCABULARY

animal model (p. 423)	Human Genome Project
gene marker (p. 422)	(p. 427)
gene therapy (p. 419)	physical map (p. 427)
germline gene therapy	somatic cell gene therapy
(p. 419)	(p. 419)

REVIEW QUESTIONS

1. How is the ability to diagnose genetic disorders related to social issues? What are some of the social issues that now confront us as a result of new gene technologies?

2. What is genetic screening? Have genetic screening programs been successful in the past? Why?

3. What is gene therapy?

4. What are some of the ethical issues associated with human gene therapy?

5. What are the two general types of somatic cell gene therapy?

6. How are vectors used in gene therapy? Which vectors have been developed for human gene therapy?

7. How are human gene therapy protocols developed?

8. How are animals used in gene therapy research? What are animal models? Why are they important in research on human diseases?

9. Describe the first experiment conducted on humans using a form of gene therapy.

10. What was the significance of the ADA gene therapy experiments?

11. Describe how the experimental gene human therapy protocol was developed and tested for treating familial hypercholesterolemia.

12. How might antisense compounds be used to treat genetic disorders? Why does stem cell gene therapy offer the prospect of curing genetic disorders?

13. What types of research will be done under the umbrella of the Human Genome Project?

14. How are the results of the Human Genome Project expected to be used?

ESSAY AND DISCUSSION QUESTIONS

1. Describe policies that you would recommend for (a) conducting germline therapy research and (b) applying the results of such research to humans.

2. Somatic cell gene therapy trials are now under way for various genetic disorders. Do you support the idea of using gene therapies to treat genetic disorders? Why? What criteria should be used to determine whether or not an individual should receive gene therapy? For example, should infants receive gene therapy? Should terminally ill patients receive an untested gene therapy if there is no other possible treatment?

3. What do you feel are the advantages and disadvantages of the Human Genome Project? Would you recommend that the project continue go forward, be modified, or be terminated? Why?

4. Genetic knowledge has been misapplied in the past in the United States (recall Chapter 19's Focus on Scientific Process, "Human Genetics and the Abuse of Science"). Are you concerned that knowledge from the Human Genome Project might be misused? Explain.

REFERENCES AND RECOMMENDED READING

Alton, E. W., and D. M. Geddes. 1995. Gene therapy for cystic fibrosis: A clinical perspective. *Gene Therapy*, 2:88–95.

Anderson, W. F. 1992. Human gene therapy. *Science,* 256: 808–813.

Baird, P. A. 1994. Altering human genes: Social, ethical, and legal implications. *Perspectives in Biology and Medicine,* 37: 566–575.

Benjamin, C. M., S. Adam, S. Wiggins, and J. L. Theilmann. 1994. Proceed with care: Direct predictive testing for Huntington disease. *American Journal of Genetics,* 55: 606–617.

Bulger, R. E., E. Heithman, and S. J. Reiser (eds.). 1993. *Ethical Dimensions of the Biological Sciences.* New York: Cambridge University Press.

Cooperative Human Linkage Center. 1994. A comprehensive human linkage map with centimorgan density. *Science,* 265: 2049–2054.

Farzanh, F. 1995. *Functional Analysis of the Human Genome.* Oxford: BIOS.

Hodson, M. E., and D. M. Geddes (eds.). 1995. *Cystic Fibrosis.* London: Chapman & Hall.

Jorde, L. B. 1995. *Medical Genetics.* St. Louis, Mo.: Mosby.

Kay, L. E. 1993. *The Molecular Vision of Life: Caltech, the Rockefeller Foundation, and the Rise of the New Biology.* New York: Oxford University Press.

Keller, E. F. 1995. *Refiguring Life: Metaphors of Twentieth-Century Biology.* New York: Columbia University Press.

Kevles, D. J., and L. Hood (eds.). 1992. *Scientific and Social Issues in the Human Genome Project.* Cambridge, Mass.: Harvard University Press.

Marteau, T., and M. Richards (eds.) 1996. *The Troubled Helix: Social and Psychological Implications of the New Human Genetics.* New York: Cambridge University Press.

McConkey, E. H. 1993. *Human Genetics: The Molecular Revolution.* Boston: Jones & Bartlett.

National Academy of Sciences. 1994. *Assessing Genetic Risks: Implications for Health and Social Policy.* Washington, D.C.: National Academy Press.

Polmin, R., M. J. Owen, and P. McGuffin. 1994. The genetic basis of complex human behaviors. *Science,* 264: 1733–1739.

Rasko, I. 1995. *Genes in Medicine.* London: Chapman & Hall.

White, R., and C. T. Caskey. 1988. The human as an experimental system in molecular genetics. *Science,* 240: 1483–1488.

Wivel, N. A., and L. Walters. 1993. Germ-line gene modification and disease prevention: Some medical and ethical perspectives. *Science,* 262: 533–538.

ANSWERS TO FIGURE QUESTIONS

Figure 22.4 The fact that it is widely viewed as being just another approach to treating human diseases.

Figure 22.6 The transfusion type. The first disorder treated was severe combined immune deficiency in 1990.

Figure 22.7 So that they cannot replicate in cells and continue to infect and kill new cells.

Figure 22.9 Humans are not experimental animals, and relatively few have specific genetic disorders. An animal model has the same disorder, with the same cause, as a disorder that occurs in humans. By studying animal models, scientists can learn more about specific disorders in a much shorter time period.

Figure 22.13 The *dystrophin* gene (see Chapter 20).

Unit Three Conclusion
The Language of Life and the Process of Science

The earliest written records reveal that humans have always been interested in understanding why offspring resemble their parents. Aristotle's erroneous ideas on the subject lasted almost 2,000 years. The scientific revolution of the seventeenth century engendered a mechanical philosophy of nature, which introduced new ideas that replaced Aristotle's outdated concepts.

During the second half of the nineteenth century, scientific perspectives on inheritance were developed through experimental studies with plants and efforts to answer questions raised by the theory of evolution. In 1900, de Vries published results from his experiments that supported the concept that discrete units were inherited from parents. He also called attention to Mendel's earlier overlooked experiments on segregation and independent assortment. De Vries's work also gave rise to a new question that drove research for over a decade: where were these discrete units located?

In 1902, Sutton articulated the chromosome theory of heredity: genes are carried on chromosomes. Between 1910 and 1915, Morgan and his coworkers conducted classic experiments on fruit flies that confirmed the chromosome theory. An interesting new question concerning the chemical identity of the genetic material then dominated research for over two decades. The answer—that DNA is the genetic material—came from celebrated experiments, using bacteria and viruses, conducted between 1928 and 1953. Two new fields of science, molecular biology and molecular genetics, trace their roots to work done during this period. What was the structure of DNA? In 1953, Watson and Crick answered that question and gave rise to new questions about DNA replication and gene expression that were quickly answered. The answers laid the groundwork for new investigations of DNA, RNA, gene expression, and tools that could be used to manipulate genes. In the past two decades, the fruits of this research have provided answers to questions related to genetic engineering, gene cloning, human genetic disorders, and gene therapy.

The importance of genetic knowledge for society is reflected by the massive amount of money that government and private agencies have invested in research. The Human Genome Project is part of a worldwide effort to increase our understanding of human heredity. Recombinant DNA technology and genetic engineering promise to revolutionize agriculture, animal husbandry, and medicine.

Despite these advances, society faces difficult questions related to new knowledge and new genetic technologies. The early history of eugenics stands as a warning of the ease with which contemporary social values can become integrated into the science of genetics. Many thoughtful individuals are troubled by the prospect that the pace of scientific discovery will surpass society's ability to deal adequately with the consequences.

The Evolution of Life

IV

2000	
Present	
1980	
	How has social behavior evolved?
1960	
	How have physiological pathways evolved?
	How have ecological units evolved?
	How do genes in a population change?
1940	
1920	
	What are the isolating mechanisms in evolution?
1900	
	Are there different patterns of evolutionary change?
	Are there different rates of evolution?
1880	Can human behavior be explained by evolution?
	What is the evolutionary significance of behavior?
1860	What is the source of variation?
	How is variation inherited?
	What accounts for biogeographical patterns?
1840	How do new species arise?
A.D.1700	What is the relationship between fossils and living forms of life?

23

Darwin and the Origin of the *Origin*

Chapter Outline

Reading Questions

1. What types of information and material did Darwin collect on his voyage?

2. What questions did Darwin ask as a result of observations that he made on his journey?

3. What is the importance of Darwin's theory of evolution?

4. How did Darwin's ideas fit in with those of the time?

As we saw in Chapter 1, the traditional goals of natural history have been to describe and classify the diversity of life on Earth. Naturalists, working in museums, have constructed detailed taxonomic systems, such as those discussed in Chapters 5 and 6. Although these systems are the best known products of natural history, naturalists have always been interested in discovering overall biological *patterns* and in explaining them. For example, scientific expeditions in the eighteenth century revealed that some species of plants and animals were found only in certain parts of the world, whereas other species inhabited most areas of Earth. What could account for such a pattern?

NATURAL HISTORY AND EVOLUTION

Until the nineteenth century, most naturalists believed that the plants and animals they described had not changed in time. Linnaeus, for example, explained that the distribution patterns discovered by eighteenth-century naturalists were the result of plants and animals having been dispersed from the original site of God's creation. These explanations did not include the concept of change through time. A few individuals, such as the famous eighteenth-century naturalist Georges-Louis Leclerc, comte de Buffon, thought that the environment could influence the appearance of living beings, but he believed that such changes were limited and reversible. Those few scientists and philosophers who did believe that significant change had occurred were not taken seriously in established scientific circles. For example, Jean-Baptiste-Pierre-Antoine de Monet de Lamarck, a French scientist of the late eighteenth century, believed that there had been a general *evolution* from simple one-celled organisms to highly complex animals such as humans. Because he could not identify any mechanisms responsible for such changes, his scientific contemporaries dismissed his ideas as idle speculation. Ironically, Lamarck is better known today for his ideas on evolution than he was in his own day. This later fame resulted from efforts by some writers to identify earlier thinkers who may have had similar ideas (for further information on this subject, see the Focus on Scientific Process, "The Convergence of Ideas in Biology" in Chapter 24).

Early in the nineteenth century, Georges Cuvier, a great comparative anatomist and one of Lamarck's colleagues at the famous Paris museum of natural history, made detailed studies on animal structures. Based on results of his research, he claimed that he had identified a comprehensive anatomical pattern that could serve as the scientific foundation for a new classification system. His "laws of comparative anatomy" stated that all parts of the animal body were interrelated, that some organ systems were more important to understanding the basic body plan than others, and that all taxonomic units could be rigorously defined and were anatomically stable. From these ideas, Cuvier then produced the most advanced and complete taxonomy of animals that had ever been accomplished.

Cuvier's ideas had important theoretical implications for any theory of evolution. He maintained that animal structures were so complex that it would be *impossible* for any major structural changes to occur in a specific group (species) over time. He argued that a significant anatomical change in any one part would result in death of the organism. Other naturalists extended his ideas to include plants. Consequently, the most rigorous and widely accepted classification system included this basic principle: significant change was not possible for any species.

The industrial revolution of the early nineteenth century had many important consequences for natural history. During this period, European powers conducted extensive explorations throughout the world, searching for new markets and sources of inexpensive raw materials. These expeditions resulted in waves of new plant and animal specimens flooding into Europe from Africa, Asia, Australia, and the Americas. As a result, museums in London, Paris, Leyden, Berlin, and Vienna increased their holdings dramatically. Cuvier's classification system again showed its value because it could accommodate the new species discovered throughout the world.

Although Cuvier and his followers could name and classify all new species brought back to the great museums of Europe, these specimens raised a set of interesting, new questions. What could account for global distribution patterns that were observed among plants and animals? Why were plants and animals from closely related parts of the world similar, yet different? What were the relationships between fossils and living forms? Why was there so much variation among individuals of certain species? Cuvier's writings held no answers to these new questions.

Some naturalists, referring to theological ideas and beliefs, advanced the idea that whatever patterns or relationships existed, they were part of "God's plan." The ultimate purpose of this plan may have been difficult to understand, but its general outline could be perceived. Zoologist Louis Agassiz (discussed later in this chapter) took this approach. Other naturalists reverted to some of the speculative ideas of the eighteenth and early nineteenth centuries and used the idea of change in time in attempting to answer the new questions. But many naturalists were still uncomfortable with that idea because they could not conceive of a mechanism to account for dramatic changes in animals and plants. There was a third approach that offered greater hope for answering the new questions: observe, compare, and seek fundamental, scientific laws that might explain the patterns. Where such an inquiry would lead was unknown, but several naturalists were committed to seeking a general solution to the problems in natural history. The most famous of these was Charles Darwin.

CHARLES DARWIN

Charles Darwin was born in Shrewsbury, England, on February 12, 1809—the same date as Abraham Lincoln. Unlike Lincoln, who was born in modest circumstances, Darwin came from a well known and wealthy English medical family, and his relatives expected him to become a physician like his father and grandfather. After two years of medical school in Edinburgh, however, Darwin decided that he did not want to practice medicine; he wanted to become a clergyman instead. To prepare for the ministry he enrolled at the University of Cambridge, but an unexpected chance to make a long voyage caused him to change his goals once again.

DARWIN'S VOYAGE

Shortly after he completed his studies in 1831 at the University of Cambridge, Charles Darwin received a letter from one of his professors, the Rev. John Stevens Henslow, informing him of an exciting opportunity. The British government was sending out a surveying ship to chart the coast of South America and to make various measurements at sea. The captain of the ship was looking for a naturalist to accompany him, and Henslow had suggested Darwin. The letter urged the young college graduate to follow up on the recommendation. Henslow knew that Darwin would be a reliable naturalist on the expedition, for he himself had helped to train him.

Darwin was excited by the prospect of the voyage, but his father Robert Darwin was initially opposed to the idea. He thought that it was about time for his son to settle down. His son had already made an unsuccessful attempt at studying medicine in Edinburgh and had subsequently gone to Cambridge to prepare for the ministry. Robert Darwin didn't think that a voyage would add to his son's qualifications for his chosen profession. Eventually, however, Charles was able to obtain his father's consent.

That voyage is very famous today. As a naturalist abord the ship, Darwin not only had the opportunity to make thousands of observations, but he also developed into a mature scientist (see Figure 23.1). More important, he began to ask himself a set of questions that ultimately led him to formulate the theory of evolution.

Figure 23.1 Charles Darwin was 22 at the time of his voyage on the HMS *Beagle*. Naturalists aboard British expedition vessels were expected to make observations and collections that were useful in several areas of science, including botany (plants), zoology (animals), paleontology (fossils), geography (exploration and mapping), ethnology (native peoples), and geology (minerals and land forms).

The journey was long, dangerous, and at times difficult (he was dreadfully seasick at first), but it was clearly one of the greatest experiences of Darwin's life. The ship, HMS *Beagle* (Figure 23.2), left England two days after Christmas in 1831 and returned in October 1836 after circling the globe. A map of the voyage is shown in Figure 23.3.

Darwin's Observations and Questions

The expedition's activities were concentrated primarily on South America, but other areas—Australia, South Africa, and many oceanic islands—were also studied. Henslow's recommendation had been accurate, for Darwin had been well trained at Cambridge to look carefully at geological formations, plants, and animals. He assembled enormous collections; it would take scientists years to examine them and publish descriptions of the new species Darwin discovered. An illustration of some beetle specimens that Darwin observed is shown in Figure 23.4. Even more significant than the contribution Darwin made to the natural history of South America and other places he visited was what he did with his observations and collections. Darwin went well beyond being a collector by asking interesting questions about his data and what they meant.

Fossils

Fossils were among the most exciting finds Darwin made in South America. He was very impressed by the large number of fossils from extinct species. What was the significance of all these fossils? They suggested to him that the number of species that had become extinct must be enormous. Many of the fossils were of unusual extinct animals such as the giant ground sloth, shown in Figure 23.5, and giant armadillos. One of the things that struck Darwin about these discoveries was that although the bones were the remains of extinct animals, they

Figure 23.2 This watercolor, owned by Darwin, shows the HMS *Beagle* in the Murray Narrow, a small channel at the southern extremity of South America.

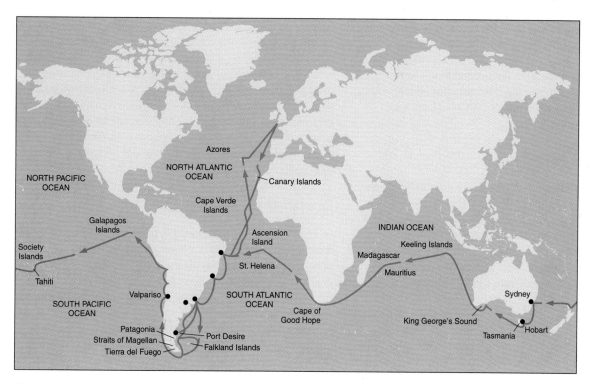

Figure 23.3   Darwin's voyage (1831–1836) proceeded from England to the coast of South America, across the Pacific to Australia, and back around the south of Africa to England.

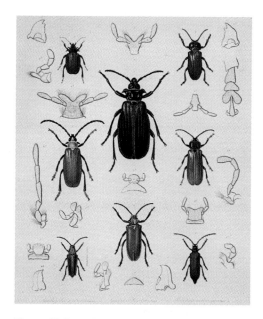

Figure 23.4   Darwin arranged to publish his observations after he returned to England. He contacted noted specialists to describe his collections and artists to draw them. This illustration contains beetles that Darwin saw in Chile.

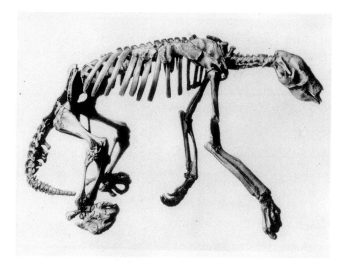

Figure 23.5   One of the fossils that impressed Darwin was the giant sloth. This fossil was found in an area that contained smaller living animals that resembled the giant fossil.

Question: *What might be the relationship between the two?*

were found in areas where similar animals still lived. Was there a relationship between the extinct forms and those presently living in the area? Other naturalists had noted similar relationships between the remains of extinct and living forms. For example, many Australian fossils were similar to contemporary Australian forms. The same relationship could be found in other areas of the world.

Biogeographical Patterns

Darwin also observed a similarity among organisms (living and extinct) in large geographical areas. Different regions of the world appeared to him to have a basic "character." Many South American animals and plants "looked" South American compared to European species. Darwin's view of a basic geographic pattern reflected a common idea in the nineteenth century: the world could be divided into several large regions, and the plants and animals of those regions have distinct characteristics. Although some plant and animal groups are found throughout the world, many are centered in only one region. These large-scale distribution patterns are now called **biogeographical realms.** What particularly struck Darwin was that a

definite pattern existed in localized regions within larger, general areas. As he traveled southward in South America, Darwin noted that closely related species were often found in adjacent areas. For example, he discovered a previously unknown rhea (a large, flightless bird) and found that the common rhea, to which the new species was closely related, inhabited the adjacent area (see Figure 23.6). This and similar examples suggested to him that there was a complex distribution of plants and animals, in both time (the fossils) and space (biogeographical patterns). What could explain these patterns?

Adaptation

In addition to being distributed in patterns, plants and animals are also remarkably well suited to the particular spot in which they are found. As a boy, Darwin had been taught that God created each species, unchangeable and "perfectly adapted to its environment." For most of his voyage, Darwin saw again and again how this teaching seemed to be borne out in the wonderful adaptations of organisms. Adaptation in Darwin's day referred to how the structure and function of organisms were perfectly "designed" for survival in the specific envi-

A

B

Figure 23.6 As Darwin traveled southward in South America, he often found closely related species in overlapping ranges. A good example of this was a previously unknown rhea he found in Argentina, (A) *Pterocnemia pennata,* which is closely related to the common rhea, (B) *Rhea americana,* also found in Argentina.

ronment in which they were found. Just as people design different machines for different jobs, God had designed different animals and plants to live in different places on the globe. For example, in what is now Argentina, Darwin visited a salt lake, like the one in Figure 23.7, and discovered that in spite of a high concentration of salt in the water, large numbers of animals survived along the shore. Similarly, he marveled at the unusual black skimmer, *Rhynchops nigra,* shown in Figure 23.8, which has a long beak flattened laterally and an extended lower mandible, giving it an uncommon ability to catch fish.

Toward the end of his voyage, however, Darwin made some observations that cast doubt on the concept that animals were perfectly adapted. In Australia, for example, he saw that some introduced species had replaced native species. This did not fit the concept of perfect adaptation, for if each species was indeed perfectly adapted to its own environment, then an introduced species should compete poorly in the "foreign" envi-

Figure 23.7 Mono Lake in California is a salt lake that resembles the one Darwin saw on his voyage. Salt-tolerant grasses grow along its shore.

Figure 23.8 Darwin marveled at the beak of this *Rhynchops nigra,* which allowed it to catch fish. In *The Voyage of the Beagle,* Darwin wrote: "I here saw a very extraordinary bird [*Rhynchops nigra*]. It has short legs, web feet, extremely long-pointed wings, and is of about the size of a tern. The beak is flattened laterally, that is, in a plane at right angles to that of a spoonbill or duck. It is as flat and elastic as an ivory paper-cutter, and the lower mandible, differently from every other bird, is an inch and a half longer than the upper. . . . I saw several of these birds, generally in small flocks, flying rapidly backwards and forwards close to the surface of the lake. They kept their bills wide open, and the lower mandible half buried in the water. Thus skimming the surface, they ploughed it in their course."

ronment, and the local species should have the competitive advantage. Although Darwin came to realize that adaptation is never perfect, he nonetheless maintained his fascination with the exquisite adaptations found in nature and realized their importance for survival.

Oceanic Islands

> The natural history of these islands is eminently curious, and well deserves attention. Most of the organic productions are aboriginal creations, found nowhere else; there is even a difference between the inhabitants of the different islands; yet all show a marked relationship with those of America, though separated from that continent by an open space of ocean, between 500 and 600 miles in width. The archipelago is a little world within itself, or rather a satellite attached to America, whence it has derived a few stray colonists, and has received the general character of its indigenous productions. Considering the small size of these islands, we feel the more astonished at the number of their aboriginal beings, and at their confined range. Seeing every height crowned with its crater, and the boundaries of most of the lava-streams still distinct, we are led to believe that within a period, geologically recent, the unbroken ocean was here spread out. Hence, both in space and time, we seem to be brought somewhat near to that great fact—that mystery of mysteries—the first appearance of new beings on this earth.

This quotation from *The Voyage of the Beagle* (1845) reflects Darwin's fascination for the life that he found on islands in the ocean. The most famous of these islands, and the ones referred to in the quotation, are the **Galápagos Islands** (Figure 23.9), located in the Pacific near the equator, about 600 miles off the coast of South America. Darwin was surprised by the number of indigenous species ("aboriginal creations") in the Galápagos; that is, the species were native, not introduced from elsewhere. It seemed to him that although they were distinct species, many of the island's plants and animals must be related to South American organisms. In descriptions of the Galápagos Islands, published when he returned to England, Darwin wrote that there were even differences in the animals among the islands. Why should there be such diversity in a cluster of small and remote islands? Moreover, what was the original source of the native plants and animals? The islands themselves had been formed by volcanoes during the present epoch, yet the flora and fauna were mostly unique to the Galápagos. In thinking about these fairly barren islands containing species found nowhere else, Darwin was led to ask, What is the origin of new species?

BEFORE YOU GO ON   Several critical questions occurred to Darwin during his voyage: What was the relationship of fossil to living forms? What was the cause of biogeographical patterns? How had animals and plants become adapted to their environment? What was the origin of plants and animals on the Galápagos Islands?

DARWIN'S COLLECTIONS AND THE CHANGE OF SPECIES

After Darwin returned from his voyage in 1836, he settled in London for several years to work on his collections and to arrange for the publication of their description. As part of this process, he consulted the leading British authorities in natural history. They were very impressed with the collections, which contained not only unusual specimens of unknown species but also interesting information on their distribution.

Questions About Darwin's Data

Discussions with preeminent naturalists led Darwin to consider some fruitful questions. For example, John Gould, a famous ornithologist (bird specialist), informed Darwin that his collection of Galápagos finches actually included several different species. Darwin was quite puzzled by this. Why should there be so many species of finches on a remote group of islands? Where had they come from?

For many months, Darwin pondered these questions and others relating to fossils and geographic distribution. Finally, he decided that if he made the unorthodox assumption that species *could* change over time, he could explain many of the unusual aspects of his data. It became evident why there were different species of finches: they were the descendants of an original species that had somehow become modified to fit new, different habitats on the islands. If species could change, many other questions could also be resolved. For example, the relationship of fossils to present-day forms could be explained by assuming that the contemporary forms were related to the fossil forms by descent. That is, existing species that had strong similarities to fossils of extinct species were actually the distant descendants of those (or related) species. By studying the fossil record, it was possible to construct a **phylogeny,** a history of the evolutionary development of a species or of a larger group of organisms. Closely related species that were found in adjacent areas might be related to a common ancestor, or one species might be an offshoot of the other. The South American character of the Galápagos Islands might be explained if the original inhabitants had come from South America. The questions that could be generated and answered were endless. Some of these are illustrated in Figure 23.10 (see page 448).

Natural Selection

How do species change? Darwin reflected on this question for over a year, trying to develop some hypothesis to explain it. He finally came up with a possible solution: **natural selection** (natural selection, which is discussed in detail in Chapter 24, is considered here briefly). Darwin was aware that *variation* existed in plant and animal populations because he had studied how domestic breeders had created new varieties by selecting offspring with certain desired characteristics. For example,

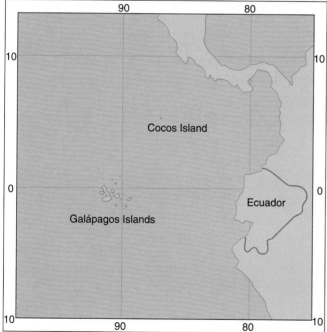

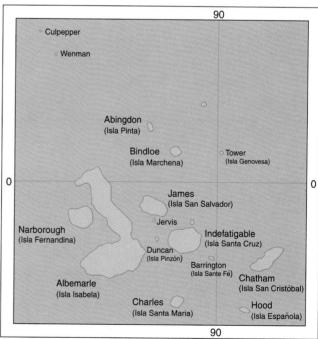

Figure 23.9 The natural history of the Galápagos Islands fascinated Darwin. As can be seen in the photo above, the main islands of the Galápagos feature a rugged terrain. Maps of the islands are shown below photo.

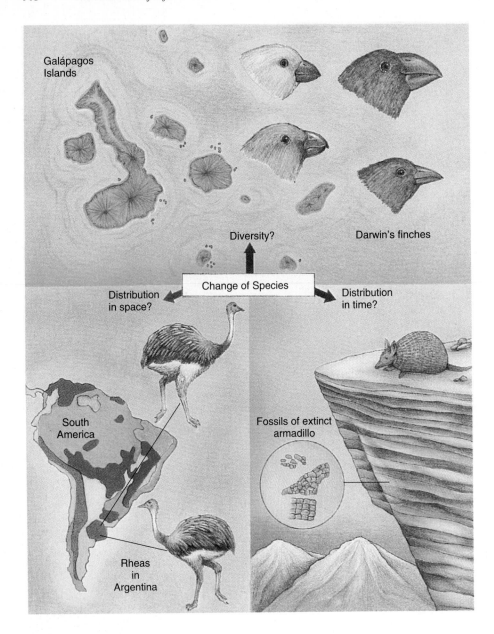

Galápagos Islands

Darwin's finches

Diversity?

Change of Species

Distribution in space?

Distribution in time?

South America

Rheas in Argentina

Fossils of extinct armadillo

Figure 23.10 Darwin came to realize that if he accepted the idea that species change, many of the puzzling observations that he had made on his voyage could be explained. He could understand the diversity of island organisms, relate fossils to living forms, and explain biogeographical distribution patterns. For example, Darwin reasoned that the different species of Galápagos finches were descendants of an ancestral species that had somehow been modified to fit different habitats on the islands (see also Figure 23.12, page 450). He also reasoned that contemporary organisms were related to fossil forms of the organisms by descent. For instance, he assumed that existing species of armadillos were actually the distant descendants of certain species or related species of fossilized armadillos. Finally, faced with different but closely related species found in adjacent areas, such as two different species of rheas (see also Figure 23.6, page 444), Darwin explained that these species might be either related to a common ancestor or offshoot species of one another.

Question: *What is phylogeny, and how did it play a role in Darwin's studies?*

he was amazed by the great number of varieties of pigeons that pigeon fanciers had produced.

In 1838, Darwin read an essay by the Rev. Thomas Malthus (see Chapter 12), who believed that the rate of growth in human populations was exponential (or "geometric"; that is, 2, 4, 16, 32, 64, . . .), whereas the rate of increase of food was arithmetic (that is, 2, 4, 6, 8, . . .). Consequently, Malthus concluded that humans would face a dramatic struggle for existence. Darwin combined these ideas of variation and population growth. He realized that animal and plant populations have tremendous potential for reproducing (see Figure 23.11 for one rather dramatic example), yet they appear to occur in roughly the same numbers over periods of time; therefore, only a few offspring must actually survive to reproduce and leave progeny. Further, he reasoned that those who were more "fit" would be more likely to survive and leave offspring than those less "fit." In other words, Darwin hypothesized that there would be a *natural selection*. The result of natural selection would be analogous to events in domestic breeding, where breeders select for mating only offspring displaying the traits that they are trying to enhance.

Natural selection differed from domestic selection in two important ways: no one was performing the selecting, and there was no predetermined goal in mind (for example, producing tomatoes that ripen early and have the diameter of hamburger buns). Also, domestic breeders had never created a new species, only new varieties of the same species. Darwin thought that given sufficient time, natural selection would pro-

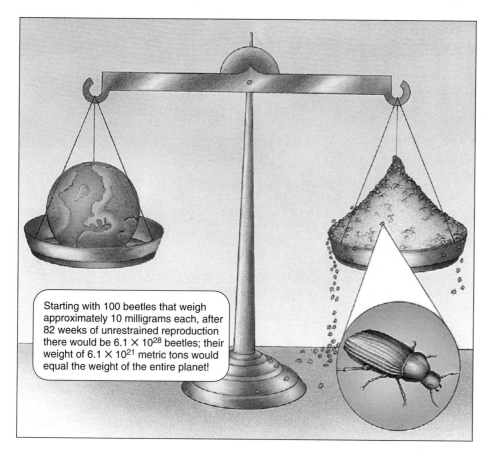

Starting with 100 beetles that weigh approximately 10 milligrams each, after 82 weeks of unrestrained reproduction there would be 6.1×10^{28} beetles; their weight of 6.1×10^{21} metric tons would equal the weight of the entire planet!

Figure 23.11 In his discussion of unrestrained growth, Malthus provided Darwin with the key to understanding how species change. If there are no checks, a population will increase and soon deplete its food sources. This drawing dramatizes how many beetles would exist if all offspring of a colony of 100 beetles lived and reproduced for 82 weeks.

duce new species. He thought that this would occur because as the environment gradually changed in time, new adaptations would be favored in the new environment. In addition, new ways of exploiting the environment might be developed by individuals with novel variations. In time, many new species could arise. Figure 23.12 (see page 450) provides a modern analysis of the Galápagos finches and the beaks that distinguish them and permit them to exploit resources from different habitats. Such models may be used to help explain how the great diversity of life on Earth originated.

Darwin's hypothesis that natural selection resulted in the development of new species provided a new perspective on species. Rather than considering a species to be a design or blueprint for an organism, a species now had to be considered a population of individual organisms. The adaptation of a species to its environment was no longer just a part of the species' "design." Instead, adaptation was a *process* that resulted in survival in surrounding conditions. From this new perspective, a species can never be "perfectly" adapted, only relatively well adapted.

> **BEFORE YOU GO ON** By hypothesizing that species changed over time, Darwin could solve all of the problems he addressed. His concept of natural selection explained how species changed.

THE THEORY OF EVOLUTION

Of course, Darwin could not test his natural selection hypothesis directly (place a population of mice on an island and watch them for 100,000 years?), so he attempted to test it indirectly by assuming that natural selection was operating. Then by considering the known facts of natural history, he tried to determine whether the expected patterns occurred or whether they were consistent with what might be expected to occur.

Darwin's Theory

Darwin was a cautious investigator with indefatigable patience. He read vast numbers of books and journals, wrote to hundreds of specialists, made many observations, and conducted numerous experiments. As a result, he broadened his perspective and was finally able to construct a theory of evolution that not only explained the origin of species but also synthesized known facts from the biological sciences (see the Focus on Scientific Process, "Levels of Scientific Inquiry"). Moreover, the new theory pointed the way to areas of research that proved to be highly productive.

Darwin's explanation about the origin of species was a bold move in his day, but one that fit a general trend toward explaining nature in physical terms. Scientists wanted studies

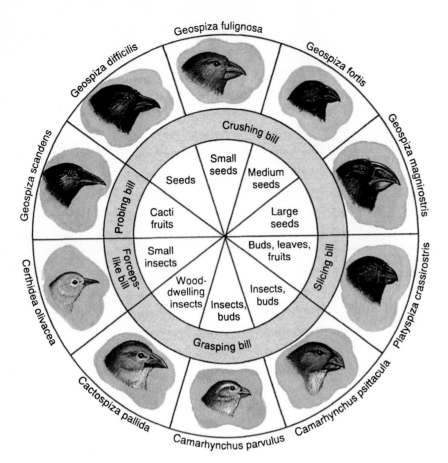

Figure 23.12   A result of natural selection is adaptation, as illustrated by the various beaks of Galápagos finch species. The diagram relates the beaks of some finches to their food supply.

of the living world to be rigorous, like those of physics, chemistry, astronomy, and geology. To Darwin and many of his contemporaries, that meant that such studies should be based on careful observations of living objects and attempts should be made to discover the laws that governed them. Practitioners of the other leading sciences (such as physics) attempted to explain nature in terms of physical concepts and underlying mechanisms. Darwin's theory of evolution was in the spirit of modern science because it used physical and mechanical explanations to account for the phenomena of life.

We should state here that Darwin was not the only naturalist of his time to develop interesting and intriguing biological theories. As you will see in the Chapter 24 Focus on Scientific Process, "The Convergence of Ideas in Biology," Alfred Russel Wallace formulated a theory of evolution that was almost identical to Darwin's. Naturalists in France and Germany also attempted to construct broad theories based on their understanding of the physical world. None of these, however, was as successful and far-reaching as Darwin's theory of evolution.

BEFORE YOU GO ON   The hypothesis of natural selection explained more than the immediate problem of how species change. By considering how natural selection and the change of species were related to the major areas of biology, Darwin was able to formulate a general theory that explained the phenomena of life.

Accepted Ideas on the Origin of Species in Darwin's Time

All organized beings exhibit in themselves all those categories of structure and of existence upon which a natural system may be founded, in such a manner that, in tracing it, the human mind is only translating into human language the Divine thought expressed in nature in living realities. . . .

The combination in time and space of all these thoughtful conceptions exhibits not only thought, it shows also premeditation, power, wisdom, greatness, prescience, omniscience, providence. In one word, all these facts in their natural connection proclaim aloud the One God, whom man may know, adore, and love; and Natural History must in good time become the analysis of the thoughts of the Creator of the Universe, as manifested in the animals and vegetable kingdoms, as well as in the inorganic world.

Louis Agassiz, *Essay on Classification* (1857)

He who believes in separate and innumerable acts of creation will say, that in these cases it has pleased the Creator to cause a being of one type to take the place of one of another type; but this seems to me only restating the fact in dignified language. He who believes in the struggle for existence and in the principle of natural selection, will acknowledge that every organic being is constantly endeavouring to increase in numbers; and that if any one being vary ever so little, either in habits or structure, and thus gain an advantage over some other inhabitant of the country, it will seize on the place of

FOCUS ON SCIENTIFIC PROCESS

Levels of Scientific Inquiry

Scientists use a variety of techniques and methods to gain knowledge of the living world. As we have seen many times in this text, there is no single, simple "scientific method." Science is a process, and what scientists do often resembles what scholars do in other areas of research. Let's consider what scientists do by looking at different *levels of inquiry.*

Observation

Observation provides much of the basic data of science. Observations, however, are not made randomly. During his voyage, Darwin made observations related to topics that interested him—indications of past geological change, unusual patterns in the distribution of organisms, and interesting adaptations of various plants and animals. On reading Darwin's account of his trip, *The Voyage of the Beagle,* one finds that he selected his material carefully. In describing South American fossils, for example, Darwin placed his observations in the context of the question "What is the relationship of fossils to living forms?" Scientists, such as Darwin, make observations related to relevant questions or interests.

Generalization

An individual observation takes on greater significance if it can be related to some broader statement. Scientists look for patterns. In reviewing his observations, Darwin made many generalizations: He noted that closely related species were found in adjacent areas, he recognized a relationship of island fauna to nearby continents (such as the "African character" of the Cape Verde Islands), and after examining over 10,000 specimens of barnacles, he stat-

ed that practically all the external characters of a species are highly variable. Figure 1 shows some of the barnacles Darwin studied.

Explanation

What makes science interesting is the attempt to explain patterns, or even single events, in nature. Initially, a calculated statement is made about the underlying reason to account for an observation or a natural pattern. These statements are called *hypotheses,* and they are formulated in such a way that either they can be tested directly or indirect evidence can be gathered to support them.

Darwin formulated the hypothesis of natural selection to explain the change of species. He could not test his hypothesis directly, so he compared what a scientist would expect to find in some different branches of natural history, if natural selection operated, with what was known to exist in those branches. For example, he reasoned that in examining the anatomy of the members of a taxon—say, a family—numerous anatomical similarities could be explained by descent from a common ancestor. The variations could be understood as the product of natural selection operating over time. His results were very convincing, and he went on to consider other areas of natural history.

Can testing or gathering more evidence make us believe that a hypothesis is true? Scientists are uncomfortable speaking about the truth of hypotheses. They have learned from experience that hypotheses are often shown to be false or that observations can be better explained in some other way. Positive test results give us confidence to accept a hypothesis provisionally, while keeping an open mind. Negative results,

however, may lead to the rejection or modification of a hypothesis. Some hypotheses turn out to be very good in that they repeatedly stand up to new tests. Natural selection, for example, has now been demonstrated in laboratory and field tests. When scientists have a great amount of confidence in a hypothesis, they call it a *law.* For example, the regularities that Gregor Mendel explained and then tested are called Mendel's laws (see Chapter 14).

Darwin was searching for a hypothesis that he hoped would become a law that explained the origin of species. As it turned out, he did far more. He constructed a theory.

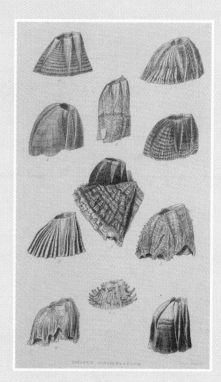

Figure 1   Darwin examined over 10,000 barnacles and came to appreciate the enormous variation that existed among them. This illustration is from Darwin's book on barnacles.

box continues

FOCUS ON SCIENTIFIC PROCESS

Theory

Darwin initially formulated a hypothesis dealing with the origin of species by natural selection. He could not test it directly, so he tested it indirectly by asking himself, "If natural selection did operate on populations, what patterns might we expect to find?" He applied this idea to the fossil record, biogeographical distribution, comparative anatomy (the comparison of animal structures), breeding experiments, embryology, classification—in short, to all the known areas of natural history at that time. In each case, his hypothesis not only seemed to hold but also served to explain related regularities. Using a few simple ideas (such as variation and natural selection), Darwin was able to provide an explanation for a vast quantity of data, and in so doing he synthesized these data in a coherent, unified formulation, as is indicated in Figure 2. A theory, like Darwin's theory of evolution, explains vast amounts of data with a relatively simple set of concepts and processes.

A theory does more than that, however. Theories are highly regarded in science, not only because their development represents an impressive intellectual achievement but also because they *direct research* by raising new questions. Much of the knowledge of genetics discussed in previous chapters was generated as scientists attempted to resolve the problems posed by Darwin's theory or its implications (see Figure 3).

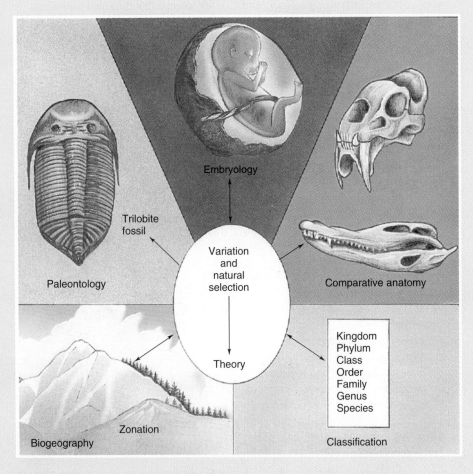

Embryology

Trilobite
fossil

Paleontology

Variation
and
natural
selection

Comparative anatomy

Theory

Kingdom
Phylum
Class
Order
Family
Genus
Species

Zonation

Biogeography

Classification

Figure 2   Darwin's hypothesis of natural selection led him to consider many areas of biology. In each case, it seemed that natural selection explained the facts involved and allowed interpretation from a new perspective. Ultimately, this led Darwin to formulate what has become the major synthesizing theory of biology—the theory of evolution.

that inhabitant, however different it may be from its own place.

Charles Darwin, *On the Origin of Species*(1859)

These quotations reflect the fresh approach that Darwin took in examining the problems of biology. In Darwin's time, the origin of species and the patterns found among living organisms were explained as part of God's plan. According to this view, each species was created at a particular time and was suited for a particular place. In 1857 Louis Agassiz (see Figure 23.13), the leading zoologist in the United States, published the *Essay on Classification,* which attempted to place all natural history data into this religious framework. He argued that God had a complex plan of continuous creation and that by studying classification, biogeographical distribution, and the fossil record, the naturalist only translates into human language God's plan as revealed in nature. Like Darwin, Agassiz had traveled and read extensively—his personal library became the core of the library at Harvard University's Museum of Comparative Zoology—and he, too, wanted an explanation for the many phenomena and patterns he observed in nature. Agassiz's solution was to attribute all patterns directly to the Creator. This was not a modern scientific theory because it did not explain the world of natural history using a few basic scientific concepts, and it did not raise any useful scientific questions. Rather, life was explained as the result of an unknowable "divine intelligence."

Darwin's theory, which he published in 1859 as *On the Origin of Species by Means of Natural Selection, or the Preservation of Favoured Races in the Struggle for Life,* argued against using this form of explanation in biology. He did not argue against a belief in God or against a view that God might have created the world and its laws. Instead, he argued that such beliefs belonged in the realm of religion and theology, not in science. Darwin's position was that it was more useful to ask questions such as "What are the causes of the origin of species?" and "How do species change?" than to attribute what was observed to a divine plan. Answering questions such as "Why are there the distribution patterns that we observe?" with "Because it was God's plan" did not encourage research to solve scientific problems. It might lead to more descriptive work filling in details of the plan, but it would not extend our knowledge of the underlying laws of nature.

By struggling with the question "How do new species come into being?" Darwin was led to the hypothesis of natural selection and ultimately to a broad evolutionary theory. That theory, in turn, raised questions that led to research on the origin of variation, the laws of heredity, and the rates of evolutionary change. The hypotheses (and laws) formulated to answer those questions have, in turn, led to dozens of other questions that have resulted in an enormous amount of valuable research.

By the third quarter of the nineteenth century, biologists had been convinced that they should limit their inquiry to questions that allowed a physical answer. The results of this inquiry have been overwhelmingly successful, and it is hardly surprising that today this bias is very strong among biological scientists.

Figure 3   Darwin, aged 72, a year before he died in 1882.

Are theories "true"? As with hypotheses, it is inappropriate to label theories as true. Rather, if theories are successful in answering questions, unifying knowledge, and guiding research, scientists come to value them highly. Because a theory directs research, it is as much an approach (a plan of action, a research program) as it is an explanation, and it is therefore *dynamic* (it changes). A theory that never changes with time is not a very interesting or useful theory. Because of the changing nature of theories, a blind, unquestioning acceptance of theory (dogmatism) is unacceptable in science.

Figure 23.13 Louis Agassiz was the leading zoologist in the United States in the nineteenth century.

SUMMARY

1. Darwin's voyage aboard HMS *Beagle* provided him with a wealth of natural history data and prompted his thinking about a set of questions that led him to formulate the theory of evolution.

2. Darwin was especially struck with his data on fossils, biogeographical patterns, adaptations, and the relationship of island life to mainland life.

3. After his return to England, Darwin decided that he could explain many biological puzzles if he assumed that species change. To explain how species change, he formulated the concept of natural selection.

4. Although his ideas were novel for his time, they gave Darwin a key to understanding life on this planet.

WORKING VOCABULARY

biogeographical realm (p. 444)   natural selection (p. 446)
Galápagos Islands (p. 446)   phylogeny (p. 446)

REVIEW QUESTIONS

1. What was the significance of Darwin's voyage?

2. What did Darwin find interesting about the fossils he saw on his trip?

3. What biogeographical patterns did Darwin observe?

4. What did adaptation mean in Darwin's day?

5. What struck Darwin about organisms living on islands?

6. How did the idea that species change answer the questions that Darwin was addressing?

7. Why is variation important for natural selection?

8. How does natural selection differ from domestic selection?

9. How did Darwin's ideas differ from Agassiz's?

ESSAY AND DISCUSSION QUESTIONS

1. It has often been pointed out that Louis Agassiz was familiar with all of the data that Darwin had and that he even later traveled and visited some of the locations that Darwin visited. Why couldn't Agassiz accept the theory of evolution?

2. In his day, Darwin was criticized for not providing "proof" of his theory. Was that a fair criticism? Why?

3. When Darwin proposed the theory of evolution, some naturalists indicated that it contradicted accepted laws in biology—for example, the law in anatomy stating that organisms were of such complexity that it was impossible for them to change. How might a scientist convinced of Darwin's ideas address this problem (a common one, it turns out, in the history of science)?

REFERENCES AND RECOMMENDED READING

Agassiz, Louis. 1857. Essay on Classification. *Contributions to the Natural History of the United States of America,* 1:1–232. Boston: Little, Brown and Company.

Appleman, P.(ed.). 1979. *Darwin.* New York: Norton.

Bowler, P. J. 1984. *Evolution: The History of an Idea.* Berkeley: University of California Press.

Browne, J. 1995. *Charles Darwin: Voyaging.* New York: Knopf.

Darwin, Charles. 1859. *On the Origin of Species by Means of Natural Selection, or the Preservation of Favoured Races in the Struggle for Life.* London: Murray.

Desmond, A., and J. Moore, 1991. *Darwin: The Life of a Tormented Evolutionist.* New York: Warner Books.

Kohn, D. (ed.). 1985. *The Darwinian Heritage.* Princeton, N.J.: Princeton University Press.

Lurie, E. 1960. *Louis Agassiz: A Life in Science.* Chicago: University of Chicago Press.

Ruse, M. 1979. *The Darwinian Revolution.* Chicago: University of Chicago Press.

Sulloway, F. 1982. Darwin and his finches: The evolution of a legend. *Journal of the History of Biology,* 5:1–53.

ANSWERS TO FIGURE QUESTIONS

Figure 23.5 One, some, or even all of the smaller, living sloths may be descendants of the fossilized giant sloth.

Figure 23.10 Phylogeny is a history of the evolutionary development of a species or of a larger group of organisms. Darwin saw that by studying the fossil record of a species, it was possible to construct a phylogeny for that species.

24

The Modern Synthesis: Genetics and Natural Selection

Chapter Outline

Reading Questions

1. What is the modern synthesis?
2. What are the sources of variation in populations?
3. What is the significance of the Hardy–Weinberg law for evolution?
4. What selective processes affect the composition of gene pools?

B y 1870, Darwin's theory of evolution had become widely accepted in the scientific community. It answered several critical questions and raised many new ones: If natural selection acted on small variations to create new species, then what regulates the inheritance of variations? What are the sources of variation? How is variation distributed in a population? For half a century after the general acceptance of Darwin's theory, much pioneering research addressed these and related questions. Results of this research led most scientists to believe that Darwin was correct about evolution in general, but that he had overstated the role of natural selection. Research in the late nineteenth and early twentieth centuries focused on alternative mechanisms or processes that might also influence evolution. By the mid-1930s, new information and concepts derived from work done in the early part of this century led several scientists to believe that a complete reformulation of the theory was in order. Ironically, that reformulation resulted in a return to Darwin's original perspective, emphasizing the role of natural selection.

The first attempt to articulate a reformulated theory was made by Theodosius Dobzhansky in *Genetics and the Origin of Species* (1937). His book was followed by Ernst Mayr's *Systematics and the Origin of Species* (1942), Julian Huxley's *Evolution: The Modern Synthesis* (1942), and George Gaylord Simpson's *Tempo and Mode in Evolution* (1944) (see Figure 24.1). All four of these books stressed the central role of natural selection and incorporated the results of new research in genetics. These authors had all worked primarily with evidence from animals. In 1950, G. Ledyard Stebbins brought plants into the new theory with his book *Variation and Evolution in Plants*. To distinguish the "new" Darwinian theory from Darwin's original theory, biologists call it the *modern synthesis*.

The **modern synthesis** is based on two assumptions: (1) that gradual evolutionary change can be explained by the action of natural selection on small genetic changes and recombina-

tion during sexual reproduction, and (2) that the processes of species formation, **speciation,** and evolutionary change in taxonomic groups higher than species are explainable in terms that are consistent with known genetic mechanisms. This chapter first considers **microevolution,** the genetic changes within a species that provide a source of variation on which natural selection operates and the forces of change that cause populations to differ. The next two chapters describe **macroevolution,** the coming into being of new species (*speciation,* Chapter 25), the dying out of species, and larger-scale evolutionary change above the species level (Chapter 26).

> **BEFORE YOU GO ON**   The modern synthesis emphasizes the action of natural selection on small genetic changes and recombinations in explaining gradual evolutionary change. It states that speciation and macroevolution can be explained in terms consistent with known genetic mechanisms.

THE GENETICS OF EVOLUTION

Darwin was acutely aware of variations in populations. After all, he had examined over 10,000 barnacles and had observed extensive variations in this invertebrate group. The modern synthesis has enhanced the appreciation of variations in the gross physical characteristics among individuals by providing an understanding of the basis of genetic variation in individuals and populations.

Genetic Variation

What are the sources of new variations? Hereditary mechanisms basically promote stability. The replication and translation of genetic messages result in offspring that resemble parents in fundamentally important ways. However, chemical processes are not always absolutely reproducible. As described in Chapters 16, 17, and 18, the enormous chemical complexity of genetic systems results in many different kinds of errors or changes (inexact duplications, mutations, transposable elements, recombinations). These errors and changes are an important source of variation. Early supporters of the modern synthesis emphasized simple gene mutations. They did not have detailed knowledge of the chemical events involved. Later research distinguished and described frameshift mutations and the more common point mutations (see Chapter 18).

Recall that a *point mutation* is a change in the sequence of nucleotide bases in DNA or RNA resulting from a base substitution. Point mutations can alter the amino acid sequence specified by the genetic code, with the result that the original message is modified. Altered sequences may have no serious effects, or they can cause a significant change in protein structure, as in the case of the base substitution that gives rise to sickle-cell anemia (see Chapter 20).

A   B
C   D

Figure 24.1   The four chief architects of the modern synthesis: (A) Theodosius Dobzhansky, (B) Ernst Mayr, (C) Julian Huxley, and (D) George Gaylord Simpson.

Frameshift mutations occur when the deletion or addition of a base pair shifts the reading of the triplet code along DNA by one unit, which ultimately leads to an entirely new amino acid sequence in a protein. Figure 24.2 illustrates point and frameshift mutations.

New genetic variation is not confined to gene mutations. Segments of DNA can "jump" or be transferred between chromosomes (and, as described in Chapter 17, perhaps even between individuals of different species!), and chromosomes can undergo a variety of changes during replication, which may result in deletions, alterations, or duplications of the genetic material. Although losses and alterations are usually harmful, duplication occasionally results in more robust forms. For example, in certain yeasts, chromosomal duplications have increased the capacity to produce particular enzymes that improve their chance of survival in environments that are low in necessary nutrients.

As discussed in Chapter 16, duplication of whole chromosome sets, polyploidy, is common among plants. One-third to one-half of all flowering plants, including many familiar crop plants (for example, wheat, potatoes, and tobacco) and numerous common wildflowers, are thought to have originated through polyploidy. The changes that result from the loss, alteration, or duplication of chromosomes during replication, together with mutations, transpositions, and recombinations of genes, are important in generating variations in populations.

BEFORE YOU GO ON New genetic variation can arise by way of point mutation, frameshift mutation, transposition, recombination, and various structural changes in chromosomes during duplication. These changes provide the basis for variation in individuals.

The Genetics of Populations

The study of genetics reveals a great potential for creating variation among the individuals of a population. Investigations of natural populations have confirmed that great genetic diversity does indeed exist in such populations. Attempts to under-

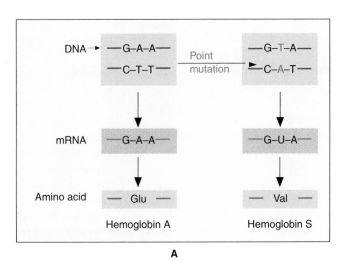

A

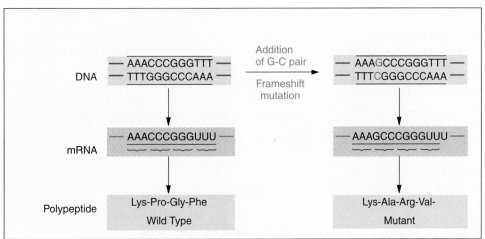

B

Figure 24.2 The modern synthesis initially stressed the two general forms of gene mutations; point mutations and frameshift mutations. (A) A base substitution (point mutation) is responsible for sickle-cell hemoglobin. (B) A frameshift mutation alters the amino acid sequence of a protein.

stand the significance of this vast storehouse of genetic differences within populations has raised fruitful questions, for example, "How are variable traits maintained, expressed, suppressed, or altered?" The study of **population genetics,** which is concerned with the genetic constitution of populations and how populations change, attempts to answer such questions. This field is of preeminent importance to the modern theory of evolution.

Populations

Populations are groups of individuals of the same species that live in the same place. Geneticists use several more specific terms to describe populations. A **Mendelian population** consists of interbreeding, sexually reproducing individuals. The genes distributed among all individuals of a population are collectively called a **gene pool.** All alleles in a gene pool are dispersed in individuals and determine the genotypes present in the population at any one time. As a result of sexual reproduction, the alleles of each generation are sorted, shuffled, and incorporated into gametes that then combine in fertilization and lead to the formation of new genotypes. Population geneticists sometimes find it more convenient to emphasize gene pools than populations of individuals.

Genotype Frequencies: The Hardy–Weinberg Law

The essence of evolution is change, and population geneticists study how the frequency of genes and genotypes of populations changes. In this endeavor, they rely on an idealized model, just as physicists rely on ideal gas laws or models of perfectly spherical balls rolling down frictionless inclined planes. The model that serves as the touchstone in population genetics is called the **Hardy–Weinberg law** (see Figure 24.3). This law starts with an initially defined distribution of genotypes and then proceeds to predict the genotypic frequencies of succeeding generations. The law is based on the following assumptions:

1. The population is very large.
2. No mutation, selection, or migration occurs.
3. Mating is random (that is, there is no selection of mates).

In other words, the model describes a large, randomly mating Mendelian population that is not subjected to any selective pressures. By a simple mathematical calculation, this model predicts that after a single generation, the frequency of each genotype will remain constant; that is, the frequencies will be in equilibrium.

This relationship was first described in nonmathematical terms in 1903 by the famous American geneticist William Castle. Castle was one of the first scientists to popularize the Mendelian laws in the United States. A few years later (1908), it was expressed independently in mathematical terms by G. H. Hardy and W. Weinberg, and since their writings called attention to the relationship, it is now called the Hardy–Weinberg law.

The Hardy–Weinberg law is useful because it defines the genetic stability in an idealized, undisturbed population. It thereby provides a standard of comparison with naturally occurring populations. By studying natural populations or experimental laboratory populations, which do change, biologists can compare them with an ideal, unchanging population and attempt to identify and analyze the factors that actually do cause changes in genotype frequencies.

> **BEFORE YOU GO ON**   The Hardy–Weinberg law provides a basis for understanding population changes. The law describes the genetic stability of an ideal, large population in which no mutation, selection, or migration occurs and mating is random. Population geneticists use the law as a baseline for evaluating changes in genotype frequencies of natural populations.

FORCES OF CHANGE

Population geneticists recognize four major, related forces of change that can alter the frequencies in a population: (1) natural selection, (2) mutation, (3) migration, and (4) random genetic drift. The remainder of this chapter describes the operation of these forces and how they cause changes to occur in populations. Chapter 25 will explore why change in populations is the central process of evolution.

Natural Selection

The main thrust of Darwinian evolution is its emphasis on natural selection. In its most general sense, natural selection refers to the different survival and reproductive rates of beings that differ in one or more ways. Natural selection is a mechanical and statistical process that does not assume any sort of conscious selector, any value judgment concerning inherent worth, or any sort of predetermined goal. Natural selection can be the result of many different factors and, contrary to popular conceptions, can be discussed without reference to "a violent struggle for existence." It is also a concept that can refer to different kinds of objects and different levels of complexity.

As applied in the modern synthesis, **natural selection** is the differential reproductive success of genotypes–that is, their fitness. **Fitness** may be defined as the *net reproductive rate* (the average number of offspring produced by individuals) multiplied by the probability that the individuals will survive to reproductive age, as described in the following simple equation:

$$w = n \times s$$

where w = reproductive success (fitness)

n = net reproductive rate (average number of offspring produced per surviving individual)

s = probability that an individual present at the start of the generation will survive to reproductive age

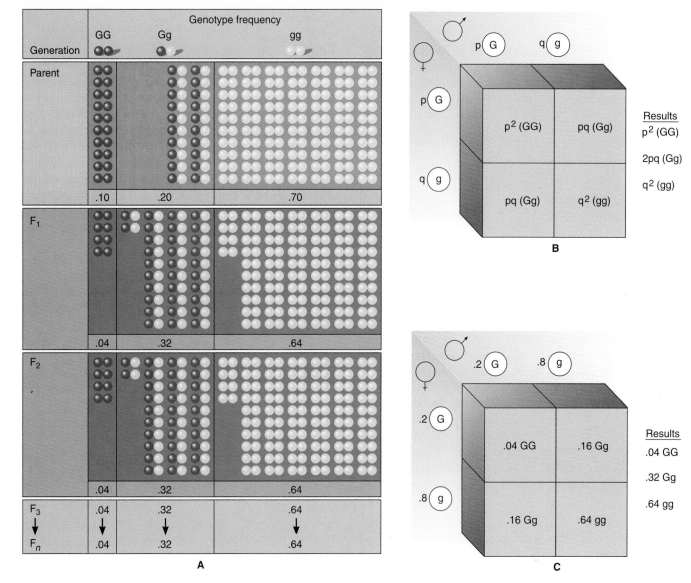

Figure 24.3 (A) The Hardy–Weinberg law predicts that after one generation in a large population where mating is random and no mutation, selection, or migration occurs, the frequencies of the genotypes will remain constant. In this illustration, the parent generation, shown in the red box, has genotype frequencies of .10 for GG, .20 for Gg, and .70 for gg. In one generation the genotype frequencies become .04 for GG, .32 for Gg, and .64 for gg.

Question: *All subsequent generations remain the same. Are such populations ever observed in nature?*

(B) We can understand how the frequencies in (A) are established in the F_1 generation and why they remain the same in subsequent generations by considering the algebraic formula that states the Hardy–Weinberg law: $p^2 + 2pq + q^2 = 1$. In this equation, p equals the frequency of the allele G and q equals the frequency of the allele g. The origin of this formula is simple for those who are comfortable with algebra. The left side of the formula is the sum of the possible genotype frequencies, and the right side reflects that the frequencies add up to all the whole (1). This relationship can be visualized with a Punnett square (B) that shows the combinations of gametes possible from the parent generation.

To understand where the frequencies of the F_1 genotypes come from, remember that allele frequency is not the same as genotype frequency. The number of G alleles in the parent generation is 40, and the number of g alleles is 160 (if it is not obvious, count them!). Because the total number of alleles in the parent generation is 200 (remember: 2 alleles for each parent), the frequency of the G allele is 40 divided by 200, or .2, and the frequency of the g allele is 160 divided by 200, or .8.

Drawing a Punnett square to see the possible genotypes that result from the parent generation (C) shows the genotype frequencies of the F_1 generation (.04GG, .32Gg, .64gg). If you calculate the allele frequencies in the F_1 generation, you will see that they are the same as the parent generation because all the alleles from one generation have been used in the next—nothing has dropped out, and nothing has been added in the ideal model. Because the gene frequencies are the same, the next set of genotype frequencies will be the same as the F_1, and on, and on, and on.

The fitness of different genotypes in a population—that is, the **relative fitness**—can be expressed by dividing the net reproductive rate by the reproductive rate of the genotype with the greatest fitness. The following equation is used by population biologists:

$$F_r = \frac{N_{ab}}{N_{xy}}$$

where F_r = relative fitness

N_{ab} = net reproductive rate of genotype *ab*

N_{xy} = net reproductive rate of the genotype with the greatest fitness (*xy*)

Using this equation, 1 is the highest possible fitness (when $N_{ab} = N_{xy}$), and 0 is the lowest (when $N_{ab} = 0$).

Natural selection modifies the composition of genotypes in a population through an indirect process. Selection operates directly on phenotypes, the expression of genotypes in individual organisms. Because the phenotypic expression of genes is complex and often related to environmental factors, discussing selection of genotypes is correspondingly complicated. To get a grasp of the way geneticists study selection, we can consider a simple case involving natural selection operating on only two alleles at a single gene locus. Then we can look at a more complicated example of natural selection, creating an equilibrium involving more than one genotype.

Selection Against Deleterious Alleles

If an allele reduces the fitness of a phenotype, we might expect it to be selected against. For example, when an allele causes infertility or death before reproductive age, the fitness of the organism is 0. If the allele is recessive and has no effect on the fitness of heterozygotes, selection will operate only on recessive homozygotes. After several generations, the number of homozygous, recessive individuals will be reduced, but the allele, as shown in Figure 24.4, will never be eliminated from the population because some will always be conserved in the heterozygous state. For this reason, the ability to remove harmful or undesirable recessive alleles from a population (human, animal stock, or plant crop) is ultimately limited. If the harmful allele is dominant, however, it will be quickly eliminated because affected individuals will not reproduce. Nevertheless, new mutations may generate the allele at a low frequency in the population.

Balancing Selection

In contrast to reducing the frequency of or eliminating an allele, natural selection can also create a state of equilibrium involving more than one genotype. Such selection is called **balancing selection,** and the resulting existence within a population of two or more genotypes for a given trait is called **polymorphism.** For example, some heterozygous genotypes are

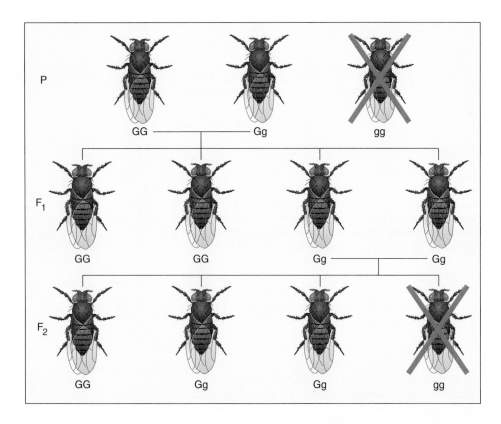

Figure 24.4 Selection against deleterious alleles may not lead to their elimination from a population if the allele is recessive and has no effect on the fitness of heterozygotes. As can be seen from this figure, the deleterious allele, g (which is lethal in the homozygous state), continues to appear in later generations because it has no harmful effect on heterozygotes (Gg).

Question: *What would happen in this diagram if the deleterious allele was dominant?*

more fit than either of the homozygous conditions; our discussion of sickle-cell anemia in Chapter 20 showed an example of balancing selection. Individuals homozygous for the allele are affected by a severe anemia that causes infant mortality. The heterozygotes have a milder form of the disorder, but

they have a greater resistance to malaria, thereby increasing their fitness in environments where malaria is a health hazard. Figure 24.5 illustrates a different example of balancing selection, where different genotypes permit adaptation to different environmental conditions.

A

B

Figure 24.5 A classic example of balancing selection involves a land snail (*Cepaea nemoralis*) found near Oxford, England. This simplified illustration shows two forms of the snail's shell: the yellow, banded shell and the brown, unbanded shell. (A) Selection by thrushes in open grasslands favors the yellow snails, which are more difficult to distinguish than the darker brown ones. (B) Similarly, the darker shells are more difficult to distinguish among the leaves of the forest.

Other Patterns of Selection

The foregoing discussion of selection concentrates on the change in frequencies of genotypes. Several patterns of selection that are recognized in populations were discovered by studies of phenotypes and their interaction with the environment. These patterns include stabilizing selection, directional selection, disruptive selection, sexual selection, and coevolution.

Stabilizing Selection

Stabilizing selection refers to selection against extreme variants in a population, with the result that a "standard phenotype" is favored. This can occur when the relationship between an organism and its environment remains constant, at least in domains that are critical for the survival and reproduction of individuals. Stabilizing selection can be considered the usual state of affairs in which a balance that promotes an optimum range of characteristics has been reached. Stabilizing selection, however, does not act to establish a set of uniformly homozygous genotypes.

Studies of variation in natural populations have shown that even where phenotypes are highly uniform, there are significant variations in the genotypes. Most populations possess a large reservoir of genetic variation. Some of the variation consists of small molecular variants that appear to be selectively neutral and cannot be detected in the phenotype. However, larger variations usually can be detected in populations, and they have potential importance for a population if, for example, the environment changes. This type of variation has been tapped by both scientists and breeders who have altered the appearance of small groups of organisms by controlling their breeding and artificially selecting for certain desired characteristics that have scientific or commercial value, such as early ripening in tomatoes or exotic eye color in fruit flies. In nature, however, average individuals often appear to have a better chance of reproducing than those on the periphery of appearances. Almost a century ago, Hermon C. Bumpus illustrated this point by studying sparrows that died during a severe storm. He argued that surviving birds tended to be of average size, whereas those that died were larger or smaller (see Figure 24.6). Although more recent analyses of his data suggest that only female sparrows are subject to stabilizing selection for average size, Bumpus's classic study remains one of the standard examples of this pattern of selection.

Directional Selection

Directional selection shifts the mean (average) in a distribution of certain characteristics in response to a change in the environment. For example, if an area is becoming progressively cooler or drier, some traits, such as density of coat, behavior, or metabolic production of water, might be favored. Similarly, a population that migrates into a new environment that has recently undergone change (for example, due to glaciers or volcanic eruption) might be subjected to directional selection. One of the most famous examples of directional selection—the changes in the colors of peppered moths as a result of industrialization—belongs to this category (see Figure 24.7).

Average-sized birds survive

Undersized birds die

Oversized birds die

Figure 24.6 An early example of stabilizing selection was recorded by H. C. Bumpus after observing house sparrows. In 1899 Bumpus published a classic paper on natural selection that described his research conducted during the previous year after "an uncommonly severe storm of snow, rain, and sleet." He studied a set of birds that was brought to the Anatomy Laboratory at Brown University after the storm. Some of the birds revived, and some died. Bumpus carefully measured and compared those that survived with those that perished. He observed that many of the birds with extreme dimensions (for example, the largest, smallest, heaviest, lightest, and those with the biggest and smallest heads) did not survive, whereas many with average dimensions did.

Figure 24.7 Perhaps the most widely known example of directional selection, often used as a basic example of natural selection, concerns peppered moths (*Biston betularia*), which have been carefully studied in England. These moths fly at night and pass the day resting on trees. (A) Until 1848 only the light form of the moth was known in England; the light-colored moths were difficult to see against the background of lichen-encrusted trees. (B) During the nineteenth century, the environment in many parts of England was radically changed by industrialization. Smoke from factories killed the light-colored lichens and darkened the trunks of the trees that the moths rested on during the day. In 1848 a dark form of the moth was seen in Manchester, England; the frequency of the dark form increased in polluted areas. Later studies showed that after installing new pollution controls, some areas having less pollution have more light-colored lichens on trees. In those areas, the frequency of the light-colored moths has increased. (C) Manchester, 1850. There are six moths on this tree trunk. (D) Thirty years later. There are six moths on this blackened tree trunk.

Question: *In (C), which are better adapted to their environment: the pale moths or the dark moths?*

Question: *In (D), which are better adapted to their environment: the pale moths or the dark moths?*

Disruptive Selection

Disruptive selection (also called *diversifying selection*) acts to favor two or more traits simultaneously. This typically occurs when a population is subjected to separate selective pressures in different occupied areas. In Africa, for example, the Mock-er swallowtail butterfly (*Papilio dardanus*) has a color pattern that mimics (that is, it has come to resemble another organism or object to which it is not related) other butterflies that have an offensive taste to their predators (see Figure 24.8). *Papilio dardanus,* however, is found over a wide range, and individu-

Mocker Swallowtail Male

Mocker Swallowtail Female

Mocker Swallowtail Mimic Female

Mocker Swallowtail Mimic Female

Amauris niavius
(Distasteful species)

A

Mocker Swallowtail females...

Papilio dardanus

Papilio dardanus

Papilio dardanus

....and the distasteful species
they mimic (Danaids)

Danaus chrysippus

Amauris echeria

Amauris niavius

B

Figure 24.8 A well-known form of mimicry, called *Batesian mimicry,* occurs when a harmless species comes to resemble a poisonous, dangerous, or distasteful species. The Mocker swallowtail butterfly displays the result of disruptive selection and mimics a number of different noxious butterflies. (A) All the (1) male Mocker swallow-tails are black and yellow. One type of female (2) resembles the male. Others (4) mimic distasteful species (5). One mimic (3) has a wing shape like the male but colors like the distasteful species. This may be a result of natural selection not having completed the change to mimic the distasteful species. (B) Mocker swallowtail females mimic numerous distasteful species.

als mimic one of several different noxious butterflies! Populations of *Papilio* often have primarily one of the six patterns reflecting the distribution of the butterflies they mimic.

Another famous example, illustrated in Figure 24.9, involves certain bentgrass populations that grow in Wales in areas that are heavily contaminated with toxic metals. Careful experimental tests have shown that the bentgrass growing on contaminated soil belongs to subpopulations that are genetically resistant to the contaminants, whereas the grass growing in adjacent uncontaminated areas has no resistance to metal toxicity. Because the resistant plants grow more slowly than nonresistant grasses, selective pressure operates on two traits: in contaminated soil, the resistant grasses are at an advantage, but in uncontaminated soil, the faster-growing nonresistant plants are at an advantage.

Sexual Selection

Sexual selection results in differences in the external appearance of males and females of the same species, a condition called **sexual dimorphism.** Often these differences involve sex-limited traits found in males that affect size, strength, or ornamentation. Darwin believed that a form of selection was operating among competing males, and he termed it *sexual selection.* Although Darwin treated sexual selection as a process supplementary to natural selection, the modern synthesis has included it as a special case of natural selection to explain characteristics that have a selective value affecting success in mating. These characteristics can be behavioral as well as physical.

A commonly cited example of sexual selection is that of polygamous mammals, where the strongest male collects and guards a group of females or a territory containing females and drives off weaker males. A well-known case involves the red deer, *Cervus elaphus,* studied in Scotland. Red deer males are almost twice the size of females, and in the fall breeding season, they fight to gain control of groups of females, or harems (see Figure 24.10). The stronger, larger, and more agile males have greater reproductive success.

Finally, the striking plumage of numerous birds, like that of the peafowl and the sage grouse (see Figure 24.11, page 466),

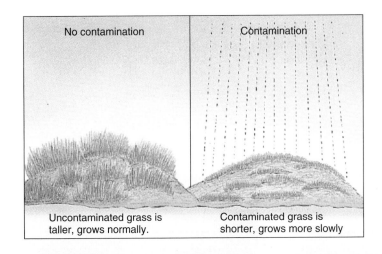

No contamination	Contamination
Uncontaminated grass is taller, grows normally.	Contaminated grass is shorter, grows more slowly

Figure 24.9 In Wales, populations of bentgrass (*Agrostis tenuis*) in the neighborhoods of old mines show a striking pattern of tolerance to toxic heavy metals (lead, copper, and zinc). A. D. Bradshaw showed that populations of bentgrass growing on contaminated soil were tolerant to the metals and that a very sharp transition occurred over a few yards between tolerant and nontolerant species.

Figure 24.10 Red deer males fight during the fall breeding season to gain control of groups of females.

Question: *This figure shows an example of one type of sexual selection. What is the other?*

Figure 24.11 Sexual selection is thought to be responsible for some of the striking plumage of male birds, such as (A) the peafowl and (B) the sage grouse.

is thought to have resulted from the preference of females for males with such fancy feathers.

Coevolution

Coevolution occurs when two or more interacting populations of different species change mutually. The term *coevolution* was coined in 1964 by ecologists Paul Ehrlich and Peter Raven in a paper that discusses how various plants have evolved protective chemical compounds for defense against insects and how insects have subsequently evolved a tolerance of those compounds. Since that classic paper, biologists have applied the concept of coevolution to the evolution of predator–prey interactions, parasite and host relationships, and related animal–plant traits involved in pollination and seed dispersal. Two examples of coevolution are shown in Figure 24.12.

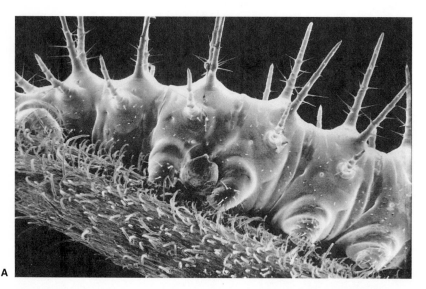

Figure 24.12 Coevolution involves mutual changes in interacting populations of different species. (A) *Heliconius* butterfly larvae feed on passion vines. They are among the few insects that can do this because passion vines contain chemicals that protect them against herbivores. Scientists believe that *Heliconius* butterflies evolved adaptations that permit them to feed on these plants. But this story continues—some species of passion vines have evolved hooked hairs (trichomes) that protect them from the *Heliconius* larvae, as this photo shows. (B) Acacia trees and ants of the genus *Pseudomyrmex* have evolved a mutual relationship that benefits both. The ants live in the tree's hollow thorns and feed on tree products that are rich in oil and protein. The ants, in turn, protect the tree from herbivores and competing plants.

Natural selection, as applied in the modern synthesis, is the differential reproductive success of genotypes in a population. Five patterns of selection that have been identified are stabilizing selection, directional selection, disruptive selection, sexual selection, and coevolution. Each of these operates on populations and results in changes in genotypic frequencies that are reflected by alterations in observed phenotypic ratios.

Natural Selection and Adaptation

Natural selection is a complex process that involves gene fitness, gene interaction, and the interaction of organisms with their environment, including other organisms. Unlike the other forces of change that will be discussed in this chapter, natural selection not only results in changing the gene pool, or genotype frequencies in populations, but also promotes **adaptation,** a change resulting in a structural, behavioral, or functional trait that promotes survival and reproduction. It accomplishes this because individuals with certain variations are more likely to survive and leave offspring than those without them. Similarly, individuals with variations that afford them the opportunity to exploit new food sources, new defenses, or new life strategies will be selected. For this reason, modern evolutionary theory emphasizes the central role of natural selection in accounting for adaptation.

Natural selection alone does not determine how a population changes. Other processes and effects, although not directly adaptive, influence population changes. These include mutation, migration, random genetic drift, founder effects, and bottleneck effects.

Mutation

Mutation and the various causes of genetic variation were discussed earlier in this chapter. Significant gene mutations are rare events, as can be seen in Table 24.1. Although mutations can be caused by various environmental factors—such as ionizing radiation, ultraviolet light, heat, and chemical mutagens—mutation rates are very low, and changes associated with them do not occur frequently.

Mutations that result in a major change are usually harmful, but occasionally they increase fitness. Are these "beneficial" mutations a response by the organism to its environment, or are mutations random; that is, do they occur independently of their usefulness?

Whether or not mutations occur randomly has been a hotly debated issue in the past. Many of the foremost American evolutionists of the nineteenth century believed that organisms could respond directly to environmental change either through "preadapted" or even "willed" mutations that increased their chances of surviving and leaving offspring. Many experiments have been performed to test the existence of such responses.

Table 24.1 Mutation Rates in Various Organisms

Organism	Mutation	Value	Units
Bacteriophage T2 (bacterial virus)	Lysis inhibition $r \rightarrow r^+$	1×10^{-8}	*Rate:* mutant genes per gene replication
	Host range $h^+ \rightarrow h$	3×10^{-9}	
Escherichia coli (bacterium)	Lactose fermentation $lac \rightarrow lac^+$	2×10^{-7}	*Rate:* mutant cells per cell division
	Histidine requirement $his^- \rightarrow his^+$	4×10^{-8}	
	$his^+ \rightarrow his^-$	2×10^{-6}	
Chlamydomonas reinhardi (alga)	Streptomycin sensitivity $str\text{-}s \rightarrow str\text{-}r$	1×10^{-6}	
Neurospora crassa (fungus)	Inositol requirement $inos^- \rightarrow inos^+$	8×10^{-8}	*Frequency* per asexual spore
	Adenine requirement $ad^- \rightarrow ad^+$	4×10^{-8}	
Drosophila melanogaster (fruit fly)	Eye color $W \rightarrow w$	4×10^{-5}	*Frequency* per gamete
Mouse	Dilution $D \rightarrow d$	3×10^{-5}	
Human (autosomal dominant)	Huntington disease	0.1×10^{-5}	
	Nail-patella syndrome	0.2×10^{-5}	
	Epiloia (predisposition to brain tumor)	$0.4 - 0.8 \times 10^{-5}$	
	Multiple polyposis of large intestine	$1 - 3 \times 10^{-5}$	
	Achondroplasia (dwarfism)	$4 - 12 \times 10^{-5}$	
	Neurofibromatosis (predisposition to tumors of nervous system)	$3 - 25 \times 10^{-5}$	

Mutation rates are extremely low.
Source: After Sager and Ryan (1961).

For example, in 1960, Jack Bennett reported data from studies on *Drosophila* resistance to the pesticide DDT (see Figure 24.13). In one experiment, he selected siblings of the most resistant flies for many generations; that is, the flies selected for breeding from each generation had *not* been exposed to DDT, but some of their siblings, used in exposure studies, had shown some resistance. By the end of 15 generations, Bennett had created a strain of DDT-resistant flies that was descended from flies that had not been exposed to DDT. He concluded that mutations related to DDT resistance must have been present in the original population and that by selection he had isolated some of their carriers, rather than concluding that mutations had arisen by some response to exposure on the part of the flies. Experiments like those of Bennett have convinced biologists that mutations arise randomly.

Thus although mutations arise randomly, the value of a mutation depends entirely on the environment. What may be neutral or deleterious in one setting can be advantageous in another.

Migration

When considering the genetic composition of a species, we often treat it as an isolated entity. In nature, however, species consist of many breeding populations of varying size and proximity. These groups are usually not genetically isolated; that is, some individuals interbreed with individuals from different populations, a process called **migration.** The resulting exchange of alleles among populations is called **gene flow** (see Figure 24.14).

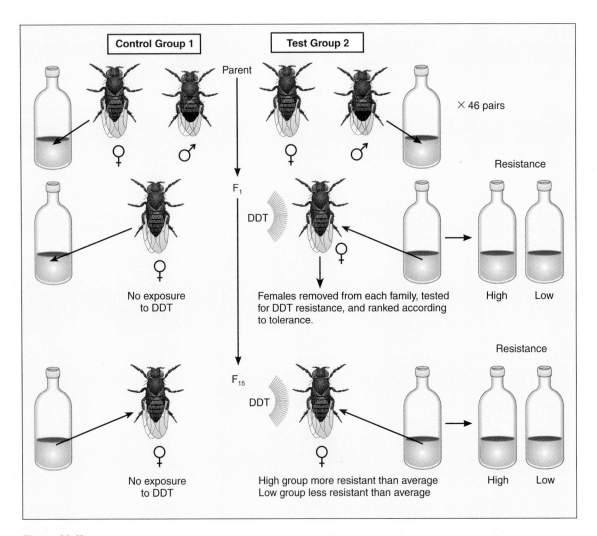

Figure 24.13 Jack Bennett performed experiments on preadaptation of DDT resistance in *Drosophila melanogaster.* In the experiment shown here, a control group was later tested for DDT resistance against an experimental group. The experimental group consisted of an initial set of 46 pairs of *Drosophila* that were placed in bottles containing food and allowed to reproduce. The offspring were tested for DDT resistance by removing a female from the offspring and exposing her to DDT. Bennett selected the families that were most resistant and least resistant and allowed them to reproduce. Continual selection produced strains of flies that were more resistant and less resistant to DDT when compared to the control group.

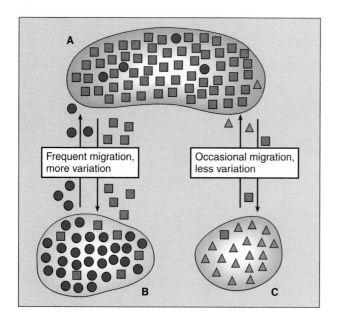

Figure 24.14 Gene flow resulting from migration can vary among populations. In this illustration, population A has frequent exchanges with population B, resulting in more variation for both populations. In contrast, population A has infrequent exchanges with population C, resulting in less variation for both populations. There is no direct gene flow between populations B and C.

Question: *In time, would you expect population B to have any variation introduced from population C?*

Migration is an important factor in evolution. The greater the extent of migration, the greater the genetic uniformity of various breeding populations. Migration also extends successful genetic combinations and increases the genetic variability of semi-isolated populations. Alternatively, restriction of migration can lead to the formation of different races or even more significant changes, which will be discussed in Chapter 25.

Random Genetic Drift

Random genetic drift refers to chance fluctuations in gene frequencies in a small population and is at least as if not more important than migration. In a large population, the Hardy–Weinberg law predicts that the frequency of an allele not subject to selective pressure will remain constant. However, this unchanging frequency actually represents a mean. If a small subset of reproducing individuals is examined, deviations from the average will be observed. This chance deviation in gene frequencies in a small population is the basis for random genetic drift (see Figure 24.15, page 470). If all individuals in a population are included in the analysis, however, the gene frequencies remain constant. Is random genetic drift important in evolution?

The significance of random genetic drift has been highly controversial within the modern synthesis. Sewall Wright, who

developed the concept and stressed its importance, argued that natural populations are composed of many small subpopulations that undergo varying amounts of migration. According to Wright, allele frequencies will be altered in these small subpopulations owing to the combined interaction of four factors: the size of the population, the selective pressure on the allele, the rate of mutation, and the amount of gene flow from migration. By chance alone, an allele could increase from a low frequency to a high frequency in a small subpopulation after several generations. Similarly, it could disappear altogether.

Many biologists believe that genetic drift may be a major evolutionary process. Because populations of a species are often spread over a wide geographical area, genetic drift can result in significant variations among the frequencies of genotypes within these subpopulations. Such differences may be enhanced by other factors, such as selective pressures, migration, and variable mutation rates. Theoretically, this could lead to a subpopulation developing advantageous genotypes that could spread through the wider population by migration. Or the advantageous genotypes could remain in a balanced polymorphic state, allowing the population to meet a wider range of geographical conditions. Although Wright's emphasis on the importance of genetic drift has been controversial, there is considerable evidence to support his position that the joint action of genetic drift and selection does operate in nature.

Founder Effects and Bottleneck Effects

Two special cases of random genetic drift are important in understanding change in populations: founder effects and bottleneck effects (see Figure 24.16, page 471). The **founder effect** occurs when a small group of individuals establishes a colony. Because the founders are only a small subset of the larger population, they may be genetically unrepresentative of the population. A certain allele may be absent in the founding population, or its frequency may be so low that it disappears before the population can increase in size.

A British study on the former inhabitants of a small island illustrates the founder effect. In 1816, British soldiers established a garrison on Tristan da Cunha, a small, isolated volcanic island in the South Atlantic midway between South Africa and Brazil. When the garrison left, one soldier and his family stayed, and they were later joined by a few other settlers. The population soon grew to 100, straining the resources of the island, and some inhabitants left. In 1961, when the population was 267, a volcanic eruption forced the evacuation of all island inhabitants to England.

British geneticist, D. F. Roberts, studied these individuals and was able to chart a reproductive history of the island population. He found that only ten of the original adult settlers had living descendants in 1961, and that of the ten, two had contributed twice as much to the gene pool as the other eight. The island population in 1961 was found to have an atypical genetic composition compared to the general European population from which it was derived. For example, there were higher fre-

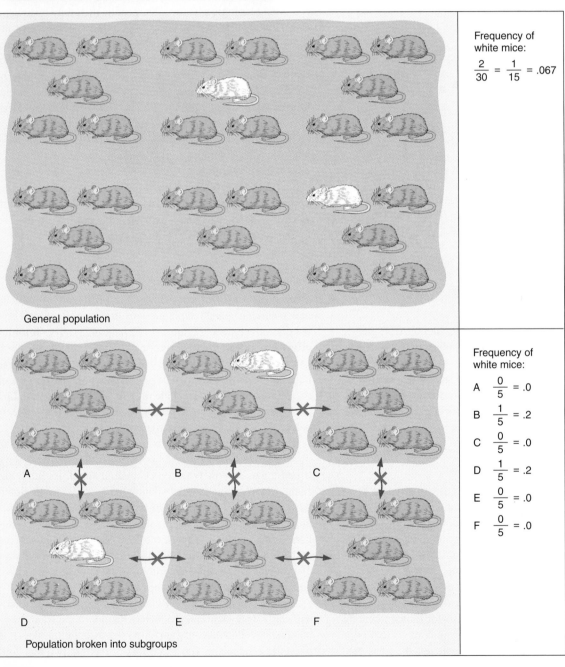

Figure 24.15 (A) Sewall Wright, a major figure in developing the modern synthesis, originated the concept of random genetic drift. (B) As an example, suppose that a population of mice has a white mouse frequency of 2 in 30. If the population is broken into smaller subpopulations, note how, by chance, two of the populations have white mice and four populations have none. Various pressures could increase or decrease the frequency of white mice in the two populations; hence the frequency could be altered by chance alone.

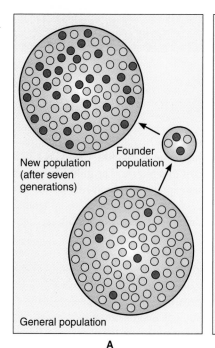

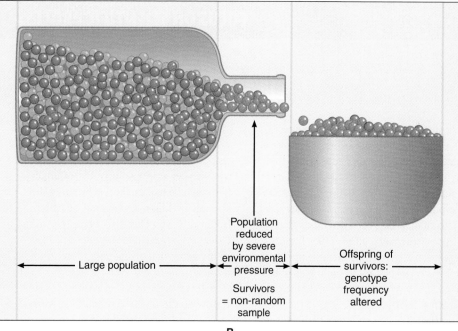

A   **B**

Figure 24.16   (A) A small subset of founders break off from a general population. Because, by chance, this subset is not genetically representative, the new population differs genetically from the original general population. (B) Similarly, when a population experiences a severe fluctuation in size, the survivors may, by chance, be unrepresentative of the original large population. When the offspring of the survivors reproduce, the resulting population may be different than the original large population.

quencies of genetic disorders, such as retinitis pigmentosa, and genetically linked mental retardation.

A second, and similar, phenomenon, the **bottleneck effect,** occurs in populations that experience severe periodic fluctuations in size. In such circumstances, when the population is at its smallest size, random changes in gene frequency can occur. When the population then attains a greater size, the altered gene frequencies may be retained. This bottleneck effect can radically alter the gene frequencies of a large population in a relatively short period of time.

> **BEFORE YOU GO ON**   In addition to natural selection, the major forces of change in gene frequencies and the genotypic frequency of a population are mutation, migration, and random genetic drift (including founder effect and bottleneck effect). Natural selection differs from these other forces of change because it alone promotes adaptation.

JOINT ACTION OF EVOLUTIONARY FORCES

In this chapter we have examined several processes that influence the genetic composition of a population: natural selection, mutation, migration, and random genetic drift. These forces act jointly to cause population changes. Proponents of the modern synthesis stress the importance of natural selection as the main agent of evolutionary change, for it alone promotes adaptation

(see the Focus on Scientific Process, "The Convergence of Ideas in Biology"). What is the significance and the result of population change? Chapters 25 and 26 will focus on the consequences of genetic change in populations.

SUMMARY

1. The modern synthesis states that gradual evolutionary change can be explained by the action of natural selection on small genetic changes and recombination, and that the processes of species formation and of evolutionary change in groups higher than species are consistent with known genetic mechanisms and processes.

2. The study of genetics reveals a great potential for creating heritable variations among individuals of a population. Population genetics is the study of gene and genotypic frequencies in populations and how they change. The touchstone of population genetics is the Hardy–Weinberg law. This law predicts that after a single generation in a large, randomly mating Mendelian population not subject to any selective pressures, the frequency of genotypes will be in equilibrium.

3. Four related forces of change that alter the genetic composition of a population are natural selection, mutation, migration, and random genetic drift. These forces act jointly to change gene and genotypic frequency in populations.

4. Often-cited patterns of selection are stabilizing selection, directional selection, disruptive selection, sexual selection, and coevolution.

The Convergence of Ideas in Biology

The history of science provides numerous examples of main ideas that were formulated at approximately the same time by different people. In the seventeenth century, Isaac Newton and Gottfried Leibniz, two of the greatest mathematicians of their day, each invented calculus at about the same time. They fought a bitter battle over who had accomplished it "first." The history of biology also has its share of what is often called "simultaneous discovery."

In 1900, three scientists independently formulated genetic principles that Gregor Mendel had published in 1866 (see Chapter 14). After Darwin had formulated his theory of evolution, but before he had published it, he received an essay from Alfred Russel Wallace, a younger naturalist with whom he occasionally corresponded (see Figure 1). In the essay, Wallace sketched out almost exactly the theory that Darwin had constructed! Subsequently, Wallace's essay was published along with some of Darwin's earlier writings, so Wallace formally received equal credit for inventing the idea of evolution by natural selection. Because it was Darwin's book, *On the Origin of Species,* that convinced the scientific world that evolution had occurred, we hear little about Wallace today, even though Wallace was a very productive scientist who did much to support Darwin.

Several formulations of the modern synthesis were written within a short period of time. Although not as startling, perhaps, as the recovery of Mendel's laws or the independent description of the theory of evolution, it raises interesting questions: What are simultaneous discoveries, and why do they occur?

A possible explanation, and one that was popular a few decades ago, is that great ideas are "in the air" at particular times—not like germs, but like pieces of a puzzle. The individual parts of an explanation are available, but it is necessary for someone to assemble them in the correct order. This explanation for simultaneous discovery places a high value on the earlier scientists who contributed individual pieces of the puzzle or who had described rough approximations of the completed version. Such people are called *precursors.* In Darwin's case, for instance, a likely precursor is Georges-Louis Leclerc, comte de Buffon (see Figure 2A), the famous eighteenth-century French naturalist who believed in a limited form of change in plants and animals as a response to their environment. Jean-Baptiste-Pierre-Antoine de Monet de Lamarck (see Figure 2B), a protégé of Buffon, later speculated that species

Figure 1   Alfred Russel Wallace was coinventor of the theory of evolution.

Figure 2   (A) Georges-Louis Leclerc, comte de Buffon, and (B) Jean-Baptiste-Pierre-Antoine de Monet de Lamarck were allegedly Darwin's precursors.

FOCUS ON SCIENTIFIC PROCESS

change and that modern species are the descendants of earlier ones.

Unfortunately, when scholars seriously analyze the writings of precursors, they usually discover that the ideas of the precursor were significantly different from those they are supposed to have foreshadowed. Their ideas, though often very interesting, are most important for understanding the science *of their own time* rather than that of a later period. What is equally problematic is that when historians try to link major scientific figures with their precursors, it turns out that connections do not exist. Darwin read Lamarck and thought that his ideas were nonsense! Later, it was a source of considerable embarrassment and annoyance to Darwin that some of his conclusions were similar to those of this earlier French writer. The search for precursors usually turns out to be a sterile exercise that obscures more than it illuminates. This is especially the case when the alleged precursor is a distant or unknown figure.

A more profitable approach to understanding why simultaneous discoveries occur lies in perceiving that much scientific investigation occurs because of an enthusiasm for answering what are considered interesting or significant questions. That is, focusing on the *problem* being addressed rather than the answer that was given leads to a better understanding of the origin of ideas. Scientists, even if they work alone, rarely work in isolation. They work on problems passed down from earlier researchers or topics that have come to be regarded as especially engaging by their contemporaries. Often these activities are aimed at explaining a certain set of facts from a shared perspective, such as the search

for a particulate theory of genetics. Thus it is not surprising that occasionally more than one person is involved in a simultaneous discovery.

In the case of Darwin and Wallace, both naturalists had traveled to South America and the Pacific, each was fascinated by the relationship between fossils and living forms, each was curious about the underlying causes of biogeographical distribution patterns, and each was struck by the enormous variation in nature. Finally, each wondered—in light of island life that was obviously related to, but distinct from, mainland life—how new species had come into existence. Darwin and Wallace were also interested in explaining these questions without reference to supernatural causes. Each realized that many of his questions could be solved if he assumed that species changed. But how did they change? Each had read Malthus, and it was in thinking about Malthus's discussion of the growth of unrestricted populations—which Malthus in no way connected with the idea of change of species—that each was led to the idea of natural selection.

The formulation of the modern synthesis also came about as a result of different scientists attempting to answer certain questions by synthesizing copious data that had become available through the enormous amount of research already conducted on evolution. Recall that Darwin by 1870 had convinced the scientific community that evolution of life had taken place and that it was the key to understanding life on Earth. He had not convinced the scientific community, however, that natural selection was the chief factor responsible for evolution. During the next 50 years, scientists struggled to construct

an alternative. Some, such as the famous American paleontologist Edward Cope (see Figure 3), thought that there must be a cosmic force driving the evolutionary process. Others, such as the pioneering ichthyologist (a scientist who studies fish), and later president of Stanford University, David Starr Jordan (see Figure 4), thought that the solution lay in an understanding of how the environment affects species. At the turn of the twentieth century, a number of publications claimed that Darwinism—the belief that natural selection was the primary agent of species change—was dead. The most promising alternative in the early years of the century was proposed by Hugo de Vries, who claimed to have experimental evidence indicating that individual plants occasionally mutate and give rise to individuals of a new species. This theory of a species appearing over a single generation did not last, but de Vries's research on genetics did lead to the recovery of Mendel's laws (see Chapter 14).

Figure 3 Edward Drinker Cope was one of America's foremost paleontologists.

box continues

FOCUS ON SCIENTIFIC PROCESS

Biologists may have been depressed by the confusion surrounding the theory of evolution at the beginning of the twentieth century, but they were very excited about its chances for explaining the living world. The expansion of biology into university and college education and growing government support signaled a new age for the biological sciences. At that time, new disciplines such as genetics, cytology, embryology, and biochemistry were generating vast quantities of information that many thought was relevant to a broad, evolutionary understanding of life. Intensive research in traditional fields such as biogeography, anatomy, and taxonomy added even more information.

By the 1920s and 1930s, many scientists felt that the time had come to apply the body of new information to the theory of evolution. But how? The individuals mentioned in this chapter—Dobzhansky, Mayr, Huxley, Simpson, and Stebbins—believed that Darwin had it right in the beginning. What was needed was a return to his original insight, so they combined natural selection with the new results from genetics, population biology, biogeography, and other areas of research. Not everyone agreed. Nevertheless, the five books written by the leaders of what came to be called the modern synthesis convinced the biological community and served as a guiding vision for research during the past half century. All five attempted a similar task: to survey all that was known in biology from the perspective of natural selection. Their highly successful accomplishment came to be known as the modern synthesis because it synthesized our knowledge of biology from a Darwinian, evolutionary perspective. It was a powerful theory because it not only explained vast amounts of data, but it also pointed to interesting research questions. It was a theory that solved the problem of explaining the basic mechanism of evolution in a way that was consistent with the known facts of biology.

Simultaneous discovery can be understood, then, by determining the questions driving research at the time and examining the assumptions made in attempting to resolve them. Examining the common background of a problem renders its ultimate solution a little less mysterious.

Figure 4   David Starr Jordan was a leading American ichthyologist and president of Stanford University.

WORKING VOCABULARY

balancing selection (p. 460)
bottleneck effect (p. 471)
coevolution (p. 466)
directional selection (p. 462)
disruptive selection (p. 464)
fitness (p. 458)
founder effect (p. 469)
gene flow (p. 468)
gene pool (p. 458)
Hardy–Weinberg law (p. 458)
macroevolution (p. 456)

Mendelian population (p. 458)
microevolution (p. 456)
migration (p. 468)
modern synthesis (p. 456)
natural selection (p. 458)
population genetics (p. 458)
random genetic drift (p. 469)
sexual dimorphism (p. 465)
sexual selection (p. 465)
speciation (p. 456)
stabilizing selection (p. 462)

REVIEW QUESTIONS

1. What basic assumptions does the modern synthesis make?
2. What is the study of population genetics?
3. What is the Hardy–Weinberg law?
4. What assumptions does the Hardy–Weinberg law make?
5. What forces alter the gene and genotypic frequencies of a population?
6. Describe the concepts of fitness and relative fitness.
7. Describe five patterns of selection and their effects on a population.
8. How does coevolution differ from other patterns of selection described in this chapter?
9. How do mutation, migration, random genetic drift, and the founder and bottleneck effects influence the genetic composition of populations?

ESSAY AND DISCUSSION QUESTIONS

1. Imagine that you and several members of your class are dropped off on a planet somewhere in space to establish a colony. In 500 years, would that population have the same characteristics as the U.S. population today? How might the genotypic and phenotypic features of the population differ from the parent population?
2. Why might farmers who tried to develop the "perfect" cattle stock in the nineteenth century have become discouraged?
3. "The theory of evolution with its emphasis on the survival of the fittest has lowered the moral fiber of modern society." Based on your understanding of the modern synthesis, comment on this statement.

REFERENCES AND RECOMMENDED READING

Ayala, F. J. 1982. *Population and Evolutionary Genetics.* Menlo Park, Calif.: Benjamin Cummings.
Bennett, J. 1960. A comparison of selective methods and a test of the preadaptation hypothesis. *Heredity,* 15: 65–77.
Dobzhansky, T. 1937. *Genetics and the Origin of Species.* New York: Columbia University Press.
Ehrlich, P. R., and P. H. Raven. 1964. Butterflies and plants: A study in coevolution. *Evolution,* 18: 586–608.
Futuyma, D. 1986. *Evolutionary Biology.* Sunderland, Mass.: Sinauer.
Gilbert, L. 1971. Butterfly-plant coevolution: Has *Pasciflora adenopoda* won the selectional race with Heliconiine butterflies? *Science,* 172: 585–586.
Huxley, J. S. 1942. *Evolution: The Modern Synthesis.* London: Allen & Unwin.
Lenski, R. E., and J. E. Mittler. 1993. The directed mutation controversy and neo-Darwmism. *Science,* 259:188–194.
Mayr, E. 1942. *Systematics and the Origin of Species.* New York: Columbia University Press.
Mayr, E., and W. B., Provine (eds.). 1980. *The Evolutionary Synthesis: Perspectives on the Unification of Biology.* Cambridge, Mass.: Harvard University Press.
Mettler, L. E., T. G. Gregg, and H. E. Schaffer. 1988. *Population Genetics and Evolution.* Englewood Cliffs, N.J.: Prentice-Hall.
Panchen, Alec. 1993. *Evolution.* New York: St. Martin's Press.
Provine, W. B. 1986. *Sewall Wright and Evolutionary Biology.* Chicago: University of Chicago Press.
Sager, R., and F. J. Ryan. 1961. *Cell Heredity.* New York: John Wiley.
Simpson, G. G. 1942. *Tempo and Mode in Evolution.* New York: Columbia University Press.
Stebbins, G. L. 1950. *Variation and Evolution in Plants.* New York: Columbia University Press.
Sutherland, W. J. 1996. *From Individual Behavior to Population Ecology.* New York: Oxford University Press.
Tempo and Mode in Evolution: Genetics and Paleontology 50 Years after Simpson. 1995. Washington, D. C.: National Academy Press.

ANSWERS TO FIGURE QUESTIONS

Figure 24.3A No. They are ideal populations.
Figure 24.4 All carriers of the harmful allele would die. Therefore, in the parent generation only the individual who is homozygous for the nonlethal allele would live, and no lethal alleles would be found in the F_1 or F_2 generations.
Figure 24.7 (C) The pale moths. (D) The dark moths.
Figure 24.10 The selection by female birds of male partners with striking plumage.
Figure 24.14 Yes.

The Origin of Species

Chapter Outline

Reading Questions

1. What is a species?

2. How does speciation occur?

3. What are reproductive isolating mechanisms?

4. What is the significance of reproductive isolating mechanisms?

The number of species that have now been identified is just under 1.4 million. E. O. Wilson, one of many biologists who is worried about the alarming rate at which humans are causing species to become extinct, believes that the total number of species that inhabit the globe may be as high as 30 million. Although that figure is astonishingly high, it is relatively small compared to the number of species of plants and animals that paleontologists believe have lived on Earth at some time in the geological past—a number they estimate to be 4 billion.

How did all of these species originate? In Darwin's youth, there were far fewer species known than now, and a widely accepted idea was that God had created each species, perfectly adapted to its environment. This view was more plausible when the number of known species was only in the thousands.

However, the swelling number of identified species was beginning to strain the idea that God had created each species individually. Darwin's observations in South America gave him firsthand evidence that a vast number of different plants and animals inhabited the globe. Further, his examination of the fossil record showed him that the number of extinct forms exceeded the number that were living. The fossil record also revealed that plants and animals in the past were very different from those now inhabiting the planet, and that most present species have not been around for long.

Darwin referred to the appearance of new species as "that mystery of mysteries," and, as noted in Chapter 23, he spent many years trying to understand from a scientific perspective how new species come into being. Later scientists have been equally fascinated with the processes that lead to the production of new species, and our current understanding of those mechanisms is at the center of the modern theory of evolution.

Chapter 24 reviewed the forces of change responsible for altering populations. Changes within species' populations are called *microevolution.* This chapter and the next will concentrate on *macroevolution.* The following pages contain a discussion of how species form and change, defined in Chapter 24 as *speciation.* Chapter 26 examines the question of evolution above the species level.

SPECIES

In the 1940s, Ernst Mayr proposed the most widely accepted definition of **species** from the perspective of the modern synthesis: *Species are groups of actually or potentially interbreeding natural populations that are reproductively isolated from other such groups.* This definition stresses reproductive isolation. A species can occupy a wide geographical area, which allows subpopulations to be subjected to different selection pressures or to other forces of change such as genetic drift. In time, some of these subpopulations may acquire different genetic compositions and varying physical appearances. However, as long as there is migration (resulting in gene flow) between them and individuals from different subpopulations that can successfully interbreed, all are considered to be members of the same species. The different subpopulations of a single species are called **races** (see Figure 25.1, page 478). If the differences in subpopulations are sufficiently great, they are called *subspecies* and given a separate taxonomic name in Latin.

According to Mayr's evolutionary definition, species are distinguished by the existence of genetic barriers to gene flow among populations—that is, **reproductive isolation**—rather than physical characteristics. The significance of reproductive isolation can be understood by considering **sibling species.** These are groups of species that are so similar they cannot be distinguished from one another by their external characteristics. Sibling species, however, are considered to represent distinct species because of isolating mechanisms that keep them from interbreeding. An example of considerable medical importance is the European *Anopheles* mosquito. Originally this mosquito was thought to represent one species, but scientists have discovered that it is actually a group of six sibling species. It is not possible to tell them apart on the basis of their external appearance. They differ, however, in habitat preferences and egg color—characteristics that permit scientists to differentiate them. The ability to identify the sibling species of *Anopheles* is of practical value in the control of malaria, for only some of these mosquitoes transmit the disease.

Species, then, are not always as easily and simply recognized as descriptions from a bird or tree guidebook might lead us to believe. Whereas earlier concepts of species stressed external physical characteristics (morphological features), Mayr's newer definition is much more difficult to apply. It requires biologists to make many difficult inferences to distinguish species. As we saw in Chapter 5, taxonomists cannot always use the evolutionary concept of reproductive isolation for classifying species, and they often fall back on morphological and behavioral features.

BEFORE YOU GO ON   Ernst Mayr's widely quoted definition of species is as follows: "Species are groups of actually or potentially interbreeding natural populations that are reproductively isolated from other such groups." This definition stresses reproductive isolation rather than external physical characteristics.

SPECIATION

Speciation, the formation and changing of species, occurs basically in one of two distinct ways: through slow, gradual change of a population (or set of populations) into a new species, a process called *phyletic evolution,* or through the splitting of a population into two or more species, a process that produces diversity and is called *cladogenesis.*

Phyletic Evolution

Phyletic evolution occurs because of the action of natural selection. The selection can act to increase the fitness of the population's genotypes in a constant environment, or it can favor certain genotypes in response to changes in the environment. In a case where phyletic evolution occurs, cumulative change in succeeding generations results in a population that is sufficiently different from the initial population, and interbreeding with the original ancestral population is deemed impossible. In other words, phyletic evolution occurs when a population over time becomes so different owing to natural selection that if it were possible to bring back an individual from the original population, it could not produce offspring by interbreeding with an individual of the current population.

Finding evidence of phyletic evolution is difficult. Some **paleontologists,** scientists who study fossils, estimate that the average life span of a species is probably about 3 million years. So it is not possible simply to give a science student the assignment of checking the genotypes of a species until it changes! The fossil record provides evidence of long-term change, and scientists have found remains that appear to represent sequences of forms that are consistent with the concept of phyletic evolution.

The formation of a fossil is actually a rare event. Nevertheless, paleontologists have been able to identify certain regions of the globe that have yielded impressive numbers of fossils, which permit a fragmented view of ancient events. Fossils of marine organisms have been the most useful objects for study. Drillings from the ocean floor have provided samples that are rich in the remains of microscopic organisms. Samples that are rich in foraminiferans have been particularly interesting. Foraminiferans are enclosed in hard external coverings that have been preserved in samples of sediments brought up from the ocean floor. A study shows a gradation of changes in foraminiferans during the Cenozoic era that started with *Globorotalia conoidea,* continued through three other species, and finally ended with *G. inflata* (see Figure 25.2, page 479).

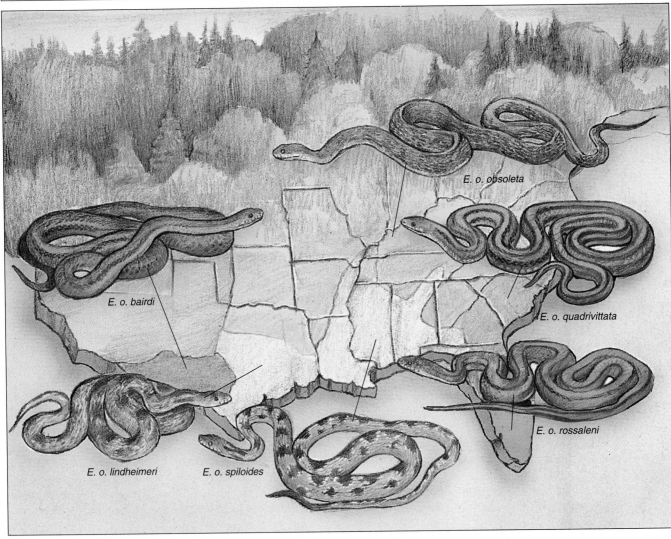

Figure 25.1 (A) The different golden whistlers (*Pachycephala pectoralis*) of the Solomon Islands constitute different races. (B) The rat snake (*Elaphe obsoleta*) of the eastern part of the United States consists of several subspecies.

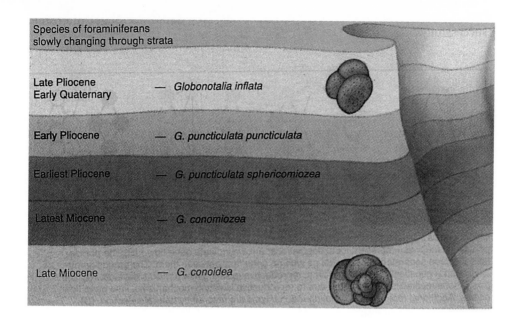

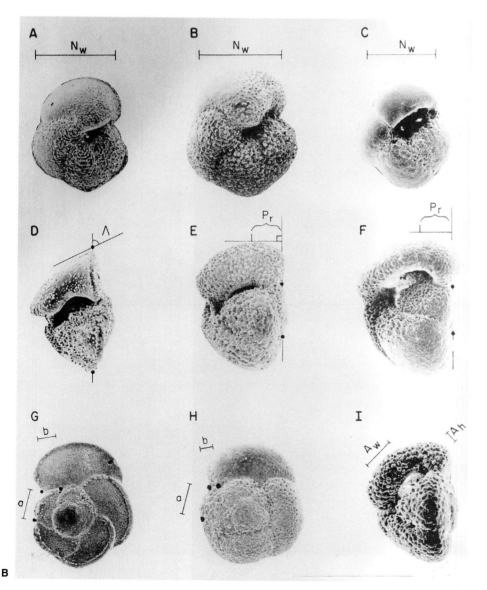

Figure 25.2 (A) Samples of sediments drilled from the ocean floor show a gradation of changes (from most recent, top, to oldest, bottom) in foraminiferans during the Cenozoic era that started with *Globorotalia conoidea* (bottom) continued through three other species, and ended with *G. inflata* (top). This transitional series of species of the genus *Globorotalia* is an example of phyletic evolution. (B) Photos of *G. conoidea* (G), *G. conomiozea* (A, D), *G. puncticulata* (B, E, H, I) and *G. inflata* (C, F).

Examples of phyletic evolution are not confined to micro-fossils. Recently, a British scientist, Peter Sheldon, examined 14,888 trilobites that had been uncovered in central Wales. These fossils, shown in Figure 25.3, are the remains of an interesting ancient group of invertebrates that have been studied extensively by numerous paleontologists. Sheldon's painstakingly exact studies showed that over a period of 3 million years, some species gradually developed an increase in pygidial ribs (ribs of the tail region), a characteristic that is used to distinguish different trilobite species. This finding nicely fits the pattern expected in phyletic evolution.

> **BEFORE YOU GO ON** Phyletic evolution is the slow, gradual change of a population or set of populations into a new species. Most of the evidence that supports the concept of phyletic evolution has come from analyzing fossils. In some cases, transitional series of fossils have been discovered that are consistent with the idea of slow, gradual change.

Cladogenesis

In contrast to the gradual transformation of one species into another over a period of time, **cladogenesis** is the splitting of one species into two or more species. Recall from Chapter 24 that many natural forces can alter the gene pool of populations.

What would happen if a population were to be divided and different selective pressures acted on each of the two separate populations? Might the two populations become so different that they could no longer interbreed successfully?

Let us consider a simple, ideal case of cladogenesis, illustrated in Figure 25.4. A single population consists of several local populations. The populations become separated into two groups. Gene flow ceases between the two groups; over time they become genetically different, and reproductive isolating mechanisms develop. Later the populations reestablish contact, but because of reproductive isolating mechanisms, little gene flow occurs. Natural selection favors the development of additional isolating mechanisms. Ultimately, the populations become two distinct species with no gene flow at all. Don't forget that this is an ideal picture! In nature, of course, the story is more complicated, but this model serves as a guide for investigating species origin and diversity.

In nature, when do populations split? A population spread over a wide range can become broken into subpopulations as a result of new geographic or ecological barriers. Directional selection, random genetic drift, the founder effect, or the bottleneck effect can then act to alter the separated populations. In time, these populations might no longer be able to interbreed. Figure 25.5 (see pages 482–483) illustrates two classic examples of this sort of speciation: the Galápagos finches and the Lake Victoria cichlids.

Figure 25.3 Sheldon's study of trilobites showed that over a period of 3 million years, new species evolved gradually from older ones. He based his conclusions on the appearance of additional pygidial ribs.

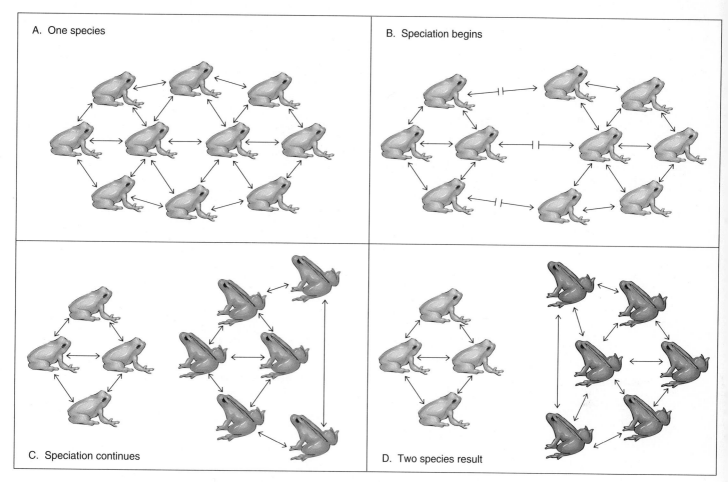

Figure 25.4   Cladogenesis is illustrated in this general model. (A) A single species consists of several local populations. (Each frog represents a local population; arrows indicate gene flow between populations.) (B) The frog populations become separated into two groups with no gene flow between them. The groups gradually become genetically different. In time, reproductive isolating mechanisms develop. (C) The populations reestablish contact, but owing to reproductive isolating mechanisms, little gene flow occurs. Additional reproductive isolating mechanisms are favored by natural selection. (D) Speciation is complete when the two sets of populations are fully isolated reproductively; that is, the two populations coexist without any gene flow between them.

Question: *What are some natural causes that could initially separate populations?*

If a species arises while remaining geographically isolated, as with the Galápagos finches, the process is called **allopatric speciation.** Some scientists propose that a species can also originate even when it exists together with the parent species through a process called **sympatric speciation.** The occurrence of sympatric speciation is a hotly debated issue among biologists. Early work done in the modern synthesis was hostile to the idea, but recent studies have documented several possible cases. One involves a treehopper (*Enchenopa binotata*) that is thought originally to have fed on several plant species (see Figure 25.6, page 484). Currently, six races of the treehopper exist, each infesting a different plant species. These races have each developed different morphological features and also show genetic differences. Female treehoppers will normally lay eggs only on their race's plant, and if raised in a cage, they mate only with individuals on the same host plant. Supporters of sympatric speciation believe that these treehopper

races are on their way to becoming new species. Although several other possible cases of sympatric speciation have been recognized, allopatric speciation is still regarded as the main avenue of cladogenic speciation.

How much time is required for speciation to occur? The answer varies. Some speciation events may occur very rapidly, in a process called **quantum speciation.** The best known form of quantum speciation is *polyploidy,* a condition often found in plants in which cells contain three or more complete sets of chromosomes. Polyploids can arise through the doubling of the same genome within a single species or when multiple sets of chromosomes result from the hybridization of individuals from different species. Polyploid individuals develop in one generation, and they are reproductively isolated from their parent population—hence they can arise by sympatric speciation. Figure 25.7 (see page 484) illustrates a classic example of polyploid evolution in present-day species of bread wheat.

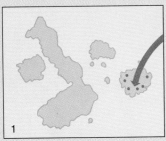
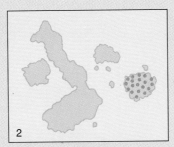

1

In the early history of the Galápagos Islands, there were no finches. Then, a few finches from the mainland of South America reached one of the islands.

2

The number of finches increased. Due to natural selection, the finches gradually became adapted to the island's environment.

3

Some finches flew to a second island, which had a different environment.

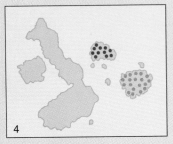

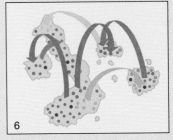

4

These finches adapted to the new environment.

5

In time, the finches on the second island became sufficiently different so that when some of them flew back to the first island, they could not interbreed with the finches there. The populations had become different species.

6

Similar events were repeated. Now there are 13 different species of finches on the islands.

A

B

C

D

E

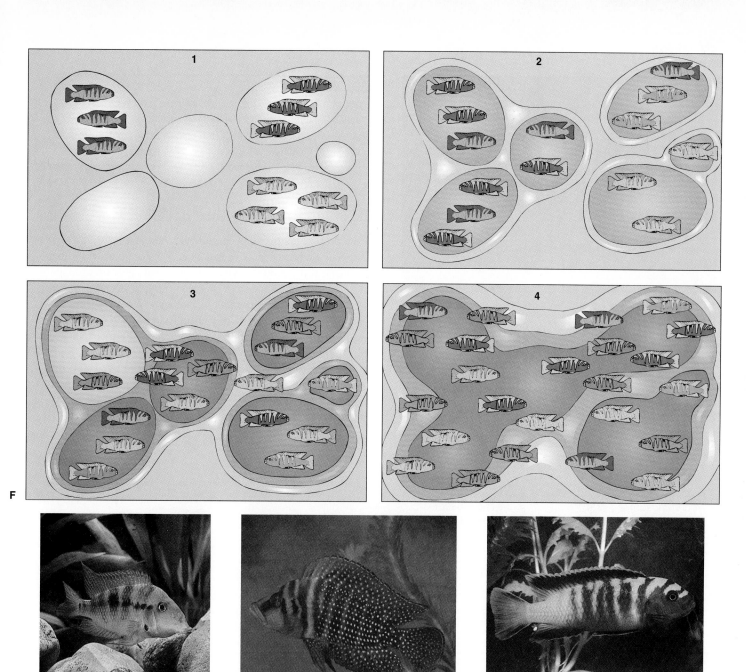

Figure 25.5 Two classic examples of speciation are the Galápagos finches and the cichlid fish of Lake Victoria. Shown here is a simplified version of what is known about the evolutionary story. (A) (1) In the early history of the Galápagos Islands there were no finches. Then, a few finches from the mainland of South America reached one of the islands. (2) The number of finches increased. Because of natural selection, the finches gradually became adapted to the island's environment. (3) Some finches flew to a second island, which had a different environment. (4) These finches adapted to the new environment. (5) In time, the finches on the second island became sufficiently different so that when some of them flew back to the first island, they could not interbreed with the finches there. The populations had become different species. (6) Similar events were repeated. Now there are 13 different species of finches on the islands. Four of the 13 species of finches are shown in the photos: (B) large ground finch, (C) medium ground finch, (D) woodpecker finch, and (E) warbler finch. (F) Lake Victoria, the largest lake in Africa, began to form about 1 million years ago as a result of movements on Earth that created some small lakes along existing rivers. (1) Initially, there were only one or a few species of cichlid fish living in the rivers and the new lakes. (2) In time, the lakes grew and some merged. Continued movement on earth altered the dimensions and borders of the lakes. In the process, many new habitats were created. The original species of cichlid fish became adapted to these new habitats. (3) When the lakes all merged, the cichlid fish were brought together, but individuals from many different populations could no longer interbreed. The populations had become different species. (4) Lake Victoria has continued to experience changes, and new habitats have continued to come into existence, providing chances for new species to evolve. There are now over 170 species of cichlid fish in this one lake. (5) Three examples of cichlid fish are shown in (G), (H), and (I).

Figure 25.6   Initially, many scientists were hostile to the idea of sympatric speciation. One possible case of such speciation, however, may involve this treehopper (*Enchenopa binotata*). Six races of the treehopper have each developed different morphological features; they also show genetic differences. Supporters of sympatric speciation believe that these treehopper races are in the process of becoming distinct new species.

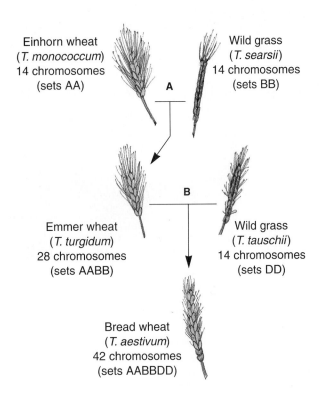

Einhorn wheat
(*T. monococcum*)
14 chromosomes
(sets AA)

Wild grass
(*T. searsii*)
14 chromosomes
(sets BB)

A

B

Emmer wheat
(*T. turgidum*)
28 chromosomes
(sets AABB)

Wild grass
(*T. tauschii*)
14 chromosomes
(sets DD)

Bread wheat
(*T. aestivum*)
42 chromosomes
(sets AABBDD)

Figure 25.7   The likely origin of the wheat plant used for bread involves two hybridization events and chromosome doubling. The first hybridization was (A) between einkorn wheat (*Triticum monococcum*) and a wild grass (*T. searsii*), followed by a doubling of chromosomes in the F₁ to produce emmer wheat (*T. turgidum*). The second hybridization was (B) between emmer wheat and wild grass (*T. tauschii*), followed by a doubling of the chromosomes in the F₁ to produce bread wheat (*T. aestivum*). Bread wheat contains sets of chromosomes derived from three diploid species.

BEFORE YOU GO ON   Cladogenesis is the splitting of one species into two or more species. This can occur when a population is split into separate subpopulations that experience different selective pressures. In time these subpopulations diverge to the point where individuals of the different populations cannot interbreed.

REPRODUCTIVE ISOLATING MECHANISMS

Slow or fast, involving geographical isolation or not, the key to speciation is reproductive isolation. A major question for the founders of the modern synthesis was what causes reproductive isolation. The answers were not simple because many mechanisms lead to reproductive isolation. When two similar species live in close proximity, usually more than one isolating mechanism is operating. In some cases, the mechanisms are often not completely effective; instead, they limit gene exchange between species. The major **reproductive isolating mechanisms** that have been identified are grouped under two main categories: *prezygotic* (premating) and *postzygotic* (postmating) (see Table 25.1).

Table 25.1   Reproductive Isolating Mechanisms

Prezygotic mechanisms (prevent or reduce successful interspecific crosses)

• **Habitat isolation.** Populations of different species live in different habitats within the same geographical region. Example: *Quercus velutina and Q. coccinea* (page 485).

• **Seasonal isolation.** Individuals of different species inhabit the same region but are sexually mature or breed at different times. Example: *Pinus radiata* and *P. muricata* (page 485).

• **Ethological isolation.** Individuals of different species do not mate because of differing behavior patterns. Example: Mating calls of Australian frogs *Crinia parinsignifera, C. sloanei,* and *C. signifera.* (Figure 25.8, page 485)

• **Mechanical isolation.** Individuals of different species have structural differences that prevent successful fertilization. Example: *Ficus* species pollinated by different wasps (Figure 25.9, page 486)

• **Gametic isolation.** Gametes from different species do not combine in fertilization. Example: sea urchins *Strongylocentrotus purpuratus,* and *S. franciscanus.* (Figure 25.10, page 486)

Postzygotic mechanisms (reduce the fitness of hybrids)

• **Hybrid inviability.** Hybrid does not reach muturity. Example: goat and sheep cross (p. 486).

• **Hybrid sterility.** Partial or complete sterility of hybrid. Example: mule (p. 486).

• **F₂ breakdown.** Hybrid is fertile but produces sterile or less fit offspring. Example: cotton hybrids from *Gossypium barbadense* and *G. hirsutum* (p. 486).

The major isolating mechanisms are either prezygotic or postzygotic.

Prezygotic Isolating Mechanisms

Prezygotic isolating mechanisms prevent or reduce hybridization among members of different species. Major prezygotic mechanisms include habitat, seasonal, ethological, mechanical, and gametic isolation. Examining these mechanisms allows us to understand how different species remain separated in nature.

Habitat isolation occurs when populations of different species occupy different habitats within the same general geographical region. Such isolation may be the result of some specialization that permits them to live in particular conditions. Habitat isolation is common in plants. For example, black oak (*Quercus velutina*) and scarlet oak (*Q. coccinea*) are found throughout the eastern United States. The black oak is found on well-drained soils, but the scarlet oak lives in swampy or poorly drained (acidic) soils. The two species, therefore, are rarely able to interbreed because they seldom grow in close enough proximity. Habitat isolation also occurs in animals. The sibling species of mosquitoes in the genus *Anopheles* discussed earlier live in different, although nearby, habitats.

Seasonal isolation (or *temporal isolation*) refers to species that live in the same region but whose populations are prevented from interbreeding because they are sexually mature at different time periods or seasons, or they breed at different times of the day. For example, two species of *Drosophila—D. pseudoobscura* and *D. persimilis*—are found in the same regions in the western United States and breed during the same season. The former, however, mates in the evening; the latter in the morning. Another classic example involves two pine species that grow on the Monterey peninsula south of San Francisco. *Pinus radiata* releases pollen in February, *P. muricata* in April; therefore, the populations remain distinct even though the individuals grow in the same area.

Ethological isolation refers to differing behavior patterns in courtship or a lack of sexual attraction between males and females of different species. It occurs only in animals and is the major prezygotic isolating mechanism in many animal groups . Courtship behavior—calls, songs, and displays—is often species specific, and ethological isolation serves to prevent males and females of different species from attempting to interbreed. Figure 25.8 shows one frog species that is similar to two other closely related species. They all inhabit the same range, but each species has a distinct mating call that keeps it from mating with the others.

Mechanical isolation and gametic isolation both refer to sexual factors that inhibit or prevent fertilization. **Mechanical isolation** in animals is usually the result of incompatible sizes or shapes of genital parts. In plants, mechanical isolation results from structural differences in flowers that do not permit cross-pollination by insects or other animals. The floral structures in two populations of plants may be adapted to facilitate pollination by different animals or to prevent an animal from transferring pollen between individuals of different species. Coevolution of plants and animals also promotes mechanical isolation: most of the more than 900 species of fig (*Ficus*) are each pollinated by a different species of wasp (see Figure 25. 9).

Figure 25.8   This frog, *Crinia signifera,* looks identical to two other *Crinia* species. They all inhabit the same range but each has a distinct mating call that serves as an isolating mechanism because it prevents interspecies mating.

Gametic isolation occurs when male and female gametes cannot combine in fertilization, or when pollen or sperm are rendered inviable in the female sexual structures of another species. Aquatic animals that release gametes into the water often depend on the attraction of sperm and eggs for fertilization. Sea urchins of different species, for example, display gametic isolation. In laboratory tests, male and female urchins can be stimulated to release their gametes into the water, but fertilization rarely occurs between gametes of different species (see Figure 25.10, page 486).

BEFORE YOU GO ON   The major prezygotic isolating mechanisms are habitat, seasonal, ethological, mechanical, and gametic isolation. Each basically ensures that the integrity of a species is maintained by preventing reproduction among members of different species.

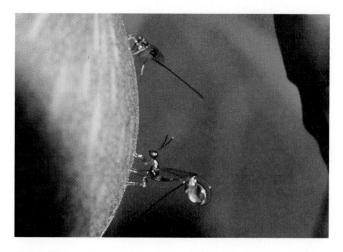

Figure 25.9   *Blastophaga psenes* is one of many hundreds of species of fig wasps. The 900 or so species of fig (*Ficus*) are pollinated by wasps of the same family. With just a few exceptions, each species of fig has its own species of wasp.

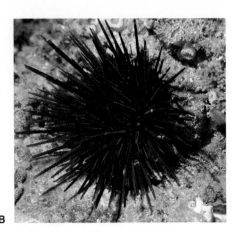

A   B

Figure 25.10 Stimulated to release gametes into the water, these specimens of two different species of sea urchins—(A) *Strongylocentrotus purpuratus* and (B) *S. franciscanus*—will not produce viable offspring because of gametic isolation.

Postzygotic Isolating Mechanisms

Prezygotic isolating mechanisms prevent or reduce hybridization between members of different species. If such mechanisms do not exist or fail to prevent hybridization, however, *postzygotic isolating mechanisms* may cause the resulting offspring to die before maturity, to be sterile, or to be less fit. **Postzygotic isolating mechanisms** are processes that reduce the fitness of hybrids. They include hybrid inviability, hybrid sterility, and F_2 breakdown.

All postzygotic isolating mechanisms depend on the inability of parental genes to function effectively together in a hybrid offspring. In its most extreme form, **hybrid inviability,** the hybrid does not reach maturity. Crossing a goat and a sheep, for example, results in an embryo that is spontaneously aborted. **Hybrid sterility,** either partial or complete, can be a consequence of hybrid inviability. It may also result from abnormal development of sex organs, incompatibility of the embryo with its food source (only in plants), abnormal segregation during meiosis, or other irregularities that prevent hybrids from reproducing. The mule is a well-known example of a sterile hybrid. Although the hybrids are vigorous, they are sterile owing to faulty germ-cell formation. Yet another way in which postzygotic events isolate populations is F_2 **breakdown.** In this case, the hybrids are fertile but produce sterile or less fit offspring. For example, hybrids of certain cotton species are hardy and fertile, but F_2 crosses produce plants that do not reach maturity or are less fit in other ways.

BEFORE YOU GO ON Major postzygotic isolating mechanisms are hybrid inviability, sterility, and F_2 breakdown. These mechanisms prevent successful reproduction among members of different species in cases where offspring are produced by matings from individuals of different species.

Selection for Isolating Mechanisms

Isolating mechanisms (especially in combination) are effective in reducing gene flow between populations and, therefore, in maintaining the separation of species. Natural populations display a wide range of isolating mechanisms. The origins of these mechanisms are correspondingly diverse. Some may be attributed to chance genetic variation enhanced by random genetic drift or geographical isolation. Postzygotic isolating mechanisms, for instance, are generally considered to have developed in these ways. Once partial reproductive isolation exists, however, natural selection may operate to increase that degree of reproductive isolation. An organism that mates with a member of another species and produces less fit offspring will obviously leave fewer of its own genes in the next generation than organisms that avoid such hybridization. Under these conditions, prezygotic isolating mechanisms might increase fitness because preventing hybridization would be an advantage to the species.

Care must be taken not to oversimplify the picture. Many species that apparently do not hybridize in nature will do so under experimental conditions. By itself, the ability or inability to form hybrids cannot alone be used as a criterion for defining species.

THE EVOLUTION OF SPECIES

The origins of species are complex and varied. Some species are distributed over a wide range, whereas others are very localized. Some are polymorphic (see Chapter 24), some are broken into many races, and some have numerous subspecies. Species may be found living adjacent to related species, or they may have a region of overlap in which some hybridization occurs. Two related species may be found in the same region with little or no hybridization. In some cases, usually in areas disturbed by people, swarms of hybrids occur, and the boundary between separate species becomes obscured. These diverse forms of species (and we have only mentioned some obvious ones) have come about in many different ways. Some forms evolve over a long period of time, and some evolve very quickly. The separation of species is maintained by a variety of mechanisms, often working together. It is little wonder that the concept of species is one of the most confusing topics among theoretical biologists. Thus the subject of species evolution has been controversial for several reasons (see the Focus on Scientific Process, "Evolution and Creation in U.S. Public High Schools").

Evolution and Creation in U.S. Public High Schools

High school science textbooks usually reflect beliefs of the scientific community. It is not surprising, therefore, that by the 1880s evolution was a topic included in new texts. By then the American scientific community had completely accepted the theory of evolution. Before the turn of the twentieth century, botany, geology, and zoology textbooks all contained serious discussions of evolution. After the turn of the century, curriculum reform resulted in the emergence of biology textbooks that combined the life sciences into one subject. They, too, continued the evolutionary thrust of earlier texts. In addition, the leading teachers' journal and the National Education Association stressed the value of teaching evolution as a unifying theme in biology.

By the 1920s, however, a strong antievolution crusade had emerged, with the goal of reversing the emphasis on evolution in biological science courses. This movement was not based on any change in the scientific community's commitment to evolutionary theory; instead, it reflected a number of social changes that had taken place in the United States.

One important factor was the tremendous expansion of the number of high school students. In 1890, only 3.8 percent of individuals 14 to 17 years of age were enrolled in high school. The number of students enrolling in high schools doubled each decade from 1890 to 1920, increasing from about 200,000 in 1890 to almost 2 million in 1920. This expansion brought many more high school students in contact with the theory of evolution and was a major factor in alarming members of the general public who opposed the theory.

A second and equally important element was the rise of Christian fundamentalism after the First World War

(1914–1918) and its drive to combat "modernism." This entailed supporting a literal reading of the Bible and combating forces that fundamentalists believed were undermining their traditional values.

William Jennings Bryan (see Figure 1) was perhaps the most ardent crusader against teaching evolution in the public schools. He believed that government had the obligation to pass legislation to improve the social well-being of its citizens, and he had taken part in the campaign for prohibition as well as for numerous liberal reforms, such as the income tax and women's suffrage.

Bryan opposed the teaching of human evolution because he thought it would lead people away from God. Furthermore, he believed that it had contributed to the causes of World War I by allegedly encouraging in Germany a justification for an armed struggle among nations. He also feared that it would promote the exploitation of people by stressing the value of competition.

Although Bryan rejected the theory of evolution in general, he and the antievolution crusaders of the 1920s focused their attack only on the teaching of *human evolution* and sought legislation

Figure 1   William Jennings Bryan, the leading advocate of the antievolution crusade of the 1920s, is shown here during his first speech at the Scopes trial in 1925.

FOCUS ON SCIENTIFIC PROCESS

that restricted it alone. Bryan's position was that church and state were separate; therefore, public schools should not teach *anything* concerning the origin of humans.

The antievolution campaign succeeded in introducing many bills in state legislatures to prohibit the teaching of human evolution. The first bill was passed in Oklahoma (1923). The best-known bill was passed in Tennessee (1925); it led to the famous "Monkey Trial" in which Bryan and the World's Christian Fundamentals Association (WCFA) confronted the American Civil Liberties Union (ACLU) and Clarence Darrow, America's most famous defense attorney. In that trial, a science teacher, John Scopes (shown with Darrow in Figure 2) was accused of violating the newly enacted law, providing the occasion for an impassioned debate between two leading masters of rhetoric. Although Darrow succeeded in making the antievolution movement look dogmatic and ignorant, the jury convicted

Scopes of violating the law. Darrow and the ACLU hoped to test the constitutionality of the law by appealing the decision to the Tennessee Supreme Court, but that court overturned the conviction on a technicality (without accepting the argument of the law's unconstitutionality). So the defense could not take its case to the Supreme Court for a final judgment.

Some historians have described the Scopes trial as a turning point in the creation–evolution controversy and claim that the antievolution crusade was discredited as ignorant, intolerant, and backward-looking. A more accurate description would be to see the trial as marking the end of a short but stormy period of confrontation and the beginning of a truce. Because the Tennessee law had not been declared unconstitutional, other states could enact antievolution laws. Few did, however. For the next 30 years, no action was taken by any state regarding antievolution laws, although some local school boards did take action. The teaching of evolution in

public school nonetheless suffered because publishers of high school texts tended to reduce or eliminate the topic of evolution in an effort to avoid controversy and to sell more books.

A turning point in the history of evolution theory in the high school curriculum was reached in 1959. The launching of a space satellite, *Sputnik,* by the Soviet Union in 1957 had shocked the American educational community and had led to a serious effort to increase the quantity and quality of science taught in the schools. In 1959, the American Institute of Biological Sciences, financed by the National Science Foundation, sponsored the Biological Science Curriculum Study (BSCS). Many leading biologists who strongly believed that evolution should be given a prominent emphasis were brought into the process of planning and writing biology texts. The materials produced by the BSCS were widely used and strongly featured evolution.

Antievolutionists were quick to respond. This time, however, their reac-

Figure 2 Clarence Darrow (left at table) was one of America's most famous defense attorneys when he defended John Scopes (second right at table) for teaching human evolution in 1925.

tion was not to ban evolution from the schools but to demand "equal time" for teaching creation based on the biblical account in the Book of Genesis. This new approach was probably inspired by rules that regulate the media, which require equal time for political candidates and opposite sides of controversial public issues. The Creation-Science Research Center, the Creation Research Society, and other groups that either prepared teaching materials or worked to influence local and state political bodies mounted a strong campaign in support of this demand.

Science and teaching organizations, both on the state and national levels, launched a dual defense: to repeal the old antievolution laws and to fend off new equal-time legislation. The American Association for the Advancement of Science (AAAS), the National Academy of Science, and the National Science Teachers Association, along with the ACLU, were active in these efforts. They were successful in overturning the old antievolution laws, but new equal-time laws were passed in some states.

In 1982, Federal Judge William Overton ruled that an Arkansas equal-time law was unconstitutional. He argued that its intent was to introduce particular religious beliefs into the curriculum and that "creation science" was not science. The same year, Judge Adrian Duplantier struck down a Louisiana equal-time law as an unconstitutional attempt to promote religion, a violation of the First Amendment of the Constitution which states, "Congress shall make no law respecting an establishment of religion." A Supreme Court decision in 1987 affirmed the Duplantier judgment, and for the present, it seems unlikely that any state law restricting the teaching of evolution will be acceptable. (For more information on the legal battle over creation and evolution, see Edward J. Larson's *Trial and Error: The American Controversy over Creation and Evolution*.)

The modern synthesis has tried to bring some order to the complexity of species by stressing underlying genetic mechanisms. This approach has proven to be very useful, but like all scientific ideas, it has limitations. For example, when we look at the fossil record, the concept of a species as a set of potentially interbreeding populations poses some formidable problems (Do fossils reproduce?). Similarly, some botanists feel that the modern synthesis stresses mechanisms important for animal species formation at the expense of those, such as polyploidy, that are more common in plants. Nevertheless, Mayr's new definition of a species has allowed scientists to explore populations in the field and in the laboratory. It has permitted them to bring biochemical analysis into the picture, and it has greatly expanded Charles Darwin's original idea.

SUMMARY

1. Species are groups of actual or potentially interbreeding natural populations that are reproductively isolated from other such groups.

2. Speciation occurs in two ways: through phyletic evolution, which is the slow, gradual change of a population or set of populations, and through cladogenesis, which is the splitting of one species into two or more species.

3. Allopatric speciation refers to speciation that occurs when the parent population is divided into geographically isolated areas. Sympatric speciation occurs when the new and parent species live in the same geographical area.

4. Speciation can be a slow or rapid process. When rapid, it is called quantum speciation.

5. Reproductive isolation is essential for speciation. Major known isolating mechanisms can be classed as either prezygotic or postzygotic.

6. Prezygotic isolating mechanisms include habitat, seasonal, ethological, mechanical, and gametic isolation. Postzygotic isolating mechanisms include hybrid inviability, hybrid sterility, and F_2 breakdown.

WORKING VOCABULARY

allopatric speciation (p. 481)
cladogenesis (p. 480)
paleontologists (p. 477)
phyletic evolution (p. 477)
races (p. 477)
reproductive isolating mechanisms (p. 484)
reproductive isolation (p. 477)
species (p. 477)
sympatric speciation (p. 481)

REVIEW QUESTIONS

1. Define a species.

2. What are the two general patterns of speciation?

3. What is phyletic evolution?

4. What is cladogenesis? What is sympatric speciation?

5. What is allopatric speciation?

6. What is the importance of reproductive isolation for speciation?

7. Describe the operation of major prezygotic isolating mechanisms.

8. Describe the operation of major postzygotic isolating mechanisms.

ESSAY AND DISCUSSION QUESTIONS

1. Some botanists have expressed dissatisfaction with Mayr's model of speciation. Can you explain why that might be?

2. The majority of pre-Darwinian naturalists in England and the United States defined *species* as an ideal plan of God's that specified a set of unchanging essential characteristics and their relationships for each animal and plant. Darwin, by contrast, used the concept of species to refer to a population that changes over time. The radically different usage caused much confusion, and even today there is controversy over the definition of species. Should scientists come up with new terms rather than redefine old ones?

3. Because it is not possible to revive fossils and run tests on them to see if individuals can interbreed, what evidence might be used to determine that two specimens belong to the same or different species?

REFERENCES AND RECOMMENDED READING

British Museum (Natural History). 1981. *Origin of Species.* London.

Dobzhansky, T., F. J. Ayala, G. L. Stebbins, and J. W. Valentine. 1977. *Evolution.* New York: Freeman.

Ford, E. B. 1975. *Ecological Genetics.* New York: John Wiley.

Futuyma, D. J. 1986. *Evolutionary Biology.* Sunderland, Mass.: Sinauer.

Larson, E. J. 1989. *Trial and Error: The American Controversy over Creation and Evolution.* Oxford: Oxford University Press.

Mayr, E. 1942. *Systematics and the Origin of Species.* New York: Columbia University Press.

Prokopy, R., and G. Bush. 1993. Evolution in an orchard. *Natural History,* 9/93: 4–10.

Ramirez, R. 1970. Host specificity of fig wasps (*Agaonidae*). *Evolution,* 24: 681–691.

Stebbins, L. G. 1982. *Darwin to DNA, Molecules to Humanity.* New York: Freeman.

White, M. J. D. 1978. *Modes of Speciation.* New York: Freeman.

Wiebes, J. T. 1979. Co-evolution of figs and their insect pollinators. *Annual Review of Ecology and Systematics,* 10:1–12.

ANSWER TO FIGURE QUESTION

Figure 25.4   New geographical or ecological barriers can separate populations.

26

The Unfinished Synthesis

Chapter Outline

Reading Questions

1. What does the fossil record tell us about past life on Earth?

2. What patterns of change does the fossil record reflect?

3. How are different areas of biology related to the theory of evolution?

4. Why is the theory of evolution an "unfinished" synthesis?

The modern synthesis brings together knowledge from genetics, biogeographical distribution, and taxonomy to explain how new species come into being. Like Darwin's theory of evolution, however, the modern synthesis goes beyond explaining just the origin of species. It attempts to unify all of the current life sciences under the theory of evolution. This unification includes subjects as diverse as paleontology and physiology. Theodosius Dobzhansky epitomized this view when he commented in 1973, "Nothing in biology makes sense except in the light of evolution." How does the modern synthesis attempt to encompass all phenomena dealing with living things?

It approaches this task through various avenues. Some areas of biology, such as **paleontology** (the study of fossils), include knowledge that is closely linked to evolutionary theory. Architects of the modern synthesis have used concepts and processes from evolutionary theory to explain the appearance and significance of the fossil record. In turn, the study of fossils has contributed to the development and extension of the modern synthesis. The modern theory of evolution also raises questions about the fossil record that direct research. For example, studies of ancient community evolution, population dynamics, and rates of speciation have been inspired by questions about evolution.

In other areas of biology, the relationship to evolutionary theory is less direct. *Physiology,* for example, is concerned with how organisms function. Physiologists are most interested in processes that can be studied at the present time, not their evolutionary development. Nonetheless, by comparing certain common biochemical compounds or metabolic pathways in different organisms, physiologists have thrown some light on possible evolutionary relationships among organisms. By showing the adaptive importance of various physiological processes, biologists have also related the chemical activities of an organism to its evolutionary history. Still, an examination of the major questions that now guide research in physiology does not reveal many that are directly concerned with evolution. Rather, the theory of evolution provides an intellectual framework that allows physiologists to relate the functions of organisms to other aspects of life and its history.

In this chapter, the relationship of the modern synthesis to paleontology and physiology are discussed in order to elucidate evolution as a unifying theory. The examination of paleontology also provides a basis for describing some of the main topics in macroevolution (evolution above the species level), which was introduced in Chapter 25.

PALEONTOLOGY AND EVOLUTION

Our consideration of paleontology begins with a general account of the fossil record. Then patterns of change revealed in the fossil record are described and explained. These topics permit an evaluation of the relationship between paleontology and the theory of evolution.

The Fossil Record

The **fossil record** is a partial chronicle of past life on this planet (see the Focus on Scientific Process, "The Fossil Record—700 Million Years of Life"). Studies of fossils were central to Darwin in formulating his theory of evolution, and they have continued to be an important source of new ideas. Studies of the fossil record have been extremely valuable, giving scientists their only direct picture of past life on Earth. Nevertheless, the fossil record is woefully incomplete. Why is the record so fragmentary?

Fossils are remnants or traces of organisms from a past geologic age that are embedded in Earth's crust. The conditions necessary for forming fossils rarely existed. Therefore, only a fraction of the species that have lived during Earth's history left any fossils (see Figure 26.1). Moreover, because fossils are most often formed under water or in wet soil, the record tells us more about shallow-water marine organisms than terrestri-

al species (refer to Figure 25.2, in the previous chapter, which shows five different species of foraminiferans found in samples of sediment drilled from the ocean floor). For the same reason, scientists have learned more about organisms that lived near lakes than those that lived on mountains. Because the soft parts of organisms have rarely been preserved, paleontologists must work mainly with the remains of substances like teeth, bones, shells, pollen, and wood. Entire groups of organisms (for example, jellyfish) have left practically no fossils, and even for groups that have left a record, the picture is incomplete.

The fossil record as a whole, then, tells us little about the specific history of individual species or the origins of major phyla. Remains from all the major animal phyla go back to the Cambrian period, the earliest period for which we have extensive fossil remains. It is the evolution within phyla, especially animal phyla, that is best revealed by the fossil record.

Despite the inherent difficulties in studying the record of past life, a great amount has been learned. For example, in Darwin's day very little was known about the evolutionary history of the angiosperms (flowering plants). Like many scientists who came after him, Darwin despaired of ever tracing the development of these plants. Since the 1970s, however, scientists studying the fossil remains of leaves and pollen from the Cretaceous period have demonstrated that in this period angiosperm leaves and pollen increased enormously in diversity and complexity. Further investigation of the fossil record has provided a clearer picture of the history of the flowering plants (see Figure 26.2, page 494).

The fossil record has allowed biologists to recognize many significant patterns, to trace specific lines of development, and to reconstruct a broad picture of the history of life on this planet.

BEFORE YOU GO ON   Fossils are the permanent remains of organisms that lived and died in a past geologic age. Although fossils yield a fragmentary picture, they provide the clearest chronicle of life that has evolved on this planet during past geologic eras. Analyses of the fossil record have provided strong support for the theory of evolution.

Patterns of Change

Studies of fossils cannot disclose much direct information about genetic changes in populations. For the most part what is revealed are *morphological* (structural) differences that have developed among species and higher taxa.

The fossil record shows that there are different patterns of evolutionary change. Some species appear to have remained constant for millions of years without any recognizable modification. These species are referred to as "living fossils," and the best-known examples are the coelacanths (an ancient marine fish species) and ginkgo trees, both shown in Figure 26.3 (see page 494). Such examples, however, are rare. More common is evidence of a gradual change without any splitting

Figure 26.1   This figure describes the steps in one possible path in the formation of a fossil—in this case, that of an icthyosaur. (A) An animal lives in a sea. (B) When it dies, its remains sink to the bottom of the sea and in time become covered by mud and silt, which prevents their destruction. Soft tissues decompose, but bones and teeth remain. The soft parts may leave impressions in the surrounding mud. (C) As more material sinks to the sea bottom, the animal remains get buried deeper and deeper. Eventually the entire area may be submerged by more water, and the mud layer may become part of a larger sediment. (D) Elements in the surrounding sediment slowly penetrate, replace, or fill spaces in the original organic material, which may become totally replaced by a different mineral. (E) Geological processes may lead to the entire region's sinking under seas to form sedimentary rock, and lakes can be uplifted to produce mountains. (F) Ultimately, the rock strata may be pushed near the surface and uncovered by erosion or by a passing traveler, research scientist, or road construction team.

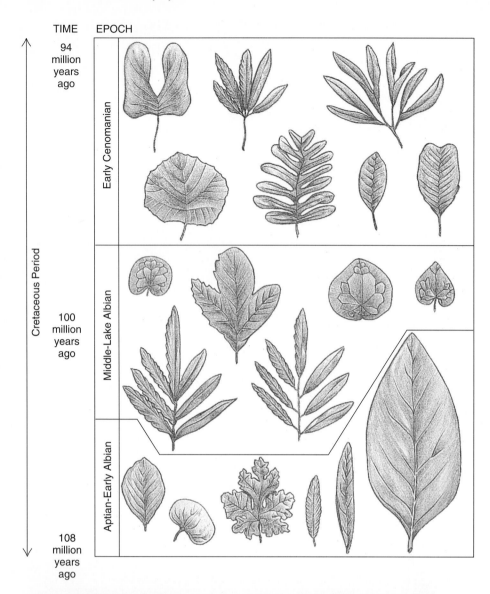

TIME
EPOCH

94 million years ago

Cretaceous Period

Early Cenomanian

100 million years ago

Middle-Lake Albian

Aptian-Early Albian

108 million years ago

Figure 26.2 The origin of major leaf types during three epochs of the Cretaceous period.

A

B

Figure 26.3 Two famous "living fossils" are (A) the coelacanth, a marine fish, and (B) *Metasequoia occidentalis*. Both have survived unchanged for millions of years. The coelacanth, once thought to be extinct, was discovered in 1938 in the waters near Madagascar. A fossil metasequoia was discovered in 1941 in Japan; a few years later it was discovered still alive in southern China.

The Fossil Record—700 Million Years of Life

Chapter 3 explained the evolution of prokaryotic and eukaryotic cells. Prokaryotes make their first appearance in the fossil record about 3.5 billion years ago, and eukaryotes appeared almost 2 billion years ago. Very few fossil remains from before the Paleozoic era, however, have ever been found. We have some evidence of invertebrate evolution in the period immediately before the Paleozoic. The first evidence consists merely of burrows left in sediment by primitive organisms; the earliest of these traces is 700 million years old. In South Australia, a deposit of fossils that date from 680 to 580 million years ago contains the remains of corals, segmented worms, and other marine invertebrates. Some of these are from modern phyla; others are not. Figure 1 shows some fossils from the different geological eras.

Paleozoic Era

The Paleozoic era consists of six periods, beginning with the Cambrian and ending with the Permian.

Cambrian Period

What do we know about life in the Cambrian period? The fossil record reveals that starting about 570 million years ago, there was a sudden burst of diversity: nearly all the modern animal phyla appeared over the relatively short period of 50 million years. The fossil record contains not only all of the basic animal body plans that exist today, but also many that have become extinct. In the Burgess Shale of British Columbia, researchers discovered an unusual collection of soft-bodied fossils that contains the remains of almost a dozen phyla that have no modern counterparts. Producers do not show the same diversity and are limited to algal forms in the Cambrian.

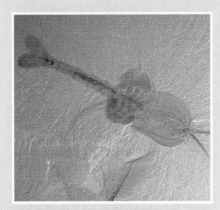

Figure 1A Cambrian period: an arthropod fossil from Burgess Shale.

Ordovician Period

Paleontologists have determined that late in the Cambrian and extending into the Ordovician period (starting about 500 million years ago), a lineage evolved that developed an effective swimming ability. It contained organisms that came to have armor plates for protection and an axlelike skeleton with vertebrae that had lateral extensions. These were the first, primitive, jawless fish, or *ostracoderms*. Animal life was still confined to the oceans, but during this period, much of what is now dry land was covered with shallow seas, which permitted primitive plant growth. A great increase in bivalves and other

Figure 1B Ordovician period: brachiopods (spherical, scallop-like structures) and crinoids.

bottom-dwelling filter-feeding invertebrates can be observed in the fossil record of this period.

Silurian Period

Fossil evidence suggests that land plants first appeared in the Silurian period, which began 441 million years ago. They were simple branching stems (up to about 25 centimeters long); some of them had small, scalelike structures that look like leaves. The adult plants developed spore cases that released airborne spores. Animals inhabited shallow, warm seas, and Silurian deposits reflect a diversification of fish, as well as the presence of dramatic-looking giant sea scorpions and bottom-dwelling invertebrates.

Figure 1C Silurian period: sea scorpion.

box continues

FOCUS ON SCIENTIFIC PROCESS

Devonian Period

Some 413 million years ago, the Devonian period—the Age of Fish—began, according to paleontologists who have studied the fossil record. As its name suggests, a great radiation of fish dates from this period. Jawed fish with paired fins came into existence. The fossil record contains cartilaginous fish (sharks and rays), which remained mostly marine, and bony fish that proliferated in fresh water. Among the bony fish were ray-finned fish, the ancestors of modern fish, and lobe-finned fish, some of which are thought to be the ancestors of amphibians, the first vertebrates that appeared on land. Ammonites, a group of externally shelled cephalopods that have been widely used by geologists to date strata, also made their appearance during this period. Land plants formed the first forests, which were dominated by ferns, club mosses, and horsetails. The first wingless insects are found in the late Devonian period.

Carboniferous Period

Beginning 365 million years ago, the fossil record shows that great portions of the Northern Hemisphere were covered with huge swamp forests that were drowned late in the period by shallow seas. This resulted in the production of massive coal deposits. Amphibians exploited many habitats during the Carboniferous period and attained large body sizes. The first reptiles, which arose from a primitive amphibian lineage, appeared, as did flying insects.

Permian Period

The Permian period, starting 290 million years ago, marks the end of the Paleozoic era. Reptiles succeeded amphibians as the dominant land animals, and new insect groups emerged. Evidence indicates that great geologic changes altered the climate and topography. Gymnosperms began to increase, and conifers came to dominate many forests. The end of the Permian period saw one of the most dramatic mass extinctions in the fossil record; nearly

Figure 1D Carboniferous period: An example of a Northern Hemisphere plant.

Figure 1E Permian period: *Lebackia piniformis,* a conifer.

half of the known families of animals disappeared. Marine invertebrates were especially hard hit.

Mesozoic Era

The Mesozoic era, which began 245 million years ago, is conventionally divided into three periods: the Triassic, the Jurassic, and the Cretaceous. The most sensational events of this era center on the reptiles, which underwent an extensive set of radiations. Dinosaurs—the group of stars from the fossil record that intrigue not only professional paleontologists but also grade school children and Hollywood movie makers—date from this era.

The dinosaurs evolved in the Triassic period and succeeded the dominant reptiles from the Permian period. The dinosaurs themselves underwent several waves of extinction, but each time they radiated from surviving lineages and dominated Earth. At the close of the Cretaceous period, they, too, entered the abyss of extinction, along with ammonites and some other widely distributed animals.

During the Mesozoic era, the supercontinent Pangaea split apart, and by the Cretaceous period, the continents were taking on their modern shapes and positions. Earth's vegetation underwent a complete change in the last part of the Mesozoic. The first flowering plants (angiosperms) appeared roughly 130 million years ago. They underwent a spectacular radiation, which led to one of the greatest success stories in the evolutionary saga—the development of over 250,000 species. Ironically, the history of that radiation is one of the least understood chapters in evolution.

Plant and reptile evolution were not the only important Mesozoic developments. Birds and mammals arose, as did the modern orders of insects and the modern fish.

Cenozoic Era

Some 65 million years ago, the "Age of Recent Life" began. It is divided by scientists into the Tertiary and Quaternary periods. The Tertiary is composed of the bulk of the era (all but the past 2 million years) and is divided into five epochs.

The Quaternary consists of two epochs: the Pleistocene and the Recent.

Depending on one's point of view, the Cenozoic era can be called the "Age of Mammals," the "Age of Birds," the "Age of Insects," or the "Age of Flowering Plants." There was enormous diversification and speciation in all these groups.

Mammals are found in the fossil record to have succeeded the dinosaurs on land. They had coexisted as small, active, often nocturnal animals, but unlike the dinosaurs, some of them (including the primates) survived the mass extinctions at the end of the Cretaceous period. From the several lineages that entered the Cenozoic, a strong set of radiations occurred, reaching a peak within the past 2 million years.

The Pleistocene witnessed the ice ages in the Northern Hemisphere as well as the rise of that most peculiar animal, *Homo sapiens*. Many of the large mammals became extinct before the Recent epoch, although there is evidence that humans coexisted with several of them.

Figure 1F   Mesozoic era: *Tyrannosaurus rex* (Jurassic period).

Figure 1G   Cenozoic era: hare.

or branching. Such **lineages** (single lines of descent) are characterized by a long series of individuals that display a continuum of change. At the ends of the continuum are different species, and there may be different species in between. This gradual evolution was discussed in Chapter 25; it is called phyletic evolution.

Even more common are fossils from lineages that split over time, and several patterns of separation have been described. Divergence, convergence, and parallelism are the best documented; they are illustrated in Figure 26.4.

Divergence

Divergence describes an evolutionary pattern in which there are increased morphological differences among separated lineages. This is most striking in **adaptive radiation,** where several lineages diverge from a common ancestor. A classic example involves the evolution of Hawaiian Island honeycreepers (Drepanididae). How can the adaptive radiation of these birds be explained? Ornithologists believe that, similar to the evolution of finches on the Galápagos Islands, a South American ancestor originally colonized the oceanic islands of Hawaii and gave rise to the entire honeycreeper family, which now consists of nine genera that are divided into two subfamilies. Naturalists hypothesize that their diversity is the result of an ancestral form that invaded a new territory, enabling its descendants to exploit new sources of food. The external characteristics of honeycreepers reflect adaptations to different types of food. Walter Bock of Columbia University has described a series of radiations that start with a nectar-eating offshoot from one of the two subfamilies of honeycreepers. In time, the different genera of the second subfamily evolved. These consisted of insect, fruit, and seed eaters, each of which have very distinctive bills (see Figure 26.5).

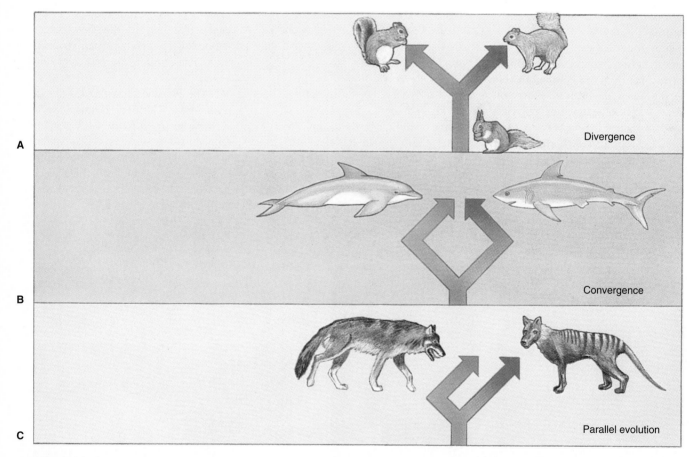

Figure 26.4 Three patterns of macroevolution as measured by morphological difference are divergence, convergence, and parallel evolution. (A) In divergence there is increased difference among separated lineages. (B) Convergence occurs when separate lineages become similar. (C) Parallel evolution resembles convergence but occurs when the separate lineages share a common ancestor and the similarity is partly a result of the shared ancestry.

Question: *Give another example of each pattern of macroevolution.*

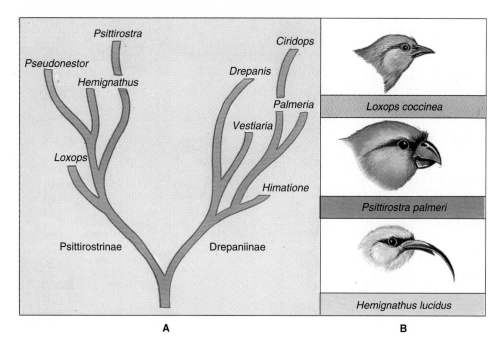

Figure 26.5 The beaks of honeycreeper birds on the Hawaiian Islands reflect great diversity and provide a classic example of the type of evolutionary divergence known as adaptive radiation. (A) A phylogenic diagram of the honeycreepers. (B) Three genera that have evolved in the Hawaiian Islands, like the six others, reflect responses to different environmental resources. *Loxops* feed on insects caught flying, *Psittirostra* feed on large seeds that need to be crushed, and *Hemignathus* feed on the nectar of flowers.

Another case of adaptive radiation involves the entire class Aves (birds). After they branched off from the reptile class during the Mesozoic period, birds underwent an initial radiation, which was then followed by successive radiations during the Paleocene and Eocene epochs. These numerous radiations were related to structural and functional advances. For example, the ability to fly not only permitted birds to escape predators but also to migrate, taking advantage of mild seasons and exploiting many different food sources.

The adaptive radiation of placental mammals at the beginning of the Tertiary period is another famous example of this type of divergence (see Figure 26.6, page 500). Placental mammals, which had an efficient form of reproduction and considerable brain size, exploded in a set of radiations. They replaced the dinosaurs, which had become extinct after dominating Earth for well over 100 million years.

Divergence explains some of the striking homologies found in the living world. **Homologous structures** share a common structural, evolutionary, and embryological background. Classic examples of homologous structures are the skeletal forelimbs of humans, dogs, whales, and birds, (as shown in Figure 26.7, page 501).

Convergence

Convergence occurs when separate lineages become morphologically similar. Environmental or functional similarities often lead to convergence over time. For example, the evolution of wings in insects, birds, and mammals (bats) reflects convergent evolution related to function (see Figure 26.8, page 501). Development of the body shapes of some aquatic mammals and fishes shows convergence that has resulted from adaptations to the same environment. The independent development of mollusk and vertebrate eyes is another quite amazing example of convergence.

Convergent evolution is responsible for many of the striking analogies that can be observed both in the fossil record and in living organisms. **Analogous structures** have similar functions but, in contrast to homologous structures, differ in their structure, development, and evolutionary history. For example, the wings of a butterfly and the wings of a bird are both used for flight but are analogous structures because they differ in structure, embryology, and evolutionary background.

Convergence can also occur with behavioral and physiological traits. The fossil record, however, yields little evidence of such traits, and biologists rely on comparative studies of living organisms to learn more about them.

Parallelism

Parallelism is similar to convergence. It is the development of similar characteristics in separate lineages that have a common ancestor, where the similarity is influenced by the characteristics of the shared ancestor. The evolution of placental and mar-

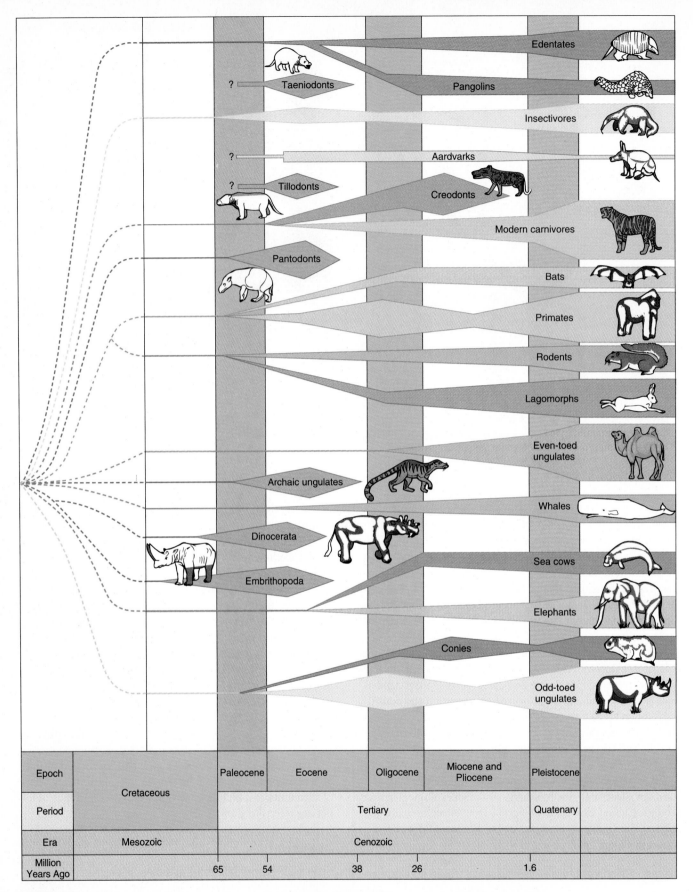

Epoch		Paleocene	Eocene	Oligocene	Miocene and Pliocene	Pleistocene	
Period	Cretaceous			Tertiary		Quaternary	
Era	Mesozoic			Cenozoic			
Million Years Ago		65	54	38	26	1.6	

Figure 26.6 Placental mammals evolved from a primitive lineage during the Cretaceous period. Their advanced brains and offspring, which were well-developed at birth, gave this group a competitive advantage that resulted in a series of adaptive radiations. Note how most of the orders of placental mammals had evolved by the early Eocene epoch.

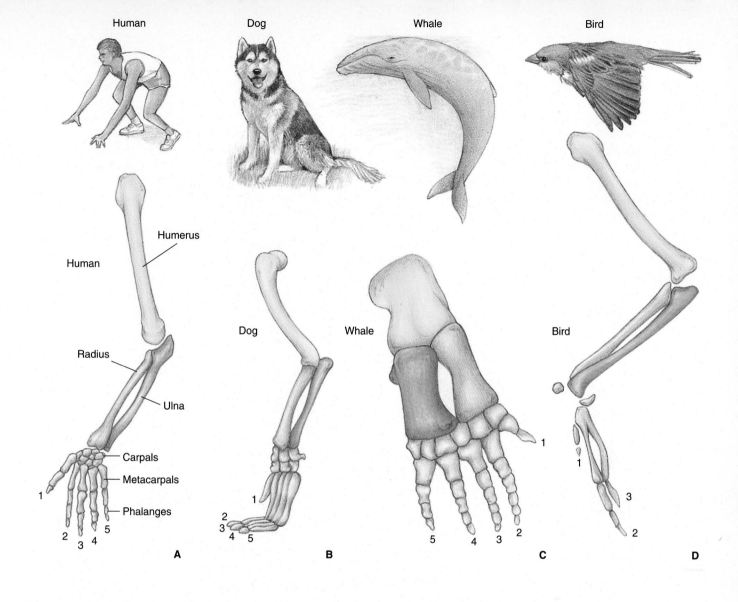

Figure 26.7 Homology among the forelimbs of (A) humans, (B) dogs, (C) whales, and (D) birds. Numbers refer to digits.

Question: *Beside structure, what do these limbs have in common?*

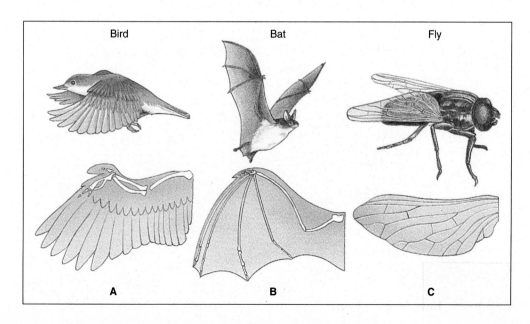

Figure 26.8 The wings of (A) birds, (B) bats, and (C) insects (fly) evolved independently and thus exemplify convergent evolution.

supial mammals (see Chapter 6) is an example of parallelism. Both evolved from the same subclass during the mid-Cretaceous period. Marsupial mammals evolved extensively in Australia (and in South America until the late Tertiary period), in the absence of competition from the placental mammals that came to dominate other continents. Because of their similarities, marsupials evolved in Australia in parallel to mammals that evolved on other continents (see Figure 26.9).

Parallel evolution sometimes results in the production of both analogous and homologous structures. For example, the true anteater and the marsupial anteater share homologous mammalian characteristics, but their anteater characteristics—long snouts and long sticky tongues—are analogous.

> **BEFORE YOU GO ON** The fossil record contains evidence of different patterns of evolutionary change. Although lineages remain constant for millions of years, and some show a slow, gradual change, the more common pattern is that of splitting into separate lineages. There are different patterns of lineage splitting. In divergence, different lineages have increased morphological differences. In convergence, separate lineages become morphologically similar. In parallelism, lineages that share a common ancestry develop similar characteristics.

Other Patterns of Change

The divergence, convergence, and parallelism of lineages are the three most common evolutionary patterns revealed by the fossil record. Scientists, however, recognize several other patterns. The most important of these are the emergence of *grades,* the appearance of *direction* in evolution, and the occurrence of *mass extinctions.*

Grades

Most evolutionary change involves the modification of existing structures and systems. Occasionally, however, a change occurs that results in a new functional ability. A group of species that have evolved to the same level of organization is called a new **grade.** Some grades are associated with dramatic evolutionary events that lead to adaptive radiations.

Direction

The recognition of grades, particularly within the histories of higher taxonomic groups, led early paleontologists to hypothesize the existence of **progressive evolution,** an evolutionary sequence displaying a constant direction, or "goal," and coming into existence independent of selective forces. Early paleontologists thought that this directed evolution was the result of some unknown "internal force" that either altered embryological development or caused the organism to acquire adaptive characteristics that could be passed on to future genera-

tions. The modern synthesis has been skeptical about such forces and about the existence of progressive evolution. Although the modern theory of evolution recognizes directional selection in a population, contemporary paleontologists have argued that a critical look at the existing evidence does not support the idea of a linear, progressive evolution of major groups. Examining the evolution of ammonites provides a good model of how analyses of data have led to challenges to the claim for progressive evolution.

The subclass Ammonoida of the class Cephalopoda (mollusks) is divided into three large orders—Goniatitida, Ceratitida, and Ammonitida—that were each dominant during three different time periods (see Figure 26.10A, page 504). The goniatites lived during the last three periods of the Paleozoic era, the ceratites followed them in the first period of the Mesozoic era, and the true ammonites flourished during the last two periods of the Mesozoic. The three orders show a clear progression in shell complexity, most notably in the suture patterns present at the junctures of the internal chambers and the outer shells (see Figure 26.10B, page 504). Although the functional significance of the shell complexity is not well understood, taxonomists rank the three orders as representing different grades. One might propose that these three grades are evidence of progressive evolution.

A careful look at the ammonite fossil record, however, dispels the notion of progressive evolution. It is true that one can arrange the three orders in a chronological order that reflects the order of shell complexity, but the "higher" ceratites did not evolve from the "lower" goniatites, nor did the "highest" grade, the true ammonites, evolve from the "middle" grade ceratites. Rather, as you can see in Figure 26.10, the ceratites arose from a different lineage, the Prolecanitida, that existed along with the "lower" goniatites and underwent a rapid radiation after the extinction of the goniatites. Similarly, the true ammonites radiated from a lineage separate from the ceratites, and did so after their extinction. To arrange the three orders of ammonites into a sequence and suggest that a linear development accounts for their evolution is to oversimplify the evidence from the fossil record.

Examples like the ammonite fossil record have made modern evolutionists very skeptical of the validity of constructing long, linear evolutionary sequences that display a continual advancement of grades. Such "evolutionary ladders" are often oversimplifications of the fossil record. A more accurate picture of the evolution of groups like the ammonites would be a branching bush.

Mass Extinctions

Mass extinction refers to the destruction of a vast number of taxa; in a sense, it is the opposite of adaptive radiation. The commonly used method of dividing the geologic record into different periods is partly a reflection of mass extinctions of the past.

Placentals

Wolf
(*Canis*)

Ocelot
(*Felis*)

Flying squirrel
(*Glaucomys*)

Ground hog
(*Marmota*)

Anteater
(*Myrmecophaga*)

Mole
(*Talpa*)

Mouse
(*Mus*)

Marsupials

Tasmanian wolf
(*Thylacinus*)

Native cat
(*Dasyurus*)

Flying phalanger
(*Petaurus*)

Wombat
(*Phascolomys*)

Anteater
(*Myrmecobius*)

Mole
(*Notoryctes*)

Mouse
(*Dasycercus*)

Figure 26.9 Placental mammals on the large continents evolved in parallel with Australian marsupial mammals.

Question: *What explains the similarities between the two groups?*

In 1982, two scientists, David Raup and John Sepkoski, Jr., of the University of Chicago, described five large, separate extinction events that occurred during the late Ordovician, Devonian, Permian, Triassic, and Cretaceous periods. Because most of these time periods coincide with known changes in Earth's surface and shifts in climate, scientists accept the view that general changes in the environment were responsible for the extinctions. Two years later, the same two scientists published results of another study concerned with families of marine vertebrates, invertebrates, and protozoans over the past

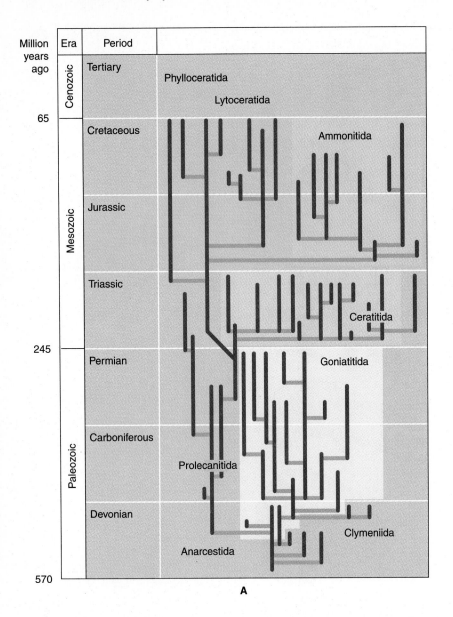

A

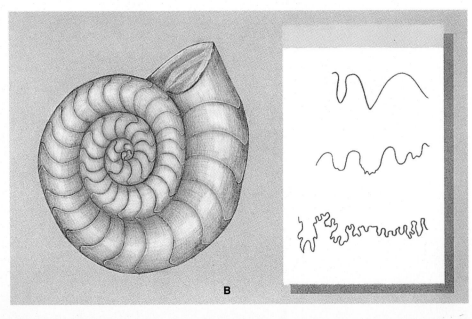

B

Figure 26.10 The fossil record of ammonite evolution shows no direct progression. (A) The three large orders of ammonites—Goniatitida, Ceratitida, and Ammonitida—appear chronologically in the fossil record. The Ceratitida, however, arose from a lineage (Prolecanitida) that existed at the same time as the Goniatitida but was not part of it. Similarly, the Ammonitida arose from a lineage that was not part of the Ceratitida. Comparing the three large orders of ammonites does reveal a progression in the complexity of the suture pattern of their shells (B), which may have misled earlier paleontologists to think that the three orders were an example of progressive evolution.

250 million years. Their new results, illustrated graphically in Figure 26.11, showed 12 peaks of extinction with a regular cycle of 26 million years between the peaks. This regularity raised some very interesting questions and suggested to some scientists that there was an extraterrestrial cause of the extinctions.

The concept of a periodic (regular, recurring) astronomical cause for extinctions on Earth was not a new idea in 1982. In 1980, Luis Alvarez, his son Walter, and his colleagues at the University of California at Berkeley proposed that extinctions at the end of the Cretaceous period were caused by the impact of an asteroid. They argued that the collision produced a cloud of dust that lasted long enough to suppress photosynthesis, which in turn resulted in a dramatic collapse of existing food webs. Their evidence was related to the presence of a layer of iridium, a rare metal found in asteroids, that was deposited in marine sediments from the late Cretaceous period, presumably as a result of the impact (see Figure 26.12). Geologists have recently identified a huge crater (180 kilometers in diameter) on the Yucatan Peninsula in Mexico, which may have been formed by a massive asteroid that slammed into Earth 65 million years ago. The Alvarez hypothesis is very controversial among paleontologists. Recent studies suggest that considerable extinction took place in the Cretaceous period *before* the iridium layer was deposited. Such evidence has weakened the Alvarez hypothesis. Nonetheless, the regularity of extinction events has led many scientists to give serious attention to nonbiological factors for explaining the fossil record.

The study of mass extinction has also suggested another feature of natural selection. Families that had a wide geographical distribution were more likely to survive than more geographically restricted families. This pattern suggests that perhaps a "rare-event selection" takes place occasionally. That is, under extreme conditions, some generally overlooked adaptations may have an increased importance that is not evident during other times.

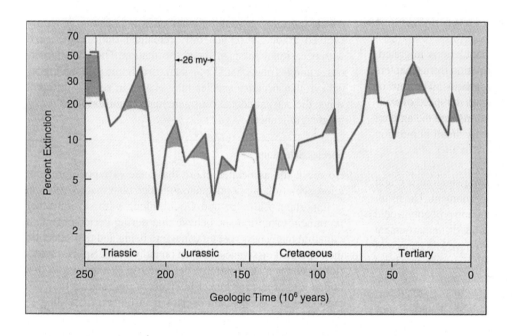

Figure 26.11 Graphing the family extinction rates of marine vertebrates, invertebrates, and protozoans for the time interval starting in the late Permian period and extending through the Pliocene period reveals a curve with 12 peaks. The lines above the curve mark 26-million-year intervals.

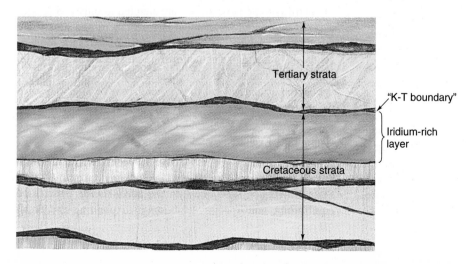

Figure 26.12 Scientists have found a layer of iridium in the boundary between the Cretaceous and Tertiary periods. This layer is considered to be evidence that a large impact, possibly caused by an asteroid, produced an enormous cloud of dust that might have significantly altered Earth's climate—and thus life on Earth—65 million years ago.

PHYSIOLOGY AND EVOLUTION

Physiology is the study of how organisms function. It depends heavily on chemical and physical methods. Historically, physiology was the first of the biological disciplines to develop experimental techniques. Unit VI of this text covers several areas of importance in human physiology.

Modern physiology is divided into cellular physiology, the physiology of special groups, and comparative physiology. At the cellular level, living organisms share many basic characteristics, so **cellular physiology** is regarded by biologists as "general physiology," or the examination of fundamental functions of living organisms. The **physiology of special groups** focuses on the functioning of specific groups or species that have a special interest. For example, human physiology (because of its value for medical research) and plant physiology (because of its agricultural importance) are topics in the physiology of special groups. **Comparative physiology** examines the mechanisms that different organisms use to perform similar functions.

Relationship of Physiology to the Theory of Evolution

The leading physiologists of the mid-nineteenth century, such as Claude Bernard in France and Rudolf Virchow in Germany, were hostile to Darwin's theory of evolution when it was first proposed. Although some physiologists were inclined to believe that organisms had evolved, they rejected Darwin's theory in part because they believed that any theory unifying the biological sciences would have to be at the cellular level (Darwin didn't even mention cells in his book *On the Origin of Species!*). Physiologists were also unreceptive because they were primarily experimental biologists, and Darwin's theory did not easily lend itself to experimental verification.

Today physiologists have a different view of evolution. The theory serves not only as an intellectual context relating physiology to other biological disciplines, but it also explains why there are similarities in basic but complex physiological processes in all cells, for example, metabolic pathways and the functions of nucleic acids (see Chapters 17 and 18). The common evolutionary history of single and multicellular organisms accounts for these similarities.

Some molecular biology research, which was done with a view toward better understanding of physiological functions, has been used to extend evolutionary theory to the biochemical level. The best-known of these studies concern the protein cytochrome *c*, which is found in the cells of all animals and plants. It plays a central role in energy-generating metabolic reactions. Because the cytochrome *c* molecules of different species perform the same function, have similar amino acid sequences, and have comparable overall structure, they are considered homologous. When the amino acid sequences of cytochrome *c* proteins of different species are placed on a chart, their arrangement is strikingly similar (see Figure 26.14).

	10	20	30	40
	1 2 3 4 5 6 7 8 9 0	1 2 3 4 5 6 7 8 9 0	1 2 3 4 5 6 7 8 9 0	1 2 3 4 5 6 7 8 9 0 1 2 3 4 5 6 7 8 9
1. Human	- - - - - - - - G D V E K G K K I	F I M K C S Q C H T V E	K G G K H K T G P N L H	G L F G R K T G
2. Rhesus monkey	- - - - - - - - G D V E K G K K I	F I M K C S Q C H T V E	K G G K H K T G P N L H	G L F G R K T G
3. Horse	- - - - - - - - G D V E K G K K I	F V Q K C A Q C H T V E	K G G K H K T G P N L H	G L F G R K T G
4. Dog	- - - - - - - - G D V E K G K K I	F V Q K C A Q C H T V E	K G G K H K T G P N L H	G L F G R K T G
5. Gray whale	- - - - - - - - G D V E K G K K I	F V Q K C A Q C H T V E	K G G K H K T G P N L H	G L F G R K T G
6. Rabbit	- - - - - - - - G D V E K G K K I	F V Q K C A Q C H T V E	K G G K H K T G P N L H	G L F G R K T G
7. Kangaroo	- - - - - - - - G D V E K G K K I	F V Q K C A Q C H T V E	K G G K H K T G P N L N	G I F G R K T G
8. Penguin	- - - - - - - - G D V E K G K K I	F V Q K C S Q C H T V E	K G G K H K T G P N L H	G I F G R K T G
9. Snapping turtle	- - - - - - - - G D V E K G K K I	F V Q K C A Q C H T V E	K G G K H K T G P N L N	G L I G R K T G
10. Bullfrog	- - - - - - - - G D V E K G K K I	F V Q K C A Q C H T C E	K G G K H K V G P N L Y	G L I G R K T G
11. Tuna	- - - - - - - - G D V A K G K K T	F V Q K C A Q C H T V E	N G G K H K V G P N L W	G L F G R K T G
12. Silkworm moth	- - - - G V P A G N A E N G K K I	F V Q R C A Q C H T V E	A G G K H K V G P N L H	G F Y G R K T G
13. Wheat	A S F S E A P P G N P D A G A K I	F K T K C A Q C H T V D	A G A G H K Q G P N L H	G L F G R Q S G
14. Fungus *(Neurospora)*	- - - - G F S A G D S K K G A N L	F K T R C A E C H G E G	G N L T Q K I G P A L H	G L F G R K T G

A Alanine	F Phenylalanine	K Lysine	P Proline	T Threonine
C Cysteine	G Glycine	L Leucine	Q Glutamine	V Valine
D Aspartic acid	H Histidine	M Methionine	R Arginine	W Tryptophan
E Glutamic acid	I Isoleucine	N Asparagine	S Serine	Y Tyrosine

Figure 26.14 A comparison of the amino acid sequences of cytochrome *c* shows a striking similarity among living organisms.

Question: *How many differences are there between the amino acid sequences of cytochrome c in humans and tuna? Between humans and dogs? What does this mean?*

Closely related species have amino acid sequences that are more alike than those of distantly related species. For example, the cytochromes *c* of humans and rhesus monkeys differ by only one amino acid, whereas those of humans and kangaroos differ by ten.

By calculating the minimum number of nucleotide substitutions in the genetic code that are necessary to produce these differences, scientists have constructed a phylogeny that agrees for the most part with those derived from the fossil record (see Figure 26.15, page 510). Other proteins have also been used in a similar manner. Not only does the comparative physiological approach verify generalizations derived from the fossil record, but it is also a valuable supplement in cases where the fossil record is weak or nonexistent.

Comparative studies of proteins and nucleic acids also suggest that base substitutions in DNA or RNA accumulate at a steady rate. Some scientists therefore believe that such macromolecules can act as "molecular clocks," and molecular comparisons can be used to date evolutionary divergences. The greater the dissimilarity, the older the evolutionary divergence.

Physiology and Paleontology

Although the theory of evolution provides a conceptual framework that relates physiology to other biological disciplines and provides an explanation for similar chemical processes in diverse organisms, it does not have as close a relationship to physiology as it does to paleontology. By analyzing, for example, the chemical basis of muscle contraction, hormonal regulation of reproduction, or the response of plants to airborne pollutants, physiologists can tell us how organisms are *currently* functioning. The frame of reference is very short from an evolutionary perspective. Although the information physiologists generate occasionally has relevance for understanding evolutionary processes, most physiologists do not design their studies with evolutionary theory in mind or with an eye toward deepening our understanding of the history of life on this planet. Rather, they explore how living things work at present.

Physiology and paleontology provide a contrast in how the theory of evolution is related to the separate disciplines of biology. In paleontology, evolution not only provides the background but is central in formulating research questions that in turn have important consequences for the theory. Physiology, in contrast, looks to evolution to explain the regularities it uncovers. Although some of the specific research questions physiology explores are derived directly from the theory of evolution, most of the field has its historical roots in medicine and agriculture. Most questions that physiologists investigate relate to the functioning (or malfunctioning) of the human body or of organisms of economic importance to humans.

```
         50           60           70           80           90           100          110
       0 1 2 3 4 5 6 7 8 9 0 1 2 3 4 5 6 7 8 9 0 1 2 3 4 5 6 7 8 9 0 1 2 3 4 5 6 7 8 9 0 1 2 3 4 5 6 7 8 9 0 1 2 3 4 5 6 7 8 9 0 1 2
 1.  Q A P G Y S Y T A A N K N K G I I W G E D T L M E Y L E N P K K Y I P G T K M I F V G I K K K E E R A D L I A Y L K K A T N E
 2.  Q A P G Y S Y T A A N K N K G I T W G E D T L M E Y L E N P K K Y I P G T K M I F V G I K K K E E R A D L I A Y L K K A T N E
 3.  Q A P G F T Y T D A N K N K G I T W K E E T L M E Y L E N P K K Y I P G T K M I F A G I K K K T E R E D L I A Y L K K A T N E
 4.  Q A P G F S Y T D A N K N K G I T W G E E T L M E Y L E N P K K Y I P G T K M I F A G I K K T G E R A D L I A Y L K K A T K E
 5.  Q A V G F S Y T D A N K N K G I T W G E E T L M E Y L E N P K K Y I P G T K M I F A G I K K K G E R A D L I A Y L K K A T N E
 6.  Q A V G F S Y T D A N K N K G I T W G E D T L M E Y L E N P K K Y I P G T K M I F A G I K K K D E R A D L I A Y L K K A T N E
 7.  Q A P G F T Y T D A N K N K G I T W G E D T L M E Y L E N P K K Y I P G T K M I F A G I K K K G E R A D L I A Y L K K A T N E
 8.  Q A E G F S Y T E A N K N K G I T W G E D T L M E Y L E N P K K Y I P G T K M I F A G I K K K S E R A D L I A Y L K D A T S K
 9.  Q A E G F S Y T E A N K N K G I T W G E E T L M E Y L E N P K K Y I P G T K M I F A G I K K K A E R A D L I A Y L K D A T S K
10.  Q A A G F S Y T D A N K N K G I T W G E D T L M E Y L E N P K K Y I P G T K M I F A G I K K K G E R Q D L I A Y L K S A C S K
11.  Q A E G Y S Y T D A N K S K G I V W N N D T L M E Y L E N P K K Y I P G T K M I F A G I K K K G E R Q D L V A Y L K S A T S -
12.  Q A P G F S Y S N A N K A K G I T W G D D T L F E Y L E N P K K Y I P G T K M V F A G L K K A N E R A D L I A Y L K E S T K -
13.  T T A G Y S Y S A A N K N K A V E W E E N T L Y D Y L L N P K K Y I P G T K M V F P G L K K P Q D R A D L I A Y L K K A T S S
14.  S V D G Y A Y T D A N K Q K G I T W D E N T L F E Y L E N P K K Y I P G T K M A F G G L K K D K D R N D I I T F M K E A T A -
```

A	Alanine	F	Phenylalanine	K	Lysine	P	Proline	T	Threonine
C	Cysteine	G	Glycine	L	Leucine	Q	Glutamine	V	Valine
D	Aspartic acid	H	Histidine	M	Methionine	R	Arginine	W	Tryptophan
E	Glutamic acid	I	Isoleucine	N	Asparagine	S	Serine	Y	Tyrosine

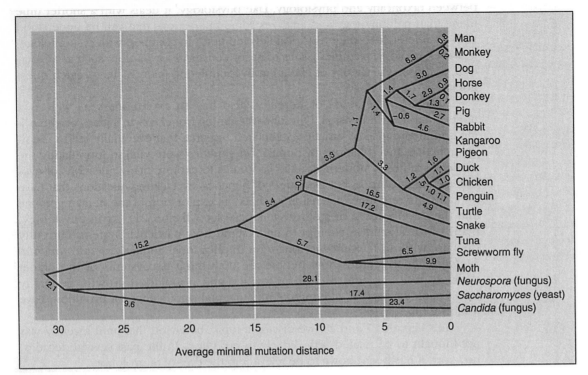

Figure 26.15 By comparing the differences among the amino acid sequences of cytochrome *c* in different species, scientists have been able to construct phylogenetic tables like this one that depict evolutionary relationships. The numbers on the lines are the number of nucleotide substitutions that have taken place.

BEFORE YOU GO ON Physiology and paleontology provide a contrast in how the modern synthesis is related to the separate disciplines of biology. For paleontology, the theory of evolution provides a foundation and is central in guiding research. Physiology has provided important insights into evolution and can be interpreted in evolutionary terms to explain the regularities it reveals. Physiologists, however, are most concerned with how organisms currently function.

OTHER FIELDS OF BIOLOGY AND EVOLUTION

Other biological disciplines fall somewhere between paleontology and physiology in their relationship to evolution, depending on how directly each is guided by questions stemming from evolution. The study of classifying organisms is close to paleontology because much of the work in taxonomy involves constructing classifications that attempt to reflect evolutionary relationships. Ecology studies the interaction of groups of organisms among themselves and with the environment. As a discipline related to evolution, it falls somewhere between taxonomy and physiology. Like physiology, it deals with a shorter time frame than paleontology and analyzes some interactions that can be understood without reference to their historical development. In common with paleontology, however, ecology examines issues of major concern to evolution, such as adaptation, and uses the theory of evolution to formulate many of its research questions.

Genetics, the study of heredity, also has an evolutionary as well as a nonevolutionary dimension. Questions stemming from Darwin's ideas on the origin and inheritance of variation stimulated research that eventually led to classical genetics. Investigations of population genetics were vital in formulating the modern synthesis. Modern genetics continues to explore many questions of evolutionary significance, but new areas of genetics are tackling questions that have a totally different time dimension, such as the exploration of the human genome and the development of medical genetics (see Chapter 22). As in physiology, many of these studies attempt to understand how particular systems currently operate, without reference to long-term implications. Similarly, environmental biology, which studies how people are disturbing ecosystems of the planet through pollution, destruction of forests, and population growth, generally involves processes that, although of monumental significance for evolution, have a shorter time period.

In Chapters 28 and 29, we discuss a topic—behavior—that traditionally was not thought to be related very directly to evolution. In the past several decades, however, the relevance of behavior to the theory of evolution has been demonstrated

THE "UNFINISHED" SYNTHESIS

The modern synthesis has been extraordinarily successful in unifying the biological sciences. No such single theory exists for the physical sciences or the social sciences. We have seen that the modern synthesis does not relate to each of the separate biology disciplines in the same way. For some fields, such as paleontology, it provides a foundation as well as many research questions. For others, such as physiology, it relates research results to other biological subjects and provides a conceptual framework in which to interpret the discipline as a whole.

The modern synthesis has expanded and changed considerably since it was first developed in the 1940s. Research in population genetics, molecular biology, and paleontology has resulted in lively debates and revisions of the theory. The dynamic nature of the theory proves its vitality and strength. There is no sense that the story has been completely told or that biologists have discovered all the laws regulating life. Rather, most disciplines in the biological sciences pose fascinating questions that await exploration. The modern synthesis is the unifying thread that ties all of these separate studies together, gives them a coherent set of relationships, and suggests avenues of investigation and integration. That the modern synthesis is an "unfinished" synthesis is not a criticism or a weakness; rather, it is an indication that the theory of evolution is one of the most exciting and dynamic products of scientific thought and that it promises to continue to be so for a long time.

SUMMARY

1. The modern synthesis attempts to unify all of the current life sciences. Some fields, such as paleontology, are closely linked to the modern synthesis; others, such as physiology, are less so.

2. The fossil record is incomplete, but it does reveal certain patterns of evolution. Some species have remained stable over long periods of time. Some lineages show gradual change without splitting (phyletic evolution); others reveal splitting. Patterns of splitting include divergence, convergence, and parallelism.

3. The fossil record also displays the emergence of grades, the appearance of direction in evolution, and the occurrence of mass extinctions.

4. The modern synthesis attempts to account for macroevolution in terms of the processes of microevolution. Challenges to that view have pointed to uneven rates of evolution and have attempted to account for them by a model called punctuated equilibrium. In addition, some evolutionists claim that some new species have more general adaptations and can give rise to many descendant species, a process they call species selection.

5. Although physiology has contributed to the study of evolution, for the most part it focuses on how current organisms function. Other biological disciplines fall somewhere between paleontology and physiology in their relationship to evolution.

6. The modern synthesis is still "unfinished" but remains an exciting, dynamic theory.

WORKING VOCABULARY

adaptive radiation (p. 498)
analogous structures (p. 499)
convergence (p. 499)
divergence (p. 498)
fossils (p. 492)

homologous structures (p. 499)
lineages (p. 498)
mass extinction (p. 502)
parallelism (p. 499)

REVIEW QUESTIONS

1. What is a fossil?

2. What limits are there on what the fossil record tells us about past life on this planet?

3. What is divergence?

4. What is convergence?

5. What is parallel evolution?

6. Why have scientists abandoned the idea of progressive evolution?

7. What is punctuated equilibrium?

8. What is the relationship of physiology to evolution theory?

9. What is meant by the "unfinished" synthesis?

ESSAY AND DISCUSSION QUESTIONS

1. What would a complete, modern synthesis of evolution theory have to encompass?

2. How might you attempt to resolve a conflict between contradictory conclusions reached by paleontologists and molecular biologists about the relationship between two taxa?

3. How would you construct a display on evolution for a presentation to the general public?

4. Many examples in this chapter relied on bird models. Why might this be?

REFERENCES AND RECOMMENDED READING

Alvarez, L. W., W. Alvarez, F. Asaro, and H. V. Michel. 1980. Extraterrestrial cause for the Cretaceous-Tertiary extinction. *Science,* 208: 1095–1108.

Alvarez, W., and F. Asaro. 1990. An extraterrestrial impact. *Scientific American,* 263: 78–84.

Angiosperm Origin, Evolution, and Phylogeny. 1995. New York: Chapman & Hall.

Bailey, J. 1995. *Genetics and Evolution: The Molecules of Inheritance.* New York: Oxford University Press.

Bock, W. J. 1970. Microevolutionary sequences as a fundamental concept in macroevolutionary models. *Evolution,* 24: 704–722.

Boucot, A. J. 1990. *Evolutionary Paleobiology of Behavior and Coevolution.* Amsterdam: Elsevier.

Cowen, R. 1995. *History of Life.* 2nd ed. Boston: Blackwell.

Dobzhansky, T. 1973. Nothing in biology makes sense except in the light of evolution. *American Biology Teacher,* 35: 125–129.

Eldredge, N. 1985. *Unfinished Synthesis: Biological Hierarchies and Modern Evolutionary Thought.* Oxford: Oxford University Press.

Eldredge, N. 1989. *Macroevolutionary Dynamics.* New York: McGraw-Hill.

Eldredge, N. 1995. Reinventing Darwin: *The Great Debate at the High Table of Evolutionary Theory.* New York: Wiley.

Eldredge, N., and S. J. Gould. 1972. Punctuated equilibrium: An alternative to phyletic gradualism. In T. J. M. Schopf (ed.), *Models in Paleobiology,* pp. 82–115. New York: Freeman.

Gordon, M. S. 1995. *Invasions of the Land: The Transitions of Organisms from Aquatic to Terrestrial Life.* New York: Columbia University Press.

Gould, S. J. 1992. *Bully for Brontosaurus. Reflections in Natural History.* New York: Norton

Hillis, D. M., J. P. Huelsenbeck, and C. W. Cunningham. 1994. Application and accuracy of molecular phylogenies. *Science,* 264: 671–677.

Long, J. A. 1995. *The Rise of Fishes: 500 Million Years of Evolution.* Sydney: USNW Press.

Mayr, E. 1988. *Toward a New Philosophy of Biology.* Cambridge, Mass.: Harvard University Press.

Mayr, E., and W. B. Provine (eds.). 1980. *The Evolutionary Synthesis: Perspectives on the Unification of Biology.* Cambridge, Mass.: Harvard University Press.

Raup, D. M., and J. J. Sepkoski, Jr. 1984. Periodicity of extinctions in the geologic past. *Proceedings of the National Academy of Science,* 81: 801–805.

Raup, D. M., and S. M. Stanley. 1978. *Principles of Paleontology.* New York: Freeman.

Stanley, S. M. 1989. *Earth and Life Through Time.* New York: Freeman.

Stebbins, G. L., and F. J. Ayala. 1981. Is a new evolutionary synthesis necessary? *Science,* 213: 967–971.

ANSWERS TO FIGURE QUESTIONS

Figure 26.4 *Divergence:* Hawaiian Island honeycreepers. *Convergence:* evolution of the mollusks and vertebrate eyes. *Parallelism:* the evolution of placental and marsupial mammals.

Figure 26.7 Function.

Figure 26.9 Common ancestors and environmental similarity.

Figure 26.14 Humans and tuna: twenty-one. Human and dog: eleven. Humans are more closely related phylogenetically to dogs than tuna.

27

Human Evolution

Chapter Outline

Reading Questions

1. What were some major trends in the evolution of primates?

2. What is known about the biological evolution of humans?

3. How does the multiregional model compare with the "Out of Africa" model of the evolution of *Homo sapiens?*

The theory of evolution is the basic organizing theory that unifies the study of life and serves as a guiding force in directing research in many areas. Chapters 23 to 26 discussed the theory of evolution and its relationship to biological disciplines. This chapter describes what evolution has to say about people. Specifically, it examines an issue that has been controversial since Darwin first published *On the Origin of Species* in 1859: the origin of humans.

HUMAN ORIGINS

Darwin didn't actually discuss human evolution in his book *On the Origin of Species*. All he wrote was a brief statement in the conclusion saying that in future research, "Light will be thrown on the origin of man and his history." A dozen years later, Darwin did publish a book on human evolution, *The Descent of Man,* in which he argued that people had evolved from some preexisting form. This idea was hotly contested, in part because many people thought that biology was overstepping its boundaries and delving into areas that had traditionally been the province of religion and philosophy.

Although well over a century has passed since Darwin's discussion of the origin of humans, the topic is still controversial. Creationists reject the theory of evolution itself and therefore its implications about humans (see Chapter 25, the Focus on Scientific Process, "Evolution and Creation in U.S. Public High Schools"). Scientists have repeatedly revised their picture of human evolution as the discovery of human fossils and other relevant new data continues.

HUMAN EVOLUTION

The story of human evolution is known in rough outline. It has been an area of intense research and public fascination. Like the histories of most individual species, the account contains many gaps. The incomplete nature of the fossil record precludes a complete history. Nevertheless, recent discoveries, supplemented with biochemical studies, have allowed for the development of a more detailed picture than ever before.

Primates

A brief look at our place in the classification system of animals provides a starting point for the study of human history. As described in Figure 27.1, we belong to the placental mammalian order of **Primates.** The primates are one of the earliest orders of placental mammals that evolved on Earth. Fossil primates date from the late Cretaceous and early Paleocene, approximately 65 million years ago. Contemporary primates

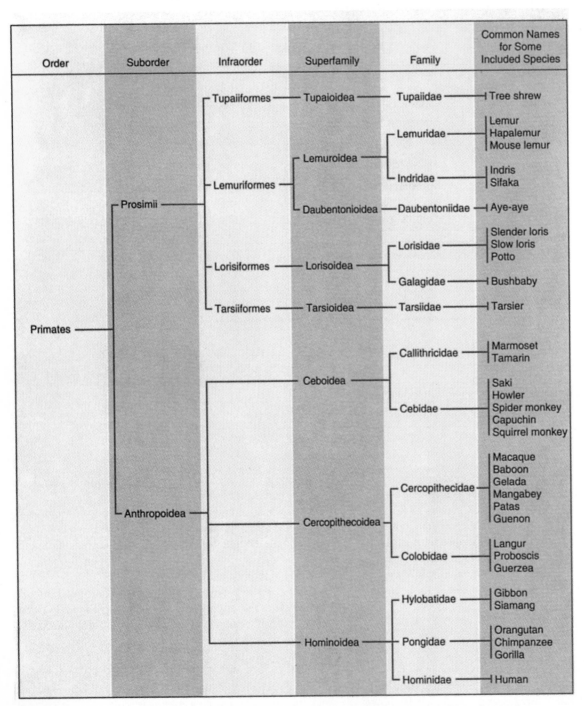

Figure 27.1   Primates are one of the 17 living orders of placental mammals. This taxonomy of the primates shows their diversity, from tree shrews to humans.

Question: *In which suborder are humans found?*

are divided into two suborders: *Prosimii,* also called the lower primates, and *Anthropoidea,* the higher primates.

The Prosimii are represented today by tree shrews, lorises, and lemurs. The Anthropoidea include monkeys, apes, and humans (see Figure 27.2). Primates share certain features, the most obvious being their adaptation for living in trees and their ability to eat many different kinds of food. The higher primates have an additional shared characteristic: they spend more time raising a smaller number of offspring.

Life in the Trees

Many animals are adapted to live in trees. Primates, however, show specific adaptations that distinguish them from other arboreal (tree-dwelling) creatures. Among the most significant primate adaptations are the *freely movable digits* of their hands and feet, an adaptation that permits grasping. In some primates, this has been further refined by the development of opposable thumbs and big toes (see Figure 27.3, page 516), which increases grasping and manipulatory abilities.

Primates have exceptional limb flexibility owing to a *generalized limb structure.* They have retained most of the individual limb bones that characterize the early placental mammals from which they evolved. Also, primates conserved the clavicle (collarbone), which was lost in many other mammals. The clavicle gives primates greater arm motion, permitting them to perform actions such as placing their hands behind their heads. Another feature retained in primates is two separate bones (the radius and the ulna) in the lower arm. These bones enhance flexibility in turning the hand.

Important changes in *sensory systems* also characterized primate evolution. Sight became the most important sense, coupled with a corresponding decline in other senses. Primates developed the ability to judge depth, to see in color, and to discriminate objects in their field of vision. The ability to grasp and manipulate objects and to see them more clearly resulted in a reduction in importance of the sense of smell. Correspondingly, the primate nose diminished in size.

Two other adaptations associated with arboreal life were the replacement of claws by *nails* and the development of an *upright posture.* Although nails are inferior to claws for gripping, they offer the advantage of having broad, flat, sensitive tactile surfaces on the underside of their digits. This touch sensitivity extends the ability to explore and examine objects. An

Figure 27.2 (A) The bush baby (*Galago senegalensis*) and (B) the brown lemur of Madagascar are representatives of the suborder Prosimii. (C) The white-faced capuchin monkey of tropical America and (D) the African lowland gorilla belong to the suborder Anthropoidea.

Figure 27.3   Thumb and toe opposability permit greater grasping power. (A) The gibbon can examine an object with one hand, whereas other mammals, such as (B) the common mouse, need to use both.

upright position made it easier for animals to hang and swing from branches and to increase their visual range. It also held particular significance for the later development in chimpanzees and people of walking on two legs.

Diet in the Trees

Primates can eat a wide variety of foods. This ability is largely related to their dental structure, combined with their ability to manipulate objects. Primates have unusual dental features. Like their retention of a generalized limb structure, they retained the primitive mammalian trait of possessing different types of teeth (see Figure 27.4). Early mammals had four types of teeth— incisors, canines, premolars, and molars—each with a different purpose. Although primate evolution shows a reduction in the number of teeth, most primates retained the different types of teeth and with them the ability to use different food sources.

Brains and Babies

Compared to other animals, higher primates invest more time raising fewer babies. As a result, offspring learn more over an

extended period of time and have a greater chance of survival. The behavioral consequence of this increase in *parental investment* was an increase in the importance of social groups. Modern higher primates are almost invariably social. The increase in parental investment is also related to a general evolution of greater intelligence in the higher primates, which is reflected by increased brain size and complexity.

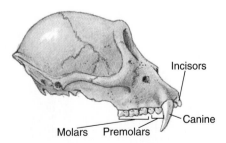

Figure 27.4   Primates have four types of teeth, each used for a different purpose.

Question: *What are large canines used for?*

BEFORE YOU GO ON The evolution of primates reflects adaptations for living in trees and for the ability to eat many different kinds of food. Among the most significant primate adaptations are freely movable digits of the hands and feet, generalized limb structure, changes in the sensory systems that enhance sight, replacement of claws by nails, the development of upright posture, and the ability to eat a wide range of food because of their dental structure and their ability to manipulate objects. Higher primates invest much time raising few offspring.

Hominids

To trace the evolution of **hominids**—that is, any species belonging to the human family—scientists usually begin with the fossil remains of the earliest known Anthropoidea (higher primates), which are approximately 35 million years old and come from Egyptian deposits dating from the Oligocene epoch. One of these fossils, belonging to the genus *Aegyptopithecus,* is thought to represent the common ancestor of apes and hominids that diversified in the Miocene (see Figure 27.5).

Higher primates underwent several radiations in the Miocene, but by the end of the epoch (approximately 5 million years ago) few descendants of these groups remained. Much of the research on these fossils has been to establish when humanlike—or ancestors immediate to humanlike—forms first appeared. One famous fossil that for many years was thought to be in a direct line with humans is *Proconsul africanus,* which was discovered by Mary Leakey in 1948. *Proconsul* species lived from 19 to 13 million years ago. Their features are both monkeylike and apelike—short arms like a monkey, but teeth more like those of an ape. Another fossil that until recently was thought to represent the first example of a hominid is *Ramapithecus,* which arose about 19 million years ago in Africa and later migrated to Europe and Asia.

Recently discovered fossils and biochemical research indicate that hominids diverged from apes much later than was previously thought. The discovery of new and more complete *Ramapithecus* fossils has led paleontologists to conclude that it was more similar to orangutans than to humans and was probably ancestral to Asian great apes, including the orangutans, but not to hominids.

Hominid Features

Three features distinguish hominids: (1) their teeth and jaws, (2) their mode of locomotion, and (3) their brain size and organization. Each involves both morphological and behavioral components.

Teeth and Jaws Teeth and jaws differ in hominids and in apes. The most obvious difference is the lack of large projecting canines in hominid males. In many primates, males have larger canines than females. Male and female hominids have canines of approximately the same size. The canines barely project above the level of the other teeth and are generally smaller in hominids than in other higher primates. They function differently as well. Hominid canines are used more like incisors for gripping, holding, and tearing rather than cutting. Figure 27.6 (see page 518) compares the teeth of hominids with those of gorillas. Hominid permanent teeth also appear in a different sequence during maturation. Canines erupt earlier, but other permanent teeth erupt later in hominids than in closely related animals such as chimpanzees.

Although all primates grind their food with their back teeth, hominids differ because they are able to apply more effective force during chewing. This enhanced capability is related to the size and orientation of the muscles that control the jaw.

Mode of Locomotion Apes and monkeys move on all fours, making extensive use of their forelimbs in locomotion. Hominids are erect bipeds; that is, they walk erect on two feet. This mode of locomotion has significant advantages. It frees the hands for other tasks such as carrying or manipulating objects. It also allows striding, a more energy-efficient mode of locomotion. As a result, hominids are able to cover longer distances without tiring.

Brain Size and Organization Intelligence, marked by changes in brain size and brain organization, is the most prominent feature of hominids. Although hominids developed larger brains than other apes or monkeys, the importance of that increase is not clear because among humans, at least, there is no correlation between brain size and intelligence. Most biologists believe that the complexity of the human brain is more important than its size, because the highly complex human brain is capable of creating language and designing tools.

Figure 27.5 The fox-sized *Aegyptopithecus zeuxis* lived in Northern Africa. This is a skull of *A. zeuxis,* which lived roughly 35 million years ago.

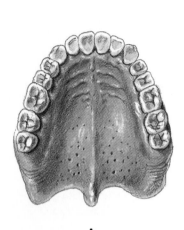

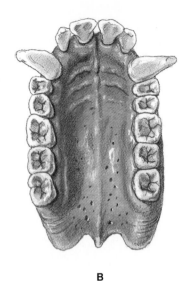

A   B

Figure 27.6   (A) Hominid teeth differ significantly from (B) gorilla teeth.

Question: *What is the function of primates' back teeth?*

Hominid Fossils

The story of hominid evolution has changed considerably in recent years as a result of some sensational fossil discoveries. Every time textbook authors think they have the story straight, another set of fossil bones is uncovered somewhere, and writers return to their word processors.

At present, the earliest known hominid fossils belong to the genus *Australopithecus.* The first *Australopithecus* fossil, described by Raymond Dart in 1925, consists of the remains of a juvenile found in South Africa. At the time of its discovery, scientists were skeptical that the fossil represented a hominid, but later studies confirmed Dart's original contention. Since then, many *Australopithecus* fossils have been found, chiefly in eastern and southern Africa. In 1994 the earliest hominid fossil, *Australopithecus ramidus,* dating from 4.4 million years ago, was reported.

The australopithecine that has received the greatest publicity is *Australopithecus afarensis,* a species that inhabited Africa 3 to 4 million years ago (see Figure 27.7A and B). The fame that surrounds this species is related to a very unusual find in Ethiopia by Donald Johanson. In 1974, Johanson discovered the fossil remains of a small female who lived approximately 3 million years ago. What made the find so unusual was that it was 40 percent complete—the most complete specimen of such an early hominid. Nicknamed "Lucy" (after the Beatles song "Lucy in the Sky with Diamonds"), this *A. afarensis* specimen is thought by many paleontologists today to be a representative of the ancestral group that gave rise to modern humans. "Lucy" lacked a skull, but in 1994 a nearly complete skull of *A. afarensis* was described, which now gives an even more complete picture of the small humanlike frame and apelike head of this species.

A. afarensis was once thought to have split into two branches: one leading to modern humans and the other, by way of the fossil discovered by Dart (*A. africanus*), to a fossil hominid with massive physical features called *A. robustus* and

to a similar hominid, *A. boisei.* The *robustus* and *boisei* lines became extinct 1.2 million years ago.

A discovery in 1985 of a 2.5-million-year-old fossil in Kenya has caused the two-branch hypothesis of *Australopithecus* evolution to be revised. The newly discovered fossil has a face that is similar to the later *A. boisei* but a more primitive cranium. Several paleontologists feel that the new fossil find, named *A. aethiopicus,* is ancestral to *A. boisei* and represents a third branch in australopithecine evolution (see Figure 27.8A, page 520).

The lineage leading to modern humans is now thought to have proceeded from *A. afarensis* to *Homo habilis,* the earliest species to have the human genus name *Homo. H. habilis* was a tool user and, like *A. afarensis,* was short, standing about 3 feet tall. A relatively brief period (in fossil record terms) of 200,000 years stands between *H. habilis* and the fossil with a distinctly modern body shape, *H. erectus. H. erectus* appeared approximately 1.6 million years ago and spread widely through Asia, Europe, and Africa. These early humans were the first hominids to migrate away from Africa. They had more sophisticated tools than *H. habilis,* and there is evidence that they used fire.

The evolution of the modern human, *H. sapiens,* has been highly controversial during the last several years because of different hypotheses that have been advanced by scientists who study the subject. Some scientists hold a view called the **multiregional model of human evolution.** According to this hypothesis, *H. sapiens* developed from populations of *H. erectus* in several different areas of the globe; subsequent human evolution was shaped by gene flow, natural selection, and genetic drift. Gene flow prevented populations of *H. erectus* from radiating into several different species (see Chapter 24). In between *H. erectus* and modern *H. sapiens,* therefore, are a number of "archaic" humans that are difficult to place in one species or the other because of their intermediate appearance.

The **"Out of Africa" model of human evolution** states that a population of the modern form of *H. sapiens* arose in

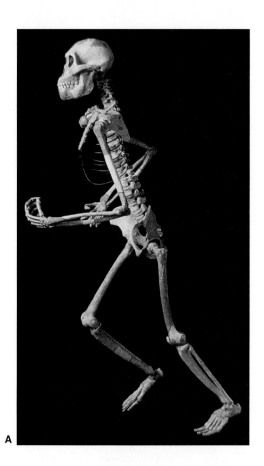

A

B

Figure 27.7 (A) This plaster reconstruction of *Australopithecus afarensis* is based on fossils that have been uncovered and on knowledge of hominid anatomy. (B) An artist's conception reflecting the fossil evidence suggests that *A. afarensis* could walk upright.

Africa at an early date and spread rapidly across Europe and Asia. The modern form is called *H. sapiens sapiens,* to distinguish it from various "archaic" forms, and it is a subspecies of *H. sapiens* that includes both modern and archaic forms.

According to the "Out of Africa" model, the many archaic forms of *H. sapiens* found in Asia and Europe (which evolved from *H. erectus*) did not evolve into *H. sapiens sapiens.* For example, a well-known archaic human called by the subspecies name *H. sapiens neanderthalensis* lived from 130,000 to 35,000 years ago in Europe. Although for many years "Neanderthal man" was considered a direct ancestor of modern humans, recent fossils found in Africa and the Middle East suggest that modern humans (*H. sapiens sapiens*) did not evolve from populations of Neanderthals. Instead, we probably evolved from *H. erectus* in Africa 100,000 years ago but did not migrate to other parts of the globe until 40,000 years ago (see Figure 27.8B1-9, page 520).

When *H. sapiens sapiens* finally did arrive on the European scene, it was with dramatically different behavior patterns than those of its predecessors. Modern people planned hunts, built hearths, and fashioned complex tools. They made elaborate jewelry and buried their dead in a ritualistic manner. Art was part of their world, and they used some form of symbolic notation. Although Neanderthals buried their dead, modern humans had more elaborate preparations, and there are clear indications that they conducted final rituals (see Figure 27.9, page 521).

The changes associated with the advent of modern humans have more to do with behavior than bones. The typical male Neanderthal would probably be difficult to distinguish in a crowd dressed in a three-piece suit or a football uniform.

Neanderthals disappeared approximately 30,000 years ago, presumably because they were less fit than modern humans. What gave modern humans the selective advantage was their culture rather than any single physical trait. Careful analyses of bones indicates that modern humans lacked the brute strength of Neanderthals, but archeological remains suggest that they had a more advanced culture. Ingenuity, social organization, and an ability to plan for the future seem to have been the critical advantages. With the arrival of modern humans, biology gives way to archaeology, anthropology, and history, which all deal with the development of human cultures.

BEFORE YOU GO ON Hominids differ from apes by their teeth and jaws, their mode of locomotion, and their brain size and organization. The earliest known hominid fossil belongs to the genus Australopithecus. The evolution of *Homo sapiens* remains controversial. Some scientists accept a multiregional model that holds that *H. sapiens* evolved from an original population of H. erectus in several areas of the globe. Other scientists accept the "Out of Africa" model, which states that a population of the modern form of *H. sapiens* arose in Africa and spread across Europe and Asia.

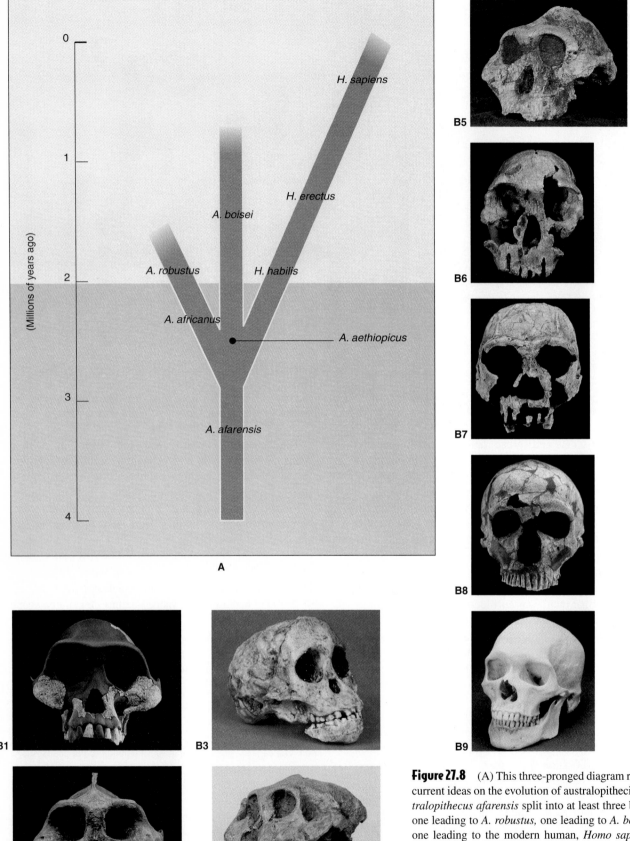

A

Figure 27.8   (A) This three-pronged diagram represents current ideas on the evolution of australopithecines. *Australopithecus afarensis* split into at least three branches: one leading to *A. robustus*, one leading to *A. boisei*, and one leading to the modern human, *Homo sapiens*. (B) Skulls of major *Australopithecus* and *Homo* species: (1) *A. afarensis*, (2) *A. aethiopicus*, (3) *A. africanus*, (4) *A. robustus*, (5) *A. boisei*, (6) *H. habilis*, (7) *H. erectus*, (8) *H. sapiens neanderthalensis*; (9) *H. sapiens sapiens*.

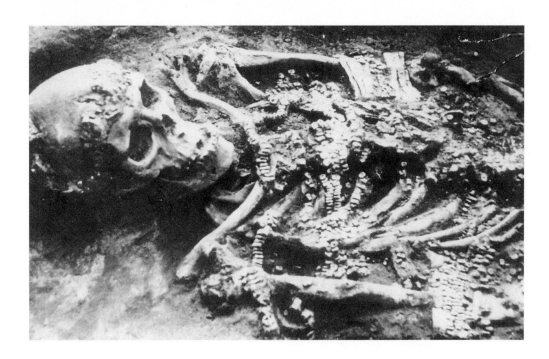

Figure 27.9 The beads adorning this early *Homo sapiens sapiens* skeleton bespeak an elaborate burial.

Hominid DNA

The "Out of Africa" model of *H. sapiens sapiens* evolution is strengthened by biochemical studies. By comparing DNA sequences among human populations, geneticists have discovered that relatively little variation occurs among humans, at least for nonfunctional DNA (see Chapter 18). In spite of the perceived great morphological variation among human groups, the genetic information based on molecular clocks (see Chapter 26) suggests that there has *not* been the great divergence one would expect if many populations of humans had evolved at the same time. Humans have relatively little genetic and morphological variation compared to many species.

A fascinating recent study of human molecular clocks has led to one of the most contentious issues in human evolution. Rebecca Cann of the University of Hawaii and Mark Stoneking and Allan Wilson of the University of California at Berkeley hypothesized from their 1987 research on *mitochondrial DNA* that the different human populations inhabiting the globe can all trace their ancestry to one particular African female who lived about 200,000 years ago. Mitochondria (small cellular organelles that provide energy to the cell) have their own unique DNA that enables them to replicate independently within the cell.

Mitochondrial DNA (mtDNA) is shorter than the DNA found in the cell's nucleus, and it is inherited only from the mother (that is, mtDNA is transmitted through the egg but not the sperm). Therefore, any changes in mtDNA must have arisen through mutations in females rather than as a result of recombination during fertilization. Because the rate of mutation defines a molecular clock, scientists contend that mtDNA is a particularly sensitive molecule for use in dating studies. Cann, Stoneking, and Wilson collected samples of placental tissue, a convenient way to obtain human tissue, from women all over the globe. (More recent research makes use of a tech-

nique that can be used to isolate a selected portion of DNA from single hairs plucked from women and to make millions of copies of the DNA segment required for analysis.) Comparison of mtDNA from these placental cells showed that the differences among them were extremely small. The researchers then calculated how long it would have taken these differences in mtDNA to have developed if humans diverged from a common ancestor. They concluded that approximately 200,000 years ago, a single common ancestor carried the mtDNA from which all mitochondria in the current human population originated.

This does not mean that only one female existed 200,000 years ago. It does suggest that of the females of her generation, only her descendants gave rise to a minimum of one female per generation and that only her mtDNA now exists in the human population. Presumably, along the way, descendants of other females did not leave daughters. Cann, Stoneking, and Wilson argue that this "mitochondrial Eve" lived in Africa and that a population of her descendants migrated from there to Asia and Europe, where they eventually became the dominant subspecies. Since the date of this common ancestral female is earlier than any recorded modern human fossils, it is possible that she belonged to an archaic human form. The research on mtDNA has caused debate because some scientists are not convinced that molecular clocks are accurate enough to provide very precise dates. Others, particularly those who accept the multiple-origin model, do not believe that the fossil record agrees sufficiently with the biochemical data. (See the Focus on Scientific Process, "Scientific Controversy and 'Mitochondrial Eve'").

Hominid Ecology

The limited number of hominid fossils makes it quite difficult to reconstruct the ecological system in which hominids

Scientific Controversy and "Mitochondrial Eve"

As we first discussed in Chapter 1, the process of science can be understood in terms of a lineage of questions. Answering one set of questions often leads to new questions or new information, or a fresh perspective may lead scientists to reexamine answers that were previously given. Scientific process is dynamic and exciting, but it may occasionally lead to controversy, which can be unsettling to the general public. After all, if science defines itself as the rational investigation of nature, why do disputes arise? Disagreements may occur for several reasons. For example:

■ Only a limited amount of information may be available, and two or more explanations may be equally plausible.

■ The use of dissimilar methodologies may cause scientists to focus attention on different dimensions of an issue and lead them to formulate different answers.

■ Alternative interpretations of facts can cause scientists to disagree.

■ Sometimes a scientist who has spent years studying a problem from a particular point of view has difficulty accepting a radically new solution. The scientist may also be concerned about the practical implications (funding, personal reputation, and so on) of admitting that she or he is wrong.

The continuing debate over "mitochondrial Eve" illustrates the complexity of scientific controversy. As discussed in this chapter, in 1987, Rebecca Cann, Mark Stoneking, and Allan Wilson concluded from results of their studies on mitochondrial DNA (mtDNA) that all modern human populations trace their ancestry to a single female who lived in Africa approxi-

mately 200,000 years ago. Their argument was based on the following assumptions: (1) mtDNA is an accurate molecular clock because it accumulated mutations at a regular rate; (2) the mtDNA mutation rate was 2 to 4 percent mutations per million years; and (3) when a new species is formed its mtDNA is the same as that of the species from which it originated, therefore, any mtDNA differences occurred since the two species became separated.

Cann, Stoneking, and Wilson compared the mtDNA of females living in five different geographical regions and claimed that the mtDNA of Africans was noticeably more diverse than the mtDNA of others. They interpreted this variation to mean that Africans appeared first (in order to have accumulated so much variation). They also analyzed their data using a computer program that builds family trees by comparing DNA sequences and relates them according to number of mtDNA mutations (see Figure 1). They concluded that the "Out of Africa" model of human evolution was strongly supported by their molecular biology evidence.

Not everyone was convinced, however, and a heated debate has developed. What is the basis of the disagreement? As is often the case with serious scientific controversy, criticism has been directed at different elements of the study. In this case, scientists have questioned the methods, assumptions, and interpretations of the "Wilson group" (Wilson was the senior scientist; he died shortly after their paper was published). Some who are opposed are committed to a different model of human origins and are hostile to what they see as an attempt to argue for the single origin model using unconventional data (that is, something other than the fossil record).

What are some criticisms of the "mitochondrial Eve" concept, and how have its supporters responded to them? The initial set of complaints tended to be technical; for example, mtDNA from African-Americans was used to represent Africa. Was this valid? The Wilson group responded with new studies involving a wider group of individuals representing more diverse geographical areas. More serious criticism was directed at the computer methods used by the Wilson group. Critics contended that their results were biased because entering the data in a different sequence would produce different conclusions. The Wilson group acknowledged that problem but maintained that their conclusion was still correct. They pointed to other studies that suggested greater diversity occurred in both the nuclear and mitochondrial DNA of Africans. For example, Stanford geneticist Luigi Luca Cavalli-Sforza, in collaboration with the Yale geneticists Judith and Kenneth Kidd, studied the nuclear DNA of different human populations and concluded that the variation among Africans is roughly twice that of peoples from other continents. Their conclusion supports the Wilson group's contention that Africa was the site from which modern humans originated.

The most strident criticism of "mitochondrial Eve" has come from supporters of the multiregional model of human evolution. Foremost among the critics is Milford Wolpoff of the University of Michigan. Wolpoff has attacked the Wilson group's methods, interpretations, and assumptions. He questions the accuracy of using mtDNA as a molecular clock. He notes that some species seem to have mtDNA that evolves at different rates than is assumed for humans by the Wilson group. Wolpoff has also suggested that because

FOCUS ON SCIENTIFIC PROCESS

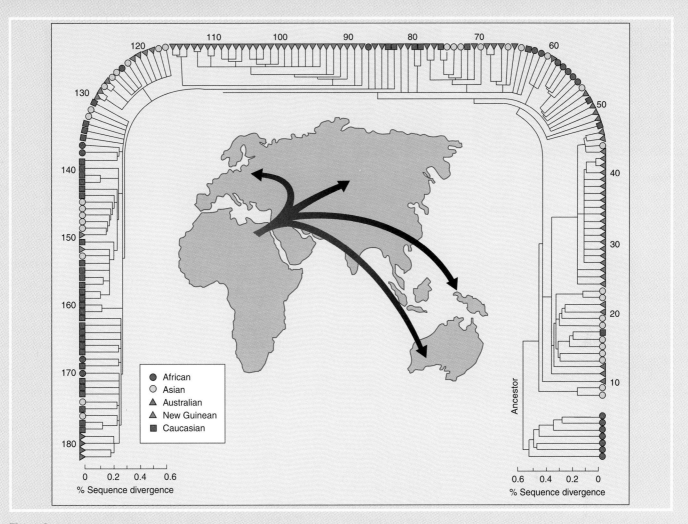

Figure 1   This genealogy is based on 241 individuals representing 182 mtDNA types. The scale at the bottom measures the percentage of differences in sequences of the mtDNA. The outer points reflect geographic groups.

mitochondrial lineages are lost in time, perhaps that derived from "Eve" is not reliable for developing an accurate picture of human evolution. His concern is based on the following logic:

1. According to the Eve model, all human populations can be traced to a single African female.

2. Other individuals who were contemporary with Eve and individuals who were descended from these contemporaries are not represented by living offspring today.

3. What if Eve was not representative of her contemporaries? For example, what if Eve was a member of a younger, atypical lineage that had more recently evolved? If that was the case, attempts to reconstruct human evolution models are destined to fail.

Wolpoff is a paleoanthropologist, a scientist who studies human fossils. His main objection to the Wilson group model rests on his interpretation of the fossil record. He claims the fossil record reveals no evidence supporting

the idea that one group of modern humans spread across the globe displacing all earlier *Homo* species 200,000 years ago. Moreover, Wolpoff feels that the fossil record flatly contradicts the "mitochondrial Eve" hypothesis and instead supports the multiregional model of human evolution. The principal evidence consists of a set of *H. erectus* skulls from different parts of the world that resemble modern human skulls from the same region. Even more compelling are a set of skulls from China that he and some others believe are intermediate between *H. erectus* and *H. sapiens* (see Figure 2).

box continues

FOCUS ON SCIENTIFIC PROCESS

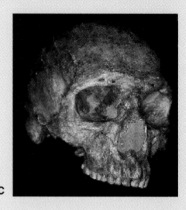

Figure 2 These three skulls found in China have been used to argue for a transition between *Homo erectus* and *Homo sapiens;* (A) is from Zhoukoudian Lower Cave, (B) is from the Dali site, and (C) is from Zhoukoudian Upper Cave.

He claims that these skulls support the view that human evolution took place in different localities throughout the world.

The fossil evidence of earliest humans, of course, has its own problems. There are few specimens, dating them is often problematic, and only a few people have studied them very carefully. Moreover, geneticists are very skeptical about the theoretical possibility of the multiregional model. Wolpoff and other paleoanthropologists believe that *H. erectus* migrated to Europe and Asia about 1 million year ago. From then on *H. erectus* continued to evolve in different parts of the world, and these populations continued to interbreed, thereby giving rise to modern humans in different geographic regions. Geneticists doubt that the characteristic genetic uniformity of modern humans could be maintained over such vast distances because the amount of gene flow would have to be very high. Instead, they argue that geographical isolation would have resulted in divergence and the development of reproductive isolation.

The debate over "mitochondrial Eve" resembles many other scientific controversies in that the parties involved disagree with certain basic assumptions underlying each position. For the Wilson group, tools of molecular biology open new possibilities for investigating questions for which the available fossil record has no answer. For Wolpoff, the fossil record is primary. Obviously, much work remains to be done before a resolution is likely. New computer models may be more powerful or reliable for generating human family trees. More research on mtDNA and nuclear DNA may clarify their rates of evolution. New fossil discoveries and further, careful examination of existing fossils may alter the argument. Extracting DNA from fossils may also throw new light on the issue. It seems likely, however, that we may be reading about this debate for several years to come.

evolved. If an African origin is accepted, the geologic changes that are known to have occurred there may provide clues about the original ecosystem. The early Miocene (25 to 5 million years ago) African environment consisted primarily of forests. Beginning approximately 16 million years ago, however, most of East Africa was altered by geologic changes that resulted in a drying trend, and with that a replacement of forests by grasslands (see Figure 27.10). In some areas, a mosaic of forest and open areas existed, and changes in the landscape created new environmental conditions. The arboreal ancestors of humans may have exploited the new opportunities and in so doing became modified. Coming down from the trees would have permitted animals to use new food sources, which might have led to the selection of altered teeth and jaws. The development of bipedal motion would have been an efficient means of moving on the grasslands. Tool use may have opened new possibilities and been an effective means of protection. New behavioral adaptations may have been dependent on a substantial amount of learned behavior, which would have resulted in more information to be learned by offspring. This emphasis on learned behavior may have added to the trend in primates toward a more complex brain.

Today, there are many different hypothetical constructions of hominid evolution. The lack of hard data makes it especially difficult to test these models, so their reliability is uncertain. More fossil finds and continued study of the geographical and geologic history of Earth may lead to a more detailed account.

Figure 27.10 The evolution of modern humans began as grasslands replaced forests in East Africa. This African savanna resembles the setting of hominid evolution.

SUMMARY

1. Humans belong to the order Primates. Contemporary primates are divided into two suborders: Prosimii and Anthropoidea. Primates share certain characteristics. The best known are their adaptation for living in trees and their ability to eat many different kinds of food. The Anthropoidea also spend more time raising a smaller number of offspring.

2. Among the significant adaptations of primates to tree life are their freely movable digits, generalized limb structure, and abilities to judge distance, see in color, and discriminate objects in their field of vision. Primates also replaced claws with nails and developed an upright posture.

3. Three features characterize hominids (species of the human family) from other primates: their teeth and jaws develop and are used differently in hominids; hominids walk erect on two feet, and hominids have more complex brains.

4. The earliest hominid fossils belong to the genus *Australopithecus. Australopithecus afarensis* inhabited Africa 3 to 4 million years ago. One widely held opinion is that *A. afarensis* split into three branches: one leading to modern humans and the other two leading to hominids that became extinct.

5. *Homo erectus,* which appeared approximately 1.6 million years ago, had a distinctly modern human body shape. Modern humans, *H. sapiens,* are thought to have evolved from *H. erectus.* After its beginnings, human culture soon became a primary factor in further human development.

WORKING VOCABULARY

hominid, (p. 517)
mitochondrial DNA, (p. 521)
multiregional model of human evolution, (p. 518)

"Out of Africa" model of human evolution, (p. 518)
primate, (p. 514)

REVIEW QUESTIONS

1. What did Darwin have to say about human evolution?

2. What adaptations favored primates for life in the trees?

3. What features distinguish higher primates?

4. What three features distinguish hominids from other primates?

5. What genus contains the earliest known hominid fossils?

6. The lineage leading to modern humans proceeded from *Australopithecus afarensis* to what species?

7. What is the multiregional model of the evolution of modern humans?

8. What is the "Out of Africa" model of the evolution of modern humans?

9. What new conclusion has research on mtDNA supported?

ESSAY AND DISCUSSION QUESTIONS

1. What paleontological evidence would seriously undermine the mtDNA hypothesis on the recent origin of humans?

2. If modern humans evolved in Africa, why are there different skin colors among humans?

3. If you were writing a grant proposal to do field research on the fossils of early humans, how might you justify the overall value of the project?

REFERENCES AND RECOMMENDED READING

Brace, C. L. 1995. *The Stages of Human Evolution.* 5th ed. Englewood Cliffs: Prentice Hall.

Bryne, R. W. 1995. *The Thinking Ape: Evolutionary Origins of Intelligence.* New York: Oxford University Press.

Cann, R. L., M. Stoneking, and A. C. Wilson. 1987. Mitochondrial DNA and human evolution. *Nature,* 325: 31–36.

Coppens, Y. 1994. East side story: the origin of humankind. *Scientific American,* 270: 88–95.

Easteal, S. 1991. The relative role of DNA evolution in primates. *Molecular Biology and Evolution,* 8: 115–127.

Easteal, S. 1995. *The Mammalian Molecular Clock.* New York: Springer-Verlag.

Fisher, A. 1988. Human origins. *Mosaic,* 19: 22–45.

Gibbons, A. 1991. Looking for the father of us all. *Science,* 251: 378–380.

Human Populations: Diversity and Adaptation. 1995. New York: Oxford University Press.

Johanson, D. C., and M. Edey. 1981. *Lucy: The Beginnings of Humankind.* New York: Simon & Schuster.

Johanson, D. C., and J. Shreeve. 1989. *Lucy's Child: The Discovery of a Human Ancestor.* New York: Morrow.

Lewin, R. 1988. *In the Age of Mankind.* Washington, D.C.: Smithsonian Books.

Milton, K. 1993. Diet and primate solution. *Scientific American,* 269: 86–93.

Morell, V. 1995. *Ancestral Passions: The Leakey Family and the Quest for Humankind's Beginnings.* New York: Simon & Schuster.

Paleoclimate and Evolution, with Emphasis on Human Origins. 1996. New Haven: Yale University Press.

Shipman, P. 1995. Climbing the family tree: What makes a hominid a hominid? *Journal of NIH Research,* 7: 50–55.

Shreeve, J. 1990. Argument over a woman. *Discover,* 11(8): 52–59.

Stringer, C. 1990. The emergence of modern humans. *Scientific American,* 263: 98–104.

Tattersall, I. 1995. *The Fossil Trail: How We Know What We Think We Know about Human Evolution.* New York: Oxford University Press.

Wenke, R. J. 1990. *Patterns in Prehistory.* Oxford: Oxford University Press.

Wolpoff, M. 1980. *Paleoanthropology.* New York: Knopf.

ANSWERS TO FIGURE QUESTIONS

Figure 27.1 *Anthropoidea.*
Figure 27.4 Cutting.
Figure 27.6 Grinding food.

28

Animal Behavior

Chapter Outline

Reading Questions

1. What are the major types of animal behavior?

2. How do various behaviors increase the fitness of animals?

3. What are the major types of learning?

4. How does the theory of evolution explain animal behaviors?

The behavior of animals has long attracted the attention of biologists. In the nineteenth century, however, the life sciences became highly specialized, and most scientists ignored the study of behavior in favor of areas such as anatomy, physiology, embryology, and biogeography. A notable exception to this trend was Charles Darwin. Not only did he continue to do research on broad general areas, but he was also interested in behavior. In his book *On the Origin of Species,* Darwin devoted a separate chapter to behavior, or "instinct," as he called it. He attempted to demonstrate that the behavior of animals could have evolved in the same ways as

their physical characteristics. Thus Darwin defined animal behavior as a field that could and should be encompassed by his new theory of evolution.

However, decades passed before animal behavior was fully integrated with evolutionary theory. Although Darwin and some of his immediate followers made observations on behavior and speculated on its evolution and development, other areas of biology continued to attract more attention. Fortunately, animal behavior was approached by scholars from other fields of science. The modern study of behavior has its roots in disciplines such as psychology, ecology, neurology, and natural history. Many of the fundamental ideas that became integrated into the discipline of behavior came from studies in these fields.

In this chapter, several basic types of animal behavior of individuals are described; social behavior is discussed in Chapter 29. In both cases, evolution serves as an important unifying theme.

INDIVIDUAL BEHAVIOR

The modern theory of evolution regards the behavior of animals as adaptive because it contributes to an individual's ability to (1) survive and (2) produce offspring. Animals do this by way of innate and learned behavior.

Innate and Learned Behavior

During the last hundred years, scientists approached the behavior of animals in two distinct ways. Some believed that animals inherit nervous systems that are organized in such a way that specific sets of reactions—**instincts**—occur in response to a proper stimulus. This group of scientists held that instincts have evolved, varied, and been subjected to selective pressures in the same way as anatomical features.

Other scientists felt that much of the behavior that animals display does not consist of such rigid behaviors; rather, they reflect what animals have learned in given situations. Therefore, **learning**—the change in behavior that occurs as a result of experience—has been of central interest to this second group of scientists. As with instincts, the ability to learn has been subjected to selective pressures and has adaptive value.

A few decades ago, debates between proponents of these two different positions generated a great amount of heat. In the history of scientific debates, when each side has a considerable amount of verifiable data, scientists eventually come to realize that both sides are at least partially correct. Today both instinct and learning are now thought to be important in understanding behavior. Types of animal behaviors that were originally thought to be purely instinctive have been shown to have a learned component, and vice versa. For some individual behaviors, one factor is more important than the other; in other cases, both are crucial.

Reflexes, Kineses, and Taxes

Among the simplest forms of behavior are **reflexes,** simple reactions to external stimulation. A stimulus such as a change in light intensity or a touch can trigger a response that is automatic, stereotyped, and involuntary. A classic example is the human knee-jerk reflex (see Figure 28.1). Another is the scratch reflex in dogs, which was studied in the early twentieth century by the famous English physiologist Charles Sherrington.

Sherrington demonstrated some of the basic characteristics of reflexes in studies on dogs (see Figure 28.2). Later

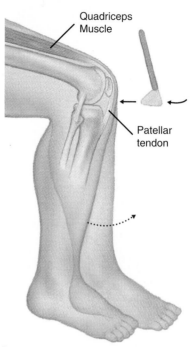

Figure 28.1   The simple knee jerk is a well-known human reflex. The reflex depends on sense organs in the patellar tendon below the kneecap and on muscles in the upper leg. A sensory nerve travels up from the tendon into the spinal cord and there makes contact with a motor nerve that goes down to the responding muscle.

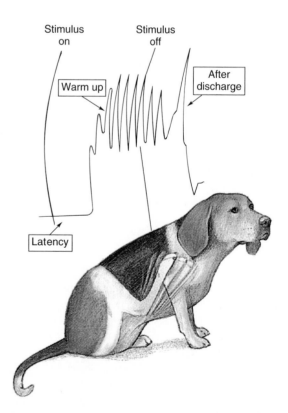

Figure 28.2   Sherrington's studies of the scratch reflex in dogs revealed various distinct stages.

investigators found that several of these features apply to learned behavior patterns as well. Sherrington found that a brief period, called a **latent period,** elapsed between a stimulus and the start of a response. Similarly, there is a small delay between removal of the stimulus and the end of the response, which is called **after-discharge.** Sherrington noted that the scratch reflex becomes stronger after a few strokes. He called this property of the reflex that does not allow the reflex to attain its maximum strength right away *warm-up.* He also noticed *fatigue:* when a stimulus is not removed, the response does not continue indefinitely. After a time, the response will diminish and stop altogether until the system has recovered.

An animal can perform some reflexes without attention to its position in space. Most animals, however, orient themselves to the source of stimulation. If the orientation is merely an undirected change in the rate of motion in response to the intensity of the stimulus, it is called a **kinesis** (the plural form of the word is, *kineses*). When an animal heads either toward or away from a source of stimulation (see Figure 28.3), the action is called a **taxis** (plural, *taxes*). Taxes are numerous. Animals can orient themselves to different stimuli such as light, chemicals, and gravity. For example, yellow jackets will fly upwind in the direction of the smell of food on a picnic table. Animals can also align themselves in different fashions, such as moving at an angle to the stimulus rather than straight at it, or comparing the amount of stimulation on each side of their bodies and moving accordingly.

Navigation

Navigation, the action of orienting toward a goal, is a more complex orientation behavior. Animals use a variety of mechanisms in navigating. Bees generally navigate by the sun, but when the sky is cloudy, they use a backup system that depends on polarized light in the sky. In both cases, they must compensate for the movement of the sun. Other animals use Earth's magnetic field and possibly other environmental cues. The impressive migration journeys of birds require the sun and the stars, which are used as compass points in guiding them.

The navigation of most species can be explained by innate (inborn) responses. European warblers, for example, make an annual trip from Europe to Africa by flying southwest for 40 days and then southeast for 20 to 30 days. The navigation of some species, however, depends on past experience. Pacific salmon hatch in streams and rivers of western North America, and after spending one or two years in the stream, they migrate downstream to the ocean, where they spend two to four years. When salmon become sexually mature, they return to the general area where their natal river enters the ocean, using the sun and other environmental cues. From there, they use their sense of smell to guide them back to the same tributary stream in which they hatched (see Figure 28.4). During the early stages of life, salmon learn the native odors of their stream. As adults, they can distinguish the odor of their native stream from that of the other waters entering the river and choose the correct direction at each fork of the river on their upstream voyage.

A

B

Figure 28.3 Kineses and taxes are simple reflexes. (A) Kinesis, an undirected change in the rate of motion in response to the intensity of a stimulus, is illustrated by these common pill bugs, which are more active in low humidity and less active in high humidity. Therefore, they become aggregated in damp places, such as under rocks. (B) Taxis, a motion toward or away from a stimulus, is illustrated by these housefly maggots, which after feeding will crawl away from a light source.

Figure 28.4 Pacific salmon navigate using past experience. Using their sense of smell, coho salmon can correctly choose among the forks in streams as they navigate back to their home stream.

BEFORE YOU GO ON   The study of animal behavior has roots in psychology, ecology, physiology, and natural history. Evolution theory focuses on the adaptive value of behavior, both learned and instinctive. Among the simplest forms of behavior is a reflex, which is a simple reaction to an external stimulus. Animals orient themselves to sources of stimulation and can direct themselves toward a goal, a process called navigation.

FIXED ACTION PATTERNS

Studies of reflexes initially led some scientists to hope that they could explain complex behavior in terms of chains of reflexes. Although complex behavior often incorporates reflexes, it is actually more complicated. Reflexes can usually be described very precisely and in terms of our detailed knowledge of the nervous system, whereas complex behavior involves so many systems and variables that a more comprehensive method of description is necessary. Early in the twentieth century, the concept of a **fixed action pattern** was developed and found to be useful. Although more complicated than reflexes, fixed action patterns often share some of the features of reflexes, such as latency and fatigue. Initially, it was also thought that fixed action patterns were entirely innate, but as we shall see later, this view has been modified.

The concept of fixed action patterns developed from the work of European **ethologists,** naturalists who are interested in observing behavior in the natural environment and attempt to explain it in an evolutionary context. Two famous ethologists, Konrad Lorenz in Germany and Niko Tinbergen in England, made careful observations of individual behavioral patterns and related them to specific biological situations.

Lorenz and Tinbergen: Fixed Action Patterns

If an egg rolls out of the nest of an incubating greylag goose and she notices it, a set of highly stereotyped actions will occur (see Figure 28.5). The goose fixes her attention on the egg,

rises slowly, reaches her neck out over the egg, rolls the egg back into the nest with the bottom of her bill, and continues the incubation. Once started, the action continues to completion in a mechanical way. Even if the egg is removed when the bird is extending her neck, she will execute the entire sequence of actions to the end.

Lorenz and Tinbergen believed that this egg-rolling behavior was innate and stereotyped. Experiments by Tinbergen showed not only that geese would return eggs to the nest but that if they saw similar objects near the nest (or even slightly similar objects, such as metal cans or baseballs), they would automatically return these objects to the nest also. Foreign objects were rejected by the geese once they were in the nest, but nonetheless they were capable of triggering the action if placed near it.

Lorenz and Tinbergen called this entire sequence of actions a "fixed action pattern." They thought that an "innate releaser" must exist that triggered such actions, and they hypothesized that an **innate releasing mechanism** could be stimulated by something egglike. The features of objects that triggered the innate releasing mechanism were called **releasers.**

Some of Tinbergen's later research identified some of these releasers. He conducted a set of classic experiments with a small freshwater fish, the three-spined stickleback, to discover which features the fish would respond to. Figure 28.6 illustrates some of his basic observations. A male stickleback that has built a nest is stimulated to engage in courtship behavior by the swollen belly of a female with eggs. A male will court a featureless dummy lacking eyes, fins, spines, and a tail, as long as it has a swelling on its bottom side.

During the mating period, the breeding male stickleback has a distinctive red underside and is very aggressive toward other males that have similar markings. Tinbergen showed that the aggressive males reacted primarily to the red marking. He did this by making models that were accurate imitations of male sticklebacks but lacked a red belly and models that barely resembled a fish but had a red belly. The breeding males reacted only to the models that had a red belly (see Figure 28.7).

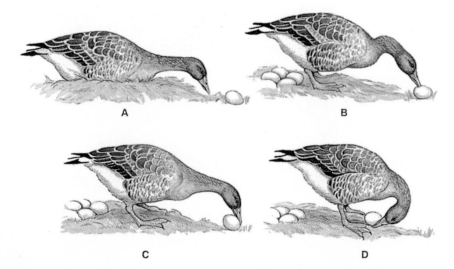

A   B

C   D

Figure 28.5   A fixed action pattern is a sequence of stereotyped behaviors that is triggered by a stimulus in the environment. The sequence of events in the egg-rolling behavior of the greylag goose is a classic example: (A) The goose looks at the egg, (B) rises and reaches beyond the egg, (C) catches the egg under her bill, and (D) pulls it back toward her into the nest. Even if the egg is removed partway through the ritual, the goose continues through all the steps.

Question: *What is the releaser in this example?*

Figure 28.6 When the male stickleback recognizes an appropriate female, he engages in a "zigzag" display. If the female responds, he leads her to his nest and shows her the entrance by poking his snout into it. The female may then enter the nest to lay eggs while the male nuzzles her tail.

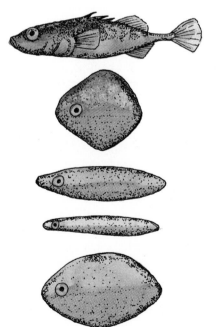

Figure 28.7 Dummy sticklebacks were used to demonstrate that the red marking of males is the releaser of aggressive behavior in breeding males.

Question: *What might be the adaptive value of this behavior?*

Drives and Clocks

We have seen that a male stickleback that has built a nest will react to the presence of another male with red markings. Males, however, do not always have red markings, and those that are not guarding a territory in preparation for mating will not react either. Similarly, the specific egg-rolling behavior of greylag geese occurs only at certain times and with particular individuals.

Why don't animals always respond in the same way to a given stimulus? Ethologists explain this by referring to the animal's "motivation." Specific motivations are called **drives;** thirst, for example, is a drive. A dog that has not had anything to drink for an extended period will search out a source of water, even if that source seems inappropriate to us—a toilet bowl, for instance. Ethologists use the concept of drive in referring to the internal state of the animal that results in a tendency to organize its behavior to achieve a certain goal. The causes of that internal state are complex mechanisms that are not very clearly understood, but they are known to involve the nervous and endocrine systems. Ethologists have found it useful to describe many stereotyped behaviors in terms of drives.

Originally, ethologists spoke of animals having a number of simple, discrete drives. However, research on these drives showed that rather than being simple forces, drives are complex states that start, stop, or modulate an animal's responsiveness to stimuli. Drives are now considered only a part of a larger group of factors, both internal and external, that can stimulate overlapping behavior patterns.

One of the important elements that determines the stimulation of behavior patterns is a **biological clock,** a biochemical mechanism that is responsible for repeated patterns—rhythms—of behavior. Some of these clocks are set by external environmental factors. For example, Atlantic fireworms that inhabit the waters off Bermuda swarm during each full moon of the summer. Fifty-five minutes after sunset, the worms form breeding swarms. The reproductive cycles of most animals are correlated with specific seasons. Sticklebacks react to the lengthening days of spring by producing reproductive hormones that result in changed behavior.

By experimentally altering the conditions under which an animal lives, scientists have been able to discover which behavioral rhythms are set by external stimuli. One can, for example, shift the time of day that the marine snail, *Nassarilus festilvus,* is active by illuminating the snails at night and leaving them in darkness during the day. Under natural conditions, these snails are active at night, but under the reversed experimental conditions, they are active by day.

Some biological clocks do not seem to be altered when the animal is placed in an experimental environment with constant conditions (for example, no change in light or temperature). Scientists consider these to be *internal clocks.* The best known are daily internal clocks, which reset every 24 hours, called **circadian rhythms.** Even these, however, are synchronized with the environment by external stimuli referred to as *time set-ters,* or **zeitgebers** (from German, meaning "time givers"). When animals are kept in isolation (that is, away from external time setters), their daily clocks will drift out of step with the environment, but if left in a normal setting, they will stay synchronized (see Figure 28.8).

Fixed Action Patterns and Learning

Early biologists who studied reflexes, kineses, taxes, and fixed action patterns assumed that these responses are inherited, unlearned, and common to all members of a species. The responses clearly depend on internal and external factors, but until recently, instinct and learning were considered distinct aspects of behavior. However, in some very clever experiments, Jack Hailman of the University of Wisconsin showed that certain stereotyped behavior patterns require subtle forms of experience for their development. In other words, at least some of the behavior normally called instinct is partly learned.

Hailman observed and experimented with the feeding behavior of sea gull chicks. In one species, the laughing gull, parent gulls lower their heads and point their beaks downward in front of young chicks. The chicks will then grasp, with a pecking motion, the parent's bill and stroke it downward. The parent responds by regurgitating some partly digested food (see Figure 28.9, page 534).

Earlier ethologists had claimed that gull chicks used an innate "picture" of a parent gull for pecking. Hailman's experiments showed that the gull chicks indeed reacted to an innate stimulus, but that learning was also important in the development of the behavior. He first demonstrated that the accuracy of the chick's pecking rapidly increases after hatching. He then compared the accuracy of chicks raised in the dark with those of normal chicks raised in the wild. Although the chicks raised in the dark improved their accuracy with time, they never achieved the "normal" accuracy of the controls. From this experiment, Hailman concluded that visual experience was necessary for full development of the fixed action pattern of gull chick feeding.

Hailman also found that the chicks do not have a complex innate picture of a parent gull as earlier ethologists, like Tinbergen, had thought. Instead, they start life by responding simply to any suitably oriented object that is shaped and moves like a parent's bill. Chicks respond to headless bills, but not to bill-less heads.

Much to Hailman's surprise, laughing gull chicks that had not been isolated from their natural parents at first responded similarly both to models that resembled laughing gull parents and to those that resembled the quite different-looking species of herring gulls. Within a week, however, the chicks responded only to models that closely resembled their parents. As with the increase in accuracy of the chicks' pecking, results from this experiment also suggest that some fixed action patterns have a component of learning and may not be totally innate. Only further research will show to what extent other "innate" behaviors have a learned component.

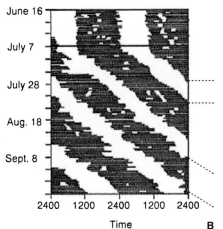

Figure 28.8   Human circadian rhythms can get out of phase. In this experiment, a woman was placed in isolation with a clock. The graph is a recording of her sleeping (light part of graph) and awake periods (dark lines) during the study. At first, she tried to keep a regular schedule (A); then (July 7) she slept when she felt like it (B). Note how her daily clock shifted.

BEFORE YOU GO ON   Ethologists call a sequence of innate, stereotyped behaviors triggered by a releaser a fixed action pattern. Internal states of animals influence how animals respond to releasers and other stimuli. Originally, fixed action patterns were thought to be totally innate; however, some of them have been shown to have a learning component.

LEARNING

As stated earlier, learned behavior occurs when animals change their responses as a result of experience. Psychologists did a considerable amount of the early work on learning. They have primarily been concerned with human learning, and even when their research has been on animals, it has been with an eye toward using animals to understand human behavior. More recently, biologists have focused directly on animal learning. Although studies on learning have been carried out on a relatively small number of species, a vast amount of information has been generated. Scientists now recognize five major categories of learning: imprinting, habituation, associative learning, latent learning, and insight.

Imprinting

Imprinting is a highly specialized form of learning. In many species, it takes place during the early stages of an animal's life, when attachment to parents, the family, or a social group is critical for survival. **Imprinting** is a process whereby a young animal forms an association or identification with another animal, object, or class of items. The best-known type of imprinting, called *filial imprinting,* concerns the behavior of young in following a "mother object" (see Figure 28.10, page 534). During a critical *sensitive period,* a young animal is susceptible to imprinting.

   Young animals are not completely indiscriminate in what they follow. For example, a mallard duckling will follow a moving object for the first two months after hatching. It will show a preference, however, for yellow-green objects (the color of its parents) over objects of different colors. Young animals may also be sensitive to sound as well as to sights.

Figure 28.9 A young gull chick stimulates its parent to regurgitate partly digested food by grasping the parent's beak and stroking it downward.

Figure 28.10 Konrad Lorenz got goslings to imprint on him as their mother object.

Wood ducks respond to a species-specific call in exiting from their nests.

Imprinting is an important form of learning, because it has both short-term effects on the immediate parent–offspring relationship and long-term effects that become evident in adult animals. For example, lack of imprinting has been shown to result in abnormal adult social behavior in some species. Also, the breeding preferences of many birds are a consequence of early imprinting experiences. As adults, they prefer to mate with birds of their imprinted parents' color or markings. This form of imprinting, which has considerable evolutionary significance as a reproductive isolating mechanism, is called **sexual imprinting.** The phenomenon can be tested by experiments that allow a bird to be raised by foster parents of a different species. When these young birds mature, they show a sexual preference for mates with the color of their foster parents. Sexual imprinting can have some unusual outcomes, as when hand-reared birds become sexually imprinted on people.

Habituation

Habituation is a simple form of learning. It occurs when an animal is repeatedly exposed to a stimulus that is not associated with any positive or negative consequence. The animal ceases to respond to that stimulus, even though it will continue to respond to other stimuli.

Another way of looking at habituation is to consider that the animal has learned not to respond to meaningless stimuli. Birds will soon ignore a scarecrow that initially caused them to avoid a garden. A snail moving along a board will withdraw into its shell if the board is tapped. After a brief time, it will emerge and continue on its course. If the board is tapped again, it will withdraw again but will emerge more quickly. After six or more responses, the snail will ignore subsequent taps. Habituation is important to animals in adjusting to their environment. If animals continued to respond to meaningless (for them) stimuli, they would not be able to function effectively.

Associative Learning

Habituation is learning that results in the loss of a response that is not relevant or useful to the animal. **Associative learning,** in contrast, is the acquisition of a response to a stimulus by associating it with another stimulus. Considerable research on associative learning has taken place in the past century. The most famous of the early scientists to study associative learning was the Russian scientist Ivan Petrovich Pavlov (see Figure 28.11).

In a number of now classic experiments, Pavlov demonstrated a form of associative learning called **classical conditioning,** which results in changes in the stimuli that elicit behavior. He did this by placing meat powder in a dog's mouth and recording the amount of saliva produced. He then repeated the procedure, only this time he rang a bell before giving the dog the powder. Although at first the dog did not respond to the bell (other than by pricking up its ears), eventually the dog salivated at just the sound of the bell. That is, it had learned to associate the stimulus of the bell with the stimulus of the meat. It had learned to respond to a new stimulus, the *conditioned stimulus* (the bell), by associating it with an old one, the *unconditioned stimulus* (the meat powder). Pavlov called the "new" response a *conditioned reflex;* now we call it a *conditioned response.* Pavlov showed that he could use almost any stimulus as a conditioned stimulus, as long as it was not too strong. Although Pavlov's experiments were elegant and have been confirmed repeatedly, it is not clear to what extent classical conditioning actually occurs in nature.

Figure 28.11 Ivan Petrovich Pavlov was one of the first investigators of associative learning, using dogs as his subjects.

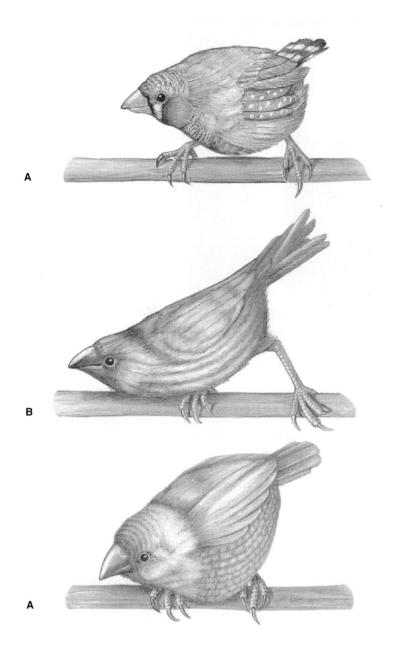

Figure 28.14 Studies of related species can lead to inferences of how specific behaviors evolved, as illustrated by this observation of beak wiping in three finches. In an unritualized movement, (A) a zebra finch prepares to wipe its beak on its perch. This movement has become ritualized in (B) the striated finch and (C) the spice finch, that hold the position for some seconds during courtship so that it resembles a bow.

would reduce the value of the well-camouflaged colored eggs. To test his hypothesis that the removal of the broken eggshells was adaptive, Tinbergen set out gull eggs in an area with broken shells nearby and gull eggs that were not close to any broken shells. Predators found two-thirds of the eggs that were near the broken shells but only one-fifth of the other eggs.

Emphasizing the adaptive significance of individual behavior closely ties the study of animal behavior to the theory of evolution by relating it to fitness. The study of behavior is also tied to the theory of evolution through the use of behavior in classification. Behavior is a trait; therefore, by comparing behavior, a taxonomist can use behavioral information in much the same way as anatomical information. Knowledge of the behavior of animals is also valuable in deciding whether two populations are actually different species. Information about behavior can determine whether the two groups interbreed in nature or whether they have different sexual behaviors that would prevent them from interbreeding.

Although a considerable amount of the study of animal behavior is done with explicit reference to evolution, a significant portion of the work is also done without it. Currently, some very exciting work concentrates on the underlying mechanisms of individual behavior. Research on neural pathways, genetic foundations, hormonal control, and learning focuses on uncovering the mechanisms involved in behavior. As with the study of physiology, evolution puts these behavioral mechanisms into a broader context and relates them to other domains of biology. (See the Focus on Scientific Process, "Organisms as Research Models.")

FOCUS ON SCIENTIFIC PROCESS

Organisms as Research Models

How do scientists select a plant or animal species for their research? If a plant biologist is interested in diseases that affect tomatoes, then tomatoes are the obvious choice. If a scientist is interested in a general problem, however, then the options are not so clear. In reading the scientific literature in a particular subject area (for example, development), we often discover that much of the research has been done on a small number of species. Why are there relatively few species used in biology research?

Ideal research organisms are both easy and inexpensive to obtain, raise, observe, and manipulate. The number of species that satisfy these criteria is relatively small compared to the number available. Moreover, scientists can compare results of different experiments if their work involves the same species. Yet these simple considerations do not explain the existence of what we might call research "superstars." A few species occupy exalted positions—they have been studied intensively by hundreds of scientists in a particular field. In Unit III, for example, numerous *Drosophila* experiments were described. More is known about the reproduction of fruit flies than about humans! The familiar bacterium, *Escherichia coli,* and the bread mold, *Neurospora* (see Figure 1), have received similar attention in molecular and biochemical studies. What accounts for such an uncommon interest in using these organisms?

A survey of highly preferred research species reveals that two critical factors are generally responsible for their popularity: practicality and specific biological characteristics that permit particular types of research. *Drosophila* originally became popular for genetics studies because they were easy to maintain. They take only two weeks to complete a generation, one pair can give rise to hundreds of offspring, and they can live on simple foods. T. H. Morgan and the "fly room" group at Columbia University demonstrated the practical value of the fruit fly (see Chapter 16). But *Drosophila* has special biological features that are critical. It was initially important that many of their physical characteristics (bristles, wing veins, eye color) were inherited according to expected Mendelian ratios. Perhaps even more important, observable mutations were also inherited in Mendelian fashion. Because *Drosophila* has only four chromosome pairs, the interesting gene linkage phenomenon was relatively easy to explore. Also, *Drosophila* have unusual, giant salivary gland chromosomes that, when stained, made it possible to study structural chromosomal mutations (see Figure 2).

Studies of *E. coli,* another research superstar, led to many of the fundamental generalizations in modern genetics. *E. coli* is easy to grow, and it is not normally virulent. Because it also has a simple structure and life cycle, and because it has molecules that can be tagged with radioactive atoms, scientists were able to use it in many pioneering physiology experiments. Finally, *E. coli* also are easily infected by phages (viruses), and this led to studies that were central to the development of gene-cloning techniques (see Chapter 21).

Currently, many animal behaviorists believe that research in neurobiology (the study of nerve cells and nerve networks) is revolutionizing our understanding of behavior. Detailed examination of the nervous systems of a few species, particularly mollusks, have been especially illuminating. The stimulus for molluscan research was the discovery that squid possess giant nerve fibers that were easy to locate and investigate. These observations led the French biologist Angelique Arvanitake to search for other animals with large nerve cells, and she concentrated her efforts on sea and land snails. She called attention to the marine snail, *Aplysia,*

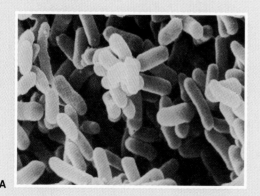

A

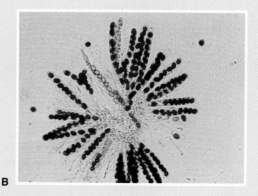

B

Figure 1 (A) Photomicrograph of *E. coli* and (B) photomicrograph of *Neuspora* (bread mold).

box continues

FOCUS ON SCIENTIFIC PROCESS

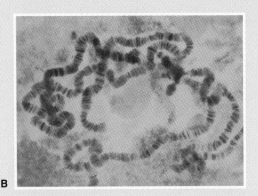

Figure 2 (A) Photo of *Drosophila* (single specimen) and (B) photomicrograph of stained *Drosophila* salivary gland chromosomes.

which has become one of the most carefully studied organisms in neurobiology (Figure 3). *Aplysia* is simple to raise in the laboratory; more important, however, specific nerve cells and nerve networks can be easily located and mapped. Furthermore, its 20,000 nerve cells are large enough to permit direct manipulation. The abdominal nerve cells are especially useful for experimental studies because they can be easily identified. This has allowed researchers to relate certain cells and specific chemical reactions to definite behavior patterns.

The meticulous neurobiology studies of organisms such as *Aplysia* have given animal behaviorists the hope of describing and understanding behavior in simple biological terms. Such knowledge is especially exciting because it provides an important link to the theory of evolution. Because the nervous system regulates behavior, genes that affect structural or chemical changes in the nervous system must also affect behavior. The *Aplysia* system is relatively simple, so neurobiologists have already been able to identify genes that influence its behavior. For example, *Aplysia*

genes have been identified that affect specific neurons located in the abdomen; these neurons influence behaviors associated with egg laying. If certain variations in egg-laying behavior are more adaptive than others, then one would expect natural selection to favor individuals carrying those genes. That is, these individuals would be more fit, and natural selection would lead to an increase in the frequency of their genes. On a deeper level, this knowledge reinforces the view that animal behavior, like animal structure, can be understood in the context of evolutionary history.

Figure 3 Photo of *Aplysia californica* (marine snail).

SUMMARY

1. During the past century, scientists have approached the behavior of animals in two distinct ways. Some have held that behavior is a function of inherited patterns of the nervous system that result in specific reactions known as instincts. Other scientists have held that behavior is learned.

2. Contemporary behavioral scientists realize that both instinct and learning are important in behavior.

3. The simplest forms of behavior are reflexes. Undirected orientation of animals to sources of stimulation are called kineses. Movement toward or away from a source of stimulation is called a taxis.

4. A complex form of orientation is navigation, the action of orienting toward a goal.

5. Fixed action patterns are sequences of actions that are innate and stereotyped.

6. Animals have specific motivations for behavior called drives.

7. Biological clocks are responsible for many rhythms of behavior. Daily internal clocks are called circadian rhythms.

8. There is evidence that some of what we usually regard as instinctive behavior is learned.

9. Learned behavior occurs when animals change their responses as a result of experience.

10. Five categories of behavior are imprinting, habituation, associative learning, latent learning, and insight.

11. Ethologists have demonstrated the adaptive significance of some individual behaviors.

WORKING VOCABULARY

associative learning (p. 535)
biological clock (p. 532)
circadian rhythm (p. 532)
displacement activity (p. 537)
fixed action pattern (p. 530)
imprinting (p. 533)

instinct (p. 528)
kinesis (p. 529)
learning (p. 528)
operant conditioning (p. 536)
reflex (p. 528)
taxis (p. 529)

REVIEW QUESTIONS

1. What are instincts?
2. What is learned behavior?
3. What is a reflex? A kinesis? A taxis? How do they differ from one another?
4. What types of environmental cues are used by animals in navigating?
5. What is a fixed action pattern? Describe one.
6. How did Lorenz and Tinbergen demonstrate the occurrence of fixed action patterns?
7. What are biological clocks? Give one example.
8. How is learning related to fixed action patterns?
9. What is imprinting? Give an example.
10. What is the significance of habituation, associative learning, latent learning, and insight?
11. How is behavior related to fitness?

ESSAY AND DISCUSSION QUESTIONS

1. What lessons came out of the battle between scientists who stressed the importance of instinct to behavior and those who stressed learning?
2. How might an understanding of individual animal behavior relate to an understanding of individual human behavior?
3. How might an understanding of the underlying mechanisms of behavior be integrated with the theory of evolution?

REFERENCES AND RECOMMENDED READING

Alcock, J. 1989. *Animal Behavior: An Evolutionary Approach.* Sunderland, Mass.: Sinauer.
Eisner, T., and E. O. Wilson. 1975. *Animal Behavior: Readings from Scientific American.* New York: Freeman.
FitzGerald, G. 1993. The reproductive behavior of the stickleback. *Scientific American,* 268: 80–85.
Gould, J. L. 1982. *Ethology: The Mechanisms and Evolution of Behavior.* New York: Norton.
Grier, J. W. 1984. *Biology of Animal Behavior.* St. Louis: Times Mirror/Mosby College Publishing.
Hinde, R. A. 1970. *Animal Behavior.* New York: McGraw-Hill.
Johnson, C. H., and J. W. Hastings. 1986. The elusive mechanism of the circadian clock. *Scientific American,* 74: 239–236.
Kelner, K., and J. Benditt. (eds.). 1994. Genes and behavior. *Science,* 264: 1685–1739.
Lederman, M. and R. Burian. (eds.), 1993. "The Right Organism for the Job," *Journal of the History of Biology,* 26(2): 235–367.
McFarland, D. 1982. *The Oxford Companion to Animal Behavior.* Oxford: Oxford University Press.
Menaker, M. (ed.). 1993. Special Topic: Circadian rhythms. *Annual Review of Physiology,* 55: 657–753.
Tinbergen, N. 1973. *The Animal in Its World: Explorations of an Ethologist.* Cambridge, Mass.: Harvard University Press.
Waterman, T. H. 1989. *Animal Navigation.* New York: Scientific American.

ANSWERS TO FIGURE QUESTIONS

Figure 28.5   The egg.
Figure 28.7   Since only males have red markings, the aggressive behavior will be released only toward other males, not females. Such behavior could help establish the males' territory and chase away any competing males.
Figure 28.12   If the animal encounters the situation again it can more quickly "solve" the problem it faces.

29

Social Behavior

Chapter Outline

Reading Questions

1. How are social interactions understood in evolutionary terms?

2. How do biologists attempt to understand social groups?

3. In what ways is behavior adaptive?

The diversity of animal forms is extraordinary. Anyone who has had to memorize part of the classification system of animals can attest to that. It will come as no surprise, then, to learn that the behaviors of animals are equally diverse. Darwin recognized this when he was formulating his theory of evolution. He treated behavioral traits as important adaptive features that had changed over time as a result of selective pressures. In Chapter 28, we discussed some basic features of individual behavior. The adaptive value of such behavior is of considerable interest for the study of evolution. So is **social behavior,** the behavior of two or more interacting animals.

SOCIAL BEHAVIOR AND EVOLUTION

Social behavior has attracted the attention of evolutionary scientists from Darwin's time to the present. Much of the recent research on this topic has been classified as **sociobiology** after the book by Edward O. Wilson of Harvard University. Wilson, shown in Figure 29.1, published *Sociobiology: The New Synthesis* in 1975 as an attempt to integrate modern work on social behavior with that done in genetics, evolution, and ecology. He hoped to establish "the systematic study of the biological basis of all social behavior."

Fitness

An evolutionary perspective currently dominates the study of social behavior. Basic to that view is the idea of relative fitness. Recall that *relative fitness* refers to an individual's chance of leaving offspring compared to other individuals in the same population (see Chapter 24). It measures the relative genetic contribution of individuals to the next generation. Much of the writing on social behavior is based on the assumption that sets of genes determine particular behavioral traits. A behavior that increases reproductive success, presumably, will result in an increase in the frequency of alleles (alternate forms of a gene) associated with that behavior. Unfortunately, it is often very difficult to gauge a specific behavior's effect on reproductive success or to separate one behavior from an animal's entire behavioral repertoire. Nonetheless, biologists have attempted to assess the consequences of a behavior and to determine whether it might enhance reproductive success.

Social Interactions

Social interactions can be classified, from an evolutionary perspective, by their effects on reproductive success of the **actor** (the individual displaying the behavior) and the **recipient** (the individual or individuals affected by the behavior). Three main behavioral categories are recognized: altruistic, cooperative, and selfish.

Altruistic Behavior

Altruistic behavior results in a benefit to the recipient at a cost to the actor. From an evolutionary point of view, you might not expect such behavior to exist because altruistic behavior seemingly reduces individual fitness, and natural selection ought to operate against it. Why should animals give alarm calls when it places them at greater risk of predation? Why should animals help raise the young of another pair or orphaned young when they could use their energies in raising their own young?

The paradox of altruistic behavior is explained by the idea of **kinship,** which is described as the degree of relatedness among organisms. Although helping another animal at the actor's own expense might decrease that individual's relative fitness, the actor's genetic contribution to the next generation could be enhanced if the recipient is related to the actor. How does this occur? An individual can increase its genes in the next generation in two ways: by producing offspring itself or by helping relatives produce offspring. If the degree of relatedness is high (that is, if the parties have many genes in common) and the cost to the actor is low, an altruistic act may well have value in promoting the relative fitness of the actor's genes. The altruistic act, seen from this perspective, is not so unselfish after all. It is a means by which an animal's genes are increased in the next generation, even if it occasionally comes at the individual's expense!

Because of the complexity of behavior, it has been difficult to document with certainty the relationship between kinship and altruistic behavior. Many cases of reported altruistic behavior, however, appear to involve kinship. In a famous study of the species of squirrel called Belding's ground squirrels, it was shown that these animals gave warning calls, at considerable danger to themselves, more often when they were near close relatives than when they were not. Adult females (such as the one shown in Figure 29.2, see page 544), which were usually found near close relatives, gave more calls than adult males, which were more widely distributed. In addition, adult females were more likely to gave a warning call if they were near many close relatives than if they were not.

The evolution of an altruistic act, because of its effects on promoting the survival and reproduction of relatives, is the result of a process termed **kin selection.** Research on kin selection has led scientists to expand the idea of fitness to **inclusive fitness,** which measures both an individual's reproductive success and its effects on the reproductive success of its relatives. Close relatives share genes. Thus, from the point of view of inclusive fitness, an altruistic individual may be enhancing the

Figure 29.1 E. O. Wilson wrote *Sociobiology: The New Synthesis* in an attempt to integrate research on social behavior with research on genetics, evolution, and ecology.

Figure 29.2 Belding's ground squirrels have been used in studies of kin selection. This photo shows a Belding's ground squirrel giving an alarm call.

Question: *Why would females make calls that benefit their close relatives even though it exposes them to considerable danger?*

survival of his or her own genes by helping a relative survive and reproduce. One can visualize this with the famous quotation attributed to J. B. S. Haldane, one of the biologists who worked on the modern synthesis: "I would lay down my life for two brothers or eight cousins." He was alluding to the fact that individuals, on the average, shares half of their genes with a brother (or sister) and one-eighth of their genes with a cousin. So if the two brothers or eight cousins each survived and reproduced, the altruistic individual would have the same number of genes passed on to the next generation as if the individual himself or herself had reproduced.

Cooperative Behavior

Although altruistic behavior has received considerable attention from researchers interested in theoretical issues, cooperative behavior is of greater social importance in organizing the behavior of animals. **Cooperative behavior** involves members of the same species and results in mutual benefits. In evolutionary terms, it is behavior whereby the fitness of both the recipient and the actor is increased. Examples of cooperative behavior among animals are numerous. Mutual vigilance is one

that has been studied extensively. Starlings in a flock, for example, can spend more time foraging, and do so more safely, than solitary birds. In experiments with caged starlings, scientists have shown that birds in flocks responded faster to the threat of a predator than solitary birds and that they spent more time foraging.

Cooperative defense can be more active than simple vigilance. Social ants, termites, bees, and wasps can launch massed attacks to defend their colonies. When attacked by wolves, Arctic musk-oxen form a defensive ring of adult females around the young, with the adult male on the outside to attack wolves that approach more closely (see Figure 29.3). Cooperative behavior also permits animals to be more successful in hunting, both in terms of capturing larger prey and in increasing the percentage of successful hunts. Lions have a higher percentage of success in capturing prey when they hunt in groups of two or more. Much of the increased success is attributable to cooperative ambush tactics (see Figure 29.4). Pack-hunting African wild dogs, shown in Figure 29.5, need four to six individuals to capture most of their regular prey.

Selfish Behavior

Selfish behavior occurs when the actor benefits and there is a cost to the recipient. Natural selection will favor such acts in the actor but will also favor recipients that avoid paying the cost of the actor's selfish behavior. A well-studied example involves the langur monkey, which travels in groups of 10 to 20 adult females with one dominant adult male. Occasionally, the dominant male is displaced by another male. When this happens, the new male will often kill all the infants in the group. This infanticide contributes to the reproductive success of the new male, for the females mate after they cease nursing. The energy they would have expended on their previous infants (fathered by the previous dominant male) will be saved for raising those of the new male. In contrast to the male, the female's reproductive success is diminished by this infanticide. As a result, female langur monkeys have developed several behaviors to counter the selfish behavior of the male. For example, after a new male has displaced the former one, females will sometimes travel alone or on the periphery of the group until their infants are older and less subject to attack (see Figure 29.6). Other females will also come to the aid of a mother in resisting attacks by the new male.

BEFORE YOU GO ON From an evolutionary perspective, social interactions may be classified by their effect on the reproductive success of the actor and the recipient. Altruistic behavior results in benefits to the recipient at a cost to the actor. The paradox of altruistic behavior is resolved by the concept of inclusive fitness, which measures both an individual's reproductive success and its effects on the reproductive success of its relatives. Cooperative behavior involves members of the same species and results in mutual benefits, whereas selfish behavior occurs when the actor benefits and there is a cost to the recipient.

Figure 29.3 Cooperative defense is used by Arctic musk-oxen for protection against wolves. When threatened, the females form a defensive circle around their calves. The adult male (right) remains outside and attacks any predators, such as wolves, that approach.

Question: *Why is it reasonable to assume that such a defense is effective for musk-oxen?*

Figure 29.4 Lions normally hunt small prey that can be killed by individual lions, but their hunting is more efficient when done in groups. Also, at certain times of the year, smaller prey are scarce, and lions must hunt larger, more dangerous prey. In hunting such animals, individual lions are much less successful than groups. This photo shows young lions cooperating in bringing down a wildebeest.

Figure 29.5 African wild dogs cooperate to kill their prey.

Question: *What are the benefits of such behavior?*

Figure 29.6 Female langur monkeys attempt to protect their infants from the selfish behavior of a new dominant male. These females and their daughters traveled apart from the main troop while the daughters were infants.

SOCIAL GROUPS

Although some animals are *solitary,* or alone, for part or most of their lives, many others live in groups. These groups are often dynamic and are difficult to classify. In very broad terms, they can be placed into three general classes: aggregations, anonymous groups, and societies.

Aggregations are groups formed as a result of individuals being attracted to a stimulus or an environmental feature such as a source of food. They have little or no internal organization and are therefore the simplest form of social group. Fruit flies gathered around a mound of rotted fruit, moths around a light, and slugs in moist places (as shown in Figure 29.7), are common examples.

Anonymous groups often exhibit coordinated movements that are more organized and may have some rudimentary division of labor. Examples are schools of fish, migratory flocks of birds, and migrating herds of caribou (see Figure 29.8). Although not highly structured, these groups still provide advantages to members in terms of mutual defense and other cooperative behaviors.

Animal *societies* have received the most attention from scientists studying animal behavior. They are very diverse and can vary with the season; therefore, they are difficult to characterize. Generally, **societies** consist of members of a species that show an attraction for one another, communicate with one another, exhibit cooperative behavior, and have synchronized activities. Table 29.1 lists various animal societies. (See also Figure 29.9.)

Features of Social Organization

The pattern of relationships among individuals within a population at a particular time is known as **social organization.** Social organization is dynamic; it can change to meet different conditions. How can social organization be understood? Biologists usually examine several key features. First, they analyze group structure. They gather population data concerning the distribution of age, population density, reproductive rates, life spans, and participation in reproduction. **Demography** is the statistical analysis of population data.

Next researchers look at how individuals and groups are distributed in the area they occupy and at the behavioral activity involved in that spacing. Animals do not use all of the available space in an area. What they do exploit on a daily basis is called their **home range.** Within the home range, individuals and groups are spaced in different *dispersion patterns* and show diverse behaviors related to those patterns. At one

Figure 29.7 Slugs form an aggregation that is based on their attraction to moist places. Gardeners in the Pacific Northwest make use of this attraction in reducing slug damage by placing slug bait in moist spots in and around their gardens.

Question: *Can you think of another everyday example?*

Figure 29.8 These caribou form an anonymous group.

Question: *What is the adaptive value in forming such a group?*

Table 29.1 Animal Societies

Type of Society	Examples
One-parent family	Polar bear (*Thalarctos maritimus*), black bear (*Ursus arctos*), hedgehog (*Erinaceus europaeus*), mallard duck, black grouse, stickleback fish, many rodents
Family	Songbirds, mute swan, golden jackal (*Canis aureus*)
Extended one-parent family	Honeybee, wasp (*Vespula vulgaris*), ants
Extended family	Termites, wolf, white-handed gibbon (*Hylobates lar*)
Harem	Wild horse (*Equus caballus*), zebra (*Equus burchelli*), vicuna (*Vicugna vicugna*), red deer, gelada baboon (*Theropithecus gelada*), patas monkey, sea lions (*Otaria*)
Female groups	African elephants (*Loxodonta africana*), wild pigs (*Sus scrofa*), red deer
Bachelor groups	Common waterbuck (*Kobus ellipsiprymnus*), Thompson's gazelle (*Gazella thomsoni*)
Multimale groups	Common baboon, rhesus monkey, spotted hyena (*Crocuta crocuta*), African hunting dogs (*Lycaon pictus*), African lion (*Panthera leo*)

extreme is the mountain gorilla, which lives in groups that have overlapping home ranges and generally tolerates its neighbors. At the other extreme, observed in robins, is a **territory,** where the home range is defended by threats, attacks, or advertisement (with scent or song) for exclusive use of the individual or group.

The study of *territorial behavior* has long attracted interest. The evolutionary significance of that behavior has also been appreciated because the occupation of a specific area gives access to resources such as food, nest sites, or mates that increase reproductive success.

In addition to studying population data and spacing, scientists have a critical interest in describing social interaction in groups. How do animals interact with other members of their group? How do they cooperate, compete, and communicate, and how do these activities develop? These are subjects of central interest for many scientists who study animals in the field and in the laboratory. Many behavioral interactions involve communication, which will be discussed shortly.

Biologists try to understand how social organization changes. It is constantly being modified in small ways to reflect the shifting relationships among individuals. It also often undergoes marked changes related to season, ecological conditions, or reproductive cycles. For instance, during the red deer mating season, males form temporary harems of females, but after the season, the deer organize into male and female groups.

A

C

B

D

Figure 29.9 Different animal societies: (A) mallard hen with young (one-parent family), (B) sea lions (harem), (C) African elephants (female group), and (D) Thompson's gazelle (bachelor group).

THE ADAPTIVE VALUE OF SOCIAL BEHAVIOR

The evolutionary perspective by which behavior is approached stresses the value of social behavior for reproductive success. A general look at the main functions of behavior supports that position. Explaining the detailed behavior of animals is not always easy, and often scientists are quite puzzled by their observations. All exciting science is like that. Without the challenge of unexplained phenomena, science would be a dull enterprise. However, inventing a good story to account for certain behavior is not considered good science. In 1902, Rudyard Kipling published a charming children's book, *Just So Stories,* in which he imaginatively explained the origin of such novel phenomena as the camel's hump, the leopard's spots, and the whale's throat.

Some of the popular books on sociobiology that attempt to explain complex behavior—especially human behavior—in terms of relative or inclusive fitness are, unfortunately, little better than Kipling's fictional stories. Yet, the abuse of sociobiology should not detract from the enormous value that it and evolution theory hold for understanding behavior. A brief review of some of the adaptive values of social behavior will illustrate this point.

An often-cited advantage of social existence is increased protection from predators. Many animals must expend energy to avoid being someone else's dinner. Cooperative behavior in a social group allows most species to benefit from mutual vigilance. Birds, for example, as mentioned earlier, gain extra foraging time by living in flocks, because they can rely on the vigilance of other flock members.

What other benefits might an individual animal gain from group living? Researchers in behavior now recognize many advantages. For example, numerous carnivores cooperatively hunt prey that an individual would be incapable of bringing down or killing. Similarly, experiments on captive birds show that groups are more efficient in finding food than single birds. Also, birds that forage over a wide area where the food is unevenly distributed benefit by belonging to a flock, which allows them to follow successful birds.

The division of labor that is possible in social insects has important adaptive value. Social insects have colonies that consist of different **castes,** which are groups of individuals that differ in structure and function. The different castes undertake distinct tasks, which increases the efficiency of the group considerably. Not only are individual tasks done more efficiently when the work is divided among different groups, but also several important tasks can be undertaken at one time, rather than in a sequence. The probability of completing a task is thereby considerably increased.

Another possible advantage of social life is the transmission of learned behavior. When generations overlap, learned behavior can be transmitted to the next generation. For example, some bird songs are partly learned. Male chaffinches raised in total isolation sing a song that is different from that of normal adults.

Learned behavior can be transmitted to peers as well. In studies on Japanese macaque monkeys, a particularly innovative female "invented" potato washing, which soon spread to other members of her troop (see Figure 29.14). She later discovered that she could wash wheat to rid it of sand, and this practical skill was also passed on to others and also to later generations. Compared to the evolution of human culture, this "animal cultural evolution" is limited. Its adaptive value, however, is nonetheless important. (See the Focus on Scientific Process, "Do Animals Think?")

Figure 29.14   This Japanese macaque "invented" potato washing for the group—a technique that was quickly learned by the other members.

FOCUS ON SCIENTIFIC PROCESS

Do Animals Think?

Scientists have investigated the human organism from a mechanistic viewpoint for over a century. Their underlying assumption has been that humans could be viewed as elegant chemical machines that follow predictable natural laws. This approach has had stunning success. We know a great deal about the human body, we can design drugs to alleviate various ailments, and we can counter numerous conditions that cause suffering or death. Although science can tell us a great deal about our physical condition by treating the human body as a machine, no one doubts that humans, unlike machines, are conscious creatures. Our own consciousness is evident. What about animals?

To those of us who have pets, such as dogs or cats, it is difficult to think of

them as machines without self-awareness—as entities more akin to our washing machines, personal computers, and blow-dryers than to our family members and friends (see Figure 1). Can it be that our clever dog, Cassie, "comforts" us when we are down, leaps with "joy" when we return from work, and has "outsmarted" the neighbor's dog that used to "steal" her food, is simply a genetically programmed automaton? Or that ZiZi, our neighbor's cat, that would seemingly "favor" starvation to dry cat food and that, if not a connoisseur of lasagna, is known to "prefer"—very definitely—smoked salmon to canned tuna, is, in her behavior, just reflecting an idiosyncratic program rather than expressing a conscious preference?

Until shortly after World War I, it seemed obvious to scientists that animals had feelings and that they could think. Charles Darwin believed that female birds showed aesthetic preferences in their choice of mates and that sexual selection was strongly influenced by it.

Many writings done in the late nineteenth century on the animal mind, however, were uncritical and highly anthropomorphic. Human desires, fears, and attitudes were attributed to animals, and numerous stories were accepted without any careful attempts at verification. It is not surprising, then, that when we read this literature today, much of it seems comical.

Psychologists in the 1920s reacted strongly to this uncritical literature and

Figure 1   (A) Can this animal think? (B) Do pet tricks reflect thinking ability?

FOCUS ON SCIENTIFIC PROCESS

took the position that it was not possible to verify whether or not animals could think. They concluded that the question of animals' thinking was not a meaningful topic for science because it could not be tested experimentally. Instead, psychologists focused on the observable *behavior* of animals. They argued that in establishing a scientific psychology, it was irrelevant whether animals thought. They intended to establish scientific laws about how animals learn and behave that could be verified by other scientists. To psychologists such as James Watson and B. F. Skinner, the private mind of the animal, if it existed, was closed to human investigation.

Ethologists who studied animal behavior, for the most part, were equally dismissive about probing the inner world of animals. A few workers were interested in how the world might "look" to animals, which have different sense organs than humans, but the primary thrust of ethology was in documenting repeatable patterns of behavior and in comparing these patterns with the object of establishing evolutionary connections.

Modern animal behavior draws on knowledge derived from psychology, ethology, and an ever-growing body of research in the fields of genetics, ecology, neurophysiology, and neuroanatomy. Until recently, all of these areas of research had been far removed from discussions of animal thought or animal awareness.

A well-known investigator of animal behavior, Donald Griffin of Rockefeller University, argues that neglecting animal awareness and thinking is

not only an overreaction to the naïve acceptance of undocumented animal stories but also a blind spot that retards advances in the scientific understanding of animal behavior. Griffin believes that mental experiences in animals could have an adaptive value—the better an animal understands its environment, the better it can adjust its behavior to survive and reproduce in it. He is also interested in animal communication, which he feels can sometimes be used to convey information about objects or events that are distant in time or space. This form of information may suggest awareness.

In support of his ideas, Griffin cites various behaviors that seem to involve accurate evaluation in complex environments. For example, he refers to a classic study on the prey selection of wagtails, a type of bird found in southern England. These birds feed on fly eggs and a number of small insects. Each day they must make several choices on where to hunt, when to move on to hunt in another area, and whether to join a flock or hunt alone. Scientists who study these wagtails have shown that they hunt with great proficiency. Although proficiency is not necessarily an indicator of awareness, Griffin argues that in cases where accurate evaluation of a changing and complex environment occurs, it is reasonable to consider that the animal is consciously thinking about what it is doing. Cooperative hunting by lions and the cultural transmission of behavior such as the potato washing done by Japanese macaques, both described in this chapter, and insight learning, described in Chapter 28, are other examples of

behaviors that suggest to Griffin and others that animals are aware and can think.

At present, Griffin and analysts who agree with him are in the minority in the scientific community. How can the question of animal thinking be resolved? One way is to attempt to design experiments that might give an indication one way or the other. Psychologists in the early twentieth century were very outspoken in their rejection of animal consciousness, claiming that testing for it was impossible. Griffin has proposed that some tests may be possible and that evidence can be gathered to support his position. He argues that once we have a better understanding of the electrical signals that are correlated with conscious thinking in humans, we could search for equivalents in animals. If none were found, that would suggest that his hypothesis of animal awareness is false. The strongest supporting evidence of Griffith's hypothesis involves cases in which animal communication is active and specialized, information is exchanged, and the receiving animal responds interactively. To Griffin, such cases are compelling examples of conscious and intentional acts.

It is too early to tell what researchers of animal behavior will conclude about animal awareness. Further research on interesting phenomena, such as animal communication, will ultimately provide the results necessary for formulating a scientific conclusion. Until then, we are confident that people will continue to discuss the world with their dogs and cats.

SOCIAL BEHAVIOR, SOCIOBIOLOGY, AND HUMANS

The study of social behavior, dominated in the 1980s by the insights of sociobiology, led to fruitful research. But the field has remained controversial. The two main issues that have attracted criticism are the emphasis on the genetic basis of social behavior and the application of sociobiology to humans.

Few experts doubt that individual and social behavior have their foundations in the genome, but believing such an idea is not the same as claiming that particular genes control a specific behavior. Likewise, some behavior does appear to improve fitness, but it is usually impossible to measure directly the impact of a particular behavior on the fitness of an allele. Even the influence of a behavior on the fitness of an individual is often difficult to assess because of the multitude of factors involved in fitness. Critics argue that although genes may be responsible for the physiological and anatomical foundations of behavior, and that behavior can evolve, behavior is too complex to be understood in terms of our current knowledge of genetics. To a large extent, the resolution of the genetic issue will depend on how successful sociobiologists are in relating behavior to genes.

The issue as it relates to humans, however, is much more nebulous, because *Homo sapiens* simply cannot be studied in the same way that scientists explore the behaviors of starlings and starfish. The issue with human behavior, moreover, raises a difficult question: To what extent is human behavior controlled by genes? The idea of human behavior being under genetic control is not attractive to those who take pride in people's freedom of action, creativity, and ethical standards.

What are the implications for social change and moral choice? A few sociobiologists have argued that the study of social behavior will permit researchers to provide a scientific foundation for human values. Such a claim is not only unlikely but also confuses different domains of thought. An understanding of the biological constraints on human action is necessary for a discussion of how humans *should* act. Philosophers often begin their investigations of morality with a discussion of human nature. Whatever light biology could shed on that issue has always been important, but understanding the biological basis of behavior will not reveal what is good or bad behavior. Some sociobiologists have attempted to argue that behaviors that increase fitness might be the basis of morality. Survival in itself, however, does not have any moral value.

The study of behavior, like other branches of the biological sciences, can be approached from numerous perspectives. The study of mechanisms of individual behavior can be profitable without reference to the theory of evolution. The theory of evolution, however, provides considerable guidance for the origin of behavior and for understanding its diversity. Studies of social behavior have been especially tied to evolution and have even contributed to the enlargement of the scope of the theory. The study of social and individual behavior may provide a background for an understanding of human values but appears to be an inappropriate guide for them. The study of morality lies in the domain of the humanities. Poets tell us much about nature, but not the same sorts of things that scientists do. Similarly, biologists tell us much about behavior, but not what philosophers have to say. Because all knowledge is ultimately connected, cooperative behavior among scholars, it is hoped, will provide the greatest benefit to all.

SUMMARY

1. Sociobiology attempts to integrate research on social behavior with knowledge from genetics, evolution, and ecology. Basic to the study of social behavior is the concept of relative fitness.

2. Social interactions are characterized as altruistic, cooperative, or selfish. Study of altruistic behavior has led to the formulation of the concept of relative fitness.

3. Although some animals are solitary, many live in groups. Animal groups fall into three broad classes: aggregations, anonymous groups, and societies.

4. The pattern of relationships among individuals in a population at a particular time is known as social organization. Among the features of social organization that scientists study are group structure, population data, distribution patterns, interactions, and change.

5. Communication is the passing of information by means of signals, resulting in a change in behavior. Animals communicate in various ways. They can use chemicals, electrical pulses, or visual, tactile, or auditory signals.

6. Animals communicate information about the environment, internal states, and their identity, as well as signals that coordinate behavior.

7. Sociobiology stresses the adaptive value of behavior. It has been a very fruitful approach, but it has been criticized because it emphasizes the genetic basis of behavior and because some sociobiologists apply their techniques to the study of human behavior.

WORKING VOCABULARY

altruistic behavior (p. 543)
castes (p. 550)
cooperative behavior (p. 544)
demography (p. 546)
inclusive fitness (p. 543)

kinship (p. 543)
pheromones (p. 548)
selfish behavior (p. 544)
sociobiology (p. 542)

REVIEW QUESTIONS

1. How is behavior classified from an evolutionary perspective?

2. What is altruistic behavior?

3. Altruistic behavior seemingly reduces individual fitness. How is this paradox resolved?

4. What is cooperative behavior? What are some of its benefits to the individual?

5. What is selfish behavior?

6. What are three general classes of social groups?

7. How are population data used in studies of social behavior?

8. What two aspects of animal spacing do scientists study?

9. What does *communication* mean in the study of animal social behavior?

10. What are pheromones? What are some of their functions in animal interaction?

11. What other simple communication signals do animals use?

12. What are some adaptive advantages that result from social existence?

ESSAY AND DISCUSSION QUESTIONS

1. A controversial issue in biology is the question of how speciation occurs. How might the study of behavior influence that discussion?

2. Some of the harshest critics of sociobiology have stated that the study of animal behavior may be of importance for ethics. What might they mean by that?

3. If it were to be established genetically that human males were more aggressive than females, would there be any implications for the formulation of social policies? What would they be? Why?

REFERENCES AND RECOMMENDED READING

Alcock, J. 1989. *Animal Behavior: An Evolutionary Approach.* Sunderland, Mass.: Sinauer.

Dawkins, M. S. 1995. *Unravelling Animal Behavior.* New York: J. Wiley & Sons.

Evolution and Ecology of Macaque Societies. 1996. New York: Cambridge University Press.

Gould, J. L. 1982. *Ethology: The Mechanisms and Evolution of Behavior.* New York: Norton.

Griffin, D. R. 1984. *Animal Thinking.* Cambridge, Mass.: Harvard University Press.

Kirchner, W. H., and W. F. Towne. 1994. The sensory basis of the honeybee's dance language. *Scientific American,* 270: 74–80.

Kitcher, P. 1985. *Vaulting Ambition: The Quest for Human Nature.* Cambridge, Mass.: MIT Press.

Ristau, C. A. 1991. *Comparative Ethology: The Minds of Other Animals.* Hillsdale, N.J.: Erlbaum.

Seven Pioneers of Psychology: Behavior and Mind. 1995. New York: Routledge.

Trivers, R. L. 1985. *Social Evolution.* Menlo Park, Calif.: Benjamin/Cummings.

Wilson, E. O. 1975. *Sociobiology: The New Synthesis.* Cambridge, Mass.: Harvard University Press.

Wilson, E. O. 1978. *On Human Nature.* Cambridge, Mass.: Harvard University Press.

Wintsch, S. 1990. Cetacean intelligence: You'd think you were thinking. *Mosaic,* 21(3): 34–48.

Wittenberger, J. F. 1981. *Animal Social Behavior.* North Scituate, Mass.: Duxbury Press.

ANSWERS TO FIGURE QUESTIONS

Figure 29.2   It might increase the animal's genes in the next generation because relatives have many genes in common.

Figure 29.3   Most vulnerable individuals have greatest protection.

Figure 29.5   The group can capture and kill prey that individuals alone could not capture.

Figure 29.7   Common pill bug.

Figure 29.8   Mutual defense.

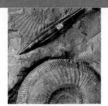

Unit Four Conclusion
The Evolution of Life and the Process of Science

The history of the theory of evolution provides one of the best examples of how the process of science generates scientific theories. Theories relate large bodies of information, provide answers to interesting related questions, and guide research by raising new questions. Theories always embody assumptions about the natural world, and they represent appropriate methods of investigation.

Early in the nineteenth century, a set of pressing scientific questions absorbed life scientists. A few of these questions, such as "What is the relationship between fossils and living forms of life," had occupied naturalists for over a century. Other questions arose because of new information about plants and animals that had been collected in previously unexplored regions of the world. This increase of knowledge was related to the social context in which science was practiced; it occurred because of explorations made by European nations that were interested in economic expansion. Similarly, the number of working scientists was greater than ever before, reflecting the emphasis placed on science by countries engaged in the industrial revolution. Among the most interesting unanswered questions of the time were these: What accounts for biogeographical patterns? What is the relationship between fossils and similar, living forms? How do new species arise?

Darwin's theory of evolution provided an elegant answer to all of those questions; the accumulation of adaptive variations could in time lead to change and ultimately to the formation of new species. The theory embodied assumptions made by a new generation of naturalists who sought to understand the laws of nature without direct reference to theology. The theory grew out of extensive observations and comparisons among organisms in the field and in museums. The theory also gave rise to new questions: What was the source of variations? How were they inherited? Were there different patterns of evolutionary change? What is the environmental significance of behavior? Can human behavior be explained by evolution?

Early in the twentieth century, research on these questions, along with discoveries in varies scientific disciplines, led to a reformation of Darwin's theory. The *modern synthesis* combined knowledge gained by researchers trying to determine how variation was inherited with results from field biology studies. The new theory was further strengthened by both observations and results from experimental methods, which led to a new set of questions. Scientists focused their attention on discovering the isolating mechanisms in evolution and studying how genes in a population change over time. After mid-century, the theory was further extended to investigate the evolution of ecological units, physiological pathways, and social behavior. Today, the theory of evolution is the major unifying theory of biology.

Plant Systems of Life

V

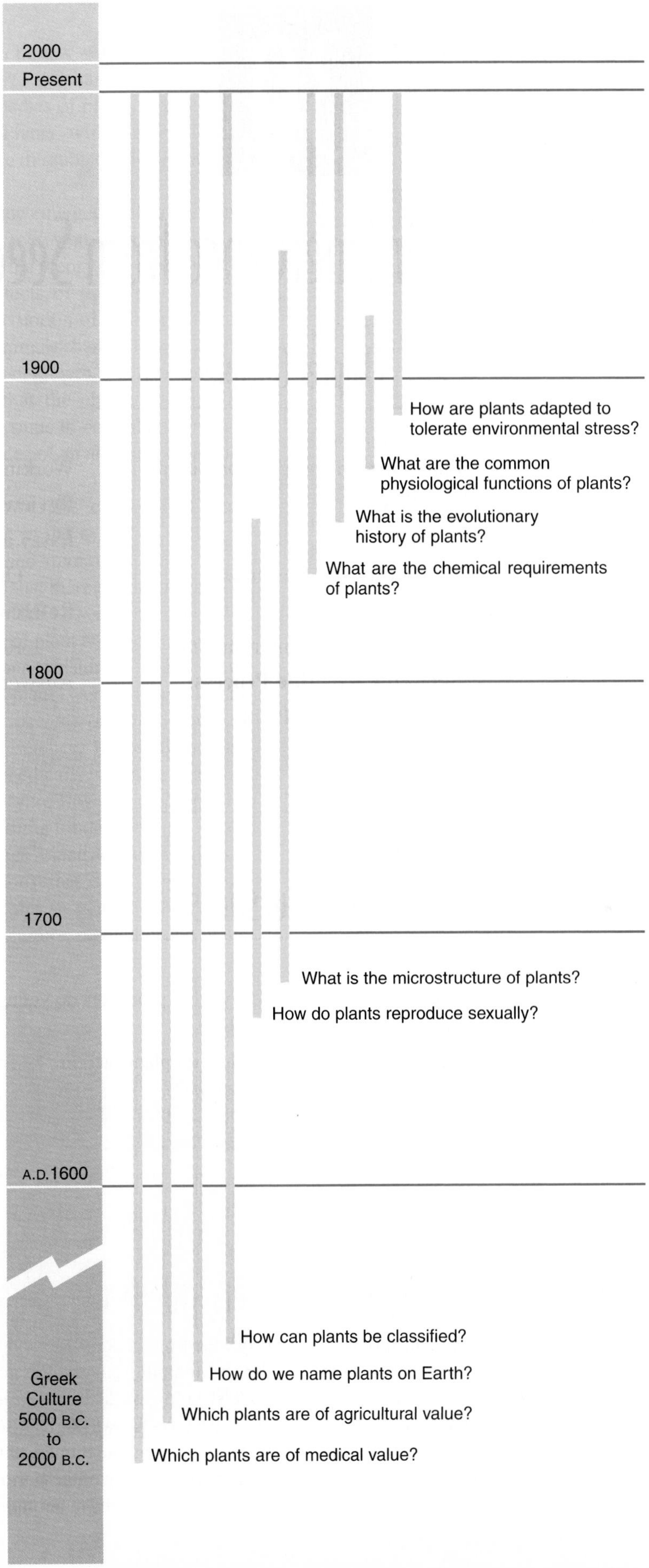

2000

Present

1900

How are plants adapted to
tolerate environmental stress?

What are the common
physiological functions of plants?

What is the evolutionary
history of plants?

What are the chemical requirements
of plants?

1800

1700

What is the microstructure of plants?

How do plants reproduce sexually?

A.D. 1600

How can plants be classified?

Greek
Culture
5000 B.C.
to
2000 B.C.

How do we name plants on Earth?

Which plants are of agricultural value?

Which plants are of medical value?

Figure 30.5   Seeds of angiosperms are contained within many types of fruits. Large fruits such as peaches (A) and grapes (B) protect seeds. Nuts, such as acorns (C), can be large and protect seeds but are also adapted for seed dispersal by animals or rolling away from the parent tree. Small fruits with wings, such as maple samaras (D), are dispersed by the wind.

Question: *How might development of a peach fruit be adaptive for seed dispersal?*

types of seeds can be dispersed by wind and transported to new areas (Figure 30.6).

Fruits range from small to large, and they vary greatly in their moisture content and fleshiness. Fruits produced by buttercups, buckwheat plants, and grasses are usually small and dry. Fruits of peach trees and date palms are medium-sized and fleshy and contain only one seed. Familiar examples of fleshy fruits that contain more than one seed are tomatoes, apples, and pears. Watermelons and squashes are examples of fruits that can reach great size.

Gymnosperm seeds generally take months or years to develop fully in cones. Once seeds mature, the cone scales of

Figure 30.6   Seeds are dispersed by various agents. Animals, such as the Eastern squirrel (A), may transport seed-containing acorns considerable distances from their origin. Wind is the principal dispersal agent for many types of small seeds, including these maple samaras (B), which have been blown from the parent tree to the ground, and for dandelion seeds, which number in the millions in the mature, dry flowers in this dandelion field in Michigan (C).

Question: *Which of these seeds is most likely to be transported the greatest distance from the parent plant? Why?*

many species dry and separate, allowing the seeds to be released. Some cones of several gymnosperm species, including lodgepole pine, will not open unless exposed to fire. (What might be the advantages of such an adaptation?) Once released, gymnosperm seeds are commonly dispersed by the wind or by animals that cache them.

Seeds of Angiosperms

Based on characteristics of their seeds, shown in Figure 30.7, angiosperms are either *monocotyledonous (monocots)* or *dicotyledonous (dicots)*. **Cotyledons** are leaves that are part of the embryo contained within the seed. In the seeds of some species, cotyledons are large and serve to store nutrients that will be used by the embryo until it begins photosynthesis. In other species, the cotyledon is a thin structure that assists in the digestion and transport of nutrients stored within the seed. **Monocot** seeds have only one cotyledon, and **dicot** seeds have two. Figure 30.7 also shows some of the other distinguishing features of monocots and dicots.

Monocots

Monocots include at least 65,000 species of plants. Monocot seeds grow into plants that have several common characteristics. These plants have numbers of flower parts, such as petals,

in multiples of three. Root systems of monocots consist of many small branches and are described as "fibrous." The *stem,* or aboveground axis, of monocots can become quite thick but does not develop the true wood, with seasonal growth rings, that is common in dicot trees. Finally, monocot plants have leaves with parallel *veins,* bundles of vascular tissues that run parallel to one another. Grasses, lilies, and palm trees are examples of monocots that have all of these characteristics.

Dicots

Approximately 170,000 dicot species are known to exist, including such plants as beans, peas, asters, roses, most shrubs, and deciduous trees. Dicots usually have flower parts in fours or fives. They typically have *taproots* (a single main root with many smaller lateral branches), and the stems of woody dicots increase in size through the addition of seasonal growth rings. Their leaves usually have midribs and a netlike arrangement of veins.

BEFORE YOU GO ON Gymnosperms and angiosperms produce seeds—structures enclosing an embryo—that can develop into new plants. Seeds are usually dispersed away from parent plants through a variety of mechanisms. An angiosperm is classified as either a monocot or a dicot, depending on characteristics of its seed and structural features of the adult plant.

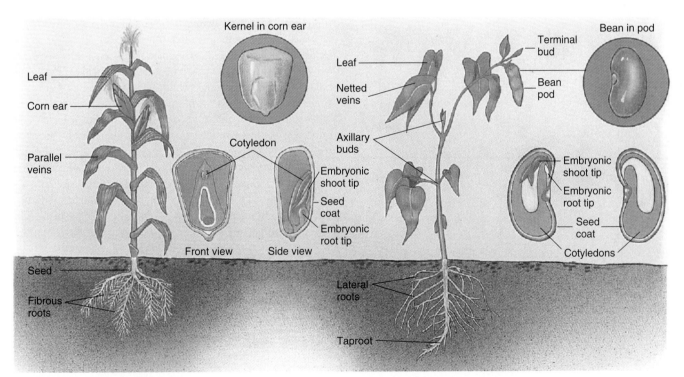

A *Zea mays*   **B** *Phaseolus vulgaris*

Figure 30.7 Angiosperms are classified as monocots or dicots on the basis of certain characteristics, including seed structure. The seed of a monocot such as corn, *Zea mays* (A), has only one cotyledon and an embryo. The seed coat protects the embryo. A dicot seed such as the bean, *Phaseolus vulgaris* (B), has a seed coat surrounding two large cotyledons and an embryo. Monocots also have fibrous roots and leaves with parallel veins. Dicots have a taproot and leaves with netted veins.

Seed Anatomy

Anatomy is the study of the structures of organisms. As described in Figure 30.8, a seed starts to take shape, and the flower begins maturing into a fruit, after sperm from the pollen fertilizes an egg. *Pollen grains* each contain two sperm cells and are formed in a *stamen,* the male reproductive organ. For fertilization to occur, the sperm must reach an egg that is located in the base of a *pistil,* the female reproductive organ.

The process of angiosperm fertilization involves both sperm contained within each pollen grain. One sperm combines with an egg to create a *zygote* that develops into an embryo. Embryonic cells ultimately grow and give rise to a seedling, which grows and matures into an adult plant. Because

flowers of many species have more than one pistil and pistils can have more than one egg, fertilization may occur many times in a single flower.

The second sperm participates in the formation of *endosperm.* **Endosperm** contains stored food that provides the energy and nutrients required for a seedling to develop. The embryo is surrounded by a tough outer layer called the *seed coat* that encloses and protects all seed parts. Both the embryo and endosperm are surrounded by membranes that play an active role in germination.

Besides cotyledons, the plant embryo also contains cells at its shoot tip that divide and develop into the stem and leaves. At the opposite end of the embryo is a group of cells that are destined to form the *radicle,* or embryonic root.

Seed Germination

Germination occurs when various structures within the seed function in an integrated way to induce the development and growth of an embryo into a new plant. During germination, the seed becomes transformed from a small, inert object into a highly organized seedling. The final stage of germination is somewhat analogous to animal birth: after embryonic development is completed, the individual plant escapes from its protective container and begins an independent life. Besides being essential for the successful reproduction of wild plants, crops, and garden plants, germination results in the development of certain products that have economic importance for human beings. For example, malt is produced from germinating barley seeds and has many important uses, including the brewing of beer.

A seed faces a demanding series of transitions before it can grow into a seedling. After formation, most seeds dry out, but there is great species variation in the events that follow. In some species, both the fruit and the enclosed seeds dry out, whereas in others, the seeds dry slowly after they have been released from a moist, fleshy fruit. Some plant species produce seeds that germinate soon after dispersal. The seeds of other species require passage through the digestive tract of an animal, where the seed coat is eroded by digestive juices, thereby freeing the embryo and allowing it to germinate soon after the animal eliminates it.

Once in a dried state, the seeds of some species become *dormant,* which means that their level of cellular activity is barely detectable. What is the advantage of seeds undergoing a period of dormancy? In most cases, dormancy enhances a seed's chances for survival and its eventual germination. Dormant seeds are extremely resistant to environmental stresses such as drought and extreme temperatures. Thus they can survive during periods when there is inadequate moisture or when it is too hot or too cold for a plant to live.

The basis of dormancy differs among plant species. Seeds of many species enter into a genetically defined period of dormancy, lasting from 3 to 16 weeks, before germination can begin. Some species produce dormant seeds that also require specific intervals of cold temperature or exposure to a certain number of hours of daylight before they can germinate. Such periods of dormancy are thought to be a mechanism that prevents seeds from germinating at inappropriate times, especial-

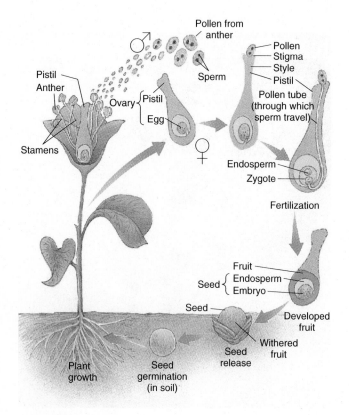

Figure 30.8   The angiosperm life cycle begins with a seed that germinates and develops into a seedling, which then grows into an adult plant. The mature plant produces a flower that contains both male *(stamens)* and female *(pistil)* reproductive organs. Meiosis occurs in the anthers and pistil to produce male and female gametes. Male gametes *(sperm)* develop within pollen grains produced by the anthers, while the female gamete *(egg)* ripens within the pistil. Pollen grains can be transported by wind, insects, or other animals to the *stigma,* a structure located on top of the pistil. Pollen grains that reach the stigma grow through the pistil, after which one sperm fertilizes the egg to form a zygote that develops into an embryo. Another sperm fuses with two pistil cell nuclei to form the *endosperm*—the nutritive tissue used by the embryo. The embryo and endosperm develop into a seed within the pistil, and the pistil subsequently matures into a fruit that encloses the seed.

Question: *Why are plants said to have a "double fertilization"?*

ly during the winter. For example, if January weather is unseasonably warm and a seed has no dormancy requirements for day length, exposure to cold, or exposure to light, it might begin to germinate, only to be killed by freezing temperatures during a February storm.

How long can dormant plant seeds survive? Botanists have found that viable dormancy periods range widely. The seeds of some species can remain alive for only a short period of time. For example, some orchid seeds are as small as dust particles and may live for only a few days because they have only a small amount of endosperm. A more typical example, however, is weed seeds, which can remain dormant and viable for up to 30 years.

Seeds of some species can remain dormant for incredibly long periods of time. In one remarkable case, a mining engineer made an amazing discovery in 1954. While working in the Yukon Territory of Canada, he uncovered ancient, preserved rodent burrows that contained the remains of lemmings, nest materials, and large seeds. He collected one lemming skull and several large seeds from one of the burrows and kept them for 12 years. Finally, several Canadian scientists gained possession of the materials and determined that the seeds were from an arctic tundra lupine (*Lupine articus*). Through other investigations, it was concluded that the lemming skull, burrow materials, and seeds were at least 10,000 years old. Would the seeds still germinate? (Do you suppose that the dormancy cold requirements had been satisfied?) About half of the seeds had been well preserved. When placed on wet paper in a dish, they germinated within 48 hours, and all grew into healthy plants!

Initial Events of Germination   Germination begins after a seed absorbs water and when other necessary conditions have been met (see Figure 30.9). The seed swells, and many of the metabolic processes essential for plant life begin. For example, in barley seeds, the cell layer located between the seed coat and

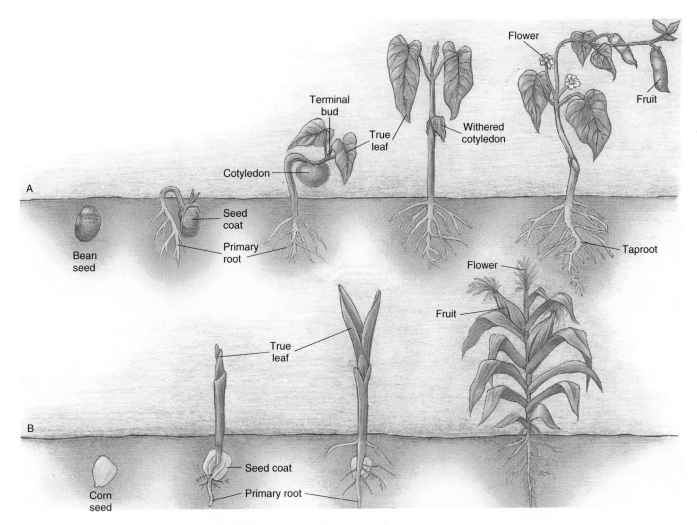

Figure 30.9   The germination of both dicot (A) and monocot (B) seeds generally follows the same course. After dormancy requirements are met, the root absorbs water and begins to swell. The primary root is the first part of the seedling to emerge from the seed. As the root grows down into the soil, the shoot elongates and the one or two cotyledons provide nutrients to the seedling. As the true leaves emerge, the seed coat is shed, and the cotyledons shrivel and fall away. At this point, the seedling must obtain water and nutrients from the soil and carry out photosynthesis to produce the carbohydrates it needs for continued growth and development.

Question: *Why is an angiosperm seed considered adaptive for surviving in terrestrial environments?*

the endosperm produces enzymes needed for growth. After being activated by tiny amounts of chemicals, known as **plant growth regulators,** these enzymes are released by the embryo and used in cellular respiration reactions and protein synthesis. Cellular respiration reactions allow the embryo to convert carbohydrates stored in the endosperm and cotyledons into energy required for growth. Protein synthesis provides the molecules that are necessary for constructing new cells and tissues during development and growth.

The first visible sign of germination is the appearance of the radicle after it expands, elongates, and then penetrates through the seed coat. As the radicle continues to elongate, it responds to gravity, turning downward and growing into the soil. As it becomes secured in the soil, the embryonic root absorbs the water and nutrients necessary for germination to continue.

Later Events of Germination Once the radicle is established, the shoot elongates and pushes up through the soil. After the shoot breaks through the soil surface, leaves form and begin to carry out photosynthesis. This allows the new plant to produce the sugars it requires for cellular respiration and growth. When the rate of photosynthesis is great enough to satisfy the needs of the developing plant, the seedling is no longer dependent on stored energy reserves, which have usually been depleted by this time.

Signs of germination can be seen by walking through a forest, meadow, or other natural habitat during the spring. It is often possible to find newly emerged seedlings in various stages of development. Some still have remnants of their seed coats attached, others appear with their newly developed leaves, and some will have already developed many of the characteristics of the adult plant.

Under some circumstances, large numbers of seeds may germinate at the same time and grow at the same rate. This is most common in fields where seeds are planted at the same time, then irrigated and fertilized. Uniform germination, growth, and maturation of crop species have obvious importance for farmers, who can consequently harvest all the plants at the same time. Seeds of wild plants may also germinate in synchrony, especially in the spring, when environmental conditions favor successful germination and seedling survival. The moist mild climate of spring normally provides the best conditions for germination and seed survival, when the plant is most sensitive to environmental stresses.

BEFORE YOU GO ON Most seeds, once formed and dispersed, enter a state of dormancy that is associated with an increased chance of survival. After the dormant period ends, germination begins when the seed absorbs water and its metabolic activities increase. The embryonic root appears first and becomes established in the soil. Subsequently, the embryonic shoot elongates, leaves form, photosynthesis begins to occur, and the new plant becomes established in its environment.

PLANT GROWTH PROCESSES

Germination marks the end of seed dormancy and the beginning of an independent existence for a new plant. Growth is an important process in the life of a seedling because new cells must develop and increase in size before a fully mature plant arises. Growth of the root system is necessary for acquiring water and nutrients from the soil. Similarly, the shoot of a new plant must elongate and new leaves must grow before photosynthesis can occur.

Patterns of Growth

Growth of plant tissues, in both the root and the shoot, often follows the pattern illustrated in Figure 30.10. Growth is initially slow, followed by a period in which plant size increases quickly. After the period of rapid growth, the rate of growth slows as the plant approaches maturity. These stages of growth are characteristic of a *sigmoidal growth curve* and can be defined by plotting a specific growth characteristic, such as plant height or weight, as a function of time. Such curves are useful for describing the growth stages of a single plant, for comparing the growth of two plants, and for evaluating the effects of various environmental factors on plant growth.

Mechanisms of Growth

The growth of plant tissues is the result of two processes, the production of new cells by mitosis and cytokinesis (that is, cell division, described in Chapter 15) and the enlargement of new or existing cells. These two mechanisms are responsible for the

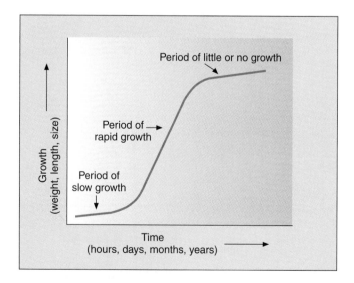

Figure 30.10 The growth of plant cells, individual plants, and populations of plants often follows a sigmoidal growth curve. Growth is initially slow, then follows a rapid phase, and finally slows or stops.

Question: *Which cellular processes are responsible for the period of rapid growth?*

growth of plant roots, stems, leaves, and reproductive tissues such as flowers.

Production of New Cells

In plants, growth resulting from cell division is generally confined to specific regions called **meristems.** Normally, plant daughter cells produced by mitosis and cytokinesis are small. The initial slow phase of the sigmoidal growth curve is thought to be associated with the production of new cells. Later, the rapid growth phase may be related to the enlargement of daughter cells produced during the initial phase. Throughout their lives, plants continue to produce new cells in meristem regions.

Enlargement of Existing Cells

Plant growth can also occur through the enlargement of existing cells, as shown in Figure 30.11. For example, some sunflower daughter cells are known to increase 15-fold in size after they are produced by cell division. Thus cell enlargement can result in substantial growth of the tissue or the entire plant.

Plant cells may enlarge because of high osmotic pressure caused by water concentrations that are greater outside the cell than on the inside (osmosis is described in Chapter 3). If moisture is abundant in the soil, water will enter the cell and cause it to enlarge. If little water is present, as in dry soils, cells will not enlarge, and the plant will be stunted and may wilt if a substantial amount of water leaves the plant cells. Thus environmental factors such as moisture can influence the growth of plants by affecting the mechanism of cell enlargement.

Plant cell enlargement depends, in part, on the stretching capacity of the *cell wall,* the tough outer layer that surrounds each plant cell. Cell walls are composed of a complex carbohydrate matrix in which fibrils composed of smaller microfibrils made of *cellulose chains* are embedded (see Figure 30.12A, page 568). **Cellulose** is a large structural polysaccharide molecule composed of linked glucose molecules. Cell walls function much like a house foundation or building walls made of reinforced concrete. In cell walls, the cellulose microfibrils correspond to steel reinforcing rods and the cell wall matrix compares to concrete. Cell wall structure varies greatly among different cells of the same plant and among cells of different species.

The *primary wall* forms first around growing daughter cells. The primary wall is composed of microfibrils that are loosely arranged in a random manner, as illustrated in Figure 30.12B (see page 568), and a matrix that has plastic properties due to the presence of *pectin,* a complex carbohydrate; these qualities allow the wall to expand as the cell grows. Actively dividing cells and most cells involved in photosynthesis, cellular respiration, and secretion have only a primary wall.

After cell growth ends and primary wall enlargement ceases, many cells form a *secondary wall* that has a denser, more uniform microfibril arrangement and a more rigid matrix because pectin is absent; these properties impart greater rigidity and strength compared to the primary wall. Some plant cells

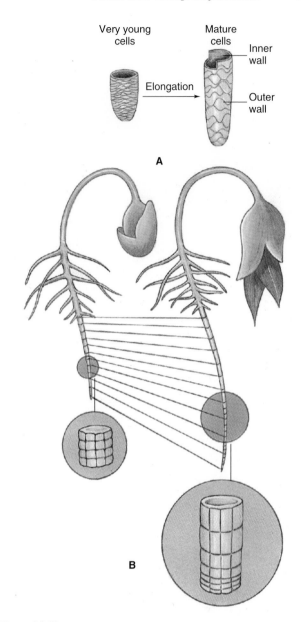

Figure 30.11 (A) Plant cells can enlarge as a result of osmotic pressure that forces elongation. (B) Cell elongation is an important component of plant growth, as shown by the stretching zones marked on a bean root.

may form several layers in the secondary wall, each having a different microfibril arrangement. Strong cell walls are necessary in vascular plants because cells must be able to withstand large osmotic pressure differences that may occur between the extracellular and intracellular fluids. Without support from cell walls, the pressure differences that occur when water concentrations are higher outside the cell may cause cells to swell and rupture. Cells that function in water transport and strengthening the plant typically die after forming a thick secondary wall. The reinforced structure of secondary walls allows plants to withstand the physical stress imposed by the enormous weight in the base of large trees.

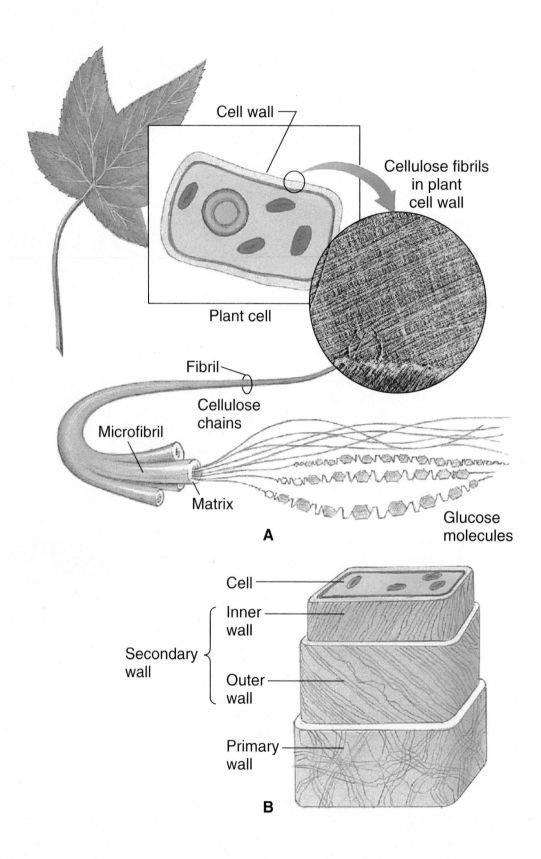

Cell wall

Cellulose fibrils
in plant
cell wall

Plant cell

Fibril

Cellulose
chains

Microfibril

Matrix

Glucose
molecules

A

Cell

Inner
wall

Secondary
wall

Outer
wall

Primary
wall

B

Figure 30.12 (A) Plant cell walls are made of fibrils containing microfibrils composed of cellulose chains embedded within a complex carbohydrate matrix. (B) Cell walls are formed sequentially. The flexible primary wall forms first in growing cells. Once growth is completed, a secondary wall forms that may consist of several layers.

Development

Cell differentiation is the process whereby newly formed cells become transformed into highly specialized cells that have specific functions. During the growth phase, one of the two daughter cells created by mitosis usually remains in the meristem, does not differentiate, and will divide again. The other daughter cell grows, differentiates, matures, and becomes capable of performing a specific function as part of a tissue.

Plants are composed of many differentiated tissues that have specific functions. Figures 30.13 and 30.14 show the anatomy of shoot and root tips. Meristems produce new

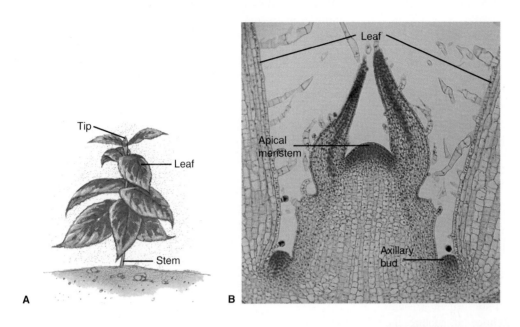

A

B

Figure 30.13   The tip of a *Coleus* shoot (A) magnified in a longitudinal section. (B) Shows the *apical meristem* where mitosis and cytokinesis occur that leads to growth and an increase in plant height. In the *axillary buds,* meristematic tissues also produce new cells by mitosis and result in the growth of lateral stems or leaves.

Question: *Where would you expect the region of elongation to be found, relative to the apical meristem?*

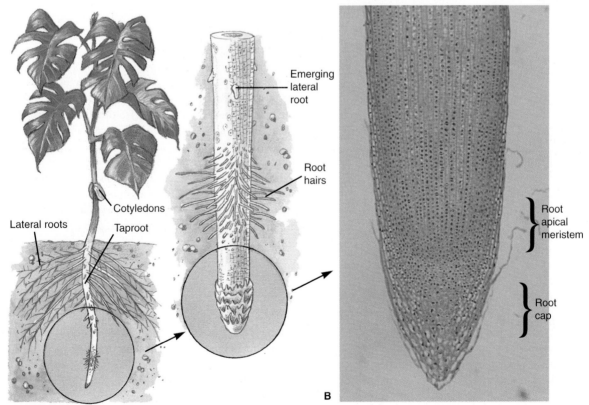

A

B

Figure 30.14   (A) This dicot root tip shows the root cap, root hairs, and emerging lateral roots. (B) The longitudinal section shows the root cap, apical meristem, and cells destined to form various root tissues.

Question: *Where else in a plant, other than the root tip, would you expect to find apical meristem?*

daughter cells that differentiate into the following major plant tissues. *Phloem* tissue transports fluids, including water, sugars, and nutrients, from photosynthetic tissues to other tissues where they are used or stored. *Xylem* tissue conducts water and minerals absorbed from the soil by the root system to the shoot. *Parenchyma* is a common tissue composed of thin-walled cells found in most leaves, stems, and roots. One type of parenchyma also makes up the photosynthetic cells found in leaves. All of these plant tissues originate from cells produced by meristem tissues.

BEFORE YOU GO ON   Plant growth is dependent on mitosis and cytokinesis in meristems to produce new daughter cells and on the enlargement of existing cells. Enlargement involves several factors, including flexible cell walls and water movement into and out of cells. New and enlarged cells eventually differentiate and form tissues that are specialized to carry out specific functions. Major plant tissues include meristem, phloem, xylem, and parenchyma.

PLANT GROWTH AND ENVIRONMENTAL FACTORS

The processes of cell division, growth, and differentiation ultimately result in the transformation of an embryo within a seed into a functional plant. The continued survival of a newly established plant depends on its abilities to acquire the resources it will need for further growth and survival. What resources does a plant require? Where do they come from? How are they obtained by the plant? Because plants are immobile, they must acquire resources from the environment in which they grow.

Energy

The sun generates energy that radiates to Earth's surface, and sunlight is the primary source of energy that sustains most life on the planet. Chlorophyll molecules in plants capture solar energy, which drives the process of photosynthesis (refer to Chapter 4). Through complex biochemical reactions, some of the energy absorbed by chlorophyll is used to construct carbohydrate molecules from water and atmospheric CO_2. These carbohydrates, including sugars and starch, can be stored in plant cells and serve as fuel that the plant uses to live and grow.

Atmospheric Resources

Plants acquire some essential chemicals from the atmosphere. Gases are the most critical, although atmospheric chemicals absorbed by dust, fog, rain, and snow can also be important sources of substances required for plant growth.

The chemistry of the atmosphere is extremely complex (see Figure 30.15). Unpolluted air contains more than 20 gases. Three gases, nitrogen (about 78 percent), oxygen (21 percent), and argon (1 percent), make up most of the volume of the atmosphere, and their concentrations remain nearly constant. Other gases are present in lower, variable concentrations. These include water vapor (0.7 percent) and carbon dioxide (0.03 percent), as well as traces of neon, helium, hydrogen, methane, ozone, ammonia, sulfur dioxide, and hydrogen sulfide. The gases of greatest biological importance are carbon dioxide (CO_2), oxygen (O_2), and nitrogen (N_2). These gases constitute environmental resources that are used by plants in photosynthesis, cellular respiration, and protein synthesis.

Nitrogen

Nitrogen is required by plants for synthesizing proteins. It is present as a gas in the atmosphere, but this form of nitrogen cannot be used by plants. Rather than acquiring nitrogen directly from the air, plants absorb nitrogen that has been converted from gaseous N_2 to other chemical forms, such as ammonia and nitrates. Recall from Chapter 9 that this transformation is carried out primarily by bacteria that live in the soil or inside nodules on some plant roots. These microbes convert atmospheric N_2 into chemical forms that can be absorbed by the roots of plants and assimilated into proteins. A few plants, in common with animals, have developed mechanisms to obtain nitrogen from food they digest (see this chapter's Focus on Scientific Process, "Carnivorous Plants").

Carbon Dioxide

Carbon dioxide is a critical biological component of Earth's atmosphere, even though it makes up only about 0.035 percent of air volume. CO_2 is absorbed by plant leaves and is used to form carbohydrate molecules during photosynthesis. Many carbohydrates, such as sugars, starch, and cellulose, are found within the plant, and each is involved with specific functions. Simple sugars can be transported from one organ, usually a leaf, to another organ, such as a root. *Starch* is made up of multiple, linked, simple sugar molecules and is stored in plants. When carbohydrates are required to provide energy for cellular maintenance and growth, starch is converted to simple sugars that can be used in cellular respiration. Other organic compounds synthesized from plant sugars include cellulose, pectin (a constituent of cell membranes and cell walls), fats, nucleic acids, and proteins.

Soil Resources

Plants are able to take root and grow in a variety of environmental substrates, as shown in Figure 30.16. Most terrestrial plants are rooted in the soil, but some can root in fallen, decaying logs known as *nurse logs*. Hemlock seedlings often grow well on nurse logs, and as they continue to grow, they develop root systems that eventually penetrate into the soil. Some plants, such as Spanish moss and other *epiphytes*—plants that grow on other plants but do not obtain nutrients from them—can grow in the tops of living trees and obtain required water and minerals from rain or other forms of atmospheric moisture.

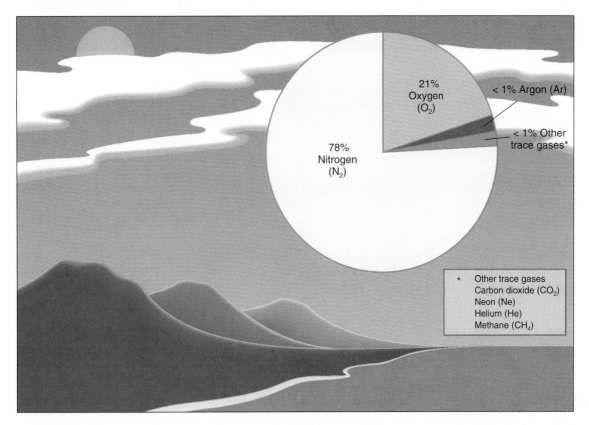

Figure 30.15 Earth's atmosphere is composed largely of nitrogen and oxygen gases. Together, they account for about 99 percent of the air. Many other gases are present in trace amounts. Carbon dioxide (CO_2) presently accounts for about 0.035 percent of Earth's atmosphere and is the source of carbon for plants. CO_2 concentrations are increasing as a result of human activities. Increasing concentrations of CO_2 and other "greenhouse" gases have the potential to cause changes in Earth's climate.

Question: *Plants extract CO_2 from the atmosphere during photosynthesis, but which plant activity adds CO_2 to the atmosphere?*

Figure 30.16 Many plants, including hemlock (A), grow in soils or decomposing logs known as nurse logs. Other species, such as Spanish moss (B), root in tree canopies using branches for their support. Dwarf mistletoe, a parasitic plant, grows in the tops of pine trees (C).

FOCUS ON SCIENTIFIC PROCESS

Carnivorous Plants

Among the most interesting plant adaptations are those of *carnivorous plants*—plants that have special organs for capturing and digesting animal prey. In a sense, carnivorous plants have turned the table on animals; rather than occupying the standard animal-eats-plant world, they function in a plant-eats-animal domain.

Carnivorous plants have fascinated scientists for centuries. In a letter written in April 1759, Governor Dobbs of North Carolina described the first known carnivorous plant, the Venus flytrap *(Dionaea)*, to amazed European botanists. Linnaeus referred to the Venus flytrap as a *Miraculum Naturae,* a "miracle of Nature." For decades, amateur exotic plant collectors coveted the Venus flytrap, and even today, it continues to be a popular houseplant. Early descriptions of the flytrap were often quite sensational; for example, in an eighteenth-century engraving, the caption noted that "each leaf is a miniature figure of a Rat trap with teeth; clos-

ing on every fly or other insect, that creeps between its lobes, and squeezing it to Death" (see Figure 1).

The roughly 600 carnivorous plant species are highly diverse and have specialized adaptations that allow them to attract and capture animals. For example, the production of glistening droplets in sundews *(Drosera),* the sweet-smelling nectar of dewy-pines *(Drosophyllum),*

and the ultraviolet-reflecting patterns of sunpitchers *(Heliamphora)* attract their prey. They can also digest the animals they capture, using acids and enzymes and, in many cases, specialized receptacles that function as stomachs.

Some carnivorous plants such as the cobra plant, *Darlingtonia,* act as passive traps. *Darlingtonia* has unusual structural features (see Figure 2A):

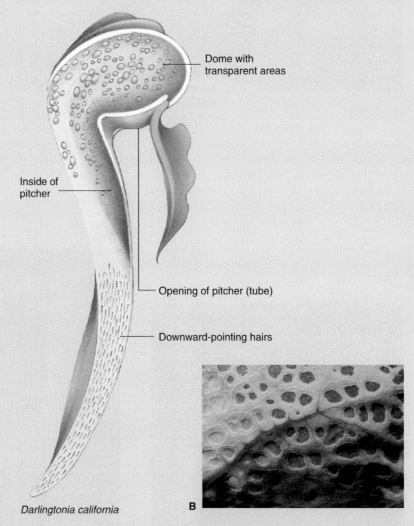

Dome with transparent areas

Inside of pitcher

Opening of pitcher (tube)

Downward-pointing hairs

Darlingtonia californica

B

Figure 1   This eighteenth-century engraving describes a Venus flytrap as having a "Rat trap with teeth" on its leaves that squeezes its prey to death!

Figure 2   (A) *Darlingtonia californica* has a hollow cylindrical tube that traps insects. Insects are attracted into it by the smell of its nectar, enter the tube, and are unable to find their way out. Flies trapped inside the pitcher apparently become confused by the transparent areas in the dome, collide with them when attempting to fly out, and fall to the bottom of the pitcher. (B) A fly's view of the dome inside the pitcher.

FOCUS ON SCIENTIFIC PROCESS

(1) the leaf is a long, hollow, cylindrical tube, which contains insect-attracting nectar; (2) the top of the leaf is a hollow dome with an opening below; and (3) the dome itself is partly transparent and may appear to be an opening to an insect (see Figure 2B). Insects can easily enter the leaf, but when they try to exit, they fly against the top of the dome and fall down the hollow tube, which has downward-pointing hairs that prevent escape.

The more dramatic active traps, like the Venus flytrap, are capable of unusual movements that contribute to capturing prey animals. The flytrap's leaves have two-lobed traps on their ends and each lobe has a set of coarse projections ("teeth") along each edge. The inner lobe surface contains two types of glands; one secretes an insect-attracting fluid, and the other both secretes an insect-digesting fluid and absorbs nutrients from digested insect tissues. Each lobe also contains three trigger hairs on their inner surface. The trap closes on insects (often ants) when they touch the trigger hairs—either one hair twice, or two hairs in succession; the trap takes about one second to close (see Figure 3).

A

B

C

Figure 3   (A) The trap of the Venus flytrap. (B) Trigger hairs are arranged in a triangular pattern on the inner lobes of the trap. When stimulated, trigger hairs cause the trap to close. (C) A green bottle fly about to trigger closure of the trap.

FOCUS ON SCIENTIFIC PROCESS

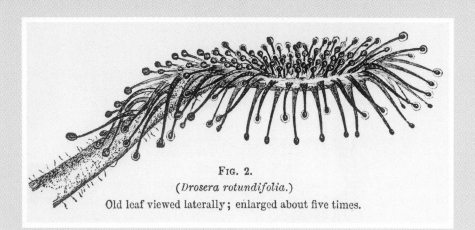

FIG. 2.
(*Drosera rotundifolia.*)
Old leaf viewed laterally; enlarged about five times.

Figure 4   An illustration from Darwin's book *Insectivorous Plants* detailing *Drosera rotundifolia,* the sundew that he studied.

What is the adaptive value of plants being able to capture and digest animals? Charles Darwin was very interested in the common sundew *(Drosera rotundifolia),* shown in Figure 4. In 1860, the year after he published *On the Origin of Species,* Darwin observed sundew plants and was struck by how many insects they caught. He noted that sundew leaves possess elongated stalks; glands inside the stalks secrete a sticky fluid that entraps small insects. Sundew stalks are flexible, and when a small insect, such as an ant, contacts several stalks on a leaf, other stalks bend toward the prey and secrete liquid, which soon smothers it. Darwin performed many experiments with the sundew. He showed that when the stalks contacted animals, they secreted acidic substances that helped digest the animal tissues. At that time, no other plant was known to have such a capability. He also showed that sundew secretions had antiseptic properties that inhibited the action of bacteria and other microbes. In 1875, Darwin published the results from his ingenious and revealing experiments in a book titled *Insectivorous Plants.*

Darwin also formulated a theoretical foundation for understanding carnivorous plants. He argued that the ability to catch and digest insects was a valuable plant adaptation. However, some botanists were initially skeptical of Darwin's claim because it had been shown that insects were not essential for the survival of some carnivorous plants; they could be raised successfully in the absence of animal prey.

The debate over Darwin's claim led to considerable research during the past century. The results indicate that the dependence of carnivorous plants on animal prey varies considerably among different species and that carnivorous plants use the nutrients they obtain from insects for different purposes. Insects are a relatively rich food source, and they contain many major nutrients required by plants—nitrogen, potassium, calcium, phosphorus, magnesium, and iron. The ability to acquire these important nutrients from animal tissue (by way of either enzymes or bacterial action) allows carnivorous plants like *Drosera* to exploit nitrogen-poor soils that are unfit for plants dependent on obtaining nitrogen from the soil. Specific insect nutrients are apparently necessary for some carnivorous plants to thrive. For example, experiments with bladderworts *(Utricularia longifolia)* suggest that insect prey is necessary for them to flower. *U. longifolia* grown in a greenhouse with no insects grew well but never flowered, whereas those raised in a greenhouse with insects produced flowers.

The development of advanced biochemical analysis and electron microscopy in the twentieth century have uncovered the details of digestion and the microstructure of carnivorous plants (see Figure 5). Careful research on *Drosera,* for example, has led to detailed knowledge of how it absorbs various nutrients, such as amino acids.

Increased understanding has not led to a decline in interest. Carnivorous plants still exert a fascination on the general and professional public. Science fiction cartoons continue to produce "man-eating" plants (Figure 6), hobbyists regularly grow over 100 different species, and research scientists actively explore the details of these unusual adaptations.

FOCUS ON SCIENTIFIC PROCESS

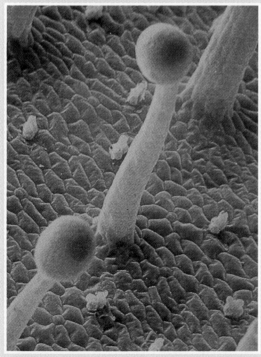

Figure 5 (A) A photo of a fly caught by a sundew, *Drosera capensis.* (B) Electron micrograph of sundew leaf stalks that trap and digest insects.

Figure 6 A comic interpretation of carnivorous plant capabilities.

Whether plants take root in logs or soils, one universal function of the rooting substrates is to provide stability so that the plant can grow in one place. Plants cannot move from one site to another, so if the rooting medium in which the seed originally falls proves unstable, the roots will be damaged, and the plant may die.

Water

Water is a major resource that plants must acquire from their rooting substrate. Even though leaves may be in direct contact with water in the form of rain, snow, or dew, most cannot absorb this water because their external surfaces are covered by a **cuticle,** which is impervious to water because it is waxy. The cuticle is an important adaptation that allows vascular plants to grow on land. Without the cuticle, these terrestrial plants would dry out and die.

Together with cell walls, water plays a central role in providing support for a plant. The osmotic pressure of water is necessary not only for cellular enlargement during growth but also for maintaining the entire structure of the plant. When plant cells are fully charged with water, they are *turgid.* When insufficient water is available for cells to remain turgid, plants lose **turgor,** the capacity to sustain their shape, and they wilt

(see Figures 30.17 and 30.18). If wilting is only temporary, or if it is not caused by extended soil dryness, many plants can recover and regain their turgor when water again becomes available.

Water is also used in a plant's metabolism. It plays a critical role in photosynthesis and is important in other reactions carried out by the plant. Water is the primary solvent in the cell; it carries nutrients into the plant and transports chemical compounds within the plant. Just as too little water is detrimental to plants, so is an excess of water. Flooding a plant results in reduced levels of oxygen around the roots and, if prolonged, may result in death. However, many plants are adapted for growing in wetlands because they have evolved mechanisms to keep air around or within their roots, even when submerged. For example, water lilies can absorb air through their leaves and transport it to their roots.

Nutrients

Studies of plant nutrients have a long and interesting history. In the early nineteenth century, many American farmers in the southern and northeastern United States had become concerned about "worn-out soil." A disturbing cycle had developed. Farmers settled on new land, exploited it until fertility decreased,

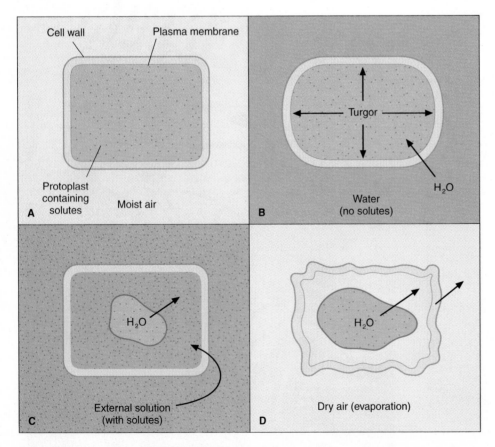

Figure 30.17 (A) Normal plant cells have regular shapes as described in this diagram. (B) High water pressure in cells results in turgor, which may cause the cell wall to expand. (C) Loss of water from the cell may occur and cause changes in the cell's profile. If water loss continues, the cell may lose enough volume that its plasma membrane pulls away from the cell wall. (D) Extreme water loss may cause collapse of both the plasma membrane and cell wall and result in the cell's death.

A

B

Figure 30.18 Absorbing sufficient amounts of water allows plant cells to maintain the pressure that causes the plant to stand upright. If water is limited, plant cells lose water pressure, and the plant wilts (A). Recovery is possible if water is applied before large numbers of cells die (B).

Question: *As the plant wilts, what change would you expect in the turgor pressure within cells?*

and then moved on to virgin lands. Consequently, vast areas had become unsuitable for farming or habitation. However, by the middle of the nineteenth century, these barren lands were once again under cultivation, and the formerly destructive cycle had been broken. In large part this was made possible by the new field of *agricultural chemistry,* which brought about fundamental changes in American agriculture.

The research of a German chemist, Justus Liebig, was central to the agricultural revolution in the mid-nineteenth century. Liebig's widely read book, *Organic Chemistry in Its Applications to Agriculture and Physiology,* published in 1840, went through several editions in the United States. Liebig stressed the importance of nitrogen for plants, and although his specific description of the nitrogen cycle was later replaced as new knowledge became available, he was the first to call attention to its importance. Of greater significance was his discussion of **essential elements.** Liebig was among the first scientists to realize that certain chemical elements were essential for proper plant growth. He showed that these elements were found in the tissues of all plants. Furthermore, plants either did not grow or were stunted in soils lacking essential elements.

Liebig advanced the idea that "worn-out soil" was in fact soil that had been depleted of essential elements. He noted that farmers could improve their fields by adding substances such as animal dung, wood ashes, ground bone, or "mineral manures" (for example, gypsum and lime). According to Liebig, these materials were effective because they replaced chemicals that had been depleted. Liebig also suggested that these chemical substances could be manufactured. This would

have the advantages of providing relatively inexpensive additives that would allow agricultural experts to apply exactly the amounts of essential elements that they determined (through testing) were necessary.

Liebig's book initiated a scientific revolution in agriculture. In the United States, it began with American scientists traveling to Germany to study in laboratories such as Liebig's. Later, the U.S. federal government (under President Lincoln) established the Department of Agriculture and funded land-grant colleges. These colleges in turn set up agricultural experiment stations that transmitted results of agricultural research to farmers throughout the country.

Plants acquire nutrient resources from soils, by root uptake, or directly from the atmosphere. Botanists now generally believe that plants require at least 20 essential elements to grow and complete their life cycles. Plants obtain carbon from the atmosphere and the other essential elements—hydrogen, oxygen, nitrogen, potassium, calcium, magnesium, phosphorus, sodium, sulfur, chlorine, iron, boron, cobalt, manganese, zinc, copper, nickel, silicon, and molybdenum—from their rooting substrate. These elements are available to plants from soil because of two processes. During *decomposition,* bacteria and fungi convert organic matter in the soil into simpler chemical forms that the root system can absorb. During *mineralization,* soil particles derived from parent rock break down and release important elements. Roots usually absorb these essential elements when the elements become dissolved in water.

As Liebig first demonstrated, nutrient deficiencies may occur in agriculture or gardening, and these deficiencies can be overcome with fertilizers. Because nutritionally deficient soils usually lack compounds that hold nitrogen (N), phosphorus (P), or potassium (K), these chemicals are most commonly contained in commercial fertilizers. Why are these chemicals most often limiting to plant growth?

In general, plants are capable of accumulating nutrients at much higher concentrations than exist in the soils they inhabit because their roots absorb essential chemicals selectively. Nutrients absorbed by root cells are distributed and retained within the plant, where they may reach high concentrations. For example, potassium concentrations in plants may be several hundred times greater than the soil concentration. As much as 95 percent of the *dry weight* (the weight after water is removed) of a plant may be composed of carbon, oxygen, and hydrogen, with all other elements contributing only about 5 percent. Of that 5 percent, nitrogen may contribute as much as 1.5 percent of a plant's dry weight, and potassium, 1 percent. The remaining essential elements are present in only trace amounts.

BEFORE YOU GO ON   To survive after germination, plants must be able to acquire essential resources from their surrounding environment. By transforming radiant energy during photosynthesis, plants are able to produce the carbohydrates they use as energy sources. Essential gases—oxygen, nitrogen, and carbon dioxide—are obtained directly or indirectly from the atmosphere. Water and essential elements are obtained from the soil through the root system and then distributed to all parts of the plant.

PLANTS AND THEIR ENVIRONMENT

In this chapter, we followed the life cycle of seed plants from the formation of a seed to its subsequent germination, then to its development into a seedling, and finally to its growth into an adult plant. These events are common to many plants that are important to humans. Plants must absorb solar radiation and chemicals from the environment in order to grow. The processes by which they obtain resources from the habitat form the basis for the relationship between plants and their environment. The dynamic nature of this relationship demonstrates that although plants remain stationary, they must respond quickly as features of the environment change over time. We explore plant–environment relations further in Chapter 31.

SUMMARY

1. Vascular plants have tissues that transport water and nutrients between their roots, stems, and leaves. Most trees, crop species, garden plants, and houseplants are vascular plants.

2. A seed is a complex structure that contains a plant embryo and nutrients to be used by the embryo during its development. Two primary classes of vascular plants produce seeds: gymnosperm seeds develop in cones, and angiosperm seeds form in fruits.

3. Most plant species have mechanisms that act to disperse seeds away from a parent plant. Seeds of different species are dispersed by various processes that are related to their size and shape.

4. Angiosperms are of two major types—monocots and dicots—that are distinguished by differences in number of cotyledons in their seeds, arrangement of veins in their leaves, number of floral parts, and type of root.

5. Seed formation in angiosperms is a complex process that begins when an egg in the flower is fertilized by a sperm contained in a pollen grain. Fertilization creates a zygote that develops into an embryo within the seed. In some species, a second sperm nucleus from the pollen grain helps form endosperm, which is also part of the seed. All of the internal seed structures are protected by a seed coat.

6. Seed germination usually occurs after the seed has dried and environmental conditions are favorable. In some plant species, seeds enter a period of dormancy that does not end until certain requirements—usually related to cold temperatures or day length—have been met.

7. During germination, the embryo constructs new cells and tissues that will be used in the development of a new plant. The radicle, or embryonic root, first emerges from the seed and becomes established in the soil. Later, the stem develops and leaves form to carry on photosynthesis. At this point, the newly formed seedling becomes independent and may eventually mature into an adult plant.

8. Growth of plant tissues depends on cell division, cell enlargement, and differentiation. Mitosis and cytokinesis take place in specialized regions, called meristems. As new cells are formed, they enlarge through the effects of osmosis and differences in osmotic pressure inside and outside the cell. Structural properties of the cell wall are also important in cell enlargement processes. Many new cells undergo differentiation and acquire the ability to perform specialized functions.

9. Plants must carry out photosynthesis, cellular respiration, and protein synthesis to survive. The energy required for these biological processes comes from solar energy that is converted into sugars by photosynthesis. Nutrient resources used by plants come from the atmosphere and the soil.

WORKING VOCABULARY

botany (p. 558)	fruit (p. 561)
cellulose (p. 567)	germination (p. 564)
cone (p. 561)	meristem (p. 567)
cuticle (p. 576)	monocot (p. 563)
dicot (p. 563)	turgor (p. 576)
endosperm (p. 564)	vascular plant (p. 559)

REVIEW QUESTIONS

1. How did the early concept of a chain of being influence early ideas about plants?

2. What is the function of vascular tissue?

3. What is the primary difference between gymnosperm and angiosperm seed production?

4. What is the significance of seed dispersal? How is seed dispersal influenced by seed shape and fruit type?

5. What are some of the differences between monocots and dicots?

6. How are seeds formed?

7. What is the function of dormancy before seed germination?

8. Describe the process of seed germination.

9. How are new plant cells produced?

10. How are cell walls formed in new cells?

11. What are the functions of phloem, xylem, and parenchyma?

12. How do plants acquire nitrogen and carbon dioxide?

13. Which resources do plants acquire from the soil? How do they obtain these substances?

ESSAY AND DISCUSSION QUESTIONS

1. Angiosperm and gymnosperm plants produce seeds during sexual reproduction. What advantages and disadvantages do seeds have as components of a plant's life cycle?

2. Does it seem likely that plants are more susceptible to environmental changes than animals? Why? What types of environmental changes might be especially harmful to plants or plant populations?

3. One plant reproductive strategy is to make lots of small seeds, and another strategy is to make a limited number of large seeds. What are the advantages and disadvantages of each of these strategies?

REFERENCES AND RECOMMENDED READING

Arteca,N. 1995. *Plant Growth Substances: Principles and Applications.* New York: Chapman & Hall.

Attenborough, D. 1995. *The Private Lives of Plants: A Natural History of Plant Behavior.* Princeton: Princeton University Press.

Clarkson, D. T. 1985. Factors affecting mineral nutrient acquisition by plants. *Annual Review of Plant Physiology and Plant Molecular Biology,* 36: 77–115.

Farago, M. E. (ed.). 1994. *Plants and the Chemical Elements: Biochemistry, Uptake, Tolerance, and Toxicity.* New York: VCH.

Francis, C. A., C. B. Flora, and L. D. King (eds.). 1990. *Sustainable Agriculture in Temperate Zones.* New York: Wiley.

Frey, K. J. (ed.). 1995. *Historical Perspectives in Plant Science.* Ames: Iowa State Univ. Press.

Galston, A., P. Davies, and R. Satter. 1980. *The Life of a Green Plant.* Englewood Cliffs, N.J.: Prentice Hall.

Gilroy, S., and T. Trewavas. 1994. A decade of plant signals. *BioEssays,* 16: 677—682.

Grace, J. B., and D. Tilman (eds.). 1990. *Perspectives on Plant Competition.* San Diego, Calif.: Academic Press.

Harlan, J. R. 1995. *The Living Fields: Our Agricultural Heritage.* New York: Cambridge University Press.

Klein, R. M. 1987. *The Green World: An Introduction to Plants and People.* New York: HarperCollins.

Raven, P. H., R. F. Evert, and S. E. Eichhorn. 1992. *Biology of Plants.* 5th ed. New York: Worth.

Villiers, T. 1975. *Dormancy and the Survival of Plants.* London: Arnold.

Young, J. A. 1991. Tumbleweed. *Scientific American,* 264: 82–87.

ANSWERS TO FIGURE QUESTIONS

Figure 30.3 The more seeds produced, the greater the chance that one or more of them will be dispersed to an environment in which they can germinate and grown into a mature plant.

Figure 30.5 The fruit may attract birds and other animals that will carry it away from the parent peach tree. After eating the fruit, the seed may be dispersed to a favorable environment where it can grow.

Figure 30.6 Dandelion seeds because they are smaller and lighter and can be transported great distances by the wind..

Figure 30.8 Each pollen grain contains two sperm—one fertilizes the egg, and the other joins with two female nuclei to form the endosperm.

Figure 30.9 Water is often the limiting factor in terrestrial environments. Seeds allow plants to survive long periods of drought and to germinate when water becomes available.

Figure 30.10 Increased cell division and cell elongation.

Figure 30.13 Below the apical meristem.

Figure 30.14 At the tip of each stem.

Figure 30.15 Plants release CO_2 during cellular respiration.

Figure 30.18 As a plant wilts, water molecules are lost from the cells, which reduces the turgor pressure.

31

Plant–Environment Relations

Chapter Outline

Reading Questions

1. How do plants acquire and use carbon? Water?

2. How do plants use sugars produced through photosynthesis?

3. What are optimal environmental conditions for plant growth and survival?

4. What types of environmental stress affect plants? How do they cope with different forms of environmental stress?

In Chapter 30, we examined the first stages in a seed-producing plant's life history. Seeds that fall into a suitable environment may develop into seedlings (see Figure 31.1). For a seedling to mature into an adult plant, it has to obtain substances and energy from the surrounding environment. An adult plant, of course, must be able to do this continuously throughout its life. In this chapter, we look deeper into the following questions: What types of mechanisms allow plants to obtain and use required environmental resources? What problems do plants face in acquiring these resources? And what types of responses enable them to overcome these problems?

As shown in Figure 31.2, plant growth can be described as a multistep process in which the following events occur: (1) **acquisition,** whereby the plant obtains needed resources from the environment; (2) **assimilation,** or modification of these resources within the plant; and (3) **allocation,** or apportionment, of the products of assimilation to specific parts of the plant according to need. For example, carbon dioxide is an extremely valuable environmental resource for plants. They must acquire CO_2 from the atmosphere, assimilate it through photosynthesis, and allocate the resulting sugars to leaves, stems, or roots for cell maintenance and growth in these tissues.

Figure 31.1 Plants produce many seeds, but only a few seeds ever develop into mature plants.

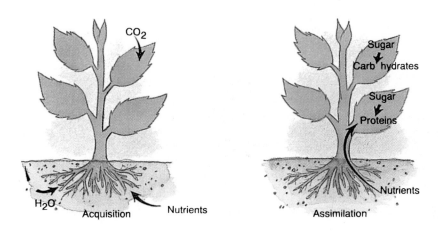
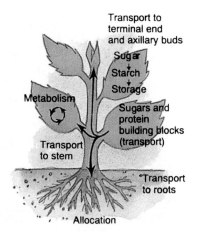

Figure 31.2 Plant growth requires nutrients, water, CO_2, and light energy from the environment. These resources are assimilated into carbohydrates, proteins, and other organic molecules and are then allocated for various uses. The products of assimilation can be used immediately for metabolism, such as respiration or protein synthesis, or they can be stored. Under some conditions, assimilation products are mobilized and transported for use in other parts of the plant.

Plants acquire nutrients and water directly from soils through their roots (see Figure 31.3, page 582). These nutrients are then transported to shoots, which ultimately allocate them to other plant tissues. Thus plants continuously obtain resources from the air and the soil and use these substances in ways that reflect the needs of the entire organism. A plant's use of environmental resources changes as a function of age, developmental state, and the availability of such resources.

The need to comprehend the nature of plant–environment relations is important because human activities and products are altering the physical and chemical climate of our planet (see Chapter 13). Climatic changes, regardless of cause, may directly affect the capacity of plants to acquire, assimilate, or allocate resources that they obtain from the environment and

thus may eventually limit plant growth and survival. Understanding resource acquisition and use in plants can help us predict how plants will respond to environmental changes. Such predictions are essential for making wise management decisions about forests, agricultural systems, and the quality of our air, soil, and water.

BEFORE YOU GO ON The establishment, growth, and survival of a plant are dependent on its success in acquiring necessary resources from the immediate environment. Carbon, nutrients, and water are examples of environmental resources that plants require. Plants use diverse mechanisms for obtaining these resources, assimilating and transporting them to different parts of the plant, and ultimately using them for growth and other activities.

Figure 31.3 Plant growth depends on acquisition of CO_2 from the air, its assimilation into carbohydrates through photosynthesis, and allocation of carbohydrates throughout the plant where these compounds are metabolized. At the same time, plants acquire nitrogen from the soil, assimilate it into amino acids, and allocate these compounds throughout the plant where they are used for protein synthesis.

Question: *What are the sources of the materials that enter the plant?*

THE CARBON STORY

Carbon dioxide constitutes a small fraction of Earth's atmosphere—about 0.035 percent by volume. However, this concentration is not constant; it changes seasonally in all areas of the world. In the Northern Hemisphere, CO_2 concentrations are highest in winter and lowest in summer. Also, atmospheric CO_2 concentrations are now increasing slightly each year, a factor that may be related to global warming (discussed in Chapter 13).

For convenience, most scientists express the amount of CO_2 (and other chemicals) in the air as *parts per million (ppm)* rather than percent volume. Thus air with 0.035 percent CO_2 is equivalent to 350 ppm; in other words, of every 1 million liters of Earth's atmosphere, 350 liters are CO_2. During an annual cycle, atmospheric CO_2 concentrations currently range between 350 and 355 ppm.

Several important processes cause seasonal variations in atmospheric CO_2 concentrations. Some seasonal variation is caused by high CO_2 emissions that occur during the winter as people use more heating fuels. Another factor causing seasonal variation is related to organisms that continue respiration in the face of declines in photosynthesis during the winter. That is, CO_2-producing cellular respiration continues during the winter, but CO_2-removing photosynthesis decreases significantly. In contrast, spring and summer peaks of photosynthesis tend to decrease atmospheric CO_2 concentrations during the growing season. Furthermore, annual mean atmospheric CO_2 concentrations increased from about 315 ppm in 1976 to about 350 ppm in 1990. This increase, and projected increases for the future, strongly influence the availability of CO_2 and the rates at which it is acquired by plants. How are plants able to control the amount of CO_2 they take into their tissues? We discuss this question next.

Acquisition and Regulation

All gases, including CO_2, are exchanged between the leaf and air through **stomata,** pores in the leaf surface that are surrounded by two *guard cells* (see Figure 31.4). Because stomata can open and close—a mechanism controlled by the guard cells—they help regulate the movement of gases into or out of leaves. The rate of CO_2 absorption is related to CO_2 concentrations in the air, leaves, and stomata.

Gas movement between the leaf and the air occurs as a result of *diffusion,* the movement of molecules, such as CO_2, from regions of high concentration to regions of low concentration. When photosynthesis removes CO_2 from the air inside a leaf, the CO_2 concentration inside the leaf is usually lower than the CO_2 concentration in the air outside the leaf. During daylight hours, when photosynthesis occurs, stomata are open, and CO_2 usually diffuses from the air into the leaf.

The magnitude of the difference in CO_2 concentration between the air and the inside of the leaf—called the *concentration gradient*—affects the rate of CO_2 acquisition by the leaf. To illustrate this effect, assume that stomata are always fully open in the following examples. If the atmospheric CO_2 concentration is 340 ppm and the concentration of CO_2 inside the leaf is 335 ppm, the concentration gradient is low, and the rate of CO_2 exchange between the leaf and the air will be slow. However, if the concentration of CO_2 in the air is 340 ppm and the concentration inside the leaf is 280 ppm, the concentration gradient is high, and the exchange rate will be much faster.

Environmental factors can influence processes that affect both the CO_2 concentrations inside the leaf and the rate of CO_2 diffusion into the leaf. Light, for example, is an environmental factor that can have profound effects on CO_2 concentrations in leaves. When the sun sets and photosynthesis stops, CO_2 concentrations in leaves increase because of cellular respiration and approach those of the air whether the stomata are opened or closed.

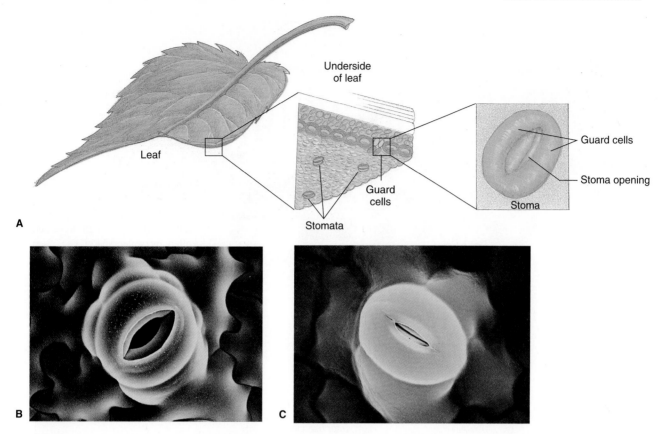

Figure 31.4 Stomata—pores in the leaf surface that help regulate the movement of gases into or out of the leaf— are surrounded by two guard cells lying next to each other on the leaf surface (A). The leaf surface is covered with a waxy cuticle that is impervious to water. Physiological processes regulated by the plant affect the guard cells and cause the stomata to open (B) or close (C). Plants can transpire only when stomata are open. Stomata open when the turgor pressure of the guard cells increases. When turgor pressure rises, the guard cells increase in length and separate.

Question: *If the stomata were closed, would the CO_2 concentration within the leaf increase or decrease?*

Water Use Efficiency

Why would a plant's stomata ever close? Because stomatal closure reduces the rate of CO_2 acquisition and therefore the quantity of photosynthetic products available to the cell, it would not seem to be beneficial to the plant. Discovering the answer to this question is essential for understanding how plants survive when environmental conditions change.

Transpiration

Water is lost from leaves through **transpiration,** the diffusion of water vapor away from the leaves and into the atmosphere. Water movement from roots to stem, branches, and leaves, and its loss through transpiration, is a crucial process for plant growth. For example, water movement helps distribute nutrients to all plant tissues.

The principles that explain CO_2 regulation by leaves also apply to transpiration. Air inside the leaf is fully saturated with water vapor, whereas water vapor concentrations in the atmosphere are usually far below saturation. Figure 31.5 (see page 584) describes stomatal regulation of CO_2 and water vapor. The rate of transpiration depends both on the magnitude of the water vapor concentration gradient between the leaf and the air and on whether the stomata are open or closed. If transpiration rates exceed water absorption rates, the plant may wilt and die.

Controlling Water Loss

Photosynthesis and transpiration rates are both affected by changes in stomata. What is the relationship between CO_2 absorption and water loss? One key factor is that water supplies are often insufficient for plants. Also, the rate of water loss is highest when stomata are open and CO_2 absorption rates are high. Thus in terms of a cost–benefit analysis, the price that plants pay for absorbing CO_2 to use in photosynthesis is the loss of water, which is an extremely valuable and often scarce resource.

It might seem that the most efficient way for plants to manage water intake and loss would be to open their stomata completely when photosynthesis can proceed most rapidly (see Figure 31.6, page 584). Ideally, such periods occur when light is high and the raw materials and enzymes used in photosynthesis are available (suitable levels of CO_2 are almost always present).

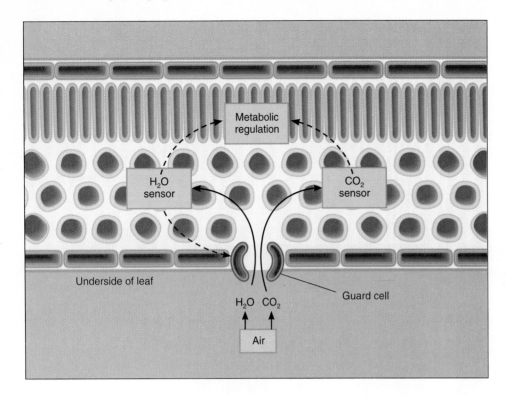

Figure 31.5 Stomata open or close in response to feedback systems that operate within the leaf. These feedback systems are thought to be controlled by water and/or CO_2 "sensors," about which little is known. When water loss rates exceed supply rates, stomata close. When environmental conditions favor high rates of photosynthesis, stomata may open.

Question: *What is the major cost for a plant attempting to maximize photosynthesis on a hot summer day?*

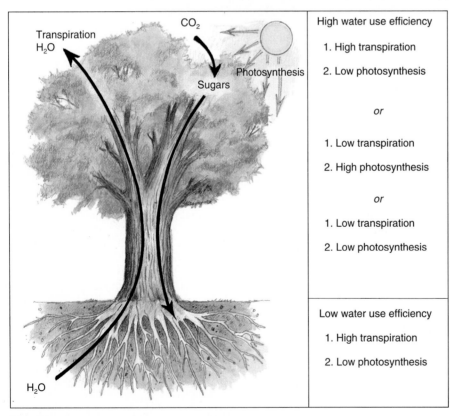

Figure 31.6 Water use efficiency—the ratio of CO_2 gain to water loss—can change with environmental conditions. High water use efficiency occurs in several circumstances. Water use efficiency is low when plants have high transpiration and low photosynthesis.

Question: *Which plants should have higher water use efficiency, those living in a desert or those living in a tropical rain forest?*

Under these conditions, the amount of carbon gained per amount of water lost is high, and plants are said to have a high **water use efficiency.** Some desert plants must operate with a high water use efficiency if they are to survive. Conditions can also exist in which plants have low water use efficiency. If water is abundant, a plant's stomata may remain open during the night, and transpiration rates will be high but no photosynthesis will occur. Water use efficiency is low because carbon gain is low and water loss is high. Plants growing on irrigated fields tend to slip into a regime of low water use efficiency.

Plants respond to changes in their environment by opening and closing their stomata, a mechanism that optimizes their water use efficiency. Accordingly, as light levels, atmospheric humidity, soil water availability, and internal activities change, plants respond by either opening or closing their stomata. This process maximizes the ratio between carbon gain and water loss.

> **BEFORE YOU GO ON** The carbon source for plants is atmospheric CO_2 that is converted to sugars during photosynthesis. CO_2 enters a plant through leaf stomata, structures that have a regulatory role in the movement of gases into and out of the plant. Quantities of CO_2 that enter a plant are influenced by several factors, including CO_2 concentration gradients, light availability, and transpiration rates. Stomata play a central role in regulating carbon gain and water loss.

Assimilation and Allocation

The assimilation of CO_2 by plants hinges on incorporating carbon atoms into sugars through the reactions of photosynthesis. As CO_2, water, and solar energy are converted to carbohydrates, plants obtain the fuel needed for their growth and development. Sugars not used in cellular respiration or growth are stored or translocated to other plant tissues.

Photosynthesis

Photosynthetic cells may produce more sugars than they can use. When conditions are optimal for photosynthesis, leaves can produce six to ten times more sugar than they use during cellular respiration. Scientists use estimates of **net photosynthesis**—the difference between the quantity of sugars produced by the leaves in photosynthesis and the quantity used in cellular respiration—as a measure of the amount available for other uses in the plant.

Cellular Respiration

Cellular respiration (described in Chapter 4) is the cellular metabolic process that releases the energy required for life-sustaining functions in the cells of all living organisms. Respiration has two aspects. **Maintenance respiration** is the sugar consumption required for simply maintaining a living cell. Maintenance functions include repair of membranes and organelles and the production of new enzymes and other essential macromolecules. **Growth respiration** is the sugar consumption needed for cell division and cell enlargement. In addition, growth requires the construction of new membranes, cell walls, and organelles within cells. All sugars that are immediately available for respiration can be used either for maintenance or for growth.

Storage

Carbohydrates not used for growth or maintenance respiration can be either allocated to storage or transported for use elsewhere in the plant. Simple sugars produced by photosynthesis can be converted to photosynthates comprising more complex sugars, cellulose, starch, and various forms of fats, all of which can be stored. Photosynthates that are allocated to storage are generally used when the respiratory demands of cells are greater than the supply provided by photosynthesis. Over the short term, some of the starch formed during the day will be used for respiration at night, when photosynthesis cannot occur. Storage of photosynthates is also important to plants for coping with long-term environmental fluctuations. For example, cloudy or foggy conditions may persist in coastal areas for days or even weeks. During that time, light levels may be too low for positive net photosynthesis, but the plant can draw on its stored reserves. Other environmental factors (for example, drought) can result in stomatal closure. This response may also suppress photosynthesis for prolonged periods and create conditions when plants must rely on stored carbohydrates. In both of these cases, plants can use stored photosynthates during periods when they cannot assimilate CO_2.

Translocation

Translocation occurs when photosynthates not immediately respired or stored are moved from one part of a plant to another (see Figure 31.7, page 586). Thus under ideal environmental conditions, a fully expanded leaf can produce excess sugars that can be exported to other tissues. Tissues that import and consume these sugars include immature leaves that are not yet capable of positive net photosynthesis and nonphotosynthetic tissues in flowers, fruits, seeds, branches, stems, and roots.

Sugars are allocated to various tissues in response to changes in the age of the plant, its developmental state, and environmental conditions. For instance, many plants produce only vegetative tissues such as roots, stems, and leaves for much of their lives. At some stage, however, a plant typically enters into a reproductive phase. At that point, most of its sugars will be allocated for use in the production of cones, flowers, pollen, fruit, or seeds.

> **BEFORE YOU GO ON** Plants use photosynthates in various ways. Some are used at once in cellular respiration reactions in leaves to provide energy for maintenance and growth. Sugars not used directly in leaf cell respiration can be handled in different ways. They may be converted to complex molecules that serve to store the remaining energy for use in the future, or they may be allocated to other parts of the plant, where they are used in respiration or growth.

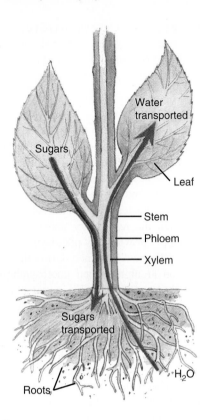

Figure 31.7   Sugars are transported in *phloem* from leaves to other tissues that are nonphotosynthetic, such as roots, or are not capable of producing all of the sugars they need for growth and maintenance, such as young leaves. Water taken up by the roots is transported to different parts of the plant through the *xylem*.

Question: *What are some possible fates of a water molecule absorbed by a root?*

OTHER NUTRIENTS

Essential elements needed by plants are commonly classified according to the amount required. *Macronutrients* are elements needed in relatively large amounts; they include carbon, hydrogen, oxygen, nitrogen, phosphorus, potassium, sulfur, calcium, and magnesium. Elements required in trace amounts are called *micronutrients;* they include boron, chlorine, cobalt, copper, iron, manganese, molybdenum, nickel, silicon, sodium, and zinc.

Acquisition of Nitrogen and Sulfur

Plants acquire the nitrogen and sulfur they require for growth directly from the environment. These elements are usually dissolved in the water between soil particles and taken in by root plant cells (see Figure 31.8). The absorptive surface of a root consists of epidermal tissue in which some cells have long extensions known as *root hairs.* Root hairs increase the surface area through which water and important minerals can be absorbed. Many plants also form symbiotic structures known

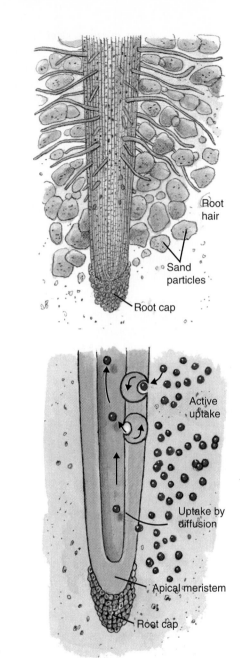

Figure 31.8   (A) Plant root tips penetrate into the soil. The root cap protects the apical meristem from being damaged by soil particles. Root hairs increase the surface area of roots and facilitate nutrient transport. (B) Uptake of nutrients by roots occurs by diffusion and by active uptake, a process in which the cell uses energy to acquire a needed molecule.

as *mycorrhizae,* which are composed of root tissues, root hairs, and fungi (see Figure 31.9). Mycorrhizae increase water and mineral absorption in plants. Just as carbon is absorbed by plants in the chemical form CO_2, nitrogen is generally absorbed by plants as nitrate (NO_3^-) or ammonium (NH_4^+), and sulfur is absorbed as sulfate (SO_4^{-2}).

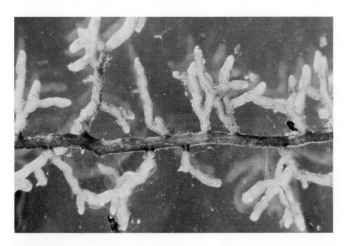

Figure 31.9 A red pine root covered with mature mycorrhizae with fully developed fungi.

Assimilation of Nitrogen and Sulfur

The patterns of nitrogen and sulfur processing reveal two common themes. First, plants acquire these elements from the environment in specific chemical forms. Second, these chemicals are modified during assimilation in leaves and other tissues, producing organic compounds of great importance to the plant. Nitrogen and sulphur compounds may become structural components, or they may be used in various cellular processes, including photosynthesis and respiration.

Following their absorption by roots, nutrients are transported through xylem to tissues that convert them to organic molecules. The processes of assimilating nitrate, ammonium, sulfate, and CO_2 are unique to plants.

ENVIRONMENTAL STRESSES

What kinds of environments are ideal for plants? Like any organism, each plant species has an optimal range of environmental conditions that provide for growth and reproduction. Given the resource requirements and mechanisms described earlier, what environmental factors would exist in a perfect plant world? Optimal conditions generally include an abundance of CO_2, water, and nutrients; adequate levels of light; high relative humidity; and temperatures that are neither too warm nor too cold. In such ideal conditions, plus a habitat free from insect herbivores, diseases, and competing plants, we might expect maximal plant growth.

Optimal conditions are closely approximated, but never achieved completely, in agricultural systems, gardens, greenhouses, and other places where environmental factors can be managed. Practices that tend to optimize environmental conditions include irrigation, application of fertilizer, planting after the last freeze, and the use of selective insecticides and herbicides to reduce predation and limit competition. Even using such expensive methods, however, may fail to provide conditions that are perfect for plant growth. Furthermore, human attempts to improve crop yields by supplementing water and nutrient resources and managing weeds, disease, and pests often result in more energy being invested than is recovered in increased plant yields.

> **BEFORE YOU GO ON** In a perfect environment, plants would be readily able to obtain and assimilate CO_2 and nutrients such as nitrogen and sulfur. There would also be an abundance of water, adequate levels of light and relative humidity, ideal temperatures, and a scarcity of herbivores, diseases, and competing plants.

The Concept of Environmental Stress

The idea that environmental conditions are never truly perfect for plant growth suggests that plants are continually subjected to stress, as depicted in Figure 31.10 (see page 588). It is helpful to characterize these stresses so that positive and negative environmental factors that affect plants can be identified and evaluated.

Natural Environmental Stress

An **environmental stress** is any component of the physical environment that reduces the capacity of plants to acquire, assimilate, or allocate resources needed for maintenance or growth. An example of a natural environmental stress is drought. An inadequate supply of water can limit plant maintenance, growth, and development.

Nutrient deficiencies can also be important environmental stresses. For example, inadequate levels of nitrogen in soils may lead to reduced levels of amino acids, structural proteins, and enzymes needed for all essential cellular processes. Researchers have found that photosynthetic capacity is often correlated with the nitrogen content of leaves. This finding indicates that when nitrogen levels in leaves are high, photosynthesis is not limited by a lack of enzymes or other molecules constructed from amino acids.

Light becomes a natural environmental stress factor if intensities are too low for sustaining high rates of photosynthesis. Light levels may be reduced by clouds; inanimate objects such as buildings, rocks, and mountains; or neighboring plants. Shading can also result from normal plant growth; as the plant canopy expands, self-shading occurs, and photosynthesis by leaves inside the canopy becomes limited due to insufficient light.

Light intensity levels are often correlated with temperature extremes, which also become a natural stressor to plants. For example, during periods of intense light in deserts, associated high temperatures cause reductions in rates of photosynthesis and increases in rates of respiration. Thus when plants are exposed to high temperatures, their rate of sugar production

Figure 31.10   Environmental stresses always limit plant growth. Stresses in deserts (A) often include drought and excessive heat. Growth in citrus orchards (B) may be affected by freezing temperatures. Stresses even exist in agricultural fields (C), where temperature, light, water, and nutrients are seldom at optimal levels.

decreases while their rate of sugar consumption increases. Low temperatures can damage plants directly. If water freezes, ice crystals formed within cells may puncture membranes and cause cell death. Low temperatures may also slow down critical biological processes such as water absorption by roots or respiration throughout the plant.

Anthropogenic Environmental Stress

How have human activities affected environments in which plants live? The industries, products, and transportation devices used by human cultures have changed the chemistry of the environment, and such changes can limit the capacity of plants to survive and grow (see Figure 31.11). Human-caused environmental stresses are called *anthropogenic*. Toxic, gaseous air pollutants such as ozone (O_3), sulfur dioxide (SO_2), and nitrous oxides (NO_x) can be absorbed through stomata and moved into leaf cells, where they cause damage or death. In addition, atmospheric acid depositions, in both dry form (dust and aerosol particles) and wet form (precipitation such as rain and snow), constitute environmental stresses. These forms of air pollution can acidify soils in some areas, create an unfavorable balance of nutrients, or add to the levels of sulfur and nitrogen already present in soils. Thus acid deposition has the potential to increase or decrease the availability of nutrients in soils and their absorption. Currently, scientists are debating the extent to which acid deposition may be affecting forest productivity in the industrialized countries (discussed in Chapter 13).

Human activities have also resulted in increased emissions of "greenhouse" gases, including CO_2 and methane, which can have direct effects on plants. Increasing atmospheric CO_2 concentrations may cause metabolic alterations that could lead to accelerated growth in some species of trees, other plants, and crops. Even if such an effect is beneficial, CO_2-caused increases in growth could ultimately be damaging if plants outgrow their water supply.

Environmental stresses are interrelated and rarely occur alone. For example, light levels may become too high or drought could result in overheating, resulting in wilting and decreased photosynthesis and growth. Interrelated multiple stresses can result in a decreased capacity for plants to acquire, assimilate, and allocate resources needed for growth.

Scientists are only beginning to determine the extent to which humans have modified Earth's environment. Researchers from many disciplines, including chemistry, physics, atmospheric science, oceanography, and biology, are turning their attention to answering critical questions about anthropogenic environmental effects.

BEFORE YOU GO ON   Plants are often subjected to environmental stresses that compromise their abilities to grow, reproduce, and survive. Deficiencies of nitrogen and other nutrients are associated with reductions in photosynthesis and growth. Extremes in light levels and abnormal temperatures also constitute environmental stresses that lead to decreased photosynthesis and other harm to plants. Anthropogenic stresses may now be limiting plant growth and survival.

Stress Compensation

Given the range of stresses to which plants are subjected, how do they survive? Plants can respond to short-term environmental stresses quickly through adjustments at the cellular level. Such responses form the foundation of a dynamic relationship between plants and their environment. Plants that compensate in response to stress have both increased survival and increased growth, compared with the survival and growth of plants that do not have such compensation mechanisms.

Stomatal Responses   Stomatal responses are among the most rapid short-term reactions to an environmental stress. When soil water is limiting, stomata close, and hence transpirational water

Figure 31.11   Air pollutants emitted from industrial sources can modify the chemical climate and thereby cause stress on plants. Gaseous pollutants can directly inhibit physiological processes that occur in leaves, or the atmospheric deposition of chemicals can alter the nutrient supply of plants.

Question: *Do anthropogenic environmental stresses affect plants locally, regionally, or globally?*

loss is reduced in affected plants. Stomata may also close if transpiration increases in response to decreased relative humidity. Plants can compensate for either insufficient water supply rates or excessive transpirational water loss rates by simply closing their stomata. Stomatal responses are considered to be compensatory because even if growth is inhibited by environmentally caused changes in plant water status, stomatal closure will help the plant survive.

Root and Shoot Responses   Compensation for environmental stresses may also be long-term and lead to permanent changes in the patterns of plant growth and development. Perhaps the best-studied example of this sort of response involves

changes in the ratio of root dry weight to shoot dry weight, or the **root–shoot ratio** illustrated in Figure 31.12 (see page 590).

The amount of carbon allocated to roots and shoots determines the extent of growth and the biomass of these tissues. Plants that allocate carbon preferentially to shoots will use photosynthates to construct more shoots, which can gain more carbon. Such a strategy results in the plant having a high growth rate and a low root–shoot ratio. In contrast, plants that allocate photosynthates preferentially to nonphotosynthetic tissues such as stems or roots have invested carbohydrates into tissues that have important roles but cannot be used to gain more carbon. The root–shoot ratio in such plants is high, and growth

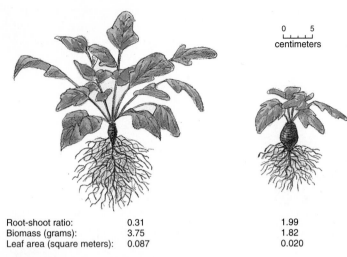

0   5
centimeters

Root-shoot ratio:	0.31	1.99
Biomass (grams):	3.75	1.82
Leaf area (square meters):	0.087	0.020

Figure 31.12 Plants with similar photosynthetic rates can have different growth rates. These two radishes are both five weeks old but show different types of growth due to different patterns of carbon use. The wild type of radish *(left)* used carbohydrates from photosynthesis to develop more foliage, which leads to more growth. The commercial, edible radish *(right)* stored carbohydrates in the root, which resulted in less growth.

rates will be low. Why don't plants always employ the shoot-favoring strategy, since it is associated with higher growth rates?

If plants were to function solely to maximize their growth rates, they would allocate all of their resources to shoots and nothing to roots. However, the success or survival of plants depends on more than a rapid growth rate. Besides providing attachment, structural support, and the capacity to absorb water and resources from soils, roots and belowground plant tissues carry out many other vital plant functions. For example, some plants store carbohydrate reserves in belowground stems called *tubers*. Potatoes are tubers that store carbohydrates until they are needed. Sweet potatoes, beets, carrots, and radishes are plants with roots that perform a similar function.

Plant Responses to Anthropogenic Environmental Stresses

A useful way to categorize plant responses to anthropogenic stresses is to distinguish between aboveground stresses and belowground stresses. One group of aboveground stresses that acts to limit carbon gain can be broadly classified as components of the atmospheric environment (see Figure 31.13). Such stress factors include gaseous air pollutants. Plants faced with these stresses will generally have less carbon for allocation to non-photosynthetic tissues. In this situation, plants can compensate for such stresses by shifting their carbon allocation patterns to favor shoots at the expense of roots, thereby decreasing their root–shoot ratio. Basically, plants compensate for atmospheric stresses that limit carbon acquisition rates by increasing the proportion of their biomass allocated to acquire carbon.

Belowground anthropogenic environmental stresses, such as the presence of toxic substances, can also limit carbon gains by plants. In these cases, plants shift carbon allocation patterns by favoring belowground tissues at the expense of shoots. The shift in carbon allocation pattern in response to belowground stresses results in an increase in the root–shoot ratio, an opposite response to that elicited by atmospheric stresses.

These stress mechanisms are also used in responding to natural environmental stresses. However, it is largely because of an increasing concern about the effects of pollution on plants that emphasis is now being placed on responses to anthropogenic stresses (see the Focus on Scientific Process, "Plant Physiological Ecology").

BEFORE YOU GO ON Plants are capable of both short-term and long-term responses to environmental stress. Stomatal reactions act to reduce water loss in plants when water becomes limiting. Carbon allocations to roots and shoots change in response to environmental stress. The net effect of this mechanism is to optimize the chances for long-term survival by increasing growth of either the shoot or the root, depending on the nature of the stress.

Initial allocation state

Leaf biomass   Photosynthetic rate   Root biomass

Carbon

Reduced carbon supply
Low light
SO_2, O_3

Reduced nitrogen
Reduced water supply
Low soil fertility

Environmental stress

New allocation state

Leaf biomass   Root biomass   Leaf biomass   Root biomass

Carbon   Carbon

Figure 31.13 Environmental stresses can reduce plant growth and influence carbon allocation to roots and shoots. Atmospheric stresses tend to shift allocations in favor of the shoot, whereas stresses in soils tend to cause shifts in favor of the roots.

Question: *Which would you expect to allocate a higher percentage of carbohydrates to its roots, a saguaro cactus or a tropical teak tree?*

FOCUS ON SCIENTIFIC PROCESS

Plant Physiological Ecology

The relationships between plants and their environment are of concern to scientists in the field of **plant physiological ecology. Plant physiology** is the study of processes, such as photosynthesis and cellular respiration, taking place within a single cell or an individual plant. **Plant ecology** is the study of the structure and functions of plant populations. Studies in plant physiological ecology are becoming increasingly important as more people realize that the global climate is changing. Consequently, plant scientists are challenged to describe, evaluate, and ultimately predict the effects that global warming will have on individual plants, plant communities, and ecosystems. To achieve this ambitious goal, extremely large-scale experiments will be conducted.

Generally, plant physiological ecologists attempt to define physiological mechanisms and describe how they are affected by various environmental stresses. Such research is important in defining a plant's range of environmental tolerances, in understanding its ecological role in a community, and in assessing the possible consequences of environmental change. The experiments described here provide some indication of the progress in this field.

Simple plant transplant experiments have provided useful information for identifying variations in stress response within a single species. These differences can be seen through studies of a single species that inhabits a wide range of elevations. In the early 1950s, scientists at Stanford University in California studied a plant species called *Potentilla glandulosa* that grew along an elevation gradient ranging from 100 meters in western California to greater than 3,000 meters in the Sierra Nevada. First, they collected seeds from plants growing at three different altitudes, designated "low elevation," "moderate elevation," and "high elevation." They

then grew these seeds in a common garden at a moderate elevation and compared the growth of seedlings and adult plants. The hypothesis being tested was that seeds of the same species would grow uniformly at the moderate elevation, regardless of their origin. Surprisingly, seeds collected from the high elevation produced small plants, even when grown at the moderate elevation. As shown in Figure 1, seeds from the low-elevation site also gave rise to plants smaller than those from the mod-

erate elevation. Therefore, this experiment demonstrated that significant genetic differences existed between populations of a single species that were growing in habitats characterized by different stresses. This was one of the first scientific studies that showed in a rigorous way that plants possess an inherited ability to respond to environmental conditions.

Studies conducted in the late 1950s began to show how environmental factors affected a plant's capacity to carry out

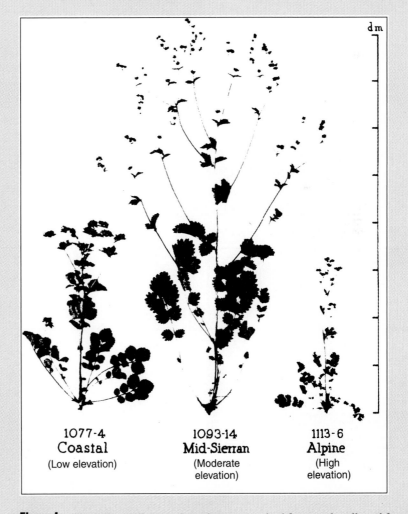

Figure 1 These *Potentilla granulosa* plants were raised from seeds collected from different habitats. The plants were raised in a single garden, and the results indicate that this species has distinct variations that represent unique adaptations to specific habitats.

box continues

FOCUS ON SCIENTIFIC PROCESS

photosynthesis and transpiration. Researchers wanted to know if (and when) plants growing above the timber-line, at a site in the Medicine Bow Mountains of Wyoming, altered their rates of photosynthesis during the day and during the growing season. A unique feature of these studies was that experiments were done at a remote field site, using plants growing in their native environment. To make the necessary measurements, a small chamber was built that could be opened and closed around a single leaf (see Figure 2). The chamber had a hole for the *petiole,* the part of the leaf that connects it to the plant, so that the leaf could remain intact and attached to the plant. The chamber was also designed to allow airflow through inlet and outlet holes. Measurements of photosynthesis were determined by measuring the differences in CO_2 concentrations between the airstreams entering and leaving the chamber. Transpiration was measured by comparing water vapor concentrations at the chamber inlet and outlet. The investigators found that leaves became photosynthetically active very early in the spring and that rates of photosynthesis increased during the day as the leaf responded to warmer temperatures and more light. These and similar experiments conducted by plant physiological ecologists established normal patterns of plants' activities and opened the door to new research that aimed to define further the effects of environmental stresses on plant physiology and growth.

Determining the extent of anthropogenic climate change and its effects on vegetation will be difficult. One of the obstacles to unraveling plant responses to human-induced environmental changes is that many natural factors change simultaneously and affect plants in the process. For example, as summer progresses toward fall, plants may shed their leaves in response to cooler temperatures, shorter days, or other factors associated with seasonal change. How do plant physiological ecologists identify which environmental factors—natural or anthropogenic—are responsible for observed or measured changes in plant physiology and growth?

Modern, highly sophisticated experimental systems not only monitor important environmental factors but also allow them to be controlled. Leaf chambers have been developed that can control temperature, CO_2, and water vapor concentrations in the air flowing into the system. Special lights have also been developed, along with shading techniques, in order to control light levels. Thus temperature around the leaf can now be held constant while light levels change, or temperature can be held constant while water vapor concentration in the air is increased or decreased. These systems can operate in the field, a greenhouse, or a laboratory and allow scientists to monitor leaf responses to systematic and controlled environmental changes. Much of the monitoring and regulation of environmental factors is done by computer systems. These new systems will be important tools for investigating the effects of changing climate on plants.

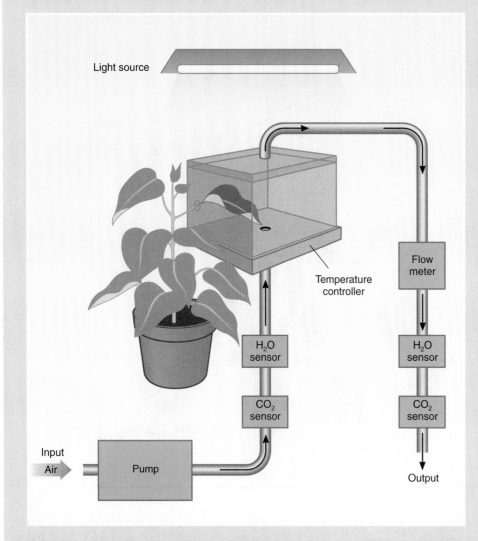

Figure 2   A chamber designed for measuring photosynthesis, respiration, and stomatal responses in single leaves has a supply of moving air and a lid that is opened to insert the leaf while it is still attached to the plant. Photosynthesis reduces the CO_2 concentration in the airstream. Transpiration increases the relative humidity of the air as it passes through the tube.

A NEW WAY OF LOOKING AT THE PLANT WORLD

As plants grow and develop over the course of a single season or over many years, the pattern of carbon allocation changes continuously. Shifts in carbon allocations can be understood from the perspective of a plant's compensation for stress, which accomplishes several purposes. Compensation results in enhanced survival and growth, at least superior to what would have occurred in the absence of compensation. Another facet of stress compensation seems to be for plants to shift structure and function so as to optimize acquisition of the most limiting resources.

Comprehending the dynamic nature of plant–environment relations provides important insights into why plants are so successful. This helps us recognize the changes in function and growth that occur in the plants we observe each day. From this perspective, the responses of plants to natural environmental changes, such as a sudden storm, a sunset, or a period of drought, take on new meaning or pose new questions. Why do certain flowers, such as tulips, close when the sun sets and open when it rises? If you live in Buffalo, New York, why do the grasses in lawns turn brown during the winter and once again become green when spring arrives? If you live in Seattle, Washington, why is grass green all winter? The consequences of anthropogenic pollution can also be analyzed in such a way that human activities may be managed more effectively.

Analyses of plant–environment relations also reveal that not all plants respond to environmental change in the same ways. In Chapter 32, we will consider the ideas that plants have many adaptations for responding to environmental change and that these adaptations are perpetuated by generations that follow.

SUMMARY

1. To grow and survive at the site where they take root, plants must be able to obtain resources from the air and the soil that surround them. Plants use a variety of processes to accomplish this task. Many of these same mechanisms are also used in adjusting to changing conditions and in coping with environmental stresses.

2. Carbon dioxide is a critical resource that is used by plants in producing sugars during photosynthesis. Plants acquire CO_2 from the atmosphere, and rates of intake are regulated primarily by the leaf stomata. Global climatic changes and increasing concentrations of CO_2 in the atmosphere may ultimately affect the ability of certain plant species to maintain and regulate rates of carbon fixation.

3. Water and nutrients obtained from the soil are transported continuously throughout the plant to tissues where they are required. Water reaching the leaf surface diffuses into the atmosphere through transpiration and is lost to the plant. The rate of transpiration and water loss is regulated by the stomata and is influenced by several environmental factors, including light, water vapor in the air, temperature, and CO_2 concentrations.

4. Under appropriate conditions, leaf cells produce excess sugars that can be transported to other plant tissues where they are metabolized or stored for future use. Stored carbohydrates represent a reserve that can be used by the plant if environmental conditions change or if the plant becomes stressed in the presence or absence of certain environmental factors.

5. Water and most nutrients required by plants, such as nitrogen and sulfur, are obtained by the roots from the soil. Once absorbed, they are assimilated and translocated to tissues where they are used for cell maintenance and growth.

6. Plants rarely inhabit an ideal environment. The degree to which they are able to compensate for environmental stresses determines whether or not they will be able to survive, grow, and reproduce. Inadequate amounts of water, insufficient levels of nutrients, extreme temperatures, inappropriate light intensities, and pollutants introduced by human activities are environmental stresses that plants must deal with.

7. Plants can make short-term adjustments to environmental stress. For example, if water becomes limiting, plants can decrease their rate of photosynthesis, or the stomata can respond to reduce water loss.

8. If subjected to continuing stress, plants can also make long-term adjustments. One long-term compensation mechanism involves regulating the flow of carbon and nutrients to different plant tissues, depending on the nature of the stress. If exposed to aboveground stresses, such as pollution, plants tend to invest resources in the development of stems and leaves at the expense of roots and other nonphotosynthetic tissues. Belowground stresses result in the opposite compensation pattern.

WORKING VOCABULARY

acquisition (p. 580)
allocation (p. 580)
assimilation (p. 580)
environmental stress (p. 587)

stomata (p. 582)
translocation (p. 585)
transpiration (p. 583)
water use efficiency (p. 585)

REVIEW QUESTIONS

1. What processes are related to plant growth?

2. How do atmospheric carbon levels fluctuate during the year? Why do these variations occur?

3. What mechanisms regulate the movement of CO_2 into and out of a plant's leaves?

4. What is transpiration? How is it regulated in plants?

5. Under what conditions do plants have a high water use efficiency?

6. What plant mechanisms operate to maximize water use efficiency?

7. What are the possible fates of photosynthates produced by a plant?

8. How do plants acquire and assimilate nitrogen and sulfur?

9. What are ideal plant conditions for plant growth and survival?

10. What is environmental stress? Under what circumstances may nutrient availability, light, temperature, and human activities cause environmental stress on plants?

11. What mechanisms do plants use to compensate for environmental stress?

ESSAY AND DISCUSSION QUESTIONS

1. In what ways do human activities affect resource availability for plants? How do your daily activities affect Earth's physical and chemical climate?

2. Tropical plants can be made to grow in a desert. How might this be done?

3. The effects on plants of an environmental stress, such as drought, are a function of an array of factors. Which plant structures and mechanisms help define their capacity to tolerate a particular stress?

REFERENCES AND RECOMMENDED READING

Bouwman, A. F. (ed.). 1990. *Soils and the Greenhouse Effect.* New York: Wiley.

Butlin, R. A. (ed.). 1995. *Ecological Relations in Historical Times: Human Impact and Adaptation.* Oxford: Blackwell.

Caldwell, M. M., and E. D. Schulze (eds.). 1994. *Ecophysiology of Photosynthesis.* Berlin: Springer-Verlag.

Farago, M. E. (ed.). 1994. *Plants and the Chemical Elements: Biochemistry, Uptake, Tolerance, and Toxicity.* New York: VCH.

Freedman, B. 1995. *Environmental Ecology: The Impacts of Pollution and Other Stresses on Ecosystem Structure and Function.* 2nd ed. San Diego, Calif.: Academic Press.

Hopkins, W. G. 1995. *Introduction to Plant Physiology.* New York: Wiley.

Jones, H. G., T. J. Flowers, and M. B. Jones. 1989. *Plants Under Stress.* New York: Cambridge University Press.

Katterman, F. 1990. *Environmental Injury to Plants.* San Diego, Calif.: Academic Press.

Kozlowski, T. T., P. J. Kramer, and S. G. Pallardy. 1991. *The Physiological Ecology of Woody Plants.* San Diego, Calif.: Academic Press.

Kuppers, M. 1989. Ecological significance of above-ground architectural patterns in woody plants: A question of cost–benefit relationships. *Trends in Ecology and Evolution,* 4: 375–379.

Larcher, W. 1995. *Physiological Plant Ecology: Ecophysiology and Stress Physiology of Function Groups.* 3d ed. New York: Springer-Verlag.

Mooney, H. A., B. G. Drake, R. J. Luxmoore, W. C. Dechel, and L. F. Pitelka. 1991. Predicting ecosystem responses to elevated CO_2 concentrations. *BioScience,* 41: 96–103.

Strain, B. R. 1987. Direct effects of increasing atmospheric CO_2 on plants and ecosystems. *Trends in Ecology and Evolution,* 2: 18–21.

Winner, W. E. 1994. Mechanistic analysis of plant responses to air pollution. *Ecological Applications,* 4: 651–662.

ANSWERS TO FIGURE QUESTIONS

Figure 31.3 CO_2 and O_2 come from the atmosphere; water and nutrients come from the soil (or the medium in which the plant is growing).

Figure 31.4 The CO_2 concentration would decrease.

Figure 31.5 The major cost would be water loss because the stomata must remain open to supply CO_2, and open stomata result in water loss through transpiration.

Figure 31.6 Generally, desert plants, because they must have high water use efficiencies to survive in areas with limited water resources. However, even though tropical plants are not water-limited, they have very high photosynthetic rates, which requires high water use efficiency.

Figure 31.7 A water molecule could be transpired directly to the atmosphere, or it could be split during photosynthesis, with the oxygen diffusing into the atmosphere and the hydrogen being fixed with carbon and oxygen to form sugar, which would be transported to the root.

Figure 31.11 Anthropogenic stresses generally affect plants locally, but concern about negative effects at the regional and global levels is increasing (see Chapter 13).

Figure 31.13 Saguaro cactus, because water is the limiting factor in most deserts.

32

Plant Adaptations and Reproduction

Chapter Outline

Reading Questions

1. What is the relationship between plant adaptations and long-term natural environmental changes?

2. What types of adaptations allow deciduous trees and conifers to survive in the environments they inhabit?

3. How do gymnosperms and angiosperms produce seeds by sexual reproduction?

Plants can respond to environmental factors in many ways and, in so doing, can increase their capacity to grow, survive, and reproduce. As described in Chapter 31, over the course of days or weeks, plants adjust their physiological and growth processes as they compensate for environmental changes or stresses (see Figure 32.1, page 596). In this chapter, we concentrate on a broader time scale. Over many generations, plants evolve adaptive characteristics as a result of natural selection. Such adaptations account for the basic shapes, structures, and physiological capacities that allow plants to survive normal environmental changes over extended time periods.

ADAPTATIONS

What is the nature of plant adaptations? Figure 32.2 shows several examples of plant adaptations to natural environmental stresses. The most familiar aspects of a plant's appearance—the size, color, and shapes of leaves; the plant's general shape, branch pattern, and reproductive structures—reflect adaptations related to survival in the environment it inhabits. Many of these structural features, along with some of their physiological capabilities, arose in response to environmental stresses in the plant's evolutionary past. Some of these stresses persist today. For example, local and regional climate changes have occurred

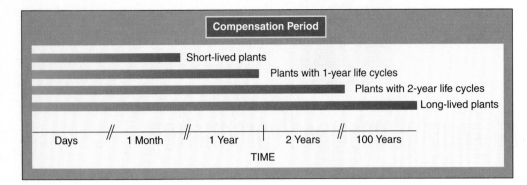

Figure 32.1 Plants use different mechanisms to respond to and compensate for natural environmental stresses. Short-term stress compensations take place during the life of an individual plant. Plants may also have adaptations that evolved over many successive generations in response to continuous environmental stresses.

slowly over centuries (see Chapter 7). How do plants respond to such changes? Generally, plant populations may evolve new adaptations for coping with a changing climate, or they may stop growing and eventually die at sites where conditions are no longer favorable for their survival. If they do not adapt, the plant population must disperse seeds to more favorable sites, or it may disappear from a geographical region. In extreme cases, species may even become extinct. For example, conifer species became widely established throughout North America after the last ice age ended. Now, after centuries of a general warming trend, the distribution of some conifer species is confined to the northern regions of the continent.

A recent study of limber pine *(Pinus flexilis)*, which grows in the Rocky Mountains at high elevations, provides a very interesting example of the relationship between long-term climate change and adaptation. Researchers measured the number of stomata per square millimeter (mm^2) of pine needle surface in pine needles that ranged in age from mere weeks to 30,000 years old. (Fossil pack rat nests were the source of the older needles.) Figure 32.3 shows that pine needle stomata, which were measured in this study, lie in recessed rows that are parallel to the long axis of the needle. Needles that grew during the last glacial period (30,000 to 15,000 years ago) had a stomatal density of 117 mm^2. During the transition period

Figure 32.2 Tropical vines (A), desert cacti (B), and other flowering plants (C) represent some of the diversity found in the plant world. The diverse shapes and forms reflect adaptations that have enabled these plants to cope with environmental pressures.

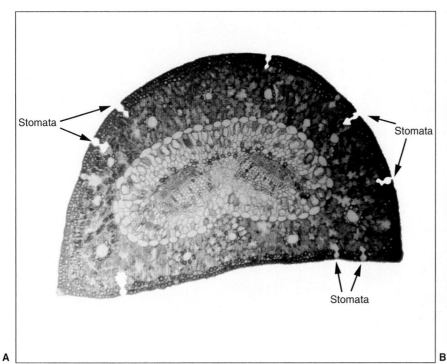

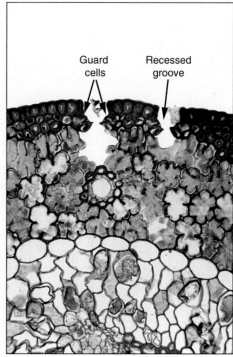

Figure 32.3 (A) This cross section of a pine needle shows numerous recessed stomata on the exterior surface. (B) At higher magnification, two stomata on the needle surface can be seen in their recessed groove. The guard cells in these two stomata are separated, indicating that they are open, as when gas exchange occurs.

when glaciers were withdrawing, 15,000 to 12,000 years ago, pine needle stomatal density was 111 mm². From the end of the glacial period to the present time, pine needle stomatal density was 97 mm². This 17 percent decrease in stomatal density correlates with a 30 percent increase in the atmospheric concentration of carbon dioxide (CO_2) over the same 30,000-year period. From these data, scientists concluded that the decrease in stomatal density was an adaptation to an increasing atmospheric concentration of CO_2. What would be the advantage to the pine species in having a smaller number of stomata in response to increasing CO_2 concentrations? Recall from Chapter 31 that plants, through their stomata, must regulate the intake of CO_2 used in photosynthesis and water lost through transpiration. Scientists hypothesize that as CO_2 levels increased over a 30,000-year period, pines with smaller numbers of stomata could acquire adequate CO_2 for photosynthesis and benefit by losing less water for having fewer stomata.

Two basic types of trees live in the temperate zones of the world, and we focus on them to illustrate the nature of plant adaptations more closely. **Deciduous trees** living in the Northern Hemisphere shed their leaves during the fall and grow new ones the following spring. Common deciduous species inhabiting temperate forests include oaks, elms, beeches, maples, ashes, and many species found in fruit orchards (for example, apple trees). In contrast, most **evergreen trees** retain most of their leaves year round, though some leaves are regularly shed. Pines, spruces, firs, cedars, and junipers are common evergreen species (see Figure 32.4, page 598).

Many deciduous and evergreen tree species are economically important. They provide timber products, such as lumber and paper; fruits; and aesthetic beauty, an asset that is difficult to price. In addition, stands of trees are ecologically important. They play central roles in water and nutrient cycling, and they influence the chemical composition of our air. The widespread distribution of deciduous and evergreen tree species indicates that they are well adapted to many environmental stresses.

BEFORE YOU GO ON Successful plants are well adapted to survive in the environments they inhabit. Adaptations are reflected in plant structures and functions. Many adaptations are related to normal environmental variations. Examining the adaptations of deciduous trees and evergreen trees provides a basis for understanding normal, long-term relationships between plants and their environment.

Deciduous Trees

Deciduous trees are widespread throughout the world and can grow up to 50 meters tall. Although scattered throughout North America, forests dominated by oak, hickory, maple, and beech are most common on the eastern part of the continent.

Leaves

With few exceptions, deciduous trees have *laminar* leaves, which are thin, flat, and wide. What is the adaptive significance of thin, flat leaves? The laminar shape favors the efficient

A

B

Figure 32.4 Deciduous trees, such as the maple (A), are usually broad-leaved and shed their leaves at the end of each growing season. Many familiar evergreen trees like the eastern white pine (B) have needles that are not shed annually.

absorption of solar energy. Specifically, photosynthetic components, including light-absorbing pigments and required biochemical compounds, are arranged in thin sheets and can intercept more light than those in leaves with other shapes. Thus the profile and structure of these leaves reflect adaptations that enhance photosynthesis.

Deciduous tree species have specific patterns of leaf arrangement, as illustrated in Figure 32.5. Most commonly, neighboring leaves are paired and positioned on opposite sides of the branch, or they are staggered and alternate on the branch.

Leaf arrangements represent an adaptation to minimize self-shading, which can result in decreased levels of photosynthesis. As branches grow, **buds** are produced that will eventually develop into new branches with their own leaves. As shown in Figure 32.6, older leaves that were once at the outside edge of the canopy, where they received abundant sunlight, become shaded by leaves that grow from new branches above them. As the growing season progresses, self-shading becomes more pronounced, although the pattern of leaf spacing on a branch tends to reduce this effect.

Branches

The location and arrangement of branches relative to other branches is an adaptation that also reduces self-shading. As Figure 32.7 indicates, this is particularly striking in *open-grown trees,* ones that grow at a distance from neighboring trees that might compete for light, which usually have extensive branching and expansive canopies. Such trees can sustain high rates of photosynthesis because of the great amount of leaf surface area available for capturing radiant energy and absorbing CO_2.

Trees that grow in dense forests usually lose their lower branches in response to low levels of light in a process called *self-pruning.* Once leaves on a branch are no longer able to produce sufficient photosynthates to support the branch or be transported to other tissues, resources are no longer allocated to that branch. Its leaves die, and the branch eventually prunes itself, drying up and falling from the tree.

A Opposite leaf orientation   **B** Alternate leaf orientation

Figure 32.5 Deciduous leaves are commonly arranged on branches in either an opposite (A) or an alternate (B) pattern of orientation.

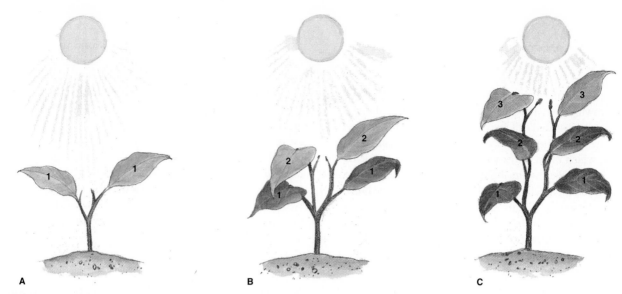

Figure 32.6   Self-shading occurs when leaves formed at the start (1) of the growing season (A) are overgrown by leaves formed later (2) during the season (B). As growth progresses and more new leaves are added (3), leaves that once formed the outside of the canopy, where light was high, will become part of the inside canopy, where light intensity is lower (C).

Question: *How might nutrient allocation change as leaves become shaded?*

Figure 32.7   Leaf distribution in the canopy of deciduous trees, such as these blue oaks, reflects a branching pattern that becomes apparent when the trees shed their foliage.

Question: *What is the adaptive advantage of this type of branching pattern?*

Nitrogen Allocation

Nitrogen is commonly used to build structural proteins and enzymes that function in photosynthesis and other forms of metabolism. Leaves and branches are well adapted to maximize the efficiency of nitrogen use in a tree. For example, the progression of self-shading during canopy growth leads to a situation in which leaves can no longer maintain high rates of photosynthesis because light levels are too low. As light levels decline, rates of enzyme synthesis decrease, and unused nitrogen accumulates. The excess nitrogen in these older, shaded leaves is then converted to a chemical form that can be translocated from an older "source" leaf, through the branch, to different plant structures such as younger, active leaves that are fully exposed to the sun. Once recycled to these new leaves, the nitrogen is again used to construct enzymes required for high levels of photosynthesis.

Loss of Leaves

What is the adaptive significance of deciduous trees shedding their leaves during the fall? Leaf loss during autumn represents an adaptation to natural environmental stresses associated with drought, short days, low temperatures, and other aspects of the progressing season. During this period, nitrogen and photosynthates in leaves are converted to soluble forms that are translocated back into the shoots and roots. These nutrients are then stored in the tree and can be reused the following spring, when the next growing season begins. Leaf loss is initiated by the development of a specialized cell layer at the site where the leaf **petiole** joins the branch (see Figure 32.8). Growth of this **abscission cell layer,** which often forms within a month in all tree leaves, first restricts and then severs the vascular tissues connecting leaf and branch and eventually causes the leaf to fall from the branch.

When leaves are finally shed from a tree's canopy, they contain very little nitrogen and sugar. At this point, leaves consist mostly of structural carbohydrates, such as cellulose, that make up cell walls (see Figure 32.9). Leaf shedding by deciduous trees in temperate regions is considered adaptive because nutrients and carbohydrates are not lost from foliage that might be destroyed by freezing or physical damage from winter ice and snowstorms.

> **BEFORE YOU GO ON**   Deciduous trees have various adaptations that enable them to survive throughout the world. Their leaf structure promotes maximum absorption of radiant energy and high rates of photosynthesis. Leaves and branches of deciduous trees are arranged in such a way that self-shading is minimized. If leaves or branches become nonproductive, they are shed by the tree. Deciduous trees routinely shed their leaves before winter to prevent the loss of valuable plant resources.

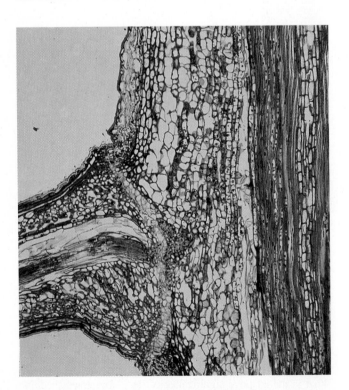

Figure 32.8   Deciduous leaves are shed after a layer of cells, the abscission layer, forms at the point where the petiole joins the branch.

Evergreen Trees

Evergreen trees live throughout the world and retain their leaves for more than one growing season. Evergreen tree species have either broad leaves or needles. Evergreens with broad leaves commonly grow in the tropics. Examples of broad-leaved evergreen trees that are found in the United States are live oak, rhododendron, and holly. A few broad-leaved evergreen species are also found in hot, dry regions.

Most evergreen tree species from temperate regions are classified as *conifers* (described in Chapter 6) and have narrow, elongated, pointed leaves called **needles** (see Figure 32.10). Needles are cylindrical, two-sided (flat), or polyhedral (having several flat surfaces). Coniferous evergreens are found throughout North America.

Needles

How do the functions and lives of needles compare with the leaves of deciduous trees? Even though most coniferous species do not shed their leaves annually, needles are not permanent parts of the tree. Some species retain their needles for only two or three years, while other species may conserve their needles for more than ten years. Needles shed from inside the conifer canopy are soon replaced by new needles produced at the outside of the canopy. Needle longevity is related to the needle's capacity to sustain positive net photosynthesis. As needles age and self-shading becomes more pronounced, the tree responds by dropping needles that are no longer self-sufficient. As with deciduous trees, nitrogen and photosynthates are reallocated to other active tissues before needles fall. As illustrated in Figure 32.11 (see page 602), branches of conifer trees also undergo self-pruning. Consequently, many conifers in dense, forested stands typically have long, straight trunks with needle-bearing branches only in the upper section of the tree, where light levels are high and photosynthesis occurs.

Summer Adaptations

Conifer needles are well adapted for conserving water during the dry summer season. Their stomata generally allow low rates of gas exchange, which results in lower water use. Also, needles are covered with a thick, prominent waxy cuticle that is impervious to water, another adaptation related to water conservation. Finally, the cylindrical shape of the needle is well suited for efficient light interception and for cooling during hot, dry summers due to air movement over the leaf surface.

Winter Adaptations

Needles are also well adapted to endure stresses imposed by winter weather and are even able to photosynthesize on warm days. Their cylindrical shape prevents the buildup of ice or wet snow that could cause branches on a broad leaf to break. This shape also offers little resistance to high winter winds. The needle's internal structure is reinforced with cellulose and other substances that make it resistant to physical damage from wind, snow, and ice.

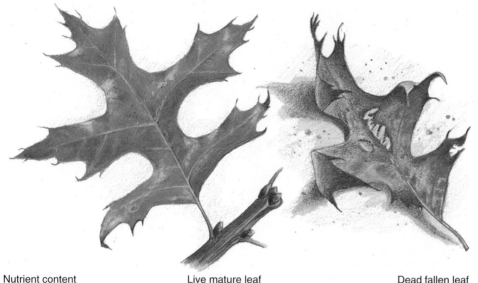

Nutrient content	Live mature leaf	Dead fallen leaf
Nitrogen	3%	0.1%
Sugar	4%	0.1%
Potassium	2%	0%
Magnesium	0.5%	0%

Figure 32.9 Mature leaves have higher concentrations of nutrients at the peak of the growing season compared to leaves that have been depleted of nutrients prior to being shed from the tree.

Question: *What is the adaptive advantage for deciduous trees to shed ther leaves each fall?*

Figure 32.10 Needles of coniferous species have a wide assortment of shapes as exemplified by the ponderosa pine (A), Douglas fir (B), blue spruce (C), and juniper (D).

Figure 32.11 In dense stands of trees, where self-pruning is common, living needles and branches occur only at the top of the tree. Needles and branches are shed from the tree after the needles' exposure to sunlight is reduced and they are no longer able to carry out a high rate of photosynthesis.

Question: *How may self-pruning be related to nutrient allocation in these Ponderosa pine trees?*

Both conifer needles and broad-leaved evergreen foliage have adaptations that prevent their freezing and protect them from cold damage. With the onset of cold fall weather, needles begin the process of "hardening," which ultimately protects them from freezing temperatures. During hardening, sugars accumulate in the needles, where they function as antifreeze; the freezing point drops because lower temperatures are required to freeze needles with a high sugar content than needles with a low sugar content. If water did freeze in the cells, the expansion that occurs when water turns to ice would damage or tear cell membranes, organelles, and cell walls. Thus needles are usually able to survive the extreme cold temperatures of winter because of biochemical adaptations that prevent water from freezing inside their cells.

The conical tree shape of conifers is also an adaptation for winter—structures with this shape easily shed heavy snow and ice. If allowed to accumulate, the increased weight from wet snow and ice could break the branches and even the main stem of a tree.

BEFORE YOU GO ON Conifer trees have highly specialized needle-shaped leaves that are not usually shed in response to seasonal variations. As a result, conifers are able to carry on photosynthesis throughout the year. Conifer needles are also adapted for conserving water during the summer and surviving the low temperatures of winter.

REPRODUCTION

Reproductive periods punctuate the life histories of plants at specific intervals. During these periods, the normal patterns of resource acquisition, assimilation, and allocation that govern the growth and development of roots, shoots, and leaves are temporarily modified. For plants to reproduce successfully, resources must be diverted to developing gametes, cones, flowers, fruits, and seeds.

Understanding plant reproduction allows humans to exploit certain plants more effectively. Desirable species can be propagated successfully, and undesirable species can be reduced or even eliminated. Many industries are totally dependent on the reproductive tissues of plants. Most agricultural activities, including the production of corn, soybeans, wheat, and apples, are directed at harvesting fruits and seeds. The landscaping and cut-flower industries rely heavily on flowering species, including roses, orchids, rhododendrons, daisies, and tulips. Products of plant reproduction provide much of the food consumed by humans, and flowers enrich our lives as well.

Growth Phases and Patterns

The life history of a plant encompasses two successive growth phases (see Figure 32.12). The phase that results in the development of roots, stems, and leaves is termed the **vegetative growth phase.** In addition, the plant normally produces and stores carbohydrates. When reproductive plant tissues begin to form, the plant enters the **reproductive phase,** and flowers, seeds, and fruits develop. The duration of these two phases and the timing of transition between them differ among species.

Three distinct patterns of vegetative growth and reproduction phases are used in classifying plants. Angiosperms have an *annual, biennial,* or *perennial* life cycle, whereas all gymnosperms are perennials. Each of these patterns has distinguishing qualities that include the common properties of growth and reproduction.

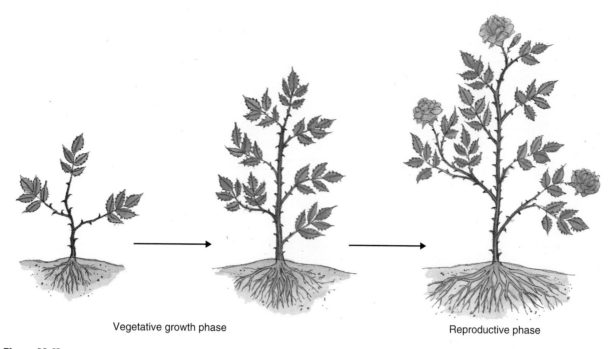

Vegetative growth phase   Reproductive phase

Figure 32.12   After completing a vegetative growth phase, plants enter a reproductive phase, during which they begin to produce cones or flowers.

Question: *What part of your own life cycle is analogous to a plant's vegetative growth phase?*

Annual Plants

As shown in Figure 32.13A (see page 604), **annual plants** complete their entire life cycle within one year or less. Annual plants include many agricultural crops, such as corn and beans; ornamental flowers, including pansies and marigolds; and weeds, such as cocklebur and vetch. Annuals commonly have small seeds and are adapted to germinate and grow on disturbed soil.

Typically, seeds of annual plants germinate in the spring and proceed through a period of vegetative growth. They enter their reproductive phase when they reach a particular size or stage of development or when they become stimulated by an environmental cue such as a specific day-to-night ratio. Once reproduction has been completed, the plant dies, leaving its seeds behind.

Some annual plants complete this life cycle very quickly. For example, desert annuals usually germinate during periods of spring or fall rains, grow rapidly, flower, and produce seeds before the weather becomes too hot and dry for them to survive. These plants frequently complete their entire life cycle within the span of two or three weeks.

Some annuals germinate in the spring, grow vegetatively, begin flowering, and persist in both growing and flowering until the plant dies from freezing in late fall. The life span of these plants may be six months or more. Familiar examples of relatively long-lived annual plants include many garden species such as beans and marigolds. Even though some of these plants do not die immediately after producing seeds and can live for long periods if protected from freezing in a greenhouse, they are still considered to have annual life cycles.

Biennial Plants

Beets, cabbage, and celery are examples of **biennial plants** that complete their lives in two growing seasons (see Figure 32.13B). During the first year, the growth of biennials is usually limited to the vegetative phase. Such plants often have a large taproot, a first-year shoot less than 2 centimeters tall, and large leaves. Much of the nitrogen and photosynthate accumulated during the vegetative growth phase is stored in the taproot. Beets and carrots are biennials that are usually harvested after their first year of growth.

During their second year of life, biennial plants usually enter into a reproductive phase. At the onset of reproduction, nutrients and photosynthates stored in the taproot, along with those being produced by leaves, are mobilized and sent to the shoot. When this happens, **bolting** occurs as the short, compressed shoot quickly grows to a height of half a meter or more. This process involves rapid stem elongation and may take one or two days or as long as two weeks. Once a plant has bolted, its resources are allocated to tissues forming flowers, seeds, and fruits for the remainder of the growing season. Biennial plants die at the end of their second year of life, after they have bolted and produced seeds.

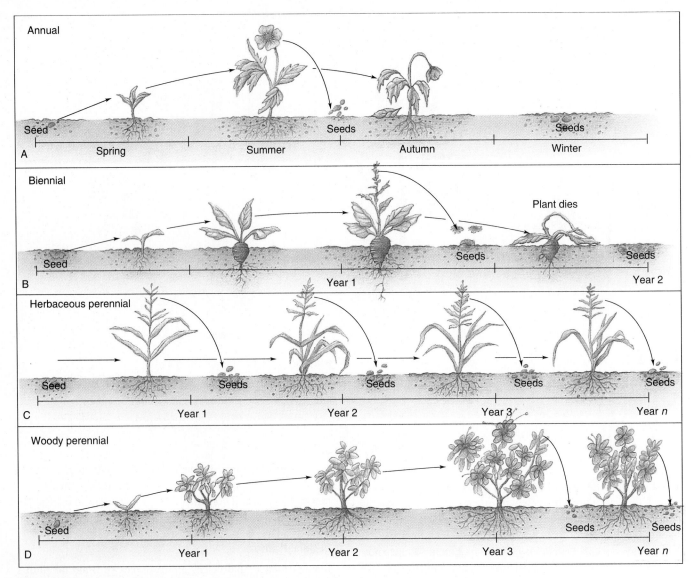

Figure 32.13   (A) Annual plants complete their life cycle in one growing season. In the temperate zone, annual plant seeds typically germinate in the spring, and flowers are produced. The plant dies after its seeds are released. (B) Biennial plants have a two-year life history. During the first year, the seed germinates, and the plant completes a vegetative growth phase. In the second year, biennial plants produce seeds during a reproductive phase and then die. (C) Perennial plants can live and produce seeds for many years. Herbaceous perennials, such as many grass species, may produce seeds during each year of growth. (D) Woody perennials, including trees and shrubs, may go through a juvenile period before they become capable of reproducing.

Question: *Which life cycle requires the longest time period to complete?*

Perennial Plants

Perennial plants live more than two years and are subdivided into two groups, which are illustrated in Figure 32.13C and D. *Herbaceous perennials* lack wood in their tissues and include asparagus, rhubarb, milkweed, lilies, and many grasses. *Woody perennials* are characterized by woody structures and include many vines (for example, grapes), shrubs (rhododendron), and trees (pines and oaks).

Many perennials have vegetative growth phases that last for extended periods of time. Some species of trees, such as walnuts, may grow in a juvenile, nonreproductive phase for up to 20 years. Once perennials have matured, they enter a reproductive phase and often reproduce during each subsequent growing sea-

son. Thus once a walnut tree produces a first crop of nuts, it can produce a nut crop every remaining year of its life.

The number and vigor of seeds produced by perennial plants differ from year to year, depending on environmental factors and plant age. Years of extreme environmental stress, such as a prolonged drought, may cause a great reduction in the seed production of affected perennials. The term *mast year* describes a year in which trees such as oaks, hickories, and beeches produce an abundance of fruits and seeds. Mast years are important because they increase the possibility of successful reproduction, and the prolific seed crop is a valuable supply of food for many animals such as wild squirrels and birds.

The life cycle of plants includes two growth phases—a vegetative growth phase when roots, shoots, and leaves are formed and a reproductive phase when reproductive structures develop. Three types of growth patterns occur in plants. Annual plants complete both phases within a single year. Biennial plants complete a vegetative phase during the first year and reproduce during the second year. Perennial plants live for many years.

Mechanisms of Vegetative Growth and Reproduction

Recall that *meristems* are active zones of growth located in plant shoots and roots. In shoots, meristems occur in both *terminal buds* and *lateral buds* and in *cork cambium* and *vascular cambium* tissues (see Figure 32.14). Root meristems are located in vascular cambium and in the root tip regions. Plant meristem cells contain a complete set of chromosomes, and some have the ability, under special conditions, to develop into a new plant. Meristem cells (described in Chapter 30) produced at regions of active plant growth are undifferentiated cells that can give rise to complete new plants. In contrast, specialized cells such as phloem or xylem are differentiated and cannot develop into other cell types.

Primary Growth

Apical meristems, located at the top of the plant and at the ends of branches and roots, give rise to all cells in the plant's shoots and roots. The production, growth, and differentiation of apical cells results in increased plant height. **Lateral buds** are located in *leaf axils,* the sites where petioles join the stem. Meristem cells of lateral buds develop and grow into branches. Growth that results in increased plant height or increased branch length is termed **primary growth.**

Terminal and lateral buds generally become active in the spring, causing branches to elongate and new leaves to emerge during a primary growth phase. As the growing season ends, branches of deciduous trees stop elongating, and their leaves eventually fall to the ground. As shown in Figure 32.15, a

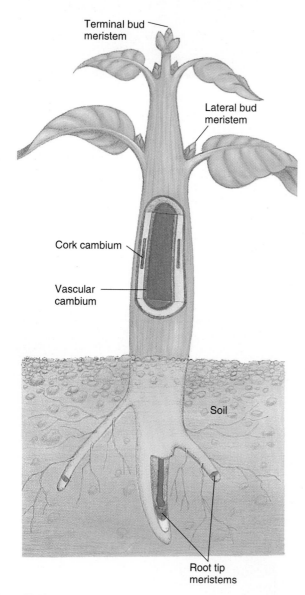

Figure 32.14 Apical buds, lateral buds, root tips, vascular cambium, and cork cambium all contain active meristem cells that divide and increase in size, resulting in growth. The cork cambium and vascular cambium lie just inside the bark of woody trees and shrubs.

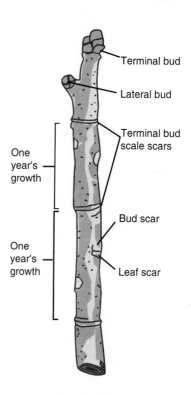

Figure 32.15 The history of leaf location, bud location, and growth during each season is recorded on branches. When apical and lateral buds open at the onset of growth, the bud scales drop, leaving a scar on the branch. The distance between apical bud scale scars represents one year of growth. Leaves also leave scars after they are shed from branches. Apical buds may be found just above the leaf scars and mark points where branches will sprout during the next growing season.

Question: *How could you determine the age of this woody sapling?*

record of these events remains on the branch in the form of leaf scars and bud scars in the former locations of leaves and buds.

Secondary Growth

In woody plants, a thin cell layer, the **vascular cambium,** forms a continuous sheath of cells near the outside of the stem. It may be only two or three cells thick and is found just inside the bark (see Figure 32.16). The vascular cambium produces cells from its inner surface that ultimately form woody xylem tissue; cells originating from its outer surface differentiate into phloem tissue, the innermost layer of the bark. The continuous activity of the vascular cambium not only produces cells specialized in

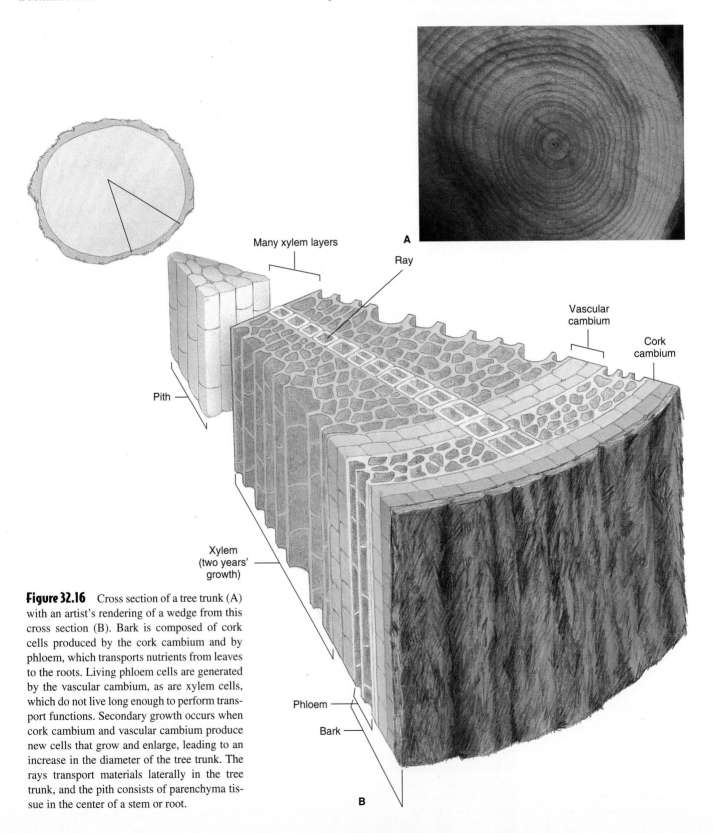

Figure 32.16   Cross section of a tree trunk (A) with an artist's rendering of a wedge from this cross section (B). Bark is composed of cork cells produced by the cork cambium and by phloem, which transports nutrients from leaves to the roots. Living phloem cells are generated by the vascular cambium, as are xylem cells, which do not live long enough to perform transport functions. Secondary growth occurs when cork cambium and vascular cambium produce new cells that grow and enlarge, leading to an increase in the diameter of the tree trunk. The rays transport materials laterally in the tree trunk, and the pith consists of parenchyma tissue in the center of a stem or root.

Many xylem layers

Ray

Vascular cambium

Cork cambium

Pith

Xylem (two years' growth)

Phloem

Bark

A

B

transporting essential resources in the plant but also causes an increase in the diameter of the stem and branches, a process known as **secondary growth.** The **cork cambium** lies outside the vascular cambium and phloem tissue and is a meristematic tissue that produces most of the bark of woody plants. It is also associated with secondary growth because it causes an increase in diameter.

Vegetative Reproduction

The list of reproductive mechanisms used by plants is long and includes various types of plant tissues. **Vegetative reproduction** refers to the origin of new plants from nonsexual tissues. Some tree species, including willow and poplar, can be vegetatively propagated by simply cutting off a young branch and planting it in the ground. Such plants have the capacity to generate roots quickly enough to allow the shoot to develop, grow, and survive. The establishment of new plants from clipped branches, or **cuttings,** depends on characteristics of the branch and environmental factors, such as sufficient warmth and water, that affect root development. Not all species of woody plants can be propagated from cuttings.

Grafting is a form of vegetative propagation in which a bud or a branch of one plant is attached to the stem or rootstock of another closely related plant. Nurseries commonly use this method to create one plant with superior features of two plants. For example, ornamental roses are produced by grafting stems from desirable roses to hardy wild rose rootstock. If stems or branches that support terminal or lateral buds become buried, cells of the vascular cambium can divide and differentiate into root tissues. Subsequently, an entirely new independent plant will arise. This type of vegetative reproduction, called **layering,** is common in many tree species with branches that grow close to the ground and are often covered by windblown soils and litter. It also occurs in tree species, such as high-elevation spruce, that grow where conditions are so harsh that sexual reproduction is rarely successful.

Plant tissues that grow belowground can also produce new plants by vegetative reproduction. Plants such as strawberries and spider plants have stems called **stolons** ("runners") that grow along the surface of the ground. Once on (or in) the soil, lateral buds on the stolons can develop into roots and shoots, which ultimately become transformed into independent plants. Stems of certain plants can grow horizontally under the soil surface. These stems, called **rhizomes,** produce lateral buds that contain cells capable of generating new plants. Bamboo, iris, and lily of the valley are examples of common plants that reproduce vegetatively from rhizomes.

Tubers are underground stems that are highly specialized for carbohydrate storage but are also capable of vegetative reproduction (see Figure 32.17). For example, the eyes of a potato are actually the lateral buds of a tuber. Potatoes are commonly propagated by planting a segment of tissue containing an eye, or lateral bud.

Finally, even roots can give rise to plant structures by vegetative reproduction and then develop into new plants. Some trees, such as willows and poplars, often reproduce in this way.

Figure 32.17 Plants such as potatoes reproduce by sprouting from underground stems, called tubers, which are also specialized to store carbohydrates.

Question: *If tubers are underground stems, what do the "eyes" of the potato represent?*

BEFORE YOU GO ON Vegetative growth is associated with increasing the size of most plant tissues. During primary growth, meristem regions located in apical and lateral buds produce new cells that develop and grow, resulting in an elongation of shoots, branches, leaves, and roots. In secondary growth, meristem tissues of the vascular cambium produce cells that form xylem and phloem, which results in an increase in the diameter of stems, branches, and roots. Cuttings, layering, and specialized structures—stolons, rhizomes, and tubers—are associated with vegetative reproduction.

Mechanisms of Sexual Reproduction in Seed Plants

The seeds of gymnosperm and angiosperm plants are produced through sexual reproduction. Within seed cells are genes that convey adaptations from previous generations of successful plants. These adaptations, encoded in the DNA of embryonic cells, may enable the seed to germinate successfully and proceed in its development from a seedling to a mature seed-bearing plant.

Gymnosperm Reproduction

Coniferous evergreen trees, including pines, firs, spruces, and hemlocks, are well-known gymnosperms. Pine trees produce seeds in cones through a series of events typical of most gymnosperms. Figure 32.18 (see page 608) describes the pine tree life cycle. Pines produce two kinds of cones, male and female, usually on the same tree. Sperm-bearing pollen grains from male cones are usually carried by the wind to egg-bearing female cones. Fertilization takes place within the female cone.

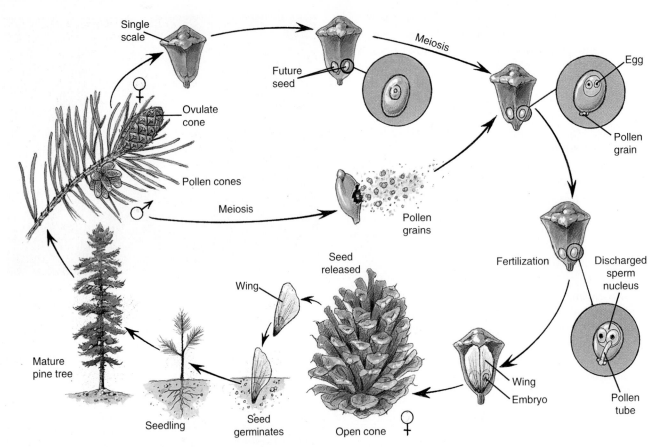

Single scale

Future seed

Meiosis

Egg

Pollen grain

Ovulate cone

♀

Pollen cones

♂

Meiosis

Pollen grains

Fertilization

Discharged sperm nucleus

Seed released

Wing

Mature pine tree

Wing

Embryo

Pollen tube

Seedling

Seed germinates

Open cone ♀

Figure 32.18 The life cycle of a gymnosperm, such as a pine, includes seed germination, growth and maturation of the tree, and the production of male and female cones. Sexual reproduction begins with the production of sperm-containing pollen grains in male cones. Pollen grains are then carried by the wind to the female ovulate cones, where the sperm fertilize eggs. Fertilization results in the formation of a zygote and, eventually, a new seed.

Question: *In some pine tree species, seeds are not released from cones until after a forest fire. How might this be adaptive for the pine life cycle?*

Male Cones In the late fall, male cones form at the ends of branches. During the spring, mature pollen grains, each containing two sperm, develop in male cones. Pine pollen grains have two small wings that aid in their dispersal by the wind. Pines and other conifers can produce pollen in such great quantities that nearby objects become covered with what appears to be yellow dust (Figure 32.19). Following the release of their pollen, male cones dry out and fall to the ground.

Female Cones Female cones are produced on the youngest branches in the canopy and are larger than male cones. During the spring, pollen released from male cones drifts through the air and lodges between scales of the female cones. The vast quantities of pollen produced ensures that most eggs in the female cones are fertilized. As a female cone grows, its scales close and prevent the entrance of any additional pollen. Over the next 12 to 14 months, pollen tubes emerge from the pollen grains in the female cone as the female reproductive cells undergo development and maturation.

Fertilization occurs within the female cone and leads to the formation of seeds. The seed contains the zygote that ultimately gives rise to an embryo, which has a radicle, an apical

Figure 32.19 A "pollen storm" from an Atlas cedar.

bud, and cotyledons (see Chapter 30), is surrounded by nutritive tissues, and is protected by a seed coat. The pine seed has a wing that enables it to be carried by the wind when it is released from the cone. Pine seeds are usually shed in the spring, about two years after the female cones first appear.

Angiosperm Reproduction

The angiosperm life cycle is illustrated in Figure 32.20. In angiosperms, pollen grains are most often transported by the wind or by insects as they move from flower to flower while foraging for nectar and pollen (see Figure 32.21, page 610). Once the pollen is in position on the stigma of the pistil, a tube grows down to the egg located in the base of the pistil. When this pollen tube contacts the egg, one sperm from the pollen grain migrates down the pollen tube and fertilizes the egg. Another sperm moves down the pollen tube and combines with a central cell within the embryo sac, which leads to the formation of endosperm that will be used by the developing embryo. These two fertilization events (double fertilization) are unique to angiosperms.

Flower Structure   Angiosperms produce an incredible variety of **flowers.** Many species, such as orchids and roses, produce large, conspicuous flowers; in contrast, the flowers of other species may be either small or short-lived and therefore not apparent. For example, grasses produce inconspicuous flowers that the casual observer seldom notices.

Angiosperms produce flowers that differ among species in the number, color, size, and arrangement of their flower parts. Figure 32.22 (see page 610) illustrates the general anatomy of a flower. Flowers are attached to the plant by a stalk known as the **peduncle.** At the end of the peduncle is the **receptacle,** to which the flower parts are attached. **Sepals** protect the flower while it is a bud, and the **petals** attract the insects, hummingbirds, or bats that are needed for pollination. The reproductive organs of a flower include the **stamens,** which produce pollen, and the **pistils,** which produce the female gametophyte, or embryo sac, located within the ovary. The pollen and embryo sac each produce gametes, sperm and eggs (see the Focus on Scientific Process, "Alternation of Generations in Plant Life Cycles").

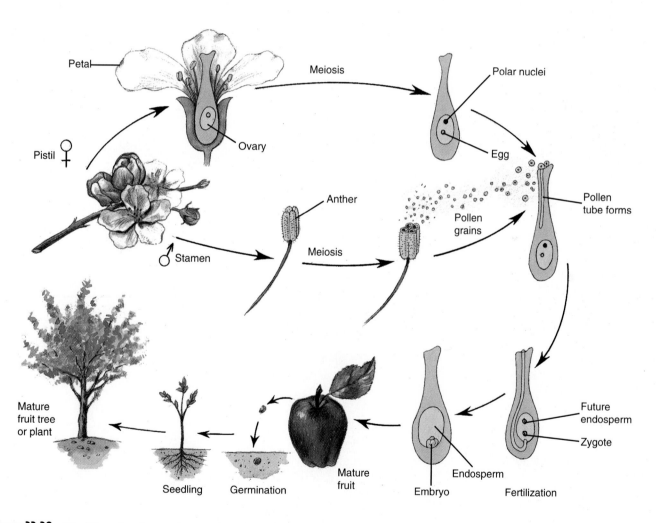

Figure 32.20   The life cycle of an angiosperm includes germination of a seed, vegetative growth, and the production of flowers that eventually bear fruit, which contains seeds. Seeds are formed when the male reproductive organs, the anthers, produce pollen that contain sperm. The pollen grains are transported by wind or animals to the stigma of the egg-bearing pistil. The sperm and egg unite to form the zygote, which becomes part of a new seed. The seed develops further within the pistil, a structure that eventually matures into a fruit.

Question: *How does apple seed dispersal differ from pine tree seed dispersal?*

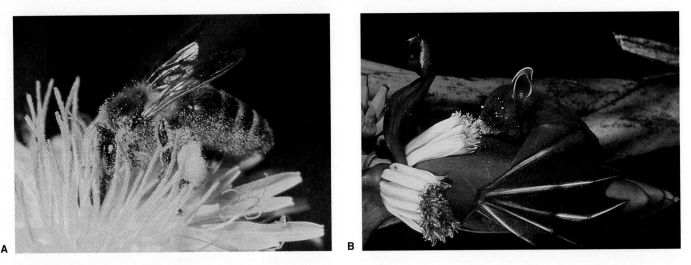

Figure 32.21 Flowers of many angiosperm species are adapted to attract animal pollinators. Certain insects, such as bees (A); birds; and mammals, such as bats that pollinate wild bananas (B), forage for nectar in flowers and consequently transfer pollen from one flower to another. The pollen of some angiosperms, such as grasses, is transported by the wind.

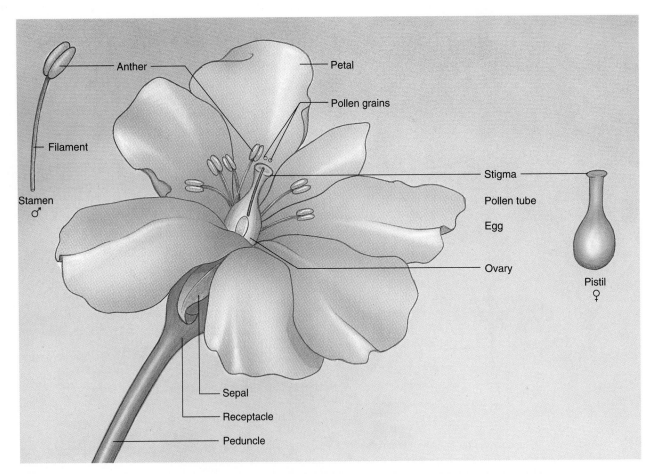

Figure 32.22 This is the structure of a typical flower. Sepals and petals are arranged in whorls around the axis of the flower. Male reproductive structures include the filament and the anther. The stigma at the top of the female reproductive structure, the pistil, is modified to receive pollen grains. The egg is located in the ovary at the base of the pistil.

FOCUS ON SCIENTIFIC PROCESS

Alternation of Generations in Plant Life Cycles

In the late 1600s, human and animal reproduction were thought to be similar to that seen in chickens, where the "female testes" (ovaries) produced vesicles called *ova*. These ova were thought to contain *germs*, fully formed, miniature embryos that grew during development. This was known as the *doctrine of germ preexistence* (see Chapter 14). Theorists who believed that these germs occurred in the ova were called *ovists*, while others, including Robert Hooke and Antoni van Leeuwenhoek, thought germs resided in male semen and were termed *spermists*. The ovists thought that the semen contained only "germinative spirit" that would stimulate the ovum to develop, while the spermists believed that the ovum provided only a place for the "animalcule" in the sperm to develop. This thinking was extended to the plant world in 1673 by Nicholas Malebranche when he wrote, "We may with some sort of certainty affirm, that all trees lie in miniature in the cicatride of their seed." In 1678, Nicolas Hatsoeker extended the spermists' view to include plant pollen grains and relegated the role of the plant ovum to that of the animal ovum, as only a place for the development of a miniature plant contained in a pollen grain.

Ovists' views dominated the learned world of the early eighteenth century, but by mid-century, Carl Linnaeus opposed the doctrine of germ preexistence when he wrote that "the offspring proceeds not from the egg alone, nor from the sperm alone." He thought that the plant anthers produced pollen that "burst and shed their seminal virtue" when they fell on the stigma. However, he could explain neither fertilization nor development. He did use the presence of sex organs in plants as the basis for their classification into 23

plant classes that he called *phanerogamia* (see Figure 1). But algae, fungi, mosses, liverworts, ferns, and fern allies had no visible sex organs, and Linnaeus called this twenty-fourth class *cryptogamia*. He thought that a union did occur in these plants but that it could not be seen.

The development of the *cell theory* by Matthias Schleiden and Theodor Schwann (see Chapter 15) and the *pollen theory* by Schleiden allowed scientists to view reproduction from a cellular perspective. Schleiden declared the reproduction of cryptogamic stem plants, like ferns, to be no different from the sexual reproduction found in the phanerogamia. He thought that the spore and the pollen were functionally identical and that only the medium where they developed into new plants differed. The sexual phanerogamic plants had female floral parts in which the pollen grain would be deposited and

the pollen tube would form, allowing the embryo to develop within the ovum. In the spore-forming ferns and other cryptogamia, germination would occur in the soil, and the new plants would develop there from the embryo carried by the spores (see Figure 2, page 612). Hence, by 1840, Schleiden had concluded that sexual reproduction was only a special type of asexual reproduction!

In 1850, a self-taught botanist from Leipzig, Germany, took exception to Schleiden's view that pollen grains and spores were functionally similar. Wilhelm Hofmeister used the term *Generationswechsel* ("alternation of generations") to describe his observations of the life cycle in ferns (see Figure 3, page 612). This was the first use of the term *alternation of generations* in botany and established a new way of looking at plant reproduction and life cycles. Hofmeister argued that fern life

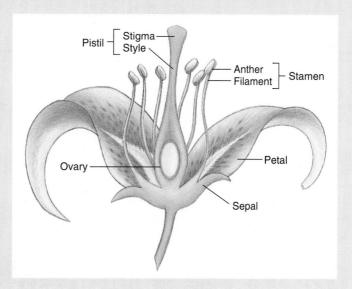

Figure 1   Linnaeus used the sex organs of flowering plants as the basis for classifying plants into 23 classes he called *phanerogamia*.

box continues

FOCUS ON SCIENTIFIC PROCESS

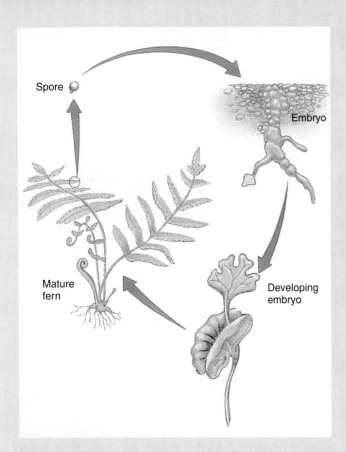

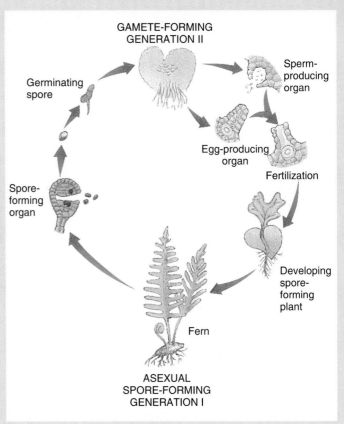

Figure 2 Schleiden thought that ferns and other cryptogamia reproduced by a special form of asexual reproduction. According to his hypothesis, mature fern plants released embryo-containing spores that entered the soil and developed into new ferns.

Figure 3 Hofmeister contended that the fern life cycle consisted of an alternation of generations in which an asexual, spore-forming multicellular generation (I) alternated with a sexual, gamete-forming generation (II).

cycles involved an asexual spore-forming multicellular generation (I) that alternated with a sexual, gamete-forming multicellular generation (II) and that *free cell formation* occurred in both generations, producing, alternatively, spores and gametes. He extended this thinking to mosses and other cryptogams (see Figure 4), and in a short review chapter in a book he published in 1851, Hofmeister wrote that "the course of development of the embryo of conifers stands intermediately between those of higher cryptogams and the phanerogams." He thought that conifers had an alternation of generations but that the gamete-producing generation

disappeared in the higher phanerogams and that these angiosperms displayed no alternation of generations.

The conceptual development of alternation of plant generations occurred concurrently with the development of better microscopes. Lenses that corrected for both spherical and color aberrations allowed for ever-finer resolution of the microscopic spores and gametes from both plants and animals. Better microscopes allowed botanists to discover the spores of angiosperms and to recognize pollen grains as the male gametophyte generation. In addition, the female spores were located in the developing pistil of

angiosperm flowers, where it produced the eight-celled female gametophyte. Double fertilization was shown to occur in angiosperms, producing both the embryo sporophyte and the endosperm that made up the seeds. These developments happened at the same time that Charles Darwin was publishing his ideas on the theory of evolution, and plant scientists began to think about the possible evolutionary significance of both sexual and asexual stages in the plant life cycles. In 1890, Frederick Bower published a paper that linked the differences seen in the alternation of generations in algae, bryophytes, ferns, conifers, and angiosperms to environ-

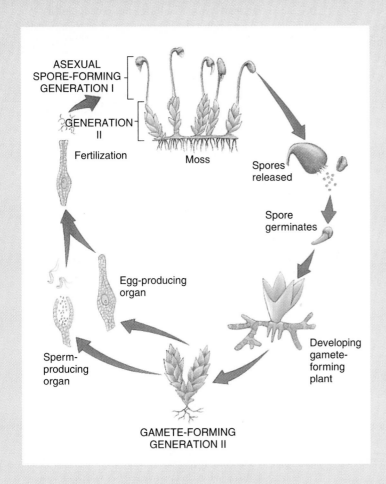

ASEXUAL
SPORE-FORMING
GENERATION I

GENERATION II

Fertilization

Moss

Spores released

Spore germinates

Egg-producing organ

Developing gamete-forming plant

Sperm-producing organ

GAMETE-FORMING
GENERATION II

Figure 4 Hofmeister later wrote that life cycles of mosses and other cryptogamia could also be explained as an alternation of generations.

mental factors that affected evolution in plants as they migrated from water to land. As plants became more terrestrial, the gametophyte generation decreased and the sporophyte became dominant. Primitive plant gametes required water for fertilization, whereas spores were adapted for wind dispersal.

Combining the ideas of Hugo de Vries and Thomas Hunt Morgan (see Chapters 14 and 16) with his own observations, Eduard Strasburger defined changes in chromosome numbers during the alternation of generations in plants. In 1906, he introduced two new terms:

It would perhaps be desirable if the terms *gametophyte* and *sporophyte,* which can be applied only to plants with single and dou-

ble chromosome numbers, should be set aside in favor of terms that apply to the animal kingdom. To this purpose, I permit myself to introduce the words *haploid* and *diploid.*

Strasburger was able to link the morphological–reproductive alternation of generations discovered by Hofmeister with the changing number of chromosomes resulting from meiosis and fertilization.

However, like many terms used in the science of botany, *gametophyte* and *sporophyte* were not set aside, as Strasburger suggested. In fact, other terms used to describe the plant life cycles have their roots in Linnaean terminology and make little sense when viewed from Strasburger's perspective. For

example, conifer cones are called male and female, but they produce asexual spores, not gametes; and angiosperm flower anthers produce spores that develop into the male gametophyte generation (pollen), while the antheridia (antherlike sperm-producing cells) of mosses and ferns produce gametes directly. In addition, pistils are often called female, but they produce an embryo sac composed of eight haploid cells. Gametes in flowering plants are produced by the greatly reduced sexual gametophytes, the male pollen grains and the female eight-celled embryo sacs. As confusing as this use of terms may be to biology students, it is nevertheless an unavoidable part of the long and venerable history of botany!

Types of Flowers Flowers that contain both stamens and pistils are called **perfect flowers; imperfect flowers** lack one of these reproductive organs. Perfect flowers may be **self-pollinating,** which means that pollen produced by the stamen of a flower can fertilize an egg within the pistil of the same flower. Stamen and pistil maturation must occur at the same time for self-pollination to take place. Some species have adaptations that prevent self-pollination. (Why would this be an advantage to a plant?) For example, pollen of a flower may mature and be released from its stamens either before or after the period when pollen can be received by the flower's pistil.

Certain species of angiosperms with imperfect flowers, such as willows and poplars, are **dioecious,** meaning that male flowers grow on one plant and female flowers on another plant. For such species to bear fruit, male plants must grow close to female plants to ensure pollination of the female flowers. Other plants with imperfect flowers, such as corn, are **monoecious,** meaning that the same plant has both male and female flowers, which are separate from each other. Pollen from the male flowers may be either compatible or incompatible with the female flowers on the same plant. Self-fertilization can occur if flowers are compatible but not if they are incompatible.

The number of floral parts varies among species. Some species have flowers with one pistil and several stamens. In other species, flowers have many pistils and many stamens. As the number, position, arrangement, and color of these structures, along with sepals and petals, change, so too does the appearance of the flower. Obviously, the number of possible combinations of these floral characteristics is enormous, as is reflected in the great diversity of flowers found in angiosperms.

BEFORE YOU GO ON Sexual reproduction in gymnosperms involves pollen production by male cones and transportation of the pollen to egg-bearing female cones, where fertilization occurs and seeds develop. Sexual reproduction in angiosperms begins after flowers are formed. Pollen is transported to female floral parts, where fertilization occurs. Subsequently, seeds develop, and the ovary becomes transformed into a fruit.

The events that lead to the production of seeds mark the point at which the life cycle—from seed to seed—has been completed. Whether reproduction occurs only once, as in the life of an annual plant, or many times, as in the life of a perennial plant, the release of viable seeds is the endpoint or product of all of the processes that occurred during the vegetative and reproductive phases of plant life.

SUMMARY

1. Plants are well adapted for surviving in the environments they inhabit. Several general types of adaptations exist in plants. These include physiological mechanisms that are used in normal activities and also in making adjustments to changing environmental conditions, specialized structures and growth forms that permit survival in different geographical areas, and life history strategies that increase the likelihood of long-term survival and successful reproduction.

2. Deciduous trees have broad, flat leaves that produce great quantities of photosynthates during spring and summer. Their leaves are arranged on branches so as to receive full sunlight. The growth form and branching pattern of deciduous trees also ensure maximum exposure of leaves to sunlight. However, their leaves cannot survive cold temperatures, and they are shed from the tree during the fall. Thus deciduous trees are characterized by high levels of photosynthesis during the growing season but very low levels of activity during the rest of the year.

3. In contrast, most evergreen trees do not shed their leaves annually. Conifers have small, needlelike leaves. Such leaves can survive low temperatures and are not shed seasonally. The growth form of these trees—an elongated pyramid with sloping branches—is favorable for shedding heavy snow and ice. Hence conifers are adapted for remaining active throughout the year.

4. Both deciduous and evergreen trees have other adaptations that are important to their survival. These include mechanisms for regulating carbohydrate and nitrogen allocation and preventing water loss.

5. Plants have two distinct growth phases during their life cycle. In the vegetative growth phase, plant resources are used in the development and growth of roots, stems, and leaves. During the reproductive phase, a plant invests in forming structures used in reproduction.

6. The timing and duration of the two growth phases varies with plant species. Three general types of life cycle pattern are used to classify plants. Annual plants complete both growth phases during a single year and then die. Biennial plants usually undergo a vegetative growth phase during their first year and a reproductive phase in their second year, after which they die. Perennial plants can live for many years, and the time required for the vegetative growth phase may last for years. Once they enter a reproductive phase, they may reproduce yearly for the rest of their lives.

7. Vegetative plant structures—stems, roots, and leaves—are produced from cells located in meristems. These cells divide and differentiate to form the various plant tissues. Primary growth leads to increases in plant length. Secondary growth—the production of additional xylem and phloem—occurs only in woody plants and results in an increase in branch and stem diameter.

8. Some plant species use specialized processes (layering) or structures (stolons, rhizomes, and tubers) to create new plants by vegetative reproduction.

9. Sexual reproduction in seed plants consists of fertilization and seed formation and development. Gymnosperms have male cones that produce pollen, which contains sperm, and female cones, which contain eggs. Fertilization leads to the formation of a zygote and a seed. The zygote develops into an embryo within the seed. Angiosperms develop flowers to use in reproduction. Although the basic process of sexual reproduction is similar to that in gymnosperms, among angiosperm species there are enormous variations in flower types, flower structure, pollinators, and fertilization mechanisms.

WORKING VOCABULARY

annual plant (p. 603)
biennial plant (p. 603)
cork cambium (p. 607)
deciduous tree (p.597)
evergreen tree (p. 597)
lateral bud (p. 605)
needle (p. 600)
perennial plant (p. 604)

primary growth (p. 605)
secondary growth (p. 607)
terminal bud (p. 605)
vascular cambium (p. 606)
vegetative growth (p. 602)
vegetative reproduction
 (p. 607)

REVIEW QUESTIONS

1. Give some examples of plant adaptations to natural environmental changes.

2. In what ways are the leaves of deciduous trees considered adaptive for survival of the entire tree?

3. What is a deciduous tree's response to self-shading? Why is it considered adaptive?

4. Why is it advantageous to deciduous trees to drop their leaves in the fall?

5. In what ways are the leaves of conifers considered adaptive for survival of the entire tree?

6. How are conifers adapted for remaining active during the winter?

7. Describe the two basic life history phases of plants.

8. Explain the primary differences in the life histories of annual plants, biennial plants, and perennial plants.

9. Describe the different mechanisms plants use to carry out vegetative reproduction. What is the difference between primary growth and secondary growth?

10. Describe the process of sexual reproduction by gymnosperms.

11. What are the functions of flowers and the various floral parts in angiosperm sexual reproduction?

ESSAY AND DISCUSSION QUESTIONS

1. The life histories of evergreen and deciduous tree species differ, yet both must be well adapted to their environment in order to survive and reproduce. How can you explain the occurrence of both types of trees in a single forest?

2. Assuming that the buildup of atmospheric greenhouse gases leads to a slow rate of global warming, what types of adaptation may evolve in plants growing in your area? What might be some of the consequences for plants of a rapid rate of warming? Why?

3. Larch is a conifer species that produces new needles each spring and sheds them each fall. What are some disadvantages to this adaptation that might explain why it is rare in conifers?

REFERENCES AND RECOMMENDED READING

Barbour, M. G., and W. D. Billings (eds.). 1988. *North American Terrestrial Vegetation*. New York: Cambridge University Press.

Doust, J. L. 1989. Plant reproductive strategies and resource allocation. *Trends in Ecology and Evolution*, 4: 230–234.

Farley, J. 1982. *Gametes and Spores: Ideas About Sexual Reproduction, 1750–1914*. Baltimore: Johns Hopkins University Press.

Feinsinger, P. 1987. Effects of plant species on each other's pollination: Is community structure influenced? *Trends in Ecology and Evolution*, 2 :123–126.

Goldberg, R. B., G. de Pavia, and R. Yadegari. 1994. Plant embryogenesis: Zygote to seed. *Science*, 226: 605–614.

Hughes, N. F. 1994. *The Enigma of Angiosperm Origins*. New York: Cambridge University Press.

Meyerowitz, E. M. 1994. The genetics of flower development. *Scientific American*, 271: 56–65.

Niklas, K. J. 1987. Aerodynamics of wind pollination. *Scientific American*, 257: 90–96.

Powell, E. 1995. *From Seed to Bloom: How to Grow over 500 Annuals, Perennials, and Herbs*. Pownal, Vt: Storey Communications.

Primack, R. B. 1987. Relationships among flowers, fruits, and seeds. *Annual Review of Ecology and Systematics*, 18: 409–430.

———. 1985. Longevity of individual flowers. *Annual Review of Ecology and Systematics*, 16: 15–38.

Scott, R. J., and A. D. Stead (eds.). 1995. *Molecular and Cellular Aspects of Plant Reproduction*. New York: Cambridge University Press.

Van de Water, P. K., S. W. Leavitt, and J. L. Betancourt. 1994. Trends in stomatal density and $^{13}C/^{12}C$ ratios of *Pinus flexilis* needles during the last glacial–interglacial cycles. *Science*, 264: 239–242.

ANSWERS TO FIGURE QUESTIONS

Figure 32.6 Nutrients would no longer be allocated to shaded leaves. Also, many nutrients within shaded leaves would be reallocated to nonshaded leaves.

Figure 32.7 It increases the leaf surface area that can be exposed to light.

Figure 32.9 If leaves were not shed, in many temperate zones they would freeze or would accumulate snow and ice, causing the branches to break. Also, shed leaves contain various waste products, conveniently removed from the tree.

Figure 32.11 Nutrients in shaded branches would be translocated to growing branches before they are pruned.

Figure 32.12 Your "vegetative growth phase" would have ended at puberty.

Figure 32.13 That of woody perennials; it takes years before they can reproduce.

Figure 32.15 By counting the number of annual rings in a cross section or the number of terminal bud scale scars on the trunk.

Figure 32.17 Buds.

Figure 32.18 It ensures that seeds will be shed on open ground rather than under mature trees, where light and nutrients would be limited.

Figure 32.20 Most pine seeds depend on wind, whereas apples attract birds and other animals to assist in seed dispersal.

Unit Five Conclusion
Plant Systems of Life and the Process of Science

Plants have been of interest throughout the great expanse of human existence. From the beginning of recorded history, through the time of the ancient Greeks, and into the present, plants have been sought for both their agricultural and medical values. Naming plants has been important for both oral and written histories in all human cultures.

In the eighteenth century, two questions arose that drove research for many decades: How do plants reproduce sexually? And what is the microstructure of plants? The first of these questions was answered after careful microscopic studies of plant structure. Unlike animals, plants undergo an alternation of generations. Many findings from these microscopic studies were also used in answering the second question, on plant microstructure.

Until the nineteenth century, studies of plants generally emphasized naming, describing, and classifying the profusion of plants found on Earth. The major purpose of these studies was to identify plants that had medical value. Plant extracts—crushed and ground leaves, bark, roots, and stems—constituted almost the entire store of products dispensed by pharmacists until the middle of the twentieth century (for example, quinine, an antimalarial drug, comes from chinchona bark). Botany matured as a scientific discipline in the nineteenth and twentieth centuries and greatly extended the range of its investigations to include questions concerned with fundamental biological processes, environmental interactions, and evolutionary history.

Botanists are very interested in learning how plants reproduce, grow, develop, function, and survive in their environment. Plant physiological ecologists have produced detailed knowledge about the processes and adaptations that account for the great evolutionary success of plants. Much of the recent research effort has centered on questions about how plants respond to various environmental stresses they encounter during their life cycle. Unlike animals, which can move away from sources of stress, plants generally remain in the same place throughout their lives. How do plants cope with environmental stresses? What types of adaptations allow them to survive in changing environments? These and related questions have driven research that has provided important information about the anatomy, physiology, and evolution of plants. Will our understanding of water use efficiency, nutrient assimilation and allocation, and anthropogenic environmental stress compensation in plants enable us to apply modern genetic engineering technologies (described in Chapter 21) to develop new agricultural products? To create plants better suited for modified environments that may result from global warming? These questions will receive much attention in the decades ahead.

Human Systems of Life

VI

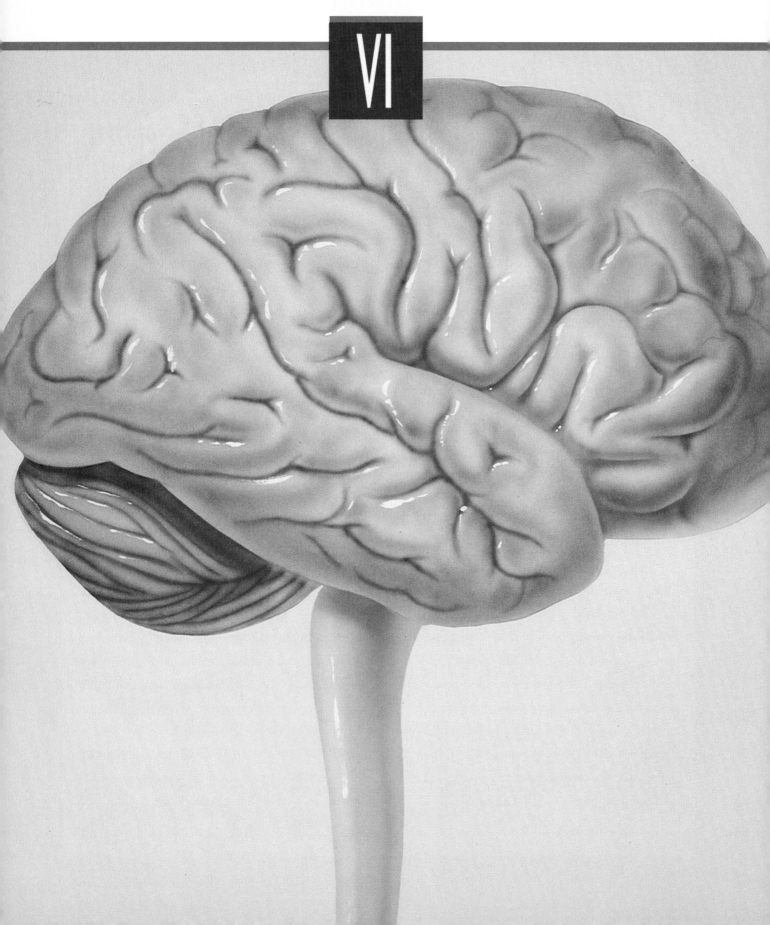

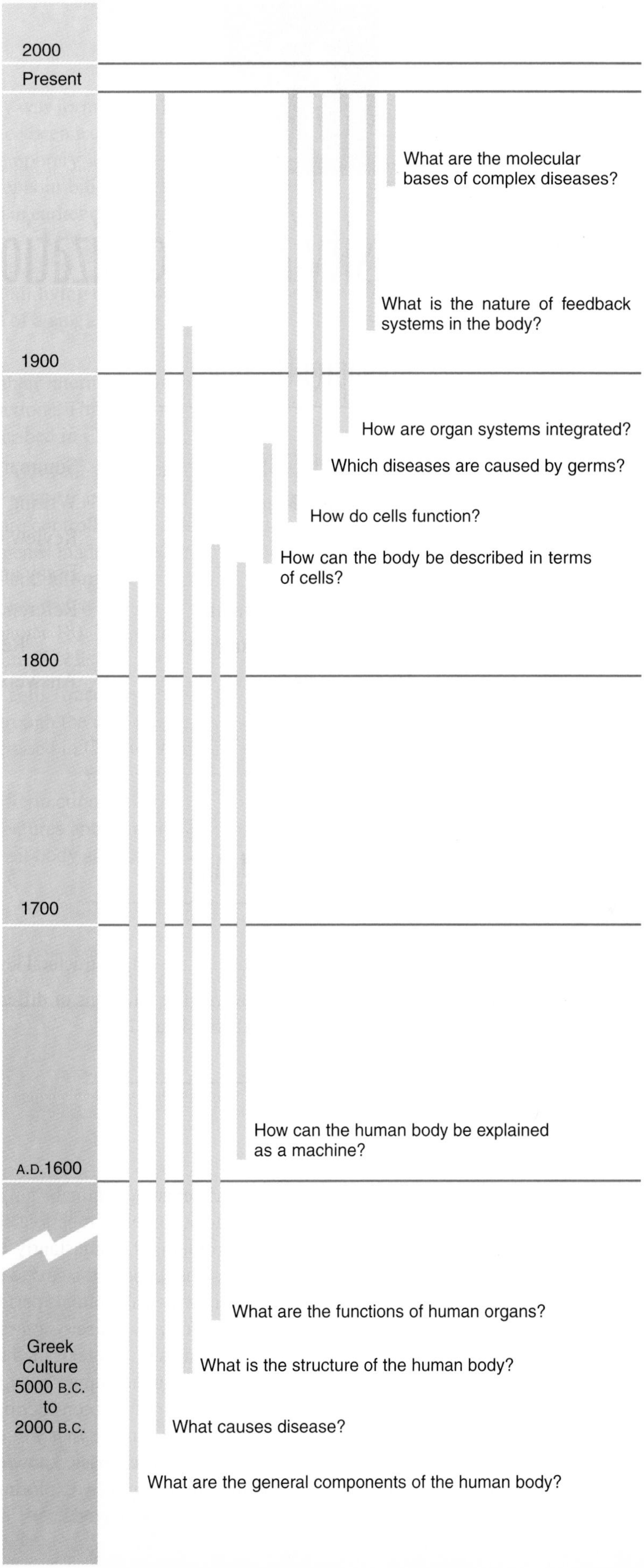

2000

Present

1900

1800

1700

A.D. 1600

Greek
Culture
5000 B.C.
to
2000 B.C.

What are the molecular
bases of complex diseases?

What is the nature of feedback
systems in the body?

How are organ systems integrated?

Which diseases are caused by germs?

How do cells function?

How can the body be described in terms
of cells?

How can the human body be explained
as a machine?

What are the functions of human organs?

What is the structure of the human body?

What causes disease?

What are the general components of the human body?

Cell Functions

All human cells are continuously engaged in normal activities that include the conversion of food materials into energy or new materials and the removal of waste products created during these transformations. However, cellular components are constantly damaged or destroyed during the cell's metabolic activities. Thus **self-maintenance** involves activities used in meeting the continuous demand for repairing or replacing ribosomes, membranes, enzymes, and other structures and molecules that are lost during normal operations. It is essential for maintaining the health of the cell and hence the health of the individual.

Cellular Reproduction

Some cells in adults are capable of undergoing **cellular reproduction** to replace cells that normally wear out and die. Before a cell can reproduce, it undergoes a precisely controlled increase in its numbers of chromosomes, mitochondria, ribosomes, plasma membranes, proteins, and other structures and molecules. Once sufficient quantities of these elements are available, a cell divides by mitosis (nuclear division) and gives rise to two new daughter cells after cytokinesis (cell division); one or both of the daughter cells can then continue to carry out the same function as the cell or cells they replaced.

Specialized Functions

Most human cells are *differentiated;* they have undergone **differentiation,** the process of changing from an unspecialized cell to a cell that has distinctive molecular and structural features that reflect a specialized function. For example, cells that actively synthesize proteins, such as active plasma cells, are packed with rough endoplasmic reticulum and ribosomes (see Figure 33.2A); cells requiring great amounts of energy, such as heart muscle cells, have numerous mitochondria (see Figure 33.2B). Some cells, however, remain *undifferentiated;* they

do not become specialized to carry out a specific function. Such cells often have the ability to continue dividing, thus producing new cells. These new cells commonly replace functional cells that are injured or die. To do this, these new cells must themselves undergo differentiation.

Some specific functions of differentiated human cells include conducting electrochemical impulses, forming bone, capturing and digesting bacteria that invade the body, secreting hormones, transporting oxygen, forming gametes, and storing fat. We will look more closely at several of these functions in the chapters that follow.

> **BEFORE YOU GO ON** Most human cells contain a nucleus and various organelles enclosed within a plasma membrane. Throughout their lives, cells engage in self-maintenance activities to repair or replace damaged molecules and organelles. Some cells are capable of producing daughter cells that replace lost or injured functional cells. These new cells become differentiated to perform specialized functions.

STRUCTURAL LEVELS OF ORGANIZATION

There are four structural levels of organization in the architecture of the human body. From simplest to most complex, they are cells, tissues, organs, and organ systems. Whereas cells are able to carry out relatively few functions, organ systems are capable of accomplishing numerous tasks within the body.

Cells

Cells represent the simplest structural level of organization in multicellular organisms. Hundreds of different cell types occur in humans. However, except for blood cells and gametes, they do not operate as independent entities. Rather, they combine with

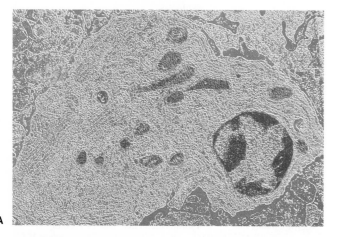

A

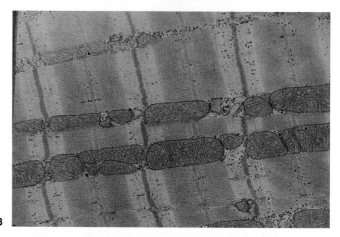

B

Figure 33.2   A cell's structure is closely related to its function. (A) Plasma cells contain an abundance of ribosomes and endoplasmic reticulum—organelles required for protein synthesis. (B) Heart muscle cells are full of mitochondria—organelles that produce energy.

Question: *What other structure is especially abundant in muscle cells?*

other cells to form more complex levels of organization that permit an efficient division of labor among different groups of cells.

Tissues

Tissues consist of a large assemblage of cells, primarily with the same type of structure, that are specialized to carry out a particular function. There are four fundamental animal tissues: epithelial, connective, muscular, and nervous. Each of these tissues is composed of cells with a characteristic appearance and functional capability.

Epithelial Tissues

Epithelial tissues (epithelium) are continuous layers of closely connected cells that cover body surfaces, line cavities and some organs inside the body, and perform various functions. Epithelial tissues are classified according to the shapes and surface specializations of the cells, the arrangement of cell layers within the tissue, and their functions. Different types of epithelium perform important tasks in the human body, some of which are shown in Figure 33.3.

Epithelium protects underlying tissues and conducts a variety of important activities. Epithelial tissues absorb nutrients from the digestive tract; secrete *mucus,* a slimy, viscous fluid, to lubricate hollow organs (for example, the lungs and the stomach); manufacture digestive enzymes and hormones; and engage in many other functions.

Connective Tissues

Connective tissues, the most abundant tissue type in the body, include a diverse group of organized cells. They support organs in their appropriate locations, bind different cells and tissues together, store energy reserves, and perform other specialized functions. Connective tissues are composed of relatively few cells held within an abundant *cellular matrix*—usually a semifluid, gel-like, or fibrous material made up of complex carbohydrates and proteins. Within the matrix are different types of **filamentous fibers** that are categorized according to size and molecular composition (see Figure 33.4A, page 624).

The three major types of connective tissue fibers found in humans are shown in Figure 33.4B–D. **Collagenous fibers** are the most abundant and are made up of bundles of smaller fibers,

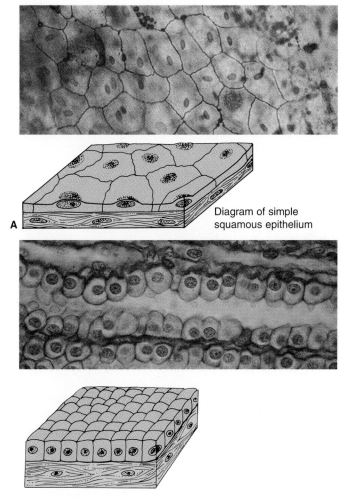

A   Diagram of simple squamous epithelium

B   Diagram of simple cuboidal epithelium

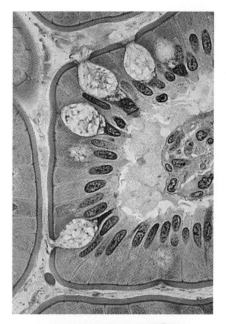

C   Diagram of simple columnar epithelium

Figure 33.3 Different types of epithelial cells cover and line tissues, organs, and body cavities: (A) simple squamous epithelium, (B) simple cuboidal epithelium, (C) simple columnar epithelium.

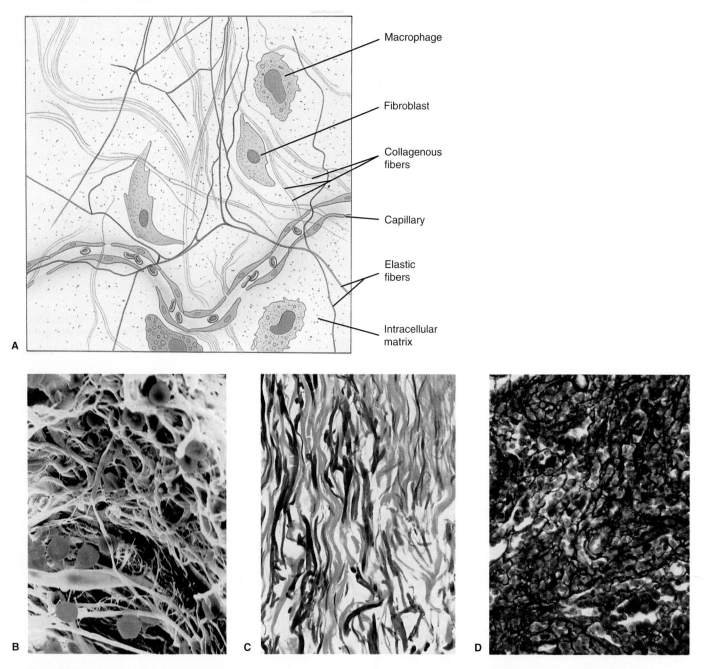

Figure 33.4 Connective tissue (A) is composed of a chemical cellular matrix and filamentous fibers that hold cells and tissues together. Collagenous fibers (B), which appear as irregular yellow strands with entangled red blood cells; elastic fibers (C); and reticular fibers (D), which are both stained dark blue or black, are composed of proteins, and each fiber type has properties related to its function. (See Table 33.2 for details.)

Question: *What is the function of a fibroblast?*

similar to the structural design of a cable. These fibers are made of a protein called *collagen* and are very strong and inelastic. **Elastic fibers** are long, thin, branching fibers made of the protein *elastin* (see Figure 33.4C). Elastin conveys the property of resilience to the fibers, allowing them to return to their original length after being stretched. Loss of this property is correlated with age and accounts for the effects of gravity on the skin and external structures of older humans. **Reticular fibers** are thin, short filaments constructed of a protein called *reticulin* (see Figure 33.4D). These fibers com-

monly form tight networks that serve as a scaffolding for the epithelial cells in organs such as glands, lymph nodes, and the spleen.

Fibroblasts and macrophages are common and important cell types associated with the connective tissues. **Fibroblasts** are large spindle-shaped cells that produce the various types of connective tissue fibers. **Macrophages** are a type of white blood cell that moves throughout the body engulfing foreign particles, such as bacteria, and debris from dead cells. They also play a major role in immunity, which is described in Chapter 39.

Connective tissues are classified according to the character of the cellular matrix, which can be liquid, fibrous, or hard, and the types and organization of fibers within the matrix. Table 33.2 provides information about the types and functions of connective tissues in the human body. Figure 33.5 illustrates some types of connective tissues in humans. Bone and blood are classified as connective tissues because they are made up of cells embedded in an intercellular matrix. In bone tissue, the matrix becomes *mineralized;* that is, minerals such as calcium are deposited in the fibrous network and form a nearly solid substance. In contrast, the surrounding matrix of blood cells is a fluid called *plasma.* Bone and blood are discussed in Chapters 34, 35, and 39.

Table 33.2 Major Types and Functions of Connective Tissues (CT)

Type	Location	Functions
Loose CT (unorganized collagen fibers)	Widely distributed throughout the body	Attaches skin to underlying tissue; fills spaces between organs and holds them in place; surrounds and supports blood vessels
Dense CT (abundance of collagen fibers)	Specific organs that require support of strong attachments	Provides capacity to withstand tension; supports underlying tissues of skin; forms tendons, ligaments, and some membranes
Elastic CT (majority of elastic fibers)	Organs that expand and return to original size	Confers strength and elasticity to walls of arteries, trachea and bronchi, vocal cords, and lungs
Reticular CT (mostly reticular fibers)	Liver, lymph nodes, and spleen	Provides a framework for support of the cells that make up these organs
Adipose CT (fat cells)	Underlying skin; in loose CT	Serves as a storage site for fats; insulates, pads, and protects certain areas of the body
Cartilage (modified collagen fibers)	Certain skeletal structures, breathing tubes	Provides strong structural support for the embryonic skeleton and ends of bones; forms flexible structures such as the ear and nose
Bone (cells in a mineralized matrix)	Major potion of the adult skeleton	Provides support and protection; skeleton serves as the attachment site for muscles
Blood (cells in a fluid matrix)	In blood vessels; propelled throughout the body by the heart	Transports oxygen, nutrients, and other materials; aids in cellular immune functions

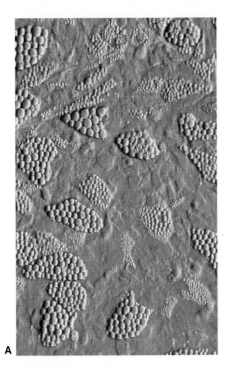
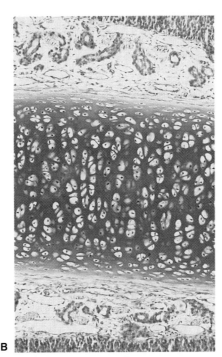
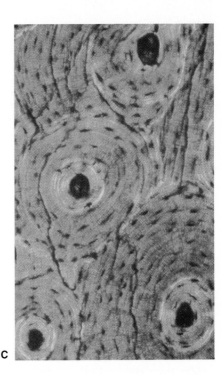

A   B   C

Figure 33.5 Adipose tissue (A), cartilage (B), and bone (C) are classified as connective tissues.

Question: *What is the function of adipose tissue?*

Muscle Tissues

Muscle tissue is composed of elongated thin cells that are commonly referred to as "fibers." Muscle "fibers" are true cells and should not be confused with connective tissue fibers. Each muscle fiber (cell) contains numerous smaller fibers, called *myofibrils,* made of specialized proteins that have the ability to contract. Contraction of muscle fibers is necessary for moving body parts, changing the diameters of hollow organs within the body, moving materials throughout the body, and removing unwanted substances from the body. In addition, muscle contractions require ATP and release energy, which results in the production of significant amounts of heat that help maintain normal body temperatures (described in Chapter 4). The human body contains three types of muscle tissue, shown in Figure 33.6, which are distinguished by their structure, function, and location.

Skeletal muscle consists of extremely long fibers, masses of which are attached to bones by special connective tissues, serving to cause movements of the skeleton. Skeletal muscle also aids in manipulating facial features and moves the eyes and tongue. Skeletal muscle cells have two unique features: they contain multiple nuclei, and they have characteristic cross-striations due to the alternation of light and dark bands along the myofibrils. The bands are associated with the mechanical contraction of muscle fibers. Skeletal muscle is under the voluntary (conscious), control of the individual.

The walls of the heart are constructed of specialized *cardiac muscle* cells, which form branching networks that impart great strength to this critical organ. Cardiac muscle cells usually contain a single centrally located nucleus and have striations that are similar to those in skeletal muscles. They also have unique cross-striations, called *intercalated disks,* that

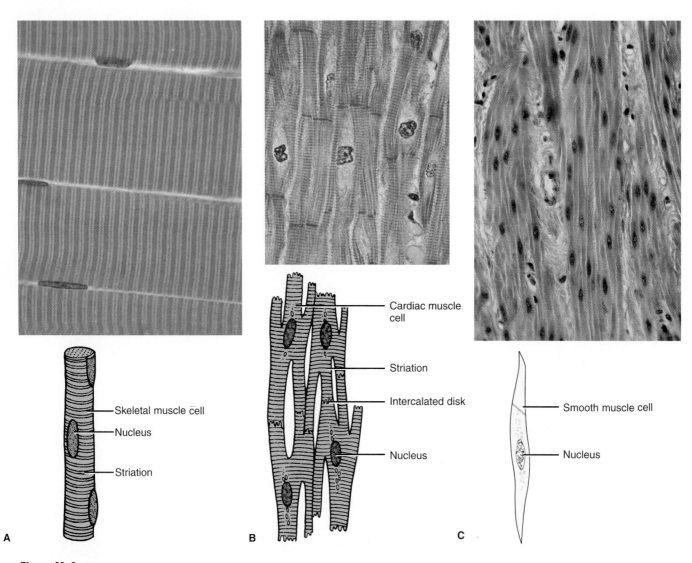

Figure 33.6 Three types of muscle tissue are found in humans: skeletal muscle (A), cardiac muscle (B), and smooth muscle (C).

Question: *What might be an advantage in cardiac muscle cells' being branched?*

occur where the ends of two cells meet, end to end, and form junctions. Cardiac muscle is under involuntary control (how would you like to have to think about when and how often your heart should beat?) and is regulated by hormones and parts of the nervous system.

Smooth muscle, made up of cells with single centrally located nuclei and no cross-striations, is present in the walls of hollow internal organs and tubes (such as the intestines and blood vessels) and in many other locations. Contractions of smooth muscle, which serve to move materials through these organs, are also under involuntary control.

Nervous Tissue

The structural unit of **nervous tissue** is the *neuron,* a cell type specialized for the rapid conduction of electrochemical impulses. As shown in Figure 33.7, neurons come in a variety of sizes and shapes, but their basic structure consists of a cell body, which contains the nucleus, and two or more slender extensions, or fibers, involved in the transmission of nerve impulses. Nervous tissues are considered in greater detail in Chapter 36.

Organs

An **organ** is a distinct structure that is made up of more than one type of tissue, has a definite form, and performs a specific function in the body. Consider a familiar example, the stomach. It is a hollow organ covered and lined with different types of epithelial tissue; its walls are composed of muscle and connective tissue; and it is regulated primarily by nervous tissue. Each of these tissues contributes to the specific function of the stomach, the breakdown of food.

Many organs are contained within one of five major body cavities (see Figure 33.8). **Body cavities** are enclosed spaces

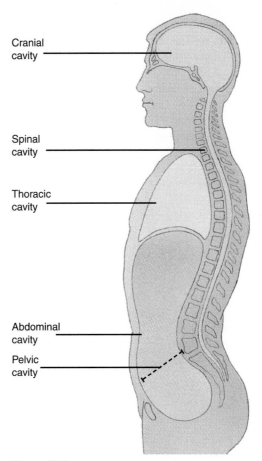

Figure 33.8 The five major cavities in the human body—cranial, spinal, thoracic, abdominal, and pelvic—each contain organs of critical importance.

Question: *Which organs are found in the thoracic cavity?*

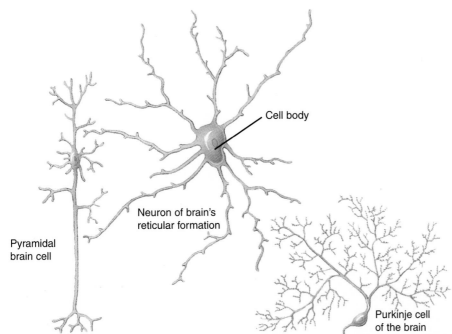

Figure 33.7 Neurons, the structural and functional unit of nervous tissue, have an amazing diversity of sizes and shapes.

Question: *How is the branched structure of neurons related to their function?*

that help support and protect organs. Cavities are separated from one other various structures, including muscles and bones. The *cranial cavity* encloses the brain, and the *spinal cavity* holds the spinal cord. The *thoracic cavity* contains two critical organs, the heart and lungs, as well as the esophagus, which relays food to the stomach. The abdominopelvic cavity is the largest human cavity and it consists of two divisions that are not separated by a distinct structure; the *abdominal cavity* holds the organs of digestion and others such as the kidneys and liver, and the *pelvic cavity* contains the internal reproductive organs and the urinary bladder. Organs such as the skin, blood vessels, most muscles, eyes, and the skeleton are not contained in body cavities. Further information on specific organs is provided in Chapters 34–39.

Organ Systems

Within individual organisms, the highest level of biological structural organization is the organ system. **Organ systems** are groups of organs operating as a unit to perform a major function or group of functions. Truly independent organ systems do not actually exist in humans. Rather, the term *organ system* is a useful way to categorize interacting structures and functions for study. The 11 defined human organ systems and their major organs, structures, and functions are summarized in Table 33.3 and illustrated in Figure 33.9.

BEFORE YOU GO ON   Cells are the basic structural unit of organization. A tissue is an organized collection of cells. The four fundamental types of tissues are epithelial, connective, muscle, and nervous. Organs are structures that consist of multiple tissues and perform a specific function. Organ systems are composed of several organs and represent the highest level of structural organization. Eleven human organ systems—integumentary, skeletal, muscular, nervous, endocrine, circulatory, lymphatic and immune, respiratory, digestive, urinary, and reproductive—perform all principal functions in the body.

ANATOMY AND PHYSIOLOGY

Anatomy is the study of the *structure* of an organism and the relationship among its parts. In general, anatomy focuses on whole organs, although microscopic anatomy involves the

Table 33.3   The Structures and Functions of Human Organ Systems

System	Major Structures	Functions
Integumentary	Skin and specialized structures such as hair, nails, and sweat glands	Covers and protects internal body structures against injury and foreign materials; prevents fluid loss; regulates temperature
Skeletal	Bones, cartilage, joints, and the ligaments connecting them	Supports and protects soft tissues and organs; provides for body movement
Muscular	Skeletal, cardiac, and smooth muscles	Provides for movement of the skeleton and internal organs; enables locomotion; propels blood through body
Nervous	Brain, spinal cord, nerves, and sense organs	Serves as the primary regulatory system; regulates activities of the body; receives and interprets information from the internal and external environment
Endocrine	Hormone-secreting glands	Regulates many body functions in conjunction with the nervous system
Circulatory	Heart, arteries, veins, capillaries lymphatic vessels, blood and lymph	Transports oxygen, nutrients, and hormones between different cells and tissues; removes cellular waste products
Lymphatic and immune	Lymph, lymph vessels, lymph nodes, thymus, and spleen	Filters body fluids; helps produce certain white blood cells; protects against disease
Respiratory	Lungs, bronchi, windpipe, mouth, and nose	Exchanges oxygen and carbon dioxide between the internal and external environment
Digestive	Mouth, esophagus, stomach, intestines, and accessory organs—liver, salivary glands, pancreas, and gallbladder	Processes foods used for energy and the synthesis of biomolecules and construction of cells and tissues
Urinary	Kidneys, bladder, and associated ducts	Regulates blood chemistry; eliminates metabolic waste products; maintains water balance
Reproductive	Testes, ovaries, and associated structures, depending on sex	Produces gametes; secretes hormones influencing growth, maturation, and other activities

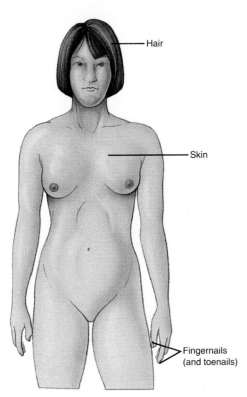

1. Integumentary System

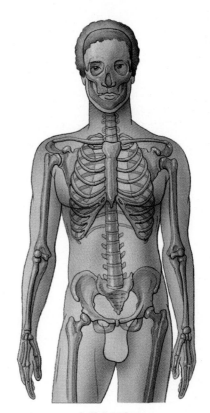

2. Skeletal System

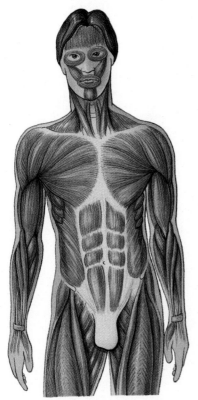

3. Muscular System

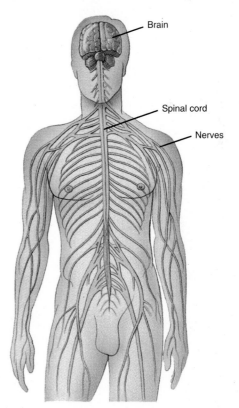

4. Nervous System

Figure 33.9 The human body has 11 major organ systems.

Question: *What are some functions of the integumentary system?*

figure continues

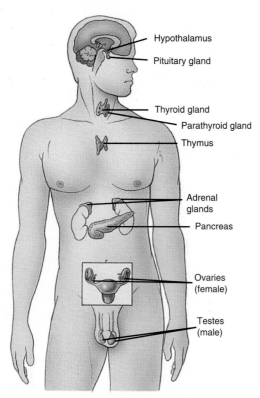

5. Endocrine System

- Hypothalamus
- Pituitary gland
- Thyroid gland
- Parathyroid gland
- Thymus
- Adrenal glands
- Pancreas
- Ovaries (female)
- Testes (male)

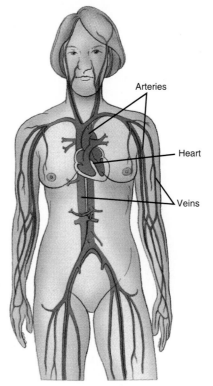

6. Circulatory System

- Arteries
- Heart
- Veins

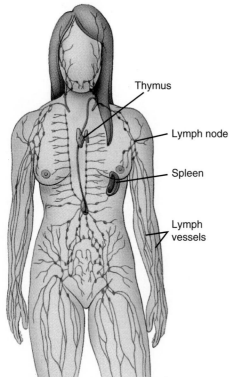

7. Lymphatic and Immune System

- Thymus
- Lymph node
- Spleen
- Lymph vessels

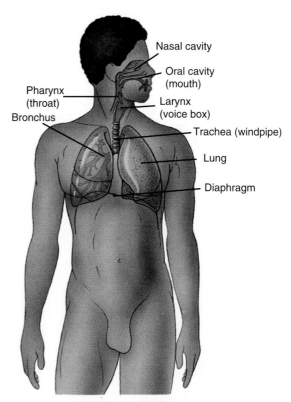

8. Respiratory System

- Nasal cavity
- Oral cavity (mouth)
- Pharynx (throat)
- Larynx (voice box)
- Bronchus
- Trachea (windpipe)
- Lung
- Diaphragm

Figure 33.9   *continued*

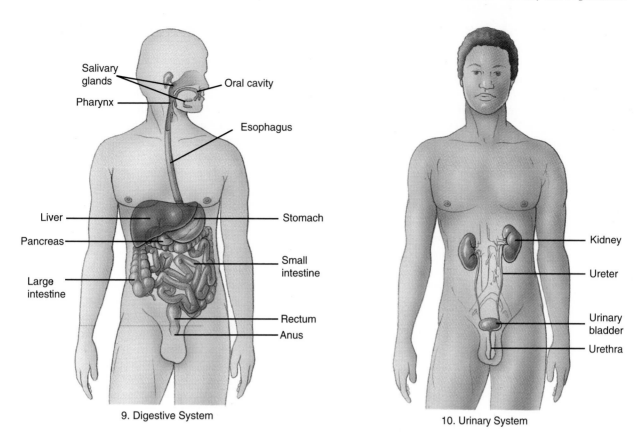

Salivary glands
Oral cavity
Pharynx
Esophagus
Liver
Stomach
Pancreas
Small intestine
Large intestine
Rectum
Anus

9. Digestive System

Kidney
Ureter
Urinary bladder
Urethra

10. Urinary System

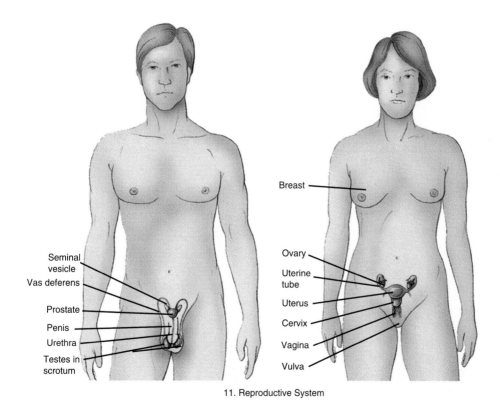

Seminal vesicle
Vas deferens
Prostate
Penis
Urethra
Testes in scrotum

Breast
Ovary
Uterine tube
Uterus
Cervix
Vagina
Vulva

11. Reproductive System

study of structures that can be seen only with the aid of a microscope. The field of anatomy has existed since the time of the ancient Greeks, when physicians in Alexandria examined dead organisms, including humans, to learn more about their internal structure. Today, anatomy still has great value in reducing the complexity of studying the human body by describing the forms, arrangements, and relationships of organs and organ systems—in other words, how we are constructed.

Physiology is the study of the *function* of an organism. It focuses on many levels of organization, ranging from functions of subcellular organelles to multicellular organs. Physiology is considered in greater detail in subsequent chapters.

A strong interdependence between structure and function is evident at all levels of organization. Every organism represents an integrated unit of structure and function. Examining the structure and function is a time-honored approach for studying humans that will be used in Chapters 34–39.

MAINTENANCE AND REGULATION

The trillions of cells making up the tissues and organs of the human body operate at high levels of activity during most of their existence. Not surprisingly, many of these cells wear out, are injured, or die after a genetically programmed period of service. Are cells of all tissues replaced? How? What are the consequences if they are not replaced? To answer these questions, it is necessary to begin by examining growth prior to adult maturity.

Cell Renewal

Cells of all tissues in the developing human embryo are capable of dividing by mitosis and cytokinesis. This permits growth and repair of all tissues during that phase of life. After birth, however, cells of certain tissues become fully differentiated and lose the ability to divide. In other words, the numbers of those cells are fixed soon after birth for the rest of the individual's life. Nervous and muscle tissues are two human tissues that consist of such cells. Once development of these tissues is completed within a year or two after birth, it is generally not possible to replace damaged or destroyed cells. Consequently, injuries to nerves and muscles are permanent and cannot be repaired by natural biological mechanisms.

In contrast, epithelial and connective tissues contain small populations of undifferentiated cells that retain the ability to divide and differentiate into new cells. This is important for two reasons. First, tissues that are injured can be repaired completely and with great accuracy. Second, many tissues consist of cells with genetically programmed life spans. For example, skin (several months), red blood cells (about 120 days), and cells lining the intestine (three to seven days) die and are replaced continuously throughout an individual's life. Other tissues and organs—including glands and certain digestive,

respiratory, and urogenital tissues—have similar replacement systems.

To illustrate the importance of cell renewal systems, consider the normal requirement for replacing red blood cells (RBCs) in humans. The average adult human contains about 25 billion RBCs, which have a life span of about 120 days. To maintain a constant population of RBCs, it is necessary for the body to produce 2,500 new cells each second of your life. It is estimated that the total number of cells produced and differentiated by the replacement systems in the human body exceeds 10 million per second!

Homeostasis

Human cells are specialized to survive and function under relatively constant internal conditions (for example, temperature, pH, and levels of glucose) that are continuously subjected to forces of change from both the external and internal environments. **Homeostasis** refers to the condition of a constant internal environment that fluctuates within a normal range. In this context, the "internal environment" refers to the fluids that surround cells in the body; these fluids contain gases, nutrients, and ions dissolved in water. Thus homeostasis exists when the internal environment is the proper temperature and contains appropriate concentrations of gases, nutrients, ions, and water. However, homeostasis is constantly subjected to *stress*—any factor that tends to disrupt the internal environment. Stress includes factors from the external environment, such as abnormal temperatures and loud noises, as well as factors from the internal environment, such as low blood sugar levels and high pH. To maintain homeostasis, the body's physiological mechanisms resist or compensate for changes that are always occurring in the dynamic internal environment (see the Focus on Scientific Process, "Homeostasis").

Homeostatic Mechanisms

Although all organ systems participate in maintaining homeostasis, the nervous and endocrine (hormone) systems are responsible for regulating homeostatic mechanisms. For example, if the internal environment becomes overheated, the nervous and endocrine systems regulate the reduction of body temperature through heat-dissipating operations of the skin and circulatory system (sweating). Other body systems also adjust to the temperature change to return to homeostasis. How does the body sense changes in the internal environment? What underlying mechanisms compensate for these disruptions?

Feedback Systems

The body uses regulatory mechanisms known as *feedback systems* to maintain stable conditions in the internal environment. In homeostasis, a **feedback system** is a cycle of events in which information about a specific condition in the internal environment is relayed to a central control region that causes change in the condition. Homeostatic feedback systems are

Homeostasis

Homeostasis, a fundamental concept in physiology, has a long and interesting history. Greek Hippocratic physicians in the fifth century B.C. believed that the human body was composed of four substances called *humors.* In a healthy body, the four humors were in balance, but an imbalance was thought to cause disease (see Chapter 40). Diet, weather, and other environmental factors could disrupt the balance, but the body actively tried to keep the humors in balance. Early physicians recommended "helping the body help itself" by prescribing rest, reasonable diet, and exercise. This approach lasted well into the seventeenth and eighteenth centuries and still has value today.

Striving to maintain a balance of humors was a powerful idea that conformed to experience—people did recover from many illnesses and temporary medical problems, such as indigestion—but it was not a concept that could provide a detailed understanding of the process by which the body "restored" itself. Partly for that reason, it was replaced during the scientific revolution of the seventeenth century by the mechanical conceptual model of the human body as a machine, like a clock. However, in that pre–computer-chip era, clocks and other machines seldom corrected themselves when they malfunctioned or ran fast or slow. Realizing this, scientists and physicians of that time eventually put aside the idea that the body somehow actively maintained an internal balance.

By the nineteenth century, many physiologists viewed the human body as a "material machine" that could be explained in terms of chemistry and physics. Recall that the cell theory was developed in this intellectual atmosphere (see Chapter 3). Cells were thought to be tiny parts of the material machine in which useful chemical reactions took place. Cells interacted with one another through the chemicals produced during these reactions. However,

this picture of the body proved inadequate: if one took the chemicals known to exist in a cell and poured them into a test tube, they would not react in ways that resembled a mechanism related to maintaining the organization of a body. In the 1860s, the French physiologist Claude Bernard claimed that the body had an internal environment that resisted change. He coined the term *internal milieu* for the concept of a stable internal environment in which individual cells existed and interacted.

Bernard was among the greatest physiologists of the nineteenth century, and he stressed the value of rigorous experimental research. Although he had an enormous impact on physiology, his internal milieu concept did not stimulate much new research. Instead, most scientists concentrated on studying basic physical and chemical processes in the cell. Two important early-twentieth-century physiologists—Lawrence J. Henderson and Walter Bradford Cannon of Harvard Medical School—finally recognized the potential in Bernard's internal milieu concept. Neither Henderson nor Cannon doubted that the body was composed of chemicals or that it could be studied using experimental methods. Like Bernard, they recognized that merely cataloging chemicals and sets of chemical reactions that occur in the body was unlikely to lead to an understanding of its ability to maintain itself. They believed that the distinguishing characteristic of the living body was its ability to regulate all of its processes through a complex system of internal self-control mechanisms.

Henderson arrived at this conclusion by studying mechanisms that allow the body to maintain a constant acid-base composition (pH). From there, he moved to pioneering studies on blood, which he described in publications from 1921 to 1931. He explained the organization of systems that helped maintain blood pH and showed how to analyze

their relationships mathematically. He emphasized that it was essential to examine chemical reactions within the context of the body in order to understand their physiological role. A chemical reaction studied in an experimental system, outside the body (*in vitro*), would not necessarily reveal its importance when studied in the context of complex interactions that took place within a living system (*in vivo*).

Cannon extended Henderson's work by applying the idea of internal self-regulation to the entire body. He hypothesized that general processes such as blood sugar level, temperature, and metabolic rate were regulated not only by cellular chemical reactions but also through interactions of the nervous and endocrine systems. In a set of classic experiments on the autonomic nervous system (the system that regulates many of the body's automatic responses, described in Chapter 36), Cannon demonstrated how this system regulates various physiological processes including endocrine functions. By removing parts of the system, he showed that animals lost the ability to control their internal environment.

Cannon's work on the nervous and endocrine systems led him to coin the term *homeostasis,* which literally means "same state or condition." In using this term, Cannon intended to communicate that the body is a self-regulating system that can maintain a stable, balanced, internal environment. If the system shifts out of balance, it can react to reestablish its stability (return it to the "same state").

The negative feedback concept described in this chapter has proved very fruitful, not only in physiology but also for understanding many forms of organization, such as populations. Equally important, the idea has been used by engineers and scientists to study nonliving systems such as computers and various communication technologies.

composed of three principal components, illustrated in Figure 33.10. The **control center** measures levels of a **regulated condition** in some part or activity of the body and acts to ensure proper functioning of that condition. Regulated conditions in the human body include body temperature, heart rate, blood pressure, blood pH, blood glucose level, and breathing rate. For most regulated conditions, the brain is the control center of the feedback system. The control center receives information ("input") about a regulated condition from a **receptor**—a cell, tissue, organ, or mechanism that monitors changes in a regulated condition. For example, exercise usually results in increased body temperature that is detected by heat receptors in the nervous system, which then sends this input to the brain. If the input received from a receptor is interpreted by the con-

trol center to require change in the regulated condition, it signals an **effector**—a cell, tissue, organ, or mechanism that receives the "output" from the control center and initiates the response necessary for change. Thus in the feedback system used to decrease body temperature that results from exercise, the brain signals sweat glands (effectors) to increase their secretions. Evaporation of sweat from the skin then causes body temperature to decrease.

Negative and Positive Feedback Mechanisms

Two general types of feedback systems operate in the body. **Negative feedback** systems cause the level of the regulated condition to change in the direction *opposite* from the original level that caused the response. In our exercise example, a negative feedback system caused the body temperature to decrease in response to an original temperature increase brought on by exercise.

Positive feedback mechanisms tend to cause the level of the variable to change in the *same* direction as the initial change. Thus they do not lead to maintenance of stable internal conditions but rather tend to increase further the level of a variable. Because this mechanism does not promote internal stability, it is not commonly found in the human body. One example of a positive feedback mechanism occurs during childbirth, when the pressure of the baby's head against the birth canal stimulates muscular contractions. In this case, positive feedback promotes expulsion of the baby at birth.

The human body has several thousand regulating systems, most of which involve negative feedback mechanisms regulated by the nervous and endocrine systems. Because of their critical role in many disease processes, these systems have become a very important area of research in human medicine.

BEFORE YOU GO ON A valuable approach for studying organisms is by concentrating on anatomy (structure) and physiology (function). The health and integrated control of organ systems depend on a number of mechanisms. Cells with finite lives are regularly replaced through the orderly operations of cell renewal systems. Homeostatic processes, primarily negative feedback mechanisms, act to regulate and coordinate the activities of organ systems.

The human is a wondrously complex organism. One of the most stimulating and worthwhile endeavors individuals can undertake is to learn about the operations of their bodies. Without such knowledge, it is difficult to acquire an adequate understanding of self. In the chapters that follow, we explore basic human structures and functions, the regulation and control of organs and selected physiological processes, and the consequences when failures in structure or function occur at different levels of organization.

SUMMARY

1. Human organs and tissues are composed of an enormous variety of cells that enable them to carry out specific functions.

2. All cells have the capacity for self-maintenance—the ability to repair damage and maintain health. A limited number of cell

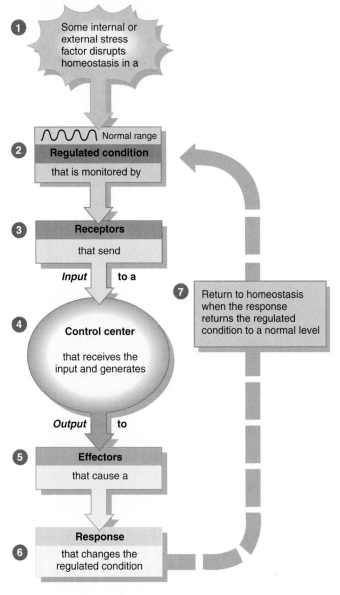

Source: Gerard J. Tortora and Sandra R. Grabowski, *Principles of Anatomy and Physiology*, 8th Edition. New York: Harper Collins College Publishers, 1996.

Figure 33.10 Homeostasis is maintained through feedback systems shown in this model.

types are able to reproduce and in so doing provide a continuous supply of cells to replace those that die in the normal course of events. Most cells are highly specialized, and some carry out unique functions such as conducting an electrochemical impulse, transporting oxygen, or storing fat.

3. There are four structural levels of organization in higher organisms. Organ systems are composed of organs, organs are made up of tissues, and tissues consist of millions of cells.

4. The fundamental types of tissues (and their primary function) are epithelial (covering body surfaces and lining cavities and some organs), connective (binding and supporting tissues and organs), muscle (moving organs), and nervous (conducting electrochemical impulses).

5. Anatomy is the study of structure, and physiology is the study of function. These two branches of biology have long been used in the systematic investigation of organisms and how they operate.

6. Mechanisms of maintenance and control in the body ensure optimal conditions for normal cell, tissue, and organ functions. For example, new cells are normally produced at the same rate at which cells are lost, and the internal environment remains relatively constant because of homeostatic mechanisms.

7. Homeostasis, the condition of a constant environment, is maintained primarily through negative feedback systems, which operate on the principle of changing the direction of the regulated condition being monitored. If temperature increases to a dangerous level, for example, negative feedback will lead to a decrease in temperature.

WORKING VOCABULARY

anatomy (p. 628)
connective tissue (p. 623)
differentiation (p. 622)
epithelium (p. 623)
feedback system (p. 632)
homeostasis (p. 632)

macrophage (p. 624)
organ (p. 627)
organ system (p. 628)
physiology (p. 632)
tissue (p. 623)

REVIEW QUESTIONS

1. What does the cell theory tell us about the structure and function of organisms?

2. What is the basic structure of eukaryotic cells?

3. What is the importance of cellular self-maintenance and cellular reproduction?

4. How do cells acquire the ability to perform specialized functions?

5. What is a tissue? Describe the basic structure and function of epithelial tissue, connective tissue, muscle tissue, and nervous tissue.

6. What is an organ? An organ system? What are the 11 major human organ systems?

7. How are anatomy and physiology related?

8. What is the function of a cell renewal system?

9. What is the role of homeostatic feedback systems?

10. What is the difference between a negative and a positive feedback system?

ESSAY AND DISCUSSION QUESTIONS

1. Do you think it is likely that the cell theory will be further modified or extended in the future? Why?

2. The endosymbiotic hypothesis explains the origin of cells (see Chapter 3). Describe a possible outline for the evolution of higher levels of structural organization in animals. Information in Chapters 3 and 4 might be useful.

3. Emphasizing anatomy and physiology is a common method for learning about organisms. What other approaches might be used in acquiring an understanding of organisms?

REFERENCES AND RECOMMENDED READING

Alberts, B. M. 1989. Introduction: On the great excitement in cell biology. *American Zoologist,* 29: 483–486.

Chiras, D. D. 1995. *Human Biology: Health, Homeostasis, and the Environment.* 2d ed. Minneapolis: West.

Chrousos, G. P., and P. W. Gold. 1992. The concepts of stress and stress system disorders: Overview of physical and behavioral homeostasis. *Journal of the American Medical Association,* 267: 1244–1252.

Crawshaw, L. I., B. P. Moffitt, D. E. Lemons, and J. A. Downey. 1981. The evolutionary development of vertebrate thermoregulation. *American Scientist,* 69: 543–550.

Diamond, J. 1994. Best size and number of human body parts. *Natural History,* 103: 78–81.

Gartner, L. P. 1994. *Color Atlas of Histology.* Baltimore: Williams & Wilkins.

Lindsay, D. T. 1995. *Functional Human Anatomy.* St. Louis: Mosby/Year Book.

Ratnoff, O. D. 1987. The evolution of homeostatic mechanisms. *Perspectives in Biology and Medicine,* 31: 4–24.

Ross, M. H. 1995. *Histology: A Text and Atlas.* Baltimore: Williams & Wilkins.

Society for Experimental Biology (Great Britain). 1964. *Homestasis and Feedback Mechanisms.* San Diego, Calif.: Academic Press.

Tortora, G. J., and S. R. Grabowski. 1996. *Principles of Anatomy and Physiology.* 8th ed. New York: HarperCollins.

ANSWERS TO FIGURE QUESTIONS

Figure 33.1 Primarily the synthesis of proteins.
Figure 33.2 Microfilaments.
Figure 33.4 To produce various types of connective tissue fibers.
Figure 33.5 To store fats, which insulate and protect parts of the body.
Figure 33.6 Gives them greater strength.
Figure 33.7 They serve to establish and operate communication networks.
Figure 33.8 Heart, lungs, and esophagus.
Figure 33.9 To cover and protect the body, to prevent fluid loss, and to help regulate temperature.

34

The Integumentary, Skeletal, and Muscular Systems

Chapter Outline

Reading Questions

1. What are the primary structures and functions of the integumentary system?

2. What are the functions of the skeletal system?

3. What is the significance of bone remodeling?

4. How is the skeletal muscle structure related to its function?

5. How do the integumentary and muscular systems help maintain temperature homeostasis?

Multicellular animals evolved hundreds of millions of years ago. The first primitive animals, such as the unadorned sponges, had only rudimentary organizations of cells that were adapted to carry out certain essential functions, such as capturing food. Over millions of years, more complex animals arose, and cells became organized into complex tissues and organs. Finally, Chordata, the taxonomic group that contains fishes, frogs, snakes, birds, and mammals (including humans), evolved. Human organs, in particular, have long been of interest; they have been studied for centuries (see the Focus on Scientific Process, "Human Anatomy").

What types of biological problems do humans face, and how do we solve these problems? In this chapter, we consider some organ systems that have functions related to the following questions: What covers and protects us from the external environment? Why are we able to counteract gravity? What enables us to move? How are we able to maintain a constant internal environment?

FOCUS ON SCIENTIFIC PROCESS

Human Anatomy

How did we come to understand structures in the human body? In the centuries before Greek civilization, when mythical worldviews prevailed, most cultures regarded illness as being caused by spirits or by people capable of using magic. Consequently, symptoms were not thought to be significant, and little attention was paid to symptoms of a disease or its effects on body structures.

Greek rational medicine of the fifth century B.C. took a dramatically different approach to disease (discussed further in Chapter 40). Early Greek doctors paid careful attention to symptoms of diseases, but cultural and religious attitudes restricted their opportunity to examine inner body structures. However, beginning in the third century B.C., Greek physicians were allowed to dissect human bodies in Alexandria, North Africa. They may have even studied internal organs in the living bodies of criminals condemned to death! Thanks to these studies, Greek physicians were able to describe the major internal organs of the body and had begun to examine smaller structures such as nerves in the head. However, their research ended abruptly when Alexandria was conquered by the Romans, who believed that the human body was sacred and should not be dissected. The growth of Christianity in the early part of the present era reinforced this Roman view.

Despite the prohibition on dissection, Galen, a Greek physician who lived in the second century A.D., produced the most accurate and detailed description of the human body done in antiquity. Galen had studied in Alexandria and was later a physician to the Roman emperor Marcus Aurelius. He synthesized all the anatomical knowledge of his day in a manuscript, *On Anatomical Procedures*. Galen's anatomy included detailed descriptions of muscles, the skeleton, blood vessels, nerves, and internal organs of the body. Because Galen lacked access to human subjects, he dissected Barbary apes, an animal whose anatomy is similar, but not identical, to that of a human. Most of the "mistakes" since identified in Galen's anatomy are actually not mistakes at all but rather accurate depictions of Barbary ape features that are slightly different from similar structures in humans (see Figure 1).

Galen's was the standard reference in human anatomy for well over 1,000 years. During that time, social, political, religious, and economic changes in the Western world led to greatly reduced interest in "higher learning." Consequently, few people studied anatomy, and those who did added little to what Galen had already described.

Beginning in the Renaissance of the fifteenth century, several physicians attempted to revive the spirit of Galenic research in anatomy. Andreas Vesalius, a physician who was born in Brussels but who taught and did his greatest work in Italy, was a prominent member of this group. Vesalius focused on the numerous "errors" in Galen's anatomy and set out to correct them. He also realized that the recent invention of printing and the associated technological innovations made it possible for him to publish descriptions with illustrations. In 1543, Vesalius published what is often described as the greatest anatomy book of all time, *De humani corporis fabrica* ("On the Fabric of the Human Body"). His book gives extensive details of all structures of the human body and is illustrated with

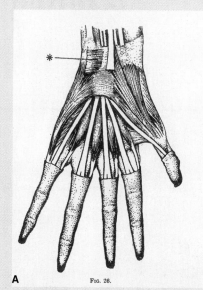 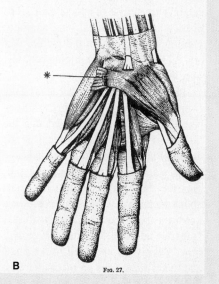

A   FIG. 26.   B   FIG. 27.

Figure 1 These drawings show partial dissections of the left hand of (A) the Barbary ape, the animal on which Galen did most of his anatomical studies, and (B) the human. Although the ape's hand contains all of the muscular structures present in the human hand, many differences are readily apparent. For example, the third (middle) finger is longest in the human, whereas the fourth is longest in the ape. Also, the human hand is broader and relatively shorter, and the human thumb is relatively much longer. Nevertheless, the similarities between the species were sufficient for Galen's work to be useful to anatomists and surgeons for hundreds of years.

box continues

FOCUS ON SCIENTIFIC PROCESS

beautiful and usually very accurate woodcuts (see Figure 2).

Finer details have since been added to the picture developed by Vesalius. Skilled surgeons in the seventeenth, eighteenth, and nineteenth centuries contributed knowledge of *morbid anatomy* (changes in anatomy caused by disease) and elaborated on lesser-known systems, particularly the nervous system. For example, Jakob Winslow, a Danish anatomist, published a four-volume set of anatomy books in 1733 that summarized and extended knowledge of the nervous system.

The development of precision microscopes and techniques for studying cells and tissues allowed studies of anatomy at the microscopic level during the nineteenth century. In the twentieth century, anatomy research has progressed to the molecular level.

This brief survey of the history of anatomy reveals a basic story in which physicians progressively described structures of the body in ever-increasing detail. There were, of course, periods of history when interest in anatomy seemed to regress, but the picture that emerges is one of a continuing progression in knowledge.

During the first few decades of the twentieth century, historians, philosophers, and scientists believed that the history of anatomy was typical of the development of knowledge in science—that is, science consists of ongoing, progressive discoveries of positive truths about the natural world. This position is known as *positivism* because it stresses "positive additions" to knowledge over time. Positivism appealed to scientists and made physicians proud of their heritage; regrettably, it was largely inaccurate.

Although knowledge in some fields of science, such as anatomy, expanded slowly and progressively, the overall process of science is more chaotic. Broad, sweeping theories change in

time, and with those changes come dramatic shifts in previously held "truths" or "relevant facts." Whole bodies of information—for example, the now-obsolete detailed descriptions of mythical animals that once graced natural history books or the once highly regarded books of anthropomorphic descriptions of animal behavior—can be rejected as unreliable by modern science.

What does history tell us about scientific knowledge? In a few areas of biology, such as human anatomy, knowledge advanced progressively over several centuries. However, the "layer cake" view of the progress of science—each generation uncovering a new layer of truth and adding it to what was inherited from the past—distorts our picture of a much more complex and interesting *process of science*.

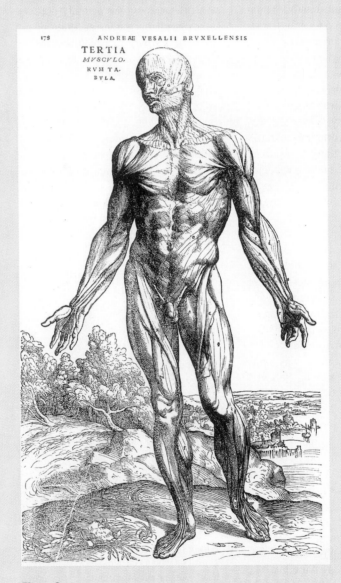

Figure 2 Vesalius's famous anatomy book of 1543, *De humani corporis fabrica* ("On the Fabric of the Human Body"), is illustrated with beautiful and accurate woodcuts such as this one. The small numbers and letters inscribed on the human figure correspond to text explanations of anatomical structures.

INTEGUMENTARY SYSTEM

The **integumentary system** includes the **skin,** the external layer that covers the body, and structures that originate in the skin such as nails, hair, certain glands, and sensory receptors that detect environmental stimuli. The skin is the largest human organ; it has a surface area of about 2 square meters and weighs about 5 kilograms. The integumentary system performs various functions related to protection and homeostasis.

Skin

The skin, shown in cross section in Figure 34.1, may be the least appreciated human organ because of its apparent simplicity.

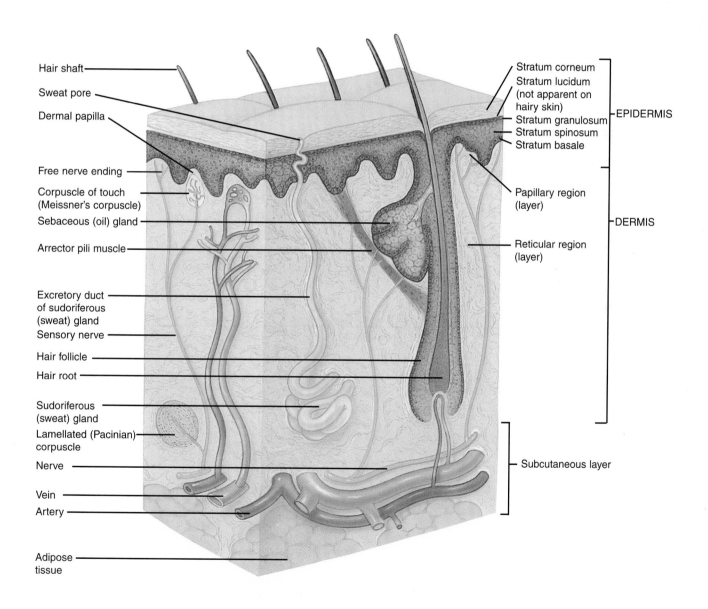

Figure 34.1   The human skin is composed of many different tissues and organs that protect the human body from harmful external environmental factors and help maintain homeostasis of the internal environment.

However, microscopic examination reveals a complex arrangement of cells, tissues, and specialized structures. Skin is composed of two primary layers: a thin, outer layer of closely packed epithelial cells called the **epidermis** and a thicker, inner layer, the **dermis**, consisting of connective tissue, various cells and tissues, and blood vessels. Connective tissue fibers from the dermis anchor the skin in a **subcutaneous layer** that contains blood vessels, nerves, and adipose (fat) tissue.

Epidermis

The epidermis is shown in Figure 34.2A. About 90 percent of cells in the epidermis progress through a simple life cycle that takes about four weeks: they originate in the basal cell layer, mature as they move toward the surface, die, and are then finally sloughed from the skin surface. The **basal cell layer** contains stem cells that reproduce rapidly and continuously by cell division to produce cells that will become **keratinocytes.** Other stem cells produce cells that enter the dermis, where they develop into glands or follicles that produce hair. The epidermis also contains sensory receptors that are sensitive to touch, temperature, and pain.

Newly formed keratinocytes mature as they begin to move toward the surface of the skin. As maturation proceeds, keratinocytes become flattened and begin to synthesize and accumulate **keratin,** a tough, fibrous protein that is flexible and impermeable to water. As keratin fills the cell, the cytoplasm, nucleus, and other organelles degenerate, and the cell becomes *keratinized*—completely filled with keratin—and dies. Dead keratinized cells are closely packed together at the skin surface, where they form a cell layer called the **stratum corneum,** a cover that protects underlying tissues from bacterial infections and physical trauma such as heat and light. Although keratinized cells are continuously sloughed from the skin's surface, the rate at which they are lost is normally equal to the rate of replacement from stem cell reproduction in the basal cell layer.

Melanocytes are epidermal cells that produce *melanin,* the brown pigment that absorbs ultraviolet (UV) light and is responsible for eye, hair, and skin color (see Figure 34.2B). Once synthesized, melanin is distributed to maturing keratinocytes. Exposure to sunlight, specifically to UV rays, initiates the following protective response: (1) melanocytes enlarge and

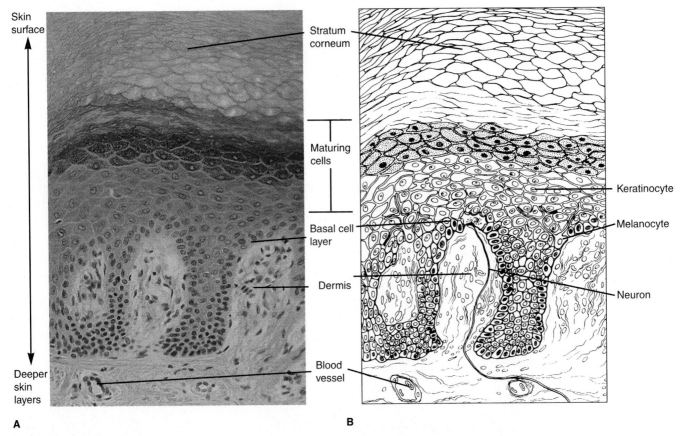

A   **B**

Figure 34.2   (A) The epidermis is composed of different types of cells. Stem cells in the basal cell layer produce new cells by mitosis and cytokinesis that become cells called keratinocytes. As keratinocytes mature, they move toward the surface of the skin, become keratinized, and finally become part of the stratum corneum. (B) Melanocytes in the epidermis produce melanin, which is transferred to keratinocytes as they mature. Melanin gives hair and skin their color and also protects skin cells from the harmful effects of ultraviolet radiation from the sun.

Question: *What is keratin? Why is it important?*

increase melanin production, (2) keratinocytes accumulate greater concentrations of melanin, and (3) melanin coalesces into granules that shield the nucleus, thus protecting DNA, which is very sensitive to UV light. Thus a suntan, an increase in skin pigmentation due to higher concentrations of melanin in keratinocytes, is actually the end result of a biological protective response because the function of melanin is to protect the skin from harmful UV radiation. Can you explain the evolutionary advantage, for human populations inhabiting equatorial regions of the world, of having darker skins? What might be the connection between UV radiation and skin cancer?

Dermis

The epidermis is firmly attached to the underlying dermis, an organ composed primarily of collagen and elastic fibers and a few fibroblasts and adipose cells. Collagen and elastic fibers in the dermis give the skin its strength and elasticity. The dermis holds hair, sweat and sebaceous glands, blood vessels, and sensory receptors, all of which are embedded in underlying connective tissues and fibers.

Hair

Like the outer epidermis, **hair** is constructed of dead, keratinized cells but traces its origin to living, rapidly dividing stem cells that exist in deeper tissue layers (see Figure 34.3). Two structural types of hair are produced by **hair follicles** located in the dermis. Fine, soft hairs cover much of the body, and coarser hair grows on everyone's scalp, eyebrows, and genital areas and on the faces of males. Hair color is caused by melanin that is incorporated into cells forming hair. Brown and black hair contain varying concentrations of typical melanin; blond and red hairs contain types of melanin with iron and sulfur; and gray and white hair lack melanin because of decreased melanocyte activity related to increasing age.

For the most part, human hair is thought to be an evolutionary relic that served as insulation for preventing heat loss in our hominid ancestors. The continuing presence of scalp hair may be partly related to that function, considering that the rate of heat loss is greatest through the scalp. Nevertheless, the importance of a heat-retaining hair covering has been diminished for humans by clothing and, more recently, by occupancy of heated dwellings. Hairs offer other types of protection in localized areas. Eyebrows and eyelashes offer protection against particles entering the eyes, and hairs in our nostrils prevent the entry of particles through the nose.

The activities of hair-producing follicles in the scalp are cyclical. Normally, follicle cells are busy dividing and constructing hair at a rate in excess of 1 centimeter per month. However, follicles periodically enter a quiescent phase after months or years of activity, and hair growth is arrested for several months. Approximately 10 percent of the hair follicles are in the resting phase at any particular time, and some people become concerned as these hairs are shed when being combed or washed. Approximately 100 scalp hairs are lost and replaced each day. However, greater hair loss occurs during periods of

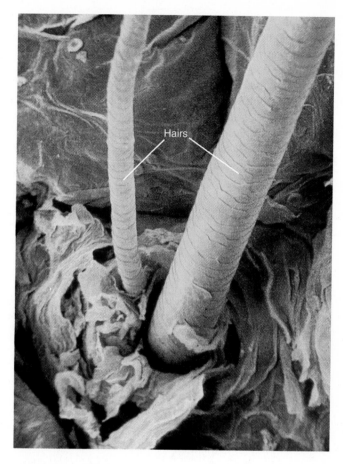

Figure 34.3 Hair projects through openings in the stratum corneum.

Question: *What determines the color of human hair?*

stress or illness, after childbirth, or after receiving radiation and drugs used in treating cancer.

Sebaceous Glands

Sebaceous glands are associated with hair follicles, and they secrete an oily substance that waterproofs the skin, prevents hair from drying, and suppresses growth of bacteria. These glands become especially active during adolescence, and when they become obstructed or infected, pimples, the universal affliction of teenagers, develop.

Sweat Glands

Sweat glands are coiled structures that secrete **sweat**—a complex solution containing water, salts, sugar, lactic acid, and small amounts of nitrogen-containing metabolic waste products—onto the skin surface. The primary function of sweat glands is related to temperature regulation.

Sensory Receptors

Maintaining homeostasis requires continuous information inputs about situations in the internal and external environment. The skin contains **sensory receptors,** which are generally specialized cells of the nervous system capable of detecting certain changes in external or internal conditions. There are different types of sensory receptors, and each is sensitive to one specific stimulus (see Figure 34.4). Sensory receptors for pain, touch, pressure, and temperature detect changes related to their function and relay this information to the appropriate control center (brain or spinal cord) responsible for maintaining homeostasis. For example, stimuli, such as fire, that are detected by pain receptors in the hand will lead to responses that will cause the person to remove the hand rapidly from the heat source.

Temperature Homeostasis

The average internal, or core, temperature of humans is 37°C (98.6°F), although it fluctuates within a normal range throughout the day. Heat results as a by-product of metabolic reactions that occur endlessly in all living cells. Except during exposure to cold temperatures, a certain amount of body heat must be dissipated to maintain a constant internal temperature.

The skin functions as a controlled radiator that regulates heat carried by circulating blood. Ninety percent of the total

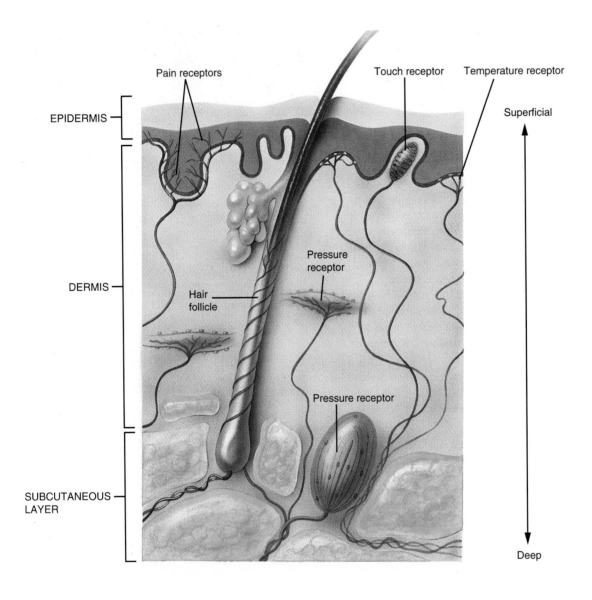

Figure 34.4   The skin contains different sensory receptors for touch, pain, temperature, and pressure. These receptors detect changes in the external environment and relay this information to the appropriate control center responsible for maintaining homeostasis.

excess heat is dissipated through about 2.5 million sweat glands. Most of the heat is lost when **perspiration,** the secretion of sweat onto the skin's surface, evaporates (converts from liquid into vapor). This change requires energy that is provided by the heat. Recall that water has a high specific heat, which means that significant amounts of heat are lost through the evaporation of sweat.

This homeostatic mechanism usually operates unnoticed except when excessive sweating occurs during periods of exercise or high external temperatures. As shown in Figure 34.5, when the core temperature increases beyond the normal range, a negative feedback system begins to operate. First, temperature receptors in the skin detect the change and send an impulse to the temperature-regulating center in the brain. The brain in turn sends signals that stimulate sweat glands to increase perspiration, which accelerates the loss of heat through evaporation until the normal temperature is restored. In cool temperatures, perspiration is reduced or does not occur. Evaporation is most efficient in a dry atmosphere and operates ineffectively in a humid environment. Special precautions are therefore necessary when living in a hot, humid environment where overheating may have potentially serious consequences. In response to increasing core temperature, the brain also stimulates the circulatory system to increase the quantity of blood flowing through blood vessels near the skin surface. This increased blood flow results in heat being lost by radiation, the transfer of heat between a warmer and cooler object. When the external temperature is cold, blood flow to the skin is reduced, and consequently, less heat is lost from the internal environment.

BEFORE YOU GO ON The skin is a large, complex organ that is the primary part of the integumentary system. It acts as a flexible armor against injury, inhibits water loss, and prevents the entrance of harmful microorganisms. The skin consists of epidermis and dermis, each having specialized functions. The basal cell layer of the epidermis generates keratinocytes that mature and develop into the outer layer of the skin. Melanocytes produce melanin, a pigment that protects skin from harmful UV radiation and gives color to skin and hair. The dermis contains hair follicles, sebaceous glands, and sweat glands, all of which have protective functions. Sweat glands help regulate body temperature, and sensory receptors have roles in maintaining homeostasis.

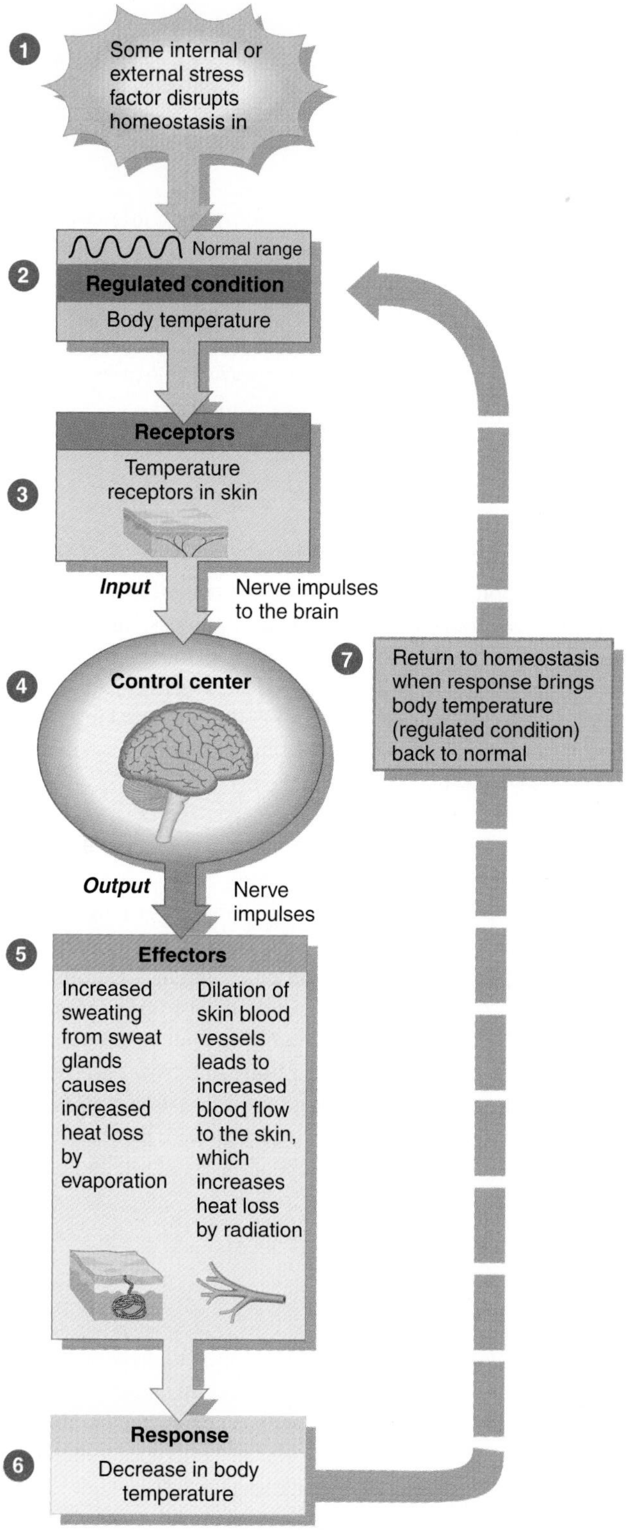

---

1. Some internal or external stress factor disrupts homeostasis in

2. **Regulated condition** — Normal range — Body temperature

3. **Receptors** — Temperature receptors in skin
   Input — Nerve impulses to the brain

4. **Control center**
   Output — Nerve impulses

5. **Effectors** — Increased sweating from sweat glands causes increased heat loss by evaporation — Dilation of skin blood vessels leads to increased blood flow to the skin, which increases heat loss by radiation

6. **Response** — Decrease in body temperature

7. Return to homeostasis when response brings body temperature (regulated condition) back to normal

Figure 34.5 One role of the skin is to help regulate body temperature. It performs this function through a classic negative feedback system. When stimulated by heat, temperature receptors in the skin transmit impulses to the brain. Subsequently, sweat glands are stimulated to produce sweat, which evaporates from the skin's surface and results in cooling. Also, blood flow increases in blood vessels near the skin, which results in increased blood flow and loss of heat to the exterior. When normal body temperature is restored, sweat production ceases and blood vessels return to their usual state.

Source: Gerard J. Tortora and Sandra R. Grabowski, *Principles of Anatomy and Physiology*, 8th Edition. New York: Harper Collins College Publishers, 1996.

Question: *How might this system respond to a* decrease *in body temperature?*

SKELETAL SYSTEM

Perhaps because of its historical association with Halloween and its traditional use in depicting Death, the misconception persists that the human skeleton is a dead structure. The living skeletal system consists of *active* tissues and organs. The structural design and architecture of bones, cartilage, joints, and ligaments, as measured by strength and flexibility, constitutes a marvel of biological engineering. To the frustration of beginning medical students who are required to know such things, the skeleton of the average human adult contains 206 bones. The general structure of the human skeleton and some specific bones are shown in Figure 34.6. Principal functions of the skeletal system are summarized in Table 34.1.

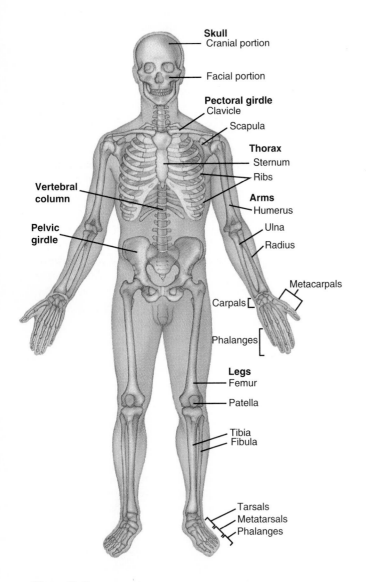

Figure 34.6 The major divisions of the human skeleton include the skull, pectoral girdle, thorax, arms, vertebral column, pelvic girdle, and legs. The names of a few of the 206 bones of the human skeleton are indicated.

Table 34.1 Some Major Functions of the Skeletal System

Function	Description
Protection	Bones protect critical internal organs, such as the brain, spinal cord, heart, lungs, and reproductive organs from harm
Support	Bone supports soft tissues and serves as attachment sites for most muscles
Movement	Skeletal muscles attached to bones help move the body
Blood cell production	Red and white blood cells are produced in marrow, a connective tissue found within certain bones
Mineral homeostasis	Bones serve as a repository for calcium, phosphorus, sodium, and potassium; through negative feedback mechanisms, they can release or take up minerals to maintain homeostasis

Bones and Bone Tissue

Bone is a connective tissue with cells embedded in a matrix of calcium compounds, which impart hardness, and collagen fibers, which provide flexibility. This chemical composition gives bones enormous strength and makes them capable of supporting massive weights without bending or breaking. Bones contain numerous spaces that are infiltrated by blood vessels that service various cells that make up bone tissue.

The architecture of different bones varies according to their shape, size, and function. There are two general types of bone tissue, compact bone and spongy bone. **Compact bone** is dense and provides protection and support for long bones of the arms and legs (see Figure 34.7). The length and strength of these bones are important for mobility and as sites for skeletal muscle attachment. **Spongy bone** makes up the inner portion of these long bones; it is much less dense than compact bone. Spongy bone surrounds the **bone marrow,** a type of connective tissue capable of storing fat (yellow marrow) or making several types of blood cells (red marrow). Bone marrow is found inside ribs, hipbones, backbones, the skull, and parts of some long bones.

Bone Transformations

Bones are dynamic organs. They continue to grow and undergo structural modifications throughout an individual's lifetime, although most skeletal growth is completed by about 20 years of age. Four connective tissue cell types function in generating new bone or remodeling old bone (see Figure 34.8). Bone **remodeling** is a normal process in which old bone tissue is continually replaced by new bone tissue after maturity.

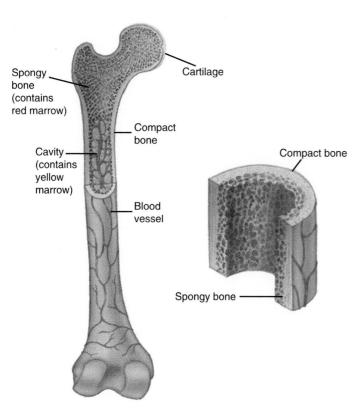

Figure 34.7 A long bone such as the humerus consists of an outer part composed of compact and spongy bone tissues that impart strength. The inner spongy bone contains red marrow, which produces red and white blood cells, and yellow marrow, which stores fat.

Cells

Stem cells in bone divide by mitosis and cytokinesis and produce cells that develop into **osteoblasts,** which build new bone tissue by secreting collagen and other organic substances that then become calcified, or impregnated with calcium salts. Once an osteoblast, which is initially found on bone surface, becomes surrounded by new bone containing collagen and calcified matrix materials, it develops into an **osteocyte,** a branched cell that sustains newly formed bone tissue by exchanging nutrients and waste materials with blood cells in nearby blood vessels (see Figure 34.9A, page 646). Bone remodeling begins when cells known as **osteoclasts** initiate bone resorption by digesting the calcified matrix.

Bone Remodeling

Bone remodeling is an active process in which bones are repaired, replaced, or strengthened. The pace of remodeling is different for different bones. Some bones, such as the ends of long bones, are replaced every four months or so; other parts of long bones are never replaced unless they are broken. The following sequence of events occurs in bone remodeling: (1) osteoclasts become attached to the bone surface and secrete enzymes and acids that digest collagen and bone minerals, such as calcium (see Figure 34.9B, page 646); (2) digested bone materials are apparently taken up by osteoclasts by phagocytosis; and (3) osteoblasts build new bone as described in the preceding paragraphs.

Several factors influence bone remodeling, and lifestyle management factors have relevance for helping you maintain a healthy skeleton. Remodeling is enhanced by mechanical stress; that is, exercise such as walking, running, or weight lifting, can help make bone tissue stronger because additional

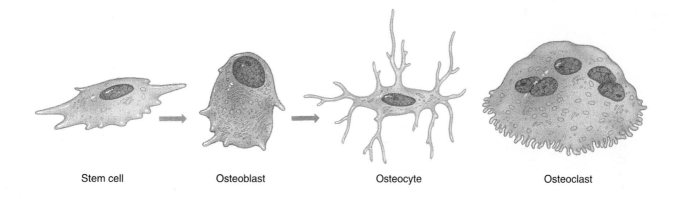

Stem cell   Osteoblast   Osteocyte   Osteoclast

Figure 34.8 Several cell types participate in building new bone and remodeling old bone. *Stem cells* divide by mitosis and cytokinesis to produce cells that develop into *osteoblasts,* which build new bone tissue. Once an osteoblast finishes building new bone, it becomes an *osteocyte,* which maintains newly formed bone by providing it with nutrients and helping remove waste products. *Osteoclasts* have multiple nuclei and function in bone remodeling.

Question: *Do all stem cells develop into osteoblasts?*

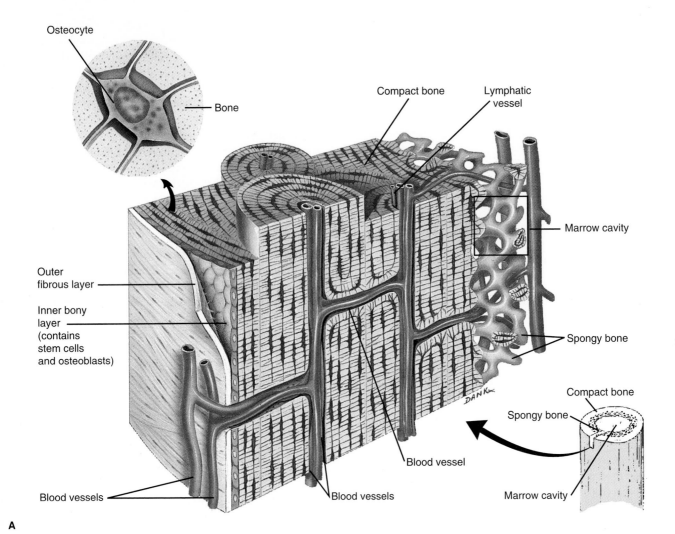

Osteocyte

Bone

Compact bone

Lymphatic vessel

Outer fibrous layer

Inner bony layer (contains stem cells and osteoblasts)

Marrow cavity

Spongy bone

Compact bone

Spongy bone

Blood vessel

Blood vessels

Blood vessels

Marrow cavity

A

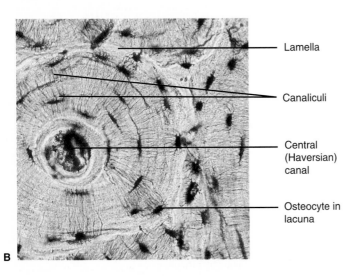

Lamella

Canaliculi

Central (Haversian) canal

Osteocyte in lacuna

B

Figure 34.9 (A) Compact bone is composed of circular layers of dense, calcified tissue that is maintained by osteocytes, which exchange materials with blood carried by blood vessels. Stem cells and osteoblasts are found on the bone surface, beneath an outer fibrous layer. (B) When bone undergoes remodeling, osteoclasts digest compact bone materials, and new bone is formed by osteoblasts to replace the bone that was removed.

Question: *Is bone remodeling a normal process? Why is it important?*

minerals and collagen are deposited in response to the stress. Conversely, a lack of exercise, or becoming bedridden, results in a weakening of bone tissue because the loss of bone materials during normal remodeling will exceed the deposition rate of new materials in bone being remodeled. Diet is also important because significant quantities of calcium, phosphorus, magnesium, other minerals, and various vitamins are used in bone remodeling.

Other Structures

Figure 34.10 shows other structures associated with the skeletal system. **Cartilage** is a connective tissue that is composed of cells, collagen, and elastic fibers embedded in a solidified chemical matrix. Cartilage is tough but flexible and is found in joints and the ends of long bones, such as those of the arms and legs, in adults. Different types of cartilage are also involved in the construction of the windpipe, backbone, and ears. **Joints** are sites of connection between different bones or bones and cartilage. These structures are linked together in such a way that skeletal movement is possible. Bones in joints are connected to one another by bands of tough connective tissue called ligaments. **Ligaments** also serve to restrict bone movement, thus preventing bones from being dislocated. *Tendons* attach muscle to bone and are described later in this chapter.

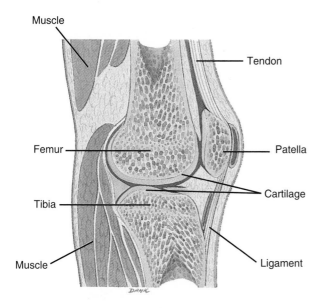

Figure 34.10 This diagram shows a knee joint and various structures that protect bones and connect bones and muscles. *Joints* are sites where bones are connected to one another. *Cartilage* is a tough, flexible connective tissue that covers the ends of long bones such as the femur and tibia. *Ligaments* join bones together, and *tendons* connect muscles to bones.

Protection and Support

Besides being a reservoir for blood cell–producing cells and mineral reserves, the skeleton has other important functions. The two most obvious are to provide support for the body and protection of critical internal organs. Different parts of the skeleton protect certain organs. The **skull** shields the brain, eyes, and ears from injury. The backbone or **vertebral column** is composed of separate small bones called **vertebrae,** a type of construction that imparts great strength but also allows great flexibility. The backbone provides primary support of the skeleton and serves as attachment sites for the skull, rib cage, and pelvic girdle. It also protects the spinal cord from injury. The lungs, heart, stomach, and liver are covered by the **rib cage,** and the **pelvic girdle** protects internal reproductive organs. Finally, the skeleton provides sites for muscle attachment. The action of muscles imparts movement to the human skeleton.

Mineral Homeostasis

The skeleton contains significant quantities of many minerals, including calcium, potassium, sodium, and phosphorus; this explains why the skeleton remains intact long after death. The importance in life, however, is that the skeleton serves as a mineral storehouse within the body. When there is a demand for a specific mineral, it can be mobilized from bones through remodeling and distributed by the blood to sites where it is required.

Calcium ions (Ca^{2+}) are essential for normal functions of muscle, nerve, and blood cells. Also, during pregnancy, some calcium from the mother's skeleton is transferred to the developing skeleton of the fetus. Consequently, Ca^{2+} levels in the blood are regulated by homeostatic control. Various hormones are involved in regulating Ca^{2+} blood levels, but the primary feedback system is described in Figure 34.11 (see page 648). If Ca^{2+} blood levels decrease below the normal range, receptors on parathyroid gland cells recognize the change and stimulate the gland cell to secrete **parathyroid hormone (PTH)** by activating the gene encoding PTH. Once released into the blood, PTH increases the rate of bone remodeling activities by osteoclasts, which results in more Ca^{2+} being released to the blood. PTH also influences kidney activity in such a way that Ca^{2+} is not eliminated in urine. Once the normal Ca^{2+} blood concentration has been restored, PTH secretion ceases and osteoclast activities decrease.

BEFORE YOU GO ON The skeleton is constructed of bones, cartilage, joints, and ligaments. Together these organs constitute an internal support system that has enormous strength and great flexibility. The principal functions of the skeletal system are to provide support and protection for internal organs, permit movement of the organism, produce blood cells, and help maintain mineral homeostasis. The skeleton is a dynamic system, with cells and tissues that remain active throughout a person's lifetime.

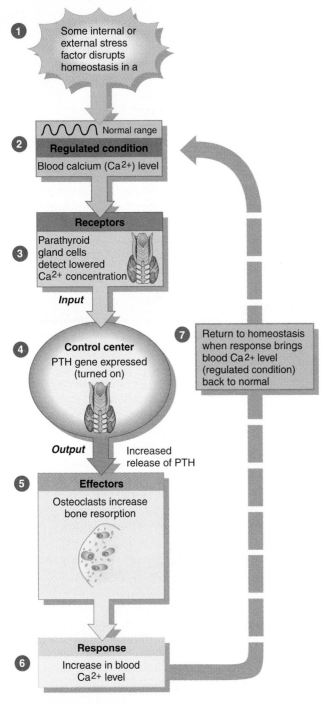

1 Some internal or external stress factor disrupts homeostasis in a

2 ⟁⟁⟁ Normal range
Regulated condition
Blood calcium (Ca^{2+}) level

3 **Receptors**
Parathyroid gland cells detect lowered Ca^{2+} concentration

Input

4 **Control center**
PTH gene expressed (turned on)

Output   Increased release of PTH

5 **Effectors**
Osteoclasts increase bone resorption

6 **Response**
Increase in blood Ca^{2+} level

7 Return to homeostasis when response brings blood Ca^{2+} level (regulated condition) back to normal

Source: Gerard J. Tortora and Sandra R. Grabowski, *Principles of Anatomy and Physiology*, 8th Edition. New York: Harper Collins College Publishers, 1996.

Figure 34.11 Bones play an important role in calcium ion (Ca^{2+}) homeostasis. If Ca^{2+} blood levels decrease below a certain point, the change is detected by receptors on parathyroid gland cells, the gene encoding parathyroid hormone (PTH) is activated, and the cells secrete PTH. PTH increases the rate of bone resorption by osteoclasts, which increases Ca^{2+} levels in the blood and eventually results in the system's being deactivated.

MUSCULAR SYSTEM

As described in Chapter 33, there are three types of human muscle tissue—skeletal, cardiac, and smooth (see Table 34.2). However, the **muscular system** includes only skeletal muscles and various supporting structures—blood vessels, connective tissues, and neurons—that together make up organs, such as the biceps muscle, involved primarily in movement. *Cardiac muscle* is considered part of the circulatory system; *smooth muscle,* part of the digestive, urinary, respiratory, and circulatory systems. Hence we will focus most of our attention on skeletal muscle.

Skeletal Muscle

Skeletal muscles function primarily in producing movement and helping in temperature homeostasis. They are under *voluntary* control, which means that a person can consciously make them contract and relax. Skeletal muscle cells (fibers) have the following features: they appear *striated* because of alternating light and dark bands that can be seen through a microscope, each cell has many nuclei, and individual cells are long and fiberlike in appearance.

Structure and Organization

The basic structure and organization of skeletal muscle tissue is illustrated in Figure 34.12. Skeletal muscles are surrounded by a **fascia,** a band of fibrous connective tissue. Within the fascia are several groups of separate muscle tissue bundles called **fascicles,** blood vessels, and *motor neurons,* which stimulate contraction. Each fascicle generally contains 10 to 100 muscle cells that are about 100 micrometers long. Three different layers of connective tissue are found in a skeletal muscle; one layer surrounds the whole muscle, one encloses fascicles, and one encircles individual muscle cells. All three connective tissue layers may extend beyond the muscle cells and, along with the fascia, form a cord of dense connective tissue called a **tendon,** which attaches the muscle to a bone.

As shown in Figure 34.13, skeletal muscles cause movements of the skeleton by applying force, transmitted through tendons, to bones. For each muscle, the **origin** refers to the tendon attached to the bone that does not move during muscle contraction, and the **insertion** is the other tendon, which is attached to the bone that does move when the muscle contracts.

Muscle Cell Structure

Muscle cells have an extremely intricate structure, as illustrated in Figure 34.14 (see page 650). The plasma membrane of a muscle cell is called the **sarcolemma,** and **sarcoplasm** refers to muscle cell cytoplasm. Each cell contains many nuclei and mitochondria that are located at the outer edge of the cell. Since muscle cells actively synthesize proteins and use great amounts of ATP during contraction, the abundance of these organelles reflect muscle cell function.

Table 34.2 Characteristics of Muscle Tissues

Characteristics	Type of Muscle Tissue		
	Skeletal	**Cardiac**	**Smooth**
Microscopic appearance	Striated, multiple nuclei, cells not branched	Straited, single nucleus, cells branched	Nonstriated, single nucleus, cells spindle-shaped
Organ Systems	Musclar system	Circulatory system	Digestive and urinary systems
Location	Most attached to bone; some to skin and muscles	Heart	Walls of digestive organs, blood vessels, and hair follicles
Nervous control	Voluntary	Involuntary	Involuntary
Contraction speed	Fast	Moderate	Slow

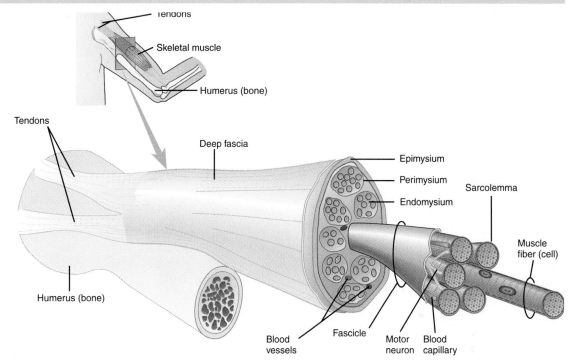

Figure 34.12 (A) Skeletal muscles are connected to bones by tendons. (B) This figure describes the structure of skeletal muscle tissue. Groups of skeletal muscle cells, which are also called "fibers," are surrounded by a connective tissue layer called a *fascicle.* In turn, bundles of fascicle are contained within *fascia,* which is composed of tough connective tissue. Layers of connective tissue underlay the fascia and fascicles and separate muscle cells from one another. Blood vessels and motor neurons are also parts of skeletal muscle tissue

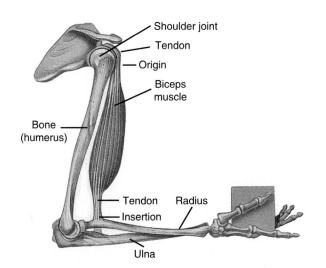

Figure 34.13 Skeletal muscles generate movements by moving bones. The *origin* refers to the tendon that connects the muscle to the bone that will not move when the biceps muscle contracts; the *insertion* is the tendon connected to the bone that will move during muscle contraction.

Question: *Which bone will move when the muscle contracts?*

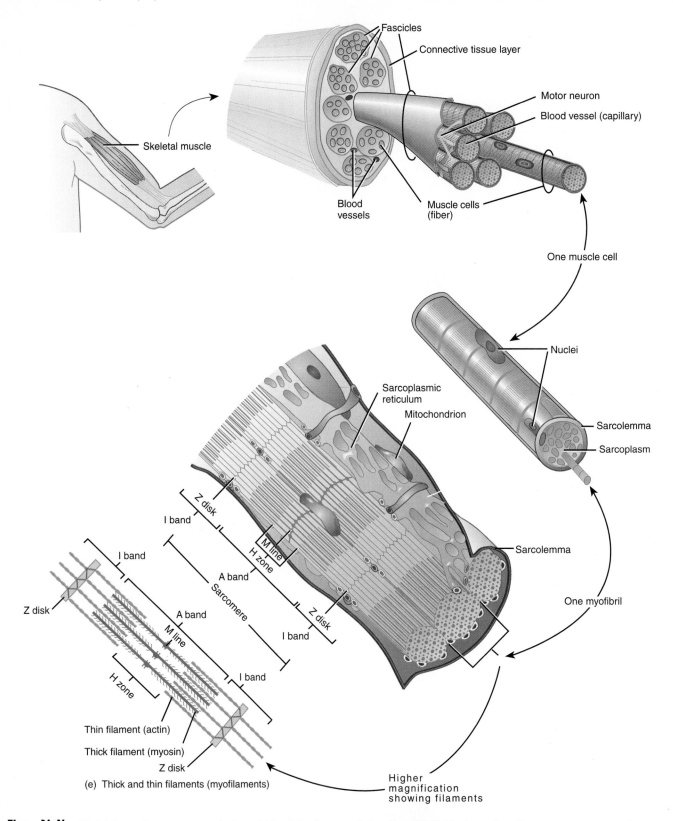

Figure 34.14   Skeletal muscles are composed of several fascicles that contain bundles of individual muscle cells. Each muscle cell has several nuclei and mitochondria and is surrounded by a *sarcolemma* (plasma membrane) that encloses the *sarcoplasm* (cytoplasm). Within the sarcoplasm are many tiny threadlike structures called *myofibrils.* These are composed of thick and thin filaments that are organized into *sarcomeres,* which have characteristic bands, disks, and zones.

Question: *What is the function of mitochondria in a muscle cell?*

The sarcoplasm of each muscle cell is filled with small threadlike structural units called **myofibrils,** that are made of proteins capable of contracting when a muscle is working. Myofibrils are composed of even smaller structures called *filaments.* There are two primary types of filaments: **thick filaments** are made primarily of the protein **myosin,** and **thin filaments** are made of the protein **actin.** The overlapping of thin and thick filaments in the myofibrils, gives skeletal muscle cells their characteristic striated appearance. Each myofibril is surrounded by a fluid-filled tubular membrane system called the **sarcoplasmic reticulum** that stores Ca^{2+}. Calcium ions play a key role in nerve impulse transmission and muscle cell contraction.

Myofibril filaments are organized in sections called **sarcomeres** (see Figure 34.15). Each sarcomere is defined by a series of bands that reflect relationships between thin and thick filaments. The various bands are as follows:

■ The *Z disk* forms the boundary between sarcomeres; it marks sites where thin filaments are fastened within a sar-

comere and thick filaments are anchored by attached elastic filaments.

■ The *A band* appears dark; it includes the two sets of thick filaments in a sarcomere and the parts of the thin filaments that overlap with thick filaments.

■ The *I band* appears light; it includes parts of the thin filaments that do not overlap with thick filaments. A Z disc appears in the center of each I band.

■ The *H zone* is narrow sarcomere segment containing the part of thick filaments that does not overlap with thin filaments.

■ The *M line* is a thin line formed by protein molecules connecting the two sets of thick filaments.

Muscle Contraction

Even though you may not realize it, skeletal muscles contract continuously, not only when you are walking or lifting weights. For example, skeletal muscle contractions to keep your head

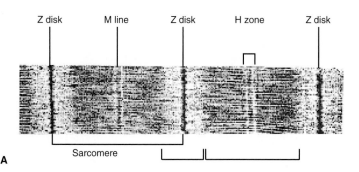

Figure 34.15 (A) An electron micrograph showing two sarcomeres. The striated appearance reflects the structure and organization of thick and thin filaments. (B) A diagram showing thick and thin filaments and the bands, disks, and line that appear as a consequence of their organization.

Question: *What accounts for the I band's appearing lighter than the A band?*

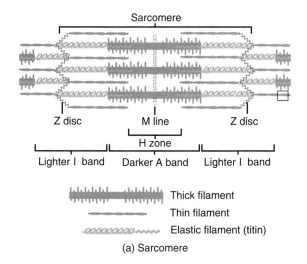

upright and maintain normal body positions. In 1954, two groups of scientists described what became known as the **sliding-filament model** of muscle contraction. According to the model, skeletal muscle contracts when thick and thin filaments slide past one another. Figure 34.16 relates the sliding-filament model to thick and thin filaments. When a muscle contracts, sections of thick-filament myosin proteins, which are attached to thin filaments, pull the thin filaments toward the H zone. As the force applied by myosin continues, the thin filaments meet, or overlap, at the center of the H zone. At the point of maximum contraction, the sarcomere appears shorter, but the lengths of thick and thin filaments actually remain unchanged. Sliding filaments and shortened sarcomeres are associated with contraction of a muscle cell and ultimately the entire skeletal muscle. Energy for muscle contraction comes from ATP produced through cellular respiration (see Chapter 4). Most of the energy used during muscle contractions is transformed to heat energy, not mechanical energy (movement). The heat generated is used to help maintain normal body temperature.

Muscles and Temperature Homeostasis

Excess heat is normally transferred from the body through the skin and lungs. What happens when there is a need to retain heat or increase the temperature of the body? Previously, we saw how the integumentary system helps maintain body temperature by stimulating or preventing heat loss through the skin. Another homeostatic feedback mechanism is illustrated in Figure 34.17. If, for some reason, body temperature declines, skeletal muscles are stimulated to contract involuntarily, a response known as **shivering.** Shivering generates

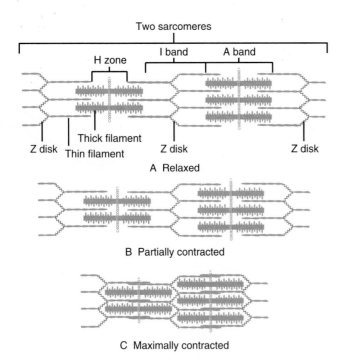

Figure 34.16 Muscle cell contraction is explained by the *sliding-filament model* shown here. (A) Sarcomeres are spaced in a relaxed muscle. (B) When contraction is initiated, segments of thick filament myosin proteins that are connected to thin filaments (represented by vertical lines through the thick filaments) pull the thin filaments inward. (C) When fully contracted, the sarcomere appears to be shorter because of the movement of sliding thick and thin filaments.

Question: *What is the source of energy for muscle cell contraction?*

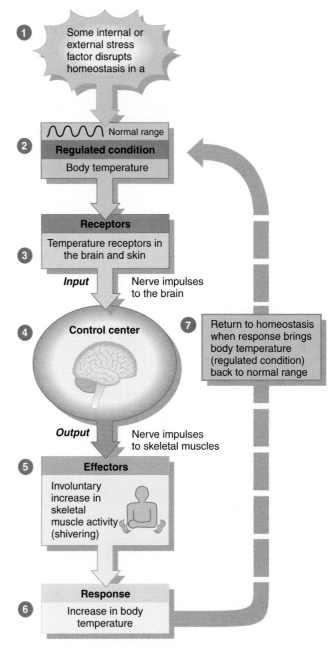

Source: Gerard J. Tortora and Sandra R. Grabowski, *Principles of Anatomy and Physiology*, 8th Edition. New York: HarperCollins College Publishers, 1996.

Figure 34.17 The muscular system helps maintain temperature homeostasis through shivering reactions. If core body temperature decreases, receptors in the skin and brain signal a control center in the brain. In response, skeletal muscles are involuntarily stimulated to contract, generating heat that increases the temperature. As body temperature increases, shivering diminishes and eventually stops.

heat, which, along with reduced heat loss through the skin, can raise body temperature back to a normal level. When that occurs, shivering ceases due to negative feedback regulation in the control center of the brain.

Cardiac Muscle

Cardiac muscle tissue occurs in only one organ, the heart wall. Cardiac muscle cells are shown in Figure 34.18. Unlike skeletal muscle cells, cardiac muscle cells are branched, have one central nucleus, have more and larger mitochondria than skeletal muscle cells, and are involuntary. Cardiac muscle cells have myofibrils and the same types of filaments, bands, zones, and Z disks as skeletal muscle cells.

Each branched cardiac muscle fiber (cell) is connected to adjoining fibers by **intercalated disks,** irregular thickened regions of the sarcolemma that contain **desmosomes**—thickened sites of attachment between two cells. Interconnected cardiac muscle fibers form two general networks—one in the ventricles (lower part) of the heart and the other in the atria (upper part). If one cell of a network is stimulated to contract, all cells in the network contract. Through this mechanism, heartbeat is normally regular and coordinated. Muscle contraction of cardiac muscle is also described by the sliding-filament model.

Smooth Muscle

Smooth muscle tissue is found in the walls of blood vessels, the stomach, the intestines, the trachea, the bronchi, the urinary bladder, and the uterus and is attached to hair follicles and to certain parts of the eye that regulate pupil diameter. When examined through the microscope, smooth muscle cells look quite different from skeletal and cardiac muscle cells (see Figure 34.19). Smooth muscle cells do contain both thick and thin filaments; however, the filaments have no regular arrangement,

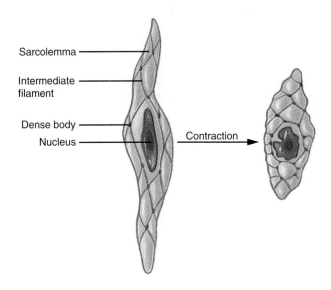

Figure 34.19 Smooth muscle cells are characterized by intermediate filaments that are attached to dense bodies. During muscle contraction, intermediate filaments shorten, and dense bodies attached to the sarcolemma are pulled inward, shortening the muscle cell.

as in sarcomeres, and so the cells appear smooth rather than striated. Smooth muscle cells contain two unique structures, **intermediate filaments** attached to **dense bodies.** Dense bodies are scattered in the sarcoplasm or attached to the sarcolemma, and bundles of intermediate filaments connect one dense body with another. Smooth muscle contraction involves a modification of the sliding-filament model. During contraction, thin and thick filaments generate force that is transmitted to the intermediate filaments. Consequently, dense bodies attached to the sarcolemma are pulled inward, shortening the muscle cell. Contraction of smooth muscle tissue is slower than contraction of skeletal or cardiac muscle tissue and lasts much longer.

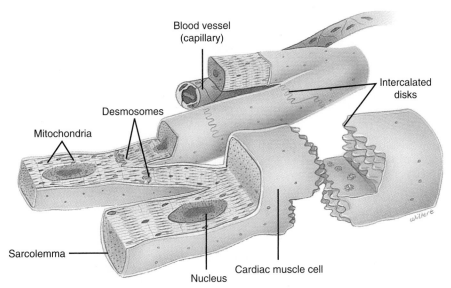

Figure 34.18 Cardiac muscle tissue is composed of branched cells that are attached to one another by intercalated disks containing desmosomes.

Question: *Which organ is composed of cardiac muscle tissue?*

BEFORE YOU GO ON There are three types of muscle tissue—skeletal, cardiac and smooth; only skeletal muscle is considered part of the muscular system. Skeletal muscle tissue has an elaborate structural architecture consisting of ever-smaller fibers, myofibrils, and filaments organized into sarcomeres. Within each sarcomere are various bands that reflect muscle cell structure and function. Skeletal muscle cell contraction is described by the sliding-filament model: during contraction, thick and thin filaments slide past one another within a sarcomere. The muscular system helps maintain temperature homeostasis through the shivering response.

In this chapter, we have seen how structures of the integumentary, skeletal, and muscular systems are related to their functions. We also found that these systems help maintain homeostasis. In Chapter 35, we continue to explore these themes in studying the respiratory, circulatory, digestive, and urinary systems.

SUMMARY

1. Humans are composed of a number of tissues, organs, and organ systems that allow them to solve different biological problems.

2. The major organ of the integumentary system is the skin, which consists of two cellular layers, the outer epidermis and the inner dermis. The life cycle of keratinocytes in the epidermis results in continuous renewal of skin tissue. Within the epidermis and dermis are various cells and tissues—melanocytes, hair, and glands—that protect underlying tissues from potentially harmful factors present in the external environment. The skin also contains sensory receptors that play important roles in maintaining temperature homeostasis.

3. The skeletal system is made up of bones and cartilage, which are connected at joints and held together by ligaments. The skeleton provides support for the body, protects vital internal organs, produces blood cells, and provides sites for muscle attachment.

4. Bones are dynamic organs that are remodeled continually throughout an individual's lifetime. Osteoblasts and osteoclasts are responsible for bone formation and remodeling; osteocytes maintain bone tissue.

5. The skeleton plays a key role in mineral homeostasis. Calcium ions are recycled between bone and other tissues through a negative feedback system.

6. The muscular system consists of skeletal muscles and supporting structures that function primarily in generating movement of the skeleton. Cardiac muscle is part of the circulatory system, and smooth muscle functions primarily in the digestive, respiratory, and urinary systems.

7. Skeletal muscle cells are composed of myofibrils and various filaments that give skeletal muscle its striated appearance. A sarcomere is characterized by various bands, disks, and lines, which are useful references for understanding muscle contraction.

8. Muscle contraction is described by the sliding-filament model in which thick and thin filaments slide past one another.

9. Skeletal muscle helps maintain temperature homeostasis through the mechanism known as shivering.

WORKING VOCABULARY

bone (p. 644)
bone marrow (p. 644)
cardiac muscle (p. 653)
cartilage (p. 647)
dermis (p. 640)
epidermis (p. 640)
hair (p. 641)
integumentary system (p. 639)

melanocyte (p. 640)
sebaceous glands (p. 641)
skeletal muscle (p. 648)
skin (p. 639)
sliding filament model (p. 652)
smooth muscle (p. 653)
sweat glands (p. 641)

REVIEW QUESTIONS

1. What are epidermis and dermis?

2. What is the function of keratinocytes? How are they involved in the renewal of skin tissue?

3. What are melanocytes? What is their function?

4. What are the protective functions of hair, sebaceous glands, and sweat glands?

5. How does the skin help regulate body temperature?

6. What are the functions of the skeletal system?

7. How is bone constructed? What is bone remodeling? Why is it important?

8. How does bone help regulate Ca^{2+} homeostasis?

9. What are the structures and functions of the muscular system?

10. Describe the structure of a skeletal muscle cell (fiber).

11. What is the sliding-filament model? What does it explain?

12. How does the muscular system function in temperature homeostasis?

ESSAY AND DISCUSSION QUESTIONS

1. What types of biological problems do the integumentary, skeletal, and muscular systems "solve"?

2. What types of adaptations involving the integumentary, skeletal, and muscular systems would you ultimately expect in humans sent to colonize the moon?

3. In *comparative anatomy* courses, students discover that various animals have anatomical structures that appear to be similar but function in different ways and, conversely, anatomical structures that function in similar ways but look quite different. What might account for these observations?

REFERENCES AND RECOMMENDED READING

Concar, D. 1992. The resistible rise in skin cancer. *New Scientist,* 134: 23–26.

Cosan, C.E. 1995. The Platonic origins of anatomy. *Perspectives in Biology and Medicine*, 38: 581–596.

Montagna, W. 1992. *Atlas of Normal Human Skin.* New York: Springer Verlag.

Parsons, K. C. 1993. *Human Thermal Environments.* Bristol, Pa.: Taylor & Francis.

Puglia, M. L. 1995. Bone: A living tissue. *Biology Digest,* 21: 10–18.

Radetsky, P. 1995. The mother of all blood cells, *Discover,* 16: 86–93.

Radinsky, L. B. 1987. *The Evolution of Vertebrate Design.* Chicago: University of Chicago Press.

Riggs, B. L., and J. L. Melton. 1992. The prevention and treatment of osteoporosis. *New England Journal of Medicine,* 327: 620–627.

Rochat, A., K. Kobayashi, and Y. Barrandon. 1994. Location of stem cells of human hair follicles by clonal analysis. *Cell,* 76: 1063–1073.

Sawday, J. 1995. *The Body Emblazoned: Dissection and the Human Body in Renaissance Culture.* London: Routledge.

Strobel, G. 1993. Minimizing molecular motor mysteries: Physics and biology work together to unravel the basis of movement. *Science News,* 144: 316--318.

Taylor, E. W. 1993. Molecular muscle. *Science,* 261: 35-36.

Tortora, G. J., and S. R. Grabowski. 1996. *Principles of Anatomy and Physiology.* 8th ed. New York: HarperCollins.

ANSWERS TO FIGURE QUESTIONS

Figure 34.2 Keratin is a tough fibrous protein. It does not allow water or other substances to enter the body.

Figure 34.3 The type and quantity of melanin present in the hair.

Figure 34.5 Sweating and evaporation would decrease, and blood vessels near the skin surface would become constricted to reduce blood flow.

Figure 34.8 No, some remain as stem cells and continue to divide.

Figure 34.9 Yes, remodeling strengthens bone and rebuilds damaged bone.

Figure 34.13 The radius (and the ulna, which is connected to it).

Figure 34.14 To provide muscle cells with energy through cellular respiration reactions (see Chapter 4).

Figure 34.15 The I bands contains only thin filaments that do not stain as darkly as the larger thick filaments.

Figure 34.16 ATP generated from cellular respiration.

Figure 34.18 Only the heart.

The Respiratory, Circulatory, Digestive, and Urinary Systems

Chapter Outline

Reading Questions

1. What organs and processes play a role in providing tissue cells with oxygen?

2. How are oxygen and carbon dioxide gas exchanges regulated?

3. How does the digestive system accomplish the task of transforming complex foods into simple compounds that can enter cells?

4. How does the urinary system help maintain internal homeostasis?

 n this chapter, we continue our exploration of organ systems that allow us to live on Earth. How do our cells receive oxygen and nutrients? How do they get rid of waste products?

RESPIRATORY SYSTEM

All cells require a steady supply of energy for carrying out their activities, and oxygen (O_2) acts as an electron acceptor in the reactions of cellular respiration that provide such energy (see Chapter 4). Consequently, most human cells die quickly if deprived of O_2 for more than a few minutes. It is also necessary that carbon dioxide (CO_2), a waste product of energy-releasing reactions, be removed from the body. The primary gas exchange mechanism involves diffusion and differences in the relative concentrations of CO_2 and O_2 at exchange sites. Air, containing O_2, is taken into the body from the environment, is moved to organs of respiration, and diffuses to red blood cells that deliver it to tissues. Carbon dioxide is then transported to the lungs by returning blood. The mechanical process of taking air into the lungs and exhaling it again is **breathing.** The exchanges of O_2 and CO_2 between an organism and the environment is called **respiration.**

Organs of the Respiratory System

The **respiratory system,** shown in Figure 35.1, consists of **lungs,** the principal organ of respiration, two treelike passages that lead to and from the lungs, and the nose, pharynx, larynx, trachea, and diaphragm. Breathing is initiated by the **diaphragm,** a large dome-shaped skeletal muscle that lies beneath the lungs and forms the floor of the thoracic cavity. When the diaphragm contracts, lung volume increases, and air

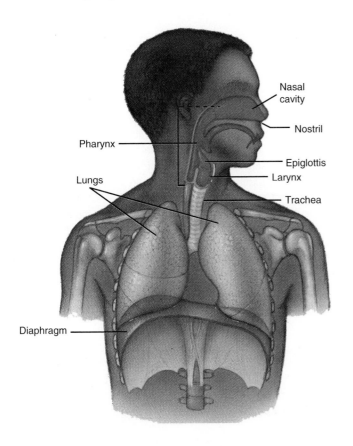

Figure 35.1 The human respiratory system provides for efficient CO_2–O_2 exchange through the organs shown in this figure.

Question: *What is the function of the diaphragm?*

enters the lung, when it relaxes, lung volume decreases, and air is forced out of the lung. Air entering through external openings, the *nostrils,* is filtered by small hairs that line the nasal surfaces. It then moves into hollow chambers called *nasal cavities,* where it is warmed, humidified, and filtered. Any small airborne particles present in the inhaled air are absorbed by the mucus layer that lines the cavity. The mucus and particles are continuously swept into the throat by ciliated epithelium that lines the nasal cavities, swallowed, and routed into the digestive tract, which ultimately disposes of them.

Air then progresses through a set of organs that divert it into a separate pathway from that taken by food and liquids. All these materials initially enter the **pharynx,** or throat. There, foods and air become separated by a flap of tissue called the **epiglottis,** which automatically seals the entry into the air passageways during swallowing. When there is a malfunction of this process and food is allowed to enter the air passage, paroxysms of uncontrolled coughing result, in an attempt to remove the obstacle. If this reflexive action fails and the air passage is blocked, choking ensues, and death results if the situation is not corrected.

From the pharynx, air moves through the **larynx,** commonly known as the "voice box," and then into the windpipe, or **trachea,** the trunk of the respiratory tree. The structure of both the larynx and the trachea reflects their function as air conduits; outer walls containing rings of cartilage hold these tubes constantly open. The larynx is a highly specialized organ composed of bone, cartilage (which forms what is commonly called the "Adam's apple"), ligaments, and skeletal muscles that moves air in such a way that sound having different pitches can be generated. Sound originates from vibrations of *vocal cords* within the larynx the mouth, tongue, nasal cavity, pharynx, and various facial muscles convert that sound to speech.

As shown in Figure 35.2A (see page 658), the trachea divides into two large **bronchi,** one advancing into each lung, that continue to branch into smaller **bronchioles.** These in turn branch repeatedly into tiny clusters of terminal thin-walled sacs called **alveoli** that are the sites of gas exchange in the lung (see Figure 35.2B, page 658). Lungs, then, are not simple bags but are complex, moist respiratory organs consisting of an enormous number of alveoli (about 750 million), with nearby capillaries, enclosed within a delicate connective tissue framework. Approximately 70 to 90 square meters of surface area is available for gas exchange in healthy lungs, an area approximately 40 times as large as the surface of the skin.

Respiratory Physiology I

The act of breathing is performed about 12 times each minute in an average adult and, in a 24-hour period, over 8,000 liters of air is inhaled and exhaled. Two organ systems—the respiratory and circulatory systems—work in partnership to carry out breathing and respiration. As with most critical functions in the human body, both breathing and respiration are involuntary activities.

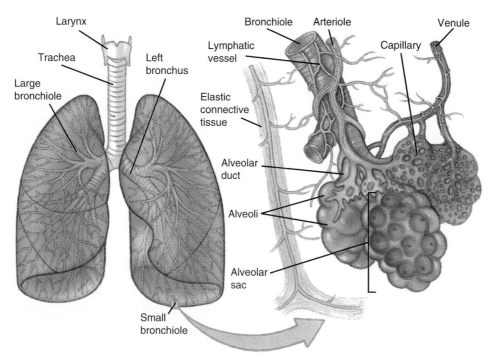

Figure 35.2   (A) Each lung contains a branching network of vessels called a *respiratory tree*. The trachea, the trunk of the tree, splits into two bronchi, which in turn branch into smaller bronchioles. (B) Bronchioles terminate in alveoli, the tissue sites where CO_2 and O_2 are exchanged. Each alveolus is serviced by lymph vessels and blood capillaries.

Question: *What is the source of the O_2 and the CO_2 exchanged in alveoli?*

Gas Exchange During Breathing

What regulates the exchange of air between the atmosphere and our lungs? The short answer to this question is that there are differences in pressure gradients. **Atmospheric pressure** is expressed as a unit of pressure that is equal to the force per unit area required to support a column of mercury (Hg) 760 mm high. Atmospheric pressure represents the weight of all gases in the atmosphere, and the normal pressure at sea level is 760 mm Hg. A difference in atmospheric pressure between two areas—that is, a difference in pressure gradients—results in air flow down a pressure gradient, from high to low pressure. For example, if atmospheric pressure at area A is 760 mm Hg and at area B it is 750 mm Hg, air will flow from A to B.

Figure 35.3 illustrates changes in atmospheric pressures that occur during breathing and their effects on air flow. The fundamental principles involved are explained by the laws of physics as they pretain to gases. Basically, a specific amount of gas enclosed in a container (the lung) exerts less pressure as the volume of the container expands; conversely, pressure increases if the volume of the container decreases. When not breathing, atmospheric pressure in both the alveoli and the atmosphere is 760 mm Hg. When the diaphragm contracts, the volume of the lung increases, which causes a slight decrease in atmospheric pressure inside the lung; consequently, air flows into the lung. When the diaphragm relaxes, lung volume is reduced, atmospheric pressure inside the lung increases, and air is exhaled.

Gas Exchange During Respiration

What controls the movement of O_2 and CO_2 between the lungs and blood? Basically, the two compartments—lung and blood—have different O_2 and CO_2 pressure gradients. **Partial pressure** (*p*) refers to the pressure exerted by a specific gas in a mixture of gases; in a sense, it is a measure of the relative concentration of a gas in a mixture. Partial pressures are also measured in mm Hg, and for O_2 and CO_2 they are expressed as pO_2 and pCO_2. In atmospheric air, $pO_2 = 160$ mm Hg and $pCO_2 = 0.3$ mm Hg. In alveoli, $pO_2 = 105$ mm Hg and $pCO_2 = 40$. Gases diffuse from an area of high partial pressure to low partial pressure; therefore, a little O_2 diffuses into the lung from the atmosphere and some CO_2 diffuses out of the lung into the atmosphere. Partial pressures are especially important in gas exchanges between alveoli and blood and between blood and tissue cells. These exchanges are described in the next section, on the circulatory system.

BEFORE YOU GO ON   The main function of the respiratory system is to supply cells with O_2 and to remove CO_2 from the body. Oxygen contained in air enters lungs, and CO_2 is exhaled, when breathing occurs. Once in the lung, air passes through a system of ever-smaller branches until it reaches the alveoli, tissues where gases are exchanged with red blood cells. Most gas exchange between the atmosphere and the lungs occurs because of pressure gradients—differences in atmospheric pressures in the atmosphere and lungs—established by breathing. Some O_2 and CO_2 exchange also occurs because of differences in partial pressures between the atmosphere and the lung.

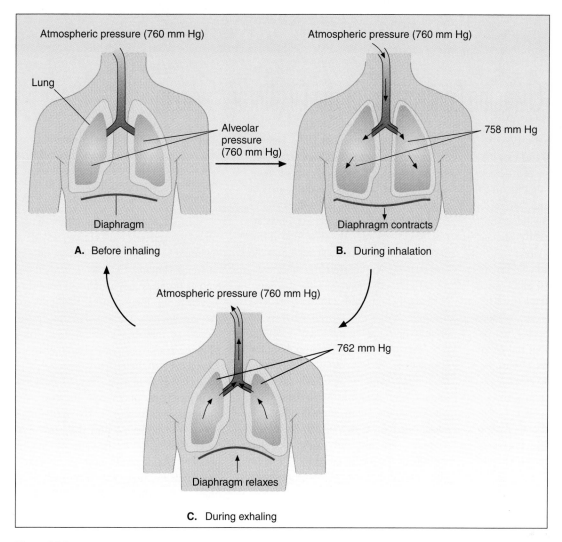

Figure 35.3 (A) Before inhaling, the diaphragm is relaxed, and atmospheric pressure both inside and outside the lungs is 760 mm Hg; consequently, air does not move in or out of the lungs. (B) During inhalation, the diaphragm contracts, causing an expansion of lung volume that reduces atmospheric pressure inside the lungs to 758 mm Hg. Because gases move down a pressure gradient, air enters the lung. (C) During exhalation, the diaphragm relaxes, lung volume is reduced, atmospheric pressure increases, and air moves out of the lungs.

CIRCULATORY SYSTEM

Cells and tissues of the human body require an incessant supply of nutrients and O_2 to survive. They also must rid themselves of CO_2 and other metabolic wastes generated continuously during their normal activities. These functions involve the **circulatory system,** an internal transport system that consists of the heart, a four-chambered muscular pump; an elaborate network of interconnecting, hollow vessels; and blood, a complex fluid containing different types of cells, cell fragments, and soluble molecules that each perform a specialized function.

The circulatory system has long been of interest to humans (see the Focus on Scientific Process, "Changing Conceptions of the Circulatory System"). Discussions of this system are usually peppered with huge numbers and exclamation marks related to such matters as the amount of blood the heart pumps in an average 70-year lifetime (18 million barrels!), the num-

ber of miles of capillaries (more than 60,000!), the number of red blood cells in the human body (around 25 billion!), and so on. More impressive than the sheer size of these numbers, however, is their indication of the magnitude of circulatory system activities within the human body.

The circulatory system functions in a range of important activities, including the transport of O_2 from the lungs to red blood cells and CO_2 from blood cells and fluids to the lungs. It is also involved in the transport of other substances. For example, the blood moves metabolic wastes generated in cells to organs that can process and eliminate them; it also transports nutrients from the digestive tract, or nutrient storage sites, to the cells where they are used. Blood cells defend the host against disease-causing agents. Finally, the circulatory system serves in certain homeostatic feedback systems such as regulating body temperature, blood pH, and fluid balance. The respiratory, digestive, and urinary systems also participate in these activities.

Changing Conceptions of the Circulatory System

To someone who casually observes the internal organs of an animal and sees the heart and some of the major blood vessels, it is not at all obvious that blood is circulating. It is not surprising, therefore, that the circulation of the blood within the body was not appreciated by ancient or even Renaissance physicians.

The first physicians and anatomists did realize that the heart and blood vessels were vital to living organisms. The ancient Egyptians, who were among the earliest people to study the structure of the human body, thought that the heart was the origin of all the "vessels" of the body and that through these vessels, various important substances, such as blood, semen, water, and air, traveled. Later, Greeks studied the internal anatomy of the human body for medical purposes and came to have a sophisticated knowledge of its structure (but not its functions). Some early Greeks ignored the connections between the heart and the blood vessels and assumed that the blood vessels were like irrigation ditches through which the nutrients from food flowed out to the various parts of the body to be consumed. Later Greek anatomists believed that the heart was the center of vitality and that from it an "animating heat" traveled through the arteries to the rest of the body. They also believed that digested food traveled from the stomach and liver through the veins to nourish the tissues. Famous anatomists like Galen, who lived in the Roman Empire during the second century A.D., provided accurate and detailed descriptions of the heart and the blood vessels of humans but had no conception that the blood circulated. Instead, most physicians and naturalists believed that the blood simply ebbed and flowed in the blood vessels.

Opinions on the nature and function of the heart and blood vessels did not begin to change until William Harvey, a distinguished English physician, published his observations in 1628. Harvey discovered that blood did not travel out to the tissues of the body to be consumed; rather, it traveled continuously in a closed loop. He was led to his discovery by anatomical studies on the heart and blood vessels. He was also influenced by the then-popular idea that the human body was a reflection of the cosmos. To him, the heart was like the "sun" of the body and fluids circulated around it, similar to the way the planets traveled around the center of our solar system.

Harvey's famous 1628 book on the circulation of the blood, *Exercitatio anatomica de motu cordis et saguinis in animalibus* ("Anatomical Studies on the Motion of the Heart and Blood"), is a classic in the history of the biological sciences. He not only described the motion of the heart and blood, but he also elegantly set out *experimental arguments* to demonstrate his points.

To show that the heart is supplied with blood from the vena cava, Harvey described observations on a living snake:

If a live snake be cut open, the heart may be seen quietly and distinctly beating for more than an hour, moving like a worm and propelling blood when it contracts longitudinally, for it is oblong. It becomes pale in systole, the reverse in diastole. . . . The vena cava enters at the lower part of the heart, the artery leaves at the upper. Now, pinching off the vena cava with a forceps or between finger and thumb, the course of blood being intercepted some distance below the heart, you will see that the space between the finger and the heart is drained at once, the blood being emptied by the heart beat. At the same time, the heart becomes much paler, even in distention, smaller from lack of blood, and beats more slowly, so that it seems to be dying. Immediately on releasing the vein, the color and size of the heart return to normal. On the other hand, leaving the vein alone, if you ligate or compress the artery a little distance above the heart, you will see the space between the compression and the heart, and

the latter also, become greatly distended and very turgid, of a purple or livid color, and choked by the blood, it will seem to suffocate. On removing the block, the normal color, size, and pulse return.

To show the one-way flow of the blood though the veins, Harvey relied on a simple but brilliant set of observations and experiments. First, he described the valves of the veins, which had been discovered by Hieronymus Fabricius of Aquapendente. These valves are positioned so that they prevent blood from flowing backward, and thus they route blood back to the heart.

Harvey then used a very elementary experiment to underscore the one-way flow of venous blood. By applying a loose tourniquet around the arm of a man, Harvey forced the veins of the arm to swell. At intervals, nodules appeared that indicated the sites of the valves of the veins, as shown in Figure 1. Then he noted:

If you will clear the blood away from the nodule or valve by pressing a thumb or finger below it, you will see that nothing can flow back, being entirely prevented by the valve, and that the part of the vein between the swelling and the finger disappears, while above the swelling or valve it is well distended. Keeping the vein thus empty of blood, if you will press downward against the valve, by a finger of the other hand on the distended upper portion, you will note that nothing can be forced through the valve. The greater effort you make, the more the vein is distended toward the valve, but you will observe that it stays empty below it.

To make this point even more dramatically, Harvey also described a related experiment:

With the arm bound as before and the veins swollen, if you will press on a vein a little below a swelling or valve and then squeeze the blood upward beyond the valve with another finger, you will see that this part of the vein stays empty, and that no back flow

FOCUS ON SCIENTIFIC PROCESS

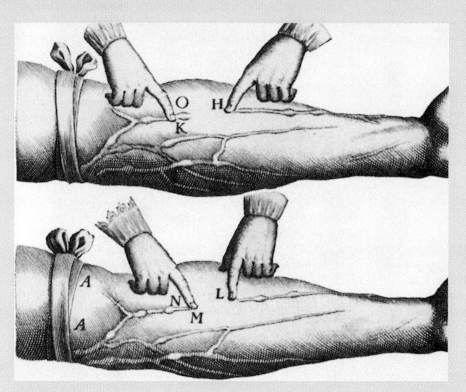

Figure 1   Harvey's experiment demonstrated blood flow in the arm.

can occur through the valve. But as soon as the finger is removed, the vein is filled from below.

Harvey provided some compelling quantitative arguments as well. By calculating the volume (not very accurately, as it turned out, but reasonably close) of blood forced out of the heart into the aorta by each beat, he easily showed that the volume of blood pumped by the heart in an hour was more blood than the entire body contained. Even though Harvey did not observe the capillary system between arteries and veins (microscopes had not yet been invented), he reasonably argued that since the flow was one-way, it must circulate through a single pathway in the body.

Harvey's discovery and description of the circulation of the blood stands as a paragon in the history of the biological sciences, but what is perhaps even more fascinating than his careful experiments and reasoned arguments was the reception of his ideas. Although a group of

English physicians at Oxford did appreciate the implications of Harvey's discovery, the medical community was slow to accept the concept of blood circulation. For decades, anatomists and physicians rejected Harvey's findings. Why? It is difficult today to read through Harvey's short book without being totally convinced. In his own day, however, Harvey's book was revolutionary in two dramatic ways, one methodological, the other theoretical.

Harvey's use of an experimental approach now seems very ordinary to us, but in the 1620s it was very unusual. The study of living beings at that time was primarily a descriptive science. Naturalists classified animals and plants; medical researchers described certain parts of the body and speculated on their functions. Experimental investigations were being done in the physical sciences, but they were still rather novel. The idea of doing experiments on living beings was so strange and new that many serious and well-educated people

found it hard to imagine that anything worthwhile could come of it. The use of powerful experiments in biology lay in the future.

The other reason that the reception of Harvey's ideas was muted was that it refuted a long-held theory. Assuming that Harvey's findings were correct meant that the accepted theory of the animal body, as it had evolved from the Greeks through the Arabs and then medieval and Renaissance Europeans, was fatally flawed. The foundation of that theory was an idea about the uptake, distribution, and fate of food in the body. It was believed that food taken into the body first traveled to the stomach, where the "nutritious portion" was separated, then to the liver, where it was further altered and supplemented, and finally to the veins, which distributed it to the rest of the body to be absorbed. A portion of the blood passed to the arterial system, where it mixed with a small amount of air taken in by the lungs. This vital mixture then flowed out via the arteries to the entire body, where it was all consumed.

Circulation of the blood didn't fit this theoretical view at all. Harvey showed that the blood was not distributed to tissues of the body to be consumed but rather traveled in a continuous circuit. What, then, was the function of this circulation? How did it connect with nutrition? If the venous and arterial systems were *not* distinct, how were they related? Since the entire edifice of practical medicine, physiology, and anatomy was based on the earlier conception of the heart and blood vessels, physicians were hesitant to accept an idea (based on a novel method) that contradicted conventional wisdom.

Harvey's ideas were eventually accepted, of course, but not until they became integrated into a new theoretical view of the body that "made sense." This case study illustrates how important theory is to science and emphasizes the reluctance one should have in thinking of science as merely a collection of facts.

Blood Components

Blood is classified as a fluid connective tissue that is transported in blood vessels and circulated by the action of the heart. It consists of a liquid matrix, called plasma, that contains cells collectively called *formed elements.*

Plasma

Plasma is mainly water (about 91 percent) that contains and transports dissolved gases, proteins, hormones, nutrients, lipids, minerals such as iron, and metabolic waste products. Plasma also contains a large variety of proteins that have specific structures and functions. Several of these proteins participate in the body's defense mechanisms and are discussed in Chapter 39.

Formed Elements

The **formed elements** of blood include **erythrocytes,** or red blood cells (RBCs), that are specialized for transporting O_2; various types of **leukocytes,** or white blood cells (WBCs), that have specialized defense functions; and **platelets,** which are very small, nonnucleated cell fragments that have a critical function in blood clotting (see Figure 35.4). Leukocytes and platelets are described in greater detail in Chapter 40.

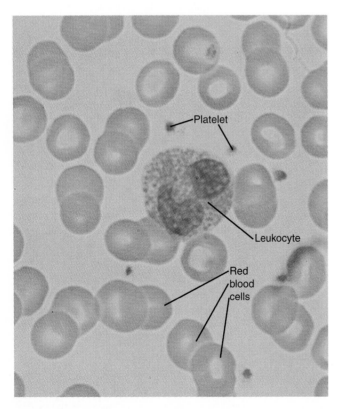

Figure 35.4 Erythrocytes, leukocytes, and platelets are the formed elements of the blood.

Question: *What is the primary function of red blood cells?*

Erythrocytes are the most abundant formed element in blood. Mature RBCs are small, circular, biconcave disks that have no nuclei. Each RBC contains up to 275 million hemoglobin molecules within its cytoplasm. **Hemoglobin** is a complex, iron-containing molecule that combines with O_2 in the lung, giving blood its characteristic bright red color. After O_2 is given up to cells, hemoglobin transports a small amount of CO_2 back to the lung. RBCs are subjected to great stress during their endless journey through blood vessels. They have an average life span of 120 days and are replaced through mitosis in the bone marrow. The body has the ability to govern the number of RBCs available; regulation is associated primarily with the body's demand for O_2. For example, individuals who exercise regularly or live at high altitudes have up to twice as many RBCs as sedentary individuals.

Vessels

Blood in the circulatory system flows in one direction, through three general types of blood vessels that are distinguished by size, structure, and function as shown Figure 35.5A. **Arteries** conduct blood *away* from the heart and move it to separate vascular territories of the body, while **veins** return blood *to* the heart. **Capillaries** connect the arterial and venous systems.

Major arteries closest to the heart are strong and thick-walled, due to large amounts of muscle and elastic fibers that allow the vessel wall to be expanded by the force of blood being pumped from the heart. When the muscular artery wall rebounds after expansion, it causes movement of blood through the vessel. If you place your fingers on the inner portion of your wrist, you will be able to feel a pulse, which represents the wave of pressure radiated outward from the heart through the arteries. Farther along the arterial tree, after considerable branching, the size and diameter of the vessel walls become reduced, and blood pressure diminishes significantly; these smaller vessels are called **arterioles.**

From arterioles, blood flows into capillaries, the smallest blood vessels, which have walls thin enough to allow the transfer of O_2, nutrients, CO_2, and wastes between cells and the blood (see Figure 35.5B). Pressure exerted from pumping by the heart has nearly dissipated by the time blood passes into capillaries. Single capillaries are about 1 millimeter long and have diameters approximately the size of RBCs. Tissues are completely infiltrated with beds of capillaries, and no cell within tissues is more than two or three cell diameters away from a capillary. Nutrients, gases, and other materials exchanged between blood cells and tissue cells move through **tissue fluid**, composed of water, salts, and proteins, that bathes the cells.

Upon leaving the capillaries, blood flows into small **venules** that unite to form veins. Veins return blood to the heart largely through the routine movements of skeletal muscles, which squeeze the vessels and move the blood forward as described in Figure 35.5C. Backflow is prevented in larger veins by the presence of **valves,** or flaps of tissues extending from the vessel wall that close when blood attempts to flow in

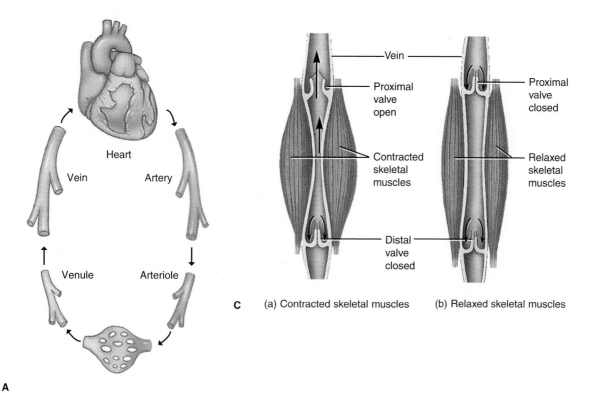

C   (a) Contracted skeletal muscles   (b) Relaxed skeletal muscles

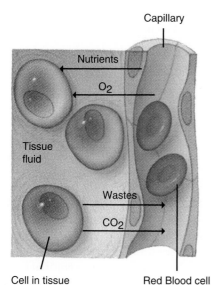

Figure 35.5 (A) The major vessels of the circulatory system are shown, and the direction of blood flow is indicated by the arrows. (B) Oxygen (O_2), carbon dioxide (CO_2), nutrients, and wastes are exchanged between tissue cells and the capillaries of the circulatory system. (C) The movement of blood through veins is governed by contraction of the surrounding muscles and valves within the veins. When muscles contract, blood is forced to flow toward the heart. When they relax, the vein expands and fills with blood. Valves in the vein prevent backflow of the blood.

the opposite direction. In the absence of movement—for example, when standing rigidly at attention—blood tends to collect in veins and in extreme cases may lead to fainting because the brain becomes deprived of an adequate supply of blood.

Heart

The **heart** is a large muscular organ capable of powerful contractions that propel blood into vessels (see Figure 35.6A, page 664). The human heart consists of four chambers, called the right and left **atria** and right and left **ventricles,** and blood flows through each chamber in a specific sequence. Chambers are separated by valves, which open and close automatically,

to ensure that blood flows only in one direction. Figure 35.6B shows the chambers of the human heart and the direction of blood flow within the heart.

Circulation of the Blood

The circulation of blood follows two specific pathways, pulmonary circulation and systemic circulation (see Figure 35.7, page 665). This design reflects a primary responsibility of transporting O_2 to the billions of cells making up the human body.

Pulmonary circulation conducts blood between the heart and the lungs. Oxygen-poor, CO_2-laden blood returns through two large veins (**venae cavae**) from tissues within the body,

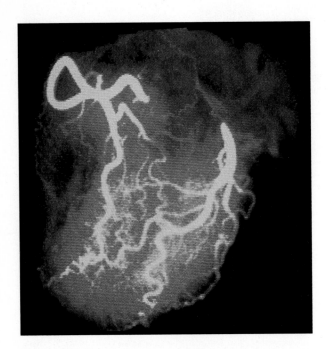

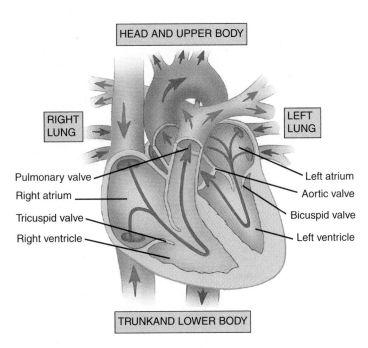

HEAD AND UPPER BODY

RIGHT LUNG

LEFT LUNG

Pulmonary valve

Left atrium

Right atrium

Aortic valve

Tricuspid valve

Bicuspid valve

Right ventricle

Left ventricle

TRUNKAND LOWER BODY

Figure 35.6   (A) An external view of the human heart as revealed by a coronary angiogram. The braches of the left and right coronary arteries appear as the network of bright white ribbons throughout the heart. (B) The four chambers of the human heart are separated by valves. Arrows indicate the direction of blood flow.

enters the right atrium, and is then moved into the right ventricle of the heart. From there, it is pumped into the pulmonary artery, which divides into two branches, each leading to one of the lungs. In the lung, the arteries undergo extensive branching, giving rise to vast networks of capillaries where gas exchange takes place, with blood becoming oxygenated while CO_2 is discharged. Oxygen-rich blood then returns to the heart via the pulmonary veins. In contrast with all other arteries and veins in the body, pulmonary arteries carry oxygen-deficient blood, and pulmonary veins convey oxygen-rich blood.

Once blood is returned from the lungs, it enters the left atrium, passes to the left ventricle, and is then distributed by **systemic circulation** to all tissues of the body. Blood emerges from the heart into the **aorta,** the largest artery in the body. The aorta branches into major arteries that lead to vascular territories that include the heart itself, the brain, the shoulder region, the intestines, the kidneys, and the legs.

Lymphatic System

The **lymphatic system** consists of a network of vessels that run near the veins and **lymph tissue**, which is a type of connective tissue. Its major function is associated with immune responses, described in Chapter 39. The other functions of the lymphatic system are generally related to draining excess fluids,

transporting fats from the digestive system to the blood, and removing foreign substances from the body. Specifically, a slight excess of tissue fluid accumulates in the body as a result of nutrient, gas, and water diffusing between tissue cells and blood cells. This fluid enters lymphatic capillaries, where it is called **lymph**, these capillaries merge to form larger vessels that return the lymph to veins near the heart (see Figure 35.8, page 666) Lymph tissue is organized into compact masses of connective tissue called **lymph nodes** that filter and remove foreign particles, such as bacteria, and also harbor some of the cells that have important functions in the immune system.

Respiratory Physiology II

Recall that *respiration* is the exchange of O_2 and CO_2 between an organism and the environment. For convenience in describing gas exchanges inside humans, **external respiration** refers to O_2 and CO_2 exchanges between alveoli and the circulatory system, and **internal respiration** to that between capillaries and tissue cells.

External and Internal Respiration

Gas exchanges are primarily regulated by O_2 and CO_2 partial pressures in the various internal environments. Table 35.1 summarizes normal partial pressures in the atmosphere, alveolar

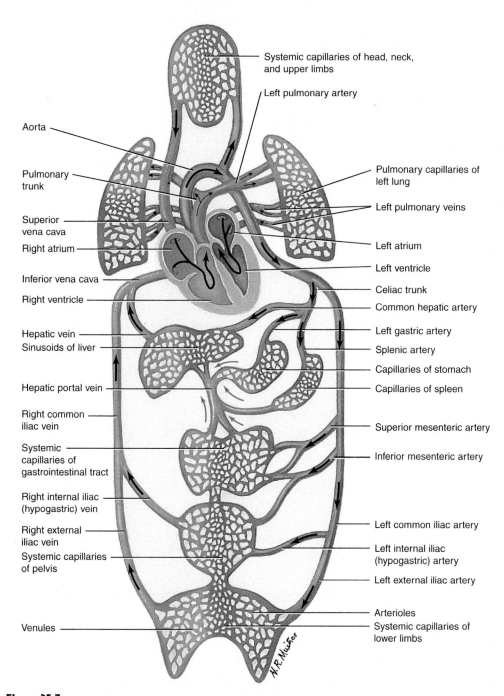

Figure 35.7   Blood circulates through two general pathways in the human body. Pulmonary circulation between the heart and the lungs loads red blood cells with O_2. Oxygen-rich blood is then distributed to the head and arms, the internal organs, and the legs by systemic circulation. Oxygen-poor blood returns to the heart through the two venae cavae.

Table 35.1   Partial Pressures (mm Hg) of Oxygen (pO_2) and Carbon Dioxide (pCO_2) in Different Environments

	Atmosphere	Alveolar Air	Blood (Oxygenated)	Tissue Cells	Blood (Deoxygenated)
pO_2	160.0	105	105	40	40
pCO_2	0.3	40	40	45	45

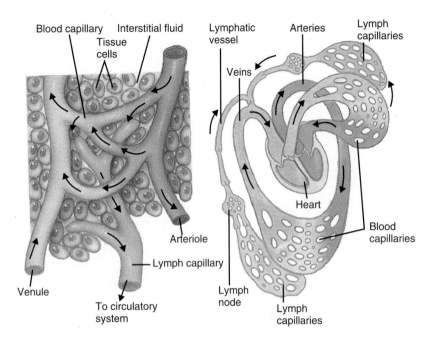

Figure 35.8 (A) Tissue fluids that surround cells diffuse into lymphatic capillaries. (B) Lymphatic capillaries flow into large lymph vessels that merge with veins near the heart, returning fluids to the blood. Lymph nodes act as filters in removing foreign particles; they also play a major role in the immune system.

air, blood leaving the lung (**oxygenated blood**), blood returning to the lung (**deoxygenated blood**), and tissue cells. Oxygen and CO_2 diffuse down a pressure gradient, that is, from a higher to a lower partial pressure.

Figure 35.9 shows the direction of O_2 and CO_2 exchanges in external and internal respiration. Blood leaving the lung (and heart) is fully oxygenated. All hemoglobin (Hb) molecules in red blood cells combine with O_2 to form **oxyhemoglobin** (HbO_2), a reaction described by the following simple formula: $Hb + O_2 \rightarrow HbO_2$. These hemoglobin–$O_2$ reactions result in a pO_2 of 105 mm Hg. From the heart, blood is pumped through arteries to arterioles to capillaries where internal respiration occurs. Oxygen is released by hemoglobin ($HbO_2 \rightarrow Hb + O_2$) and diffuses rapidly into cells, where pO_2 is only 40 mm Hg. As a result, oxygenated blood becomes deoxygenated, pO_2 in the capillary is reduced to 40 mm Hg, and diffusion no longer occurs. At the same time, CO_2 diffuses from cells, which have a pCO_2 of 45 mm Hg, into blood, where pCO_2 is 40 mm Hg. Carbon dioxide diffusion continues until the pCO_2 is 45 mm Hg in deoxygenated blood. Deoxygenated blood returns to the heart and is then pumped to the lungs, where CO_2 diffuses into the alveoli and is exhaled, and O_2 diffuses rapidly into the blood until it once again is fully oxygenated. The blood then returns to the heart and is pumped out to make another tour of the body.

Carbon dioxide is transported by the circulatory system in various ways. Some of it (about 5 percent) is dissolved in plasma, and about 25 percent is carried by hemoglobin. The rest of it (about 70 percent) is carried in plasma as **bicarbonate**

ions (HCO_3^-) that are formed when **carbonic anhydrase,** an enzyme in red blood cells, catalyzes the following chemical reaction: $CO_2 + H_2O \rightarrow H_2CO_3$ (carbonic acid). Once formed inside red blood cells, H_2CO_3 dissociates into H^+ and HCO_3^-; the HCO_3^- then diffuses into the plasma, which tends to lower blood pH, and is transported to the lungs. In the presence of normal oxygen concentrations in the lungs, H^+ and HCO_3^- recombine to form carbonic acid, which splits into H_2O and CO_2; the CO_2 is then exhaled.

Regulation of Blood Chemistry

Significant differences occur in our breathing rates, depending on whether we are exercising, reading, or attempting to perform some activity in front of an audience. Stressful activities make demands on the respiratory and circulatory systems that can alter normal blood chemistry. **Blood chemistry** generally refers to pO_2, pCO_2, and pH; recall that pH is a measure of hydrogen ion (H^+) concentration of a solution (discussed in Chapter 2). Normal blood pH is about 7.4. What regulates blood chemistry? Within the brain, a collection of neurons form a **respiratory center** that regulates breathing rate. The respiratory center monitors blood chemistry and responds instantly to significant changes in CO_2 and O_2 concentrations and pH by increasing the rate and depth of breathing. Specialized nerve receptors that detect chemical changes in the blood also exist in major arteries leaving the heart, and they help regulate blood chemistry as well.

Figure 35.10 depicts the most important homeostatic mechanisms that act to regulate blood pH, pCO_2, and pO_2. In

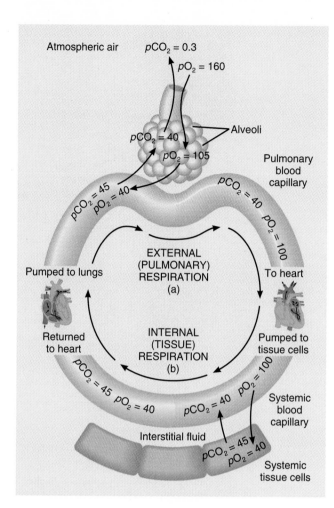

Figure 35.9 Partial pressures of carbon dioxide (pCO_2) and oxygen (pO_2) vary among the atmosphere, lung alveoli, capillaries, and tissue cells. Differences in partial pressures are responsible for determining the direction of CO_2 and O_2 exchanges between atmospheric air and alveoli, alveoli and blood in capillaries, and capillary blood and tissue cells. Gases move down a pressure gradient from an area of higher partial pressure to an area of lower partial pressure. Arrows indicate the direction of O_2 and CO_2 exchange in various parts of the body.

Question: *In which direction does O_2 diffuse between a capillary and a cell? Why does it diffuse in that direction?*

Figure 35.10 Changes in blood chemistry—an increase in pCO_2 or a decrease in pH or pO_2—are detected by specialized neurons located in the brain and in arteries leaving the heart. Nerve impulses are then transmitted to the brain's respirator center, and the diaphragm and other muscles are stimulated to contract more forcefully, which leads to deeper, more frequent breathing. As a result, more CO_2 is exhaled, more O_2 is inhaled, blood pH is increased, and normal blood chemistry is restored.

Question: *Why are these homeostatic mechanisms identified as a negative feedback system?*

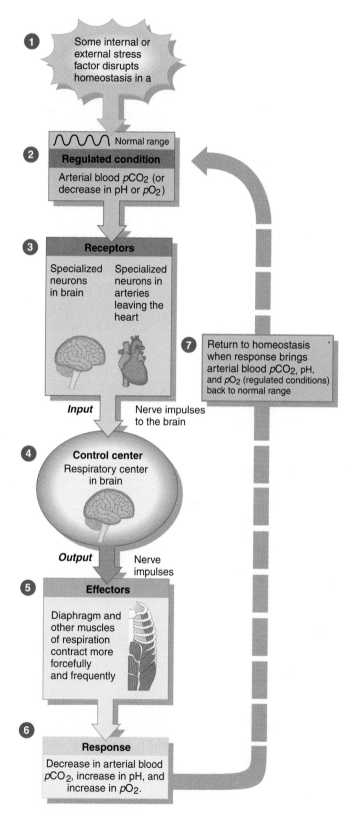

Source: Gerard J. Tortora and Sandra R. Grabowski,
Principles of Anatomy and Physiology, 8th Edition.
New York: Harper Collins College Publishers, 1996.

general, any increase in blood pCO_2 causes an increase in H+, which reduces blood pH, making it more acidic. Elevated pCO_2 and lower blood pH may result from a higher rate of cellular respiration during heavy exercise or from a reduced breathing rate. As blood pH decreases, receptors in the brain detect the change in H+ concentration and stimulate the part of the respiratory center that regulates breathing rate. Nerve impulses are then sent to the diaphragm, and the rate and depth of breathing increase, which leads to a lower pCO_2 and an increase in blood pH. The same reaction occurs in response to significant reductions in pO_2.

What if blood pH increases or pCO_2 is abnormally low? The same negative feedback system operates to *reduce* the breathing rate, which will result in blood pH being lowered and pCO_2 being elevated.

BEFORE YOU GO ON The circulatory system functions by moving blood throughout the body and working with other systems to maintain homeostasis. Its organs include a the heart (a pump) and various types of blood vessels. There are two elaborate networks of blood vessels, one associated with pulmonary circulation and the other with systemic circulation. Blood is a complex fluid composed of plasma and several formed elements, each with specific functions. The lymphatic system consists of a separate network of vessels that transport lymph and remove foreign substances from the body. Gas exchanges are largely regulated by partial pressures of O_2 and CO_2 in the lungs, blood, and tissue cells. Blood CO_2 and O_2 concentrations and pH are regulated by a respiratory center in the brain.

DIGESTIVE SYSTEM

The **digestive system** has many functions, all of which are generally related to delivering the nutrients contained in food to every cell in the body that requires them for energy, building new structures or molecules, or repairing damage. **Digestion** refers to the breakdown of complex food substances into simple molecules that can be absorbed into the body.

The digestive system is analogous to a "disassembly line" where large molecules are broken down into smaller subunits. As described in Figure 35.11, it accomplishes this task by moving foods progressively through a tube called the **gastrointestinal (GI) tract** that begins at the mouth and terminates at the **anus,** the opening through which digestive waste materials are expelled. Food is sequentially modified, first mechanically and then chemically, to achieve the reduction of complex molecules into simpler forms. Along the way, various organs and glands empty their products into the digestive tube at certain regions where specialized processes occur. All of these activities are regulated by the nervous and endocrine systems.

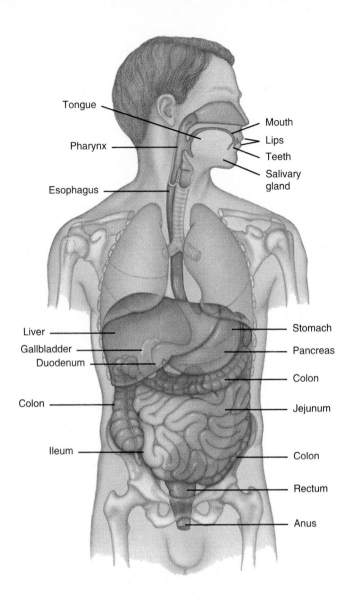

Figure 35.11   The human digestive system breaks down and delivers nutrients to all cells of the body.

Mouth, Pharynx, and Esophagous

The **mouth** is the opening to the exterior that receives food and begins the process of digestion using both mechanical and chemical devices. The tongue, teeth, and jaw are used to chew or grind the food into smaller pieces, which results in a greater amount of surface area being exposed to the action of enzymes. The presence of food stimulates the release of saliva through ducts from three pairs of **salivary glands** in the mouth. **Saliva** is a watery secretion consisting of mucus, which protects the lining of the mouth and lubricates the food for easier movement, and **salivary amylase**, a specific digestive enzyme that begins the breakdown of starch.

Once the food has been chewed, softened, and lubricated, it passes into the pharynx, a thick-walled, muscular tube. In the pharynx, the ball of pasty food is swallowed and moved to the entrance of the esophagus. The **esophagus** is an elastic tube, about 25 centimeters in length and 2.5 centimeters in diameter, that conducts food from the pharynx to the stomach. From the esophagus onward, food is forced through the GI tract by **peristalsis**, a series of contraction waves that ripple along the smooth muscles surrounding the tubular organs of the digestive system. From the esophagus, the food **bolus**—the rounded, softened mass of food material—moves into the stomach.

Stomach

The **stomach** is a relatively large muscular bag in the upper part of the abdomen that can stretch to accommodate about 2 liters of food or liquid. A number of chemical assaults are made on food once it enters the stomach. The epithelium that lines the interior of the stomach secretes *gastric juice,* a highly acidic fluid with a pH of about 1, that is actually strong enough to dissolve many metals. This powerful acid acts to dissolve the tough tissues that bind together the plant and animal cells present in food. Gastric juice also contains an enzyme called **pepsin** that initiates the breakdown of food proteins.

Given the potency of the digestive chemicals inside the stomach, why isn't the stomach lining also broken down? First, a coating of mucus secreted by epithelial cells lining the stomach offers substantial protection against the digestive actions of gastric juice. A second mechanism involves the release of an enzyme in an inactive form that becomes activated only when exposed to acid. For example, pepsin is secreted by epithelial cells in an inert form called *pepsinogen.* Upon exposure to the acid in gastric juice, it is converted to the active enzyme, pepsin. Many other digestive enzymes are restricted

in the same manner. Despite these protective mechanisms, epithelial cells of the stomach have life spans measured in days, and they are replaced by new cells arising from cell division in the basal epithelial layers. After spending two to six hours in the stomach, the food mass is reduced to a semifluid mixture of partially digested food, called **chyme,** that passes next into the small intestine.

Despite the complexity of its functions, the stomach is classified as a "nonessential" organ because life can be sustained without it. In drastic circumstances, such as cancer development or severe cases of ulceration, the stomach can be partially or totally removed by surgery. Although the rest of the digestive system can digest food without the stomach, eating modifications are necessary—for example, consuming many small meals each day rather than the three that are customary in our society.

Small Intestine

The major organ of digestion, the **small intestine,** is about 6 meters long. It consists of an upper region called the **duodenum;** a middle region, the **jejunum***;* and a lower sector, the **ileum.** Proteins, fats, and carbohydrates are finally broken down to simpler molecules in the duodenum due to the actions of numerous enzymes secreted by intestinal epithelial cells. Other chemical substances released from accessory organs (to be described shortly) also aid in digesting food in the small intestine.

Following digestion, the nutrients, along with water, are absorbed into the bloodstream through the intestinal walls of the jejunum and ileum and distributed throughout the body. The architecture of the ileum is ideally suited for its absorptive responsibilities (see Figure 35.12). Folds of the inner surface are covered with small, fingerlike projections called **villi,**

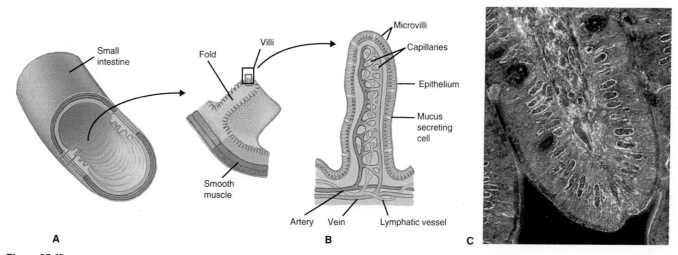

Figure 35.12 (A) The small intestine is extensively wrinkled. (B) Millions of small projections called *villi* line each wrinkle and are the site of nutrient transfer between the small intestine and the circulatory system. (C) This micrograph of villi of the small intestine shows details of the epithelial cells.

Question: *What is the principal advantage of this type of "wrinkled" surface?*

and the epithelial cells lining the villi are covered with thousands of **microvilli,** microscopic folds in the plasma membranes. This structural design provides an enormous amount of surface area through which molecules such as glucose and amino acids can pass into capillaries surrounding the intestine.

Accessory Digestive Organs

Two other organs—the pancreas and the liver—have major roles in the breakdown of food in the duodenum. They synthesize or store chemical substances that are released into the small intestine during digestion.

Pancreas

The **pancreas** carries out functions in the digestive system and also the endocrine system. It produces *pancreatic juice,* which contains an assortment of digestive enzymes and releases a bicarbonate solution that serves to neutralize the strong acids contained in the chyme received from the stomach.

Liver

The **liver** is an enormously complex organ that performs a wide variety of critical operations in the body. In addition to removing nutrients from circulation, converting glucose to **glycogen,** a storage form of glucose, and detoxifying harmful compounds, it has other major digestive functions. These include breaking down hemoglobin, storing iron and some vitamins, synthesizing certain proteins, and other metabolic conversions of amino acids, carbohydrates, and fats. The liver also synthesizes **bile,** a mixture of substances that breaks down large fat globules into smaller particles. Bile is stored in the **gallbladder,** another accessory organ, and released into the small intestine as required during digestion.

Large Intestine

By the time chyme reaches the **large intestine,** little remains except for indigestible materials. The primary functions of the colon are, first, to reabsorb sodium and the remaining water that was released into the digestive tract to facilitate digestion. Reclaiming water is necessary to prevent dehydration that would occur if the water were not recycled. Second, the colon eliminates **feces,** compacted solid wastes consisting mostly of undigested food, dead intestinal cells, and bacteria that inhabit the large intestine. The colon hosts a rich variety of bacteria that are generally harmless, including the molecular biologists' favorite, *Escherichia coli.* Bacteria are able to survive in the large intestine by extracting nutrients from certain food materials, such as cellulose, that are of no value to humans. In return, they produce small quantities of valuable vitamin K, which are absorbed by the host. Less welcome bacterial byproducts—odiferous gases such as hydrogen sulfide—are also produced in the large intestine. After one to three days, feces reach the **rectum,** the lower portion of the large intestine, where they are stored until defecation, or expulsion through the anus, occurs.

Chemical Digestion and Absorption

Table 35.2 summarizes the major enzymes and their effects on various macromolecules that pass through the GI tract. By the time foods reach the small intestine, they have been broken down into simple molecules—monosaccharides (for example, glucose), amino acids, glycerol and fatty acids, and nucleotides—that can pass through the digestive tract into the blood or lymph, a process called **absorption.** Absorption of specific nutrients through the small intestine occurs by diffusion, facilitated diffusion, or active transport. About 90 percent of digested substances are absorbed through the small intestine, and the remaining 10 percent, through the stomach or large intestine.

Figure 35.13 shows how molecules are absorbed in the small intestine and then through the villi into capillaries or lymphatic vessels. Glucose is absorbed by active transport and enters a capillary by facilitated diffusion. Single amino acids, small peptides, and nucleotides are absorbed into villus cells by active transport and then diffuse into a capillary. Peptides are broken down to amino acids inside villus cells. Lipid

Table 35.2 Some Digestive Enzymes, Their Source, and Their Effect on Products Formed When they Act on Their Substrate.

Enzyme	Source	Substrate	Product
Saliva			
Salivary amylase	Salivary glands	Starch	Maltose (disaccharide)
Gastric juice			
Pepsin	Stomach cells	Proteins	Peptides
Pancreatic juice			
Trypsin	Pancreas	Proteins	Peptides
Lipase	Pancreas	Triglycerides	Glycerol and fatty acids
Nucleases	Pancreas	Nucleic acids	Nucleotides
Intestinal enzymes			
Maltase	Small intestine	Maltose	Glucose
Peptidases	Small intestine	Peptides	Animo acids

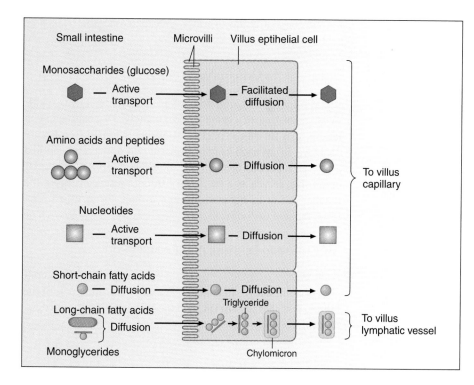

Figure 35.13 Various macromolecules are broken down to smaller structural units (carbohydrates to monosaccharides, proteins to amino acids or small peptides, nucleic acids to nucleotides, lipids to fatty acids and monoglycerides) in the stomach and small intestine. The smaller chemicals enter villus cells by diffusion or active transport. From a villus cell, glucose, amino acids, nucleotides, and short-chain fatty acids enter capillaries. Long-chain fatty acids and monoglycerides form chylomicrons—triglycerides and other lipids enclosed by a protein coat—that leave the villus cell and enter a lymphatic vessel.

Question: *Which of the transport processes does not require energy for moving substances through microvilli?*

absorption is more complicated. **Triglycerides,** the most common form of lipid, are broken down into fatty acid molecules of varying length, and **monoglycerides,** compounds composed of a glycerol molecule with one attached fatty acid. All fatty acids and monoglycerides are absorbed by simple diffusion. Once inside a villus cell, lipase breaks down monoglycerides to glycerol and fatty acids. Next, one glycerol and three fatty

acids recombine to form a triglyceride. Triglycerides and other lipid molecules clump together and become surrounded by a protein coat to form a spherical structure called a **chylomicron** that leaves the cell and enters a lymphatic vessel in the villus.

As described in Figure 35.14, after leaving the villus, absorbed molecules are transported to the liver, through the hepatic portal vein or a lymphatic vessel, where they are

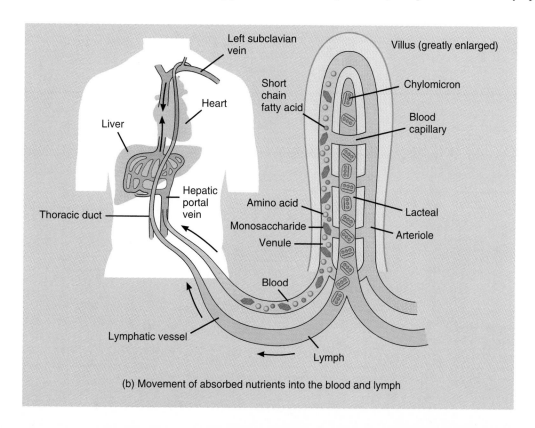

(b) Movement of absorbed nutrients into the blood and lymph

Figure 35.14 Once absorbed through the small intestine, monosaccharides, amino acids, nucleotides, and lipids are transported to the liver. From the liver, they pass through the heart and are distributed throughout the body by the circulatory system.

"processed." For example, glucose may be converted to glycogen. If present, toxic chemicals are also extracted and broken down or stored by the liver. From the liver, they are transported elsewhere in the body by the circulatory system.

BEFORE YOU GO ON   The basic function of the digestive system is to reduce food to small molecules that can be absorbed by the circulatory system and transported to all cells in the body. In the mouth, food is mechanically broken down and first exposed to digestive enzymes. Food is further degraded in the stomach by gastric juice and numerous enzymes. The final phase of digestion occurs in the small intestine where other enzymes complete the breakdown process and absorption occurs. The pancreas, liver, and large intestine, also have roles in digesting food.

URINARY SYSTEM

As we have seen, a number of human organ systems are involved in disposing of wastes produced as a result of various activities, as summarized in Figure 35.15. The **urinary system** (see Figure 35.16) is made up of organs that filter the blood and produce **urine,** a solution containing varying amounts of water and waste products that is excreted from the body. The wastes normally removed by the urinary system are excess salts, nitrogen-containing compounds formed during the chemical modification or breakdown of amino acids, proteins or nucleic acids, and other unwanted products that may be present in the blood. The primary metabolic waste that must be removed is **urea,** a small nitrogen-containing product.

In addition to filtration and waste removal, the urinary system performs critical homeostatic functions. It helps maintain a constant internal environment by continuously adjusting the concentrations of water, salts, and other materials present in blood.

Organs of the Urinary System

Blood filtration and urine formation are carried out by two **kidneys,** bean-shaped organs approximately 10 centimeters in length, that lie on the back wall of the abdomen (see Figure 35.17A). From each kidney, urine moves down a tube called a **ureter** and into the **urinary bladder,** a hollow muscular organ that serves as a temporary storage reservoir. From the bladder, urine is excreted through a duct, the **urethra,** that leads outside the body.

Structure and Function of the Kidney

Each kidney is enclosed within a **renal capsule**, which consists of a thick layer of connective tissue. The outer part of the kidney is called the **cortex,** and the inner portion is called the **medulla.**

Nephrons

Each kidney contains more than 1 million functional units called **nephrons,** which are long, curved, tubular structures of exquisite design. Nephrons regulate the composition of blood through the production of urine (see Figure 35.17B). Each nephron can be divided into two functional compartments. First, the **glomerulus,** a mass of capillaries, and associated structures filter the blood. Second, a complex of **tubules** removes valuable materials, such as glucose, and transfers them back into the bloodstream. Unneeded substances, including some excess water, proceed and are eventually excreted as urine.

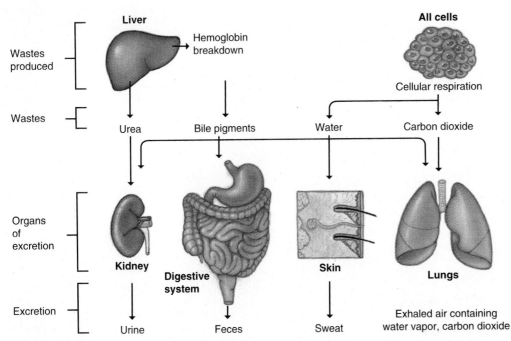

Figure 35.15 Various waste products are disposed of by several systems.

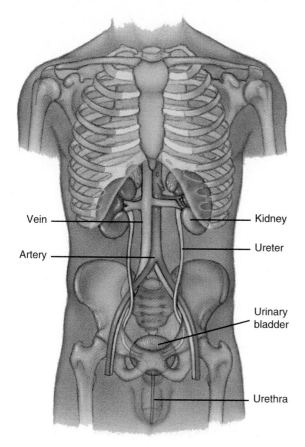

Figure 35.16 The human urinary system eliminates water-soluble waste products.

Question: *What are some water-soluble waste products eliminated by the urinary system?*

Formation of Urine

To carry out the functions of removing nitrogenous wastes from the bloodstream and regulating concentrations of water and salts in the body, the kidney conducts three operations: filtration, reabsorption, and secretion. First, wastes, salts, amino acids, glucose, some water, urea, various ions, and other substances contained in blood enter the glomerulus and are forced out of the bloodstream into the tubules, where they form a fluid called the **filtrate.** Note the unusual composition of this filtrate. Some of the materials are harmful and require elimination from the body, while others, such as glucose, are too valuable to be lost and are normally reclaimed by healthy kidneys. Thus a second step, which occurs in the tubules, involves selective **reabsorption** in which most of the water and other substances of value to the body are returned to the blood from the filtrate. A third process involves the **secretion** of unwanted ions and other substances from the blood to the filtrate in various regions of the nephron. Unwanted substances pass through the tubules without being reabsorbed and become part of the urine. Table 35.3 summarizes nephron activities in forming urine.

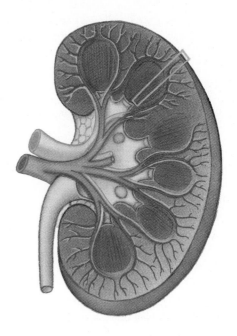

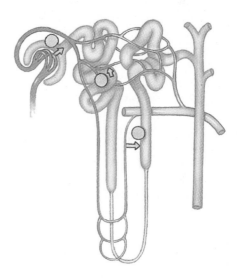

Figure 35.17 (A) The human kidney is composed of more than 1 million nephrons that filter wastes from the blood and produce urine. (B) Each nephron consists of a glomerulus and various tubules that produce urine. In the glomerulus, substances such as wastes, salts, glucose, urea, and various ions are filtered from the blood, enter the capsule, and pass into the tubule. As substances move through the tubule network, some are reabsorbed by blood in surrounding capillaries. Substances not reabsorbed from the tubule move the collecting tubule, where they become incorporated into urine. Unwanted substances may also pass from capillaries directly into tubules to become part of the urine.

Table 35.3 Summary of Major Filtration, Reabsorption, and Secretion Activities in a Nephron

Region of Nephron	Filtration	Reabsorption	Secretion
Glomerular capsule, glomerulus	Blood in glomerular capillaries; forms a filtrate containing water, glucose, amino acids, Na^+, Cl^-, K^+, HCO_3^-, urea, and other solutes	—	—
Proximal convoluted tubule	—	Glucose, amino acids, Na^+, Cl^-, K^+, HCO_3^-, urea, and water	H^+
Descending limb of the loop of Henle	—	Water, Na^+, and glucose	Urea
Ascending limb of the loop of Henle	—	Na^+, Cl^-, and K^+	—
Distal convoluted tubule	—	Water, glucose, Na^+, Cl^-, and HCO_3^-	—
Collecting duct	—	Water, Na^+, HCO_3^-, and urea	H^+ and K^+

The workload of the kidneys is astonishing. About 1,200 liters of blood is processed daily, and almost 110 liters of filtrate is formed. More than 99 percent of the filtrate is reabsorbed from the tubules, and an average of 1.5 liters of urine is produced and excreted per day.

Although the process is immensely complex, the end product, urine, contains about 96 percent water and 1 to 2 percent each of nitrogenous waste products, principally urea, and salts. However, urine composition depends on the state of the internal environment. For example, if an individual has consumed excessive quantities of fluids, the urine will be more dilute, because greater than average amounts of water will be excreted. In contrast, the urine of someone who has run 10 kilometers on a hot day will contain only the minimal amount of water necessary to form urine. Concentrations of salts also fluctuate, depending on intake and internal conditions. Some compounds, such as glucose, are reabsorbed entirely, and their appearance in urine usually signals some type of health problem. Through the precise regulation of substances reabsorbed or excreted, the kidney plays a pivotal role in maintaining internal homeostasis.

BEFORE YOU GO ON The urinary system plays a major role in maintaining internal homeostasis. Nephrons, the functional units of the kidneys, remove salts, nitrogen-containing waste products, and excess water from the blood. These substances are then incorporated into urine, which is excreted from the body.

The major structures and functions of several human organ systems have been described in Chapters 34 and 35. Though each has specific and distinctive responsibilities, all are interrelated and interdependent. Homeostasis is a unifying concept that is valuable for reflecting on the roles of various organ systems in the human body. Organ systems allow us to maintain the relatively constant internal environment that permits the highest level of cellular specialization. Organ systems enable us to operate with great efficiency and survive in a demanding environment. What controls or regulates the activities of each of the different organ systems? This question is addressed in the next two chapters.

SUMMARY

1. Lungs are the primary organ of the respiratory system, which functions to obtain O_2 from the air and get rid of CO_2 produced during cellular respiration. Gas exchange occurs in lung alveoli.

2. Gas exchange during breathing is primarily a function of differences in atmospheric pressure inside and outside the lung. Gas exchange between the lungs and blood is dependent on partial pressures of O_2 and CO_2.

3. The major functions of the circulatory system are to transport O_2, metabolic wastes, and nutrients from one area of the body to another; to assist in defense reactions against disease agents; and to regulate temperature and fluid balance. To accomplish these tasks, the heart pumps blood through an elaborately organized system of vessels that reaches all the tissues and cells of the body.

4. Oxygen is transported from the lungs to tissue cells by hemoglobin molecules in red blood cells. Carbon dioxide is transported from cells to lungs in various ways—dissolved in plasma, by hemoglobin, or in plasma as HCO_3^-.

5. The respiratory center maintains blood pH, primarily by monitoring H^+ concentrations in the blood and regulating breathing rates to increase or decrease CO_2 exchanges.

6. The mouth, pharynx, esophagus, stomach, small intestine, various accessory organs, and large intestine constitute the digestive system, which functions in the preparation and delivery of nutrients throughout the body.

7. Carbohydrates, proteins, nucleic acids, and lipids are broken down to their component parts—monosaccharides, amino acids, nucleotides, and glycerol and fatty acids, through the

small intestine by diffusion or active transport and then moved into the blood or the lymph. From the small intestine, nutrients are transported first to the liver and then throughout the entire body.

8. To rid the body of certain wastes, the kidneys filter blood to remove unneeded substances, which are later excreted in urine. The urinary system also plays a role in the regulation of water, salts, and other substances in the blood.

WORKING VOCABULARY

absorption (p. 670)
artery (p. 662)
blood (p. 662)
capillary (p. 662)
digestion (p. 668)
erythrocyte (p. 662)
esophagus (p. 669)
heart (p. 663)
kidneys (p. 672)
large intestine (p. 670)
leukocyte (p. 662)
liver (p. 670)

lung (p. 657)
lymphatic system (p. 664)
nephron (p. 672)
pancreas (p. 670)
pharynx (p. 657)
plasma (p. 662)
pulmonary circulation (p. 663)
respiration (p. 657)
small intestine (p. 669)
systemic circulation (p. 664)
vein (p. 662)

REVIEW QUESTIONS

1. Describe the pathway that air (oxygen) follows, beginning with the atmosphere and ending with alveoli.

2. How is breathing regulated?

3. What regulates gas exchange during respiration?

4. What are the functions of the circulatory system?

5. Describe the composition of blood.

6. What are the functions of various types of blood vessels? Of the heart?

7. What is the function of pulmonary circulation? Of systemic circulation?

8. What are the structures and functions of the lymphatic system?

9. How are oxygen and carbon dioxide transported in the circulatory system?

10. How does the respiratory center help regulate respiration?

11. Which organs are part of the GI tract? What is the function of each?

12. How do the pancreas and the liver contribute to digestion?

13. How are different classes of chemicals digested and absorbed?

14. What is the function of the urinary system?

15. What is the functional unit of the urinary system? Describe its structure.

16. How is urine formed?

ESSAY AND DISCUSSION QUESTIONS

1. What sort of adaptive changes might you predict would occur in humans if Earth's environment changed from its present state to become significantly hotter and dryer?

2. Several human organs are considered "essential" (required for survival). Disease or trauma can destroy any of these organs, and efforts are being made to develop artificial essential organs. How would you rank the lung, heart, liver, and kidney in terms of difficulty in development, from easiest to most difficult? Why?

3. Why do you suppose the heart became the symbol for romantic love?

4. The lymphatic system was the last major organ system to be discovered and described. What might have accounted for this delay?

REFERENCES AND RECOMMENDED READINGS

Bastian, G. F. 1994. *An Illustrated Review of the Respiratory System.* New York: HarperCollins.

Bramble, D. M., and D. R. Carrier. 1983. Running and breathing in mammals. *Science,* 219: 251–256.

Chien, K. R. 1993. Molecular advances in cardiovascular biology. *Science,* 260: 916–917.

Cooper, K. E. 1994. Some responses of the cardiovascular system to heat and fever. *Canadian Journal of Cardiology,* 10: 444–452.

French, R. K. 1994. *William Harvey's Natural Philosophy.* New York: Cambridge University Press.

Harvey, W. 1941. *Exercitatio anatomica de motu cordis et saguinis in animalibus.* Tran. C. D. Leake. 3rd ed. Springfield, Ill.: Thomas. (Originally published 1628.)

Rowell, L. B. 1993. *Human Cardiovascular Control.* New York: Oxford University Press.

Tortora, G. J., and S. R. Grabowski. 1996. *Principles of Anatomy and Physiology.* 8th ed. New York: HarperCollins.

West, J. B. 1995. *Respiratory Physiology: The Essentials.* 5th ed. Baltimore: Williams & Wilkins.

ANSWERS TO FIGURE QUESTIONS

Figure 35.1   The diaphragm is a large muscle that initiates breathing by contracting, which causes lung volume to increase.

Figure 35.2   Oxygen is in the atmosphere; CO_2 is generated during cellular respiration reactions.

Figure 35.4   To carry oxygen from the lung to tissue cells.

Figure 35.9.   From the capillary into the cell, because pO_2 is higher in the capillary (105 mm Hg) than in the cell (40 mm Hg) and O_2 moves down a pressure gradient (from high to low pressure).

Figure 35.10   An increase in pCO_2 (or a *decrease* in pH or pO_2) results in changes that lead to the opposite condition—a decrease in pCO_2 (or an *increase* in pH or pO_2).

Figure 35.12   It increases the amount of surface area through which nutrients can be absorbed.

Figure 35.13   Diffusion.

Figure 35.16   Urea, H^+, and K^+.

36

Control and Regulation: The Nervous System

Chapter Outline

Reading Questions

1. How is the structure of a neuron related to its function?

2. How are nerve impulses transmitted across a synapse?

3. How is knowledge of synaptic transmission being used in modern medicine?

4. What are the major divisions of the nervous system and the function of each?

A ll organisms, including humans, are constantly exposed to great quantities of information being transmitted from both the internal and external environments. Endless streams of signals are conveyed to internal regulatory centers from all tissues and organs in our bodies. Never-ending stimuli from the external environment—in the form of light, noise, heat, pressure, odor, color, shape, and movement—are detected and processed during every moment of our lives. Occasionally, a proper response to this incoming information must be made instantly if we are to survive or continue func-

tioning in a normal fashion. In other cases, a more leisurely, protracted reaction may be appropriate, and sometimes no response is required.

As we noted in Chapter 33, the nervous and endocrine systems play the most significant roles of all the body systems in maintaining *homeostasis*. Central to communications that occur between cells, tissues, and organs within our bodies (internal stimuli), the nervous and endocrine systems also receive, process, and respond to stimuli of widely different varieties and intensities from outside our bodies (external stimuli). Although we consider the nervous and endocrine systems in separate chapters, they are linked in governing activities that are essential to our well-being. In general, our nervous system, with its extraordinary control center, the brain, allows us to formulate prompt responses of short duration with pinpoint control. The nervous system is thus the first system to respond to stress. In contrast, our endocrine system is characterized by slower reaction times, responses of greater duration, and more diffuse control. A common feature of both systems is that they use **chemical messengers,** specialized molecules capable of transmitting signals, in their communications.

In Chapters 36 and 37, we describe the anatomical features of these complicated systems and examine a number of functions and physiological mechanisms involved in their operation. For the most part, we concentrate on processes that can be readily visualized and that occupy the spotlight in contemporary research efforts.

The nervous system coordinates and regulates most functions of the body. This system consists of billions of highly specialized cells, called *neurons,* that are interconnected to form vast communication networks. The basic "wiring" of the human nervous system has been known for centuries, and more recently we have come to understand parts of its structure and function at the molecular level. Nevertheless, a great deal remains unknown about its more sophisticated operations, particularly those centered in the brain (for example, learning).

The human nervous system is often compared to a computer because both are viewed as operating in fundamentally similar ways: information is first put into the system, some form of integration occurs, and a message is then formulated. However, such correlations have only limited applications. Even the largest mainframe computers have primitive processing capabilities in comparison with our nervous system because the nervous system operates as a parallel processor (two or more items can be manipulated at a time) and has unlimited storage capacity. What enables the human nervous system to operate with such great flexibility and with such a high degree of precision? The answers lie within the structure, organization, and functional capabilities of neurons.

BEFORE YOU GO ON | The general function of the nervous system is to receive, process, and integrate nerve impulses caused by stimuli received from the external and internal environments. Neurons are the functional units of the nervous system. Billions of neurons are concentrated in specialized organs, such as the brain and the spinal cord, and arranged in specific networks that reach all parts of the body. This organization allows for precise control of all organ systems.

NEURONS

The nervous system of all animals is composed of two principal cell types. **Neurons,** shown in Figure 36.1A (see page 678), are the functional units of the nervous system. They are highly specialized cells capable of conducting electrochemical nerve impulses. The human nervous system consists of over 100 billion neurons, most of which are located in the brain. They are supported, protected, and insulated by surrounding **neuroglia (or glial) cells,** the other major cell type of the nervous system (see Figure 36.1B, page 678). Neuroglia cells are smaller than neurons and five to ten times as abundant. There are several kinds of neuroglia cells, and they perform different functions. Certain neuroglia cells form supporting networks by being wrapped around neurons, others bind neurons to blood vessels, and some tie nervous tissues together.

Structure of Neurons

Neurons come in hundreds of sizes and shapes, but they all have a common structure that reflects their function (see Figure 36.1C). The **cell body** contains the nucleus, cytoplasm, and organelles responsible for maintenance and for synthesizing the specific proteins required by a neuron. The most prominent features of neurons are two types of stringlike cytoplasmic fibers, emanating from the cell body, that are used in conducting impulses and communicating with other cells. Most neurons contain a single **axon,** a fiber that conducts the nerve impulse away from the cell body. The length of axons in human neurons ranges from less than 1 millimeter to greater than 1 meter. The axon separates near its end into multiple small branches that each end in a tiny knob called the **synaptic bulb**. The ends of the axon's branches are called **synaptic terminals**. **Dendrites,** the second type of extension, are thinner than axons and radiate out from the cell body. Dendrites receive messages from adjacent cells or directly from the external environment and conduct impulses toward the cell body. A neuron may have up to 200 dendrites. The cell body, axons, and dendrites contain numerous **neurofibrils**, threadlike protein structures that form delicate supporting networks within a neuron.

Axons of most large neurons are covered by an insulating layer of lipid-rich whitish material called **myelin** that is produced through the actions of specialized **Schwann cells** (see Figure 36.2, page 679). During embryonic development and early childhood, Schwann cells grow in such a way that they encircle an axon several times and form a covering of overlapping membranes that is called a **myelin sheath. Nodes of Ranvier** are spaces along a myelinated neuron that are gaps between individual Schwann cells. Myelinated neurons are better insulated and conduct impulses faster than nonmyelinated neurons (see Figure 36.3, page 679). If the myelin covering of an axon becomes eroded or is injured and is replaced by scar tissue, nerve function becomes impaired, and serious consequences follow. For example, multiple sclerosis (MS) is a disease characterized by progressive deterioration of myelin sheaths covering axons in the brain and spinal cord. Affected neurons become covered with scars, resulting in the disruption of nerve conduction pathways. Early symptoms of MS are muscular weakness, double vision, and lack of coordination. Later symptoms depend on the extent and rate of damage to other neurons. The cause of MS is not yet known, although increasing evidence suggests that a virus may be involved.

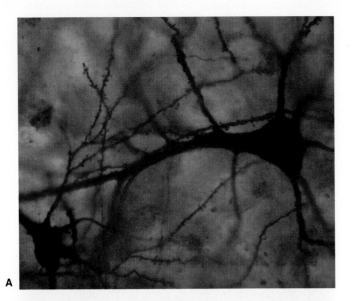

A

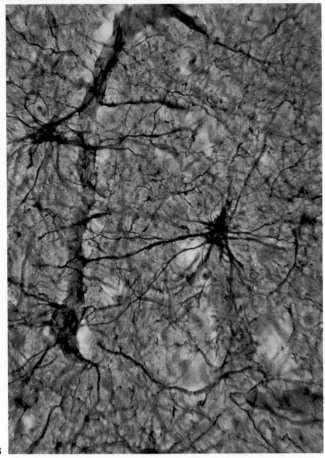

B

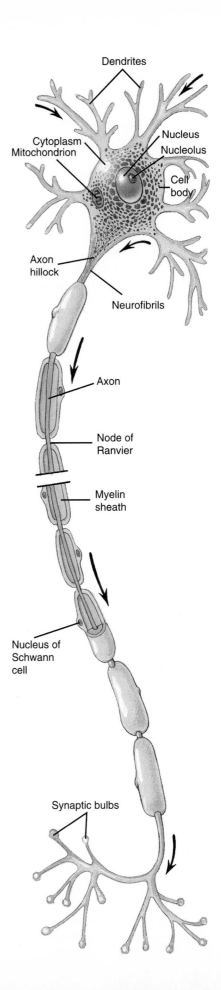

C

Figure 36.1 (A) Neurons are highly specialized cells that are the functional units of the nervous system. (B) Billions of small neuroglia cells, such as these astrocytes, nourish, protect, and support neurons. (C) There are many types of neurons, but all have a common structure. The basic components of a neuron are a cell body, an axon, and multiple dendrites. Arrows indicate the direction of impulse travel in a neuron.

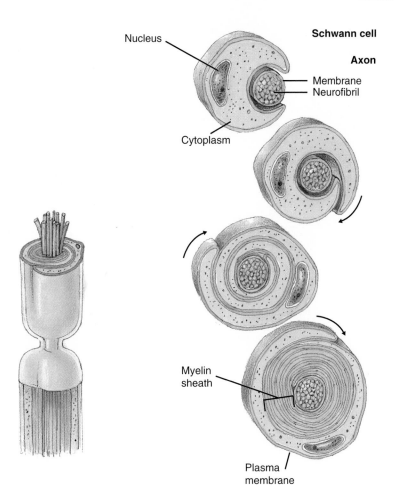

Schwann cell

Axon

Nucleus

Membrane
Neurofibril

Cytoplasm

Myelin
sheath

Plasma
membrane

Figure 36.2 (A) Axons of major neurons are enclosed within a myelin sheath formed by Schwann cells. (B) The myelin sheath is formed by Schwann cells located near an axon. As they grow, they become wrapped around themselves and the axon. Myelinated neurons conduct impulses more rapidly than nonmyelinated neurons, and they are also better protected from injury.

Question: *What effect does multiple sclerosis have on myelinated neurons?*

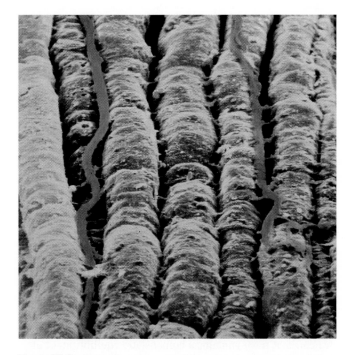

Figure 36.3 An SEM of neuron bundles. The connective tissue binding the neurons together has been removed, revealing the parallel axons (yellow) and associated blood vessels (blue). These neurons are myelinated and the external surfaces of the Schwan cells that surround myelinated nerves appear as yellow-colored structures.

Types of Neurons

Neurons are commonly classified according to their functions, which are in turn related to the routes of information flow within the nervous system (see Figure 36.4, page 680). Cells, tissues, and organs that are controlled by the nervous system through the actions of neurons are said to be "innervated."

Sensory (afferent) neurons receive information from the external and internal environments and conduct it to the brain or spinal cord, where it is processed. **Motor (efferent) neurons** transmit impulses away from the brain or spinal cord, where the cell body is situated, toward other cells, muscles, glands, or other neurons involved in making a response. Typically, motor neurons have single, long, myelinated axons. **Association neurons,** the most abundant type (more than 99 percent of all neurons), are confined to the brain and spinal cord. There, they process incoming information by transmitting it between sensory and motor neurons. Such processing may involve transmission among a small number of neurons if a simple response occurs or millions of neurons if learning or language is involved. Brain association neurons commonly have a profusion of branched dendrites that are essential for establishing connections with other neurons.

A **nerve** is a bundle of parallel neuron fibers, both axons and dendrites, and their associated blood vessels and supporting cells, all of which are enclosed by protective connective tis-

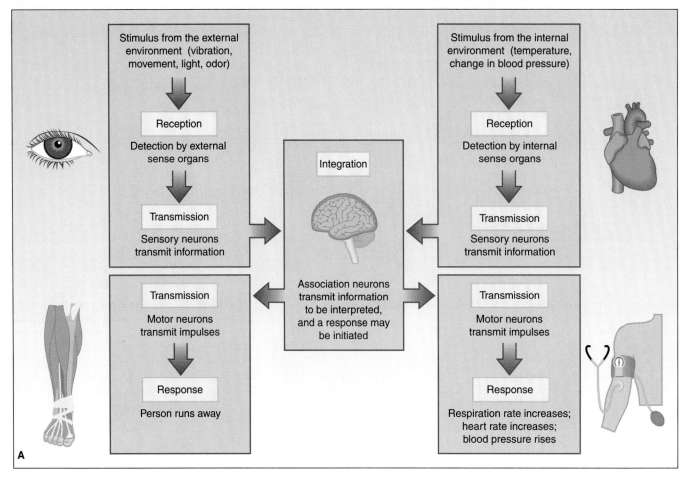

Figure 36.4 (A) Information flows through the nervous system along neural pathways. Stimuli are received from the external and internal environments and transmitted through sensory neurons to association neurons in the brain and spinal cord, where they are integrated and interpreted. If a reaction is initiated, nerve impulses are transmitted to muscle and gland cells by motor neurons. (B) The three functional types of neurons differ in subtle ways.

Question: *Where are association neurons found in the nervous system?*

sue (see Figure 36.5). Each fiber within a nerve operates independently, and most nerves have both sensory and motor components. By definition, nerves are found outside the brain and spinal cord. Structures in the brain and spinal cord that have the same structural organization are called **nerve tracts**.

Neuron Function

The basic function of a neuron is to communicate with other neurons and with other cells in the body. Neurons have two basic characteristics related to this function: they are **excitable**, meaning that they have the capacity to respond to stimuli; and they are also **conductive**, which means that they are able to transmit an electrical signal (a *nerve impulse*) along their axons. Generating and transmitting a nerve impulse is dependent on special properties of the neuronal plasma membrane that allow it to regulate movements of ions with a positive electrical charge, most notably sodium ions (Na^+) and potassium ions (K^+). Nerve impulses have the following characteristics:

- They travel rather rapidly, about 100 meters per second.
- They travel in only one direction.
- They follow an "all or none" law: a neuron conducts an impulse like a pistol fires a bullet, either at full power or not at all.

How do neurons generate and transmit electrical signals? To answer this question, we will first examine the electrical features of a neuron at rest (when not conducting an electrical impulse) and then explore changes that occur when an electrical impulse is generated and transmitted by a neuron at work.

Neurons at Rest

A neuron at rest maintains a homeostatic equilibrium of electrically charged ions on each side of the plasma membrane. As described in Figure 36.6A (see page 682), Na^+ and K^+ constitute positive electrical charges, and negative electrical charges come from chloride ions (Cl^-), phosphate ions, (PO_4^{3-}), and

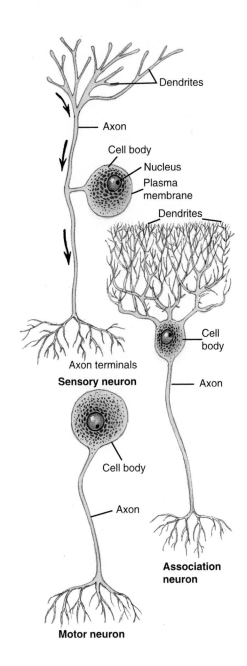

Sensory neuron

Dendrites

Axon

Cell body

Nucleus

Plasma membrane

Dendrites

Axon terminals

Cell body

Cell body

Axon

Cell body

Association neuron

Axon

Motor neuron

A

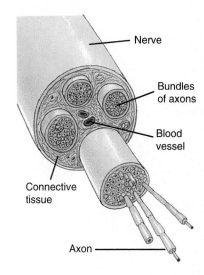

Nerve

Bundles of axons

Blood vessel

Connective tissue

Axon

B

Figure 36.5 (A) Nerves consist of bundles of neurons and blood vessels, which are enclosed within connective tissue.(B) Three types of neurons.

potential is said to be **polarized.** The potential energy is converted to electrical energy when **depolarization** occurs as Na^+ and K^+ flow through the neuron's plasma membrane and the membrane potential becomes less negative.

What regulates Na^+ and K^+ flow through the neuron when an electrical nerve impulse is to be generated and transmitted? Certain *ion channels* are formed by channel proteins embedded in the plasma membrane that contain a pore with a "gate" capable of opening and closing. When open, an ion channel allows a specific ion—Na^+, K^+, calcium (Ca^{2+}), or Cl^-—to diffuse across the membrane in either direction, depending on ion concentrations on either side of the membrane. Neuronal transmission of an electrical impulse involves channels that are regulated—so-called **gated ion channels** ("nongated" ion channels are always open). Ion channels that play a critical role in conducting a nerve impulse—gated Na^+ and K^+ channels—are **voltage-gated ion channels,** which means they open or close in response to changes in membrane potential (voltage) (see Figure 36.7, page 682). Neurons contain thousands of Na^+ and K^+ channels. As described in the next section, voltage-gated Na^+ channels have two gates, one for activation and one for inactivation. In resting neurons, voltage-gated Na^+ and K^+ channels are closed.

Neurons at Work

Have you ever been awakened in the middle of the night by a loud, sharp noise? What changes occurred in resting neurons that led to your becoming alert? A nerve impulse, technically known as an **action potential,** is generated when a neuron becomes excited in response to a stimulus. Scientists could not explain the mechanism of action potential until the 1940s and 1950s, when giant nerve axons found in squid were first identified and studied. Later research using other species confirmed that action potentials are conducted in the same way in all animals. The following model of action potential, described in

amino acids in protein. However, there is normally an uneven balance between total positive and negative electrical charges on each side of the resting membrane. The outer membrane has a slight positive electrical charge and the inner membrane a slightly greater negative charge. Consequently, there is a voltage (electrical potential energy) difference across the plasma membrane because of the uneven distribution of electrical charges. This voltage difference is called the **resting membrane potential (RMP),** and in neurons, this potential energy is measured in millivolts (mV). The RMP of a neuron is usually expressed as −70 millivolts, which means that the area inside the membrane has more negative charges relative to the number of positive charges outside the membrane (see Figure 36.6B, page 682). A neuron (or any cell) with a membrane

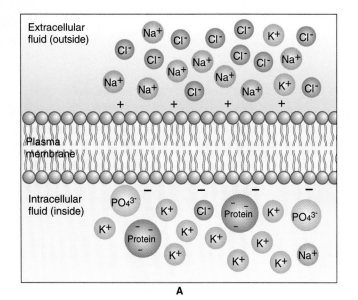

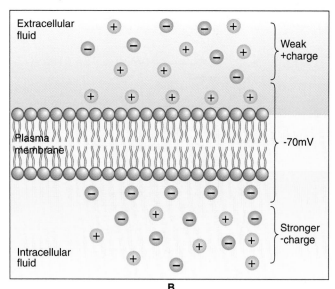

Figure 36.6   (A) Various charged ions and molecules are found on both sides of a neuron's plasma membrane. Chloride ions (Cl^-), phosphate groups (PO_4^{3-}), and amino acids in proteins have negative charges, while sodium ions (Na^+) and potassium ions (K^+) have positive charges. The inner plasma membrane has a negative electrical charge relative to the outer membrane.

Question: What type of energy does this difference in electrical charge between the inner and outer membrane represent?

(B) The resting membrane potential (RMP) is a measure of the difference in electrical charges between the inner and outer membrane. An RMP of −70 millivolts indicates that the inside of the membrane is negative relative to the outer membrane.

Question: What would be the effect on RMP if positively charged ions flowed into the membrane?

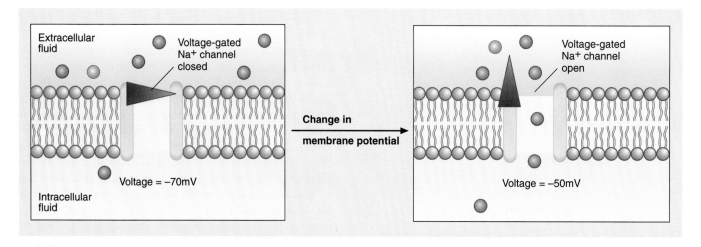

Figure 36.7   Voltage-gated Na^+ ion channels regulate the movement of Na^+ through the plasma membrane. Neurons at rest have a membrane potential of −70 millivolts and Na^+ channels remain closed at that steady-state voltage. When membrane potential increases to −50 millivolts, voltage-gated Na^+ channels respond by opening.

Question: What might cause a voltage-gated Na^+ channel to close?

Figure 36.8, emerged from those studies. An action potential takes only a few milliseconds to complete.

An action potential begins in one area of an excited axon when the membrane is briefly depolarized. In response, voltage-gated Na^+ activation channels open, Na^+ ions rush inward, and the membrane potential changes from −70 millivolts to 30 millivolts; that is, the flow of positively charged ions into the membrane results in an electrical current or change in the steady-state electrical balance. The sudden influx of Na^+ triggers further depolarization in the neighboring area of the axon,

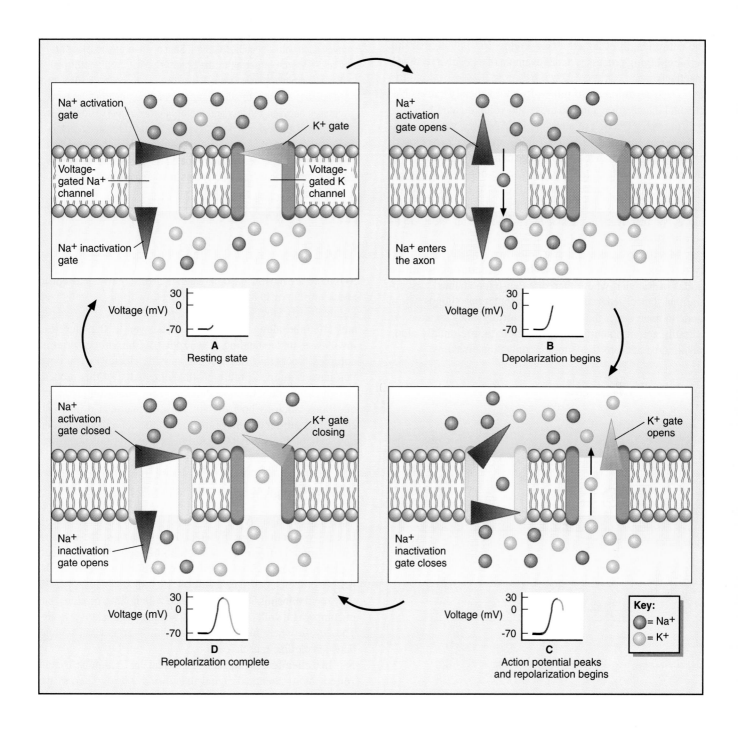

Figure 36.8 (A) In the resting state, Na⁺ activation gates and voltage-gated K⁺ channels remain closed and Na⁺ inactivation gates are open. The resting membrane potential remains at −70 millivolts. (B) When an action potential is initiated, depolarization begins as Na⁺ channel activation gates open, allowing Na⁺ to enter the axon, which causes the membrane potential to become less negative. (C) When the action potential peaks at 30 millivolts, repolarization begins as Na⁺ channel inactivation gates close automatically and voltage-gated K⁺ gates open, leading to a massive outflow of K⁺. (D) Once repolarization is complete, Na⁺ activation gates and K+ gates are closed and Na⁺ inactivation gates open. The neuron is now ready to initiate another action potential if necessary.

Question: *How long does this sequence of events take to complete?*

causing more Na$^+$ channels to open, and this continues sequentially, resulting in the transmission of an action potential along the entire length of an axon (see Figure 36.9). Note that this self-amplifying process is a rare example of a positive feedback mechanism.

Depolarization to 30 millivolts automatically causes Na$^+$ channel inactivation gates to close and voltage-gated K$^+$ channels to open a fraction of a second later. No new action potentials can be initiated when Na$^+$ inactivation gates are closed. Once these gate changes occur, large numbers of K$^+$ ions flow out of the cell, which leads to repolarization—a return to the normal polarized state and an RMP of −70 millivolts. Once repolarization is complete, Na$^+$ activation gates and K$^+$ gates close, Na$^+$ inactivation gates open, and the neuron returns to its resting state.

Myelinated neurons convey electrical signals in a slightly different way. All of the electrical activities occur in the nodes of Ranvier, where there are dense ion channel concentrations. Once generated, the action potential "jumps" from node to node rather than flowing uniformly along an axon. As a result, action potentials are transmitted more rapidly along myelinated neurons than along nonmyelinated axons.

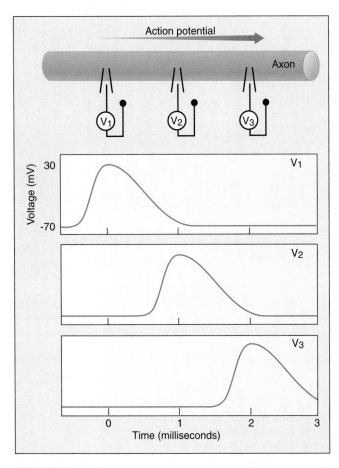

Figure 36.9   An action potential flows sequentially along an axon until it reaches a synapse. The gauges indicate the voltages that would be recorded if measured in sequence ($V_1 \rightarrow V_2 \rightarrow V_3$) along the axon, and those voltages are described in the graphs.

BEFORE YOU GO ON   Neurons are specialized cells that can respond to stimuli and conduct an impulse. There are three basic neuron types, each with a specific function. Sensory neurons receive stimuli and transmit impulses to the brain or spinal cord; motor neurons transmit impulses from these organs to cells, tissues, or organs that respond to stimuli; and association neurons transmit impulses between sensory and motor neurons. A nerve impulse is transmitted as an action potential along an axon. An action potential is generated by a change in resting membrane potential that results from the actions of voltage-gated Na$^+$ and K$^+$ channels in the plasma membrane. Action potentials are completed when repolarization occurs.

SYNAPSES

Synapses are narrow gaps between neurons, between a neuron and a muscle or gland cell, or between two muscle cells (see Figure 36.10). How is a nerve impulse able to cross the synapse and continue along its route? The answer to that question involves an understanding of *synaptic transmission,* a series of complex biochemical events that may be clearer to you once you have examined Figure 36.11. Knowledge of this process also opens the door to understanding some interesting research questions, such as how certain poisons, medical drugs, and controlled substances alter functions of the nervous system and how certain drugs may be useful in treating various neurological disorders.

Types of Synapses

There are two basic types of synapses, electrical and chemical. In cells communicating by **electrical synapses**, the impulse travels directly from one cell to another cell. In effect, such synapses can be viewed as "direct wiring" between two cells, neither of which has to be a neuron. Electrical synapses permit very rapid transmission of an impulse with no interruptions, but in comparison with chemical synapses, they are relatively rare. They occur where high speed of transmission is vital, such as between cardiac muscle cells.

In a **chemical synapse,** an electrical nerve impulse is converted into a chemical signal that forms a bridge across the synapse between neurons or between a neuron and a muscle or gland cell. This bridge allows the chemical signal to pass to the adjacent cell, where it may initiate another action potential. Let us examine how this occurs and what happens if the process is disrupted or fails.

Transmission of a Nerve Impulse Across a Chemical Synapse

Figure 36.11 summarizes the events in chemical synaptic transmission. One-way transmission of an impulse is a key feature of this process. When a nerve impulse travels along the axon of a **presynaptic neuron** (the one carrying the impulse) toward

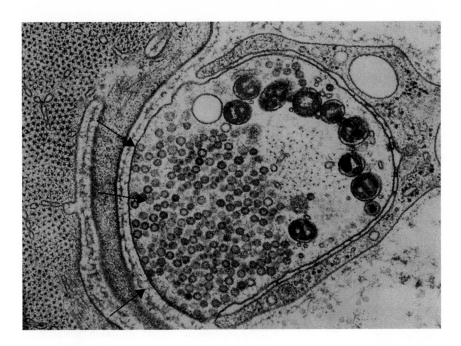

Figure 36.10 This electron micrograph shows a chemical synapse (arrows)—the space between two neurons or between a neuron and a muscle or gland cell.

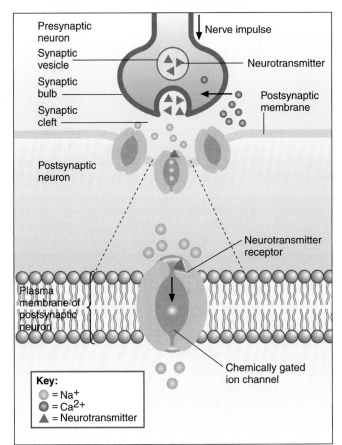

Key:
○ = Na⁺
● = Ca²⁺
▲ = Neurotransmitter

Figure 36.11 Electrical nerve impulses are transmitted from a presynaptic neuron to a postsynaptic neuron through a chemical synapse. When an impulse reaches a synaptic bulb, it triggers an influx of Ca^{2+} into the presynaptic neuron, which induces a synaptic vesicle to fuse with the presynaptic membrane and release its neurotransmitters into the synaptic cleft. The neurotransmitters diffuse across the cleft and bind with specific receptors on chemical-gated ion channels. As a result, positively charged ions can enter the postsynaptic neuron. If a sufficient number of channels become involved, a new action potential will be created, and transmission of the original nerve impulse will continue.

Question: *What might happen to the neurotransmitters released into the synaptic cleft after their mission is completed?*

the synapse, it initiates a series of events in the *synaptic bulbs* located at the ends of axon terminals. Within each synaptic bulb are thousands of **synaptic vesicles,** small membrane-bound sacs filled with **neurotransmitters,** specific chemical messengers involved in nerve impulse transmission across a chemical synapse. Each synaptic vesicle contains 10,000 to 100,000 neurotransmitter molecules. There are many different neurotransmitters, but each vesicle is thought to contain only a single type. However, a synaptic terminal can harbor two or more types of vesicles that contain different neurotransmitters.

As the electrical nerve impulse reaches the synapse, the depolarizing events open voltage-gated Ca^{2+} channels, causing

Ca^{2+} to flow inward. As a result, vesicles are stimulated to fuse with the **presynaptic membrane** and release their neurotransmitter into the **synaptic cleft,** the narrow space between cells. The neurotransmitters diffuse across the cleft and bind to specific receptors that are part of **chemically gated ion channels** that regulate Na^+, K^+, and Ca^{2+}. When neurotransmitters bind with the receptors, these channels allow positively charged ions to pass through, which may help initiate a new action potential in a postsynaptic cell, thus perpetuating transmission of the original impulse.

Not all nerve impulses, however, continue to be transmitted. Neuron cell bodies, muscle cells, and gland cells are typically contacted by numerous synaptic bulbs (see Figure 36.12). Some of these belong to excitatory neurons and are likely to induce a response. Other synaptic bulbs, however, are from inhibitory neurons that tend to make a response less like-ly. Whether or not an impulse is transmitted depends on a final accounting by the postsynaptic cell in making a yes-or-no decision about firing. If the balance sheet tilts toward excitation, the impulse is transmitted; if toward inhibition, no impulse conduction occurs.

Return to Normal

The action of neurotransmitters takes only a fraction of a second. What happens to them after the impulse is transmitted? How is the signaling terminated? These are critical questions because if the synaptic cleft is not cleared of neurotransmitters, continuous firing of the excited neuron will result, which can lead to severe problems (this is the mode of action for some poisons). There are three primary means for purging the synaptic cleft of neurotransmitters: (1) specific enzymes degrade some transmitters, (2) some transmitters are taken back up by the presynaptic membrane and recycled into vesicles, and (3) some transmitters are taken back up by nearby neuroglia cells. Whatever the process, it occurs within a fraction of a second, and calm is restored at the synapse.

The Current Picture

The mechanism of chemical synaptic transmission is well understood, but many questions remain open. The list of chemicals that serve as neurotransmitters is now lengthy but still incomplete. Also, full details about how neurotransmitters are released and interact with postsynaptic cells are known for only a handful of such substances. **Acetylcholine (ACh),** a neurotransmitter occurring at many neuron–neuron and neuron–muscle cell synapses, has been the most extensively studied. A detailed model has been developed that describes its synthesis, storage in synaptic vesicles, and release into the synaptic cleft (see Figure 36.13). Scientists have also isolated the ACh receptor found on postsynaptic chemically gated ion channels. The receptor is a protein, and researchers have cloned it and determined its amino acid sequence. After transmission, ACh is instantly split into choline and acetate, its two components, by the enzyme called *acetylcholinesterase.* Choline and acetate are recycled back into the synaptic bulb, recombined into ACh, and reincorporated into a synaptic vesicle. This type of detailed information will likely become available for other neurotransmitters in the future and will undoubtedly lead to new pharmaceutical drugs and other compounds that can modify action at the synapse.

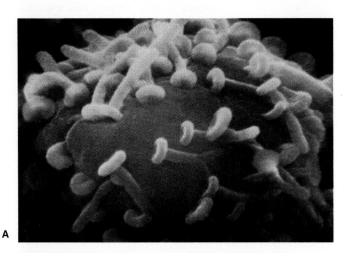

A

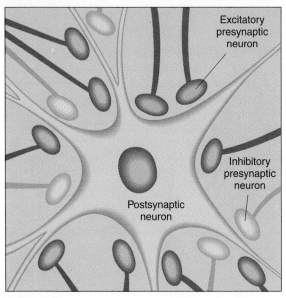

Excitatory
presynaptic
neuron

Inhibitory
presynaptic
neuron

Postsynaptic
neuron

B

Figure 36.12   (A) This scanning electron micrograph shows numerous synaptic bulbs in contact with a neuron cell body. (B) Cells are often contacted by both excitatory and inhibitory neurons. By integrating impulses from both types of neurons, the target cell determines whether or not to fire.

BEFORE YOU GO ON   Most nerve impulses are transmitted from cell to cell across a chemical synapse. Impulses traveling down presynaptic neurons trigger release of neurotransmitters from synaptic vesicles into a synaptic cleft. Neurotransmitters bind to receptors on ion channels in the postsynaptic cell and initiate changes that may trigger an impulse, depending on whether the cell becomes excited or inhibited. Once an impulse is transmitted, neurotransmitters are quickly removed from the synapse.

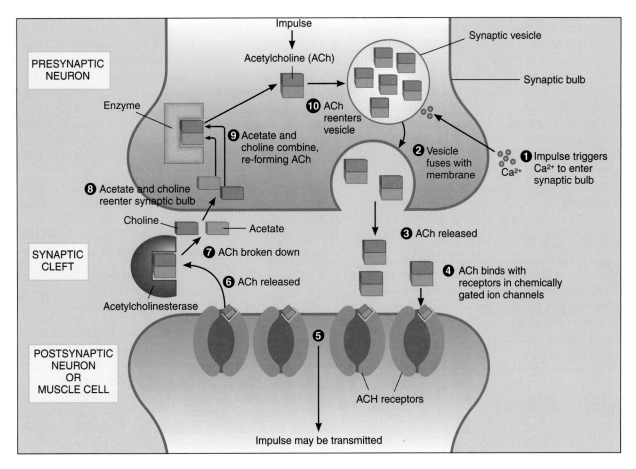

Figure 36.13 An incoming electrical nerve impulse causes calcium ions (Ca^{2+}) to enter the synaptic bulb and trigger release of acetylcholine (ACh) from synaptic vesicles into the synaptic cleft. ACh diffuses across the cleft to the postsynaptic cell, where it binds with specific ACh receptors on chemically gated channels in the membrane, causing ion fluxes that may lead to the generation of an impulse. Once this occurs, ACh is released from the receptor and broken down into acetate and choline by the enzyme acetylcholinesterase. Acetate and choline are taken up by the presynaptic cell, where they recombine to form ACh.

ALTERATIONS OF NORMAL SYNAPTIC TRANSMISSION

Understanding the process of normal synaptic transmission of nerve impulses enables us to recognize the potential for modifying the mechanism at different points. A large number of **drugs**–chemical agents that bring about functional or structural change in living tissues—exert their effects at the synapse. In humans, these effects include altering neurological processes involved in thinking, learning, memory, and complex behaviors. Perhaps the best-known drugs are those with questionable or no legitimate medical use. For example, **nicotine,** the active ingredient in tobacco, is a stimulant that mimics the action of ACh and causes a short-term excitatory effect. **Cocaine,** an addictive compound obtained from coca leaves, is now thought to prevent or reduce the uptake of neurotransmitters at the synapse and results in euphoria and an increased perception of alertness. **Crack,** a potent form of cocaine, disrupts the normal inactivation of neurotransmitters involved in mood and muscle coordination. In the short term, this can cause convulsions, high blood pressure, and weight loss. Chronic use may lead to a depletion of specific neurotransmitters and result in severe psychological effects such as persistent depression.

A more positive use of drugs is in treating serious neurological disorders. For example, **Parkinson's disease** is characterized by late age of onset (typically 50 years or older) and severe symptoms such as drooling, tremors, and nonfunctioning muscles. Although the cause of the disease is not known, a leading hypothesis is that exposure of certain brain neurons to toxic substances may play a role. This view was strengthened by a discovery involving young addicts who had taken heroin contaminated with a toxic compound known as MTTP. After entry into the body, MTTP was found to alter the same part of the brain that is affected by Parkinson's, and it caused the same effects. A major challenge to scientists is to devise ways of treating such disorders. How can the symptoms be eliminated or reduced?

The first step, always, is to identify the underlying biological mechanism that has been altered. In Parkinson's, the cause was found to be due to depletion of a neurotransmitter called **dopamine** in the brain. Consequently, neurons that require certain levels of dopamine for normal nerve transmission function abnormally, resulting in uncontrolled muscular actions. Although dopamine-requiring brain cells cannot transmit impulses normally, their dopamine receptors remain intact. Unfortunately, dopamine cannot be introduced via injection or other common procedures. Would it be possible to identify a drug structurally similar to dopamine that could be introduced into the body, bind to dopamine receptors, and allow messages to be transmitted in the normal way? After exhaustive research and testing, the drug *L-dopa* was found to meet those criteria. As a result, it is now possible to treat the disease symptoms, although such treatment does not constitute a cure. Currently, great efforts being made in the field of "synaptic chemistry" may lead to successful treatments of other neurological disorders.

BEFORE YOU GO ON Intense efforts are being made to understand how specific drugs affect synaptic transmission events. There are two aspects to this research: first, to determine how harmful substances modify normal transmission mechanisms and, second, to develop new drugs that may counteract or prevent the effects of neurological disorders.

ORGANIZATION OF THE HUMAN NERVOUS SYSTEM

Schematic wiring diagrams of the human nervous system would fill several manuals and probably astound even the most experienced electrician. At the simplest level, sensory neurons, association neurons, and motor neurons are organized into circuits, as described earlier. At more complex levels, the nervous system is organized and functions in such a way that great numbers of signals can be integrated, or "averaged," to form an appropriate response that serves to coordinate the entire organism. At the highest, and least understood, levels are the unique human abilities usually expressed with such terms as intelligence, psychology, philosophy, and problem solving.

Figure 36.14 diagrams the human nervous system, which consists of two primary divisions: the **central nervous system (CNS),** which is basically the brain and the spinal cord, and the outlying **peripheral nervous system (PNS),** which innervates all parts of the body and transmits information to and from the CNS.

Peripheral Nervous System (PNS)

The PNS consists of all neurons that exist outside of the CNS, and it contains 43 major nerves and a vast network of smaller nerves that connect the brain and spinal cord with all other structures in the body. Compared to the CNS, the PNS has the major responsibility for maintaining homeostasis.

Of the 43 major nerves of the PNS, 12 are **cranial nerves** that originate from the lower brain, and 31 are **spinal nerves** that emerge from the spinal cord (see Figure 36.15A). Through these nerves, information is transmitted continuously between every organ and tissue and the CNS.

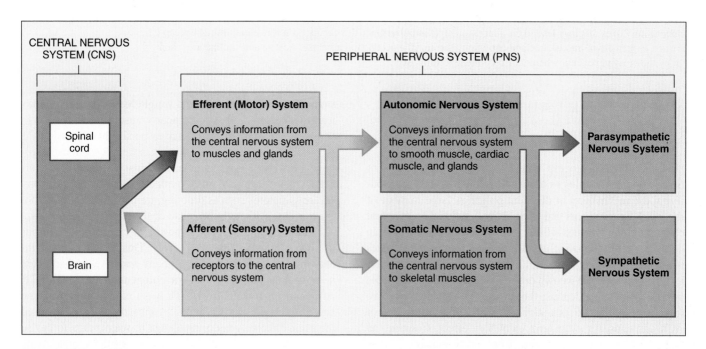

Figure 36.14 The human nervous system is composed of two principal parts: the central nervous system (CNS) and the peripheral nervous system (PNS). The PNS includes the somatic nervous system and the autonomic nervous system, which is further subdivided into the parasympathetic and sympathetic nervous systems. Each of the various divisions has specific functions.

Question: *What is the general function of the autonomic nervous system?*

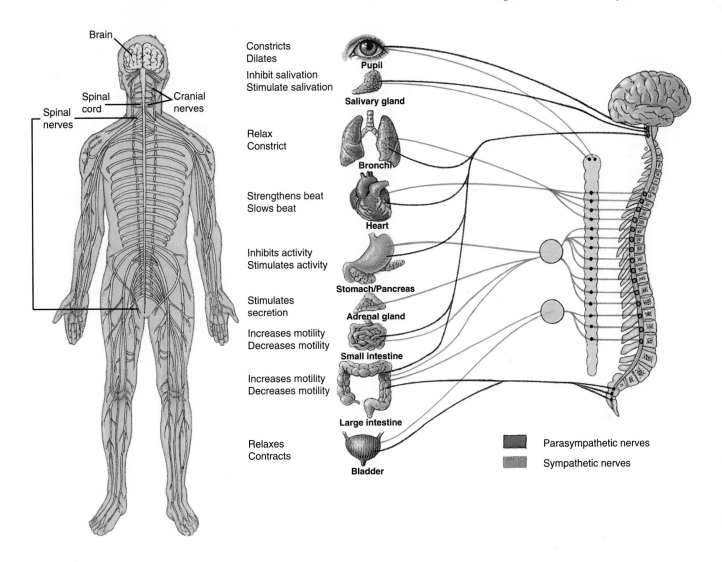

Figure 36.15 (A) The 43 nerve pairs of the peripheral nervous system emerge from the brain and spinal cord and, through branching and interconnections, reach every organ in the body. (B) The autonomic nervous system of the PNS has two divisions: the parasympathetic and the sympathetic nervous systems. Most organs are innervated by both of these systems, and they generally cause opposite effects.

Cranial and Spinal Nerves

The cranial nerves lead to major **sense organs**—the eyes, ears, nose, and tongue—and to muscles located in the head. Only one cranial nerve, the critically important **vagus nerve,** serves a region outside the head. It extends to, and regulates functions of, the heart, lungs, and digestive organs. Spinal nerves innervate all other regions of the body below the neck.

Functional Classification of the PNS

The PNS is organized according to functions performed. It is first divided into the **afferent (sensory) division,** which includes nerves that relay impulses from all areas of the body to the CNS, and the **efferent (motor) division,** which is in turn subdivided into the following two parts: the **somatic nervous system,** composed of motor nerves that run directly from the CNS to skeletal muscles, and the **autonomic nervous system,** shown in Figure 36.15B, which connects motor nerves from the CNS with different internal organs and is primarily concerned with automatic functions such as the beating of the heart. Finally, there are two divisions of the autonomic nervous system, the **sympathetic** and **parasympathetic nervous systems.** Many organs of the body are affected by both of these divisions, and the two systems generally tend to cause opposite responses. Most often, the sympathetic system prepares a person for vigorous activity, for example, by increasing the heart rate. The parasympathetic system is concerned with maintaining the body under relatively calm conditions and allows us to lead tranquil lives without worrying about when it is necessary to breathe or for our hearts to beat. The dual effects of the autonomic system allow for remarkably precise control of internal functions.

Central Nervous System (CNS)

The CNS is the integrative and control center that links the sensory and motor functions of the nervous system. Sensory neurons transmit information to the spinal cord. From there, it is relayed to the brain, the coordination center of the nervous system, where thorough processing and integration take place in the association neurons and an appropriate response, involving stimulation of specific muscles or glands, occurs. The brain also analyzes sensory information, a highly complex activity that involves thousands of association neurons. Although many motor neurons originate in the brain and numerous sensory neurons terminate there, the vast majority of brain neurons are the association neurons essential to processing information and formulating complex responses.

Spinal Cord

The spinal cord has two principal functions. First, it acts as a two-way relay system that transmits impulses between the brain and the peripheral nervous system through sensory and motor neurons. Second, it controls simple reflex actions. A reflex is an unconscious, programmed response to a specific stimulus. Each reflex is controlled by a **spinal reflex arc,** an inborn neural pathway by which impulses from sensory neurons reach motor neurons without first traveling to the brain (see Figure 36.16).

A familiar (and painful) example of a reflex will occur if you accidentally place your hand on a hot object. Pain receptors in the skin covering the hand send an impulse along a sen-sory nerve leading to the spinal cord. This message is switched directly, through association neurons in the spinal cord, to motor neurons controlling muscles of the arm, and the hand is instantly withdrawn. At the same time, the brain also receives the message, and more complex reactions, such as vocalization, follow instantly after the reflex action. Such reflex actions are important because they permit a rapid, automatic corrective response to a potentially dangerous stimulus.

Brain

The three geographical regions of the human brain and their major structures and functions are presented in Figure 36.17 and Table 36.1. The three primary structures of the **hindbrain**—cerebellum, medulla, and pons—are involved in maintaining homeostasis and coordinating large-scale body movements such as walking or jumping. They are also part of the **reticular formation,** a loosely organized network of nerves that receives input from several other areas of the brain and spinal cord and maintains the state of consciousness. It accomplishes this task by filtering the massive number of impulses received from the surrounding environment and relaying only a portion of them to the *cerebral cortex,* the conscious center of the brain. Because the reticular formation controls what reaches the cortex, it is considered to be an activating system for maintaining wakefulness and alertness and is often referred to as the **reticular activating system (RAS).** Damage to the RAS often results in coma.

The **midbrain,** a small area between the forebrain and the hindbrain, receives and integrates several types of sensory

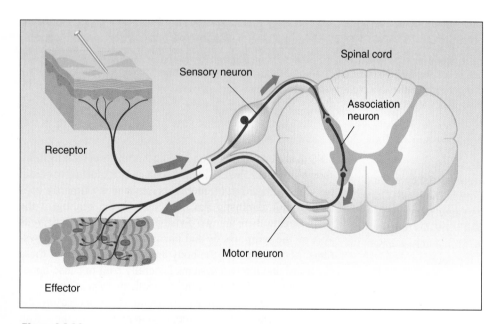

Figure 36.16 In a spinal reflex arc, reflexes occur automatically because they involve fixed neural pathways that travel directly from a receptor to a muscle or a gland.

Question: *Why are reflexes important?*

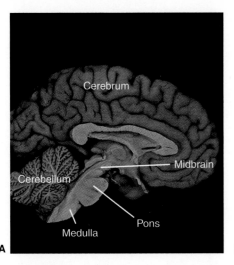

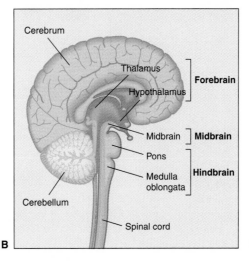

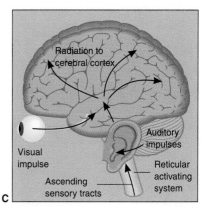

Figure 36.17 (A) The human brain contains 100 billion neurons. (B) The brain is divided into three main sections: hindbrain, midbrain, and forebrain. (C) The reticular formation is a diffuse network of nerves that receives and processes signals from the internal and external environments. Only a small number of these signals are transferred to the cerebral cortex through the reticular activating system.

Table 36.1 Major Functions of Different Structures and Regions of the Human Brain

Area	Structure	Function
Hindbrain	Cerebellum	Coordinates complex muscular movements; maintains a sense of balance
	Medulla oblongata	Regulates several automatic funtions such as breathing, heartbeat, swallowing, and digestion
	Pons	Conducts impulses between different parts of the brain; helps regulate breathing activities
Midbrain		Serves to connect the hindbrain and forebrain; contributes to coordination and consciousness
Forebrain	Cerebrum	Processes and integrates information received from several sources; is essential to thought, memory, consciousness, emotion, and higher mental processes
	Thalamus	Serves as the center for relaying and coordinating sensory information; directs information to specific anatomical areas of the cerebum
	Hypothalamus	Serves as the main coordinating center for parts of the peripheral nervous system; for example, regulates body temperature, appetite, and emotions; also regulates the pituitary gland, thus linking the nervous and endocrine systems

information and relays it to specific regions of the forebrain for processing. Certain cells of the midbrain are also integrated with the reticular formation.

The most complex neural processing and integration occur in the **forebrain,** or **cerebrum.** This is the largest part of the brain and is more highly developed in humans than in any other species. This portion of the brain is essential to higher mental functions—thought, memory, consciousness, and complex behaviors associated with humans. The cerebrum consists of two halves known as **cerebral hemispheres.** These structures are covered with a delicate layer of **gray matter** (nonmyelinated nervous tissue) called the **cerebral cortex,** the area of the forebrain that has become so highly specialized and enlarged in humans. To accommodate this increase in size, our cerebral cortex consists of a number of folds and deep grooves that allow it to fit inside the skull. The two cerebral hemispheres contain four major **lobes,** each of which carries out one or more specific functions, as indicated in Figure 36.18. In summary, the human cerebral cortex is the site of the following activities:

Control of voluntary skeletal muscle

Receipt of sensory information involving touch, pain, temperature, sight, hearing, smell, and taste

Sensory relay and integration

Emotions and emotional responses

Integration and interpretation

Intellectual functions

The human brain is perhaps the most remarkable organ to have evolved in any organism. Its operations encompass an extensive range of activities, from the simple, automatic control of muscles to the greatest intellectual achievements. Mas-

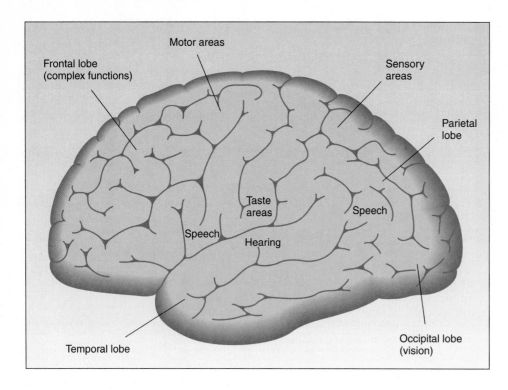

Frontal lobe
(complex functions)

Motor areas

Sensory areas

Parietal lobe

Taste areas

Speech

Speech

Hearing

Temporal lobe

Occipital lobe
(vision)

Figure 36.18 This diagram shows the four major lobes and other regions of the cerebral cortex, where major functional areas such as those for vision, speech, and hearing are centered.

sive textbooks have been written about the structure and functional domains of the human brain. However, much remains to be learned about the most wondrous aspects, which are related to intellect, and the exploration of brain structure and function represents a frontier field of research.

BEFORE YOU GO ON The nervous system has an organizational hierarchy that is related to function. The brain and spinal cord form the central nervous system, which regulates most body activities. The peripheral nervous system links the CNS with all other organs. The PNS is partitioned into an afferent division, which directs impulses to the CNS, and an efferent division, which is subdivided into two parts, the somatic nervous system, composed of motor nerves, and the autonomic nervous system, which regulates automatic functions.

FUTURE FRONTIERS

Interest in systems that regulate and control operations within the human body has a long history. Studies of the nervous system can be traced back to the ancient Greek and Roman cultures. As in other fields of biological research, progress accelerated during the early twentieth century and erupted in parallel with advances in molecular biology during the past four decades. As described in this chapter, much is known about the highly specialized cells, chemical messengers, and

underlying mechanisms that govern our activities. Nevertheless, great voids remain in our understanding of the nervous system.

What problems remain for scientists to confront in the decade that lies ahead? Perhaps first and foremost is studying how the human brain works. Our interpretations of the outside world and how we respond to it at any given moment are dependent on brain functions. What about other processes, such as emotion, memory (see the Focus on Scientific Process, "Memory"), information storage, intelligence, thought, and imagination, that make us human? How does the brain develop? How do the billions of brain neurons become so highly organized? What are the links between genetics, the brain, and behavior? Neurobiologists are making headway in some of these areas, and it should be fascinating to follow their progress in the years ahead. Recognizing the importance of research on the brain, Congress designated the 1990s the Decade of the Brain. As Francis Crick (of DNA fame) stated in a 1984 article, "There is no scientific study more vital to [humans] than the study of [our] own brain. Our entire view of the universe depends on it."

SUMMARY

1. The nervous system receives, integrates, and responds quickly to stimuli from the internal and external environments.

2. The functional unit of the nervous system is the neuron, a highly specialized cell that forms communication networks with other neurons.

Memory

Memory can be described as the store of items learned and retained as the result of an individual's activities or experience. Knowledge of a specific past event—a memory—can be recalled from storage and reproduced mentally with varying degrees of accuracy. Why are we able to remember some things and not others? Why can certain incidents in our lives, such as our first blind date, be recalled with crystal clarity or in agonizing detail? Why do sensory stimuli, especially odors—such as the fragrance of a certain perfume, new-mown grass, or a musty attic—frequently evoke detailed memories of events in the distant past?

Scientists are very interested in answering such questions as part of their efforts to understand the brain. As much as any other trait, the ability to remember events and integrate them in intelligent ways is a hallmark characteristic of our species. Neurobiologists are concentrating on answering these and other questions. Are there different types of memory? How is information processed and transformed into a memory? Where is a memory stored? How is it accessed? How are memories formed?

As outside observers, perhaps we should ask, How can such questions even be addressed? What types of studies or experiments can be done to provide data that can be used to answer these questions? Several approaches have been used in memory research. Much of the early effort (in the 1950s and 1960s) was directed at studying patients with various types of *amnesia,* loss of memory due to a brain injury or malfunction. Certain affected individuals, usually identified by their initials, provided investigators with insights that later led to carefully designed experiments. When amnesia patients participating in a study died, their brains were

carefully analyzed to determine which parts appeared damaged or abnormal and may therefore have been associated with the memory disorders.

The first and perhaps most famous case was a man referred to as H.M., an individual who had a severe form of epilepsy that could not be treated successfully in 1953. As a result, doctors made a decision to remove parts of his brain associated with the seizures. After recovery, it was discovered that the seizures were much less severe, but tragically, H.M. was no longer able to learn new facts. He was unable to remember the names of individuals he saw every day or recall what he had eaten a few moments earlier. Interestingly, he was quite proficient at solving puzzles but was unable to remember that he had done so.

What was learned by studying H.M. and other individuals with amnesia? First, it became clear that there are different types of memory linked with specific regions of the brain. Memory of past events, facts, names, dates, and places is called **declarative memory. Procedural memory,** or "habit," is acquired by repetition or continuous practice. Riding a bicycle, swinging a

golf club, playing a piano, and solving puzzles are examples of procedural memory. H.M. retained procedural memories but was unable to acquire declarative memories after his operation. Second, the parts of the brain that are important in processing and transforming information into memories were identified. They included the hippocampus, the amygdala, and parts of the cortex (see Figure 1).

To identify more precisely the key parts of the brain involved in declarative memory formation and storage, investigators turned to studying memory in animal models, primarily monkeys. The usual experimental approach was to teach the animals how to play different types of games in which they received a reward for playing correctly. Once they learned how to play the game, small parts of their brains would be surgically removed or modified. Their memories of the game they had learned would then be evaluated using different tests.

Results from these experiments extended the knowledge gained from studies of amnesia patients. The areas of the brain that were found to be critical for memory formation and recall indicated in Figure 1 were determined

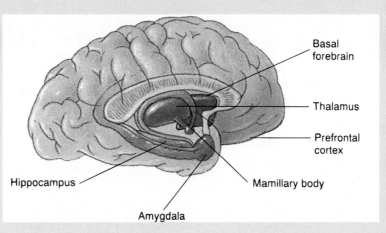

Figure 1   The indicated parts of the brain were found to be involved in memory formation.

box continues

to be related as shown in Figure 2. The most interesting findings were related to discoveries of the neuroanatomical pathways used in memory processing. From these experiments, conducted over a 20-year period, scientists hypothesized that perceptions are formed in areas of the brain that receive sensory information (for example, a vision, a smell, or a noise). Once this occurs, two parallel circuits, one leading to the amygdala and the other to the hippocampus, are activated. Information flow from both of these structures travels to the diencephalon (mammillary body and thalamus), the prefrontal cortex, and the basal forebrain.

How do these structures collaborate in creating a memory? This is a very complicated question, and the answer is not yet complete. Available evidence indicates that specific sensory areas of the brain not only process incoming signals but are also the most likely sites for memory storage. The amygdala and hippocampus are thought to be critical in determining whether or not a memory will be stored. Damage to either of these areas results in total amnesia. The amygdala is a region known to be involved in processing emotionally charged information. Thus sensory perceptions travel from sensory areas to the amygdala, which evaluates and weighs their "emotional content." "Heavy" information, such as events surrounding the first blind date, are "tagged" by the amygdala and move forward to other parts of the brain, where conscious processing ("Why did I accept this blind date?") occurs. Such events are replayed constantly ("rehearsed") in the brain and are the most likely to be stored as memories. Sensory events with no emotional tag (for example, the color of the first car you saw today) are not likely to be remembered. This model seems consistent with experience in that most of our clearest memories have strong emotional content.

The hippocampus appears to play a key role in organizing neural connections between itself and the sensory areas and perhaps other parts of the brain. The establishment of new neural circuits between the sensory areas and the hippocampus is thought to be one basis for memory formation and storage. These new pathways become permanent after an appropriate number of rehearsals. This process is analogous to the creation of a network of trails in a new campground. At first, there is nothing distinctive, but as more and more people walk from campsites to different areas of the campground, a well-worn trail system develops. According to this model, when a familiar stimulus (the perfume or cologne worn by the blind date) enters the sensory area, the memory, stored as a neural circuit (the trail network), is recalled to consciousness. The exact functions of the other areas of the brain in this model are more dimly understood. However, they are also known to be critical in information processing, as damage to each of them alone disrupts memory formation.

Many new questions have arisen from these studies and the hypothesized physical model of memory processing and storage. For example, what biochemical events are required for creating new neuronal circuits? What is the molecular basis for information storage and memory formation? These and other questions have guided research during the past decade, and new hypotheses about the finer aspects of memory formation have begun to emerge. Such research is intrinsically fascinating to scientists, and it is also fundamentally important. An ultimate hope is that memory can be understood at the molecular level so that it may become possible to treat conditions, such as Alzheimer's disease, that cause impairment of this extraordinary human ability.

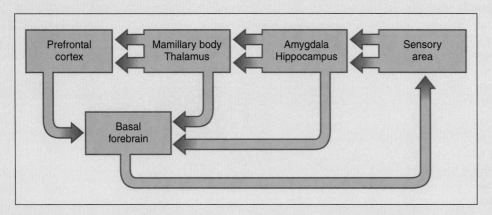

Figure 2   Experimental studies of monkeys led to this diagram of the memory system.

3. Neurons are classified according to their function. Sensory neurons receive information and transmit it to the central nervous system, motor neurons transmit impulses away from the central nervous system, and association neurons relay signals between sensory and motor neurons.

4. A nerve impulse or action potential is generated when depolarization occurs in an axon as a result of Na^+ inflow through voltage-gated ion channels. Once depolarization occurs at a single point, similar reactions are triggered along the neuron, and the action potential is transmitted along the length of the axon. Repolarization occurs when K^+ ions flow out of the cell and the resting membrane potential is reestablished.

5. An electrical nerve impulse can be transmitted from one neuron to another through a chemical synapse. The nerve impulse stimulates release of neurotransmitters from a presynaptic neuron that can interact with chemically gated ion channels in the postsynaptic membrane. If a sufficient number of ion channels are opened, a new action potential will be generated, and transmission of the nerve impulse will continue.

6. Once the nerve impulse has passed between two cells, neurotransmitters are removed from the synapse by different mechanisms. Drugs may disrupt normal neurotransmitter synthesis, prevent their removal from the synapse, or mimic their action. Though most drugs are ultimately harmful, several may be useful in treating neurological disorders.

7. The human nervous system is composed of the central nervous system—the brain and spinal cord—and the peripheral nervous system, which conveys information to and from the central nervous system.

8. The peripheral nervous system is organized according to function. Each of the divisions performs different functions, some of which are under conscious control (for example, movement of skeletal muscles) and some of which are not (for example, heartbeat). The general function of the peripheral nervous system is to maintain constant body conditions.

9. The central nervous system acts as the control center of the body. Different parts of the human brain are responsible for specific operations, such as helping maintain homeostasis, maintaining consciousness, integrating and processing sensory information, and engaging in intellectual activities.

WORKING VOCABULARY

action potential (p. 681)
association neuron (p. 679)
autonomic nervous system (p. 689)
axon (p. 677)
central nervous system (p. 688)
cerebral cortex (p. 691)
chemical messenger (p. 677)
chemically gated ion channel (p. 686)
dendrite (p. 677)

motor neuron (p. 679)
nerve (p. 679)
neuron (p. 677)
neurotransmitter (p. 685)
peripheral nervous system (p. 688)
sensory neuron (p. 679)
somatic nervous system (p. 689)
spinal reflex arc (p. 690)
synapse (p. 684)

REVIEW QUESTIONS

1. What are the basic functions of the nervous system?

2. What are neurons? Describe their structure.

3. What are the three major types of neurons? What is the general responsibility of each type?

4. How is an action potential generated and transmitted?

5. What is a synapse? What are the two types of synapses?

6. Describe the events of chemical synaptic transmission.

7. How do different drugs affect synaptic transmission? How is this knowledge being used?

8. What is the function of the peripheral nervous system? What types of nerves does it contain?

9. Describe the functional classification of the peripheral nervous system.

10. Describe the components of a spinal reflex arc. Why are they important?

11. What are the three major regions of the brain and the function of each?

12. What are the functions of the cerebral cortex?

ESSAY AND DISCUSSION QUESTIONS

1. What might be the goals of research conducted on the nervous system during the late 1990s? Beyond the twentieth century?

2. What might be the adaptive value of the following: (a) pain, (b) memory, and (c) the separation of the somatic and autonomic nervous systems?

3. What do you consider the ethical limits of doing experiments on living animals in order to obtain information and data about the nervous system?

REFERENCES AND RECOMMENDED READING

Asimov, I. 1994. *The Human Brain: Its Capacities and Functions.* New York: Viking Penguin.

Baddeley, A. 1992. Working memory. *Science,* 255: 556–559.

Chauvet, G. 1995. *Theoretical Systems in Biology: Hierarchical and Functional Integration.* New York: Pergamon.

Cohen, N. J., and H. Eichenbaum. 1993. *Memory, Amnesia, and the Hippocampal System.* Cambridge, Mass.: MIT Press.

Crick, F. 1994. *The Astonishing Hypothesis: The Scientific Search for the Soul.* New York: Scribner.

Changeux, J.-P. 1993. Chemical signaling in the brain. *Scientific American,* 269: 58–65.

Horgan, J. 1994. Can science explain consciousness? *Scientific American,* 271: 88–94.

Jennes, L. 1995. *Atlas of the Human Brain.* Philadelphia: Lippincott.

Kalin, N. H. 1993. The neurobiology of fear. *Scientific American,* 268: 94–101.

Le Doux, J. E. 1994. Emotion, memory and the brain. *Scientific American,* 270: 50–57.

Mishkin, M., and T. Appenzeller. 1987. The anatomy of memory. *Scientific American,* 256: 80–90.

O'Connor, V., G. J. Augustine, and H. Betz. 1994. Synaptic vesicle exocytosis: Molecules and models. *Cell,* 76: 785–787.

Parent, A. 1995. *Human Neuroanatomy.* Baltimore: Williams & Wilkins.

Raichle, M. E. 1994. Visualizing the mind. *Scientific American,* 270: 58–65.

Scientific American. 1992. Mind and the brain (special issue). *Scientific American,* 267.

Simmers, J., P. Meyrand, and M. Moulins. 1995. Dynamic networks of neurons. *American Scientist.* 83: 262–268

ANSWERS TO FIGURE QUESTIONS

Figure 36.2 MS causes a progressive demyelinization of affected axons.

Figure 36.4 In the brain and spinal cord.

Figure 36.6 (A) Potential energy; (B) It would become less negative or even positive.

Figure 36.7 A return of the membrane potential to −70 millivolts.

Figure 36.8 A few milliseconds.

Figure 36.11 They will be broken down, taken back into the presynaptic neuron, or taken up by nearby supporting cells.

Figure 36.14 Regulating automatic body functions such as events in digestion and the beating of the heart.

Figure 36.16 They allow an instant, programmed response that may have survival value.

37

Control and Regulation: The Endocrine System

Chapter Outline

Reading Questions

1. How do modern concepts dealing with the endocrine system differ from those of classical endocrinology?

2. What are some general functions of human hormones?

3. How do hormones carry out their tasks?

4. What are some of the functions of the pituitary gland? How is it regulated?

The endocrine system, an exquisitely controlled assemblage of tissues and organs, has moved into the spotlight of biological research during the past two decades. Progress made in understanding the endocrine system reflects the process of science, in which new knowledge leads to new questions.

THE CLASSICAL VIEW OF THE ENDOCRINE SYSTEM

The earlier classical view of the endocrine system was based largely on studies conducted during the first half of the twentieth century. The classical concept held that the **endocrine system** consists of discrete organs called **glands,** which are made up of specialized cells capable of secreting unique chemical messengers called *hormones*. Hormones are secreted into the circulatory system and then transported to a set of distant "target cells," which they affect in some way. *Endocrinology* is the study of glands and hormones and their effects on the organism.

Figure 37.1 identifies the glands of the classic endocrine system. The endocrine and nervous systems operate together in regulating an organism's functions. The common denominator linking the two systems is the use of chemical messengers in their communications. However, in comparison to the instant responses and pinpoint control provided by neurons and neurotransmitters, the effects of hormones differ in the following ways. First, they are slow-acting. From the time they are released from cells, their effects may not occur for minutes, hours, days, or even longer. In some cases, they mediate responses measured in years (for example, sexual maturation). Second, they circulate rather randomly (and slowly) throughout the body until they locate appropriate target cells. Third, their control is somewhat diffuse. Often several cells or tissues may be affected by a single hormone.

In pioneering studies during the first half of the twentieth century, many important glands and hormones and their effects

in the human endocrine system were described; they are summarized in Table 37.1. Hormones were found to affect development, growth, maturation, behavior, and a variety of other processes by activating cellular mechanisms that were either idle or operating at a greatly reduced level or by decreasing a cell's activities. What types of studies influenced formation of the classical view of endocrinology?

In the eighteenth and nineteenth centuries, physicians were familiar with human anatomy, including locations of the major glands, although little was known about the functions of these glands. There was considerable interest in determining experimentally the effects on animals when a certain tissue or organ was removed. This is not a sparkling imaginative approach by modern scientific standards, but it led to some interesting discoveries.

Descriptions of the first such experiments on record were published in 1849 by A. A. Berthold. He found that if he removed the testes from immature male chickens, they never developed structures characteristic of mature roosters, nor did they ever crow or show any sexual interest in hens. However, if a single testis was later transplanted into the castrated male, it developed normal male structures and behaviors. What could be concluded from such experiments? Clearly, testes

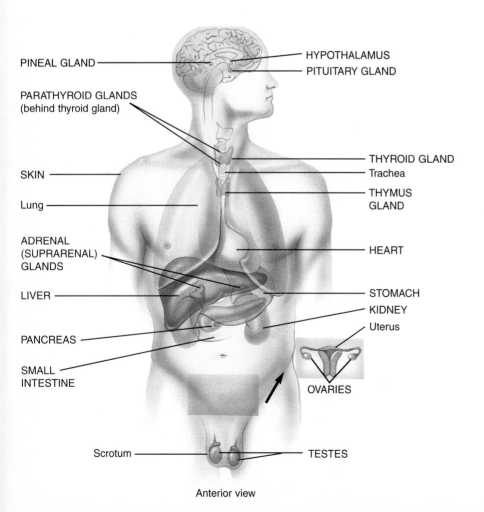

Anterior view

Figure 37.1   This drawing shows the major glands of the human endocrine system and their location relative to other organs.

Table 37.1 Major Glands, Hormones, and Actions of the Classical Endocrine System

Gland	Hormone[a]	Target	Primary Functions
Pineal	Melatonin	Hypothalamus	Block LH and FSH releasing factors
Pituitary			
Anterior	ACTH	Adrenal cortex	Synthesis of various hormones
	FSH	Ovary	Female development and hormones
		Testis	Male development and hormones
	hGH	Bones, tissues	General growth stimulation
	LH	Ovary	Stimulates ovulation, synthesis of female hormones
		Testis	Stimulates synthesis of male hormones
	Prolactin	Mammary gland	Stimulates milk synthesis
	TSH	Thyroid	Regulates thyroid gland
Posterior	ADH	Kidney	Water conservation
	Oxytocin	Uterus	Increases muscle contractions during childbirth
		Mammary gland	Stimulates milk secretion
Thyroid	Thyroxine	Most cells	Increases metabolic activities
	TRH	Various glands	Stimulates release of hGH, thyroxine, and prolactin
	Calcitonin	Bones	Stimulates calcium uptake
Parathyroid	PTH	Bones	Increases calcium levels in blood
Thymus	Thymosin	Leucocytes	Stimulates differentiation
Adrenal			
Medulla	Epinephrine, norepinephrine	Circulatory system	Prepares the body for physical action
Cortex	Steroids		
	GC	Various cells	Increases carbohydrate metabolism
	MC	Kidney	Regulates water and mineral balance
Pancreas	Insulin	Many cells	Stimulates glucose uptake from blood
	Glucagon	Many cells	Stimulates glucose release from cells into blood
	Somatostatin	Pancreas	Inhibits secretion of insulin and glucagon
Kidney	Erythropoietin	Erythrocytes	Stimulates red blood cell production
Testis	Testosterone	Many cells	Stimulates male sexual development and maturation of sex organs
Ovary	Estrogens	Many cells	Stimulates female sexual development and maturation of sex organs
	Progesterone	Uterus	Stimulates growth of uterine lining

[a]Abbreviations: ACTH = adenocorticotropin; FSH = follicle-stimulating hormone; hGH = human growth hormone; LH = luteinizing hormone; TSH = thyroid-stimulating hormone; ADH = antidiuretic hormone, TRH = thyroid-releasing hormone; PTH = parathyroid hormone; GC = glucocorticoid; MC = mineralocorticoid

were necessary for the development of male characteristics in chickens. It is now known, of course, that the testes secrete a hormone, testosterone, that influences the development of males in all vertebrate species, including humans.

Little progress was made in endocrinology during the next 50 years. In 1855, the French physiologist Claude Bernard introduced the term *internal secretion* to describe, in a general way, a hypothesized mechanism responsible for controlling concentrations of glucose in the blood. However, it was not until the turn of the century that the concept of glands secreting substances captured the attention of scientists. The term *hormone* (Greek for "arouse to activity") was introduced in 1905 by Ernest H. Starling after a series of experiments estab-

lished that pancreatic secretions were stimulated by a substance released from the small intestine. In 1889, it was reported that dogs developed a condition similar to human *diabetes,* a disease characterized by excess glucose in the blood, when their pancreases were removed. In a reciprocal experiment, diabetes did not occur if those dogs then received a graft of pancreas tissue. What could these results mean? Surely, the pancreas must be required for the proper regulation of glucose levels in the blood. How did the pancreas influence activities in other areas of the body? Perhaps some secretion was involved? That was confirmed in studies conducted during the next 35 years, and the pancreatic secretion (by now called a hormone) was given the name *insulin.*

After the successes of early workers, "the theory of internal secretions" continued to evolve and served as a springboard for studies on glands and hormones in the twentieth century. Similar to investigations in the field of genetics during this same period, the results achieved were stunning. As mentioned previously, many of the major hormones were isolated and identified, their functions were largely determined, a classical view of hormones and target cells emerged, and scientists in this field won a host of Nobel Prizes.

THE MODERN VIEW OF THE ENDOCRINE SYSTEM

Regrettably for those of us who are fond of simple pictures, recent discoveries have made it necessary to extend the classical gland–hormone view of the endocrine system. What is known now that was not understood during the formation of classical endocrinology? Four general conclusions emerged from new research. First, we now know that organs other than classical endocrine glands secrete hormones. The gastrointestinal tract, heart, and brain are now known to secrete hormones, although their effects are not yet entirely understood. Neurons were also found to secrete hormones that regulate activities of many organs. Second, nonhormonal substances were discovered that act as hormones. For example, some classic neurotransmitters, such as dopamine, may function as hormones in certain cells. Third, it was determined that not all hormones are transported to their target cells by the circulatory system. Fourth, hormonal regulatory systems were found to be enormously complex and fundamentally important in understanding normal human physiology. All of these general conclusions have generated new questions for which answers are being pursued. Which tissues and cells produce hormones? Which chemicals can act like hormones, and what are their effects on the cell and on the organism? How are hormones regulated?

As a result of the findings, new, broader definitions have been formulated that provide greater accuracy and flexibility. A **hormone** is now considered to be any substance released by one cell that acts on another cell, near or far, regardless of the means of transport. A **neurohormone** is a hormone produced by a neuron, and a *neurotransmitter* is a neurohormone that acts at the synapse. The term *chemical messenger* can be used as a synonym for all of these terms. **Endocrinology** is the study of hormones derived from classical endocrine glands or from other cells or tissues, such as the brain or heart, and their functions.

> **BEFORE YOU GO ON** Studies conducted in the nineteenth century and the first half of the twentieth gave rise to a classic model of endocrine system function. It explained that distinct glands secreted specific hormones, which acted on target cells and influenced processes such as development and growth. More recent research has made it clear that hormones can be produced by other organs and that substances other than classic hormones can affect the activities of target cells.

ENDOCRINE GLANDS AND HORMONES

A large and varied collection of hormones is required to regulate human physiological activities. Most hormones and their effects are probably now known. However, no one is any longer surprised by reports of a new hormone or of a hormone being produced by a previously unsuspected tissue or organ. The classical glands, hormones, and effects still provide a useful framework for studying the endocrine system. Nevertheless, further discoveries in the decade ahead will continue to move the field well beyond the classical concept.

Major Classes of Hormones

There are two principal chemical classes of hormones (see Figure 37.2). **Peptide hormones** are mostly small proteins composed of linear strings of 3 to 200 amino acids. They are manufactured through the protein-synthesizing pathways and processes described in Chapter 18. Most hormones listed in Table 37.1 (see page 699) belong to this class.

Steroid hormones are synthesized from a large lipid molecule called **cholesterol.** Two major types of steroids are produced by the **adrenal glands,** which are located on top of the kidneys. **Glucocorticoids** play important roles in sugar metabolism, and **mineralocorticoids** are critical in maintaining water and mineral balances. The gonads produce several **sex hormones** that are essential for reproductive development and function. The principal male sex hormones produced by the testes are *androgens,* and the female ovaries synthesize *estrogens.*

There are other types of hormones that do not fit into either major class (for example, thyroid hormones) and "candidate hormones," substances whose exact roles are unknown but that appear to function as hormones. In this latter category are chemical messengers such as dopamine, which functions as a neurotransmitter but may also act as a classical hormone in regulating the pituitary gland. Scientists have also discovered numerous **peptide growth-stimulating factors** that promote growth or development of certain cells and tissues. Two of these include **nerve growth factor,** a chemical messenger responsible for neuron maturation during embryonic development, and **erythropoietin,** which stimulates erythrocyte (red blood cell) production in the bone marrow. **Prostaglandins** are an important family of small molecules that influence a broad range of activities, including smooth muscle contraction, blood flow, reproduction, respiration, and nerve impulse transmission. They also have key roles in maintaining blood pressure, stimulating steroid secretions, and many other physiological processes.

> **BEFORE YOU GO ON** The two major classes of hormones are peptide hormones and steroid hormones. Peptide hormones are synthesized by several glands and regulate many different activities. Steroid hormones are synthesized by the adrenal glands and gonads and have a major role in reproductive organ development and function. There are also other types of substances that act as chemical messengers.

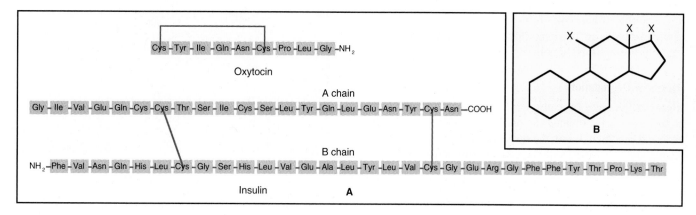

Figure 37.2   (A) Peptide hormones are small protein molecules composed of linear chains of amino acids, which are represented here by rectangles. Oxytocin has only nine amino acids, whereas insulin consists of two chains—an A chain of 21 amino acids and a B chain of 30 amino acids—held together by two chemical bonds. (B) All steroid hormones are made with part of a cholesterol molecule (shown in blue), but each has a variable chemical structure at sites indicated by the letter *X*.

Question: *Which organs synthesize most steroid hormones?*

The Lives of Hormones

Figure 37.3. summarizes the general life cycle of a hormone. All hormones are synthesized within cells, and most are enclosed within secretory vesicles until released from the cell, similar to neurotransmitters of the nervous system. Once stimulated by an appropriate external or internal environmental cue, the vesicle fuses with the plasma membrane and empties the hormone outside the cell. Steroids and thyroid hormones, however, are never stored; they are secreted directly into the bloodstream as they are produced. After release, hormones reach target cells in which they cause a response. Once the target cell's activities are completed, the hormone is removed and eliminated. Most hormone secretions are regulated by negative feedback systems.

Hormone Secretion

What stimulates a hormone-secreting cell to liberate its contents? External stimuli such as light, sound, and temperature may influence the release of hormones through their effects on the nervous system. For example, a loud noise interpreted as threat-related by the nervous system may provoke the release of the hormone epinephrine from adrenal gland cells. Internal stimuli consist of signals from either the nervous system or the endocrine system. Most frequently, other hormones stimulate

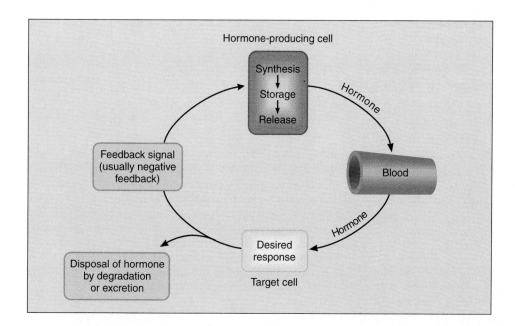

Figure 37.3 Most hormones are synthesized in cells, released into the bloodstream, and travel to target cells, where they elicit a response. Once the target cell completes its activity, the hormone is eliminated. The synthesis and release of most hormones are precisely regulated by negative feedback systems involving the endocrine and nervous systems.

Question: *Under what conditions are negative feedback systems activated?*

the secretion of hormones from specific glands. For example, follicle-stimulating hormone (FSH) stimulates release of certain hormones from the gonads.

Hormone Degradation

Because hormones can cause subtle or profound changes in the activities of cells, organs, and organisms, there must be ways to inactivate them. If not, the responses they induce would not cease, and the consequences could be extremely serious. Various mechanisms ensure that the life of a hormone does not exceed the time required for its beneficial effect.

Peptide hormones have rather short lives once they are secreted from the cell. Generally, they remain intact for less than an hour before they are inactivated, usually in the liver or kidney. A smaller number of peptide hormones are broken down at the target cell area, perhaps by actions of the target cell itself.

Steroid and thyroid hormones live for considerably longer periods, some for up to several weeks. Steroids are ultimately modified and inactivated in the liver, filtered from the blood in the kidneys, and then excreted in the urine. Thyroid hormones may also follow this route.

Feedback Control of Hormone Secretion

Levels of hormones are controlled through complex feedback systems, negative feedback being the most common mechanism (see Chapter 33). An example is shown in Figure 37.4. **Thyroxine** is a hormone secreted by the thyroid gland. It helps regulate metabolism and tissue growth and development, especially in infants and children. When blood thyroxine concentrations fall below the normal range or metabolism is significantly reduced, the hypothalamus, a part of the brain, detects the change and secretes thyrotropin-releasing hormone (TRH) in response. TRH travels to the pituitary gland, where it stimulates release of thyroid-stimulating hormone (TSH), which in turn causes the thyroid gland to release more thyroxine into the blood. Once metabolism and thyroxine blood levels return to normal, the hypothalamus releases a thyrotropin-inhibiting factor that ultimately causes a reduction in thyroxine secretion. Such complicated feedback systems (and this example is relatively simple!), involving both the endocrine and nervous systems, result in exquisite control of related functions within the body.

> **BEFORE YOU GO ON** Hormones are released from cells in response to signals from the nervous system or from other hormones. Hormone concentrations in the body are precisely regulated, usually by negative feedback systems. Once their actions are completed, hormones are broken down or inactivated and removed.

Example: Regulation of Blood Glucose Levels

To conduct normal daily activities, your cells require a constant supply of energy. Most cells use glucose, a simple sugar, as an energy source. Regulating blood glucose levels is critical for maintaining homeostasis, as abnormally high or low levels of

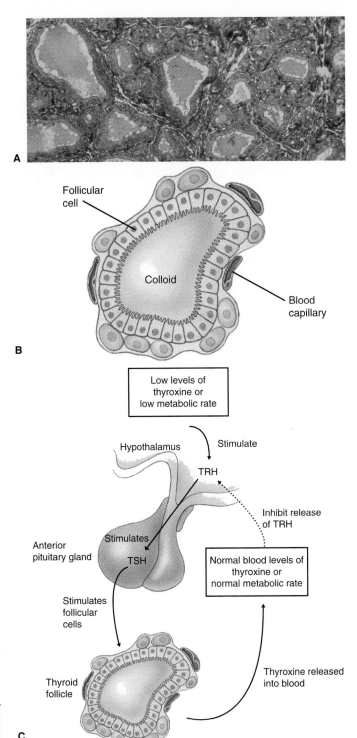

Figure 37.4 (A) The thyroid gland consists of hundreds of follicles where hormones, including thyroxine, are produced. The interior (colloid) material contains stored hormones and other substances. (B) A single follicle has the general structure shown here. (C) When blood thyroxine levels are lower than normal, the hypothalamus releases thyroxine-releasing hormone (TRH), which acts on the anterior pituitary gland, causing it to release thyroid-stimulating hormone (TSH). TSH stimulates the release of thyroxine from the thyroid gland. Increased levels of thyroxine in the blood inhibit further release of TRH from the hypothalamus.

Question: *What is this type of regulatory system called?*

glucose in the blood can cause serious problems. **Diabetes mellitus** is a name applied to a variety of disorders characterized by elevated concentrations of glucose in the blood, a condition known as **hyperglycemia.** How are blood glucose levels normally regulated? Figure 37.5 describes the organs, cells, hormones, and feedback systems involved.

Two hormones, **insulin** and **glucagon,** which are both secreted by the pancreas, are responsible for glucose homeostasis. Insulin acts to reduce the levels of blood glucose, and glucagon has the opposite effect, causing an increase in blood glucose. The pancreas contains different populations of secretory cells. Approximately 1 million cell clusters called **islets of Langerhans** exist in the pancreas. Each islet is composed of **alpha cells** that produce glucagon, **beta cells** that produce insulin, and **delta cells** that produce **somatostatin,** a hormone

that inhibits secretion of glucagon and insulin. Digestive enzymes are secreted by pancreas cells that surround the islets.

After eating a meal, blood glucose levels increase, and within minutes, insulin is released from beta cells in the pancreas. The immediate effect (in seconds or minutes) of insulin is to increase the rate of glucose uptake from the blood into target cells—primarily liver and muscle cells. Once inside cells, glucose molecules can be metabolized or linked together to produce *glycogen,* a storage form of glucose. As blood glucose levels fall below the normal range, glucagon is released by alpha cells of the pancreas in response. This hormone primarily affects liver cells and stimulates the breakdown of glycogen and release of glucose into circulation. Normally, the feedback regulation between these two hormones maintains blood glucose levels at optimal physiological levels.

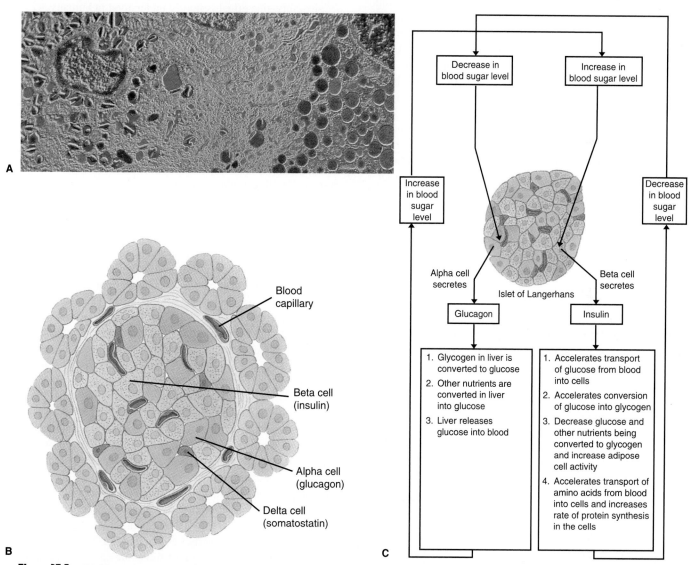

Figure 37.5 (A) Blood glucose levels are regulated by hormones released from various pancreatic cells. The pancreas contains approximately 1 million islets of Langerhans, one of which is shown in this EM. Each islet is composed of alpha (red), beta (green, yellow, and brown), and delta cells (bottom, orange); each cell type secretes a different hormone. (B) Each islet of Langerhans has the general structure shown here. (C) When blood glucose levels increase, insulin is released from beta cells. Insulin causes several changes that all serve to remove glucose from the blood. When glucose levels are low, glucagon is released from alpha cells and stimulates activities that release glucose into the bloodstream. The complete regulatory system ensures that every cell in the body will always receive appropriate amounts of glucose.

As described in Figure 37.5C, pancreatic hormones also act in routing proteins, fats, and sugars to cells where they can be stored. For example, adipose cells, take up excess glucose and convert it to fat for storage. They are also involved in liberating stored substances from cells if conditions warrant, as during dieting. These activities play an important role in glucose homeostasis over longer time periods.

What are the causes and consequences of abnormal glucose homeostasis? Too much insulin reduces blood sugar, and serious effects are generally related to brain damage, as this organ uses glucose almost exclusively as an energy source. **Diabetes mellitus**, characterized by hyperglycemia, is caused by too little insulin (Type I, or insulin-dependent diabetes) or by changes in muscle, adipose, and liver cells that make them resistant to the action of insulin (Type II, or non-insulin-dependent diabetes). Continuous high blood sugar concentrations may result in the following sequence of effects: (1) water diffuses out of tissues, (2) kidneys excrete the surplus water in urine along with excess salt molecules, (3) dehydration and salt loss can lead to coma and death. Certain chronic effects—heart problems, blindness, kidney failure—are associated with milder hyperglycemia.

Diabetes mellitus is estimated to affect about 2 percent of the United States population. Type I diabetes generally develops in children and accounts for 10 to 25 percent of diabetes cases. The more common Type II diabetes usually occurs later in life and affects about 5 percent of all people in the United States who are more than 40 years old. Type I diabetes is caused by an abnormal immune reaction that leads to the destruction of pancreatic beta cells. The underlying cause of type II diabetes has not yet been determined.

General Activities of Hormones

Like most matters of the endocrine system, describing the general activities of hormones is complicated by the great diversity of these substances. Some hormones may never be secreted or are produced only rarely. For example, oxytocin is a hormone that functions only at the end of pregnancy and during the period of nursing. Other hormones, such as insulin, are secreted in irregular on–off cycles that are related to homeostatic demands. Some, such as FSH in females, influence events for decades, but after a certain age, they play no further role in the life of the individual.

The following list attempts to categorize the vast number of physiological activities of hormones into a manageable number of general processes.

■ Hormones influence the secretions of most substances by cells in the body. These substances include, but are not limited to, other hormones, neurohormones, milk from mammary glands, enzymes and other products of the digestive tract, and sweat.

■ Hormones control both the synthesis and breakdown of fats, carbohydrates, and proteins in cells of the body.

■ Hormones stimulate or inhibit cell division (mitosis and cytokinesis) and thus direct the growth of tissues and the organism.

■ Hormones regulate activities associated with sexual development, sexual behaviors, and the production of gametes.

■ Hormones regulate levels of minerals and other chemicals, including sodium, calcium, and potassium, in the body.

■ Hormones help maintain homeostasis in emergency situations such as trauma, starvation, infection, severe emotional stress, and temperature extremes.

■ Hormones are thought to have a significant influence on various behaviors; however, except for certain aspects of sexuality, this effect is poorly understood in humans.

MECHANISMS OF HORMONE ACTION

After a gland secretes a hormone into the bloodstream, the chemical messenger reaches a target cell that then engages in some physiological process in response. Each hormone can regulate only specific target cells. How are hormones able to identify their specific target cells? What determines whether or not a cell will respond to a hormone?

Receptors

Scientists have found that each target cell possesses different **receptors,** which are complex protein molecules embedded in the plasma membrane that "recognize" hormones. Like a neuron, gland, or muscle receptor that recognizes only a single type of neurotransmitter, a specific hormone receptor has a unique structure that matches only one hormone, and when contact occurs, the hormone and receptor fit together like a lock and key. Intense research efforts are being directed at determining the structure and three-dimensional organization of hormone receptors to clarify the nature of hormone-receptor interactions (see Figure 37.6).

Each cell is equipped with receptors only for hormones that act on that cell. If a cell's activities are regulated by, say, two hormones, it will have one type of receptor for each hormone. The number and even the type of receptors a cell possesses can change at different times, depending on the function of the cell and other circumstances. In general, the total number of cellular receptors is quite large. For example, target cells for steroid hormones contain 10,000 to 100,000 steroid receptors; each receptor molecule can bind with one hormone molecule. Finally, only a fraction of the receptors have to be occupied to cause a maximum physiological response by the cell. When hormone molecules bind to their receptors, what occurs that leads to changes in the cell's activities?

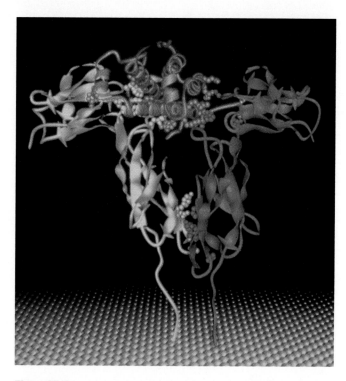

Figure 37.6   A model describing the structure of human growth hormone (hGH) and the part of an hGH receptor molecule that lies outside the plasma membrane. One hGH molecule, shown in red, is bound to two hGH receptor domains (red and blue) extending through the plasma membrane (orange). Such specific binding is hypothesized to be necessary for triggering events inside the cell that lead to effects associated with hGH.

Question: *What is the primary effect of hGH?*

Peptide Hormones

Target cells for peptide hormones have specific receptor molecules located on the surfaces of their plasma membranes. The exact nature and molecular structure of several hormone receptors has recently been determined. Although the attraction of peptide hormones to the surfaces of target cells was discovered relatively early, the mechanism for action long remained unclear. How could a hormone that became attached to the outside of a cell trigger chemical reactions inside the cell?

First and Second Messengers

The first answers came through research conducted in the 1950s by Earl Sutherland and colleagues. They discovered that a previously unknown chemical called **cyclic adenosine monophosphate (cAMP)** accumulated in "activated" target cells. The substance is similar to ATP, but it has only one phosphate group rather than three phosphates, and it is synthesized from ATP by an enzyme called *adenylyl cyclase*. For discovering cAMP, Sutherland was awarded the Nobel Prize in physiology and medicine in 1971.

What was the role of cAMP in peptide hormone actions? Building on the pioneering studies of the 1950s, researchers hypothesized that after binding to surface receptors, peptide hormones somehow triggered the synthesis of cAMP inside the cell. In turn, cAMP initiated a cascade of chemical reactions in the cell that would enable it to perform a certain function, such as synthesizing an enzyme that could break down glycogen into glucose. Subsequent research in the 1960s confirmed the basic accuracy of this hypothesis for a variety of peptide hormones.

The general cAMP mechanism, described in Figure 37.7, is known as a second-messenger system. The first messenger is the peptide hormone that acts with its receptor on the plasma membrane to initiate the synthesis of cAMP; the second messenger is the cAMP synthesized inside the cell. Increased levels of cAMP then trigger a specific molecular activity inside

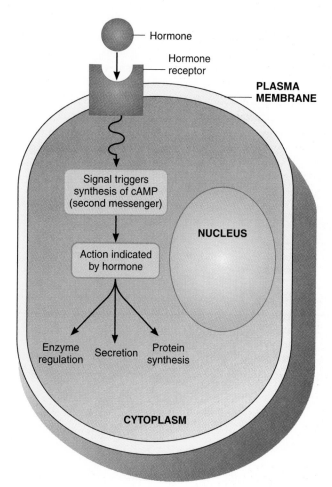

Figure 37.7   A peptide hormone—the first messenger—interacts with its target cells by binding with receptors located on the cell's surface. Binding results in the production of cAMP—the second messenger—which causes the cell to carry out the activity specified by the hormone.

Question: *What transmits the signal from the hormone–receptor to the second-messenger production system?*

the cell. Other second messengers are now known to exist, but cAMP is the most common in the human endocrine system. Once the basic second-messenger mechanism was understood, a new question arose: exactly how was the hormone–receptor signal from the plasma membrane converted into a command for initiating cAMP synthesis in the cytoplasm?

G Proteins

In the late 1970s, it was discovered that intracellular actions dictated by many peptide hormones depend on a class of molecules called **G proteins**—so named because they bind to guanine (G) nucleotides inside the cell. Basically, G proteins, which are attached to the inner plasma membrane, relay the hormone–receptor signal to an *effector,* usually a membrane-bound enzyme that converts an inactive molecule into an active second messenger (for example, adenyl cyclase, the enzyme that produces cAMP from ATP, is an effector).

Figure 37.8 describes the operation of one of the most carefully studied hormone–G protein systems, and it illustrates the level of understanding now being achieved in endocrinology. *Epinephrine,* synthesized in adrenal glands, prepares the body for physical action by increasing the rates of heartbeat, respiration, metabolism, and force of muscle contraction. These energy-dependent effects require an increase in blood sugar levels, which occurs as a result of epinephrine trigger-ing the release of glucose from liver cells. Here is the sequence of events related to this process:

1. Epinephrine, the first messenger, binds to its receptor, located in the plasma membrane.

2. Binding stimulates the specific G protein involved in this pathway (G_s) to activate the effector, adenylyl cyclase.

3. Adenylyl cyclase converts ATP into cAMP, the second messenger.

4. cAMP initiates a cascade of enzymatic reactions that results in glycogen being broken down to glucose, which is then released from the cell.

5. After functioning for a few seconds, G_s is deactivated, and the glucose export operation ceases in the liver cell.

In the past decade, over 25 distinct G proteins have been identified. Most are quite versatile, being able to participate in several different first- and second-messenger interactions. Although much is now known about G proteins, there are many open questions. For example, if a specific G protein can take part in various reactions, how are all of the options and signals coordinated within the cell? G proteins will undoubtedly continue to be intensely studied now that it is known that they are so heavily involved in normal cell functions.

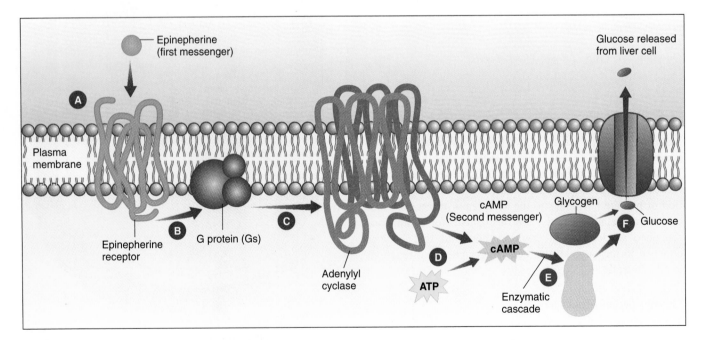

Figure 37.8 Epinephrine prepares the body for physical activities that require glucose to be released from the liver. This model describes the hypothesized sequence of events triggered by epinephrine. The first messenger, epinephrine, binds with one of thousands of epinephrine receptors in the plasma membrane of a liver cell (A). Epinephrine receptor binding stimulates G_s, a G protein (B), to activate the enzyme adenylyl cyclase (C). Activated adenylyl cyclase synthesizes cAMP, the second messenger, from ATP (D). cAMP triggers an enzymatic cascade (E) that results in glycogen being converted to glucose molecules, which are released from the cell (F).

Question: *How long will this process continue?*

Steroid Hormones

Steroids have the ability to pass through plasma membranes and enter most cells. Nevertheless, their effects occur only in cells that contain appropriate *internal* receptors. Once inside a target cell, they combine with a receptor to form a receptor–steroid complex. However, it still has not been determined whether unoccupied steroid receptors are found in the cytoplasm or in the nucleus. Figure 37.9 illustrates the model for steroid action assuming that the unoccupied receptor is in the cytoplasm.

In either case, once the hormone joins its receptor, the hormone–receptor complex migrates to chromatin within the nucleus. There it binds to a gene (DNA with a specific nucleotide sequence) that codes for a certain protein (or proteins). Once that occurs, there is an accelerated synthesis of the protein "requested" by the hormone. Increased levels of this protein commonly have some effect, usually of long duration, on the growth or maturation of specific tissues. For example, the steroid sex hormones estrogen and progesterone affect the development of secondary

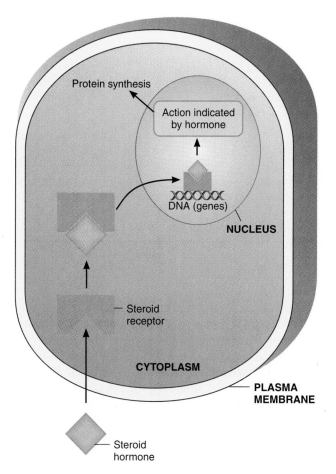

Figure 37.9 Steroid hormones bind to receptors located inside the target cell, creating a hormone–receptor complex that enters the nucleus and becomes attached to a gene. The gene becomes activated and directs the synthesis of a protein that usually affects growth or development of some tissue in the organism.

sex characteristics, such as breast development in females and facial hair in males. They also influence a variety of other activities, depending on the target cells.

> **BEFORE YOU GO ON** Target cells have receptors for specific hormones. Two mechanisms are used by most hormones for exerting their effect in a target cell. In the second-messenger system, peptide hormones are the first messenger. Once bound to a receptor, they trigger synthesis of the second messenger, cAMP, which initiates an activity in the target cell. Target cells for steroid hormones have internal receptors. Once formed, the steroid–receptor complex interacts with DNA and stimulates production of a specific protein that usually affects growth or development.

PITUITARY GLAND

Details about the **pituitary gland** illustrate the types of activities and regulatory mechanisms that characterize the human endocrine system. The pituitary gland has two lobes—anterior (front) and posterior (back)—and lies at the base of the brain, where it is attached to the hypothalamus region by a short stalk (see Figure 37.10, page 708). As you will see, the pituitary gland's two lobes are quite distinct in both form and function.

The pituitary gland was known to the ancients almost 2,000 years ago. Galen, a famous Greek physician living during the Roman period, thought that the pituitary gland served to conduct waste products from the brain into the nose, where they became incorporated into nasal mucus, which was then discharged. In the early nineteenth century, some 1,800 years later—and after extensive, detailed anatomical and physiological studies—a more enlightened view emerged. The pituitary was found to be an amazing organ that regulates many other glands of the endocrine system and affects diverse body activities through the secretion of several hormones. It was also discovered to have a very close relationship with the nervous system. Specialized neurons in the hypothalamus are linked to the pituitary. Furthermore, blood flows from the hypothalamus directly to the anterior pituitary through a complex network of blood vessels called the **hypophyseal portal system** (see Figure 37.10B).

Posterior Pituitary

The posterior pituitary releases two peptide hormones, **antidiuretic hormone (ADH)** and **oxytocin,** that are both composed of nine amino acids. ADH, also known as *vasopressin,* helps regulate water balance within the body by promoting reabsorption of water in the kidneys. In females, oxytocin stimulates muscles of the uterus during childbirth and triggers secretion of milk from the mammary glands during nursing. Its function, if any, in males has not yet been determined.

Although both ADH and oxytocin are released from the posterior pituitary, these hormones are actually synthesized by specialized neurons in the hypothalamus called **neurosecre-**

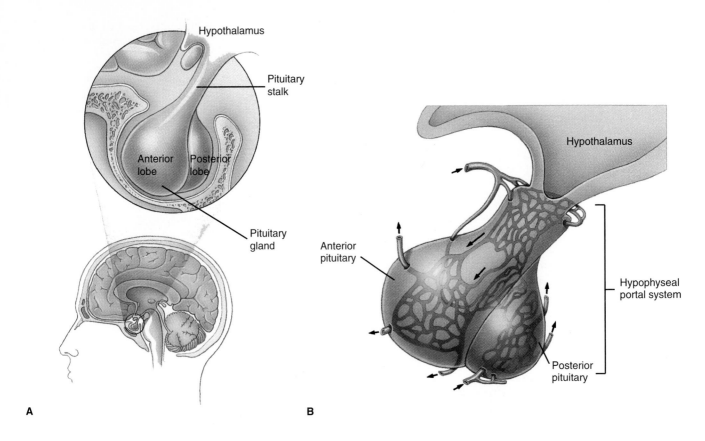

Figure 37.10 (A) The human pituitary gland is located adjacent to the brain. It has two lobes, anterior and posterior. (B) The pituitary gland is connected to the hypothalamus by a complex system of blood vessels called the *hypophyseal portal system.*

tory cells (see Figure 37.11). After being synthesized in the hypothalamus, ADH and oxytocin travel through axons of the neurosecretory cells to the pituitary, where they are released. These neurosecretory cells can also receive and conduct nerve impulses. The release of oxytocin is regulated by nerve impulses from the body. Pressure on the uterus during childbirth and stimulation of the nipples during nursing trigger the release of oxytocin. ADH release is stimulated by chemical changes in the blood and is regulated by a negative feedback system.

Anterior Pituitary

The anterior pituitary produces numerous hormones that affect many activities in the body (see Figure 37.12, page 710). For this reason, and because of the neurovascular hypothesis (see the Focus on Scientific Process, "The Neurovascular Hypothesis"), the anterior pituitary has been studied extensively.

Trophic Hormones

Hormones that affect the release of other hormones are called **trophic hormones. Follicle-stimulating hormone (FSH)** and

luteinizing hormone (LH) are classified as **gonadotrophins** because they affect endocrine functions of the gonads (ovaries and testes). In females, FSH stimulates the functions of the ovary and, along with LH, induces the production of female sex hormones. In males, FSH stimulates the development and functions of sperm-producing cells. LH stimulates the testes to produce male sex hormones and also plays a role in sperm production.

Thyroid-stimulating hormone (TSH), also known as *thyrotropin,* stimulates the synthesis and release of hormones from the thyroid gland. FSH, LH, and TSH are all glycoproteins, which are peptides combined with carbohydrate molecules. **Adrenocorticotropin (ACTH)** is a peptide hormone of 39 amino acids that stimulates the release of hormones important in the metabolism of carbohydrates from the adrenal gland.

Human growth hormone (hGH) influences a number of metabolic activities that directly or indirectly influence growth, especially of the skeleton. Its most striking effects occur during periods of rapid growth and development, but it is present throughout life. **Prolactin,** like hGH, is a relatively large peptide (191 amino acids). Its role in males is uncertain, but it functions in the production of milk in females.

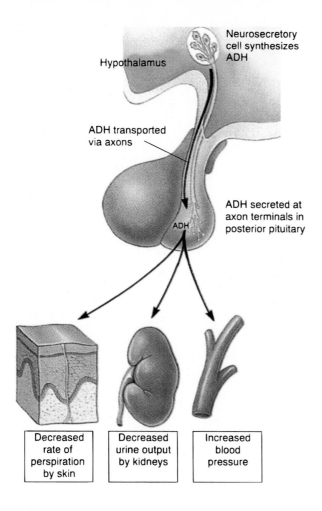

The pituitary gland consists of an anterior and a posterior lobe. The posterior pituitary releases ADH and oxytocin, which are synthesized by neurosecretory cells in the hypothalamus. The anterior pituitary produces many hormones, including several trophic hormones. Hormone secretions from the anterior pituitary are regulated by the hypothalamus through the actions of releasing and inhibiting hormones.

FUTURE FRONTIERS

In the past decade, endocrinologists have made enormous advances in understanding the mechanisms involved in the actions of chemical messengers. Like molecular genetics, and for many of the same reasons, endocrinology is a dynamic field in which new, exciting, and important discoveries are made almost daily. These two disciplines employ state-of-the-art techniques that enable investigators to address questions related to the molecular foundations on which our existence is constructed and maintained.

Hormone receptors and G proteins are very hot research topics at the present time. It has become clear that a large number of human diseases are due to defects in hormone–receptor interactions. Some of these can be traced to genetic errors that may lead to mistakes in synthesizing normal proteins. Because hormones, receptors, and G proteins are all largely or entirely constructed of specific proteins, such errors result in defective hormone or receptor structure, insufficient numbers of receptors or quantities of hormones, defects in receptor signals, or abnormal G protein function. In this regard, three recent molecular studies reported specific receptor defects as causes of different types of diabetes. The root causes are genetic defects that interfere with normal receptor structure and processing because of defective protein synthesis.

The effects of specific diseases caused by biological agents such as viruses and bacteria may also be directed at hormone–receptor–G protein interactions. For example, cholera is a disease caused by bacteria that may result in death because of dehydration. In this case, a toxin secreted by the bacteria prevents the G_s protein from being inactivated. Consequently, cAMP is produced continuously, causing such large amounts of fluids to be lost from the intestine that dehydration and death can result.

Understanding the exact molecular cause of such diseases is an essential requirement in devising an effective strategy for curing them. For hormone-related diseases, the future prospects for cures or treatment are extremely promising. Furthermore, recent studies indicate that deeper knowledge of the endocrine system will lead to exciting discoveries related to human health and diseases. For example, new research results indicate that altered relationships between hormones, receptors, and G proteins may cause certain types of cancers. Startling discoveries like this typify the enormously exciting field of endocrinology.

Figure 37.11 ADH (shown here) and oxytocin are synthesized in neurosecretory cells of the hypothalamus and travel to the posterior lobe of the pituitary gland, where they are released. ADH affects several organs in the body.

Question: *What type of hormone is produced by a neuron?*

Regulation of Anterior Pituitary Hormones

Hormones from the anterior pituitary regulate the activities of many other tissues and glands, but what governs the anterior pituitary? It is now known that neurosecretory cells in the hypothalamus manufacture and secrete substances called **releasing** and **inhibiting hormones** or **factors** (*hormone* if the chemical structure is known, *factor* if it is not). As described in Figure 37.12 (see page 710), these substances travel to the pituitary through the hypophyseal portal system, where they stimulate or inhibit the release of hormones produced by the anterior pituitary. Secretion of releasing or inhibiting factors is under the control of the peripheral nervous system and the information it relays to the brain.

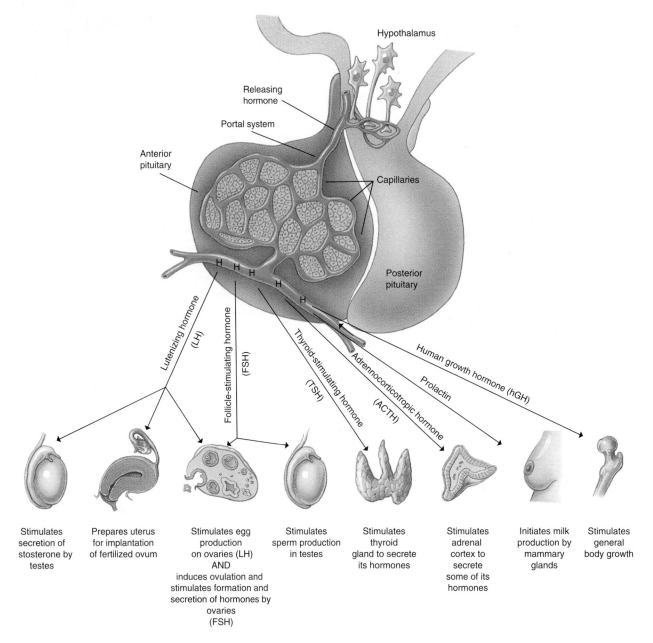

Figure 37.12   The anterior pituitary secretes numerous hormones that affect many organs and processes in the human body. Endocrine activities of the anterior pituitary are regulated by the hypothalamus through the production of releasing hormones that stimulate hormone secretion. The hypothalamus also produces inhibiting hormones that suppress secretion from the anterior pituitary.

SUMMARY

1. The classical view of the endocrine system is centered on the idea that glands secrete hormones into circulation and that these chemical messengers are then carried to target cells, where they exert their influence. In early studies, hormones were found to affect normal development, growth, sexual maturation, and behavior.

2. Because of recent research findings, modern endocrinology has found it necessary to enlarge the simple gland–hormone picture. It is now known that many organs besides glands secrete substances that act in the same way as classical hormones. Neurons secrete neurohormones, and growth factors control the growth and development of certain cells.

3. The two major types of hormones are peptide hormones and steroid hormones. Other types of hormones include thyroid hormones, growth-stimulating factors, and prostaglandins.

4. The production and release of hormones are usually regulated by complex negative feedback systems that involve both the endocrine and nervous systems. After release, peptide

The Neurovascular Hypothesis

Confirmation of a scientific hypothesis in biology often involves an extended period of time, brilliantly conceived experiments, and fragile resources. Verification of the existence and role of hypothalamic releasing and inhibiting hormones followed a somewhat different path that serves to illustrate the flexibility of scientists in answering interesting questions.

During the nineteenth century, experimental evidence indicated that the pituitary gland secreted several hormones essential for numerous body functions; hence the pituitary became known as the "master gland." The view persisted that regulation of hormone secretions from the pituitary involved classical negative feedback systems (that is, high levels in the body eventually led to a reduction in secretion from the pituitary). In the twentieth century, however, it became clear, from many types of studies, that the nervous system was somehow involved in regulating secretions from this gland. For example, stimuli from the external environment were found to cause the release of certain hormones from the pituitary, even though this gland is incapable of detecting such stimuli.

What part of the nervous system was involved in this regulation? Several studies indicated that it was the hypothalamus. In 1894, it was discovered that axons from a group of prominent neurons ran from the hypothalamus to the posterior pituitary. Subsequently, researchers found that these neurons synthesized ADH and oxytocin and transported them to the pituitary, where they were stored until appropriate stimuli triggered their release. These results were, of course, tremendously exciting because they offered clear evidence of integration between the nervous and endocrine systems.

But what regulated the anterior pituitary, the portion of the gland that releases the most hormones and was *not* linked to the hypothalamus by neurosecretory cells? Several lines of experimental evidence indicated that the hypothalamus was also responsible for controlling this part of the pituitary. For example, electrical stimulation of the hypothalamus caused the release of anterior-pituitary hormones, but direct stimulation of the anterior pituitary had no effect. Also, damage to the hypothalamus was found to cause decreases or increases in the secretion of different anterior-pituitary hormones. Finally, in the early 1930s, scientists discovered that the hypothalamus was directly connected to the anterior pituitary by a set of blood vessels called the hypophyseal portal system.

Thus a hypothesis emerged in the 1950s: neurons in the hypothalamus synthesize substances that travel straight to the anterior pituitary, via the portal system, and stimulate or inhibit secretion of hormones from this gland. Like most notable hypotheses, this new *neurovascular hypothesis* satisfied the following criteria: (1) it was consistent with existing knowledge (no neuron connections), (2) it offered a plausible interpretation of the role of a poorly understood but obviously significant element (the portal system), (3) it made predictions (the nervous system controlled the pituitary through stimulation of the hypothalamus), and (4) it served as a guide for further research. In certain cases, scientists competed to identify the releasing factors. Some of these races lasted for several decades.

How could the neurovascular hypothesis be confirmed? There seemed to be two possible lines of investigation. The first was to demonstrate that regulation of the anterior pituitary by the hypothalamus depended on the portal system. This was accomplished through several animal experiments in which surgical procedures were used. For example, the portal system between the hypothalamus and the pituitary was disconnected and hormone levels were then measured (they decreased) or hormone-dependent functions were monitored (they failed). After reconnection, normal levels and functions returned. All of these results strongly supported the idea that hypothalamic control was applied through the portal system.

The second and by far most difficult task was to isolate and identify the specific releasing factors hypothesized to regulate the pituitary. That is, did releasing hormones actually exist? How could essential details be obtained for these unique but unknown substances? What were the problems?

One of the primary problems in the 1950s was that only ACTH quantities in tissues or fluids could be measured reliably; methods for analyzing other anterior-pituitary hormones were inadequate or did not exist. Thus the first efforts were aimed at isolating the ACTH-releasing hormone (termed *corticotropin-releasing hormone,* or *CRH,* because ACTH belongs to a class of hormones called *corticotropins*). For several years, numerous scientists endeavored independently to isolate and identify CRH. The key experiments involved growing pituitary cells, obtained from rats and dogs, in test tubes. At first the cells produced ACTH, but after several days no additional ACTH was synthesized, even though the cells remained healthy. What should be attempted next? Tissue from the hypothalamus was then added to the test tubes containing the pituitary cells, and once again they produced ACTH! This observation indicated that the

box continues

FOCUS ON SCIENTIFIC PROCESS

hypothalamus released some factor (CRH?) that stimulated ACTH production. Results of these experiments were published in 1955, but despite serious efforts, overwhelming problems delayed the isolation and identity of CRH until 1981. The most stubborn obstacle was being able to obtain sufficient quantities of pure CRH from experimental animals. Eventually, investigators were forced to turn to slaughterhouses to acquire the hundreds of thousands of glands they needed to obtain adequate amounts of the desired substance.

Following the frustrating delay in identifying CRH, these same scientists (primarily two groups headed by Andrew Schally and Roger Guillemin, both of whom were to receive Nobel Prizes for their research on releasing factors) turned their attention to identifying *thyrotropin-releasing hormone (TRH)*. To overcome the fundamental problem of having pure substances to study, both investigators used enormous quantities of animal tissues. Guillemin is reported to have processed over 5 tons of hypothalamic tissue from 500,000 sheep while in Paris and the hypothalami of almost 2 million sheep after he moved his research program to Texas. Not to be outdone, Schally established a relationship with the Oscar Mayer meat-packing company that netted him donations of more than 1 million pig hypothalami. These industrial-scale quantities are not a feature shared with other studies described in earlier chapters of this book. After several false starts, some ingenious methodological developments, a few hundred thousand additional hypothalami, and seven years, both groups identified the structure of TRH, the first hypothalamus releasing hormone to be described. Thinking it had won the contest, Guillemin's group was to publish its findings on November 12, 1969. However, on November 6, the same results of Schally and his colleagues' work were described in another journal; they had won the race by a matter of days!

The competition between these two investigators (and others) to identify new releasing hormones continued. The next hunt, centered on describing *luteinizing-hormone-releasing hormone (LHRH)*, was also won by Schally's group in 1971. Guillemin's group finally scored in 1973 when it identified the hormone that inhibits release of growth hormone. Finally, in 1981, another group finally determined the structure of CRH.

The structures of all the hypothalamic releasing and inhibiting hormones are now known, and the neurovascular hypothesis has been confirmed. While less well known than the dramatic investigations of DNA, confirmation of this hypothesis was certainly as challenging and exciting.

---

hormones are degraded within hours, whereas steroid and thyroid hormones may remain in the body for several weeks.

5. Peptide hormones interact with their target cell through complex protein receptor molecules and G proteins found in the plasma membrane. Each receptor is specific for a single hormone. Hormone–receptor signals activate a G protein to initiate synthesis of cAMP, and this leads to a cascade of chemical reactions that result in the cell's performing a certain function. Steroid hormones enter the cell and bind to an internal receptor, forming a complex that attaches to DNA and causes the synthesis of specific proteins.

6. The pituitary gland is the major gland of the endocrine system in terms of the number of hormones it produces and the effects its hormones have on body activities. The pituitary gland is regulated by neurosecretory cells in the hypothalamus region of the brain that release substances that either stimulate or inhibit hormone synthesis in the pituitary.

7. Research on the endocrine system, specifically on the hormone–receptor and G protein systems, holds great promise for being able to treat a number of important diseases effectively.

WORKING VOCABULARY

cAMP (p. 705)	inhibiting hormone (p. 709)
diabetes mellitus (p. 703)	insulin (p. 703)
endocrinology (p. 700)	neurohormone (p. 700)
G protein (p. 706)	peptide hormones (p. 700)
gland (p. 697)	releasing hormone (p. 709)
hormone (p. 700)	steroid hormones (p. 700)

REVIEW QUESTIONS

1. What is the classical view of the endocrine system?

2. How did the theory of internal secretions advance the field of endocrinology?

3. What is the modern view of the endocrine system?

4. What are two major classes of hormones? How do they differ?

5. Describe the general life cycle of a hormone.

6. Which organs, cells, and hormones play roles in regulating blood glucose levels? How is blood glucose regulated?

7. What are some general activities of hormones in the human body?

8. What are hormone receptors? What is their function?

9. What is the second-messenger system used by peptide hormones? How do G proteins fit into this system?

10. How are steroid hormones able to carry out their functions?

11. Which hormones are produced by the posterior and anterior lobes of the pituitary gland? How is the anterior pituitary regulated?

12. What types of endocrinology studies are now being conducted? What is the potential importance of such studies?

ESSAY AND DISCUSSION QUESTIONS

1. There are several types of diabetes, each with a specific cause. What might be some of the causes of diabetes? Consider the hormones involved, how they act on cells, and the general life cycle of hormones.

2. Select one of the hormones listed in Table 37.1 and devise a probable feedback system that may be involved in its regulation.

3. Which control system, nervous or endocrine, do you think evolved first? Why?

REFERENCES AND RECOMMENDED READING

Archer, R. 1980. Molecular evolution of biologically active polypeptides. *Proceedings of the Royal Society of London,* B210: 21–43.

Bolander, F. F. 1994. *Molecular Endocrinology.* 2d ed. San Diego, Calif.: Academic Press.

Brown, R. E. 1994. *An Introduction to Endocrinology.* New York: Cambridge University Press.

Charron, M. J., and B. B. Kahn. 1990. Divergent molecular mechanisms for insulin-resistant glucose transport in muscle and adipose cells in vivo. *Journal of Biological Chemistry,* 265: 7994–7999.

Crapo, L. 1985. *Hormones: Messengers of Life.* New York: Freeman.

Greenstein, B. 1994. *Endocrinology at a Glance.* Boston: Blackwell.

Guillemin, R., and R. Burgus. 1972. The hormones of the hypothalamus. *Scientific American,* 227: 24–33.

Hoberman, J. M., and C. E. Yesalis. 1995. The history of synthetic testoterone. *Scientific American,* 272: 76–81.

Kaltenbach, J. C. 1988. Endocrine aspects of homeostasis. *American Zoologist,* 28: 761–773.

Linder, M. E., and A. G. Gilman. 1992. G proteins. *Scientific American,* 267: 56–65.

McDermott, M. T. 1994. *Endocrinology Secrets.* St. Louis: Hanley & Belfus.

McGarry, J. D. 1992. What if Minkowski had been ageusic? An alternative angle on diabetes. *Science,* 258: 766–770.

Wade, N. 1978. Guillemin and Schally. *Science,* 200: 279–282.

Watkins, P.J. 1996. *Diabetes and its Management.* 5th ed. Cambridge: Blackwell Science.

ANSWERS TO FIGURE QUESTIONS

Figure 37.2 Adrenal glands and gonads (ovaries and testes).
Figure 37.3 When hormone levels are higher or lower than required for homeostasis.
Figure 37.4 A negative feedback system.
Figure 37.6 It influences many metabolic activities associated with growth, especially of the skeleton.
Figure 37.7 G proteins.
Figure 37.8 A few seconds.
Figure 37.11 A neurohormone.

38

Human Reproduction and Development

Chapter Outline

Reading Questions

1. What are the major structures and functions of the male and female reproductive systems?

2. How are the activities of the male and female reproductive systems regulated?

3. How does a fertilized egg become transformed into a baby in nine months?

4. What types of questions are now guiding research in reproduction and development?

Reproduction is the process by which organisms produce new individuals. Most people are extremely interested in the way humans accomplish this task. Biologically, human reproduction mostly depends on organs, controlling systems, and strategies that first evolved in our distant vertebrate ancestors, the reptiles (see Chapter 26), and our more modern mammalian ancestors.

What are some of these inventions from evolutionary history? Like our earliest animal ancestors, we carry out **sexual reproduction,** which involves the fusion of haploid gametes to create offspring (see Chapter 15). We use **internal fertilization,** the process in which gametes fuse within the body of the female. This ancient strategy enabled reptiles to occupy terrestrial environments hundreds of millions of years ago.

Finally, from our mammalian ancestors, we inherited abilities to retain the developing embryo inside the female's body and to provide it with nourishment from milk-secreting glands, practices that greatly enhanced survival of the offspring.

Humans have, of course, added a new wrinkle or two through the course of our evolutionary history. Perhaps the most interesting is that unlike all other mammalian species, which reproduce only during a specific time of the year as determined by the female's **estrous cycle** (a regular period of sexual receptivity that occurs when an egg is available to be fertilized), humans can reproduce year-round. They also engage in sexual activities when fertilization is unlikely to occur. These unique features of human reproduction have been analyzed extensively. What would be the advantage of a strategy in which sexual behavior is not necessarily linked to reproductive success or a greater chance of producing offspring? Though not convincing to all scientists, one hypothesis proposes that in early humans, this strategy may have been important in establishing long-term pair bonds. In other words, the continuous availability of pleasurable sex was the glue that held a relationship together. A durable human male–female relationship could then provide the justification for individuals to invest major amounts of time in caring for intensely dependent human offspring. Early human couples who engaged in such behavior would have been more likely to leave behind more offspring than those who did not.

The origins and evolutionary significance of human sexual behaviors will continue to be debated. However, there is little argument that human reproduction has been enormously successful throughout the course of history.

THE HUMAN REPRODUCTIVE SYSTEM

Males and females possess **gonads,** the **primary sex organs** that create gametes and produce hormones that influence development and functions related to reproduction. The primary sex organs of males are called **testes** and produce **spermatozoa,** or **sperm,** by the billions as well as male sex hormones called **androgens. Ovaries** are the primary sex organs of females. They produce a small number of **ova,** or **eggs,** during their reproductive lifetime. The ovaries also produce sex hormones, including **progesterone** and a number of **estrogens. Accessory reproductive organs** include all other organs involved in gamete protection, storage, and transport.

BEFORE YOU GO ON   Humans can produce progeny year-round through sexual reproduction, internal fertilization, and embryonic development. All of these processes result in a high offspring survival rate. The primary sex organs of males are testes, which produce sperm. Female primary sex organs are ovaries, which produce ova. Reproductive organs and gamete production are regulated primarily by hormones.

THE MALE REPRODUCTIVE SYSTEM

Organs of the male reproductive system are illustrated in Figure 38.1. The two major functions of the testes are to manufacture sperm and to secrete androgens. The primary androgen is **testosterone,** a hormone that influences many activities in the male reproductive system. Various accessory organs are responsible for storing and nourishing sperm and for delivering them to the reproductive tract of a female. Regulation of all reproductive functions is carried out by pituitary hormones, with secretions managed by the hypothalamus of the brain.

Sperm Production

The testes lie outside the abdominal cavity, contained within a pouch of skin called the **scrotum.** The external location of the testes is significant because normal sperm development requires a slightly cooler temperature than that maintained in the internal environment. Each testis is surrounded and protected by a capsule of connective tissue, and inward extensions of this tissue divide the testes into about 250 compartments. Each compartment contains sperm-producing **seminiferous tubules,** highly convoluted structures that harbor male gametes in various stages of development (see Figure 38.2A, page 716). Testosterone is produced by clusters of **interstitial (Leydig) cells** located within the connective tissue between the seminiferous tubules.

Production of spermatozoa within the testes is called **spermatogenesis.** It begins in males at **puberty,** the time of life

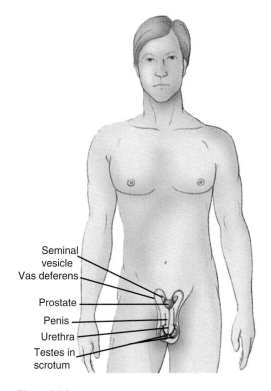

Seminal
vesicle
Vas deferens
Prostate
Penis
Urethra
Testes in
scrotum

Figure 38.1   The external organs of the male reproductive system are the penis and the scrotum. The other organs are internal.

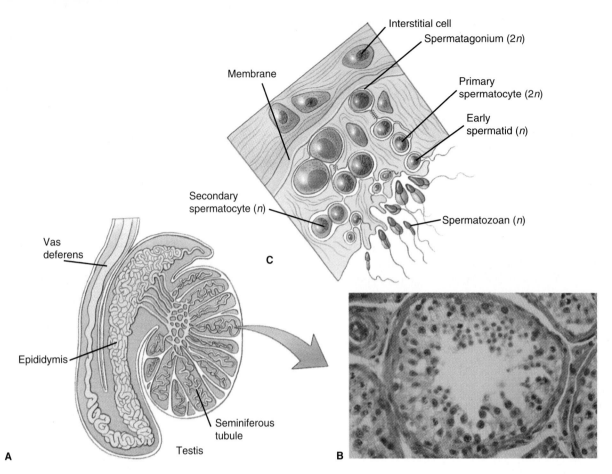

Figure 38.2   (A) Sperm are produced within seminiferous tubules in the testes. (B) This light micrograph shows sperm in various stages of production. (C) Diploid (2*n*) spermatogonia proliferate by mitosis and produce cells that develop into primary spermatocytes. Primary spermatocytes undergo meiosis, giving rise to secondary spermatocytes that each contain a haploid *(n)* number of chromosomes. Secondary spermatocytes then develop into spermatids, which mature into functional spermatozoa.

Question: *When does spermatogenesis begin in human males? How long does it continue?*

when the gonads mature and reproduction becomes possible. After puberty, spermatogenesis continues throughout life.

The manufacture of mature sperm begins in germinal cells of the seminiferous tubules shown in Figure 38.2B and C. These cells, called **spermatogonia,** divide by mitosis and cytokinesis to give rise to large numbers of identical cells. Spermatogonia contain the diploid number of 23 pairs of chromosomes (44 autosomes and an X and a Y chromosome). Each day, millions of spermatogonia move toward the center of the tubule, begin to grow, and change into large developing cells known as **primary spermatocytes.** Ultimately, it is essential for each sperm to possess a haploid number of chromosomes ($n = 23$). How does this occur? The process is described in Figure 38.3. First, each primary spermatocyte undergoes meiosis. During the first meiotic division, two **secondary spermatocytes** are produced that each have 22 paired autosomes plus either a pair of X or Y chromosomes. Subsequently, during the second meiotic division, each secondary spermatocyte gives rise to two **spermatids,** each of which contains 22 autosomes

and one sex chromosome. Each spermatid then undergoes a sequence of changes that leads to its final destiny, becoming a mature spermatozoon.

Sperm have a structure that reveals a perfect relationship with their function (see Figure 38.4). The head of a sperm contains a nucleus with 23 chromosomes and is covered by a caplike, fluid-filled structure called an **acrosome.** The acrosomal fluid contains enzymes that are released when the sperm encounters an ovum in the female reproductive tract. Apparently, these enzymes assist the sperm in dissolving membranes that surround the ovum. Another striking feature of sperm is the "tail," a single whiplike *flagellum* that provides the cell with mobility.

Sperm Storage and Transport

Once spermatozoa are formed, they move through a system of ducts, shown in Figure 38.5 (see page 718), until they are eventually released from the body. From the seminiferous tubules,

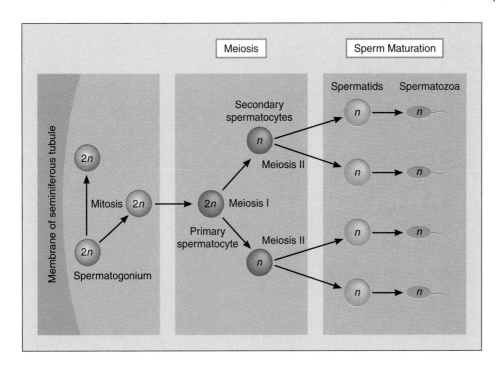

Figure 38.3   The maturation sequence from spermatogonium to spermatozoa includes a change in the chromosome number from 2*n* (diploid) to *n* (haploid) after meiosis.

Question: *What are the diploid and haploid number of human chromosomes?*

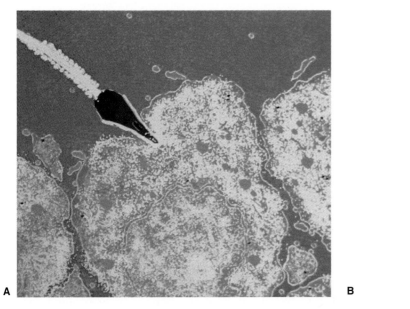

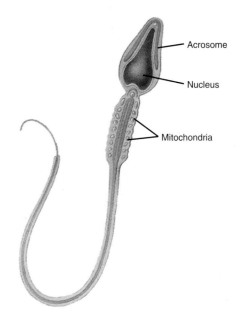

A   B

Figure 38.4   (A) A functional human sperm looks like this as it begins to penetrate an ovum. (B) The structure of a sperm is related to its function—to fertilize an ovum. Chromosomes are carried in the nucleus, and the acrosome contains fluid filled with enzymes that break down ovum membranes. Mobility is provided directly by the flagellum and indirectly by mitochondria that provide the energy required to move the sperm's "tail."

sperm pass into highly coiled tubules of the **epididymis.** There they undergo final maturation and are stored until **ejaculation,** or discharge from the body, takes place. This occurs when a male becomes sexually stimulated and a series of physiological events leads to the sperm's being propelled through the **penis,** the male organ that conveys sperm to the external environment.

Between the epididymis and the penis are a number of ducts (tubelike structures) and glands that play important roles

in the sperm's journey. During ejaculation, sperm from both testes are propelled from each epididymis into the **vasa deferentia.** These two ducts are surrounded by walls of smooth muscle that contract rhythmically during ejaculation, transporting the sperm forward. A vas deferens ascends from each testis and merges with a duct from a **seminal vesicle** to form a short **ejaculatory duct** that is located behind the urinary bladder. These ducts then pass through the **prostate gland** and

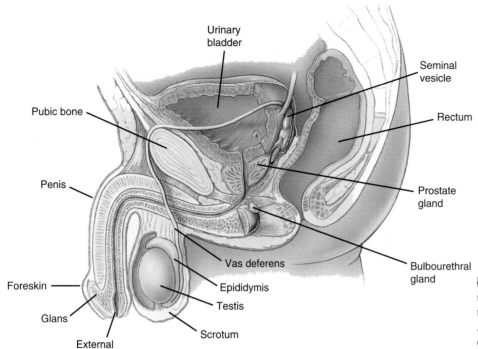

Figure 38.5 This median section (from the side) shows the organs involved in the transport of sperm and adjacent organs.

Question: *What is the function of the seminal vesicles?*

enter the *urethra,* the tube that drains both the urinary and reproductive systems.

Secretions from three organs combine with sperm to form **semen,** the fluid that is ejaculated. The two seminal vesicles secrete an alkaline, sticky fluid that contains *fructose,* a sugar that is the major energy source for ejaculated sperm, and *prostaglandins,* substances thought to enhance the movement of sperm when deposited in the female reproductive tract. The alkaline nature of the fluid helps counteract the acid environment within the female reproductive tract, which can inactivate sperm if not neutralized. The prostate gland secretes a milky fluid containing enzymes that contribute to sperm motility and survival. Finally, secretions from two **bulbourethral glands** precede ejaculation and are thought to play some role in cleaning and reducing the acidity of the urethra.

Hormonal Regulation of the Male Reproductive System

Prior to the onset of puberty, which typically occurs at about 14 years of age, males are unable to produce mature sperm. After puberty, through the actions of testosterone, males are able to produce sperm by the billions. In addition, this hormone radically transforms several body features. Changes in the body related to sexual maturation are known as **secondary sex characteristics.** In males, these include a deepening of the voice, increased development of skeletal muscle tissue and a decrease in body fat, growth of hair on the face and other areas of the body, and enlargement of the penis, scrotum, and testes.

What happens at puberty to cause these profound changes that enable males to become capable of producing offspring? The leading hypothesis to explain the onset of puberty in humans focuses on maturation of the part of the brain's neuroendocrine control system involved in secreting releasing hormones from the hypothalamus. Recall that specific releasing hormones from the hypothalamus are required to stimulate production of certain hormones from the anterior pituitary gland. However, the factors responsible for the onset of puberty in males are not totally understood. During childhood, the endocrine system produces only meager amounts of male sex hormones. These hormones, which are required for sexual maturation and function, consist of testosterone and two *gonadotropins* called *follicle-stimulating hormone (FSH)* and *luteinizing hormone (LH).* At puberty, the hypothalamus, in reaction to a signal from an undefined area of the brain, begins to secrete increased levels of a **gonadotropin-releasing hormone (GnRH).** It is still not known whether a single GnRH stimulates the production of both FSH and LH or if two different releasing hormones are necessary. All these hormones are believed to interact in classical negative feedback cycles.

Figure 38.6 indicates that GnRH stimulates production and release of FSH and LH, which are then transported by the circulatory system to the testes. FSH activates sperm production, and LH stimulates interstitial cells to secrete testosterone, which in turn triggers spermatogenesis and the development and maintenance of secondary sex characteristics.

How are levels of these various hormones regulated by control systems? The self-regulating pathways are generally understood, although some uncertainties linger. High levels of testosterone inhibit LH release through negative feedback that is thought to be directed at the hypothalamus and perhaps the pituitary. In other words, as testosterone concentrations in the blood increase, the pituitary responds by reducing production of LH.

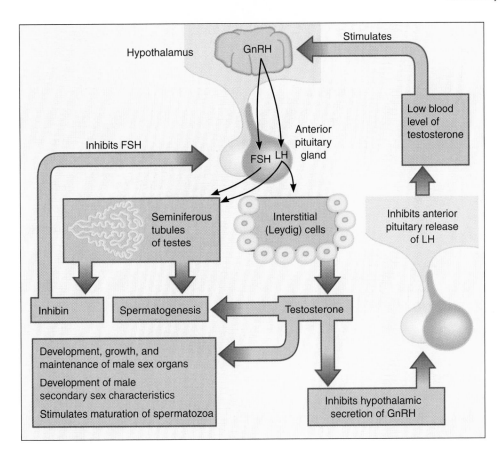

Figure 38.6 The onset of puberty and continuous hormonal regulation of sperm production in males involve several hormones. GnRH released from the hypothalamus stimulates production and release of FSH and LH from the pituitary gland. FSH activates and sustains sperm production, and LH stimulates production of testosterone, the hormone that plays a central role in spermatogenesis and the development and maintenance of secondary sex characteristics.

Question: *What seems to trigger the beginning of puberty and sexual maturation?*

FSH production in both sexes is hypothesized to be regulated by a hormone called **inhibin.** In males, this hormone is thought to be produced by cells in the seminiferous tubules and to exert negative feedback effects on the pituitary, the hypothalamus, or both. As testosterone levels decrease, the pituitary once again produces increased levels of FSH and LH. This hormonal cycle operates continuously throughout the lifetime of a normal male.

> **BEFORE YOU GO ON** Testosterone, a sex hormone, plays a major role in regulating the maturation and function of male reproductive organs and is also responsible for development of secondary sex characteristics. Spermatogenesis begins in the testes at puberty and continues throughout life. Two pituitary hormones are important in maintaining this process: FSH initiates sperm production, and LH stimulates testosterone secretion. All these hormones are regulated by negative feedback mechanisms.

THE FEMALE REPRODUCTIVE SYSTEM

Though far from simple, the functions of the male reproductive system are relatively straightforward and linear and can be summarized as producing constant quantities of sperm and testosterone. By contrast, the tasks of the female reproductive system are multiple and involve two overlapping 28-day cycles. During the **ovarian cycle,** ova of human females mature according to a precise, genetically programmed series of events. In the **menstrual cycle,** parts of the reproductive tract also undergo a hormonally controlled cycle of changes that occurs about every 28 days.

Ovum Production and Transport

The primary sex organs of the female reproductive system are two ovaries that produce ova and sex hormones (see Figure 38.7, page 720). Major accessory reproductive organs include the **uterine tubes** *(oviducts* or *Fallopian tubes),* which transport an ovum from the ovary to the **uterus** *(womb),* a thick-walled, muscular organ where an embryo develops and grows; the **vagina,** which receives sperm from a male and also serves as the birth canal during childbirth; and accessory glands that secrete a lubricating fluid.

In addition to the internal organs, females have several external genital organs that are referred to collectively as the **vulva.** The general functions of most of these organs are related to protection of sensitive tissues and facilitation of sexual intercourse. **Mammary glands** *(breasts),* found also in males, develop fully only in females. Their function is to provide milk for nourishing offspring after birth.

Oogenesis

In females, puberty generally begins about a year earlier than in males. Available evidence suggests that the onset of puberty in females is related to attaining a critical body weight or body composition. The beginning of puberty is marked by the

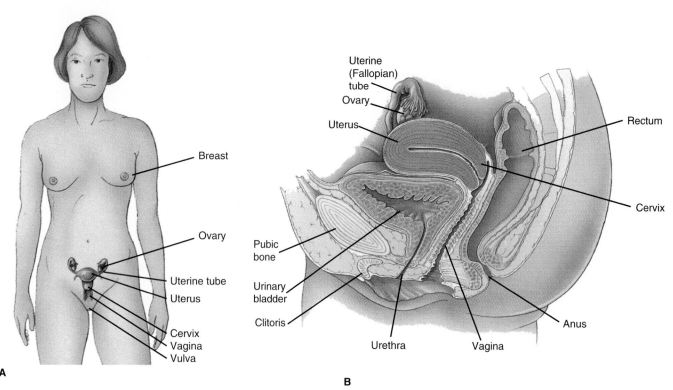

Figure 38.7   (A) The primary sex organs of the female reproductive system. (B) This median section shows the structure of the major female sex organs.

Question: *What is the function of the uterus?*

first incidence of **menstruation,** a periodic (approximately monthly) loss of blood and other tissue from the uterus. This event signifies that the ovaries have begun to produce mature ova, a process called **oogenesis.**

The gametogenic potential of the ovary is actually determined during embryonic development. Before birth, **oogonia,** cells in the female embryo's ovary, divide by mitosis and cytokinesis, giving rise to about 3 million **primary oocytes** (immature ova). At birth, about 1 million oocytes remain, and by puberty, this number has been further reduced to about 250,000, of which 400 to 500 will ultimately develop into mature ova. Prior to birth, primary oocytes enter the first phase of meiosis (prophase), where they remain in arrested development until later in life.

Development of Follicles

Beginning at puberty, **ovulation,** the release of a single mature egg from one of the ovaries, occurs approximately once each month (the average cycle is 28 days). At the start of the cycle, primary oocytes are surrounded by a layer of supporting follicle cells that, together with the primary oocyte, are called a **primary follicle** (see Figure 38.8A). Events in the ovarian cycle begin when, under hormonal stimulation, 6 to 12 primary follicles began to grow and develop into **secondary follicles** (see Figure 38.8B). Several events occur during the maturation of a secondary follicle.

1. The first meiotic division is completed, producing two haploid cells of unequal size, as described in Figure 38.9 (see page 722). The largest, which may become a mature ovum, is called a **secondary oocyte,** and the smaller is referred to as the **first polar body,** a cell that has no functional future. If fertilization occurs, each of these cells may undergo a second meiotic division, giving rise to two additional polar bodies.

2. Follicle cells proliferate and form glandular cells that secrete a fluid around the developing oocyte.

3. As the oocyte approaches maturity, it becomes surrounded by a fluid-filled cavity.

4. At maturity, which occurs 10 to 14 days into the ovarian cycle, the follicle moves toward the surface of the ovary. Although several follicles undergo initial development each month, generally only one reaches the final stage of maturity. The rest simply degenerate.

Ovulation

Upon receiving the appropriate hormonal cue, ovulation occurs as the mature follicle ruptures and the ovum, along with a surrounding layer of follicle cells, moves into the uterine tube. Once there, it is swept toward the uterus by the action of cilia that line the tube.

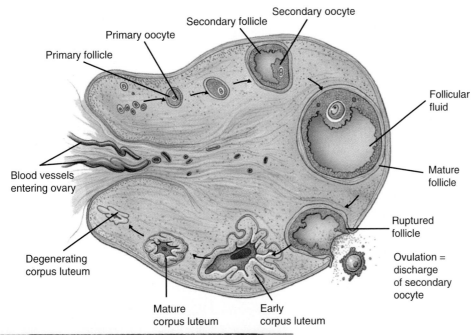

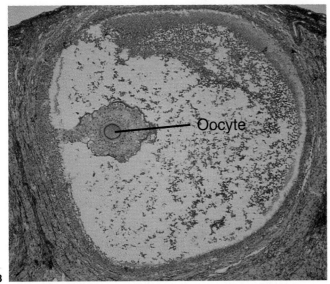

Figure 38.8 (A) Once each month, in one of the two ovaries, a primary follicle containing an oocyte develops into a secondary follicle, which releases the mature oocyte during ovulation. The follicle then becomes transformed into a corpus luteum that will degenerate if fertilization of the ovum does not occur. (B) This secondary follicle contains a maturing oocyte.

Question: *When are primary follicles formed in the life of a female?*

Formation of the Corpus Luteum

Soon after ovulation occurs, the ruptured follicle undergoes a series of changes, eventually becoming a cholesterol-filled tissue called a **corpus luteum.** The fate of this structure depends on whether or not the female becomes pregnant. If she does not, it degenerates about 14 days after ovulation. If pregnancy occurs, the corpus luteum undergoes further development, forming a temporary endocrine gland that regulates several important functions during pregnancy.

Hormonal Regulation of the Female Reproductive System

Female hormonal regulating mechanisms are described in Figure 38.10 (see page 722). At puberty, the GnRH–FSH–LH feedback system begins to function, and **estrogens,** female sex hormones produced by cells of the ovaries, induce the development of secondary sex characteristics. During this period, the vagina, uterus, and uterine tubes grow and develop; fat becomes deposited in certain areas such as the hips and the breasts; and mammary growth and development occur. The female reproductive tract is prepared to receive a fertilized egg and to maintain the uterine lining if pregnancy occurs. All of these actions are induced by estrogens and a number of other hormones.

The regular ovarian and menstrual cycles continue to occur in a woman until **menopause,** the time in her life when reproduction is no longer possible. The onset of menopause, usually around age 50, has been thought to be associated with the failure of the ovary to respond any longer to FSH and LH. Consequently, the ovarian hormonal cycles cease, estrogen

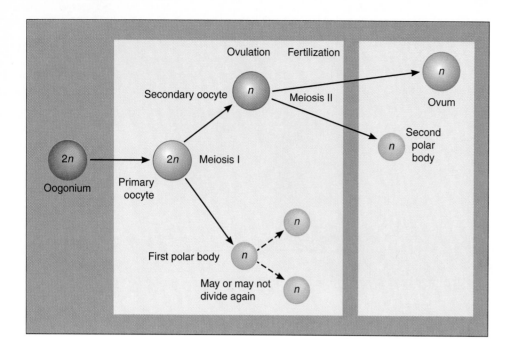

Figure 38.9 During embryonic development, diploid oogonia (*2n*) proliferate by mitosis and produce primary oocytes (*2n*). Each month after puberty, a primary oocyte undergoes the first meiotic division, giving rise to one haploid secondary oocyte (*n*) and a haploid polar body (*n*). The first polar body, and the secondary oocyte if fertilized, may complete a second meiotic division.

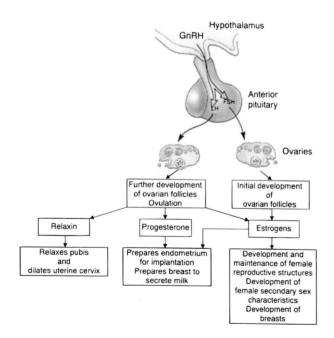

Figure 38.10 The female hormones GnRH, FSH, LH, progesterone, and estrogen are all involved in regulating the monthly ovarian and menstrual cycles in females after puberty. They also, along with relaxin, play important roles if pregnancy occurs.

Question: *When and why do menstrual cycles cease in women?*

levels become greatly diminished, and this leads to a number of physiological and structural changes in menopausal women. Recent studies indicate that alterations in GnRH release patterns may also contribute to the onset of menopause.

The Ovarian Cycle

The events and timing of the ovarian and menstrual cycles are described in Figure 38.11A. The concentrations of various hor-

mones in the blood during a typical 28-day period are shown in Figure 38.11B. The effects of GnRH, LH, and FSH are similar to those described for males. In females, FSH stimulates the development of ova, and an examination of Figure 38.11 reveals the relationships between hormone concentrations and events during development of the follicle. FSH levels decrease until midway through the cycle, probably as a result of increasing estrogen (and perhaps inhibin) concentrations. LH production is also suppressed during this period by the negative feedback

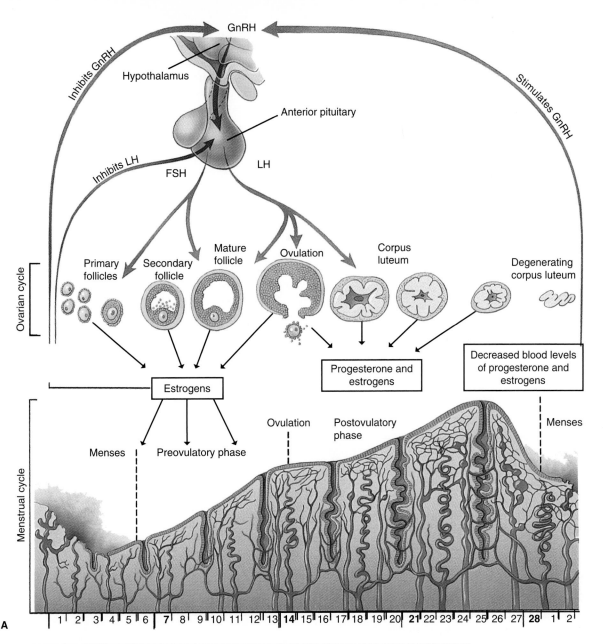

GnRH

Hypothalamus

Inhibits GnRH

Stimulates GnRH

Anterior pituitary

Inhibits LH

FSH

LH

Ovarian cycle

Primary follicles

Secondary follicle

Mature follicle

Ovulation

Corpus luteum

Degenerating corpus luteum

Estrogens

Progesterone and estrogens

Decreased blood levels of progesterone and estrogens

Ovulation

Postovulatory phase

Menses

Menses

Preovulatory phase

Menstrual cycle

1 2 3 4 5 6 7 8 9 10 11 12 13 14 15 16 17 18 19 20 21 22 23 24 25 26 27 28 1 2

A

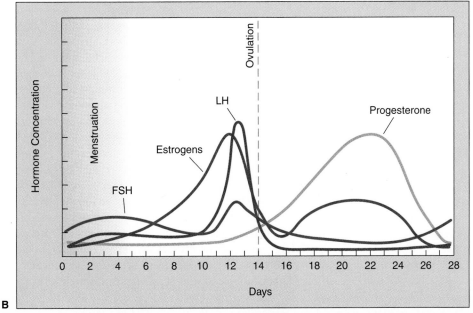

Hormone Concentration

Menstruation

Ovulation

LH

Estrogens

FSH

Progesterone

0 2 4 6 8 10 12 14 16 18 20 22 24 26 28

Days

B

Figure 38.11 (A) This diagram traces the changes that occur during the 28-day ovarian and menstrual cycles, the hormones involved in these changes, and the relationships between the two cycles. (B) Hormone concentrations vary during a normal 28-day cycle. Inhibin concentrations are not shown, but they are apparently very low.

Question: *What is the role of progesterone in this complicated negative feedback system?*

inhibition of estrogens on the pituitary. At mid-cycle, however, just prior to ovulation, the high estrogen levels exert a positive feedback that triggers a surge in the levels of LH and FSH. The high LH levels stimulate release of the mature ovum, which in turn results in an elevation of progesterone levels. Progesterone—which continues to increase during the latter part of the cycle—and inhibin exert negative feedback on the hypothalamus, and further FSH and LH secretion is inhibited during this time. If pregnancy does not occur, the corpus luteum degenerates, and levels of all hormones begin to fall. Consequently, negative feedback inhibition of FSH and LH secretion by estrogens and progesterone no longer occurs. As a result, FSH and LH levels begin to increase once again, new follicles will begin to develop, and the ovarian cycle is repeated.

The Menstrual Cycle

The menstrual, or uterine, cycle involves a sequence of changes that occur primarily in the **endometrium,** the mucous glandular membrane that lines the uterus. It overlaps with the ovarian cycle because estrogens and progesterone also regulate the menstrual cycle (see Figure 38.11). During the developmental phase, which is first stimulated by estrogens, blood vessels invade the epithelium, glandular cells increase, and the endometrium becomes thicker. After ovulation, the corpus luteum secretes both estrogens and progesterone, which continue to stimulate further development of the endometrium. Essentially, it is being prepared to receive a fertilized egg. If pregnancy does not occur, hormone levels decline as the corpus luteum degenerates, and as a result, maintenance of the endometrium is discontinued. When that happens, the membrane deteriorates, and cells, blood vessels, and blood are sloughed and discharged from the uterus as the menstrual flow. The endometrium is quickly repaired, and before the next cycle begins, it can again undergo the cyclical changes that characterize the menstrual cycle.

BEFORE YOU GO ON | In females, mature ova are produced in a monthly cycle between puberty and menopause. Estrogens induce development of secondary sex characteristics and also play key roles in all female reproductive functions. The ovarian and menstrual cycles are connected by negative feedback mechanisms involving FSH, LH, estrogens, progesterone, and perhaps inhibin. Fluctuating levels of these hormones are responsible for the cyclical maturation and release of ova and the development and decline of the corpus luteum and the endometrium.

SEXUAL RESPONSES

Males and females are capable of responding to many types of stimuli that have sexual connotations. In males, sexual stimuli, be they visual, tactile, or psychological, may lead to **erection** (enlarging and firming) of the penis. This is brought about by engorgement of the highly vascular tissues of the penis with blood, and it enables the penis to be inserted into the vagina. As sexual stimulation intensifies, pulse, blood pressure, and rate of breathing all

increase, and **emission,** movement of the contents of sexual ducts and glands to the urethra, occurs. At this point, small amounts of semen may be emitted from the penis. Ultimately, sexual excitement reaches its peak, and sympathetic nerve impulses lead to ejaculation of semen from the urethra. Ejaculation is accompanied by a series of brief, widespread muscular contractions and pleasurable sensations that are collectively referred to as **orgasm.** Following ejaculation, regular patterns of blood flow, heart rate, and breathing are restored, and the penis returns to its normal, unaroused state. For several minutes or hours after ejaculation, most males are unresponsive to further sexual stimulation.

Sexual stimulation in women leads to many physiological responses that are similar to those in men, which is not surprising in that many of the same nervous pathways are involved. The **clitoris,** a small sensitive organ that has the same embryological origin as the male penis, becomes engorged with blood, as do the breasts and the area surrounding the vaginal opening. Copious quantities of fluids lubricate the vagina, and further arousal leads to rhythmic stimulatory responses of all external organs involved. The pleasurable sensations and physiological changes that characterize a female's orgasm are similar to those of the male, except for the absence of ejaculation.

PREGNANCY

If all of the reproductive and hormonal functions of both sexes are operational, the various sexual responses occur, and no contraceptive devices are used, an ovum may be fertilized in one of the female's uterine tubes (see the Focus on Scientific Process, "Contraception"). The resulting **zygote** (fertilized egg) now faces a precarious series of tasks that must be accomplished if it is to emerge nine months later as a normal human being. The time during which the zygote develops inside the female is referred to as **pregnancy.**

Fertilization

A human ovum can be fertilized only within 24 to 48 hours after ovulation, a limited window of opportunity for a sperm cell. Of the 300 to 400 million sperm deposited in the vagina during ejaculation, only a few hundred are able to make their way to the uterine tube, where the ovum is usually located. For fertilization to occur, one of the sperm must penetrate the membrane surrounding the egg and pass through, entering into its cytoplasm (see Figure 38.12A, page 728). Once a sperm has penetrated the egg's membrane, these next events happen quickly: (1) the ovum completes the second meiotic division and becomes a true haploid cell, (2) the nuclei of the sperm and egg (called pronuclei at this stage) combine to form one nucleus and reestablish the diploid number of chromosomes (see Figure 38.12B), and (3) the outer membranes of the fertilized egg undergo changes that prevent the entry of any additional sperm. The zygote begins to undergo cell division, development, and relocation and is now called an **embryo.**

Contraception

Contraception, or birth control, refers to any device or strategy that prevents fertilization of the ovum or growth of the zygote. According to this definition, *abstinence*—voluntarily avoiding sexual contact—qualifies as a contraceptive method. Although this is undoubtedly the most effective method of preventing pregnancy, many people do not consider it an acceptable option in making decisions about their sexuality. Hence through the course of history, humans have devised other methods for birth control. Contraceptive methods currently in use are summarized in Table 1 (see page 726). Note that of all the methods listed there, only condoms offer any significant protection against *sexually transmitted diseases (STDs),* which we discuss in Chapter 40.

Contraceptive methods depend on behavioral, mechanical, hormonal, or surgical maneuvers that can be categorized as follows: (1) preventing the sperm from contacting the egg in the female reproductive tract, (2) suppressing the production or release of eggs or sperm, and (3) preventing implantation. In the last category, an intrauterine device (IUD) was commonly used by women during the 1970s and 1980s to prevent implantation of the blastocyst. The IUD is a small spiral, ring, or loop made of various materials that requires insertion into the uterus by a physician. Many women suffered negative health effects from using the IUD (for example, increased menstrual bleeding, uterine infections, and allergic-type reactions to the IUD). As a result, it is no longer recommended as a method of contraception for young women who may wish to have children in the future.

Withdrawal *(coitus interruptus)* and the rhythm method are not reliable because they require behaviors that are not always practical during passionate moments. Furthermore, the effectiveness of the rhythm method depends on accurate prediction of ovulation, which can often be irregular, and that of withdrawal depends on nondeposition of sperm, which in fact seep out before completion of the sex act. Consequently, reliance on these methods is problematic if avoidance of pregnancy is a serious concern.

Condoms and diaphragms (used with a spermicide) are very effective barriers when used properly but may be useless if torn or defective. Another barrier device, the cervical cap, was approved for use by the U.S. Food and Drug Administration (FDA) in 1989. It is made of latex or plastic, fits over the cervix, and is used with a spermicide. Its advantages over the diaphragm are that it can be worn longer and it is tighter fitting and rarely leaks. Both the diaphragm and the cervical cap must be fitted by a physician.

Oral contraceptives that are designed to alter regulation of hormonal cycles are currently limited to use by women. The pill commonly contains estrogens or progesterone (or both). Pills are usually taken daily for 21 days, from the fifth to twenty-fifth days of the menstrual cycle, after which their use is discontinued for about seven days. During that time, menses (uterine bleeding) occurs, and then the menstrual cycle begins again. What is the basis of this hormonal strategy, which is currently used by over 8 million women in the United States? Essentially, the pill's hormones exert negative feedback to the hypothalamus, and perhaps the pituitary, that suppresses both LH and FSH secretions. Thus there is no LH–FSH surge, and follicle maturation and ovulation are prevented. Oral contraceptives have been linked to an increased risk in certain vascular (blood clotting) disorders and perhaps other side effects. However, in women under 35, the risk is about 25 percent or less than the risk of death due to complications of pregnancy and birth.

The two surgical methods, vasectomy in males and tubal ligation in females, are considered 100 percent effective, although a rare pregnancy may result in cases where the tubes are not completely separated or reattach spontaneously. The procedures have no effect on sexual desire or intercourse, and both are now considered relatively risk free. The major drawback is that they are usually irreversible. However, new surgical techniques have had modest success in restoring the severed ducts.

Another form of birth control is induced abortion, which entails removing the embryo from the uterus. Written records indicate that abortion has been used to terminate pregnancies for over 4,000 years. The use of this technique as a birth control measure is very controversial, even though in some parts of the world it is the predominant method of birth control. The moral issues surrounding abortion continue to be debated, and in the United States and other countries, no consensus is expected soon.

Several questions are relevant to individuals or couples in evaluating the various methods of contraception available. Is the method effective in preventing pregnancy? How effective? Is it safe for the individual using the method, or are risks involved? What are the risks? Is it reversible? Is it convenient, or does it interfere with the enjoyment and spontaneity of sexual activities for one or both partners? No existing method provides the desired answers for all these questions. Thus individuals must use personal criteria, balancing advantages and disadvantages against

box continues

Table 1 Methods of Contraception Most Commonly Used in Developed Countries

PREVENTION OF SPERM–EGG CONTACT

Method	Procedure	Effectiveness[a]	Advantages	Disadvantages	Protection Against STDs?[b]
Withdrawal	Male withdraws penis from vagina prior to ejaculation	75–80%	No costs; no side effects	Unnatural action required; ineffective	No
Rhythm	No intercourse for several days around time of ovulation	65–85%	No costs; no side effects; no action required during intercourse	Requires careful studies and record keeping to establish pattern of ovulation; periods of abstinence required; ineffective	No
Condom	Rubber or plastic sheath fitted over penis; sperm trapped, cannot enter vagina	85–90%	No side effects; simple to use; prevents spread of sexually transmitted disease; reversible	Moderately expensive; must be applied before intercourse; disrupts activities; somewhat ineffective	Yes[c]
Female condom	Polyurethane sheath with two rings—one that fits over the cervix and one that remains outisde the vagina, covering the external area; sperm trapped in sheath	85–90%	No side effects; relatively simple to use; reversible	Must be applied before intercourse; disrupts sexual activities, somewhat ineffective	Yes[c]
Diaphragm (with spermicide)	Plastic or rubber cup inserted to cover entrance to uterus; sperm unable to enter uterus; kills sperm	90–95	No side effects; reversible	Must be applied before intercourse; may disrupt sexual activities; must be inserted correctly	No
Cervical cap (with spermicide)	Small latex or plastic thimble-shaped cup that fits over the cervix; sperm unable to enter uterus; kills sperm	90–95%	No side effects; reversible; can be worn up to 48 hours; fits tightly and rarely leaks	Must be applied before intercourse; may disrupt sexual activities; must be inserted correctly	No
Vaginal spermicides alone	Kills most sperm	60–80%	No side effects; simple to use	Must be inserted just before intercourse; may disrupt sexual activities; ineffective	No
Vaginal douche	Kills or inhibits sperm	50–70%	Inexpensive; simple; can be done after intercourse	Must be done immediately following intercourse; ineffective	No

SUPPRESSION OF SPERM OR EGG PRODUCTION OR RELEASE

Method	Procedure	Effectiveness[a]	Advantages	Disadvantages	Protection Against STDs?[b]
Oral contraceptive (pill)	Prevents follicle maturation; suppresses ovulation	95–100%	Effective; reversible; simple	Expensive; increased risk of blood clots, heart-disease; must take daily; possible short-term effects including weight gain, nausea	No
Norplant	Prevents follicle maturation; suppresses ovulation	99+%	Effective; long-term; simple; reversible	Relatively expensive; must be implanted	No
Vasectomy	Vasa deferentia cut and tied	99–100%	Effective; simple	Permanent sterility	No
Tubal ligation	Uterine tubes cut and tied	99–100%	Effective	Permanent sterility	No

[a]The percentage of 100 sexually active women who did not become pregnant when using the method for one year.
[b]STDs = sexually transmitted diseases, including AIDS, herpes, syphilis, gonorrhea, and chlamydia (see Chapter 40).
[c]Condoms reduce but do not eliminate the risk of acquiring STDS.

FOCUS ON SCIENTIFIC PROCESS

the possibility of an unwanted pregnancy, in selecting an appropriate method for their own use.

As described in an important report from the National Academy of Sciences (NAS), the consensus among health officials is that the United States has remained in the Stone Age when it comes to the development and use of advanced birth control technologies. In countries throughout the world, new forms of contraception are continually being developed and used. For example, women in many European countries have long had access to a process in which a contraceptive (hormone-releasing) device is implanted under the skin of the upper arm. Not until 1991 was a similar under-skin implant system (Norplant) finally approved by the FDA for use in the United States. The implant method protects against pregnancy for up to five years. The French, British, and Chinese have RU-486, which is described in the next paragraph. The NAS report also stressed that more contraceptive choices for Americans would result in fewer abortions. Available data indicate that contraceptive failure now accounts for at least half of all abortions in the United States each year.

In 1986, a new birth control substance called *RU-486* (also known as *mitepristone*) was introduced in France. It was discovered in 1980, not from investigations related to birth control but rather through cancer research. Subsequent studies indicated that the drug might be effective in treating breast cancer, brain cancer, pituitary gland tumors, ovarian tumors, and infertility. The development and use of this compound as an abortifacient (abortion-inducing agent) dramatizes how advances in biomedical research can create new social problems. The standard treatment, which usually begins within ten days after a missed menstrual period, consists of a woman taking three RU-486 pills, followed by a small amount of progesterone 48 hours later. RU-486 binds to progesterone receptors of the uterus. Without progesterone, the endometrium deteriorates and, along with the embryo, is expelled. Because studies indicated that RU-486 was very effective, caused few side effects, and did not impair future fertility, its use was approved in France. Considerable controversy has erupted over its use in France and in other countries, but the product remains available in France, Great Britain, and China.

The primary social issue associated with the use of RU-486 is a moral one. In general, people who oppose abortion did not support the drug's approval for general use under any circumstances, while people who consider abortion an option for terminating pregnancy favored allowing access to RU-486 as quickly as possible. Both groups exerted considerable pressure on the product's manufacturer and also on researchers interested in studying potential beneficial uses of the drug. Consequently, the use of RU-486 in the United States for any purpose was mired in political controversy for several years, and prior to 1993, it was illegal to bring the product into the country, except for use in a limited number of studies. However, political changes in the United States and new research results from various studies conducted mostly in other countries led to a reconsideration of existing government policies. Two-year clinical trials, involving 4,000 women, were begun in the United States in 1994 to establish the safety of using RU-486 as an abortion pill. Medical researchers will also have access to RU-486 for use in various health studies. It remains to be seen whether or not the FDA will authorize the sale of RU-486 after the safety trials are completed.

Contraceptive technologies, old and new, will always be subject to moral discussion. And even among people who approve of them, their use will depend on knowledge of the available alternatives and how to gain access to them. The United States has the highest rate of unplanned pregnancies among industrialized countries of the West. It also lags behind those other countries in the dissemination of relevant information in a formal way, through the educational system.

Implantation

Figure 38.13 summarizes the events that occur between fertilization and implantation of the embryo in the endometrium. Within three to four days after fertilization, the developing embryo has migrated down the uterine tube and entered the uterus. By this time, it has developed into a microscopic hollow ball of cells called a **blastocyst.** One week after fertilization, the blastocyst becomes implanted in the endometrium and begins to secrete enzymes that digest endometrial cells. This serves two purposes: the blastocyst creates a "nest" for itself in the endometrium, and the digested cells provide nutrients for its continuing development.

The Endometrium During Pregnancy

Figure 38.14 (see page 729) summarizes hormonal actions during pregnancy. The endometrium provides nourishment to the developing embryo for about two months. Therefore, it cannot be allowed to degenerate as it normally does during the regular menstrual cycle. Can you hypothesize about how this might be accomplished, knowing that hormones play a key role? High levels of estrogens and progesterone are required to maintain the endometrium and also to suppress FSH and LH production, which prevents development of any new follicles during pregnancy. For the first few months, supplemental quantities of

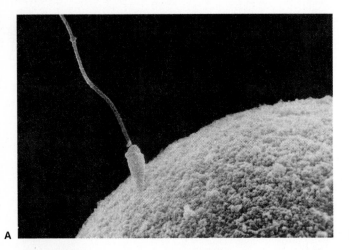

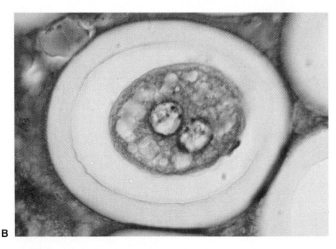

Figure 38.12 (A) This scanning electron micrograph shows a sperm penetrating the membrane of an ovum, the first step in fertilization. (B) The male and female pronuclei in this fertilized ovum are just about to fuse to form a zygote.

Question: *How many chromosomes does each pronucleus contain?*

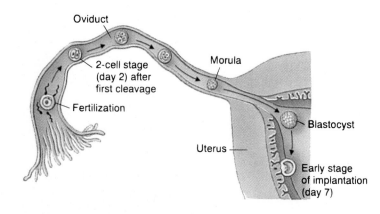

Figure 38.13 After fertilization, the zygote travels down the uterine tube while being transformed into a morula and then a blastocyst through a series of cell divisions. Seven days after fertilization, the blastocyst becomes implanted in the endometrium of the uterus.

estrogens and progesterone are secreted by the corpus luteum, a temporary endocrine gland that remains intact if pregnancy occurs. Two other unique, temporary tissues also secrete hormones that are critical during pregnancy. The **chorion** is the outer membrane of the embryo, and it secretes **human chorionic gonadotropin (hCG),** a hormone that helps sustain the corpus luteum during the first months of pregnancy. The presence of this hormone in urine can be detected and is used in simple tests to confirm pregnancy. After three months, the *placeuta* produces estrogens and progesterone until the end of pregnancy.

The Placenta

The **placenta** is a vascular organ that connects mother and fetus and secretes estrogens and progesterone throughout pregnancy (see Figure 38.15, page 730). (An embryo becomes a *fetus* after the second month of development.) It is a complex organ that serves as a lifeline for the developing fetus from the third month of pregnancy until birth. The placenta is a combination of maternal and fetal tissues. During placental forma-

tion, the chorion develops numerous fingerlike extensions called **chorionic villi.** As these villi develop, they erode limited areas of the endometrium, creating microscopic lagoons of maternal blood. Eventually, tiny arteries and veins from the fetus grow into the villi, each of which is surrounded by a pool of maternal blood. The structure containing fetal arteries and veins that connects the fetus with the placenta is called the **umbilical cord.** Blood vessels from the mother are also connected with these blood pools in the placenta, but they are not in direct contact with the blood vessels of the fetus. The pools serve as common reservoirs through which waste substances from the fetus, such as carbon dioxide, and nutrients and oxygen contained in the mother's blood can be exchanged.

During pregnancy, the placenta produces progesterone, which is important in maintaining the uterus, inhibiting GnRH secretion, and stimulating mammary development. The placenta also produces estrogen, but its synthesis and secretion are exceedingly complex, requiring substances from both the fetus and the mother. The full details about this process are not yet known.

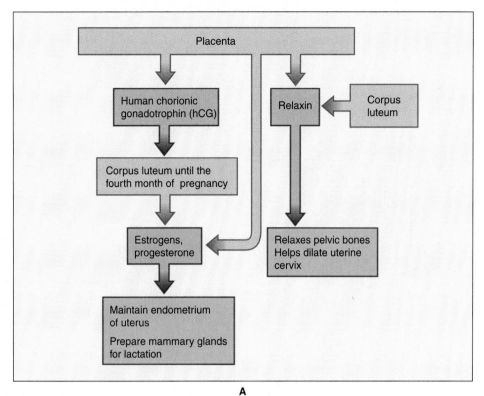

A

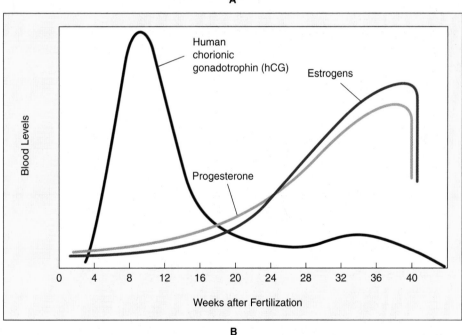

B

Figure 38.14 (A) Hormones have various functions during pregnancy. (B) High levels of estrogens and progesterone are necessary to maintain the endometrium and the placenta and to prevent the maturation of new follicles. Early in pregnancy, hCG helps maintain the corpus luteum, a source of the other two hormones. Relaxin affects certain organs just prior to childbirth.

Question: *Which tissues might be classified as transient endocrine glands?*

BEFORE YOU GO ON After fertilization takes place, the resulting zygote begins a series of transformations that ends nine months later when the final product, a baby, is born. Within one week, the zygote develops into a blastocyst that becomes implanted in the endometrium, where it will remain for two months and develop into an embryo. After two months, the fetus becomes connected with the placenta (and hence the mother) through the umbilical cord. Hormones control the development and maintenance of all these structures.

DEVELOPMENT

Development is a general concept that encompasses all irreversible changes that occur throughout an individual's lifetime. The major developmental phases in a human life are commonly classified as embryonic development, infancy, childhood, and adulthood, this last of which is a lengthy stage with few profound changes that ends with the death of the individual.

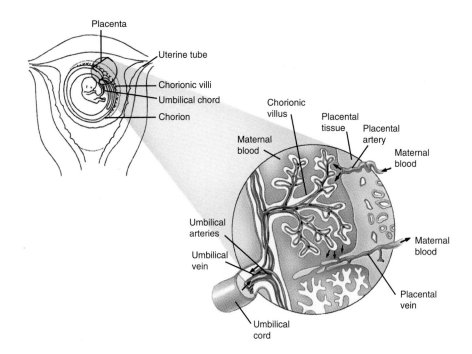

Figure 38.15 (A) The umbilical cord contains arteries and veins to and from the fetus and connects the fetus with the placenta. (B) In the placenta, nutrients, waste products, carbon dioxide, and oxygen are exchanged between blood in vessels from the mother and fetal vessels in the umbilical cord.

In this chapter, however, *development* specifically refers to the systematic increase in complexity that occurs from the time of fertilization until birth. How does the zygote, a single cell, give rise to a new human being consisting of about 200 billion highly organized cells within a period of only nine months? This question can be broken into three parts. First, what cellular events occur during development? Second, how are these events initiated and regulated at the molecular level? Third, how do cell aggregates form specialized tissues and organs? The answers to the first and third questions are quite well understood. Extensive, ongoing research is slowly enabling scientists to close in on the answers to the second question.

Embryonic Development

The first two months of development constitute the *embryonic period,* when the developing human is referred to as an **embryo.** Events that occur during this period are staggeringly complex. At the most fundamental level, a single-celled zygote undergoes cell divisions (by mitosis and cytokinesis), a process that eventually produces millions of cells. Through controlled—but poorly understood—organizational processes, these cells become the normal tissues and organs that make up a human being. There are three stages of embryonic development, referred to as *cleavage, gastrulation,* and *organogenesis.*

Cleavage

Cleavage is a series of rapid cell divisions that gives rise to a sequence of structures (see Figure 38.16). Cleavage begins almost immediately after the zygote is formed and continues during its passage down the uterine tube to the uterus. During cleavage, the embryo does not increase in size; rather, the cells produced simply do not grow after division. Thus after mitosis and cytokinesis, the resulting two daughter cells are half the size of the mother cell.

By day 4 after fertilization, a solid ball of cells called the **morula** is formed. As further divisions occur, additional cells accumulate in the morula. By the sixth day, some of these cells become rearranged, forming a hollow, fluid-filled *blastocyst* consisting of an **inner cell mass,** a cluster of cells at one end, and a single layer of surrounding cells called the **trophoblast.** After one week, the trophoblast of the developing embryo burrows into the wall of the uterus, where it ultimately forms the embryo's portion of the placenta.

Gastrulation

By the third week, **gastrulation** occurs, and the cells undergo a series of migrations that lead to a dramatic remodeling of the embryo, described in Figure 38.17 on page 732. The inner cell mass becomes the **embryonic disk,** which will form three layers of cells, the *ectoderm, mesoderm,* and *endoderm.* These cell layers are called the **primordial germ layers,** and during the next few weeks, each gives rise to a specific set of tissues and organs.

During this period, before the placenta has developed, four **extraembryonic membranes** form in the developing embryo. The membrane surrounding the cavity between the trophoblast and the inner cell mass is the fluid-filled **amnion.** This structure eventually surrounds the fetus, and the fluid it encloses (the "water" released during birth) cushions the developing fetus. The **yolk sac** emerges as a small cavity below the embryonic disk and eventually becomes incorporated into the

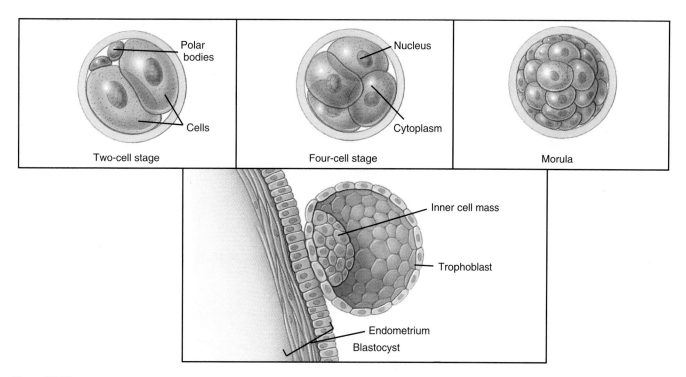

Figure 38.16   During cleavage, the embryo undergoes a series of cell divisions and transformations that ultimately lead to implantation in the endometrium.

Question: *What are the processes by which embryonic cells divide?*

umbilical cord. Despite its name, it provides no nutrients to the developing human embryo (although this is its major function in nonmammalian vertebrates). This membrane contains germ cells that will later migrate to the developing gonad and give rise to gametes. The **allantois,** a third embryonic membrane, is also found in the umbilical cord and may be involved in the transport of wastes from the embryo and oxygen and nutrients from the mother. The fourth membrane, the *chorion,* develops from the trophoblast and completely surrounds the embryo. As described previously, this membrane secretes hormones essential for maintaining the corpus luteum.

Organogenesis

The process through which primary organs are formed in the developing embryo is called **organogenesis.** During this period, which lasts for about five weeks, or until the embryo reaches its eighth week, cells of the primordial germ layers differentiate, becoming specialized cells that form major tissues and organs (see Figure 38.18, page 733). Perhaps most dramatically, the sex of the offspring is established about six or seven weeks after fertilization. In humans and other mammals, current evidence indicates that the embryo develops into a male due to the action of a gene called *SRY* located on the Y chromosome (see Chapter 19). This gene is thought to encode a protein that transforms sexually neutral embryonic tissues into the testes and other organs that characterize the male phe-

notype. If this protein is not present, the embryo follows an alternate, longer pathway that leads to the development of a female phenotype between the thirteenth and sixteenth weeks. By the sixteenth week, the brain, spinal cord, eyes, limbs, heart, and most other organs can be identified (see Figure 38.19, page 733).

Fetal Development

The months after the second month of pregnancy are called the *fetal period,* and the developing human is now referred to as a **fetus.** During this time, organs established during the embryonic period grow rapidly, and the fetus takes on a human appearance. By the fourth month, a mother can detect movements of the fetus. The respiratory and circulatory systems have appeared, although they will not be functional for another month or two, which severely limits the chances of survival if birth were to occur before the sixth month. By the sixth month, the fetus is about 5 centimeters long and weighs 700 grams. Scalp hair is visible, the fetus may suck its thumb, and the first bones (ribs) develop. By the seventh month, the fetus weighs around 1 kilogram. The nervous system has developed to the point where controlled breathing can occur if birth takes place, although the chances of survival would still be low. During the eighth and ninth months, all essential organs become fully developed, and the weight of the fetus doubles.

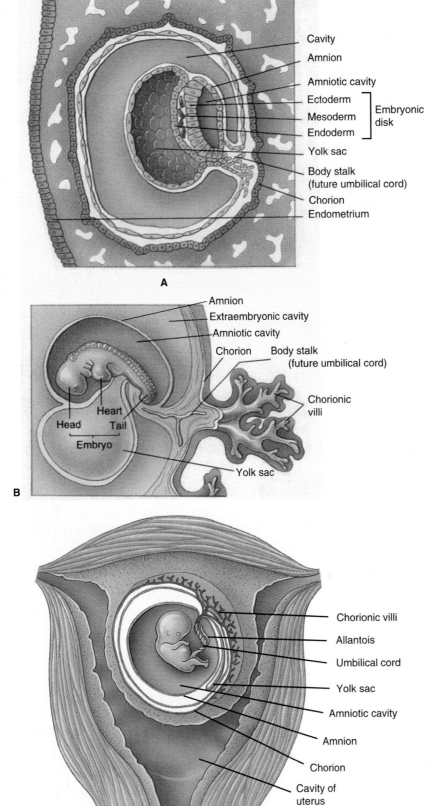

A

Cavity
Amnion
Amniotic cavity
Ectoderm   } Embryonic
Mesoderm   } disk
Endoderm
Yolk sac
Body stalk
(future umbilical cord)
Chorion
Endometrium

B

Amnion
Extraembryonic cavity
Amniotic cavity
Chorion   Body stalk
(future umbilical cord)
Head   Heart   Tail
Embryo
Chorionic
villi
Yolk sac

C

Chorionic villi
Allantois
Umbilical cord
Yolk sac
Amniotic cavity
Amnion
Chorion
Cavity of
uterus

Figure 38.17   (A) By the third week of development, gastrulation has taken place, and three primordial germ layers—the ectoderm, endoderm, and mesoderm—have originated in the embryo. Each of these three germ layers will give rise to specific tissues and organs as organogenesis continues during the next few weeks. (B) The developing embryo looks animal-like three to four weeks after fertilization. (C) Four extraembryonic membranes—the amnion, yolk sac, allantois, and chorion—develop during gastrulation.

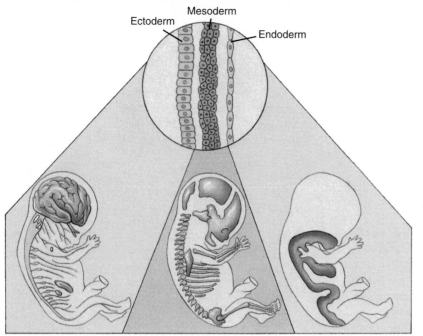

Figure 38.18 The three embryonic cell layers—ectoderm, mesoderm, and endoderm—each give rise to different tissues and organs during development.

Question: *When during pregnancy does this tissue and organ development occur?*

ECTODERM:
Skin, sweat glands, hair and nails; the entire nervous system, some endocrine gland cells, tooth enamel, lining of nose, mouth, and anus

MESODERM:
Bones, muscles, blood, blood vessels, connective tissue, reproductive organs, and kidneys

ENDODERM:
Lining of digestive and breathing passages, liver, pancreas, most endocrine glands, bladder, and urethra

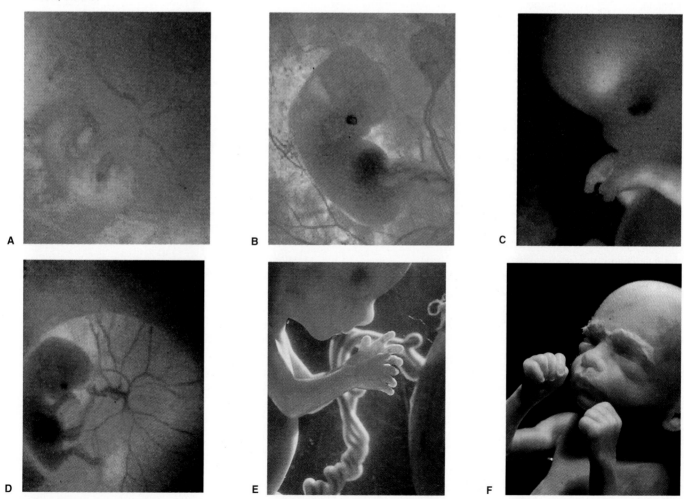

Figure 38.19 The sequence of human development during the first six months of pregnancy, showing an embryo 4 weeks after fertilization (A) and at 6–7 weeks (B) and a fetus at 10 weeks (C), 12 weeks (D), 4 months (E), and 6 months (F).

BEFORE YOU GO ON Embryonic development occurs during the first three months of life. In the first three weeks, the fertilized egg undergoes cleavage. Gastrulation then occurs, characterized by the formation of ectoderm, mesoderm, and endoderm, the three cell layers that ultimately give rise to all tissues and organs by the third month of pregnancy. During the last six months of pregnancy, all fetal organs undergo rapid growth.

BIRTH

Parturition, the process by which the fetus is ousted from the uterus about 265 days after fertilization, is a complicated process. As in other reproductive events, a number of hormones influence this dramatic separation of the fetus from the mother.

The onset of **labor,** the period when strong uterine contractions ultimately lead to expulsion of the fetus from the uterus, is thought to be triggered by three hormones. The fetus initiates the "contraction cascade" by releasing the hormone *oxytocin,* which travels through the umbilical cord and binds to receptors on the placenta and uterus. Oxytocin induces the placenta to produce prostaglandins that act to stimulate muscle contractions in the uterus. As uterine contractions continue, moving the fetus down the birth canal, nerve impulses are transmitted from the uterus to the hypothalamus. These impulses lead to the release of more oxytocin, this time from the mother's pituitary gland, which further increases uterine contractions. The cycle continues to escalate until the fetus is finally ejected from the birth canal. This cyclical activity, involving oxytocin, uterine contractions, and hypothalamus–pituitary stimulation, is a rare example of a positive feedback system.

Before birth, **relaxin,** a hormone whose origin is unclear, acts to increase the flexibility of joints between bones around the mother's birth canal. This allows enlargement of the birth canal and enables the baby to pass through, usually head first. Because relaxin is present in the blood as early as the fourth week of pregnancy, it has been hypothesized that relaxin is produced by the corpus luteum and then stored within certain blood cells until release near the time of parturition.

At birth, several changes take place in the fetus that enable it to make a successful transition from living in an aquatic environment to becoming an air-breathing animal. Most notably, blood enters capillaries in the newborn infant's lungs and causes them to expand and begin functioning (confirmed by a crying response). A few minutes after delivery, the uterus contracts, and the placenta separates from the uterine wall and is expelled as the **afterbirth.** The uterus and its membranes then undergo repair, and if the mother does not nurse the child, menstrual cycles usually resume after six weeks. Nursing inhibits resumption of the menstrual cycle through the actions of *prolactin,* a pituitary hormone secreted during pregnancy that has diverse reproductive roles. This hormone stimulates mammary gland growth and development during pregnancy, and after birth, it induces **lactation,** the secretion of milk. In addition, it inhibits release of LH and probably FSH, which in turn leads to repression of ovulation and the menstrual cycle. In certain cultures, women continue to nurse infants for two years or longer, a practice that tends to spread the ages of siblings, especially if no other form of birth control is used. However, this natural method of birth control is unreliable because the length of time that the menstrual cycle remains repressed varies greatly among individuals.

BEFORE YOU GO ON Parturition begins when oxytocin and prostaglandins initiate uterine contractions, which first cause the fetus to move into the birth canal and then force its entry into the outside world. After birth, the placenta is expelled, and the uterus and other maternal tissues are restored. The ovarian and menstrual cycles begin six weeks later if the mother does not nurse the child. Otherwise, they will resume soon after nursing is discontinued.

CELLS AND DEVELOPMENT

As in many fields of the life sciences, the cardinal problems of developmental biology are now being successfully investigated using modern molecular methods. To date, much of what has been learned has come from studies of animal models including fruit flies *(Drosophila),* frogs, and rats. For much of the past century, the central question has been, How does a single-celled zygote give rise to an organism that consists of billions of cells, each with distinct capabilities, which are organized perfectly into different tissues and organs?

Relatively little is known about the development of **cell specificity,** or exactly how a cell ultimately becomes specialized to perform a precise activity or function. The processes by which a single cell, the fertilized egg, ultimately produces an entire organism by mitosis and cell specialization are understood only in outline form. For example, genes that determine the course of development of certain groups of cells have been identified in some animals. The pathways of embryonic development, including the genes responsible for creating an adult organism, have been determined in *Drosophila.* However, important questions about molecular events during embryonic development remain open. How are signals transmitted between embryonic cells during development? How do genes or gene products control the development of cells with defined sizes, shapes, and capabilities?

At the other end of the life span, why do cells ultimately fail, causing death? What causes aging? Some evidence supports a *genetic clock* hypothesis, the view that all species contain unique genetic instructions that dictate its course of aging. Every carefully studied species has been found to have a defined average life span, and all appear to follow an unalterable, programmed timetable that leads to death. For example, mice normally live about 2 years, beagle dogs 10 to 15 years, humans 75 to 100 years, and Galápagos turtles perhaps hundreds of years. Why does the effectiveness of our tissues and organs begin to decline at about age 30? During the autumn of our lives, do "aging genes" become activated? Do they produce a protein, an aging substance, that disrupts cell function or prevents cell division? Is normal cell division affected by some other, age-related process? Cell development and mechanisms of aging are fascinating, challenging fields that will continue to occupy scientists well into the next century.

SUMMARY

1. Humans produce offspring by sexual reproduction. Distinguishing features of human reproduction include internal fertilization, year-round sexual receptivity, unique behaviors, and complex endocrine mechanisms that are responsible for sexual maturity and the regulation of gamete production.

2. Testes are the primary sex organ of males. Once a male becomes sexually mature, the testes continuously produce sperm and androgens (hormones). Testosterone, the most important androgen, initiates the development of male secondary sex characteristics and sustains continuous production of sperm. Spermatogenesis is regulated by releasing factors (hormones) from the hypothalamus and hormones from the pituitary gland.

3. Ovaries are the primary sex organs of females. At puberty, the ovaries begin releasing mature ova and various sex hormones that influence the development of secondary sex characteristics. Gamete production in females is characterized by monthly cycles. Ovulation occurs about once each month and is regulated by hormones secreted by the ovary and numerous releasing factors (hormones) from the hypothalamus and hormones from the pituitary gland.

4. If the egg is fertilized and pregnancy occurs, the resulting embryo becomes implanted in the endometrium. By the third month of pregnancy, the fetus becomes linked with the mother via the placenta. From the third month of pregnancy until birth, various substances are exchanged between the mother and the fetus through the placenta. The placenta is maintained by the actions of various hormones.

5. During the first three months of pregnancy, major organs arise in the developing embryo through the processes of cleavage, gastrulation, and organogenesis. During the last six months, organs undergo rapid growth and become part of fully functional organ systems.

6. Pregnancy ends about nine months after fertilization. Parturition, and associated events such as labor, are controlled and coordinated by various hormones.

WORKING VOCABULARY

androgens (p. 715)
cleavage (p. 730)
embryo (p. 724)
estrogens (p. 715)
gastrulation (p. 730)
menopause (p. 721)
menstrual cycle (p. 719)
menstruation (p. 720)
oogenesis (p. 720)

organogenesis (p. 731)
ova (p. 715)
ovarian cycle (p. 719)
ovaries (p. 715)
ovulation (p. 720)
placenta (p. 728)
spermatogenesis (p. 715)
testosterone (p. 715)

REVIEW QUESTIONS

1. What are the significant features of human reproduction?
2. What are the major organs of the male reproductive system?
3. How are sperm produced in the male?
4. How is the male reproductive system regulated by hormones?
5. What are the major organs of the female reproductive system?
6. How are mature ova produced by the female?
7. Describe the major events in the ovarian and menstrual cycles. How are these two cycles regulated by hormones?
8. List the events that occur between fertilization and the embryo's becoming connected with the uterus.
9. What occurs during the following stages of embryonic development: cleavage, gastrulation, and organogenesis?
10. Which hormones are involved in parturition? What are their functions?
11. What questions are now guiding research in developmental biology?

ESSAY AND DISCUSSION QUESTIONS

1. Women often stop menstruating when placed under great stress or when not receiving adequate amounts of food. What might be the adaptive significance of this response?
2. In certain cultures (ancient and modern), people do not believe that sexual intercourse is responsible for producing children. Can you identify any biological facts that might account for such a belief?
3. If you could redesign the human body, what would you do to improve the human reproductive system?
4. Which human type would probably be most harmed by being exposed to mutagens (chemicals or physical agents that cause mutations): adult male, adult female, embryo, or fetus? The least sensitive? Why?

REFERENCES AND RECOMMENDED READING

Baulieu, E.-E. 1995. Foundations and principles of human reproduction. *Perspectives in Biology and Medicine,* 38: 640–658.

Bayertz, K. 1995. *GenEthics: Technological Intervention in Human Reproduction as a Philosophical Problem.* New York: Cambridge University Press.

Bogan, J. S., and D. C. Page. 1994. Ovary? Testis? A mammalian dilemma. *Cell,* 76: 603–607.

Carlson, B. M. 1994. *Human Embryology and Developmental Biology.* St. Louis: Mosby.

Crenshaw, T. L. 1996. *The Alchemy of Love and Lust: How Hormones Dictate Our Sexuality.* New York: Putnam.

Crews, D. 1994. Animal sexuality. *Scientific American,* 270: 108–116.

Editors of *Science.* 1994. Frontiers in biology: Development. *Science,* 226: 56–603.

Galton Institute. 1995. *Human Reproductive Decisions: Biological and Social Perspectives.* New York: St. Martin's Press.

McGinnis, W., and M. Kuziora. 1994. The molecular architects of body design. *Scientific American,* 270: 58–67.

National Academy of Sciences. 1990. *Developing New Contraceptives: Obstacles and Opportunities.* Washington, D.C.: NAS Press.

National Institutes of Health. 1994. Special Section: Sex and reproduction. *Journal of NIH Research,* 6: 10–63.

Pollard, I. 1994. *A Guide to Reproduction: Social and Human Concerns.* Cambridge: Cambridge University Press.

Rowe, J. W., and R. L. Kahn. 1987. Human aging: Usual and successful. *Science,* 238: 143–149.

Spira, A. 1994. Contraception by the end of the 20th century. *Human Reproduction,* 9: 1–125.

Yu, H. S. 1994. *Human Reproductive Biology.* Boca Raton, Fla.: CRC Press.

ANSWERS TO FIGURE QUESTIONS

Figure 38.2   It begins at puberty, about 14 years of age, and continues throughout life.

Figure 38.3   Diploid = 46; haploid = 23.

Figure 38.5   They secrete a fluid that contains fructose, prostaglandins, and substances that help neutralize the acid environment within the female reproductive tract.

Figure 38.6   Increased GnRH concentrations.

Figure 38.7   It is the site of menstruation and, if pregnancy occurs, embryonic and fetal development.

Figure 38.8   During her embryonic development.

Figure 38.10   Around age 50, when the ovaries no longer respond to FSH and LH.

Figure 38.11   It inhibits further FSH and LH secretions after ovulation.

Figure 38.12   23.

Figure 38.14   The corpus luteum and the placenta.

Figure 38.16   Mitosis and cytokinesis.

Figure 38.18   From the fourth through the eighth week after fertilization.

39

Human Defense Systems

Chapter Outline

Reading Questions

1. How do nonspecific defense mechanisms prevent infection and respond to injury?

2. What is the significance of self and nonself in immunity?

3. What is the function of cellular and humoral defense responses? When do they occur? How do they operate?

4. What problems associated with the immune system are of current interest to researchers?

Throughout their lives, all organisms are constantly threatened by biological, chemical, and physical agents from the surrounding environment. Many of these potentially harmful agents are called *microbes,* a term that commonly includes viruses, bacteria, protozoans, and fungi. To survive, or complete their life cycle, microbes must colonize an environment that provides them with nutrition and a suitable place to grow and multiply. The cells and tissues in a multicellular organism's body provide a nearly ideal habitat in which many microbes and parasitic worms can thrive. In this chapter, we examine the human defense systems that usually keep us safe from such unwelcome guests.

DEFENSE SYSTEMS

Human defense systems are separated into two general categories that are summarized in Table 39.1 (see page 738) and

Table 39.1 Human Defense Systems

Classification of System	Components	Functions
Nonspecific Defense Mechanisms		
Body surfaces	Skin	Barrier to invasion by external agents
	Sweat and sebaceous glands	Secrete substances toxic to many bacteria
	Mucous membranes	Barrier to invasion by external agents
	Secreted products	Mucus traps small particles; tears in eyes contain an enzyme with antimicrobial activity; gastric juice in stomach destroys bacteria and bacterial toxins
Inflammation	Leukocytes:	
	Basophils (mast cells)	Phagocytosis of foreign materials; release histamine
	Eosinophils	Phagocytosis; important in parasite infections and allergic responses
	Macrophages (monocytes)	Phagocytosis; release pyrogens
	Neutrophils	Phagocytosis; produce factors that attract macrophages; release enzymes that digest foreign materials
	Natural killer cells	Kill nonself cells or modified self cells
	Chemical mediators:	
	Histamine	Causes dilation of blood vessels
	Complement system	Mediate events in inflammation
	Chemical factors:	
	Pyrogens	Increase body temperature
Specific Immune Responses		
Cellular Immunity	Macrophages	Initate immune responses
	T lymphocytes:	
	Helper T cells	Assist in activating various T and B cells
	Cytotoxic T cells	Kill infected self cells
	Suppressor T cells	Diminish an immune response
	Memory T cells	Possess "memory" of a specific antigen
	Chemical mediators:	
	Interleukin-1	Attract and activate T cells
	Interleukin-2	Stimulate proliferation of T_H cells
Humoral Immunity	B lymphocytes:	
	Plasma cells	Produce antibodies
	Memory B cells	Possess "memory" of a specific antigen
	T lymphocytes:	
	Helper T cells	Activate naive B cells
	Suppressor T cells	Diminish a humoral immune respone
	Antibodies	Molecules that react with antigens
	Chemical mediators:	
	Interleukin-4	Activate B cells
	Interleukin-5	Stimulate activated B cells to proliferate
	Interleukin-6	Cause B cells to mature into plasma cells

Figure 39.1. **Nonspecific defense mechanisms** always operate in the same way when presented with a challenge. They include mechanical barriers, secreted products, and the inflammation response. In contrast, **specific immune responses** involve a complicated adaptive system composed of various cells, cell products, and control mechanisms; it is characterized by specificity, self-recognition, and memory. The specific immune system has great flexibility and can respond in a specific way to each unique provocateur that enters the body.

The immune system is fascinating to explore. It embodies intricacies that offer a unique perspective on the development and success of an advanced system that evolved in higher organisms in response to natural selection pressures. Of all animals, only birds and mammals have the elaborate specific immune system described in this chapter.

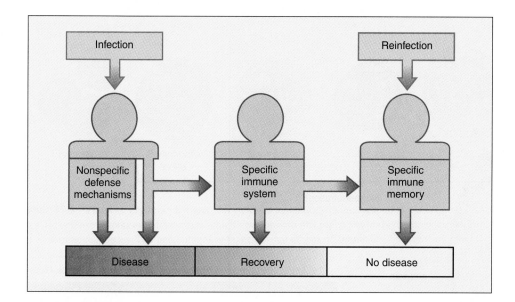

Figure 39.1 Nonspecific defense mechanisms generally prevent infections by blocking the entry of disease-causing agents into the body. If this first line of defense fails, disease results; the specific immune system is then activated and begins to respond about five days after the original infection. Specific immune responses ultimately eliminate foreign invaders, and recovery occurs. If the same agent reinfects the body, an immune memory process ensures that it will not cause disease a second time.

Question: *Which defense system do you think evolved first? Why?*

BEFORE YOU GO ON Humans have nonspecific defense mechanisms that are inborn and conventional, reacting in the same way to all initiating agents, and specific immune responses that can create a customized reply to each challenge.

NONSPECIFIC DEFENSE MECHANISMS

Figure 39.2 (see page 740) identifies the major nonspecific defense mechanisms. Most materials (microbes, chemicals, and physical agents such as splinters) that pose a threat to an individual's internal homeostatic systems gain access to the body from the external environment.

Body Surfaces

Routes of entry by which potentially dangerous substances may pass from the external to an internal environment include penetration or absorption through the skin, eyes, or ears and the digestive, respiratory, urinary, and reproductive systems. Our first line of defense consists of organs, tissues, and cell products that form barriers and prevent or limit access at these portals.

Skin and Mucous Membranes

The basic structure of the skin was described in Chapter 34. The skin is a complex organ that covers the external surface of the body. The outer epithelial layer of the skin, the epidermis, is a formidable barrier to the entrance of microbes. *Keratin,* the tough protein found in the skin's outer cell layers, is resistant to water, some chemicals, and enzymes secreted by certain bacteria. Even though it is subjected to considerable abuse (for example, exposure to sun, soaps, and cosmetics) and is frequently damaged, it has a remarkable ability to repair itself. Because of this property, it generally remains intact and functions as a highly effective, dependable, defense barrier.

Like the skin, **mucous membranes** are composed of an outer epithelial cell layer and an underlying connective tissue layer. Many of the epithelial cells secrete **mucus,** a substance that covers the surfaces of mucous membranes. Mucous membranes line body cavities and the surface of organs or tissues that open to the external environment. The digestive, reproductive, respiratory, and urinary systems and the eyes are all lined with these moist membranes. Like the skin, they form a barrier between the external and internal environments. The mucous membrane of the nose has mucus-coated *hairs* that filter and trap pollutants from the air. The mucous membrane of the upper respiratory tract contains *cilia,* microscopic hairlike projections of the epithelial cells that propel inhaled particles trapped in mucus toward the throat for expulsion; coughing and sneezing accelerate movement of this mucus.

Secreted Products

The skin contains glands that secrete substances that play a role in defense. *Sebaceous glands* secrete an oily substance *(sebum)* that prevents hair and skin from drying and is also toxic to many types of bacteria. *Sweat glands,* in addition to producing sweat, also secrete antimicrobial substances such as **lysozyme,** an enzyme that can break down the outer cell wall of many bacterial species.

Mucus entraps particles that can then be swept away by cilia or removed by white blood cells capable of engulfing them. Saliva rinses foreign materials from surfaces in the mouth and also contains antimicrobial substances. Enzymes (lysozyme) are present in tears, which bathe the eyes, and in the gastric juices of the digestive tract; they destroy millions of bacteria every day of our lives.

Despite the remarkable effectiveness of the skin, mucous membranes, and special substances in their defense roles, they are often breached by unwanted materials. What protects us when that happens?

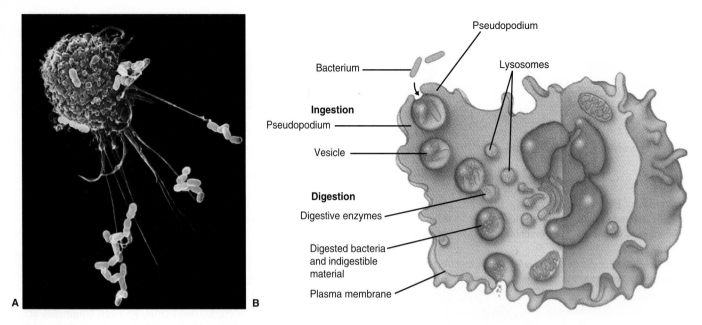

Figure 39.5   (A) A macrophage's capture of a bacterium (green) is the first step in phagocytosis. (B) Once ingested after capture, the bacterium is enclosed within a membrane vesicle that fuses with a lysosome. The lysosome releases enzymes into the vesicle that digest the bacterium but not the vesicle. The remaining material may then be released if the vesicle empties its contents to the exterior.

Question: *Which other types of blood cells can carry out phagocytosis?*

2. Fluid accumulates at the site of injury (swelling).

3. Neutrophils invade the affected site (pain).

4. Lymphocytes and macrophages infiltrate the area.

5. The normal architecture of the affected tissue is restored (wound repair), if possible.

All of the inflammatory processes are controlled by interrelated feedback systems. The magnitude of the responses varies according to the severity of the trauma. Hence only increased blood flow may occur, for example, if the skin is lightly scratched, or the entire inflammatory response may occur if the damage is serious.

The first three events on the list are part of the **acute inflammation response.** There are two key features of acute inflammation: it occurs in exactly the same way in all humans, no matter what the cause, and it happens quickly, beginning within seconds of the initiating event. Although numerous chemical and cellular activities are involved in inflammation, our discussion will focus on the major events of an inflammatory response.

The Inflammatory Response: Minutes to Hours

Immediately following injury, blood vessels in the surrounding area enlarge, and their permeability increases (see Figure 39.6A). As a result, blood cells flow quickly into the affected area.

When a serious injury occurs, several mechanisms, collectively called the **coagulation system,** are activated by chemical mediators to arrest bleeding. **Chemical mediators** are small molecules that either attract other molecules or cells to a site or signal them to perform some function. Platelets arrive on the scene and begin to plug leaks in small vessels of the vascular system. Blood **clotting** is a very complex process that requires many proteins and other factors present in the plasma. The most important mechanism involves two proteins; the first, **fibrinogen,** is converted into the second, **fibrin.** Fibrin, a threadlike protein, creates a network that entraps erythrocytes, platelets, and other materials to create a blood clot. Formation of a blood clot, shown in Figure 39.6B, halts bleeding and walls off the damaged area. Later, the fibrin complex is dissolved by enzymes so that the area can be repaired.

Hours to Days

Increased blood flow results in sharp increases in the delivery of blood cells, plasma, fluids, and other circulating substances to the site of injury. Also, damaged and dead cells release their contents into this mixture. The **complement system** consists of a set of blood proteins—chemical mediators—that function during acute inflammation and in various defense reactions. They help in breaking down invading organisms and dead cells, dilating blood vessels, attracting various leukocytes, and enhancing phagocytosis by leukocytes.

Neutrophils are generally the first leukocytes to migrate into damaged tissues. They are attracted by chemical mediators released from basophils at the site and also by substances released from dead or dying tissue cells or bacterial cells. Neutrophils phagocytose foreign substances and the remains of dead cells. In so doing, they also release enzymes that degrade

Tissue injury

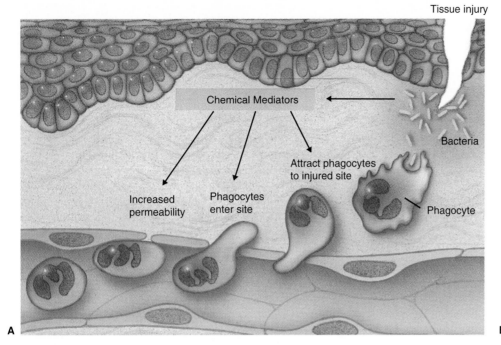

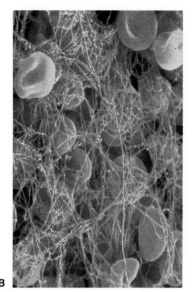

A   B

Figure 39.6 (A) Immediately after injury, affected tissues and bacteria, if present, release chemical mediators that increase blood flow to the area and induce phagocytic cells to migrate to the damaged site. First neutrophils and later macrophages phagocytose bacteria and cell debris and prepare the site for repair. (B) If necessary, early in the inflammatory process, a fibrin network (pink threads) is thrown up to trap red blood cells to form a clot that prevents further loss of blood.

dead cells or tissues. Neutrophils carry out these processes by the thousands or even millions, and many are killed or incapacitated as a result.

Macrophages are large phagocytic cells that are critically important in inflammation and also in specific immune responses. Because of their size, they are slower-moving than the smaller neutrophils and arrive at the inflammatory site a few hours later, attracted by various chemical mediators. Macrophages (and some other inflammatory cells) release substances called **pyrogens,** which cause *fever,* or elevation of the body temperature. What does this accomplish? The elevated temperature increases the efficiency of phagocytic cells and also retards multiplication of certain viruses or bacteria that may be present. Macrophages put the finishing touches on an acute inflammatory response. They phagocytose dead cells and tissue fragments, eliminate any remaining microbes, and clear the site of other inflammatory products.

Days to Weeks

The goal of an acute inflammatory response is to limit injury, prevent invasion by microorganisms, clear the inflamed area of debris, and allow the tissue to return its normal state (**resolution**). Sometimes, however, tissue damage is too extensive, and the inflammatory response must be extended in time (become **chronic**) in order to repair the tissue. When resolution cannot be accomplished, a wound repair response will occur, with damaged tissues being replaced by collagen, the tough, fibrous protein produced by fibroblasts. This results in the formation of a scar and loss of function in the tissue replaced.

BEFORE YOU GO ON Inflammation is a nonspecific defense mechanism that occurs when cells and tissues are damaged. During an acute inflammatory response, blood flow into the injured site increases, the coagulation and complement systems are activated, and neutrophils and macrophages carry out phagocytosis and perform other functions that lead to resolution. In cases of extensive injury, damaged tissues are replaced by collagen, forming a scar.

SPECIFIC IMMUNE RESPONSES

All animals, from the simplest invertebrates to the most complicated mammals, use inflammation and wound repair responses in maintaining the integrity of their bodies when injury or infection occurs. However, phagocytosis, the primary antimicrobial response of inflammation, is relatively sluggish and inefficient. A highly sophisticated protective system evolved in birds and mammals that is much more effective in the continual campaign to prevent being successfully colonized by microbes and multicellular parasites. Our understanding of specific immune responses is derived largely from studies conducted during the past 20 years. However, many questions about immune system functions remain unanswered.

Overview

The **immune system** consists of organs and several different classes of lymphocytes and other leukocytes that work together to inactivate or destroy foreign organisms or substances. Where do infectious agents live after they enter the body? Generally, infective microbes are either *intracellular* (living inside cells) or *extracellular* (living in areas outside cells). Bacteria and protozoan disease agents cause both intracellular and extracellular infections; viruses are always intracellular. **Antigens** are substances that activate lymphocytes and induce the formation of specialized protein molecules called *antibodies* as part of complex responses that lead to *immunity*. **Immunity** is the state of being able to resist a particular disease-causing agent or the substances that such an agent may produce. **Immunology** is the study of the immune system, immune responses, and immunity. Table 39.2 defines terms used frequently in immunology.

A critical feature of immune responses is that the immune system is able to recognize *self* molecules and does not normally react against them. Thus the goal of an immune response is to recognize and eliminate any foreign, or *nonself*, substance (antigen) present in the body. The critical feature of modern immunology is that immune cells are able to communicate with one another through chemical mediators and specific cell-surface **receptors** and self molecules.

When challenged by an antigen, the immune system responds in two distinct but interrelated ways. **Cellular immunity** involves the actions of several classes of *T lymphocytes (T cells),* which participate in reactions directed toward the destruction of nonself cells or self cells that are infected with microbes. **Humoral immunity** consists of responses characterized by the formation of antibodies against extracellular nonself antigens. Antibodies are manufactured by *plasma cells,* which are derived from *B lymphocytes (B cells).* A key feature of antibodies is their **specificity.** Each antibody can recognize and bind to only one type of antigen; however, the immune system is capable of generating millions of different antibodies, each with a different specificity.

Where are B and T cell populations found in the body? Several organs, collectively referred to as **lymphoid organs,** contain dense populations of lymphocytes and have critical functions relative to the immune system (see Figure 39.7). **Lymph nodes** are complex structures containing specific tissue territories composed of lymphocytes, with a few phagocytic leukocytes mixed in. They are found in areas where lymph drains in the body and act as filters for fluids in lymphatic vessels. Lymph nodes are the sites where most activities of the immune response actually occur, and each gram of tissue contains about 1 billion lymphocytes. The **spleen** filters circulating blood and also contains populations of T cells, B cells, and macrophages. It is especially important in clearing the blood of harmful microbes. The **thymus gland** is involved in the production and maturation of T cells. Local collections of diffuse, unencapsulated **lymphoid tissues** are found in the appendix and in tissues underlying the gastrointestinal tract, the airways leading to the lung (tonsils), and the urinary and reproductive tracts. Lymphocytes also circulate throughout the body, which increases the likelihood that they may contact any antigen that may be present.

BEFORE YOU GO ON Specific immunity involves a set of cellular and humoral responses to an antigen. Underlying immune system actions is an ability to distinguish between self and nonself substances. Cellular immunity consists of coordinated responses by T cell populations that are directed at destroying nonself cells or infected self cells. Humoral immunity is based on plasma cells producing antibodies that react against extracellular antigens. Most T and B cells are found in lymphoid organs and tissues.

Table 39.2 Terms Used in Immunology

Term	Definition
Antibodies	Specialized protein molecules (immunoglobulins) produced by plasma cells in response to exposure to an antigen
Antigen	A substance that includes a specific immune response
Immunity	State of protection from a foreign agent or organism, due to previous exposure
Cellular immunity	Reactions mediated by populations of lymphoid cells called T lymphocytes
Humoral immunity	Reactions mediated by antibodies produced by plasma cells, a type of B lymphocyte
Lymphocytes	A population of leukocytes with similar appearance but different immune functions
T cells	Thymus-derived lymphocytes consisting of several cell subsets that participate in cellular immune responses
B cells	Lymphocytes that develop into plasma cells that produce antibodies or into memory cells; associated with humoral immune responses
Major Histocompatibility Complex (MHC) molecules	Self molecules that are unique for each individual's cells
MHC-I molecules	MHC molecules embedded in the plasma membranes of most cells; important in helping activate T_C cells
MHC-II molecules	MHC molecules found primarily on macrophages and B cells; important in activating T_H cells and B cells
Self/nonself concept	
Self	A person's own cells as determined by the presence of specific MHC surface molecules
Nonself	Foreign antigens, or cells whose MHC molecules differ from those of self cells

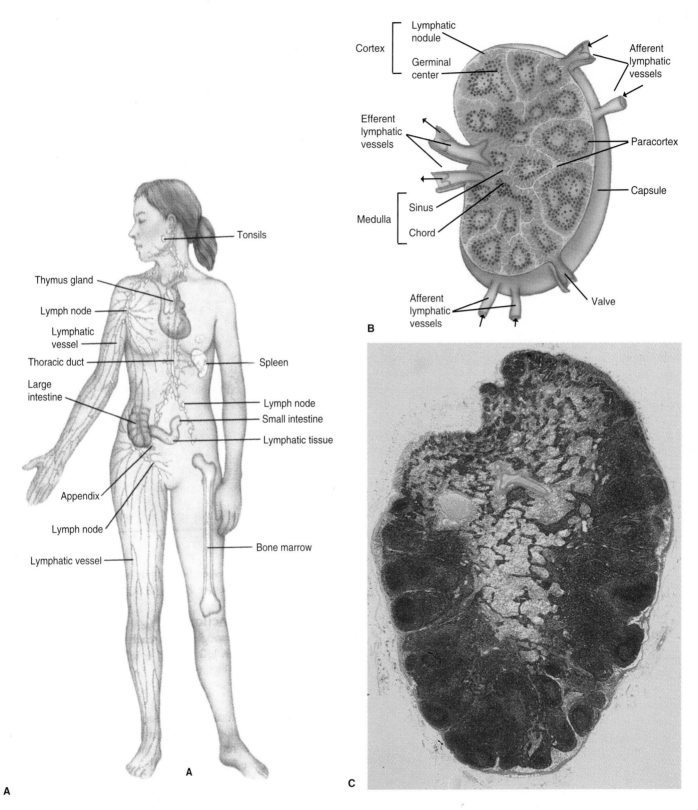

Figure 39.7 (A) The human lymphoid system consists of various organs and tissues in which immune reactions occur. (B) Each lymph node is enclosed by a thick collagen capsule and contains clusters of two types of specialized functional regions, the cortex and the medulla. Each outermost cortex cluster contains a germinal center where B lymphocytes proliferate and a lymphatic nodule that contains T lymphocytes and macrophages. Inner medulla clusters each consist of a sinus lined by phagocytic cells that remove and process any foreign materials present in lymph entering from the afferent vessels and a cord area that contains antibody-secreting plasma cells. Lymph and lymphocytes leave the lymph node through efferent vessels. (C) This photomicrograph shows a cross section of a lymph node.

Antigens

Because most scientists believe that the immune system evolved in response to selection pressures exerted by **pathogens** (organisms that cause disease), antigens are classically considered to consist of proteins and polysaccharides that are recognized as nonself. In general, an immune response is directed at a limited region of an antigen called the **antigenic determinant,** rather than at the entire antigen (see Figure 39.8A). A specific pathogen may have many different antigenic determinants on its surface, but individual antibody molecules can recognize only one.

The immune system can also react against self cells that come to possess antigens on their surface that are recognized as nonself. Certain cancer cells, for example, may be killed by the immune system if they have nonself ("tumor") antigens on their surface. If the immune system fails to recognize and kill such cells, a cancer may then develop.

Antibodies

Antibodies are types of specialized proteins, known as **immunoglobulins,** that have the capacity to bind specifically to an antigen or antigenic determinant (see Figure 39.8B). Antibodies may be either attached to the surfaces of B cells, where they act as receptors, or unattached when circulating in the blood, lymph, and tissue fluids. An antibody is able to combine with a specific antigen in lock-and-key fashion to form an **antigen–antibody complex** (see Figure 39.8C). Once bound to antibodies, antigens are inactivated or targeted for removal by phagocytic cells, such as macrophages.

Antibody molecules have a fundamental structure that is described in Figure 39.9. They are Y- or T-shaped and consist of two identical long protein chains, called **heavy (H) chains,** and two identical shorter protein chains, called **light (L) chains.** The amino acid sequence of much of the H and L chains does not vary, and these parts of the antibody chains are referred to as the **constant (C) region.** At the end of each chain are small **variable (V) regions** with amino acid sequences that are unique for each antibody; distinctive variable regions are responsible for antibody specificity. The variable amino acid sequences for each antibody result from the expression of assorted gene fragments involved in producing antibody molecules. A different set of gene fragments is expressed for each antibody produced; consequently, over 100 million unique antibody molecules can be generated. The V portions of one H and one L chain that form an arm of the antibody interact with, and bind to, a specific antigen.

Major Histocompatibility Complex Molecules

How does our immune system recognize self cells? Major pieces of the puzzle were fit into place in the 1980s when interesting types of self molecules were identified and their three-dimensional structures were elucidated. We now know that cells of every individual (or pair of identical twins) have a unique set of self **major histocompatibility complex (MHC) molecules** of protein embedded in their plasma membranes. Cells possessing MHC molecules on their surface will not normally be destroyed by the immune system. What is the basis for each person having a different set of MHC molecules? In humans, MHC molecules are encoded in a cluster of at least seven gene loci—the *MHC complex*—located on chromosome 6, and each locus has a remarkably large number of alleles (some have more than 100). If we assume that each locus contributes one allele for producing an MHC molecule, and only

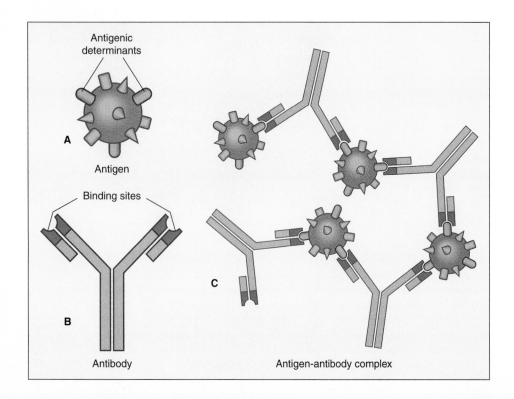

Figure 39.8 (A) Antigens have a three-dimensional structure in which specific regions, known as antigenic determinants, are the target for antibodies. (B) Antibodies have sites that bind with an antigenic determinant much as a key fits a lock. (C) Binding results in the formation of an antigen–antibody complex.

Question: *How many antigenic determinants can this antibody recognize?*

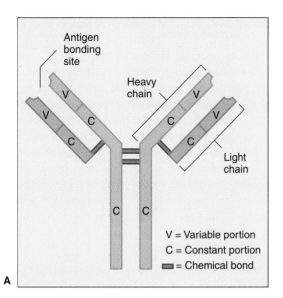

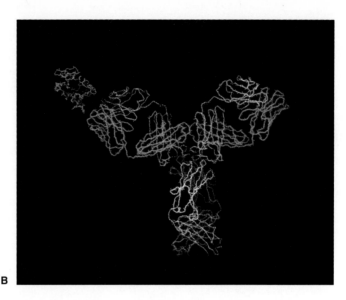

A   **B**

Figure 39.9   (A) In an antibody molecule, both heavy (H) chains and both light (L) chains have identical amino acid sequences, but the sequences of H and L chains vary. The V (variable) regions of antibody chains are responsible for binding to an antigen. Each V region of an antibody has a different amino acid sequence. The hinge portion provides flexibility and allows the two binding sites to act independently. (B) A computer-generated antibody–antigen (bright green structure, upperleft) model.

Question: *Which regions account for antibody specificity?*

one allele is expressed from a homologous chromosome pair, the *minimum* number of different MHC molecules possible in the human population is roughly 10^{14} (100 trillion)!

There are two principal types of MHC molecules, *class I (MHC-I)* and *class II (MHC-II);* their general structures are shown in Figure 39.10. There are slight differences in the extra-cellular structure of the two MHC molecules, but the primary difference is that MHC-I have one segment that spans the plasma membrane and MHC-II molecules have two. MHC-I molecules are normally found in the plasma membranes of all human cells. Self cells without MHC-I molecules or with only a small number of such molecules are destroyed by NK cells.

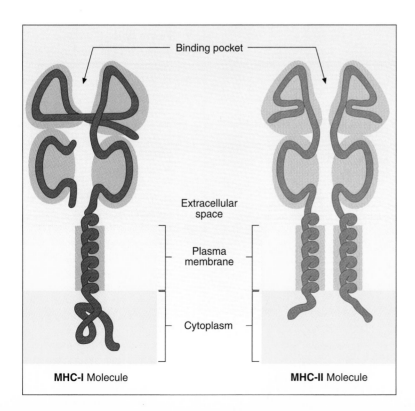

Figure 39.10   Humans have two types of major histocompatibility complex (MHC) protein molecules, MHC-I and MHC-II, that have slightly different structures: MHC-I molecules have one segment that spans the plasma membrane, and MHC-II have two such segments. MHC-I molecules are found on all cells; MHC-II molecules are found primarily on macrophages and B cells. Antigens fit into MHC binding pockets.

MHC-II molecules occur only in the plasma membranes of macrophages, B lymphocytes, and a few other cells involved in an immune response. The critical feature of MHC molecules is a special extracellular binding pocket in which nonself antigens can fit. As we shall see, antigen–MHC binding is a key process in initiating an immune reaction.

BEFORE YOU GO ON Antigens are nonself molecules or substances that induce specific immune responses. Antibodies consist of two H chains and two L chains. Each chain has constant structural regions and a variable, antigen-binding region. Self cells have MHC molecules that are recognized by certain immune cells.

Cells of the Immune System

The immune system contains billions of lymphocytes, and although all lymphocytes have a similar appearance, several different functional classes are known to exist. In healthy adults, about 20 percent of the lymphocytes, mostly T cells, circulate in the blood, while the remainder reside in lymph nodes, the spleen, the thymus gland, the tonsils, and the appendix. Lymphocytes arise from stem cells, but until they mature, they have no special immune capabilities. After maturation, they are capable of performing a specific function when challenged by an appropriate antigen. The production, maturation, and general functions of lymphocytes are described in Figure 39.11.

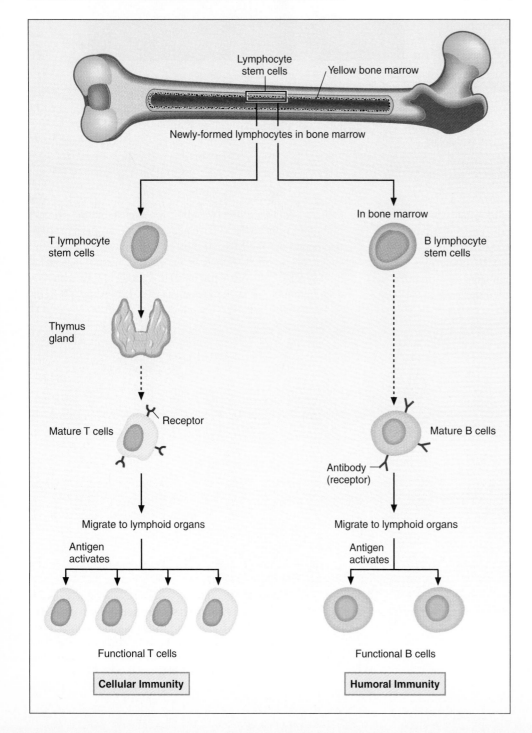

Figure 39.11 T and B cells are produced from stem cells that originate in the bone marrow. Stem cells that generate T cells then move to the thymus gland, where they divide to produce new T cells. While in the thymus, new T cells acquire receptors that can recognize a single antigenic determinant. If they have appropriate receptors that allow them to function in an immune response, they migrate from the thymus to various lymphoid organs. B cells are born and begin to mature in the bone marrow. Once they acquire antibody receptors capable of recognizing an antigen, they, too, migrate to lymphoid organs. If activated by an antigen, various T cells develop and participate in cellular immune reactions. By contrast, most B cells develop into plasma cells, which are the basis of humoral immunity.

T Lymphocytes

T lymphocytes, or **T cells**, *function in cellular immune responses.* They are produced from stem cells in the thymus gland that originally migrated from the bone marrow. After birth in the thymus, each new T cell acquires unique antigen-binding protein molecules, called *T cell receptors,* that span the plasma membrane. What is the nature and function of T cell receptors? As illustrated in Figure 39.12, the general structure of T cell receptors is similar to that of MHC molecules and antibodies, and like those molecules, they can bind with an antigen. However, receptors on a T cell can recognize and react with a single type of antigen and can do so only when it is "presented" on the surface of another self cell, usually a macrophage, along with a self MHC molecule. **Antigen presentation** is a process by which certain self cells—**antigen-presenting cells**—display a foreign antigen on their surface in a form that can be recognized by a T cell receptor. The nonself antigen must be bound to an MHC (self) molecule before it will be recognized by a T cell (see Figure 39.13A, page 750). Note the functional utility of this response: T cells react primarily against intracellular infectious agents, and through this mechanism, in which an antigen is combined with an MHC molecule, they are able to recognize infected self cells. Figure 39.13B reveals that a nonself antigen becomes bound to a MHC molecule because it fits into the "binding pocket" of the MHC protein.

During maturation in the thymus, only about 5 percent of newly formed T cells survive to become functional T cells. What happens to the other 95 percent? Evidence suggests that new T cells are subjected to a brutal selection process in the thymus. Only T cells that recognize MHC molecules *and* pos-sess receptors that do not bind to self molecules survive and mature to become functional cells. T cells that recognize MHC-I molecules mature into *cytotoxic T cells,* and those that recognize MHC-II molecules become *T helper cells;* both T cell types are discussed in the next section. New T cells that do not recognize an MHC molecule or that have receptors that bind with self molecules die by **apoptosis,** a complex, genetically programmed series of events involving intracellular metabolic pathways that leads to cell death. Mature T cells leave the thymus and migrate to lymphoid tissues, where they may have a chance to participate in an immune response.

T Cell Subpopulations

Two distinct T cell subpopulations function in normal cellular immune responses. **Cytotoxic T cells** (T_C cells, also known by the more colorful name *killer T cells*) primarily destroy self cells infected with viruses or intracellular bacteria or cancer cells that possess molecules recognized as nonself (see Figure 39.14, page 750). **Helper T cells** (T_H cells) help regulate the activities of T_C cells, B cells, and macrophages. Two other T cell types, hypothesized to be derived from T_H cells, function during later stages of a cellular immune response. **Suppressor T cells** (T_S cells) are thought to diminish or terminate ongoing cellular immune responses by reducing the number of T_H cells that can stimulate T_C cells and B cells. **Memory T cells** (T_M cells) remain after completion of a cellular immune response. T_M cells are long-lived and can react against previously encountered antigens.

B Lymphocytes

B lymphocytes, or **B cells**, *function in humoral immune responses*; they arise from stem cells and develop in the bone marrow. During maturation, B cells acquire antibodies on their surface that act as receptors. All antibody receptors on a single B cell are identical, which means that each B cell can recognize only a single antigen or antigenic determinant. Because B cells are specialized to react against extracellular antigens, they do not need to recognize self cells, as T cells do. After acquiring antibody receptors, B cells migrate to lymph nodes and other lymphoid organs and tissues. If one of the surface antibody receptors contacts its matching antigen, the B cell can become activated and give rise to **plasma cells** that are capable of manufacturing and secreting the same type of antibody, and to **memory B cells** (see Figure 39.15, page 751). Memory B cells also develop during humoral immune responses. They are long-lived and can react against previously encountered extracellular antigens.

Only about half of B cells produced in bone marrow live to become functional cells. Like self-reactive T cells, B cells with receptors that bind with self molecules die by apoptosis before they are released from the bone marrow.

Millions of different T and B cells are produced daily, each with unique receptors and recognition capabilities. These cells live only a few days, and only a small fraction of them ever come in contact with their corresponding antigens. However, because of the great diversity and the continuous production of unique lymphocytes, an antigen has little chance of escaping detection for any length of time.

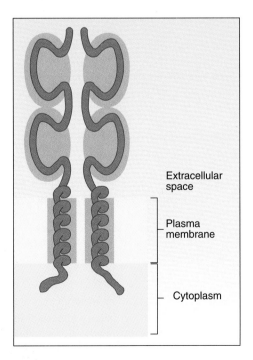

Extracellular space

Plasma membrane

Cytoplasm

Figure 39.12 T cells have receptors with this general structure. Each T cell has a receptor that can recognize and react with a single antigen when it is displayed by an antigen-presenting cell in association with an MHC molecule.

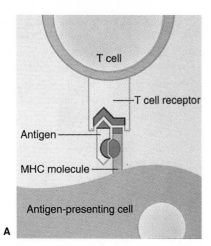

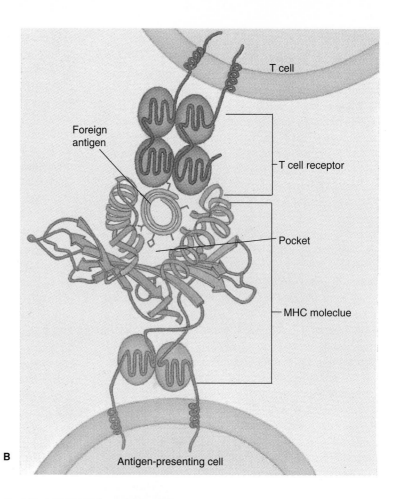

Figure 39.13 (A) T cells are able to recognize foreign (nonself) antigens when they are combined with a self MHC molecule and presented on the surface of an antigen-presenting cell. The basis for recognition is the presence of a specific receptor on the T cell that matches the shape of the nonself antigen when combined with a self MHC molecule. (B) MHC-presenting molecules have a structure that matches their function. Foreign antigens fit into the binding pocket on MHC proteins.

Question: *What is an example of an antigen-presenting cell?*

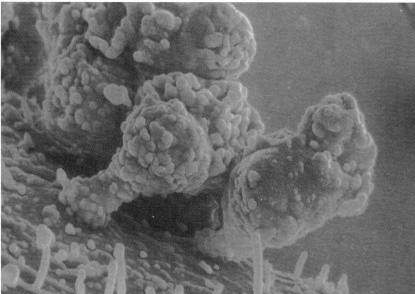

Figure 39.14 Cytotoxic T cells constitute the primary natural defense mechanism against virus-infected self cells and certain types of cancer cells. In this EM, a T_C cell (green) is shown attacking a much larger cancer cell (pink and yellow).

Antigen-presenting Cells

There are two principal types of antigen-presenting cells, *follicular dendritic cells,* which are described in Chapter 41, and macrophages. Antigen-presenting cells are found in lymphoid organs and tissues, and they have significant functions in spe-cific antibody responses. As described earlier, macrophages are the primary phagocytic cells and are particularly important in cleaning up damaged tissues by ingesting dead cell fragments and any bacteria or other microorganisms present. They can also become infected by certain microbes. The uptake of (or

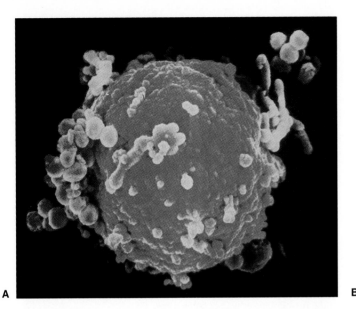

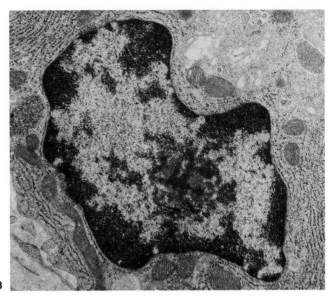

Figure 39.15   (A) A B cell (orange) can be directly activated to make a specific antibody by contact with its matching antigen. In this photograph, B cell antibody receptors have contacted chlamydia bacteria (green) that contain nonself antigens. (B) Once activated, the B cell proliferates and gives rise to identical cells that develop into plasma cells (shown here) that produce antibodies against the initiating antigen.

Question: *Which organelle is especially abundant in plasma cells? What explains its abundance?*

infection by) nonself antigens by macrophages usually constitutes the first step leading to an immune response. In a sense, at least one member from an invading group has been recognized and captured. What happens next?

> **BEFORE YOU GO ON** T cells are produced and develop in the thymus gland. They survive only if they have receptors capable of recognizing a foreign antigen in combination with an MHC molecule. B cells are produced and develop in the bone marrow, where they acquire antibody receptors. After maturing, T and B cells migrate to lymphoid organs and tissues. Four classes of T cells—T_C, T_H, T_S, and T_M—participate in cellular immune responses. B cell types involved in humoral immune responses are plasma cells and B memory cells. Antigen-presenting cells, such as macrophages, initiate specific immune reactions by processing foreign antigens and presenting them to other cells.

Primary Immune Responses

Specific immune reactions are commonly divided into cellular responses and humoral responses. Although such a separation of functions facilitates understanding, it must be emphasized that both types of immune response may occur simultaneously, in the same tissues, and are regulated by overlapping mechanisms. The first response to an antigen not previously encountered by the immune system is called a **primary immune response.**

T Cells

Recall that T cell reactions are generally directed against infectious agents, such as viruses, that live inside self cells. A primary immune response is initiated after a nonself antigen is first "processed" by an antigen-presenting cell, which is often a macrophage (see Figure 39.16, page 752). Macrophages themselves may be infected by certain disease agents, or they can acquire an antigen through phagocytosing other infected self cells. During antigen processing, a nonself protein from the infectious agent is first broken down to smaller segments (10 to 20 amino acids), at least one of which acts as an antigenic determinant. Once this task is completed, the antigen fragment is bound to an MHC molecule and transported to the surface membrane of the antigen-presenting cell. Recall that T_H cells and T_C cells recognize different MHC molecules, which reflects their different functions in an immune response.

T_H cells most commonly become activated in lymphoid tissues after contacting matching antigens presented by macrophages in combination with MHC-II molecules (see Figure 39.17, page 752). An antigen-presenting macrophage secretes a substance called *interleukin-1* that attracts and activates T cells. **Interleukins** are hormones that act on lymphocytes, and they are critical for the activation, maturation, proliferation, and function of all cells involved in immune responses. An activated T_H cell will secrete *interleukin-2,* which stimulates it to divide (an example of a positive feedback system) and also induces the proliferation, growth, and development of functional T_C cells.

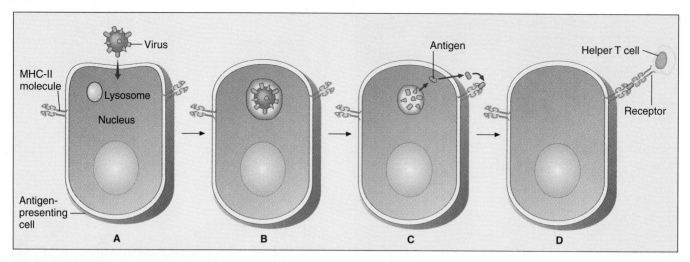

Figure 39.16 (A) Protein antigens are taken into antigen-presenting cells, such as macrophages, by phagocytosis. (B) Once inside the cell, they combine with a lysosome and are partially broken down into smaller fragments. (C) Fragments that act as antigens (or antigenic determinants) move through the membrane surface, become associated with an MHC-II molecule, and are displayed by the antigen-presenting cell. (D) Helper T cells with matching receptors bind with the antigen being presented.

Question: *Can T_C cells react with this particular antigen-presenting cell?*

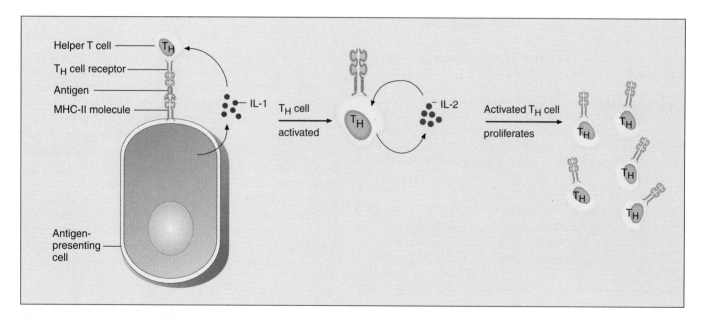

Figure 39.17 T_H cells become activated when they encounter their antigen being displayed with an MHC-II molecule by an antigen-presenting cell, most commonly a macrophage. Once contact is established, the macrophage releases interleukin-1 (IL-1), a hormone that activates the T_H cell and causes it to secrete interleukin-2 (IL-2). Interleukin-2 stimulates proliferation of the activated T_H cell, which leads to the production of identical T_H cells.

Question: *How many antigens can the newly formed T_H cells react against?*

As shown in Figure 39.18, T_C cells recognize antigens—most commonly, an antigenic determinant from a virus that has infected a self cell—presented in association with MHC-I molecules. Because T_C cells interact with MHC-I molecules, which are found on all cells, they can destroy any self cell that becomes infected with a virus. Once exposed to a viral antigen, the T_C cell becomes activated by T_H cells to proliferate and form additional T_C cells that can react against the antigen. Mature T_C cells attack virus-infected self cells by releasing a protein called **perforin,** which forms holes in the plasma membrane that drain the cell's contents, causing it to die.

Following activation, an immune response could cause the body to become overloaded with immune cells, antibodies, interleukins, and other immune substances that could result in damage. What stops T and B cell proliferation and antibody production once the quantity of an antigen is reduced or eliminated? The complete answer to that question is still uncertain. Complex experiments have produced results that support the following model: (1) T_H cells become activated through the presence of an antigen, as just described. (2) Some of the activated, antigen-specific T_H cells become nonfunctional as the immune response continues; these cells may then act to suppress immune reactions, hence they are called suppressor T (T_S) cells. (3) T_S cells reduce the immune response through two processes: by binding to sites on antigen-presenting cells,

which are then unable to activate as many new T cells, and by competing for interleukin-2, which reduces the amount available for stimulating newly activated T cells to proliferate. Together, these two forms of competition act to suppress the ongoing immune response. Further research is necessary to confirm the accuracy of this T_S cell model.

B Cells

Whereas T cells generally react against infected self cells, B cells produce antibodies that are most important in reacting against extracellular antigens. Helper T cells are required for activating B cell responses to most antigens. However, a few antigens—mostly polysaccharide carbohydrates that make up the capsules surrounding certain types of bacteria (for example, the S strain *Diplococcus* bacterium described in Chapter 17) can activate B cells directly. In the latter case, many antibody receptors on a B cell will bind to capsule antigens, which leads to activation of the cell. Once activated, the B cell proliferates, giving rise to daughter cells that mature into plasma cells capable of producing antibodies against the initiating antigen. All daughter cells produced from an activated B cell are *clones* (identical copies) of the parent cell. Each clone is capable of producing only a single type of antibody—the type that was responsible for activation of the parent B cell (see the Focus on Scientific Process, "The Clonal Selection Theory").

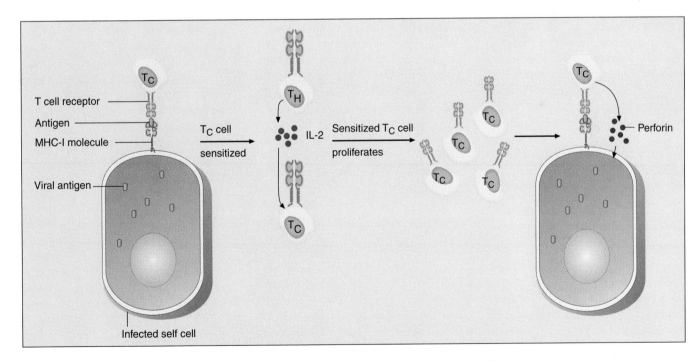

Figure 39.18 T_C cell activation occurs in two stages. First, a T_C cell encounters its antigen being displayed by an infected self cell and becomes sensitized. Second, the T_H cell, activated by the same antigen, secretes interleukin-2 (IL-2), which causes the sensitized T_C cell to proliferate and form additional T_C cells. Mature T_C cells bind with infected self cells and release perforin, which creates holes in the plasma membrane of an infected cell and leads to its death.

Question: *What would be the consequence if T_C cells interacted with MHC-II molecules instead of MHC-I molecules?*

The Clonal Selection Theory

We now know that the human immune system can recognize and react against an almost indefinite number of antigens. We also understand much about the underlying cellular and genetic mechanisms that provide us with this capability. Like many major intellectual advances in biology, this knowledge was derived from creating hypotheses and then designing and conducting experiments to test them. Early hypotheses about antigen recognition were formulated and provided a foundation for further advances. As new information became available, they were modified or gave way to new hypotheses. The evolution of hypotheses about antibody formation ultimately led to the formation of a major theory and illustrates the dynamic nature of scientific process.

In common with other fields of physiology, much of the experimental work in immunology was (and still is) done on animals, especially mice. Because these fellow mammals have a similar evolutionary history, it is presumed that results and conclusions from such studies can be extrapolated to humans. Except for minor details, this has generally proved to be the case. Modern immunology terms are used to simplify the story.

By the late nineteenth century, scientists recognized that animals were capable of reacting against foreign materials. If an animal was infected with bacteria that produced a toxin, "antitoxins" would soon appear in the blood serum of the host. This was illustrated by the formation of clumps when fresh serum from the host was mixed with the bacterial toxin. E. A. von Behring, a German bacteriologist, referred to the host's clumping substances as *antibodies*. Paul Ehrlich, a German physician, was an active researcher in immunology. During the 1890s, he developed a method for measuring amounts of antibodies produced in response to a bacterial infection. This quantitative technique was to prove especially valuable in forming early pictures of the antibody response in animals. Its use made clear that exposure to an infectious agent resulted in an irruptive increase in responding antibodies. How did these agents cause such a reaction?

In 1897, Ehrlich proposed the first coherent hypothesis about antibody formation (see Figure 1). He postulated that white blood cells had special chemical groups ("side chains") on their surface that normally functioned in the cell's metabolic processes. However, these groups could also serve as receptors for antigens that entered the body. Once a side chain was linked to an antigen, though, the side chain's normal functioning would be disrupted. To counteract this effect, the cell would be stimulated to synthesize new side chains to replace those that no longer functioned. Ehrlich claimed that "the antitoxins represent nothing more than the side-chains reproduced in excess during regeneration and are therefore pushed off from the protoplasm—thus to exist in a free state." Note that this early hypothesis was essentially correct in two respects: antigens are captured by surface receptors, and cells do produce antibodies that appear in the circulation after exposure to an antigen. However, the hypothesis was too vague and was based in part on incorrect assumptions. For example, Ehrlich felt that cells naturally made side chains that could bind to all antigens. Thus specificity could not be explained by this hypothesis.

Karl Landsteiner, an immunologist who spent most of his career at the Rockefeller Institute for Medical Research, demonstrated that specific antibodies were formed in response to specific antigens. In his 1945 book *The Specificity of Serologic Reactions,* he summarized the view of immunology at that time:

> The immune antibodies all have in common the property of specificity, that is, they react as a rule only with the antigens that were used for immunizing or with similar ones, for instance, with proteins or blood cells of one species, or particular bacteria, and closely related species. . . . [Further,] the specificity of antibodies . . . constitutes one of the two chief theoretical problems, the other being the formation of antibodies.

In other words, although the concept of specificity had become clearly established, neither the basis for specificity nor the mechanisms of antibody formation were known.

How could an antigen induce the formation of a specific antibody? In the early 1930s, several hypotheses were advanced to answer this question, but one claimed the attention of most scientists for the next 25 years. The *direct-template* or *instructive* hypothesis of antibody formation maintained that an antigen entered the antibody-forming cell and acted as a template against which the antibody molecule would shape itself in a way complementary to the antigen, as shown in Figure 2. Most of the arguments in favor of this hypothesis were based more on conceptual ideas than on experimental data. Professor Linus Pauling published a detailed paper in 1940 ("A Theory of the Structure and Process of Formation of Antibodies") that attempted to explain how the antibody might fold, or be shaped, by the antigen. The basis of complementary folding was related to chemical bonding phenomena, one of his fields of expertise. However, one of

FOCUS ON SCIENTIFIC PROCESS

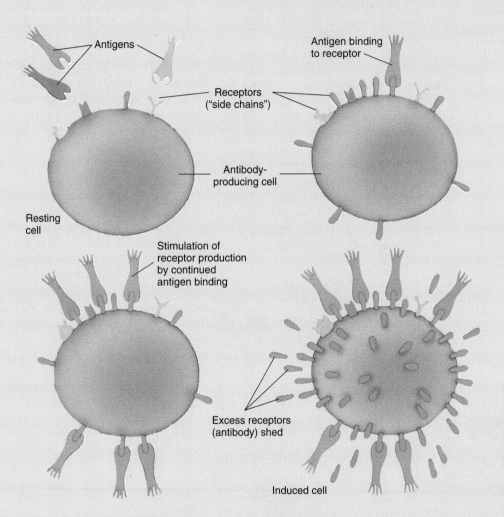

Figure 1 Ehrlich's illustration of his side-chain hypothesis, published in 1897. Using modern terms, Ehrlich thought that a foreign antigen, indicated in orange, became bound to a cell receptor and stimulated the cell to synthesize and release identical receptors, which acted as antibodies.

his assumptions was that all antibody molecules contained the same amino acid structure, and hence specificity could not be related to variations in an antibody's protein chains. This, of course, turned out to be incorrect, but that was not determined until years later. Nevertheless, Pauling was correct about amino acid sequences in the constant part of antibody molecules.

By 1950, the template hypothesis was crumbling. A number of scientists, including F. Macfarlane Burnet, an Australian immunologist, identified several shortcomings of the hypothesis.

For example, it could not account for the early rapid rise in antibody concentrations following infection. How could antibodies so quickly outnumber the antigen templates? The hypothesis also failed to explain secondary responses (well understood by the 1930s), continuing antibody production even in the absence of an antigen, or immunological tolerance (the concept of self-toleration became known in the 1940s). Obviously, the fall of the template hypothesis could be anticipated as soon as a new hypothesis arose that could address these criticisms.

A 1955 paper published by a Danish immunologist, Niels Jerne, provided a foundation for a new hypothesis that was to emerge in 1957. In this paper, Jerne formulated a *natural-selection* or *selective* hypothesis of antibody formation that proposed the following scheme:

1. Animals contain at least one circulating antibody that can react to every antigen.

2. After an antigen becomes bound to the antibody, it will interact with a lymphoid cell to trigger an immune response.

box continues

FOCUS ON SCIENTIFIC PROCESS

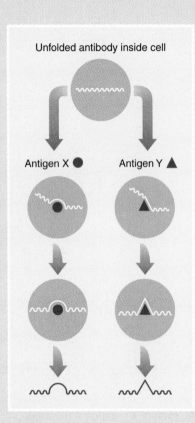

Figure 2   According to the instructive hypothesis, nonself antigens (X and Y) served as templates for antibodies that would then react against them.

3. It does this by stimulating the cell to generate and release large quantities of the specific antibody.

Note the similarity of this theory with Ehrlich's earlier side-chain theory. The primary flaw in Jerne's theory was his assumption that the body contained pre-existing antibodies for every conceivable antigen.

In 1957, David W. Talmage, of the University of Colorado, added to Jerne's ideas. He proposed that in addition to circulating antibodies, replicating cells possessing antibodies (receptors) were involved in the immune response. Such cells would then multiply after their antibody (receptor) was

"selected" by the antigen. He further suggested that single cells might produce only one type of antibody. In the same year, Burnet, independent of Talmage, published his **clonal selection theory** of antibody formation. Though his theory was very similar to that proposed by Talmage, Burnet is usually given credit for its development.

The clonal selection theory, described in Figure 3, has been called one of the great intellectual constructs of modern biology. It explained how antigens stimulated antibody production and why there had to be an enormous number of potential antibody-producing B cells in the body. The theory made the following assumptions:

1. Naive B cells are produced continuously in the body.

2. Each B cell has only a single type of antibody receptor.

3. An antigen is capable of reacting with ("selecting") only a B cell that has the "correct" receptor.

4. Once selected, the B cell proliferates and creates a population of identical cells ("cloning").

5. Each of the clones secretes the same antibody.

This description of immune events remains valid today, even though scientists now recognize that numerous other factors are involved.

Thus from Ehrlich's first revolutionary hypothesis, the modern theory of antibody formation evolved. As with many endeavors in the sciences, there were many false starts, blind alleys, and nonproductive ideas that have not been described here. And the story is still not completed. Much remains to be learned about the induction and preservation of antibody diversity as well as the intricate regulating features of the entire immune process. All of the participants in this story, except the great Linus Pauling, were later awarded Nobel Prizes for their studies in immunology. (Pauling went on to receive two Nobel Prizes for his work in other areas.)

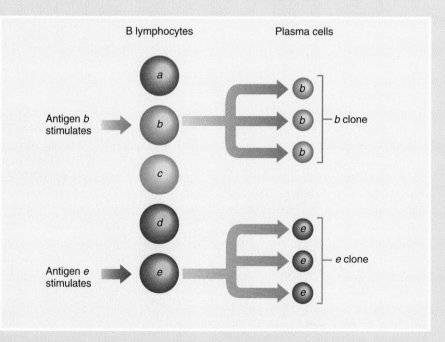

Figure 3   The clonal selection theory, formulated in 1957, has proved essentially correct and is used today to account for the humoral immune response.

B cell activation by T_H cells is a two-step process that is described in Figure 39.19. First, a receptor on a "naive" B cell (one that has not yet been activated) binds with its matching antigen, which is then taken into the cell, processed, and presented on the surface of the B cell in combination with an MHC-II molecule. Second, a T_H cell binds with the antigen being presented by the B cell. Consequently, the T_H cell secretes three interleukins that activate presenting B cells *(interleukin-4)*, causes them to proliferate *(interleukin-5)*, and causes them develop into plasma cells *(interleukin-6)*. Plasma cells remain in lymphoid tissues, but the antibodies produced are secreted and are present in plasma and tissue fluids. Most antigen–antibody reactions occur in lymphoid organs and tissues.

Memory Cells

In a primary immune response, summarized in Figure 39.20 (see page 758), activated T and B cells proliferate and develop into cells that have the capacity to react directly against a specific, initiating nonself antigen. Once the antigen is cleared, reactive T and B cells die within a few days and are eliminated from the body. The reason for the rapid disappearance of these functional immune cells is still unclear, although some hypotheses have been advanced. For example, prolonged interleukin signaling may lead to apoptosis, or cell death may occur if cells are deprived of continuous contact with interleukins resulting from the actions of T_S cells. Though most antigen-reactive T and B cells are removed at the end of an immune response, some antigen-specific cells do survive and become long-lived memory cells. Memory T cells are thought to be derived from T_H cells but the source of memory B cells has not yet been clearly established. Memory T and B cells are specific for (retain a memory of) the antigenic determinant that provoked their development. That is, like their antecedents, they possess receptors that bind with only one antigen. Memory cells may live for decades, and they do not become active immune cells unless they contact the same antigen that led to their creation.

Secondary Immune Responses

As illustrated in Figure 39.21 (see page 759), if the immune system encounters an antigen for the second time, a more intense **secondary immune response** occurs quickly because it is not necessary to activate naive cells. Memory cells can give rise to active B and T cells directly. Memory cells and secondary immune responses provide the basis for long-lasting immunity. After having been exposed to a foreign antigen and mounting a successful immune response, an individual will usually not suffer from the effects of that antigen at a later time. If the original antigen is encountered again, memory cells ensure that it will be eliminated before it can cause any noticeable symptoms.

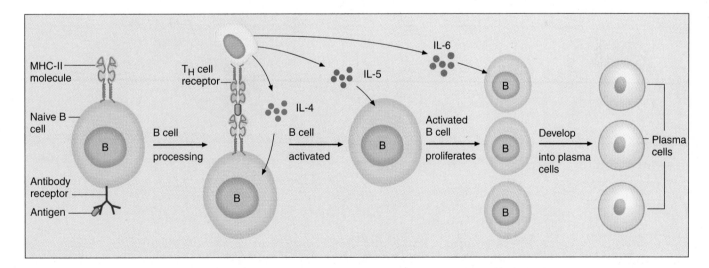

Figure 39.19 B cell activation by T_H cells is a two-step process. First, a B cell antibody receptor contacts its matching antigen, processes it, and then displays it along with an MHC-II molecule on its surface. Second, a T_H cell, activated by the same antigen, binds with the B cell antigen being displayed and secretes three interleukins. Interleukin-4 (IL-4) activates the presenting B cell, interleukin-5 (IL-5) stimulates the B cell to proliferate, and interleukin-6 (IL-6) promotes development of the new B cells into plasma cells.

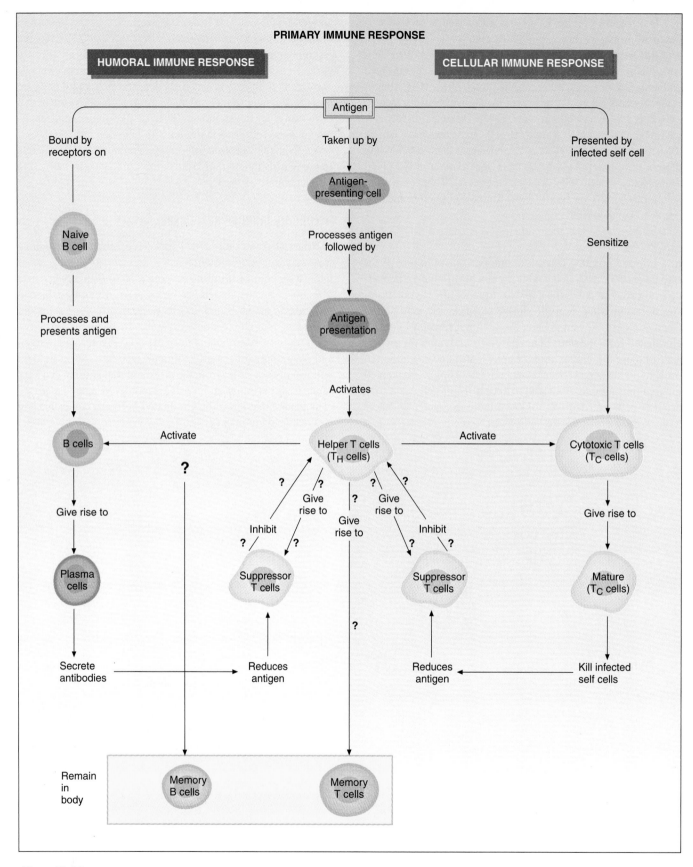

Figure 39.20   A summary of cells and reactions in a primary immune response. Question marks indicate a hypothesized pathway or mechanism that is not yet fully understood.

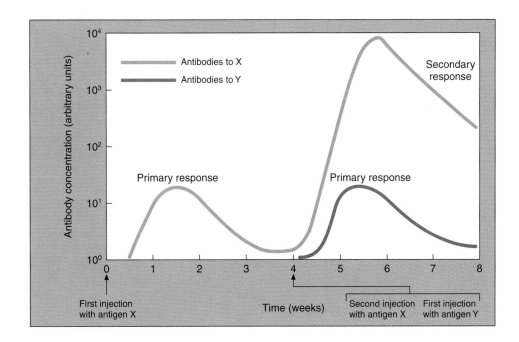

Figure 39.21 The primary immune response to antigen X leads to antibody production, resulting in a certain antibody concentration. If reinfection to antigen X occurs, a rapid secondary immune response results in much higher antibody concentrations.

Question: *Which cells are primarily responsible for the rapid secondary response?*

BEFORE YOU GO ON A primary cellular immune response begins when a T cell contacts its antigen being displayed on the surface of an antigen-presenting cell. T_H cells are activated by antigens displayed with MHC-II molecules. T_H cells induce production and maturation of T_C cell populations and B cells by secreting interleukins. T_C cells are activated by antigens displayed in combination with MHC-I molecules. Once activated, they kill self cells infected with intracellular agents. B cells can be activated by direct contact with an antigen or by T_H cells. They then proliferate and develop into antibody-secreting plasma cells. Ultimately, the initiating foreign antigen is eliminated, and the immune reactions are halted by T_S cells. Memory B and T cells remain in the body and can trigger a secondary response if the antigen reappears at a later time.

THE OTHER EDGE OF THE SWORD

Given the enormous complexity of immune responses, the number of chemical and genetic factors involved, and the coordination of events required for success, it is perhaps not surprising that many problems can and do occur. What happens if the immune system fails to distinguish between self and nonself molecules?

Tolerance

Immunological **tolerance** is the process whereby lymphocytes do not react against self molecules. It is quite obvious that it would not be beneficial to you if your immune system produced many antibodies or T cells that reacted against your own proteins or cells. What prevents or limits our immune system from reacting against self molecules? In 1959, Joshua Lederberg formulated the *clonal deletion* hypothesis, which proposed that immature lymphocytes with receptors capable of

binding to self molecules would die before they became activated. Experiments conducted since then strongly support Lederberg's hypothesis. As described previously, new self-reactive T and B cells die in the thymus and bone marrow by apoptosis before they can migrate from these organs.

More recently, it has become clear that other tolerance mechanisms must exist because newly formed lymphocytes cannot possibly be exposed to all self molecules while they remain in the thymus or bone marrow. A second hypothesis, known as *clonal anergy* ("inaction"), proposes that mature B or T lymphocytes with antiself potential receive some sort of "signal" that retards or abolishes their ability to react against a self molecule. The exact nature of the anergic signal is now being extensively investigated.

Autoimmune responses and allergies are examples of abnormal immune responses. In some cases, the normal responses of our immune system cause frustrating problems for the medical community. For example, it is extremely difficult to transplant organs from one individual into another if the immune system is operating normally. What accounts for autoimmune diseases, allergies, and organ transplant rejections?

Autoimmune Diseases

In 5 to 7 percent of humans, tolerance mechanisms fail, and an **autoimmune response** occurs in which T cells attack normal self cells or antibodies react against self molecules. More than 40 diseases have been characterized by the presence of antibodies that react against self molecules; these include multiple sclerosis, rheumatoid arthritis, and insulin-dependent diabetes mellitus. In some cases, the autoimmune response is directed against specific self molecules; in others, the response is directed at molecules present in the nucleus or cytoplasm of self cells. Autoimmunity can, of course, have disastrous consequences. Why do these harmful immune responses occur?

Their causes are poorly understood, although genetic mutations, chemicals, viruses, bacteria, and drugs have been implicated in some cases.

Intense research is currently being conducted to identify the molecular targets of autoimmune responses and to determine the underlying causes of the reactions. Are self-reactive T cells involved? Is there a failure in the mechanisms that normally suppress self-reactive T cells? Some evidence suggests that autoimmune diseases involve genes encoding MHC proteins. How does this fit into the autoimmune picture? Answering these questions presents a major challenge to immunologists.

Allergies

Hypersensitivity is a heightened immune response that may be harmful to the body. If a person is *allergic* to a substance (called an **allergen**), exposure to that allergen will cause a hypersensitive reaction. About 35 to 40 million Americans suffer from allergies. Common allergens include dust, hair, feathers, pollens, penicillin, strawberries, and lobster. One type of hypersensitivity is mediated by a type of antibody called *immunoglobulin E (IgE)* (see Figure 39.22A). During the first exposure to an allergen, the susceptible individual becomes sensitized, and his or her B cells produce IgE antibodies that bind to mast cells. In subsequent exposures, the allergen binds to these IgE antibodies in such a way that it causes mast cells to release great quantities of chemical mediators such as histamine, which activates various defense reactions (see Figure 39.22B). When this occurs, breathing becomes difficult because smooth muscles lining the air passages constrict, excess mucus is released (the nose runs), and there is much coughing, sneezing, and sometimes itching. Drugs called **antihistamines** can sometimes be effective in neutralizing histamine and relieving some of these effects. In the allergies just described, the reactions are concentrated in the nasal area. In some allergies, particularly those involving injected substances, such as insect stings or penicillin shots, the immune reactions are more generalized and affect the whole body. Severe reactions involving the respiratory and circulatory systems may occur and can be fatal if not treated immediately. Epinephrine, a hormone, is commonly injected in affected individuals to counteract the effects of histamine and other mediators. Attempts are now being made to develop new types of therapies for preventing or treating allergies.

Rejection of Transplanted Organs

Beginning in the 1960s, it became technically possible to transplant organs from one individual to another. There was initially great enthusiasm over the possibility of providing new organs to

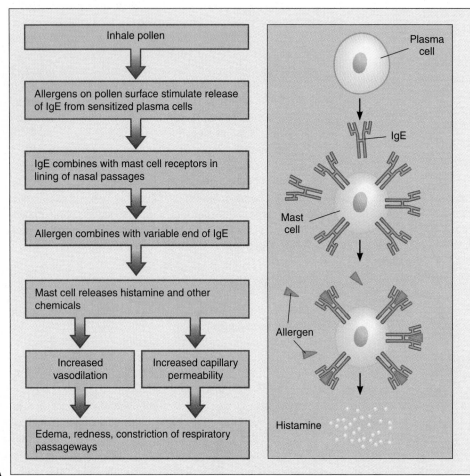

A

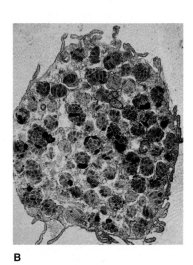

B

Figure 39.22 (A) In allergic responses, basophils are stimulated to release great quantities of histamine when an appropriate antigen, such as pollen, is encountered. Histamine is a chemical mediator that triggers allergic reactions. (B) This mast cell contains numerous granules (stained dark) that are filled with histamine and other immune system mediators.

patients with hopeless heart or kidney ailments. However, disappointment followed a short time later when it became clear that transplanted organs tended to be rejected because of immune system reactions against nonself MHC molecules.

Since then, higher success rates have been achieved through the use of precise tissue typing and drugs that suppress immune responses. Great efforts are made to match the MHC molecules between patient and donor as closely as possible. In most cases, close relatives now provide the transplanted organ or tissue. The use of drugs to suppress the immune system is problematic, for it may lead to serious infections from pathogenic microorganisms. Drugs are usually used for only a short period of time or in quantities that reduce but do not abolish the immune response. New approaches may soon be developed. For example, it may be possible to prevent nonself antigens from binding to self MHC presentation molecules, thereby preventing a full-blown immune response to the transplanted tissue.

BEFORE YOU GO ON Many important questions about immunity remain unanswered. The basis of immunological tolerance has not yet been fully defined, although two attractive hypotheses—clonal deletion and clonal anergy—have been supported. If tolerance mechanisms fail, autoimmune diseases may result, but their causes are generally unknown. Being able to diminish immune responses may be desirable in two cases: extreme immune responses that occur in allergic reactions and organ transplants that have poor success rates because of immune reactions.

Given its great importance in maintaining the health of humans, research on the immune system is one of the most active fields of investigation. In Chapters 40 and 41, we consider what happens when this marvelous system fails to repel disease agents or to eliminate self cells that become transformed into nonself cells.

SUMMARY

1. Humans have two general defense systems that are used against pathogens and other harmful agents. Nonspecific defense mechanisms are innate and operate in the same inflexible way against any harmful agent and the damage it may cause. The specific immune system is capable of highly precise responses directed against a single agent.

2. Nonspecific defense mechanisms consist of body surfaces—the skin and mucous membranes—that act as barriers against entry, secreted products that inhibit the entry of foreign agents, and blood cells that react against foreign substances or, in some cases, organisms that gain entry into the body. Granular leukocytes and macrophages remove various materials, including microorganisms and dead or dying cells, by phagocytosis. Natural killer cells destroy host cells that have been infected by viruses or have become cancerous.

3. Inflammation is a nonspecific two-phase process that occurs after injury. Immediately after cells, tissues, or organs are damaged, various blood cells invade the site and, through the action of chemical mediators, wall off the affected area. Subsequently, destroyed cells and tissues are removed, and resolution or wound repair takes place.

4. The specific immune system employs two basic mechanisms to react against nonself antigens. In cellular immune reactions, various T lymphocytes react directly against nonself cells or infected self cells. In humoral immune reactions, plasma cells, a type of B lymphocyte, produce specific antibodies that respond to extracellular antigens or antigenic determinants. Self cells are protected from being destroyed by the immune system because of the presence of MHC molecules on their surfaces.

5. T lymphocytes are produced and develop in the thymus gland and B lymphocytes in the bone marrow. After acquiring appropriate receptors, mature T and B cells migrate to lymphoid organs and tissues, where they carry out their immune functions.

6. In a primary immune response, an antigen is first captured and processed by an antigen-presenting cell, frequently a macrophage. Subsequently, a nonself antigenic fragment is joined with a self MHC molecule, and both are presented on the surface of the antigen-presenting cell. A T_H cell becomes activated when its surface receptors bind with specific nonself antigenic determinant that is combined with an MHC-II molecule present on an antigen-presenting cell. Once activated, the T_H cell proliferates and stimulates the development of other types of T cells and, in some cases, B cells. A T_C cell recognizes a specific nonself antigenic determinant combined with a MHC-I molecule on the surface of an infected self cell; they are then activated to destroy self cells that are infected with the triggering antigen. B cells can be activated directly by contact with certain types of antigens or by T_H cells. Activated B cells proliferate, giving rise to plasma cells that secrete antibodies against the initiating antigen.

7. A legacy of the primary immune response is the presence of memory T and B cells—cells that have specific receptors for the initiating antigen. These cells have very long lives, and if the original antigen reappears, a secondary immune response quickly results in its elimination.

8. Autoimmune diseases, allergies, and organ transplant rejections all involve undesirable immune reactions. They constitute major challenges for immunology researchers.

WORKING VOCABULARY

antibody (p. 746)
antigen (p. 744)
antigenic determinant (p. 746)
B cell (p. 749)
cellular immunity (p. 744)
chemical mediator (p. 742)
complement system (p. 742)
cytotoxic T cell (p. 749)
helper T cell (p. 749)
humoral immunity (p. 744)
inflammation (p. 741)
interleukin (p. 751)
lymphocyte (p. 741)
macrophage (p. 743)

memory cell (p. 749)
neutrophil (p. 741)
nonspecific defense
   mechanisms (p. 738)
plasma cell (p. 749)
platelet (p. 740)
primary immune response
   (p. 751)
secondary immune response
   (p. 757)
specific immune response
   (p. 738)
tolerance (p. 759)

REVIEW QUESTIONS

1. What are some fundamental differences between the two general human defense systems?

2. What are the defense functions of skin, mucous membranes, and various secretions?

3. Describe the types of blood cells that participate in nonspecific defense responses. What is the role of each?

4. What is inflammation? Why is it important? What are the major events in an inflammatory response?

5. How are the concepts of self and nonself used in immunology? Why are they important?

6. What is an antigen? An antibody? How are they related? How do their structures reflect this relationship?

7. Describe the life cycle of a T cell and a B cell. What type of immune response is associated with each cell?

8. What are the general functions of T cells, B cells, and macrophages in immune responses?

9. How does a primary immune response occur? What does this type of immune response accomplish?

10. What is the significance of a secondary immune response? How does it differ from a primary immune response?

11. What explains the development of immune tolerance?

12. What is the relationship between immune responses and autoimmune diseases? Allergies? Organ transplants?

ESSAY AND DISCUSSION QUESTIONS

1. What might be some of the consequences of removing the thymus gland at birth?

2. Some physicians who did pioneering immunology research believed that understanding immune processes could lead to curing all human disease. What are some inherent limits to this idea?

3. Explain the possible evolutionary significance of complicated specific immune systems being limited to birds and mammals.

4. Allergy sufferers rely on drugs such as antihistamines to control symptoms. What other strategies might be used to control allergies in the future?

REFERENCES AND RECOMMENDED READING

Abbas, A. K., A. M. Lichtman, and J. S. Pober. 1994. *Cellular and Molecular Immunology*. 2d ed. Philadelphia: Saunders.

Ada, G. L., and G. Nossal. 1987. The clonal-selection theory. *Scientific American*, 257: 62—69.

Burnet, F. M. 1959. *The Clonal Selection Theory of Immunity*. London: Vanderbilt/Cambridge University Press.

Engelhard, V. H. 1994. How cells possess antigens. *Scientific American*, 271: 54–61.

Flajnik, M. F. 1994. Advances in immunology. *BioEssays*, 16:671–675.

Kärre, K. 1995. Express yourself or die: Peptides, MHC molecules, and NK cells. *Science*, 267: 978–979.

Labro, M. T. 1994. *Host Defense and Infection*. New York: Hoechst/Dekker.

Lombardi, G., S. Sidhu, R. Batchelor, and R. Lechler. 1994. Anergic T cells as suppressor cells in vitro. *Science*, 264: 1587–1589.

Miller, J. 1994. The thymus: Maestro of the immune system. *BioEssays*, 16:509–513,

Mossalayi, M. D., F. Mentz, A. H. Dalloul, A. H. Blanc, and H. Merle-Beral. 1994. Functional analysis of human bone-marrow- and thymus-derived early T cells. *Research in Immunology*, 145: 134–139.

Roitt, I. M. 1994. *Essential Immunology*. 8th ed. Boston: Blackwell.

Schwartz, R. H. 1993. T cell anergy. *Scientific American*, 269: 62–71.

Scientific American. 1993. Life, death, and the immune system. (special issue). *Scientific American*, 269.

Silverstein, A. M. 1989. *A History of Immunity*. San Diego, Calif.: Academic Press.

Sprent, J. 1994. T and B memory cells. *Cell*, 76: 315–322.

Tauber, A. I. 1995. *The Immune Self: Theory or Metaphor? A Philosophical Inquiry*. New York: Cambridge University Press.

Tizard, I. R. 1995. *Immunology: An Introduction*. 4th ed. Philadelphia: Saunders..

ANSWERS TO FIGURE QUESTIONS

Figure 39.1 Nonspecific defense mechanisms. They exist in all animals, whereas specific immune responses occur only in birds and mammals, which indicates that they evolved later.

Figure 39.3 Help in blood-clotting reactions.

Figure 39.5 All granulocytes (basophils, eosinophils, and neutrophils).

Figure 39.8 One.

Figure 39.9 The variable (V) regions.

Figure 39.13 Macrophages, B cells.

Figure 39.15 Rough endoplasmic reticulum (RER). RER functions in protein synthesis, and plasma cells synthesize antibodies, which are proteins.

Figure 39.16 No, they react with MHC-I molecules.

Figure 39.17 One.

Figure 39.18 Only infected self macrophages and B cells (and a few others) could be killed. Because different viruses infect almost all cell types, this would not be particularly beneficial in the long run.

Figure 39.21 Memory T and B cells.

40

Human Diseases

Chapter Outline

Reading Questions

1. Which human diseases are of greatest importance in the United States?

2. Which sexually transmitted diseases have become a major public health problem in the United States?

3. What is cancer? What are vascular diseases?

4. Why are individuals being asked to assume control in managing disease risk factors in their lives?

As a result of impressive strides made in producing antibiotics during the first half of the twentieth century, progress in understanding the causes of human diseases actually slowed. Many people came to believe that all diseases were due to invasions of the body by **microbes** (bacteria and certain protozoans and fungi). They also thought that such disease agents could be controlled by antibiotics, vaccines, and drugs. Consequently, there was considerable hope in the 1950s and 1960s that all human disease would soon be conquered and that it was not necessary to spend vast amounts of money studying human diseases. However, further research revealed that such optimism was misguided. The discovery of numerous human viruses revealed a class of disease agents that was much more difficult to control than bacteria. It also became clear that many human diseases were not caused by microbes or viruses but by other factors.

DISEASES AND MODERN MEDICINE

Concepts of human disease have changed with time (see the Focus on Scientific Process, "Changing Concepts of Disease"). Modern medicine is based on the fundamental concept that disease impairs normal physiological function. This approach reflects the mechanical view of the body that has been the foundation of Western medicine for over 100 years. Today, we define **disease** as any abnormal condition of the body that impairs normal functioning. This broad definition includes mental disease, social maladaptions, and trauma, as well as more traditional diseases—heart disease, cancer, and **infectious diseases** that are caused by microbes and parasitic worms. Other causes of disease are also becoming better understood (for example, the genetic disorders discussed in Chapter 20). Nutritional diseases have been recognized for nearly a century, and our current national preoccupation with vitamins and "health foods" reflects a widespread concern. Nevertheless, malnutrition remains a serious problem for parts of our society.

The Role of Lifestyles

The medical community has become increasingly aware of the influences of lifestyle and environment on the individual. Chemical and psychological stresses have been found to cause

harmful effects on the body (see Figure 40.1) Public health officials consider cigarette smoking the single greatest health risk factor, accounting for nearly one third of all deaths due to vascular diseases, cancer, and respiratory diseases. Other environmental factors such as preservatives, charcoal-broiled food, drinking water contaminated with chemicals, fouled air, food additives, pesticide residues, and a host of other agents may be involved in some diseases, although their contributions have not been established definitively.

The Future

A lesson from the nineteenth century may have relevance for controlling diseases of modern civilization. Before the discovery of antibiotics, several of the most potent infectious diseases, such as cholera, were effectively controlled by the development of public health programs. Sanitation and clean water ended some of the worst epidemics in industrial Europe well before the real causes (bacteria) were discovered. Most scientists consider the development of sewage systems and water treatment facilities the most important events in the history of disease control and eradication. So it is not altogether novel that control of environmental and lifestyle diseases should involve engineering, social workers, nutritionists, education, and modification of our attitudes toward exercise.

Modern medicine also appreciates the extent to which different categories of disease often affect the same individual. Invading organisms, genetics, lifestyle, and environmental conditions are often different facets of problems that the modern physician sees. The patient who complains of a chronic sinus infection is likely infected with a microbe. However, other factors may be involved if the patient smokes heavily, is a perfectionist who works 80 hours a week, or thinks that a balanced meal is a hamburger with fries and a milkshake.

Figure 40.1 Until the middle of the twentieth century, it was thought that infectious agents accounted for most diseases. Now it is clear that many other factors, such as stress, nutrition, and lifestyle, contribute to disease.

FOCUS ON SCIENTIFIC PROCESS

Changing Concepts of Disease

Early Human Cultures

The concept of disease has a long history and has changed along with prevailing cultures. In very early cultures, the human worldview was essentially magical. No distinctions were made between living and nonliving things, and all objects were thought to have a potential for affecting other objects. In such a setting, illnesses were believed to be a consequence of someone's anger or spite. They could be brought on by a spell cast by a jealous rival, a person who had been mistreated, or a demon or spirit that was offended or was seeking punishment for some offense. The symptoms of any illness were of little use in determining its cause. The primary concerns were discovering who had caused the problem and why. All early human societies had individuals whose special function was to investigate the causes of illness. These medicine men and women, or *shamans,* used various methods of divination and careful interviews to uncover the force, and perhaps the motive, behind the problem. Then the shaman would attempt to resolve the issue through exhortation, prayers, and magic, an approach with a low degree of success.

Rational Medicine

The early magical worldview was widespread, and even today it exists in many cultures that are classified as "primitive." The Greeks were the first to formulate a system of rational medicine (as well as the first systematic rational worldview). The early Greeks' view of disease was expressed in the fifth century B.C. and survives in a set of books called the *Hippocratic Corpus.*

Greek doctors thought that the body was composed of substances they termed *humors.* There were four humors: black bile, yellow bile, phlegm, and blood. Health was the optimal body state when the four humors were in balance, and illness was caused by an imbalance, most often an excess of one of the humors. The body also had an internal force known as its *vital heat* that maintained the balance of humors. If there was an excess of a humor, the vital heat "cooked" it, which allowed the body to expel it.

Many factors could influence the balance of humors. Diet, age, weather, daily habits, and season were among the most important. To correct humors imbalances (disease), rest and proper diet were recommended to help the body heal itself. Greek doctors recognized many specific diseases and carefully described their symptoms. They believed that humor imbalances could be best treated at certain periods and that many diseases had critical periods when the patient would either recover or die.

The Greek humor theory endured for centuries. The Romans inherited the humor theory and it passed to the Arabs, who in turn transmitted it back to Western Europe in the Middle Ages. The humor theory remained dominant until the end of the eighteenth century.

Nineteenth-Century Medicine

Starting at the end of the eighteenth century, Europe was transformed by the industrial revolution. As part of that revolution, science came to play a greater role in all aspects of society. Medicine was particularly affected, and the concept of disease was fundamentally revised. Rather than being caused by an altered balance of the body humors, disease came to be seen as a specific entity in itself. The new concept of disease arose from three developments: localism, cell theory, and germ theory.

A group of physicians in Paris began to perform careful autopsies with an eye toward correlating internal changes in the body with external symptoms of disease. Their work gave rise to the concept of **localism,** the idea that local disturbances in specific organs were responsible for disease.

Localism was extended by the cell theory in the 1830s. As described in Chapter 3, the **cell theory** claimed that the body was composed of cells, which were organized into tissues and organs. Rudolf Virchow, the great theoretician of the cell theory, maintained that a local disturbance in a cell or group of cells could spread through the body and affect other cells. The cell theory replaced the humor theory as the major conception of the body and of disease. Now disease was considered to be caused by a local disturbance of a group of cells that could undermine their proper functioning.

The great French chemist Louis Pasteur (see Figure 1) further extended the cell theory. In a set of brilliant experiments, he demonstrated that many diseases were caused by microbes, such as bacteria, that invaded healthy bodies; his results gave rise to the **germ theory.** Pasteur's ideas came from studying germs that spoiled beer, milk, and other natural products. To counteract these germs, he invented the process of *pasteurization* (partial sterilization) to keep milk and beer from spoiling. Obviously, humans can't be gently boiled to keep germs from growing in them. Fortunately, Pasteur also discovered that some diseases could be combated by **vaccination.** In modern terms, the basic approach involves exposing a potential host to an appropriate disease-agent antigen (a **vaccine**). This will lead to an immune response and result in permanent immunity because of memory cells (explained in Chapter 39). Vaccinations were pow-

box continues

FOCUS ON SCIENTIFIC PROCESS

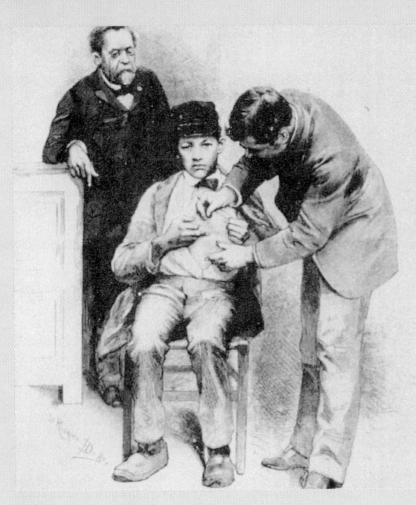

Antibiotics and Vaccines

An even more dramatic discovery occurred early in the twentieth century, when it was found that certain microbes produce chemicals that inhibit the growth of other microorganisms. These chemicals are called **antibiotics,** and they revolutionized medicine because they made it possible to treat bacterial infections after they occurred. Antibiotics disrupt vital metabolic pathways in the microbe. For example, the most famous antibiotic is produced by a mold, *Penicillium.* When a *Penicillium* extract is injected into the human body, it prevents bacterial cell wall synthesis in susceptible species (and thus kills them) but does not damage human cells.

Viruses cannot be treated with antibiotics because they live inside cells and use the host cell's metabolic machinery. To control modern viral diseases, emphasis was first placed on developing new vaccines. Highly effective virus vaccines have been developed for measles, polio, mumps, and rabies, and in one of the great triumphs of modern medicine, smallpox was eradicated by a vaccine.

Although the development of new vaccines remains an important goal, much attention is now being directed toward identifying or creating antiviral drugs. **Drugs** are chemical substances used to control diseases. Antiviral drugs are targeted at a specific feature of the virus life cycle. By disrupting or abolishing an essential step in viral synthesis, the disease may never develop. For example, the drug acyclovir (Zovirax) is a specific inhibitor of replication by certain herpesviruses. It prevents the normal synthesis of virus-directed DNA, and hence new virions cannot be produced. It does not, however, eliminate the virus. Despite these advances in treating many diseases, there are several important diseases for which no effective antibiotic or vaccine is available (for example, malaria and herpes). Treatment of these diseases remains a challenge for the future.

Figure 1   Louis Pasteur developed the *germ theory*—the idea that diseases were caused by microbes that invaded the body. This drawing, from a nineteenth-century French magazine, shows Pasteur (with glasses) supervising a rabies vaccination.

erful preventives but of little value after an individual had contracted a disease. Consequently, much of the medical research in the second half on the nineteenth century involved attempts to identify and try to kill microbes.

Joseph Lister appreciated the finding that germs spoiled organic matter. He quickly realized that germs were probably responsible for causing wound infections. Shortly thereafter, he found that carbolic acid could be used to kill germs that caused infections in wounds and surgical incisions. Through his use of antiseptic surgery and dressings, the mortality rate in surgery was drastically reduced. Indeed, before Lister, surgery

was a last-resort procedure because the survival rate was abysmally low.

Finding a means for combating infectious diseases was more difficult. Paul Ehrlich, the German physician mentioned in Chapter 39, had the idea of trying to find chemicals that would adhere to microbes but not to human cells. Ehrlich reasoned that if one could attach a poison, like arsenic, to such a chemical, it would kill the microbe but not the host. He referred to these hypothetical agents as "magic bullets." After many discouraging years of research, such a drug was found. Called *salvarsan,* it was the first of the sulfa drugs that were effective in treating certain diseases.

DISEASES OF CONTEMPORARY IMPORTANCE IN DEVELOPED COUNTRIES

Advances in modern medicine have greatly benefited citizens living in developed countries. Antibiotics, effective sanitation practices, public health measures, clean living environments, and education have diminished the importance of infectious diseases as a cause of death. For example, deaths due to cholera, typhus, scarlet fever, and many other diseases of historical importance are now rare. Much of that success is due to childhood immunizations against dangerous infectious diseases. The picture is less cheerful for inhabitants of developing countries, where infectious diseases still claim a frightful toll. Table 40.1 summarizes data on fatal infectious diseases in developing countries of the world.

Although mortalities have diminished, the frequency of infectious diseases in all citizens of the world has probably not decreased significantly in the twentieth century. Rather, some new diseases and some diseases that have been around for a long time have become more common because of new opportunities for infecting dense, mobile human populations that now exist on Earth.

Two consequences of the reduction in mortalities caused by infectious diseases are that we now live to older ages, and we die of other diseases. What are the primary causes of death

Table 40.1 Infectious Diseases Responsible for Mortality in Developing Countries of the World

Cause of Death	Primary Disease Agents	Number of Deaths Each Year
Respiratory infections	Various bacteria and viruses	6.9 million
Diarrheal diseases	Rotaviruses; various bacteria	4.2 million
Tuberculosis	*Mycobacterium tuberculosis*	3.3 million
Malaria	Protozoan, *Plasmodium* species	1 to 2 million
Hepatitis	Hepatitis A and B viruses	1 to 2 million

Source: World Health Organization, 1992.

in the United States? Figure 40.2 indicates that heart diseases are the leading cause of death, and cancer is second. Although infectious diseases (except for pneumonia and influenza) do not cause significant mortality, they cause serious problems for certain parts of our society.

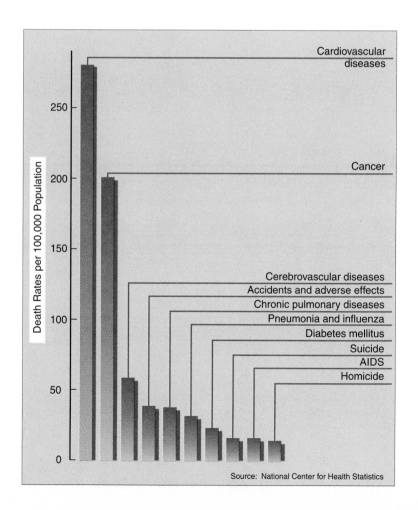

Source: National Center for Health Statistics

Figure 40.2 The ten leading causes of death in the United States are described in this figure. Each year, approximately 2 million people die of these conditions.

Question: *Which one of these became a leading cause of death only in the past decade?*

It once seemed possible that human diseases could be eradicated by the use of antibiotics. It is now recognized that many diseases have causes that cannot be addressed with antibiotics. In developed countries, infectious agents still cause serious diseases and heavy mortality. However, nutrition and lifestyle factors are associated with circulatory diseases and cancer, the major causes of death in the United States.

INFECTIOUS DISEASE

Humans live in an environment filled with microorganisms. Vast numbers of beneficial or harmless bacteria normally inhabit our skin, our mouth, and all of our body openings. Also, different viruses routinely enter and exit our body during the course of our lives. Fortunately, we serve as hosts to only a few bacteria, viruses, and other microorganisms that cause disease.

Infectious Disease-causing Agents

Diseases can be caused by microbes representing several levels of classification, but we shall confine our discussion to bacteria (kingdom Prokaryotae, subkingdom Eubacteria) and viruses. Viruses are not considered to be living by many scientists because they are unable to carry out independently the

activities that define living cells or organisms. Table 40.2 contains information about a few important human bacterial and viral diseases. More than 400 viruses and bacteria are known to infect humans. Many protozoans (kingdom Protoctista), fungi (kingdom Fungi), and parasitic worms (kingdom Animalia) also cause important human diseases.

Bacteria

Bacteria are a diverse group of single-celled organisms of varying shapes and sizes whose most important biological roles are decomposition, nitrogen fixation, and disease agents (see Figure 40.3A). As illustrated in Figure 40.3B, bacterial cells are surrounded by a *cell wall,* composed of protein and carbohydrate molecules, that provides rigidity and protection. Many bacteria have a surrounding **capsule** that is gelatinous and lies outside the cell wall. The capsule provides protection, aids in attaching bacteria to cells they can infect, and helps them resist phagocytosis. Molecules of the cell wall and capsule often act as antigens in initiating a host immune response. Bacteria may also have external flagella and pili, structures that originate inside the cell. Flagella permit movement, and **pili** are used to attach bacteria to host cells and in bacterial reproduction. Finally, a few bacteria, such as *Clostridium,* can produce **spores** that are resistant to most environmental stresses (for example, heat, cold, and drying). New bacteria develop from spores when favorable conditions occur.

Table 40.2 Some Bacteria and Viruses That Cause Diseases in Humans

Organism	Diseases	Features
Bacteria[a]		
Staphylococcus	Boils	Common inhabitant of skin; may infect skin cells, causing pimples or boils
Streptococcus	Dental cavities, pneumonia	Regular inhabitant of mouth; causes plaque; normal inhabitant of upper respiratory tract; causes infections in damaged lungs
Neisseria	Gonorrhea	Sexually transmitted disease; infects epithelium lining urogenital tract
Clostridium	Tetanus, gangrene, botulism	Widely distributed in soil and intestines; opportunistic infections cause disease
Salmonella	Food poisoning	Caused by an exotoxin that affects cells of the intestinal epithelium
Pseudomonas	Urinary tract and wound infections	Common human intestinal bacteria; resistant to many antibiotics; opportunistic infections
Legionella	Legionnaire's disease	Respiratory pathogen; often acquired from contaminated air-conditioning units
Viruses		
Herpesvirus	Herpes	Two forms: type 1 infects salivary glands; type 2, the genital tract. Both are persistent and cannot be eliminated
Papovavirus	Warts	Infects skin cells
Orthomyxovirus	Influenza types A, B, and C	Respiratory infections
Coronavirus	Common cold	More than 100 antigenetically distinct cold viruses
Retrovirus	Leukemia and other cancers	A retrovirus (HIV-1) causes AIDS

[a]For bacteria, the genus name is indicated in this table. There are several species of most of these bacteria that cause disease.

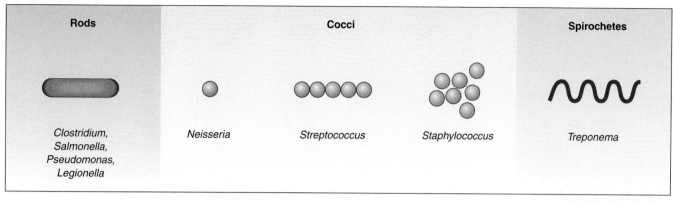

Rods	Cocci			Spirochetes

Clostridium, Salmonella, Pseudomonas, Legionella   *Neisseria*   *Streptococcus*   *Staphylococcus*   *Treponema*

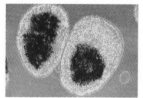
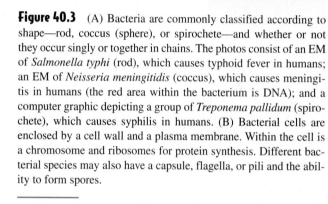

A

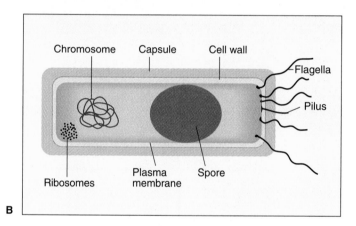

Chromosome   Capsule   Cell wall   Flagella   Pilus

Ribosomes   Plasma membrane   Spore

B

Figure 40.3 (A) Bacteria are commonly classified according to shape—rod, coccus (sphere), or spirochete—and whether or not they occur singly or together in chains. The photos consist of an EM of *Salmonella typhi* (rod), which causes typhoid fever in humans; an EM of *Neisseria meningitidis* (coccus), which causes meningitis in humans (the red area within the bacterium is DNA); and a computer graphic depicting a group of *Treponema pallidum* (spirochete), which causes syphilis in humans. (B) Bacterial cells are enclosed by a cell wall and a plasma membrane. Within the cell is a chromosome and ribosomes for protein synthesis. Different bacterial species may also have a capsule, flagella, or pili and the ability to form spores.

Question: *Which one of the bacteria shown here cause a sexually transmitted disease?*

One outstanding characteristic of many bacteria is that they can reproduce by cell division (binary fission) every few minutes when they inhabit a favorable environment. Thus a single bacterium can give rise to a population that numbers in the millions in less than 24 hours. Bacteria that normally colonize external areas of the human body cause little or no harm, other than distinctive odors from inhabited areas such as feet, mouth, and underarms, unless they have an opportunity to gain entry through wounds. It is impossible to eradicate these normal occupants, but proper personal hygiene serves to keep their numbers within a tolerable range. *Pathogenic* (disease-causing) bacteria, by contrast, may cause disease if they evade the immune system and become established on or in the human body. The mechanisms by which they cause damage vary, depending on the bacteria and tissues affected. Bacterial infections can generally be treated with antibiotics.

Viruses

Viruses occupy a world between living and nonliving. They can be replicated only inside a host cell, for they are unable to carry on metabolic activities by themselves. The structure of different virus particles varies (see Figure 40.4A, page 770). In general, a virus is a submicroscopic particle consisting of a nucleic acid genome—DNA or RNA—surrounded by a protein coat called a **capsid.** Some viruses, such as the influenza **virion** (an infectious virus particle) have an outer **envelope** consisting of lipids and viral proteins. Unlike all living cells, which contain both DNA and RNA, viruses contain either RNA or DNA but not both. The genome of viruses is limited in size, ranging between 3 and 200 genes. The origin of viruses is steeped in mystery, but they are amazingly successful pathogens that cause many serious human diseases.

A general viral growth cycle is described in Figure 40.4B. A viral infection begins when a virion comes in contact with a host cell, becomes attached to a specific receptor site on the plasma membrane, penetrates the plasma membrane, and introduces its DNA or RNA into the cytoplasm. Receptors used by viruses are normal membrane molecules; for example, a specific virus may become attached to a hormone receptor in the plasma membrane of a host cell. A DNA virion's nucleic acid then enters the nucleus, where it takes over the host cell's protein- and DNA-synthesizing machinery. RNA viruses usually operate in the cytoplasm, where conditions for RNA synthesis are optimal. Regardless of location, the virion nucleic acid

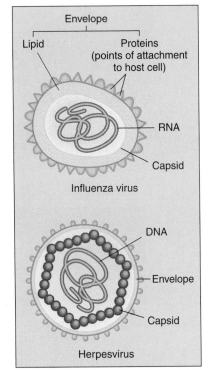

Envelope
Lipid
Proteins (points of attachment to host cell)
RNA
Capsid
Influenza virus

DNA
Envelope
Capsid
Herpesvirus

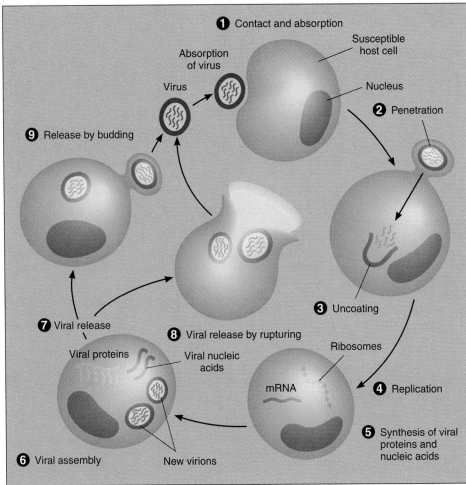

❶ Contact and absorption

Susceptible host cell

Absorption of virus

Virus

Nucleus

❷ Penetration

❾ Release by budding

❸ Uncoating

Ribosomes

❼ Viral release

❽ Viral release by rupturing

mRNA

❹ Replication

Viral proteins

Viral nucleic acids

❺ Synthesis of viral proteins and nucleic acids

❻ Viral assembly

New virions

B

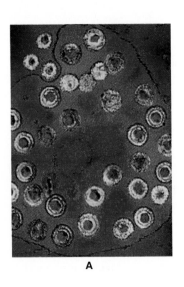

A

Figure 40.4 (A) Viruses are extremely small particles that infect cells and cause many human diseases. Their basic structure includes an outer envelope composed of lipid and protein, a protein capsid, and genetic material—either DNA or RNA—enclosed within the capsid. The photo shows herpesviruses (yellow and green spheres). (B) Viral replication can take place only within host cells and occurs in a stepwise fashion. A virion first contacts a host cell and enters the cytoplasm, where it sheds its envelope and capsid and releases its genetic material. The viral DNA (or RNA) directs the host cell's protein and DNA-synthesizing machinery to produce viral proteins and DNA. These viral molecules are then used in assembling new virions. Once formed, they are released after rupturing the host cell or by budding from the cell's surface. Released virions may then infect other cells.

"orders" the host cell's different RNAs, enzymes, and ribosomes to synthesize virion proteins and nucleic acids. These viral proteins and nucleic acids are then used in assembling new virions. The time required for most viruses to produce 100 to 1,000 new virions in a cell is less than one hour. The normal activities of the host cell often cease once the virion takes control. When the infected cell becomes full of new virions, it may rupture, releasing its contents all at once; it may simply disintegrate and release the virions more slowly; or the virions may "bud" from the cell's membrane. The effect of these processes is to kill the host cell and create new virions that can go on to infect fresh cells.

Other viruses, including some DNA "tumor viruses" that cause cancer and the extremely important RNA retroviruses that cause cancer and AIDS (see Chapter 41), use a different strategy. When retrovirus virions infect a cell, their nucleic acid can be stably integrated into the host's DNA molecules, where it remains "silent" until turned on by a certain signal.

The treatment of viral diseases is generally complex. Many of the early symptoms of viral disease—fever and inflammation—are the direct result of nonspecific defense responses to damaged cells. However, much to everyone's frustration, there is no quick cure for many viral diseases.

Why can't the symptoms of a common cold or flu be alleviated by getting a shot or taking a pill? The viral life cycle holds the answer. Drugs or antibiotics intended to kill a virus directly would have to kill host cells, an unacceptable approach. Thus the strategy commonly employed ("get lots of rest and drink as much fluid as possible") is simply to wait until the immune system catches up with the virus. This eventually occurs during the virus's passage from one host cell to another, when appropriate immune cells are able to capture antigenic virions. For several of the most important viral diseases, such as polio and smallpox, effective vaccines have been developed.

BEFORE YOU GO ON Bacteria, viruses, fungi, protozoans, and parasitic worms cause infectious diseases. Both harmless and pathogenic bacteria may cause disease if they enter the body. Bacterial infections can usually be treated with antibiotics. Viruses have unique life cycles. Host cells are required to synthesize viral genetic material and proteins and to assemble new virions. Because they live inside host cells, viruses cannot be killed with antibiotics.

Evolutionary Considerations

What are the basic requirements for a microbe that causes disease? They must include infection of a host, survival, successful reproduction, and also, at some point, transmission to a new host. If any of these is not accomplished, the microbe will not be an evolutionary success.

At the heart of a microbial pathogen's success is an extraordinary rate of evolution in comparison with that of the human hosts. As discussed in Chapter 39, people have developed highly effective immune defenses against most infective disease agents. However, the rate at which new generations of bacteria and viruses are produced is measured in minutes or hours, not decades. Hence there are many more chances for microbes to benefit from the genetic events (mutations, recombination) that may lead to new opportunities. As a result, successful disease microbes are usually a step ahead in the race with the immune system.

Microbe Survival

To survive and cause disease, an individual microbe must, at some stage of its life cycle, be able to finesse the human immune system. Different microbes accomplish this in a number of distinct ways. Some microbes are only weakly antigenic, so the immune system fails to mount an effective response against them. Others prevent or suppress immune reactions (to their antigens) in the infected host. For example, bacteria use several mechanisms to avoid being phagocytosed and processed by macrophages (see Figure 40.5). Intracellular infectious agents, such as the **herpesviruses,** which live inside nerve cells, are not usually exposed to immune cells or circulating antibodies. Also, a few disease agents, especially certain viruses, frequently undergo changes in the molecular composition of their antigens (see Figure 40.6, page 772). As a result, the immune system does not initially recognize them when infection occurs, and they are able to become established in the body. Finally, many microbes (for example, those causing colds and sexually transmitted diseases) are successful because their "offspring" pass quickly and efficiently to a new host before an effective immune response that kills them in the original host can be initiated.

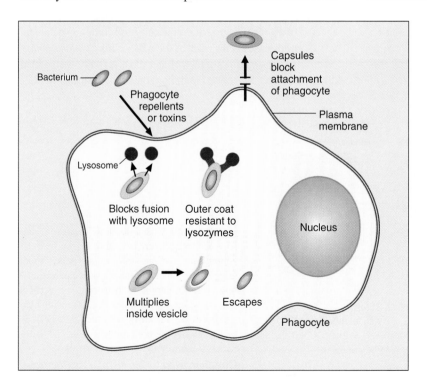

Figure 40.5 Some bacterial species have mechanisms to avoid being killed by phagocytes, among them secreting substances that kill or deflect phagocytes, capsules that block phagocyte attachment or fusion with lysosomes, and capsules that are resistant to the action of lysozymes. Also, some bacteria can multiply inside phagocytes without being destroyed.

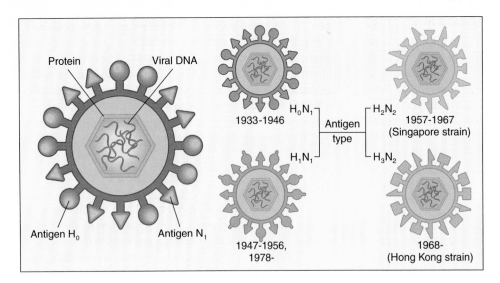

Figure 40.6 Influenza viruses occasionally undergo changes in their envelope proteins (antigens) that make them resistant to host immune responses, at least for a while. Strain A of the influenza virus has two major antigenic proteins on its envelope—antigen H is associated with cell attachment, antigen N with cell penetration. The "standard" strain A antigens are designated H_0N_1. In 1947, a new strain, H_1N_1, with different H antigens appeared. This strain remained in the population for ten years and then subsided, only to reappear in 1978. Other changes are indicated in the figure. The Hong Kong strain was particularly virulent and remains in the population today.

Question: *Why do changes in envelope proteins make influenza viruses resistant to an immune response?*

Disease Transmission

Assuming successful reproduction, how are microbes that require transmission passed on to other human hosts? Two general types of transmission are described in Figure 40.7. **Horizontal transmission** occurs when one individual infects another, directly or indirectly. **Vertical transmission** occurs when the microbe is transmitted from parent to offspring via infected sperm, ova, or milk or through the placenta or the birth canal. Certain cancer-causing viruses are known to be vertically transmitted, and several other diseases (for example, multiple sclerosis) are suspected of being spread by this route.

In developed countries, horizontal transfers via infected feces, food, and urine have decreased in significance over the past century, while two other routes have become increasingly important. Viruses that spread by *respiratory transmission* (sneezing, coughing, talking, kissing) and *sexual transmission* have been described as having entered their "golden age"—the first because of increasing population size and mobility, which allow individuals to contract infections from all parts of the world, and the second largely because of a greater degree of promiscuity and complex social factors. Some disease agents, such as the AIDS virus *[human immunodeficiency virus, type 1 (HIV-1)]* and most sexually transmitted diseases, can be transmitted both horizontally and vertically.

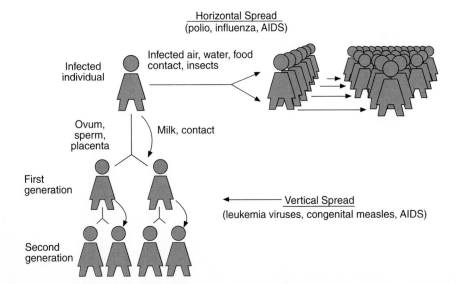

Figure 40.7 Most infectious disease organisms are spread from person to person by horizontal transmission—through air, water, food, or personal contact. Some diseases can be transmitted from parent to offspring by vertical transmission—through the placenta or infected sperm or ova. A few diseases, such as AIDS, can be transmitted by both routes.

Question: *What other diseases can be transmitted by both routes?*

Microbial pathogens must be able to infect hosts, survive in the host's environment, reproduce, and be transmitted to new hosts. To accomplish these tasks, they depend on different mechanisms for eluding the immune system and being transferred to a new host. Diseases can be transmitted horizontally, from one person to another; vertically, from parent to offspring; or both.

The Life of a Disease Organism

Bacteria, viruses, and other disease-causing agents face many challenges to be successful. They must gain entry into a human host, survive once they arrive, find a suitable place to grow and reproduce, and succeed in being transmitted to another human.

In Search of a Host

Disease-causing microbes commonly enter and leave the host through openings in the respiratory, gastrointestinal, or urogenital (urinary and reproductive) tracts. The portal of exit is generally related to the means by which the pathogen will be transmitted to and infect a new host.

Microbes in the respiratory passages and mouth can become incorporated into fluid droplets (aerosols) and expelled to the external environment by coughing, sneezing, talking, or kissing. For example, a person with a common cold can emit thousands of virus-containing droplets during one sneeze. Given the crowded environments in which most people spend at least part of the day (classrooms, buses, trains, subways, lunch lines, sporting events, offices, and so on), little can be done to prevent the spread of diseases via respiratory transmission.

Microbes that infect organs of the digestive tract are expelled in feces, and for many diseases, the process is accelerated by diarrhea. Numerous pathogens can be transmitted indirectly from feces to the mouth through contaminated drinking water. Many of the great disease epidemics in our history (for example, cholera) involve this route of transmission. Waterborne diseases still cause millions of deaths each year, primarily of children, in developing countries.

Most disease organisms of the urogenital tract are transmitted to new hosts through direct mucous membrane contact during sexual activities. In the past two decades, there has been a tremendous increase in the **prevalence** (percentage of the population infected) of sexually transmitted diseases such as gonorrhea, syphilis, and genital herpes. The underlying causes of this continuing increase are complex but include changes in social and sexual customs. Sexually transmitted diseases are described in a subsequent section.

Entry into the Body

Some microbes, such as *Staphylococcus* bacteria, can invade the body only through breaks in the surfaces of the skin or in mucous membranes lining the urogenital tract, respiratory tract, or digestive tract (see Figure 40.8A). However, many microorganisms, through evolution, have developed mechanisms for entering into or between cells of mucous membranes lining specific organs (see Figure 40.8B). For example, flu and cold viruses enter cells lining the respiratory tract, *Neisseria*

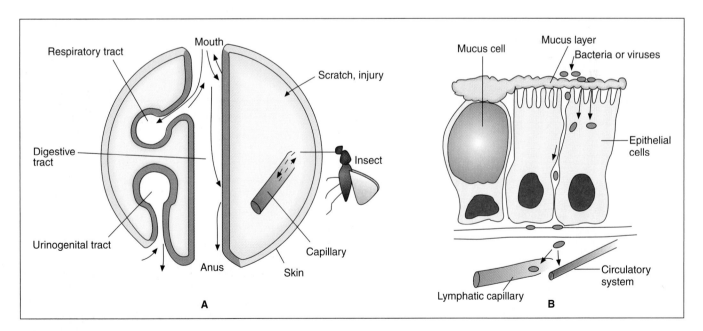

Figure 40.8 (A) Many pathogens enter and leave the human body through the digestive, respiratory, and urogenital tracts. Others enter through wounds in the skin or are injected into the body by mosquitoes or other insects. (B) Viruses and bacteria that cause disease often invade through or between cells of mucous membranes that line the major tracts.

Question: *Which disease transmitted by a mosquito kills more than 1 million people each year?*

(gonorrhea bacteria) infect cells of the urogenital tract, and *Salmonella* (food poisoning bacteria) pass through cells lining the intestinal tract. Finally, for certain diseases such as malaria, the agent (a protozoan) is injected through the skin by a mosquito, an insect that carries and transmits the agent.

There are also organisms—yeast, fungi, and, especially, bacteria—that are **opportunistic pathogens;** such agents can invade tissues and cause disease only under special conditions, most commonly involving diminished defense or immune capabilities. Opportunistic pathogens may normally live in the body where they do not cause disease in healthy humans, or they may live in water or soil but do not normally inhabit the human body unless defense systems fail. For example, *Pseudomonas aeruginosa* thrives in natural moist environments. Unfortunately, these bacteria have also become widely established in hospitals (showers, baths, wet kitchen sponges, and so on) and can cause infections in patients with weakened

defense systems. It is very difficult to control *P. aeruginosa* in hospitals because they are resistant to many antibiotics.

Infection of Cells and Tissues

Figure 40.9 describes the routes taken by different pathogenic microbes once they enter the body. Many disease microbes cause their effects by multiplying in the epithelial surface at the site of their establishment. In these cases, there is no invasion of underlying tissues. For example, respiratory flu viruses infect epithelial cells of the respiratory tract and produce new virions that quickly infect nearby cells. This hit-and-run process continues for a few days, with new viruses being shed, until the immune system catches up or until all available cells have been infected. In contrast, gonorrhea bacteria infect the epithelial surface of the urogenital tract and underlying tissues, which causes an inflammatory response. This leads to a char-

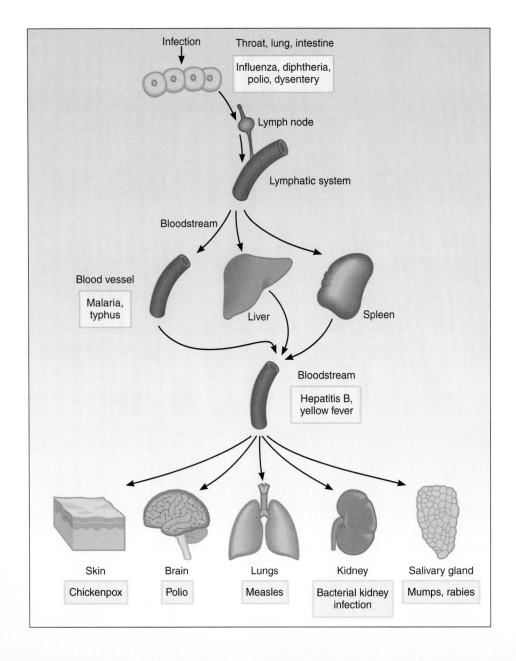

Figure 40.9 Once inside the body, pathogenic viruses and bacteria usually infect a specific cell population or tissue, where they become established, multiply, and cause damage.

acteristic yellowish discharge consisting of epithelial cells, leukocytes, bacteria, and inflammatory substances.

Once they enter the internal environment, most microbes are filtered out, inactivated, or destroyed in lymph nodes. This is the reason why swollen, sore nodes indicate that a disease agent–immune system encounter is under way. Microbes that escape destruction by this system can spread to other parts of the body in plasma, blood cells, or both. Components of the immune system continue to remove such microbes, but if they avoid these defenses or replicate more rapidly than they are removed and reach their target cells or tissues where they can grow and multiply, disease occurs.

Cell and Tissue Damage

All human cells, tissues, and organs can be occupied and damaged by specific disease-causing microbes. There is an obvious relationship between the tissue affected and the severity of a disease. Hence infections of the heart and brain that cause harmful changes and disease are usually much more serious than those of the liver or intestine. Pathogenic microbes can cause damage by several different mechanisms.

Viral replication usually results in death of the infected host cell. For example, one of the common cold viruses infects epithelial cells lining the nasal cavity. Infected cells die, become detached, and are carried away in a stream of fluid (a "runny nose"). Bacteria that infect cells also destroy their host cell. Tissue damage occurs after many cells have been killed.

Dental cavities are perhaps the most common disease of developed countries. They occur as a result of direct damage caused by a complex of bacteria, which colonize the surfaces of teeth, cause plaque (a film of bacteria and saliva) to form, and ultimately, as shown in Figure 40.10, create a cavity in the

Figure 40.10   Bacteria that colonize the teeth can cause cavities through the release of acids that penetrate the tooth enamel.

tooth. These bacteria metabolize sugars in the mouth as an energy source, producing acids that act to decalcify the tooth, and a cavity results. This process can be reduced by removing or preventing plaque by flossing and brushing teeth and reducing the dietary intake of sugars.

Exotoxins are compounds released by certain bacteria while they are multiplying. These substances are mostly proteins, and some are among the most powerful biological poisons known. For example, tetanus and botulism are often fatal diseases caused by bacterial exotoxins. Many bacteria, including species of *Streptococcus, Staphylococcus,* and *Legionella,* secrete exotoxins that kill or damage cells. Toxic shock syndrome is a dangerous disease that occurs most commonly in menstruating women. In some cases, tampons have apparently been contaminated by an exotoxin-secreting *Staphylococcus.* Alternatively, considering that menstrual fluids are a good growth medium for bacteria, tampons may scratch the vaginal lining, allowing *Staphylococcus* direct entry into the body.

Susceptibility and Risk Factors

Infectious diseases, like most diseases, do not tend to strike individuals randomly. Many factors influence the chances of an individual's suffering from a certain disease. A **risk factor** is anything that increases the probability of developing a disease. For example, some risk factors for sexually transmitted diseases are fairly obvious: sexual intercourse with multiple partners, intercourse with an infected person, intercourse with no condom, and so on. There are two general types of risk factors: **modifiable risk factors**—those that are subject to some degree of individual control (use of a condom, cigarette smoking)—and **nonmodifiable risk factors**—those that are not subject to individual control (genetic disorders, age). Scientists attempt to identify specific risk factors for a disease and assign a quantitative value to them that reflects their relative importance. This approach has emerged as an extremely valuable tool that can be used by enlightened individuals in making decisions about their personal behaviors or lifestyle relative to disease.

Some general risk factors for infectious diseases are immune depression, age and sex of the host, malnutrition, hormonal factors, fatigue, and stress. Which of those are modifiable, and which are not?

BEFORE YOU GO ON   Microbial infections usually depend on delivery to a new host through the respiratory, digestive, or urogenital tract. Once inside the body, a microbe often lives in specific cells or tissues, which may be damaged or killed. Through knowledge and management of risk factors, individuals can greatly reduce their chances of getting some infectious diseases.

SEXUALLY TRANSMITTED DISEASES

A **sexually transmitted disease (STD)** is any infectious disease that is spread by sexual contact. Over 50 STDs are caused by parasites, bacteria, viruses, and fungi. Some STDs of great importance, in terms of numbers of people infected and severity of effects, include **gonorrhea, syphilis,** and **chlamydia,** which are all caused by bacteria, and **genital herpes** and **genital warts,** which are caused by viruses. AIDS is a virus-caused STD, and it is described in Chapter 41.

STDs in the United States

Information on STDs and data for the United States are given in Table 40.3. Most developed countries—Japan, Australia and New Zealand, the European Union—have a low prevalence of STDs because of effective public health programs. In the United States, however, STD cases have increased at astonishing rates in the past decade. For example, **chlamydia**, which was rarely reported ten years ago, is now the most common STD, with 4 million new cases being reported yearly. The U.S. Centers for Disease Control and Prevention (CDC) estimates that there are over 12 million new STD infections each year and that at least 25 percent of Americans will develop an STD in their lifetime. The CDC contributes funds and technical expertise to state and local agencies to conduct diagnostic tests for STDs, conducts research related to STDs, and supports programs on risk reduction education, epidemiology, and training.

Chlamydia

Chlamydia trachomatis can serve as a model to describe infectious agents that cause STDs. Questions that guide research are related to structure of the disease agent, its life cycle, cells and tissues affected, and harmful effects produced.

Chlamydia species are classified as bacteria because they have a cell wall, RNA and DNA, and ribosomes. However, they were once considered to be viruses because they can replicate

only inside a host cell. Also, *Chlamydia* are "energy parasites"; they depend on host cells for energy because they are unable to synthesize ATP. *C. trachomatis* uses two specialized forms in its life cycle—an *elementary body (EB),* which attaches to a target cell and promotes entry into the cell, and a *reticulate body (RB),* which reproduces inside the cell. EBs infect epithelial cells lining mucous membranes of the urogenital tract, and they have two interesting adaptations that allow them to enter and survive inside cells (see Figure 40.11). First, they secrete a protein that induces phagocytosis by epithelial cells, a cell type that does not normally have phagocytic capabilities. Second, once inside the cell, the phagocytic vesicle is prevented from fusing with a lysosome, which prevents the EB from being digested. The complete *C. trachomatis* life cycle is described in Figure 40.12 (see page 778).

Individuals infected with *C. trachomatis* may not develop any early symptoms, or they may have mild symptoms that go undiagnosed, depending on the number of cells affected. When symptoms do occur, they are related to inflammatory and cellular immune responses directed against infected cells or damaged cells and tissues. In women, infection begins in the cervix and can spread from there to the uterus, uterine tubes, and ovaries. In men, infections occur in the urethra and may spread to the epididymis. The inflammatory and immune responses cause pus and fluid accumulations and discharges, itching, pain, and in women, scars in affected organs and, in some cases, sterility.

Effects of STDs

Cells and tissues affected by various STD agents are summarized in Table 40.3, along with typical symptoms in both sexes. Certain features about these STDs require further elaboration. There are no cures for viral-caused STDs, although symptoms can be

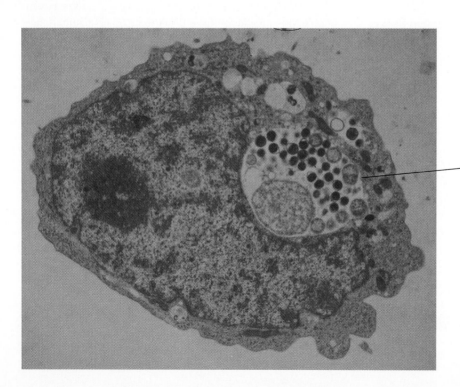

— *Chlamydia*

Figure 40.11   A *Chlamydia* bacterium (arrow) within an infected cell.

Question: *Why were* Chlamydia *once thought to be viruses?*

Table 40.3 Information on Sexually Transmitted Diseases (STDs) in the United States

Disease Characteristic	Sexually Transmitted Disease Agent				
	Gonorrhea	**Syphilis**	**Chlamydia**	**Genital Herpes**	**Genital Warts**
Type of disease agent	Bacterium (coccus)	Bacterium (spirochete)	Bacterium (coccus)	Virus	Virus
Name of disease agent	*Neisseria gonorrhoeae*	*Treponema pallidum*	*Chlamydia trachomatis*	Herpes simplex, type II	Human papillomavirus
New cases each year	1 million	100,000	4 million	500,000	1 million
Site of infection	Extracellular; lives among epithelial cells of mucous membranes lining the urogenital tract	Extracellular; lives among epithelial cells of mucous membranes lining the urogenital tract; spreads rapidly to other tissues	Intracellular; epithelial cells of mucous membranes lining the urogenital tract	Intracellular; epithelial cells of mucous membranes lining the urogenital tract; later spreads to neurons, where it replicates	Intracellular; epithelial cells of mucous membranes lining the urogenital tract and rectum
Tissues generally affected	Mucous membranes of the urogenital tract, rectum, mouth; eyes of newborn infants	Initially, mucous membranes of urogenital tract; after weeks to years, spreads to almost every tissue	Mucous membranes of the urogenital tract	Mucous membranes of the urogenital tract and rectum	Mucous membranes of the urogenital tract and rectum
Transmission	Contact with mucous membranes of an infected person	Contact with mucous membranes of an infected person	Contact with mucous membranes of an infected person	Contact with mucous membranes of an infected person	Contact with mucous membranes of an infected person
Symptoms in females	Inflammation of the vagina, cervix, or urethra; may lead to pelvic inflammatory disease (PID)[a]	Initially, open sores on cervix; after 6 to 24 weeks, skin rash, muscle and joint aches; after years, cardiovascular and nervous system effects	Inflammation of the cervix; vaginal discharge, itching burning, and bleeding; may lead to PID	Formation of painful blisters or sores in the cervix and vagina; sores may reappear throughout life	Warty growths around genital and anal area; some people have no symptoms
Symptoms in males	Inflammation of the urethra; painful urination	Same as in females, except sores on penis	Inflammation of the urethra painful, frequent urination	Same as in females except sores on penis	Same as in females
Transmission from mother to infant	In the birth canal during birth	Through placenta (50 percent of infected fetuses are aborted or stillborn)	During birth	During birth	During birth
Cure	Antibiotics	Antibiotics	Antibiotics (not penicillin)	None	None

[a]Pelvic inflammatory disease (PID) occurs when a cervical infection spreads to the uterus, uterine tube, or ovaries. It is estimated that 100,000 to 150,000 women become sterile each year because of an STD infection that led to a PID.

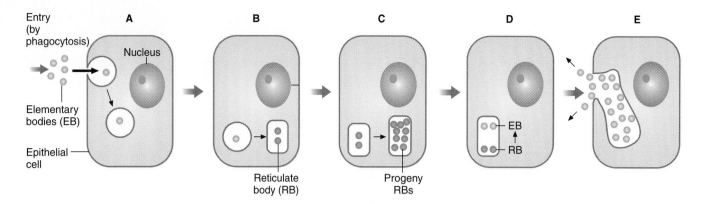

Figure 40.12 The *Chlamydia* life cycle consists of the following sequence of events: (A) An elementary body (EB) attaches to and is phagocytized into an epithelial cell lining the urogenital tract. (B) Once inside the cell, elementary bodies develop into reticulate bodies (RB) within 24 hours. (C) RBs reproduce by binary fission during the next 24 hours, forming 100 to 500 progeny RBs. (D) RBs develop into EBs. (E) Some 72 to 96 hours after infection, new EBs move to the plasma membrane, where they are released (infected cells may or may not be killed by this process). The new EBs then go on to infect new cells. EBs are relatively resistant, inactive particles that can withstand conditions outside of a cell; they are specialized for survival and infection. RBs are fragile and metabolically active. They are specialized for reproduction and are totally dependent on host cells for providing them with energy to perform this function.

treated when they become apparent. The CDC estimates that over 50 million Americans are now infected with incurable viral STDs, exclusive of HIV-1 (the AIDS virus); over 1 million people in the United States have been infected with HIV-1. Bacterial STDs can usually be cured once they are detected. However, the number of antibiotic-resistant sexually transmitted bacterial strains has increased sharply in the past decade. Also, not all infections become obvious until serious symptoms develop.

Women are more susceptible to STD infections than men because the female reproductive tract is more suitable for microbial growth. Women are also at much greater risk than men for suffering severe effects if an STD is not detected and treated soon after infection occurs. Each year, an estimated 1 million women develop **pelvic inflammatory disease (PID),** which occurs when a gonorrhea or chlamydia infection spreads from the cervix to the uterus, uterine tubes, or ovaries. PIDs cause sterility in over 100,000 women each year. Women infected with viral STDs also have an increased risk for developing genital cancers. That is, women infected with the herpes simplex virus, type II, and the human papilloma virus are more likely to develop cancers of the cervix, uterine tubes, or uterus than women who are not infected. Infected women can also transmit STDs to their offspring through the placenta or at birth when the child descends through the birth canal (genital tract). Some of these infections can have serious consequences for newborn infants. For example, *Chlamydia* can cause pneumonia, and 25 percent of babies infected with syphilis die in early childhood.

Risk Factors for STDs

The greatest risk for acquiring a STD is obvious—having sex with an infected partner. However, there are two separate processes to consider, each with a different set of risk factors.

Reducing risk by, *not contacting an infected partner* focuses on *sexual behavior; reducing the chance of getting an STD* involves *health care behaviors.* For bacterial-caused STDs (gonorrhea, syphilis, and chlamydia), both processes are relevant because the diseases can be cured by antibiotics after infection occurs. For viral-caused STDs (genital herpes and genital warts), the first process is especially important because these diseases cannot be cured after infection occurs.

Sexual behavior risk factors are correlated with the probability of encountering an infected partner. Major sexual behavior risk factors include many sex partners, high rates of acquiring new sexual partners and changing partners, contact with casual sexual partners, and sex with a prostitute. **Health care behaviors** that reduce the risk of developing an STD include use of condoms for protection, early diagnosis and treatment if infection occurs, and informing partners if infected. All of these are classified as modifiable behaviors. However, to be effective, individuals must be educated about their importance and use *before* they become sexually active.

Society and STDs

Who acquires STDs in the United States? This subject has generated controversy because of factors associated with the recent dramatic increase in STDs. STD data are relatively imprecise because infected people may not seek treatment or, if treated by private physicians or in private clinics, may not be reported to the CDC. Nevertheless, certain STD trends have been identified:

■ Gonorrhea and syphilis are five to ten times more common in African Americans than in whites. However, much of the difference involves young, inner-city black populations, which have the highest infection rates for gonorrhea, syphilis, and herpesvirus. Young, inner-city Latinos also have high infection rates for these STDs. These extreme infection rates for inner-city areas have been attributed to high population densities, the high percentage of teenagers and young adults in the populations, urban poverty, social disintegration, lack of economic opportunities, widespread prostitution, and drug abuse.

■ *Chlamydia* infections occur in all racial and ethnic groups and in all socioeconomic classes.

■ *Chlamydia* and syphilis are increasing at epidemic rates in teenagers of all incomes, races, and ethnic groups.

■ More than 3 million teenagers—one of every eight people between the ages of 13 and 19—acquire an STD each year.

What is the solution to the enormously difficult problem posed by STDs? How can our society begin to address and reduce the epidemic rates of STDs in poor, inner-city populations and *Chlamydia* in all racial and ethnic populations and socioeconomic groups? Some proposals have been generated, which are summarized in Table 40.4, but no political consensus has yet been reached, and funding remains a problem.

Table 40.4 Proposals for Reducing STD Infections

• Increased emphasis should be given to diagnosing and treating *Chlamydia* because of the huge number of people infected and the serious reproductive health problems that it causes. *Chlamydia* is the leading cause of sterility that can be prevented by providing diagnostic tests and treating infected women.

• Increased educational efforts to change individual sexual behaviors and help individuals reduce their risk of being infected. Special attempts should be made to provide young people with up-to-date education on STDs before they become sexually active.

• *Design programs that emphasize prevention rather than treatment.*

• Increase public health care capacities, especially in inner-city communities.

Source: Centers for Disease Control and Prevention.

None of these proposals, of course, addresses the fundamental societal problems related to poor, inner-city populations. Many health officials believe that educational approaches, without significant economic and social programs to improve inner-city life, are not likely to be effective. Also, there is resistance in some communities to educating children about STDs. All of the strategies cited require money, yet the amount now spent on STD programs—approximately $100 million—is only 75 percent of that spent in 1950, when adjusted for inflation. Ultimately, society will decide whether to continue to fund STD programs and if so, to what extent. Should state and federal governments and private institutions be doing more? What can be done?

BEFORE YOU GO ON In the past decade, sexually transmitted diseases have become an increasing public health problem in the United States. Young, poor, inner-city minority populations have high rates of gonorrhea, syphilis, and genital herpesvirus infections. Chlamydia and genital warts are common in all ethnic groups and all socioeconomic classes. Risk factors for avoiding STDs have been identified. Programs for reducing STDs in the United States have been proposed, but society has yet to commit to attacking underlying problems.

CANCER

The word *cancer* evokes powerful feelings in many individuals. Fear is one of the most common responses, perhaps because cancer is perceived to strike individuals at random and to have no cure. However, cancer is generally a disease of old age. The major reason for this age association is that cancer takes a long time to develop, but there are also other considerations. The ravages of a lifetime of exposure to cancer-causing substances, the age-related decline in immune system effectiveness, and an inability to repair cell damage may explain, in part, why cancer most often affects older individuals.

Because of the importance of this disease in terms of numbers of people affected directly and indirectly, society has mandated that enormous amounts of money and research effort be directed toward solving the cancer problem. Certain efforts, such as the "war against cancer" proclaimed in the 1960s, were based on political concerns and naive assumptions. Underlying this effort was a simplistic belief that cancer was a distinct disease and that a cure could be found, given enough time and money.

It is now recognized that understanding cancer represented an extremely difficult challenge. Yet phenomenal progress was made in cancer research during the 1980s, and certain facets of this disease have rapidly come into focus. At least partial answers are now available for many fundamental questions. What is cancer? What causes cancer? Who gets cancer? Can cancer be cured? What does the future hold?

Cancer in the United States

The American Cancer Society estimates that approximately 1 million new cases of cancer will be diagnosed each year in the 1990s. About 30 percent of all Americans will develop cancer during their lifetime, and, projecting current rates, 40 percent will be treated successfully (survive for at least five years after treatment). Each year, about 500,000 Americans die of cancer. Cancers of various organs have declined or remained nearly constant for over 50 years; the exception is lung cancer, which has increased. Figure 40.13 (see page 780) provides information about the most common cancers occurring in the American population.

The direct causes of cancer involve changes in gene structure or function (see Chapter 41). However, the occurrence of many cancers is thought to be related to an individual's envi-

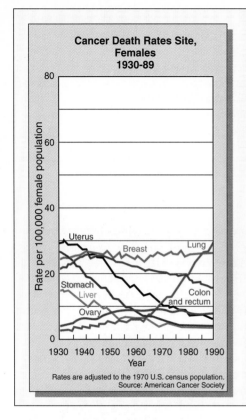

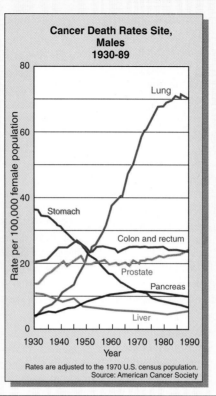

Figure 40.13   Death rates for various cancers in females and males from 1930 to 1989.

Question: *What accounts for the increases in lung cancers during this period?*

ronment, personal habits, or lifestyle. Risk factors for cancer include tobacco and alcohol use, improper diet, and occupational or environmental exposures to **carcinogens** (cancer-causing agents). The dominant cancer risk factor is the use of tobacco; it is estimated to cause at least 30 percent of all cancers. However, it is impossible to quantify the level of risk associated with most other factors or even to identify specific risk factors with certainty. The exact relationship between diet and cancer is unknown, although diet is suspected to be associated with 20 to 35 percent of all cancer deaths.

Normal and Abnormal Cell Growth

Tissues and organs are composed of cells that carry out specific functions. In many tissues, cells carry out their activities for a genetically prescribed length of time, die, and are replaced by new cells, which are produced at a constant pace. The rate of cell replacement varies among organs. For example, intestinal cells and most leukocytes live only a few days; red blood cells live about three to four months; liver cells seldom die yet can be rapidly replaced by cell division if serious loss occurs; and cells of the central nervous system are never replaced. In adults, steady-state homeostasis exists between cell birth and cell death. To maintain this balance, cell growth and reproduction are exquisitely regulated, primarily by hormones and growth factors. Different regulatory substances exert their control either by inhibiting or stimulating cell growth or proliferation. What happens if regulation is lost? On

rare occasion, one or more cells escapes from the regulatory system and grows and divides at an accelerated rate, well beyond the normal replacement requirements of a tissue. If such a cell gives rise to a clone of cells that also does not respond to regulatory substances, a cell mass known as a **tumor** eventually develops.

Tumors

Benign tumors grow slowly, become surrounded by connective tissue, and remain localized; they can usually be removed by a surgeon if necessary. Tumors that consist of rapidly reproducing cells that invade surrounding tissues and spread throughout the body are called **malignant tumors. Cancer** is a large group of diseases that are characterized by malignant tumors.

Malignant Cells

Cells of a malignant tumor are known as **malignant cells** or *cancer cells.* Why do malignant cells fail to respond to normal regulatory mechanisms? A simple (and incomplete) answer is that these mechanisms become permanently altered through some process and consequently no longer receive, recognize, or respond to regulating signals. Such changes cause alterations in the growth and developmental properties of a cell, a process called **transformation.** Transformed cells, such as those shown in Figure 40.14, specialize in proliferation, and this is reflected in their structure; they have an undifferentiated

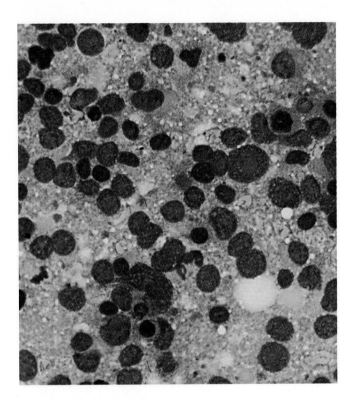

Figure 40.14 Cancer cells arise from transformed normal cells. Once transformation occurs, malignant cells are able to proliferate and grow quickly because they are not regulated by host control mechanisms. This ability is reflected in their structure. Breast cancer cells, shown here, are poorly differentiated, are abnormally large and vary in size, and have large nuclei.

appearance that is characteristic of rapidly reproducing cells. The frequency at which cells are transformed is not known, but it is thought to occur repeatedly. Most transformed cells are destroyed by the natural killer cells and cytotoxic T cells of the immune system. Under circumstances of immune suppression (and others as yet unknown), malignant cells that cause a cancer escape such detection and elimination.

Malignant cells have other distinguishing characteristics. They may have abnormal numbers of chromosomes, and they can penetrate and invade surrounding tissues. Most important, they can undergo **metastasis,** a process in which cells from the original *primary tumor* can be transported through the circulatory or lymphatic system to other tissues in the body, where they become established and create new malignant growths called *secondary tumors* (see Figure 40.15). Most secondary tumors develop in the liver, lungs, or bone marrow (see Figure 40.16, page 782).

Effects on Untreated Patients

If unchecked, the hallmark characteristics of malignant cancer—uncontrolled proliferation, invasiveness, and metastasis—lead to death of the individual. Because their rate of growth vastly exceeds replacement requirements, malignant cells displace or overwhelm normal cells. Invasiveness and metastasis result in the same effect, uncontrolled growth and displacement of normal tissues over a widespread area (see Figure 40.17, page 782). Death caused by cancer is related to depletion of the host's nutritional reserves due to the growth requirements of malignant cells or loss of function in the organs or tissues affected.

Understanding these effects enables one to appreciate the critical importance of early treatment with drugs, radiation, surgery, or a combination of these therapies. Rarely does cancer regress without some form of medical intervention. The earlier this whole progression of events—loss of regulation, transformation, proliferation, malignant tumor formation—is halted, the better the chance for survival. Once invasion and metastasis have occurred, the probability of a successful treatment is very low.

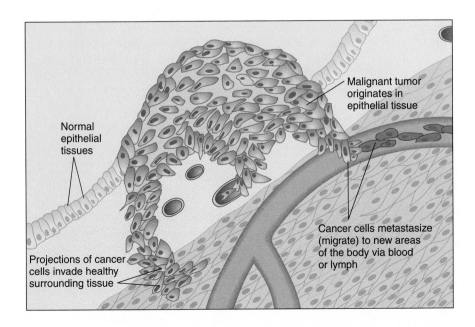

Normal epithelial tissues

Projections of cancer cells invade healthy surrounding tissue

Malignant tumor originates in epithelial tissue

Cancer cells metastasize (migrate) to new areas of the body via blood or lymph

Figure 40.15 Malignant cells can invade surrounding tissues and metastasize—travel to new tissues or organs via the circulatory or lymphatic systems, where they become established and form secondary tumors.

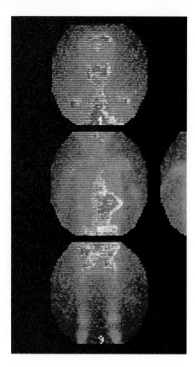

Figure 40.16   Cancer cells often metastasize to new organs. Specialized imaging technology shows that cancerous growths (in red) have become established in the brain, lung, liver, and bone marrow of the pelvic girdle.

Question: *Why are the lungs the principal organ where secondary tumors form after metastasis?*

Causes of Cancer

To inquire about the cause of cancer is to ask, "What causes cell transformation?" Until very recently, most efforts to answer this question were directed at identifying specific types of cancer-causing agents. From research conducted over the past half century, certain viruses, chemicals, and radiation have been implicated in causing human cancers, although there is no clear understanding of the numbers of cancers attributable to each. Causes of cancer are considered in greater detail in Chapter 41.

BEFORE YOU GO ON Cancer occurs when normal cells become transformed into malignant cells that are not killed by cells of the immune system. The growth of malignant cell populations cannot be regulated by the host; consequently, they increase rapidly and may metastasize to new sites. If treatment is delayed, death usually occurs. The major risk factor for cancer is tobacco products.

VASCULAR DISEASES

Cardiovascular diseases are malfunctions of the heart and blood vessels. **Cerebrovascular diseases** occur when the blood supply to the brain is either temporarily or permanently interrupted. Given the critical functions of the circulatory system and these two organs, it is not surprising that failures of this system have serious, often fatal consequences. Approximately one million Americans die of vascular diseases each year.

Cardiovascular Disease

The major cause of coronary heart disease is **atherosclerosis,** the buildup of fibrous, fatty deposits, called **plaques,** on the inner walls of major arteries. The development of plaques consists of a complex series of interactions in the artery lining involving cholesterol, macrophages, and cell growth factors (see Figure 40.18). Atherosclerosis results in loss of artery elasticity and a thickening of arterial walls, which can diminish or block blood flow to major organs, including the heart and brain. Blood flow dynamics around plaques are altered in such a way that blood clots may also form on the arterial wall. If these break loose, they may block a smaller artery downstream.

When the flow of blood to the heart is hindered because of plaques or blocked by a blood clot, muscle cells of the heart are deprived of oxygen and glucose. When this occurs, the cells die, causing some loss of heart function, and a heart attack follows. The degree of damage is related to the location of the obstruction and the number of heart cells lost. If the block occurs at the beginning of a coronary artery, the heart attack will be severe and most likely fatal (see Figure 40.19). If the

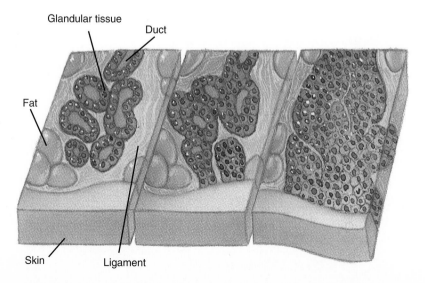

Figure 40.17   The increase in malignant cell populations is characterized by two processes—a rapid rate of cell division and invasiveness, in which cells grow and penetrate into surrounding tissues. This drawing illustrates a breast tumor that arose in the glandular tissue (A), grew rapidly (B), and invaded the surrounding tissues (C). A lump would first be detectable when a tumor had formed (B).

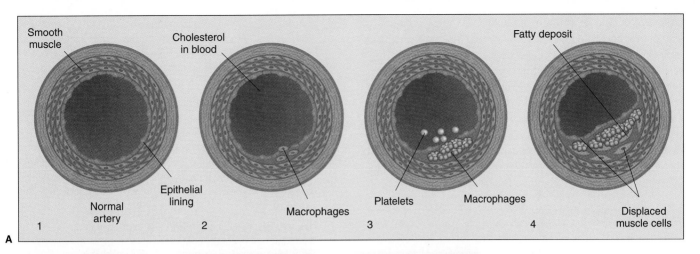

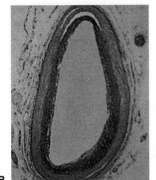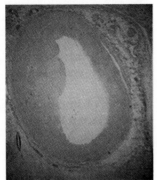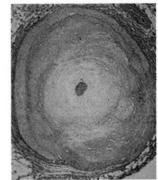

Figure 40.18 (A) Plaque formation in blood vessels begins when excess cholesterol becomes incorporated into the lining of blood vessels and macrophages attempt but fail to remove it. Subsequently, fat deposits at the site increase, underlying muscle and connective tissue cells proliferate, and the entire complex eventually becomes a plaque. (B) The photographs show a normal artery (left), a partially obstructed artery (center), and an artery with advanced atherosclerosis (right).

Cerebrovascular Disease

A **stroke** is a sudden, severe event causing irreversible damage to the brain due to interference in normal blood circulation. Such interruptions occur if a blood clot blocks one of the arteries leading to the brain (**cerebral thrombosis**) or if an artery in the brain ruptures (**cerebral hemorrhage**). The severity of the stroke generally depends on the part of the brain affected and the extent of the damaged area. There is a broad range of possible effects, from relatively minor (a speech impediment) to major (coma, death). The common causes of stroke are atherosclerosis and **hypertension** (high blood pressure).

Risk Factors for Cardiovascular and Cerebrovascular Diseases

Risk factors for vascular diseases as determined through various studies are tobacco use, elevated serum cholesterol, high blood pressure, obesity, diabetes, and a sedentary lifestyle. The basic approach used in identifying risk factors is to collect and analyze relevant data on individuals who were affected by a cardiovascular disease. Two basic questions are asked: Is there a correlation between a risk factor of interest and the disease under study, and what is the strength of the relationship?

Typically, results from hundreds of studies are analyzed by scientists representing agencies associated with different diseases (for example, the National Cancer Institute; the National Heart, Lung and Blood Institute; and the American Heart Association), and a consensus is formed about the degree of

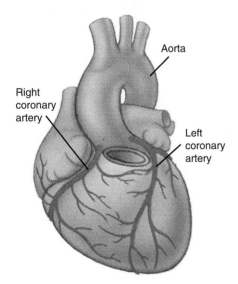

Figure 40.19 The severity of heart attacks is related to the site where circulation is blocked. If blockage occurs at the beginning of a major coronary artery, death usually results.

blood supply is reduced or lost in a more distant part of the artery, the effect may be less severe.

A number of important advances in preventing and treating heart attacks have been made during the past decade. New drugs, such as blood clot–dissolving enzymes, and new surgical procedures (to remove or bypass plaques) are now available that are highly effective in extending the lives of individuals suffering from heart disease.

risk. *Major* risk factors are those that are most commonly found to be correlated with the disease and that most scientists agree are likely to be associated with the disease.

It must be emphasized, however, that these do not represent *proof* in the classic scientific use of the term. What would it take to prove that a specific risk factor caused a disease? The simplest study required would be to compare the occurrence of the disease in two groups of people that differed only in the risk factor being investigated (for instance, one group smoked and the other did not). Because of the great diversity of individual lifestyles, such a study would be impossible.

> **BEFORE YOU GO ON** Circulatory diseases of the heart and brain are the number one cause of death in the United States. Atherosclerosis is the principal cause of cardiovascular disease and, along with hypertension, cerebrovascular disease. Major risk factors for these diseases are tobacco use, diet, and lack of exercise.

RISK FACTORS AND LIFESTYLE MANAGEMENT

Each year, about 11 percent of the U.S. gross domestic product ($2,500 to $3,000 for every man, woman, and child) is spent on health, with the largest share by far going for treatment—not prevention—of disease. Because it is estimated that more than 60 percent of all Americans die prematurely (not of "old age"), it is widely believed that much more of that amount should be directed toward preventing disease.

Detection and treatment are important for all diseases, but preventive actions addressed at changing individual behaviors could be much more effective in terms of personal and social costs. Researchers estimate that more effective management of major risk factors for vascular diseases and cancer (poor diet, use of tobacco, abuse of alcohol, lack of exercise) could prevent between 40 and 70 percent of all premature deaths. In comparison, the use of advanced medical treatments is not likely to reduce premature death by more than 10 to 15 percent.

Understanding risk factors is helpful to individuals wishing to optimize their chances of not getting a particular disease. A straightforward analysis suggests that a person who makes decisions to not smoke, to eat foods with little or no fat and ample amounts of fiber, and to lower blood pressure through exercise and careful nutrition would greatly reduce the risk of coronary heart disease and stroke. In fact, data indicate that such lifestyle changes reduce the chances of suffering from one of these diseases by a factor of 5 to 10.

This approach recognizes the complexities of modern diseases. It also enables individuals to take primary responsibility in making decisions about managing their lives in such a way that they reduce the likelihood of falling victim to a disease.

SUMMARY

1. Until the 1950s, it was widely believed that most diseases were caused by infectious agents that could be controlled if the proper antibiotic was discovered. Since then, other disease-causing agents have been found that modified this view of disease. Today, disease is considered to be any abnormal condition that impairs functioning. Causes of disease include infectious microbes, nutritional deficiencies, abnormal genes, and other agents, as well as lifestyle factors.

2. Many infectious diseases of historical importance have largely been eliminated in developed countries thanks to antibiotics, sanitation practices, and public health measures. As a result, citizens in these countries generally die from diseases commonly associated with old age—cancer and cardiovascular disease. However, some infectious diseases and various sexually transmitted diseases are becoming reestablished in North America.

3. Bacteria and pathogenic viruses can gain entry to the interior of body through openings in the external surface or through the digestive, respiratory, and urogenital systems. If they gain entry, they can become established, grow, and reproduce in various cells and tissues. As a result, they may kill or damage the cells or tissues they infect and other tissues if they are able to spread to different areas of the body. Bacterial infections can often be treated with antibiotics without harming host cells. Viruses infect cells and cannot be killed by antibiotics without affecting host cells. Most bacterial and viral infections are eventually eliminated through actions of the immune system.

4. Sexually transmitted diseases—gonorrhea, syphilis, chlamydia, genital herpes, and genital warts—have become a major public health problem in the United States. Gonorrhea, syphilis, and genital herpes are especially prevalent in young, poor, inner-city African-American and Latino populations; chlamydia and genital warts are common in all ethnic groups and socioeconomic classes. Modifiable risk factors for STDs have been identified, but complex social problems have hindered their implementation. Prospects for reducing the epidemic rates of STD infections are uncertain.

5. Cancer includes a large number of diseases characterized by the unregulated and excessive growth of a malignant cell population that arises from the transformation of a normal cell. If not brought under control, the growth of malignant cell populations will usually deplete host resources, displace normal cells, disrupt normal organ function, and cause death of the host.

6. Vascular diseases are a group of circulatory system diseases that cause the deaths of about 1 million Americans each year. The major cause of coronary heart disease and primary cause of death is atherosclerosis, a complex process that leads to a failure of blood vessels leading to the heart, which results in heart attacks. Strokes occur if blood flow to the brain is disrupted.

7. Risk factors for various diseases and other causes of death have been identified and can be used for preventive purposes by informed individuals.

WORKING VOCABULARY

atherosclerosis (p. 782)
benign tumor (p. 780)
cancer (p. 779)
carcinogen (p. 780)
cardiovascular disease (p. 782)
cerebrovascular disease (p. 782)
chlamydia (p. 776)
disease (p. 764)
infectious disease (p. 764)

REVIEW QUESTIONS

1. What is disease? What are some of the causes of disease?

2. Why are infectious diseases less important in developed countries than in undeveloped countries? What types of diseases are most serious in developed countries?

3. How do viruses cause disease?

4. What are the two general processes of disease transmission?

5. How do bacteria and viruses enter the body?

6. How do bacteria and viruses affect cells and tissues after infection?

7. What are risk factors? How can they be used in health protection?

8. What are sexually transmitted diseases (STDs)? Which STDs are presently of great concern in the United States?

9. What are the risk factors for sexually transmitted diseases?

10. What is cancer? How does cell transformation occur? What is the significance of cell transformation?

11. Why is prompt treatment of cancer necessary for increasing the chances of survival?

12. What is cardiovascular disease? Cerebrovascular disease? What are the major risk factors for these diseases?

13. Why is there increasing emphasis on lifestyle management in reducing disease?

ESSAY AND DISCUSSION QUESTIONS

1. If you were the minister of health in a developing country, what recommendations would you make when preparing your budget to address the nation's health problems? How would they differ from those made for a developed country?

2. Assume that you are asked to create a lifestyle management program for yourself. Which diseases would concern you most? What risk factors would you emphasize? Why?

3. Drugs, antibiotics, and lifestyle management are now used to combat diseases in the United States. What new approaches might be used in the future? Explain.

4. A significant number of people die from disease. What factors, other than disease, are important in determining the theoretical maximum life expectancy of humans?

REFERENCES AND RECOMMENDED READING

Amábile-Cuevas, C., M. Cárdenas-Garcia, and M. Ludgar. 1995. Antibiotic resistance. *American Scientist*, 83:320–329.

Amler, R. W., and H. B. Dull. 1987. *Closing the Gap: The Burden of Unnecessary Illness.* New York: Oxford University Press.

Bartecchi, C. E., T. D. MacKenzie, and R. W. Schrier. 1995. The global tobacco epidemic. *Scientific American*, 272:44–51.

Brandt, A. M. 1987. *No Magic Bullet: A Social History of Venereal Disease in the United States Since 1880.* New York: Oxford University Press.

Epstein, J. 1994. *Altered Conditions: Disease, Medicine, and Storytelling.* New York: Routledge.

Ewald, P. W. 1994. *Evolution of Infectious Diseases.* New York: Oxford University Press.

Hamann, B. P. 1994. *Disease: Identification, Prevention, and Control.* St Louis: Mosby.

Kraut, A. M. 1994. *Silent Travelers: Germs, Genes, and the "Immigrant Menace."* New York: Basic Books.

Levins, R., T. Awerbuch, U. Brinkmann, I. Eckhardt, and P. Epstein. 1994. The emergence of new diseases. *American Scientist*, 82: 52–60.

Magner, L. 1992. *A History of Medicine.* New York: Dekker.

Morse, S. S. 1993. *Emerging Viruses.* New York: Oxford University Press.

Mulvihill, M. L. 1995. *Human Diseases: A Systemic Approach.* 4th ed. Norwalk, Conn.: Appleton & Lange.

National Academy of Sciences. 1995. *Infectious Diseases in an Age of Change: The Impact of Human Ecology and Behavior on Disease Transmission.* Washington, D.C.: National Academy Press.

Paul, W. E. 1993. Infectious diseases and the immune system. *Scientific American*, 269: 91–97.

Stanbury, J. B. 1995. On the patterns of disease: a nosography. *Perspectives in Biology and Medicine*, 38:521–534.

Tamparo, C. D. 1994. *Diseases of the Human Body.* 2nd ed. Philadelphia: Davis.

Willett, W. C. 1994. Diet and health: What should we eat? *Science*, 264: 532–537.

ANSWERS TO FIGURE QUESTIONS

Figure 40.2 AIDS.

Figure 40.3 *Treponema* (syphilis).

Figure 40.6 Envelope proteins act as antigenic determinants in a primary immune response, and antibodies are formed against them, along with memory cells that prevent reinfection. Viruses with new envelope proteins (antigenic determinants) are not recognized by memory cells, and so no rapid immune response is initiated.

Figure 40.7 Most sexually transmitted diseases.

Figure 40.8 Malaria.

Figure 40.11 They are intracellular and depend on the host cell for energy.

Figure 40.13 Use of tobacco products (smoking).

Figure 40.16 All cells carried by the circulatory system normally pass though the lungs.

41

Contemporary Research on Human Diseases

Chapter Outline

Reading Questions

1. How did the existence of a new disease, AIDS, become known?

2. How does HIV-1 cause AIDS?

3. What does the future hold for AIDS in the United States? The world?

4. What is the concept of cellular homeostasis? How is it related to cancer?

5. How are oncogenes and tumor suppressor genes related to cancer?

B eginning in the early 1950s, rapid progress was made in learning about all aspects of human disease, primarily through the use of powerful tools developed in molecular biology. New knowledge was obtained about host biology, the immune system, and the underlying molecular causes of "simple" human diseases—those caused by a single disease agent (such as polio).

STUDYING COMPLEX HUMAN DISEASES

As we approach the twenty-first century, the principal focus of human disease research has shifted to "complex" diseases— those with complicated or multiple causes, including AIDS, cancer, vascular diseases, and mental disorders such as schizophrenia. Investigators are concentrating on the molecular level because complex disorders often involve sequences of molecular changes. Also, understanding diseases at the molecular level may lead to the development of treatments or cures.

Several relevant questions guide research on complex diseases. What causes the disease? Which cellular processes are disrupted and why? What approaches might be used to treat or cure the disease? In this chapter, we describe research on two of the most important human diseases—AIDS and cancer—to illustrate the complexities and the power of modern disease research. We emphasize that great gaps remain in our understanding of AIDS and cancer. The intent of this chapter is to present the most current information available and to provide a background that will enable you to follow progress in the years ahead.

AIDS: A NEW DISEASE

Beginning in 1981, it slowly became evident that a new disease, which became known as *AIDS,* was causing mortalities in the United States, Europe, and Africa. This new disease was to become one of two great **pandemics** (diseases that occur worldwide) of the twentieth century (the other was the great flu pandemic of 1918 that killed over 20 million people). How did scientists determine that a new disease was affecting people and that it was caused by an infectious agent?

Discovery of the Causative Agent

Epidemiology is the field of science concerned with the relationships of various factors that influence the frequencies and distributions of a disease in a human community or population. Epidemiologists conduct investigations designed to accomplish the following:

- Determine the occurrence of different diseases
- Identify patterns of mortality in different geographical areas of a country or the world
- Define methods of transmission
- Identify specific groups that may be at risk
- Determine risk factors
- Create strategies to prevent and control disease

In the United States, primary responsibility for conducting such investigations rests with the Centers for Disease Control and Prevention (CDC), located in Atlanta. Scientists at the CDC rely on data forwarded from local and state health departments. Such data provided the first indications that a new disease was present in the United States and also contributed clues about the nature of the disease.

Early Epidemiological Investigations

In 1981, workers at the CDC became aware that over an eight-month period, five new cases of an extremely rare type of pneumonia caused by a protozoan *(Pneumocystis carinii)* had occurred in young homosexual men from Los Angeles. At approximately the same time, more than 25 cases of an unusual type of cancer, Kaposi's sarcoma, were reported in young homosexual men in New York and California. Many other severe opportunistic infections were also reported in the early 1980s. Recall that opportunistic diseases occur in individuals whose immune systems have become impaired for some reason.

Occurrence and Transmission

Most of the early AIDS cases in the United States occurred in male homosexuals living in New York and California. Was the disease actually caused by an infectious agent? Or was some lifestyle factor responsible?

In comprehensive studies of homosexuals with AIDS, CDC investigators established a strong link between the occurrence of AIDS and the pattern of sexual contacts. Patients with the disease tended to have frequent sex with multiple partners in comparison with control subjects (homosexuals without AIDS). Such a finding was consistent with the view that AIDS was caused by an infectious agent and that it was transmitted through sexual contacts between males.

By 1982, AIDS cases were being reported in individuals who were not homosexuals. People with hemophilia and recipients of blood transfusions had contracted AIDS, apparently from injections of blood or blood products. AIDS was also reported in people who injected themselves intravenously with controlled drugs using shared hypodermic needles. Finally, AIDS was reported in two females who were sexual partners of male intravenous drug abusers. Clearly, AIDS could also be transmitted sexually through heterosexual intercourse. Some time later, babies of mothers with AIDS were shown to develop the disease (see Figure 41.1, page 788). All of these findings emphasized the crucial importance of developing tests to identify infected individuals and to detect the AIDS agent, whatever it was, in blood.

The results of these various studies led scientists to conclude that the AIDS agent was being transmitted by the injection of blood from an affected person or from body fluids exchanged during sexual contacts. Further, because the disease could be transmitted, it must be caused by an infectious agent. In 1982, the CDC identified individuals with the new disease syndrome (a *syndrome* is a set of symptoms or characteristics that occur together) as having **acquired immune deficiency syndrome (AIDS).** Thus the combination of rare opportunistic infections along with a seriously deficient immune system became known as AIDS. Further, there appeared to be a **latent period** of several years between the time of initial infection and the appearance of AIDS. Therefore, infections in AIDS patients living in the United States had actually occurred in the 1970s.

A New Retrovirus

What relevant information about a possible AIDS-causing agent was available to researchers in 1982, when the search for such an agent began in earnest? AIDS was hypothesized to be a new disease, apparently caused by an agent that was present in different body fluids including blood, plasma, and semen.

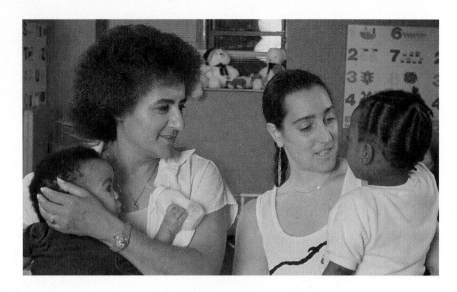

Figure 41.1 Volunteers at the Birk Childcare Center in Brooklyn, New York, caring for infants with AIDS.

Question: *Why do babies develop AIDS?*

Transmission could occur through sexual contact, blood transfusions, or injections using shared needles containing blood or plasma. Two additional clues were particularly significant. First, before plasma is transfused into individuals with hemophilia, it is filtered, a process that removes bacteria and other microbes but not the much smaller viruses. Second, immunodeficient patients with AIDS were found to have very low levels of helper **T4 lymphocytes** *(T4 cells),* a class of helper T cell. T4 cells regulate a variety of immune functions and appeared to be the primary cells affected by the AIDS agent. Also, one type of leukemia (T lymphocyte cancer) was already known to be caused by a retrovirus.

The problem of identifying the AIDS agent had attracted top scientists in Europe and the United States, and they accomplished their goal in 1983 and 1984. Researchers in France and the United States reported that a new **retrovirus** was responsible for causing AIDS. During this same period, diagnostic blood tests were also developed for detecting the presence of the virus. These tests were important for identifying infected individuals and also for screening blood donated for transfusions. Use of these tests enhanced the safety of blood supplies.

BEFORE YOU GO ON In the early 1980s, epidemiologists discovered that a new disease, characterized by opportunistic infections, was causing mortalities. The people initially affected in the United States were male homosexuals and recipients of blood transfusions. Later, other population groups were found to have the disease. In 1982, the CDC designated the new disease syndrome as *AIDS.* By 1984, the cause of AIDS had been determined to be a new retrovirus.

HIV-1

Progress in learning more about the new retrovirus continued at a rapid pace following its discovery. In 1985, a virology nomenclature committee named it *human immunodeficiency virus, type 1,* now known as **HIV-1** (see Figure 41.2A). Throughout the rest of this chapter, we use HIV-1 in reference

to the established AIDS virus. To many people, HIV-1 has become synonymous with AIDS; however, HIV-1 causes a broad range of disease effects. The syndrome of AIDS, as defined by the CDC, occurs late in the disease progression caused by an HIV-1 infection.

Molecular Structure and Genome

The major molecular structures of HIV-1 are described in Figure 41.2B and C. The *virion,* or virus particle, is enclosed by an *envelope* composed of two layers of lipid molecules that are organized in the same way as in a plasma membrane. In fact, the viral envelope is derived (comes) from plasma membrane molecules provided by an infected human cell. The envelope is studded with knobs made of two glycoproteins (proteins containing carbohydrate side chains) called *gp120* and *gp41.* A protein called *p17* forms a layer inside the envelope that surrounds the core, or *capsid,* which is composed of *p24.* Within the *p24* capsid are two identical strands of RNA, the genetic information of retroviruses. The RNA is surrounded by a structural protein *(p7)* and **reverse transcriptase,** the enzyme that enables HIV-1 to make DNA corresponding to its RNA sequence. This retro (backward) flow of genetic information—RNA converted to DNA—is the distinguishing characteristic of retroviruses. Three other enzymes—*ribonuclease, integrase, and protease*—are also found inside the capsid.

Scientists have successfully determined the entire molecular structure of HIV-1. The HIV-1 genome consists of 9,749 nucleotides and only nine genes, which are described in Figure 41.3 (see page 790). Three of the genes *(gag, pol,* and *env)* encode viral structural proteins. The other six genes have roles in different phases of the HIV-1 life cycle. Remarkably, these six genes guide all HIV-1 molecular processes concerned with regulation of the HIV-1 life cycle within the infected cell, construction of new HIV-1 virions, and the escape and infection of fresh cells. The nine viral genes are bounded at each end by stretches of DNA known as long terminal repeats *(LTRs).* The LTRs do not encode proteins but rather initiate replication of new HIV-1 virions within an infected cell.

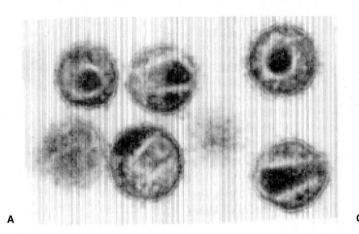

A

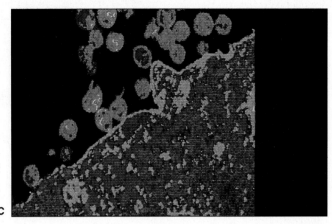

C

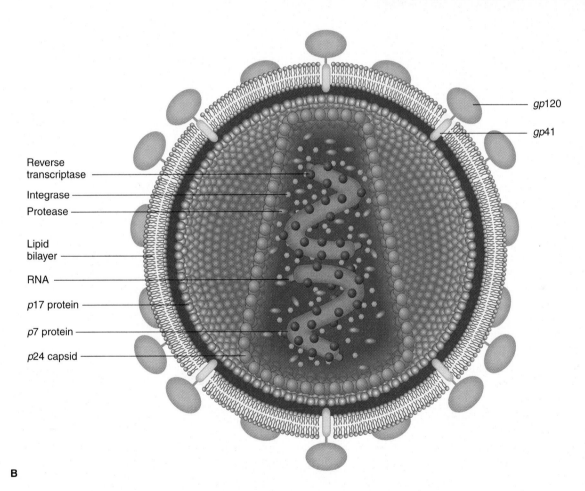

gp120

gp41

Reverse transcriptase

Integrase

Protease

Lipid bilayer

RNA

p17 protein

p7 protein

p24 capsid

B

Figure 41.2   (A) This electron micrograph shows HIV-1 virions. (B) This cross section shows the structure of HIV-1. The knobs are composed two glycoproteins; *gp*120 attaches to CD4 receptors of T4 lymphocytes prior to infection of the cell, and *gp*41 is embedded in the viral envelope. The *p*17 protein forms a layer that surrounds the capsid, which contains RNA and enzymes. The protein (*p*24) capsid contains genetic material—RNA—that is surrounded by *p*7 protein and four enzymes. The enzyme reverse transcriptase transcribes viral RNA into DNA, which becomes integrated into a host cell chromosome; ribonuclease degrades viral RNA; integrase integrates viral DNA into the host chromosome; and protease helps make new HIV-1 virions. (C) This scanning electron micrograph (SEM) image shows HIV-1 RNA (red), glycoproteins (blue), and proteins (green).

Question: *What is unique about the genetic material of HIV-1?*

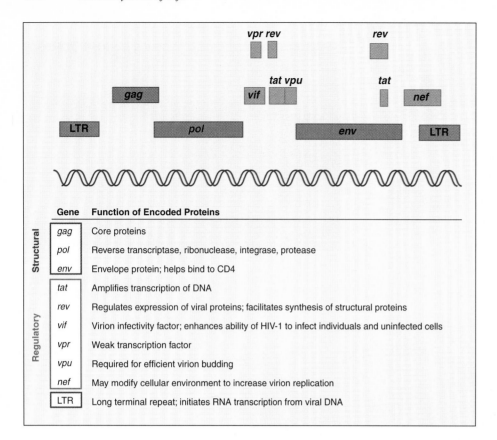

Gene	Function of Encoded Proteins
Structural	
gag	Core proteins
pol	Reverse transcriptase, ribonuclease, integrase, protease
env	Envelope protein; helps bind to CD4
Regulatory	
tat	Amplifies transcription of DNA
rev	Regulates expression of viral proteins; facilitates synthesis of structural proteins
vif	Virion infectivity factor; enhances ability of HIV-1 to infect individuals and uninfected cells
vpr	Weak transcription factor
vpu	Required for efficient virion budding
nef	May modify cellular environment to increase virion replication
LTR	Long terminal repeat; initiates RNA transcription from viral DNA

Figure 41.3 The genome of HIV-1 consists of nine genes. Three are structural genes that encode the core and envelope proteins and enzymes. The other six genes encode regulatory proteins that influence the production and assembly of HIV-1 virions within the infected cell.

Life Cycle

Like all viruses, HIV-1 cannot be replicated or cause harmful effects unless it enters an appropriate host cell. What happens once HIV-1 virions gain entry into the body through sexual contact, injection, or other means? How does HIV-1 find and enter an appropriate cell? What events follow that lead to the cell's being killed? The HIV-1 life cycle is described in Figure 41.4.

Entering a Host Cell

After HIV-1 virions enter the circulatory system, they eventually contact, bind to, and then enter specific host ("target") cells. The HIV-1 *gp*120 molecule binds to host cells that have *CD4 (T4)* molecules, or receptors, protruding from their surface membrane. CD4 is found on T4 cells, where its functional role is related to interactions between immune cells or molecules involved in immune responses. CD4 is also found on monocytes, macrophages, and follicular dendritic cells, which are found in lymph nodes, the spleen, and tonsils, where they function as antigen-presenting cells for T4 cells and help filter out various infectious microbes. Following the initial binding event, the viral envelope and plasma membrane fuse through the action of *gp*41, and the viral capsid is injected into the host cell.

RNA Transcription and Provirus Integration

Once inside the host cell, a series of events may take place that ultimately leads to the production of new HIV-1 virions. First, the capsid is uncoated, and the viral RNA genome is copied by reverse transcriptase into a complementary DNA strand, using host cell nucleotides. The RNA is then degraded by viral ribonuclease, and a double-stranded DNA molecule is produced using the new DNA strand as a template. Once synthesized, *integration* occurs—the viral DNA enters the nucleus and becomes inserted into a host cell chromosome by the viral integrase enzyme. In this state, the integrated viral DNA is known as a **provirus.** Recent research indicates that integration apparently does not take place in all T4 cells. If integration does not occur, the viral DNA breaks down a few days after being synthesized.

Replication

The replication of new HIV-1 virions begins when an infected T4 cell becomes activated by contact with an antigen. Replication is initiated by a molecular signal from the LTRs that activates proviral DNA and leads to host cell enzymes' transcribing the integrated HIV-1 DNA into messenger RNA, which can be translated by the host cell's protein-synthesizing system, and HIV-1 genomic RNA. Specific viral proteins are then synthesized by the host cell from the newly formed messenger RNA. The first viral proteins synthesized are regulatory proteins that carry out their functions in the nucleus of the host cell; HIV-1 structural proteins—core proteins, enzymes, and envelope proteins—are produced later in the replication cycle. Newly-formed HIV-1 proteins are then converted by the viral protease enzyme into their functional form.

In the final phase of HIV-1 replication, viral proteins, enzymes, and genomic RNA are assembled at the plasma membrane (see Figure 41.5A, page 792). The plasma membrane, which becomes the lipid component of the viral envelope, begins to constrict around a new HIV-1 particle and

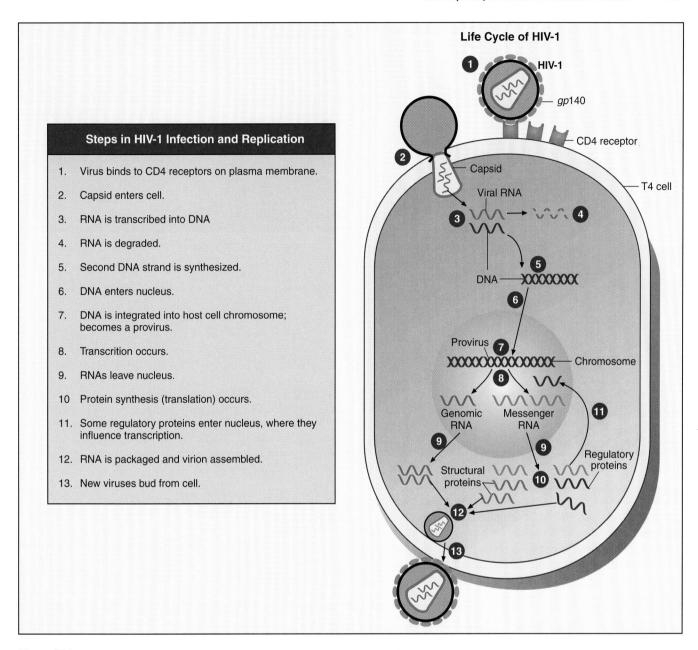

Life Cycle of HIV-1

Steps in HIV-1 Infection and Replication

1. Virus binds to CD4 receptors on plasma membrane.

2. Capsid enters cell.

3. RNA is transcribed into DNA

4. RNA is degraded.

5. Second DNA strand is synthesized.

6. DNA enters nucleus.

7. DNA is integrated into host cell chromosome; becomes a provirus.

8. Transcrition occurs.

9. RNAs leave nucleus.

10 Protein synthesis (translation) occurs.

11. Some regulatory proteins enter nucleus, where they influence transcription.

12. RNA is packaged and virion assembled.

13. New viruses bud from cell.

Figure 41.4 The HIV-1 life cycle consists of many steps. Infection occurs when *gp*120 knobs of an HIV-1 virion attach to CD4 receptors; they fuse with the plasma membrane, and the capsid is inserted into the cell. Once inside the cell, the protein coat is shed, and reverse transcription follows when viral RNA is copied into DNA by reverse transcriptase. The new single-stranded DNA then serves as a template for making a double-stranded molecule that migrates into the nucleus. The viral DNA is integrated into a host chromosome and becomes a provirus. Once activated by exposure to an antigen, the infected cell's metabolic machinery transcribes viral DNA into messenger RNA encoding viral proteins and genomic RNA—the virus's genetic material. Viral proteins and RNA are then assembled into new virions, which are enclosed within host cell membrane material and then released by budding through the plasma membrane.

forms a bud at the surface of the host cell. When the virus is completely assembled, the membrane pinches off, and a new HIV-1 buds from the cell as shown in Figure 41.5B and C.

When virions bud off from the cell, holes may be torn in the plasma membrane, which causes the loss of structural integrity and death of the cell.

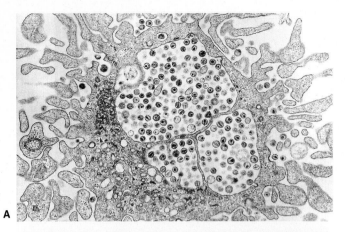

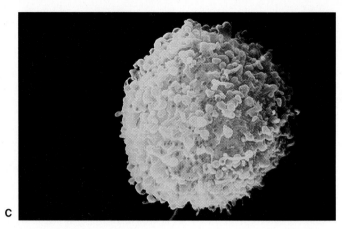

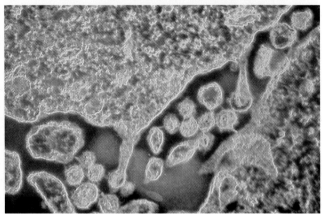

Figure 41.5   (A) This electron micrograph shows HIV-1 assembled within an infected cell. (B) This scanning electron micrograph shows HIV-1 virions (orange, green) in different stages of budding from the plasma membrane of an infected cell. (C) This SEM shows HIV-1 virions (green) as they bud from a T4 lymphocyte.

BEFORE YOU GO ON   HIV-1 virions consist of an outer envelope and an inner core containing RNA, its genetic material, and various proteins and enzymes. During infection, the RNA is inserted into a host cell bearing T4 receptors. Reverse transcriptase then converts viral RNA into DNA that may become integrated into a host cell chromosome. After integration, proviral genes can be expressed and directed to produce HIV-1 virions. The new virions leave the original host cell and infect fresh cells.

HIV-1 and Human Diseases

The consequences of HIV-1 infection are numerous and varied. The disease syndrome called AIDS represents the progressive effects of immune system destruction that become evident only during the terminal phase, years after infection. What happens during the period between infection and the appearance of AIDS? New hypotheses related to this question have recently emerged.

General Aspects of HIV-1-induced Diseases

HIV-1-related diseases are associated with a gradual depletion of the T4 cell population. As these helper T cells are lost, other viruses and various opportunistic bacteria, fungi, and protozoans may cause infections. Individuals may also develop cancers and disorders of the central nervous system. There is considerable variation in diseases and disease progression among infected patients.

Most infected individuals follow a predictable progression of disease once infection occurs and for convenience, effects are categorized as *early stage, middle stages,* and *late stages.* The number of T4 cells per microliter of blood, the functional capabilities of the patient's immune system, and abundance of HIV-1 virions in blood are important elements in distinguishing between the three stages. The different stages are summarized here, and Figure 41.6 describes the progressive phases of HIV-1 infection.

Early Stage.   T4 cell counts in circulating blood are normal, greater than 500 cells per microliter. Within a few weeks after an individual becomes infected with HIV-1, great quantities of the virus are found in infected blood cells. Then a vigorous, antibody immune response is mounted against HIV-1, and virus concentrations in the blood decline rapidly (HIV-1 antibody presence is the basis of the test used to identify infected individuals). Two to three months after infection, T4 cell counts usually return to normal (800 to 1,000 cells per microliter of blood), and there are relatively few infected T4 cells found in circulating blood.

Middle Stages.   The middle stages of infection are characterized by generally good health; there may or may not be any disease symptoms—fatigue, fever, swollen lymph nodes, diar-

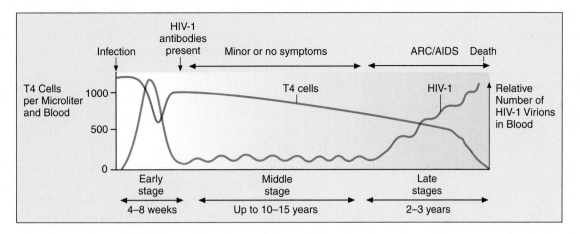

Figure 41.6   Shortly after infection with HIV-1, virus replication proceeds at a rapid pace, and billions of virions are found circulating in the blood. An immune response is quickly mounted against the virus, and virion counts in the blood decrease rapidly. From about two months after infection to up to 10–15 years, T4 cell counts progressively decline, and the infected patient may or may not have disease symptoms. Once T4 cell counts drop below about 500 cells per microliter of blood, virion counts in the blood begin to increase sharply, disease symptoms become evident, and AIDS develops, which usually leads to death two to three years later.

Question: *When do opportunistic diseases become a problem in HIV-1-infected people? Why do such infections occur?*

rhea, headaches—for a period of up to 10–15 years. T4 cell counts are initially normal but gradually decline as a function of disease progression. Cellular and humoral (antibody) immune systems appear to be effective in controlling HIV-1 replication for several years in most cases.

Late Stages.   When T4 cell counts drop to between 200 to 500 cells per microliter, there are usually clear indications of immune impairment—poor response to special antigens injected under the skin—which signal the impending loss of humoral immunity. Consequently, more severe disease symptoms develop that are often referred to as an **AIDS-related complex (ARC).** ARC symptoms include fever, weight loss, diarrhea, fatigue, and swollen lymph nodes. When T4 cell counts drop below 200 cells per microliter, the patient has AIDS, the immune system functions progressively deteriorate until there is no measurable immune response, and HIV-1 numbers escalate in the blood. Patients frequently develop an uncontrolled infection in the mouth, a disease called **thrush,** caused by the fungus *Candida,* which marks the onset of AIDS. Other serious opportunistic viral and bacterial pathogens infect the skin and mucous membranes. Some opportunistic diseases that commonly occur during AIDS include *Pneumocystis carinii* and other protozoan infections; fungal infections of the nervous system, liver, bone, and other tissues; bacterial infections of the respiratory and digestive systems; and viral infections of the digestive tract. Many AIDS patients develop significant neurological disorders, the causes of which are poorly understood. Opportunistic infections and complete collapse of the immune system usually occur 5 to 15 years after infection with HIV-1. Most patients die within two years after T4 cell counts drop below 200 cells per microliter.

BEFORE YOU GO ON   HIV-1 infection causes a progressive destruction of the immune system that is associated with a variety of diseases. Health generally remains good for up to 10 to 15 years after infection. Then there is a loss of humoral and cellular immunity, and opportunistic infections occur. In the final stage, known as AIDS, many diseases may develop.

Immune System Destruction

During the first decade after HIV-1 had been identified, studies of infected people revealed a consistent pattern of T4 cell decline that took place over several years. This delay between time of infection and immune system dysfunction was the basis for believing that there was a latent period, after infection occurred, in which the virus did not damage the immune system. However, during the middle stages, T4 cell counts decrease continuously, and during the final stages, when AIDS develops, T4 cell counts decline to less than 100 cells per microliter. What could explain these observations?

An Early Hypothesis   A reasonable hypothesis emerged from these observations: T4 cell populations gradually decreased because cells died after becoming infected with HIV-1 and this decline finally led to loss of immune system function. However, certain data did not seem to support this hypothesis. For example, during the early stages of infection, fewer than 0.0001 percent of circulating T4 cells become infected with HIV-1. In the final stages, when T4 cell counts are very low, only about 1 percent of T4 cells are infected. These observations suggested that HIV-1 infection of T4 cells alone could not account for the T4 cell population decline. How, then, does

HIV-1 infection cause destruction of the T4 cell population and loss of immune system function?

New Hypotheses   New research has given rise to more complex hypotheses that attempt to explain how immune system function is lost. The hypothesized sequence of events, from first (HIV-1 infection) to last (AIDS), is summarized below here.

1. Soon after infection, there is an enormous number of cells infected; virion concentrations in circulating blood are commonly about 1 million per microliter of blood.

2. Lymph nodes act as virus filters during this early HIV-1 replication phase.

3. An immune response is rapidly initiated, and blood virion concentrations are sharply reduced to about 10,000 to 50,000 per microliter.

4. HIV-1 primarily infects the tissues in lymph nodes and the lymphoid tissue associated with the digestive tract. The principal targets are T4 cells, and once infected, they are destroyed continuously as they become activated.

5. Within the lymph nodes, large numbers of HIV-1 virions become attached to the surfaces of **follicular dendritic cells (FDCs)** (see Figure 41.7). FDCs form a dense network of interconnected cells that acts as a "trapping system," which filters antibody-bound HIV-1 antigens in circulating lymph and presents them to nearby immune cells for processing and destruction.

6. T4 cells and other immune cells that circulate through lymph nodes and tissues may become infected through the trapping and presentation process (see Figure 41.8). The number of immune cells infected during the early stage has been estimated to be as high as 100 billion. Fewer than 1 percent of those infected cells produce new HIV-1 virions; the remainder serve as *reservoirs* for the virus.

7. By some unknown mechanism, FDCs and the trapping system are slowly destroyed as T4 cell counts decline. Thus there is no latent period in which HIV-1 is not actively replicating and damaging the immune system. However, individual cells may be latently infected; that is, HIV-1 DNA can be integrated into a host cell chromosome, but new virions may not be produced for weeks, months, or years after the cell is infected.

8. A relatively small percentage of infected T4 cells are killed directly when they become activated and produce new virions. What happens to the billions of other infected T4 cells? Infected T4 cells may be killed by other processes. Several hypotheses explain T4 cell loss: infected T4 cells may be killed by natural killer (NK) cells or cytotoxic T cells of the immune system; antibodies against viral *gp*120 and *gp*41 may also react against CD4 molecules on T4 cells, which leads to their being inactivated or killed; HIV-1 may destroy developing T4 cells in the thymus or perhaps thymus cells that are necessary for normal T cell processing; or T4 cells may be killed by the abnormal use of mechanisms that normally remove new T cells in the thymus during T cell processing. One or more of these mechanisms may account for T4 cell loss, along with direct killing by HIV-1.

9. Ultimately, the lymph nodes are destroyed, FDCs perish, and HIV-1 is no longer filtered from the lymphatic system. These events, along with the loss of T4 cells, signal the onset of AIDS.

Much additional research will be necessary to test the various hypotheses. Despite the uncertainties, these new hypotheses are consistent with observations and may help guide research on developing new treatments for HIV-1-infected people (see the Focus on Scientific Process, "Can We Prevent or Cure HIV-1 Infections and Disease?").

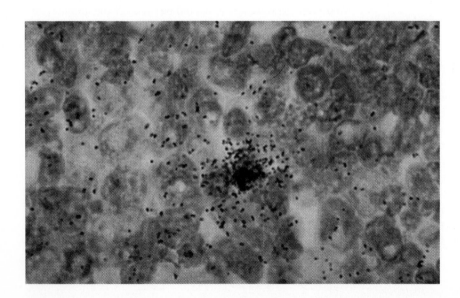

Figure 41.7   Follicular dendritic cells (purple and red) trap HIV-1 virions (viral RNA appears as black dots) in lymph nodes.

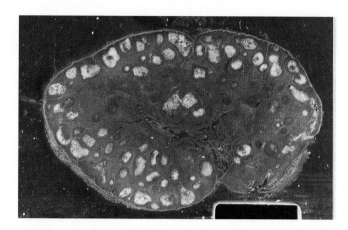

Figure 41.8 During the middle stages of infection, HIV-1 is concentrated and stored within lymph nodes. This photo shows virions (white particles) that fill parts of a lymph node.

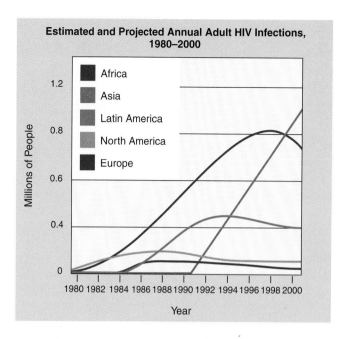

Figure 41.9 This graph shows the number of HIV-1 infections in different parts of the world from 1980 to 2000; the data for 1994 to 2000 are projections based on analyses by the World Health Organization.

Question: *Which region will have the greatest number of AIDS cases in the year 2000? In 2010?*

Epidemiology of HIV-1

Before AIDS, the last great pandemic occurred just after World War I, when influenza killed more than 20 million people. A complete AIDS accounting, in terms of mortalities and economic costs, will not be known for decades. However, it is already evident that the final costs of HIV-1 will be enormous, and for some countries and population groups, they will be absolutely disastrous.

Worldwide

The World Health Organization (WHO) has primary responsibility for studying HIV-1 and AIDS outside the United States. The WHO's global projections of HIV-1 infection are shown in Figure 41.9. Data from developed countries are generally considered reliable, but information from developing nations is often inadequate.

According to WHO estimates, by 1995, approximately 20 million people had become infected with HIV-1 since the start of the pandemic in 1981. Over 2 million HIV-1-infected people have died of AIDS. As shown in Figure 41.9, more than 10 million infections have occurred in Africa, where the HIV-1 pandemic first appeared. Infections are increasing in Latin America but are starting to level off or decline in developed countries. However, growth of the epidemic in Asia has been explosive in the past few years. In Thailand, for example, it is estimated that over 5 percent of the general population is now infected. In South Asia, there are at least 1 million infected adults, most of them in India. The HIV-1 epidemic in Asia may ultimately dwarf all other regions because of its huge population.

To date, Africa has been the continent most seriously affected by HIV-1 infections. Here HIV-1 is spread primarily through heterosexual intercourse. AIDS is now the leading health problem in many central and eastern African countries. The WHO estimates that between 10 and 30 percent of sexually active people in urban centers in these countries are now infected, and that figure is expected to increase. In some cities,

AIDS patients occupy up to 80 percent of hospital beds. Given the lack of health and economic resources in these areas, the current picture in Africa is devastating in terms of both direct and indirect costs (lost income and decreased workforce output). The situation in South and Southeast Asia is also appalling. The government of Thailand estimates that by the year 2000, the direct and indirect costs of AIDS will reach $10 billion.

How is this global pandemic going to be controlled? The WHO has proposed a comprehensive AIDS prevention program for the developing world. It emphasizes promotion and distribution of condoms in the general population, treatment of conventional sexually transmitted diseases, providing AIDS information in schools and through the mass media, promotion of condom use by prostitutes and their clients, maintenance of a safe blood supply, and needle-exchange programs for injecting drug users. The cost of the program is estimated to be $1.5 to $3 billion per year. Where is the money going to come from? Given the magnitude of the problem, can the world afford to delay implementing this or some other type of program?

United States

What is currently known about the status of HIV-1 in the United States? How many people have been or will be infected? Who are they? How did they become infected with HIV-1? The answers to these questions are based on data collected mostly by the CDC. These data have allowed scientists to for-

Can We Prevent or Cure HIV-1 Infections and Disease?

In 1984, after AIDS was shown to be caused by HIV-1, investigators in laboratories around the world began to search for ways to prevent infections and treat or cure the disease. Two basic approaches are used to counteract disease agents: drug therapy and vaccines. **Drug therapy** involves creating drugs that attack the disease agent once it has infected the host. **Vaccines** are produced from inactivated (killed) or live-attenuated (altered or weakened) disease-causing viruses, bacteria, or protozoa that, when introduced into the body, stimulate an immune response.

HIV-1 Drug Therapies

How does a therapeutic agent, or drug, act to treat or cure a disease? To be effective, it must either kill the disease-causing agent or prevent it from reproducing (in the case of bacteria and protozoa) or replicating (in the case of viruses) without irreparably harming host cells. To carry out its task, a drug targets a biochemical pathway that is unique to the agent and disrupts its function. Consequently, some critical substance may not be available to the pathogen, or it may not be able to complete its life cycle.

Because viruses live inside host cells, they represent a difficult target for a drug to reach. This problem becomes even more severe when dealing with integrated HIV-1 proviruses. The fundamental problem is how to eliminate the virus without also killing the host cell. Unfortunately, to some degree, all drugs are harmful to at least some cells. Hence a fine balance must be maintained between effects on the pathogen and effects on the host. For nonthreatening diseases such as colds, a conservative approach is used, and drugs having little effect on the host are prescribed. In contrast, for life-threatening diseases such as AIDS, radical therapies that have substantial side effects may be employed.

How do scientists identify possible drug targets of opportunity in the case of HIV-1? A general approach is to study the virus's life cycle and identify vulnerable stages. Examination of the HIV-1 life cycle indicates several possible points of attack by drugs. Research is being conducted to identify drugs capable of acting in these various stages. We describe here the only useful drug therapy developed so far.

In 1984, when HIV-1 was identified as the causal agent of AIDS, scientists at the National Cancer Institute began testing known antiviral drugs to determine if any of them might react against the new virus. One, *azidothymidine (AZT, or zidovudine)*, was shown to be a powerful inhibitor of HIV-1 growth in T4 cells maintained in culture. Initially, AZT did not appear to be harmful to cells, although it was later shown to be toxic to bone marrow cells at high doses. Intensive efforts led to the development of an AZT drug that could be tested in HIV-1 infected patients. By 1986, small studies suggested that AZT extended the survival of patients and improved the quality of their lives. The U.S. Food and Drug Administration (FDA) approved (licensed) AZT for use in 1987. Subsequently, two similar drugs, *dideoxycytidine (ddC)* and *dideoxyinosine (ddI)*, were also licensed for use.

How do these drugs react against HIV-1 or prevent T4 cells from being affected by the virus? AZT mimics thymidine (T), one of the nucleotides used to construct DNA. When reverse transcriptase converts viral RNA into DNA, it may mistakenly incorporate AZT instead of T into the chain being synthesized. When that happens, the next DNA nucleotide base cannot be added, viral DNA synthesis is aborted, and no provirus is manufactured. The drugs ddC and ddI are also reverse transcriptase inhibitors. Unfortunately, recent results from two large studies led to a conclusion that AZT neither prolongs the life nor delays the onset of AIDS in people infected with HIV-1. However, AZT will still be used because it seems to help in treating certain patients. Further research is now being conducted using combinations of the three reverse transcriptase inhibitors. Early results have not been encouraging.

Other drugs are now being developed that may interfere with other parts of the HIV-1 life cycle, but they are in early stages of testing. New ideas are also being pursued. For example, efforts are being made to develop a vaginal chemical that will kill HIV-1 and other STD agents. Gene therapy methods involving mutant proteins that may inhibit HIV-1 are also being investigated. Given the large number of viral targets, the enormous effort being made by researchers, and the support provided by various funding agencies, there is hope that more effective drugs will soon be developed.

HIV-1 Vaccines

In contrast to the celebrated successes in developing live-attenuated viral vaccines against the viruses that cause smallpox, polio, mumps, and measles and inactivated viral vaccines against influenza, no useful HIV-1 vaccines have yet been produced, and the prospects are not promising at this time. Useful vaccines must meet three criteria: *safety*, which must be established in animal model tests before human trials can be approved; *efficacy* (extreme effectiveness); and *availability*, which usually implies affordability.

FOCUS ON SCIENTIFIC PROCESS

A common question asked by concerned citizens is, Why, given the economic and human costs of AIDS, can't scientists create a vaccine that works on HIV-1? In 1984, after HIV-1 had been identified as the AIDS causal agent, the secretary of the Department of Health and Human Services announced that "an AIDS vaccine would be available in two years." This statement was widely publicized and helped raise hopes that the AIDS problem would be quickly solved with a new vaccine. Unfortunately, it reflected complete ignorance about the biological puzzles posed by HIV-1 that soon became clear. First, there is no definite understanding of the precise immune protective mechanism that must be stimulated to destroy or inhibit HIV-1. Second, once HIV-1 becomes integrated as a provirus, the immune system is helpless in locating and destroying it. Thus even if a vaccine led to the development of HIV-1 memory cells, integration would likely occur before they could act. Third, and most important, key HIV-1 proteins that might serve as antigenic determinants change continuously because of **hypermutation** (extraordinarily rapid mutation rates) of the viral genome. Much of this variation is due to the inaccuracy of reverse transcriptase—it makes about one mistake per 5,000 nucleotides added during DNA synthesis. Thus an infected individual may have several different *strains* of HIV-1, varieties that differ only slightly in their protein structure or genetic content. (As many as 100 strains have been isolated from a single individual.) Consequently, virion molecules represent an elusive target for vaccines, which by their nature are designed to represent a precise, unchanging part of the viral structure.

Despite these and other difficulties, efforts continue to develop a successful vaccine against HIV-1. Which part of the virion might serve as an antigenic determinant for initiating and sustaining an immune response sufficient for conferring resistance? Each viral protein or glycoprotein represents a potential antigenic determinant, and most attention has been directed at the envelope glycoproteins, especially *gp*120. The rationale for this approach is that if viral *gp*120 could be prevented from binding to CD4 receptors, cells with that receptor would not become infected by HIV-1. However, results from various studies of test vaccines have not been encouraging, perhaps because of hypermutations now known to occur in the *env* gene. Thus any *gp*120 vaccines may become obsolete soon after their production.

Given the molecular problems presented by HIV-1 hypermutation, debate has begun over the question of whether or not to develop and test a live-attenuated virus vaccine against HIV-1. Information about this issue is summarized in Table 1. Whatever the outcome of this

Table 1   Developing and Using a Live-Attenuated-Virus Vaccine Against HIV-1

Strategy for Developing A Live-Attenuated HIV-1 Vaccine

- Create a strain that can replicate at a low level but does not cause disease.
- To create such a strain, remove viral genes that are not essential for replication but play a role in causing disease.
- Likely basis for attenuation is deleting genes *nef, vpr*, and/or *vpu*.
- Do first attenuation studies on SIV.
- Test live-attenuated SIV vaccine on normal monkey hosts.
- If successful, proceed to develop and test a live-attenuated HIV-1 vaccine.

Should A Live-Attenuated HIV-1 Vaccine Be Developed and Tested?

Some Arguments in Favor	Some Arguments Against
• No current vaccines seem to work.	• Information about safety is lacking. No live-attenuated virus vaccine is 100 percent safe.
• Whole attenuated viruses are more likely to elicit an immune response.	• Introducing a live retrovirus into the human population is an enormous conceptual risk.
• Replication-competent strains can be made safe for the infected host.	• Once integrated, live-attenuated HIV-1 could disrupt normal host genes.
• Risks are worthwhile because nothing else is on the horizon.	• Infections will be permanent; this could be dangerous because the diseases they cause generally occur years after infection.
• Costs will be much lower than for a product created through genetic technologies.	• It will take at least a decade to develop an acceptable live-attenuated vaccine; new developments could lead to a safe vaccine.
• Early studies using attenuated SIV and tests on rhesus monkeys have been promising.	• There is no adequate animal model for studying HIV-1 and AIDS.

box continues

FOCUS ON SCIENTIFIC PROCESS

debate, a live-attenuated HIV-1 vaccine is not expected for at least a decade.

The Future

Biological obstacles are not the only reason why progress has been slow in developing HIV-1 vaccines. Few major companies in the developed world have made a serious commitment to develop a vaccine due to economic and political factors. The market for an HIV-1 vaccine was originally projected to be substantial but now appears to be much smaller. Many North American compa-

nies are not willing to make an expensive investment for little, if any, return on that investment. There are also fears of governmental regulation and concerns about lawsuits by people claiming to have been injured by vaccines. What is the answer to this problem? Should the government fund the development of an HIV-1 vaccine? Will taxpayers support such an effort?

What would happen if a successful vaccine is developed? Would HIV-1 infections be reduced or eradicated? Clearly, that would depend on whether or not the vaccine was widely dissemi-

nated, affordable, and used effectively in mass vaccination campaigns. Most analysts believe that risk behavior change will also be critical for success. If the availability of an effective vaccine leads to an increase in risk behaviors, overall reductions in HIV-1 infections may be slight, and the magnitude of the epidemic might even increase. Relative to the next decade, at least, there is universal agreement that the most effective approach for preventing HIV-1 infections and AIDS involves behavior modification and education, not the appearance of a wonder drug or vaccine.

mulate some important conclusions about infection rates, mortality rates, transmission vectors, and populations at risk of becoming infected.

Numbers of Infected People   In 1995, the CDC estimated that more than 1 million people were infected with HIV-1 in the United States. Since the start of the epidemic in 1981, more than 300,000 cases of AIDS have been reported to CDC, and more than 200,000 people have died. It is expected that 50,000 to 100,000 new cases of AIDS will occur each year until the year 2000.

Who Gets AIDS   Figure 41.10 gives the annual AIDS incidence for the population groups that have accounted for most

of the AIDS cases in the United States. The data for 1990 to 1995 were based on projections from earlier data. About 90 percent of the AIDS cases were associated with two risk groups, homosexual or bisexual men and intravenous drug users. Most heterosexual transmission involved sexual contact with a partner in one of the two primary risk groups. At present, the greatest rate of increase in AIDS cases is occurring in newborn children. Children born to mothers infected with HIV-1 have about a 50 percent chance of being infected. The infections in these cases occur during pregnancy, birth, or nursing, and most are due intravenous drug use by one or both parents.

The CDC recently conducted an epidemiological study in 22 states that suggests future changes relative to HIV-1 infec-

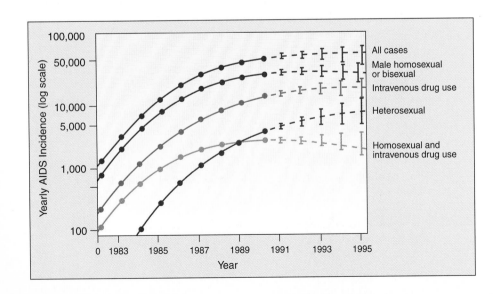

Figure 41.10   This graph shows projections (dashed part of the lines) of annual AIDS incidence for the major United States risk groups.

tions and AIDS. For example, of 24,000 AIDS cases, 90 percent were in men and 10 percent were in women. Of HIV-infected people, 82 percent were men and 18 percent were women. By race, of the AIDS cases, 63 percent were among whites, 31 percent among African Americans, and the remainder among Latinos, Asians, and other groups. Among HIV-1-infected people, 48 percent were whites, 44 percent African Americans, and the remainder members of other groups. What do these data seem to forecast? The CDC data suggest that AIDS cases will increase more rapidly among younger people than in the past and that the mode of HIV-1 transmission is shifting slightly from homosexual to heterosexual intercourse. Poor women and young African Americans and Latinos will have a greater percentage of infections, and heterosexual intercourse will be increasingly important for transmitting HIV-1 in these groups.

Risk Factors   What are the major risk factors for becoming infected with HIV-1? It is now well established that HIV-1 is transmitted predominantly through sexual contact, exposure (usually via needles) to contaminated blood or blood products, and from mother to offspring during pregnancy or while nursing. Risk factors for homosexual men include number of sexual partners, frequency of unprotected anal intercourse, and the presence of sexually transmitted diseases (STDs), especially syphilis and genital herpes, in either partner. Available evidence indicates that use of a condom significantly reduces the risk of contracting HIV-1 from an infected sexual partner. The major risk factor for drug users is injecting HIV-1-contaminated blood or plasma directly into the body, using a needle shared with someone infected with HIV-1. From the data that are available, the CDC has identified the major risk factors associated with heterosexual intercourse, with no other risk factors involved. They are number of sexual partners, the chance of the partner's being infected with HIV-1, and not using a condom during intercourse.

Who Doesn't Get AIDS   A great deal of research has been directed at determining the following: the risk of transmission between members of a household when one of the members is infected with HIV-1, the risk of contracting the virus from an infected individual at the workplace or school, and the possibility of contracting the virus from an insect vector. Except for sexual partners of an infected individual and a few workers accidentally injected in hospitals or research labs, no one has yet been infected with HIV-1 in any of these circumstances. Convincing evidence also indicates that HIV-1 is not transmitted by insects or from food, coughing, sneezing, toilet seats, holding hands, or kissing.

CDC studies have also been conducted to determine if HIV-1-infected doctors, surgeons, dentists, or health care workers can transmit the virus to patients. Except for one dentist who apparently transmitted HIV-1 to several patients, no other infections have been recorded. The CDC has concluded that the probability of a patient's being infected by an HIV-1-positive health specialist is extremely low.

BEFORE YOU GO ON   Epidemiological data have enabled scientists to identify risk groups and risk factors and to determine approximate numbers of people infected. In underdeveloped countries, HIV-1 infections are spread primarily through heterosexual intercourse. Several African populations have very high rates of HIV-1 infection and numbers of AIDS cases. HIV-1 infection rates have soared in South and Southeast Asia in recent years. More than 1 million people in the United States are now estimated to be infected with HIV-1. Most are homosexual or bisexual men or intravenous drug users, but the numbers of infected women, children, and young people are increasing. Major risk factors involve unsafe sex practices and drug abuse.

Origin of HIV-1

How did HIV-1 originate? Where did it come from? Most new human infectious diseases arise by transmission of a disease agent from a different animal species. The agent may be relatively harmless to its natural animal host, but it may cause a serious or even lethal disease in the new human host. For example, the major flu pandemics of the past have been caused by influenza viruses that normally infect pigs and waterfowl such as ducks.

It is now generally accepted that HIV-1 evolved from a **simian immunodeficiency virus (SIV)** that infects an African monkey species. Over 30 distinct SIV types have been identified in their African monkey hosts. In general, SIVs do not cause harmful effects in their normal monkey host, which suggests a long evolutionary relationship between virus and host. Analyses of viral genomes indicates that HIV-1 is most likely to have evolved from an SIV strain found in African chimpanzees (SIV_{CPZ}); that is, the molecular composition of HIV-1 is more like SIV_{CPZ} than any other SIV.

It is not clear how or when the virus may have been transmitted to humans. Analyses of blood stored in different countries for decades revealed that a sample from Zaire (Africa), collected in 1956, contained antibodies against HIV-1. Thus the virus may have infected humans in that country at least 20 years before it appeared elsewhere. Epidemiologists hypothesize that HIV-1 existed in isolated pockets in Central Africa until the 1950s. At that time, rural Africans began migrating into large cities and transported HIV-1 with them. In the cities, the virus was transmitted to larger numbers of individuals, presumably through heterosexual intercourse and transfusion of contaminated blood. By the 1960s and 1970s, technology in the form of passenger jets and global distribution of blood for medical uses had led to the dissemination of HIV-1 throughout the world. AIDS then appeared in countries to which the virus had been conveyed.

Society and AIDS

Scientists can provide the type of technical information contained in this chapter. However, it falls to all educated citizens not only to understand the nature of these biological problems but also to plot a course of action that may lead to solutions of related social and cultural problems. How can the spread of AIDS be stopped? Why has our society been reluctant to confront the issues—sexual behaviors and practices of homosexual men and drug abuse—historically associated with HIV-1 infections and AIDS? What about the next decade, in which poor women and urban youth of color will be heavily affected? Health specialists have concluded that education about safe sex and risk factors will be critical. However, what is the best means of educating people, especially those that are difficult to reach, about HIV-1 and AIDS? Also, whereas scientists can describe risk groups and risk behaviors, only individuals can modify their own risky behaviors.

Finally, how are we going to pay for the medical bills of AIDS patients, now expected to cost tens of billions of dollars by the end of the 1990s? Perhaps the most important consideration for all citizens is to recognize that AIDS has not disappeared and will not disappear within our lifetimes. The AIDS epidemic will continue and is likely to escalate unless appropriate programs are developed and implemented to combat it.

> **BEFORE YOU GO ON** Available evidence indicates that HIV-1 evolved from SIV in African monkeys. It first affected people in rural areas and then spread to large African cities. Subsequently, it was dispersed throughout the world. For the heterosexual population, major risk factors for becoming infected with HIV-1 are multiple partners, infected partners, and failure to use condoms. Trying to prevent the spread of HIV-1 presents major social and cultural problems.

A UNIFYING HYPOTHESIS OF CANCER CAUSATION

Cancer, the disease, was described in Chapter 40. It begins when fundamental molecular changes lead to unregulated cellular proliferation and loss of cellular homeostasis. Before the 1970s, cancer researchers placed great emphasis on identifying the causes of cancer in experimental animals and evaluating the behavior of transformed, malignant cells.

In the past 20 years, however, new discoveries have opened doors to deeper, molecular levels of understanding cancer and its underlying causes. Currently, there is great excitement in the cancer research community because of new conceptual breakthroughs. Scientists are now attempting to develop a **unifying hypothesis**—a synthesis of existing knowledge to create a single series of statements or ideas—that will explain the causes of cancer. The general principle that gene mutations are a prerequisite for transforming normal cells into cancer cells led to several new and crucial questions that guided related research during the 1980s. Exactly what types of genetic damage lead to the transformation of normal cells? Why is cancer the consequence of cell transformation?

The Concept of Cellular Homeostasis

Figure 41.11 illustrates the general concept of **cellular homeostasis.** A relatively constant number of cells makes up tissues and organs in the body, and this number is maintained throughout an individual's life. The four major types of tissues have two different patterns of cellular homeostasis. In general, cells of nervous and muscle tissues cannot be replaced if they are damaged or destroyed. Thus their numbers normally decline slowly as a function of age. In contrast, cells of most epithelial and connective tissues are systematically and continuously replaced through the operations of highly regulated **cell renewal systems.** These systems consist of the following four "compartments," each of which describes a sequential stage in the life of a cell: *proliferation* (mitosis and cytokinesis), *growth and differentiation, function,* and *senescence* through genetically programmed cell death (apoptosis). Complex regulatory mechanisms operate in each compartment of the system. For example, the frequency of mitosis, the rate of growth and differentiation, the length of time a cell will function, and the life span of a cell are all controlled by positive and negative regulatory proteins produced from genes; that is, all of these processes are genetically regulated. What is the nature of these genes? What happens if they malfunction? The answers to these questions have their roots in a famous discovery made early in the twentieth century.

Oncogenes

In 1910, Peyton Rous demonstrated that he could transmit cancer (specifically, a **sarcoma,** a type of connective cell cancer) from an affected chicken to one that was not affected. He did this by removing cancer cells, crushing them, and then filtering the material to remove all cellular fragments, bacteria, and other substances. When the filtered fluid was injected into healthy chickens, they developed sarcomas! What was the agent that caused these cancers? Like the revelations of Mendel (Chapter 14) and Garrod (Chapter 18), Rous's discovery was scientifically premature—the agent and its significance were not understood until decades later. It is now known that Rous had discovered the first oncogenic (cancer-causing) retrovirus (now called the *Rous sarcoma virus*). Further, this virus was found to have only four genes. Three of them help replicate the virus in infected cells, and the fourth is a gene that causes cancer—an **oncogene** known as *src*. For his discovery, Rous finally received a Nobel Prize in 1966, when he was 87 years old.

Retroviruses proved to be fascinating to study. Beginning in the 1950s, they were shown to cause cancers in several types of animals. (More than 20 retrovirus oncogenes have since been discovered.) As a result, in 1969, two researchers at the National Cancer Institute proposed the *oncogene hypothesis* to explain how cancer developed in humans. Their idea was that because of infections that occurred early in the evolutionary history of humans, viral oncogenes existed within the chromosomes of our normal cells, and cancer developed when they became activated by chemical carcinogens or radiation. In other words, viral oncogenes were integrated in the normal genome, and malignant growths arose in humans when they were activated. What type of research finding would support or confirm the oncogene hypothesis? Investigators began by searching for the *src* gene in various cells. These efforts led to a very surprising discovery.

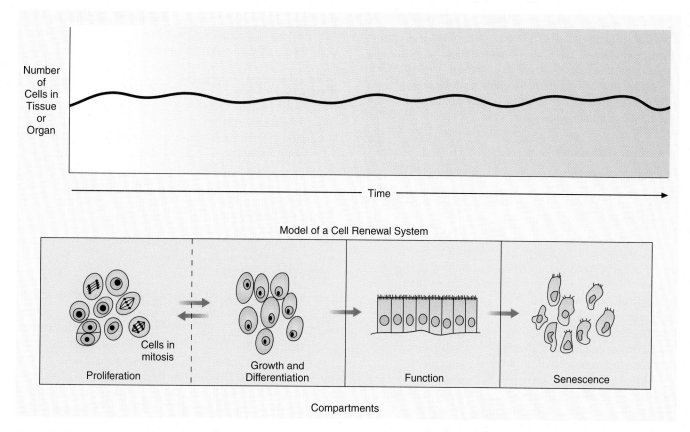

Figure 41.11 This is a conceptual respresentation of cellular homeostasis and the components of a normal cell renewal system. Cells of different tissues arise through mitosis and cytokinesis. They then undergo a period of differentiation when they become specialized to perform specific functions in the body. Such cells are functional for a genetically defined period of time, after which they die. In normal tissues, the number of new cells produced by mitosis is approximately equal to the number lost. If cells fail to differentiate, they may continue to proliferate— a common characteristic of cancerous growth.

Question: *When cancer occurs, which compartment contains malignant cells?*

Proto-oncogenes

In 1976, J. Michael Bishop and Harold Varmus found *src* sequences in the normal DNA of birds and later in mammals, including humans. However, the *src* gene in these cells is *not* a viral oncogene! Rather, in normal DNA, it is a gene that encodes an ordinary protein found in the plasma membrane. Thus in normal cells, *src* is not an oncogene; consequently, it was classified as a **proto-oncogene**—a gene that has a specific function but when damaged or mutated has the potential to become a cellular **oncogene.** For their crucial discovery that normal cells contain proto-oncogenes that can cause cancer if they become transformed into oncogenes, Bishop and Varmus were awarded a Nobel Prize in 1989.

Proto-oncogenes and Cancer

How are retroviral oncogenes related to normal cellular proto-oncogenes and oncogenes? Recall that the retroviral genome becomes integrated as a provirus into cellular chromosomes after infection. According to current theory, at some point in their evolutionary history, a provirus picked up part of a cellular proto-oncogene, most likely during a replication phase, and incorporated it into its genome. The frequency of such an event over many million years of evolutionary history must have been extremely rare. Evidence from molecular studies indicates that pirated proto-oncogenes underwent modification in the provirus replication process and eventually became established as viral oncogenes that can cause certain cancers in susceptible bird and animal species.

By the early 1980s, it had been established that some cancers could be a consequence of two distinct processes. First, viral oncogenes could cause cancer in certain birds and animals (but not humans) if they became activated. Second, carcinogens could mutate normal cellular proto-oncogenes into oncogenes, an event that could lead to cancer. Thus the original oncogene hypothesis was incorrect—the ultimate cause of human cancer (and most cancers of other animal species) does not involve retroviral oncogenes. Which proteins are encoded by cellular proto-oncogenes and oncogenes, and what do they do in the cell?

By the mid-1990s, over 70 proto-oncogenes had been identified. All known proto-oncogenes and oncogenes encode either growth factors, their cell surface receptors, or factors that regulate gene transcription or cell development. These products all function in the regulation of cellular proliferation, growth, differentiation, or senescence; that is, they are all involved in maintaining cellular homeostasis.

Conversion to Oncogenes

How are proto-oncogenes converted to oncogenes? Different types of damage have been identified—translocations between chromosomes or within the same chromosome, deletions of tiny segments of specific chromosomes, mutations within proto-oncogenes, and abnormal increases in portions of a chromosome, a process called **gene amplification.** When a proto-oncogene becomes an oncogene, three types of malfunctions can occur. First, the oncogene or its product cannot be regulated, and increased, uncontrolled cell activity may result. Second, gene amplification may lead to **overexpression,** the production of excessive quantities of the encoded protein or the production of an abnormal protein. Third, *mutations* can alter the action of the protein produced.

Any of these events may result in a loss of ability to regulate cell differentiation and proliferation, a step that can lead to cancer. For example, the first oncogene protein identified is encoded by a member of the *ras* oncogene "family," (one of a number of genes with similar structure and function) and has a role in regulating cell growth. The difference between the normal *ras* proto-oncogene protein and the *ras* oncogene protein usually involves a change in only a single nucleotide base. Yet the consequence of this change is that the oncogene encodes a protein that is associated with uncontrolled proliferation of the affected cell. Active *ras* genes are the most common oncogenes isolated from human cancers.

Thus from research conducted during the 1980s, it became known that there were links between oncogenes and certain cancers in humans. In general, oncogenes produce active products that lead to a loss of control at some point in a normal cell renewal system. New questions continued to arise. Can single oncogenes cause cancer, or are many active oncogenes required? Are there other genes involved?

Tumor Suppressor Genes

Throughout the 1980s, evidence mounted that another type of regulatory gene played a role in the genesis of many cancers. Such genes became known as **tumor suppressor (TS) genes.** In contrast to oncogenes, they influenced cancer induction if they were inactivated or lost from chromosomes. TS genes have been implicated in many cancers. At present, the picture of TS genes is not fully developed, but it is known that their encoded proteins tend to inhibit cellular renewal processes rather than driving them forward as oncogenes do. In addition, some TS genes regulate DNA repair—an essential mechanism for maintaining chromosome stability. How are these genes lost or inactivated? By the same processes involved in proto-oncogene transforma-

tion—translocations, deletions, and mutations. Figure 41.12 illustrates the two general types of mutations linked to loss of normal cell regulation and the development of cancer.

> **BEFORE YOU GO ON** A unifying hypothesis on cancer causation is now evolving. Cancer develops when there is a loss of regulated cellular homeostasis. This can occur when proto-oncogenes encoding regulatory proteins are converted to oncogenes. Consequently, the oncogenes may not be regulated, or their encoded proteins may be abnormal or overexpressed. Any of these outcomes may lead to cancer. Proteins encoded by tumor suppressor genes are also involved in cellular homeostasis. The loss or inactivation of these genes has also been implicated in some cancers.

A Model: Colorectal Cancer

One model will illustrate the type of research now being conducted. Some very exciting research results may have great relevance in developing effective treatments for colorectal cancer.

Figure 41.13 shows a model for colorectal tumor formation that has emerged in the 1990s. Colon and rectal cancers develop in about 150,000 people in the United States each year, killing more than one third of them. Colorectal cancers develop relatively slowly, taking years or decades to become malignant. The evolution of this cancer, beginning with benign growth and progressing to malignancy and metastasis, requires a sequence of genetic alterations involving both oncogenes and tumor suppressor genes. According to this model, tumor suppressor genes are sequentially lost from chromosomes 5, 18, and 17, and perhaps others just prior to metastasis. Activated *ras* oncogenes are apparently the only oncogenes involved in the progressive development of colorectal cancer. Normal *ras* genes apparently help regulate cell growth and differentiation. What is the identity and function of the normal tumor suppressor genes? At present, the tumor suppressor genes identified in the colorectal model are a gene identified as *MCC* (*mutated in colorectal cancer*) located on chromosome 5, the *DCC gene* (*deleted in colon cancer*) on chromosome 18, and the *p53 gene* on chromosome 17. Loss or mutation of the *p53* gene also occurs in many other types of human cancer. For that reason, it has been intensely studied, and its regulatory function in human cells has finally been established. The *p53* protein induces cell death by apoptosis. It is now hypothesized that *p53* acts as a tumor suppressor gene by eliminating cells generated from proliferation induced by oncogene activation. The *DCC* gene may encode a cell-adhesion protein that holds cells together in a tissue, and *MCC* is similar to genes that encode G proteins.

This model is extremely useful for several purposes. First, it explains the relatively late age of onset of colorectal cancer—it takes years or decades for the multiple mutations to occur. Second, it may be possible to detect the tumor in an early, benign stage and remove the growth before additional mutations accumulate and it becomes malignant and metastasizes. Finally, it provides points of reference for devising treatments, or even cures, for colorectal cancer in the future.

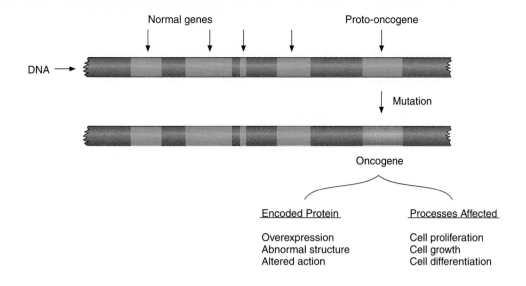

Overexpression   Cell proliferation
Abnormal structure   Cell growth
Altered action   Cell differentiation

A

Tumor suppressor gene

Incactivated or lost

Encoded Protein   Processes Affected

Not synthesized   DNA repair (?)
                           Cell proliferation
                           Cell gowth
                           Cell differentiation

Figure 41.12 Two general types of mutations lead to loss of cell regulation and to cancer. (A) A proto-oncogene is converted to an oncogene, a change that leads to alterations in levels or types of proteins involved in normal cell regulation. (B) Loss or inactivation of a tumor suppressor gene results in abnormal cell regulation because of the absence of an essential protein.

B

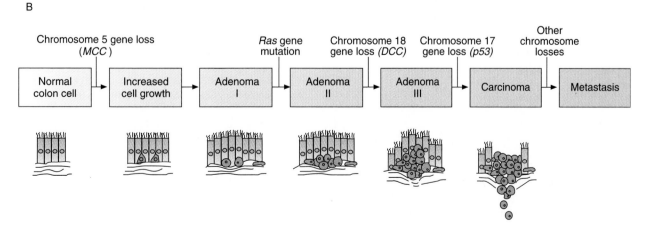

Figure 41.13 Colorectal cancer develops over many years or decades and requires a sequence of mutations or other types of genetic damage. Each of these events is associated with a change in the dynamics of the cell renewal system. Initially, there is increased growth, followed by the development of a series of benign tumors (adenomas). Eventually, cells acquire the ability to grow invasively into underlying tissues (carcinoma) and metastasize to other sites in the body.

BEFORE YOU GO ON Colorectal cancer serves to illustrate the power of a developing hypothesis on cancer causation. This cancer develops as a result of sequential changes in a proto-oncogene and several tumor suppressor genes. Understanding cancer at the molecular level may lead to effective preventive measures in the future.

Current Ideas About Cancer Causation

The emerging hypothesis of cancer causation suggests that cancer occurs when some part of a complex cell renewal regulatory system fails because of an accumulation of genetic changes within a single cell. Consequently, the affected cell continues to proliferate but no longer differentiates or becomes functional, or dies, according to a genetically defined timetable. These cells may eventually invade surrounding tissues and metastasize to other sites in the body. The underlying elements that lead to this loss of control include active oncogenes and their proteins and lost tumor suppressor genes and their missing proteins.

Many human cancers have now been studied, and most are thought to occur because of a combination of activated oncogenes and the loss of tumor suppressor genes. In some cases, a specific oncogene has been linked to several different cancers. The same is also thought to be true for tumor suppressor genes; many cancers are characterized by the loss of a specific TS gene. The model for colorectal cancer—beginning with normal epithelium and progressing to benign tumors (adenomas) and finally to the malignant, invasive stage of the disease—strongly supports the emerging hypothesis of cancer causation. As more becomes known about the specific functions of the various genes, the terms *oncogene* and *tumor suppressor gene* may be replaced by terms that are more descriptive of their proteins and how they function in normal cells.

Much remains unknown, and years of additional research will be required to put the complete puzzle together. Nevertheless, understanding cancer at the level described in this chapter offers great hope for eventually being able to prevent or retard its development. For example, it may become possible to disrupt the sequence of genetic events that leads to the final, malignant stage of the disease. Some day it may become practical to use gene therapy approaches (see Chapter 22) for replacing defective or missing genes or proteins involved in cancer progression.

LOOKING AHEAD

This chapter has focused on HIV-1 and cancer research being conducted at the molecular level of biological organization. It is worth noting that the speed with which many of the puzzles associated with HIV-1 were solved is simply phenomenal. Equally rapid headway is now being made in unraveling the causes of cancer. You will recognize that this progress was due in large part to the advances made in molecular biology discussed in earlier chapters.

Despite this progress, much remains to be learned about HIV-1 and cancer. It should be very exciting to follow research advances in AIDS and cancer into the twenty-first century.

SUMMARY

1. Complex human diseases such as AIDS and cancer are being investigated at the molecular level. This approach will enable scientists to determine the basis for pathogenic effects caused by these diseases and to establish rationales for diagnosis, therapy, and eventual cures for these disorders.

2. AIDS was identified as a significant disease in the United States and other areas of the world in 1981. AIDS is characterized by loss of immune function and the inability to control opportunistic infections.

3. Through epidemiological investigations conducted by the CDC and other agencies, it became obvious by 1982 that AIDS was caused by an infectious agent, not some unique lifestyle factor. In 1983 and 1984, the cause of AIDS was discovered to be a new retrovirus called HIV-1.

4. HIV-1 infects helper T4 lymphocytes and becomes integrated into host cell chromosomes. When the T4 cell becomes activated by contact with a foreign antigen, it replicates new virions rather than substances and cells required for a normal immune response. Subsequently, the virions bud from the cell, enter circulation, and infect other T4 cells. The original T4 cell dies during the budding process.

5. The eventual depletion of T4 cell populations accounts for the characteristic immune deficiency of AIDS and is the indirect cause of death of infected individuals.

6. The major groups infected with HIV-1 in the United States are homosexual and bisexual men, intravenous drug users, sexual partners of individuals in these two groups, people who received blood transfusions, and babies of infected mothers.

7. In the future, rates of HIV-1 infection are expected to increase in poor women and young, urban African Americans and Latinos.

8. It is now hypothesized that cancer arises because of sequential mutations in genes associated with cellular homeostasis. These changes modify the structure and function of proto-oncogenes, converting them into oncogenes, and cause the loss of tumor suppressor genes.

9. Both proto-oncogenes and tumor suppressor genes play key roles in the complex regulating mechanisms of cell renewal systems. The combination of active oncogenes and lost tumor suppressor genes results in an inability to regulate normal cell proliferation and differentiation, which leads to cancer.

WORKING VOCABULARY

AIDS (p. 787)
cell renewal system (p. 800)
cellular homeostasis (p. 800)
epidemiology (p. 787)
HIV-1 (p. 788)
oncogene (p. 800)

proto-oncogene (p. 801)
provirus (p. 790)
retrovirus (p. 787)
reverse transcriptase (p. 788)
tumor suppressor gene (p. 802)

REVIEW QUESTIONS

1. Why is modern disease research emphasizing changes at the molecular level?

2. What is epidemiology? What are the objectives of epidemiological research?

3. What types of results from early studies indicated that AIDS was caused by an infectious agent?

4. What is AIDS? What causes AIDS?

5. Describe the HIV-1 life cycle.

6. What events precede the development of AIDS?

7. Who is most likely to be infected with HIV-1 in the United States? Why?

8. Why is the AIDS picture so bleak for Africa and Asia?

9. What are proto-oncogenes? Oncogenes? How are they related to cellular homeostasis?

10. What are tumor suppressor genes? What is their normal function?

11. Explain one cause of colorectal cancer.

12. Summarize the developing hypothesis of cancer causation.

ESSAY AND DISCUSSION QUESTIONS

1. Based on current epidemiological information in the United States, what would you recommend to halt the spread of HIV-1? What problems might affect implementation of your recommendations? How could they be overcome?

2. New data indicate that AIDS is spreading in the heterosexual population, the rise in teenage AIDS cases is alarming, and high school and college students have not modified their sexual behaviors. What might be done to change the attitudes and behaviors of this risk group? Should anything be done? Why?

3. Whether or not HIV-1 tests should be required for all marriage license applicants remains an issue in the United States. An analysis published in 1987 (in the *Journal of the American Medical Association* (258: 1757–1762) included the following data: applicants for marriage licenses, 3,800,000; true positives (actual cases identified), 1,219; false positives (cases misdiagnosed), 382; false negatives (cases missed), 129; HIV-infected births prevented, 250; cost of screening, $100 million. Do you favor marriage applicant screening? Why? What arguments support or do not support HIV-1 screening?

4. Describe the types of studies and data that you think will be necessary to confirm the unifying hypothesis of cancer causation. Do you believe that this hypothesis will explain the origin of all cancers? Why? How do the risk factors described in Chapter 40 fit into the hypothesis?

REFERENCES AND RECOMMENDED READING

Aggleton, P., K. O'Reilly, G. Slutkin, and P. Davies. 1994. Risking everything? Risk behavior, behavior change, and AIDS. *Science,* 265: 341–345.

Beardsley, T. 1994. A war not won. *Scientific American,* 270: 130–137.

Bethel, E.R. 1995. *AIDS: Readings on a Global Crisis.* New York: Macmillan.

Cavenee, W. K., and R. L. White. 1995. The genetic basis of cancer. *Scientific American,* 272: 72–79.

Coffin, J. M. 1995. HIV population dynamics in vivo: Implications for genetic variation, pathogenesis, and therapy. *Science,* 267: 483–489.

Des Jarlais, D. C., and S. R. Friedman. 1994. AIDS and the use of injected drugs. *Scientific American,* 270:82–89.

Des Jarlais, D. C., S. R. Friedman, and J. L. Sotheran. 1994.Continuity and change within an HIV epidemic. *Journal of the American Medical Association,* 271: 121–127.

Grady, C. 1995. *The Search for an AIDS Vaccine: Ethical Issues in the Development and Testing of a Preventive HIV Vaccine.* Bloomington: Indiana University Press.

Greene, W. C. 1993. AIDS and the immune system. *Scientific American,* 269: 99–105.

Grmek, M. D. 1990. *History of AIDS: Emergence and Origin of a Modern Pandemic.* Princeton, N.J.: Princeton University Press.

Hoffman, M. 1994. AIDS: Solving the molecular puzzle. *American Scientist,* 82: 171–177.

Mendelsohn, J., P. M. Howky, M. A. Israel, and L. A. Liotta (eds.). 1995. *The Molecular Basis of Cancer.* Philadelphia: Saunders.

Murphy, J. S. 1995. *AIDS, Reproductive Technology, and Ethics.* Albany: State University of New York Press.

Nowak, M. A., and A. J. McMichael. 1995. How HIV defeats the immune system. *Scientific American,* 273:58–65.

Silin, J. G. 1995. *Sex, Death, and the Education of Children: Our Passion for Ignorance in the Age of AIDS.* New York: Teachers College Press.

Ruddon, R. W. 1995. *Cancer Biology.* New York: Oxford University Press.

Tarantola, D., and J. Mann. 1993. Coming to terms with the AIDS pandemic. *Issues in Science and Technology,* 9: 41–48.

ANSWERS TO FIGURE QUESTIONS

Figure 41.1   They are infected by their mothers during pregnancy, birth, or nursing.

Figure 41.2   It is RNA, not DNA, which is the genetic material of all life forms other than viruses.

Figure 41.6.   Generally, opportunistic diseases become a problem when cell counts drop below 200 to 500 T4 cells per microliter of blood, at which point the immune system can no longer mount an effective reaction against opportunistic infectious agents.

Figure 41.9   In 2000, Africa (because AIDS develops several years after HIV-1 infection); in 2010, Asia.

Figure 41.11   Proliferation; malignant cells generally do not differentiate, function, or die according to a genetically programmed schedule.

Biology in Future Decades

Chapter Outline

Reading Questions

1. Why is the integration of biological knowledge considered an important goal for the coming decades?

2. Of what scientific or practical value is the study of individual taxonomic groups?

3. What are some ways in which our biological knowledge may be applied in the near future?

4. What are currently active areas of biological research?

Throughout this book, we have emphasized the dynamic nature of biology, discovering that studies of life have changed significantly over the past century. Biology has shifted focus on its central questions, expanded its domain to incorporate what were formerly separate disciplines, and developed new methods, technologies, concepts, and theories for understanding nature. There is little reason to doubt that biology will continue to evolve in the coming decades. What might we look forward to?

At best, crystal ball gazing is a perilous task. One of the primary lessons from the history of science is that biology has often veered in unexpected directions. New ideas, discoveries, and challenges have altered the course of biological research in unpredictable ways. Nevertheless, it is possible to reflect on probable directions, likely areas of development, or desirable paths that biology might follow in the coming years.

How does one attempt to answer the question, What direction will, or should, biology take in coming decades? One obvious method is to consider the fastest-growing and most exciting areas in biology. Throughout this book, we have described developments in rapidly progressing fields. These "hot areas" are recognized by funding agencies, experts in the field, and, through the media, the general public. Some of these prominent areas of research are described here.

There are also broad, general ways in which the biological sciences might change or in which serious researchers would like it to move. These general changes include the overall manner in which scientific ideas relate to one another, the basic perspectives that guide research, and the makeup and structure of the scientific community. Forecasting such changes is an uncomfortable activity for scientists, but what Ernst Mayr, one of the architects of the modern theory of evolution and an active participant in evolutionary thinking for several decades, calls "wishful hoping" is not. Many authorities, like Mayr, have given serious thought to how they would like to see biology develop in the coming decades. We shall describe several major themes, related to the future of biology, that leading scientists, historians of science, and philosophers of science have identified.

The cost of scientific research will also be a subject of continuing discussion. According to National Science Foundation statistics, federal funding for research in the life sciences increased from $1.5 billion in 1970 to $4.2 billion in 1980 and to $8.8 billion in 1990. These funds were used to support research in government agencies, such as the Environmental Protection Agency, and to finance projects at universities, private corporations, and state agencies. But as the cost of research has increased, so too has the number of critics who question the size of the expenditure. Recent cuts in the science budget may reflect public dissatisfaction with the level of scientific funding. Defenders of substantial government support cite the dramatic and economically valuable advances in medicine, in agriculture, and in our understanding of environmental problems to justify the investment.

ACTIVE AREAS OF BASIC RESEARCH

Several active areas of basic research, because of their importance and momentum, will remain prominent in the coming decade. Neurobiology is one of these. Research on the immune system and human diseases is another field where considerable work will be done.

Many active areas of research were described in earlier chapters. From these we developed a list of "hot questions" for the next decade. We challenge you to formulate other questions.

How are different ecosystems integrated?

How is the human population affecting ecosystems?

What are the causes of global climate change? Can global warming be reversed?

What is the molecular basis for various human genetic disorders?

What is the best use of genetic engineering of microbes, plants, and animals?

How can genetic technology be used in treating or curing human disorders?

Is the process of macroevolution different from that of microevolution?

What is the genetic basis of behavior?

Will there be a new synthesis of the modern theory of evolution?

How is the endocrine system related to other systems?

What is the molecular basis of complex human diseases?

GENERAL TRENDS

The list of ways in which biology might change is much shorter than the inventory of active areas for research. Three major items on the list of potential changes are (1) integrating knowledge, (2) shifting from studying life in terms of levels of biological organization to studying taxonomic groups of organization, and (3) applying biological knowledge.

Integrating Knowledge

The most commonly expressed concern about change in biology involves integrating scientific knowledge. During the past century, biology has progressively become more specialized. There once was a time when scientists could envision writing a set of volumes that would encompass all knowledge of the living world. That time is long past. Now many societies and journals concentrate on finely defined areas of research, and scientists find it increasingly difficult to keep abreast of significant research in their own area of interest. As a consequence, more and more information is becoming available, but it is not being connected or integrated into existing bodies of knowledge.

Why is this a problem? One reason is related to efficiency. On a practical level, the proliferation of information—what has often been referred to as an information explosion—creates a situation where similar problems arise in different fields but are attacked independently rather than jointly. Also, time and money may be lost in "reinventing the wheel" because scientists are often ignorant of relevant developments in other fields.

The enormous number of new research articles published each year tells us a great deal about pieces of the story of life, but it leaves great gaps in developing the overall picture. Ernst Mayr has stated:

> My own attitude is perfectly clear. In science, we have to do analysis first so that we know what the pieces are with which we are working. However, in many areas we have now reached the place where we have to try to put the systems together that we have analyzed. The two areas—and almost everybody seems to agree in this—that are in particular need of this are the genotype as a whole and the functioning of the central nervous system.[1]

Mayr is referring to the great concern among biologists about our current lack of understanding the functional genotype—how it becomes expressed in forming a phenotype, how it has evolved, and its inherent constraints. His comment on the nervous system points to a hot area of research, neurobiology. Many scientists are now attempting to understand the brain and the nervous system. Notable figures like Francis Crick, who pioneered our understanding of DNA's structure and function, are now exploring the

---

[1] All quotations in this chapter are from personal communications.

brain. Although scientists have uncovered much information about genotype and the nervous system, they have not produced comprehensive explanations of either (see Figure 42.1).

Other examples illustrate the need to integrate knowledge in addressing scientific issues. Margaret Davis, of the University of Minnesota and past president of the Ecological Society of America, recently stated that she was convinced that global change can be understood only if research is coordinated at all levels of ecological organization. The scientific problems of global change, she contends, touch every subfield of the discipline and present the most exciting intellectual challenge that has confronted ecologists in decades. But the answers we seek can be reached only by coordinating and integrating research.

In Chapter 26, we considered why the theory of evolution is the broadest integrating theory in the biological sciences and how it is more closely related to some fields than others. For many biologists, the intellectual challenge of the future is to integrate all biological disciplines within an evolutionary perspective. Ledyard Stebbins, a leading evolutionary botanist associated with applying the modern synthesis to studying plants, feels this way and agrees with Theodosius Dobzhansky's dictum that "nothing makes sense except in the light of evolution." Stebbins also recognizes that other unifying concepts, such as energy transfer and biological organization, may be useful in drawing together all information about the living world. David Hull, of Northwestern University and one of the foremost philosophers of biology in the United States, hopes to see advances in our understanding of the general process of evolution. We know a lot about a few processes, but when it comes to understanding

Figure 42.1   Much research is now being done on the human brain. At center, left, are flask-shaped Purkinje neurons (dark red). A network of Purkinje cell dendrites occur in upper layer of the cerebellum (red, upper left). Other nerve cell bodies (orange) and dendrites (yellow) are seen at the lower right.

how different areas in biology relate evolutionarily, we are still mostly in the dark. We do not yet understand how ecology and population biology go together.

Biodiversity and the New Natural History

Biologists have questioned the current focus on studying distinct levels of organization. E. O. Wilson, father of sociobiology, the world's authority on ants, and curator of entomology at the Museum of Comparative Zoology at Harvard University, has recently argued that it is time to shift from a biology that concentrates on levels of organization to a biology that concentrates on different taxonomic groups. Since the 1950s, biological research has concentrated on searching for generalizations on different levels: organism, cell, molecule. Wilson argues that it is time to emphasize "the study of particular groups of organisms across all levels of organization." He is not advocating a return to old-fashioned descriptive natural history but rather, he stresses, in this new natural history, the value of studying particular taxa with tools that have been developed in molecular and cell biology.

Why pursue the study of individual taxa? In part, Wilson says, the search for new and interesting generalizations that hold true across all the species at any one level of organization is yielding fewer and fewer successes. Although we continue to discover many interesting truths, they often turn out to apply to just the group in which they were discovered, not all groups. The deep study of different taxa will, it is hoped, reveal new unifying generalizations.

Of equal importance is an appreciation of the inherent interest and value of individual groups. Another way of stating this idea is that future emphasis will be placed on exploring diversity. The theory of evolution provides a basic framework for understanding the phenomenon of diversity itself. Wilson and others hope to expand on that general understanding. The diversity of life, *biodiversity,* is the storehouse of all that exists in the biological world. A detailed understanding of diversity can provide information about which organisms are of benefit to humans, which are endangered, and which are useful for different types of research. This approach is particularly relevant at the present time because scientists have documented the alarming rate at which the biodiversity of Earth is being dissipated by destruction of natural habitats (see Figure 42.2).

Applications of Biological Knowledge

Another common theme that will run through discussions of biology in the coming decades concerns the enormous potential for greater practical applications of knowledge. Which areas of active research will lead to applications in the 1990s and beyond? Recent discoveries associated with medicine and medical technology will undoubtedly lead the way (see Figure 42.3). Marvin Druger, a geneticist and science educator at Syracuse University, states:

> This is an especially exciting time for biology. Technological advances have given us the tools to explore living things in ways that were not possible before. Biologists can use these

Figure 42.2 Many species in this tropical rain forest of eastern Ecuador might be endangered due to the loss of their native habitat.

Figure 42.3 These flasks in a National Cancer Institute laboratory contain interferon, an antiviral substance produced in large quantities through genetic engineering.

tools in clever and creative ways that benefit humankind. New technologies will continue to be developed and refined, and they will be applied to making the environment cleaner, curing diseases, predicting disease risks, improving agriculture, and learning more about organism and cell functions. The entire pharmaceutical industry will be transformed by genetic engineering technology, and most major body chemicals will be produced by microorganisms. Also, all sorts of new creatures will be designed by genetic engineers, combining the desirable qualities of different species. Organ transplants will become more common, and we may have centers where arms, legs, eyes, kidneys, and other parts are generated from cell cultures *in vitro*. Cures will be found for AIDS, cancer, and other menacing diseases. . . . Chemical substances will be available to prevent deterioration that occurs with aging.

Linus Pauling, a distinguished biochemist and twice a Nobel laureate, was optimistic right up to his death in 1994 about what future research may hold for *human health*. For example, he felt that research on nutrition will be particularly fruitful, especially in determining

> the properties of vitamins and other orthomolecular substances [substances normally present in the human body, usually required for life]. I foresee that much information will be gathered about the optimum levels of these substances in the blood, or perhaps I should say optimum levels in the body. . . . I think that it will be recognized that the greatest progress in the control of disease, improvement of health and well-being, and extension of the life span will be made during the coming decade by studies in this direction.

A concern for the *health of life on Earth* also occupies a prominent position in biological research and will continue to be important in the coming decade. Scientists are exploring the effects of pollution on various environments and attempting to understand alterations in ecosystems that will follow if predicted global climate changes occur (see Figure 42.4). Biologists are also concerned with the management of natural resources, such as forests, and with restoration of areas that have been severely altered by humans.

Scientists now understand that processes affecting the environment operate over different time scales. Long-term effects are sometimes "invisible" in the short term and may not be noticed until substantial damage has occurred. Until recently, much ecological research was focused on relatively short time scales. The current work on global climate change is concerned with very long periods. Recently, the National Science Foundation established a new program, the Long-Term Ecological Research Program (LTERP), that bridges the gap between short and very long term effects and also attempts to look at problems on an intermediate spatial scale.

The Ecological Society of America, in a report titled *The Sustainable Biosphere Initiative: An Ecological Research Agenda for the Nineties,* urges scientists to investigate topics that are important for science and society, such as preparing for global change, protecting biodiversity, and maintaining a sustainable biological ecosystem.

Figure 42.4 The short-term effects of this oil spill near Galveston, Texas, were very dramatic. Long-term effects, which are often less obvious, are now being studied by scientists.

ORGANIZATION OF BIOLOGY

Numerous scientists deplore the overspecialization—the narrow attention of researchers to their own area—that seems to characterize biology today. One way to reduce or eliminate overspecialization is to reform education and have it become broader and more integrative. Another way is to encourage more group research and collaborative efforts. Much research is now being done by teams, especially in some fields. For example, the problems associated with global climate change are so complex that teams of atmospheric scientists, ecologists, physiologists, geographers, and oceanographers will be required to provide some answers.

BIOLOGY AND SOCIETY

If the nineteenth century was the "century of physics," the twentieth may be labeled by historians as the "century of biology." It has been a period of enormous expansion in the quantity and complexity of the questions asked, the solutions developed, and the impact of biology on our lives. Biology is now a part of big business, the educational establishment, the recreation industry, health care, government policy, and government budgets (see Figure 42.5). We see or hear about biology on comics pages, on television, and even during congressional debates. In the coming decade, this interest and significance will increase. The number of scientists studying the living world will also increase, but probably in a somewhat different way. Women and minorities, who now constitute a small fraction of the biological research community, will increase proportionately to the white males who now make up the vast

majority of biologists. The shift will come about because of a concerted effort by the scientific and educational communities to recruit and retain women and minorities who are interested in pursuing careers in biology (see the Focus on Scientific Process, "Diversity and the Future of Biology"). Biology will see many new faces in the coming decade in more than one way, for it will interface increasingly with other intellectual disciplines, such as philosophy and medical ethics. New insights will provide us with a deeper understanding of humans, relationships among all forms of life, and the planet Earth. Because biology is and will continue to be concerned with problems that affect so many aspects of human life, it will increasingly have a political dimension. Issues concerning land use, the protection of endangered species, and the direction of medical research are examples of topics that have scientific, ethical, and political facets. Scientists will have to work closely with other specialists in the coming decades to face the challenges that now confront us.

Figure 42.5 This genetically engineered cantaloupe plant is being field-tested to see if it is resistant to two common viruses that often damage cantaloupe crops. It is hoped that such genetic engineering will replace the need for chemicals currently used to control infestations.

Diversity and the Future of Biology

The Current Lack of Diversity

Today, after two decades of good intentions, hard work, and billions of dollars of public and private support, women and especially, ethnic minorities are still underrepresented in biology. That is, the proportion of women and ethnic minorities in the life sciences is significantly lower than their percentage in the overall population. Many scientists are concerned by this continuing imbalance and have called for renewed efforts to achieve a more representative distribution of women and ethnic minorities. Others, however, have asked why we should worry about who does science, especially in light of the spectacular accomplishments in biological research that are reported almost daily. What can diversity add to such a seemingly efficient and successful enterprise?

As might be expected, emotions run high in exchanges between those who argue that science is independent of social context and those who maintain that cultural representation is an issue in biology because all knowledge is influenced by social, political, and historical factors within a cultural context.

Why Increase Diversity?

Two different types of reasons—pragmatic and ameliorative (improvement-oriented)—are generally raised in discussions about the importance of diversity to science. The *pragmatic argument* is based on the changing demographics of the United States, which reflect greater numbers of "minorities" in the population (in 10 to 20 years, some may no longer be minorities in certain regions of the country, and women, of course, are not currently a minority of the population). If the scientific community hopes to

recruit future scientists to sustain the scientific enterprise, it will have to draw from this expanding pool or become diminished (see Figure 1). The *ameliorative argument* takes a completely different approach. It assumes that science is always done in a social context and that by enlarging that context, we can enlarge our scope of vision because people from different backgrounds will bring new perspectives and fresh ideas about important areas of biological research.

Figure 1 Efforts are being made to ensure that the diversity of our society will be reflected in the future diversity of the scientific community.

Social Influences on Biology

About 20 years ago, Steven Weinberg, a Nobel Prize–winning physicist, wrote that "the laws of nature are as impersonal and free of human value as the rules of arithmetic. We didn't want it to come out that way, but it did." He was expressing a widely held view that science is objective and independent of any social influence. However, recent studies by science historians, philosophers, and sociologists have shown the limitations of such a belief. The modern view that has emerged from these studies does *not* imply that a scientific understanding of nature can be constructed on personal beliefs, hopes, or interpretations. Science *is* objective in the sense that it is based on community evaluations and agreements. Observations and experiments that confirm scientific hypotheses must be repeatable and openly communicated before they become accepted. However, the historical record clearly reveals that there is an important social dimension to all sciences, especially the life sciences. Most notably, the selection of research topics in science is strongly influenced by social factors. For example, Charles Davenport was a prominent early-twentieth-century American geneticist who spent the late stages of his career studying human genetics. He was convinced that "inferior" foreign immigrants and "mentally defective" individuals in the population would cause the decline of the American people unless their reproduction could somehow be restricted. His work was central to the early eugenics movement, which was discussed in Chapter 19.

A more recent example involved justifying the enormous cost of the Human Genome Project on the basis of discovering causes of "genetic disease" and potentially eliminating them from the human population. One government report from the Office of Technology Assessment quotes the statement that "individuals have a paramount right to be born with a normal, adequate hereditary endowment." But who will decide what is "normal"? How is the "right" to be born with a "normal hereditary endowment" to be guaranteed? Will prospective parents have to be certified as being capable of providing a "normal, adequate hereditary endowment" before they can have children?

If social context is an important element of science, what might be expected if the scientific community was more diverse? Will our vision of nature be expanded? Might new questions be asked that lead to fruitful investigations?

Toward an "Ethnic- and Gender-Diverse" Biology?

Not only is broadening the composition of the scientific community a realistic response to an increasingly diverse American population, but it can also bring exciting new visions to research in biology. For example, Mary-Clair King decided very early in her scientific career to pursue research on the genetics of breast cancer. (Breast cancer and ovarian cancer are responsible for one fourth of cancer-related deaths in women.) When she began her work in 1974, little was known about the subject, although one set of tantalizing clues was available—epidemiological studies had shown that a woman's risk of developing breast cancer was significantly increased if she had a mother or a sister who had died of the disease before age 50 or if more than one sister had been affected. Nevertheless, because of technological limitations and despite intense efforts, advances were slow until a dramatic breakthrough was announced in 1990—her research group found a gene that was linked to a marker on chromosome 17 in families with early-onset breast cancer. The specific gene, now known as *BRCA1* (*Br*east *Ca*ncer *1*), was identified in 1994. Soon, it will be possible to diagnose women at risk and, it is hoped, to develop treatments or a cure for this genetically caused type of breast cancer. The gender-specific nature of the disease gave King a strong incentive for studying this important biological problem at a time when little commitment had been made to learn more about its underlying cause (see Figure 2).

Figure 2   Mary-Clair King (right) has been a leader in breast cancer research for 20 years.

Earlier and broader diversity in the biological community might have helped reduce the magnitude of suffering and deaths for many African Americans that was caused by sickle-cell anemia. Even though the medical and research communities understood the basis of sickle-cell anemia for almost 50 years, little was done until the 1960s, when affected individuals called attention to the problem and sought knowledge to deal with it. How many ethnic- and gender-specific issues are ignored because of the narrow representation of the contemporary biological community? Will a more diverse community of scientists doing basic research avoid such blind spots in the future? One disturbing aspect of the current redirection in science funding is the possible impact on expanding diversity. If the scientific enterprise is diminished, there may be fewer opportunities to increase the number of women and ethnic minorities who wish to enter the sciences.

Does the call for increased diversity mean that finding the cause of breast cancer or a cure for sickle-cell anemia requires a "different science" with separate departments of "women's medicine" or "black genetics"? The answer is an unequivocal no. Biology has undergone major transformations in the past century. We are heirs to a flourishing, expanding, and exciting set of disciplines that has been marvelously successful in explaining nature. Broadening the scope of questions asked, investigating biological problems that concern specific ethnic groups, and evaluating certain medical issues from a different perspective will allow us to respond more equitably to a changing society and to enrich the existing, dynamic world of the modern life sciences.

REFERENCES AND RECOMMENDED READING

Biodiversity Loss: Economic and Ecological Issues. 1995. New York: Cambridge University Press.

"Breast Cancer Research: A Special Report." 1993. *Science,* 259: 616--638.

"Comparisons Across Cultures: Women in Science '94." 1994. *Science,* 263: 1468–1496.

Crick, F. 1994. *The Astonishing Hypothesis: The Scientific Search for the Soul.* New York: Scribner.

Davis, M. 1989. Insights from paleoecology on global change. *Bulletin of the Ecological Society of America,* 70: 222–228.

Etzkowitz, H., C. Kemelgor, M. Neuschatz, B. Uzzi, and J. Alonzo. 1994. The paradox of critical mass for women in science. *Science,* 266: 51–54.

Farber, P. L. 1994. *The Temptations of Evolutionary Ethics.* Berkeley: University of California Press.

Greenspan, R. J. 1995. Understanding the genetic construction of behavior. *Scientific American,* 272: 72–78.

Holloway, M. 1993. A lab of her own. *Scientific American,* 269: 94–97.

Hull, D. 1988. *Science as a Process: An Evolutionary Account of the Social and Conceptual Development of Science.* Chicago: University of Chicago Press.

Longino, H. 1990. *Science as Social Knowledge: Values and Objectivity in Scientific Inquiry.* Princeton, N.J.: Princeton University Press.

Maksie, Z. B. 1991. *Molecules in Natural Science and Medicine: An Encomium for Linus Pauling.* New York: Horwood.

Mayr, E. 1988. *Toward a New Philosophy of Biology: Observations of an Evolutionist.* Cambridge, Mass.: Harvard University Press.

"Minorities in Science '93." 1993. *Science,* 262: 1089–1135.

Science. 1995. The future of the Ph.D. *Science,* 270: 121-146.

Wilson, E. O. 1992. *The Diversity of Life.* Cambridge, Mass.: Harvard University Press.

Wilson, E. O. 1989. The coming pluralization of biology and the stewardship of systematics. *BioScience,* 39: 242-245.

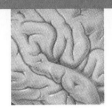

Unit Six Conclusion
Human Systems of Life and the Process of Science

Scientific investigators have studied the structure and function of the human body since at least the fifth century B.C. Early medical writers were concerned about the causes of disease. But they also asked questions about the general components of the human body, its structure, and the functions of its primary organs. It took several centuries before they developed detailed answers to these questions. In the second century, Galen combined the results of his research with knowledge acquired by past medical researchers to produce the first detailed book on human anatomy. His description of the human body endured until the sixteenth century, when anatomists began to raise questions. In an effort to correct some of Galen's minor errors, anatomy was reborn during the sixteenth century in Europe. For example, Vesalius produced a magnificent new book on human anatomy in 1543.

In the seventeenth century, physicians and other scientists attempted to describe the human body as if it were a machine. This ultimately led, in the nineteenth century, to the formation of cell theory, which described the body as being composed of cells. Scientists then turned to new questions: How do cells function? How do they reproduce? How is cell action coordinated? How do groups of cells respond to environmental factors? These questions are still being addressed today.

The cell theory gave scientists a new way to think about disease. What happened to cells when a person had a disease? Did cells from outside the body cause disease? Pasteur thought so, and he demonstrated to the scientific world that germs caused many diseases. If germs caused disease, how could they be controlled or killed? Research has focused on this question for over 100 years.

As scientists explored the structure and function of the body, they realized that cells were part of complex systems. They then asked new, broader questions: How are organ systems integrated? What is the nature of feedback systems in the body? Subsequently, new technologies have been used to increase our understanding of the structure and function of the human body. Today, because of new hypotheses about the causes of cancer and the discovery of HIV-1, the virus that causes AIDS, scientists are pursuing answers to old and new questions about normal and abnormal structure and function at the molecular level.

Appendix A
Chemistry

Matter

Matter: Anything that has mass and occupies space.
Mass: The amount of matter in an object

Atomic Structure [*Refer to Figure 1*]

All matter, living and nonliving, is composed of atoms. Atoms are the smallest units of matter that enter into chemical reactions. Structurally, they have two main parts, a single nucleus surrounded by one or more electrons.

Nucleus

The mass of an atom is concentrated in a tiny nucleus at the center of the atom. The nucleus consists of two types of subatomic particles: positively charged protons and neutrons, which are electrically neutral. Protons and neutrons are approximately equal in mass and weigh about 2,000 times more than an electron.

Electrons

The space occupied by an atom is determined by the negatively charged electrons, which move at enormous speeds in pathways around the nucleus. The electrons are arrayed around the nucleus in regions called energy levels; the farther out an electron is from the nucleus, the higher its energy level.

Related Terms

Atomic number: The number of protons in the nucleus of an atom.
Atomic weight (atomic mass): The number of protons plus neutrons in a nucleus.
Ions: Atoms that have a positive or negative charge.
Isotopes: Atoms of the same element that have different numbers of neutrons in their nuclei. Carbon-12 (^{12}C) has 6 protons and 6 neutrons; carbon-14 (^{14}C) has 6 protons and 8 neutrons.

Chemical Structures and Formulas

Element	Atomic Number	Atomic Weight	Number of Electrons in Each Energy Level[a]	Electron-Dot Diagram	Sample Molecular Formula with the Element	Structural Formula
Hydrogen	1	1	1 1	H ·	Water H_2O	H – O – H
Carbon	6	12	6 2 4	· Ċ ·	Methane CH_4	H – C – H with H above and H below
Nitrogen	7	14	7 2 5	· N :	Ammonia NH_3	H – N – with H above and H below
Oxygen	8	16	8 2 6	: Ö ·	Oxygen O_2	O = O
Phosphorus	15	31	15 2 8 5	· P̈ ·	Phosphoric acid H_3PO_4	O (double bond) H – O – P – O – H with O – H below
Sulfur	16	32	16 2 8 6	: S̈ :	Hydrogen sulfide H_2S	H – S – H

Figure 1 Chemical structures and formulas.

Organization of Matter

Elements

Element: A single type of matter that cannot be broken down into simpler substances by ordinary chemical processes.

Compounds

Compound: A substance that consists of two or more atoms of different elements that are bonded together. The compound water (H_2O) has 2 hydrogen atoms and one oxygen atom.

Molecule: A substance that contains two or more atoms of the same or different elements. An oxygen molecule (O_2) consists of 2 oxygen atoms; a molecule of water (H_2O) consists of 2 hydrogen atoms and 1 oxygen atom.

Classification of Compounds:

Inorganic compounds: Acids, bases, salts, and water.

Organic compounds: Carbohydrates, lipids, proteins, and nucleic acids.

Formulas:

Electron-dot diagrams: Indicate the location of the electrons in the outer shells.

Molecular formulas: Indicate the elements in a molecule and their proportions (H_2O).

Structural formulas: Indicate the elements in a molecule and the location of covalent bonds.

Chemical Bonds

Electrons

An electron is a fundamental and stable atomic particle that possesses one negative electric charge. Electrons are found in shells surrounding the nuclei of atoms; their interactions with the electrons of neighboring atoms create the chemical bonds that link atoms together as molecules.

Electron interactions: The basis of all chemical reactions.

Inert elements: Elements that do not take part in chemical reactions because their outermost electron shells are filled to the maximum. Examples are helium and neon.

Chemical stability: To achieve chemical stability, an atom must reach a configuration of 8 electrons in its outermost energy level. This is achieved by losing, gaining, or sharing electrons.

Valence electrons: Electrons in the outermost energy level of an atom.

Types of Chemical Bonds [*Refer to Figures 2, 3, and 4*]

Covalent Bonds (*Shared Electrons*)

Chemical bonds that result when two atoms share valence electrons.

Nonpolar: Covalent bonds formed between two atoms that do not result in molecules with regions of unbalanced electrical charges.

Polar: Covalent bonds formed between atoms that result in a molecule with regions of unbalanced electrical charges. Water is a polar molecule.

Ionic Bonds (*Electron Transfer*)

Chemical bonds that form as a result of attraction between ions when one or more electrons are transferred from one atom to another.

Electron acceptor: The atom that accepts one or more electrons and becomes negatively charged.

Electron donor: The atom that donates one or more electrons and becomes positively charged.

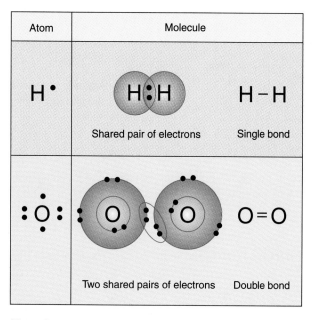

Figure 2   Covalent bonds.

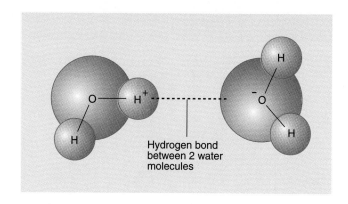

Figure 4   Hydrogen bond.

Hydrogen Bonds (*Between Electrically Charged Atoms*)

Very delicate bonds formed by the weak electrical attraction between positively charged, covalently bonded hydrogen atoms of one molecule and negatively charged atoms, usually oxygen or nitrogen, of another molecule. Hydrogen bonds exist between adjacent water molecules and two strands of a DNA molecule.

Chemical Reactions

Chemical Reactions in Cells

Chemical reactions involve breaking chemical bonds and forming new chemical bonds.

Anabolism (*Synthesis Reactions*)

Anabolic reactions are synthesis reactions—reactants combine to form a new product. Anabolic reactions require energy. Examples:

Carbohydrate synthesis: Monosaccharides combine to form a polysaccharide.

Lipid synthesis: 3 fatty acids and 1 glycerol combine to form a triglyceride.

Protein synthesis: Amino acids combine to form a protein.

Nucleic acid synthesis: Nucleotides combine to form DNA or RNA.

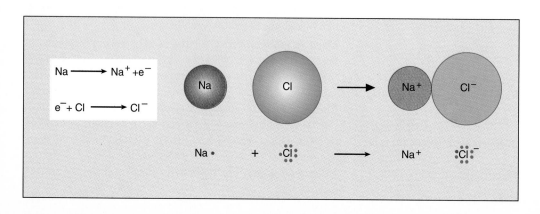

Figure 3   Ionic bond.

Catabolism *(Decomposition Reactions)*

Catabolic reactions are decomposition reactions—a molecule breaks into smaller parts. Catabolic reactions release energy. Examples:

Carbohydrate decomposition: A disaccharide splits, forming 2 monosaccharides.

Cellular respiration: A series of 19 reactions that break glu-close into CO_2 and H_2O.

Metabolic Pathways *(Series of Reactions)*

A metabolic pathway is a complete sequence of chemical reactions in which a molecule called a substrate becomes progressively modified, a substrate is transformed into products, or two or more substrates combine to form a product.

General types of metabolic pathways:

Synthesis: Substrates are combined to form a final product (for example, $A + B \longrightarrow X$, where substrates A and B are combined to form product X).

Modification: Stepwise changes in a single substrate to form a new product (for example, $M \longrightarrow N \longrightarrow O \longrightarrow P$, where substrate M is ultimately modified to product P).

Breakdown (decomposition): A substrate is broken down into different products (for example, $X \longrightarrow Y + Z$, where substrate X is broken down to products Y and Z).

Oxidation–Reduction Reactions [*Refer to Figure 5*]

Oxidation–reduction reactions involve electrons (e^-) in atoms or molecules.

Oxidation: Occurs when electrons or hydrogen atoms are removed from a molecule. Oxidation reactions cause a decrease in the energy content of a molecule, which means that energy is liberated.

Reduction: Occurs when electrons or hydrogen atoms are added to a molecule. Reduction reactions increase the energy content of a molecule, which means that energy is stored.

Oxidation–reduction reactions are always *coupled*—whenever one molecule is oxidized, electrons and energy are transferred to another molecule that is then reduced.

Related Terms

Catalyst: A substance that alters the rate of a chemical reaction without being changed in the process.

Enzyme: A substance that alters the rate of a chemical reaction in a cell; an organic catalyst.

Metabolism: All of the chemical reactions that occur in cells (anabolic and catabolic reactions).

Metabolite: A molecule that participates in metabolism.

Product: A molecule formed in an enzyme-mediated reaction.

Substrate: A reactant in an enzyme-mediated reaction.

Water and Electrolytes

Water

Body Water: Weight and Volume

The most abundant compound in the body is water. It is responsible for about 70 percent of total body weight, and it has a volume of 42 liters. For a body cell to stay alive, it must have a thin layer of water surrounding it. The outer layer of skin cells is not surrounded by fluid. It consists of dead cells completely filled with keratin.

Properties of Water [*Refer to Figure 6*]

The properties of water that make it so important for the survival of cells result from its molecular structure: Although it

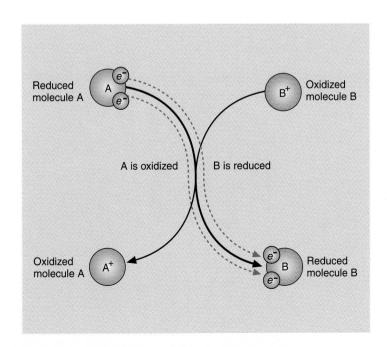

Figure 5 Oxidation-reduction reactions.

is an electrically neutral molecule, it is relatively negative at one end and positive at the other. This is because the oxygen atom has a stronger attraction for electrons than the hydrogen atoms. Consequently, the electrons in the covalent bonds are pulled closer to the oxygen atom, hence the bonds are polar covalent bonds. Because of the orientation of these polar bonds, the water molecule is polar, positive at one end and negative at the other. The following list summarizes the important properties of water:

Polarity: Water molecules have oppositely charged poles.

Hydrogen bonds: Weak attractions between nearby water molecules due to polarity.

Surface tension, cohesion, and adhesion: Attractive forces due to hydrogen bonds.

High specific heat: 1 calorie of heat is required to raise 1 gram of water 1°C.

High heat of vaporization: Heat required to evaporate water (cooling of sweat).

Solvent: A good solvent for salts and polar molecules.

Dissociation: The separation of a compound into ions of water.

Functions of Body Water

Temporary moderation: Due to high specific heat and high heat of vaporization.

Transport and exchange: For nutrients, wastes, hormones, antibodies, gases, and other substances.

Medium for chemical reactions: Molecules must be dissolved in water to react.

Hydrolysis: Water is inserted during the digestive breakdown of foods.

Body lubricant: Water is the base for mucus, saliva, tears, and serous and synovial fluids.

Protection: Cerebrospinal fluid (in brain and spinal cord); amniotic fluid (surrounding fetus).

Electrolytes [*Refer to Figure 7*]

Electrolytes are substances that dissociate when dissolved in water, releasing charged particles called ions (see Figure 6B).

Common Electrolytes That Dissociate in Body Fluids:

Sodium chloride: $NaCl$
Calcium carbonate: $CaCO_3$
Potassium chloride: KCl
Calcium phosphate: $Ca_3(PO_4)_2$
Calcium chloride: $CaCl_2$
Sodium sulfate: Na_2SO_4
Magnesium chloride: $MgCl_2$
Sodium bicarbonate: $NaHCO_3$

Common Ions Found in Body Fluids

Cations (positive ions): sodium (Na^+), potassium (K^+), calcium (Ca^{2+}), magnesium (Mg^{2+})

Anions (negative ions): chloride (Cl^-), carbonate (CO_3^-), phosphate (PO_4^{3-}), sulfate (SO_4^{2-})

Passive and Active Transport Mechanisms

Cells exist in environments containing hundreds or thousands of different molecules and ions. Plasma membranes are selectively permeable: they allow some substances to pass directly into the cell but not other substances. Whether or not molecules can pass through a plasma membrane depends on their size, polarity, and electrical charge.

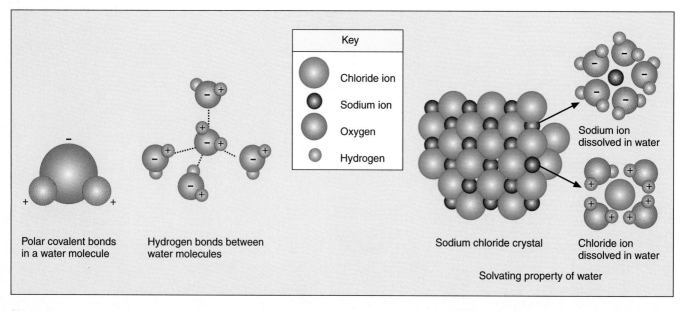

Key

Chloride ion
Sodium ion
Oxygen
Hydrogen

Polar covalent bonds in a water molecule

Hydrogen bonds between water molecules

Sodium chloride crystal

Sodium ion dissolved in water

Chloride ion dissolved in water

Solvating property of water

Figure 6   Water molecules are polar and can dissolve some substances.

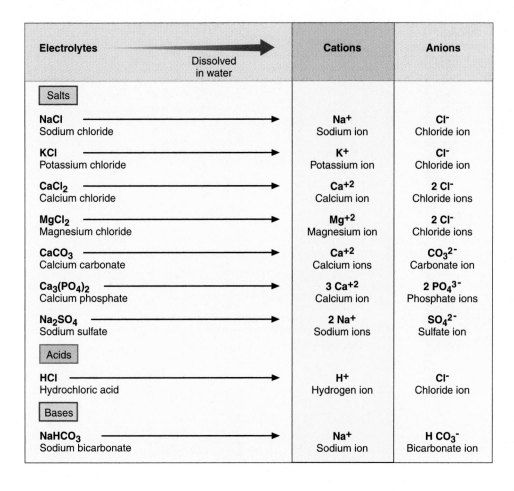

Figure 7 Electrolytes.

Passive Transport [*Refer to Figure 8*]

In passive transport, which does not require energy, the movement of a substance is related to its concentration gradient—the difference in the number of molecules, atoms, or ions—between two points or areas.

Simple Diffusion

The movement of molecules down a concentration gradient—that is, from an area where they are highly concentrated to an area where they are less concentrated. Gases such as carbon dioxide and oxygen diffuse through plasma membranes until the concentrations on both sides are equal, a point that represents the equilibrium level.

Facilitated Diffusion

A process in which essential molecules, such as glucose, that are too large to diffuse through the plasma membrane combine with carrier proteins within the membrane and are then transported through the membrane and released on the other side.

Osmosis

The diffusion of water molecules across the plasma membrane, from the side where it is more concentrated to the side where it is less concentrated.

Active Transport [*Refer to Figure 8*]

In active transport, membrane proteins use energy to transport important ions or molecules against their concentration gradient. In general, active transport is used to move needed molecules and ions into the cell and waste molecules out of the cell. Membrane proteins participate in active transport.

Related Terms

Concentration gradient: The difference in the number of molecules, atoms, or ions between two points or areas.

Solution: The mixture formed when some substance is dissolved in water or another liquid.

Solvent: The water or liquid in which a substance is dissolved.

Solute: The substance dissolved in a solvent.

Isotonic solution: A solution with the same solute concentration as the inside of a cell; no osmosis occurs.

Hypertonic solution: A solution with a higher solute concentration than exists inside a cell; therefore, water diffuses out of the cell in response.

Hypotonic solution: A solution with a lower solvent concentration than exists inside a cell; therefore, water diffuses into the cell, which may cause it to rupture.

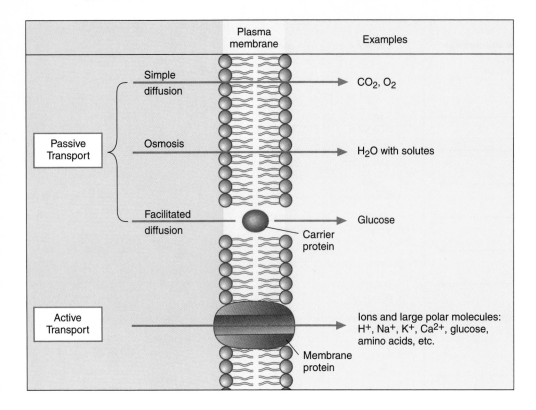

Figure 8 Passive and active transport.

Organic Compounds

Carbohydrates and *lipids* consist mostly of carbon, hydrogen, and oxygen atoms.

Polysaccharides are polymers made up of chains of simple sugars such as glucose. The main type of *lipids* are triglycerides, phospholipids, steroids, and lipoproteins. *Proteins* and *nucleic acids* contain nitrogen atoms as well as carbon, hydrogen, and oxygen atoms. *Proteins* consist of chains of amino acids, and *nucleic acids* consist of chains of nucleotides.

Elements: All organic compounds contain carbon (C) and hydrogen (H). Many also contain nitrogen (N) and oxygen (O); some contain phosphorus (P) and sulfur (S).

Carbon framework: Carbon atoms react with other carbon atoms to form chains and rings. Thousands of different molecules can be formed from these carbon skeletons.

Polymers and monomers: Large molecules (macromolecules) formed by linking many subunits together are called *polymers*; the subunits of polymers are called *monomers*.

Carbohydrates [*Refer to Figure 9*]

Carbohydrates are sugars or starches; they provide most of the energy used by the cells.

Glucose

$$
\begin{array}{c}
H \\
| \\
C = O \\
H - C - OH \\
HO - C - H \\
H - C - OH \\
H - C - OH \\
H - C - OH \\
| \\
H
\end{array}
$$

Figure 9   A carbohydrate (monosaccharide).

Monosaccharides

Simple sugars, including glucose, fructose, and galactose.

Polysaccharides

Plant starch, animal starch (glycogen), and cellulose; polymers of glucose.

Lipids [*Refer to Figure 10*]

Lipids are a diverse group of compounds that are nonpolar and therefore insoluble in water.

Triglycerides

The most common lipids in the body. A triglyceride molecule consists of 1 glycerol and 3 fatty acids. All excess food is converted into triglycerides and stored in fat cells (adipose tissue) as an energy reserve. Adipose tissue insulates and protects the organs.

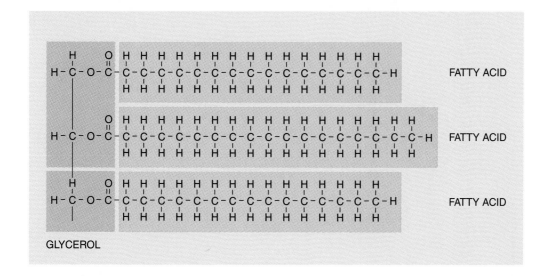

Figure 10 A triglyceride.

Phospholipids

Consist primarily of 1 glycerol, 2 fatty acids, or 1 phosphate; membrane component.

Carotenes

Yellow pigments found in egg yolk, carrots, and tomatoes; needed for vitamin A.

Proteins [*Refer to Figures 11 and 12*]

Amino Acids

The building blocks (monomers) of proteins. There are some 20 different amino acids, which can be linked together in various sequences from thousands of different protein molecules. Each amino acid has four main parts: a central carbon atom, an amino group (—NH_2), a carboxl group (—COOH), and an R group (a variable side chain). Each of the 20 amino acids has a different R group. A peptide bond is a covalent bond linking the carboxyl group of one amino acid to the amino group of another amino acid.

Polypeptides

A polymer consisting of 10 to more than 2,000 amino acids.

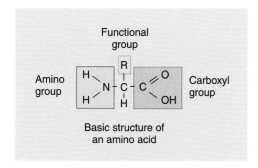

Figure 11 Amino acid structure.

Structure

Consists of one to several polypeptides.

Nucleic Acids [*Refer to Figures 13 and 14*]

Nucleotides

The building blocks (monomers) of nucleic acids. There are four different types of nucleotides in a given nucleic acid. Each nucleotide has three main parts: a nitrogenous base, a phosphate group, and a sugar.

DNA (*Deoxyribonucleic Acid*)

The nucleic acid found in the nucleus of cells. The hereditary information for the entire organism is stored in the DNA of every cell. A gene is a sequence of DNA nucleotides that serves as a blueprint for the synthesis of a particular polypeptide (portion of a protein).

RNA (*Ribonucleic Acid*)

There are three types of RNA. Messenger RNA (mRNA) carries information needed for polypeptide synthesis from the nucleus to ribosomes (cell structures that provide the sites for protein synthesis) mRNAs also serve as templates (patterns) for building specific amino acid sequences. Transfer RNA (tRNA) carries specific amino acids to the messenger RNA attached to ribosomes. Ribosomes are made of ribosomal RNA (rRNA) and proteins.

Proteins

Amino Acids, Polypeptides, and Proteins [*Refer to Figures 11 and 12*]

Amino acids: The building blocks of proteins; there are 20 different types.

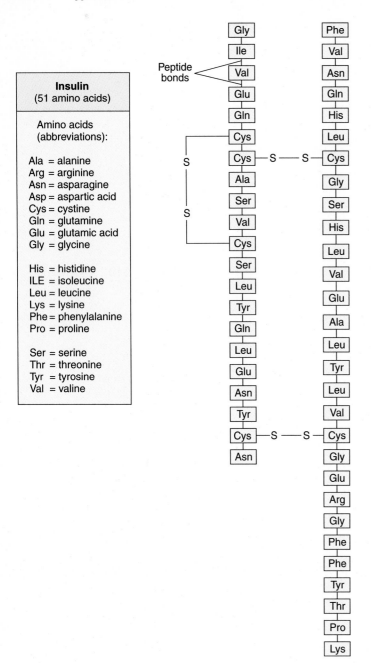

Figure 12 Insulin, a protein.

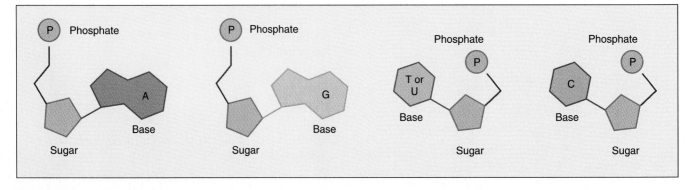

Figure 13 Structure of nucleotides.

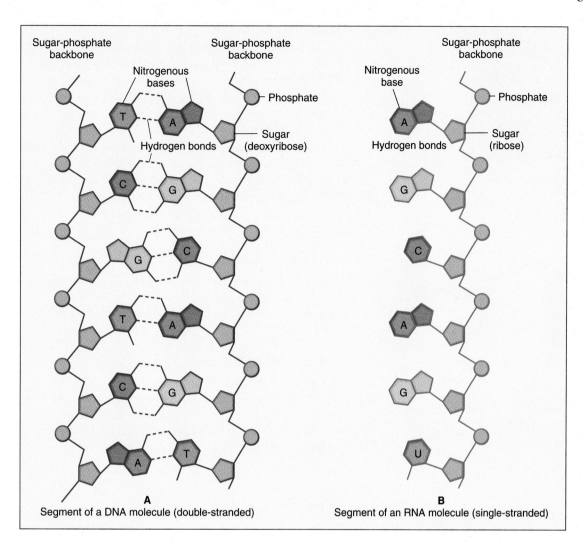

Figure 14 Structure of DNA (A) and RNA (B).

A
Segment of a DNA molecule (double-stranded)

B
Segment of an RNA molecule (single-stranded)

Polypeptides: 10 to 2,000 amino acids linked by covalent bonds.

Proteins: One or several polypeptides linked by covalent bonds. Examples:

Lysozyme (enzyme): 1 polypeptide, consisting of 129 amino acids.

Insulin (hormone): 2 polypeptides; a total of 51 amino acids.

Collagen (fiber): 3 polypeptides; a total of about 1,000 amino acids.

Hemoglobin: 4 polypeptides; a total of about 600 amino acids.

Levels of Structural Organization

Primary: The sequence of amino acids in a polypeptide.

Secondary: The spiral or pleated shape of a polypeptide (hydrogen bonding; regular intervals).

Tertiary: The three-dimensional shape of a polypeptide (interactions of side chains).

Quaternary: The shape resulting from the interactions of more than one polypeptide.

Protein Functions

Structural: Collagen fibers are found in connective tissues; keratin is in hair and skin.

Regulatory: Many hormones with a wide variety of regulatory functions are proteins.

Contractile: Actin and myosin are protein filaments found in all muscle cells.

Immunological: Antibodies and interleukins are proteins.

Transport: Hemoglobin carries oxygen and carbon dioxide; lipoproteins carry lipids.

Catalytic: Enzymes; catalysts alter the rate of a reaction without changing themselves.

Enzymes   [*Refer to Figure 15*]

Parts

Protein portion (apoenzyme): The main part of the enzyme molecule.

Nonprotein portion (cofactor): Metal ions or coenzymes (vitamin derivatives).

Active site: Region of an enzyme that has a shape that fits a portion of the substrate molecule.

Shape

Specificity: Each type of enzyme has a unique three-dimensional shape.

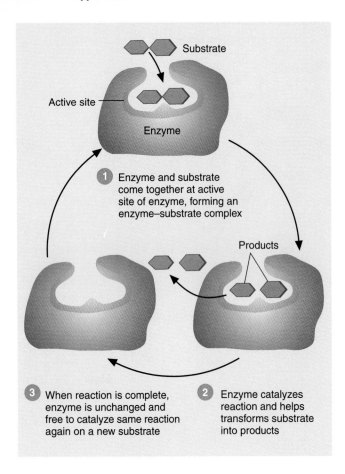

Figure 15   Enzyme function.

Induced fit model: Active sites are flexible and are able to close tightly around a substrate molecule so that the molecule becomes properly aligned for the reaction to occur.

Denaturation: Changes in pH or temperature can change enzyme shape, affecting function.

Function

Catalyst: A substance that alters the rate of a chemical reaction without being changed in the process.

Naming Enzymes

The names of most enzymes end in -ase. Enzymes discovered before this system of naming was established do not end in -ase; an example is the digestive enzyme pepsin.

Acids and Bases

Acid

A compound that dissociates into one or more hydrogen ions (H⁺) and one or more anions.

Base

A compound that dissociates into one or more hydroxide ions (OH⁻) and one or more cations. For example, when sodium hydroxide dissociates, it forms hydroxide ions and sodium cations (Na^+).

pH [*Refer to Figure 16*]

pH measures the concentration of hydrogen ions in a solution (p stands for the power of 10, and H stands for hydrogen ions). For example, pH × 2 = 1×10^{-2} moles per liter of hydrogen ions (or 1/100 moles per liter). A solution with pH 2 has 10 times as many hydrogen ions as a solution with pH 3.

pH scale (pH 0–pH 14)

Above 7 = base (pH 7.35–7.45 is normal for blood plasma).
pH 7 = neutral (pure water).
Below 7 = acid (pH 2 is normal for gastric juices in the stomach).

Effect of pH on Enzymes

Most reactions in the body are regulated by enzymes. A change in pH can alter the shape of an enzyme, which alters its effectiveness as a catalyst. Digestive enzymes in the stomach function best when the pH is 2, while those in the small intestine require a pH of about 8.

Energy

The capacity to do work.

Kinds of Energy

Potential energy: Stored or inactive energy that is capable of doing work later.

Kinetic energy: Energy of action or motion.

Forms of Energy

Radiant energy (electromagnetic energy): Energy from the sun; includes gamma rays, X rays, ultraviolet rays, visible light rays, infrared rays, microwaves, radio, and television waves. All chemical energy is derived from radiant energy by the process of photosynthesis; the energy from certain wavelengths of visible light is trapped by the chlorophyll molecules in green plants and is used to make the C—H bonds of glucose.

Chemical energy: Potential energy stored in chemical bonds, especially C—H bonds.

Electrical energy: Energy resulting from the flow of electrons.

Thermal energy (heat): Kinetic energy of molecular motion that can be measured with a thermometer.

Mechanical energy: Energy expressed as motion.

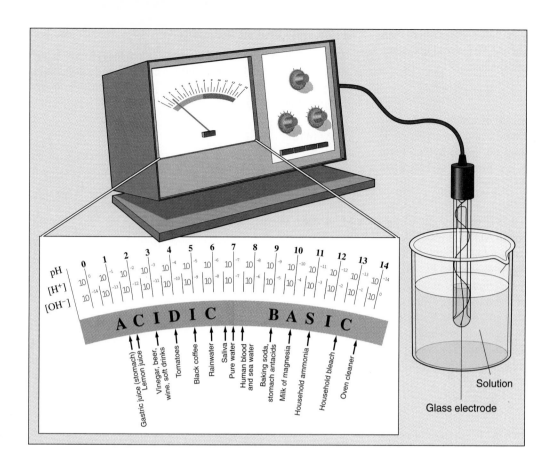

Figure 16   pH scale.

Sources of Chemical Energy (Fuels)

Hydrocarbons

Fuels such as natural gas, gasoline, oil, wood (cellulose), and organic compounds are all called hydrocarbons because they have many C—H bonds. When natural gas, gasoline, oil, or wood is burned, C—H bonds are broken, and energy is released as heat.

Cellular Fuels

In cells, the major fuels are organic compounds, which are also hydrocarbons. Cells oxidize (catabolize) organic compounds in mitochondria (cell structures where nearly all of the ATP for the cell is produced). Some 60 percent of the energy released is in the form of heat; 40 percent is used to synthesize ATP. Carbohydrates are the chief sources of fuel for ATP synthesis. When there are insufficient carbohydrate reserves, lipids (triglycerides) are catabolized. Excess dietary proteins are also catabolized for ATP synthesis. When both carbohydrate and lipid reserves are exhausted, the body will catabolize tissue proteins (muscle) for ATP synthesis.

ATP (Adenosine Triphosphate)   [*Refer to Figure 17*]

Energy-dependent activities of cells depend on the energy stored in the high-energy phosphate bonds of ATP. These bonds

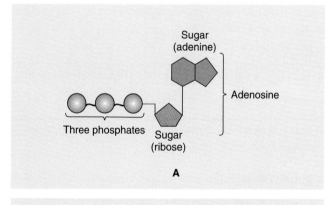

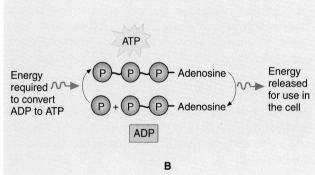

Figure 17   ATP structure (A) and function (B).

are broken as energy is required by the cell. Catabolism of fuels provides energy for the regeneration of ATP from ADP (adenosine diphosphate) and inorganic phosphates.

Structure

ATP is a large molecule that consists of a nitrogenous base (adenine), a five-carbon sugar (ribose), and three phosphate groups. The three phosphate groups are linked by two high-energy bonds; when these bonds are broken (hydrolysis of ATP), energy is released.

ATP–ADP Cycle

Energy is continuously cycled through ATP molecules. A typical ATP molecule may exist for only a few seconds before it is broken down into ADP and inorganic phosphate; the released energy is used to perform a cell function. The total energy stored in all the ATP molecules of a cell can supply the energy requirements of that cell for less than a minute. Using energy derived from the catabolism of organic compounds, ADP and inorganic phosphate are quickly converted back into ATP. ATP is synthesized by a series of reactions called *cellular respiration.*

Uses

ATP is the cell's chief energy "currency." It is a form of energy that is immediately available—like cash or a credit card. It is used by cells to carry out their energy-requiring functions, such as the movement of cilia, the contraction of muscle cells, the active transport of molecules across a plasma membrane, and the synthesis of organic molecules.

Energy Transformation in Cells [*Refer to*

Figure 18]

A reaction or set of reactions within a cell in which energy is changed from one form to another.

Photosynthesis

The energy transformation process by which radiant (solar) energy is captured by plant pigments, such as chlorophyll, and converted into high-energy organic compounds, such as the sugar glucose. Photosynthesis occurs in chloroplasts and consists of light-dependent reactions and light-independent reactions.

Light-dependent Reactions

The first set of chemical reactions, in which light energy is transformed into ATP or high-energy electrons that are captured by the coenzyme NADP+, which then becomes converted to NADPH.

Light-independent Reactions

The second set of chemical reactions in photosynthesis, which use energy provided by ATP and NADPH formed during light-dependent reactions to produce glucose or other high-energy carbohydrates.

Glycolysis

A set of reactions in which glucose is broken down to pyruvate, with a net gain of 2 ATP molecules. Occurs in the cytoplasm of all cells.

Cellular Respiration

Two sets of chemical reactions—Krebs cycle reactions and electron transport chain reactions—that occur in the mitochondria of eukaryotic cells and result in the production of ATP from glucose and other organic molecules.

Krebs Cycle Reactions

A stepwise set of reactions in which an acetyl-coA molecule is completely broken down, yielding a variety of products, some of which (electron carriers) enter into electron transport chain reactions.

Electron Transport Chain Reactions

High-energy electrons carried by NADH and $FADH_2$ are converted to ATP by chemiosmosis.

Related Terms

Pigment: Protein molecules capable of absorbing photons and transfering energy by rearranging their own molecular structures to create an energized state. Chlorophyll is the most common pigment.

Electron carriers: Coenzymes that are able to accept high-energy electrons generated during photosynthesis and cellular respiration. Three coenzymes are important in cellular energy transformations: When oxidized (not carrying electrons), they are identified as NAPD+, NAD+, and FAD; when reduced (carrying electrons), they are NADPH, NADH, and $FADH_2$.

Coenzyme: A nonprotein organic molecule that is required for enzyme function; coenzymes usually serve as donors or acceptors of electrons during chemical reactions.

ATP synthetase: An enzyme in inner mitochondrial and chloroplast membranes that uses energy from proton flow in chemiosmosis to synthesize ATP.

Chemiosmosis: The process that powers ATP synthesis. Protons pumped across the inner mitochondrial membrane during electron transport reactions flow back through a protein channel and provide the enzyme ATP synthetase with energy to generate ATP.

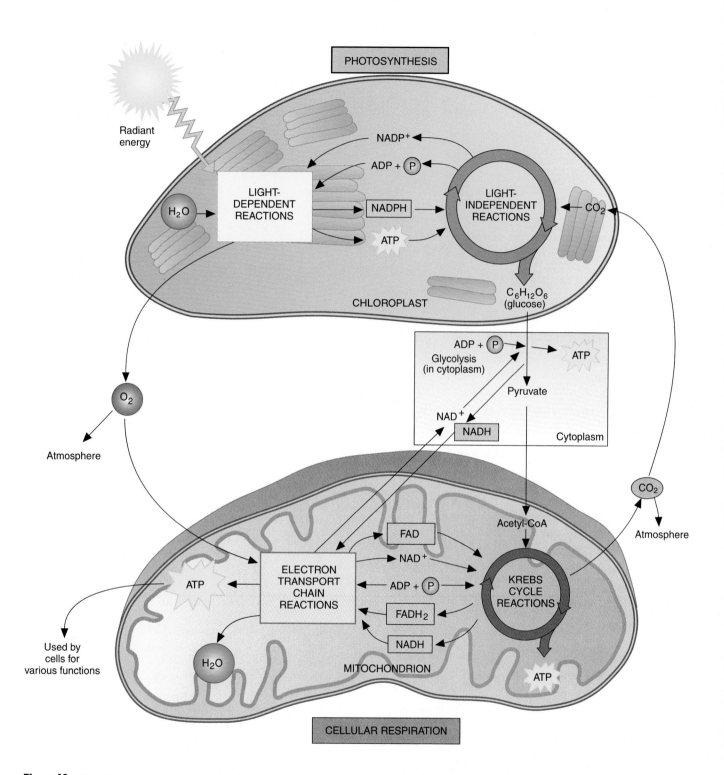

Figure 18 Cellular energy transformation reactions.

Appendix B
Classification of Organisms: The Five Kingdoms, Their Phyla, and One Genus of Each

	Genera	*Common Names*
KINGDOM: PROKARYOTAE		
Subkingdom: Archaebacteria		
Division: Mendosicutes (methanogenic bacteria)		
Methanocreatrices	*Methanobacterium*	Methanogenic bacteria
Unnamed (1988)	*Thermoplasma*	Hot springs bacteria
Subkingdom: Eubacteria		
Division: Tenericutes (lack of rigid cell walls)		
Aphragmabacteria	*Mycoplasma*	Pneumonia bacteria
Division: Gracilicutes (Gram-negative bacteria)		
Spirochaetae	*Treponema*	Syphilis bacteria
Thiopneutes	*Desulfovibrio*	Sulfate bacteria
Unnamed (1988)	*Rhodomicrobium*	Purple bacteria
Cyanobacteria	*Anabaena*	Blue-green algae
Chloroxybacteria	*Prochloron*	Prochlorophytes
Unnamed (1988)	*Azotobacter*	N_2-fixing aerobes
Pseudomonads	*Pseudomonas*	Pseudomonas
Omnibacteria	*Escherichia*	Escherichia
Unnamed (1988)	*Nitrobacter*	Chemoautotrophs
Myxobacteria	*Simonsiella*	Mucus bacteria
Division: Firmicutes (Gram-positive bacteria)		
Unnamed (1988)	*Streptococcus*	Lactic-acid bacteria
Aeroendospora	*Bacillus*	Bacillus
Micrococci	*Micrococcus*	Micrococcus
Actinobacteria	*Streptomyces*	Strep bacteria
KINGDOM: PROTOCTISTA		
Caryoblastea	*Pelomyxa*	Giant amoeba
Dinoflagellata	*Gonyaulax*	Gonyaulax
Rhizopoda	*Amoeba*	Amoeba
Chrysophyta	*Dinobryon*	Golden algae
Haptophyta	*Prymnesium*	Marine heterophytes
Euglenophyta	*Euglena*	Euglena
Cryptophyta	*Cryptomonas*	Cryptophytes
Zoomastigina	*Trichonympha*	Flagellates
Xanthophyta	*Ophiocytium*	Zoospore algae
Eustigmatophyta	*Vischeria*	Eustigs

	Genera	*Common Names*
Bacillariophyta	*Diploneis*	Diatoms
Phaeophyta	*Postelsia*	Sea palm
Rhodophyta	*Corallina*	Corallin algae
Gamophyta	*Spirogyra*	Spirogyra
Chlorophyta	*Codium*	Dead man's fingers
Actinopoda	*Acanthocystis*	Actinopods
Foraminifera	*Globigerina*	Forams
Ciliophora	*Paramecium*	Parameca
Apicomplexa	*Plasmodium*	Malaria sporozoa
Cnidosporidia	*Ichthyosporidium*	Fish microsporidians
Labyrinthulomycota	*Labyrinthula*	Slime nets
Acrasiomycota	*Dictyostelium*	Cellular slime molds
Myxomycota	*Echinostelium*	Slime molds
Plasmodiophoromycota	*Plasmodiophora*	Plasmodia
Hypochytridiomycota	*Hypochytrium*	Hypochytrids
Chytridiomycota	*Blastocladiella*	Chytrids
Oomycota	*Saprolegnia*	Fish fungus

KINGDOM: FUNGI

Zygomycota	*Rhizopus*	Black bread mold
Ascomycota	*Saccharomyces*	Yeast
Basidiomycota	*Amanita*	Fly agaric
Deuteromycota	*Penicillium*	Penicillium
Mycophycophyta	*Leparia*	Lungwort lichen

KINGDOM: PLANTAE

Bryophyta	*Sphagnum*	Peat moss
Psilophyta	*Psilotum*	Whisk ferns
Lycopodophyta	*Lycopodium*	Club moss
Sphenophyta	*Equisetum*	Horsetail
Filicinophyta	*Pteridium*	Bracken fern
Cycadophyta	*Zamia*	Cycads
Ginkgophyta	*Ginkgo*	Ginkgo
Coniferophyta	*Pinus*	Pine
Gnetophyta	*Welwitschia*	Welwitschia
Angiospermophyta	*Rosa*	Roses

KINGDOM: ANIMALIA

Placozoa	*Trichoplax*	Trichoplax
Porifera	*Leucilla*	Urn-shaped sponge
Cnidaria	*Hydra*	Hydra
Ctenophora	*Pleurobrachia*	Cat's-eye comb jelly

Acoelomate Phyla

Mesozoa	*Dicyema*	Mesozoans
Platyhelminthes	*Planaria*	Planaria
Nemertina	*Nemertea*	Ribbon worms
Gnathostomulida	*Gnathostomaria*	Gnathostomulids
Gastrotricha	*Tetranchyroderma*	Gastrotrichs

Pseudocoelomate Phyla

Rotifera	*Brachionus*	Rotifers
Kinorhyncha	*Kinorhynchus*	Kinorhynchs

	Genera	*Common Names*
Loricifera	*Nanaloricus*	Loriciferans
Acanthocephala	*Acanthocephalus*	Spiny-headed worms
Entoprocta	*Barentisa*	Entoprocts
Nematoda	*Trichinella*	Trichinosis worms
Nematomorpha	*Gordius*	Horsehair worms

Protostome Coelomate Phyla

Ectoprocta	*Plumatella*	Ectoprocts
Phoronida	*Phoronis*	Phoronid worms
Brachiopoda	*Terebraulina*	Lamp shells
Mollusca	*Octopus*	Octopus
Priapulida	*Tubiluchus*	Priapulids
Sipuncula	*Phascolosoma*	Peanut worms
Echiura	*Urechis*	Spoon worms
Annelida	*Lumbricus*	Earthworms
Tardigrada	*Echiniscus*	Water bears
Pentastoma	*Linguatula*	Tongue worms
Onychophora	*Peripatopis*	Velvet worms
Arthropoda	*Homarus*	Lobsters

Deutrostome Coelomate Phyla

Pogonophora	*Riftia*	Beard worms
Echinodermata	*Pisaster*	Sea star
Chaetognatha	*Sagitta*	Arrow worms
Hemichordata	*Saccoglossus*	Acorn worms
Chordata	*Odocoileus*	Deer

Glossary

A

AAET See *average annual evapotranspiration.*

abdominal (ab-DOM-i-nul) **cavity** The largest human body cavity; contains the organs of digestion and the kidneys, liver, spleen, and pancreas.

abiotic (ā′bī-OT-ik) **factors** All environmental factors not directly associated with living organisms; physical, chemical, and temporal (time) components of the environment.

aboveground net primary production (ANPP) The amount of plant biomass (leaves and stems) produced after subtracting that used in plant respiration per unit area per unit of time.

abscission (ab-SIZH-un) **cell layer** Layer of cells connecting the leaf to the branch; changes in it in autumn cause the leaf to fall.

absorption A process in which simple molecules pass through the digestive tract into the blood or lymph.

abyssal (uh-BIS-ul) **zone** Ocean region 1,500 to 6,000 meters below the surface.

accessory reproductive organs All organs other than testes or ovaries involved in gamete protection, storage, and transport.

acetylcholine (uh-set′ul-KŌ-lēn) **(ACh)** A neurotransmitter released at many neuron-neuron and neuron-muscle cell synapses.

acid A compound that dissociates into one or more hydrogen ions (H^+) and one or more anions; a proton donor.

acid deposition Process in which acidic substances are delivered from the atmosphere to Earth's surface.

acid neutralizing capacity (ANC) The capacity of alkaline substances in water to neutralize a certain quantity of acid.

acoelomates (ā′SĒ-luh-māts) The most primitive bilateral phyla, which do not develop a coelom.

acquired immune deficiency syndrome See *AIDS.*

acquisition Process whereby a plant obtains required resources such as carbon dioxide from the environment.

acrosome (AK-ruh-sōm) A caplike, fluid-filled structure that covers the head of a sperm.

ACTH See *adrenocorticotropin.*

actin A contractile protein that makes up thin filaments of muscle cells.

action potential A rapid change in membrane potential involving depolarization at one site in the plasma membrane of an axon or muscle cell followed by repolarization; also called *nerve impulse.*

active site A small surface region of an enzyme that recognizes and reacts with a substrate.

active transport The energy-dependent movement of molecules or ions across a plasma membrane against a concentration gradient.

acute inflammation response An initial response to injury, characterized by redness, heat, swelling, and pain.

adaptation (1) In ecology, the process through which an individual organism adjusts to environmental stresses. (2) In evolution, a change that results in a structural, functional, or behavioral trait that promotes survival and reproduction.

adaptive radiation Evolutionary divergence in which several lineages arise from a common ancestor.

adenosine triphosphate (uh-DEN-uh-sēn trī-FOS-fāt) **(ATP)** An energy-liberating molecule used by plant and animal cells, composed of the molecule adenosine and three linked phosphate groups.

adenoviruses (ad′uh-nō-VĪ-rus-iz) Viruses that cause colds and viral pneumonia; have been used in gene therapy protocols for treating certain genetic disorders such as cystic fibrosis.

ADH See *antidiuretic hormone.*

adrenal (uh-DRĒN-ul) **glands** Two glands that produce certain steroid hormones; located on top of the kidneys.

adrenocorticotropin (uh-drē′nō-kor′ti-kō-TRŌ-pin) **(ACTH)** A peptide hormone of 39 amino acids that stimulates the release from the adrenal gland of other hormones that are important in carbohydrate metabolism.

aerobic (e-RŌ-bik) **cellular respiration** A set of oxygen-requiring chemical reactions that release energy stored in sugar for use by an organism.

afferent (AF-uh-runt) **(sensory) division** Part of the peripheral nervous system that includes nerves that relay impulses from all areas of the body to the central nervous system.

afferent (AF-uh-runt) **neurons** See *sensory neuron.*

afterbirth The placenta expelled after childbirth.

after-discharge The brief delay between removal of a stimulus and the end of a response.

aggregation (ag′ruh-GĀ-shun) Group formed by individuals that are attracted to a stimulus or some feature of the environment.

agranulocytes (ā-GRAN-yoo-lō-sīts) White blood cells without cytoplasmic granules; macrophages and lymphocytes.

Agrobacterium tumefaciens (ag′rō-bak-TĒ-rē-um too′muh-FĀ-shuns) Bacterium used as a vector to insert genes into plants in genetic engineering.

AIDS Acquired immune deficiency syndrome, caused by the virus HIV-1; characterized by a combination of opportunistic infections and a seriously deficient immune system.

AIDS-related complex (ARC) Various symptoms, including fever, diarrhea, and weight loss, that occur in AIDS patients.

albinism (AL-buh-niz′um) The condition of having no pigment in certain cells.

alchemists (AL-kuh-mists) People who studied or attempted transmutations such as turning base metals into gold.

alimentary (al′uh-MEN-tuh-rē) **canal** See *gastrointestinal tract.*

alkaptonuria (al-kap′tuh-NYOOR-ē-uh) A metabolic disorder characterized in children by urine turning black after it is exposed briefly to air; may cause arthritis in adults.

allantois (uh-LAN-tuh-wis) A membrane involved in the transport of wastes from the embryo and oxygen and nutrients from the mother.

alleles (uh-LĒLZ) Alternate forms of a gene; one of two or more forms that can exist at a single gene locus.

allergen (AL-ur-jin) A substance (an antigen) to which some people have a hypersensitive reaction.

allocation Process in which plant metabolism products are apportioned to specific tissues and organs according to need.

allopatric speciation (al′uh-PAT-rik spē′shē-Ā-shun) Process by which a species arises in geographical isolation.

alpha cells Pancreas cells that produce glucagon.

alpine zones Habitats on mountains located between timberline and snow line, with environmental features similar to those of tundra biomes.

alternation of generations Plant life cycle feature characterized by two different structural forms, *gametophytes* and *sporophytes,* produced alternatively.

alternative hypothesis A statement that is accepted if a null hypothesis is rejected.

altitudinal zonation A change in community structure as a function of changing elevation.

altruistic (al′troo-IS-tik) **behavior** Behavior that results in a benefit to the recipient at a cost to the actor.

alveoli (al-VĒ-uh-lī) Tiny clusters of thin-walled air sacs that are sites of gas exchange in the lung.

amino (uh-MĒ-nō) **acids** Small building-block molecules that make up protein.

ammonification (uh-mon′uh-fuh-KĀ-shun) Process in which fungi and bacteria convert organic nitrogen from plants and animals into ammonia; the release of ammonia from nitrogenous organic matter.

amniocentesis (am′nē-ō-sen-TĒ-sus) Procedure in which a long, thin needle is inserted through the mother's abdominal wall and into the amniotic cavity to withdraw cells and fluids that are analyzed to determine if genetic abnormalities are present in a fetus.

amnion (AM-nē-on) Fluid-filled membrane that first surrounds the cavity between the trophoblast and the inner cell mass and later becomes a membranous sac holding the fetus.

amniote (AM-nē-ōt) **egg** Type of leathery or hard-shelled egg produced by reptiles or birds that is waterproof and contains a fluid-filled membrane.

amoebocytes (uh-MĒ-buh-sīts) Blood cells that can move through the cell or tissue layers with their pseudopodia.

anabolism (uh-NAB-uh-liz-um) Chemical reactions involved in the synthesis of molecules.

anaerobic (an′e-RŌ-bic) **exercise** Type of exercise in which not enough oxygen is available in muscle cells for all pyruvate to move into the pathway of cellular respiration.

anaerobic (an′e-RŌ-bic) **reaction** A reaction that does not require oxygen.

analogous (uh-NAL-uh-gus) **structures** Structures that have similar functions but which differ in their structural, embryological, and evolutionary background.

anaphase (AN-uh-fāz) The fourth stage of mitosis, in which paired chromatids separate at their centromeres and appear to be pulled toward the poles of the parent cell.

anaphase I Stage of meiosis in which homologous chromosomes are separated.

anaphase II Stage of meiosis in which sister chromatids are separated.

anatomy (uh-NAT-uh-mē) Study of the structure of an organism and the relationship among its parts.

ANC See *acid neutralizing capacity.*

androgens (AN-druh-jinz) Principal male sex hormones produced by the testes; stimulate development and maintenance of the male reproductive system and secondary sex characteristics.

anemia (uh-NĒ-mē-uh) A low number of red blood cells.

aneuploid (AN-yoo-ploid) Containing an abnormal number of chromosomes.

angiosperms (AN-jē-uh-spurmz) Plants that produce seed-containing fruits from flowers.

animal model An animal species with a disease that serves as a model for learning about the nature of a similar or identical condition in humans.

anion (AN-ī-on) A negatively charged ion.

annual plants Plants such as tomatoes, beans, pansies, and lettuce that complete their entire life cycle within one calendar year.

anonymous group A loosely structured group that sometimes exhibits coordinated movements and cooperative behaviors.

ANPP See *aboveground net primary production.*

anthropogenic (an′thruh-puh-JEN-ik) Tracing its origin to humans.

antibiotics (an′tē-bī-OT-iks) Chemicals that inhibit the growth of microorganisms.

antibodies (AN-ti-bod′ēz) Specialized proteins that have the capacity to bind specifically to an antigen or antigenic determinant, leading to neutralization or destruction of the antigen.

anticodon (an′te-KŌ-don) A particular three-nucleotide sequence in transfer RNA that is complementary to the three base pairs of a specific codon.

antidiuretic (an′tē-dī-yuh-RET-ik) **hormone (ADH)** A hormone that helps regulate water balance in the body by promoting reabsorption of water in the kidney; also called *vasopressin.*

antigen (AN-ti-jin) A foreign macromolecule that induces an immune response.

antigen-antibody complex The structure formed when an antibody combines with a specific antigen in lock-and-key fashion.

antigenic determinant Region of an antigen that is the site of antibody attachment.

antigen presentation A process by which certain self cells, usually macrophages, display a foreign antigen on their surfaces in a form that is recognized by T cell receptors.

antigen-presenting cell Macrophages and follicular dendritic self cells that can display a foreign antigen on their surface in a form that can be recognized by a T cell receptor.

antihistamine (an′tē-HIS-tuh-mēn) A drug used in neutralizing histamine and relieving some of the effects of allergic reactions.

antisense compounds Usually single-stranded DNA molecules that bind to complementary mRNA molecules encoding abnormal proteins and prevent them from being translated.

anus (Ā-nus) The opening through which digestive waste materials are expelled.

aorta (ā-OR-tuh) The largest artery in the body; emerges from the left ventricle of the heart.

apical meristem (Ā-pi-kul MER-i-stem) Plant cells at the tips of branches and roots that give rise to all new cells in shoots and roots.

apoptosis (uh-POP-tu-sis) A complex, genetically programmed series of events, involving intracellular metabolic pathways, that leads to cell death.

aquatic (uh-KWAT-ik) Living in or on water; includes organisms living in both fresh water and seawater.

aquatic (uh-KWAT-ik) **ecosystem** An ecosystem that occurs in water; includes both fresh and marine bodies of water.

ARC See *AIDS-related complex.*

archaebacteria (ar'kē-bak-TĒR-ē-uh) The oldest of two major lineages of bacteria; the subkingdom Archaebacteria in the kingdom Prokaryotae includes the methanogenic, halophilic, and thermoacidophilic bacteria.

arteries Vessels that conduct blood away from the heart.

arterioles (ar-TĒR-ē-ōlz) Small blood vessels that branch off arteries and deliver blood to capillaries.

artificial system A classification system that is used primarily for retrieving information and makes no claims about the relationships among the objects classified.

asci (AS-kī) Microscopic reproductive structures in fungi that appear as tubular spore sacs filled with ascospores.

asexual reproduction Reproduction in which a single parent cell or individual gives rise to two or more identical (or nearly identical) offspring.

assimilation (uh-sim'uh-LĀ-shun) Modification of molecules by means of cellular metabolism.

association neuron Neuron in the brain and spinal cord that processes incoming information by transmitting it between sensory and motor neurons.

associative learning Learning in which the subject acquires a response to a stimulus by associating it with another stimulus.

astronomy (uh-STRON-uh-mē) The branch of science that studies the world beyond Earth.

atherosclerosis (ath'uh-rō-skluh-RŌ-sus) Blockage of arteries that supply the heart due to the buildup of fibrous, fatty deposits on their inner walls.

atmosphere The gaseous envelope surrounding Earth.

atmospheric cells Circular zones where air rises and moves north or south, then descends in cooler regions.

atmospheric pressure The unit of pressure that is equal to the force per unit area required to support a column of mercury 760 millimeters high.

atom The fundamental unit of a chemical element; consists of a nucleus, which may contain protons and neutrons, as well as electrons, which occupy shells outside the nucleus.

atomic mass For a single isotope, the number of protons plus the number of neutrons in the atom.

atomic number For each element, the number of protons in the nucleus of an atom.

atomic theory The concept that all matter is composed of atoms.

ATP See *adenosine triphosphate.*

ATP synthetase An enzyme in inner mitochondrial and chloroplast membranes that uses the energy from proton flow in chemiosmosis to synthesize ATP.

atrium (Ā-trē-um) An upper chamber of the heart.

autoimmune response Response in which T cells attack self cells or antibodies react against self molecules.

autonomic (ot'uh-NOM-ik) **nervous system** Part of the nervous system that connects motor neurons from the central nervous system to various internal organs; primarily concerned with automatic functions such as heartbeat.

autosomes (OT-uh-sōmz) The first 22 pairs of the chromosomes that are identical in human males and females; chromosomes not involved in sex determination.

autotroph (OT-uh-trōf) An organism that synthesizes organic molecules from inorganic molecules using sunlight (photosynthesis) or the energy released from the oxidation of chemically reduced inorganic compounds (chemosynthesis). Also known as a *producer.*

autotrophic (ot'uh-TRŌ-fik) **bacteria** Self-feeding bacteria, including the photosynthetic blue-green algae and the chemosynthetic hydrogen sulfide and methane bacteria.

average annual evapotranspiration (AAET) The average amount of water that evaporates from soil plus that transpired from vegetation per year.

axon (AK-son) A process of a neuron that conducts a nerve impulse away from the cell body.

B

bacteria (bak-TĒR-ē-uh) Organisms composed of a single prokaryotic cell without a nucleus or membranous organelles.

bacterial chromosome A single circular DNA molecule inside a bacterial cell.

bacteriophage (bak-TĒR-ē-uh-fāj) A virus that infects bacteria.

balanced pathogenicity (path'uh-juh-NIS-uh-tē) The idea that a balance exists between disease agents and their hosts; the agent is able to survive, reproduce, and be transmitted without the host's suffering serious harm.

balancing selection Selection that favors an equilibrium based on more than one genotype for a given trait.

Barr body A densely staining structure in the nucleus that represents an inactivated X chromosome.

basal cell layer Deep cell layer in the epidermis of the skin, containing undifferentiated cells that reproduce rapidly and continuously by mitosis in order to replace cells that are sloughed from the skin's surface.

base A substance that dissociates into one or more hydroxide ions (OH⁻) and one or more cations; a proton acceptor.

base substitution A gene mutation in which one nucleotide or base pair is replaced by another; for example, a T replaces an A or a T-A base pair becomes a G-C pair.

basophils (BĀ-suh-filz) Granulocytes containing factors that react against parasites and in allergic responses initiated by mast cells.

bathyal (BATH-ē-ul) **zone** Ocean region between 200 and 1,500 meters below the surface.

B cells See *B lymphocytes.*

benign (bē-NĪN) **tumor** A tumor that remains localized, grows slowly, and becomes surrounded by connective tissue.

benthic zone The bottom of lakes, where sediments accumulate and most bacterial decomposition occurs, releasing nutrients required by surface producers.

beta cells Pancreas cells that produce insulin.

bicarbonate ion HCO_3^-; transports carbon dioxide in plasma.

biennial (bī-EN-ē-ul) **plant** A plant that completes its life cycle in two growing seasons or two calendar years.

big bang theory Model of cosmic history in which the universe begins in a state of high density and temperature, both of which decrease as the universe expands; less a theory than a set of theories that attempt to trace how the universe evolved.

bile Secretion of the liver containing a mixture of substances that break down large fat globules.

binary fission (BĪ-nuh-rē FISH-un) A type of cell division that occurs in prokaryotes.

binomial (bī-NŌ-mē-ul) **system** The use of a genus and species name for identifying each kind of organism.

biogeochemical cycles Cyclical systems in which important chemical elements such as nitrogen, phosphorus, and carbon are transferred between abiotic and biotic components of the biosphere.

of animal cell membranes and is also used in the synthesis of steroid hormones.

cholinesterase (kō′luh-NES-tur-ās) An enzyme that splits acetylcholine into choline and acetate.

Chondrichthyes (kon′DRIK-thē-ēz) Phylum of the cartilaginous sharks and rays.

Chordata (kor-DĀ-tuh) Phylum containing the most advanced animals.

chorea (kuh-RĒ-uh) Constant and uncontrollable body movements.

chorion (KOR-ē-on) The outer membrane surrounding an embryo.

chorionic villi (kor′ē-ON-ik VIL-ī) Finger-like extensions of the chorion that contain fetal blood vessels.

chorionic villus sampling (CVS) Removing a plug of tissue from a developing fetus with a small tube inserted through the mother's cervix and analyzing the cells removed to determine if genetic abnormalities are present in the fetus.

chromatin (KRŌ-muh-tin) The dispersed substance of chromosomes between cell divisions; consists of DNA and proteins.

chromosomal (krō′muh-SŌ-mul) **mutations** Changes in chromosome structure, function, or number.

chromosome (KRŌ-muh-sōm) **bands** Sections of the chromosome that differ from other areas because of their lighter or darker staining intensity.

chromosome (KRŌ-muh-sōm) **maps** Diagrams that show the location of gene loci and their exact linear order along the length of a chromosome.

chromosomes (KRŌ-muh-sōmz) Filamentous structures composed of protein and tightly coiled DNA; the genetic material.

chromosome (KRŌ-muh-sōm) **theory** The theory that inheritance patterns can be explained by assuming that genes are located on chromosomes.

chronic (KRON-ik) Extended in time, as of an inflammation or a disease.

chylomicron (kī′lō-MĪ-kron) A protein-coated structure containing triglycerides and other lipids that is absorbed into the lymphatic system from villus cells of the small intestine.

chyme (kīm) The mass of partially digested food that passes from the stomach into the small intestine.

cilia (SIL-ē-uh) Short hairlike structures used for motility in some protozoans and for the movement of particles or fluids in certain cells of more advanced eukaryotic organisms.

circadian (sur-KĀ-dē-un) **rhythms** Daily internal clocks that are synchronized to the environment.

circulatory (SUR-kyuh-luh-tor′ē) **system** An internal transport system consisting of the heart, a network of blood vessels, and blood.

cladistic (klad-IS-tik) **classification** A method of classification, based on phylogenetic relationships, that uses recency of common ancestry as the criterion for taxonomic groupings.

cladogenesis (klad′uh-JEN-uh-sus) The splitting of one species into two or more species.

class A taxon that includes similar and related orders; the major subdivision of a phylum.

classical conditioning Associative learning that results in changes in the stimuli that elicit behavior.

cleavage (KLĒ-vij) A series of rapid cell divisions that gives rise to a sequence of structures in a developing embryo.

climax community A permanent community of organisms that is stable through time; the final stage resulting from ecological succession in a given biotic community.

clitoris (KLIT-uh-rus) A small, sensitive organ in the female that has the same embryological origin as the penis in the male.

clonal selection theory An attempt to explain how antigens stimulate antibody production by the process of clonal selection.

cloning vector Usually a plasmid or a virus that can enter a living cell where replication can occur under appropriate conditions; used to transfer a DNA fragment from a test tube into a living cell.

closed-canopy forests Forests where mature trees shade understories of small shrubs, herbs, and moss-covered ground, limiting the penetration of sunlight in the lower strata and on the forest floor.

clotting A change in plasma from a liquid to a fibrous mass that entraps blood cells; requires a number of proteins and other factors present in the plasma.

CNS See *central nervous system.*

coagulation (kō-ag′yuh-LĀ-shun) **system** Several mechanisms that are activated by chemical mediators to arrest bleeding.

cocaine (kō-KĀN) An addictive compound obtained from coca leaves.

codominance (kō′DOM-uh-nuns) Equal expression of both alleles in a heterozygote.

codon (KŌ-don) A section of DNA, three nucleotides in length, that codes for a single amino acid or acts as a start or stop signal in protein synthesis.

coelom (SĒ-lum) A body cavity that develops within the mesoderm of more advanced animals (coelomates).

coelomates (SĒ-luh-māts) Animals in the most advanced bilateral phyla in which a cavity (coelom) forms within the mesoderm, where most internal organ systems develop and are suspended.

coenzymes (kō′EN-zīmz) Protein molecules that are required for the functioning of enzymes; many are able to accept energetic electrons emitted from chemical reactions in cells.

coevolution The mutual evolution of two or more interacting populations of different species.

cold deserts Deserts distinguished by their low average yearly temperatures, usually occurring at high latitudes in the rain shadows of mountain ranges.

collagenous (kuh-LAJ-uh-nus) **fibers** The most abundant connective tissue in humans, made of bundles of smaller fibers composed of a strong, inelastic protein called *collagen.*

collectors Consumer guilds in lotic ecosystems that feed on fine particulate organic matter.

colon (KŌ-lun) A division of the large intestine.

communication Transmission of information by means of signals from one animal to another that results in a change in behavior.

compact bone A solid, strong outer layer of bone found in the long bones of the arms and legs.

comparative method Looking at similarities and differences to discover regularities.

complementary base pairs In DNA, the pairs of bases that are held together by weak hydrogen bonds; adenine (A) bonds with thymine (T), and guanine (G) bonds with cytosine (C).

complementary DNA (cDNA) DNA that is synthesized from messenger RNA.

complement system A set of chemical mediators that function during acute inflammation and other defense reactions.

compound A substance composed of two or more elements.

compound microscope A viewing system consisting of objective and ocular lenses and a lighting system; used to magnify objects up to 1,000 times actual size.

computerized thermal imagery A technique used to create colorized visual images based on temperature variations that result in emissions of different heat patterns.

concentration gradient The difference in the number of atoms, ions, or molecules between two points or areas.

conductive Able to conduct a nerve impulse when excited.

cones Specialized reproductive structures of gymnosperms, consisting of modified leaves.

congenital (kun-JEN-uh-tul) **malformations** Deformities that appear during fetal development.

conidia (kuh-NID-ē-uh) Sexual spores produced in fruiting bodies of mature fungi from which new fungi arise.

coniferous (kuh-NIF-uh-rus) **forest biome** Forest region in which coniferous trees predominate, including the boreal forests and the forests of the Rocky Mountains and the Pacific Northwest.

conifers (KON-uh-furz) Evergreen trees that have narrow, elongated, pointed leaves called *needles;* include fir, pine, spruce, cedar, and larch.

conjugation (kon′juh-GĀ-shun) Process of reproduction in ciliated protozoans that involves meiosis and the exchange of haploid nuclei to produce new individuals with unique genetic combinations.

connective tissue The most abundant of four tissue types in the body, consisting of cells and intercellular substances that bind and support.

constant region Portion of antibody heavy and light chains with similar amino acid sequences.

consumers Organisms that obtain their nutritional requirements by feeding on producers or their products.

continental upwelling The movement of cold, nutrient-laden, deep ocean water up to the surface along some western continental coasts where the combined effects of surface winds and Earth's rotation toward the east cause a net offshore movement of surface water.

contraception (kon′truh-SEP-shun) Prevention of fertilization of an ovum or growth of a zygote.

contractile vacuoles (kun-TRAK-tul VAK-yoo-ōlz) Specialized structures used by ciliated protozoans for expelling excess water from their cells.

control A sample that is not manipulated—a basic feature of experiments.

control site A section of DNA that includes a promoter and an operator.

convergence (kun-VUR-juns) An evolutionary pattern in which separate lineages become morphologically similar over time.

cooperative behavior Behavior among members of the same species that results in mutual benefit.

Coriolis (kor-ē-Ō-lus) **forces** Forces that result from Earth's rotation and cause circular air and water currents to rotate clockwise in the Northern Hemisphere and counterclockwise in the Southern Hemisphere.

cork cambium (KAM-bē-um) A living tissue that lies outside the vascular cambium and produces cells that form bark.

corpuscular (kor-PUS-kyuh-lur) **theory** An erroneous early theory that certain particles of matter were more "elastic" or active than others.

corpus luteum (KOR-pus LOO-tē-um) A temporary cholesterol-filled tissue formed from a ruptured ovarian follicle that acts as a gland, secreting estrogens, progesterone, and relaxin.

cortex (KOR-teks) The outer part of an organ; commonly used in reference to the brain and the kidney.

cosmic background radiation Microwave radiation peaking at a wavelength of 1 millimeter, which is visible at the same intensity all over the sky; interpreted as the cooled remains of the fireball of the big bang that started the universe.

cosmology (koz′MOL-uh-jē) The science concerned with discerning the structure and composition of the universe; combines astronomy, astrophysics, particle physics, and a variety of mathematical approaches.

cotyledon (kot′uh-LĒ-dun) An embryonic seed leaf that contains stored food for young angiosperms; monocots have one and dicots have two.

covalent (kō′VĀ-lunt) **bond** The type of bond formed when electrons are shared by two atomic nuclei.

crack A potent form of cocaine.

cranial (KRĀ-nē-ul) **cavity** The cavity formed by the skull that encloses the brain.

cranial (KRĀ-nē-ul) **nerves** Twelve pairs of nerves that originate from the lower brain and innervate the head, neck, and limb attachments.

crossing-over The exchange of corresponding chromosome parts between duplicated homologs; a process that can introduce genetic variation in sexually reproducing organisms.

cultural eutrophic (yoo-TRŌ-fik) **lakes** Lentic systems where excessive nutrients generated by human activity are introduced into the water, resulting in production that exceeds 1,000 grams of carbon per square meter per year; decomposition of the excessive production results in oxygen depletion and fish mortality.

culture Attributes of human societies that include language, social structure, value systems, and the development of tools and their use in agriculture.

cuticle (KYOO-ti-kul) A waxy, waterproof covering found on upper and lower leaf surfaces.

cuttings Clipped branches that can be used to propagate new plants.

CVS See *chorionic villus sampling.*

cyanobacteria (sī′uh-nō-bak-TĒR-ē-uh) Prokaryotic, photosynthetic microorganisms.

cyclic adenosine monophosphate (SIK-lik uh-DEN-uh-sēn mon′uh-FOS-fāt) **(cAMP)** A second messenger in initiating the actions of peptide hormones in cells.

cystic fibrosis A recessive genetic disorder caused by abnormal chloride ion channel functions related to mutations of the *CF* gene and protein; many organs are affected, but lung problems are the leading cause of death.

cystic fibrosis transmembrane-conductance regulator (CFTR) A huge protein that functions as a chloride ion channel in various cells; *CFTR* gene mutations cause cystic fibrosis.

cysts (sists) Tough, enclosed resting stages formed by single-celled amoebas during some stage of their life cycles.

cytokinesis (sī′tuh-kuh-NĒ-sus) The division of the cytoplasm of a parent cell into two daughter cells.

cytology (sī-TOL-uh-jē) The study of cells.

cytoplasm (SĪ-tuh-plaz′um) A viscous fluid enclosed within the plasma membrane that surrounds submicroscopic organelles and the cell nucleus.

cytoskeleton (sī′tuh-SKEL-uh-tun) A delicate weblike structure within the cytoplasm composed of microfilaments and microtubules.

cytotoxic (sī′tuh-TOK-sik) **T cell** A T cell that destroys cells possessing foreign antigens on their surface.

D

DAP See *dystrophin-associated proteins.*

daughter cells Two cells produced by cell division of a parent cell.

deciduous (dē-SID-yuh-wus) **forests** Forests found in three temperate regions of the Northern Hemisphere in which deciduous trees predominate.

deciduous (dē-SID-yuh-wus) **trees** Trees that shed their leaves at the end of a growing season before entering a dormant period.

declarative memory Memory of past events, facts, names, dates, and places.

decomposers (dē′kum-PŌ-zurz) Certain fungi and bacteria that obtain food by breaking down nonliving organic materials from any source.

deforestation (dē-for′uh-STĀ-shun) Clearance of vast areas of forest for lumber, planting subsistence crops, or grazing cattle.

deletion The loss of a chromosome segment.

delta cells Pancreas cells that produce somatostatin.

deme (dēm) An interbreeding population that exists in a limited geographical area; a tier in the biotic hierarchy between populations and individuals.

deme density The number of individuals of the same species in a unit area or volume.

demography (di-MOG-ruh-fē) The study of populations, especially of growth rates and age structure.

dendrites (DEN-drīts) Processes of neurons that transmit nerve impulses toward the cell body.

denitrification (dē-nī′truh-fuh-KĀ-shun) The process in which anaerobic bacteria convert nitrates (NO_3) to nitrites and ammonia.

dense bodies Structures in smooth muscle cells to which intermediate filaments are attached.

density-dependent factors Environmental factors that affect populations as a function of changes in deme density; generally retard population growth as density increases or enhance growth as density decreases.

density-independent factors Environmental factors, usually abiotic, that cause changes in the number of individuals per unit area or in a deme.

deoxygenated (dē-OK-suh-juh-nā′tid) **blood** Blood returning to the lungs from systemic circulation.

deoxyribonucleic (dē-ok′sē-rī′bō-noo-KLĒ-ik) **acid** See *DNA*.

depolarization Reduction of voltage across a plasma membrane; expressed as a movement toward less negative or more positive voltages on the inside of the membrane.

dermis (DUR-mus) The inner layer of skin, consisting of dense connective tissue and various other cells and tissues.

desert Regions of environmental extremes with dry areas usually located in the doldrums of 30° north and south latitude and receiving less than 25 centimeters of annual rainfall.

desmosome (DEZ-muh-sōm) The thickened site of attachment between two cells.

deuterostomes (DYOO-tuh-ruh-stōmz) Animal phyla in which the blastopore becomes the anus.

development All irreversible changes that occur throughout an individual's lifetime.

diabetes mellitus (dī-uh-BĒ-tēz MEL-uh-tus) A variety of hereditary disorders characterized by elevated concentrations of glucose in the blood.

diaphragm (DĪ-uh-fram) The large, dome-shaped skeletal muscle that lies beneath the lungs and initiates breathing.

diatoms (DĪ-uh-tomz) Microscopic algae enclosed in a two-part siliceous capsule that are very important in aquatic food chains.

dicots (DĪ-kots) Angiosperms that develop from germinated seeds with two cotyledons and grow into plants with netted leaf venation and concentric vascular systems that produce flowers with parts in fours or fives or multiples of four or five.

differentiated (dif′uh-REN-chē-ā-tid) Of cells, having acquired distinctive molecular and structural features associated with a specific function.

differentiation (dif′uh-ren′chē-Ā-shun) The process of changing from an unspecialized cell to a cell that has a specific structure and function.

diffusion (dif-YOO-zhun) The process in which molecules in a gas or a solution move from regions of high concentration to regions of low concentration.

digestion The breakdown of complex food substances into simple molecules that can be absorbed into the body.

digestive system The system that performs numerous functions related to delivering nutrients to every cell in the body.

dihybrid cross (1) In Mendelian genetics, a cross involving individuals with two different pairs of traits. (2) A cross that involves differences at two gene loci.

dinitrogen molecules The most stable form of nitrogen (N_2).

dioecious (dī′E-shus) Having male flowers on one plant and female flowers on another plant.

diploid (DIP-loid) Of a cell, having two complete sets of homologous chromosomes; of an organism, having two chromosome sets in each of its cells.

directional selection Selection that shifts the mean (average) in a distribution of certain characteristics in response to a change in the environment.

disease Any abnormal condition of the body that impairs normal functioning.

displacement activities Behavioral activities characterized by irrelevance to the situation, usually in a response to frustration.

disruptive selection Selection that acts to favor two or more traits simultaneously; typically occurs when a population is subjected to separate selective pressures in different occupied areas.

dissociation The process in which a molecule separates into ions in solution.

divergence In evolution, a pattern in which there are increased morphological differences among separated lineages.

division A taxon in the kingdom Prokaryotae that divides the subkingdom Eubacteria into three different groups.

DNA Deoxyribonucleic acid, the material of which genes are composed; a double chain of linked nucleotides with deoxyribose sugars.

DNA insert A foreign DNA segment that is placed into vector DNA.

DNA ligase (LĪ-gās) An enzyme that connects two DNA molecules end to end.

DNA polymerase (POL-uh-mur-ās) An enzyme that replicates DNA by joining complementary bases to a parental strand, which serves as a template.

dominant trait (1) In Mendelian genetics, the trait that dominates in the first generation of a monohybrid cross. (2) The allele in a heterozygous genotype that is expressed in the phenotype.

dopamine (DŌ-puh-mēn) A neurotransmitter in the brain; depletion may lead to Parkinson's disease.

dorsal (DOR-sul) Located near the back or upper surface of most animals; opposite the ventral surface.

double helix The shape of a DNA molecule, much like a ladder twisted about its long axis.

doubling time The time it takes a population growing at a given rate to double in size.

Down syndrome (DS) An abnormal human phenotype usually caused by a trisomy of chromosome 21, characterized by mental retardation and various physical abnormalities; more common in babies born to older mothers.

drift Dissolved materials, suspended particulates, living insects, and other organisms that move seaward with the flowing water in lotic ecosystems.

drive In classical ethology, the internal state of an animal that results in a tendency to organize its behavior to achieve a certain goal.

drugs (1) Chemical agents that bring about a functional or structural change in living tissues. (2) Chemical substances used to control disease.

drug therapy The use of drugs to attack a disease agent after it infects a host.

dry deposition Acid deposition from atmospheric particle fallout.

DS See *Down syndrome.*

Duchenne muscular dystrophy (DMD) (doo-SHEN MUS-kyuh-lur DIS-truh-fē) A genetic disorder associated with an X-linked, recessive allele that encodes a protein, dystrophin, necessary for normal muscle cell development; occurs in at least one in 3500 males born.

duodenum (dyoo′uh-DĒ-num) The upper region of the small intestine.

duplication An increase in the number of genes carried by a chromosome.

dystrophin (dis-TRŌ-fin) A large protein necessary for normal muscle cell development and function; may interact with dystrophin-associated proteins in regulating the flow of calcium ions into muscle cells.

dystrophin-associated proteins (DAP) A complex of six protein molecules that may form ion channels in muscle cell membranes.

E

ECC See *environmental carrying capacity.*

ecological pyramids Models of community trophic structure in which the base of the pyramid represents the number, biomass, or energy of the producers and the successive upper layers represent higher trophic levels.

ecological succession The predictable, gradual transformation of a site into different communities with the passage of time; ends with formation of a climax community.

ecology (ē-KOL-uh-jē) The scientific study of interrelationships that exist between organisms and their environments.

ecosystem (Ē-kō-sis′tum) (1) All organisms and physical factors that form an ecological unit. (2) A system of physical and biological processes in a space-time unit of any magnitude, such as the biotic community plus its abiotic environment.

ecosystem school An informal group of early ecologists who considered ecosystems, consisting of both biotic and abiotic factors, to represent the fundamental ecological level of organization. Followers were often placed in one of two groups—those who agreed with F. E. Clements or those who agreed with H. C. Cowles.

ectoderm (EK-tuh-durm) The outermost of three primordial germ layers in a developing animal embryo; gives rise to the nervous system and epidermis.

effector In homeostasis, a cell, tissue, or organ that receives the output from the control center and initiates the response necessary for change.

efferent (EF-uh-runt) **(motor) division** Division of the peripheral nervous system composed of the somatic nervous system and the autonomic nervous system.

efferent (EF-uh-runt) **neurons** See *motor neuron.*

eggs Haploid gametes produced by females.

ejaculation (ē-jak′yuh-LĀ-shun) The reflexive discharge of semen from the penis.

ejaculatory (ē-JAK-yuh-luh-tor′ē) **duct** A tube that transmits sperm from the vas deferens to the urethra.

elastic fibers Long, thin, branching fibers made of the protein elastin that are found in certain connective tissue cells.

electrical energy A form of energy produced by a flow of electrons (negatively charged particles).

electrical synapse Site where a nerve impulse travels directly from one cell to another cell.

electromagnetic radiation Particles of light (photons) that have different wavelengths.

electromagnetic spectrum The range of energy with different wavelengths emitted from the sun.

electron A fundamental and stable atomic particle that possesses one negative electric charge. Electrons surround the nuclei of atoms; their interactions with the electrons of neighboring atoms create the chemical bonds that link atoms together as molecules.

electron carriers Coenzymes that accept energetic electrons.

electron micrograph A photograph taken through an electron microscope.

electron transport chain reactions Sets of stepwise chemical reactions in which high-energy electrons from NADH and $FADH_2$ pass through a series of electron carriers located in mitochondrial membranes, producing large quantities of ATP.

element A substance composed entirely of one type of atom.

elementary body The form in a *Chlamydia* life cycle that infects a host cell.

elementary particle A particle that assumedly cannot be subdivided.

EM See *electron micrograph.*

embryo (EM-brē-ō) A diploid organism in its early stages of development—in humans, the first two months after fertilization.

embryonic (em′brē-ON-ik) **disk** Part of the developing embryo that gives rise to the three primordial germ layers.

emigration (em′uh-GRĀ-shun) The movement of individuals out of a deme, usually to a deme in a different area.

emission Movement of the contents of sexual ducts and glands to the urethra in males.

endocrine (EN-duh-krin) **system** A system consisting of discrete organs called glands, made up of specialized cells capable of secreting hormones, as well as other organs that secrete hormones and nonhormonal substances that act as hormones.

endocrinology (en′duh-kruh-NOL-uh-jē) The study of hormones derived from classical endocrine glands or from other cells or tissues, such as the brain or heart, and their functions.

endocytosis (en′duh-sī-TŌ-sus) The process by which cells engulf and take in particles or other material.

endoderm (EN-duh-durm) The innermost of the three primordial germ layers in a developing animal embryo; gives rise to the stomach, intestines, urinary bladder, and respiratory tract.

endometrium (en-duh-MĒ-trē-um) The mucous glandular membrane that lines the uterus.

endoplasmic reticulum (en-duh-PLAZ-mik rē-TIK-yuh-lum) **(ER)** An intracellular network of interconnected membranous tubules where proteins and lipids are synthesized in the cell.

endosperm (EN-duh-spurm) A tissue containing stored food in seeds, which provides the energy and nutrients required for seedling development.

endosymbiosis (en′dō-sim′bī-Ō-sis) **hypothesis** A hypothesis stating that modern eukaryotic cells arose as a result of unions between at least two types of eubacteria (prokaryotic cells) that came to live within ancestral eukaryotic cells.

energetics The study of energy transformations in an ecosystem.

energy The capacity to do work or cause change; occurs in two general states, potential and kinetic.

energy flow The movement of energy into or out of organisms or through trophic levels, ecosystems, or the biosphere.

energy levels Zones around an atom's nucleus in which electrons reside.

energy transformation A change from one form of energy to another.

entropy (EN-truh-pē) The useless energy of any system, whether the universe or a cell; a quantitative measure of randomness or disorder.

envelope A structure surrounding some viruses, composed of lipids and proteins.

environmental carrying capacity (ECC) The density of an animal deme that can be sustained in each stage of ecosystem succession.

environmental resistance The effect of unfavorable environmental conditions on the reproductive potential of a population.

environmental stress Any component of the physical environment that reduces the capacity of plants to acquire, assimilate, or allocate resources needed for growth.

enzymes (EN-zīmz) Catalysts that participate in most chemical reactions within cells; usually composed of proteins.

eosinophil (ē′uh-SIN-uh-fil) A granulocyte whose granules hold various substances used in defense reactions.

epidemiology (ep′uh-dē′mē-OL-uh-jē) Field of science dealing with the relationships of various factors that influence the frequencies and distributions of a disease in a human community or population.

epidermis (ep′uh-DUR-mus) The skin's thin, outer layer, composed of closely packed epithelial cells.

epididymis (ep′uh-DID-uh-mus) Small mass of tubules in which sperm undergo final maturation.

epiglottis (ep′uh-GLOT-us) A flap of tissue in the throat that automatically seals off the air passageways during swallowing.

epiphyte (EP-uh-fīt) A plant that grows on another plant but does not obtain any nutrients from it.

epithelial (ep′uh-THĒ-lē-ul) **tissue** Continuous layers of closely connected cells that cover body surfaces, form glands and part of the skin, and line cavities inside the body.

ER See *endoplasmic reticulum.*

erection The enlarged, firm temporary state of the penis or the clitoris.

erythrocytes (ē-RITH-ruh-sīts) Red blood cells; specialized for transporting oxygen.

erythropoietin (ē-rith′-rō-POI-uh-tin) A chemical messenger that stimulates erythrocyte production in red bone marrow.

esophagus (i-SOF-uh-gus) The tube that conducts food from the pharynx to the stomach.

essential elements Chemical elements essential for normal plant growth.

estrogens (ES-truh-jinz) Female sex hormones produced by the ovaries that affect development and maintenance of female reproductive organs and secondary sex characteristics.

estrous (ES-trus) **cycle** A regular period of sexual receptivity in female mammals (except higher primates); occurs when an egg is available to be fertilized.

ethological (ē′thuh-LOJ-i-kul) **isolation** An isolating mechanism that occurs due to differing behavior patterns in courtship or a lack of sexual attraction between males and females of different species.

ethologist (ē-THOL-uh-jist) A scientist who studies animal behavior.

ethology (ē-THOL-uh-jē) The scientific study of animal behavior.

eubacteria (yoo′bak-TĒR-ē-uh) Prokaryotic cells that can live in diverse environments, including most common groups of contemporary bacteria; the more recent of two main lineages of bacteria.

eugenicists (yoo-JEN-uh-sists) Scientists concerned with the implications of evolution for human heredity; they believe that knowledge from both genetics and evolution can be used for medical and social improvement.

eukaryotes (yoo-KER-ē-ōts) Organisms composed of one or more eukaryotic cells.

eukaryotic (yoo′ker-ē-OT-ik) **cell** The more complex of two cell types, with complicated internal structures including a nucleus defined by a nuclear envelope and numerous membrane-bound organelles; all protoctists, fungi, plants, and animals are composed of eukaryotic cells.

euphotic (yoo-FŌ-tik) **zone** Lake and ocean region to a maximum depth of about 200 meters, the limit of light penetration.

eutrophic (yoo-TRŌ-fik) **lakes** See *natural eutrophic lakes.*

evapotranspiration The total water loss from direct evaporation and plant transpiration.

evergreen rain forests Forests containing a variety of conifer trees, such as spruce, fir, pine, and hemlock, occurring in the coastal plains and mountains of the Pacific Northwest.

evergreen trees Broad-leaved and coniferous trees and shrubs that retain leaves year round, although the leaves are regularly being shed and replaced.

evolution (ev′uh-LOO-shun) Changes that occur in populations of living organisms over long periods of time. See *theory of evolution and modern synthesis.*

excitable Able to respond to a stimulus.

exons (EK-sonz) The protein-coding sequences of a gene; the parts of a gene that are transcribed as RNA.

exotoxins (ek′sō-TOK-sinz) Toxic substances released from multiplying bacteria.

experimental method The use of experiments as a tool of investigation.

exponential (ek′spuh-NEN-chul) **growth** Growth that can be described mathematically by the formula 10^x, where x (the exponent) is some computed number that defines the rate of growth; also called *geometric growth.*

expression vector A plasmid designed to permit expression (transcription) of a DNA insert.

external respiration Oxygen and carbon dioxide exchanges between alveoli and the circulatory system.

extraembryonic membranes Four membranes formed early in embryonic development that later become the amnion, yolk sac, allantois, and chorion.

f

facilitated diffusion Transport of a substance across a membrane by a carrier protein or channel; it does not require energy because it proceeds down a concentration gradient.

fallopian tube See *uterine tube.*

familial hypercholesterolemia (fuh-MIL-yul hī′pur-kuh-les′tuh-ruh-LĒ-mē-uh) **(FH)** A genetic disorder characterized by high levels of cholesterol in the blood; the most frequent cause of inherited heart disease.

family A taxon that includes similar and related genera; the major subdivision of an order.

fascia (FASH-uh) A fibrous connective tissue that covers, supports, and separates muscle.

fascicle (FAS-i-kul) A bundle of muscle cells.

FDC See *follicular dendritic cell.*

feces (FĒ-sēz) Compacted solid digestive wastes discharged through the anus.

feedback system   A cycle of events in which information about a regulated condition in the internal environment is relayed to a central control region that can then cause a change in the condition. Thousands of such systems maintain homeostasis in the human body.

fermentation   (fur′men-TĀ-shun)   The process whereby microorganisms synthesize useful products from nutrients and raw materials in the absence of oxygen; specifically, an anaerobic reaction in which pyruvate is converted to a different three-carbon compound, either alcohol or lactate, depending on cell type, producing a small amount of ATP.

fertilization   (fur′tuh-luh-ZĀ-shun)   The process in which haploid gametes unite to form a diploid zygote.

fetoscopy (fē-TOS-kuh-pē)   A technique for viewing an embryo that requires insertion of a fetoscope into the amniotic sac through a small surgical opening in the abdomen.

fetus (FĒ-tus)   A developing human from the third month of pregnancy to birth.

FH   See *familial hypercholesterolemia.*

fibrin (FĪ-brin)   A threadlike protein that forms a network entrapping erythrocytes, platelets, and other materials to create a blood clot.

fibrinogen (fi-BRIN-uh-jin)   A protein that is converted to fibrin during blood-clotting reactions.

fibroblasts (FĪ-bruh-blasts)   Spindle-shaped connective tissue cells that produce collagenous and elastic fibers and intracellular substances.

filamentous (fil′uh-MEN-tus) **fibers**   Structures that make up part of the intercellular matrix.

filopodia   (fil′uh-PŌ-dē-uh)   Needlelike pseudopodia that protrude from pores in the tests of foraminifera and are used in movement and feeding.

filtrate   Substances, including salts, amino acids, glucose, water, and urea, that are removed from blood as it enters the kidneys.

first law of thermodynamics   A law stating that energy can be converted from one form to another but can never be created or destroyed; also called the *law of conservation of energy.*

first-order consumers   Herbivores or herbivore guilds.

first-order streams   The smallest headwater streams.

first polar body   The smaller of two haploid cells formed during the first meiotic division of an oocyte.

fitness   A measure of success among genotypes that is calculated as the net reproductive rate (the average number of offspring produced by individuals) times the probability that the individuals will survive to reproductive age.

fixed action pattern   A species-specific set of stereotyped, coordinated behavioral activities.

flagella (fluh-JEL-uh)   Long, thin, whiplike structures used in eukaryotic cell mobility.

flowers   Plant reproductive structures containing male and female reproductive parts that produce male and female gametes.

fluid mosaic model   A model of plasma membrane structure that describes the arrangement of phospholipid molecules relative to the watery internal and external cellular environment.

flux   The movement of material between trophic levels of an ecosystem in a unit of time; the amount of any substance that moves from one place to another in a unit of time.

foliar (FŌ-lē-ur) **uptake**   The process by which trees take up substances from the air through their leaves or needles.

follicle-stimulating hormone (FSH)   The hormone that in females initiates development of ova and stimulates the ovaries to secrete estrogens and in males initiates sperm production.

follicular dendritic (fuh-LIK-yuh-lur den-DRIT-ik) **cell (FDC)**   A type of cell present in lymph nodes that traps and filters infectious agents from circulating lymph.

food chain   The sequence of energy or food transfers from one trophic level to another.

food cycles   Early conceptual diagrams that incorporated both energy flow and biogeochemical cycles in studies of communities.

food web   The complex, interlocking series of food chains in a community.

forcing   The tendency of a specific gas to increase or decrease atmospheric temperature.

forebrain   The part of the brain where most complex neural processing and integration occur; also called the *cerebrum.*

forest decline   A condition characterized by trees' developing yellow leaves or needles and by a decrease in tree growth per unit of ground area; often associated with acid deposition.

forest dieback   The mortality of select species within a forest stand or even of an entire stand; often associated with acid deposition.

formed elements   Solids in blood, including erythrocytes, various types of leukocytes, and platelets.

fossil record   All fossils, which collectively form a chronicle of past life on Earth.

fossils   Remnants or traces of organisms from past geologic ages embedded in Earth's crust.

founder effect   A special case of random genetic drift in which a small group of individuals establishes a genetically unique colony.

fourth-order consumers   Trophic guilds that feed on third-order consumers; in lentic ecosystems, larger trout, bass, or pike.

fourth-order streams   Streams that form at the junction of two third-order streams.

frameshift mutation   In genetics, the insertion or deletion of nucleotides, which disrupts translation.

freshwater ecosystems   Aquatic systems other than marine ocean ecosystems, including lotic systems characterized by free-flowing streams and rivers and lentic systems with lakes and ponds.

fruit   A seed-containing structure of angiosperms that originates from the female parts of fertilized flowers.

FSH   See *follicle-stimulating hormone.*

F_2 breakdown   An isolating mechanism that results in a hybrid producing sterile or less fit offspring.

functional classification system   A way of classifying organisms in the five kingdoms of life in which their mode of nutrient acquisition is used to categorize them as producers, consumers, or decomposers.

functional ecosystem models   Models based on the rate at which energy and materials move through the trophic guilds of an ecosystem.

fusion   A process in which small atomic nuclei combine to form large, more stable nuclei, resulting in the release of energy.

fusion gene   A gene constructed by combining DNA segments from different organisms.

G

Gaia (GĪ-uh) **hypothesis**   The controversial idea that the entire biosphere is a self-regulating system capable of maintaining Earth's environment within the narrow range of conditions required for life; organisms have the major role in regulating the system.

Galápagos (guh-LOP-uh-gus) **Islands**   A group of Pacific islands near the equator about 600 miles off the coast of South America.

galaxy (GAL-ik-sē)   A large aggregation of dust, gas, and stars bound together by gravity.

gallbladder   An accessory digestive organ that stores bile.

gametes (guh-MĒTS)   Specialized cells that contain the haploid number of chromosomes; in humans, sperm and egg cells.

gametic (guh-MET-ik) **isolation**   Mechanism that occurs when male and female gametes cannot combine in fertilization or when pollen or sperm are rendered inviable in the female sexual structures of another species.

gametophyte (guh-MĒ-tuh-fīt)   The gamete-producing generation of plants; plants that grow from germinated haploid spores.

gaseous cycles   Biogeochemical element cycles in which the atmosphere or hydrosphere is a major reservoir.

gastrointestinal (gas′trō-in-TES-ti-nul) **(GI) tract**   The digestive tube that begins at the mouth and terminates at the anus; also called the *alimentary canal.*

gastrula (GAS-truh-luh)   Developmental stage of an animal with two cell layers and a blastopore that develops from a blastula.

gastrulation (gas′truh-LĀ-shun)   A process during which cells undergo a series of migrations that lead to a dramatic remodeling of the embryo and the establishment of primordial germ layers.

gated ion channel   An ion channel capable of opening and closing.

GCMs   See *general circulation models.*

gene   The fundamental physical and functional hereditary unit, composed of a DNA segment.

gene amplification   Abnormal increases in portions of a chromosome.

gene cloning   The process used in genetic engineering to create unlimited numbers of genes.

gene dosage compensation   The process in which X chromosome inactivation results in both males' and females' producing the same amount of X chromosome gene proteins.

gene expression   The process in which genes are converted into structures; all the steps necessary to transpose a genotype to a phenotype.

gene families   Sets of genes descended from an ancestral gene; genes that encode similar proteins.

gene flow   See *migration.*

gene marker   (1) A unique allele or DNA segment on a chromosome that is used to locate a gene of interest. (2) A foreign allele encoding an easily detected protein that is inserted into a vector containing a gene of interest.

gene mutation   A change in a single gene from one allelic form to another.

gene pool   All genes that are distributed among all individuals of a population.

general circulation models (GCMs)   Computer models used to make predictions about long-term changes in global atmospheric conditions.

gene therapy   The substitution of a normal allele for a mutant allele.

genetic code   The sequence of nucleotides, coded in mRNA triplets, that determines the amino acid sequence of a protein.

genetic counseling   Communication concerning the risks of occurrence of a genetic disorder in a family; involves an attempt to help the person or family comprehend the medical facts, appreciate the hereditary nature and recurrence risks in specific relatives, understand the options for dealing with the risk, choose the most appropriate course of action, and make the best possible personal decision.

genetic disorder   A predictable consequence or set of consequences associated with a specific gene or chromosomal mutation.

genetic engineering   The manipulation of genetic information to alter the characteristics of an organism.

genetic recombination   New combinations of genes or chromosomes that are created by any process.

genetics (juh-NET-iks)   The scientific study of heredity and the transmission of characteristics from parents to offspring.

genetic screening   Testing programs generally used to identify people who are heterozygous for a genetic disorder caused by a single gene.

genital herpes (HUR-pēz)   A sexually transmitted disease caused by the human papillomavirus.

genital warts   A sexually transmitted disease caused by the herpes simplex virus, type II.

genome (JĒ-nōm)   The complete collection of genes in one organism or in a chromosome set.

genomic (juh-NŌ-mik) **DNA**   All the DNA sequences of an organism; the total genetic information carried by a cell or an organism.

genotype (JĒ-nuh-tīp)   The specific alleles contained in a cell or an individual.

genus (JĒ-nus)   A taxon that includes similar and related species; the major subdivision in a family.

geological time scale   The various eons, eras, periods, and epochs that occurred in Earth's 4.5-billion-year history.

geology (jē-OL-uh-jē)   The study of Earth's history.

geometric growth   See *exponential growth.*

germination (jur′muh-NĀ-shun)   The growth of an embryo within a seed that develops into a new plant.

germline gene therapy   A form of gene therapy in which genes are inserted into cells that produce sperm and eggs or in early embryonic cells.

germ theory   The theory that many diseases are caused by microbes, such as bacteria, that invade healthy bodies.

glands   Organs composed of specialized epithelial cells that secrete substances such as hormones.

glial (GLĒ-ul) **cells**   See *neuroglia cells.*

glomerulus (gluh-MER-yuh-lus)   A rounded mass of capillaries (or nerves) involved in blood filtration, especially in the kidneys.

glucagon (GLOO-kuh-gon)   A hormone secreted by the pancreas involved in glucose homeostasis; acts to increase glucose levels in blood.

glucocorticoids (gloo′kō-KOR-tuh-koidz)   Steroid hormones that play an important role in sugar metabolism.

glucose (GLOO-kōs)   A monosaccharide, the principal carbohydrate used for energy in most organisms; a building block of many polysaccharides.

glycogen (GLĪ-kuh-jin)   A storage form of glucose in animals.

glycolysis (glī-KOL-uh-sus)   The first chemical reaction in the step-by-step breakdown of glucose to water and carbon dioxide; occurs in all living cells.

GnRH   See *gonadotrophin-releasing hormone.*

Golgi (GOL-jē) **complex**   An organelle in which proteins from the endoplasmic reticulum are modified and stored prior to export from the cell.

gonadotrophin-releasing hormone (GnRH)   A hormone (possibly more than one) that stimulates production of FSH and LH.

gonadotrophins (gō-nad′uh-TRŌ-finz)   Hormones that regulate the endocrine functions of the gonads (ovaries and testes).

gonads (GŌ-nadz)   The reproductive organs of animals.

gonorrhea (gon′uh-RĒ-uh)   A sexually transmitted disease caused by a bacterium, *Neisseria gonorrhoeae.*

G proteins A family of membrane proteins that regulate second-messenger systems.

grade An evolutionary change resulting in a new functional ability.

gradient variations Changes in the rate of variables over distance; for example, increasing or decreasing air temperature with changing elevation or latitude.

grafting A form of vegetative reproduction in which a bud or a branch of one plant is attached to the stem or rootstock of a closely related plant.

Gram stain A specific biological staining agent used to distinguish different types of bacteria; named after the Danish microbiologist Hans Christian Gram.

granules Microscopic organelles in granulocytes that contain substances related to specific defense functions.

granulocytes (GRAN-yuh-lō-sīts) Several types of white blood cells that possess granules in their cytoplasm.

grassland biomes Areas dominated primarily by grasses and grasslike plants where about one fourth of the total area is covered by vegetation, found in a number of different latitudinal zones and, depending on elevation, interspersed within tropical and temperate forest biomes; the largest of four major natural vegetation formations covering Earth and the largest biome type in North America.

gravity The universal mutual attraction of all massive objects for one another.

gray matter Nonmyelinated nervous tissue of the type covering the cerebrum.

grazers Insect guilds in lotic ecosystems that feed on algae attached to benthic surfaces.

greenhouse effect Warming of the atmosphere that occurs when concentrations of CO_2, water vapor, and other gases increase and absorb more of the sun's longer (infrared) wavelengths radiated from Earth's surface.

green revolution An agricultural movement built on the use of chemical fertilizers and insecticides and on the selective propagation of high-yielding varieties of cereal grains such as wheat and rice.

gross production The total assimilation of organic matter by an organism, population, or trophic unit per unit time per unit volume or area.

growth respiration The amount of metabolic energy used for mitosis and cell enlargement in plants.

gymnosperms (JIM-nuh-spurmz) Tracheophytes such as conifer trees that produce seeds within cones.

H

habitat (HAB-i-tat) The local environment occupied by an organism.

habitat isolation An isolating mechanism that occurs when populations of different species occupy different habitats within the same general geographical region.

habituation (huh-bich'uh-WĀ-shun) A simple form of learning in which an animal is repeatedly exposed to a stimulus not associated with any positive or negative consequence and eventually ceases to respond.

hadal (HĀ-dul) **zone** Deep region of the ocean between 6000 and 11,000 meters below the surface.

hair A threadlike structure composed of keratinized cells that is produced by hair follicles in the dermis.

hair follicles Structures in the dermis that produce hairs.

half-life The length of time required for one half of a radioisotope's atoms to decay.

halophilic (hal-uh-FIL-ik) **bacteria** "Salt-loving" bacteria that can thrive in saline habitats.

haploid (HAP-loid) Having one chromosome set; having only one homolog of each chromosome type.

Hardy-Weinberg law A law that predicts genotypic frequencies of succeeding generations of a population on the basis of an initially defined distribution of genotypes.

HB See *hydrological balance.*

hCG See *human chorionic gonadotrophin.*

HD See *Huntington disease.*

health care behaviors Actions that reduce the risk of developing a sexually transmitted disease, including condom use, early diagnosis and treatment, and informing sexual partners.

heart A large hollow muscular organ capable of powerful contractions that propel blood through arteries.

heat See *thermal energy.*

heavy chains Two identical long protein chains that form part of an antibody molecule.

helper T cell A T cell that stimulates B cells and other T cells involved in the total immune response.

hemoglobin (HĒ-muh-glō-bin) An iron-containing protein molecule of red blood cells that carries oxygen from the lungs to the internal tissues.

herbaceous perennials (hur-BĀ-shus puh-REN-ē-ulz) Plants that lack wood in their tissues and live more than two years, such as asparagus, rhubarb, milkweed, lilies, and many grasses.

herbicides (HUR-buh-sīdz) Pesticides that kill plants.

herbivores (HUR-buh-vorz) Members of trophic guilds that feed on primary producers, plants, and algae.

hereditary factor Mendel's term for the hereditary material responsible for individual traits.

herpesviruses (hur'pēz-VĪ-ruh-siz) Viruses that normally live inside nerve cells.

heterotrophic (het'uh-ruh-TRŌ-fik) **bacteria** Bacteria that obtain their nutrients from organic molecules formed by autotrophs and other heterotrophs.

heterotrophs (HET-uh-ruh-trōfs) Organisms that must obtain energy and nutrients from the molecules produced and stored in autotrophs.

hGH See *human growth hormone.*

High Arctic Tundra at higher latitudes north of the Low Arctic; considered a desert because annual precipitation is low.

hindbrain The part of the brain that includes the cerebellum, medulla, and pons; largely involved in maintaining homeostasis and coordinating large-scale body movements.

histone A type of protein around which DNA is coiled in eukaryotic chromosomes.

HIV-1 Human immunodeficiency virus, type 1; the retrovirus that causes AIDS.

holdfasts Structures that anchor kelp to rocky shores and allow it to survive in surf zones.

homeostasis (hō'mē-ō-STĀ-sus) The condition of maintaining a relatively constant internal environment.

homeothermic (hō'mē-ō-THUR-mik) Having internal mechanisms to control body temperature.

home range The space or area that animals exploit on a daily basis.

hominid (HOM-uh-nid) A family of the Primate order that includes the human species.

homolog (HŌ-muh-log) A member of a homologous chromosome pair.

homologous (huh-MOL-uh-gus) **chromosome pair** A pair of chromosomes that are identical in size, shape, and gene composition, one homolog from each parent.

homologous (huh-MOL-uh-gus) **structures** Structures that share a common evolutionary, embryological, and structural background.

homology (huh-MOL-uh-jē) (1) Similarity between different genes. (2) Similarity of embryological and evolutionary history between structures.

horizontal transmission   Direct or indirect transfer of a disease agent from one individual to another.

hormone   (HOR-mōn) Any substance released by one cell that acts on another cell anywhere in the body.

host   The organism upon or within which a parasite lives and feeds.

hot deserts   Deserts distinguished by a high average yearly temperature.

human chorionic gonadotrophin (kor′ē-ON-ik gō-nad′uh-TRŌ-fin) **(hCG)**   A hormone that helps sustain the corpus luteum during the first months of pregnancy.

Human Genome Project   A monumental research project attempting to determine the location of all human genes, identify the complete nucleotide sequence of the entire human genome, and sequence the genomes of important bacteria, plants, and animals used in genetics research.

human growth hormone (hGH)   A hormone that influences a number of metabolic activities that directly or indirectly influence growth, especially of the skeleton.

humoral immunity   (HYOO-muh-rul im-YOO-nuh-tē) Part of the immune response, characterized by the formation of antibodies.

Huntingtin   The gene, discovered in 1993, thought to be responsible for Huntington disease.

Huntington disease (HD)   An inherited genetic disorder caused by a late-acting dominant allele that results in harmful nervous system effects.

hybrid   (HĪ-brid) (1) The heterozygous offspring in crosses between parents having different forms of a single trait. (2) The offspring of a cross between individuals of different species or different varieties.

hybrid inviability   An isolating mechanism that results in a hybrid's failure to reach maturity.

hybrid sterility   An isolating mechanism that results in partial or complete sterility of a hybrid.

hydrogen   (HĪ-druh-jin) **bond**   A very delicate bond formed by the weak electrical attraction between positively charged, covalently bonded hydrogen atoms of one molecule and a negatively charged atom, usually oxygen or nitrogen, of another molecule.

hydrogen   (HĪ-druh-jin) **ion**   A proton (H^+).

hydrological   (hī′druh-LOJ-i-kul) **balance (HB)**   The amount of precipitation that penetrates a canopy and reaches the ground surface minus the total water loss from evaporation and from transpiration by the foliage.

hydrologic (hī′druh-LOJ-ik) **cycle**   The total water cycle of the biosphere, including evaporation, condensation, precipitation, and the elevational gradient-induced flow of water.

hydrosphere   (HĪ-drus-fēr) All the water on Earth's surface.

hydroxide (hī-DROK-sīd) **ion**   An OH^- ion.

hyperglycemia   (hī′pur-glī-SĒ-mē-uh) An elevated level of sugar in the blood.

hypermutation   (hī′pur-myoo-TĀ-shun) An extraordinarily high rate of change in DNA or genomic RNA.

hypersensitivity   A heightened immune response that may be harmful to the body.

hypertension   High blood pressure.

hypertonic   (hī′pur-TON-ik) **solution**   A solution surrounding a cell that has a higher solute concentration than the interior of the cell; causes water to move out of the cell.

hyphae (HĪ-fē)   Fungal structures that develop from conidia and usually become multicellular.

hypophyseal portal   (hī-pof′uh-SĒ-ul POR-tul) **system**   A complex network of blood vessels through which blood flows from the hypothalamus directly to the pituitary gland.

hypothesis   (hī-POTH-uh-sus) An explanation derived by scientists from careful observations and supported by results from experiments and other evidence; less certain than a theory or a law.

hypotonic (hī′puh-TON-ik) **solution**   A solution surrounding a cell that has a lower solute concentration than the interior of the cell; causes water to move into the cell.

I

ileum   (IL-ē-um) The lower section of the small intestine.

immigration   The movement of individuals into a deme, usually from a deme in another area.

immune system   System composed of many types of cells and organs that participate in highly coordinated processes to inactivate or destroy abnormal self cells, foreign organisms, or antigens.

immunity   (im-YOO-nuh-tē) The state of being able to resist a particular disease-causing agent or the antigens of such an agent.

immunoglobulins   (im′yuh-nō-GLOB-yuh-linz)   A family of proteins that includes antibodies.

immunology   (im′yuh-NOL-uh-jē) The study of the immune system, immune responses, and immunity.

imperfect flowers   Flowers that lack one of the reproductive organs (stamens or pistils).

imprinting   The process whereby a young animal forms an association or identification with another animal or an object.

inclusive fitness   The measure of an individual's reproductive success and the effects of that success on the reproductive success of its relatives.

incomplete dominance   The situation when offspring have a range of intermediate phenotypes from crosses between different homozygous parents.

incomplete penetrance   The situation when a phenotypic trait is not evident in all individuals who have the relevant genotype.

induced-fit model   The idea that the active site of an enzyme has a flexibility that enables it to close tightly around a substrate molecule so that the molecule becomes properly aligned for the reaction to occur.

inducer   An agent that activates transcription from an operon.

infectious diseases   Diseases caused by viruses or microorganisms such as bacteria and protozoans, and by fungi and parasitic worms.

inflammation   (in′fluh-MĀ-shun) A nonspecific defense mechanism that eliminates or prevents the spread of foreign materials at a site of injury and prepares the damaged site for repair.

inhibin   (in-HIB-in) A male sex hormone thought to be produced by cells in the seminiferous tubules that exerts negative feedback effects on the pituitary, the hypothalamus, or both and inhibits FSH release.

inhibiting hormones (inhibiting factors)   Substances that travel through the hypophyseal portal system and inhibit the release of hormones produced by the anterior pituitary.

initiation site   The site in a DNA molecule where replication begins.

innate releasing mechanism   In classical ethology, a hypothesized mechanism responsible for triggering a fixed action pattern in response to a stimulus.

inner cell mass   A cluster of cells at one end of a blastula.

inorganic (in′or-GAN-ik) **compound**   Any chemical containing no carbon.

insecticides   (in-SEK-tuh-sīdz) Pesticides that kill insects.

insertion (1) Relocation of one or more genes to a different section of the same chromosome or to a different chromosome; a type of translocation. (2) The attachment of a muscle tendon to a movable bone.

insight The ability of some animals to devise new behaviors based on past experiences.

instinct An innate set of reactions that occurs in response to a stimulus.

insulin (IN-suh-lin) A hormone secreted by the pancreas, required by humans for normal processing of sugar in the blood.

integrase A retroviral enzyme that integrates viral DNA into a host cell chromosome.

integumentary (in-teg′yuh-MEN-tuh-rē) **system** System that includes the skin and structures such as nails, hair, and certain glands.

intercalated (in-TUR-kuh-lā′tid) **disks** Unique cross-striations in cardiac muscle cells that occur where two cells meet end to end and form junctions.

interkinesis (in′tur-kuh-Nē-sus) A period in meiosis when chromosomes fade and new nuclear envelopes form.

interleukins (in′tur-LOO-kinz) Hormones that act on lymphocytes, affecting activation, maturation, proliferation, and functioning of all cells involved in immune responses.

intermediate filament A protein filament found in smooth muscle cells; connected to dense bodies.

internal fertilization The process in which gametes fuse inside the body of the female.

internal respiration Exchanges of oxygen and carbon dioxide between capillaries and tissue cells.

interphase The first and longest stage of cell division, in which chromosomes are functionally active and genes are transcribed and translated; also, the stage in which each chromosome is duplicated and the cell acquires enough materials to survive division into two daughter cells.

interspecies interactions Interactions between demes of two or more different species.

interstitial (in′tur-STISH-ul) **cells** Cells that produce testosterone, located within the connective tissue between seminiferous tubules in the testes; also called *Leydig cells.*

intraspecies interactions Interactions among individuals of a species within a deme or between members of the same species from different demes.

intrinsic control Factors that regulate parameters affecting individuals, groups, or systems; for example, biotic factors in a deme that affect the number of individuals per unit area or volume.

introns (IN-tronz) The intervening sequences between exons; noncoding, nontranscribed sections of a gene.

inversion The situation when portions of chromosomes break, rotate 180°, and become reinserted at the same position in the chromosome.

ion A particle that has a positive or negative electrical charge.

ion channel A protein channel complex in a membrane that allows ions of a specific type to diffuse through the membrane.

ionic bond The attraction between ions formed when one or more electrons are transferred from one atom to another.

islets of Langerhans (LONG-ur-honz) Clusters of alpha, beta, and delta cells in the pancreas.

isopleth (Ī-suh-pleth) A line on a graph or map that connects points of equal or corresponding values.

isotonic (ī′suh-TON-ik) **solution** A solution surrounding a cell that has the same solute concentration as the interior of the cell; thus, no osmosis occurs.

isotope (Ī-suh-tōp) An atom of an element that differs in the number of neutrons in its nucleus.

J

jejunum (ji-JOO-num) The middle region of the small intestine.

joints Sites of connection between bones or between bone and cartilage.

K

karyotype (KER-ē-uh-tīp) A complete set of chromosomes, each with a characteristic size, shape, and staining pattern.

kelp Large forms of brown algae, some of which attain a length of nearly 100 meters.

keratin (KER-uh-tin) A tough, fibrous protein that is flexible and impermeable to water, found in the hair, nails, and epidermis.

keratinocytes (KER-uh-tin-uh-sīts) Epidermal cells that produce keratin; after becoming keratinized, they move to the skin surface, where they form the stratum corneum.

kidneys Two bean-shaped organs on the back wall of the abdomen that regulate the composition of the blood and produce urine.

kinesis (kuh-NĒ-sus) An undirected change in the rate of motion in response to the intensity of a stimulus.

kinetic (kuh-NET-ik) **energy** The energy of action or motion that results in change.

kingdom The most inclusive taxon of living organisms in the modern classification system; includes similar and related phyla.

kin selection Selection of behavior that promotes the survival and reproduction of relatives.

kinship Degree of relatedness among organisms.

Krebs cycle reactions A stepwise set of chemical reactions in which an acetyl-CoA molecule is completely dismantled, yielding a variety of products that can enter into electron transport chain reactions.

krill Small crustaceans that are abundant in some marine ecosystems, especially in arctic and antarctic waters.

L

labor The period when strong uterine contractions ultimately lead to expulsion of the fetus from the uterus.

lac **operon** (LAK OP-uh-ron) In bacteria, three consecutive regions of DNA consisting of structural genes (Z, Y, and A) that code for enzymes needed to metabolize lactose.

lactation (lak-TĀ-shun) The secretion of milk from mammary glands.

LAI See *leaf area index.*

landmarks Consistent features—position of the centromere, position of major bands, or the ends of chromosome arms—used in identifying a specific chromosome.

large intestine The digestive organ between the small intestine and the anus that reabsorbs sodium and water.

larynx (LAR-inks) The passageway between the pharynx and the trachea; the Adam's apple or voice box.

latent learning Learning in which there is no obvious reward at the time of learning and what is learned remains latent until put to use.

latent period (1) The brief period of time between a stimulus and the start of a response. (2) A span of time in which a disease agent is inactive or its effects are not apparent.

lateral buds Meristematic zones where leaf petioles join the stem and grow to form branches.

law A generalization that has stood the test of time and is continuously confirmed by new evidence; laws have the highest level of certainty.

law of independent assortment   Mendel's second law, stating that traits are transmitted independently of one another.

law of segregation   Mendel's first law, stating that hereditary traits are caused by a pair of factors and that during gamete production these factors separate, with only one member of the pair going to any single gamete.

law of toleration   The concept that limiting factors have both maximum and minimum levels in their effect on an organism and that organisms do not always live under optimum conditions.

layering   A form of vegetative reproduction in which meristems, stems, or branches give rise to a new and independent plant.

LDL   See *low-density lipoproteins.*

leaf area index (LAI)   The ratio of total plant leaf surface area to the total ground surface area covered by the leaves.

learning   A change in behavior that occurs as a result of experience.

lentic ecosystems (LEN-tik Ē-kō-sis'tumz)   Ecosystems that occur in standing bodies of fresh water, including lakes and ponds.

lethal mutation   A change in a gene that results in the death of the organism.

leukocytes (LOO-kuh-sīts)   White blood cells; of two general types, granulocytes and agranulocytes.

Leydig (LĪ-dig) **cells**   See *interstitial cells.*

LH   See *luteinizing hormone.*

lichens (LĪ-kinz)   A phylum of fungi consisting of two different species, one a fungus and the other a cyanobacterium or chlorophytic alga, living symbiotically.

life zone   A habitat and the plants and animals that live there.

ligaments (LIG-uh-mints)   Tough tissues that connect bones in joints.

light chains   Two identical short protein chains that form part of an antibody molecule.

light-dependent reactions   The first set of chemical reactions in photosynthesis in which light energy is transformed into ATP or high-energy electrons that are captured by the coenzyme NADP$^+$, which then becomes converted to NADPH.

light-independent reactions   The second set of chemical reactions in photosynthesis in which energy from the ATP and NADPH formed during light-dependent reactions is used to produce glucose and other sugars.

limiting factors   Environmental factors at suboptimal levels that prevent organisms

from reaching their optimal biotic potential.

limnetic (lim-NET-ik) **zone**   The part of a lake, excluding the littoral zone, in which sufficient light energy penetrates to depths where photosynthesis can occur.

lineage (LIN-ē-ij)   A single line of evolutionary descent.

linkage   The proximity of two or more genes or markers on a chromosome; the closer together the markers are, the lower the probability that they will be separated during meiosis and the greater the probability that they will be inherited together.

linkage group   All the genes present on a single chromosome.

lipids (LIP-idz)   Various fat molecules; key components of cellular membranes.

lithosphere (LITH-us-fēr)   The rigid crustal plates of Earth; distinguished from the atmosphere and the hydrosphere.

littoral (LIT-uh-rul) **zone**   An area of transition between riparian zones along the shore and the water, extending to a depth of about 10 meters in lentic ecosystems.

liver   A complex organ that removes nutrients from circulation, converts glucose to glycogen, and detoxifies harmful compounds.

lobe   One of four main areas in the cerebral hemispheres; each carries out one or more specific functions.

localism   An early idea that local disturbances in specific organs were responsible for disease.

lock-and-key hypothesis   A discredited idea that explained enzyme action by portraying the active site's being the "lock" and the substrate's the "key"; when the key was "turned," a reaction occurred.

locus (LŌ-kus)   Gene locus; the specific site on a chromosome where a gene is located.

lotic ecosystems (LŌ-tik Ē-kō-sis'tumz)   Ecosystems that occur in flowing bodies of fresh water, streams, and rivers.

Low Arctic   The area south of the High Arctic; a treeless zone at the edge of boreal woodlands that extends north to areas where low temperature and available moisture limit the growth of vegetation.

low-density lipoproteins (LĪ-pō-prō'tēns) **(LDL)**   Particles composed of lipid and protein molecules that transport cholesterol in the blood.

lungs   The principal organs of the respiratory system in which oxygen and carbon dioxide are exchanged between the internal and external environments.

luteinizing (LOO-tē-uh-nī'zing) **hormone (LH)**   A hormone that stimulates ovu-

lation and progesterone secretion by the corpus luteum in females and testosterone secretion in males.

lymph   The fluid in lymph vessels, composed of water and substances collected from various body tissues.

lymphatic (lim-FAT-ik) **system**   A system composed of networks of vessels that run near veins and lymph tissue.

lymph nodes   Compact masses of connective tissue organized from lymph tissue; they filter and remove foreign particles and certain B and T cells and macrophages that have important functions in the immune system.

lymphocytes (LIM-fuh-sīts)   Types of leukocytes; two general classes, T cells and B cells, function in specific immune responses.

lymphoid organs   Organs that contain dense populations of lymphocytes; important in immune responses.

lymphoid tissues   Tissues containing lymphocytes; underlie the gastrointestinal tract, airways leading to the lungs, and the urinary and reproductive tracts.

lymph tissue   Connective tissue organized into nodes containing cells that remove foreign particles and participate in immune responses.

Lyon hypothesis   An explanation of the process of X chromosome activation in mammals.

lysosome (LĪ-suh-zōm)   A membrane-bound organelle that contains enzymes that participate in intracellular digestion.

lysozyme (LĪ-suh-zīm)   An enzyme that can break down the cell wall of many bacterial species.

Ⓜ

macroevolution (mak'rō-ev-uh-LOO-shun)   Evolutionary change in taxonomic groups higher than species.

macromolecule (mak'rō-MOL-uh-kyool)   A large molecule of living matter, composed of subunits.

macronucleus (mak'rō-NYOO-klē-us)   The larger of two nucleus types present in ciliate protozoans.

macronutrients   Elements required by plants in relatively large amounts; include carbon, hydrogen, oxygen, nitrogen, phosphorus, potassium, sulfur, calcium, and magnesium.

macrophages (MAK-ruh-fāj-iz)   Phagocytic cells that engulf foreign particles, such as bacteria, and debris from dead cells and also play a major role in immunity in processing foreign antigens and activating T helper cells.

macroproducers Plants, mosses, and some algae in freshwater ecosystems and large macroalgae in marine ecosystems.

maintenance respiration The energy from carbohydrates that is used in maintaining a living cell.

major histocompatibility complex (MHC) proteins A unique set of self protein molecules found on cells of an individual that are the basis for immune tolerance. There are two main types: MHC-I occur on all human cells except erythrocytes, and MHC-II occur on macrophages, B lymphocytes, and few other antigen-presenting cells.

malignant cells Cells of malignant tumors.

malignant tumors Tumors composed of rapidly reproducing cells that invade surrounding tissues and spread throughout the body.

mammary glands Breasts; develop fully only in females to provide milk for nourishing offspring after birth.

marine ecosystems Ecosystems in open ocean systems, intertidal systems, deep ocean systems, and bays and estuaries.

marsh A wetland covered by 15 to 300 centimeters of fresh or salt water and supporting grasses, sedges, rushes, cattails, reeds, and a variety of aquatic plants.

marsupials (mar-SOO-pē-ulz) Animals such as opossums, koalas, and kangaroos in which fertilization and early development occur internally but later development takes place in an external pouch where the embryo feeds on milk provided through the nipple of a mammary gland.

mass A measure of the amount of matter in an object; measured roughly by weight.

mass extinction The sudden destruction of a vast number of taxa.

mast year A year in which trees such as oaks, hickories, and beeches produce an unusually large seed crop.

matter Anything that has physical substance and occupies space.

mechanical energy Energy expressed as motion.

mechanical isolation An isolating mechanism that occurs due to incompatible reproductive structures, such as incompatible genital parts in animals or features that prevent cross-pollination in plants.

mechanical philosophy The attempt to explain all phenomena in terms of matter following the laws of physics.

medulla (muh-DUL-uh) The inner portion of an organ such as the kidney.

meiosis (mī-Ō-sus) A process consisting of two consecutive cell divisions that result in the formation of four gametes in animals and spores in plants and fungi.

melanin (MEL-uh-nin) A brownish pigment that gives color to skin, hair, and eyes.

melanocytes (muh-LAN-uh-sīts) Epidermal cells of the skin that synthesize melanin.

memory The store of information learned and retained somewhere in the brain as a result of an individual's activities or experience.

memory B cells B lymphocytes that are specific for (retain a memory of) the antigenic determinant that provoked their development.

memory T cells T lymphocytes that are specific for (retain a memory of) the antigenic determinant that provoked their development.

Mendelian (men-DĒ-lē-un) **population** A population of interbreeding, sexually reproducing individuals.

menopause The time in a woman's life after which reproduction is no longer possible, marked by the cessation of menstrual cycles.

menstrual (MEN-stroo-ul) **cycle** A hormonally controlled cycle of changes in the endometrium that occurs about every 28 days.

menstruation (men'stroo-Ā-shun) A loss of blood, mucus, and cells from the uterus that usually lasts about five days and recurs monthly.

meristems (MER-i-stemz) Tissues in plant shoots and roots that produce unspecialized cells that differentiate and become new tissues such as phloem, xylem, and epidermis.

mesenchyme (MEZ-un-kīm) A gelatinous layer in which amoebocytes and spicules can be found in sponges.

mesoderm (MEZ-uh-durm) The middle layer of the primordial germ layers in animal embryos; gives rise to blood and blood vessels, muscles, and connective tissues.

mesotrophic (mez'uh-TRŌ-fik) **lakes** Lentic ecosystems in which the production ranges between 10 and 70 grams of carbon per square meter per year.

messenger RNA (mRNA) The type of RNA used to transmit information from the DNA of a gene to a ribosome, where the information is used to make a protein.

metabolic (met'uh-BOL-ik) **pathway** A complete sequence of chemical reactions in which a molecule becomes progressively modified.

metabolism (muh-TAB-uh-liz'um) All chemical reactions involved in the synthesis or breakdown of molecules within a living cell.

metabolites (muh-TAB-uh-līts) The chemical products or by-products of metabolism.

metallothionine (muh-tal'uh-THĪ-uh-nēn) A protein that binds to metals and prevents metal poisoning; expressed in all tissues, especially the liver.

metaphase (MET-uh-fāz) The third stage of mitosis, in which sister chromatids are aligned at the center of the cell.

metaphase I The stage of meiosis in which paired homologous chromosomes become aligned at the equatorial plane of the cell.

metaphase II A static phase of meiosis in which sister chromatid pairs are aligned at the equatorial plane of each daughter cell.

metastasis (muh-TAS-tuh-sus) The process in which malignant cells are transported through the circulatory system to distant sites in the body, where they create new malignant growths.

methanogenic (muh-than'uh-JEN-ik) **bacteria** Organisms that use simple organic molecules as food sources; most use CO_2 and hydrogen as an energy source and produce methane and water as by-products.

MHC proteins See *major histocompatibility complex proteins*.

microbes (MĪ-krōbz) Single-celled microorganisms.

microevolution The changes within species' populations.

microfibrils (mī'-krō-FĪ-brilz) Elongated structures made of cellulose chains within the fibrils of a plant cell wall.

microfilaments Components of the cytoskeleton; thin fibers composed of globular protein subunits that function in moving organelles around the cell and in contraction movements in cells specialized for that activity.

microfossils Fossils of single-celled or simple multicellular organisms.

microhabitat A small, specialized habitat; the immediate surrounding environment of an organism.

microinjection A technique of inserting cloned genes or DNA strands generated by recombinant DNA techniques into fertilized eggs or cells of an embryo using finely engineered instruments.

micronucleus The smaller of two nucleus types in ciliated protozoans.

micronutrients Elements required by plants in trace amounts; include boron, chlorine, cobalt, copper, iron, manganese, molybdenum, nickel, silicon, sodium, and zinc.

microproducers Diatoms, desmids, and other small species of algae in aquatic ecosystems.

microtubules Components of the cytoskeleton; made of globular, beadlike subunits organized into hollow, cylindrical tubes that form a skeletal fiber network.

microvilli (mī′krō-VIL-ī) Microscopic folds in the plasma membranes of small intestinal cells.

midbrain A small area between the forebrain and the hindbrain that receives and integrates several types of sensory information and relays it to specific regions of the forebrain for processing.

migration (1) The exchange of genes among populations; also called *gene flow*. (2) The cyclic annual movement of populations between different distant regions.

Milky Way galaxy The spiral galaxy in which the sun is located; appears as a hazy band of light crossing the entire sky in both the Northern and the Southern Hemisphere; consists of myriad stars, dust, and gases, lying in a central plane.

mineralization The process by which soil particles formed from parent rock break down and release important chemical elements.

mineralocorticoids (min′uh-ruh-lō-KOR-tu-koidz) Steroid hormones involved in maintaining water and mineral balances.

mitochondria (mī′tuh-KON-drē-uh) Membranous organelles found in all eukaryotic cells that supply energy to the cell through reactions of cellular respiration.

mitochondrial (mī′tuh-KON-drē-ul) **DNA (mtDNA)** DNA found in mitochondria and transmitted to offspring through an egg but not a sperm.

mitosis (mī-TŌ-sus) The process in nuclear division in which each daughter cell receives the same number and kind of chromosomes as the parent cell.

mixed-grass prairies Grasslands composed of both shortgrass and tallgrass vegetation.

model A simplified representation of some natural phenomenon; can be expressed in various forms, including written description, picture, graph, mathematical equation, and computer program.

modern synthesis The modern theory of evolution stating that gradual evolutionary change can be explained by the action of natural selection on small genetic changes and that the processes of species formation and evolutionary changes in higher taxonomic groups are explainable in terms that are consistent with known genetic mechanisms.

modifiable risk factor A risk factor that is subject to some degree of individual control; for example, use of a condom or cigarette smoking.

modification A stepwise change in chemical structure ($M \rightarrow N \rightarrow O \rightarrow P$).

molecular biology The field of study based on physics, biochemistry, microbiology, and genetics.

molecular genetics The study of molecules involved in heredity.

molecule The smallest unit of a pure substance that will retain its composition and hence its chemical properties; composed of two or more atoms, linked by interactions of their electrons.

Monera (muh-NĒ-ruh) The name Ernst Haeckel used in 1866 for the blue-green algae and the bacteria (prokaryotes), which he placed in his kingdom Protista.

monocots (MON-uh-kots) Angiosperms that develop from germinated seeds having a single cotyledon and grow into plants with parallel leaf venation, scattered vascular bundles, and floral parts that occur in threes or multiples of three.

monoculture Growth of a single-species crop.

monocyte (MON-uh-sīt) An agranulocyte that develops into a macrophage.

monoecious (muh-NĒ-shus) Having both male and female reproductive structures.

monoglyceride (mon′ō-GLIS-uh-rīd) A compound composed of one glycerol molecule and one attached fatty acid.

monohybrid cross (1) In Mendelian genetics, a cross that involves two forms of a single trait. (2) A cross involving different alleles at a single gene locus.

monoploid (MON-uh-ploid) Having the haploid number of chromosomes in any cell.

monosaccharide (mon′ō-SAK-uh-rīd) The simplest carbohydrate; a single sugar molecule.

monosomy (MON-uh-sō′mē) The lack of one chromosome from a homologous pair.

montane (mon′TĀN) The cool, moist, mountainous habitat where evergreen trees grow.

montane forests Forests in the mountains of western North America.

mortality The death rate of a population expressed as a percentage or fraction.

morula (MOR-yuh-luh) A solid ball of cells that forms a few days after fertilization.

motor neuron A neuron that transmits impulses away from the brain or spinal cord toward other cells, muscles, glands, or other neurons involved in making a response.

mouth The opening to the exterior that receives food and begins the process of digestion using both mechanical and chemical processes.

mRNA See *messenger RNA*.

mtDNA See *mitochondrial DNA*.

mucous (MYOO-kus) **membranes** Membranes composed of an outer epithelial cell layer and an inner connective tissue layer that line body cavities and the surface of organs or tissues that open to the external environment.

mucus (MYOO-kus) A viscous fluid secreted by mucous glands that covers the surfaces of mucous membranes.

multiple alleles Several alleles that can occur at a single gene locus.

multiple cropping An agricultural approach in which several different crops are grown within a given area.

multiregional model of human evolution The hypothesis that *Homo sapiens* developed from populations of *H. erectus* in several areas of the globe and that subsequent human evolution has been shaped by gene flow, natural selection, and genetic drift.

muscle tissue Specialized tissue that produces motion, composed of elongated, thin cells commonly referred to as *fibers*.

muscular system Skeletal muscle and the other structures associated with muscle contraction and movement of the skeleton.

mutant An organism possessing a trait resulting from a mutation.

mutation (myoo-TĀ-shun) A permanent structural change in a DNA molecule that may result in a new form of an expressed trait.

mutualism (MYOO-choo-ul-iz′um) A relationship in which two organisms of different species live together in an intimate association that benefits both.

mycelia (mī-SĒ-lē-uh) Mats of tissue formed by fungal hyphae.

myelin (MĪ-uh-lin) A lipid-rich, whitish material that covers and insulates the axons of most large neurons; produced by neuroglia cells.

myelin (MĪ-uh-lin) **sheath** Overlapping membranes covering an axon.

myofibrils (mī′ō-FĪ-brilz) Threadlike structures found in muscle cells; composed of thick and thin filaments.

myosin (MĪ-ō-sin) A contractile protein that makes up thick filaments of muscle cells.

N

nasal cavities Hollow chambers inside the nose where air is warmed, humidified, and filtered.

natality (nā-TAL-uh-tē) Birth rate; the number of live births per female in a deme or population per unit of time.

natural eutrophic (yoo-TRŌ-fik) **lakes** Lentic ecosystems in which productivity ranges between 70 and 400 grams of carbon per square meter per year; lakes where excessive organic production, resulting from nutrients of natural origin, is decomposed by bacteria, depleting oxygen concentrations to levels that cause fish mortality.

natural history The general study of nature, stressing description, classification, and interrelationships.

natural killer (NK) cells Lymphocytes that play an active role in destroying virus-infected cells, fungi and other parasites, and cancer cells.

natural selection (1) In modern evolution theory, the nonrandom, differential reproductive success of genotypes. (2) More generally, the nonrandom, differential survival or reproduction of individuals or sets of individuals that possess advantageous characteristics for survival in specific environments.

natural system A classification system that reflects an existing order in nature.

nature (natural world) The world or the universe perceived as intelligible in a fashion distinct from poetic or religious myths or stories.

navigation The action of orienting toward a goal.

needles The narrow, elongated, pointed leaves of conifers.

negative feedback The process in which feedback from a controlled system is monitored by a receptor that relays information to a processing center, which causes the level of the variable to change in the direction opposite that of the original change if it deviates from a set point value; governs most actions of the endocrine and nervous systems.

nephrons (NEF-ronz) Functional units of the kidney that regulate the composition of blood through the production of urine.

neritic (nuh-RIT-ik) **waters** Portions of oceans that include the littoral zone in near-shore waters adjacent to coasts, bays, estuaries, and waters extending over part of the continental shelf.

nerve A bundle of parallel neuron fibers and their associated blood vessels and supporting cells, all of which are enclosed by protective connective tissue.

nerve growth factor A chemical messenger responsible for neuron maturation during embryonic development.

nerve impulse See *action potential.*

nerve tracts Collections of nerves in a bundle; found in the brain and spinal cord.

nervous tissue Tissue composed of neurons specialized for initiating and conducting electrochemical impulses.

net photosynthesis The difference between total gross photosynthesis and cellular respiration.

net plant production The gross production of an organism, trophic guild, or community less that consumed during cellular respiration.

net productivity The amount of organic material produced by plant photosynthesis less that used during plant respiration (P_n).

neural (NYOO-rul) **tube defect** An abnormal development of the neural tube, the structure that gives rise to the nervous system in vertebrate embryos.

neurofibrils (nyoo′rō-FĪ-brilz) Threadlike proteins that form delicate supporting networks within a neuron.

neuroglia (nyoo-RŌ-glē-uh) **cells** One of two major cell types of the nervous system; most are specialized to function as connective tissue cells.

neurohormone (nyoo′rō-HOR-mōn) A hormone produced by a neuron.

neuron (NYOO-ron) One of two major cell types of the nervous system, consisting of a cell body, dendrites, and an axon and specialized for the rapid conduction of electrochemical impulses.

neurosecretory (nyoo′rō-si-KRĒ-tuh-rē) **cells** Specialized neurons in the hypothalamus that produce hormones.

neurotransmitters (nyoo′rō-tranz-MIT-urz) Chemical messengers that allow nerve impulse transmission across a chemical synapse.

neurovascular (nyoo′rō-VAS-kyuh-lur) **hypothesis** The hypothesis that neurosecretory cells in the hypothalamus produce substances that regulate hormone secretions of the anterior pituitary gland.

neutron A fundamental particle that has no electrical charge but a mass just a little greater than a proton; a constituent of most atomic nuclei.

neutrophils (NYOO-truh-filz) Granulocytes whose granules contain enzymes used to digest microorganisms.

niche (nich) The ecological role that a species plays within a community.

nicotine The highly addictive active ingredient of tobacco; mimics acetylcholine.

nitrification (nī′truh-fuh-KĀ-shun) The conversion of ammonia (NH_4^+) into nitrite (NO_2^-) and nitrate (NO_3^-).

nitrogen fixation The process of converting atmospheric nitrogen (N_2) into ammonia (NH_4^+).

nitrogen oxides Molecules, such as nitrites and nitrates, composed of nitrogen and oxygen.

NK cells See *natural killer cells.*

nodes of Ranvier (RON-vē-ā′) Spaces along a myelinated neuron that represent gaps between individual Schwann cells.

nondisjunction The failure of paired homologous chromosomes to separate during anaphase of meiosis I or meiosis II.

nonmodifiable risk factor A risk factor that is not subject to individual control, such as age.

nonspecific defense mechanisms Defense mechanisms that always operate in the same way when presented with a challenge; include mechanical barriers, secreted products, and the inflammation response.

nonstructural genes Genes that code for tRNA and recombinant RNA.

nonvascular plants Algae, fungi, mosses, and lichens, which lack true phloem and xylem tissues.

notochord (NOT-uh-kord) An internal, cartilaginous rod present in all chordates during some developmental stage; it is replaced by the developing vertebral column in higher vertebrates.

nuclear (NOO-klē-ur) **envelope** The double membrane surrounding the nucleus in eukaryotic cells.

nuclear (NOO-klē-ur) **fusion** An interaction in which protons and neutrons are forged together, creating new atomic nuclei and releasing energy.

nuclear (NOO-klē-ur) **pores** Openings in the surface of a nuclear envelope.

nucleic (noo-KLĒ-ik) **acids** Large molecules such as DNA and RNA that are composed of nucleotides.

nuclein (NOO-klē-in) A nineteenth-century term for what turned out to be DNA.

nucleoid (NOO-klē-oid) The region in a prokaryotic cell, mitochondrion, or chloroplast where DNA is concentrated.

nucleolus (noo-KLĒ-uh-lus) A prominent structure in the nucleus of most eukaryotic cells that is active in the synthesis of ribosomes.

nucleoplasm (NOO-klē-uh-plaz′um) The substance in the interior of a nucleus.

nucleosome (NOO-klē-uh-sōm) The basic structural unit of eukaryotic chromosomes; consists of a central core of eight histones around which the DNA molecule is wrapped twice.

nucleotide (NOO-klē-uh-tīd) A subunit of DNA or RNA that consists of a base (adenine, guanine, thymine, or cytosine in DNA; uracil instead of thymine in RNA), a phosphate molecule, and a sugar molecule (deoxyribose in DNA, ribose in RNA); thousands of nucleotides are linked to form a DNA or RNA molecule.

nucleus (NOO-klē-us) A membrane-enclosed structure that contains genetic material and regulates many cell activities in eukaryotic cells.

null hypothesis A statement formulated before performing an experiment that the experimental condition is due to chance alone; to be tested and accepted or rejected.

numerical changes Mutations that increase or decrease the number of whole chromosomes without changing the structure of individual chromosomes.

nutrient deserts Areas in oceans where there are no mechanisms to cycle nutrients from the bottom back into the euphotic zone.

O

occult deposition Direct acid deposition from fog or cloud droplets onto surfaces.

oceanic waters All areas of ocean zones, excluding neritic waters.

oligotrophic (OL-i-go-trō'fik) **lakes** Lakes in which productivity is low, ranging between 0.1 and 10 grams of carbon per square meter per year.

omnivore (OM-ni-vor) A consumer of both plants and animals; an organism that feeds in both carnivore and herbivore guilds.

oncogene (ON-kuh-jēn) A gene that can cause cancer.

one-gene, one-enzyme hypothesis A hypothesis that originally held that each biochemical reaction was catalyzed by a single enzyme and that each enzyme was specified by one gene; as more became known about genes and proteins and their structure, the hypothesis was extended to become the *one-gene, one-polypeptide hypothesis* and then the *one-gene, one-protein* (or *one-RNA-product*) *hypothesis.*

one-gene, one-polypeptide hypothesis See *one-gene, one-enzyme hypothesis.*

one-gene, one-protein hypothesis See *one-gene, one-enzyme hypothesis.*

one gene, one-RNA-product hypothesis See *one-gene, one-enzyme hypothesis.*

oogenesis (ō'uh-JEN-uh-sus) The process by which ovaries produce mature ova.

oogonia (ō'uh-GŌ-nē-uh) Cells in the ovaries of a female embryo that ultimately give rise to oocytes.

open-grown trees Trees growing far from neighboring trees that might compete for light and other needed resources.

open woodlands Areas where short trees fail to form a closed canopy.

operant conditioning A form of associative learning in which an animal is conditioned to repeat a behavior by reinforcement.

operator A DNA segment capable of interacting with a repressor in controlling the function of an adjacent region.

operculum (ō-PUR-kyuh-lum) A structure that covers and protects the gill slits of bony fishes.

operon (OP-uh-ron) A set of adjacent structural genes whose mRNA is synthesized in one piece, plus the adjacent regulating regions that affect transcription of the structural genes.

opportunistic pathogens Agents (bacteria or fungi) that can invade tissues and cause disease only under special conditions, such as diminished defense or immune capabilities.

order A taxon that includes similar and related families; the major subdivision of a class.

organ A distinct structure that is made up of more than one type of tissue, has a definite form, and performs a specific function in the body.

organelles (or'guh-NELZ) Microscopic membranous structures in a eukaryotic cell that are responsible for carrying out specific functions; include mitochondria, ribosomes, endoplasmic reticula, and Golgi complexes in animal cells.

organic compound A chemical substance that contains carbon and hydrogen atoms held together by covalent bonds.

organism An individual living being; a whole made up of functionally interrelated parts or organs.

organogenesis (or'guh-nō-JEN-uh-sus) The formation of primary organs during embryonic development.

organ system A group of organs integrated into a unit for performing a major function or group of functions.

orgasm A series of brief, widespread muscular contractions and pleasurable sensations that occur in sexual intercourse; accompanied by ejaculation in males.

origin The attachment of a muscle tendon to a stationary bone.

osmosis (os-MŌ-sus) The diffusion of water molecules across a selectively permeable membrane.

osteoblast (OS-tē-ō-blast) A type of bone cell that builds new bone.

osteoclast (OS-tē-ō-klast) A large, multinucleated cell that digests and reabsorbs bone tissue.

osteocyte (OS-tē-ō-sīt) A type of bone cell that maintains bone tissue.

out of Africa model The hypothesis that a population of the modern form of *Homo Sapiens* arose in Africa at an early date and spread rapidly across Europe and Asia.

ovarian (ō-VER-ē-un) **cycle** A monthly cycle in human females in which an ovum matures.

ovary (Ō-vuh-rē) Primary sex organ of females that produces ova and hormones, including estrogens, progesterone, and relaxin.

overexpression The production of excess quantities of a protein or an abnormal protein.

oviduct See *uterine tube.*

ovulation (ov'yuh-LĀ-shun) The release of a single mature egg from one of the ovaries.

ovum A mature egg cell.

oxidation (ok'suh-DĀ-shun) The process in which an atom or a molecule loses one or more electrons.

oxygenated (OK-suh-juh-nā'tid) **blood** Blood leaving the lung.

oxyhemoglobin (ok'si-HĒ-muh-glō-bin) Hemoglobin combined with oxygen.

oxytocin (ok'si-TŌ-sin) A hormone that stimulates muscles of the uterus during childbirth and triggers secretion of milk from the mammary glands during nursing.

ozone (Ō-zōn) A gas composed of three oxygen atoms (O_3) that forms a layer in the outer atmosphere that absorbs ultraviolet radiation emitted by the sun and protects life from the harmful effects of this highly reactive form of energy.

P

paleoclimatic (pā'lē-ō-klī-MAT-ik) **indicators** Records that provide information about ancient climates.

paleontologists (pā'lē-on-TOL-uh-jists) Scientists who study the fossil record.

paleontology (pā'lē-on-TOL-uh-jē) The study of fossils and ancient life.

pancreas (PAN-krē-us) An organ that carries out functions in the digestive system by secreting pancreatic juice and in the endocrine system by secreting the hormones insulin, glucagon, and somatostatin.

pandemic (pan-DEM-ik) A disease that occurs worldwide.

parallelism The development of similar characteristics in separate lineages that have a common ancestor.

parasites (PAR-uh-sīts) Organisms such as protozoans and worms that obtain substances from a host.

parasitism (PAR-uh-suh-tiz′um) A relationship in which two organisms of different species live together in an intimate association in which only one of the organisms benefits while the other is harmed.

parasympathetic nervous system The part of the nervous system that maintains the body under relatively calm conditions.

parathyroid hormone (PTH) A hormone secreted by the parathyroid glands that increases calcium levels.

parent cell A cell that divides, giving rise to two daughter cells.

Parkinson's disease A disease of the nervous system, characterized by late age of onset and severe symptoms such as drooling, tremors, and nonfunctioning muscles.

partial pressure The pressure exerted by a specific gas in a mixture of gases.

parturition The process in which a human fetus is expelled from the uterus about 265 days after fertilization.

pathogens (PATH-uh-jinz) Organisms or biological agents that cause disease.

peat Partly decomposed plant matter found in ancient bogs and swamps.

pectin A complex carbohydrate present in the matrix of a primary cell wall.

pedigree (PED-uh-grē) A simple genetic diagram used in tracing the inheritance of genetic traits and disorders in families.

peduncle (PĒ-dun-kul) The stalk that attaches a flower to a plant.

pelvic girdle The portion of the skeleton that protects the internal reproductive organs.

pelvic inflammatory disease (PID) A scar-forming process that occurs when a cervical gonorrhea or chlamydia infection spreads to the uterus, uterine tube, or ovaries.

penis (PĒ-nis) The male copulatory organ that conveys sperm into the female vagina.

pepsin (PEP-sin) An enzyme in gastric juice that initiates the breakdown of food proteins.

peptide bond The type of chemical linkage that joins two amino acids.

peptide growth-stimulating factors Substances that promote the growth or development of certain cells and tissues.

peptide hormones Mostly small proteins composed of linear strings of 3 to 200 amino acids that affect target cells through the operation of a second messenger system.

perennial (puh-REN-ē-ul) **plants** Plants that may live more than two years, grow continuously, and produce seeds once a year on several occasions; examples are cattails, rhododendrons, and redwood trees.

perfect flowers Flowers that have both stamens and pistils.

perforin (PUR-fuh-rin) A protein secreted by cytotoxic T cells that creates holes in the plasma membranes of virus-infected self cells and causes them to die.

peripheral nervous system (PNS) The part of the nervous system that innervates all parts of the body and transmits information to and from the central nervous system.

periphyton (puh-RIF-uh-ton) Plants or algae adhering to rock substrates in streams and rivers.

peristalsis (per′uh-STOL-sus) A series of contraction waves that ripple along the smooth muscles surrounding the tubular organs of the digestive system.

permafrost The subsurface layer of tundra ground that is permanently frozen to a depth of 400 to 600 meters.

permeability A measure of the ease with which specific types of ions or molecules can pass through a membrane.

perspiration A substance produced by sweat glands that is secreted onto the skin's surface; its evaporation helps maintain body temperature.

pesticides (PES-tuh-sīdz) Substances that kill unwanted or harmful organisms.

petals The parts of a flower that are usually colored to attract insects, hummingbirds, or bats needed for pollination.

petiole (PET-ē-ōl) The stalk that connects a leaf to a plant.

phage (fāj) See *bacteriophage.*

phagocytosis (fag′uh-suh-TŌ-sus) The process in which cells engulf and destroy foreign particles.

pharyngeal (fuh-RIN-jē-ul) **gill slits** Openings that allow water to flow through the mouth, over the gills, and out the gill slits of fishes.

pharynx (FĀ-rinks) The throat; a thick-walled, muscular tube.

phenetic (fi-NET-ik) **system** A classification system based on computer tabulations and analyses of the number and degree of similarities that exist among organisms.

phenotype (FĒ-nuh-tīp) The physical trait or traits that occur as a result of a specific genotype.

phenylketonuria (fen′ul-kē-tuh-NYOO-rē-uh) **(PKU)** A disorder distinguished by abnormally high levels of an amino acid, phenylalanine, in the blood of newborn babies; can result in severe mental retardation.

pheromones (FER-uh-mōnz) Chemical signals that convey information between animals.

phlogiston (flō-JIS-tun) **theory** An erroneous early theory contending that when a substance was burned, something called *phlogiston,* the "material and principle of fire," was expelled.

phospholipid (fos′fō-LIP-id) A lipid compound formed from one glycerol molecule, two fatty acid molecules, and a molecule with a negatively charged phosphate group (PO_4^{3-}) linked with a positively charged nitrogen-containing group; the principal structural component of biological membranes.

photons (FŌ-tonz) The fundamental units of energy of the electromagnetic spectrum.

photoperiod The length of the day between sunrise and sunset.

photosynthates (fō′tō-SIN-thātz) All sugars or other carbohydrates made during photosynthesis.

photosynthesis (fō′tō-SIN-thuh-sis) The process in which photosynthetic cells in plants, algae, and certain cyanobacteria use radiant energy from the sun to convert carbon dioxide and water into energy-rich sugars that are the source of energy for most organisms on Earth, releasing oxygen as a by-product.

photosynthetic bacteria Autotrophic prokaryotes that use light as an energy source for converting simple molecules into complex organic molecules.

photosynthetic biomass The fraction of energy in an ecosystem captured by producers during photosynthesis and converted to their growth.

pH scale A measure of hydrogen ion (H^+) concentration in a solution; expressed mathematically as $pH = -\log H^+$.

phyletic (fi-LET-ik) **evolution** The slow, gradual change of a population, or set of populations, into a new species.

phylogeny (fī-LOJ-uh-nē) The evolutionary development of a species or of a larger taxonomic group of organisms.

phylum (FĪ-lum) A taxon that includes similar and related orders; the major subdivision of a kingdom.

physical factors Gravity and light and heat energy; all environmental factors that are not biotic.

physical map A type of genetic map showing the actual location of genes on a chromosome and also the distances between genes as determined by constant, identifiable landmarks.

physics The scientific study of the interactions of matter and energy.

physiognomy (fiz'ē-OG-nuh-mē) The appearance or characteristic features of vegetation present in a community.

physiology (fiz'ē-OL-uh-jē) The study of functions of an organism.

PID See *pelvic inflammatory disease.*

pigments Protein molecules capable of absorbing photons and transferring energy by rearranging their own molecular structures to create an energized state.

pili (PĪ-lī) Bacterial structures used in attachment to host cells and in bacterial reproduction.

pioneer community The first collection of species that colonize or recolonize a barren or disturbed site, initiating ecological succession.

pistil (PIS-tul) The plant reproductive organ that produces female gametes.

pituitary (pi-TOO-uh-ter'ē) **gland** The major gland of the endocrine system, located at the base of the brain, producing numerous hormones that effect various body activities; regulated by neurosecretory cells in the hypothalamus region of the brain.

PKU See *phenylketonuria.*

placenta (pluh-SEN-tuh) The structure through which materials are exchanged between the mother and the embryo or fetus; secretes estrogens and progesterone throughout pregnancy.

placental (pluh-SEN-tul) **mammals** Animals that complete their development in the female uterus.

plant ecology The study of the structure and functions of plant populations and their environments.

plant growth regulators Substances that are produced in one tissue, transported to another part of the plant, and have a physiological effect at a different time.

plant physiological ecology The study of the relationships between individual plants and their environment.

plant physiology The study of processes, such as photosynthesis and respiration, taking place in a single cell or in an individual plant.

plant respiratory biomass The amount of stored energy required by plants to conduct their normal activities.

plaques (plaks) Fibrous, fatty deposits in blood vessels.

plasma (PLAZ-muh) The extracellular fluid that surrounds blood cells, containing and transporting dissolved gases, proteins, hormones, nutrients, lipids, certain minerals such as iron, and metabolic waste products; blood without the formed elements.

plasma cell A B lymphocyte that secretes antibodies.

plasma membrane A complex structure surrounding a cell, composed of two layers of highly organized phospholipid molecules in which various proteins are embedded; responsible for regulating substances entering and leaving the cell.

plasmids (PLAZ-midz) Small, circular DNA molecules found inside a bacterial cell.

platelets (PLAT-lits) Very small, nonnucleated cell fragments that have a critical function in blood clotting.

plate tectonics The field of science that studies the continuous, slow movement of plates on which Earth's landmasses and ocean basins rest.

PNS See *peripheral nervous system.*

poikilothermic (poi'ki-lō-THUR-mik) Having no internal mechanisms for controlling body temperature.

point mutation A change in the sequence of nucleotide bases in DNA or RNA due to a base substitution or to the addition or deletion of a single base pair.

polarized In a condition in which opposite states exist at the same time. In neurophysiology, having a plasma membrane positively charged on the outside and negatively charged on the inside.

polar molecule A covalent molecule with regions of unbalanced positive and negative electrical charges.

polygenic inheritance A trait determined by many genes at different loci; the numbers, chromosomal locations, and degree of expression of all the different genes are usually not known.

polymorphism (pol'i-MOR-fiz-um) The existence within a population of two or more genotypes for a given trait that results from balancing selection.

polypeptide A chain of amino acids that may be smaller than a complete protein.

polyploid (POL-i-ploid) An organism that has three or more complete sets of chromosomes.

polysaccharide (pol'i-SAK-uh-rīd) A carbohydrate composed of three or more monosaccharides.

population A group of individuals of the same species that live in the same place.

population-community school An informal group of early ecologists who considered populations to represent fundamental levels of organization. They identified factors such as competition and predation and studied their influence on populations.

population genetics The study of the genetic constitution of populations and how it changes.

positive feedback A process in which the level of a variable changes in the same direction as the initial change.

postsynaptic cell The part of a cell that receives a nerve impulse from another cell; transmits impulses away from the cell.

postsynaptic receptors Proteins on the membranes of a postsynaptic cell; of many types, each with a unique structure enabling it to bind with one type of neurotransmitter.

postzygotic isolating mechanisms Mechanisms that reduce the fitness of hybrids; they include hybrid inviability, sterility, and F_2 breakdown.

potential energy Stored or inactive energy that is capable of doing work or creating change at a later time.

preformation An erroneous explanation of reproduction that held that successive generations were preformed and encapsulated in previous generations like nested boxes.

pregnancy The period of development between fertilization and birth.

prenatal diagnosis A process used to determine the genetic status of a developing fetus.

presynaptic membrane The membrane of a presynaptic neuron at the site of a synapse.

presynaptic neuron The part of a neuron that transmits a nerve impulse toward the synapse.

prevalence (PREV-uh-luns) The percentage of a population affected by a disorder.

prezygotic isolating mechanisms Mechanisms that prevent or reduce hybridization between members of different species, including habitat, seasonal, ethological, mechanical, and gametic isolation.

primary follicle A primary oocyte and surrounding follicle cells.

primary growth Growth from apical meristems that results in increased length of roots, stems, and branches.

primary immune response The first specific immune response to an antigen not previously encountered.

primary oocytes (Ō-uh-sīts) Immature ova.

primary sex organs Organs that produce gametes and secrete hormones involved in development and functions related to reproduction.

primary spermatocytes (spur-MAT-uh-sīts) Large developing cells that give rise to mature sperm.

primary succession The progression of different species that come to inhabit new ground where life forms have not previously existed.

primary wall The part of a plant cell wall that forms first around growing cells; composed of microfibrils embedded in a matrix containing pectin, which provides the plasticity required for cell enlargement.

primates (PRĪ-māts) The order of placental mammals that includes tree shrews, lorises, lemurs, monkeys, apes, and humans.

primordial (prī-MOR-dē-ul) **germ layers** Three embryonic cell layers—ectoderm, mesoderm, and endoderm—that give rise to all tissues and organs during embryonic development.

procedural memory Memory acquired by repetition or continuous practice; habit.

producer See *autotroph.*

product A substance that results from a chemical reaction.

profundal (prō-FUN-dul) **zone** The deep part of lakes where light does not penetrate.

progenote (PRŌ-juh-nōt) The hypothesized precellular stage of chemical organization that gave rise to both prokaryotic and eukaryotic cells.

progesterone (prō-JES-tuh-rōn) A sex hormone produced by the ovaries that helps prepare the endometrium for implantation of the embryo.

progressive evolution An evolutionary sequence displaying a constant direction, or *goal,* and coming into existence independently of selective forces.

Prokaryotae (prō′ker-ē-Ō-tā) The kingdom name for all organisms composed of cells without nuclei, including the blue-green algae and the bacteria.

prokaryotes (prō-KER-ē-ōts) Simple organisms, mostly bacteria, composed of a single prokaryotic cell.

prokaryotic (prō′ker-ē-OT-ik) **cell** A cell that lacks a membrane-bound nucleus and has no membranous organelles; a bacterium.

prolactin (prō-LAK-tin) A pituitary hormone secreted during pregnancy that stimulates mammary growth and development and after birth induces lactation; its role in males, if any, is not known.

promoter A short DNA nucleotide sequence where RNA polymerase binds and initiates transcription.

pronucleus The nucleus of a sperm or egg after the sperm enters the egg but before they combine to form a zygote.

prophase (PRŌ-fāz) The second stage of mitosis, in which chromosomes become shorter and thicker.

prophase I The stage of meiosis in which chromatin coalesces and sister chromatids become visible.

prophase II The stage of meiosis in which newly formed nuclear envelopes of the two daughter cells disappear as chromosomes re-form.

prostaglandins (pros′tuh-GLAN-dinz) Substances thought to enhance the movement of sperm when deposited in the female reproductive tract.

prostate (PROS-tāt) **gland** A gland that empties into the urethra in males and secretes a substance that contributes to sperm mobility.

protease An enzyme that breaks down protein molecules.

proteins (PRŌ-tēnz) A diverse group of organic molecules composed of chains of amino acid molecules.

Protista The kingdom name that Ernst Haeckel used in 1866 for free-living microscopic eukaryotic organisms; includes single-celled protozoans, unicellular algae, multicellular algae, and slime molds; now known as the kingdom *Protoctista.*

protocol The detailed plan of a scientific experiment or medical treatment.

Protoctista Kingdom name first proposed by John Hogg in 1861 for free-living microscopic eukaryotic organisms; includes single-celled protozoans, unicellular algae, multicellular algae, and slime molds.

proton (PRŌ-ton) A particle with a positive electrical charge, found in the nuclei of atoms.

proto-oncogene (prō′tō-ON-kuh-jēn) A gene that has a specific function but when damaged or malfunctioning has the potential to become an oncogene.

protoplast fusion The process in which two protoplasts from different plant species may be stimulated to fuse.

protoplasts Plant cells that have had their outer cell wall removed through digestion by certain enzymes.

protostomes Animal phyla in which the blastopore becomes the mouth.

provirus The state in which viral DNA is integrated into a host chromosome.

pseudocoelomates (soo′dō-SĒ-luh-mātz) Animal phyla that develop a body cavity between the mesoderm and the endoderm that is often called a *false coelom.*

pseudogenes (SOO-dō-jēnz) Genes that are no longer functional (not transcribed); derived from an ancestral cell.

pseudopodia (soo′dō-PŌ-dē-uh) Specialized, temporary membrane extensions formed by single-celled amoebalike cells and used in locomotion or engulfing food.

PTH See *parathyroid hormone.*

puberty (PYOO-bur-tē) The time of life when gonads mature and reproduction becomes possible.

pulmonary circulation The flow of deoxygenated blood between the right ventricle and the lungs and the return of oxygenated blood to the left atrium.

pulse A wave of pressure radiated outward from the heart through the arteries.

punctuated equilibrium Evolutionary sequences in which macroevolution occurs in short bursts of change, followed by long stable periods.

Punnett (PUN-it) **square** A diagram for illustrating all possible combinations of gametes in crosses.

pure-breeding lines Lines that when self-pollinated or cross-pollinated with a member from the same line produce only offspring that are identical to the parents.

purine (PYOO-rēn) A base; adenine and guanine in DNA and RNA.

pyramids of biomass Representations of the amount of living mass in each trophic level of an ecosystem, usually expressed as grams of dry weight per area or volume.

pyramids of energy Representations of the amount of energy in each trophic level of an ecosystem, usually presented as kilocalories per unit area or volume.

pyramids of numbers Representations of the density of organisms in each trophic level, expressed as numbers per unit area or volume.

pyrimidine (pī-RIM-uh-dēn) A base; cytosine and thymine in DNA, cytosine and uracil in RNA.

pyrogens (PĪ-ruh-jinz) Substances that cause an elevation of body temperature.

pyruvate (pī-ROO-vāt) A molecule containing only three carbon atoms that is formed by the splitting of a glucose molecule during glycolysis.

Q

quantitative methods Methods scientists use to measure, count, and calculate results.

quantum speciation Speciation events that occur very rapidly.

R

races Different subpopulations of a single species.

radiant energy Energy emitted from the sun in the form of light waves or photons.

radioisotope (rā′dē-ō-Ī-suh-tōp) An unstable atom that "decays" or undergoes nuclear changes that result in the emission of subatomic particles and the release of energy from the atom.

rainshadow The warm, dry area on the leeward side of a mountain range that results from air losing its moisture as it rises up and over the windward side and then becoming warmer as it descends.

random genetic drift Chance fluctuations in the gene frequencies of a small population.

reabsorption The process in which most of the water and other substances of value to the body are returned to the blood from the filtrate in a nephron.

reactant A starting substance in a chemical reaction.

receptacle The part of a flower attached to the petals and sepals.

receptor Any of various classes of complex protein molecules that recognize hormones, neurotransmitters, or antigens; found on the surface of the plasma membrane or inside the cell.

receptor-mediated endocytosis (en′dō-sī-TŌ-sus) The import of specific proteins into a cell by their binding to receptors in the plasma membrane and their inclusion in vesicles.

recessive trait (1) In Mendelian genetics, a trait that disappears in the first generation in a monohybrid cross. (2) The allele in a heterozygous genotype that is not expressed in the phenotype.

recombinant (rē-KOM-buh-nunt) An individual with traits determined by genes on different homologous chromosomes.

recombinant DNA A DNA molecule containing two or more segments of DNA from two different genes or species.

recombination See *genetic recombination.*

rectum The lower portion of the large intestine.

reduced nitrogen Molecules such as ammonia, ammonium, and various forms of organic nitrogen that are composed of nitrogen and hydrogen.

reduction The process in which an atom or a molecule gains one or more electrons.

reflex (RĒ-fleks) An unconscious, rapid response of the nervous system to a specific stimulus.

regulated condition Activity or part of the body that is maintained at some level by a feedback system.

regulatory gene A gene encoding a repressor protein that controls transcription of structural genes.

regulatory sequences Specific DNA sequences that control the production of a protein.

relative fitness A comparative measure of the fitness of different genotypes in a population, calculated by dividing the net reproductive rate of a genotype by the reproductive rate of the genotype with the greatest fitness.

relaxin (ri-LAK-sin) A female hormone that acts to increase the flexibility of joints between bones around the birth canal, facilitating birth.

releaser In classical ethology, any aspect of an external stimulus that triggers an innate releasing mechanism.

releasing hormones (releasing factors) Substances that travel from the hypothalamus to the pituitary gland through the hypophyseal portal system and stimulate release of hormones from the anterior pituitary.

remodeling Replacement of old or damaged bone tissue by new bone tissue.

renal capsule A thick layer of connective tissue that encloses a kidney.

replication (rep′luh-KĀ-shun) DNA synthesis.

repolarization A return to the original polarized state.

repressor A protein molecule that binds to an operator and prevents transcription of an operon.

reproduction The origination of new cells or organisms from preexisting ones; also, the formation of cells used in growth, repair, and replacement.

reproductive isolation Genetic barriers to gene flow among populations.

reproductive phase Part of the plant life cycle in which reproductive tissues begin to form, leading to the formation of seeds.

reservoirs (1) Places where large amounts of a chemical element or compound are found. (2) Sites in the body from which viruses may spread to uninfected cells.

resistance The ability of survivors of pesticide applications to give rise to new populations that can tolerate pesticide applications in ever-increasing concentrations; also used in reference to bacteria and antibiotics.

resolution The final stage of inflammation, in which an injured tissue returns to its normal state.

respiration The exchange of oxygen and carbon dioxide between an organism and its environment.

respiratory (RES-pur-uh-tor-ē) **center** A special group of nerve cells that responds to levels of carbon dioxide in the blood and regulates the rate of respiration.

respiratory (RES-pur-uh-tor-ē) **system** The lungs and various other organs that function in exchanging oxygen and carbon dioxide between the internal and external environments.

respiratory (RES-pur-uh-tor-ē) **transmission** Transmission of a disease agent by sneezing, coughing, talking, or kissing.

resting membrane potential (RMP) The voltage that exists between the inside and outside of a neuron's plasma membrane when it is at rest; typically, −70 millivolts.

restriction enzymes Enzymes that cut DNA molecules at specific nucleotide sequences.

reticular (ri-TIK-yuh-lur) **activating system** A network of branched neurons in the brain; when stimulated, the individual becomes awake and alert.

reticular (ri-TIK-yuh-lur) **fibers** Thin, short filaments constructed of the protein *reticulin.*

reticular (ri-TIK-yuh-lur) **formation** A loosely organized network of nerves that receives input from several other areas of the brain and spinal cord and maintains the state of consciousness.

retroviruses (ret′ruh-VĪ-rus-iz) A class of viruses that can direct the synthesis of DNA from RNA.

reverse transcriptase (tran-SKRIP-tās) An enzyme that synthesizes DNA from RNA.

rhizoids (RĪ-zoidz) Rootlike structures of bryophytes and ferns that are composed of elongated single cells or multicellular filaments that anchor the plant and absorb water and mineral nutrients.

rhizomes (RĪ-zōmz) Underground plant stems from which new plants can sprout.

rib cage The part of the skeleton that covers and protects the lungs, heart, stomach, and liver.

ribonuclease An enzyme that breaks down RNA molecules.

ribonucleic (rī′bō-noo-KLĒ-ik) **acid** See *RNA.*

ribosomal (rī′buh-SŌ-mul) **RNA (rRNA)** RNA molecules used in ribosomal structure and function.

ribosomes (RĪ-buh-sōmz) Structures composed of rRNA and proteins that appear as small particles in the cytoplasm or attached to rough endoplasmic reticulum; sites where mRNA is translated into an amino acid sequence.

riparian (ri-PER-ē-in) **zones** Areas along streams or riverbanks that are influenced by flowing water.

risk factor Anything that increases the probability of developing a disease or suffering harm.

ritualization The modification of behavior patterns to serve a new function, usually involving communication.

RMP See *receptor-mediated endocytosis.*

RNA Ribonucleic acid; single-chain molecules formed by nucleotide links; different RNAs include messenger RNA (mRNA), ribosomal RNA (rRNA), and transfer RNA (tRNA); retroviruses use RNA as their genetic material.

RNA polymerase (POL-uh-mur-ās) An enzyme that attaches to a specific nucleotide sequence on the DNA molecule, causes it to separate partly into two strands, and then moves through a gene, making an RNA molecule complementary to the gene DNA.

root-shoot ratio The ratio of root dry weight to shoot dry weight.

rough ER The part of the endoplasmic reticulum that is studded with ribosomes, where proteins are synthesized.

rRNA See *ribosomal RNA.*

S

saliva (suh-LĪ-vuh) A watery secretion from the salivary glands that consists mainly of mucus, which protects the lining of the mouth and lubricates the food for easier movement, plus salts and enzymes.

salivary amylase (SAL-uh-ver′ē AM-uh-lās) A specific enzyme in saliva that begins the breakdown of starch.

salivary (SAL-uh-ver′ē) **glands** Three pairs of glands near the mouth that secrete saliva.

salt A compound that dissociates into anions and cations, other than H⁺ or OH⁻.

saprophytic (sap′ruh-FIT-ik) Obtaining nutrients by absorbing breakdown products from dead plants and animals.

sarcolemma (sar′kuh-LEM-muh) The plasma membrane of skeletal and cardiac muscle cells.

sarcoma (sar-KŌ-muh) A cancer involving connective tissue cells.

sarcomere (SAR-kuh-mir) The contractile unit in a striated muscle cell; extends from one Z disk to the next Z disk.

sarcoplasm (SAR-kuh-plaz′um) Cytoplasm in skeletal and cardiac muscle cells.

sarcoplasmic reticulum (sar-kuh-PLAZ-mik ri-TIK-yuh-lum) A tubular network surrounding myofibrils that contains calcium ions used in muscle contractions.

savannas (suh-VAN-uz) Modified tropical grasslands characterized by alternating wet and dry seasons.

scale of analysis The relative size of different biotic hierarchies in the biosphere; for example, in moving from biome to ecosystem to community to population to deme to individual, the total number of variables that may influence an observed pattern becomes greater.

scanning electron microscope (SEM) A microscope that makes it possible to see the three-dimensional appearance of a cell at high magnifications.

Schwann (shwon) **cells** Neuroglia cells that produce myelin.

scientific revolution The movement, begun in the seventeenth century, emphasizing studying the world as natural system and explaining observed regularities as laws of nature; gave rise to rational methods—careful observation, comparisons with existing information, experimentation—for obtaining knowledge of the natural world.

scrotum (SKRŌ-tum) A pouch of skin containing the testes, located outside the abdominal cavity.

scrub forest Shrublands that develop between coniferous forests and deserts in the Rocky Mountain region, where dominant species include scrub oak, mountain mahogany, and Gambel's oak, depending on latitude and elevation.

seasonal isolation An isolating mechanism that occurs when species that live in the same region have populations that are prevented from interbreeding because they are sexually mature at different time periods or seasons.

sebaceous (si-BĀ-shus) **glands** Glands in the dermis that secrete an oily substance (*sebum*) that prevents hair and skin from drying and is also toxic to many types of bacteria.

secondary follicle An intermediate developmental stage in the formation of a mature ovum.

secondary growth An increase in stem and branch diameters due to new cells produced by the vascular and cork cambiums.

secondary immune response An immune response to a previously encountered antigen that occurs quickly because of the presence of memory cells specific for the antigen.

secondary oocytes (Ō-uh-sīts) The larger of the two types of haploid cells formed during the first meiotic division of a secondary follicle.

secondary sex characteristics Changes in the body that are related to sexual maturation, including changes in distribution of body hair and body fat, voice pitch, and muscle development.

secondary spermatocytes (spur-MAT-uh-sīts) Cells produced during the first meiotic division of a primary spermatocyte that give rise to spermatids.

secondary succession The progression of a community, after disruption, through a number of predictable types (*seres*) that eventually culminates in a new climax community.

secondary wall The part of a plant cell wall that is formed after cell growth is completed; composed of microfibrils embedded in a rigid matrix, which provides strength.

second law of thermodynamics A law stating that every energy transformation results in a reduction in the total usable energy of the system.

second-order consumers Large crustacean zooplankton, insect nymphs, and other small arthropods in guilds that feed on herbivores in lentic ecosystems.

second-order streams Streams that originate when two first-order streams meet.

secretion (suh-KRĒ-shun) A regulated stage in which ions and other substances are transferred from capillaries to nephrons in the kidneys during urine formation.

secretory vesicle (SĒ-kruh-tor′ē VES-i-kul) A membrane-bound collection of cell products that are exported from the cell.

sedimentary (sed′uh-MEN-tuh-rē) **cycles** Biogeochemical cycles in which rocks, soil, or sediments act as primary reservoirs.

seeds The reproductive structures formed following fertilization of a cone or a flower that contain a protective coat, stored nutrients, and an embryo.

selectively permeable Permitting entry of certain substances and not others.

selfish behavior Behavior that benefits the actor at a cost to the recipient.

self-maintenance The set of activities used in meeting the continuous demand for repairing or replacing ribosomes, membranes, enzymes, and other cellular structures and molecules that are lost during normal functioning.

self-pollination Fertilization of an egg by pollen from the same flower.

self-pruning The shedding of unproductive branches by trees.

SEM See *scanning electron microscope.*

semen (SĒ-min) Fluid ejaculated by a male that consists of sperm and secretions from various glands.

semiconservative replication DNA replication in which each new double-stranded molecule is composed of one parental strand and one newly formed daughter strand.

seminal vesicles (SEM-uh-nul VES-i-kulz) A pair of structures that secrete a component of semen into ejaculatory ducts.

seminiferous tubules (sem'uh-NIF-uh-rus TOO-byoolz) Highly convoluted structures in the testes that harbor male gametes in various stages of development.

sense organs The eyes, ears, nose, skin, and tongue.

sensory neuron A neuron that receives information from the external and internal environments and conducts it to the brain or spinal cord, where it is processed.

sensory receptors Specialized cells of the nervous system that are capable of detecting certain changes in external or internal conditions. Each sensory receptor is sensitive to one specific stimulus, such as touch, pressure, or temperature.

sepals (SĒ-pulz) Flower parts that surround a bud.

septa Connecting cross walls in fungi.

sequence hypothesis The hypothesis that the sequence of bases in DNA and RNA molecules specifies the sequence of amino acids in proteins.

sex chromosomes Chromosomes that determine the sex of an individual.

sex hormones Steroid hormones produced by the gonads that are essential for reproductive development and function.

sexual behaviors Actions that reduce the probability of having sex with an infected partner; risk factors include having multiple partners, frequent changing of partners, casual sex, and sex with a prostitute.

sexual dimorphism (di-MOR-fiz-um) Differences in the external appearance of males and females of the same species.

sexual imprinting The breeding preferences of many birds as determined by early imprinting experiences.

sexually transmitted disease (STD) Any infectious disease that is spread by sexual contact.

sexual reproduction The fusion of haploid gametes from two parents to create offspring.

sexual selection Selection for characteristics that have value for success in mating.

sexual transmission The transfer of disease agents through sexual intercourse.

shift The insertion of portions of a chromosome in a different region of the same chromosome.

shivering Involuntary contraction of skeletal muscles that generates heat.

shortgrass prairies Plains grasslands in which drought-resistant shortgrass species constitute the climax community; occur over vast areas of the Great Plains of North America.

shredders Lotic insect guilds that consume coarse particulate organic matter and obtain most of their nutrients from digesting bacteria and fungi that colonize leaves and other organic debris entering streams.

sibling species Morphologically similar species that cannot be distinguished from one another by external characteristics but have isolating mechanisms that keep them from interbreeding.

sickle-cell anemia A disorder in which red blood cells become sickle-shaped, resulting in a loss of function; caused by an autosomal recessive gene.

sickle-cell disorder A disorder that occurs in individuals who are homozygous recessive for a particular gene, characterized by severe anemia.

sickle-cell trait A disorder that occurs in individuals who are heterozygous for a particular gene, characterized by mild anemia.

simian immunodeficiency virus (SIV) A group of viruses that cause mild diseases in monkeys; HIV-1 is hypothesized to have evolved from an SIV strain that infects chimpanzees.

simple diffusion The movement of molecules down a concentration gradient—that is, from an area where they are highly concentrated to an area where they are less concentrated.

sister chromatids (KRŌ-muh-tidz) Replicated homologs joined by a centromere that exist in mitosis and meiosis.

SIV See *simian immunodeficiency virus.*

skeletal muscle A tissue specialized for contraction that consists of extremely long fibers, masses of which are attached to bones by special connective tissues; causes movements of the skeleton.

skin A complex organ forming the external surface layer of the body.

skull The bones of the skeleton that protect the brain, eyes, and ears.

sliding-filament model An explanation of muscle contraction based on the movements of thick and thin filaments past one another.

small intestine The major organ of digestion, consisting of a long tube from which nutrients are absorbed into the blood and lymph.

smooth ER The part of the endoplasmic reticulum where lipids are synthesized; has no ribosomes on its surface.

smooth muscle A tissue specialized for contraction that is made up of cells with single nuclei and no cross-striations, present in the walls of hollow internal organs and tubes.

social behavior Behavior involving two or more animals.

social organization The pattern of relationships among individuals within a population at a particular time.

societies Groups consisting of members of a species that are attracted to one another, communicate with one another, exhibit cooperative behavior, and engage in synchronized activities.

sociobiology The study of the biological basis of social behavior.

softwood Lumber from conifers.

solar system The sun and the planets, asteroids, comets and other bodies that orbit it.

solute (SOL-yoot) The substance dissolved in a liquid (the *solvent*) to make a solution.

solution A chemical made by dissolving one substance (the *solute*) in a liquid (the *solvent*).

solvent A liquid in which other substances are dissolved.

somatic (sō-MAT-ik) **cell gene therapy** A type of gene therapy in which genes in somatic cells are replaced or modified.

somatic (sō-MAT-ik) **nervous system** The part of the nervous system composed of motor nerves that run directly from the central nervous system to skeletal muscles.

somatostatin (sō-mat'uh-STAT-in) A hormone secreted by the pancreas, involved in glucose homeostasis; inhibits secretion of glucagon and insulin.

speciation (spē'shē-Ā-shun) The processes of species formation.

species (SPĒ-shēz) Groups of actual or potentially interbreeding natural populations that are reproductively isolated from other such groups; the major subdivision of a genus.

species diversity The number of different species present in an area.

species selection Evolutionary change in which species are the units of selection.

specific heat The amount of heat required to raise the temperature of any substance one unit; the standard is the amount of heat required to raise the temperature of 1 gram of water 1°C.

specific immune responses A complicated adaptive defense system that functions in response to nonself antigens, involving various cells, cell products, and control mechanisms.

specificity (spes'uh-FIS-uh-tē) The ability of an antibody to recognize and bind to only one type of antigen.

sperm Haploid gametes produced by males; spermatozoa.

spermatids (SPUR-muh-tidz) Haploid cells arising from secondary spermatocytes; in humans, each contains 22 autosomes and one sex chromosome.

spermatogenesis (spur-mat'uh-JEN-uh-sus) Production of spermatozoa in the testes.

spermatogonia (spur-mat'uh-GŌ-nē-uh) Diploid germinal cells of the seminiferous tubules that divide by mitosis to produce cells that ultimately develop into mature sperm.

spermatozoon (spur-mat'uh-ZŌ-on) A mature sperm cell.

spicules (SPIK-yoolz) Skeletal elements of sponges, radiolarians, and sea cucumbers composed of calcium carbonate or silicates produced by amoebocytes.

spina bifida (SPĪ-nuh BĪ-fuh-duh) A developmental anomaly characterized by a defect in the bony encasement of the spinal cord.

spinal nerves Thirty-one pairs of nerves that emerge from the spinal cord and innervate most organs below the neck.

spinal reflex arc An inborn neural pathway by which impulses from sensory neurons reach motor neurons without first traveling to the brain.

spleen An organ that filters circulating blood and also contains populations of B cells, T cells, and macrophages.

spongy bone The network of inner bone surrounding the marrow.

spontaneous generation The notion that living organisms can arise from nonliving substances.

spore (1) A haploid structure that can germinate directly into the gametophyte stage of a plant. (2) An inert form of a bacterium or a fungus enclosed in a thick protective wall that is resistant to most environmental stresses.

sporophyte (SPOR-uh-fit) The spore-producing generation in plant life cycles; a diploid plant stage that produces haploid spores through meiosis.

stabilizing selection Selection against extreme variants in a population, with the result that a standard phenotype is favored.

stamens (STĀ-menz) Male reproductive parts that produce pollen in flowers.

standing crop The number or amount of organisms present per unit of area or volume at any one time.

starch A polysaccharide composed of hundreds of glucose molecules; the energy storage molecule in plants.

statistically significant Shown by mathematical analysis to be associated with a low probability of error.

statistical methods The mathematical analyses and interpretations of data (numerical facts) that enable scientists to make an objective appraisal about the reliability of conclusions based on the data.

STD See *sexually transmitted disease.*

stem cell A type of cell that divides continuously to produce daughter cells that either mature to become functional cells or continue to act as stem cells.

steroid (STĒ-roid) **hormones** Hormones synthesized from cholesterol that usually cause target cells to manufacture proteins through a mechanism involving internal receptors.

stolons (STŌ-lunz) Stems that extend to the soil, from which lateral buds can develop into roots and shoots.

stomach A relatively large, muscular organ of the digestive system in the upper abdomen that can stretch to accommodate about 2 liters of food or liquid.

stomata (STŌ-muh-tuh) Pores in the leaf surface that can open and close in response to metabolic control mechanisms.

strains Varieties of a specific virus or bacterium that differ only slightly in their protein structure or genetic content.

stratum corneum (STRĀ-tum KOR-nē-um) A layer of dead skin cells that forms the outer protective layer of the skin and is continuously sloughed.

stream order A numerical scale from 1 to 12 used to classify streams in a flowing-water ecosystem; based on the streams' size and location in a drainage basin.

stroke A sudden, severe trauma causing temporary or permanent loss of consciousness or sensation, resulting from an interference in normal blood circulation.

stromatolites (strō-MAT-uh-līts) Mound-shaped, layered, rocky structures formed by surface communities of different prokaryotic cells.

structural genes Genes that code for proteins.

structural rearrangement The loss or relocation of genes or sets of genes along specific chromosomes.

subatomic particles Subunits of atoms; include neutrons, protons, and electrons.

subcutaneous (sub'kyoo-TĀ-nē-us) **layer** A sheet of connective tissue and adipose tissue that lies between the dermis and muscles.

subduction The downward movement of oceanic plates when continental plates ride up and over them.

subspecies (SUB-spē'shēz) A subpopulation of a single species that is sufficiently distinct to be given a separate taxonomic name.

substrate See *reactant.*

subtropical forests Forests found in northern and southern portions of the Torrid Zone and at higher elevations on mountains within rain forests, resulting from a decrease in total available moisture or a seasonal distribution of rain that is less than that received by tropical forests.

succession See *ecological succession.*

supercooling Chilling water rapidly below 0°C without causing freezing.

suppressor T cells T cells that inhibit production of certain immune cells and antibodies and help control and limit humoral immune responses.

swamp A wetland intermittently inundated by standing water, dominated by trees and other woody plants.

sweat A complex solution containing salts and small amounts of nitrogen-containing metabolic waste products.

sweat glands Glands in the skin that secrete sweat and antimicrobial substances such as lysozyme.

symbionts (SIM-bī-onts) Organisms of different species participating in a close living association *(symbiosis);* for example, algae and fungi in a lichen.

sympathetic nervous system The part of the nervous system that prepares for vigorous activity, for example, by increasing the heart rate.

sympatric speciation The process by which a species originates in the same geographical region as its parent species.

synapse (SIN-aps) A narrow gap between neurons or between a neuron and a muscle or gland cell.

synapsis (suh-NAP-sus) A close pairing of homologs during prophase I of meiosis.

synaptic (suh-NAP-tik) **bulb** The knob at the branch at the end of an axon.

synaptic (suh-NAP-tik) **cleft** The narrow space between two cells at a synapse.

synaptic (suh-NAP-tik) **terminals** The ends of an axon's branches.

synaptic vesicles (suh-NAP-tik VES-i-kulz) Small membrane-bound sacs filled with neurotransmitters that are found in presynaptic neurons.

synaptonemal (suh-nap′tuh-NĒ-mul) **complex** A zipperlike protein structure that binds synapsed homologs together during prophase I of meiosis.

syndrome (SIN-drōm) A set of symptoms or characteristics that occur together.

synthesis (SIN-thuh-sus) The combining of simple chemical building blocks into a larger molecule.

syphilis A sexually transmitted disease caused by a bacterium, *Treponema pallidum.*

system Any whole or unit resulting from the organization of parts or subunits.

systematics See *taxonomy.*

systemic (sis-TEM-ik) **circulation** The circulation of oxygenated blood from the left ventricle to all tissues of the body and the return of deoxygenated blood to the right atrium.

T

taiga (TĪ-guh) The northern coniferous forest biome located about 50° north latitude that ends at the lower limit of the permafrost.

tallgrass prairies Plains grasslands receiving ample rainfall, once covered with tall grasses but now used to grow corn, soybeans, and other crops.

taxa (TAK-suh) The various levels in the classification system.

taxis (TAK-sis) A motion toward or away from a source of stimulation.

taxonomy (tak-SON-uh-mē) The science of classifying organisms; also called *systematics.*

T cells See *T lymphocytes.*

T-DNA The DNA in Ti plasmids that contains genes encoding enzymes that cause

tumors in plants infected with *Agrobacterium tumefaciens.*

telophase (TEL-uh-fāz) The fifth and last stage of mitosis, in which daughter nuclei form, nuclear envelopes and nucleoli appear, and chromosomes disappear.

telophase I The stage of meiosis in which the separated homologous chromosomes cluster at each pole and cytokinesis takes place.

telophase II The stage of meiosis in which each pole receives one set of chromosomes.

TEM See *transmission electron microscope.*

temperate forest biome A forest biome that occurs in the Temperate Zones, where increasing latitude results in greater seasonal extremes, with lower average temperatures and less precipitation than in biomes of the Torrid Zone.

temperate forests Forests in latitudes of the westerlies, between 5° and 10° north and south of the 40th parallel.

temporal factors (1) Normal changes that occur throughout the life of an organism. (2) Gradual environmental changes in a site over long periods of time.

tendon Connective tissue that attaches muscle to bone.

terminal buds Buds located at the top of the plant and at the ends of branches and roots that give rise to all cells in the plant's shoots and roots.

territory A home range that is defended with threats, attacks, or advertisement (via scent or song) for exclusive use by an individual or group.

test cross A cross between an individual with a dominant trait and one with a recessive phenotype to determine the genotype of the dominant individual.

testis The male gonad or reproductive organ that produces sperm and sex hormones.

testosterone (tuh-STOS-tuh-rōn) The male sex hormone that influences the growth and development of male sex organs, secondary sex characteristics, and sperm.

tests Hard coverings of foraminiferans composed of mineral and organic molecular complexes; the internal skeletons of echinoderms.

tetranucleotide (tet′ruh-NOO-klē-uh-tīd) **hypothesis** The erroneous belief that DNA was a static molecule, composed of equal amounts of adenine, guanine, thymine, and cytosine.

T4 lymphocytes (LIM-fuh-sīts) T cells that regulate a variety of immune functions; the cell type infected with HIV-1, the AIDS virus.

theory A systematic set of concepts that explain and relate data, answer questions, and guide research.

thermal energy Kinetic energy of molecular motion that can be measured with a thermometer; heat.

theory of evolution and modern synthesis The theory that gradual changes in a species or population can be explained by the action of natural selection on small genetic changes, and further, that the processes of species formation and of evolutionary change in groups higher than species can be understood in terms of known genetic mechanisms.

theory of galaxy evolution The belief that hydrogen and helium nuclei were created by the big bang and became mixed into galactic gas clouds that eventually condensed, forming first-generation stars. Within these stars, fusion reactions gave rise to heavy elements that became scattered through space by supernova explosions. These heavy elements became incorporated into later generations of stars and other celestial bodies.

thermoacidophilic (thur′mō-uh-sid′uh-FIL-ik) **bacteria** Literally, "heat-and-acid-loving" prokaryotes; bacteria that thrive in habitats with high temperatures and high concentrations of acids.

thermodynamics The study of relationships between heat and other forms of energy.

thick filament Muscle filament composed of myosin.

thin filament Muscle filament composed of actin.

third-order consumers Lentic guilds of minnows and young trout, bass, or pike that feed on second-order consumers and are in turn consumed by fourth-order consumers.

third-order streams Streams that arise where two second-order streams meet.

thoracic (thuh-RAS-ik) **cavity** The body cavity that contains the heart, lungs, and esophagus.

thrush An infection in the mouth, caused by the fungus *Candida,* that is common in AIDS patients.

thymus (THĪ-mus) **gland** A gland involved in the production and maturation of T cells.

thyroid-stimulating hormone A hormone that stimulates synthesis and release of hormones from the thyroid gland; also called *thyrotropin.*

thyrotropin (thī′ruh-TRŌ-pin) See *thyroid-stimulating hormone.*

thyroxine (thī-ROK-sēn) A hormone secreted by the thyroid gland that helps regulate metabolism and tissue growth and development.

TIL cells See *tumor-infiltrating lymphocytes.*

timberline The upper limit of tree growth; varies from 4200 meters in the southwestern forests to about 3500 meters in Glacier National Park, in Montana, near the Canadian border.

Ti plasmid (TĪ PLAZ-mid) A segment of DNA that disrupts normal plant cell function, resulting in the formation of a tumor; exists in the bacterium *Agrobacterium tumefaciens.*

tissue fluid The fluid that bathes cells, composed of water, salts, and proteins.

tissues Large assemblages of similar cells that are specialized to carry out a particular function.

T lymphocytes (LIM-fuh-sīts) One of two general types of lymphocytes that circulate to the thymus gland, where they are processed before being released to the circulation and distributed to lymphoid organs and tissues; develop into T helper cells, cytotoxic T cells, T suppressor cells, or T memory cells.

tolerance The capacity of lymphocytes not to respond to self molecules.

top carnivores Flesh-eating organisms in the final consumer guild of a food chain.

topography (tuh-POG-ruh-fē) The surface features of a geographical area.

Torrid Zone The area of Earth between the Tropics of Cancer and Capricorn; divided by the equator.

trachea (TRĀ-kē-uh) The windpipe; the major trunk of the respiratory tree.

tracheophytes (TRĀ-kē-uh-fīts) Plants with vascular tissues (xylem and phloem), including advanced sport-forming species and the seed-forming gymnosperms and angiosperms.

transcription Synthesis of mRNA from a sequence of DNA; so called because information coded in the sequence of DNA nucleotides is copied (transcribed) into RNA nucleotides; the first step in gene expression.

transfer RNA (tRNA) RNA molecules with triplet nucleotide sequences that are complementary to the triplet coding sequences of mRNA; tRNAs bond to amino acids and transfer them to ribosomes.

transformation Alteration in the normal growth and developmental properties of a cell.

transforming principle An unknown substance that changed one form of *Diplococcus* bacteria into a different form; later determined to be DNA.

transgenic animal An animal that developed from an embryo in which chromosomes, genes, or DNA from a different species had been integrated.

transgenic plant A plant with cells containing genes from a different species.

translation The process of producing a protein whose linear amino acid sequence is derived from the mRNA codon specified by a gene.

translocation (1) The insertion of genes or portions of chromosomes in nonhomologous chromosomes. (2) The movement of substances from one part of a plant to another.

transmembrane Spanning a membrane; usually applied to a protein channel or proton pump that spans a mitochondrial, chloroplast, or plasma membrane.

transmission electron microscope (TEM) A microscope that can magnify objects up to 300,000 times.

transmutation theory An erroneous early theory that one substance could be transformed into another.

transpiration The loss of water vapor through a membrane or pore of an organism; usually associated with water loss from plants during photosynthesis or cooling.

transposons (trans-PŌ-zonz) Short DNA sequences capable of migrating and becoming inserted into a different chromosomal site.

trichocysts (TRIK-uh-sists) Specialized structures used by ciliated protozoans for capturing food and for defense.

triglyceride (trī-GLIS-uh-rīd) A lipid compound formed from one glycerol molecule and three fatty acid molecules.

trisomy (TRĪ-sō-mē) The presence of a third copy of a homologous chromosome in cells.

tRNA See *transfer RNA.*

trophic dynamics The transfer of energy within ecosystems.

trophic guilds Groups of species that exploit the same class of trophic resources in a similar way.

trophic hormone A hormone that stimulates the release of other hormones.

trophic levels Successive steps of a food chain, each of which has less energy available than the previous level; the levels are often referred to as producers; primary, secondary, tertiary (and higher) levels of consumers; and decomposers.

trophoblast (TRŌ-fuh-blast) A single layer of cells surrounding a blastocyst.

tropical forests Forests that occur in the equatorial portion of the Torrid Zone, where more than 240 centimeters of annual rainfall combined with an average annual temperature greater than 17°C has resulted in the most productive forests on Earth.

TS genes See *tumor suppressor genes.*

tubers (TOO-burz) Large, fleshy underground stems, like that of the potato, that are highly specialized for carbohydrate storage and are also capable of vegetative reproduction.

tubule (TOO-byool) A small tube.

tumor (TOO-mur) A cell mass that develops when cells no longer respond to substances that normally regulate their growth, differentiation, or reproduction.

tumor-infiltrating lymphocytes (TIL cells) Certain leukocytes that attack cancer cells.

tumor suppressor (TS) genes Genes that normally suppress rates of cell growth and development; may play a role in cancer induction if they become inactivated or removed from a chromosome.

tundra (TUN-druh) A northern biome that begins at the southern limit of the permafrost and extends north to permanent ice fields.

turgor (TUR-gur) The capacity of plants to sustain their shape with water pressure.

turnover The process by which cold surface water sinks and is replaced by deep water from near the lake bottom; occurs when the surface water reaches its greatest density at 4°C.

U

ultrasound scanning A visualization procedure that employs high-frequency sound waves to provide a profile of the structural features of the fetus.

umbilical (um-BIL-i-kul) **cord** The structure containing fetal arteries and veins that connects the fetus with the placenta.

undifferentiated Not specialized to carry out a specific function.

uniformitarianism The theory that geologic phenomena can be explained from observable processes that have occurred in a uniform way; first proposed by James Hutton in 1788 and further developed by Charles Lyell in the early 1830s.

unifying hypothesis A synthesis of existing knowledge to create a single explanation that can be tested.

unique-sequence DNA A DNA sequence with one or more copies per genome; includes single genes, certain multiple genes and pseudogenes, and sequences with no known function.

upwelling The vertical transport of nutrients from ocean depths to the surface.

urea (yoo-RĒ-uh) The main nitrogen-containing waste product excreted in urine.

ureter (YOO-ruh-tur) One of two tubes from the kidney to the urinary bladder.

urethra (yoo-RĒ-thruh) A tube from the urinary bladder to the exterior; transports urine in females and urine and semen in males.

urinary (YOO-ruh-ner-ē) **bladder** A hollow muscular organ that serves as a temporary storage reservoir for urine.

urinary (YOO-ruh-ner-ē) **system** A system composed of organs that function in filtering blood and producing urine.

urine (YOO-rin) A fluid containing varying amounts of water and waste products that is excreted from the body.

urkaryote (oor′KER-ē-ōt) The hypothesized first eukaryote to have evolved from the progenote.

uterine (YOO-tuh-rīn) **tube** One of two tubes that transport an ovum from the ovary to the uterus; also called the *fallopian tube* or *oviduct*.

uterus (YOO-tuh-rus) A thick-walled, muscular organ in females where an embryo develops and grows.

V

vaccination (vak′suh-NĀ-shun) The process of exposing a potential host to a disease-agent antigen to stimulate immunity to the antigen.

vaccine (vak-SĒN) A substance produced from a dead or modified form of a disease-causing virus, bacterium, or protozoan that stimulates an immune response after vaccination.

vacuole (VAK-yuh-wōl) An intracellular cavity in plant cells that is surrounded by a single membrane and stores water, sugars, proteins, or waste products.

vagina (vuh-JĪ-nuh) The female organ that receives sperm from a male and also serves as the birth canal during childbirth.

vagus (VĀ-gus) **nerve** The cranial nerve that extends to and regulates the functions of the heart, lungs, and digestive organs.

valence electrons Electrons in the outermost energy level of an atom.

valves Flaps of tissues in blood vessels that prevent blood from flowing in the opposite direction; in the heart, they regulate blood flow between chambers.

variable expressivity The appearance of a trait differently in people who apparently have the same genotype.

variable regions Amino acid sequences in antibody protein chains that are unique for each antibody.

vascular cambium (VAS-kyuh-lur KAM-bē-um) A cylinder of meristematic cells that surround the stem, branches, and roots of woody plants and give rise to phloem and xylem cells.

vascular (VAS-kyuh-lur) **plants** Plants with phloem and xylem to transport substances such as water and nutrients.

vas deferens (VAS DEF-ur-enz) Either of two male reproductive ducts surrounded by walls of smooth muscle that contract rhythmically during ejaculation to transport sperm forward.

vasopressin (vā′zō-PRES-in) See *antidiuretic hormone.*

vector Any plasmid, phage, or DNA segment that carries a DNA insert.

vegetative growth phase The part of the plant life cycle in which roots, stems, and leaves develop, but not reproductive tissues.

vegetative reproduction Reproduction by nonsexual plant tissues.

vein (vān) A blood vessel that returns blood from tissues to the heart.

vena cava (VĒ-nuh KĀ-vuh) Either of two large veins through which blood returns to the right atrium of the heart from tissues within the body.

ventricle A lower chamber of the heart.

venules (VEN-yoolz) Small blood vessels that transport blood from capillaries to veins.

vertebrae (VUR-tuh-brā) Small bones that make up the backbone or vertebral column.

vertebral (vur-TĒ-brul) **column** The part of the skeleton that encloses and protects the spinal cord and serves as the site of attachment for the rib cage and back muscles.

vertical transmission The transfer of a disease agent from parent to offspring by infected sperm, ova, or milk or through the placenta.

vesicles (VES-i-kulz) Small membrane-bound sacs in which proteins are stored.

villi (VIL-ī) Small, fingerlike projections located in the folds of the inner surface of the ileum that contain blood vessels and a lymphatic vessel; sites where food products of digestion are absorbed.

viral oncogene A viral gene that can cause cancers in certain animals.

virion (VĪ-rē-on) An infectious virus particle.

viruses (VĪ-rus-iz) Nonliving disease agents that can replicate only inside host cells.

visible light spectrum The range of photons with wavelengths between 400 and 700 nanometers that are detectable as various colors by the human eye.

voltage-gated ion channel An ion channel that opens in response to a change in membrane potential.

vulva (VUL-vuh) The external genital organs of females.

W

water use efficiency The amount of carbon gained per amount of water lost.

wet deposition Acid deposition on surfaces from precipitation.

woody perennials Plants that live for two or more years and are characterized by woody structures; include vines (such as grapes), shrubs (rhododendrons), and trees (pines and oaks).

X

X chromosome inactivation The process in which one of two X chromosomes in cells of females becomes inactivated during embryonic development.

X inactivation center A gene locus where events occur that regulate the X chromosome inactivation processes.

X-ray crystallography (kris′tuh-LOG-ruh-fē) A technique used in determining molecular structure.

Y

yolk sac A small cavity below the embryonic disk; nonfunctional in humans' development.

Z

zeitgeber (TSĪT-gā-bur) An external stimulus that sets a biological clock.

zygote (ZĪ-gōt) A fertilized egg formed from the union of haploid female and male gametes (egg and sperm).

zymotechnology (ZĪ-mō-tek-nol′uh-jē) A strategy developed in the late nineteenth century that aimed at improving all types of industrial fermentation, from the manufacture of lactic acid to curing leather, through engineering and applied chemistry, bacteriology, and botany.

Credits

Unlimited. Page (108)BR/©T. E. Adams/Visuals Unlimited. Page (109)T/©Joel Arrington/Visuals Unlimited. Page (109)CL/ ©R. DeGoursey/Visuals Unlimited. Page (109)CR/©Glenn M. Oliver/Visuals Unlimited. Page (109)BL/©Brian Parker/Tom Stack & Associates. Page (109)BR/©Zig Leszczynski/ANIMALS ANIMALS. Page (110)TL/©Don & Esther Phillips/Tom Stack & Associates. Page (110)TR/©Dr. William H. Amos. Page (110)CL/ ©C. P. Hiekiman/Visuals Unlimited. Page (110)CR/©J. H. Robinson/ANIMALS ANIMALS. Page (110)BL/©A. Desbonnet/Visuals Unlimited. Page (110)BR/©B. G. Murray, Jr./ANIMALS ANIMALS. Page (111)TL/©Jeff Rotman/Peter Arnold, Inc. Page (111)TR/©Denise Tackett/Tom Stack & Associates. Page (111)BL/©Daniel Gotshall/Visuals Unlimited. Page (111)BR/ ©William C. Jorgensen/Visuals Unlimited. Page (113)T/©Patrice/ Visuals Unlimited. Page (113)TC/©Gary Milburn/Tom Stack & Associates. Page (113)CL/©Breck Kent/ANIMALS ANIMALS. Page (113)BL/©Breck Kent/ANIMALS ANIMALS. Page (113)R/ ©Leonard Lee Rue III/ANIMALS ANIMALS. Page (114)T/ ©Allen G. Nelson/ANIMALS ANIMALS. Page (114)C/©Kjell Sandved/Visuals Unlimited. Page (114)B/©Allan Roberts/Visuals Unlimited. Page (115)TL/©Dave Watts/Tom Stack & Associates. Page (115)TC/©Dr. William H. Amos. Page (115)TR/©John D. Cunningham/Visuals Unlimited. Page (115)CL/©Allen Morgan/ Peter Arnold, Inc. Page (115)CR/©Martyn Colbeck/OSF/ANIMALS ANIMALS. Page (115)BL/©John Cancalosi/Peter Arnold, Inc. Page (115)BR/©Thomas Kitchin/Tom Stack & Associates. Page (117)T/©Ray Coleman/Photo Researchers. Page (117)C/ ©Donald Specker/Earth Scenes. Page (117)B/©John S. Flannery/ Visuals Unlimited. Page (121)/©NOAO.

Unit Two
Page (122)/©NASA.

Chapter 7
Page (127)TL/©Ed Wolff/Earth Scenes. Page (127)TC/©Keith L. King. Page (127)TR/©Keith L. King. Page (128)/©Keith L. King. Page (130)TL/©Martin G. Miller/Visuals Unlimited. Page (132)L/©Link/Visuals Unlimited. Page (132)R/©Greg Vaughn/Tom Stack & Associates. Page (133)T/©Keith L. King. Page (134)L/©Donovan Reese/Visuals Unlimited. Page (134)R/ ©Bob Newman/Visuals Unlimited. Page (135)/©Keith L. King.

Chapter 8
Page (141)/©Michael Fogden/Earth Scenes. Page (142)ALL/©Bill Beatty/Visuals Unlimited. Page (143)L/©Greg Vaughn/Tom Stack & Associates. Page (143)R/©Glenn Oliver/Visuals Unlimited. Page (144)/©Steve Kaufman/Peter Arnold, Inc. Page (145)T/ ©Doug Sokell/Visuals Unlimited. Page (145)BL/©Mickey Gibson/ANIMALS ANIMALS. Page (145)BR/©W. J. Weber/Visuals Unlimited. Page (146)/©Joe McDonald/Visuals Unlimited. Page (147)TL/©Keith L. King. Page (147)TR/©John S. Flannery/Visuals Unlimited. Page (147)B/©Steve McCutcheon/Visuals Unlimited. Page (148)L/©Tom Alrich/Visuals Unlimited. Page (148)R/ ©Kutley Perkins/Visuals Unlimited. Page (149)L/©Breck P. Kent/Earth Scenes. Page (149)R/©Gary Schultz/The Wallace Collection. Page (150)ALL/©NASA. Page (151)L/©Sandy Lovelock. Page (151)R/©Courtesy of Lynn Margulis. Page (152)T/©Steve McCutcheon/Visuals Unlimited. Page (152)C/©Tom Ulrich/Visuals Unlimited. Page (152)B/©Steve McCutcheon/Visuals Unlimited. Page (153)ALL/©Keith L. King. Page (154)T/©E. R. Degginger. Page (154)C/©Doug

Sokill/Visuals Unlimited. Page (154)B/ ©Keith L. King. Page (155)T/©Bruce Davidson/ANIMALS ANIMALS. Page (155)B/©Dr. Nigel Smith/Earth Scenes.

Chapter 9
Page (159)/©T. Kitchin/Tom Stack & Associates. Page (163)B/ ©E. C. Williams/Visuals Unlimited. Page (166)/©Glenn Oliver/ Visuals Unlimited. Page (168)B/©Will Troyer/Visuals Unlimited. Page (170)/©Science VU/Visuals Unlimited. Page (171)TL/©J. Burton/ VU. Page (173)/©John D. Cunningham/Visuals Unlimited.

Chapter 10
Page (178)/©Jeff Foott/Tom Stack & Associates. Page (182)/©Patti Murray/Earth Scenes. Page (183)L/©Barbara Gerlach/Visuals Unlimited. Page (183)R/©Ron Spomer/Visuals Unlimited. Page (185)L/©John D. Cunningham/Visuals Unlimited. Page (185)R/ ©Keith L. King. Page (188)/©Jeff Henry/Peter Arnold, Inc. Page (189)T/©Richard Thom/Visuals Unlimited. Page (191)ALL/ ©Keith L. King. Page (192)L/©John Gerlach/Visuals Unlimited. Page (192)TR/©Joe McDonald/ANIMALS ANIMALS. Page (192)BR/©Mickey Gibson/ANIMALS ANIMALS.

Chapter 11
Page (196)L/©David Matherly/Visuals Unlimited. Page (196)R/ ©Greg Vaughn/Tom Stack & Associates. Page (200)B/©NASA. Page (202)L/©J. R.Williams/ANIMALS ANIMALS. Page (202)TR/©A.Gurmankin/Visuals Unlimited. Page (202)BR/©Doug Sokell/Tom Stack & Associates. Page (203)/©Kutley-Perkins/Visuals Unlimited. Page (205)T/©Paul Gier/Visuals Unlimited. Page (205)B/©Keith L. King. Page (208)/©Denise Tackette/Tom Stack & Associates. Page (211)T/©Jack Swenson/The Wallace Collection. Page (211)CL/©Richard Hermann/The Wallace Collection. Page (211)CR/©Thomas Kitchin/Tom Stack & Associates. Page (211)B/©Keith L. King. Page (212)B/©Brian Parker/Tom Stack & Associates. Page (213)T/©Tom Ulrich/Visuals Unlimited. Page (213)/©David M. Doody/Tom Stack & Associates. Page (216)T/ ©Patti Murray/Earth Scenes. Page (216)BL/©Keith L. King. Page (216)BC/©Dwight R. Kuhn. Page (216)BR/©Keith L. King. Page (217)/©SuperStock, Inc.

Chapter 12
Page (221)/©Joe McDonald/Tom Stack & Associates. Page (223)/©Scala/Art Resource, New York. Page (227)/©Science VU/Visuals Unlimited. Page (228)L/©Stephen W. Kress/Visuals Unlimited. Page (228)R/©Ted Levin/ANIMALS ANIMALS. Page (232)T/©NASA. Page (232)B/©Jill Isengart/Tom Stack & Associates.

Chapter 13
Page (240)/©Frank T. Aubrey/Visuals Unlimited. Page (241)L/ ©Keith L. King. Page (241)C/©Jack Swenson/Tom Stack & Associates. Page (241)R/©Andrew Holbrooke/Gamma-Liaison. Page (245)/©Collier/Stock Boston. Page (249)TL/©William Campbell/ TIME Magazine/© Time Warner, Inc. Page (249)TR/©Chip & Jill Isenhart/Tom Stack & Associates. Page (249)CL/©NASA. Page (251)/©NASA. Page (253)/©Keith L. King. Page (259)/©NASA.

Unit Three
Page (260)/©10(bl)Lionel I. Rebhun, University of Virginia, Charlottesville.

Chapter 14
Page (263)/©From MICROGRAPHIA by Robert Hooke/The Royal Society (1665). Page (264)/©Lord/The Image Works. Page

(265)T/©Bettmann Archive. Page (265)B/©D. Cavagnaro/Visuals Unlimited. Page (266)ALL/©Mendelianum of the Moravian Museum, Brno, Czechoslovakia. Page (273)/©Courtesy of The Newberry Library, Chicago. Page (274)/©MEMOIRON HEAT by Messeurs Oavoisier &. Dela Place.

Chapter 15
Page (277)T/©Biophoto Assoc./SS/Photo Researchers. Page (277)B/©Courtesy of S. M. Gollin & W. Wray. Page (281)ALL/©Conly L. Rieder/Biological Photo Service. Page (282)CL/©Conly L. Rieder/Biological Photo Service. Page (282)TL/©Tui de Roy/Bruce Coleman Inc. Page (282)TR/©William E. Ferguson Photography. Page (282)BR/©William E. Ferguson Photography. Page(288)TL/©Don Fawcett/Photo Researchers.

Chapter 16
Page (297)/©AP/Wide World. Page (299)/©Runk/Schoenberger/Grant Heilman Photography. Page (308)/©Barry L. Runk/Grant Heilman Photography.

Chapter 17
Page (315)L/©Custom Medical Stock Photo. Page (315)R/©A. B. Dowcett/SPL/Photo Researchers. Page (318)/©/Ed Reschke. Page (321)/©Les Simon/Stammers/SS/Photo Researchers. Page (324)L/©From *THE DOUBLE HELIX* by J. D. Watson 1958, Atheneum, New York a. d. Barrington Brown photo. Page (324)R/ ©Cold Spring Harbor Laboratories Archives. Page (325)/©David M. Dennis/Tom Stack & Associates. Page (327)/©Ann Sayre. Page (330)/©Douglas Struthers/Tony Stone Images.

Chapter 18
Page (333)/©Joe McDonald/Visuals Unlimited. Page (334)T/©James W. Richardson/Visuals Unlimited.

Chapter 19
Page (356)CL/©D. W. Fawcett/Visuals Unlimited. Page (356)CR/©James King-Holmes/ICRF/SPL/Photo Researchers. Page (357)TC/©Courtesy Children's Memorial Hospital. Page (359)/©Custom Medical Stock Photo. Page (360)/©AP/Wide World. Page (361)T/©Custom Medical Stock Photo. Page (361)C/©Charles Gupton/Tony Stone Images. Page (361)BL/ ©Carolina Biological Supply Co./Phototake. Page (361)BR/©Carolina Biological Supply Co./Phototake. Page (364)B/©Courtesy FBI. Page (365)T/©G. Argent/Camera Press/Retna Ltd., NY.

Chapter 20
Page (378)/©Jerry Wachter/Focus On Sports. Page (380)ALL/©Stanley Fiegler/Visuals Unlimited. Page (384)/©Steve Uzzell. Page (387)ALL/©National Heart, Lung and Blood Institute/National Institutes of Health. Page (389)/©Howard Sochurek.

Chapter 21
Page (395)/©Bettmann Archive. Page (400)/©M.Wiertz/Biozentrum/Univ. of Basel/SPL/Photo Researcher. Page (401)L/©David M. Phillips/Visuals Unlimited. Page (401)R/©Elmer Koneman/Visuals Unlimited. Page (403)/ ©Dan McCoy/Rainbow. Page (404)L/©Dr. Jeremy Burgess/Photo Researchers. Page (404)R/©C. P. Vana/Visuals Unlimited. Page (405)BL/©Sinclair Stammers/SPL/Photo Researchers. Page (405)BR/©Plantek/Photo Researchers. Page (406)/©Dan McCoy/ Rainbow. Page (407)/©Dr. Jeremy Burgess/SPL/Photo Researchers. Page (408)/©University of Pennsylvania/R. L. Brinster, Laboratory of Reproductive Physiology. Page (411)/©Science VU/Jackson Laboratory/Visuals Unlimited.

Chapter 22
Page (417)/©Ed Horowitz/Tony Stone Images. Page (418)/©Suzanna Arms/The Image Works. Page (419)/©UPI/Bettmann. Page (421)L/©CNRI/SPL/Photo Researchers. Page (421)R/©M. Wurtz, Biozentrum/Univ. of Basel/SPL/Photo Researchers. Page (425)/©Matt Meadows/Peter Arnold, Inc.

Unit Four
Page (438)/©Chip Clark/Smithsonian Institution.

Chapter 23
Page (442)T/©Bettmann Archive. Page (442)B/©Courtesy of Mr. G. P. Darwin/By permission of the Darwin Museum, Down House, Royal College of Surgeons. Page (443)L/©From *HISTORICA FISCIA Y POLITICA DE CHILE.* Page (443)R/ ©UPI/Bettmann. Page (444)L/©Michael Dick/ANIMALS ANIMALS. Page (444)R/©Jany Sauwanet/Photo Researchers. Page (445)T/©William E. Ferguson Photography. Page (445)B/©Kjell Sandved/Visuals Unlimited. Page (447)/©Breck P. Kent/Earth Scenes. Page (453)/©Bettmann Archive. Page (454)/©Bettmann Archive.

Chapter 24
Page (456)TR/©AP/Wide World. Page (456)BL/©Bettmann Archive. Page (456)/©AP/Wide World. Page (463)TL/©Tweedie/Bruce Coleman Inc. Page (463)R/©Breck B. Kent/ANIMALS ANIMALS. Page (465)/©Warren & Genny Garst/Tom Stack & Associates. Page (466)TL/©M. Long/Visuals Unlimited. Page (466)TR/©Diana L. Stratton/Tom Stack & Associates. Page (466)BL/©Lawrence Gilbert, Dept. of Zoology, University of Texas, Austin). Page (466)BR/©D. Wilder/Tom Stack & Associates. Page (470)/©James F. Crow, Laboratory of Genetics/University of Wisconsin. Page (472)L/©Bettmann Archive. Page (472)C/©Farber Collection. Page (472)R/©Farber Collection. Page (473)/©Bettmann Archive. Page (474)/©UPI/Bettmann.

Chapter 25
Page (479)/©James P. Kennett, Marine Sciences Institute, University of California, Santa Barbara. Page (480)/©John Cancalosi/Tom Stack & Associates. Page (482)BL/©Jack Couffer/Bruce Coleman Inc. Page (483)L/©Jane Burton/Bruce Coleman Inc. Page (483)C/©E. R. Degginger/ANIMALS ANIMALS. Page (483)R/©E. R. Degginger. Page (484)/©Dwight R. Kuhn. Page (485)T/©C. A. Henley. Page (485)B/©Kjell B. Sandved/Visuals Unlimited. Page (486)L/©Brian Parker/Tom Stack & Associates. Page (486)R/©Daniel W. Gotshall/Visuals Unlimited. Page (487)/©Bettmann Archive. Page (488)/©Bettmann Archive.

Chapter 26
Page (494)L/©L. L. T. Rhodes/ANIMALS ANIMALS. Page (494)R/©Cabisco/Visuals Unlimited. Page (495)TL/©William E. Ferguson Photography. Page (495)TR/©Sinclair Stammers/ SPL/Photo Researchers. Page (495)B/©John S. Flannery/Visuals Unlimited. Page (496)BL/©Biological Photo Service. Page (496)T/ ©Alfred Pasieka/SPL/Photo Researchers. Page (496)BR/©William E. Ferguson Photography. Page (497)L/©Tom McHugh/Photo Researchers. Page (497)R/©John Cancalosi/Tom Stack & Associates.

Chapter 36
Page (678)T/©Dan McCoy & Arnold Scheibel/Rainbow. Page (678)B/©John D. Cunningham/Visuals Unlimited. Page (679)/©P. Motla/SPL/Photo Researchers. Page (685)/©T .Reese & D. Fawcett/Visuals Unlimited. Page (686)/©Omikron/SS/Photo Researchers. Page (691)/©Dr. Colin Chumbley/SPL/Photo Researchers.

Chapter 37
Page (702)/©Astrid & Hanns-Frieder Michler/SPL/Photo Researchers. Page (703)/©Secchi-Lecaque/Roussel-UCLAF/SPL/Photo Researchers. Page (705)/©T. Hynes & A. M. de Vos with MIDAS-plus software, Genetitech, Inc.

Chapter 38
Page (716)/©SIU/Visuals Unlimited. Page (717)/©C. Edelmann/LaVillette/Photo Researchers. Page (721)/©Biophoto Associates/Photo Researchers. Page (728)L/©Don W. Fawcett/Photo Researchers. Page (728)R/©Carolina Biological Supply/Phototake. Page (733)TL,TC,BL,BR/©Petit Format/Nestle/SS/Photo Researchers. Page (733)TR/©*A Child Is Born,* Dell Publishing Company/Lennart Nilsson/Bonnier Fakta.

Chapter 39
Page (741)/©Ken Eward/SS/Photo Researchers. Page (742)/©Boehringer Ingleheim Zentrale,GmbH. Page (743)/ ©CNRI/SPL/Photo Researchers. Page (745)/©John D. Cunningham/Visuals Unlimited. Page (747)/©Petit Format/I.P.R.P./Photo Researchers. Page (750)/©Dr. Andrejs Liepirs/SPL/Photo Researchers. Page (751)L/©Boehringer Ingelheim Zentrale GmbH. Page (751)R/

©David M. Phillips/Visuals Unlimited. Page (760)/©Dr. Rosalind King/SPL/Photo Researchers.

Chapter 40
Page (764)/©Kim Newton/Woodfin Camp & Associates. Page (766)/©Pasteur Institute, Paris. Page (769)L/©CNRI/SPL/Custom Medical Stock Photo. Page (769)C/©A. B. Dowsett/SPL/Photo Researchers. Page (769)R/©Alfred Pasieka/SPL/Photo Researchers. Page (770)T/©CDCI/SS/Photo Researchers. Page (770)B/©K. G. Murti/Visuals Unlimited. Page (775)/©KROPPENS FORSVAR/Lennart Nilsson/Bonnier Fakta. Page (776)/©Fred Hossler/Visuals Unlimited. Page (781)/©Cecil Fox/SS/Photo Researchers.

Chapter 41
Page (788)/©Hank Morgan/Photo Researchers. Page (789)L/©Hans Gelderblom, Robert Koch Inst., Berlin. Page (789)TR/©Wagner Herbert Stock/Phototake. Page (792)TL/©Hans Gelderblom, Robert Koch Inst., Berlin. Page (792)TR/©NIBSC/SPL/Photo Researchers. Page (792)B/©CNRI/SPL/Photo Researchers. Page (794)/©Courtesy Cecil Fox, Molecular Histology, Inc. Page (795)/©Courtesy Cecil Fox, Molecular Histology, Inc.

Chapter 42
Page (808)/©Alfred Pasieka/Custom Medical Stock Photo. Page (809)T/©Dr. Morley Read/SPL/Photo Researchers. Page (809)B/©National Cancer Institute/National Institutes of Health. Page (810)/©Matt Bradley/Tom Stack & Associates. Page (810)B/©The Upjohn Company. Page (811)T/©Peter Correz/Tony Stone Images.

Index

Page numbers in italic type indicate the entry appears in a figure.

Oparin-Haldane hypothesis, 30
Open-grown trees, 598
Open woodlands, 147
Operant conditioning, 536, *536, 537*
Operator, 346
Operculum, 112
Operons, 346, 348
Opossum, *115*
Opportunistic pathogens, 774
Oral contraceptives, 725, 726t
Orbital factors, and climate changes, 237-238, *237*
Order, 81
Ordovician period, 495, *495*, 503
Organ systems, 628, 628t, *629-631. See also* specific systems
Organelles
    definition of, 37, 621
    division of, 280, 283, *284*
    in eukaryotic cells, 40, 44-46, *45, 46*
Organic Chemistry in Its Applications to Agriculture and Physiology *(Liebig), 577*
Organic compounds, 27-28, 32-33, *32-34*
Organismic ecology, 160
Organized cell aggregation, 93
Organogenesis, 731, *733*
Organs, 627-628
Orgasm, 724
Origin, of muscle, 648, *649*
Origin of species. *See also* Origins
    cladogenesis, 480-481, *481-484*
    Darwin on, 476
    definition of races, 477, *478*
    definition of species, 477
    evolution of species, 486-489
    and number of species, 476
    phyletic evolution, 477, *479*, 480, *480*
    postzygotic isolating mechanisms, 484, 484t, 486
    prezygotic isolating mechanisms, 484-485, 484t, *485, 486*
    reproductive isolating mechanisms, 484-486, 484t, *485, 486*
    selection for isolating mechanisms, 486
    speciation, 477, *479-484*, 480-484
Origin of Species, The (Darwin), 159, 450, 453, 472, *507*, 508, 513
Origins. *See also* Evolution; Human evolution; Origin of species
    ancient worldviews of, 14-16, *15*
    big bang theory, 17-18, 21, *21*
    of cells, 49-56, *50-52, 54, 55*
    of chemical elements, 21, *21*
    chemical evolution hypothesis, 29-31
    of complex molecules, 26-33
    different levels of certainty for theories and hypotheses on, 16, *16*
    of Earth, 16, 17, 23-25, *23, 24, 25*
    endosymbiosis hypothesis, 52
    of humans, 513-525

    of life, 29-31, 49-56, *50-52, 54, 55*, 348-351, *350, 351*
    of multicellular organisms, 53
    of solar system, 16, 22-23, *22, 23*
    spontaneous generation, 29
    of universe, 16, 17-18
    vent origin of life hypothesis, 31
Orthomyxovirus, 768t
OSHA. *See* Occupational Safety and Health Administration (OSHA)
Osmosis, 42-43, *44*
Ospreys, 228, *228*
Osteichthyes, 112, 112t, *113*
Osteoblasts, 645, *645, 646*
Osteoclasts, 645, *646*
Osteocytes, 645, *645*
Ostracoderms, 495
Otters, *211*
"Out of Africa" model of human evolution, 518-519, 521, 522
Outbreeding, 289
Outer core, of Earth, 24, *24*
Ova
    fertilization of human ova, *717, 724, 728*
    in human reproductive system, 715
    oogenesis in humans, 720
    production and transport of, in humans, 719, *720*
Ovarian cycle, 719, 722, *723*, 724
Ovaries
    of angiosperms, 102
    hormones produced by, 699t
    of humans, 699t, 700, 715
Overexpression, 802
Overpopulation, 230-234, *231, 232*
Overton, William, 489
Oviducts, 719, *720*
Ovis musimon, 85, 85t
Ovism, 264, 611
Ovulation, 720, *723*
Oxidation, 63, *63*
Oxidation-reduction reactions, 63, *63*
Oxygen
    carbon-oxygen cycle, 167, 169-170, *169, 170*
    evolutionary importance of, 51-52
    in human circulatory system, 662, *663*, 664, 665t, 666, *667*, 668
    human respiratory system and, 656-658, *657-659*
    plants and, 570, *571*
Oxygenated blood, 665t, 666
Oxyhemoglobin, 666
Oxytocin, 699t, 707-708, 711, 734
Ozone, 51-52, 232, 238, 239t, 248, *248*, 250, *251*, 588

P

p7, 788, *789*
p17, 788, *789*

p24, 788, *789*
P/R. *See* Production to respiration ratio (P/R)
Painter, T. S., 355
Paleoclimatic indicators, 237
Paleontologists, definition of, 477
Paleontology. *See also* Fossils
    community evolution, 506-507
    convergence, *498, 499, 501*
    definition of, 477, 491
    divergence, 498-499, *498-501*, 506, *507*
    ecological collapse, 507
    evolution and, 492-507
    and expansion of modern synthesis, 506-507
    fossil record, 492, *493-497*, 495-497
    fossils and modern synthesis, 506, *507*
    grades, 502
    parallelism, *498*, 502, *503*
    and patterns of change, 492, *497-505*, 498-505
    physiology and, 509
    progressive evolution, 502, *504*
    punctuated equilibrium, 506
    species selection, 506
Paleozoic era, 495-497, *495, 496*, 502
Palmiter, Richard, 409
Palo verde, 191, *192*
Pancreas, 670, 699t, 703, *703*
Pancreatic juice, 670, 670t
Pandemics, 787
Pangaea, 125
Papilio dardanus, 464-465, *464*
Papovavirus, 768t
Parallelism, *498*, 502, *503*
Parasites, 90-92, 108, 111
Parasitic diseases, 91-92, *91*
Parasitism, 91
Parasympathetic nervous system, *688*, 689
Parathyroid, 699t
Parathyroid hormone (PTH), 647, *648*, 699t
Pardee, J. T., 129
Parenchyma, 570
Parental investment, of primates, 516
Parkinson's disease, 687
Partial pressure, 658, 665t
Particles, 18, 18t
Parturition, 734
Passive transport, 42-43, *43*, 43t, *44*
Pasteur, Louis, 4, 29, *30*, 765, *766*
Pasteurization, 765
Patau syndrome, 376
Pathogenic bacteria, 769
Pathogens, 746
Pauling, Linus, 380, 754-755, 756, 809
Pavlov, Ivan Petrovich, 535, *535*
Pea plants, genetic experiments with, 266-272, *266-272*, 290-291
Peafowl, 465-466, *466*
Pectin, 567

A BIOLOGICAL LEXICON

A list of Greek and Latin prefixes,
suffixes, and word roots commonly
used in biological terms.

a-, an- [Gk. *an-*, not, without, lacking]: anaerobic, atom, abiotic, anonymous, anaphrodisiac, anemia

ad- [L. *ad-*, toward, to]: adhesion, admixture, adopt, adrenaline

amphi- [Gk. *amphi-*, two, both, both sides of]: amphibian, Amphineura

ana- [Gk. *ana-*, up, up against]: anaphase, analogy, anabolic

andro- [Gk. *andros*, an old man]: androecium, androgen

anti- [Gk. *anti-*, against, opposite, opposed to]: antibiotic, antibody, antigen, antidiuretic hormone

archeo- [Gk. *archaios*, beginning]: archegonium, archenteron, archaic, menarche

arthro- [Gk. *arthron*, a joint]: arthropod, arthritis, arthrodire, condylarthra

auto- [Gk. *auto-*, self, same]: autoimmune, autotroph, autosome, autonomic

bi-, bin- [L. *bis*, twice; *bini*, two-by-two]: binary fission, binocular vision, biennial, bicarbonate, bilateral, binomial, bipolar

bio- [Gk. *bios*, life]: biology, biomass, biome, biosphere, biosynthesis, biotic

blasto-, -blast [Gr. *blastos*, sprout; now "pertaining to the embryo"]: blastoderm, blastodisc, blastopore, blastula, trophoblast, osteoblast

brachi- [Gk. *brachion*, arm]: brachiation, brachiopod

broncho- [Gk. *bronchos*, windpipe]: bronchus, bronchi, bronchiole, bronchitis

carcino- [Gk. *karkin*, a crab, cancer]: carcinogen, carcinoma

cardio- [Gk. *kardia*, heart]: cardiac, myocardium, electrocardiogram

cephalo- [Gk. *kephale*, head]: cephalization, cephalochordate, cephalopod, cephalothorax

chloro- [Gk. *chloros*, green]: chlorophyll, chloroplast, chlorine

chromo- [Gk. *chroma*, color]: chromosome, chromoplast, chromatin

coelo-, -coel [Gk. *koilos*, hollow, cavity]: coelacanth, coelenteron, coelenterate, coelom

com-, con-, col-, cor-, co- [L. *cum*, with, together]: coenzyme, commensal, conjugation, convergence, covalent

cranio- [Gk. *kranios*, L. *cranium*, skull]: cranial, cranium

cuti- [L. *cutis*, skin]: cutaneous, cuticle, cutin

cyclo-, -cycle [Gk. *kyklos*, circle, ring, cycle]: cyclostome, pericycle

cyto-, -cyte [Gk. *kytos*, vessel or container; now, "cell"]: cytoplasm, cytology, cytochrome, cytokinesis, erythrocyte, leucocyte

de- [L. *de-*, "away, off"; deprivation, removal, separation, negation]: deciduous, decomposer, decarboxylase, dehydration

derm-, dermato- [Gk. *derma*, skin]: dermis, epidermis, ectoderm, endoderm, mesoderm

di- [Gk. *dis*, twice]: dicaryon, dicotyledon, dioxide, dipole

dia- [Gk. through, passing through, thorough, thoroughly]: diabetes, dialysis, diaphragm

diplo- [Gk. *diploos*, two-fold]: diploid, diploblastic

eco- [Gk. *oikos*, house, home]: ecology, androecium, ecosphere, ecosystem, economy

ecto- [Gk. *ektos*, outside]: ectoderm, ectoplasm, ectoparasite

endo- [Gk. *endon*, within]: endocrine, endoderm, endodermis, endometrium, endoskeleton

epi- [Gk. *epi*, on, upon, over]: epicotyl, epidermis, epididymis, epiglottis, epiphyte, epithelium

eu- [Gk. *eus*, good; *eu*, well; now "true"]: eubacterium, eucaryote

ex-, exo-, ec, e- [Gk., L. out, out of, from, beyond]: emission, ejaculation, excretion, exergonic, exhale, exocytosis, exon, exoskeleton

extra- [L. outside of, beyond]: extracellular, extraembryonic

-fer [L. *ferre*, to bear]: fertile, fertilization, conifer, rotifer

galacto- [Gk. *galakt-*, milk]: galactose, galactic [Milky Way]

gam-, gameto- [Gk. *gamos*, marriage; now usually in reference to gametes (sex cells)]: gamete, polygamy, isogamete, cryptogam

gastro- [Gk. *gaster*, stomach]: gastric, gastrula, gastrin, gastrovascular cavity, gastropod

gen- [Gk. *gen*, born, produced by; Gk. *genos*, race, kind; L. *genus*, *generare*, to beget]: polygenic, genotype, geneology, glycogen, florigen, pyrogen, estrogen, heterogenous

gluco, glyco- [Gk. *glykys*, sweet; now pertaining to sugar]: glucose, glycogen, glycolysis, glycoprotein

gyn-, gyno-, gyneco- [Gk. *gyne*, woman]: gynecology, gynoecium, polygyny, misogyny

hemo-, hemato-, -hemia, -emia [Gk. *haima*, blood]: hematology, hemodialysis, hemoglobin, hemophilia, anemia, leukemia

hepato- [Gk. *hepar, hepat-*, liver]: hepatitis, hepatic portal system

hetero- [Gk. *heteros*, other, different]: heterogeneous, heterogamous, heterozygote

histo- [Gk. *histos*, web of a loom, tissue; now pertaining to biological tissues]: histology, histone, histamine, antihistamine

homo-, homeo- [Gk. *homos*, same; Gk. *homios*, similar]: homeostasis, homeothermy, homogeneous, homologous, homozygote